WORKED SOLUTIONS

Roger Dixon
James Foley
Michael Haese
Robert Haese
Sandra Haese
Mark Humphries

Haese & Harris Publications

MATHEMATICS FOR THE INTERNATIONAL STUDENT
Mathematics SL second edition – WORKED SOLUTIONS
IB Diploma Programme

Roger Dixon	B.Ed.
James Foley	B.Ma.Comp.Sc.(Hons.)
Michael Haese	B.Sc.(Hons.), Ph.D.
Robert Haese	B.Sc.
Sandra Haese	B.Sc.
Mark Humphries	B.Sc.(Hons.)

Haese & Harris Publications
3 Frank Collopy Court, Adelaide Airport, SA 5950, AUSTRALIA
Telephone: +61 8 8355 9444, Fax: + 61 8 8355 9471
Email: info@haeseandharris.com.au
Web: www.haeseandharris.com.au

National Library of Australia Card Number & ISBN 978-1-921500-10-7

Published by Raksar Nominees Pty Ltd
3 Frank Collopy Court, Adelaide Airport, SA 5950, AUSTRALIA

First Edition	2005
Second Edition	2009

Artwork and cover design by Piotr Poturaj

Typeset in Australia by Susan Haese and Charlotte Sabel (Raksar Nominees).
Typeset in Times Roman $8\frac{1}{2}/10$

Acknowledgements: While every attempt has been made to trace and acknowledge copyright, the authors and publishers apologise for any accidental infringement where copyright has proved untraceable. They would be pleased to come to a suitable agreement with the rightful owner.

FOREWORD

This book gives you fully worked solutions for every question in each chapter of our textbook *Mathematics SL second edition* which is one of the textbooks in our series **Mathematics for the International Student** intended for use with IB Diploma and Middle Years courses.

Correct answers can sometimes be obtained by different methods. In this book, where applicable, each worked solution is modelled on the worked example in the textbook.

Be aware of the limitations of calculators and computer modelling packages. Understand that when your calculator gives an answer that is different from the answer you find in the book, you have not necessarily made a mistake, but the book may not be wrong either.

We have a list of errata for our books on our website. Please contact us if you notice any errors in this book.

RLD JMF PMH
RCH SHH MAH

e-mail: info@haeseandharris.com.au
web: www.haeseandharris.com.au

TABLE OF CONTENTS

Chapter 1	FUNCTIONS	5
Chapter 2	SEQUENCES AND SERIES	25
Chapter 3	EXPONENTIALS	49
Chapter 4	LOGARITHMS	69
Chapter 5	GRAPHING AND TRANSFORMING FUNCTIONS	92
Chapter 6	QUADRATIC EQUATIONS AND FUNCTIONS	105
Chapter 7	THE BINOMIAL EXPANSION	144
Chapter 8	THE UNIT CIRCLE AND RADIAN MEASURE	151
Chapter 9	NON-RIGHT ANGLED TRIANGLE TRIGONOMETRY	165
Chapter 10	ADVANCED TRIGONOMETRY	177
Chapter 11	MATRICES	203
Chapter 12	VECTORS IN 2 AND 3 DIMENSIONS	236
Chapter 13	LINES AND PLANES IN SPACE	270
Chapter 14	DESCRIPTIVE STATISTICS	290
Chapter 15	PROBABILITY	312
Chapter 16	INTRODUCTION TO CALCULUS	334
Chapter 17	DIFFERENTIAL CALCULUS	340
Chapter 18	APPLICATIONS OF DIFFERENTIAL CALCULUS	373
Chapter 19	DERIVATIVES OF EXPONENTIAL AND LOGARITHMIC FUNCTIONS	411
Chapter 20	DERIVATIVES OF TRIGONOMETRIC FUNCTIONS	433
Chapter 21	INTEGRATION	444
Chapter 22	APPLICATIONS OF INTEGRATION	464
Chapter 23	STATISTICAL DISTRIBUTIONS OF DISCRETE RANDOM VARIABLES	489
Chapter 24	STATISTICAL DISTRIBUTIONS OF CONTINUOUS RANDOM VARIABLES	500
Chapter 25	MISCELLANEOUS QUESTIONS	514

Chapter 1
FUNCTIONS

EXERCISE 1A

1 **a** $\{(1, 3), (2, 4), (3, 5), (4, 6)\}$ is a function since no two ordered pairs have the same x-coordinate.

b $\{(1, 3), (3, 2), (1, 7), (-1, 4)\}$ is not a function since two of the ordered pairs, $(1, 3)$ and $(1, 7)$, have the same x-coordinate 1.

c $\{(2, -1), (2, 0), (2, 3), (2, 11)\}$ is not a function since each ordered pair has the same x-coordinate 2.

d $\{(7, 6), (5, 6), (3, 6), (-4, 6)\}$ is a function since no two ordered pairs have the same x-coordinate.

e $\{(0, 0), (1, 0), (3, 0), (5, 0)\}$ is a function since no two ordered pairs have the same x-coordinate.

f $\{(0, 0), (0, -2), (0, 2), (0, 4)\}$ is not a function since each ordered pair has the same x-coordinate 0.

2 **a**

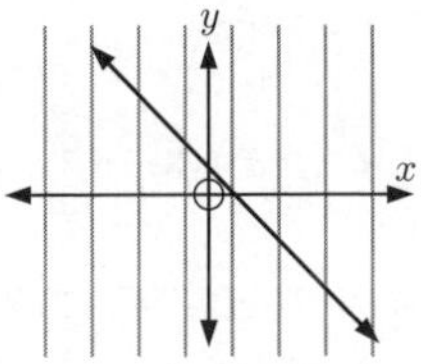

Each line cuts the graph no more than once, so it is a function.

b

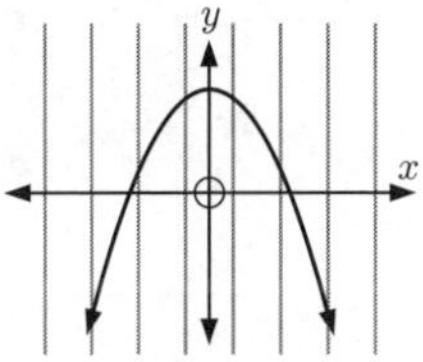

Each line cuts the graph no more than once, so it is a function.

c

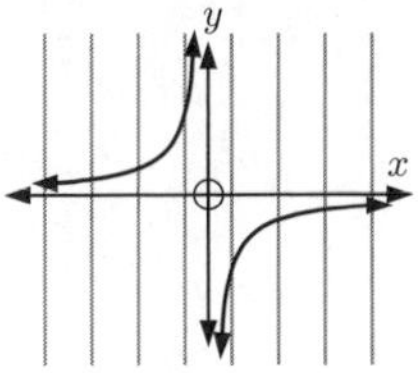

Each line cuts the graph no more than once, so it is a function.

d

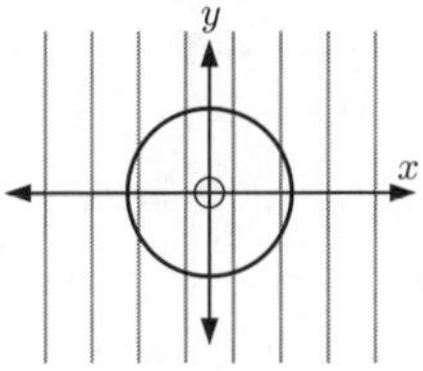

Some lines cut the graph more than once, so it is not a function.

e

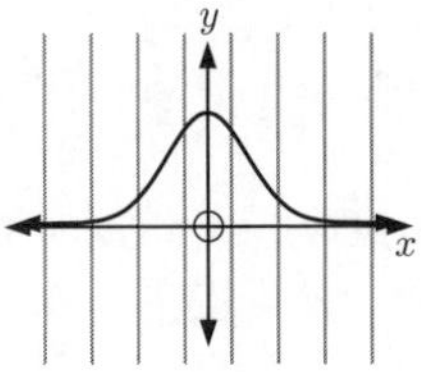

Each line cuts the graph no more than once, so it is a function.

f

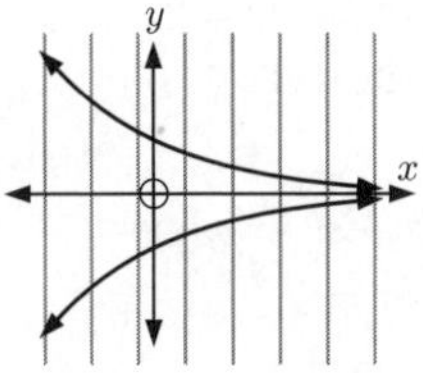

The lines cut the graph more than once, so it is not a function.

g

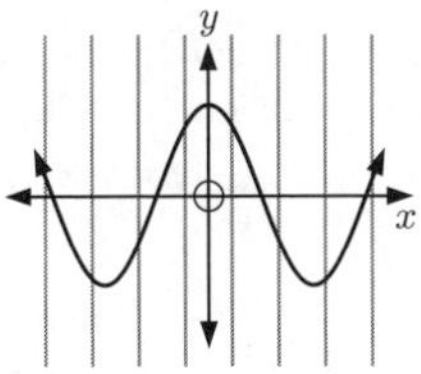

Each line cuts the graph no more than once, so it is a function.

h

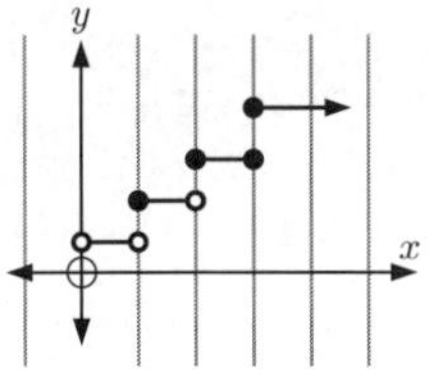

One line cuts the graph more than once, so it is not a function.

3 The graph of a straight line is not a function if the graph is a vertical line. So, it is not a function if it has the form $x = a$ for some constant a.
The vertical line through $x = a$ cuts the graph at every point, so it is not a function.

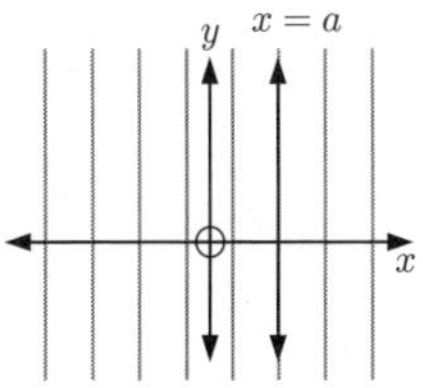

4 $x^2 + y^2 = 9$ is the equation of a circle, centre $(0, 0)$ and radius 3.

Now $x^2 + y^2 = 9$

$\therefore \quad y^2 = 9 - x^2$

$\therefore \quad y = \pm\sqrt{9 - x^2}$

For any value of x where $-3 < x < 3$, y has two real values. Hence $x^2 + y^2 = 9$ is not a function.

EXERCISE 1B

1 **a** $f(0) = 3(0) + 2 = 2$ **b** $f(2) = 3(2) + 2 = 8$ **c** $f(-1) = 3(-1) + 2 = -1$

d $f(-5) = 3(-5) + 2 = -13$ **e** $f(-\frac{1}{3}) = 3(-\frac{1}{3}) + 2 = 1$

2 **a** $f(0) = 3(0) - 0^2 + 2$
$= 2$

b $f(3) = 3(3) - 3^2 + 2$
$= 9 - 9 + 2$
$= 2$

c $f(-3) = 3(-3) - (-3)^2 + 2$
$= -9 - 9 + 2$
$= -16$

d $f(-7) = 3(-7) - (-7)^2 + 2$
$= -21 - 49 + 2$
$= -68$

e $f(\frac{3}{2}) = 3(\frac{3}{2}) - (\frac{3}{2})^2 + 2$
$= \frac{9}{2} - \frac{9}{4} + 2$
$= \frac{17}{4}$

3 **a** $g(1) = 1 - \frac{4}{1} = -3$ **b** $g(4) = 4 - \frac{4}{4} = 3$ **c** $g(-1) = -1 - \frac{4}{(-1)} = 3$

d $g(-4) = -4 - \frac{4}{(-4)} = -3$ **e** $g(-\frac{1}{2}) = -\frac{1}{2} - \dfrac{4}{(-\frac{1}{2})} = -\frac{1}{2} + 8 = \frac{15}{2}$

4 **a** $f(a) = 7 - 3a$

b $f(-a) = 7 - 3(-a)$
$= 7 + 3a$

c $f(a+3) = 7 - 3(a+3)$
$= 7 - 3a - 9$
$= -3a - 2$

d $f(b-1) = 7 - 3(b-1)$
$= 7 - 3b + 3$
$= 10 - 3b$

e $f(x+2) = 7 - 3(x+2)$
$= 7 - 3x - 6$
$= 1 - 3x$

f $f(x+h) = 7 - 3(x+h)$
$= 7 - 3x - 3h$

5 **a** $F(x+4)$
$= 2(x+4)^2 + 3(x+4) - 1$
$= 2(x^2 + 8x + 16) + 3x + 12 - 1$
$= 2x^2 + 16x + 32 + 3x + 11$
$= 2x^2 + 19x + 43$

b $F(2-x)$
$= 2(2-x)^2 + 3(2-x) - 1$
$= 2(4 - 4x + x^2) + 6 - 3x - 1$
$= 8 - 8x + 2x^2 + 5 - 3x$
$= 2x^2 - 11x + 13$

c $F(-x)$
$= 2(-x)^2 + 3(-x) - 1$
$= 2x^2 - 3x - 1$

d $F(x^2)$
$= 2(x^2)^2 + 3(x^2) - 1$
$= 2x^4 + 3x^2 - 1$

e $F(x^2 - 1)$
$= 2(x^2 - 1)^2 + 3(x^2 - 1) - 1$
$= 2(x^4 - 2x^2 + 1) + 3x^2 - 3 - 1$
$= 2x^4 - 4x^2 + 2 + 3x^2 - 4$
$= 2x^4 - x^2 - 2$

f $F(x+h)$
$= 2(x+h)^2 + 3(x+h) - 1$
$= 2(x^2 + 2xh + h^2) + 3x + 3h - 1$
$= 2x^2 + 4xh + 2h^2 + 3x + 3h - 1$

6 **a** **i** $G(2) = \dfrac{2(2)+3}{2-4}$
$= \frac{7}{-2}$
$= -\frac{7}{2}$

ii $G(0) = \dfrac{2(0)+3}{0-4}$
$= \frac{3}{-4}$
$= -\frac{3}{4}$

iii $G(-\frac{1}{2}) = \dfrac{2(-\frac{1}{2})+3}{-\frac{1}{2}-4}$
$= \dfrac{-1+3}{(-\frac{9}{2})}$
$= \dfrac{2}{(-\frac{9}{2})}$
$= -\frac{4}{9}$

b $G(x) = \dfrac{2x+3}{x-4}$ is undefined when $x - 4 = 0$
$\therefore \quad x = 4$
So, when $x = 4$, $G(x)$ does not exist.

c $G(x+2) = \dfrac{2(x+2)+3}{(x+2)-4} = \dfrac{2x+4+3}{x+2-4} = \dfrac{2x+7}{x-2}$

d $G(x) = -3$, so $\dfrac{2x+3}{x-4} = -3$ $\quad\therefore\ 2x + 3 = -3(x-4)$

$\therefore\ 2x + 3 = -3x + 12$

$\therefore\ 5x = 9$ and so $x = \frac{9}{5}$

7 f is the function which converts x into $f(x)$ whereas $f(x)$ is the value of the function at any value of x.

8 **a** $V(4) = 9650 - 860(4)$
$= 9650 - 3440$
$= 6210$

The value of the photocopier 4 years after purchase is 6210 euros.

b If $V(t) = 5780$,
then $9650 - 860t = 5780$
$\therefore\ 860t = 3870$
$\therefore\ t = 4.5$

The value of the photocopier is 5780 euros after $4\frac{1}{2}$ years.

c Original purchase price is when $t = 0$,
$V(0) = 9650 - 860(0)$
$= 9650$

The original purchase price was 9650 euros.

9

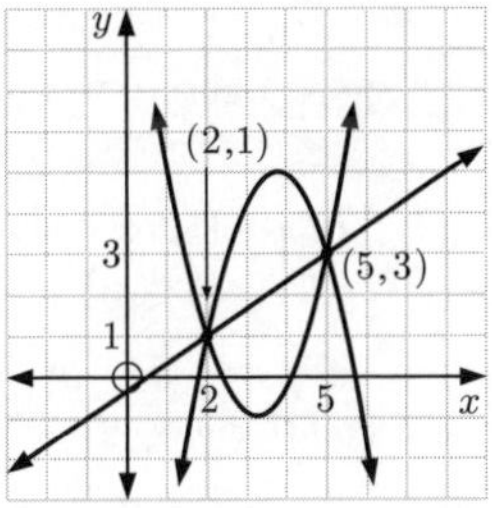

First sketch the linear function which passes through the two points (2, 1) and (5, 3).

Then sketch two quadratic functions which also pass through the two points.

10 $f(x) = ax + b$ where $f(2) = 1$ and $f(-3) = 11$

So, $a(2) + b = 1$ and $a(-3) + b = 11$

$\therefore\ 2a + b = 1$ $\qquad \therefore\ -3a + b = 11$

$\therefore\ b = 1 - 2a$ .... (1) $\qquad \therefore\ b = 11 + 3a$ .... (2)

Solving (1) and (2) simultaneously, $1 - 2a = 11 + 3a$

$\therefore\ 5a = -10$

$\therefore\ a = -2$

Substituting $a = -2$ into (1) gives $b = 1 - 2(-2) = 5$. So, $a = -2$, $b = 5$

Hence $f(x) = -2x + 5$

11 $f(x) = ax + \dfrac{b}{x}$ where $f(1) = 1$ and $f(2) = 5$

So, $a(1) + \dfrac{b}{1} = 1$ and $a(2) + \dfrac{b}{2} = 5$

$\therefore\ a + b = 1$ $\qquad \therefore\ 2a + \dfrac{b}{2} = 5$ .... (2)

$\therefore\ a = 1 - b$ .... (1)

Substituting (1) into (2), $2(1-b) + \dfrac{b}{2} = 5$

$\therefore\ 2 - 2b + \dfrac{b}{2} = 5$

$\therefore\ -\dfrac{3b}{2} = 3$

$\therefore\ b = -2$

Substituting $b = -2$ into (1) gives $a = 1 - (-2) = 3$.

So, $a = 3$, $b = -2$.

12 $T(x) = ax^2 + bx + c$ where $T(0) = -4$, $T(1) = -2$ and $T(2) = 6$

So, $a(0)^2 + b(0) + c = -4$

$\therefore\ c = -4$

Also, $a(1)^2 + b(1) + c = -2$ and $a(2)^2 + b(2) + c = 6$

$\therefore\ a + b + c = -2$ and $\therefore\ 4a + 2b + c = 6$

Substituting $c = -4$ into both equations gives

$a + b + (-4) = -2$

$\therefore\ a + b = 2$

$\therefore\ a = 2 - b$ .... (1)

and $4a + 2b + (-4) = 6$

$\therefore\ 4a + 2b = 10$ .... (2)

Substituting (1) into (2) gives $4(2 - b) + 2b = 10$ $\therefore\ 8 - 4b + 2b = 10$

$\therefore\ -2b = 2$

$\therefore\ b = -1$

Substituting $b = -1$ into (1) gives $a = 2 - (-1) = 3$.

So, $a = 3$, $b = -1$, $c = -4$, and $T(x) = 3x^2 - x - 4$.

EXERCISE 1C

1 **a** Domain is $\{x \mid x \geqslant -1\}$
Range is $\{y \mid y \leqslant 3\}$

b Domain is $\{x \mid -1 < x \leqslant 5\}$
Range is $\{y \mid 1 < y \leqslant 3\}$

c Domain is $\{x \mid x \neq 2\}$
Range is $\{y \mid y \neq -1\}$

d Domain is $\{x \mid x \in \mathbb{R}\}$
Range is $\{y \mid 0 < y \leqslant 2\}$

e Domain is $\{x \mid x \in \mathbb{R}\}$
Range is $\{y \mid y \geqslant -1\}$

f Domain is $\{x \mid x \in \mathbb{R}\}$
Range is $\{y \mid y \leqslant 6\frac{1}{4}\}$ *or* $\{y \mid y \leqslant \frac{25}{4}\}$

g Domain is $\{x \mid x \geqslant -4\}$
Range is $\{y \mid y \geqslant -3\}$

h Domain is $\{x \mid x \in \mathbb{R}\}$
Range is $\{y \mid y > -2\}$

i Domain is $\{x \mid x \neq \pm 2\}$
Range is $\{y \mid y \leqslant -1 \text{ or } y > 0\}$

2 **a** $f(x)$ is undefined when $x + 6 < 0$ $\therefore$ $f(x)$ is undefined for $\{x \mid x < -6\}$.

b $f(x)$ is undefined when $x^2 = 0$ $\therefore$ $f(x)$ is undefined for $x = 0$.

c $f(x)$ is undefined when $3 - 2x \leqslant 0$ $\therefore$ $f(x)$ is undefined for $\{x \mid x \geqslant \frac{3}{2}\}$.

3 **a** $y = 2x - 1$ can take any x-value and any y-value.
$\therefore$ the domain is $\{x \mid x \in \mathbb{R}\}$ and the range is $\{y \mid y \in \mathbb{R}\}$.

b $y = 3$ can take any value of x, but the only permissible value for y is 3
$\therefore$ the domain is $\{x \mid x \in \mathbb{R}\}$ and the range is $\{3\}$.

c $y = \sqrt{x^2 + 4}$ can take any value of x, as $x^2 + 4 \geqslant 0$ for all real x.
In fact, $x^2 + 4 \geqslant 4$, so $\sqrt{x^2 + 4} \geqslant 2$
$\therefore$ the domain is $\{x \mid x \in \mathbb{R}\}$ and the range is $\{y \mid y \geqslant 2\}$.

d $y = \sqrt{x^2 - 4}$ is defined when $x^2 - 4 \geqslant 0$,
which is when $x \leqslant -2$ or $x \geqslant 2$.
Also, $\sqrt{x^2 - 4} \geqslant 0$
$\therefore$ the domain is $\{x \mid x \leqslant -2,\ x \geqslant 2\}$
and the range is $\{y \mid y \geqslant 0\}$.

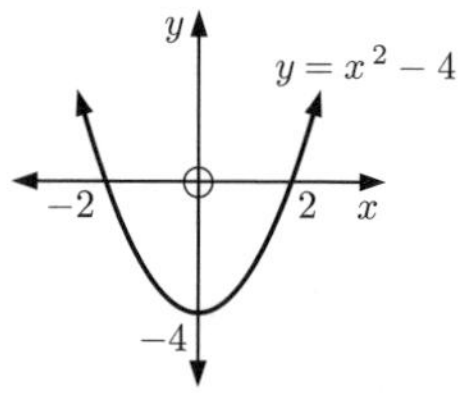

e $y = \dfrac{5}{x-2}$ is undefined when $x - 2 = 0$, or when $x = 2$

$y = \dfrac{5}{x-2}$ cannot be 0 for any value of x.

$\therefore$ the domain is $\{x \mid x \neq 2\}$ and the range is $\{y \mid y \neq 0\}$.

f $y = \sqrt{2-x}$ is defined when $2 - x \geqslant 0$, or when $x \leqslant 2$

Also, $\sqrt{2-x} \geqslant 0$ for all $x \leqslant 2$.

$\therefore$ the domain is $\{x \mid x \leqslant 2\}$ and the range is $\{y \mid y \geqslant 0\}$.

g $y = \dfrac{3}{\sqrt{2x-5}}$ is defined when $2x - 5 > 0$, or when $x > \frac{5}{2}$

Since $\sqrt{2x-5} > 0$, $\dfrac{3}{\sqrt{2x-5}} > 0$

$\therefore$ the domain is $\{x \mid x > \frac{5}{2}\}$ and the range is $\{y \mid y > 0\}$.

h $y = 2 + \dfrac{3}{5-x}$ is undefined when $5 - x = 0$, or when $x = 5$

Since $\dfrac{3}{5-x} \neq 0$, $2 + \dfrac{3}{5-x} \neq 2$

$\therefore$ the domain is $\{x \mid x \neq 5\}$ and the range is $\{y \mid y \neq 2\}$.

4 **a**

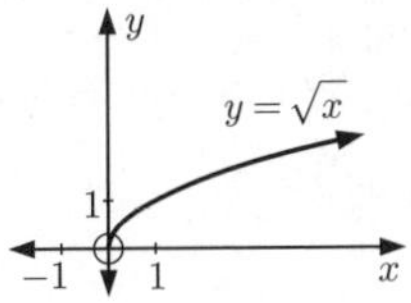

Domain is $\{x \mid x \geqslant 0\}$
Range is $\{y \mid y \geqslant 0\}$

b

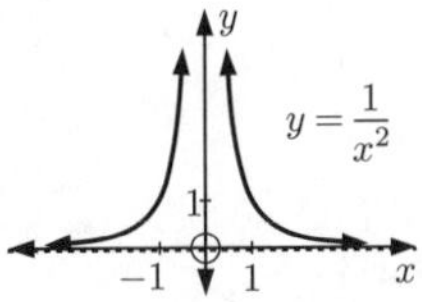

Domain is $\{x \mid x \neq 0\}$
Range is $\{y \mid y > 0\}$

c

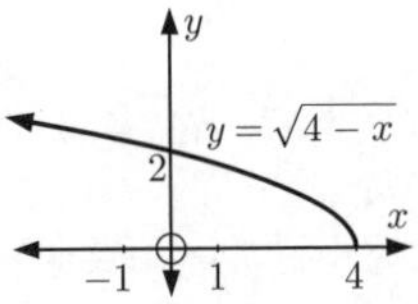

Domain is $\{x \mid x \leqslant 4\}$
Range is $\{y \mid y \geqslant 0\}$

d

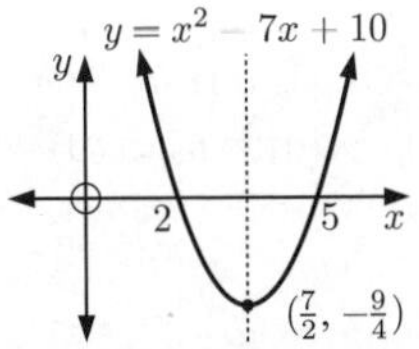

Domain is $\{x \mid x \in \mathbb{R}\}$
Range is $\{y \mid y \geqslant -2\frac{1}{4}\}$

e

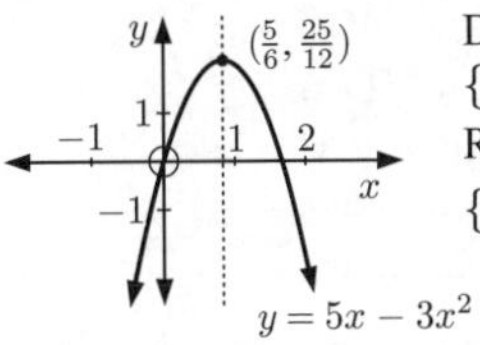

Domain is $\{x \mid x \in \mathbb{R}\}$
Range is $\{y \mid y \leqslant \frac{25}{12}\}$

f

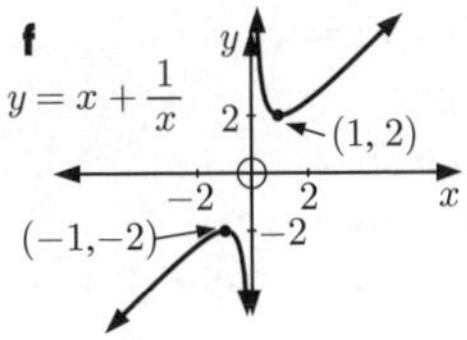

Domain is $\{x \mid x \neq 0\}$
Range is $\{y \mid y \leqslant -2 \text{ or } y \geqslant 2\}$

g

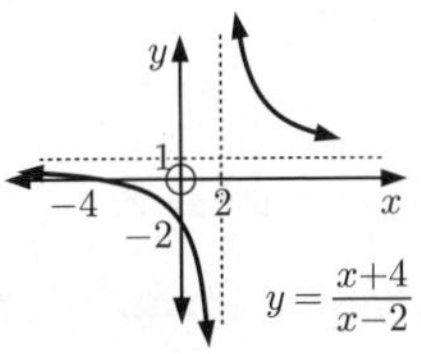

Domain is $\{x \mid x \neq 2\}$
Range is $\{y \mid y \neq 1\}$

h

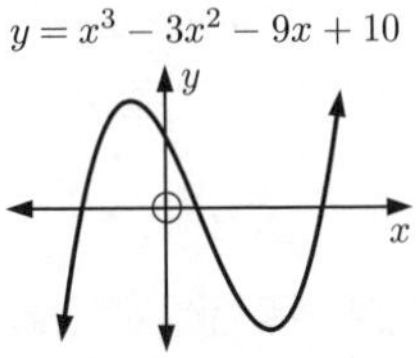

Domain is $\{x \mid x \in \mathbb{R}\}$
Range is $\{y \mid y \in \mathbb{R}\}$

i

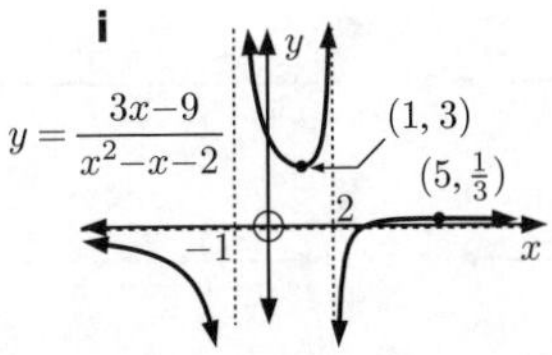

Domain is $\{x \mid x \neq -1, \ x \neq 2\}$
Range is $\{y \mid y \leqslant \frac{1}{3} \text{ or } y \geqslant 3\}$

j

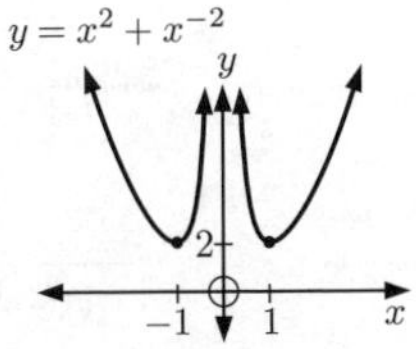

Domain is $\{x \mid x \neq 0\}$
Range is $\{y \mid y \geqslant 2\}$

k

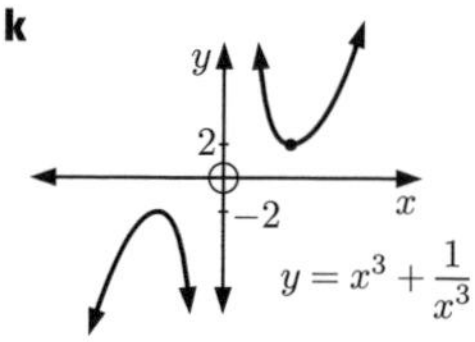

Domain is $\{x \mid x \neq 0\}$
Range is $\{y \mid y \leqslant -2 \text{ or } y \geqslant 2\}$

l

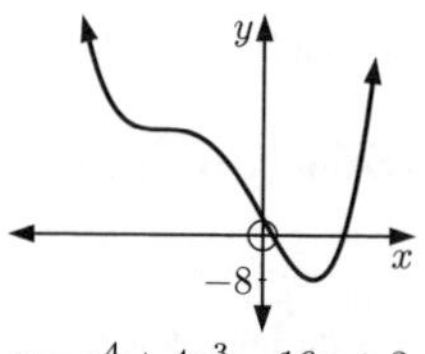

Domain is $\{x \mid x \in \mathbb{R}\}$
Range is $\{y \mid y \geqslant -8\}$

EXERCISE 1D

1 **a** $(f \circ g)(x)$
$= f(g(x))$
$= f(1-x)$
$= 2(1-x)+3$
$= 2-2x+3$
$= 5-2x$

b $(g \circ f)(x)$
$= g(f(x))$
$= g(2x+3)$
$= 1-(2x+3)$
$= 1-2x-3$
$= -2x-2$

c $(f \circ g)(-3)$
$= 5-2(-3)$ {from **a**}
$= 11$

2 $(f \circ g)(x) = f(g(x))$
$= f(2-x)$
$= (2-x)^2$

Domain is $\{x \mid x \in \mathbb{R}\}$
Range is $\{y \mid y \geqslant 0\}$

$(g \circ f)(x) = g(f(x))$
$= g(x^2)$
$= 2-x^2$

Domain is $\{x \mid x \in \mathbb{R}\}$
Range is $\{y \mid y \leqslant 2\}$

3 **a** $(f \circ g)(x)$
$= f(g(x))$
$= f(3-x)$
$= (3-x)^2+1$
$= 9-6x+x^2+1$
$= x^2-6x+10$

b $(g \circ f)(x)$
$= g(f(x))$
$= g(x^2+1)$
$= 3-(x^2+1)$
$= 3-x^2-1$
$= 2-x^2$

c $(g \circ f)(x) = f(x)$
$\therefore\ 2-x^2 = f(x)$ {from **b**}
$\therefore\ 2-x^2 = x^2+1$
$\therefore\ 2x^2 = 1$
$\therefore\ x^2 = \frac{1}{2}$
$\therefore\ x = \pm\frac{1}{\sqrt{2}}$

4 **a** $ax+b = cx+d$ is true for all x {given}
When $x=0$, $a(0)+b = c(0)+d$
$\therefore\ b = d$ .... $(*)$

When $x=1$, $a(1)+b = c(1)+d$
$\therefore\ a+b = c+d$
But from $(*)$, $b=d$, so $a+d = c+d$
$\therefore\ a = c$

b $(f \circ g)(x) = x$ for all x {given}
$\therefore\ f(g(x)) = x$
$\therefore\ f(ax+b) = x$
$\therefore\ 2(ax+b)+3 = x$
$\therefore\ 2ax+2b+3 = x$ for all x
$\therefore\ 2a = 1$ and $2b+3 = 0$ {using **a**}
$\therefore\ a = \frac{1}{2}$ and $2b = -3$
So, $a = \frac{1}{2}$ and $b = -\frac{3}{2}$ as required.

c If $(g \circ f)(x) = x$
then $g(f(x)) = x$
$\therefore\ g(2x+3) = x$
$\therefore\ a(2x+3)+b = x$
$\therefore\ 2ax+3a+b = x$
$\therefore\ 2a = 1$ and $3a+b = 0$ {using **a**}
$\therefore\ a = \frac{1}{2}$ and $b = -3a$
So, $a = \frac{1}{2}$ and $b = -\frac{3}{2}$
$\therefore$ the result in **b** is also true if $(g \circ f)(x) = x$ for all x.

EXERCISE 1E

1 **a**

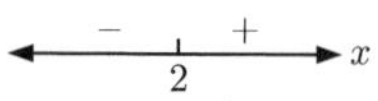

b

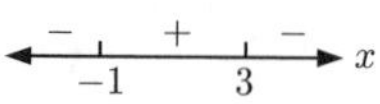

c

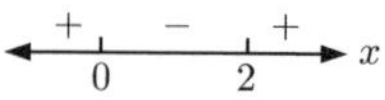

d

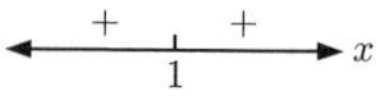

e

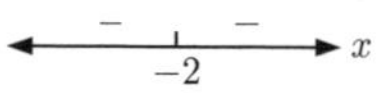

f

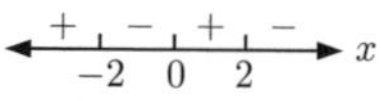

g

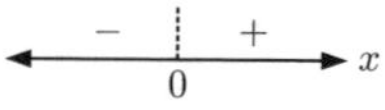

h

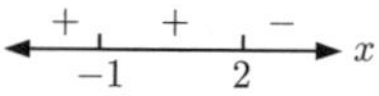

i

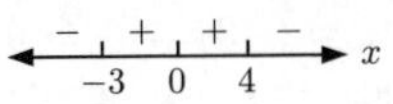

j

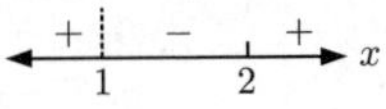

k

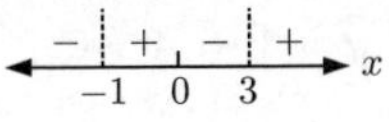

l

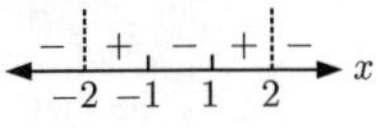

2 **a** $y = (x+4)(x-2)$ is zero when $x = -4$ or 2.
When $x = 0$, $y = (4)(-2) = -8 < 0$.
The factors are single, so the signs alternate.
$\therefore$ sign diagram is:

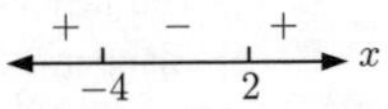

b $y = x(x-3)$ is zero when $x = 0$ or 3.
When $x = 10$, $y = 10(7) = 70 > 0$.
The factors are single, so the signs alternate.
$\therefore$ sign diagram is:

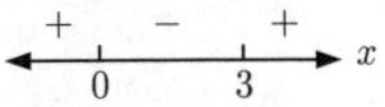

c $y = x(x+2)$ is zero when $x = -2$ or 0.
When $x = 10$, $y = 10(12) = 120 > 0$.
The factors are single, so the signs alternate.
$\therefore$ sign diagram is:

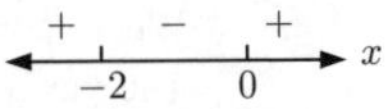

d $y = -(x+1)(x-3)$ is zero when $x = -1$ or 3.
When $x = 0$, $y = -(1)(-3) = 3 > 0$.
The factors are single, so the signs alternate.
$\therefore$ sign diagram is:

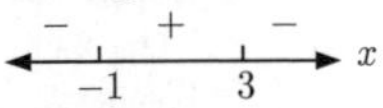

e $y = (2x-1)(3-x)$ is zero when $x = \frac{1}{2}$ or 3.
When $x = 0$, $y = (-1)(3) = -3 < 0$.
The factors are single, so the signs alternate.
$\therefore$ sign diagram is:

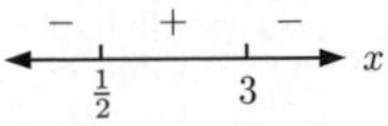

f $y = (5-x)(1-2x)$ is zero when $x = \frac{1}{2}$ or 5.
When $x = 0$, $y = (5)(1) = 5 > 0$.
The factors are single, so the signs alternate.
$\therefore$ sign diagram is:

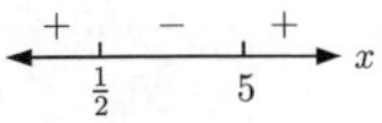

g $y = x^2 - 9 = (x+3)(x-3)$ is zero when $x = -3$ or 3.
When $x = 0$, $y = (3)(-3) = -9 < 0$.
The factors are single, so the signs alternate.
$\therefore$ sign diagram is:

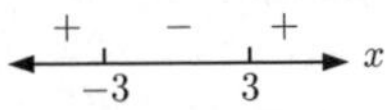

h $y = 4 - x^2 = (2+x)(2-x)$ is zero when $x = -2$ or 2.
When $x = 0$, $y = (2)(2) = 4 > 0$.
The factors are single, so the signs alternate.
$\therefore$ sign diagram is:

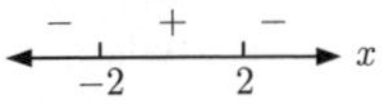

i $y = 5x - x^2 = x(5-x)$ is zero when $x = 0$ or 5.
When $x = 10$, $y = 10(-5) = -50 < 0$.
The factors are single, so the signs alternate.
$\therefore$ sign diagram is:

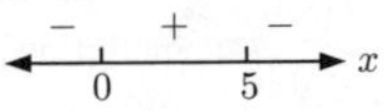

j $y = x^2 - 3x + 2 = (x-1)(x-2)$ is zero when $x = 1$ or 2.
When $x = 0$, $y = (-1)(-2) = 2 > 0$.
The factors are single, so the signs alternate.
$\therefore$ sign diagram is:

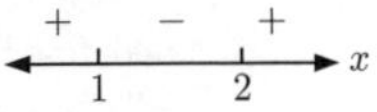

k $y = 2 - 8x^2 = 2(1+2x)(1-2x)$ is zero when $x = -\frac{1}{2}$ or $\frac{1}{2}$.
When $x = 0$, $y = 2(1)(1) = 2 > 0$.
The factors are single, so the signs alternate.
$\therefore$ sign diagram is:

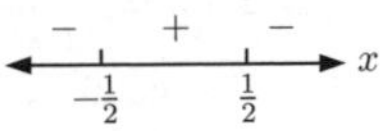

l $y = 6x^2 + x - 2 = (3x+2)(2x-1)$ is zero when $x = -\frac{2}{3}$ or $\frac{1}{2}$.
When $x = 0$, $y = (2)(-1) = -2 < 0$.
The factors are single, so the signs alternate.
$\therefore$ sign diagram is:

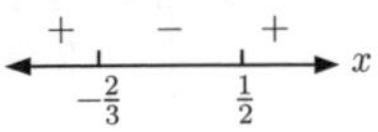

m $y = 6 - 16x - 6x^2 = 2(3+x)(1-3x)$ is zero when $x = -3$ or $\frac{1}{3}$.
When $x = 0$, $y = 2(3)(1) = 6 > 0$.
The factors are single, so the signs alternate.
$\therefore$ sign diagram is:

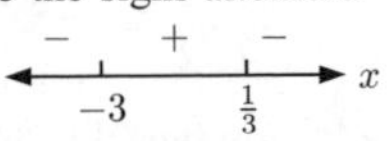

n $y = -2x^2 + 9x + 5 = (2x+1)(5-x)$ is zero when $x = -\frac{1}{2}$ or 5.
When $x = 0$, $y = (1)(5) = 5 > 0$.
The factors are single, so the signs alternate.
$\therefore$ sign diagram is:

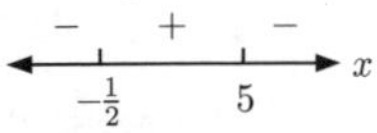

o $y = -15x^2 - x + 2 = (5x+2)(1-3x)$ is zero when $x = -\frac{2}{5}$ or $\frac{1}{3}$.
When $x = 0$, $y = (2)(1) = 2 > 0$.
The factors are single, so the signs alternate.
$\therefore$ sign diagram is:

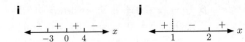

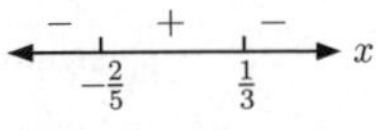

3 **a** $y=(x+2)^2$ is zero when $x=-2$.
When $x=0$, $y=2^2=4>0$.
The factor is squared, so the sign does not change.
$\therefore$ sign diagram is:

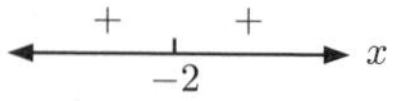

b $y=(x-3)^2$ is zero when $x=3$.
When $x=0$, $y=(-3)^2=9>0$.
The factor is squared, so the sign does not change.
$\therefore$ sign diagram is:

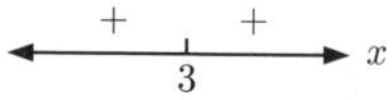

c $y=-(x+2)^2$ is zero when $x=-2$.
When $x=0$, $y=-(2^2)=-4<0$.
The factor is squared, so the sign does not change.
$\therefore$ sign diagram is:

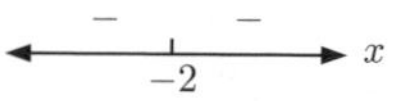

d $y=-(x-4)^2$ is zero when $x=4$.
When $x=0$, $y=-(-4)^2=-16<0$.
The factor is squared, so the sign does not change.
$\therefore$ sign diagram is:

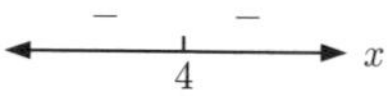

e $y=x^2-2x+1=(x-1)^2$ is zero when $x=1$.
When $x=0$, $y=(-1)^2=1>0$.
The factor is squared, so the sign does not change.
$\therefore$ sign diagram is:

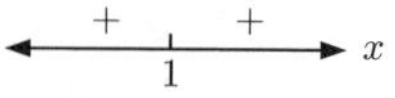

f $y=-x^2+4x-4=-(x-2)^2$ is zero when $x=2$.
When $x=0$, $y=-(-2)^2=-4<0$.
The factor is squared, so the sign does not change.
$\therefore$ sign diagram is:

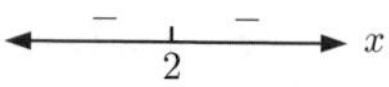

g $y=4x^2-4x+1=(2x-1)^2$ is zero when $x=\frac{1}{2}$.
When $x=0$, $y=(-1)^2=1>0$.
The factor is squared, so the sign does not change.
$\therefore$ sign diagram is:

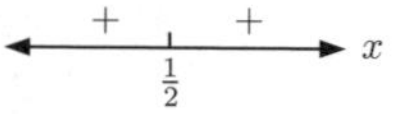

h $y=-x^2-6x-9=-(x+3)^2$ is zero when $x=-3$.
When $x=0$, $y=-(3^2)=-9<0$.
The factor is squared, so the sign does not change.
$\therefore$ sign diagram is:

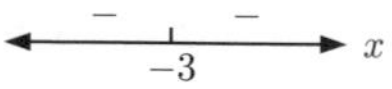

i $y=-4x^2+12x-9=-(2x-3)^2$ is zero when $x=\frac{3}{2}$.
When $x=0$, $y=-(-3)^2=-9<0$.
The factor is squared, so the sign does not change.

$\therefore$ sign diagram is:

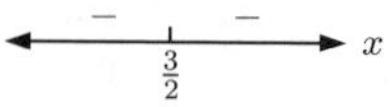

4 **a** $y=\dfrac{x+2}{x-1}$ is zero when $x=-2$ and undefined when $x=1$.
When $x=0$, $y=\frac{2}{-1}=-2<0$.
Since the factors are single, the signs alternate.

$\therefore$ sign diagram is:

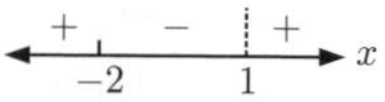

b $y=\dfrac{x}{x+3}$ is zero when $x=0$ and undefined when $x=-3$.
When $x=10$, $y=\frac{10}{13}>0$.
Since the factors are single, the signs alternate.

$\therefore$ sign diagram is:

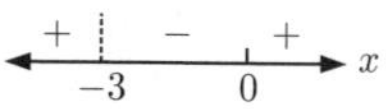

c $y=\dfrac{2x+3}{4-x}$ is zero when $x=-\frac{3}{2}$ and undefined when $x=4$.
When $x=0$, $y=\frac{3}{4}>0$.
Since the factors are single, the signs alternate.

$\therefore$ sign diagram is:

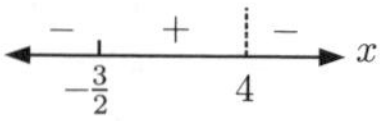

d $y=\dfrac{4x-1}{2-x}$ is zero when $x=\frac{1}{4}$ and undefined when $x=2$.
When $x=0$, $y=\frac{-1}{2}=-\frac{1}{2}<0$.
Since the factors are single, the signs alternate.

$\therefore$ sign diagram is:

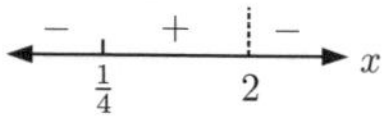

e $y=\dfrac{3x}{x-2}$ is zero when $x=0$ and undefined when $x=2$.
When $x=5$, $y=\frac{15}{3}=5>0$.
Since the factors are single, the signs alternate.

$\therefore$ sign diagram is:

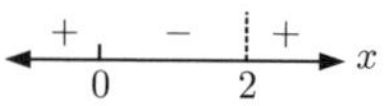

f $y=\dfrac{-8x}{3-x}$ is zero when $x=0$ and undefined when $x=3$.
When $x=5$, $y=\frac{-40}{-2}=20>0$.
Since the factors are single, the signs alternate.

$\therefore$ sign diagram is:

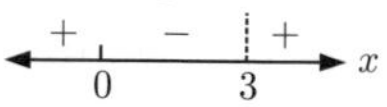

g $y = \dfrac{(x-1)^2}{x}$ is zero when $x = 1$ and undefined when $x = 0$.
When $x = 2$, $y = \frac{1^2}{2} = \frac{1}{2} > 0$.
Since the $(x-1)$ factor is squared, the sign does not change at $x = 1$.
$\therefore$ sign diagram is:

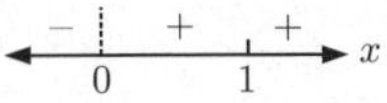

h $y = \dfrac{4x}{(x+1)^2}$ is zero when $x = 0$ and undefined when $x = -1$.
When $x = 1$, $y = \frac{4}{2^2} = 1 > 0$.
Since the $(x+1)$ factor is squared, the sign does not change at $x = -1$.
$\therefore$ sign diagram is:

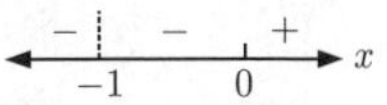

i $y = \dfrac{(x+2)(x-1)}{3-x}$ is zero when $x = -2$ or 1 and undefined when $x = 3$.
When $x = 0$, $y = \frac{(2)(-1)}{3} = -\frac{2}{3} < 0$.
Since the factors are single, the signs alternate.
$\therefore$ sign diagram is:

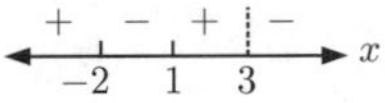

j $y = \dfrac{x(x-1)}{2-x}$ is zero when $x = 0$ or 1 and undefined when $x = 2$
When $x = 3$, $y = \frac{3(2)}{-1} = -6 < 0$.
Since the factors are single, the signs alternate.
$\therefore$ sign diagram is:

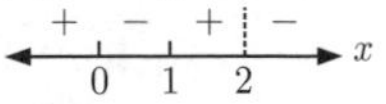

k $y = \dfrac{x^2-4}{-x} = \dfrac{(x-2)(x+2)}{-x}$ is zero when $x = \pm 2$ and undefined when $x = 0$.
When $x = 1$, $y = \dfrac{(-1)(3)}{-1} = 3 > 0$.
Since the factors are single, the signs alternate.
$\therefore$ sign diagram is:

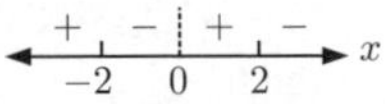

l $y = \dfrac{3-x}{2x^2-x-6} = \dfrac{3-x}{(2x+3)(x-2)}$ is zero when $x = 3$ and undefined when $x = -\frac{3}{2}$ or 2.
When $x = 0$, $y = \frac{3}{-6} = -\frac{1}{2} < 0$.
Since the factors are single, the signs alternate.
$\therefore$ sign diagram is:

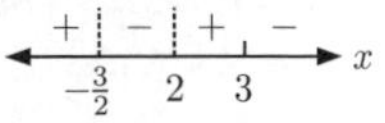

EXERCISE 1F

1

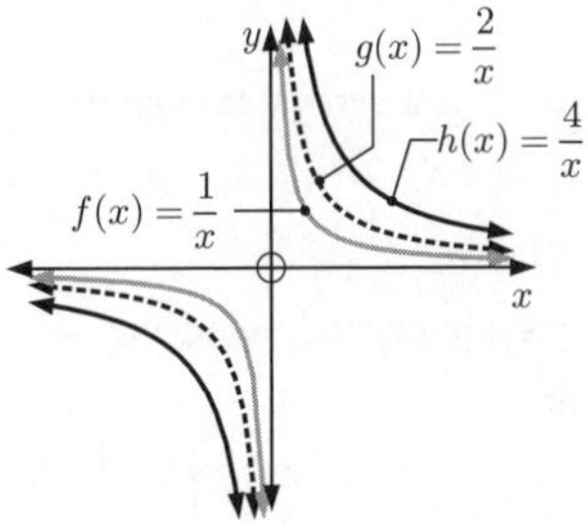

$f(x)$, $g(x)$ and $h(x)$ are all reciprocal functions which are all asymptotic about the x- and y-axes.
The graphs all lie in the 1st and 3rd quadrants.
The smaller the numerator, the closer is the graph to the axes.
Thus the graph of $f(x) = \dfrac{1}{x}$ is closer to the axes than $g(x) = \dfrac{2}{x}$ for corresponding values of x, and $g(x) = \dfrac{2}{x}$ is closer to the axes than $h(x) = \dfrac{4}{x}$.

2

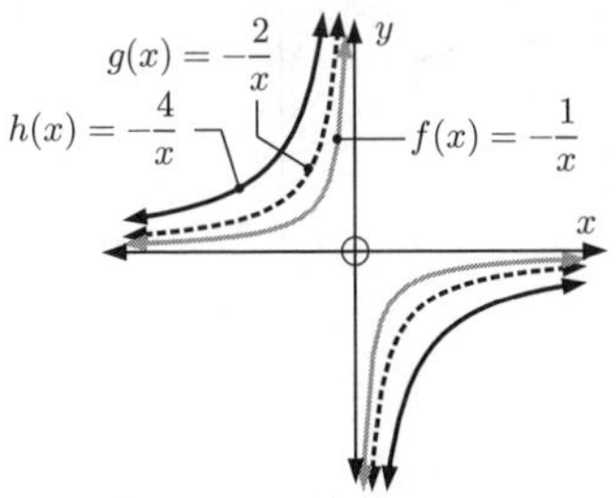

$f(x)$, $g(x)$ and $h(x)$ are all reciprocal functions which are all asymptotic about the x- and y-axes.
The graphs all lie in the 2nd and 4th quadrants.
The smaller the numerator, the closer is the graph to the axes.
Thus the graph of $f(x) = -\dfrac{1}{x}$ is closer to the axes than $g(x) = -\dfrac{2}{x}$ for corresponding values of x, and $g(x) = -\dfrac{2}{x}$ is closer to the axes than $h(x) = -\dfrac{4}{x}$.

EXERCISE 1G

1 **a** **i** $f : x \mapsto \dfrac{3}{x-2}$ is undefined when $x = 2$, so $x = 2$ is a vertical asymptote.
As $|x| \to \infty$, $f(x) \to 0$, so $y = 0$ is a horizontal asymptote.

ii As $x \to 2^-$, $y \to -\infty$.
As $x \to 2^+$, $y \to \infty$.
As $x \to \infty$, $y \to 0^+$.
As $x \to -\infty$, $y \to 0^-$.

iii $f(0) = \dfrac{3}{0-2} = -\frac{3}{2}$

So, the y-intercept is $-\frac{3}{2}$.

$f(x) = 0$ when $\dfrac{3}{x-2} = 0$, which has no solutions.

$\therefore$ there is no x-intercept.

iv

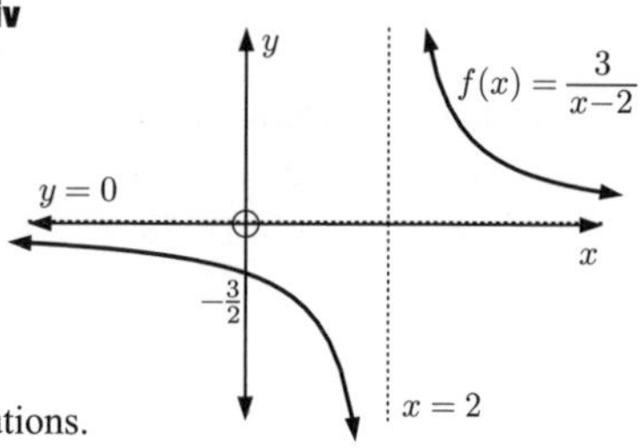

b **i** $f(x) = 2 - \dfrac{3}{x+1}$ is undefined when $x = -1$, so $x = -1$ is a vertical asymptote.

As $|x| \to \infty$, $\dfrac{3}{x+1} \to 0$, so $f(x) \to 2$ $\quad\therefore$ $y = 2$ is a horizontal asymptote.

ii As $x \to -1^-$, $y \to \infty$.
As $x \to -1^+$, $y \to -\infty$.
As $x \to \infty$, $y \to 2^-$.
As $x \to -\infty$, $y \to 2^+$.

iii $f(0) = 2 - \dfrac{3}{0+1} = -1$

So, the y-intercept is -1

$f(x) = 0$ when $2 - \dfrac{3}{x+1} = 0$

$\therefore \dfrac{3}{x+1} = 2$

$\therefore x + 1 = \frac{3}{2}$

$\therefore x = \frac{1}{2}$ So, the x-intercept is $\frac{1}{2}$.

iv

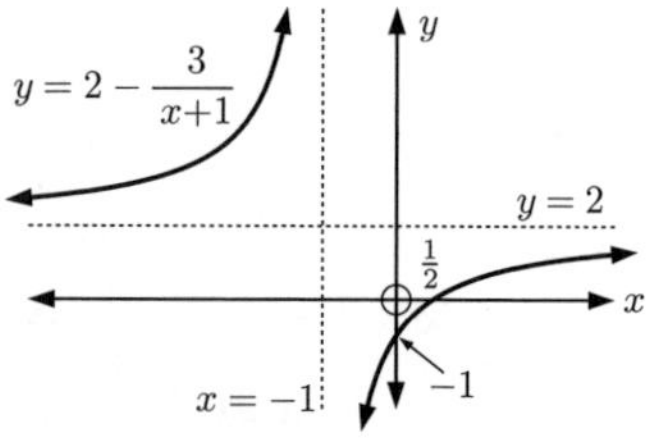

c **i** $f : x \mapsto \dfrac{x+3}{x-2}$ is undefined when $x = 2$, so $x = 2$ is a vertical asymptote.

Now $f(x) = \dfrac{x+3}{x-2} = \dfrac{1 + \frac{3}{x}}{1 - \frac{2}{x}}$

$\therefore$ as $|x| \to \infty$, $f(x) \to \frac{1}{1} = 1$, and so $y = 1$ is a horizontal asymptote.

ii As $x \to 2^-$, $y \to -\infty$.
As $x \to 2^+$, $y \to \infty$.
As $x \to \infty$, $y \to 1^+$.
As $x \to -\infty$, $y \to 1^-$.

iii $f(0) = \dfrac{0+3}{0-2} = -\frac{3}{2}$

So, the y-intercept is $-\frac{3}{2}$.

$f(x) = 0$ when $\dfrac{x+3}{x-2} = 0$

$\therefore x + 3 = 0$

$\therefore x = -3$

So, the x-intercept is -3.

iv

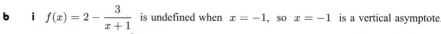

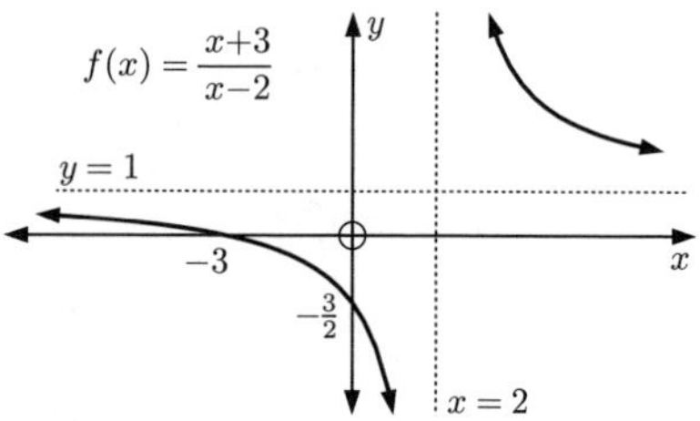

d **i** $f(x) = \dfrac{3x-1}{x+2}$ is undefined when $x = -2$, so $x = -2$ is a vertical asymptote.

$f(x) = \dfrac{3x-1}{x+2} = \dfrac{3 - \frac{1}{x}}{1 + \frac{2}{x}}$

As $|x| \to \infty$, $f(x) \to \frac{3}{1} = 3$ and so $y = 3$ is a horizontal asymptote.

ii As $x \to -2^-$, $y \to \infty$.
As $x \to -2^+$, $y \to -\infty$.
As $x \to \infty$, $y \to 3^-$.
As $x \to -\infty$, $y \to 3^+$.

iii $f(0) = \dfrac{3(0)-1}{0+2} = -\frac{1}{2}$

So, the y-intercept is $-\frac{1}{2}$.

$f(x) = 0$ when $\dfrac{3x-1}{x+2} = 0$

$\therefore \quad 3x - 1 = 0$

$\therefore \quad x = \frac{1}{3}$ So, the x-intercept is $\frac{1}{3}$.

iv

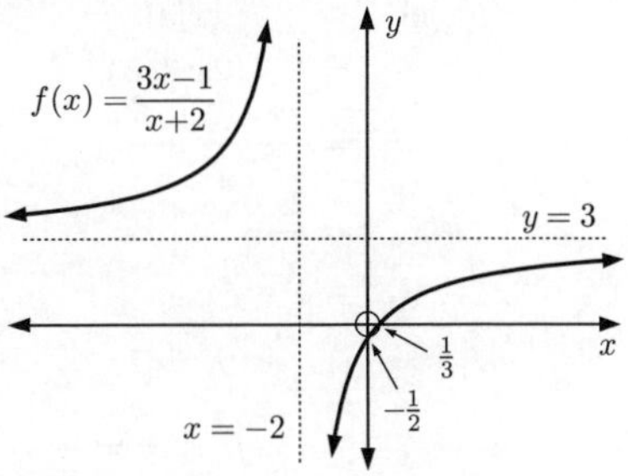

EXERCISE 1H

1 a i

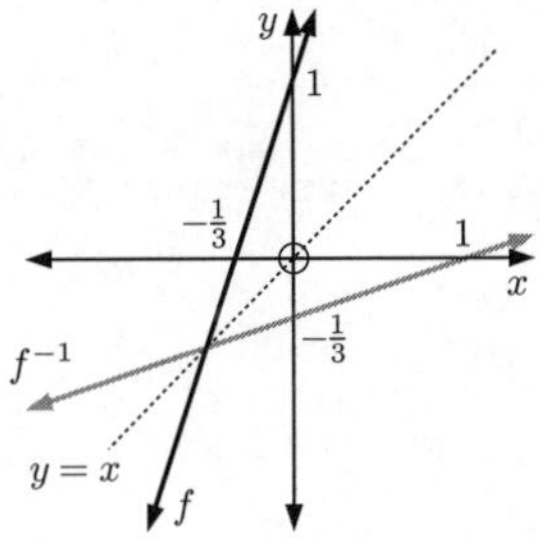

ii $f(x)$ passes through $(0, 1)$ and $(-\frac{1}{3}, 0)$

$\therefore \quad f^{-1}(x)$ passes through $(1, 0)$ and $(0, -\frac{1}{3})$

$f^{-1}(x)$ has gradient $\dfrac{-\frac{1}{3}-0}{0-1} = \dfrac{-\frac{1}{3}}{-1} = \frac{1}{3}$

So, its equation is $\dfrac{y-0}{x-1} = \frac{1}{3}$

which is $y = \dfrac{x-1}{3}$.

So, $f^{-1}(x) = \dfrac{x-1}{3}$

iii f is $y = 3x + 1$

so f^{-1} is $x = 3y + 1$

$\therefore \quad x - 1 = 3y$

$\therefore \quad y = \dfrac{x-1}{3}$. So, $f^{-1}(x) = \dfrac{x-1}{3}$

b i

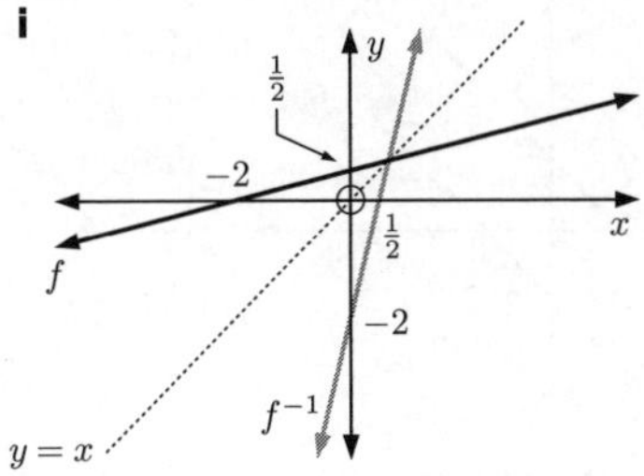

ii $f(x)$ passes through $(0, \frac{1}{2})$ and $(-2, 0)$

$\therefore \quad f^{-1}(x)$ passes through $(\frac{1}{2}, 0)$ and $(0, -2)$

$f^{-1}(x)$ has gradient $\dfrac{-2-0}{0-\frac{1}{2}} = \dfrac{-2}{-\frac{1}{2}} = 4$

So, its equation is $\dfrac{y-0}{x-\frac{1}{2}} = 4$

which is $y = 4x - 2$.

So, $f^{-1}(x) = 4x - 2$

iii f is $y = \dfrac{x+2}{4}$

so f^{-1} is $x = \dfrac{y+2}{4}$

$\therefore \quad 4x = y + 2$

$\therefore \quad y = 4x - 2$. So, $f^{-1}(x) = 4x - 2$

2 a i f is $y = 2x + 5$

so f^{-1} is $x = 2y + 5$

$\therefore \quad x - 5 = 2y$

$\therefore \quad y = \dfrac{x-5}{2}$

So, $f^{-1}(x) = \dfrac{x-5}{2}$

ii

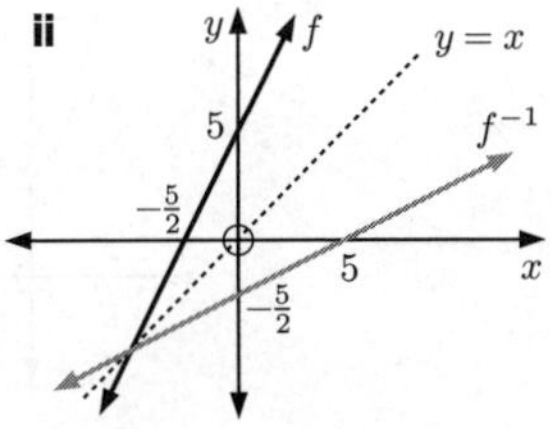

$f(x)$ passes through $(0, 5)$ and $(-\frac{5}{2}, 0)$

$\therefore \quad f^{-1}(x)$ passes through $(5, 0)$ and $(0, -\frac{5}{2})$.

iii $(f^{-1} \circ f)(x)$
$= f^{-1}(2x+5)$
$= \dfrac{2x+5-5}{2}$
$= \dfrac{2x}{2}$
$= x$

and

$(f \circ f^{-1})(x)$
$= f(f^{-1}(x))$
$= f\left(\dfrac{x-5}{2}\right)$
$= 2\left(\dfrac{x-5}{2}\right)+5$
$= x-5+5$
$= x$

b **i** f is $y = \dfrac{3-2x}{4}$
so f^{-1} is $x = \dfrac{3-2y}{4}$
$\therefore\ 4x = 3-2y$
$\therefore\ 4x-3 = -2y$
$\therefore\ y = -2x + \frac{3}{2}$
So, $f^{-1}(x) = -2x + \frac{3}{2}$

ii

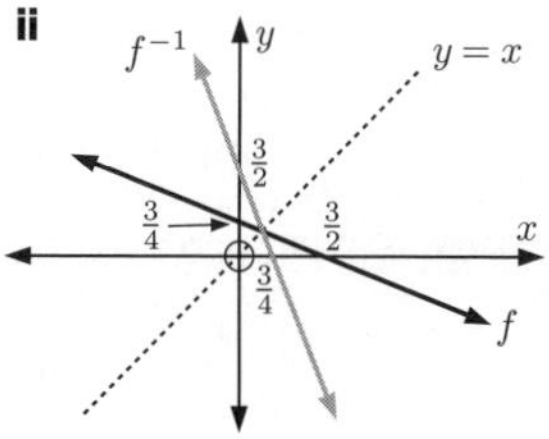

$f(x)$ passes through $(0, \frac{3}{4})$ and $(\frac{3}{2}, 0)$
$\therefore\ f^{-1}(x)$ passes through $(\frac{3}{4}, 0)$ and $(0, \frac{3}{2})$.

iii $(f^{-1} \circ f)(x)$
$= f^{-1}(f(x))$
$= f^{-1}\left(\dfrac{3-2x}{4}\right)$
$= -2\left(\dfrac{3-2x}{4}\right) + \frac{3}{2}$
$= \dfrac{3-2x}{-2} + \frac{3}{2}$
$= -\frac{3}{2} + x + \frac{3}{2}$
$= x$

and

$(f \circ f^{-1})(x)$
$= f(f^{-1}(x))$
$= f\left(-2x + \frac{3}{2}\right)$
$= \dfrac{3-2(-2x+\frac{3}{2})}{4}$
$= \dfrac{3+4x-3}{4}$
$= \dfrac{4x}{4}$
$= x$

c **i** f is $y = x+3$
so f^{-1} is $x = y+3$
$\therefore\ y = x-3$
So, $f^{-1}(x) = x-3$

ii

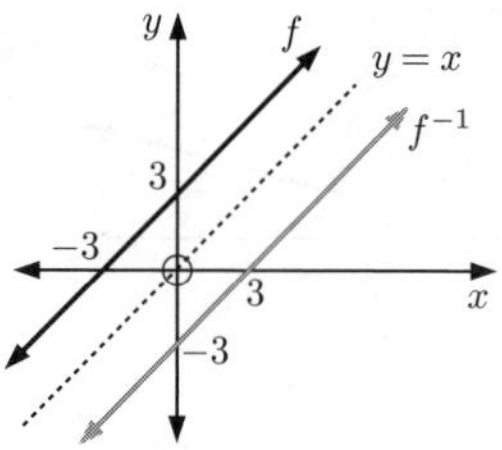

$f(x)$ passes through $(0, 3)$ and $(-3, 0)$
$\therefore\ f^{-1}(x)$ passes through $(3, 0)$ and $(0, -3)$.

iii $(f^{-1} \circ f)(x) = f^{-1}(f(x))$
$= f^{-1}(x+3)$
$= (x+3)-3$
$= x$

and

$(f \circ f^{-1})(x) = f(f^{-1}(x))$
$= f(x-3)$
$= (x-3)+3$
$= x$

3 **a**

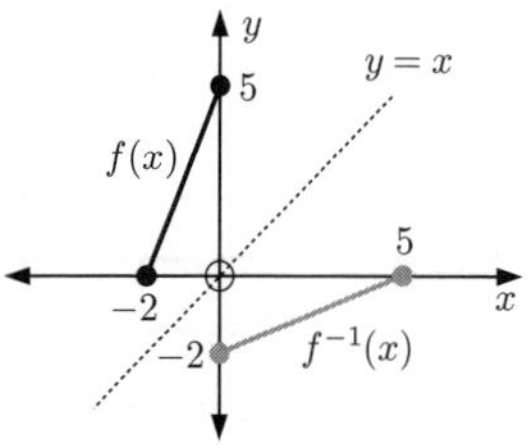

b

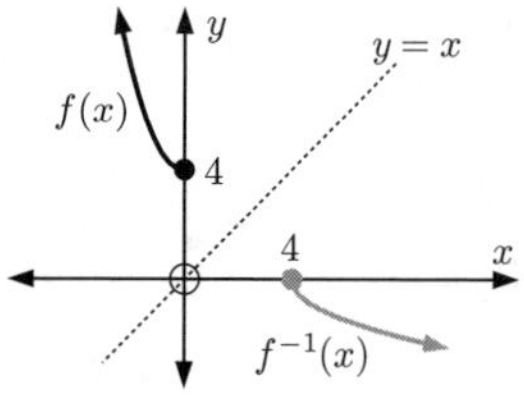

c

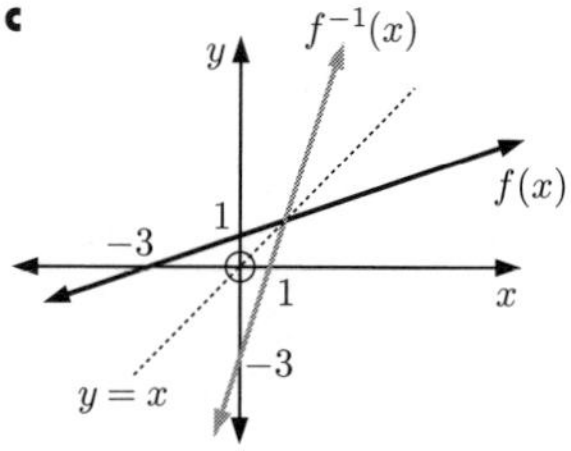

d

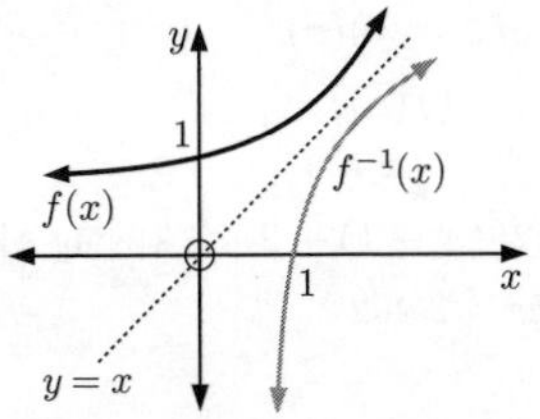

e

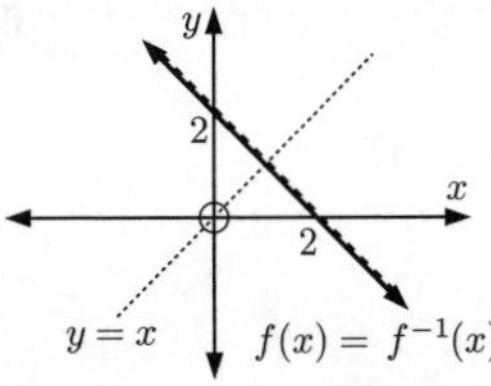

f

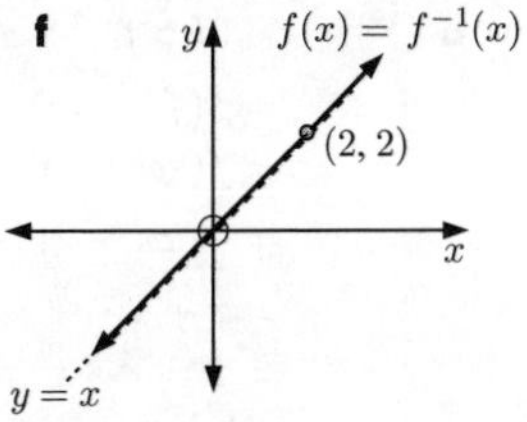

4 **a** Domain of $f(x)$ is $\{x \mid -2 \leqslant x \leqslant 0\}$ **b** Range of $f(x)$ is $\{y \mid 0 \leqslant y \leqslant 5\}$

c Domain of $f^{-1}(x)$ is $\{x \mid 0 \leqslant x \leqslant 5\}$ **d** Range of $f^{-1}(x)$ is $\{y \mid -2 \leqslant y \leqslant 0\}$

5 Range of $H^{-1}(x)$ is $\{y \mid -2 \leqslant y < 3\}$

6

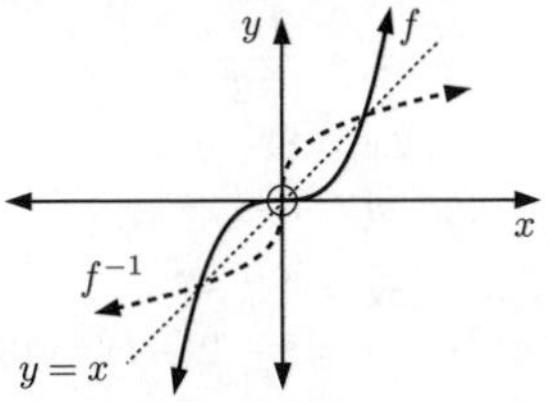

7 **a** $f(x) = \dfrac{1}{x}$ has graph

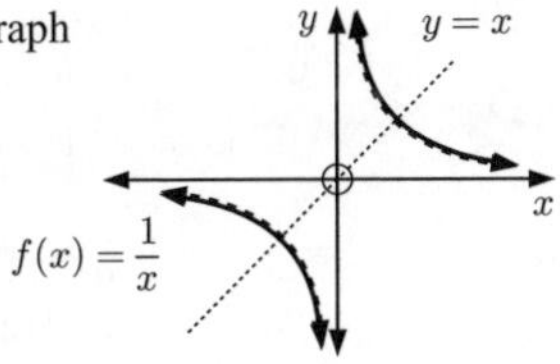

No vertical line cuts the graph more than once, so it is a function. Its reflection in the line $y = x$ also passes the vertical line test. Hence, $f(x) = \dfrac{1}{x}$, $x \neq 0$ has an inverse function.

b $f(x) = \dfrac{1}{x}$ has inverse function $x = \dfrac{1}{y}$ or $y = \dfrac{1}{x}$

So, $f^{-1}(x) = \dfrac{1}{x}$, which means $f(x)$ is a self-inverse function.

8 **a** $f(x) = \dfrac{3x-8}{x-3}$ has graph

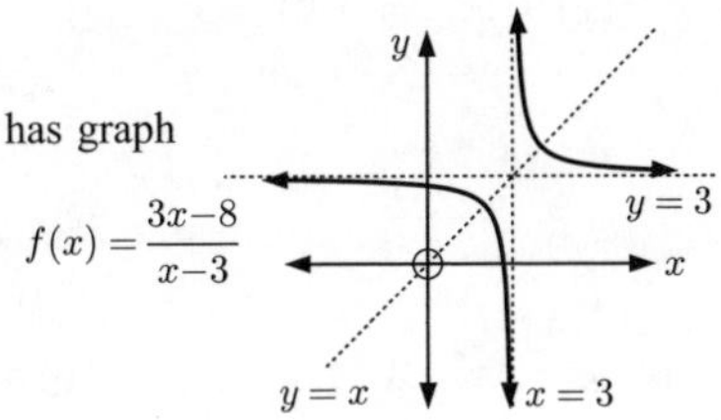

The vertical line test shows it to be a function. Symmetry about $y = x$ shows it is a self-inverse function.

b $f(x) = \dfrac{3x-8}{x-3}$ has inverse function $x = \dfrac{3y-8}{y-3}$

$$\therefore \quad x(y-3) = 3y - 8$$
$$\therefore \quad xy - 3x = 3y - 8$$
$$\therefore \quad xy - 3y = 3x - 8$$
$$\therefore \quad y(x-3) = 3x - 8$$
$$\therefore \quad y = \frac{3x-8}{x-3}$$
$$\therefore \quad f^{-1}(x) = \frac{3x-8}{x-3}$$

So, $f(x) = f^{-1}(x)$, which means $f(x)$ is a self-inverse function.

9 **a** $f(x) = \frac{1}{2}x - 1$ has inverse function $x = \frac{1}{2}y - 1$

$$\therefore \quad x + 1 = \tfrac{1}{2}y$$
$$\therefore \quad y = 2x + 2$$

So, $f^{-1}(x) = 2x + 2$

b **i**
$$\begin{aligned}&(f\circ f^{-1})(x)\\&=f(f^{-1}(x))\\&=f(2x+2)\quad\{\text{using } \mathbf{a}\}\\&=\tfrac{1}{2}(2x+2)-1\\&=x+1-1\\&=x\end{aligned}$$

ii
$$\begin{aligned}&(f^{-1}\circ f)(x)\\&=f^{-1}(f(x))\\&=f^{-1}(\tfrac{1}{2}x-1)\\&=2(\tfrac{1}{2}x-1)+2\quad\{\text{using } \mathbf{a}\}\\&=x-2+2\\&=x\end{aligned}$$

10 **a**
$$\begin{aligned}g \text{ is } y&=\frac{8-x}{2}\\ \text{so } g^{-1} \text{ is } x&=\frac{8-y}{2}\\ \therefore\ 2x&=8-y\\ \therefore\ y&=8-2x\\ \text{So, } g^{-1}(x)&=8-2x\\ \therefore\ g^{-1}(-1)&=8-2(-1)=10\end{aligned}$$

b
$$\begin{aligned}(f\circ g^{-1})(x)&=9\\ \therefore\ f(g^{-1}(x))&=9\\ \therefore\ f(8-2x)&=9\\ \therefore\ 2(8-2x)+5&=9\\ \therefore\ 16-4x+5&=9\\ \therefore\ -4x&=-12\\ \therefore\ x&=3\end{aligned}$$

11 **a** **i**
$$\begin{aligned}f \text{ is } y&=5^x\\ \text{so } f(2)&=5^2\\&=25\end{aligned}$$

ii
$$\begin{aligned}g \text{ is } y&=\sqrt{x} \text{ where } y\geqslant 0\\ \text{so } g^{-1} \text{ is } x&=\sqrt{y} \text{ where } x\geqslant 0\\ \therefore\ y&=x^2\\ \therefore\ g^{-1}(x)&=x^2,\ x\geqslant 0\\ \therefore\ g^{-1}(4)&=4^2\\&=16\end{aligned}$$

b
$$\begin{aligned}(g^{-1}\circ f)(x)&=25\\ \therefore\ g^{-1}(f(x))&=25\\ \therefore\ g^{-1}(5^x)&=25\\ \therefore\ (5^x)^2&=25\quad\{\text{as } g^{-1}(x)=x^2,\ x\geqslant 0\}\\ \therefore\ 5^{2x}&=5^2\\ \therefore\ 2x&=2\\ \therefore\ x&=1\end{aligned}$$

12
$$\begin{aligned}f \text{ is } y&=2x\\ \text{so } f^{-1} \text{ is } x&=2y\\ \therefore\ y&=\frac{x}{2}\\ \therefore\ f^{-1}(x)&=\frac{x}{2}\end{aligned}$$

$$\begin{aligned}g \text{ is } y&=4x-3\\ \text{so } g^{-1} \text{ is } x&=4y-3\\ \therefore\ 4y&=x+3\\ \therefore\ y&=\frac{x+3}{4}\\ \therefore\ g^{-1}(x)&=\frac{x+3}{4}\end{aligned}$$

$$\begin{aligned}(g\circ f)(x)&=g(f(x))\\&=g(2x)\\&=4(2x)-3\\ \therefore\ (g\circ f)(x)&=8x-3\\ \therefore\ g\circ f \text{ is } y&=8x-3\\ \text{so } (g\circ f)^{-1} \text{ is } x&=8y-3\\ \therefore\ y&=\frac{x+3}{8}\\ \text{So, } (g\circ f)^{-1}(x)&=\frac{x+3}{8}\end{aligned}$$

$$\begin{aligned}\text{Now } (f^{-1}\circ g^{-1})(x)&=f^{-1}(g^{-1}(x))\\&=f^{-1}\left(\frac{x+3}{4}\right)\\&=\frac{\left(\frac{x+3}{4}\right)}{2}\\ \therefore\ (f^{-1}\circ g^{-1})(x)&=\frac{x+3}{8}=(g\circ f)^{-1}(x)\quad\text{as required}\end{aligned}$$

13 **a** f is $y = 2x$
so f^{-1} is $x = 2y$
$\therefore\ y = \dfrac{x}{2}$
so $f^{-1}(x) = \dfrac{x}{2} \neq 2x$
So, $f^{-1}(x) \neq f(x)$

b f is $y = x$
so f^{-1} is $x = y$
$\therefore\ y = x$
so $f^{-1}(x) = x$
So, $f^{-1}(x) = f(x)$

c f is $y = -x$
so f^{-1} is $x = -y$
$\therefore\ y = -x$
so $f^{-1}(x) = -x$
So, $f^{-1}(x) = f(x)$

d f is $y = \dfrac{2}{x}$
so f^{-1} is $x = \dfrac{2}{y}$
$\therefore\ y = \dfrac{2}{x}$
so $f^{-1}(x) = \dfrac{2}{x}$
So, $f^{-1}(x) = f(x)$

e f is $y = -\dfrac{6}{x}$
so f^{-1} is $x = -\dfrac{6}{y}$
$\therefore\ y = -\dfrac{6}{x}$
so $f^{-1}(x) = -\dfrac{6}{x}$
So, $f^{-1}(x) = f(x)$

So, $f^{-1}(x) = f(x)$ is true for parts **b**, **c**, **d** and **e**.

14 **a** If $y = f(x)$ has an inverse function, then the inverse function must also be a function. It must satisfy the 'vertical line test', so no vertical line can cut it more than once. This condition for the inverse function cannot be satisfied if the original function does not satisfy the 'horizontal line test'. Thus, the 'horizontal line test' is a valid test for the existence of an inverse function.

b **i** This graph satisfies the 'horizontal line test' and therefore has an inverse function.
ii, iii These graphs both fail the 'horizontal line test' so neither of these have inverse functions.

i

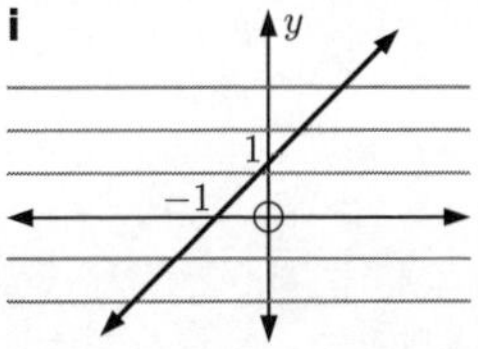

ii

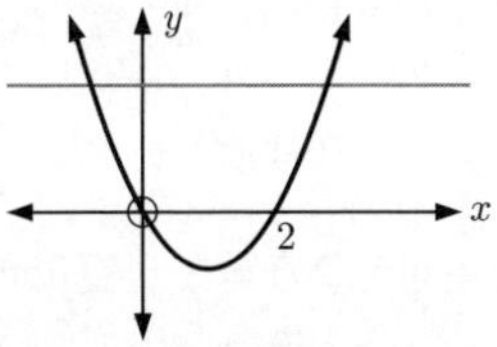

iii

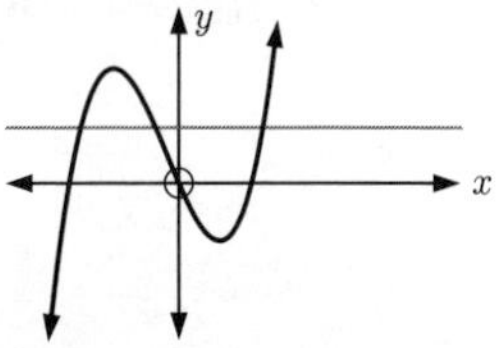

REVIEW SET 1A

1 **a** $f(x) = 2x - x^2$
$f(2) = 2(2) - 2^2$
$= 0$

b $f(-3) = 2(-3) - (-3)^2$
$= -6 - 9$
$= -15$

c $f(-\frac{1}{2}) = 2(-\frac{1}{2}) - (-\frac{1}{2})^2$
$= -1 - \frac{1}{4}$
$= -\frac{5}{4}$

2 $f(x) = ax + b$, where $f(1) = 7$ and $f(3) = -5$

When $f(1) = 7$,
$7 = a(1) + b$
$\therefore\ 7 = a + b$
$\therefore\ a = 7 - b$ .... (1)

When $f(3) = -5$,
$-5 = a(3) + b$
$\therefore\ -5 = 3a + b$
$\therefore\ -5 = 3(7 - b) + b$ {using (1)}
$\therefore\ -5 = 21 - 3b + b$
$\therefore\ 2b = 26$ and so $b = 13$

Substituting $b = 13$ into (1), $a = 7 - 13 = -6$

$\therefore\ a = -6$ and $b = 13$

3 $g(x) = x^2 - 3x$

a $g(x+1) = (x+1)^2 - 3(x+1)$
$= x^2 + 2x + 1 - 3x - 3$
$= x^2 - x - 2$

b $g(x^2 - 2) = (x^2 - 2)^2 - 3(x^2 - 2)$
$= x^4 - 4x^2 + 4 - 3x^2 + 6$
$= x^4 - 7x^2 + 10$

4 **a** **i** Range is $\{y \mid y \geqslant -5\}$. Domain is $\{x \mid x \in \mathbb{R}\}$.

ii x-intercepts are -1 and 5, y-intercept is $-\frac{25}{9}$

iii The graph passes the 'vertical line test' so is therefore a function.

b **i** Range is $\{y \mid y = 1 \text{ or } -3\}$. Domain is $\{x \mid x \in \mathbb{R}\}$.

ii there are no x-intercepts, y-intercept is 1

iii The graph passes the 'vertical line test' so is therefore a function.

5 **a** $y = (3x+2)(4-x)$ is zero when $x = -\frac{2}{3}$ or 4.

When $x = 0$, $y = (2)(4) = 8 > 0$.

Since the factors are single, the signs alternate.

$\therefore$ sign diagram is 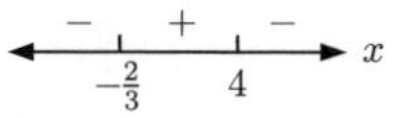

b $y = \dfrac{x-3}{x^2+4x+4} = \dfrac{x-3}{(x+2)^2}$ is zero when $x = 3$ and undefined when $x = -2$.

When $x = 0$, $y = \dfrac{-3}{2^2} = -\frac{3}{4} < 0$.

Since the $(x+2)$ factor is squared, the sign does not change at $x = -2$

$\therefore$ sign diagram is 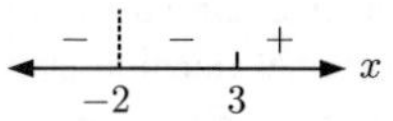

6 $f(x) = ax + b$

Now $f(2) = 1$, so $a(2) + b = 1$

$\therefore\ b = 1 - 2a$ .... $(*)$

Now $f^{-1}(3) = 4$, so $f(4) = 3$

$\therefore\ a(4) + b = 3$

$\therefore\ 4a + (1 - 2a) = 3$ {from $(*)$}

$\therefore\ 2a = 2$

$\therefore\ a = 1$

Substituting $a = 1$ into $(*)$, $b = 1 - 2(1) = -1$

So, $a = 1$ and $b = -1$.

7 **a**

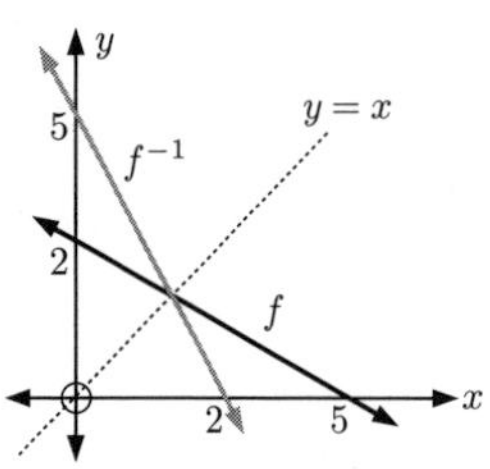

b

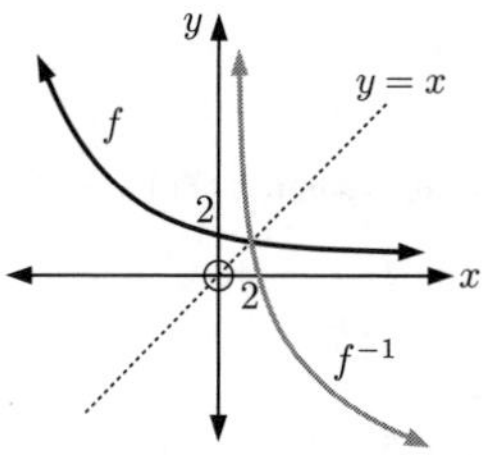

8 **a** f is $y = 4x + 2$

$\therefore\ f^{-1}(x)$ is $x = 4y + 2$

$\therefore\ y = \dfrac{x-2}{4}$

$\therefore\ f^{-1}(x) = \dfrac{x-2}{4}$

b f is $y = \dfrac{3-5x}{4}$

so $f^{-1}(x)$ is $x = \dfrac{3-5y}{4}$

$\therefore\ 4x = 3 - 5y$

$\therefore\ y = \dfrac{3-4x}{5}$

$\therefore\ f^{-1}(x) = \dfrac{3-4x}{5}$

9 f is $y = 3x + 6$
so $f^{-1}(x)$ is $x = 3y + 6$
$\therefore\ y = \dfrac{x-6}{3}$
$\therefore\ f^{-1}(x) = \dfrac{x-6}{3}$

h is $y = \dfrac{x}{3}$
so $h^{-1}(x)$ is $x = \dfrac{y}{3}$
$\therefore\ y = 3x$
$\therefore\ h^{-1}(x) = 3x$

Now $(f^{-1} \circ h^{-1})(x) = f^{-1}(h^{-1}(x))$
$= f^{-1}(3x)$
$= \dfrac{3x-6}{3}$
$= x - 2$

$(h \circ f)(x) = h(f(x))$
$= h(3x+6)$
$= \dfrac{3x+6}{3}$
$\therefore\ h \circ f$ is $y = x + 2$
so $(h \circ f)^{-1}(x)$ is $x = y + 2$
$\therefore\ y = x - 2$
$\therefore\ (h \circ f)^{-1}(x) = x - 2$

$\therefore\ (f^{-1} \circ h^{-1})(x) = (h \circ f)^{-1}(x)$ as required.

REVIEW SET 1B

1 **a** $y = (x-1)(x-5)$
$\therefore$ the x-intercepts are $x = 1$ and 5
The vertex is at $x = 3$, with $y = (3-1)(3-5) = 2 \times (-2) = -4$
$\therefore$ the vertex is at $(3, -4)$
The domain is $\{x \mid x \in \mathbb{R}\}$. The range is $\{y \mid y \geqslant -4\}$.

b From the graph, the domain is $\{x \mid x \neq 0, 2\}$ and the range is $\{y \mid y \leqslant -1 \text{ or } y > 0\}$.

2 **a** $(f \circ g)(x) = f(g(x))$
$= f(x^2 + 2)$
$= 2(x^2+2) - 3$
$= 2x^2 + 4 - 3$
$= 2x^2 + 1$

b $(g \circ f)(x) = g(f(x))$
$= g(2x-3)$
$= (2x-3)^2 + 2$
$= 4x^2 - 12x + 9 + 2$
$= 4x^2 - 12x + 11$

3 **a** $y = \dfrac{x^2 - 6x - 16}{x-3} = \dfrac{(x+2)(x-8)}{x-3}$ is zero when $x = -2$ or 8 and undefined when $x = 3$.
When $x = 0$, $y = \frac{-16}{-3} > 0$.
Since the factors are single, the signs alternate. So, the sign diagram is:

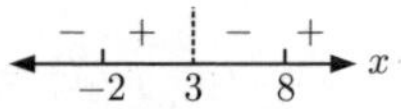

b $y = \dfrac{x+9}{x+5} + x\left(\dfrac{x+5}{x+5}\right) = \dfrac{x^2+6x+9}{x+5} = \dfrac{(x+3)^2}{x+5}$ is zero when $x = -3$
and undefined when $x = -5$.
When $x = 0$, $y = \dfrac{3^2}{5} > 0$. The $(x+3)$ factor is squared, so the sign does not change at $x = -3$.

So, the sign diagram is:

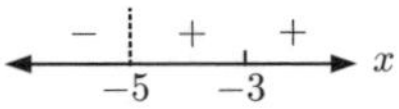

4 **a** $f(x) = \dfrac{1}{x^2}$ is meaningless when $x = 0$.

b

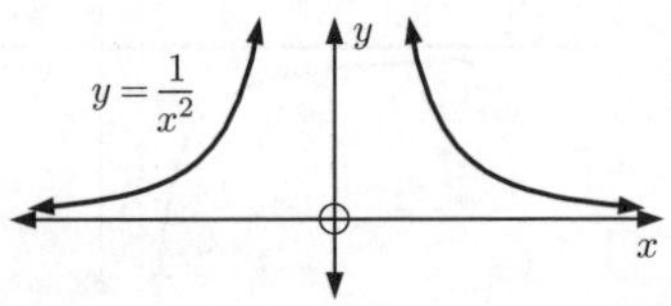

c Domain of $f(x)$ is $\{x \mid x \neq 0\}$.
Range of $f(x)$ is $\{y \mid y > 0\}$.

5 **a** $f(x) = \dfrac{ax+3}{x-b}$ has asymptotes $x=-1$, $y=2$ .... $(*)$

$f(x)$ is undefined when $x - b = 0$

$\therefore \; x = b$

But $x=-1$ is the vertical asymptote, so $b=-1$.

So, $f(x) = \dfrac{ax+3}{x-(-1)} = \dfrac{ax+3}{x+1} = \dfrac{a+\frac{3}{x}}{1+\frac{1}{x}}$

As $|x| \to \infty$, $f(x) \to \dfrac{a}{1} = a$ so the horizontal asymptote is $y=a$.

$\therefore$ using $(*)$, $a=2$.

So, $a=2$ and $b=-1$.

b Domain of f is $\{x \mid x \neq -1\}$ and range of f is $\{y \mid y \neq 2\}$.

$\therefore$ domain of f^{-1} is $\{x \mid x \neq 2\}$ and range of f^{-1} is $\{y \mid y \neq -1\}$.

6 **a** $f : x \mapsto \dfrac{4x+1}{2-x}$ is undefined when $x=2$

$\therefore \; x=2$ is a vertical asymptote.

Now $f(x) = \dfrac{4x+1}{2-x} = \dfrac{4+\frac{1}{x}}{-1+\frac{2}{x}}$

$\therefore$ as $|x| \to \infty$, $f(x) \to \frac{4}{-1} = -4$, and so $y=-4$ is a horizontal asymptote.

b As $x \to 2^-$, $y \to \infty$.
As $x \to 2^+$, $y \to -\infty$.
As $x \to \infty$, $y \to -4^-$.
As $x \to -\infty$, $y \to -4^+$.

c $f(0) = \dfrac{4(0)+1}{2-0} = \frac{1}{2}$

So, the y-intercept is $\frac{1}{2}$.

$f(x) = 0$ when $\dfrac{4x+1}{2-x} = 0$

$\therefore \; 4x+1 = 0$

$\therefore \; x = -\frac{1}{4}$

So, the x-intercept is $-\frac{1}{4}$.

d

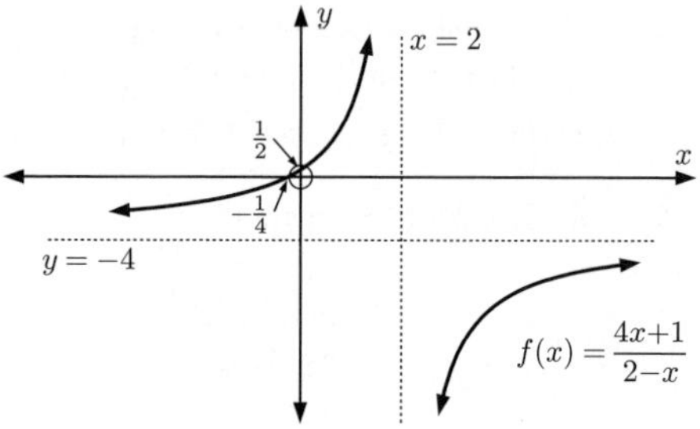

7 **a** $(g \circ f)(x) = g(f(x))$
$= g(3x+1)$
$= \dfrac{2}{3x+1}$

b $(g \circ f)(x) = -4$

$\therefore \; \dfrac{2}{3x+1} = -4$

$\therefore \; -4(3x+1) = 2$

$\therefore \; -12x - 4 = 2$

$\therefore \; 12x = -6$

$\therefore \; x = -\frac{1}{2}$

c **i** $h(x) = \dfrac{2}{3x+1}$ is undefined when

$3x+1=0$ or $x = -\frac{1}{3}$.

So, $x=-\frac{1}{3}$ is a vertical asymptote.

As $|x| \to \infty$, $h(x) \to 0$

$\therefore \; y=0$ is a horizontal asymptote.

ii

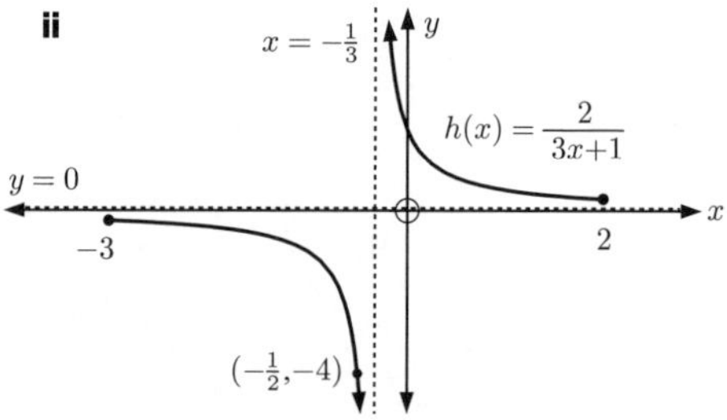

iii Domain of h is $\{x \mid x \neq -\frac{1}{3}\}$.

Range of h is $\{y \mid y \neq 0\}$.

8 **a**

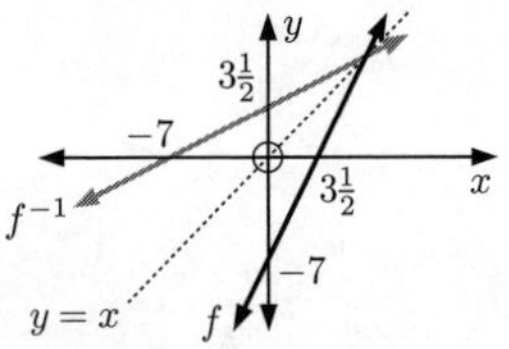

b The function f is $y = 2x - 7$

so f^{-1} is $x = 2y - 7$

$\therefore \quad y = \dfrac{x+7}{2}$

So, $f^{-1}(x) = \dfrac{x+7}{2}$

c

$$f \circ f^{-1} = f\left(f^{-1}(x)\right) = f\left(\frac{x+7}{2}\right) = 2\left(\frac{x+7}{2}\right) - 7 = x + 7 - 7 = x$$

and

$$f^{-1} \circ f = f^{-1}(f(x)) = f^{-1}(2x-7) = \frac{2x-7+7}{2} = \frac{2x}{2} = x$$

So, $f \circ f^{-1} = f^{-1} \circ f = x$

9 **a**

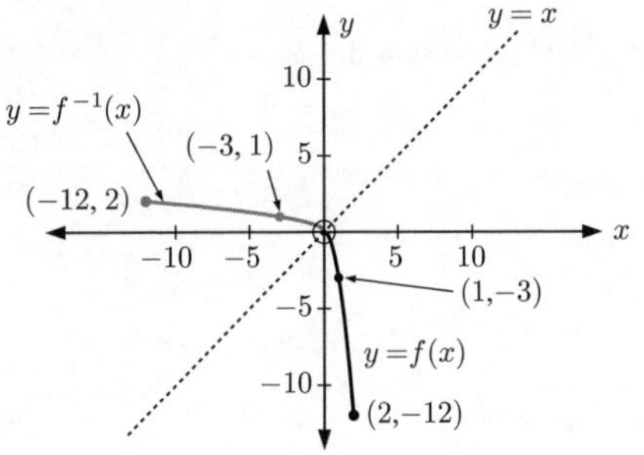

b Range of f^{-1} is $\{y \mid 0 \leqslant y \leqslant 2\}$.

c **i** $f(x) = -10$

$\therefore \quad -3x^2 = -10, \quad 0 \leqslant x \leqslant 2$

$\therefore \quad x^2 = \frac{10}{3}$

$\therefore \quad x = \sqrt{\frac{10}{3}} \quad \{0 \leqslant x \leqslant 2\}$

ii $f^{-1}(x) = 1$ so $f(1) = x$

The graph shows that $f(1) = -3$

$\therefore \quad x = -3$

REVIEW SET 1C

1 **a** Domain is $\{x \mid x \geqslant -2\}$. Range is $\{y \mid 1 \leqslant y < 3\}$.

b Domain is $\{x \mid x \in \mathbb{R}\}$. Range is $\{y \mid y = -1, 1 \text{ or } 2\}$.

2 **a**

$$h(x) = 7 - 3x$$
$$h(2x-1) = 7 - 3(2x-1) = 7 - 6x + 3 = 10 - 6x$$

b

$h(2x-1) = -2$

$\therefore \quad 10 - 6x = -2 \quad$ {using **a**}

$\therefore \quad -6x = -12$

$\therefore \quad x = 2$

3 **a** $(f \circ g)(x) = f(g(x)) = f(\sqrt{x}) = 1 - 2\sqrt{x}$

b $(g \circ f)(x) = g(f(x)) = g(1-2x) = \sqrt{1-2x}$

4 $f(x) = ax^2 + bx + c$, where $f(0) = 5$, $f(-2) = 21$ and $f(3) = -4$

When $f(0) = 5$,

$5 = a(0)^2 + b(0) + c$

$\therefore \quad 5 = c$

$\therefore \quad c = 5 \quad .... (1)$

When $f(-2) = 21$,

$21 = a(-2)^2 + b(-2) + c$

$= 4a - 2b + c$

$= 4a - 2b + 5 \quad$ {using (1)}

$\therefore \quad 4a - 2b = 16$

$\therefore \quad 2a - b = 8$ and so $b = 2a - 8 \quad .... (2)$

When $f(3) = -4$, $\quad -4 = a(3)^2 + b(3) + c$

$\therefore \quad -4 = 9a + 3b + c$

$\therefore \quad -4 = 9a + 3b + 5 \quad$ {using (1)}

$\therefore \quad -4 = 9a + 3(2a-8) + 5 \quad$ {using (2)}

$\therefore \quad -4 = 9a + 6a - 24 + 5$

$\therefore \quad 15 = 15a$ and so $a = 1$

Now, substituting $a = 1$ into (2) gives $b = 2(1) - 8 = -6$

So, $a = 1, \quad b = -6, \quad c = 5.$

5 **a**

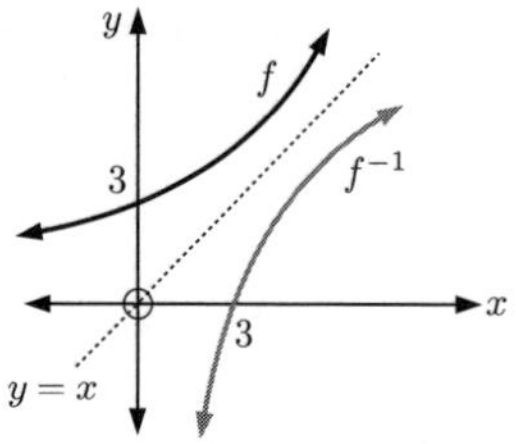

b

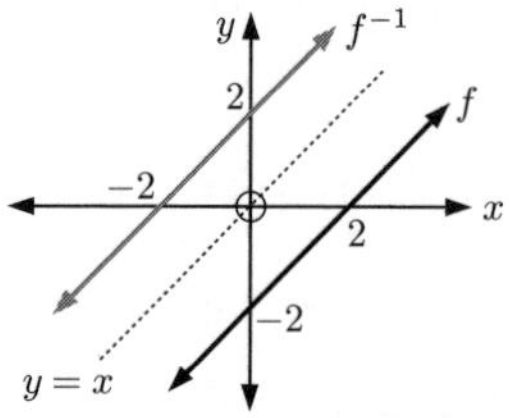

6 **a** f is $y = 7 - 4x$

$\therefore \; f^{-1}$ is $x = 7 - 4y$

$\therefore \; y = \dfrac{7-x}{4}$

So, $f^{-1}(x) = \dfrac{7-x}{4}$

b f is $y = \dfrac{3+2x}{5}$

$\therefore \; f^{-1}$ is $x = \dfrac{3+2y}{5}$

$\therefore \; 5x = 3 + 2y$

$\therefore \; y = \dfrac{5x-3}{2}$ So, $f^{-1}(x) = \dfrac{5x-3}{2}$

7 f is $y = 5x - 2$

$\therefore \; f^{-1}$ is $x = 5y - 2$

$\therefore \; y = \dfrac{x+2}{5}$

$\therefore \; f^{-1}(x) = \dfrac{x+2}{5}$

h is $y = \dfrac{3x}{4}$

$\therefore \; h^{-1}$ is $x = \dfrac{3y}{4}$

$\therefore \; y = \dfrac{4x}{3}$

$\therefore \; h^{-1}(x) = \dfrac{4x}{3}$

Now
$$\begin{aligned}(f^{-1} \circ h^{-1})(x) &= f^{-1}(h^{-1}(x)) \\ &= f^{-1}\left(\frac{4x}{3}\right) \\ &= \frac{\frac{4x}{3}+2}{5} \\ &= \frac{4x+6}{15}\end{aligned}$$

and
$$\begin{aligned}(h \circ f)(x) &= h(f(x)) \\ &= h(5x-2) \\ &= \frac{3(5x-2)}{4}\end{aligned}$$

So, $y = \dfrac{15x-6}{4}$

$\therefore \; (h \circ f)^{-1}(x)$ is $x = \dfrac{15y-6}{4}$

$\therefore \; 4x = 15y - 6$

$\therefore \; y = \dfrac{4x+6}{15}$

$\therefore \; (h \circ f)^{-1}(x) = \dfrac{4x+6}{15}$

Hence, $(f^{-1} \circ h^{-1})(x) = (h \circ f)^{-1}(x)$ as required.

8 f is $y = 2x + 11$

so $f^{-1}(x)$ is $x = 2y + 11$

$\therefore \; y = \dfrac{x-11}{2}$

$\therefore \; f^{-1}(x) = \dfrac{x-11}{2}$

$g(x) = x^2$

$$\begin{aligned}(g \circ f^{-1})(x) &= g(f^{-1}(x)) \\ &= g\left(\frac{x-11}{2}\right) \\ &= \left(\frac{x-11}{2}\right)^2\end{aligned}$$

$$\begin{aligned}\therefore \; (g \circ f^{-1})(3) &= \left(\frac{3-11}{2}\right)^2 \\ &= (-4)^2 = 16\end{aligned}$$

Chapter 2
SEQUENCES AND SERIES

EXERCISE 2A

1 **a** 4, 13, 22, 31 **b** 45, 39, 33, 27 **c** 2, 6, 18, 54 **d** 96, 48, 24, 12

2 **a** The sequence starts at 8 and each term is 8 more than the previous term. The next two terms are 40 and 48.

b The sequence starts at 2 and each term is 3 more than the previous term. The next two terms are 14 and 17.

c The sequence starts at 36 and each term is 5 less than the previous term. The next two terms are 16 and 11.

d The sequence starts at 96 and each term is 7 less than the previous term. The next two terms are 68 and 61.

e The sequence starts at 1 and each term is 4 times the previous term. The next two terms are 256 and 1024.

f The sequence starts at 2 and each term is 3 times the previous term. The next two terms are 162 and 486.

g The sequence starts at 480 and each term is half the previous term. The next two terms are 30 and 15.

h The sequence starts at 243 and each term is one third of the previous term. The next two terms are 3 and 1.

i The sequence starts at 50 000 and each term is one fifth of the previous term. The next two terms are 80 and 16.

3 **a** Each term is the square of the number of the term. The next three terms are 25, 36 and 49.

b Each term is the cube of the number of the term. The next three terms are 125, 216 and 343.

c Each term is $n \times (n+1)$ where n is the number of the term. The next three terms are 30, 42 and 56.

4 **a** 79, 75 (subtracting 4 each time) **b** 1280, 5120 (multiplying by 4 each time)

c 625, 1296 (1^4, 2^4, 3^4, 4^4, ...) **d** 13, 17 (prime numbers)

e 16, 22 (the difference between terms increases by 1) **f** 14, 18 (prime numbers $+1$)

EXERCISE 2B

1 **a** The sequence $\{2n\}$ begins 2, 4, 6, 8, 10 (letting $n = 1, 2, 3, 4, 5, ...$).

b The sequence $\{2n+2\}$ begins 4, 6, 8, 10, 12 (letting $n = 1, 2, 3, 4, 5, ...$).

c The sequence $\{2n-1\}$ begins 1, 3, 5, 7, 9 (letting $n = 1, 2, 3, 4, 5, ...$).

d The sequence $\{2n-3\}$ begins -1, 1, 3, 5, 7 (letting $n = 1, 2, 3, 4, 5, ...$).

e The sequence $\{2n+3\}$ begins 5, 7, 9, 11, 13 (letting $n = 1, 2, 3, 4, 5, ...$).

f The sequence $\{2n+11\}$ begins 13, 15, 17, 19, 21 (letting $n = 1, 2, 3, 4, 5, ...$).

g The sequence $\{3n+1\}$ begins 4, 7, 10, 13, 16 (letting $n = 1, 2, 3, 4, 5, ...$).

h The sequence $\{4n-3\}$ begins 1, 5, 9, 13, 17 (letting $n = 1, 2, 3, 4, 5, ...$).

2 **a** The sequence $\{2^n\}$ begins 2, 4, 8, 16, 32 (letting $n = 1, 2, 3, 4, 5, ...$).

b The sequence $\{3 \times 2^n\}$ begins 6, 12, 24, 48, 96 (letting $n = 1, 2, 3, 4, 5, ...$).

c The sequence $\{6 \times (\frac{1}{2})^n\}$ begins 3, $\frac{3}{2}$, $\frac{3}{4}$, $\frac{3}{8}$, $\frac{3}{16}$ (letting $n = 1, 2, 3, 4, 5, ...$).

d The sequence $\{(-2)^n\}$ begins -2, 4, -8, 16, -32 (letting $n = 1, 2, 3, 4, 5, ...$).

3 $\{15-(-2)^n\}$ generates the sequence with first five terms:
$t_1 = 15-(-2)^1 = 17$, $t_2 = 15-(-2)^2 = 11$, $t_3 = 15-(-2)^3 = 23$,
$t_4 = 15-(-2)^4 = -1$, $t_5 = 15-(-2)^5 = 47$

EXERCISE 2C

1 **a** The sequence begins with 19 and the common difference is $25-19=6$.

$$\begin{aligned} u_n &= u_1+(n-1)d \\ \therefore\ u_n &= 19+6(n-1) \\ \text{So, } u_{10} &= 19+6(10-1) \\ &= 19+6\times 9 \\ &= 73 \end{aligned}$$

b The sequence begins with 101 and the common difference is $97-101=-4$.

$$\begin{aligned} u_n &= u_1+(n-1)d \\ \therefore\ u_n &= 101+(-4)(n-1) \\ \text{So, } u_{10} &= 101+(-4)(10-1) \\ &= 101-4\times 9 \\ &= 65 \end{aligned}$$

c The sequence begins with 8 and the common difference is $9\frac{1}{2}-8=\frac{3}{2}$.

$$\begin{aligned} u_n &= u_1+(n-1)d \\ \therefore\ u_n &= 8+\tfrac{3}{2}(n-1) \\ \text{So, } u_{10} &= 8+\tfrac{3}{2}(10-1) \\ &= 8+\tfrac{3}{2}\times 9 \\ &= 21\tfrac{1}{2} \end{aligned}$$

2 **a** The first term of the arithmetic sequence is 31 and the common difference is $36-31=5$.

$$\begin{aligned} u_n &= u_1+(n-1)d \\ \therefore\ u_n &= 31+5(n-1) \\ \text{So, } u_{15} &= 31+5(15-1) \\ &= 31+5\times 14 \\ &= 101 \end{aligned}$$

b The first term of the arithmetic sequence is 5 and the common difference is $-3-5=-8$.

$$\begin{aligned} u_n &= u_1+(n-1)d \\ \therefore\ u_n &= 5+(-8)(n-1) \\ \text{So, } u_{15} &= 5+(-8)(15-1) \\ &= 5-8\times 14 \\ &= -107 \end{aligned}$$

c The first term of the arithmetic sequence is a and the common difference is $a+d-a=d$.

$$\begin{aligned} u_n &= u_1+(n-1)d \\ \therefore\ u_n &= a+(n-1)d \\ \text{So, } u_{15} &= a+d(15-1) \\ &= a+14d \end{aligned}$$

3 **a** $17-6=11$
$28-17=11$
$39-28=11$
$50-39=11$

Assuming that the pattern continues, consecutive terms differ by 11.
$\therefore$ the sequence is arithmetic with $u_1=6$, $d=11$.

b $$\begin{aligned} u_n &= u_1+(n-1)d \\ &= 6+(n-1)11 \\ &= 11n-5 \end{aligned}$$

c $$\begin{aligned} u_{50} &= 11(50)-5 \\ &= 545 \end{aligned}$$

d Let $u_n = 325 = 11n-5$
$\therefore\ 330 = 11n$
$\therefore\ n = 30$
So, 325 is the 30th member.

e Let $u_n = 761 = 11n-5$
$\therefore\ 766 = 11n$
$\therefore\ n = 69\frac{7}{11}$, but n must be an integer, so 761 is not a member of the sequence.

4 **a** $83-87=-4$
$79-83=-4$
$75-79=-4$

Assuming that the pattern continues, consecutive terms differ by -4.
$\therefore$ the sequence is arithmetic with $u_1=87$, $d=-4$.

b $u_n = u_1 + (n-1)d$
$= 87 + (n-1)(-4)$
$= 87 - 4n + 4$
$= 91 - 4n$

c $u_{40} = 91 - 4(40)$
$= 91 - 160$
$= -69$

d Let $u_n = -297 = 91 - 4n$
$\therefore\ 4n = 388$
$\therefore\ n = 97$
So, -297 is the 97th term of the sequence.

5 **a** $u_n = 3n - 2,\quad u_{n+1} = 3(n+1) - 2 = 3n + 1$
$u_{n+1} - u_n = (3n+1) - (3n-2)$
$= 3$, a constant
Consecutive terms differ by 3.
$\therefore$ the sequence is arithmetic.

b $u_1 = 3(1) - 2 = 1,\quad d = 3$

c $u_{57} = 3(57) - 2 = 169$

d Let $u_n = 450 = 3n - 2$, so $3n = 452$ and hence $n = 150\frac{2}{3}$.
We try the two values on either side of $n = 150\frac{2}{3}$, which are $n = 150$ and $n = 151$:
$u_{150} = 3(150) - 2 = 448$ and $u_{151} = 3(151) - 2 = 451$
So, $u_{151} = 451$ is the least term which is greater than 450.

6 **a** $u_n = \dfrac{71 - 7n}{2} = 35\frac{1}{2} - \frac{7}{2}n$
$u_{n+1} = \dfrac{71 - 7(n+1)}{2} = \dfrac{71 - 7n - 7}{2} = \dfrac{64 - 7n}{2} = 32 - \frac{7}{2}n$
$u_{n+1} - u_n = (32 - \frac{7}{2}n) - (35\frac{1}{2} - \frac{7}{2}n) = -\frac{7}{2}$, a constant
So, consecutive terms differ by $-\frac{7}{2}$.
$\therefore$ the sequence is arithmetic.

b $u_1 = \dfrac{71 - 7(1)}{2} = 32,\quad d = -\frac{7}{2}$

c $u_{75} = \dfrac{71 - 7(75)}{2} = -227$

d Let $u_n = -200 = \dfrac{71 - 7n}{2}$ so $-400 = 71 - 7n$ $\therefore\ 7n = 471$
$\therefore\ n = 67\frac{2}{7}$
We try the two values on either side of $n = 67\frac{2}{7}$, which are $n = 67$ and $n = 68$:
$u_{67} = \dfrac{71 - 7(67)}{2} = -199$ and $u_{68} = \dfrac{71 - 7(68)}{2} = -202\frac{1}{2}$
So, the terms of the sequence are less than -200 for $n \geqslant 68$.

7 **a** The terms are consecutive, so we equate common differences:
$k - 32 = 3 - k$
$\therefore\ 2k = 35$
$\therefore\ k = 17\frac{1}{2}$

b The terms are consecutive, so we equate common differences:
$7 - k = 10 - 7$
$\therefore\ 7 - k = 3$
$\therefore\ k = 4$

c The terms are consecutive, so we equate common differences:
$(2k+1) - (k+1) = 13 - (2k+1)$
$\therefore\ k = 12 - 2k$
$\therefore\ 3k = 12$
$\therefore\ k = 4$

d The terms are consecutive, so we equate common differences:
$(2k+3) - (k-1) = (7-k) - (2k+3)$
$\therefore\ k + 4 = 4 - 3k$
$\therefore\ 4k = 0$
$\therefore\ k = 0$

e The terms are consecutive, so we equate common differences:
$k^2 - k = (k^2 + 6) - k^2$
$\therefore\ k^2 - k - 6 = 0$
$\therefore\ (k+2)(k-3) = 0$
$\therefore\ k = -2$ or 3

f The terms are consecutive, so we equate common differences:
$k - 5 = k^2 - 8 - k$
$\therefore\ k^2 - 2k - 3 = 0$
$\therefore\ (k-3)(k+1) = 0$
$\therefore\ k = -1$ or 3

8 **a** $u_7 = 41 \quad \therefore\ u_1 + 6d = 41$ .... (1)
$u_{13} = 77 \quad \therefore\ u_1 + 12d = 77$ .... (2)
Solving simultaneously,
$-u_1 - 6d = -41$
$u_1 + 12d = 77$
$\therefore\ 6d = 36$ {adding the equations}
$\therefore\ d = 6$
So in (1), $u_1 + 6(6) = 41$
$\therefore\ u_1 + 36 = 41$
$\therefore\ u_1 = 5$
Now $u_n = u_1 + (n-1)d$
$\therefore\ u_n = 5 + (n-1)6$
$\therefore\ u_n = 6n - 1$

b $u_5 = -2 \quad \therefore\ u_1 + 4d = -2$ .... (1)
$u_{12} = -12\frac{1}{2} \quad \therefore\ u_1 + 11d = -12\frac{1}{2}$ .... (2)
Solving simultaneously,
$-u_1 - 4d = 2$
$u_1 + 11d = -12\frac{1}{2}$
$\therefore\ 7d = -10\frac{1}{2}$ {adding the equations}
$\therefore\ d = -\frac{3}{2}$
So in (1), $u_1 + 4(-\frac{3}{2}) = -2$
$\therefore\ u_1 = 4$
Now $u_n = u_1 + (n-1)d$
$\therefore\ u_n = 4 + (n-1)(-\frac{3}{2})$
$\therefore\ u_n = -\frac{3}{2}n + \frac{11}{2}$

c $u_7 = 1 \quad \therefore\ u_1 + 6d = 1$ .... (1)
$u_{15} = -39 \quad \therefore\ u_1 + 14d = -39$ .... (2)
Solving simultaneously,
$-u_1 - 6d = -1$
$u_1 + 14d = -39$
$\therefore\ 8d = -40$ {adding the equations}
$\therefore\ d = -5$
So in (1), $u_1 + 6(-5) = 1$
$\therefore\ u_1 - 30 = 1$
$\therefore\ u_1 = 31$
Now $u_n = u_1 + (n-1)d$
$\therefore\ u_n = 31 + (n-1)(-5)$
$\therefore\ u_n = 31 - 5n + 5$
$\therefore\ u_n = -5n + 36$

d $u_{11} = -16 \quad \therefore\ u_1 + 10d = -16$ .... (1)
$u_8 = -11\frac{1}{2} \quad \therefore\ u_1 + 7d = -11\frac{1}{2}$ .... (2)
Solving simultaneously,
$-u_1 - 10d = 16$
$u_1 + 7d = -11\frac{1}{2}$
$\therefore\ -3d = 4\frac{1}{2}$ {adding the equations}
$\therefore\ d = -\frac{3}{2}$
So in (1), $u_1 + 10(-\frac{3}{2}) = -16$
$\therefore\ u_1 - 15 = -16$
$\therefore\ u_1 = -1$
Now $u_n = u_1 + (n-1)d$
$\therefore\ u_n = -1 + (n-1)(-\frac{3}{2})$
$\therefore\ u_n = -\frac{3}{2}n + \frac{1}{2}$

9 **a** Let the numbers be $5,\ 5+d,\ 5+2d,\ 5+3d,\ 10$.
Then $5 + 4d = 10$
$\therefore\ 4d = 5$
$\therefore\ d = \frac{5}{4} = 1\frac{1}{4}$ So, the numbers are $5, 6\frac{1}{4}, 7\frac{1}{2}, 8\frac{3}{4}, 10$.

b Let the numbers be $-1,\ -1+d,\ -1+2d,\ -1+3d,\ -1+4d,\ -1+5d,\ -1+6d,\ 32$.
Then $-1 + 7d = 32$
$\therefore\ 7d = 33$
$\therefore\ d = \frac{33}{7} = 4\frac{5}{7}$
So, the numbers are $-1,\ 3\frac{5}{7},\ 8\frac{3}{7},\ 13\frac{1}{7},\ 17\frac{6}{7},\ 22\frac{4}{7},\ 27\frac{2}{7},\ 32$.

10 **a** $u_1 = 36, \quad 35\frac{1}{3} - 36 = -\frac{2}{3}, \quad 34\frac{2}{3} - 35\frac{1}{3} = -\frac{2}{3}$, so $d = -\frac{2}{3}$

b $u_n = u_1 + (n-1)d$
$\therefore\ -30 = 36 + (n-1)(-\frac{2}{3})$ {letting $u_n = -30$, the last term of the sequence}
$\therefore\ -66 = -\frac{2}{3}n + \frac{2}{3}$
$\therefore\ \frac{2}{3}n = 66\frac{2}{3}$
$\therefore\ n = 100$ So, the sequence has 100 terms.

11 $u_1 = 23$, $36 - 23 = 13$
$49 - 36 = 13$
$62 - 49 = 13$, so $d = 13$

$\therefore\ u_n = u_1 + (n-1)d$
$\therefore\ u_n = 23 + (n-1)13$
$= 23 + 13n - 13$
$\therefore\ u_n = 13n + 10$

Let $u_n = 100\,000 = 13n + 10$

$\therefore \quad 99\,990 = 13n$

$\therefore \quad n = 7691\frac{7}{13}$

We try the two values on either side of $n = 7691\frac{7}{13}$, which are $n = 7691$ and $n = 7692$:

$u_{7691} = 13(7691) + 10 = 99\,993$ and $u_{7692} = 13(7692) + 10 = 100\,006$

So, the first term to exceed 100 000 is $u_{7692} = 100\,006$.

EXERCISE 2D.1

1 **a** $\frac{6}{2} = 3 \quad \therefore \ r = 3, \ u_1 = 2 \quad \therefore \ b = 6 \times 3 = 18$ and $c = 18 \times 3 = 54$

b $\frac{5}{10} = \frac{1}{2} \quad \therefore \ r = \frac{1}{2}, \ u_1 = 10 \quad \therefore \ b = 5 \times \frac{1}{2} = 2\frac{1}{2}$ and $c = 2\frac{1}{2} \times \frac{1}{2} = 1\frac{1}{4}$

c $\frac{-6}{12} = -\frac{1}{2} \quad \therefore \ r = -\frac{1}{2}, \ u_1 = 12 \quad \therefore \ b = -6 \times -\frac{1}{2} = 3$ and $c = 3 \times -\frac{1}{2} = -1\frac{1}{2}$

2 **a** $\frac{6}{3} = 2 \quad \therefore \ r = 2, \ u_1 = 3 \quad \therefore \ u_6 = 3 \times 2^5 = 96$

b $\frac{10}{2} = 5 \quad \therefore \ r = 5, \ u_1 = 2 \quad \therefore \ u_6 = 2 \times 5^5 = 6250$

c $\frac{256}{512} = \frac{1}{2} \quad \therefore \ r = \frac{1}{2}, \ u_1 = 512 \quad \therefore \ u_6 = 512 \times (\frac{1}{2})^5 = 16$

3 **a** $\frac{3}{1} = 3 \quad \therefore \ r = 3, \ u_1 = 1 \quad \therefore \ u_9 = 1 \times 3^8 = 6561$

b $\frac{18}{12} = \frac{3}{2} \quad \therefore \ r = \frac{3}{2}, \ u_1 = 12 \quad \therefore \ u_9 = 12 \times (\frac{3}{2})^8 = 307\frac{35}{64}$ or $\frac{19\,683}{64}$

c $\dfrac{-\frac{1}{8}}{\frac{1}{16}} = -2 \quad \therefore \ r = -2, \ u_1 = \frac{1}{16} \quad \therefore \ u_9 = \frac{1}{16} \times (-2)^8 = 16$

d $\dfrac{ar}{a} = r \quad \therefore \ r = r, \ u_1 = a \quad \therefore \ u_9 = a \times r^8 = ar^8$

4 **a** $\frac{10}{5} = \frac{20}{10} = \frac{40}{20} = 2$

Assuming the pattern continues, consecutive terms have a common ratio of 2.

$\therefore$ the sequence is geometric with $u_1 = 5$ and $r = 2$.

b $u_n = u_1 r^{n-1}$

$\therefore \quad u_n = 5 \times 2^{n-1}$

so $\quad u_{15} = 5 \times 2^{14} = 81\,920$

5 **a** $\frac{-6}{12} = -\frac{1}{2}$

$\frac{3}{-6} = -\frac{1}{2}$

$\dfrac{(-\frac{3}{2})}{3} = -\frac{1}{2}$

Assuming the pattern continues, consecutive terms have a common ratio of $-\frac{1}{2}$.

$\therefore$ the sequence is geometric with $u_1 = 12$ and $r = -\frac{1}{2}$.

b $u_n = u_1 r^{n-1}$

$\therefore \quad u_n = 12 \times \left(-\frac{1}{2}\right)^{n-1}$

so $\quad u_{13} = 12 \times (-\frac{1}{2})^{13-1}$

$= 12 \times (-\frac{1}{2})^{12}$

$= 12 \times \frac{1}{4096}$

$= 3 \times \frac{1}{1024} = \frac{3}{1024}$

6 $\frac{-6}{8} = -\frac{3}{4}$

$\dfrac{4.5}{-6} = -\dfrac{(\frac{9}{2})}{6} = -\frac{3}{4}$

$\dfrac{-3.375}{4.5} = \dfrac{(-\frac{27}{8})}{(\frac{9}{2})} = -\frac{3}{4}$

Assuming the pattern continues, consecutive terms have a common ratio of $-\frac{3}{4}$.

$\therefore$ the sequence is geometric with $u_1 = 8$ and $r = -\frac{3}{4}$.

$u_n = u_1 r^{n-1} = 8 \times (-\frac{3}{4})^{n-1}$

So, $\quad u_{10} = 8 \times (-\frac{3}{4})^9 = -0.600\,677\,49$

≈ -0.601

7 $\frac{4\sqrt{2}}{8} = \frac{\sqrt{2}}{2} = \frac{1}{\sqrt{2}}$

$\frac{4}{4\sqrt{2}} = \frac{1}{\sqrt{2}}$

$\frac{2\sqrt{2}}{4} = \frac{\sqrt{2}}{2} = \frac{1}{\sqrt{2}}$

Assuming the pattern continues, consecutive terms have a common ratio of $\frac{1}{\sqrt{2}}$.

$\therefore$ the sequence is geometric with $u_1 = 8$ and $r = \frac{1}{\sqrt{2}}$.

$u_n = u_1 r^{n-1} = 8\left(\frac{1}{\sqrt{2}}\right)^{n-1} = 2^3 \times \left(2^{-\frac{1}{2}}\right)^{n-1} = 2^3 \times 2^{-\frac{1}{2}n+\frac{1}{2}}$

So, $u_n = 2^{\frac{7}{2}-\frac{n}{2}}$

8 **a** Since the terms are geometric,

$\frac{k}{7} = \frac{28}{k}$ $\therefore$ $k^2 = 196$ $\therefore$ $k = \pm 14$

b Since the terms are geometric,

$\frac{3k}{k} = \frac{20-k}{3k} = 3$

$\therefore$ $20 - k = 9k$

$\therefore$ $20 = 10k$

$\therefore$ $k = 2$

c Since the terms are geometric,

$\frac{k+8}{k} = \frac{9k}{k+8}$

$\therefore$ $(k+8)^2 = 9k^2$

$\therefore$ $k^2 + 16k + 64 = 9k^2$

$\therefore$ $8k^2 - 16k - 64 = 0$

$\therefore$ $8(k^2 - 2k - 8) = 0$

$\therefore$ $8(k+2)(k-4) = 0$ and so $k = -2$ or 4

9 **a** $u_4 = 24$ $\therefore$ $u_1 \times r^3 = 24$ .... (1)

$u_7 = 192$ $\therefore$ $u_1 \times r^6 = 192$ .... (2)

So, $\frac{u_1 r^6}{u_1 r^3} = \frac{192}{24}$ {(2) ÷ (1)}

$\therefore$ $r^3 = 8$ $\therefore$ $r = 2$

So in (1), $u_1 \times 2^3 = 24$

$\therefore$ $u_1 \times 8 = 24$

$\therefore$ $u_1 = 3$

$\therefore$ $u_n = 3 \times 2^{n-1}$

b $u_3 = 8$ $\therefore$ $u_1 \times r^2 = 8$ .... (1)

$u_6 = -1$ $\therefore$ $u_1 \times r^5 = -1$ .... (2)

So, $\frac{u_1 r^5}{u_1 r^2} = -\frac{1}{8}$ {(2) ÷ (1)}

$\therefore$ $r^3 = -\frac{1}{8}$ $\therefore$ $r = -\frac{1}{2}$

So in (1), $u_1 \times (-\frac{1}{2})^2 = 8$

$\therefore$ $u_1 \times \frac{1}{4} = 8$

$\therefore$ $u_1 = 32$

$\therefore$ $u_n = 32 \times (-\frac{1}{2})^{n-1}$

c $u_7 = 24$ $\therefore$ $u_1 \times r^6 = 24$ .... (1)

$u_{15} = 384$ $\therefore$ $u_1 \times r^{14} = 384$ .... (2)

So, $\frac{u_1 r^{14}}{u_1 r^6} = \frac{384}{24}$ {(2) ÷ (1)}

$\therefore$ $r^8 = 16$ $\therefore$ $r = \pm\sqrt{2}$

So in (1), $u_1 \times (\pm\sqrt{2})^6 = 24$

$\therefore$ $u_1 \times 8 = 24$

$\therefore$ $u_1 = 3$

Now $u_n = u_1 r^{n-1}$

$\therefore$ $u_n = 3 \times (\sqrt{2})^{n-1}$

or $u_n = 3 \times (-\sqrt{2})^{n-1}$

d $u_3 = 5$ $\therefore$ $u_1 \times r^2 = 5$ .... (1)

$u_7 = \frac{5}{4}$ $\therefore$ $u_1 \times r^6 = \frac{5}{4}$ .... (2)

So, $\frac{u_1 r^6}{u_1 r^2} = \frac{(\frac{5}{4})}{5}$ {(2) ÷ (1)}

$\therefore$ $r^4 = \frac{1}{4}$ $\therefore$ $r = \pm\frac{1}{\sqrt{2}}$

So in (1), $u_1 \times (\pm\frac{1}{\sqrt{2}})^2 = 5$

$\therefore$ $u_1 \times \frac{1}{2} = 5$

$\therefore$ $u_1 = 10$

Now $u_n = u_1 r^{n-1}$

$\therefore$ $u_n = 10 \times (\frac{1}{\sqrt{2}})^{n-1}$

$= 10 \times (\sqrt{2})^{1-n}$

or $u_n = 10 \times (-\frac{1}{\sqrt{2}})^{n-1}$

$= 10 \times (-\sqrt{2})^{1-n}$

10 **a** 2, 6, 18, 54, ... has $u_1 = 2$ and $r = 3$

$u_n = u_1 r^{n-1}$ $\therefore$ $u_n = 2 \times 3^{n-1}$

Let $u_n = 10\,000 = 2 \times 3^{n-1}$, so $5000 = 3^{n-1}$

$\therefore$ $n \approx 8.7527$ {using technology}

We try the two values on either side of $n = 8.7527$, which are $n = 8$ and $n = 9$:

$u_8 = 2 \times 3^7$ and $u_9 = 2 \times 3^8$

$= 4374$ $= 13\,122$

So, the first term to exceed 10 000 is $u_9 = 13\,122$.

b $4, 4\sqrt{3}, 12, 12\sqrt{3}, ...$ has $u_1 = 4$ and $r = \sqrt{3}$

$u_n = u_1 r^{n-1}$ $\therefore$ $u_n = 4 \times \left(\sqrt{3}\right)^{n-1}$

Let $u_n = 4800 = 4 \times \left(\sqrt{3}\right)^{n-1}$, so $1200 = \left(\sqrt{3}\right)^{n-1}$

$\therefore$ $n \approx 13.91$ {using technology}

We try the two values on either side of $n \approx 13.91$, which are $n = 13$ and $n = 14$:

$u_{13} = 4 \times \left(\sqrt{3}\right)^{12}$ and $u_{14} = 4 \times \left(\sqrt{3}\right)^{13}$

$= 2916$ $= 2916\sqrt{3} \approx 5050.66$

So, the first term to exceed 4800 is $u_{14} \approx 5050$.

c 12, 6, 3, 1.5, ... has $u_1 = 12$ and $r = \frac{1}{2}$

$u_n = u_1 r^{n-1}$ $\therefore$ $u_n = 12 \times (\frac{1}{2})^{n-1}$

Let $0.0001 = u_n = 12 \times \left(\frac{1}{2}\right)^{n-1}$

$\therefore$ $0.000\,008\bar{3} = \left(\frac{1}{2}\right)^{n-1}$

$\therefore$ $n \approx 17.87$ {using technology}

We try the two values on either side of $n \approx 17.87$, which are $n = 17$ and $n = 18$:

$u_{17} = 12 \times \left(\frac{1}{2}\right)^{16}$ and $u_{18} = 12 \times \left(\frac{1}{2}\right)^{17}$

$\approx 0.000\,1831$ $\approx 0.000\,091\,55$

So, the first term of the sequence which is less than 0.0001 is $u_{18} \approx 0.000\,091\,6$.

EXERCISE 2D.2

1 **a** $u_{n+1} = u_1 \times r^n$

where $u_1 = 3000$, $r = 1.1$, $n = 3$

$\therefore$ $u_4 = 3000 \times (1.1)^3$

$= 3993$

So it amounts to \$3993.

b Interest = amount after 3 years – initial amount

= \$3993 – \$3000

= \$993

2 $u_{n+1} = u_1 \times r^n$ where $u_1 = 20\,000$, $r = 1.12$, $n = 4$

$\therefore$ $u_5 = 20\,000 \times (1.12)^4$

$\approx 31\,470.39$

Interest = €31 470.39 – €20 000

= €11 470.39

3 **a** $u_{n+1} = u_1 \times r^n$ where $u_1 = 30\,000$, $r = 1.1$, $n = 4$

$\therefore$ $u_5 = 30\,000 \times (1.1)^4$

$= 43\,923$ The investment amounts to ¥43 923.

b Interest = amount after 4 years – initial amount

= ¥43 923 – ¥30 000

= ¥13 923

4 $u_{n+1} = u_1 \times r^n$ where $u_1 = 80\,000$, $r = 1.09$, $n = 3$

$\therefore \quad u_4 = 80\,000 \times (1.09)^3$

$= 103\,602.32$

Interest = amount after 3 years − initial amount

$= \$103\,602.32 - \$80\,000$

$= \$23\,602.32$

5 $u_{n+1} = u_1 \times r^n$ where $u_1 = 100\,000$, $r = 1 + \dfrac{0.08}{2} = 1.04$, $n = 10$

$\therefore \quad u_{11} = 100\,000 \times (1.04)^{10}$

$\approx 148\,024.43$ It amounts to ¥148 024.43.

6 $u_{n+1} = u_1 \times r^n$ where $u_1 = 45\,000$, $r = 1 + \dfrac{0.075}{4} = 1.018\,75$,

$n = 7$ {21 months = 7 'quarters'}

$\therefore \quad u_{10} = 45\,000 \times (1.018\,75)^7$

$\approx 51\,249.06$ It amounts to £51 249.06.

7 $u_{n+1} = u_1 \times r^n$ where $u_{n+1} = 20\,000$, $r = 1.075$, $n = 4$

$\therefore \quad 20\,000 = u_1 \times (1.075)^4$

$\therefore \quad u_1 \approx 14\,976.01$ \$14 976.01 should be invested now.

8 $u_{n+1} = u_1 \times r^n$ where $u_{n+1} = 15\,000$, $r = 1.055$, $n = \frac{60}{12} = 5$

$\therefore \quad 15\,000 = u_1 \times (1.055)^5$

$\therefore \quad u_1 \approx 11\,477.02$ The initial investment required is £11 477.02.

9 $u_{n+1} = u_1 \times r^n$ where $u_{n+1} = 25\,000$, $r = 1 + \dfrac{0.08}{4} = 1.02$, $n = 3 \times 4 = 12$

$\therefore \quad 25\,000 = u_1 \times (1.02)^{12}$

$\therefore \quad u_1 \approx 19\,712.33$ I should invest €19 712.33 now.

10 $u_{n+1} = u_1 \times r^n$ where $u_{n+1} = 40\,000$, $r = 1 + \dfrac{0.09}{12} = 1.0075$, $n = 8 \times 12 = 96$

$\therefore \quad 40\,000 = u_1 \times (1.0075)^{96}$

$\therefore \quad u_1 \approx 19\,522.47$ The initial investment should be ¥19 522.47.

EXERCISE 2D.3

1 $u_{n+1} = u_1 \times r^n$, where $u_1 = 500$, $r = 1.12$

a **i** $u_{11} = 500 \times (1.12)^{10}$

≈ 1552.92

There will be approximately 1550 ants.

ii $u_{21} = 500 \times (1.12)^{20}$

≈ 4823.15

There will be approximately 4820 ants.

b For the population to reach 2000, $u_{n+1} = 500 \times (1.12)^n = 2000$

$\therefore \quad (1.12)^n = 4$

$\therefore \quad n \approx 12.23$ {using technology}

It will take approximately 12.2 weeks.

2 $u_{n+1} = u_1 \times r^n$, where $u_1 = 555$, $r = 0.955$

a $u_{16} = 555 \times (0.955)^{15}$

≈ 278.19 The population is approximately 278 animals in the year 2007.

b For the population to have declined to 50,

$u_{n+1} = 555 \times (0.955)^n = 50$

$\therefore \quad (0.955)^n = 0.0\overline{900}$

$\therefore \quad n \approx 52.3$ {using technology}

So, after approximately 52 years the population is 50. This is the year 2044.

EXERCISE 2E.1

1 **a** **i** 3, 11, 19, 27, ... is arithmetic with $u_1 = 3$, $d = 8$, so $u_n = 3 + (n-1)8 = 8n - 5$

$S_n = 3 + 11 + 19 + 27 + ... + (8n - 5)$

ii $S_5 = 3 + 11 + 19 + 27 + 35 = 95$

b **i** 42, 37, 32, 27, ... is arithmetic with $u_1 = 42$, $d = -5$,

so $u_n = 42 + (n-1)(-5) = 47 - 5n$

$\therefore \quad S_n = 42 + 37 + 32 + 27 + ... + (47 - 5n)$

ii $S_5 = 42 + 37 + 32 + 27 + 22 = 160$

c **i** 12, 6, 3, $1\frac{1}{2}$, ... is geometric with $u_1 = 12$, $r = \frac{1}{2}$, so $u_n = 12 \times (\frac{1}{2})^{n-1}$

$S_n = 12 + 6 + 3 + 1\frac{1}{2} + ... + 12(\frac{1}{2})^{n-1}$

ii $S_5 = 12 + 6 + 3 + 1\frac{1}{2} + \frac{3}{4} = 23\frac{1}{4}$

d **i** 2, 3, $4\frac{1}{2}$, $6\frac{3}{4}$, ... is geometric with $u_1 = 2$, $r = \frac{3}{2}$, so $u_n = 2 \times (\frac{3}{2})^{n-1}$

$S_n = 2 + 3 + 4\frac{1}{2} + 6\frac{3}{4} + ... + 2(\frac{3}{2})^{n-1}$

ii $S_5 = 2 + 3 + 4\frac{1}{2} + 6\frac{3}{4} + 10\frac{1}{8} = 26\frac{3}{8}$

e **i** 1, $\frac{1}{2}$, $\frac{1}{4}$, $\frac{1}{8}$, ... is geometric with $u_1 = 1$, $r = \frac{1}{2}$, so $u_n = 1 \times (\frac{1}{2})^{n-1} = \dfrac{1}{2^{n-1}}$

$S_n = 1 + \frac{1}{2} + \frac{1}{4} + \frac{1}{8} + ... + \dfrac{1}{2^{n-1}}$

ii $S_5 = 1 + \frac{1}{2} + \frac{1}{4} + \frac{1}{8} + \frac{1}{16} = 1\frac{15}{16}$

f **i** 1, 8, 27, 64, ...

$S_n = 1 + 8 + 27 + 64 + ... + n^3$ {since $1 = 1^3$, $8 = 2^3$, $27 = 3^3$, $64 = 4^3$}

ii $S_5 = 1 + 8 + 27 + 64 + 125 = 225$

2 **a** $\displaystyle\sum_{k=1}^{4}(3k-5) = -2 + 1 + 4 + 7 = 10$

b $\displaystyle\sum_{k=1}^{5}(11-2k) = 9 + 7 + 5 + 3 + 1 = 25$

c $\displaystyle\sum_{k=1}^{7}k(k+1) = 2 + 6 + 12 + 20 + 30 + 42 + 56 = 168$

d $\displaystyle\sum_{k=1}^{5}10 \times 2^{k-1} = 10 + 20 + 40 + 80 + 160 = 310$

3 $u_n = 3n - 1$

$$\begin{aligned}\therefore \quad u_1 + u_2 + u_3 + ... + u_{20} &= \sum_{n=1}^{20}(3n-1)\\ &= 2 + 5 + 8 + 11 + 14 + 17 + 20 + 23 + 26 + 29 + 32 + 35\\ &\quad + 38 + 41 + 44 + 47 + 50 + 53 + 56 + 59\\ &= 610\end{aligned}$$

4 **a** $\displaystyle\sum_{k=1}^{n}c = \underbrace{c + c + c + ... + c}_{n \text{ times}} = cn$

b $$\begin{aligned}\sum_{k=1}^{n}ca_k &= ca_1 + ca_2 + ... + ca_n\\ &= c(a_1 + a_2 + ... + a_n)\\ &= c\sum_{k=1}^{n}a_k\end{aligned}$$

c $$\begin{aligned}\sum_{k=1}^{n}(a_k + b_k) &= (a_1 + b_1) + (a_2 + b_2) + ... + (a_n + b_n)\\ &= (a_1 + a_2 + ... + a_n) + (b_1 + b_2 + ... + b_n)\\ &= \sum_{k=1}^{n}a_k + \sum_{k=1}^{n}b_k\end{aligned}$$

EXERCISE 2E.2

1 **a** The series is arithmetic with
$u_1 = 3, \quad d = 4, \quad n = 20$

$$S_n = \frac{n}{2}(2u_1 + (n-1)d)$$

So, $S_{20} = \frac{20}{2}(2 \times 3 + 19 \times 4)$
$= 10(6 + 76)$
$= 820$

b The series is arithmetic with
$u_1 = \frac{1}{2}, \quad d = \frac{5}{2}, \quad n = 50$

$$S_n = \frac{n}{2}(2u_1 + (n-1)d)$$

So, $S_{50} = \frac{50}{2}\left(2 \times \frac{1}{2} + 49 \times \frac{5}{2}\right)$
$= 25(1 + 122\frac{1}{2})$
$= 3087\frac{1}{2}$

c The series is arithmetic with
$u_1 = 100, \quad d = -7, \quad n = 40$

$$S_n = \frac{n}{2}(2u_1 + (n-1)d)$$

So, $S_{40} = \frac{40}{2}(2 \times 100 + 39 \times (-7))$
$= 20(200 - 273)$
$= -1460$

d The series is arithmetic with
$u_1 = 50, \quad d = -\frac{3}{2}, \quad n = 80$

$$S_n = \frac{n}{2}(2u_1 + (n-1)d)$$

So, $S_{80} = \frac{80}{2}\left(2 \times 50 + 79 \times (-\frac{3}{2})\right)$
$= 40(100 - \frac{237}{2})$
$= -740$

2 **a** The series is arithmetic with
$u_1 = 5, \quad d = 3, \quad u_n = 101$

Since $u_n = 101$,
then $u_1 + (n-1)d = 101$
$\therefore \; 5 + 3(n-1) = 101$
$\therefore \; 5 + 3n - 3 = 101$
$\therefore \; 3n = 99$
$\therefore \; n = 33$

So, $S_n = \frac{n}{2}(u_1 + u_n)$
$= \frac{33}{2}(5 + 101)$
$= 1749$

b The series is arithmetic with
$u_1 = 50, \quad d = -\frac{1}{2}, \quad u_n = -20$

Since $u_n = -20$,
then $u_1 + (n-1)d = -20$
$\therefore \; 50 + (-\frac{1}{2})(n-1) = -20$
$\therefore \; -\frac{1}{2}n + \frac{1}{2} = -70$
$\therefore \; -\frac{1}{2}n = -\frac{141}{2}$
$\therefore \; n = 141$

So, $S_n = \frac{n}{2}(u_1 + u_n)$
$= \frac{141}{2}(50 + (-20))$
$= 2115$

c The series is arithmetic with
$u_1 = 8, \quad d = \frac{5}{2}, \quad u_n = 83$

Since $u_n = 83$,
then $u_1 + (n-1)d = 83$
$\therefore \; 8 + \frac{5}{2}(n-1) = 83$
$\therefore \; \frac{5}{2}n - \frac{5}{2} = 75$
$\therefore \; \frac{5}{2}n = \frac{155}{2}$
$\therefore \; n = 31$

So, $S_n = \frac{n}{2}(u_1 + u_n)$
$= \frac{31}{2}(8 + 83)$
$= 1410\frac{1}{2}$

3 **a** $\sum_{k=1}^{10}(2k+5) = 7 + 9 + 11 + ... + 25$

This series is arithmetic with $u_1 = 7, \quad d = 2$ and $n = 10$.

$\therefore$ sum $= \frac{n}{2}[2u_1 + (n-1)d] = \frac{10}{2}[14 + 9 \times 2] = 160$

b $\sum_{k=1}^{15}(k - 50) = (-49) + (-48) + (-47) + ... + (-35)$

This series is arithmetic with $u_1 = -49, \quad d = 1$ and $n = 15$.

$\therefore$ sum $= \frac{n}{2}[2u_1 + (n-1)d] = \frac{15}{2}[-98 + 14 \times 1] = -630$

c $\sum_{k=1}^{20}\left(\frac{k+3}{2}\right) = 2 + \frac{5}{2} + 3 + ... + \frac{23}{2}$

This series is arithmetic with $u_1 = 2$, $r = \frac{1}{2}$ and $n = 20$.

$\therefore$ sum $= \frac{n}{2}[2u_1 + (n-1)d] = \frac{20}{2}[4 + 19 \times \frac{1}{2}] = 135$

4 $u_1 = 5, \quad n = 7, \quad u_n = 53$

$$S_n = \frac{n}{2}(u_1 + u_n)$$
$$= \tfrac{7}{2}(5 + 53)$$
$$= 203$$

5 $u_1 = 6, \quad n = 11, \quad u_n = -27$

$$S_n = \frac{n}{2}(u_1 + u_n)$$
$$= \tfrac{11}{2}(6 + (-27))$$
$$= -115\tfrac{1}{2}$$

6 The total number of bricks can be expressed as an arithmetic series:
$1 + 2 + 3 + 4 + ... + n$
We know that the total number of bricks is 171, so $S_n = 171$.
Also, $u_1 = 1$, $d = 1$ and we need to find n, the number of members (layers) of the series.

$$S_n = \frac{n}{2}(2u_1 + (n-1)d) = 171$$
$$\therefore \quad \frac{n}{2}(2 \times 1 + (n-1) \times 1) = 171$$
$$\therefore \quad n(2 + n - 1) = 342$$
$$\therefore \quad n(n+1) = 342$$
$$\therefore \quad n^2 + n - 342 = 0$$
$$\therefore \quad (n-18)(n+19) = 0$$
$$\therefore \quad n = -19 \text{ or } 18$$

But $n > 0$, so $n = 18$. So, there are 18 layers placed.

7 The total number of seats in n rows can be expressed as an arithmetic series:
$22 + 23 + 24 + ... + u_n$
Row 1 has 22 seats, so $u_1 = 22$. Row 2 has 23 seats, so $d = 1$.

$$S_n = \frac{n}{2}(2u_1 + (n-1)d)$$
$$= \frac{n}{2}(2 \times 22 + 1(n-1))$$
$$= \frac{n}{2}(44 + n - 1)$$

$\therefore \quad S_n = \frac{n}{2}(n + 43)$ which is the total number of seats in n rows.

a Number of seats in row 44 = total no. of seats in every row – no. of seats in the first 43 rows
$$= S_{44} - S_{43}$$
$$= \tfrac{44}{2}(44 + 43) - \tfrac{43}{2}(43 + 43)$$
$$= 1914 - 1849$$
$$= 65$$

b Number of seats in a section $= S_{44} = 1914$ (from **a**)

c Number of seats in 25 sections $= S_{44} \times 25 = 1914 \times 25 = 47\,850$

8 **a** The first 50 multiples of 11 can be expressed as an arithmetic series:
$11 + 22 + 33 + ... + u_{50}$ where $u_1 = 11$, $d = 11$, $n = 50$

$S_n = \frac{n}{2}(2u_1 + (n-1)d)$ $\quad \therefore \quad S_{50} = \frac{50}{2}(2 \times 11 + 11(50 - 1))$
$$= 25(22 + 539)$$
$$= 14\,025$$

b The multiples of 7 between 0 and 1000 can be expressed as an arithmetic series:
$7 + 14 + 21 + 28 + ... + u_n$ where $u_1 = 7$, $d = 7$
To find u_n, we need to find the largest multiple of 7 less than 1000.
Now $\frac{1000}{7} \approx 142.9$, so $u_n = 142 \times 7 = 994$

But $u_n = u_1 + (n-1)d$
$\therefore\ 994 = 7 + 7(n-1)$
$\therefore\ 987 = 7n - 7$
$\therefore\ 7n = 994$
$\therefore\ n = 142$

So, $S_{142} = \frac{142}{2}(7 + 994) = 71\,071$

c The integers between 1 and 100 which are not divisible by 3 can be expressed as:
1, 2, 4, 5, 7, 8, ..., 100 where $u_1 = 1$, $u_n = 100$.
Alternatively, these integers can be expressed as two separate arithmetic series:

$S_A = 1 + 4 + 7 + ... + 97 + 100$ where $u_1 = 1$, $d = 3$, $u_n = 100$
and $S_B = 2 + 5 + 8 + ... + 95 + 98$ where $u_1 = 2$, $d = 3$, $u_n = 98$

Now for S_A, $u_n = u_1 + (n-1)d$
$\therefore\ 100 = 1 + 3(n-1)$
$\therefore\ 99 = 3n - 3$
$\therefore\ 3n = 102$
$\therefore\ n = 34$

and for S_B, $u_n = u_1 + (n-1)d$
$\therefore\ 98 = 2 + 3(n-1)$
$\therefore\ 96 = 3n - 3$
$\therefore\ 3n = 99$
$\therefore\ n = 33$

So, $S_A = \frac{34}{2}(1 + 100) = 1717$ and $S_B = \frac{33}{2}(2 + 98) = 1650$

The total sum $= S_A + S_B = 1717 + 1650 = 3367$

9 We need to show that $1 + 2 + 3 + 4 + ... + n = \dfrac{n(n+1)}{2}$

The sum of the first n positive integers can be expressed as an arithmetic series:
$1 + 2 + 3 + 4 + ... + n$, where $u_1 = 1$, $d = 1$, $u_n = n$.

So the sum of the series is $S_n = \dfrac{n}{2}(u_1 + u_n)$

$= \dfrac{n}{2}(1 + n)$ Hence $S_n = \dfrac{n(n+1)}{2}$ as required.

10 The series of odd numbers can be expressed as an arithmetic series:
$1 + 3 + 5 + 7 + ...$ where $u_1 = 1$, $d = 2$

a Now $u_n = u_1 + (n-1)d = 1 + 2(n-1)$
$\therefore\ u_n = 2n - 1$

b We need to show that S_n is n^2.
The sum of the first n odd numbers can be expressed as an arithmetic series:
$1 + 3 + 5 + 7 + ... + (2n-1)$ {using $u_n = 2n - 1$ from **a**}

So, $S_n = \dfrac{n}{2}(u_1 + u_n)$

$= \dfrac{n}{2}(1 + (2n-1))$

$= \dfrac{n}{2}(2n)$ Hence $S_n = n^2$ as required.

c $S_1 = 1 = 1 = 1^2 = n^2$ for $n = 1$ ✓

$S_2 = 1 + 3 = 4 = 2^2 = n^2$ for $n = 2$ ✓

$S_3 = 1 + 3 + 5 = 9 = 3^2 = n^2$ for $n = 3$ ✓

$S_4 = 1 + 3 + 5 + 7 = 16 = 4^2 = n^2$ for $n = 4$ ✓

11 $u_6 = 21$, $S_{17} = 0$. We need to find u_1 and u_2.

$S_n = \frac{n}{2}(2u_1 + (n-1)d)$

$\therefore\ S_{17} = \frac{17}{2}(2u_1 + 16d) = 0$

$\therefore\ u_1 + 8d = 0$

$\therefore\ u_1 = -8d$ .... (1)

Also, $u_n = u_1 + (n-1)d$

$\therefore\ u_6 = u_1 + 5d$

$\therefore\ 21 = -8d + 5d$ {using (1)}

$\therefore\ 21 = -3d$

$\therefore\ d = -7$

So, $u_1 = -8(-7) = 56$ and $u_2 = 56 - 7 = 49$

The first two terms are 56 and 49.

12 Let the three consecutive terms be $x-d$, x and $x+d$.

Now, sum of terms = 12

$\therefore\ (x-d) + x + (x+d) = 12$

$\therefore\ 3x = 12$

$\therefore\ x = 4$

So, the terms are $4-d$, 4, $4+d$

Also, product of terms = -80

$\therefore\ (4-d)4(4+d) = -80$

$\therefore\ 4(4^2 - d^2) = -80$

$\therefore\ 16 - d^2 = -20$

$\therefore\ d^2 = 36$

$\therefore\ d = \pm 6$

So, the three terms could be $4-6$, 4, $4+6$, which are -2, 4, 10
or $4-(-6)$, 4, $4+(-6)$, which are 10, 4, -2.

13 Let the five consecutive terms be $n-2d$, $n-d$, n, $n+d$, $n+2d$.

Now, sum of terms = 40

$\therefore\ (n-2d) + (n-d) + n + (n+d) + (n+2d) = 40$

$\therefore\ 5n = 40$

$\therefore\ n = 8$

So the terms are $8-2d$, $8-d$, 8, $8+d$, $8+2d$

Also, the product of the first, middle and last terms $= (8-2d) \times 8 \times (8+2d) = 224$

$\therefore\ 8(8^2 - 4d^2) = 224$

$\therefore\ 64 - 4d^2 = 28$

$\therefore\ 4d^2 = 36$

$\therefore\ d^2 = 9$

$\therefore\ d = \pm 3$

So, the five terms could be $8-2(3)$, $8-3$, 8, $8+3$, $8+2(3)$, which are 2, 5, 8, 11, 14
or $8-2(-3)$, $8-(-3)$, 8, $8+(-3)$, $8+2(-3)$, which are 14, 11, 8, 5, 2.

EXERCISE 2E.3

1 **a** The series is geometric with $u_1 = 12$, $r = \frac{1}{2}$, $n = 10$

Now $S_n = \dfrac{u_1(1-r^n)}{1-r}$

$\therefore\ S_{10} = \dfrac{12\left(1-(\frac{1}{2})^{10}\right)}{1-\frac{1}{2}}$

$\approx 23.9766 \approx 24.0$

b The series is geometric with $u_1 = \sqrt{7}$, $r = \sqrt{7}$, $n = 12$

Now $S_n = \dfrac{u_1(r^n - 1)}{r-1}$

$\therefore\ S_{12} = \dfrac{\sqrt{7}\left((\sqrt{7})^{12}-1\right)}{\sqrt{7}-1}$

$\approx 189\,134 \approx 189\,000$

c The series is geometric with $u_1 = 6$, $r = -\frac{1}{2}$, $n = 15$

Now $S_n = \dfrac{u_1(1-r^n)}{1-r}$

$\therefore\ S_{15} = \dfrac{6\left(1-(-\frac{1}{2})^{15}\right)}{1-(-\frac{1}{2})} \approx 4.00$

d The series is geometric with $u_1 = 1$, $r = -\frac{1}{\sqrt{2}}$, $n = 20$

Now $S_n = \dfrac{u_1(1-r^n)}{1-r}$

$\therefore\ S_{20} = \dfrac{1\left(1-(-\frac{1}{\sqrt{2}})^{20}\right)}{1-(-\frac{1}{\sqrt{2}})} \approx 0.585$

2 **a** The series is geometric with

$u_1 = \sqrt{3}, \quad r = \sqrt{3}$

$$S_n = \frac{u_1(r^n - 1)}{r - 1}$$

$$= \frac{\sqrt{3}\left((\sqrt{3})^n - 1\right)}{\sqrt{3} - 1} \times \left(\frac{\sqrt{3}+1}{\sqrt{3}+1}\right)$$

$$= \frac{\left(3+\sqrt{3}\right)\left((\sqrt{3})^n - 1\right)}{3-1}$$

$$= \frac{3+\sqrt{3}}{2}\left((\sqrt{3})^n - 1\right)$$

b The series is geometric with

$u_1 = 12, \quad r = \frac{1}{2}$

$$S_n = \frac{u_1(1 - r^n)}{1 - r}$$

$$= \frac{12\left(1 - (\frac{1}{2})^n\right)}{1 - \frac{1}{2}}$$

$$= 24\left(1 - (\tfrac{1}{2})^n\right)$$

c The series is geometric with

$u_1 = 0.9, \quad r = 0.1$

$$S_n = \frac{u_1(1 - r^n)}{1 - r}$$

$$= \frac{0.9\,(1 - (0.1)^n)}{1 - 0.1}$$

$$= 1 - (0.1)^n$$

d The series is geometric with

$u_1 = 20, \quad r = -\frac{1}{2}$

$$S_n = \frac{u_1(1 - r^n)}{1 - r} = \frac{20\left(1 - (-\frac{1}{2})^n\right)}{1 - (-\frac{1}{2})}$$

$$= \frac{20\left(1 - (-\frac{1}{2})^n\right)}{(\frac{3}{2})}$$

$$= \frac{40}{3}\left(1 - (-\tfrac{1}{2})^n\right)$$

3 **a** $S_1 = u_1 \quad \therefore \quad u_1 = 3$

b $u_2 = S_2 - S_1$

$= 4 - 3 = 1$

So, $r = \frac{1}{3}$

c $u_1 = 3, \quad r = \frac{1}{3}$

so $u_n = 3 \times (\frac{1}{3})^{n-1}$

$\therefore \quad u_5 = 3 \times (\frac{1}{3})^4 = \frac{1}{27}$

4 **a** $\displaystyle\sum_{k=1}^{10} 3 \times 2^{k-1} = 3 + 6 + 12 + ... + 384 + 768 + 1536$

This series is geometric with $u_1 = 3, \quad r = 2 \quad$ and $\quad n = 10.$

$\therefore \quad \text{sum} = \dfrac{u_1(r^n - 1)}{r - 1} = \dfrac{3(2^{10} - 1)}{1} = 3069$

b $\displaystyle\sum_{k=1}^{12} \left(\tfrac{1}{2}\right)^{k-2} = 2 + 1 + \tfrac{1}{2} + ... + \tfrac{1}{256} + \tfrac{1}{512} + \tfrac{1}{1024}$

This series is geometric with $u_1 = 2, \quad r = \frac{1}{2} \quad$ and $\quad n = 12.$

$\therefore \quad \text{sum} = \dfrac{u_1(1 - r^n)}{1 - r} = \dfrac{2\left(1 - (\frac{1}{2})^{12}\right)}{\frac{1}{2}} = 4\left(1 - \left(\tfrac{1}{2}\right)^{12}\right) = \dfrac{2^{12} - 1}{2^{10}}$

$\therefore \quad \text{sum} = \frac{4095}{1024} \approx 4.00$

c $\displaystyle\sum_{k=1}^{25} 6 \times (-2)^k = -12 + 24 + (-48) + ... + 100\,663\,296 + (-201\,326\,592)$

This series is geometric with $u_1 = -12, \quad r = -2 \quad$ and $\quad n = 25.$

$\therefore \quad \text{sum} = \dfrac{u_1(1 - r^n)}{1 - r} = \dfrac{-12\,(1 - (-2)^{25})}{1 + 2} = -4\left(1 - (-2)^{25}\right)$

$\therefore \quad \text{sum} = -134\,217\,732$

5 **a** $A_2 = A_1 \times 1.06 + 2000$

$= (A_0 \times 1.06 + 2000) \times 1.06 + 2000$

$= (2000 \times 1.06 + 2000) \times 1.06 + 2000$

$\therefore \quad A_2 = 2000 + 2000 \times 1.06 + 2000 \times (1.06)^2 \quad$ as required.

b $A_3 = A_2 \times 1.06 + 2000$
$= \left[2000 + 2000 \times 1.06 + 2000 \times (1.06)^2\right] \times 1.06 + 2000$ {from **a**}
$\therefore$ $A_3 = 2000\left[1 + 1.06 + (1.06)^2 + (1.06)^3\right]$ as required.

c $A_9 = 2000[1 + 1.06 + (1.06)^2 + (1.06)^3 + (1.06)^4 + (1.06)^5 + (1.06)^6 + (1.06)^7 + (1.06)^8 + (1.06)^9]$
$\therefore$ $A_9 \approx 26\,361.59$
$\therefore$ the total bank balance after 10 years is \$26 361.59

6 **a** $S_1 = \frac{1}{2}$, $S_2 = \frac{1}{2} + \frac{1}{4} = \frac{3}{4}$, $S_3 = \frac{1}{2} + \frac{1}{4} + \frac{1}{8} = \frac{7}{8}$, $S_4 = \frac{1}{2} + \frac{1}{4} + \frac{1}{8} + \frac{1}{16} = \frac{15}{16}$,
$S_5 = \frac{1}{2} + \frac{1}{4} + \frac{1}{8} + \frac{1}{16} + \frac{1}{32} = \frac{31}{32}$

b $S_n = \dfrac{2^n - 1}{2^n}$

c $S_n = \dfrac{u_1(1 - r^n)}{1 - r}$, where $u_1 = \frac{1}{2}$, $r = \frac{1}{2}$
$= \dfrac{\frac{1}{2}\left(1 - (\frac{1}{2})^n\right)}{1 - \frac{1}{2}}$
$\therefore$ $S_n = 1 - (\frac{1}{2})^n = 1 - \dfrac{1}{2^n} = \dfrac{2^n - 1}{2^n}$

d As $n \to \infty$, $\left(\frac{1}{2}\right)^n \to 0$, and so $S_n \to 1$ (from below)

e The diagram represents one whole unit divided into smaller and smaller fractions.
As $n \to \infty$, the area which the fraction represents becomes smaller and smaller, and the total area approaches one whole unit.

EXERCISE 2E.4

1 **a** **i** $u_1 = \frac{3}{10}$

ii $r = \dfrac{(\frac{3}{100})}{(\frac{3}{10})} = \frac{1}{10}$

b We need to show that $0.\overline{3} = \frac{1}{3}$.
Now $0.\overline{3} = \frac{3}{10} + \frac{3}{100} + \frac{3}{1000} + ...$
So, let $S_n = \frac{3}{10} + \frac{3}{100} + \frac{3}{1000} + ...$
Since $n \to \infty$, then $S = \dfrac{u_1}{1 - r} = \dfrac{\frac{3}{10}}{1 - (\frac{1}{10})} = \frac{1}{3}$
So, $0.\overline{3} = \frac{1}{3}$ as required.

2 **a** $0.\overline{4} = 0.444444....$
$= \frac{4}{10} + \frac{4}{100} + \frac{4}{1000} + ...$
which is a geometric series with
$u_1 = 0.4$, $r = 0.1$
$\therefore$ $S = \dfrac{u_1}{1 - r} = \dfrac{0.4}{1 - 0.1} = \dfrac{0.4}{0.9}$
$= \frac{4}{9}$
So, $0.\overline{4} = \frac{4}{9}$

b $0.\overline{16} = 0.161616....$
$= \frac{16}{10^2} + \frac{16}{10^4} + \frac{16}{10^6} + ...$
which is a geometric series with
$u_1 = 0.16$, $r = 0.01$
$\therefore$ $S = \dfrac{u_1}{1 - r} = \dfrac{0.16}{0.99} = \dfrac{16}{99}$
So, $0.\overline{16} = \frac{16}{99}$

c $0.\overline{312} = 0.312\,312\,312\,....$
$= \frac{312}{10^3} + \frac{312}{10^6} + \frac{312}{10^9} + ...$
which is a geometric series with $u_1 = 0.312$, $r = 0.001$
$\therefore$ $S = \dfrac{u_1}{1 - r} = \dfrac{0.312}{0.999} = \dfrac{312}{999} = \dfrac{104}{333}$ So, $0.\overline{312} = \frac{104}{333}$

3 Checking **Exercise 2E.3 6d**: $S = \dfrac{u_1}{1-r} = \dfrac{\frac{1}{2}}{1-\frac{1}{2}} = 1$ ✓

4 **a** $18 + 12 + 8 + ...$ is an infinite geometric series with $u_1 = 18$, $r = \frac{2}{3}$.

$\therefore\ S = \dfrac{u_1}{1-r} = \dfrac{18}{\frac{1}{3}} = 54$

b $18.9 - 6.3 + 2.1 - ....$ is an infinite geometric series with $u_1 = 18.9$, $r = -\frac{1}{3}$.

$\therefore\ S = \dfrac{u_1}{1-r} = \dfrac{18.9}{\frac{4}{3}} = 14.175$

5 **a** $\displaystyle\sum_{k=1}^{\infty} \frac{3}{4^k} = \frac{3}{4} + \frac{3}{16} + \frac{3}{64} + ...$

is an infinite geometric series with $u_1 = \frac{3}{4}$, $r = \frac{1}{4}$.

$\therefore\ S = \dfrac{u_1}{1-r} = \dfrac{\frac{3}{4}}{\frac{3}{4}} = 1$

b $\displaystyle\sum_{k=0}^{\infty} 6(-\tfrac{2}{5})^k = 6 - 6 \times (\tfrac{2}{5}) + 6 \times (\tfrac{2}{5})^2 - ....$

is an infinite geometric series with $u_1 = 6$, $r = -\frac{2}{5}$.

$\therefore\ S = \dfrac{u_1}{1-r} = \dfrac{6}{\frac{7}{5}} = \frac{30}{7}$

6 Let the terms of the geometric series be $u_1,\ u_1r,\ u_1r^2, ...$

Then $u_1 + u_1r + u_1r^2 = 19$

$\therefore\ u_1(1 + r + r^2) = 19$

$\therefore\ u_1 = \dfrac{19}{1 + r + r^2}$ .... (1)

and $\dfrac{u_1}{1-r} = 27$

$\therefore\ u_1 = 27(1-r)$ .... (2)

Equating (1) and (2), $\dfrac{19}{1+r+r^2} = 27(1-r)$

$\therefore\ \frac{19}{27} = (1-r)(1+r+r^2)$

$\therefore\ \frac{19}{27} = 1 + r + r^2 - r - r^2 - r^3$

$\therefore\ \frac{19}{27} = 1 - r^3$

$\therefore\ r^3 = \frac{8}{27}$

$\therefore\ r = \frac{2}{3}$

Substituting $r = \frac{2}{3}$ into (2) gives $u_1 = 27(1 - \frac{2}{3}) = 9$

$\therefore$ the first term is 9 and the common ratio is $\frac{2}{3}$.

7 Let the terms of the geometric series be $u_1,\ u_1r,\ u_1r^2, ...$

Then $u_1r = \frac{8}{5}$

$\therefore\ u_1 = \dfrac{8}{5r}$ .... (1)

and $\dfrac{u_1}{1-r} = 10$

$\therefore\ u_1 = 10 - 10r$ .... (2)

Equating (1) and (2), $\dfrac{8}{5r} = 10 - 10r$

$\therefore\ 8 = 50r - 50r^2$

$\therefore\ 50r^2 - 50r + 8 = 0$

$\therefore\ 2(25r^2 - 25r + 4) = 0$

$\therefore\ 2(5r-1)(5r-4) = 0$

$\therefore\ r = \frac{1}{5}$ or $\frac{4}{5}$

Using (2), if $r = \frac{1}{5}$, $u_1 = 10 - 10(\frac{1}{5}) = 8$

if $r = \frac{4}{5}$, $u_1 = 10 - 10(\frac{4}{5}) = 2$

$\therefore$ either $u_1 = 8$, $r = \frac{1}{5}$ or $u_1 = 2$, $r = \frac{4}{5}$

8 **a** Total time of motion $= 1 + (90\% \times 1) + (90\% \times 1) + (90\% \times 90\% \times 1)$

$+ (90\% \times 90\% \times 1) + (90\% \times 90\% \times 90\% \times 1) + ...$

$= 1 + 0.9 + 0.9 + (0.9)^2 + (0.9)^2 + (0.9)^3 + ...$

$= 1 + 2(0.9) + 2(0.9)^2 + 2(0.9)^3 + ...$ as required.

b $S_n = 1 + \dfrac{u_1(1-r^n)}{1-r}$, where $u_1 = 2(0.9)$, $r = 0.9$, 'n' $= n-1$

{since the term $u_1 = 2(0.9)$ is the second term of the series, not the first}

$$\therefore\ S_n = 1 + \frac{2(0.9)\,(1-(0.9)^{n-1})}{1-0.9}$$
$$= 1 + \frac{1.8\,(1-(0.9)^{n-1})}{0.1}$$
$$= 1 + 18\left(1-(0.9)^{n-1}\right)$$

c For the ball to come to rest, n must approach infinity

$\therefore\ (0.9)^{n-1} \to 0$ and so $(1-(0.9)^{n-1}) \to 1$ (from below)

$\therefore\ S_n \to 1 + 18(1) = 19$

So, it takes 19 seconds for the ball to come to rest.

REVIEW SET 2A

1 **a** arithmetic **b** both arithmetic and geometric **c** geometric **d** neither

e 4, 8, 12, 16, ... arithmetic

2 Since the terms are consecutive,

$(k-2) - 3k = k + 7 - (k-2)$ {equating common differences}

$\therefore\ k - 2 - 3k = k + 7 - k + 2$

$\therefore\ -2 - 2k = 9$

$\therefore\ 2k = -11$

$\therefore\ k = -\frac{11}{2}$

3 28, 23, 18, 13, ...

$23 - 28 = -5$

$18 - 23 = -5$

$13 - 18 = -5$

Assuming that the pattern continues, consecutive terms differ by -5.

$\therefore$ the sequence is arithmetic with $u_1 = 28$, $d = -5$.

$$u_n = u_1 + (n-1)d$$
$$= 28 + (n-1)(-5)$$
$$= 28 - 5n + 5$$
$$= -5n + 33$$

$$S_n = \frac{n}{2}(2u_1 + (n-1)d)$$
$$= \frac{n}{2}(2 \times 28 + (n-1)(-5))$$
$$= \frac{n}{2}(56 - 5n + 5)$$
$$= \frac{n}{2}(61 - 5n)$$

4 The terms are geometric, so $\dfrac{k}{4} = \dfrac{k^2-1}{k}$

$\therefore\ k^2 = 4(k^2-1)$

$\therefore\ 3k^2 = 4$

$\therefore\ k^2 = \frac{4}{3}$

$\therefore\ k = \pm\frac{2}{\sqrt{3}} = \pm\frac{2\sqrt{3}}{3}$

5 $u_6 = \frac{16}{3}$ $\therefore\ u_1 \times r^5 = \frac{16}{3}$ .... (1)

$u_{10} = \frac{256}{3}$ $\therefore\ u_1 \times r^9 = \frac{256}{3}$ .... (2)

So, $\dfrac{u_1 r^9}{u_1 r^5} = \dfrac{(\frac{256}{3})}{(\frac{16}{3})}$ {(2) ÷ (1)}

$\therefore\ r^4 = 16$

$\therefore\ r = \pm 2$

Substituting $r = 2$ into (1) gives

$$u_1 \times 2^5 = \tfrac{16}{3}$$
$$\therefore \quad u_1 \times 32 = \tfrac{16}{3}$$
$$\therefore \quad u_1 = \tfrac{1}{6}$$

Substituting $r = -2$ into (1) gives

$$u_1 \times (-2)^5 = \tfrac{16}{3}$$
$$\therefore \quad u_1 \times (-32) = \tfrac{16}{3}$$
$$\therefore \quad u_1 = -\tfrac{1}{6}$$

Now $u_n = u_1 r^{n-1}$ $\quad \therefore \quad u_n = \frac{1}{6} \times 2^{n-1}$ or $-\frac{1}{6} \times (-2)^{n-1}$

6 Let the numbers be $23,\ 23+d,\ 23+2d,\ 23+3d,\ 23+4d,\ 23+5d,\ 23+6d,\ 9$

Then $23 + 7d = 9$

$\therefore \quad 7d = -14$

$\therefore \quad d = -2$ $\qquad$ So, the numbers are 23, 21, 19, 17, 15, 13, 11, 9.

7 **a** The sequence 86, 83, 80, 77, ... is arithmetic with $u_1 = 86$, $d = -3$.

$$u_n = u_1 + (n-1)d$$
$$\therefore \quad u_n = 86 + (n-1)(-3) = 86 - 3n + 3$$
$$\therefore \quad u_n = 89 - 3n$$

b $\frac{3}{4}, 1, \frac{7}{6}, \frac{9}{7}, ...$ can also be written as $\frac{3}{4}, \frac{5}{5}, \frac{7}{6}, \frac{9}{7}, ...$

So, the numerator starts at 3 and increases by 2 each time,
whilst the denominator starts at 4 and increases by 1 each time.

The nth term is $\dfrac{2n+1}{n+3}$, and so $u_n = \dfrac{2n+1}{n+3}$

c The sequence 100, 90, 81, 72.9, ... is geometric with $u_1 = 100$, $r = \frac{90}{100} = 0.9$

$$u_n = u_1 r^{n-1}$$
$$\therefore \quad u_n = 100(0.9)^{n-1}$$

8 **a**
$$\sum_{k=1}^{7} k^2 = 1^2 + 2^2 + 3^2 + 4^2 + 5^2 + 6^2 + 7^2$$
$$= 1 + 4 + 9 + 16 + 25 + 36 + 49$$

b
$$\sum_{k=1}^{8} \frac{k+3}{k+2} = \tfrac{4}{3} + \tfrac{5}{4} + \tfrac{6}{5} + \tfrac{7}{6} + \tfrac{8}{7} + \tfrac{9}{8} + \tfrac{10}{9} + \tfrac{11}{10}$$

9 **a** $18 - 12 + 8 - ....$

The series is geometric with

$u_1 = 18$, $r = -\frac{2}{3}$

$$\therefore \quad S = \frac{u_1}{1-r}$$
$$= \frac{18}{\frac{5}{3}}$$
$$= \tfrac{54}{5} \text{ or } 10\tfrac{4}{5}$$

b $8 + 4\sqrt{2} + 4 + ...$

The series is geometric with

$u_1 = 8$, $r = \frac{1}{\sqrt{2}}$

$$\therefore \quad S = \frac{u_1}{1-r} = \frac{8}{(1 - \frac{1}{\sqrt{2}})} \times \frac{(1 + \frac{1}{\sqrt{2}})}{(1 + \frac{1}{\sqrt{2}})}$$
$$= \frac{8 + \frac{8}{\sqrt{2}}}{1 - \frac{1}{2}}$$
$$= \frac{8 + 4\sqrt{2}}{\frac{1}{2}}$$
$$= 16 + 8\sqrt{2}$$

10

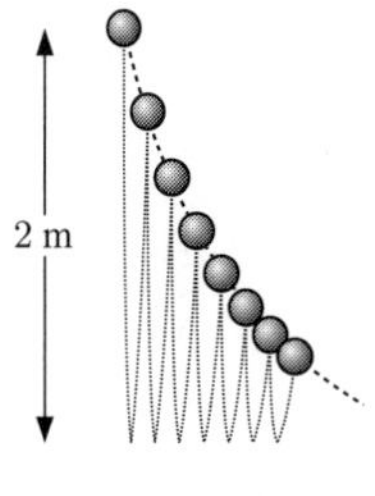

Total distance travelled

$$= 2 + 2 \times 0.8 \times 2 + 2 \times (0.8)^2 \times 2 + 2 \times (0.8)^3 \times 2 + ...$$
$$= 2 + 2 \times 0.8 \times 2 \left[1 + 0.8 + (0.8)^2 + (0.8)^3 + ...\right]$$
$$= 2 + 3.2 \times \frac{1}{1 - 0.8} \qquad \left\{\text{as } r = 0.8,\ |r| < 1 \text{ so converges to } \frac{u_1}{1-r}\right\}$$
$$= 2 + \tfrac{3.2}{0.2}$$
$$= 2 + 16 = 18 \text{ metres}$$

11 **a** $S_n = \dfrac{3n^2+5n}{2}$

$\therefore \quad u_n = S_n - S_{n-1}$

$= \dfrac{3n^2+5n}{2} - \dfrac{3(n-1)^2+5(n-1)}{2}$

$= \dfrac{3n^2+5n-3(n^2-2n+1)-5(n-1)}{2}$

$= \dfrac{3n^2+5n-3n^2+6n-3-5n+5}{2}$

$= \dfrac{6n+2}{2}$

$\therefore \quad u_n = 3n+1$

b Using part **a**,

$u_n - u_{n-1} = [3n+1]-[3(n-1)+1]$

$= 3n+1-3n+3-1$

$= 3$

The difference between consecutive terms is constant for all n, so the sequence is arithmetic.

REVIEW SET 2B

1 $u_n = 6\left(\frac{1}{2}\right)^{n-1}$

a $\dfrac{u_{n+1}}{u_n} = \dfrac{6\left(\frac{1}{2}\right)^{n+1-1}}{6\left(\frac{1}{2}\right)^{n-1}} = \frac{1}{2}$ for all n

$\therefore$ $\{u_n\}$ is a geometric sequence.

b $u_1 = 6,$

$r = \frac{1}{2}$

c $u_{16} = 6\left(\frac{1}{2}\right)^{15}$

$\approx 0.000\,183$

2 **a** Given $24, 23\frac{1}{4}, 22\frac{1}{2}, ..., -36$ we have $u_1 = 24,$ $u_n = -36,$ and we need to find n.

The sequence is arithmetic with $d = -\frac{3}{4}$.

Now $u_n = u_1 + (n-1)d$

$\therefore \quad -36 = 24 + (n-1)(-\frac{3}{4})$

$\therefore \quad -60 = -\frac{3}{4}n + \frac{3}{4}$

$\therefore \quad \frac{3}{4}n = \frac{243}{4}$

$\therefore \quad n = 81$ So, there are 81 terms in the sequence.

b $u_{35} = 24 + (35-1)(-\frac{3}{4})$

$= 24 - \frac{102}{4}$

$= -\frac{3}{2}$

c $S_n = \dfrac{n}{2}(u_1 + u_n)$

$\therefore \quad S_{81} = \frac{81}{2}(24 + (-36))$

$= -486$

3 **a** $3 + 9 + 15 + 21 + ...$

The series is arithmetic with

$u_1 = 3, \quad d = 6, \quad n = 23$

Now $S_n = \dfrac{n}{2}(2u_1 + (n-1)d)$

$\therefore \quad S_{23} = \frac{23}{2}(2 \times 3 + 6(23-1))$

$\therefore \quad S_{23} = \frac{23}{2}(6 + 132)$

$= 1587$

b $24 + 12 + 6 + 3 + ...$

The series is geometric with

$u_1 = 24, \quad r = \frac{1}{2}, \quad n = 12$

$S_n = \dfrac{u_1(1-r^n)}{1-r}$

$\therefore \quad S_{12} = \dfrac{24\left(1-(\frac{1}{2})^{12}\right)}{1-\frac{1}{2}}$

$= 48\left(1-(\frac{1}{2})^{12}\right)$

$= 47\frac{253}{256} \approx 48.0$

4 5, 10, 20, 40, ... The sequence is geometric with $u_1 = 5, \quad r = 2$

$u_n = u_1 r^{n-1} = 5 \times 2^{n-1}$

Let $u_n = 10\,000 = 5 \times 2^{n-1}$

$\therefore \quad 2000 = 2^{n-1}$

$\therefore \quad n \approx 11.97$ {using technology}

We try the two values on either side of $n \approx 11.97$, which are $n = 11$ and $n = 12$:

$u_{11} = 5 \times 2^{10} = 5120$ and $u_{12} = 5 \times 2^{11} = 10\,240$

So, the first term to exceed 10 000 is $u_{12} = 10\,240$.

5 **a** $u_6 = u_1 \times r^5$ is the amount after 5 years, where $u_1 = 6000$, $r = 1.07$

$= 6000 \times (1.07)^5$

≈ 8415.31 So, the value of the investment will be €8415.31.

b If interest is compounded quarterly, then $r = 1 + \dfrac{0.07}{4} = 1.0175$

and $n = 5 \times 4 = 20$

$\therefore\ u_{21} = u_1 \times r^{20}$

$= 6000 \times (1.0175)^{20}$

≈ 8488.67 So, the value of the investment will be €8488.67.

c If interest is compounded monthly, then $r = 1 + \dfrac{0.07}{12} = 1.005\,8\bar{3}$

and $n = 5 \times 12 = 60$

$\therefore\ u_{61} = u_1 \times r^{60}$

$= 6000 \times (1.005\,8\bar{3})^{60}$

≈ 8505.75 So, the value of the investment will be €8505.75

6 **a** $u_n = 5n - 8$

$\therefore\ u_{10} = 5 \times 10 - 8 = 42$

b $u_{n+1} - u_n = (5(n+1) - 8) - (5n - 8)$

$= 5n + 5 - 8 - 5n + 8$

$= 5$

c The difference between consecutive terms u_n and u_{n+1} is constant for all n, so the sequence is arithmetic.

d $S_n = \dfrac{n}{2}(2u_1 + (n-1)d)$ where $d = 5$ and $u_1 = 5 \times 1 - 8 = -3$

Now, $u_{15} + u_{16} + u_{17} + ... + u_{30} = S_{30} - S_{14}$

$= \frac{30}{2}(2(-3) + (30-1) \times 5) - \frac{14}{2}(2(-3) + (14-1) \times 5)$

$= 1672$

7 $u_6 = 24$ $\therefore\ u_1 \times r^5 = 24$ .... (1)

$u_{11} = 768$ $\therefore\ u_1 \times r^{10} = 768$ .... (2)

So $\dfrac{u_1 r^{10}}{u_1 r^5} = \dfrac{768}{24}$ {(2) ÷ (1)}

$\therefore\ r^5 = 32$

$\therefore\ r = 2$

Substituting $r = 2$ into (1) gives $u_1 \times 2^5 = 24$

$\therefore\ u_1 = \frac{24}{32} = \frac{3}{4}$

$u_n = u_1 r^{n-1} = \left(\frac{3}{4}\right) 2^{n-1}$

a $u_{17} = \left(\frac{3}{4}\right) 2^{17-1}$

$= 49\,152$

b $S_n = \dfrac{u_1(r^n - 1)}{r - 1} = \dfrac{\frac{3}{4}(2^n - 1)}{2 - 1}$

$= \frac{3}{4}(2^n - 1)$

$\therefore\ S_{15} = \frac{3}{4}(2^{15} - 1) = 24\,575.25$

8 24, 8, $\frac{8}{3}$, $\frac{8}{9}$, ... is geometric with $u_1 = 24$, $r = \frac{1}{3}$

$u_n = u_1 r^{n-1} = 24\left(\frac{1}{3}\right)^{n-1}$

Given $u_n = 0.001$, we need to find n, so $u_n = 24\left(\frac{1}{3}\right)^{n-1} = 0.001$

$\therefore \left(\frac{1}{3}\right)^{n-1} = \frac{0.001}{24}$

$\therefore n \approx 10.18$ {using technology}

We try the two values on either side of $n \approx 10.18$, which are $n = 10$ and $n = 11$:

$u_{10} = 24\left(\frac{1}{3}\right)^9 = \frac{8}{6561} \approx 0.001\,22$ and $u_{11} = 24\left(\frac{1}{3}\right)^{10} = \frac{8}{19\,683} \approx 0.000\,406$

$\therefore u_{11} \approx 0.000\,406$ is the first term of the sequence which is less than 0.001.

9 **a** $128, 64, 32, 16, ..., \frac{1}{512}$ is geometric with:

$u_1 = 128,\ r = \frac{1}{2},\ u_n = \frac{1}{512}$

$$\begin{aligned} u_n &= u_1 r^{n-1} \\ &= 128\left(\tfrac{1}{2}\right)^{n-1} \\ &= 2^7 \times 2^{1-n} \end{aligned}$$

$\therefore \frac{1}{512} = 2^7 \times 2^{1-n}$

$\therefore 2^{-9} = 2^{8-n}$

$\therefore -9 = 8 - n$

$\therefore n = 17$ So, there are 17 terms in the sequence.

b $S_n = \frac{u_1(1 - r^n)}{1 - r}$

$\therefore S_{17} = \frac{128\left(1 - (\frac{1}{2})^{17}\right)}{1 - \frac{1}{2}}$

$= 255\frac{511}{512}$

≈ 255.998

≈ 256

10 **a** $1.21 - 1.1 + 1 - ....$ is an infinite geometric series with $u_1 = 1.21$, $r = -\frac{10}{11}$.

$\therefore S = \frac{u_1}{1 - r} = \frac{1.21}{\frac{21}{11}} = \frac{1331}{2100}$

$\therefore S \approx 0.634$

b $\frac{14}{3} + \frac{4}{3} + \frac{8}{21} + ...$ is an infinite geometric series with $u_1 = \frac{14}{3}$, $r = \frac{2}{7}$.

$\therefore S = \frac{u_1}{1 - r} = \frac{\frac{14}{3}}{\frac{5}{7}}$

$\therefore S = \frac{98}{15}$ or $6\frac{8}{15}$

11 $u_{n+1} = u_1 \times r^n$ where $u_{n+1} = 20\,000$, $r = 1 + \frac{0.09}{12} = 1.0075$, $n = 4 \times 12 = 48$

$\therefore 20\,000 = u_1 \times (1.0075)^{48}$

$\therefore u_1 \approx 13\,972.28$ So, \$13 972.28 should be invested.

12 **a** $u_{n+1} = u_1 \times r^n$ where $u_1 = 3000$, $r = 1.05$, $n = 3$

$\therefore u_{n+1} = 3000 \times (1.05)^3$

$= 3472.875$ There were approximately 3470 koalas.

b $u_{n+1} = u_1 \times r^n$ where $u_1 = 3000$, $u_{n+1} = 5000$, $r = 1.05$

$\therefore 5000 = 3000 \times (1.05)^n$

$\therefore n \approx 10.47$ {using technology}

After 10.47 years the population will exceed 5000. This is during the year 2014.

REVIEW SET 2C

1 $u_n = 68 - 5n$

a $u_{n+1} - u_n = [68 - 5(n + 1)] - [68 - 5n]$

$= 68 - 5n - 5 - 68 + 5n$

$= -5$ for all n

$\therefore$ the sequence is arithmetic with common difference $d = -5$.

b $u_1 = 68 - 5(1) = 63$, $d = -5$

c $u_{37} = 68 - 5(37) = -117$

d Let $u_n = -200$, and we need to find n.

$u_n = 68 - 5n = -200$

$\therefore \quad 5n = 268$

$\therefore \quad n = 53\frac{3}{5}$

We try the two values on either side of $n = 53\frac{3}{5}$, which are $n = 53$ and $n = 54$:

$u_{53} = 68 - 5(53) = -197$ and $u_{54} = 68 - 5(54) = -202$

So, the first term of the sequence less than -200 is $u_{54} = -202$.

2 **a** 3, 12, 48, 192, ...

$\frac{12}{3} = 4 \quad \frac{48}{12} = 4 \quad \frac{192}{48} = 4$

Assuming the pattern continues, consecutive terms have a common ratio of 4.

$\therefore$ the sequence is geometric with $u_1 = 3$ and $r = 4$.

b $u_n = u_1 r^{n-1}$

$\therefore \quad u_n = 3 \times 4^{n-1}$

$\therefore \quad u_9 = 3 \times 4^8 = 196\,608$

3 $u_7 = 31 \quad \therefore \quad u_1 + 6d = 31 \quad .... (1)$

$u_{15} = -17 \quad \therefore \quad u_1 + 14d = -17 \quad .... (2)$

So, $(u_1 + 14d) - (u_1 + 6d) = -17 - 31 \quad \{(2) - (1)\}$

$\therefore \quad 8d = -48$

$\therefore \quad d = -6$

So in (1), $u_1 + 6(-6) = 31$

$\therefore \quad u_1 - 36 = 31$

$\therefore \quad u_1 = 67$

Now $u_n = u_1 + (n-1)d$

$\therefore \quad u_n = 67 + (n-1)(-6)$

$\therefore \quad u_n = 67 - 6n + 6$

$\therefore \quad u_n = -6n + 73$

So, $u_{34} = -6(34) + 73 = -131$

4 **a** $4 + 11 + 18 + 25 + ...$

The series is arithmetic with $u_1 = 4$, $d = 7$, $u_k = u_1 + (k-1)d = 4 + 7(k-1) = 7k - 3$

So, the series is $\sum_{k=1}^{n} (7k - 3)$

b $\frac{1}{4} + \frac{1}{8} + \frac{1}{16} + \frac{1}{32} + ...$

The series is geometric with $u_1 = \frac{1}{4}$, $r = \frac{1}{2}$,

$$u_k = u_1 r^{k-1} = \frac{1}{4} \times \left(\frac{1}{2}\right)^{k-1} = \left(\frac{1}{2}\right)^2 \left(\frac{1}{2}\right)^{k-1} = \left(\frac{1}{2}\right)^{k+1}$$

So, the series is $\sum_{k=1}^{n} \left(\frac{1}{2}\right)^{k+1}$

5 **a** $\sum_{k=1}^{8} \left(\frac{31 - 3k}{2}\right) = 14 + 12\frac{1}{2} + 11 + 9\frac{1}{2} + 8 + 6\frac{1}{2} + 5 + 3\frac{1}{2}$

This series is arithmetic with $u_1 = 14$, $n = 8$ and $u_n = 3\frac{1}{2}$.

$\therefore$ the sum is $\frac{8}{2}(14 + 3\frac{1}{2}) = 70$

b $\sum_{k=1}^{15} 50(0.8)^{k-1} \approx 50 + 40 + 32 + ... + 3.436 + 2.749 + 2.199$

This series is geometric with $u_1 = 50$, $r = 0.8$ and $n = 15$.

$\therefore$ the sum is $\dfrac{50\left[1-(0.8)^{15}\right]}{1-0.8} \approx 241$

6 $\sum\limits_{k=7}^{\infty} 5(\frac{2}{5})^{k-1} = 5(\frac{2}{5})^6 + 5(\frac{2}{5})^7 + 5(\frac{2}{5})^8 + ...$

The series is an infinite geometric series with $u_1 = 5(\frac{2}{5})^6$, $r = \frac{2}{5}$

$$\therefore \quad S = \frac{u_1}{1-r} = \frac{5(\frac{2}{5})^6}{\frac{3}{5}} = \frac{2^6}{3 \times 5^4} = \tfrac{64}{1875}$$

7 $11 + 16 + 21 + 26 + ...$ is arithmetic with $u_1 = 11$, $d = 5$

$$\therefore \quad S_n = \frac{n}{2}(2u_1 + (n-1)d) = \frac{n}{2}(2 \times 11 + 5(n-1)) = \frac{n}{2}(22 + 5n - 5) = \frac{n}{2}(5n+17)$$

Given $S_n = 450$, we need to find n,

so $S_n = \frac{n}{2}(5n+17) = 450$

$\therefore$ $\frac{5}{2}n^2 + \frac{17}{2}n - 450 = 0$

$\therefore$ $5n^2 + 17n - 900 = 0$

$\therefore$ $n \approx -15.2,\ 11.8$ {using technology}

But $n > 0$, so $n \approx 11.8$

We try the two values on either side of $n \approx 11.8$, which are $n = 11$ and $n = 12$:

$S_{11} = \frac{11}{2}(5(11) + 17) = 396$ and $S_{12} = \frac{12}{2}(5(12) + 17) = 462$

$\therefore$ 12 terms of the series are required to exceed a sum of 450.

8 **a** $u_{n+1} = u_1 \times r^n$ where $u_1 = 12\,500$, $r = 1 + \dfrac{0.0825}{2} = 1.041\,25$, $n = 5 \times 2 = 10$

So, $u_{n+1} = 12\,500 \times (1.041\,25)^{10}$
$\approx 18\,726.65$ The value of the investment is £18 726.65.

b $u_{n+1} = u_1 \times r^n$ where $u_1 = 12\,500$, $r = 1 + \dfrac{0.0825}{12} = 1.006\,875$, $n = 5 \times 12 = 60$

So, $u_{n+1} = 12\,500 \times (1.006\,875)^{60}$
$\approx 18\,855.74$ The value of the investment is £18 855.74.

9 **a** Let the terms of the geometric series be $u_1, u_1r, u_1r^2, ...$

Then $u_1 + u_1 r = 90$ and $u_1 r^2 = 24$

$\therefore$ $u_1(1+r) = 90$

$\therefore$ $u_1 = \dfrac{90}{1+r}$ .... (1)

$\therefore$ $u_1 = \dfrac{24}{r^2}$ .... (2)

Equating (1) and (2) gives $\dfrac{90}{1+r} = \dfrac{24}{r^2}$

$\therefore$ $90r^2 = 24r + 24$

$\therefore$ $90r^2 - 24r - 24 = 0$

$\therefore$ $6(15r^2 - 4r - 4) = 0$

$\therefore$ $6(5r+2)(3r-2) = 0$

$\therefore$ $r = -\frac{2}{5}$ or $\frac{2}{3}$

Using (2), if $r = -\frac{2}{5}$ then $u_1 = \dfrac{24}{(-\frac{2}{5})^2} = \dfrac{24}{\frac{4}{25}} = 150$

if $r = \frac{2}{3}$ then $u_1 = \dfrac{24}{(\frac{2}{3})^2} = \dfrac{24}{\frac{4}{9}} = 54$

$\therefore$ either $u_1 = 150$, $r = -\frac{2}{5}$ or $u_1 = 54$, $r = \frac{2}{3}$

b Since $|r| < 1$ in each case, both series converge.

When $u_1 = 150,\ r = -\frac{2}{5}$

$$\therefore\ S = \frac{u_1}{1-r} = \frac{150}{\frac{7}{5}} = \frac{750}{7} \text{ or } 107\tfrac{1}{7}$$

When $u_1 = 54,\ r = \frac{2}{3}$

$$\therefore\ S = \frac{u_1}{1-r} = \frac{54}{\frac{1}{3}} = 162$$

10 Since Seve walks an additional 500 m $= 0.5$ km each week, we have an arithmetic sequence with $u_1 = 10$ and constant difference $d = 0.5$.

$$u_n = u_1 + (n-1)d$$
$$\therefore\ u_n = 10 + (n-1)0.5$$

a $u_{52} = 10 + (52-1)0.5$ {52 weeks in a year}
$= 35.5$

$\therefore$ Seve walks 35.5 km in the last week.

b In total, Seve walks $10 + 10.5 + 11 + ... + 35.5$, which is an arithmetic series.

$$S_n = \frac{n}{2}(u_1 + u_n)$$
$$\therefore\ S_{52} = \tfrac{52}{2}(10 + 35.5) = 1183$$

$\therefore$ Seve walks 1183 km in total.

11 **a** $\sum_{k=1}^{\infty} 50\,(2x-1)^{k-1}$ is a geometric series with $r = 2x - 1$ and converges if $-1 < r < 1$

$$\therefore\ -1 < 2x - 1 < 1$$
$$\therefore\ 0 < 2x < 2$$
$$\therefore\ 0 < x < 1$$

b When $x = 0.3,\ 2x - 1 = 0.6 - 1 = -0.4$

and $\sum_{k=1}^{\infty} 50\,(2x-1)^{k-1} = 50(-0.4)^0 + 50(-0.4)^1 + 50(-0.4)^2 + ...$

which is geometric with $u_1 = 50,\ r = -0.4$

Now as $0 < 0.3 < 1$, the series converges and $S = \dfrac{u_1}{1-r} = \dfrac{50}{1+0.4} = \dfrac{50}{\frac{7}{5}} = 35\tfrac{5}{7}$

Chapter 3
EXPONENTIALS

EXERCISE 3A

1 **a** $2^1 = 2,\ 2^2 = 4,\ 2^3 = 8,\ 2^4 = 16,\ 2^5 = 32,\ 2^6 = 64$
b $3^1 = 3,\ 3^2 = 9,\ 3^3 = 27,\ 3^4 = 81,\ 3^5 = 243,\ 3^6 = 729$
c $4^1 = 4,\ 4^2 = 16,\ 4^3 = 64,\ 4^4 = 256,\ 4^5 = 1024,\ 4^6 = 4096$

2 **a** $5^1 = 5,\ 5^2 = 25,\ 5^3 = 125,\ 5^4 = 625$ **b** $6^1 = 6,\ 6^2 = 36,\ 6^3 = 216,\ 6^4 = 1296$
c $7^1 = 7,\ 7^2 = 49,\ 7^3 = 343,\ 7^4 = 2401$

EXERCISE 3B

1 **a** $(-1)^5$
$= (-1) \times (-1) \times (-1) \times (-1) \times (-1)$
$= 1 \times 1 \times (-1)$
$= -1$

b $(-1)^6$
$= (-1)^5 \times (-1)$
$= (-1) \times (-1)$
$= 1$

c $(-1)^{14}$
$= 1$

d $(-1)^{19}$
$= -1$

e $(-1)^8$
$= 1$

f -1^8
$= -(1^8)$
$= -1$

g $-(-1)^8$
$= -(1)$
$= -1$

h $(-2)^5$
$= (-2) \times (-2) \times (-2) \times (-2) \times (-2)$
$= 4 \times 4 \times (-2)$
$= -32$

i -2^5
$= -(2^5)$
$= -32$

j $-(-2)^6$
$= -(-2)^5 \times (-2)$
$= 32 \times (-2)$
$= -64$

k $(-5)^4$
$= (-5) \times (-5) \times (-5) \times (-5)$
$= 25 \times 25$
$= 625$

l $-(-5)^4$
$= -(-5) \times (-5) \times (-5) \times (-5)$
$= -25 \times 25$
$= -625$

2 **a** $4^7 = 16\,384$ **b** $7^4 = 2401$ **c** $-5^5 = -3125$ **d** $(-5)^5 = -3125$
e $8^6 = 262\,144$ **f** $(-8)^6 = 262\,144$ **g** $-8^6 = -262\,144$
h $2.13^9 \approx 902.436\,039\,6$ **i** $-2.13^9 \approx -902.436\,039\,6$ **j** $(-2.13)^9 \approx -902.436\,039\,6$

3 **a** $9^{-1} = 0.\overline{1}$ **b** $\dfrac{1}{9^1} = 0.\overline{1}$ **c** $6^{-2} = 0.02\overline{7}$ **d** $\dfrac{1}{6^2} = 0.02\overline{7}$
e $3^{-4} \approx 0.012\,345\,679$ **f** $\dfrac{1}{3^4} \approx 0.012\,345\,679$ **g** $17^0 = 1$ **h** $(0.366)^0 = 1$

We notice that $a^{-n} = \dfrac{1}{a^n}$ and $a^0 = 1$ for $a \neq 0$.

4 $3^{101} = \underbrace{3^4 \times 3^4 \times 3^4 \times .... \times 3^4}_{\text{25 of these}} \times 3^1$

But $3^4 = 81$ which ends in a 1
$\therefore \underbrace{3^4 \times 3^4 \times 3^4 \times .... \times 3^4}_{\text{25 of these}}$ ends in a 1
$\therefore 3^{101}$ ends in a 3

5 $7^1 = 7,\ \ 7^2 = 49,\ \ 7^3 = 343,\ \ 7^4 = 2401,\ \ 7^5 = 16\,807$
Now $7^{217} = \underbrace{7^4 \times 7^4 \times 7^4 \times .... \times 7^4}_{\text{54 of these so this part ends in a 1}} \times 7^1$

$\therefore 7^{217}$ ends in $1 \times 7 = 7$.

6 **a** On the first square, there is $1 = 2^0$ grains of rice.
On the second square, there are $2 = 2^1$ grains of rice.
On the third square, there are $4 = 2^2$ grains of rice.
$\therefore$ in general, on the sth square, there are $N = 2^{s-1}$ grains of rice.

b When $s = 40$, $N = 2^{40-1} = 2^{39}$
$\therefore$ on the 40th square, there are $2^{39} \approx 5.50 \times 10^{11}$ grains of rice.

c Total number of grains of rice $= 2^0 + 2^1 + 2^2 + 2^3 + ... + 2^{63}$, which is a geometric series with $u_1 = 1$, $r = 2$, $n = 64$.

$$\therefore \text{ sum} = \frac{u_1(r^n - 1)}{r - 1} = \frac{1(2^{64} - 1)}{2 - 1} = 2^{64} - 1$$

$\therefore$ the king owes a total of $2^{64} - 1 \approx 1.84 \times 10^{19}$ grains of rice.

EXERCISE 3C

1 **a** $5^4 \times 5^7 = 5^{4+7} = 5^{11}$

b $d^2 \times d^6 = d^{2+6} = d^8$

c $\dfrac{k^8}{k^3} = k^{8-3} = k^5$

d $\dfrac{7^5}{7^6} = 7^{5-6} = 7^{-1} = \frac{1}{7}$

e $(x^2)^5 = x^{2\times 5} = x^{10}$

f $(3^4)^4 = 3^{4\times 4} = 3^{16}$

g $\dfrac{p^3}{p^7} = p^{3-7} = p^{-4}$ or $\dfrac{1}{p^4}$

h $n^3 \times n^9 = n^{3+9} = n^{12}$

i $(5^t)^3 = 5^{t\times 3} = 5^{3t}$

j $7^x \times 7^2 = 7^{x+2}$

k $\dfrac{10^3}{10^q} = 10^{3-q}$

l $(c^4)^m = c^{4\times m} = c^{4m}$

2 **a** $4 = 2 \times 2 = 2^2$

b $\dfrac{1}{4} = \dfrac{1}{2^2} = 2^{-2}$

c $8 = 2 \times 2 \times 2 = 2^3$

d $\dfrac{1}{8} = \dfrac{1}{2^3} = 2^{-3}$

e $32 = 2 \times 2 \times 2 \times 2 \times 2 = 2^5$

f $\dfrac{1}{32} = \dfrac{1}{2^5} = 2^{-5}$

g $2 = 2^1$

h $\dfrac{1}{2} = \dfrac{1}{2^1} = 2^{-1}$

i $64 = 32 \times 2 = 2^5 \times 2^1 = 2^6$

j $\dfrac{1}{64} = \dfrac{1}{2^6} = 2^{-6}$

k $128 = 64 \times 2 = 2^6 \times 2^1 = 2^7$

l $\dfrac{1}{128} = \dfrac{1}{2^7} = 2^{-7}$

3 **a** $9 = 3 \times 3 = 3^2$

b $\dfrac{1}{9} = \dfrac{1}{3^2} = 3^{-2}$

c $27 = 3 \times 3 \times 3 = 3^3$

d $\dfrac{1}{27} = \dfrac{1}{3^3} = 3^{-3}$

e $3 = 3^1$

f $\dfrac{1}{3} = \dfrac{1}{3^1} = 3^{-1}$

g $81 = 3 \times 3 \times 3 \times 3 = 3^4$

h $\dfrac{1}{81} = \dfrac{1}{3^4} = 3^{-4}$

i $1 = 3^0$

j $243 = 81 \times 3 = 3^4 \times 3^1 = 3^5$

k $\dfrac{1}{243} = \dfrac{1}{3^5} = 3^{-5}$

4 **a** $2 \times 2^a = 2^1 \times 2^a$
$= 2^{a+1}$

b $4 \times 2^b = 2^2 \times 2^b$
$= 2^{b+2}$

c $8 \times 2^t = 2^3 \times 2^t$
$= 2^{t+3}$

d $(2^{x+1})^2 = 2^{2(x+1)}$
$= 2^{2x+2}$

e $(2^{1-n})^{-1} = 2^{-(1-n)}$
$= 2^{n-1}$

f $\dfrac{2^c}{4} = \dfrac{2^c}{2^2} = 2^{c-2}$

g $\dfrac{2^m}{2^{-m}} = 2^{m-(-m)}$
$= 2^{2m}$

h $\dfrac{4}{2^{1-n}} = \dfrac{2^2}{2^{1-n}}$
$= 2^{2-(1-n)}$
$= 2^{n+1}$

i $\dfrac{2^{x+1}}{2^x} = 2^{x+1-x}$
$= 2^1$

j $\dfrac{4^x}{2^{1-x}} = \dfrac{(2^2)^x}{2^{1-x}}$
$= 2^{2x-(1-x)}$
$= 2^{3x-1}$

5 **a** $9 \times 3^p = 3^2 \times 3^p$
$= 3^{p+2}$

b $27^a = (3^3)^a$
$= 3^{3a}$

c $3 \times 9^n = 3^1 \times (3^2)^n$
$= 3^{2n+1}$

d $27 \times 3^d = 3^3 \times 3^d$
$= 3^{d+3}$

e $9 \times 27^t = 3^2 \times (3^3)^t$
$= 3^{3t+2}$

f $\dfrac{3^y}{3} = \dfrac{3^y}{3^1} = 3^{y-1}$

g $\dfrac{3}{3^y} = \dfrac{3^1}{3^y}$
$= 3^{1-y}$

h $\dfrac{9}{27^t} = \dfrac{3^2}{(3^3)^t}$
$= 3^{2-3t}$

i $\dfrac{9^a}{3^{1-a}} = \dfrac{(3^2)^a}{3^{1-a}}$
$= 3^{2a-(1-a)}$
$= 3^{3a-1}$

j $\dfrac{9^{n+1}}{3^{2n-1}} = \dfrac{(3^2)^{n+1}}{3^{2n-1}}$
$= 3^{2n+2-(2n-1)}$
$= 3^3$

6 **a** $(2a)^2 = 2^2 \times a^2$
$= 4a^2$

b $(3b)^3 = 3^3 \times b^3$
$= 27b^3$

c $(ab)^4 = a^4 \times b^4$
$= a^4b^4$

d $(pq)^3 = p^3 \times q^3$
$= p^3q^3$

e $\left(\dfrac{m}{n}\right)^2 = \dfrac{m^2}{n^2}$

f $\left(\dfrac{a}{3}\right)^3 = \dfrac{a^3}{3^3} = \dfrac{a^3}{27}$

g $\left(\dfrac{b}{c}\right)^4 = \dfrac{b^4}{c^4}$

h $\left(\dfrac{2a}{b}\right)^0 = 1$

i $\left(\dfrac{m}{3n}\right)^4 = \dfrac{m^4}{3^4 \times n^4} = \dfrac{m^4}{81n^4}$

j $\left(\dfrac{xy}{2}\right)^3 = \dfrac{x^3y^3}{2^3} = \dfrac{x^3y^3}{8}$

7 **a** $(-2a)^2$
$= (-2)^2a^2$
$= 4a^2$

b $(-6b^2)^2$
$= (-6)^2b^4$
$= 36b^4$

c $(-2a)^3$
$= (-2)^3a^3$
$= -8a^3$

d $(-3m^2n^2)^3$
$= (-3)^3m^6n^6$
$= -27m^6n^6$

e $(-2ab^4)^4$
$= (-2)^4a^4b^{16}$
$= 16a^4b^{16}$

f $\left(\dfrac{-2a^2}{b^2}\right)^3$
$= \dfrac{(-2)^3a^6}{b^6}$
$= -\dfrac{8a^6}{b^6}$

g $\left(\dfrac{-4a^3}{b}\right)^2$
$= \dfrac{(-4)^2a^6}{b^2}$
$= \dfrac{16a^6}{b^2}$

h $\left(\dfrac{-3p^2}{q^3}\right)^2$
$= \dfrac{(-3)^2p^4}{q^6}$
$= \dfrac{9p^4}{q^6}$

8 **a** $ab^{-2} = \dfrac{a}{b^2}$

b $(ab)^{-2} = \dfrac{1}{(ab)^2}$
$= \dfrac{1}{a^2b^2}$

c $(2ab^{-1})^2 = 2^2a^2b^{-2}$
$= \dfrac{4a^2}{b^2}$

d $(3a^{-2}b)^2 = 3^2a^{-4}b^2$
$= \dfrac{9b^2}{a^4}$

e $\dfrac{a^2b^{-1}}{c^2} = \dfrac{a^2}{bc^2}$

f $\dfrac{a^2b^{-1}}{c^{-2}} = \dfrac{a^2c^2}{b}$

g $\dfrac{1}{a^{-3}} = a^3$

h $\dfrac{a^{-2}}{b^{-3}} = \dfrac{b^3}{a^2}$

i $\dfrac{2a^{-1}}{d^2} = \dfrac{2}{ad^2}$

j $\dfrac{12a}{m^{-3}} = 12am^3$

9 **a** $\dfrac{1}{a^n} = a^{-n}$ **b** $\dfrac{1}{b^{-n}} = b^n$ **c** $\dfrac{1}{3^{2-n}} = 3^{n-2}$ **d** $\dfrac{a^n}{b^{-m}} = a^n b^m$

e $\dfrac{a^{-n}}{a^{2+n}} = a^{-n-(2+n)}$
$= a^{-2n-2}$

10 **a** $(\frac{5}{3})^0 = 1$ **b** $(\frac{7}{4})^{-1} = \frac{4}{7}$ **c** $(\frac{1}{6})^{-1} = \frac{6}{1} = 6$ **d** $\dfrac{3^3}{3^0} = \frac{27}{1} = 27$

e $(\frac{4}{3})^{-2} = \dfrac{3^2}{4^2}$
$= \frac{9}{16}$

f $2^1 + 2^{-1} = 2 + \frac{1}{2}$
$= \frac{5}{2}$

g $(1\frac{2}{3})^{-3} = (\frac{5}{3})^{-3}$
$= \dfrac{3^3}{5^3}$
$= \frac{27}{125}$

h $5^2 + 5^1 + 5^{-1}$
$= 25 + 5 + \frac{1}{5}$
$= \frac{151}{5}$

11 **a** $\dfrac{1}{9} = \dfrac{1}{3^2}$
$= 3^{-2}$

b $\dfrac{1}{16} = \dfrac{1}{2^4}$
$= 2^{-4}$

c $\dfrac{1}{125} = \dfrac{1}{5^3}$
$= 5^{-3}$

d $\frac{3}{5} = 3 \times \frac{1}{5}$
$= 3 \times 5^{-1}$

e $\dfrac{4}{27} = \dfrac{2^2}{3^3}$
$= 2^2 \times 3^{-3}$

f $\dfrac{2^c}{8 \times 9} = \dfrac{2^c}{2^3 \times 3^2}$
$= 2^{c-3} \times 3^{-2}$

g $\dfrac{9^k}{10} = \dfrac{(3^2)^k}{2 \times 5}$
$= 3^{2k} \times 2^{-1} \times 5^{-1}$

h $\dfrac{6^p}{75} = \dfrac{(2 \times 3)^p}{3 \times 5^2}$
$= \dfrac{2^p \times 3^p}{3 \times 5^2}$
$= 2^p \times 3^{p-1} \times 5^{-2}$

12 **a** $5^3 = 21 + 23 + 25 + 27 + 29$ **b** $7^3 = 43 + 45 + 47 + 49 + 51 + 53 + 55$

c $12^3 = 133 + 135 + 137 + 139 + 141 + 143 + 145 + 147 + 149 + 151 + 153 + 155$

EXERCISE 3D

1 **a** $\sqrt[5]{2} = 2^{\frac{1}{5}}$

b $\dfrac{1}{\sqrt[5]{2}} = \dfrac{1}{2^{\frac{1}{5}}}$
$= 2^{-\frac{1}{5}}$

c $2\sqrt{2} = 2^1 \times 2^{\frac{1}{2}}$
$= 2^{\frac{3}{2}}$

d $4\sqrt{2} = 2^2 \times 2^{\frac{1}{2}}$
$= 2^{\frac{5}{2}}$

e $\dfrac{1}{\sqrt[3]{2}} = \dfrac{1}{2^{\frac{1}{3}}}$
$= 2^{-\frac{1}{3}}$

f $2 \times \sqrt[3]{2} = 2^1 \times 2^{\frac{1}{3}}$
$= 2^{\frac{4}{3}}$

g $\dfrac{4}{\sqrt{2}} = \dfrac{2^2}{2^{\frac{1}{2}}}$
$= 2^{\frac{3}{2}}$

h $(\sqrt{2})^3 = (2^{\frac{1}{2}})^3$
$= 2^{\frac{3}{2}}$

i $\dfrac{1}{\sqrt[3]{16}} = \dfrac{1}{\sqrt[3]{2^4}}$
$= \dfrac{1}{2^{\frac{4}{3}}}$
$= 2^{-\frac{4}{3}}$

j $\dfrac{1}{\sqrt{8}} = \dfrac{1}{\sqrt{2^3}}$
$= \dfrac{1}{2^{\frac{3}{2}}}$
$= 2^{-\frac{3}{2}}$

2 **a** $\sqrt[3]{3} = 3^{\frac{1}{3}}$

b $\dfrac{1}{\sqrt[3]{3}} = \dfrac{1}{3^{\frac{1}{3}}}$
$= 3^{-\frac{1}{3}}$

c $\sqrt[4]{3} = 3^{\frac{1}{4}}$

d $3\sqrt{3} = 3^1 \times 3^{\frac{1}{2}}$
$= 3^{\frac{3}{2}}$

e $\dfrac{1}{9\sqrt{3}} = \dfrac{1}{3^2 \times 3^{\frac{1}{2}}} = \dfrac{1}{3^{\frac{5}{2}}} = 3^{-\frac{5}{2}}$

3 **a** $\sqrt[3]{7} = 7^{\frac{1}{3}}$

b $\sqrt[4]{27} = \sqrt[4]{3^3}$
$= 3^{\frac{3}{4}}$

c $\sqrt[5]{16} = \sqrt[5]{2^4}$
$= 2^{\frac{4}{5}}$

d $\sqrt[3]{32} = \sqrt[3]{2^5}$
$= 2^{\frac{5}{3}}$

e $\sqrt[7]{49} = \sqrt[7]{7^2}$
$= 7^{\frac{2}{7}}$

f $\dfrac{1}{\sqrt[3]{7}} = \dfrac{1}{7^{\frac{1}{3}}}$
$= 7^{-\frac{1}{3}}$

g $\dfrac{1}{\sqrt[4]{27}} = \dfrac{1}{3^{\frac{3}{4}}}$
$= 3^{-\frac{3}{4}}$

h $\dfrac{1}{\sqrt[5]{16}} = \dfrac{1}{2^{\frac{4}{5}}}$
$= 2^{-\frac{4}{5}}$

i $\dfrac{1}{\sqrt[3]{32}} = \dfrac{1}{2^{\frac{5}{3}}}$
$= 2^{-\frac{5}{3}}$

j $\dfrac{1}{\sqrt[7]{49}} = \dfrac{1}{7^{\frac{2}{7}}}$
$= 7^{-\frac{2}{7}}$

4 **a** $3^{\frac{3}{4}} \approx 2.28$ **b** $2^{\frac{7}{8}} \approx 1.83$ **c** $2^{-\frac{1}{3}} \approx 0.794$ **d** $4^{-\frac{3}{5}} \approx 0.435$

e $\sqrt[4]{8} \approx 1.68$ **f** $\sqrt[5]{27} \approx 1.93$ **g** $\dfrac{1}{\sqrt[3]{7}} \approx 0.523$

5 **a** $4^{\frac{3}{2}} = (2^2)^{\frac{3}{2}}$
$= 2^3$
$= 8$

b $8^{\frac{5}{3}} = (2^3)^{\frac{5}{3}}$
$= 2^5$
$= 32$

c $16^{\frac{3}{4}} = (2^4)^{\frac{3}{4}}$
$= 2^3$
$= 8$

d $25^{\frac{3}{2}} = (5^2)^{\frac{3}{2}}$
$= 5^3$
$= 125$

e $32^{\frac{2}{5}} = (2^5)^{\frac{2}{5}}$
$= 2^2$
$= 4$

f $4^{-\frac{1}{2}} = (2^2)^{-\frac{1}{2}}$
$= 2^{-1}$
$= \frac{1}{2}$

g $9^{-\frac{3}{2}} = (3^2)^{-\frac{3}{2}}$
$= 3^{-3}$
$= \frac{1}{27}$

h $8^{-\frac{4}{3}} = (2^3)^{-\frac{4}{3}}$
$= 2^{-4}$
$= \frac{1}{16}$

i $27^{-\frac{4}{3}} = (3^3)^{-\frac{4}{3}}$
$= 3^{-4}$
$= \frac{1}{81}$

j $125^{-\frac{2}{3}} = (5^3)^{-\frac{2}{3}}$
$= 5^{-2}$
$= \frac{1}{25}$

EXERCISE 3E.1

1 **a** $x^2(x^3 + 2x^2 + 1)$
$= x^2 \times x^3 + x^2 \times 2x^2 + x^2 \times 1$
$= x^5 + 2x^4 + x^2$

b $2^x(2^x + 1)$
$= 2^x \times 2^x + 2^x \times 1$
$= 2^{2x} + 2^x$

c $x^{\frac{1}{2}}(x^{\frac{1}{2}} + x^{-\frac{1}{2}})$
$= x^{\frac{1}{2}} \times x^{\frac{1}{2}} + x^{\frac{1}{2}} \times x^{-\frac{1}{2}}$
$= x^1 + x^0$
$= x + 1$

d $7^x(7^x + 2)$
$= 7^x \times 7^x + 7^x \times 2$
$= 7^{2x} + 2(7^x)$

e $3^x(2 - 3^{-x})$
$= 3^x \times 2 - 3^x \times 3^{-x}$
$= 2(3^x) - 3^0$
$= 2(3^x) - 1$

f $x^{\frac{1}{2}}(x^{\frac{3}{2}} + 2x^{\frac{1}{2}} + 3x^{-\frac{1}{2}})$
$= x^{\frac{1}{2}} \times x^{\frac{3}{2}} + x^{\frac{1}{2}} \times 2x^{\frac{1}{2}} + x^{\frac{1}{2}} \times 3x^{-\frac{1}{2}}$
$= x^2 + 2x^1 + 3x^0$
$= x^2 + 2x + 3$

g $2^{-x}(2^x + 5)$
$= 2^{-x} \times 2^x + 2^{-x} \times 5$
$= 2^0 + 5(2^{-x})$
$= 1 + 5(2^{-x})$

h $5^{-x}(5^{2x} + 5^x)$
$= 5^{-x} \times 5^{2x} + 5^{-x} \times 5^x$
$= 5^x + 5^0$
$= 5^x + 1$

i $x^{-\frac{1}{2}}(x^2 + x + x^{\frac{1}{2}})$
$= x^{-\frac{1}{2}} \times x^2 + x^{-\frac{1}{2}} \times x^1 + x^{-\frac{1}{2}} \times x^{\frac{1}{2}}$
$= x^{\frac{3}{2}} + x^{\frac{1}{2}} + x^0$
$= x^{\frac{3}{2}} + x^{\frac{1}{2}} + 1$

2 **a** $(2^x + 1)(2^x + 3)$
$= 2^x \times 2^x + 2^x \times 3 + 1 \times 2^x + 3$
$= 2^{2x} + 4(2^x) + 3$
$= 4^x + 2^{2+x} + 3$

b $(3^x + 2)(3^x + 5)$
$= 3^x \times 3^x + 3^x \times 5 + 2 \times 3^x + 10$
$= 3^{2x} + 7(3^x) + 10$
$= 9^x + 7(3^x) + 10$

c $(5^x-2)(5^x-4)$
$= 5^x \times 5^x - 5^x \times 4 - 2 \times 5^x + 8$
$= 5^{2x} - 6(5^x) + 8$
$= 25^x - 6(5^x) + 8$

d $(2^x+3)^2$
$= (2^x)^2 + 2 \times 2^x \times 3 + 3^2$
$= 2^{2x} + 6(2^x) + 9$
$= 4^x + 6(2^x) + 9$

e $(3^x-1)^2$
$= (3^x)^2 - 2 \times 3^x \times 1 + 1^2$
$= 3^{2x} - 2(3^x) + 1$
$= 9^x - 2(3^x) + 1$

f $(4^x+7)^2$
$= (4^x)^2 + 2 \times 4^x \times 7 + 7^2$
$= 4^{2x} + 14(4^x) + 49$
$= 16^x + 14(4^x) + 49$

g $(x^{\frac{1}{2}}+2)(x^{\frac{1}{2}}-2)$
$= (x^{\frac{1}{2}})^2 - 2^2$
$= x - 4$

h $(2^x+3)(2^x-3)$
$= (2^x)^2 - 3^2$
$= 2^{2x} - 9$
$= 4^x - 9$

i $(x^{\frac{1}{2}}+x^{-\frac{1}{2}})(x^{\frac{1}{2}}-x^{-\frac{1}{2}})$
$= (x^{\frac{1}{2}})^2 - (x^{-\frac{1}{2}})^2$
$= x^1 - x^{-1}$
$= x - x^{-1}$

j $\left(x+\frac{2}{x}\right)^2$
$= x^2 + 2 \times x \times \left(\frac{2}{x}\right) + \left(\frac{2}{x}\right)^2$
$= x^2 + 4 + \frac{4}{x^2}$

k $(7^x-7^{-x})^2$
$= (7^x)^2 - 2 \times 7^x \times 7^{-x} + (7^{-x})^2$
$= 7^{2x} - 2 \times 7^0 + 7^{-2x}$
$= 7^{2x} - 2 + 7^{-2x}$

l $(5-2^{-x})^2$
$= 5^2 - 2 \times 5 \times 2^{-x} + (2^{-x})^2$
$= 25 - 10(2^{-x}) + 2^{-2x}$
$= 25 - 10(2^{-x}) + 4^{-x}$

EXERCISE 3E.2

1 **a** $5^{2x}+5^x$
$= 5^x \times 5^x + 5^x$
$= 5^x(5^x+1)$

b $3^{n+2}+3^n$
$= 3^n \times 3^2 + 3^n$
$= 3^n(3^2+1)$
$= 10(3^n)$

c 7^n+7^{3n}
$= 7^n + 7^n \times 7^{2n}$
$= 7^n(1+7^{2n})$

d $5^{n+1}-5$
$= 5 \times 5^n - 5$
$= 5(5^n-1)$

e $6^{n+2}-6$
$= 6 \times 6^{n+1} - 6$
$= 6(6^{n+1}-1)$

f $4^{n+2}-16$
$= 4^2 \times 4^n - 16$
$= 16 \times 4^n - 16$
$= 16(4^n-1)$

2 **a** 9^x-4
$= (3^x)^2 - 2^2$
$= (3^x+2)(3^x-2)$

b 4^x-25
$= (2^x)^2 - 5^2$
$= (2^x+5)(2^x-5)$

c $16-9^x$
$= 4^2 - (3^x)^2$
$= (4+3^x)(4-3^x)$

d $25-4^x$
$= 5^2 - (2^x)^2$
$= (5+2^x)(5-2^x)$

e 9^x-4^x
$= (3^x)^2 - (2^x)^2$
$= (3^x+2^x)(3^x-2^x)$

f $4^x+6(2^x)+9$
$= (2^x)^2 + 6(2^x) + 9$
$= (2^x+3)^2$
$\{a^2+6a+9=(a+3)^2\}$

g $9^x+10(3^x)+25$
$= (3^x)^2 + 10(3^x) + 25$
$= (3^x+5)^2$
$\{a^2+10a+25=(a+5)^2\}$

h $4^x-14(2^x)+49$
$= (2^x)^2 - 14(2^x) + 49$
$= (2^x-7)^2$
$\{a^2-14a+49=(a-7)^2\}$

i $25^x-4(5^x)+4$
$= (5^x)^2 - 4(5^x) + 4$
$= (5^x-2)^2$
$\{a^2-4a+4=(a-2)^2\}$

3 **a** $4^x + 9(2^x) + 18$
$= (2^x)^2 + 9(2^x) + 18$
$= (2^x + 3)(2^x + 6)$
$\{a^2 + 9a + 18 = (a+3)(a+6)\}$

b $4^x - 2^x - 20$
$= (2^x)^2 - 2^x - 20$
$= (2^x + 4)(2^x - 5)$
$\{a^2 - a - 20 = (a+4)(a-5)\}$

c $9^x + 9(3^x) + 14$
$= (3^x)^2 + 9(3^x) + 14$
$= (3^x + 2)(3^x + 7)$
$\{a^2 + 9a + 14 = (a+2)(a+7)\}$

d $9^x + 4(3^x) - 5$
$= (3^x)^2 + 4(3^x) - 5$
$= (3^x + 5)(3^x - 1)$
$\{a^2 + 4a - 5 = (a+5)(a-1)\}$

e $25^x + 5^x - 2$
$= (5^x)^2 + 5^x - 2$
$= (5^x + 2)(5^x - 1)$
$\{a^2 + a - 2 = (a+2)(a-1)\}$

f $49^x - 7^{x+1} + 12$
$= (7^x)^2 - 7(7^x) + 12$
$= (7^x - 4)(7^x - 3)$
$\{a^2 - 7a + 12 = (a-4)(a-3)\}$

4 **a** $\dfrac{12^n}{6^n} = \left(\dfrac{12}{6}\right)^n = 2^n$

b $\dfrac{20^a}{2^a} = \left(\dfrac{20}{2}\right)^a = 10^a$

c $\dfrac{6^b}{2^b} = \left(\dfrac{6}{2}\right)^b = 3^b$

d $\dfrac{4^n}{20^n} = \left(\dfrac{4}{20}\right)^n = \left(\frac{1}{5}\right)^n = \dfrac{1}{5^n}$

e $\dfrac{35^x}{7^x} = \left(\dfrac{35}{7}\right)^x = 5^x$

f $\dfrac{6^a}{8^a} = \left(\dfrac{6}{8}\right)^a = \left(\frac{3}{4}\right)^a$

g $\dfrac{5^{n+1}}{5^n} = \dfrac{5 \times \cancel{5^n}}{\cancel{5^n}_1} = 5$

h $\dfrac{5^{n+1}}{5} = \dfrac{\cancel{5} \times 5^n}{\cancel{5}_1} = 5^n$

5 **a** $\dfrac{6^m + 2^m}{2^m}$
$= \dfrac{2^m 3^m + 2^m}{2^m}$
$= \dfrac{\cancel{2^m}(3^m + 1)}{\cancel{2^m}_1}$
$= 3^m + 1$

b $\dfrac{2^n + 12^n}{2^n}$
$= \dfrac{2^n + 2^n 6^n}{2^n}$
$= \dfrac{\cancel{2^n}(1 + 6^n)}{\cancel{2^n}_1}$
$= 1 + 6^n$

c $\dfrac{8^n + 4^n}{2^n}$
$= \dfrac{2^n 4^n + 2^n 2^n}{2^n}$
$= \dfrac{\cancel{2^n}(4^n + 2^n)}{\cancel{2^n}_1}$
$= 4^n + 2^n$

d $\dfrac{12^x - 3^x}{3^x}$
$= \dfrac{3^x 4^x - 3^x}{3^x}$
$= \dfrac{\cancel{3^x}(4^x - 1)}{\cancel{3^x}_1}$
$= 4^x - 1$

e $\dfrac{6^n + 12^n}{1 + 2^n}$
$= \dfrac{6^n + 6^n 2^n}{1 + 2^n}$
$= \dfrac{6^n \cancel{(1 + 2^n)}}{\cancel{1 + 2^n}_1}$
$= 6^n$

f $\dfrac{5^{n+1} - 5^n}{4}$
$= \dfrac{5^n \times 5 - 5^n}{4}$
$= \dfrac{5^n \cancel{(5-1)}}{\cancel{4}_1}$
$= 5^n$

g $\dfrac{5^{n+1} - 5^n}{5^n}$
$= \dfrac{5^n \times 5 - 5^n}{5^n}$
$= \dfrac{\cancel{5^n}(5-1)}{\cancel{5^n}_1}$
$= 4$

h $\dfrac{4^n - 2^n}{2^n}$
$= \dfrac{2^n 2^n - 2^n}{2^n}$
$= \dfrac{\cancel{2^n}(2^n - 1)}{\cancel{2^n}_1}$
$= 2^n - 1$

i $\dfrac{2^n - 2^{n-1}}{2^n}$
$= \dfrac{2^{n-1} \times 2 - 2^{n-1}}{2^{n-1} \times 2}$
$= \dfrac{\cancel{2^{n-1}}(2-1)}{\cancel{2^{n-1}} \times 2}$
$= \frac{1}{2}$

6 **a** $2^n(n+1)+2^n(n-1)$
$=2^n(n+1+n-1)$
$=2^n(2n)$
$=n2^{n+1}$

b $3^n\left(\frac{n-1}{6}\right)-3^n\left(\frac{n+1}{6}\right)$
$=3^n\left(\frac{n-1}{6}-\frac{n+1}{6}\right)$
$=3^n(-\frac{1}{3})$
$=3^n\times -3^{-1}$
$=-3^{n-1}$

EXERCISE 3F

1 **a** $2^x=8$
$\therefore\ 2^x=2^3$
$\therefore\ x=3$

b $5^x=25$
$\therefore\ 5^x=5^2$
$\therefore\ x=2$

c $3^x=81$
$\therefore\ 3^x=3^4$
$\therefore\ x=4$

d $7^x=1$
$\therefore\ 7^x=7^0$
$\therefore\ x=0$

e $3^x=\frac{1}{3}$
$\therefore\ 3^x=3^{-1}$
$\therefore\ x=-1$

f $4^x=\frac{1}{16}$
$\therefore\ 4^x=4^{-2}$
$\therefore\ x=-2$

g $5^x=\frac{1}{125}$
$\therefore\ 5^x=5^{-3}$
$\therefore\ x=-3$

h $4^{x+1}=64$
$\therefore\ 4^{x+1}=4^3$
$\therefore\ x+1=3$
$\therefore\ x=2$

i $2^{x-2}=\frac{1}{32}$
$\therefore\ 2^{x-2}=2^{-5}$
$\therefore\ x-2=-5$
$\therefore\ x=-3$

j $3^{x+1}=\frac{1}{27}$
$\therefore\ 3^{x+1}=3^{-3}$
$\therefore\ x+1=-3$
$\therefore\ x=-4$

k $7^{x+1}=343$
$\therefore\ 7^{x+1}=7^3$
$\therefore\ x+1=3$
$\therefore\ x=2$

l $5^{1-2x}=\frac{1}{5}$
$\therefore\ 5^{1-2x}=5^{-1}$
$\therefore\ 1-2x=-1$
$\therefore\ -2x=-2$
$\therefore\ x=1$

2 **a** $8^x=32$
$\therefore\ 2^{3x}=2^5$
$\therefore\ 3x=5$
$\therefore\ x=\frac{5}{3}$

b $4^x=\frac{1}{8}$
$\therefore\ 2^{2x}=2^{-3}$
$\therefore\ 2x=-3$
$\therefore\ x=-\frac{3}{2}$

c $9^x=\frac{1}{27}$
$\therefore\ 3^{2x}=3^{-3}$
$\therefore\ 2x=-3$
$\therefore\ x=-\frac{3}{2}$

d $25^x=\frac{1}{5}$
$\therefore\ 5^{2x}=5^{-1}$
$\therefore\ 2x=-1$
$\therefore\ x=-\frac{1}{2}$

e $27^x=\frac{1}{9}$
$\therefore\ 3^{3x}=3^{-2}$
$\therefore\ 3x=-2$
$\therefore\ x=-\frac{2}{3}$

f $16^x=\frac{1}{32}$
$\therefore\ 2^{4x}=2^{-5}$
$\therefore\ 4x=-5$
$\therefore\ x=-\frac{5}{4}$

g $4^{x+2}=128$
$\therefore\ 2^{2(x+2)}=2^7$
$\therefore\ 2x+4=7$
$\therefore\ 2x=3$
$\therefore\ x=\frac{3}{2}$

h $25^{1-x}=\frac{1}{125}$
$\therefore\ 5^{2(1-x)}=5^{-3}$
$\therefore\ 2-2x=-3$
$\therefore\ -2x=-5$
$\therefore\ x=\frac{5}{2}$

i $4^{4x-1}=\frac{1}{2}$
$\therefore\ 2^{2(4x-1)}=2^{-1}$
$\therefore\ 8x-2=-1$
$\therefore\ 8x=1$
$\therefore\ x=\frac{1}{8}$

j $9^{x-3}=27$
$\therefore\ 3^{2(x-3)}=3^3$
$\therefore\ 2x-6=3$
$\therefore\ 2x=9$
$\therefore\ x=\frac{9}{2}$

k $\left(\frac{1}{2}\right)^{x+1}=8$
$\therefore\ \left(2^{-1}\right)^{x+1}=2^3$
$\therefore\ -x-1=3$
$\therefore\ -x=4$
$\therefore\ x=-4$

l $\left(\frac{1}{3}\right)^{x+2}=9$
$\therefore\ \left(3^{-1}\right)^{x+2}=3^2$
$\therefore\ -x-2=2$
$\therefore\ -x=4$
$\therefore\ x=-4$

m $81^x=27^{-x}$
$\therefore\ 3^{4x}=3^{-3x}$
$\therefore\ 4x=-3x$
$\therefore\ 7x=0$
$\therefore\ x=0$

n
$$\left(\tfrac{1}{4}\right)^{1-x} = 32$$
$$\therefore \left(2^{-2}\right)^{1-x} = 2^5$$
$$\therefore -2+2x = 5$$
$$\therefore 2x = 7$$
$$\therefore x = \tfrac{7}{2}$$

o
$$\left(\tfrac{1}{7}\right)^{x} = 49$$
$$\therefore 7^{-x} = 7^2$$
$$\therefore -x = 2$$
$$\therefore x = -2$$

p
$$\left(\tfrac{1}{3}\right)^{x+1} = 243$$
$$\therefore \left(3^{-1}\right)^{x+1} = 3^5$$
$$\therefore -x-1 = 5$$
$$\therefore -x = 6$$
$$\therefore x = -6$$

3 **a**
$$4^{2x+1} = 8^{1-x}$$
$$\therefore \left(2^2\right)^{2x+1} = \left(2^3\right)^{1-x}$$
$$\therefore 4x+2 = 3-3x$$
$$\therefore 7x = 1$$
$$\therefore x = \tfrac{1}{7}$$

b
$$9^{2-x} = \left(\tfrac{1}{3}\right)^{2x+1}$$
$$\therefore \left(3^2\right)^{2-x} = \left(3^{-1}\right)^{2x+1}$$
$$\therefore 4-2x = -2x-1$$
$$\therefore 4 = -1$$
This is clearly false, so no solutions exist.

c
$$2^x \times 8^{1-x} = \tfrac{1}{4}$$
$$\therefore 2^x \times \left(2^3\right)^{1-x} = 2^{-2}$$
$$\therefore x+3-3x = -2$$
$$\therefore -2x = -5$$
$$\therefore x = \tfrac{5}{2}$$

4 **a**
$$3 \times 2^x = 24$$
$$\therefore 2^x = 8$$
$$\therefore 2^x = 2^3$$
$$\therefore x = 3$$

b
$$7 \times 2^x = 56$$
$$\therefore 2^x = 8$$
$$\therefore 2^x = 2^3$$
$$\therefore x = 3$$

c
$$3 \times 2^{x+1} = 24$$
$$\therefore 2^{x+1} = 8$$
$$\therefore 2^{x+1} = 2^3$$
$$\therefore x+1 = 3$$
$$\therefore x = 2$$

d
$$12 \times 3^{-x} = \tfrac{4}{3}$$
$$\therefore 3^{-x} = \tfrac{4}{3} \div 12$$
$$\therefore 3^{-x} = \tfrac{4}{3} \times \tfrac{1}{12}$$
$$\therefore 3^{-x} = \tfrac{1}{9}$$
$$\therefore 3^{-x} = 3^{-2}$$
$$\therefore x = 2$$

e
$$4 \times \left(\tfrac{1}{3}\right)^x = 36$$
$$\therefore \left(\tfrac{1}{3}\right)^x = 9$$
$$\therefore \left(3^{-1}\right)^x = 3^2$$
$$\therefore 3^{-x} = 3^2$$
$$\therefore -x = 2$$
$$\therefore x = -2$$

f
$$5 \times \left(\tfrac{1}{2}\right)^x = 20$$
$$\therefore \left(\tfrac{1}{2}\right)^x = 4$$
$$\therefore \left(2^{-1}\right)^x = 2^2$$
$$\therefore -x = 2$$
$$\therefore x = -2$$

5 **a**
$$4^x - 6(2^x) + 8 = 0$$
$$\therefore (2^x)^2 - 6(2^x) + 8 = 0$$
$$\therefore (2^x-2)(2^x-4) = 0 \qquad \{a^2-6a+8 = (a-2)(a-4)\}$$
$$\therefore 2^x = 2 \text{ or } 4$$
$$\therefore 2^x = 2^1 \text{ or } 2^2$$
$$\therefore x = 1 \text{ or } 2$$

b
$$4^x - 2^x - 2 = 0$$
$$\therefore (2^x)^2 - 2^x - 2 = 0$$
$$\therefore (2^x-2)(2^x+1) = 0 \qquad \{a^2-a-2 = (a-2)(a+1)\}$$
$$\therefore 2^x = 2 \text{ or } -1$$
$$\therefore 2^x = 2^1 \qquad \{\text{since } 2^x \text{ cannot be negative}\}$$
$$\therefore x = 1$$

c
$$9^x - 12(3^x) + 27 = 0$$
$$\therefore (3^x)^2 - 12(3^x) + 27 = 0$$
$$\therefore (3^x-3)(3^x-9) = 0 \qquad \{a^2-12a+27 = (a-3)(a-9)\}$$
$$\therefore 3^x = 3 \text{ or } 9$$
$$\therefore 3^x = 3^1 \text{ or } 3^2$$
$$\therefore x = 1 \text{ or } 2$$

d $9^x = 3^x + 6$

$\therefore\ (3^x)^2 - 3^x - 6 = 0$

$\therefore\ (3^x - 3)(3^x + 2) = 0$ $\{a^2 - a - 6 = (a-3)(a+2)\}$

$\therefore\ 3^x = 3$ or -2

$\therefore\ 3^x = 3^1$ {since 3^x cannot be negative}

$\therefore\ x = 1$

e $25^x - 23(5^x) - 50 = 0$

$\therefore\ (5^x)^2 - 23(5^x) - 50 = 0$

$\therefore\ (5^x - 25)(5^x + 2) = 0$ $\{a^2 - 23a - 50 = (a-25)(a+2)\}$

$\therefore\ 5^x = 25$ or -2

$\therefore\ 5^x = 5^2$ {since 5^x cannot be negative}

$\therefore\ x = 2$

f $49^x + 1 = 2(7^x)$

$\therefore\ (7^x)^2 - 2(7^x) + 1 = 0$

$\therefore\ (7^x - 1)^2 = 0$ $\{a^2 - 2a + 1 = (a-1)^2\}$

$\therefore\ 7^x = 1$

$\therefore\ 7^x = 7^0$

$\therefore\ x = 0$

EXERCISE 3G

1 **a** When $x = \frac{1}{2}$, $y = 2^{\frac{1}{2}}$

From point A, $y \approx 1.4$

$\therefore\ 2^{\frac{1}{2}} \approx 1.4$

b When $x = 0.8$, $y = 2^{0.8}$

From point B, $y \approx 1.7$

$\therefore\ 2^{0.8} \approx 1.7$

c When $x = 1.5$, $y = 2^{1.5}$

From point C, $y \approx 2.8$

$\therefore\ 2^{1.5} \approx 2.8$

d When $x = -1.6$, $y = 2^{-1.6}$

From point D, $y \approx 0.3$

$\therefore\ 2^{-1.6} \approx 0.3$

e When $x = \sqrt{2}$, $y = 2^{\sqrt{2}}$

Using **a** we know $x \approx 1.4$

From point E, $y \approx 2.7$

$\therefore\ 2^{\sqrt{2}} \approx 2.7$

f When $x = -\sqrt{2}$, $y = 2^{-\sqrt{2}}$

Using **a** we know $x \approx -1.4$

From point F, $y \approx 0.4$

$\therefore\ 2^{-\sqrt{2}} \approx 0.4$

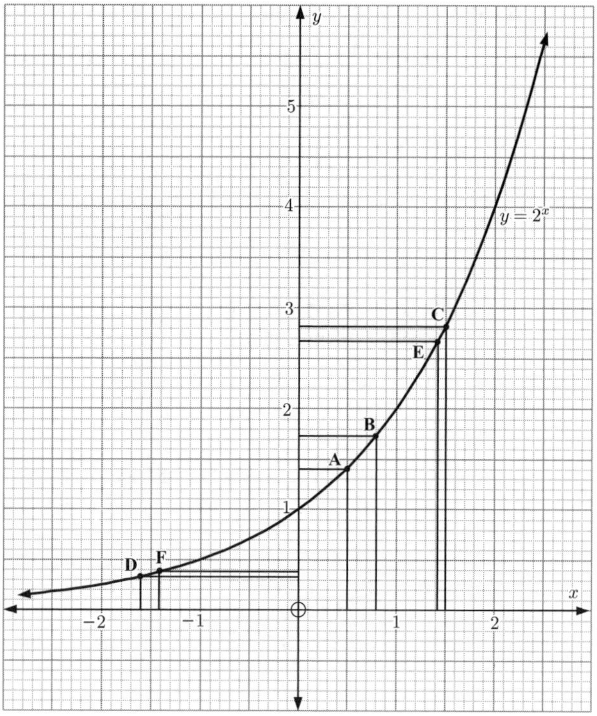

2 **a**

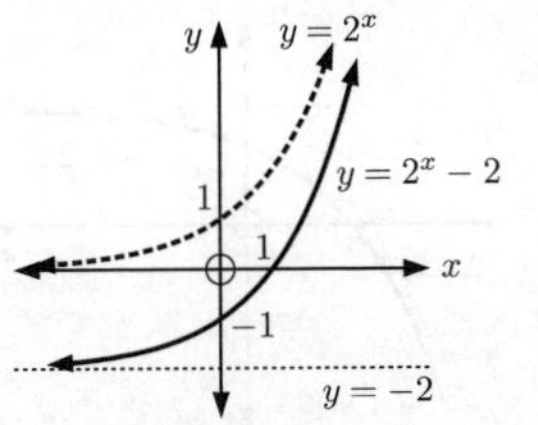

a vertical translation of 2 units downwards

$y = -2$ is the H.A.

b

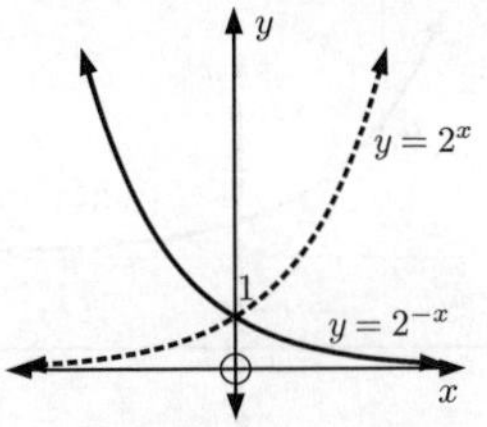

a reflection in the y-axis

c

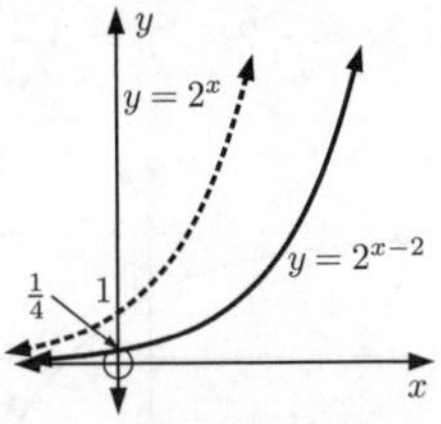

a horizontal translation of 2 units right

d

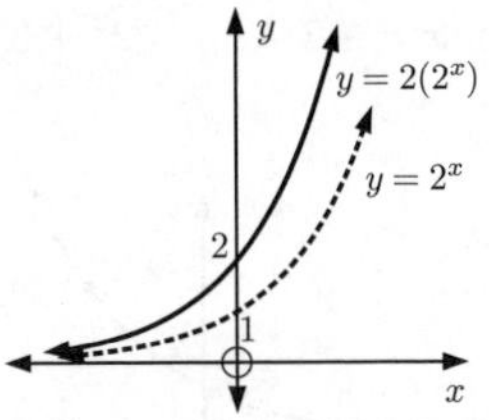

a vertical stretch of factor 2

3 **a**

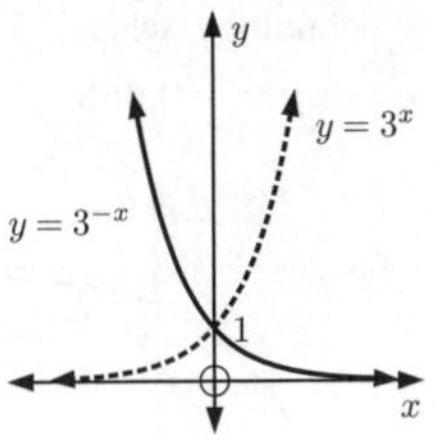

a reflection in the y-axis

b

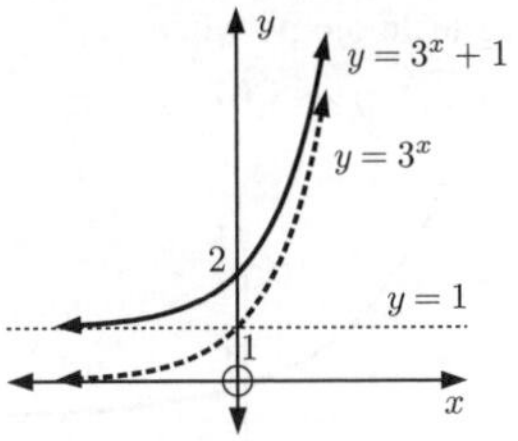

a vertical translation of 1 unit upwards

$y = 1$ is the H.A.

c

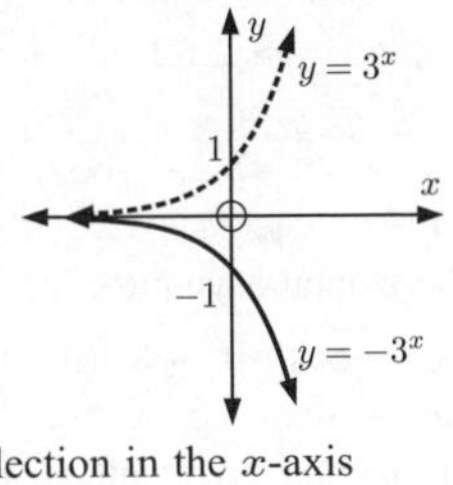

a reflection in the x-axis

d

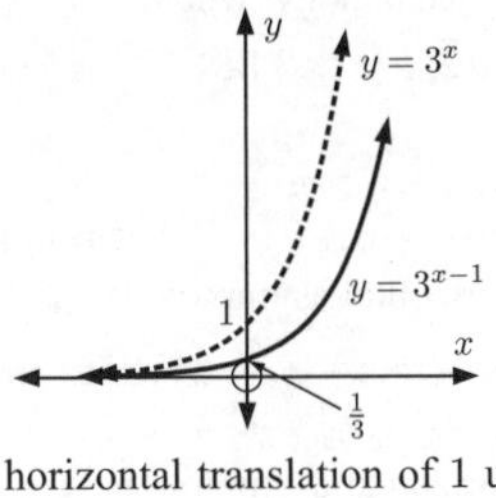

a horizontal translation of 1 unit right

4 **a**

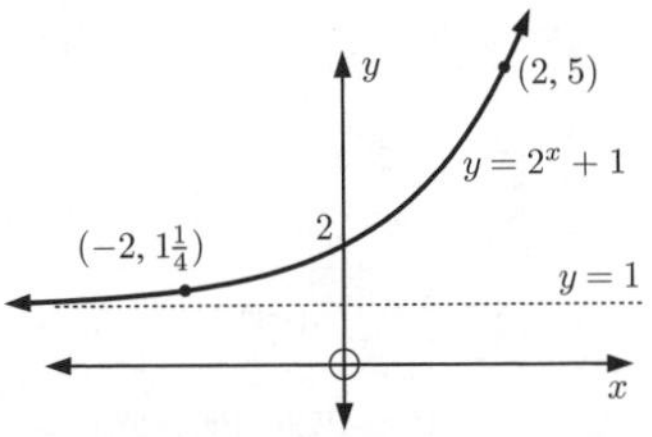

a vertical translation of 1 unit upwards

When $x = 2$, $y = 4 + 1 = 5$

When $x = -2$, $y = \frac{1}{4} + 1 = 1\frac{1}{4}$

b

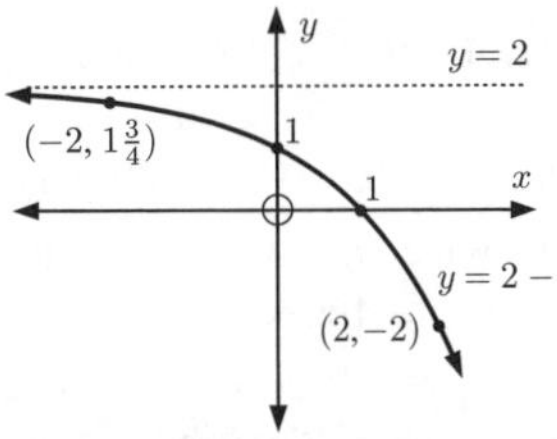

When $x = 0$, $y = 2 - 2^0 = 2 - 1 = 1$

$\therefore$ the y-intercept is 1

When $x = 1$, $y = 2 - 2 = 0$

When $x = 2$, $y = 2 - 4 = -2$

When $x = -2$, $y = 2 - \frac{1}{4} = 1\frac{3}{4}$

c

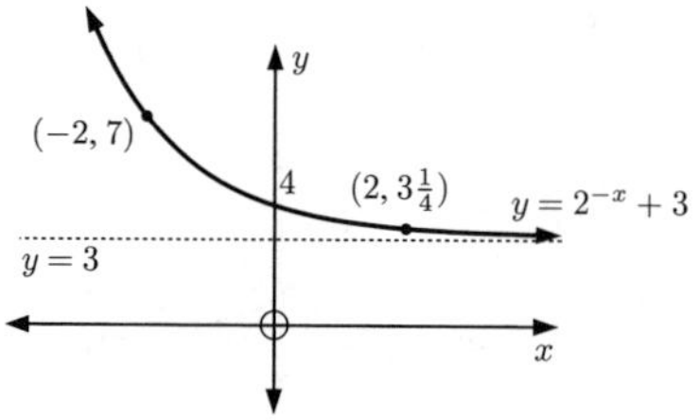

When $x = 0$, $y = 1 + 3 = 4$

When $x = 2$, $y = \frac{1}{4} + 3 = 3\frac{1}{4}$

When $x = -2$, $y = 2^2 + 3 = 7$

d

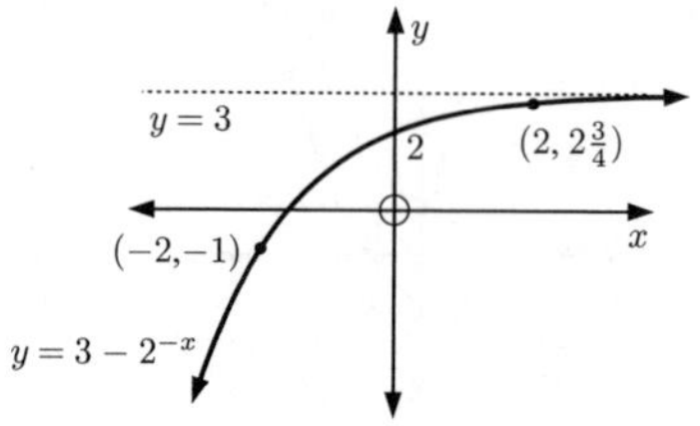

When $x = 0$, $y = 3 - 1 = 2$

When $x = 2$, $y = 3 - \frac{1}{4} = 2\frac{3}{4}$

When $x = -2$, $y = 3 - 4 = -1$

5 **a**

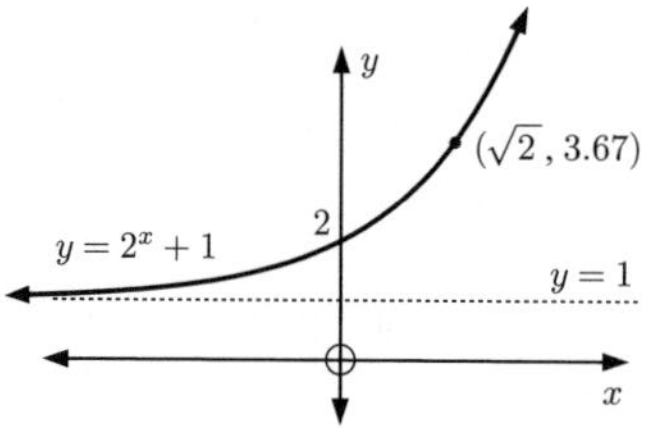

Using technology, when
$x = \sqrt{2}$, $y \approx 3.67$

b

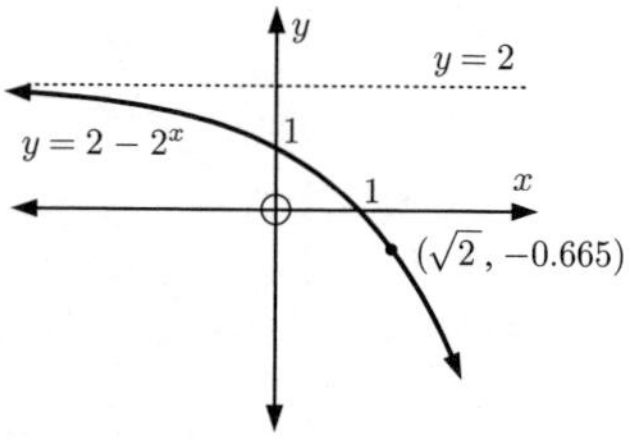

Using technology, when
$x = \sqrt{2}$, $y \approx -0.665$

c

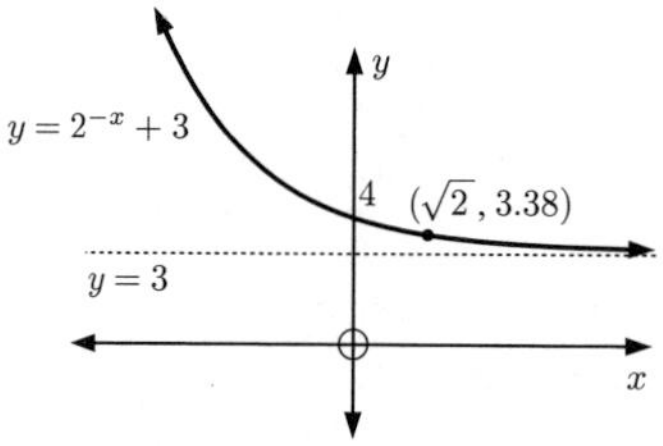

Using technology, when
$x = \sqrt{2}$, $y \approx 3.38$

d

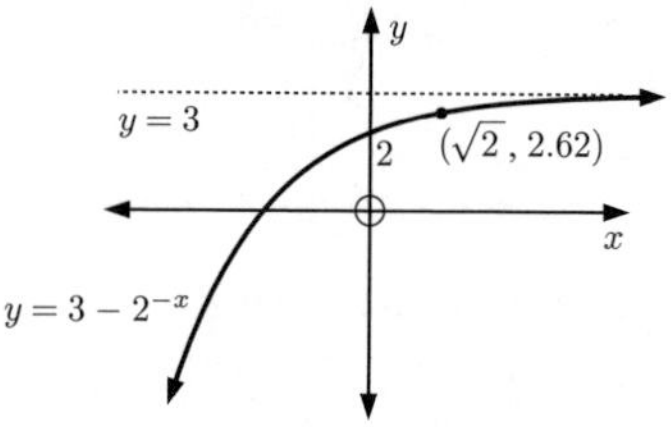

Using technology, when
$x = \sqrt{2}$, $y \approx 2.62$

6 **a** As $x \to \infty$, $y \to \infty$
As $x \to -\infty$, $y \to 1$ (from above)
the horizontal asymptote is $y = 1$

b As $x \to \infty$, $y \to -\infty$
As $x \to -\infty$, $y \to 2$ (from below)
the horizontal asymptote is $y = 2$

c As $x \to \infty$, $y \to 3$ (from above)
As $x \to -\infty$, $y \to \infty$
the horizontal asymptote is $y = 3$

d As $x \to \infty$, $y \to 3$ (from below)
As $x \to -\infty$, $y \to -\infty$
the horizontal asymptote is $y = 3$

EXERCISE 3H.1

1 **a** When $t = 0$, $W_0 = 100$ grams = the initial weight

b **i** When $t = 4$,
$W_4 = 100 \times 2^{0.1 \times 4}$
$= 100 \times 2^{0.4}$
≈ 132 grams

ii When $t = 10$,
$W_{10} = 100 \times 2^1$
$= 200$ grams

iii When $t = 24$, $W_{24} = 100 \times 2^{0.1 \times 24}$
$= 100 \times 2^{2.4}$
≈ 528 grams

c

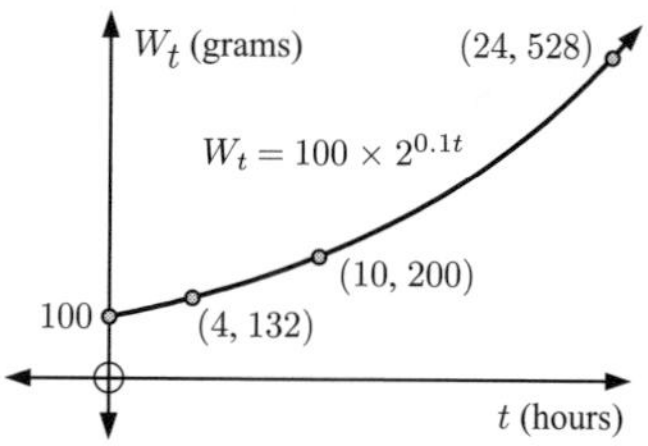

2 **a** $P_0 = 50$ (the initial population)

b **i** When $n = 2$,
$P_2 = 50 \times 2^{0.3\times2}$
$= 50 \times 2^{0.6}$
≈ 75.785
So, the expected population is 76 possums.

ii When $n = 5$,
$P_5 = 50 \times 2^{0.3\times5}$
$= 50 \times 2^{1.5}$
≈ 141.421
So, the expected population is 141 possums.

iii When $n = 10$, $P_{10} = 50 \times 2^{0.3\times10}$
$= 50 \times 2^3 = 400$ So, the expected population is 400 possums.

c

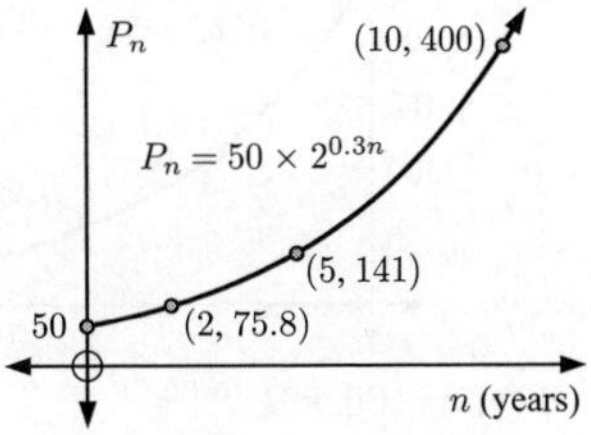

3 **a** When $t = 0$,
$V_0 = V_0 \times 2^0$
$= V_0$
So, the speed is V_0.

b When $t = 20$,
$V_{20} = V_0 \times 2^{0.05\times20}$
$= V_0 \times 2^1$
$= 2V_0$
So, the speed is $2V_0$.

c V_0 becomes $2V_0$
So, there was a 100% increase in speed.

d
$$\left(\frac{V_{50} - V_{20}}{V_{20}}\right) \times 100\%$$
$$= \left(\frac{V_0 \times 2^{2.5} - V_0 \times 2^1}{V_0 \times 2^1}\right) \times 100\%$$
$$= \left(\frac{2^{2.5} - 2^1}{2^1}\right) \times 100\%$$
$$\approx 183\%$$

This expression is the percentage increase in the speed from the speed at 20°C to the speed at 50°C.
($V_{50} - V_{20}$ is the increase in speed.)

4 **a** $B_0 = 6$ pairs $= 12$ bears

b In 2018, $t = 20$
$\therefore\ B_{20} = 12 \times 2^{0.18\times20}$
$= 12 \times 2^{3.6}$
≈ 145.508
≈ 146 bears

c In 2008, $t = 10$
$$\therefore\ \%\text{ increase} = \left(\frac{B_{20} - B_{10}}{B_{10}}\right) \times 100\%$$
$$= \left(\frac{12 \times 2^{3.6} - 12 \times 2^{1.8}}{12 \times 2^{1.8}}\right) \times 100\%$$
$$= \left(\frac{2^{3.6} - 2^{1.8}}{2^{1.8}}\right) \times 100\%$$
$$\approx 248\%$$

EXERCISE 3H.2

1 $W(t) = 250 \times (0.998)^t$ grams

a $W(0) = 250 \times (0.998)^0$
$= 250 \times 1$
$= 250$ grams $\therefore$ 250 g of radioactive substance was put aside.

b **i** When $t = 400$
$W(400)$
$= 250 \times (0.998)^{400}$
≈ 112 grams

ii When $t = 800$
$W(800)$
$= 250 \times (0.998)^{800}$
≈ 50.4 grams

iii When $t = 1200$
$W(1200)$
$= 250 \times (0.998)^{1200}$
≈ 22.6 grams

c

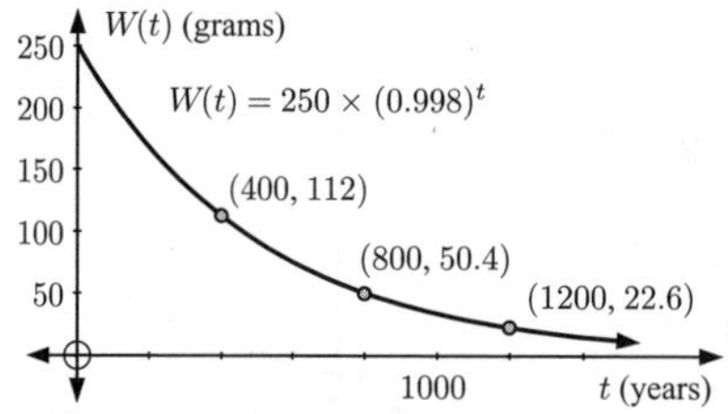

d When $W(t) = 125$

$250 \times (0.998)^t = 125$

$\therefore \quad (0.998)^t = 0.5$

$\therefore \quad t \approx 346.2$ {using technology}

It takes approximately 346 years

2 $T(t) = 100 \times 2^{-0.02t}$

a $T(0) = 100 \times 2^0$
$= 100 \times 1$
$= 100°C$

b **i** $T(15) = 100 \times 2^{-0.02\times15}$
$= 100 \times 2^{-0.3}$
$\approx 81.2°C$

ii $T(20) = 100 \times 2^{-0.02\times20}$
$= 100 \times 2^{-0.4}$
$\approx 75.8°C$

iii $T(78) = 100 \times 2^{-0.02\times78}$
$= 100 \times 2^{-1.56}$
$\approx 33.9°C$

c

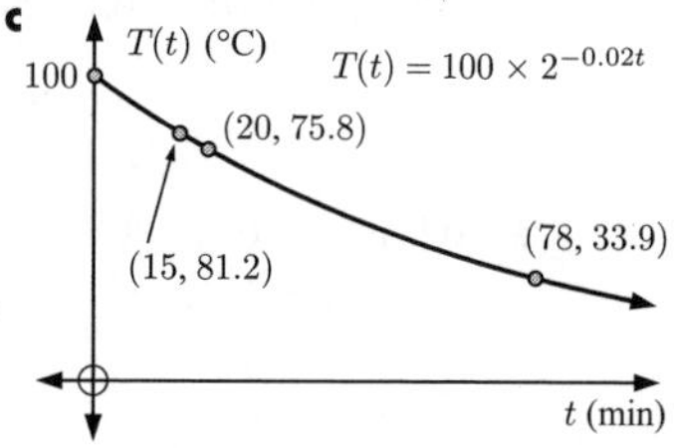

3 $W_t = 1000 \times 2^{-0.03t}$

a $W_0 = 1000 \times 2^0$
$= 1000 \times 1$
$= 1000$ g

b **i** W_{10}
$= 1000 \times 2^{-0.3}$
≈ 812 g

ii W_{100}
$= 1000 \times 2^{-3}$
$= 125$ g

iii W_{1000}
$= 1000 \times 2^{-30}$
$\approx 9.31 \times 10^{-7}$ g

c

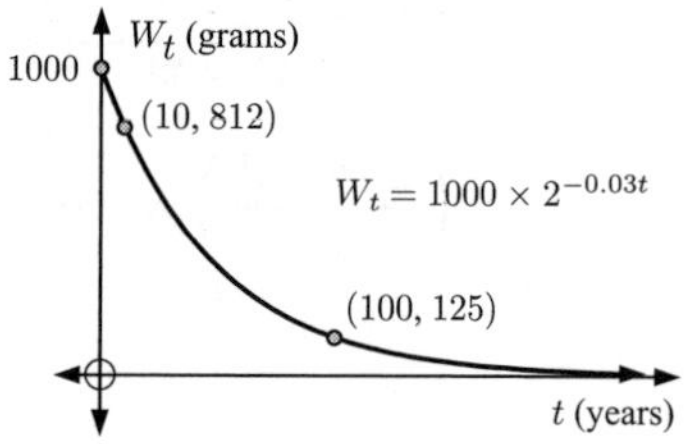

4 **a** When $t = 0$, $W_0 = W_0 2^0$
$= W_0$ grams

$\therefore$ the original weight was W_0 grams.

b $\%\text{ change} = \left(\dfrac{W_{1000} - W_0}{W_0}\right) \times 100\%$

$= \left(\dfrac{W_0 \times 2^{-0.2} - W_0}{W_0}\right) \times 100\%$

$= (2^{-0.2} - 1) \times 100\%$

$\approx -12.9\%$

The weight loss was about 12.9%.

EXERCISE 3I

1 $e^1 \approx 2.718\,281\,828$

2

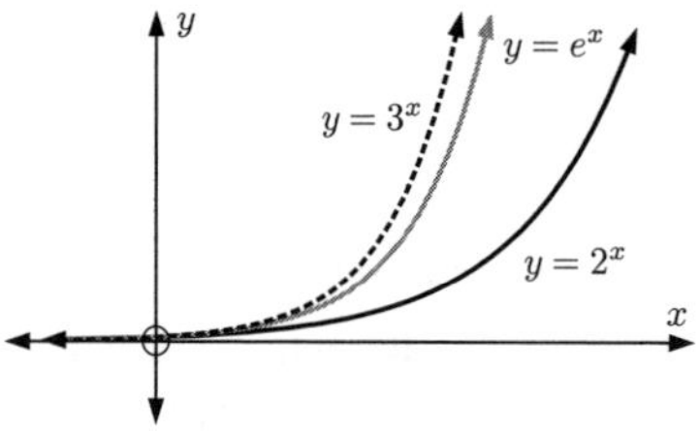

The graph of $y = e^x$ lies between $y = 2^x$ and $y = 3^x$.

3

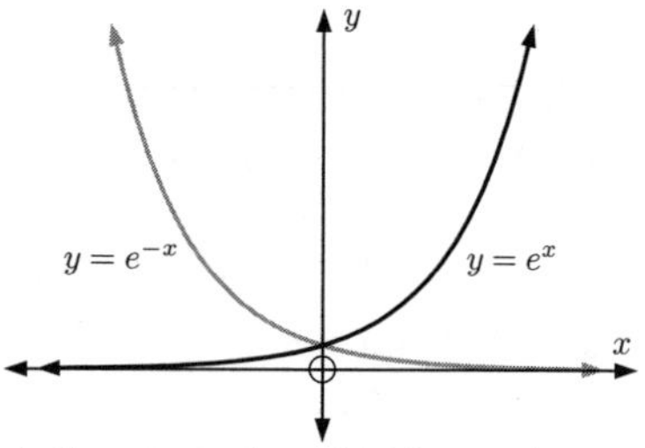

One is the other reflected in the y-axis.

4 When $x = 0$, $y = ae^0 = a \times 1 = a$ $\therefore$ the y-intercept is a.

5 **a** The graph of $y = e^x$ is entirely above the x-axis.

$y > 0$ for all x

$\therefore$ $e^x > 0$ for all x

$\therefore$ $2e^x > 0$ for all x

$\therefore$ $y = 2e^x$ cannot be negative.

b **i** When $x = -20$, $y = 2e^{-20} \approx 4.12 \times 10^{-9}$

ii When $x = 20$, $y = 2e^{20} \approx 9.70 \times 10^8$

6 **a** $e^2 \approx 7.39$ **b** $e^3 \approx 20.1$ **c** $e^{0.7} \approx 2.01$ **d** $\sqrt{e} \approx 1.65$ **e** $e^{-1} \approx 0.368$

7 **a** $\sqrt{e} = e^{\frac{1}{2}}$

b $e\sqrt{e} = e^1 e^{\frac{1}{2}} = e^{\frac{3}{2}}$

c $\dfrac{1}{\sqrt{e}} = \dfrac{1}{e^{\frac{1}{2}}} = e^{-\frac{1}{2}}$

d $\dfrac{1}{e^2} = e^{-2}$

8 **a** $\left(e^{0.36}\right)^{\frac{t}{2}} = e^{0.36 \times \frac{t}{2}} = e^{0.18t}$

b $\left(e^{0.064}\right)^{\frac{t}{16}} = e^{0.064 \times \frac{t}{16}} = e^{0.004t}$

c $\left(e^{-0.04}\right)^{\frac{t}{8}} = e^{-0.04 \times \frac{t}{8}} = e^{-0.005t}$

d $\left(e^{-0.836}\right)^{\frac{t}{5}} = e^{-0.836 \times \frac{t}{5}} \approx e^{-0.167t}$

9 **a** ≈ 10.074 **b** $\approx 0.099\,261$ **c** ≈ 125.09 **d** $\approx 0.007\,994\,5$

e ≈ 41.914 **f** ≈ 42.429 **g** ≈ 3540.3 **h** $\approx 0.006\,342\,4$

10

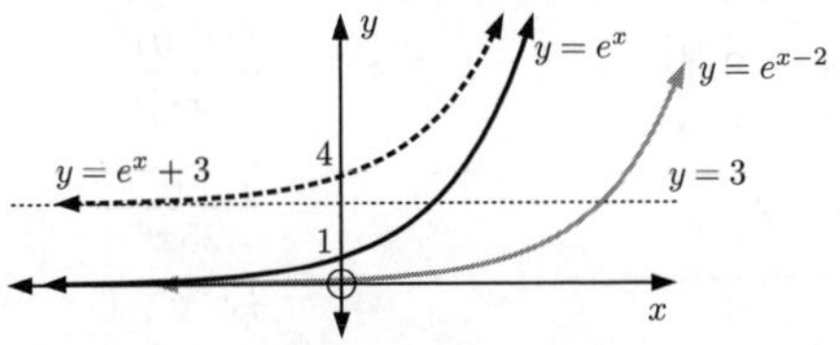

Domain of f, g and h is $\{x \mid x \in \mathbb{R}\}$
Range of f is $\{y \mid y > 0\}$
Range of g is $\{y \mid y > 0\}$
Range of h is $\{y \mid y > 3\}$

11

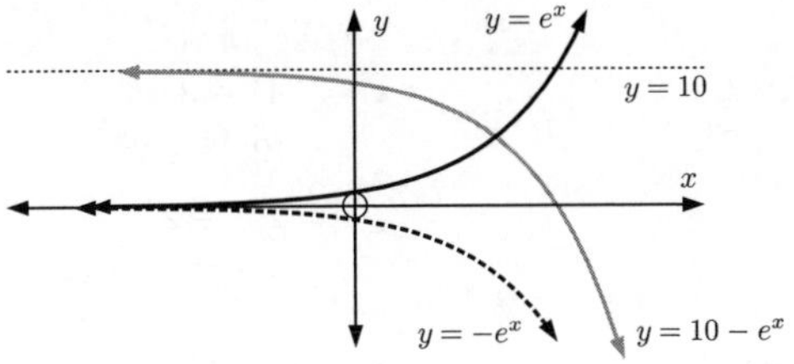

Domain of f, g and h is $\{x \mid x \in \mathbb{R}\}$
Range of f is $\{y \mid y > 0\}$
Range of g is $\{y \mid y < 0\}$
Range of h is $\{y \mid y < 10\}$

12 $W(t) = 2e^{\frac{t}{2}}$ grams

a **i** $W(0) = 2e^0 = 2 \times 1 = 2$ g

ii $W(\frac{1}{2}) = 2e^{\frac{1}{4}} \approx 2.57$ g

iii $W(1\frac{1}{2}) = 2e^{\frac{3}{4}} \approx 4.23$ g

iv $W(6) = 2e^3 \approx 40.2$ g

b

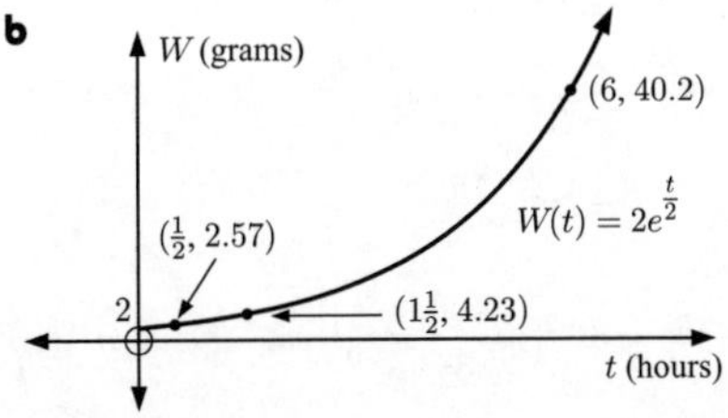

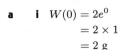

13 $I(t) = 75e^{-0.15t}$

a **i** $I(1) = 75e^{-0.15}$
≈ 64.6 amps

ii $I(10) = 75e^{-1.5}$
≈ 16.7 amps

b

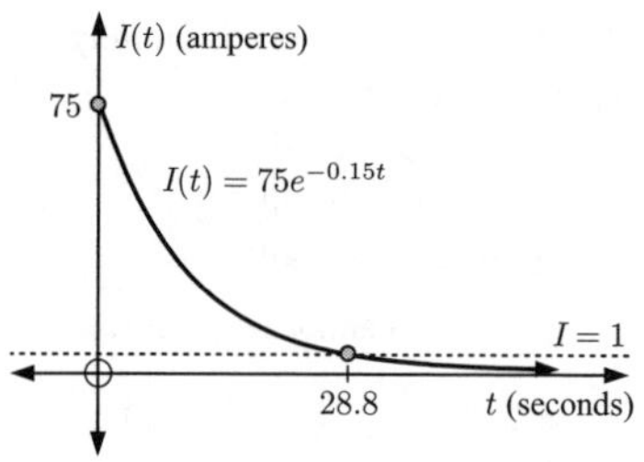

c We need to solve $75e^{-0.15t} = 1$.
Using technology, $t \approx 28.8$ s

14 $f(x) = e^x$

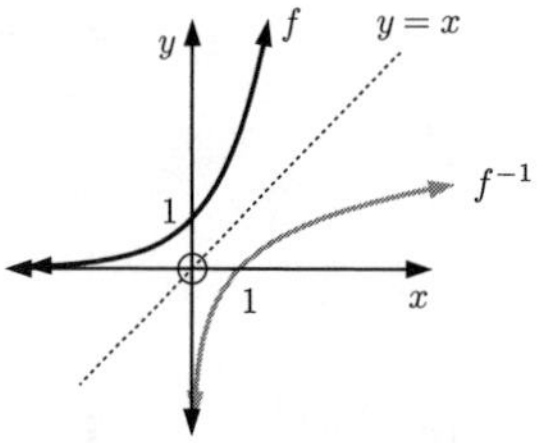

REVIEW SET 3A

1 **a** $-(-1)^{10}$
$= -1$

b $-(-3)^3$
$= -(-27)$
$= 27$

c $3^0 - 3^{-1}$
$= 1 - \frac{1}{3}$
$= \frac{2}{3}$

2 **a** $a^4b^5 \times a^2b^2$
$= a^{4+2} \times b^{5+2}$
$= a^6b^7$

b $6xy^5 \div 9x^2y^5$
$= \frac{6}{9}x^{1-2}y^{5-5}$
$= \frac{2}{3}x^{-1}y^0$
$= \dfrac{2}{3x}$

c $\dfrac{5(x^2y)^2}{(5x^2)^2}$
$= \dfrac{5 \times x^4y^2}{25x^4}$
$= \frac{1}{5}x^0y^2$
$= \dfrac{y^2}{5}$

3 **a** $\frac{1}{16}$
$= \dfrac{1}{2^4}$
$= 2^{-4}$

b $2^x \times 4$
$= 2^x \times 2^2$
$= 2^{x+2}$

c $4^x \div 8$
$= (2^2)^x \div 2^3$
$= 2^{2x} \div 2^3$
$= 2^{2x-3}$

4 **a** $x^{-2} \times x^{-3}$
$= x^{-2+(-3)}$
$= x^{-5}$
$= \dfrac{1}{x^5}$

b $2(ab)^{-2}$
$= 2 \times \dfrac{1}{(ab)^2}$
$= \dfrac{2}{a^2b^2}$

c $2ab^{-2}$
$= 2a \times \left(\dfrac{1}{b^2}\right)$
$= \dfrac{2a}{b^2}$

5 **a** $\dfrac{27}{9^a} = \dfrac{3^3}{(3^2)^a}$
$= 3^{3-2a}$

b $\left(\sqrt{3}\right)^{1-x} \times 9^{1-2x} = (3^{\frac{1}{2}})^{1-x} \times (3^2)^{1-2x}$
$= 3^{\frac{1}{2}-\frac{1}{2}x+2-4x}$
$= 3^{\frac{5}{2}-\frac{9}{2}x}$

6 **a** $8^{\frac{2}{3}} = (2^3)^{\frac{2}{3}} = 2^2 = 4$

b $27^{-\frac{2}{3}} = (3^3)^{-\frac{2}{3}} = 3^{-2} = \dfrac{1}{3^2} = \frac{1}{9}$

7 **a** mn^{-2}
$= m \times \frac{1}{n^2}$
$= \frac{m}{n^2}$

b $(mn)^{-3}$
$= \frac{1}{(mn)^3}$
$= \frac{1}{m^3n^3}$

c $\frac{m^2n^{-1}}{p^{-2}}$
$= m^2\left(\frac{1}{n}\right)p^2$
$= \frac{m^2p^2}{n}$

d $(4m^{-1}n)^2$
$= 4^2m^{-2}n^2$
$= \frac{16n^2}{m^2}$

8 **a** $(3-2^a)^2$
$= 3^2 - 2\times 3\times 2^a + (2^a)^2$
$= 9 - 3\times 2^{a+1} + 2^{2a}$
$\{or\ \ 9-6(2^a)+2^{2a}\}$

b $(\sqrt{x}+2)(\sqrt{x}-2)$
$= (\sqrt{x})^2 - 2^2$
$= x-4$

c $2^{-x}(2^{2x}+2^x)$
$= 2^{-x+2x} + 2^{-x+x}$
$= 2^x + 2^0$
$= 2^x + 1$

9 **a** $2^{x-3} = \frac{1}{32}$
$\therefore\ 2^{x-3} = 2^{-5}$
$\therefore\ x-3 = -5$
$\therefore\ x = -2$

b $9^x = 27^{2-2x}$
$\therefore\ (3^2)^x = (3^3)^{2-2x}$
$\therefore\ 2x = 6-6x$
$\therefore\ 8x = 6$
$\therefore\ x = \frac{6}{8} = \frac{3}{4}$

10 **a** $27^x = 3$
$\therefore\ (3^3)^x = 3^1$
$\therefore\ 3x = 1$
$\therefore\ x = \frac{1}{3}$

b $9^{1-x} = 27^{x+2}$
$\therefore\ (3^2)^{1-x} = (3^3)^{x+2}$
$\therefore\ 2-2x = 3x+6$
$\therefore\ -5x = 4$ and so $x = -\frac{4}{5}$

REVIEW SET 3B

1 **a** 4×2^n
$= 2^2\times 2^n$
$= 2^{n+2}$

b $7^{-1} - 7^0$
$= \frac{1}{7} - 1$
$= -\frac{6}{7}$

c $\left(\frac{2}{3}\right)^{-3}$
$= \left(\frac{3}{2}\right)^3$
$= \frac{27}{8}$
$= 3\frac{3}{8}$

d $\left(\frac{2a^{-1}}{b^2}\right)^2$
$= \frac{2^2a^{-2}}{b^4}$
$= \frac{4}{a^2b^4}$

2 **a** $3^{\frac{3}{4}} \approx 2.28$ **b** $27^{-\frac{1}{5}} \approx 0.517$ **c** $\sqrt[4]{100} \approx 3.16$

3 $f(x) = 3\times 2^x$

a $f(0) = 3\times 2^0$
$= 3\times 1$
$= 3$

b $f(3) = 3\times 2^3$
$= 3\times 8$
$= 24$

c $f(-2) = 3\times 2^{-2}$
$= 3\times\frac{1}{2^2} = \frac{3}{4}$

4 **a** $81 = 3^4$ **b** $1 = 3^0$ **c** $\frac{1}{27} = \frac{1}{3^3} = 3^{-3}$ **d** $\frac{1}{243} = \frac{1}{3^5} = 3^{-5}$

5

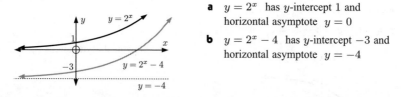

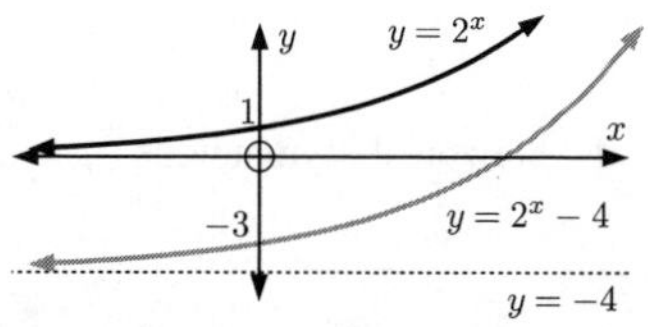

a $y = 2^x$ has y-intercept 1 and horizontal asymptote $y = 0$

b $y = 2^x - 4$ has y-intercept -3 and horizontal asymptote $y = -4$

6 $T = 80\times(0.913)^t\ ^o\text{C}$

a When $t = 0$, $T = 80\times(0.913)^0$
$= 80\times 1$
$= 80$ $\therefore$ the initial temperature was 80^oC.

b **i** When $t = 12$,
$T = 80 \times (0.913)^{12}$
$\approx 26.8^o\text{C}$

ii When $t = 24$,
$T = 80 \times (0.913)^{24}$
$\approx 9.00^o\text{C}$

iii When $t = 36$,
$T = 80 \times (0.913)^{36}$
$\approx 3.02^o\text{C}$

c

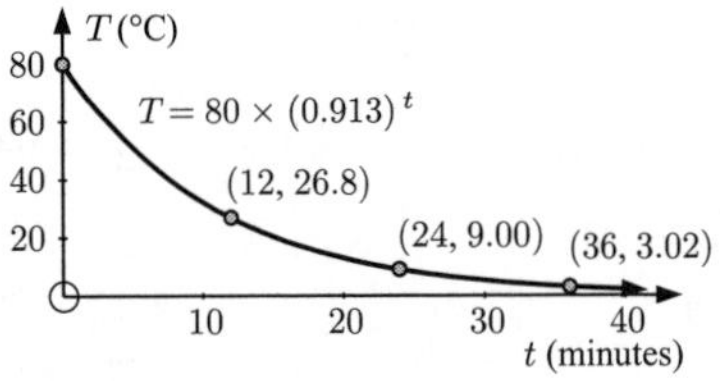

d When $T = 25$
$80 \times (0.913)^t = 25$
$\therefore\ 0.913^t = 0.3125$
$\therefore\ t \approx 12.8$ min {using technology}

7 **a** When $x = 0$, $y = 3^0 - 5 = 1 - 5 = -4$
When $x = 1$, $y = 3^1 - 5 = 3 - 5 = -2$
When $x = 2$, $y = 3^2 - 5 = 9 - 5 = 4$
When $x = -1$, $y = 3^{-1} - 5 = \frac{1}{3} - 5 = -4\frac{2}{3}$
When $x = -2$, $y = 3^{-2} - 5 = \frac{1}{9} - 5 = -4\frac{8}{9}$

b As $x \to \infty$, $3^x \to \infty$
and so $y \to \infty$
As $x \to -\infty$, $3^x \to 0$
and so $y \to -5$ (from above)

c

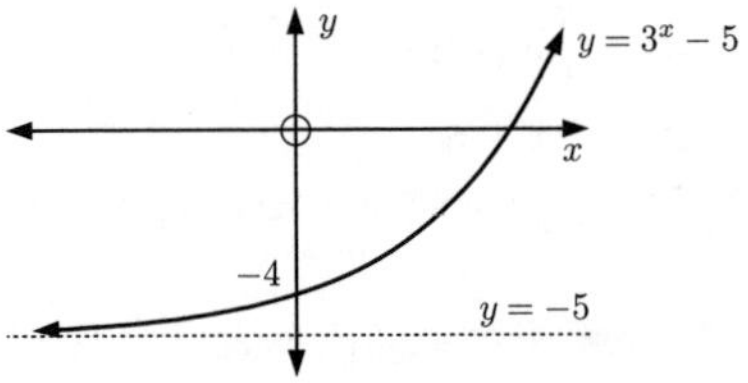

d $y = -5$ is the horizontal asymptote.

8

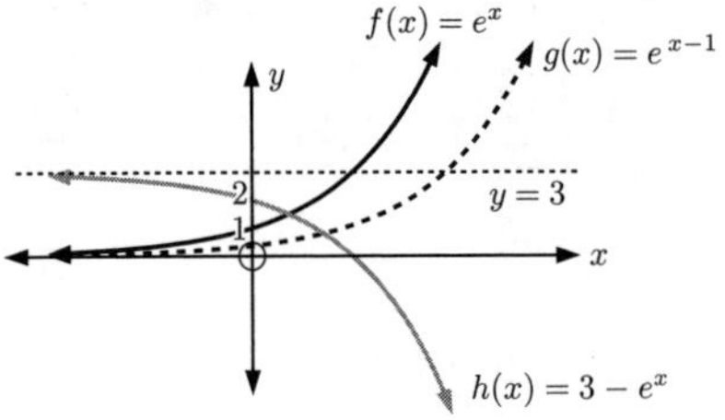

Domain of f, g and h is $\{x \mid x \in \mathbb{R}\}$
Range of f is $\{y \mid y > 0\}$
Range of g is $\{y \mid y > 0\}$
Range of h is $\{y \mid y < 3\}$

9 **a** When $x = 0$, $y = 3 - 2^0 = 3 - 1 = 2$
When $x = 1$, $y = 3 - 2^{-1} = 3 - \frac{1}{2} = 2\frac{1}{2}$
When $x = 2$, $y = 3 - 2^{-2} = 3 - \frac{1}{4} = 2\frac{3}{4}$
When $x = -1$, $y = 3 - 2^1 = 3 - 2 = 1$
When $x = -2$, $y = 3 - 2^2 = 3 - 4 = -1$

b as $x \to \infty$, $2^{-x} \to 0$,
$\therefore\ y \to 3$ (below)
as $x \to -\infty$, $2^{-x} \to \infty$,
$\therefore\ y \to -\infty$

c

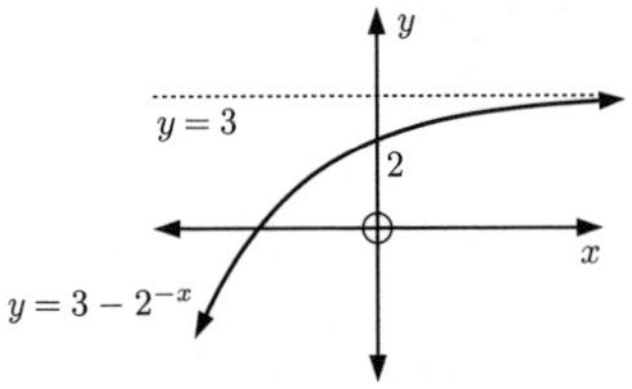

d horizontal asymptote is $y = 3$

10 $W = 1500 \times (0.993)^t$ grams

a When $t = 0$,
$W = 1500 \times (0.993)^0$
$= 1500 \times 1$
$= 1500$ grams

b **i** When $t = 400$,
$W = 1500 \times (0.993)^{400}$
≈ 90.3 grams

ii When $t = 800$,
$W = 1500 \times (0.993)^{800}$
≈ 5.44 grams

c

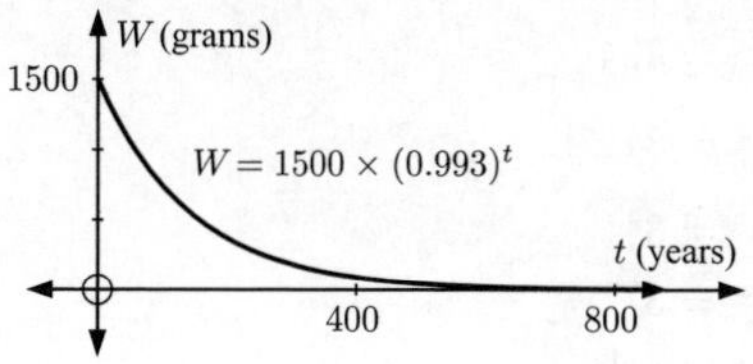

d When $W = 100$,
$1500 \times (0.993)^t = 100$
$\therefore \quad (0.993)^t \approx 0.0667$
$\therefore \quad t \approx 385.5$ {using technology}
So, it will take about 386 years.

REVIEW SET 3C

1 **a** $-(-2)^3$
$= -(-8)$
$= 8$

b $5^{-1} - 5^0$
$= \frac{1}{5} - 1$
$= -\frac{4}{5}$

2 **a** $(a^7)^3$
$= a^{7\times3}$
$= a^{21}$

b $pq^2 \times p^3q^4$
$= p^{1+3}q^{2+4}$
$= p^4q^6$

c $\dfrac{8ab^5}{2a^4b^4}$
$= \frac{8}{2}a^{1-4}b^{5-4}$
$= 4a^{-3}b^1$
$= \dfrac{4b}{a^3}$

3 **a** 2×2^{-4}
$= 2^1 \times 2^{-4}$
$= 2^{1+(-4)}$
$= 2^{-3}$

b $16 \div 2^{-3}$
$= 2^4 \div 2^{-3}$
$= 2^{4-(-3)}$
$= 2^7$

c 8^4
$= (2^3)^4$
$= 2^{12}$

4 **a** $b^{-3} = \dfrac{1}{b^3}$

b $(ab)^{-1}$
$= a^{-1}b^{-1}$
$= \dfrac{1}{ab}$

c ab^{-1}
$= a \times \dfrac{1}{b}$
$= \dfrac{a}{b}$

5 $\dfrac{2^{x+1}}{2^{1-x}} = 2^{x+1-(1-x)}$
$= 2^{x+1-1+x}$
$= 2^{2x}$

6 **a** $1 = 5^0$

b $5\sqrt{5}$
$= 5^1 \times 5^{\frac{1}{2}}$
$= 5^{\frac{3}{2}}$

c $\dfrac{1}{\sqrt[4]{5}}$
$= \dfrac{1}{5^{\frac{1}{4}}}$
$= 5^{-\frac{1}{4}}$

d 25^{a+3}
$= (5^2)^{a+3}$
$= 5^{2a+6}$

7 **a** $e^x(e^{-x}+e^x)$
$= e^0 + e^{2x}$
$= 1 + e^{2x}$

b $(2^x+5)^2$
$= (2^x)^2 + 2\times 2^x \times 5 + 5^2$
$= 2^{2x} + 5\times 2^{x+1} + 25$
$= 4^x + 5\times 2^{x+1} + 25$
$\{or\ \ 2^{2x} + 10(2^x) + 25\}$

c $(x^{\frac{1}{2}}-7)(x^{\frac{1}{2}}+7)$
$= (x^{\frac{1}{2}})^2 - 7^2$
$= x^1 - 49$
$= x - 49$

8 **a** $6\times 2^x = 192$
$\therefore\ 2^x = 32$
$\therefore\ 2^x = 2^5$
$\therefore\ x = 5$

b $4\times(\frac{1}{3})^x = 324$
$\therefore\ (\frac{1}{3})^x = 81$
$\therefore\ (3^{-1})^x = 3^4$
$\therefore\ 3^{-x} = 3^4$
$\therefore\ x = -4$

9 **a** $2^{x+1} = 32$
$\therefore\ 2^{x+1} = 2^5$
$\therefore\ x+1 = 5$
$\therefore\ x = 4$

b $4^{x+1} = \left(\frac{1}{8}\right)^x$
$\therefore\ (2^2)^{x+1} = (2^{-3})^x$
$\therefore\ 2x+2 = -3x$
$\therefore\ 5x = -2$
$\therefore\ x = -\frac{2}{5}$

10 **a** When $x=0$, $y = 2e^{-0}+1 = 3$
When $x=1$, $y = 2e^{-1}+1 \approx 1.74$
When $x=2$, $y = 2e^{-2}+1 \approx 1.27$
When $x=-1$, $y = 2e^{1}+1 \approx 6.44$
When $x=-2$, $y = 2e^{2}+1 \approx 15.8$

b As $x\to\infty$, $y\to 1$ (from above)
As $x\to-\infty$, $y\to\infty$

c

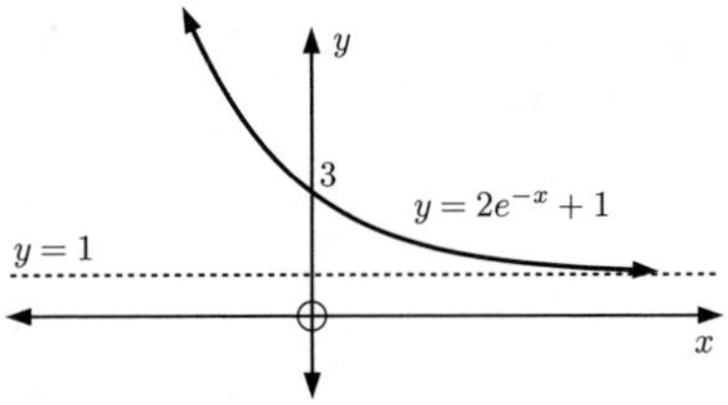

d $y = 1$ is a horizontal asymptote.

Chapter 4
LOGARITHMS

EXERCISE 4A

1 **a** $10^4 = 10\,000$ **b** $10^{-1} = 0.1$ **c** $10^{\frac{1}{2}} = \sqrt{10}$
d $2^3 = 8$ **e** $2^{-2} = \frac{1}{4}$ **f** $3^{1.5} = \sqrt{27}$

2 **a** $\log_2 4 = 2$ **b** $\log_2(\frac{1}{8}) = -3$ **c** $\log_{10}(0.01) = -2$
d $\log_7 49 = 2$ **e** $\log_2 64 = 6$ **f** $\log_3(\frac{1}{27}) = -3$

3 **a** $10^5 = 100\,000$ so $\log_{10}(100\,000) = 5$
b $10^{-2} = 0.01$ so $\log_{10}(0.01) = -2$
c $3^{\frac{1}{2}} = \sqrt{3}$ so $\log_3(\sqrt{3}) = \frac{1}{2}$
d $2^3 = 8$ so $\log_2 8 = 3$
e $2^6 = 64$ so $\log_2 64 = 6$
f $2^7 = 128$ so $\log_2 128 = 7$
g $5^2 = 25$ so $\log_5 25 = 2$
h $5^3 = 125$ so $\log_5 125 = 3$
i $2^{-3} = \frac{1}{8} = 0.125$ so $\log_2(0.125) = -3$
j $9^{\frac{1}{2}} = \sqrt{9} = 3$ so $\log_9 3 = \frac{1}{2}$
k $4^2 = 16$ so $\log_4 16 = 2$
l $36^{\frac{1}{2}} = \sqrt{36} = 6$ so $\log_{36} 6 = \frac{1}{2}$
m $243 = 3^5$ so $\log_3 243 = 5$
n $\sqrt[3]{2} = 2^{\frac{1}{3}}$ so $\log_2 \sqrt[3]{2} = \frac{1}{3}$
o $\log_a a^n = n$
p $2 = 8^{\frac{1}{3}}$ so $\log_8 2 = \frac{1}{3}$
q $\dfrac{1}{t} = t^{-1}$ so $\log_t \left(\dfrac{1}{t}\right) = -1$
r $6\sqrt{6} = 6^1 \times 6^{\frac{1}{2}} = 6^{\frac{3}{2}}$ so $\log_6(6\sqrt{6}) = \frac{3}{2}$
s $1 = 4^0$ so $\log_4 1 = 0$
t $9 = 9^1$ so $\log_9 9 = 1$

4 **a** $\log_{10} 152 \approx 2.18$ **b** $\log_{10} 25 \approx 1.40$ **c** $\log_{10} 74 \approx 1.87$ **d** $\log_{10} 0.8 \approx -0.0969$

5 **a** $\log_2 x = 3$
$\therefore\ x = 2^3$
$\therefore\ x = 8$

b $\log_4 x = \frac{1}{2}$
$\therefore\ x = 4^{\frac{1}{2}}$
$\therefore\ x = 2$

c $\log_x 81 = 4$
$\therefore\ 81 = x^4$
$\therefore\ x = \pm\sqrt[4]{81}$
$\therefore\ x = \pm 3$
$\therefore\ x = 3$ {as $x > 0$}

d $\log_2(x-6) = 3$
$\therefore\ x - 6 = 2^3$
$\therefore\ x - 6 = 8$
$\therefore\ x = 14$

6 **a** $\log_4 16$
$= \log_4 4^2$
$= 2$

b $\log_2 4$
$= \log_2 2^2$
$= 2$

c $\log_3(\frac{1}{3})$
$= \log_3 3^{-1}$
$= -1$

d $\log_{10} \sqrt[4]{1000}$
$= \log_{10}(10^3)^{\frac{1}{4}}$
$= \log_{10} 10^{\frac{3}{4}}$
$= \frac{3}{4}$

e $\log_7\left(\frac{1}{\sqrt{7}}\right)$
$= \log_7 7^{-\frac{1}{2}}$
$= -\frac{1}{2}$

f $\log_5(25\sqrt{5})$
$= \log_5\left(5^2 5^{\frac{1}{2}}\right)$
$= \log_5 5^{\frac{5}{2}}$
$= \frac{5}{2}$

g $\log_3\left(\frac{1}{\sqrt{27}}\right)$
$= \log_3\left(\dfrac{1}{(3^3)^{\frac{1}{2}}}\right)$
$= \log_3 3^{-\frac{3}{2}}$
$= -\frac{3}{2}$

h $\log_4\left(\frac{1}{2\sqrt{2}}\right)$
$= \log_4\left(2^{-\frac{3}{2}}\right)$
$= \log_4\left((2^2)^{-\frac{3}{4}}\right)$
$= \log_4 4^{-\frac{3}{4}}$
$= -\frac{3}{4}$

EXERCISE 4B

1 **a** $\log 10\,000 = \log_{10} 10^4 = 4$

b $\log 0.001 = \log_{10} 10^{-3} = -3$

c $\log 10 = \log_{10} 10^1 = 1$

d $\log 1 = \log_{10} 10^0 = 0$

e $\log \sqrt{10} = \log_{10} 10^{\frac{1}{2}} = \frac{1}{2}$

f $\log \sqrt[3]{10} = \log_{10} 10^{\frac{1}{3}} = \frac{1}{3}$

g $\log \left(\frac{1}{\sqrt[4]{10}}\right) = \log_{10} 10^{-\frac{1}{4}} = -\frac{1}{4}$

h $\log 10\sqrt{10} = \log_{10} 10^{\frac{3}{2}} = \frac{3}{2}$

i $\log \sqrt[3]{100} = \log_{10}(10^2)^{\frac{1}{3}} = \log_{10} 10^{\frac{2}{3}} = \frac{2}{3}$

j $\log\left(\frac{100}{\sqrt{10}}\right) = \log_{10}\left(\frac{10^2}{10^{\frac{1}{2}}}\right) = \log_{10} 10^{\frac{3}{2}} = \frac{3}{2}$

k $\log\left(10 \times \sqrt[3]{10}\right) = \log_{10}(10^1 \times 10^{\frac{1}{3}}) = \log_{10} 10^{\frac{4}{3}} = \frac{4}{3}$

l $\log 1000\sqrt{10} = \log_{10}(10^3 \times 10^{\frac{1}{2}}) = \log_{10} 10^{\frac{7}{2}} = \frac{7}{2}$

m $\log 10^n = \log_{10} 10^n = n$

n $\log(10^a \times 100) = \log_{10}\left(10^a \times 10^2\right) = \log_{10}\left(10^{a+2}\right) = a+2$

o $\log\left(\frac{10}{10^m}\right) = \log_{10}(10^{1-m}) = 1-m$

p $\log\left(\frac{10^a}{10^b}\right) = \log_{10}(10^{a-b}) = a-b$

2 Instructions given for a TI-84 plus:

a [log] 10 000 [ENTER], 4

b [log] 0.001 [ENTER], −3

c [log] [2nd] [√] 10 [)] [)] [ENTER], 0.5

d [log] 10 [^] [(] 1 [÷] 3 [)] [)] [ENTER], $0.\overline{3}$

e [log] 100 [^] [(] 1 [÷] 3 [)] [)] [ENTER], $0.\overline{6}$

f [log] 10 [×] [2nd] [√] 10 [)] [)] [ENTER], 1.5

g [log] 1 [÷] [2nd] [√] 10 [)] [)] [ENTER], −0.5

h [log] 1 [÷] 10 [^] 0.25 [)] [ENTER], −0.25

3 **a** $6 = 10^{\log 6} \approx 10^{0.7782}$

b $60 = 10^{\log 60} \approx 10^{1.7782}$

c $6000 = 10^{\log 6000} \approx 10^{3.7782}$

d $0.6 = 10^{\log(0.6)} \approx 10^{-0.2218}$

e $0.006 = 10^{\log(0.006)} \approx 10^{-2.2218}$

f $15 = 10^{\log 15} \approx 10^{1.1761}$

g $1500 = 10^{\log 1500} \approx 10^{3.1761}$

h $1.5 = 10^{\log 1.5} \approx 10^{0.1761}$

i $0.15 = 10^{\log(0.15)} \approx 10^{-0.8239}$

j $0.000\,15 = 10^{\log(0.000\,15)} \approx 10^{-3.8239}$

4 **a** **i** $\log 3 \approx 0.477$ **ii** $\log 300 \approx 2.477$

b
$$300 = 3 \times 10^2 = 10^{\log 3} \times 10^2 = 10^{\log 3 + 2}$$
$$\therefore \log 300 = \log 3 + 2$$

5 **a** **i** $\log 5 \approx 0.699$

ii $\log 0.05 \approx -1.301$

b $0.05 = 5 \times 10^{-2} = 10^{\log 5} \times 10^{-2} = 10^{\log 5 - 2}$

$\therefore \log 0.05 = \log 5 - 2$

6 **a** $\log x = 2$ $\therefore x = 10^2$ $\therefore x = 100$

b $\log x = 1$ $\therefore x = 10^1$ $\therefore x = 10$

c $\log x = 0$ $\therefore x = 10^0$ $\therefore x = 1$

d $\log x = -1$ $\therefore x = 10^{-1}$ $\therefore x = \frac{1}{10}$

e $\log x = \frac{1}{2}$ $\therefore x = 10^{\frac{1}{2}}$ $\therefore x = \sqrt{10}$

f $\log x = -\frac{1}{2}$ $\therefore x = 10^{-\frac{1}{2}}$ $\therefore x = \dfrac{1}{10^{\frac{1}{2}}}$ $\therefore x = \dfrac{1}{\sqrt{10}}$

g $\log x = 4$ $\therefore x = 10^4$ $\therefore x = 10\,000$

h $\log x = -5$ $\therefore x = 10^{-5}$ $\therefore x = 0.000\,01$

i $\log x \approx 0.8351$ $\therefore x \approx 10^{0.8351}$ $\therefore x \approx 6.84$

j $\log x \approx 2.1457$ $\therefore x \approx 10^{2.1457}$ $\therefore x \approx 140$

k $\log x \approx -1.378$ $\therefore x \approx 10^{-1.378}$ $\therefore x \approx 0.0419$

l $\log x \approx -3.1997$ $\therefore x \approx 10^{-3.1997}$ $\therefore x \approx 0.000\,631$

EXERCISE 4C.1

1 **a** $\log 8 + \log 2 = \log(8 \times 2) = \log 16$

b $\log 8 - \log 2 = \log(\frac{8}{2}) = \log 4$

c $\log 40 - \log 5 = \log(\frac{40}{5}) = \log 8$

d $\log 4 + \log 5 = \log(4 \times 5) = \log 20$

e $\log 5 + \log(0.4) = \log(5 \times 0.4) = \log 2$

f $\log 2 + \log 3 + \log 4 = \log(2 \times 3 \times 4) = \log 24$

g $1 + \log 3 = \log 10^1 + \log 3 = \log(10 \times 3) = \log 30$

h $\log 4 - 1 = \log 4 - \log 10^1 = \log(\frac{4}{10}) = \log 0.4$

i $\log 5 + \log 4 - \log 2 = \log\left(\frac{5 \times 4}{2}\right) = \log 10$

j $2 + \log 2 = \log 10^2 + \log 2 = \log(100 \times 2) = \log 200$

k $\log 40 - 2 = \log 40 - \log 10^2 = \log\left(\frac{40}{100}\right) = \log 0.4$

l $\log 6 - \log 2 - \log 3 = \log(6 \div 2 \div 3) = \log 1$ $(= 0)$

m $\log 50 - 4 = \log 50 - \log 10^4 = \log\left(\frac{50}{10^4}\right) = \log 0.005$

n $3 - \log 50 = \log 10^3 - \log 50 = \log\left(\frac{1000}{50}\right) = \log 20$

o $\log(\frac{4}{3}) + \log 3 + \log 7 = \log\left(\frac{4}{3} \times 3 \times 7\right) = \log 28$

2 **a** $5 \log 2 + \log 3 = \log 2^5 + \log 3 = \log(2^5 \times 3) = \log 96$

b $2 \log 3 + 3 \log 2 = \log 3^2 + \log 2^3 = \log(9 \times 8) = \log 72$

c $3 \log 4 - \log 8 = \log 4^3 - \log 8 = \log(\frac{64}{8}) = \log 8$

d $2\log 5 - 3\log 2$
$= \log 5^2 - \log 2^3$
$= \log\left(\frac{25}{8}\right)$

e $\frac{1}{2}\log 4 + \log 3$
$= \log 4^{\frac{1}{2}} + \log 3$
$= \log(2 \times 3)$
$= \log 6$

f $\frac{1}{3}\log\left(\frac{1}{8}\right)$
$= \log\left(\frac{1}{8}\right)^{\frac{1}{3}}$
$= \log\left(2^{-3}\right)^{\frac{1}{3}}$
$= \log 2^{-1}$
$= \log(\frac{1}{2})$ or $-\log 2$

g $3 - \log 2 - 2\log 5$
$= \log 10^3 - \log 2 - \log 5^2$
$= \log(1000 \div 2 \div 25)$
$= \log 20$

h $1 - 3\log 2 + \log 20$
$= \log 10^1 - \log 2^3 + \log 20$
$= \log(10 \div 8 \times 20)$
$= \log 25$

i $2 - \frac{1}{2}\log 4 - \log 5$
$= \log 10^2 - \log 4^{\frac{1}{2}} - \log 5$
$= \log(100 \div 2 \div 5)$
$= \log 10$
$= 1$

3 **a** $\dfrac{\log 4}{\log 2} = \dfrac{\log 2^2}{\log 2} = \dfrac{2\log 2}{\log 2} = 2$

b $\dfrac{\log 27}{\log 9} = \dfrac{\log 3^3}{\log 3^2} = \dfrac{3\log 3}{2\log 3} = \frac{3}{2}$

c $\dfrac{\log 8}{\log 2} = \dfrac{\log 2^3}{\log 2} = \dfrac{3\log 2}{\log 2} = 3$

d $\dfrac{\log 3}{\log 9} = \dfrac{\log 3}{\log 3^2} = \dfrac{\log 3}{2\log 3} = \frac{1}{2}$

e $\dfrac{\log 25}{\log(0.2)} = \dfrac{\log 5^2}{\log 5^{-1}} = \dfrac{2\log 5}{-1\log 5} = -2$

f $\dfrac{\log 8}{\log(0.25)} = \dfrac{\log 2^3}{\log 2^{-2}}$ $\{0.25 = \frac{1}{4} = \frac{1}{2^2}\}$
$= \dfrac{3\log 2}{-2\log 2} = -\frac{3}{2}$

4 **a** $\log 9 = \log 3^2$
$= 2\log 3$

b $\log\sqrt{2} = \log 2^{\frac{1}{2}}$
$= \frac{1}{2}\log 2$

c $\log\left(\frac{1}{8}\right) = \log\left(\dfrac{1}{2^3}\right)$
$= \log 2^{-3}$
$= -3\log 2$

d $\log\left(\frac{1}{5}\right) = \log 5^{-1}$
$= -1\log 5$
$= -\log 5$

e $\log 5 = \log\left(\frac{10}{2}\right)$
$= \log 10^1 - \log 2$
$= 1 - \log 2$

f $\log 5000 = \log\left(\frac{10\,000}{2}\right)$
$= \log 10^4 - \log 2$
$= 4 - \log 2$

5 **a** $\log_b 6$
$= \log_b(2 \times 3)$
$= \log_b 2 + \log_b 3$
$= p + q$

b $\log_b 108$
$= \log_b(2^2 3^3)$
$= 2\log_b 2 + 3\log_b 3$
$= 2p + 3q$

c $\log_b 45$
$= \log_b(3^2 5)$
$= 2\log_b 3 + \log_b 5$
$= 2q + r$

d $\log_b\left(\frac{5\sqrt{3}}{2}\right)$
$= \log_b(5 \times 3^{\frac{1}{2}}) - \log_b 2$
$= \log_b 5 + \frac{1}{2}\log_b 3 - \log_b 2$
$= r + \frac{1}{2}q - p$

e $\log_b(\frac{5}{32})$
$= \log_b 5 - \log_b 2^5$
$= \log_b 5 - 5\log_b 2$
$= r - 5p$

f $\log_b(0.\overline{2})$
$= \log_b(\frac{2}{9})$
$= \log_b 2 - \log_b 3^2$
$= p - 2q$

6 **a** $\log_2(PR)$
$= \log_2 P + \log_2 R$
$= x + z$

b $\log_2(RQ^2)$
$= \log_2 R + \log_2 Q^2$
$= \log_2 R + 2\log_2 Q$
$= z + 2y$

c $\log_2\left(\dfrac{PR}{Q}\right)$
$= \log_2(PR) - \log_2 Q$
$= \log_2 P + \log_2 R - \log_2 Q$
$= x + z - y$

d $\log_2\left(P^2\sqrt{Q}\right)$
$= \log_2 P^2 + \log_2 Q^{\frac{1}{2}}$
$= 2\log_2 P + \frac{1}{2}\log_2 Q$
$= 2x + \frac{1}{2}y$

e $\log_2\left(\dfrac{Q^3}{\sqrt{R}}\right)$
$= \log_2 Q^3 - \log_2 R^{\frac{1}{2}}$
$= 3\log_2 Q - \frac{1}{2}\log_2 R$
$= 3y - \frac{1}{2}z$

f $\log_2\left(\dfrac{R^2\sqrt{Q}}{P^3}\right)$
$= \log_2 R^2 + \log_2 Q^{\frac{1}{2}} - \log_2 P^3$
$= 2\log_2 R + \frac{1}{2}\log_2 Q - 3\log_2 P$
$= 2z + \frac{1}{2}y - 3x$

7 **a** $\log_t N^2 = 1.72$
$\therefore\ 2\log_t N = 1.72$
$\therefore\ \log_t N = 1.72 \div 2$
$= 0.86$

b $\log_t(MN)$
$= \log_t M + \log_t N$
$= 1.29 + 0.86$
$= 2.15$

c $\log_t\left(\dfrac{N^2}{\sqrt{M}}\right)$
$= \log_t N^2 - \log_t M^{\frac{1}{2}}$
$= 1.72 - \frac{1}{2}\log_t M$
$= 1.72 - \frac{1}{2}(1.29)$
$= 1.075$

EXERCISE 4C.2

1 **a** $y = 2^x$
$\therefore\ \log y = \log 2^x$
$\therefore\ \log y = x\log 2$

b $y = 20b^3$
$\therefore\ \log y = \log(20b^3)$
$\therefore\ \log y = \log 20 + \log b^3$
$\therefore\ \log y \approx 1.30 + 3\log b$

c $M = ad^4$
$\therefore\ \log M = \log(ad^4)$
$\therefore\ \log M = \log a + \log d^4$
$\therefore\ \log M = \log a + 4\log d$

d $T = 5\sqrt{d} = 5d^{\frac{1}{2}}$
$\therefore\ \log T = \log(5d^{\frac{1}{2}})$
$\therefore\ \log T = \log 5 + \log d^{\frac{1}{2}}$
$\therefore\ \log T \approx 0.699 + \frac{1}{2}\log d$

e $R = b\sqrt{l} = bl^{\frac{1}{2}}$
$\therefore\ \log R = \log(bl^{\frac{1}{2}})$
$\therefore\ \log R = \log b + \log l^{\frac{1}{2}}$
$\therefore\ \log R = \log b + \frac{1}{2}\log l$

f $Q = \dfrac{a}{b^n}$
$\therefore\ \log Q = \log\left(\dfrac{a}{b^n}\right)$
$\therefore\ \log Q = \log a - \log b^n$
$\therefore\ \log Q = \log a - n\log b$

g $y = ab^x$
$\therefore\ \log y = \log(ab^x)$
$\therefore\ \log y = \log a + \log b^x$
$\therefore\ \log y = \log a + x\log b$

h $F = \dfrac{20}{\sqrt{n}} = \dfrac{20}{n^{\frac{1}{2}}}$
$\therefore\ \log F = \log\left(\dfrac{20}{n^{\frac{1}{2}}}\right)$
$\therefore\ \log F = \log 20 - \log n^{\frac{1}{2}}$
$\therefore\ \log F \approx 1.30 - \frac{1}{2}\log n$

i $L = \dfrac{ab}{c}$
$\therefore\ \log L = \log\left(\dfrac{ab}{c}\right)$
$\therefore\ \log L = \log ab - \log c$
$\therefore\ \log L = \log a + \log b - \log c$

j $N = \sqrt{\dfrac{a}{b}}$
$\therefore\ N = \left(\dfrac{a}{b}\right)^{\frac{1}{2}}$
$\therefore\ \log N = \log\left(\dfrac{a}{b}\right)^{\frac{1}{2}}$
$\therefore\ \log N = \frac{1}{2}\log\left(\dfrac{a}{b}\right)$
$\therefore\ \log N = \frac{1}{2}\log a - \frac{1}{2}\log b$

k $S = 200 \times 2^t$
$\therefore\ \log S = \log(200 \times 2^t)$
$\therefore\ \log S = \log 200 + \log 2^t$
$\therefore\ \log S = \log 200 + t\log 2$
$\therefore\ \log S \approx 2.30 + t\log 2$

l $y = \dfrac{a^m}{b^n}$
$\therefore\ \log y = \log\left(\dfrac{a^m}{b^n}\right)$
$\therefore\ \log y = \log a^m - \log b^n$
$\therefore\ \log y = m\log a - n\log b$

2 **a** $\log D = \log e + \log 2$
$= \log(e \times 2)$
$\therefore \quad D = 2e$

b $\log F = \log 5 - \log t$
$= \log\left(\frac{5}{t}\right)$
$\therefore \quad F = \frac{5}{t}$

c $\log P = \frac{1}{2}\log x$
$= \log x^{\frac{1}{2}}$
$\therefore \quad P = \sqrt{x}$

d $\log M = 2\log b + \log c$
$= \log b^2 + \log c$
$= \log(b^2 c)$
$\therefore \quad M = b^2 c$

e $\log B = 3\log m - 2\log n$
$= \log m^3 - \log n^2$
$= \log\left(\frac{m^3}{n^2}\right)$
$\therefore \quad B = \frac{m^3}{n^2}$

f $\log N = -\frac{1}{3}\log p$
$= \log p^{-\frac{1}{3}}$
$= \log\left(\frac{1}{\sqrt[3]{p}}\right)$
$\therefore \quad N = \frac{1}{\sqrt[3]{p}}$

g $\log P = 3\log x + 1$
$= \log x^3 + \log 10^1$
$= \log(10x^3)$
$\therefore \quad P = 10x^3$

h $\log Q = 2 - \log x$
$= \log 10^2 - \log x$
$= \log\left(\frac{100}{x}\right)$
$\therefore \quad Q = \frac{100}{x}$

3 **a** $\log_3 27 + \log_3(\frac{1}{3}) = \log_3 x$
$\therefore \quad \log_3(27 \times \frac{1}{3}) = \log_3 x$
$\therefore \quad \log_3 9 = \log_3 x$
$\therefore \quad x = 9$

b $\log_5 x = \log_5 8 - \log_5(6 - x)$
$\therefore \quad \log_5 x = \log_5\left(\frac{8}{6-x}\right)$
$\therefore \quad x = \frac{8}{6-x}$
$\therefore \quad 6x - x^2 = 8$
$\therefore \quad x^2 - 6x + 8 = 0$
$\therefore \quad (x-2)(x-4) = 0$
$\therefore \quad x = 2$ or 4

Note: $x > 0$
and $6 - x > 0$
so $0 < x < 6$

c $\log_5 125 - \log_5 \sqrt{5} = \log_5 x$
$\therefore \quad \log_5\left(\frac{125}{\sqrt{5}}\right) = \log_5 x$
$\therefore \quad x = \frac{125}{\sqrt{5}}$ or $25\sqrt{5}$

d $\log_{20} x = 1 + \log_{20} 10$
$\therefore \quad \log_{20} x = \log_{20} 20^1 + \log_{20} 10$
$= \log_{20} 200$
$\therefore \quad x = 200$

e $\log x + \log(x+1) = \log 30$
$\therefore \quad \log[x(x+1)] = \log 30$
$\therefore \quad x^2 + x = 30$
$\therefore \quad x^2 + x - 30 = 0$
$\therefore \quad (x+6)(x-5) = 0$
$\therefore \quad x = -6$ or 5
but $x > 0$ for $\log x$ to exist
$\therefore \quad x = 5$

f $\log(x+2) - \log(x-2) = \log 5$
$\therefore \quad \log\left(\frac{x+2}{x-2}\right) = \log 5$
$\therefore \quad \frac{x+2}{x-2} = 5$
$\therefore \quad x + 2 = 5x - 10$
$\therefore \quad -4x = -12$
$\therefore \quad x = 3$

Note: $x + 2 > 0$ and $x - 2 > 0$
$\therefore \quad x > 2$ ✓

EXERCISE 4D.1

1 **a** $\ln e^3$
$= 3 \quad \{\ln e^a = a\}$

b $\ln 1$
$= \ln e^0$
$= 0$

c $\ln \sqrt[3]{e}$
$= \ln e^{\frac{1}{3}}$
$= \frac{1}{3}$

d $\ln\left(\frac{1}{e^2}\right)$
$= \ln e^{-2}$
$= -2$

3 $\ln x$ exists only when $x > 0$.
$\therefore \quad \ln(-2)$ and $\ln(0)$ do not exist.

Note: If $\ln(-2) = a$ then $-2 = e^a$
and $e^a = -2$ has no solutions as $e^a > 0$ for all a.

4 **a** $\ln e^a$
$= a$

b $\ln(e \times e^a)$
$= \ln e^{1+a}$
$= a + 1$

c $\ln(e^a \times e^b)$
$= \ln(e^{a+b})$
$= a + b$

d $\ln(e^a)^b$
$= \ln e^{ab}$
$= ab$

e $\ln\left(\dfrac{e^a}{e^b}\right)$
$= \ln(e^{a-b})$
$= a - b$

5 **a** $e^{1.7918}$ **b** $e^{4.0943}$ **c** $e^{8.6995}$ **d** $e^{-0.5108}$ **e** $e^{-5.1160}$
f $e^{2.7081}$ **g** $e^{7.3132}$ **h** $e^{0.4055}$ **i** $e^{-1.8971}$ **j** $e^{-8.8049}$

6 **a** $\ln x = 3$
$\therefore\ x = e^3$
$\therefore\ x \approx 20.1$

b $\ln x = 1$
$\therefore\ x = e^1$
$\therefore\ x = e \approx 2.72$

c $\ln x = 0$
$\therefore\ x = e^0$
$\therefore\ x = 1$

d $\ln x = -1$
$\therefore\ x = e^{-1}$
$\therefore\ x \approx 0.368$

e $\ln x = -5$
$\therefore\ x = e^{-5}$
$\therefore\ x \approx 0.006\,74$

f $\ln x \approx 0.835$
$\therefore\ x \approx e^{0.835}$
$\therefore\ x \approx 2.30$

g $\ln x \approx 2.145$
$\therefore\ x \approx e^{2.145}$
$\therefore\ x \approx 8.54$

h $\ln x \approx -3.2971$
$\therefore\ x \approx e^{-3.2971}$
$\therefore\ x \approx 0.0370$

EXERCISE 4D.2

1 **a** $\ln 15 + \ln 3$
$= \ln(15 \times 3)$
$= \ln 45$

b $\ln 15 - \ln 3$
$= \ln(\frac{15}{3})$
$= \ln 5$

c $\ln 20 - \ln 5$
$= \ln(\frac{20}{5})$
$= \ln 4$

d $\ln 4 + \ln 6$
$= \ln(4 \times 6)$
$= \ln 24$

e $\ln 5 + \ln(0.2)$
$= \ln(5 \times 0.2)$
$= \ln 1 = 0$

f $\ln 2 + \ln 3 + \ln 5$
$= \ln(2 \times 3 \times 5)$
$= \ln 30$

g $1 + \ln 4$
$= \ln e^1 + \ln 4$
$= \ln(e \times 4)$
$= \ln 4e$

h $\ln 6 - 1$
$= \ln 6 - \ln e^1$
$= \ln(\frac{6}{e})$

i $\ln 5 + \ln 8 - \ln 2$
$= \ln(5 \times 8 \div 2)$
$= \ln 20$

j $2 + \ln 4$
$= \ln e^2 + \ln 4$
$= \ln(e^2 \times 4)$
$= \ln 4e^2$

k $\ln 20 - 2$
$= \ln 20 - \ln e^2$
$= \ln\left(\dfrac{20}{e^2}\right)$

l $\ln 12 - \ln 4 - \ln 3$
$= \ln(12 \div 4 \div 3)$
$= \ln 1$
$= 0$

2 **a** $5\ln 3 + \ln 4$
$= \ln(3^5) + \ln 4$
$= \ln(3^5 \times 4)$
$= \ln 972$

b $3\ln 2 + 2\ln 5$
$= \ln(2^3) + \ln(5^2)$
$= \ln(2^3 \times 5^2)$
$= \ln 200$

c $3\ln 2 - \ln 8$
$= \ln(2^3) - \ln 8$
$= \ln(\frac{2^3}{8})$
$= \ln 1 = 0$

d $3\ln 4 - 2\ln 2$
$= \ln(4^3) - \ln(2^2)$
$= \ln(\frac{64}{4})$
$= \ln 16$

e $\frac{1}{3}\ln 8 + \ln 3$
$= \ln(8^{\frac{1}{3}}) + \ln 3$
$= \ln(8^{\frac{1}{3}} \times 3)$
$= \ln 6$

f $\frac{1}{3}\ln(\frac{1}{27})$
$= \ln\left((\frac{1}{27})^{\frac{1}{3}}\right)$
$= \ln\left(\dfrac{1}{27^{\frac{1}{3}}}\right)$
$= \ln(\frac{1}{3})$

g $-\ln 2$
$= \ln(2^{-1})$
$= \ln(\frac{1}{2})$

h $-\ln(\frac{1}{2})$
$= \ln((\frac{1}{2})^{-1})$
$= \ln 2$

i $-2\ln(\frac{1}{4})$
$= \ln((\frac{1}{4})^{-2})$
$= \ln(4^2)$
$= \ln 16$

3 **a** $\ln 27$
$= \ln 3^3$
$= 3\ln 3$

b $\ln\sqrt{3}$
$= \ln 3^{\frac{1}{2}}$
$= \frac{1}{2}\ln 3$

c $\ln(\frac{1}{16})$
$= \ln\left(\frac{1}{2^4}\right)$
$= \ln(2^{-4})$
$= -4\ln 2$

d $\ln(\frac{1}{6})$
$= \ln 6^{-1}$
$= -1\ln 6$
$= -\ln 6$

e $\ln\left(\frac{1}{\sqrt{2}}\right)$
$= \ln 2^{-\frac{1}{2}}$
$= -\frac{1}{2}\ln 2$

f $\ln\left(\frac{e}{5}\right)$
$= \ln e^1 - \ln 5$
$= 1 - \ln 5$

g $\ln\sqrt[3]{5}$
$= \ln 5^{\frac{1}{3}}$
$= \frac{1}{3}\ln 5$

h $\ln(\frac{1}{32})$
$= \ln 2^{-5}$
$= -5\ln 2$

i $\ln\left(\frac{1}{\sqrt[5]{2}}\right)$
$= \ln\left(\frac{1}{2^{\frac{1}{5}}}\right)$
$= \ln 2^{-\frac{1}{5}}$
$= -\frac{1}{5}\ln 2$

4 $\ln\left(\frac{e^2}{8}\right)$
$= \ln e^2 - \ln 8$
$= 2 - \ln 2^3$
$= 2 - 3\ln 2$

5 **a** $\ln D = \ln x + 1$
$\therefore\ \ln D - \ln x = 1$
$\therefore\ \ln\left(\frac{D}{x}\right) = 1$
$\therefore\ \frac{D}{x} = e^1$
$\therefore\ D = ex$

b $\ln F = -\ln p + 2$
$\therefore\ \ln F + \ln p = 2$
$\therefore\ \ln(Fp) = 2$
$\therefore\ Fp = e^2$
$\therefore\ F = \frac{e^2}{p}$

c $\ln P = \frac{1}{2}\ln x$
$\therefore\ \ln P = \ln x^{\frac{1}{2}}$
$\therefore\ P = \sqrt{x}$

d $\ln M = 2\ln y + 3$
$\therefore\ \ln M - 2\ln y = 3$
$\therefore\ \ln\left(\frac{M}{y^2}\right) = 3$
$\therefore\ \frac{M}{y^2} = e^3$
$\therefore\ M = e^3y^2$

e $\ln B = 3\ln t - 1$
$\therefore\ \ln B - \ln t^3 = -1$
$\therefore\ \ln\left(\frac{B}{t^3}\right) = -1$
$\therefore\ \frac{B}{t^3} = e^{-1}$
$\therefore\ B = \frac{t^3}{e}$

f $\ln N = -\frac{1}{3}\ln g$
$\therefore\ \ln N = \ln g^{-\frac{1}{3}}$
$\therefore\ N = g^{-\frac{1}{3}}$
$\therefore\ N = \frac{1}{\sqrt[3]{g}}$

g $\ln Q \approx 3\ln x + 2.159$
$\therefore\ \ln Q - 3\ln x \approx 2.159$
$\therefore\ \ln\left(\frac{Q}{x^3}\right) \approx 2.159$
$\therefore\ \frac{Q}{x^3} \approx e^{2.159}$
$\therefore\ \frac{Q}{x^3} \approx 8.66$
$\therefore\ Q \approx 8.66x^3$

h $\ln D \approx 0.4\ln n - 0.6582$
$\therefore\ \ln D - \ln n^{0.4} \approx -0.6582$
$\therefore\ \ln\left(\frac{D}{n^{0.4}}\right) \approx -0.6582$
$\therefore\ \frac{D}{n^{0.4}} \approx e^{-0.6582}$
$\therefore\ \frac{D}{n^{0.4}} \approx 0.518$
$\therefore\ D \approx 0.518n^{0.4}$

EXERCISE 4E

1 **a** $2^x = 10$
$\therefore\ \log 2^x = \log 10$
$\therefore\ x\log 2 = \log 10$
$\therefore\ x = \frac{\log 10}{\log 2} \approx 3.32$

b $3^x = 20$
$\therefore\ \log 3^x = \log 20$
$\therefore\ x\log 3 = \log 20$
$\therefore\ x = \frac{\log 20}{\log 3} \approx 2.73$

c $4^x = 100$
$\therefore\ \log 4^x = \log 100$
$\therefore\ x\log 4 = \log 100$
$\therefore\ x = \frac{\log 100}{\log 4}$
≈ 3.32

d
$$\begin{aligned} (1.2)^x &= 1000 \\ \therefore\ x\log(1.2) &= \log 1000 \\ \therefore\ x &= \frac{\log 1000}{\log(1.2)} \\ \therefore\ x &\approx 37.9 \end{aligned}$$

e
$$\begin{aligned} 2^x &= 0.08 \\ \therefore\ \log 2^x &= \log(0.08) \\ \therefore\ x\log 2 &= \log(0.08) \\ \therefore\ x &= \frac{\log(0.08)}{\log 2} \\ \therefore\ x &\approx -3.64 \end{aligned}$$

f
$$\begin{aligned} 3^x &= 0.000\,25 \\ \therefore\ \log 3^x &= \log(0.000\,25) \\ \therefore\ x\log 3 &= \log(0.000\,25) \\ \therefore\ x &= \frac{\log(0.000\,25)}{\log 3} \\ \therefore\ x &\approx -7.55 \end{aligned}$$

g
$$\begin{aligned} \left(\tfrac{1}{2}\right)^x &= 0.005 \\ \therefore\ \log(0.5)^x &= \log(0.005) \\ \therefore\ x\log(0.5) &= \log(0.005) \\ \therefore\ x &= \frac{\log(0.005)}{\log(0.5)} \\ \therefore\ x &\approx 7.64 \end{aligned}$$

h
$$\begin{aligned} \left(\tfrac{3}{4}\right)^x &= 10^{-4} \\ \therefore\ \log(0.75)^x &= -4 \\ \therefore\ x\log(0.75) &= -4 \\ \therefore\ x &= \frac{-4}{\log(0.75)} \\ \therefore\ x &\approx 32.0 \end{aligned}$$

i
$$\begin{aligned} (0.99)^x &= 0.000\,01 \\ \therefore\ \log(0.99)^x &= \log(0.000\,01) \\ \therefore\ x\log(0.99) &= \log(0.000\,01) \\ \therefore\ x &= \frac{\log(0.000\,01)}{\log(0.99)} \\ \therefore\ x &\approx 1145.5 \\ \therefore\ x &\approx 1150 \ \text{(3 s.f.)} \end{aligned}$$

2 **a**
$$\begin{aligned} 200\times 2^{0.25t} &= 600 \\ \therefore\ 2^{0.25t} &= 3 \\ \therefore\ \log(2^{0.25t}) &= \log 3 \\ \therefore\ 0.25t\log 2 &= \log 3 \\ \therefore\ t &= \frac{\log 3}{0.25\times\log 2} \\ \therefore\ t &\approx 6.340 \end{aligned}$$

b
$$\begin{aligned} 20\times 2^{0.06t} &= 450 \\ \therefore\ 2^{0.06t} &= 22.5 \\ \therefore\ \log(2^{0.06t}) &= \log(22.5) \\ \therefore\ 0.06t\log 2 &= \log(22.5) \\ \therefore\ t &= \frac{\log(22.5)}{0.06\times\log 2} \\ \therefore\ t &\approx 74.86 \end{aligned}$$

c
$$\begin{aligned} 30\times 3^{-0.25t} &= 3 \\ \therefore\ 3^{-0.25t} &= \tfrac{1}{10} \\ \therefore\ \log 3^{-0.25t} &= \log(0.1) \\ \therefore\ -0.25t\log 3 &= \log(0.1) \\ \therefore\ t &= \frac{\log(0.1)}{-0.25\times\log 3} \\ \therefore\ t &\approx 8.384 \end{aligned}$$

d
$$\begin{aligned} 12\times 2^{-0.05t} &= 0.12 \\ \therefore\ 2^{-0.05t} &= \tfrac{1}{100} \\ \therefore\ \log 2^{-0.05t} &= \log(0.01) \\ \therefore\ -0.05t\log 2 &= \log(0.01) \\ \therefore\ t &= \frac{\log(0.01)}{-0.05\times\log 2} \\ \therefore\ t &\approx 132.9 \end{aligned}$$

e
$$\begin{aligned} 50\times 5^{-0.02t} &= 1 \\ \therefore\ 5^{-0.02t} &= \tfrac{1}{50} = 0.02 \\ \therefore\ \log 5^{-0.02t} &= \log(0.02) \\ \therefore\ -0.02t\log 5 &= \log(0.02) \\ \therefore\ t &= \frac{\log(0.02)}{-0.02\times\log 5} \\ \therefore\ t &\approx 121.5 \end{aligned}$$

f
$$\begin{aligned} 300\times 2^{0.005t} &= 1000 \\ \therefore\ 2^{0.005t} &= \tfrac{10}{3} \\ \therefore\ \log 2^{0.005t} &= \log(\tfrac{10}{3}) \\ \therefore\ 0.005t\log 2 &= \log(\tfrac{10}{3}) \\ \therefore\ t &= \frac{\log(\frac{10}{3})}{0.005\times\log 2} \\ \therefore\ t &\approx 347.4 \end{aligned}$$

3 **a**
$$\begin{aligned} e^x &= 10 \\ \therefore\ x &= \ln 10 \\ \therefore\ x &\approx 2.303 \end{aligned}$$

b
$$\begin{aligned} e^x &= 1000 \\ \therefore\ x &= \ln 1000 \\ \therefore\ x &\approx 6.908 \end{aligned}$$

c
$$\begin{aligned} e^x &= 0.008\,62 \\ \therefore\ x &= \ln(0.008\,62) \\ \therefore\ x &\approx -4.754 \end{aligned}$$

d
$$e^{\frac{x}{2}} = 5$$
$$\therefore \frac{x}{2} = \ln 5$$
$$\therefore x = 2\ln 5$$
$$\therefore x \approx 3.219$$

e
$$e^{\frac{x}{3}} = 157.8$$
$$\therefore \frac{x}{3} = \ln(157.8)$$
$$\therefore x = 3\ln(157.8)$$
$$\therefore x \approx 15.18$$

f
$$e^{\frac{x}{10}} = 0.016\,82$$
$$\therefore \frac{x}{10} = \ln(0.016\,82)$$
$$\therefore x = 10\ln(0.016\,82)$$
$$\therefore x \approx -40.85$$

g
$$20 \times e^{0.06x} = 8.312$$
$$\therefore e^{0.06x} = 0.4156$$
$$\therefore 0.06x = \ln(0.4156)$$
$$\therefore x = \frac{\ln(0.4156)}{0.06}$$
$$\therefore x \approx -14.63$$

h
$$50e^{-0.03x} = 0.816$$
$$\therefore e^{-0.03x} = 0.016\,32$$
$$\therefore -0.03x = \ln(0.016\,32)$$
$$\therefore x = \frac{\ln(0.016\,32)}{-0.03}$$
$$\therefore x \approx 137.2$$

i
$$41.83e^{0.652x} = 1000$$
$$\therefore e^{0.652x} \approx 23.91$$
$$\therefore 0.652x \approx \ln(23.91)$$
$$\therefore x \approx \frac{\ln(23.91)}{0.652}$$
$$\therefore x \approx 4.868$$

EXERCISE 4F

1 **a**
$$\log_3 12 = \frac{\log_{10} 12}{\log_{10} 3} \approx 2.26$$

b
$$\log_{\frac{1}{2}} 1250 = \frac{\log_{10} 1250}{\log_{10}(0.5)} \approx -10.3$$

c
$$\log_3(0.067) = \frac{\log_{10}(0.067)}{\log_{10} 3} \approx -2.46$$

d
$$\log_{0.4}(0.006\,984) = \frac{\log_{10}(0.006\,984)}{\log_{10}(0.4)} \approx 5.42$$

2 **a**
$$2^x = 0.051$$
$$\therefore x = \log_2(0.051)$$
$$\therefore x = \frac{\ln(0.051)}{\ln 2}$$
$$\therefore x \approx -4.29$$

b
$$4^x = 213.8$$
$$\therefore x = \log_4 213.8$$
$$\therefore x = \frac{\ln(213.8)}{\ln 4}$$
$$\therefore x \approx 3.87$$

c
$$3^{2x+1} = 4.069$$
$$\therefore 2x+1 = \log_3(4.069)$$
$$\therefore 2x+1 = \frac{\ln(4.069)}{\ln 3}$$
$$\therefore 2x+1 \approx 1.2774$$
$$\therefore 2x \approx 0.2774$$
$$\therefore x \approx 0.139$$

3 **a**
$$25^x - 3(5^x) = 0$$
$$\therefore 5^{2x} - 3(5^x) = 0$$
$$\therefore 5^x(5^x - 3) = 0$$
$$\therefore 5^x = 3$$
$$\{\text{as } 5^x > 0 \text{ for all } x\}$$
$$\therefore x = \log_5 3$$
$$\therefore x = \frac{\log 3}{\log 5}$$
$$\therefore x \approx 0.683$$

b
$$8(9^x) - 3^x = 0$$
$$\therefore 8 \times 3^{2x} - 3^x = 0$$
$$\therefore 3^x(8 \times 3^x - 1) = 0$$
$$\therefore 8 \times 3^x - 1 = 0$$
$$\{\text{as } 3^x > 0 \text{ for all } x\}$$
$$\therefore 3^x = \tfrac{1}{8}$$
$$\therefore x = \log_3(\tfrac{1}{8})$$
$$\therefore x = \frac{\log(\frac{1}{8})}{\log 3} \approx -1.89$$

4 **a**
$$\log_4 x^3 + \log_2 \sqrt{x} = 8$$
$$\therefore \frac{\log x^3}{\log 4} + \frac{\log x^{\frac{1}{2}}}{\log 2} = 8$$
$$\therefore \frac{3\log x}{2\log 2} + \frac{\frac{1}{2}\log x}{\log 2} = 8$$
$$\therefore \frac{3\log x}{2\log 2} + \frac{\log x}{2\log 2} = 8$$
$$\therefore \frac{4\log x}{2\log 2} = 8$$
$$\therefore \log x = 4\log 2$$
$$\therefore \log x = \log 2^4$$
$$\therefore x = 16$$

b
$$\log_{16} x^5 = \log_{64} 125 - \log_4 \sqrt{x}$$
$$\therefore \frac{\log x^5}{\log 16} = \frac{\log 125}{\log 64} - \frac{\log x^{\frac{1}{2}}}{\log 4}$$
$$\therefore \frac{5\log x}{4\log 2} = \frac{\log 125}{6\log 2} - \frac{\frac{1}{2}\log x}{2\log 2}$$
$$\therefore \frac{15\log x}{12\log 2} = \frac{2\log 125}{12\log 2} - \frac{3\log x}{12\log 2}$$
$$\therefore 15\log x = 2\log 125 - 3\log x$$
$$\therefore 18\log x = 2\log 125$$
$$\therefore \log x = \tfrac{1}{9}\log 5^3$$
$$\therefore \log x = \log(5^3)^{\frac{1}{9}}$$
$$\therefore x = 5^{\frac{1}{3}} \approx 1.71$$

5 $4^x \times 5^{4x+3} = 10^{2x+3}$

$\therefore\ 4^x \times \left(5^4\right)^x \times 5^3 = \left(10^2\right)^x \times 10^3$

$\therefore\ 125 \times \left(4 \times 5^4\right)^x = 1000 \times 100^x$

$\therefore\ 2500^x = 8 \times 100^x$

$\therefore\ 25^x = 8$

$\therefore\ x = \log_{25} 8$ or $\dfrac{\log 8}{\log 25}$

EXERCISE 4G

1 **a** $f(x) = \log_3(x+1)$

i We require $x + 1 > 0$ $\therefore$ $x > -1$ So, the domain is $\{x \mid x > -1\}$ and the range is $y \in \mathbb{R}$.

ii As $x \to -1$ (from the right), $y \to -\infty$,
so $x = -1$ is a vertical asymptote.
As $x \to \infty$, $y \to \infty$.
When $x = 0$, $y = \log_3 1 = 0$
$\therefore$ y-intercept is 0.
When $y = 0$, $\log_3(x+1) = 0$
$\therefore$ $x + 1 = 3^0$, so $x + 1 = 1$, $x = 0$
$\therefore$ x-intercept is 0.

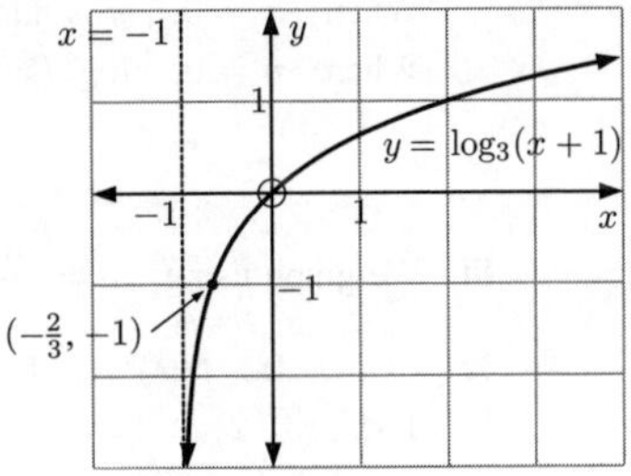

iii We graph, using $y = \dfrac{\log(x+1)}{\log 3}$

iv If $f(x) = -1$
then $\log_3(x+1) = -1$
$\therefore\ x + 1 = 3^{-1}$
$\therefore\ x = \frac{1}{3} - 1$
$\therefore\ x = -\frac{2}{3}$
which checks with the graph

v f is defined by $y = \log_3(x+1)$
$\therefore\ f^{-1}$ is defined by $x = \log_3(y+1)$
$\therefore\ y + 1 = 3^x$
$\therefore\ y = 3^x - 1$
$\therefore\ f^{-1}(x) = 3^x - 1$
and has H.A. $y = -1$
Its domain is $x \in \mathbb{R}$
and range is $\{y \mid y > -1\}$.

We can verify the inverse function by checking that $(f^{-1} \circ f)(x) = x$.

b $f(x) = 1 - \log_3(x+1)$

i We require $x + 1 > 0$ $\therefore$ $x > -1$ So, domain is $\{x \mid x > -1\}$ and range is $y \in \mathbb{R}$.

ii As $x \to -1$ (right), $y \to \infty$,
so $x = -1$ is a vertical asymptote.
As $x \to \infty$, $y \to -\infty$.
When $x = 0$, $y = 1 - \log_3 1 = 1 - 0 = 1$
$\therefore$ y-intercept is 1.
When $y = 0$, $1 - \log_3(x+1) = 0$
$\therefore\ \log_3(x+1) = 1$
$\therefore\ x + 1 = 3^1 = 3$
$\therefore\ x = 2$

So, the x-intercept is 2.

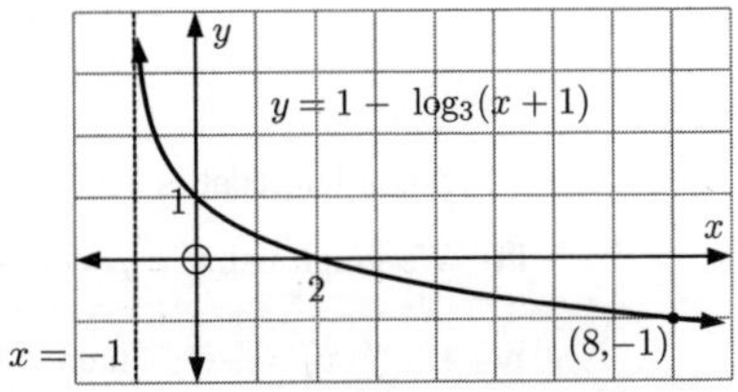

iii We graph using $y = 1 - \dfrac{\log(x+1)}{\log 3}$

iv If $f(x) = -1$, $1 - \log_3(x+1) = -1$

$\therefore \log_3(x+1) = 2$

$\therefore x + 1 = 3^2$

$\therefore x = 8$

v f is defined by $y = 1 - \log_3(x+1)$

$\therefore$ f^{-1} is defined by $x = 1 - \log_3(y+1)$

$\therefore \log_3(y+1) = 1 - x$

$\therefore y + 1 = 3^{1-x}$

$\therefore y = 3^{1-x} - 1$

$\therefore f^{-1}(x) = 3^{1-x} - 1$

Horizontal asymptote is $y = -1$.
Domain is $x \in \mathbb{R}$.
Range is $\{y \mid y > -1\}$.

c $f(x) = \log_5(x-2) - 2$

i We require $x - 2 > 0$ $\therefore$ $x > 2$. So, domain is $\{x \mid x > 2\}$ and range is $y \in \mathbb{R}$.

ii As $x \to 2$ (right), $y \to -\infty$. So, $x = 2$ is a vertical asymptote.
As $x \to \infty$, $y \to \infty$.
When $x = 0$, y is undefined $\therefore$ no y-intercept.
When $y = 0$, $\log_5(x-2) = 2$ $\therefore$ $x - 2 = 5^2 = 25$

$\therefore x = 27$

$\therefore$ x-intercept is 27

iii We graph using $y = \dfrac{\log(x-2)}{\log 5} - 2$

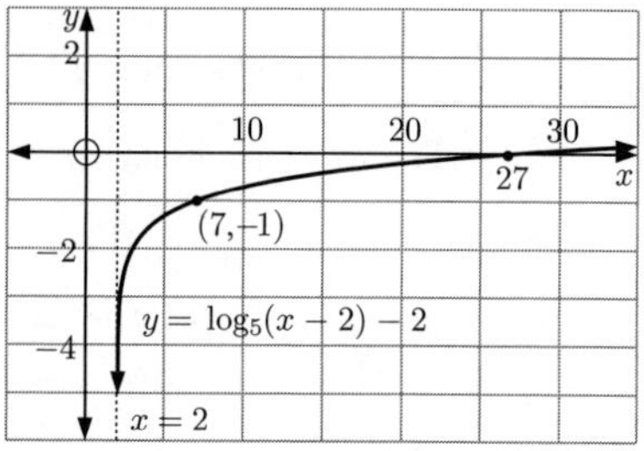

iv If $f(x) = -1$

$\log_5(x-2) - 2 = -1$

$\therefore \log_5(x-2) = 1$

$\therefore x - 2 = 5^1$

$\therefore x = 5 + 2 = 7$

v f is defined by $y = \log_5(x-2) - 2$

$\therefore$ f^{-1} is defined by $x = \log_5(y-2) - 2$

$\therefore x + 2 = \log_5(y-2)$

$\therefore y - 2 = 5^{x+2}$

$\therefore y = 5^{x+2} + 2$

$\therefore f^{-1}(x) = 5^{x+2} + 2$

Horizontal asymptote is $y = 2$.
Domain is $x \in \mathbb{R}$.
Range is $\{y \mid y > 2\}$.

d $f(x) = 1 - \log_5(x-2)$

i We require $x - 2 > 0$ $\therefore$ $x > 2$ So, domain is $\{x \mid x > 2\}$ and range is $y \in \mathbb{R}$.

ii As $x \to 2$ (right), $y \to \infty$. $\therefore$ $x = 2$ is a vertical asymptote.
As $x \to \infty$, $y \to -\infty$.
When $x = 0$, y is undefined $\therefore$ no y-intercept.
When $y = 0$, $1 - \log_5(x-2) = 0$

$\therefore \log_5(x-2) = 1$

$\therefore x - 2 = 5^1$

$\therefore x = 7$

So, x-intercept is 7.

iii We graph using $y = 1 - \dfrac{\log(x-2)}{\log 5}$

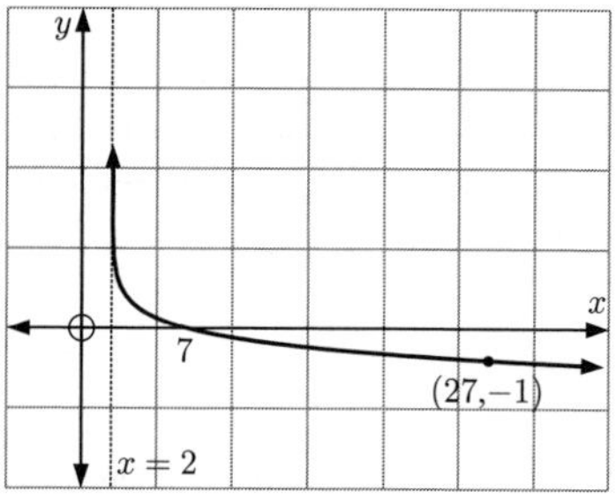

iv If $f(x) = -1$, then $1 - \log_5(x-2) = -1$

$\therefore \log_5(x-2) = 2$

$\therefore x - 2 = 5^2$

$\therefore x = 27$

v f is defined by $y = 1 - \log_5(x-2)$

$\therefore$ f^{-1} is defined by $x = 1 - \log_5(y-2)$

$\therefore$ $\log_5(y-2) = 1 - x$

$\therefore$ $y - 2 = 5^{1-x}$

$\therefore$ $y = 5^{1-x} + 2$

$\therefore$ $f^{-1}(x) = 5^{1-x} + 2$

Horizontal asymptote is $y = 2$.
Domain is $x \in \mathbb{R}$.
Range is $\{y \mid y > 2\}$.

e $f(x) = 1 - \log_2 x^2$

i We require $x^2 > 0$ which is true for all x except $x = 0$
$\therefore$ domain is $\{x \mid x \neq 0\}$, range is $y \in \mathbb{R}$.

ii As $x \to 0$ (right or left), $y \to \infty$
$\therefore$ $x = 0$ is a vertical asymptote.
As $x \to \infty$, $y \to -\infty$,
as $x \to -\infty$, $y \to -\infty$.
When $x = 0$, $y = 1 - \log_2 0$
which is undefined $\therefore$ no y-intercept.
When $y = 0$, $\log_2 x^2 = 1$

$\therefore$ $x^2 = 2^1 = 2$

$\therefore$ $x = \pm\sqrt{2}$

$\therefore$ x-intercepts are $\sqrt{2}$ and $-\sqrt{2}$

iii We graph using $y = 1 - \dfrac{\log x^2}{\log 2}$

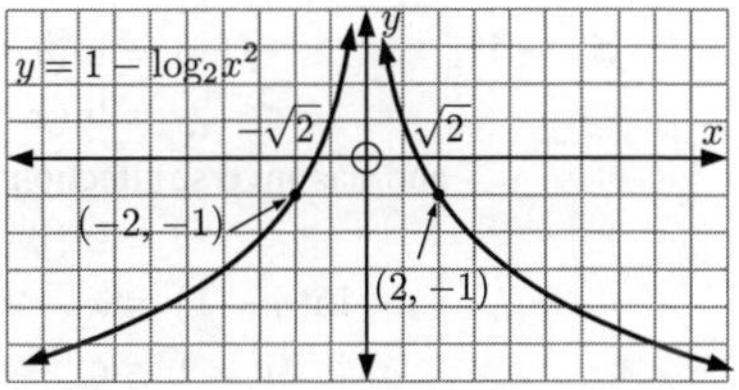

iv When $f(x) = -1$ $\therefore$ $x^2 = 2^2$
$1 - \log_2 x^2 = -1$ $\therefore$ $x^2 = 4$
$\therefore$ $\log_2 x^2 = 2$ $\therefore$ $x = \pm 2$

v If $f(x) = 1 - \log_2 x^2$, $x > 0$
then $f^{-1}(x)$ exists and is defined
by $x = 1 - \log_2 y^2$, $y > 0$

$\therefore$ $\log_2 y^2 = 1 - x$

$\therefore$ $y^2 = 2^{1-x}$

$\therefore$ $y = 2^{\frac{1-x}{2}}$ as $y > 0$

So, $f^{-1}(x) = 2^{\frac{1-x}{2}}$

If $f(x) = 1 - \log_2 x^2$, $x < 0$
then $f^{-1}(x)$ also exists and is defined
by $x = 1 - \log_2 y^2$, $y < 0$

$\therefore$ $\log_2 y^2 = 1 - x$

$\therefore$ $y^2 = 2^{1-x}$

$\therefore$ $y = -2^{\frac{1-x}{2}}$ as $y < 0$

So, $f^{-1}(x) = -2^{\frac{1-x}{2}}$

2 **a** **i** $f(x) = e^x + 5$
or $y = e^x + 5$
has inverse function
$x = e^y + 5$

$\therefore$ $x - 5 = e^y$

$\therefore$ $y = \ln(x-5)$

$\therefore$ $f^{-1}(x) = \ln(x-5)$

ii

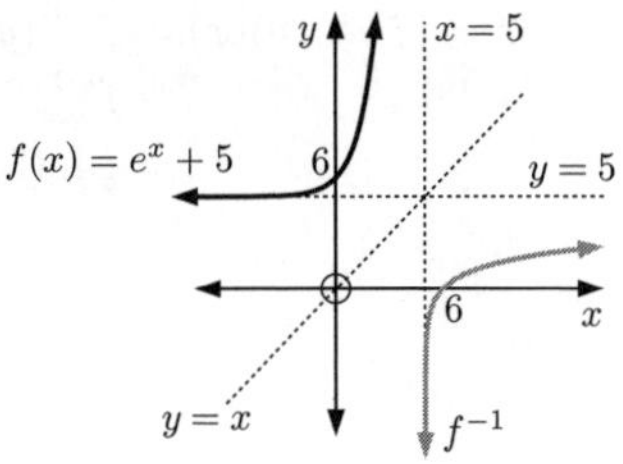

iii Domain of f is $\{x \mid x \in \mathbb{R}\}$, range is $\{y \mid y > 5\}$.
Domain of f^{-1} is $\{x \mid x > 5\}$, range is $\{y \mid y \in \mathbb{R}\}$.

iv f has a H.A. $y = 5$. f^{-1} has a V.A. $x = 5$.

b **i** $f(x) = e^{x+1} - 3$
or $y = e^{x+1} - 3$
has inverse function
$x = e^{y+1} - 3$

$\therefore$ $x + 3 = e^{y+1}$

$\therefore$ $y + 1 = \ln(x+3)$

$\therefore$ $f^{-1}(x) = \ln(x+3) - 1$

ii

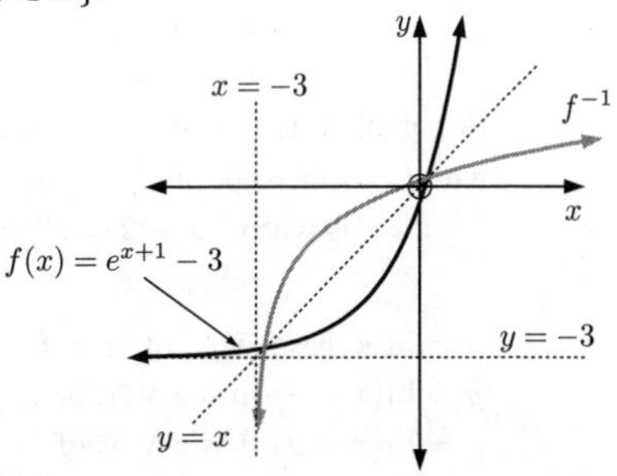

iii Domain of f is $\{x \mid x \in \mathbb{R}\}$, range is $\{y \mid y > -3\}$.
Domain of f^{-1} is $\{x \mid x > -3\}$, range is $\{y \mid y \in \mathbb{R}\}$.

iv f has a H.A. $y = -3$. f^{-1} has a V.A. $x = -3$.

c **i**
$$f(x) = \ln x - 4$$
$$\therefore \quad y = \ln x - 4$$
and has inverse function
$$x = \ln y - 4$$
$$\therefore \quad x + 4 = \ln y$$
$$\therefore \quad y = e^{x+4}$$
$$\therefore \quad f^{-1}(x) = e^{x+4}$$

ii

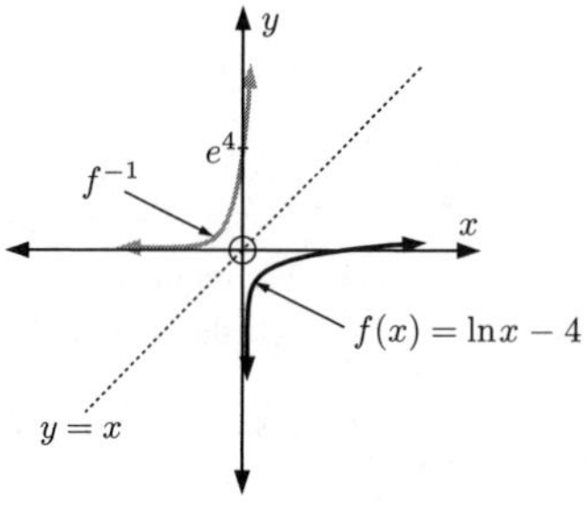

iii Domain of f is $\{x \mid x > 0\}$, range is $\{y \mid y \in \mathbb{R}\}$.
Domain of f^{-1} is $\{x \mid x \in \mathbb{R}\}$, range is $\{y \mid y > 0\}$.

iv f has a V.A. $x = 0$. f^{-1} has a H.A. $y = 0$.

d **i**
$$f(x) = \ln(x-1) + 2, \ x > 1$$
$$\therefore \quad y = \ln(x-1) + 2$$
and has inverse function
$$x = \ln(y-1) + 2$$
$$\therefore \quad \ln(y-1) = x - 2$$
$$\therefore \quad y - 1 = e^{x-2}$$
$$\therefore \quad y = e^{x-2} + 1$$
$$\therefore \quad f^{-1}(x) = e^{x-2} + 1$$

ii

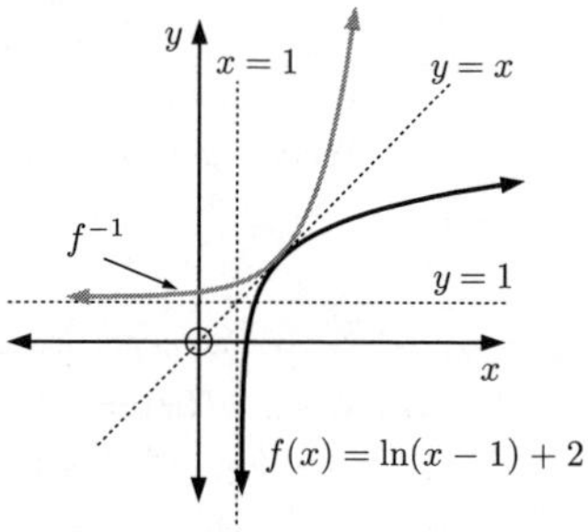

iii Domain of f is $\{x \mid x > 1\}$, range is $\{y \mid y \in \mathbb{R}\}$.
Domain of f^{-1} is $\{x \mid x \in \mathbb{R}\}$, range is $\{y \mid y > 1\}$.

iv f has a V.A. $x = 1$. f^{-1} has a H.A. $y = 1$.

3 **a** f is $y = e^{2x}$
so the inverse function f^{-1} is
$$x = e^{2y}$$
$$\therefore \quad 2y = \ln x$$
$$\therefore \quad y = \tfrac{1}{2}\ln x$$
$$\therefore \quad f^{-1}(x) = \tfrac{1}{2}\ln x$$
$$\therefore \quad (f^{-1} \circ g)(x) = f^{-1}(g(x)) = f^{-1}(2x-1) = \tfrac{1}{2}\ln(2x-1)$$

b
$$(g \circ f)(x) = g(f(x)) = g(e^{2x}) = 2(e^{2x}) - 1$$
So, $y = 2e^{2x} - 1$ which has inverse
$$x = 2e^{2y} - 1$$
$$\therefore \quad x + 1 = 2e^{2y}$$
$$\therefore \quad \tfrac{1}{2}(x+1) = e^{2y}$$
$$\therefore \quad 2y = \ln\left(\frac{x+1}{2}\right)$$
$$\therefore \quad y = \tfrac{1}{2}\ln\left(\frac{x+1}{2}\right)$$
$$\therefore \quad (g \circ f)^{-1}(x) = \tfrac{1}{2}\ln\left(\frac{x+1}{2}\right)$$

4 **a** $y = \ln x$ cuts the x-axis when $y = 0$
$$\therefore \quad \ln x = 0$$
$$\therefore \quad x = e^0 = 1$$
So, graph A is that of $y = \ln x$.

Note: x-intercept of $y = \ln(x-2)$ is when $x - 2 = e^0 = 1$
$$\therefore \quad x = 3$$

b The x-intercept of $y = \ln(x+2)$ occurs when $x + 2 = e^0 = 1$
$$\therefore \quad x = -1$$

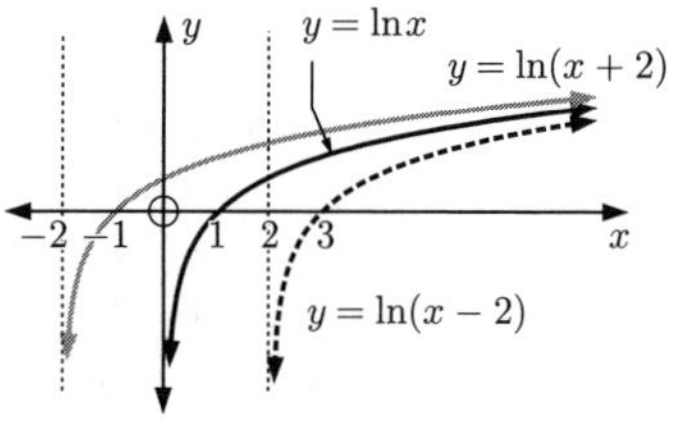

c $y = \ln x$ has a V.A. of $x = 0$.
$y = \ln(x-2)$ has a V.A. of $x = 2$.
$y = \ln(x+2)$ has a V.A. of $x = -2$.

5 Since $y = \ln x^2$, $y = 2\ln x$ {log law}

$\therefore$ the new y-values are $2 \times$ old y-values. So, she is partly correct.

However, unlike $y = \ln x$, $y = \ln x^2$ is also defined for $x < 0$, so she must also find $\ln x^2$ for $x < 0$.

6 **a** $f(x) = e^{x+3} + 2$

or $y = e^{x+3} + 2$ has inverse function

$x = e^{y+3} + 2$

$\therefore$ $x - 2 = e^{y+3}$

$\therefore$ $\ln(x-2) = y + 3$

$\therefore$ $y = \ln(x-2) - 3$

So, $f^{-1}(x) = \ln(x-2) - 3$

b **i** $f(x) < 2.1$ when $e^{x+3} + 2 < 2.1$

$\therefore$ $e^{x+3} < 0.1$

$\therefore$ $x + 3 < \ln(0.1)$

$\therefore$ $x < \ln(0.1) - 3$

$\therefore$ $x < -5.30$

ii Similarly, $f(x) < 2.01$ when

$x < \ln(0.01) - 3$

$\therefore$ $x < -7.61$

iii $f(x) < 2.001$ when

$x < \ln(0.001) - 3$

$\therefore$ $x < -9.91$

iv $f(x) < 2.0001$ when

$x < \ln(0.0001) - 3$

$\therefore$ $x < -12.2$

We conjecture that the H.A. is $y = 2$.

c As $x \to \infty$, $y \to \infty$

As $x \to -\infty$, $e^{x+3} \to 0$ $\therefore$ $y \to 2$

$\therefore$ H.A. is $y = 2$.

d f has a H.A. $y = 2$ and range $\{y \mid y > 2\}$

$\therefore$ f^{-1} has a V.A. $x = 2$ and domain $\{x \mid x > 2\}$

EXERCISE 4H

1 $W_t = 20 \times 2^{0.15t}$ grams

a When $W_t = 30$,

$20 \times 2^{0.15t} = 30$

$\therefore$ $2^{0.15t} = 1.5$

$\therefore$ $\log 2^{0.15t} = \log(1.5)$

$\therefore$ $0.15t \log 2 = \log(1.5)$

$\therefore$ $t = \dfrac{\log(1.5)}{0.15 \times \log 2}$

$\therefore$ $t \approx 3.90$ hours

$\therefore$ it takes about 3.90 hours to reach 30 g.

b When $W_t = 100$,

$20 \times 2^{0.15t} = 100$

$\therefore$ $2^{0.15t} = 5$

$\therefore$ $\log 2^{0.15t} = \log 5$

$\therefore$ $0.15t \log 2 = \log 5$

$\therefore$ $t = \dfrac{\log 5}{0.15 \times \log 2}$

$\therefore$ $t \approx 15.5$ hours

$\therefore$ it takes about 15.5 hours to reach 100 g.

2 **a** When $M_t = 50$, $25 \times e^{0.1t} = 50$

$\therefore$ $e^{0.1t} = 2$

$\therefore$ $\ln e^{0.1t} = \ln 2$

$\therefore$ $0.1t = \ln 2$

$\therefore$ $t = 10 \times \ln 2$

$\therefore$ $t \approx 6.93$

$\therefore$ it takes about 6.93 hours to reach 50 g.

b When $M_t = 100$, $25 \times e^{0.1t} = 100$

$\therefore$ $e^{0.1t} = 4$

$\therefore$ $\ln e^{0.1t} = \ln 4$

$\therefore$ $0.1t = \ln 4$

$\therefore$ $t = 10 \times \ln 4$

$\therefore$ $t \approx 13.9$

$\therefore$ it takes about 13.9 hours to reach 100 g.

3 **a**

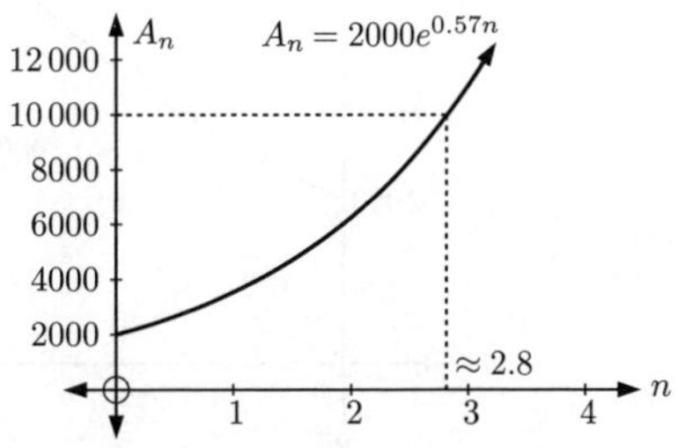

When $A_n = 10\,000$, $t \approx 2.8$

$\therefore$ we estimate that it will take 2.8 weeks for the infested area to reach 10 000 ha.

b When $A_n = 10\,000$, $2000 \times e^{0.57n} = 10\,000$

$$\therefore \quad e^{0.57n} = 5$$
$$\therefore \quad \ln e^{0.57n} = \ln 5$$
$$\therefore \quad 0.57n = \ln 5$$
$$\therefore \quad n = \frac{\ln 5}{0.57}$$
$$\therefore \quad n \approx 2.82$$

$\therefore$ it takes about 2.82 weeks for the infested area to reach 10 000 hectares.

4 $r = 107.5\% = 1.075$,
$u_1 = 160\,000$,
$u_{n+1} = 250\,000$

$$u_{n+1} = u_1 \times r^n$$
$$\therefore \quad 250\,000 = 160\,000 \times (1.075)^n$$
$$\therefore \quad (1.075)^n = \tfrac{25}{16}$$
$$\therefore \quad \log(1.075)^n = \log(\tfrac{25}{16})$$
$$\therefore \quad n\log(1.075) = \log(\tfrac{25}{16})$$
$$\therefore \quad n = \frac{\log(\frac{25}{16})}{\log(1.075)} \approx 6.1709$$

$\therefore$ it would take 6.17 years or 6 years 62 days.

5 $u_1 = 10\,000$,
$u_{n+1} = 15\,000$,
$r = 104.8\% = 1.048$

$$u_{n+1} = u_1 \times r^n$$
$$\therefore \quad 15\,000 = 10\,000 \times (1.048)^n$$
$$\therefore \quad (1.048)^n = 1.5$$
$$\therefore \quad \log(1.048)^n = \log(1.5)$$
$$\therefore \quad n\log(1.048) = \log(1.5)$$
$$\therefore \quad n = \frac{\log(1.5)}{\log(1.048)}$$
$$\therefore \quad n \approx 8.648$$

$\therefore$ it would take 9 years.
{interest compounded annually}

6 **a** 8.4% p.a. compounded monthly

is $\dfrac{8.4\%}{12} = 0.7\%$ a month
$= 0.007$

So $r = 1 + 0.007$
$= 1.007$

b $u_1 = 15\,000$ and $u_{n+1} = 25\,000$

$$u_{n+1} = u_1 \times r^n$$
$$\therefore \quad 25\,000 = 15\,000 \times (1.007)^n$$
$$\therefore \quad (1.007)^n = \tfrac{25}{15} = \tfrac{5}{3}$$
$$\therefore \quad \log(1.007)^n = \log(\tfrac{5}{3})$$
$$\therefore \quad n\log(1.007) = \log(\tfrac{5}{3})$$
$$\therefore \quad n = \frac{\log(\frac{5}{3})}{\log(1.007)} \approx 73.23$$

$\therefore$ he can withdraw the money after 74 months.

7 **a** $A_{n+1} = 12\,000 \times (1.0835)^n$

$\therefore$ when $n = 0$, $A_1 = 12\,000 \times (1.0835)^0$

$\therefore \quad A_1 = 12\,000$ euros

This is the initial investment.

b When $n = 5$, $A_6 = 12\,000 \times (1.0835)^5$
$\approx 17\,919.50$

$\therefore$ the value of the investment after 5 years is €17 919.50

c $A_{3.25}$ is the value of the investment after 2.25 years, or after 2 years 3 months.

d When $A_{n+1} = 24\,000$,

$$12\,000 \times (1.0835)^n = 24\,000$$
$$\therefore \quad (1.0835)^n = 2$$
$$\therefore \quad \log(1.0835)^n = \log 2$$
$$\therefore \quad n\log 1.0835 = \log 2$$
$$\therefore \quad n = \frac{\log 2}{\log 1.0835}$$
$$\therefore \quad n \approx 8.64$$

$\therefore$ it takes 9 years for the investment to double in value.

e

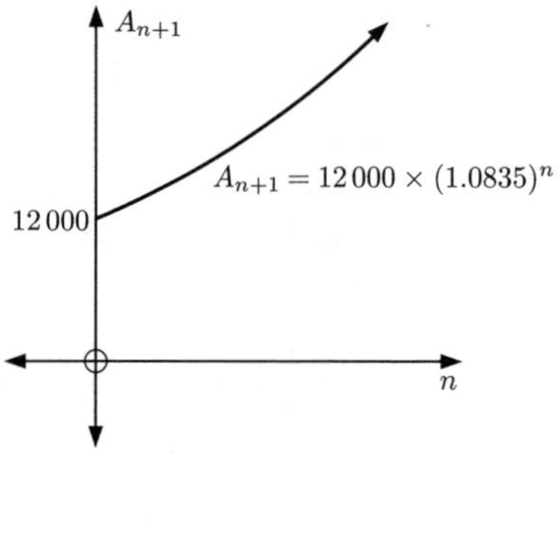

8 $M_t = 1000e^{-0.04t}$ $\quad \therefore \quad M_0 = 1000e^0 = 1000$ g

a For M_t to halve,

$$\begin{aligned} M_t &= 500 \\ \therefore \quad 1000e^{-0.04t} &= 500 \\ \therefore \quad e^{-0.04t} &= 0.5 \\ \therefore \quad -0.04t &= \ln(0.5) \\ \therefore \quad t &= \frac{\ln(0.5)}{-0.04} \\ \therefore \quad t &\approx 17.3 \text{ years} \end{aligned}$$

b For $M_t = 25$ g,

$$\begin{aligned} \therefore \quad 1000e^{-0.04t} &= 25 \\ \therefore \quad e^{-0.04t} &= 0.025 \\ \therefore \quad -0.04t &= \ln(0.025) \\ \therefore \quad t &= \frac{\ln(0.025)}{-0.04} \\ \therefore \quad t &\approx 92.2 \text{ years} \end{aligned}$$

c For $M_t = 1\%$ of M_0

$$\begin{aligned} \therefore \quad 1000e^{-0.04t} &= 0.01 \times 1000 \\ \therefore \quad e^{-0.04t} &= 0.01 \\ \therefore \quad -0.04t &= \ln(0.01) \\ \therefore \quad t &= \frac{\ln(0.01)}{-0.04} \\ \therefore \quad t &\approx 115 \text{ years} \end{aligned}$$

9 $V = 50(1 - e^{-0.2t})$ m s^{-1}

So, when $V = 40$, $50(1 - e^{-0.2t}) = 40$

$$\begin{aligned} \therefore \quad 1 - e^{-0.2t} &= 0.8 \\ \therefore \quad e^{-0.2t} &= 0.2 \\ \therefore \quad -0.2t &= \ln(0.2) \\ \therefore \quad t &= \frac{\ln(0.2)}{-0.2} \\ \therefore \quad t &\approx 8.05 \text{ s} \end{aligned}$$

So, it would take 8.05 s.

10 $T = 4 + 96 \times e^{-0.03t}$ ^{o}C

a When $T = 25$,

$$\begin{aligned} 4 + 96 \times e^{-0.03t} &= 25 \\ \therefore \quad 96 \times e^{-0.03t} &= 21 \\ \therefore \quad e^{-0.03t} &= \tfrac{21}{96} \\ \therefore \quad -0.03t &= \ln(\tfrac{21}{96}) \\ \therefore \quad t &= \frac{\ln(\frac{21}{96})}{-0.03} \\ \therefore \quad t &\approx 50.7 \text{ minutes} \end{aligned}$$

b When $T = 5$,

$$\begin{aligned} 4 + 96 \times e^{-0.03t} &= 5 \\ \therefore \quad 96 \times e^{-0.03t} &= 1 \\ \therefore \quad e^{-0.03t} &= \tfrac{1}{96} \\ \therefore \quad -0.03t &= \ln(\tfrac{1}{96}) \\ \therefore \quad t &= \frac{\ln(\frac{1}{96})}{-0.03} \\ \therefore \quad t &\approx 152 \text{ minutes} \end{aligned}$$

11 $W_t = 1000 \times 2^{-0.04t}$ has $W_0 = 1000 \times 2^0 = 1000$ grams

a For the weight to halve,

$$\begin{aligned} W_t &= 500 \\ \therefore \quad 1000 \times 2^{-0.04t} &= 500 \\ \therefore \quad 2^{-0.04t} &= \tfrac{1}{2} = 2^{-1} \\ \therefore \quad -0.04t &= -1 \\ \therefore \quad t &= \frac{1}{0.04} \\ \therefore \quad t &= 25 \text{ years} \end{aligned}$$

b For $W_t = 20$,

$$\begin{aligned} 1000 \times 2^{-0.04t} &= 20 \\ \therefore \quad 2^{-0.04t} &= 0.02 \\ \therefore \quad \log 2^{-0.04t} &= \log(0.02) \\ \therefore \quad -0.04t \log 2 &= \log(0.02) \\ \therefore \quad t &= \frac{\log(0.02)}{-0.04 \times \log 2} \\ \therefore \quad t &\approx 141 \text{ years} \end{aligned}$$

c When $W_t = 1\%$ of 1000 grams $= 10$ g,

$$\begin{aligned} 1000 \times 2^{-0.04t} &= 10 \\ \therefore\ 2^{-0.04t} &= 0.01 \\ \therefore\ \log 2^{-0.04t} &= \log(0.01) \\ \therefore\ -0.04t\log 2 &= \log(0.01) \\ \therefore\ t &= \frac{\log(0.01)}{-0.04 \times \log 2} \\ \therefore\ t &\approx 166 \text{ years} \end{aligned}$$

12 $W = W_0 \times 2^{-0.0002t}$ grams

a When W is 25% of original,

$$\begin{aligned} W &= \tfrac{1}{4} \text{ of } W_0 \\ \therefore\ W_0 \times 2^{-0.0002t} &= \tfrac{1}{4} \times W_0 \\ \therefore\ 2^{-0.0002t} &= 2^{-2} \\ \therefore\ 0.0002t &= 2 \\ \therefore\ t &= \frac{2}{0.0002} \\ \therefore\ t &= 10\,000 \end{aligned}$$

$\therefore$ it would take 10 000 years.

b When W is 0.1% of original,

$$\begin{aligned} W &= \tfrac{0.1}{100} \text{ of } W_0 \\ \therefore\ W_0 \times 2^{-0.0002t} &= \tfrac{1}{1000} \times W_0 \\ \therefore\ \log 2^{-0.0002t} &= \log(0.001) \\ \therefore\ -0.0002t\log 2 &= \log(0.001) \\ \therefore\ t &= \frac{\log(0.001)}{-0.0002 \times \log 2} \\ \therefore\ t &\approx 49\,829 \end{aligned}$$

$\therefore$ it would take about 49 800 years.

13 $I = I_0 \times 2^{-0.02t}$ amps

When I is 10% of its original value,

$$\begin{aligned} I &= 10\% \text{ of } I_0 \\ \therefore\ I_0 \times 2^{-0.02t} &= 0.1 \times I_0 \\ \therefore\ 2^{-0.02t} &= 0.1 \\ \therefore\ \log 2^{-0.02t} &= \log(0.1) \\ \therefore\ -0.02t\log 2 &= \log(0.1) \\ \therefore\ t &= \frac{\log(0.1)}{-0.02 \times \log 2} \\ \therefore\ t &\approx 166 \text{ seconds} \end{aligned}$$

14 $V = 60(1 - 2^{-0.2t})$ m s^{-1}

When $V = 50$, $60(1 - 2^{-0.2t}) = 50$

$$\begin{aligned} \therefore\ 1 - 2^{-0.2t} &= 0.8\overline{3} \\ \therefore\ 2^{-0.2t} &= 0.1\overline{6} \\ \therefore\ \log 2^{-0.2t} &= \log 0.1\overline{6} \\ \therefore\ -0.2t\log 2 &= \log 0.1\overline{6} \\ \therefore\ t &= \frac{\log 0.1\overline{6}}{-0.2 \times \log 2} \\ \therefore\ t &\approx 12.9 \text{ seconds} \end{aligned}$$

REVIEW SET 4A

1

a $\log_4 64 = \log_4 4^3 = 3$

b $\log_2 256 = \log_2 2^8 = 8$

c $\log_2(0.25) = \log_2(\tfrac{1}{4}) = \log_2 2^{-2} = -2$

d $\log_{25} 5 = \log_{25} 25^{\frac{1}{2}} = \tfrac{1}{2}$

e $\log_8 1 = \log_8 8^0 = 0$

f $\log_6 6 = \log_6 6^1 = 1$

g $\log_{81} 3 = \log_{81} 81^{\frac{1}{4}} = \tfrac{1}{4}$

h $\log_9(0.\overline{1}) = \log_9(\tfrac{1}{9}) = \log_9 9^{-1} = -1$

i $\log_{27} 3 = \log_{27} 27^{\frac{1}{3}} = \tfrac{1}{3}$

j $\log_k \sqrt{k} = \log_k k^{\frac{1}{2}} = \tfrac{1}{2}$ provided $k > 0$, $k \neq 1$

2 **a** $\log\sqrt{10}$
$= \log 10^{\frac{1}{2}}$
$= \frac{1}{2}$

b $\log\left(\frac{1}{\sqrt[3]{10}}\right)$
$= \log 10^{-\frac{1}{3}}$
$= -\frac{1}{3}$

c $\log(10^a \times 10^{b+1})$
$= \log 10^{a+b+1}$
$= a+b+1$

3 **a** $4\ln 2 + 2\ln 3$
$= \ln 2^4 + \ln 3^2$
$= \ln(16 \times 9)$
$= \ln 144$

b $\frac{1}{2}\ln 9 - \ln 2$
$= \ln 9^{\frac{1}{2}} - \ln 2$
$= \ln 3 - \ln 2$
$= \ln(\frac{3}{2})$

c $2\ln 5 - 1$
$= \ln 5^2 - \ln e^1$
$= \ln\left(\frac{25}{e}\right)$

d $\frac{1}{4}\ln 81$
$= \ln\left(3^4\right)^{\frac{1}{4}}$
$= \ln 3^1$
$= \ln 3$

4 **a** $\ln\left(e\sqrt{e}\right)$
$= \ln(e^1 e^{\frac{1}{2}})$
$= \ln e^{\frac{3}{2}}$
$= \frac{3}{2}$

b $\ln\left(\frac{1}{e^3}\right)$
$= \ln e^{-3}$
$= -3$

c $\ln\left(\frac{e}{\sqrt{e^5}}\right)$
$= \ln\left(\frac{e^1}{e^{\frac{5}{2}}}\right)$
$= \ln(e^{1-\frac{5}{2}})$
$= \ln e^{-\frac{3}{2}}$
$= -\frac{3}{2}$

5 **a** $\log 16 + 2\log 3$
$= \log 16 + \log 3^2$
$= \log(16 \times 9)$
$= \log 144$

b $\log_2 16 - 2\log_2 3$
$= \log_2 16 - \log_2 3^2$
$= \log_2(\frac{16}{9})$

c $2 + \log_4 5$
$= \log_4 4^2 + \log_4 5$
$= \log_4(16 \times 5)$
$= \log_4 80$

6 **a** $P = 3 \times b^x$
$\therefore\ \log P = \log(3 \times b^x)$
$\therefore\ \log P = \log 3 + \log b^x$
$\therefore\ \log P = \log 3 + x\log b$

b $m = \frac{n^3}{p^2}$
$\therefore\ \log m = \log\left(\frac{n^3}{p^2}\right)$
$\therefore\ \log m = \log n^3 - \log p^2$
$\therefore\ \log m = 3\log n - 2\log p$

7 **a** $\ln(e^{2x}) = 2x$

b $\ln(e^2 e^x) = \ln(e^{2+x})$
$= 2 + x$

c $\ln\left(\frac{e}{e^x}\right) = \ln(e^{1-x})$
$= 1 - x$

8 **a** $\log T = 2\log x - \log y$
$\therefore\ \log T = \log x^2 - \log y$
$\therefore\ \log T = \log\left(\frac{x^2}{y}\right)$
$\therefore\ T = \frac{x^2}{y}$

b $\log_2 K = \log_2 n + \frac{1}{2}\log_2 t$
$\therefore\ \log_2 K = \log_2 n + \log_2 t^{\frac{1}{2}}$
$\therefore\ \log_2 K = \log_2(n \times \sqrt{t})$
$\therefore\ K = n\sqrt{t}$

9 **a** $\ln 32 = \ln 2^5$
$= 5\ln 2$

b $\ln 125 = \ln 5^3$
$= 3\ln 5$

c $\ln 729 = \ln 3^6$
$= 6\ln 3$

10 **a** $\log_5 36$
$= \log_5(2^2 \times 3^2)$
$= \log_5 2^2 + \log_5 3^2$
$= 2\log_5 2 + 2\log_5 3$
$= 2A + 2B$

b $\log_5 54$
$= \log_5(2 \times 3^3)$
$= \log_5 2 + \log_5 3^3$
$= \log_5 2 + 3\log_5 3$
$= A + 3B$

c $\log_5(8\sqrt{3})$
$= \log_5(2^3 \times 3^{\frac{1}{2}})$
$= \log_5 2^3 + \log_5 3^{\frac{1}{2}}$
$= 3\log_5 2 + \frac{1}{2}\log_5 3$
$= 3A + \frac{1}{2}B$

d
$$\begin{aligned}&\log_5(20.25)\\ &= \log_5(\tfrac{81}{4})\\ &= \log_5\left(\frac{3^4}{2^2}\right)\\ &= \log_5 3^4 - \log_5 2^2\\ &= 4\log_5 3 - 2\log_5 2\\ &= 4B - 2A\end{aligned}$$

e
$$\begin{aligned}&\log_5(0.\overline{8})\\ &= \log_5(\tfrac{8}{9})\\ &= \log_5\left(\frac{2^3}{3^2}\right)\\ &= \log_5 2^3 - \log_5 3^2\\ &= 3\log_5 2 - 2\log_5 3\\ &= 3A - 2B\end{aligned}$$

REVIEW SET 4B

1 **a** $32 = 10^{\log 32}$ $\approx 10^{1.51}$

b 0.0013 $= 10^{\log(0.0013)}$ $\approx 10^{-2.89}$

c 8.963×10^{-5} $= 10^{\log(8.963)} \times 10^{-5}$ $\approx 10^{0.952} \times 10^{-5}$ $\approx 10^{-4.05}$

2 **a** $\log_2 x = -3$ $\therefore\ x = 2^{-3}$ $\therefore\ x = \frac{1}{8}$

b $\log_5 x \approx 2.743$ $\therefore\ x \approx 5^{2.743}$ $\therefore\ x \approx 82.7$

c $\log_3 x \approx -3.145$ $\therefore\ x \approx 3^{-3.145}$ $\therefore\ x \approx 0.0316$

3 **a** $\log_2 k \approx 1.699 + x$ $\therefore\ k \approx 2^{1.699+x}$ $\therefore\ k \approx 2^{1.699} \times 2^x$ $\therefore\ k \approx 3.25 \times 2^x$

b $\log_a Q = 3\log_a P + \log_a R$ $= \log_a P^3 + \log_a R$ $= \log_a(P^3 \times R)$ $\therefore\ Q = P^3 R$

c $\log A \approx 5\log B - 2.602$ $\therefore\ \log A - \log B^5 \approx -2.602$ $\therefore\ \log\left(\frac{A}{B^5}\right) \approx -2.602$ $\therefore\ \frac{A}{B^5} \approx 10^{-2.602} \approx 0.0025$ $\therefore\ A \approx \frac{B^5}{400}$

4 **a** $5^x = 7$ $\therefore\ \log 5^x = \log 7$ $\therefore\ x\log 5 = \log 7$ $\therefore\ x = \frac{\log 7}{\log 5}$ $\therefore\ x \approx 1.209$

b $20 \times 2^{2x+1} = 500$ $\therefore\ 2^{2x+1} = 25$ $\therefore\ \log 2^{2x+1} = \log 25$ $\therefore\ (2x+1)\log 2 = \log 25$ $\therefore\ 2x + 1 = \frac{\log 25}{\log 2} \approx 4.6439$ $\therefore\ 2x \approx 3.6439$ $\therefore\ x \approx 1.822$

5 $W_t = 2500 \times 3^{-\frac{t}{3000}}$ grams

a $W_0 = 2500 \times 3^0$ $= 2500 \times 1$ $= 2500$ grams

b We need t when $W_t = 30\%$ of 2500 g
$\therefore\ 2500 \times 3^{-\frac{t}{3000}} = 0.3 \times 2500$
$\therefore\ \log 3^{-\frac{t}{3000}} = \log(0.3)$
$\therefore\ -\frac{t}{3000} \times \log 3 = \log(0.3)$
$\therefore\ t = \frac{-\log(0.3) \times 3000}{\log 3}$
$\therefore\ t \approx 3287.7$
$\therefore$ about 3290 years

c % change

$= \left(\frac{W_{1500} - W_0}{W_0}\right) \times 100\%$

$= \left(\frac{2500 \times 3^{-\frac{1500}{3000}} - 2500}{2500}\right) \times 100\%$

$= (3^{-\frac{1}{2}} - 1) \times 100\%$

$\approx -42.3\%$

So, a loss of 42.3%.

d

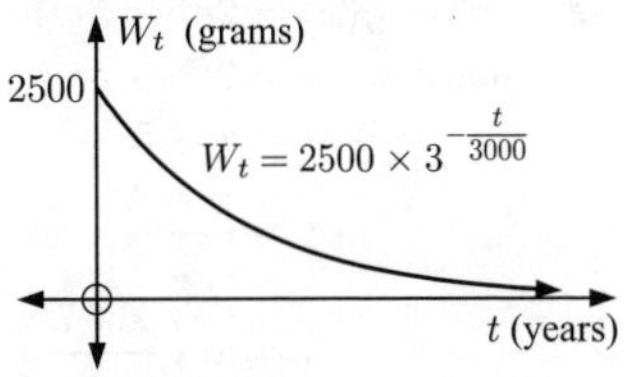

6 $16^x - 5 \times 8^x = 0$

$\therefore\ 2^x \times 8^x - 5 \times 8^x = 0$

$\therefore\ 8^x(2^x - 5) = 0$

$\therefore\ 2^x = 5$ as $8^x > 0$ for all x

$\therefore\ x = \log_2 5 = \frac{\log 5}{\log 2} \approx 2.32$

7 **a** $\ln x = 5$

$\therefore\ x = e^5$

$\therefore\ x \approx 148$

b $3\ln x + 2 = 0$

$\therefore\ 3\ln x = -2$

$\therefore\ \ln x = -\frac{2}{3}$

$\therefore\ x = e^{-\frac{2}{3}}$

$\therefore\ x \approx 0.513$

8 **a** $e^x = 400$

$\therefore\ x = \ln 400$

$\therefore\ x \approx 5.99$

b $e^{2x+1} = 11$

$\therefore\ 2x + 1 = \ln 11$

$\therefore\ 2x = \ln 11 - 1$

$\therefore\ x = \frac{\ln 11 - 1}{2}$

$\therefore\ x \approx 0.699$

c $25e^{\frac{x}{2}} = 750$

$\therefore\ e^{\frac{x}{2}} = 30$

$\therefore\ \frac{x}{2} = \ln 30$

$\therefore\ x = 2\ln 30$

$\therefore\ x \approx 6.80$

d $e^{2x} = 7e^x - 12$

$\therefore\ e^{2x} - 7e^x + 12 = 0$

$\therefore\ (e^x - 3)(e^x - 4) = 0$

$\therefore\ e^x = 3$ or $e^x = 4$

$\therefore\ x = \ln 3$ or $\ln 4$

$\therefore\ x \approx 1.10$ or 1.39

9 $P_t = P_0 \times 2^{\frac{t}{3}},\ \ t \geqslant 0$

When $t = 0$, $P_0 = P_0 \times 2^0 = P_0$. So the initial population was P_0.

a If P_t doubles, $P_t = 2P_0$

$\therefore\ \cancel{P_0}2^{\frac{t}{3}} = 2\cancel{P_0}$

$\therefore\ 2^{\frac{t}{3}} = 2^1$

$\therefore\ \frac{t}{3} = 1$

$\therefore\ t = 3$

So, it would take 3 years.

b % increase $= \left(\frac{P_4 - P_0}{P_0}\right) \times 100\%$

$= \left(\frac{P_0 \times 2^{\frac{4}{3}} - P_0}{P_0}\right) \times 100\%$

$= (2^{\frac{4}{3}} - 1) \times 100\%$

$\approx 151.98\%$

So, the % increase is 152%.

or $P_4 = P_0 \times 2^{\frac{4}{3}}$

$\approx P_0 \times 2.52$

$\approx 252\%$ of P_0

So, an increase of 152%.

10 **a** $g(x) = 2e^x - 5$ has inverse function $x = 2e^y - 5$

$\therefore\ 2e^y = x + 5$

$\therefore\ e^y = \dfrac{x+5}{2}$

$\therefore\ y = \ln\left(\dfrac{x+5}{2}\right)$

$\therefore\ g^{-1}(x) = \ln\left(\dfrac{x+5}{2}\right)$

b

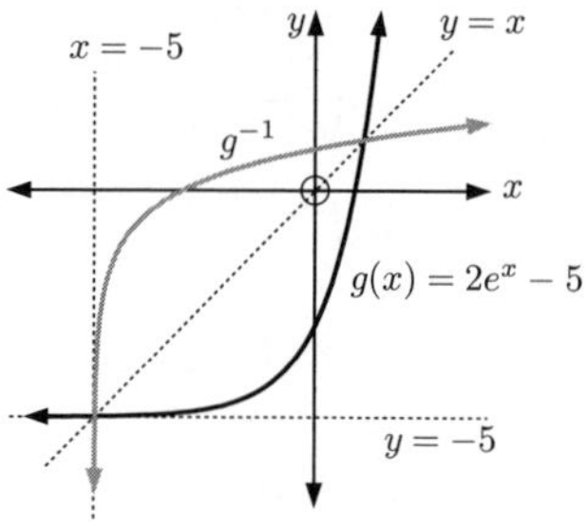

c domain of g is $\{x \mid x \in \mathbb{R}\}$, range is $\{y \mid y > -5\}$
domain of g^{-1} is $\{x \mid x > -5\}$, range is $\{y \mid y \in \mathbb{R}\}$

REVIEW SET 4C

1 **a** $\log\sqrt{1000}$
$= \log\left(10^3\right)^{\frac{1}{2}}$
$= \log 10^{\frac{3}{2}}$
$= \frac{3}{2}$

b $\log\left(\dfrac{10}{\sqrt[3]{10}}\right)$
$= \log\left(\dfrac{10^1}{10^{\frac{1}{3}}}\right)$
$= \log 10^{\frac{2}{3}} = \frac{2}{3}$

c $\log\left(\dfrac{10^a}{10^{-b}}\right)$
$= \log\left(10^{a-(-b)}\right)$
$= \log 10^{a+b}$
$= a + b$

2 **a** $\ln(e^5) = 5$
$\{$as $\ln e^a = a\}$

b $\ln\left(\sqrt{e}\right) = \ln e^{\frac{1}{2}}$
$= \frac{1}{2}$

c $\ln\left(\frac{1}{e}\right) = \ln e^{-1}$
$= -1$

3 **a** $20 = e^{\ln 20}$
$\approx e^{3.00}$

b $3000 = e^{\ln 3000}$
$\approx e^{8.01}$

c $0.075 = e^{\ln(0.075)}$
$\approx e^{-2.59}$

4 **a** $\log x = 3$
$\therefore\ x = 10^3$
$\therefore\ x = 1000$

b $\log_3(x+2) = 1.732$
$\therefore\ x + 2 = 3^{1.732}$
$\therefore\ x + 2 \approx 6.7046$
$\therefore\ x \approx 4.7046$
$\therefore\ x \approx 4.70$

c $\log_2\left(\dfrac{x}{10}\right) = -0.671$
$\therefore\ \dfrac{x}{10} = 2^{-0.671}$
$\therefore\ \dfrac{x}{10} \approx 0.628\,07$
$\therefore\ x \approx 6.28$

5 **a** $\ln 6 + \ln 4$
$= \ln(6 \times 4)$
$= \ln 24$

b $\ln 60 - \ln 20$
$= \ln(\frac{60}{20})$
$= \ln 3$

c $\ln 4 + \ln 1$
$= \ln 4 + 0$
$= \ln 4$

d $\ln 200 - \ln 8 + \ln 5$
$= \ln(\frac{200}{8}) + \ln 5$
$= \ln\left(\frac{200}{8} \times 5\right)$
$= \ln 125$

6 **a** $M = ab^n$
$\therefore\ \log M = \log(ab^n)$
$\therefore\ \log M = \log a + \log b^n$
$\therefore\ \log M = \log a + n\log b$

b $T = \dfrac{5}{\sqrt{l}}$
$\therefore\ \log T = \log\left(\dfrac{5}{l^{\frac{1}{2}}}\right)$
$\therefore\ \log T = \log 5 - \log l^{\frac{1}{2}}$
$\therefore\ \log T = \log 5 - \frac{1}{2}\log l$

c $G = \dfrac{a^2 b}{c}$
$\therefore\ \log G = \log\left(\dfrac{a^2 b}{c}\right)$
$\therefore\ \log G = \log(a^2 b) - \log c$
$\therefore\ \log G = \log a^2 + \log b - \log c$
$\therefore\ \log G = 2\log a + \log b - \log c$

7 **a**
$$3^x = 300$$
$$\therefore\ \log 3^x = \log 300$$
$$\therefore\ x\log 3 = \log 300$$
$$\therefore\ x = \frac{\log 300}{\log 3}$$
$$\therefore\ x \approx 5.19$$

b
$$30 \times 5^{1-x} = 0.15$$
$$\therefore\ 5^{1-x} = 0.005$$
$$\therefore\ \log 5^{1-x} = \log(0.005)$$
$$\therefore\ (1-x)\log 5 = \log(0.005)$$
$$\therefore\ 1-x = \frac{\log(0.005)}{\log 5}$$
$$\therefore\ 1-x \approx -3.292$$
$$\therefore\ x \approx 4.29$$

c
$$3^{x+2} = 2^{1-x}$$
$$\therefore\ \log 3^{x+2} = \log 2^{1-x}$$
$$\therefore\ (x+2)\log 3 = (1-x)\log 2$$
$$\therefore\ x\log 3 + 2\log 3 = \log 2 - x\log 2$$
$$\therefore\ x(\log 3 + \log 2) = \log 2 - 2\log 3$$
$$\therefore\ x\log 6 = \log(\tfrac{2}{9})$$
$$\therefore\ x = \frac{\log(\frac{2}{9})}{\log 6} \approx -0.839$$

8 **a**
$$\ln P = 1.5\ln Q + \ln T$$
$$\therefore\ \ln P = \ln Q^{1.5} + \ln T$$
$$= \ln(TQ^{1.5})$$
$$\therefore\ P = TQ^{1.5}$$

b
$$\ln M = 1.2 - 0.5\ln N$$
$$\therefore\ \ln M + \ln N^{\frac{1}{2}} = 1.2$$
$$\therefore\ \ln\left(M\sqrt{N}\right) = 1.2$$
$$\therefore\ M\sqrt{N} = e^{1.2}$$
$$\therefore\ M = \frac{e^{1.2}}{\sqrt{N}}$$

9 $g(x) = \log_3(x+2) - 2$

a We require $x + 2 > 0$, so $x > -2$

$\therefore$ the domain is $\{x \mid x > -2\}$ and the range is $y \in \mathbb{R}$.

b If $x \to -2$ (right), $y \to -\infty$ $\therefore$ V.A. is $x = -2$.

As $x \to \infty$, $y \to \infty$.

When $x = 0$, $g(0) = \log_3 2 - 2 \approx -1.37$ So, the y-intercept ≈ -1.37

When $y = 0$, $\log_3(x+2) = 2$ $\therefore$ $x + 2 = 3^2$

$\therefore$ $x = 7$ So, the x-intercept is 7.

c, e

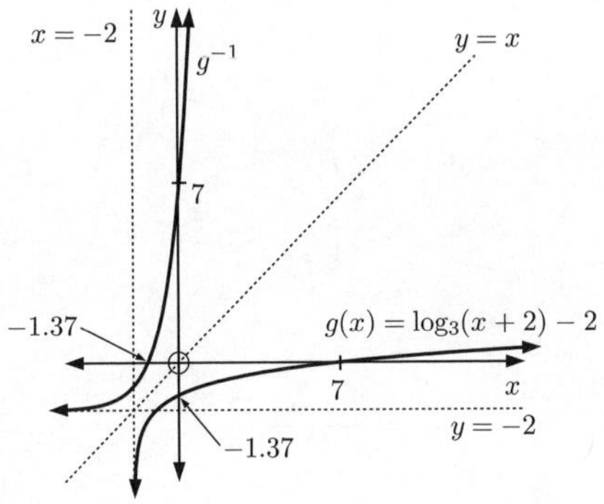

d g^{-1} is defined by $x = \log_3(y+2) - 2$
$$\therefore\ \log_3(y+2) = x + 2$$
$$\therefore\ y + 2 = 3^{x+2}$$
$$\therefore\ y = 3^{x+2} - 2$$
$$\therefore\ g^{-1}(x) = 3^{x+2} - 2$$

We can verify this by checking that $(g^{-1} \circ g)(x) = (g \circ g^{-1})(x) = x$.

10 $W_t = 8000 \times e^{-\frac{t}{20}}$ grams
$$W_0 = 8000e^0$$
$$= 8000 \times 1$$
$$= 8000 \text{ grams}$$

a When $W_t = \frac{1}{2} \times 8000$ grams, $8000e^{-\frac{t}{20}} = 4000$
$$\therefore\ e^{-\frac{t}{20}} = 0.5$$
$$\therefore\ -\frac{t}{20} = \ln(0.5)$$
$$\therefore\ t = -20\ln(0.5) \approx 13.9 \text{ weeks}$$

b When $W_t = 1000$ g,
$$8000e^{-\frac{t}{20}} = 1000$$
$$\therefore\ e^{-\frac{t}{20}} = \tfrac{1}{8}$$
$$\therefore\ -\frac{t}{20} = \ln(\tfrac{1}{8})$$
$$\therefore\ t = -20\ln(\tfrac{1}{8})$$
$$\therefore\ t \approx 41.6 \text{ weeks}$$

c When $W_t = 0.1\%$ of W_0
$$= \tfrac{1}{1000} \times 8000 = 8 \text{ g},$$
$$8000e^{-\frac{t}{20}} = 8$$
$$\therefore\ e^{-\frac{t}{20}} = 0.001$$
$$\therefore\ -\frac{t}{20} = \ln(0.001)$$
$$\therefore\ t = -20\ln(0.001) \approx 138 \text{ weeks}$$

Chapter 5

GRAPHING AND TRANSFORMING FUNCTIONS

EXERCISE 5A

1 $f(x) = x$

a $f(2x) = 2x$

b $f(x) + 2$
$= x + 2$

c $\frac{1}{2} f(x) = \dfrac{x}{2}$

d $2f(x) + 3$
$= 2x + 3$

2 $f(x) = x^2$

a $f(3x) = (3x)^2$
$= 9x^2$

b $f\left(\dfrac{x}{2}\right) = \left(\dfrac{x}{2}\right)^2$
$= \dfrac{x^2}{4}$

c $3f(x) = 3x^2$

d $2f(x-1) + 5$
$= 2(x-1)^2 + 5$
$= 2(x^2 - 2x + 1) + 5$
$= 2x^2 - 4x + 7$

3 $f(x) = x^3$

a $f(4x)$
$= (4x)^3$
$= 64x^3$

b $\frac{1}{2} f(2x)$
$= \frac{1}{2}(2x)^3$
$= \frac{1}{2} \times 8x^3$
$= 4x^3$

c $f(x+1)$
$= (x+1)^3$
$= x^3 + 3x^2 + 3x + 1$

d $2f(x+1) - 3$
$= 2(x+1)^3 - 3$
$= 2(x^3 + 3x^2 + 3x + 1) - 3$
$= 2x^3 + 6x^2 + 6x - 1$

4 $f(x) = 2^x$

a $f(2x) = 2^{2x}$
$= 4^x$

b $f(-x) + 1$
$= 2^{-x} + 1$

c $f(x-2) + 3$
$= 2^{x-2} + 3$

d $2f(x) + 3$
$= 2 \times 2^x + 3$
$= 2^{x+1} + 3$

5 $f(x) = \dfrac{1}{x}$

a $f(-x)$
$= \dfrac{1}{(-x)}$
$= -\dfrac{1}{x}$

b $f(\frac{1}{2}x)$
$= \dfrac{1}{\frac{1}{2}x}$
$= \dfrac{2}{x}$

c $2f(x) + 3$
$= 2\left(\dfrac{1}{x}\right) + 3$
$= \dfrac{2}{x} + 3$
$= \dfrac{2 + 3x}{x}$

d $3f(x-1) + 2$
$= 3\left(\dfrac{1}{x-1}\right) + 2$
$= \dfrac{3 + 2(x-1)}{x-1}$
$= \dfrac{2x+1}{x-1}$

6 **a**

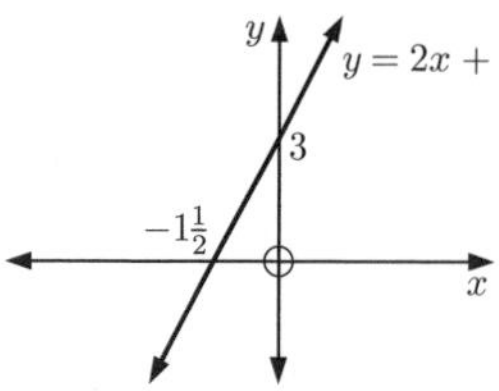

b **i** When $y = 0$, $2x + 3 = 0$ $\therefore$ $x = -\frac{3}{2}$
$\therefore$ x-intercept is $-1\frac{1}{2}$

ii When $x = 0$, $y = 0 + 3 = 3$
$\therefore$ y-intercept is 3

iii As $y = 2x + 3$, the gradient is 2 {the coefficient of x}

7 **a**

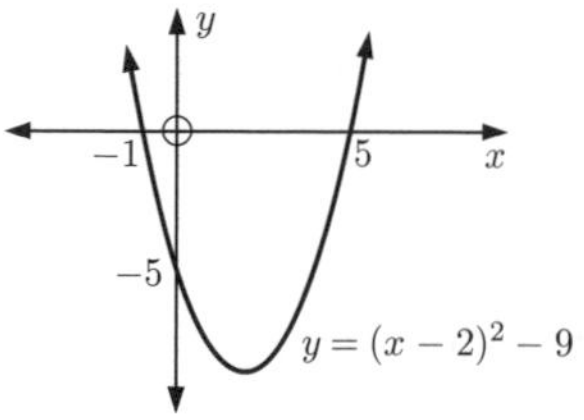

b When $y = 0$,
$(x-2)^2 - 9 = 0$
$\therefore$ $(x-2)^2 = 9$
$\therefore$ $x - 2 = \pm 3$
$\therefore$ $x = 2 + 3$ or $2 - 3$
$\therefore$ $x = 5$ or -1
$\therefore$ x-intercepts are -1 and 5

When $x = 0$,
$y = (-2)^2 - 9$
$= 4 - 9$
$= -5$
$\therefore$ y-intercept is -5

8 a, b

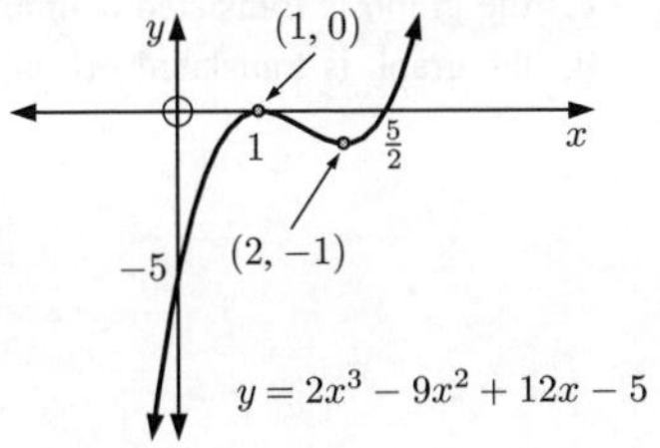

9

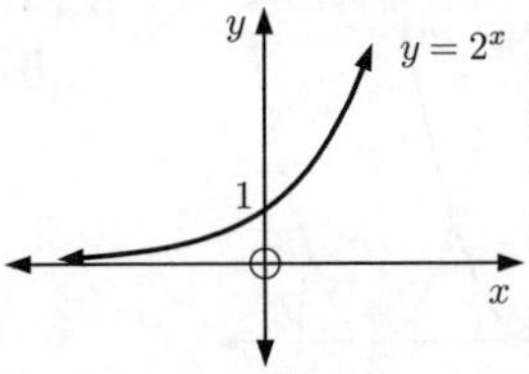

c When $x = 0$,
$y = 2^0 = 1$ ✓

d $2^x > 0$ for all x as the graph is always above the x-axis. ✓

10

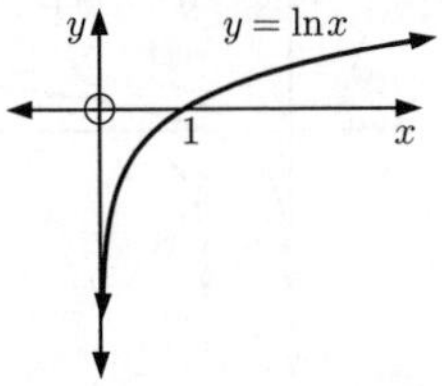

d When $y = 0$,
$\ln x = 0$
$\therefore\ x = e^0 = 1$ ✓

EXERCISE 5B.1

1 a, b

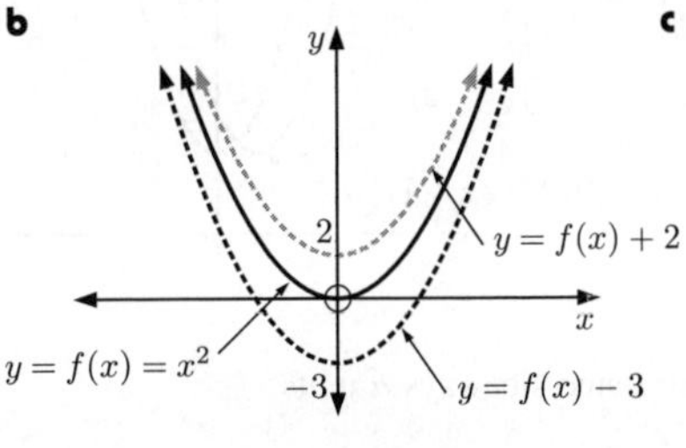

c **i** If $b > 0$, the function is translated vertically upwards through b units.

ii If $b < 0$, the function is translated vertically downwards $|b|$ units.

2 a

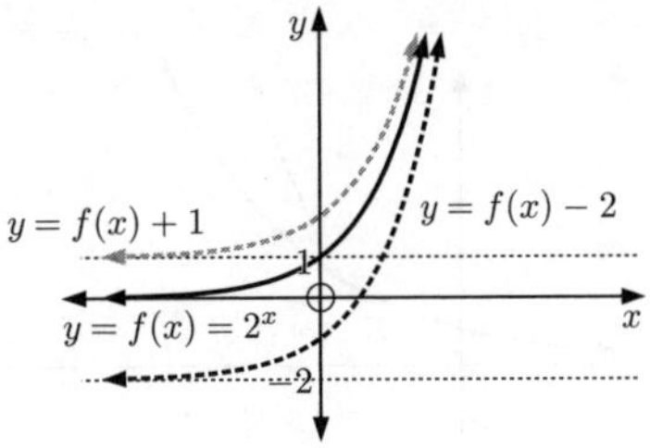

b

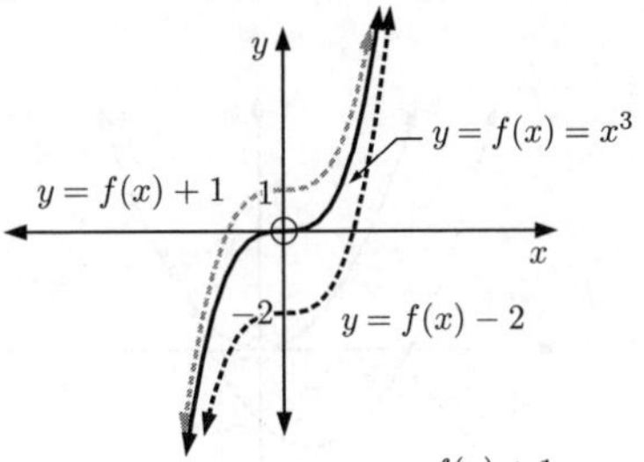

c

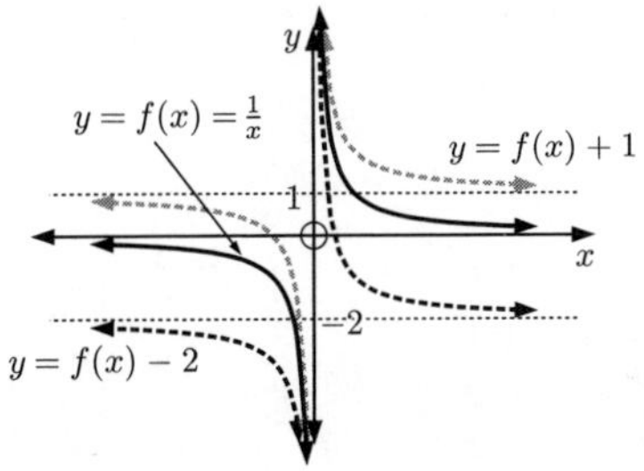

d

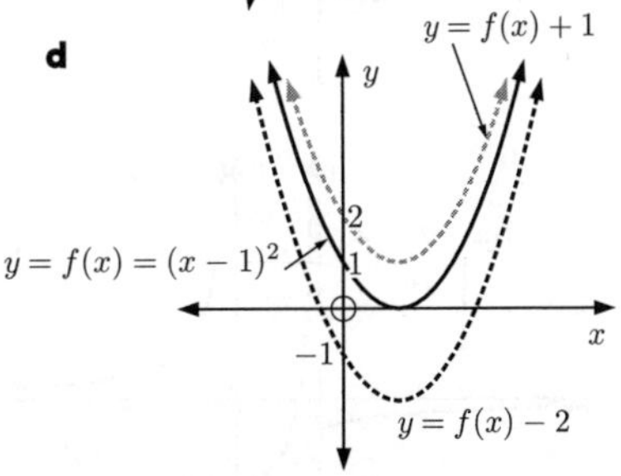

Summary: For $y = f(x) + b$, $y = f(x)$ is translated upwards b units.
If $b > 0$ movement is vertically upwards.
If $b < 0$ movement is vertically downwards.

3 **a**

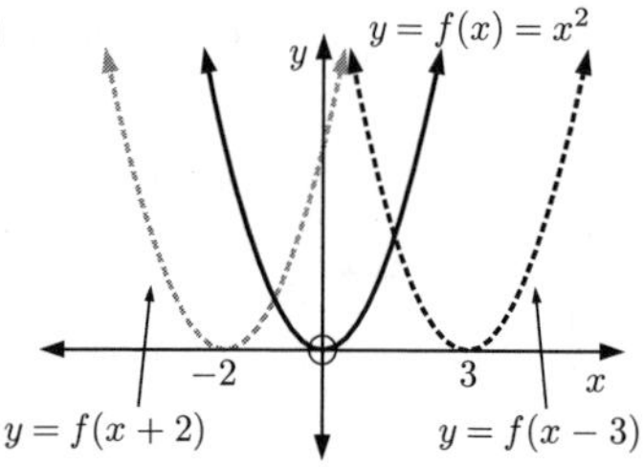

b **i** If $a > 0$, the graph is translated a units right.
ii If $a < 0$, the graph is translated $|a|$ units left.

4 **a**

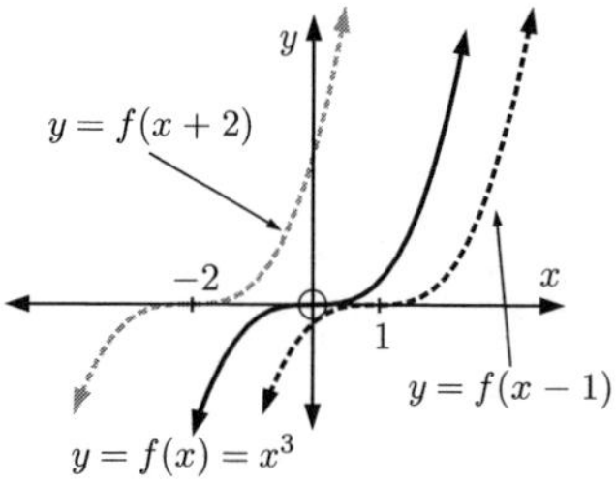

b

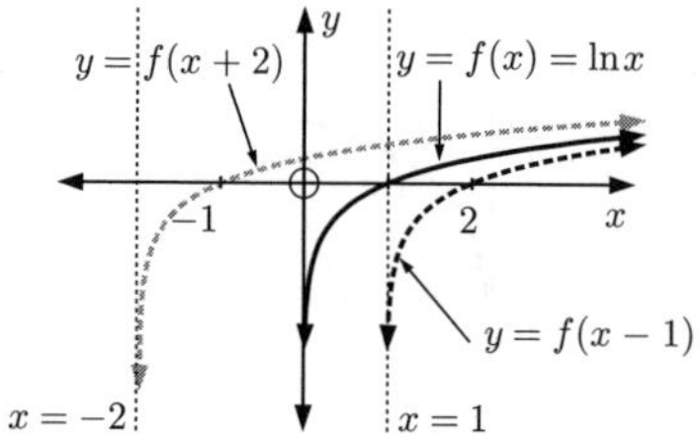

c

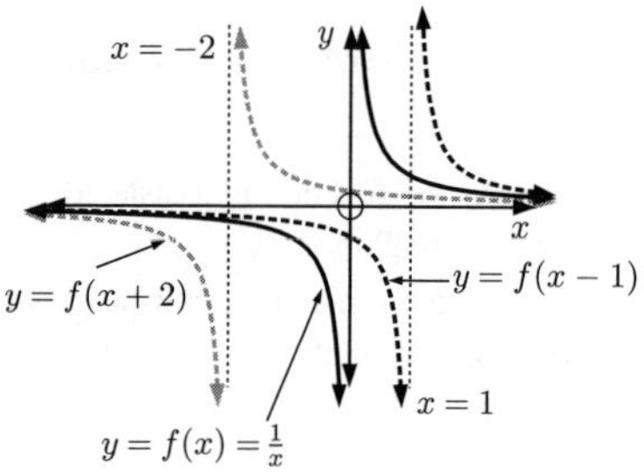

d

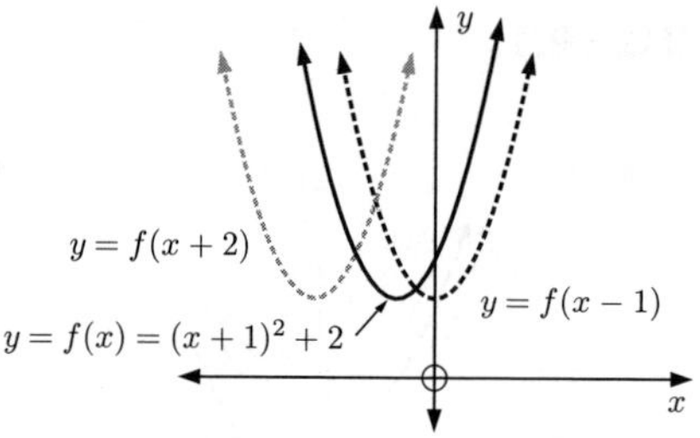

Summary: For $y = f(x-a)$, $y = f(x)$ is translated sideways a units.
If $a > 0$ movement is to the right.
If $a < 0$ movement is to the left.

5 **a**

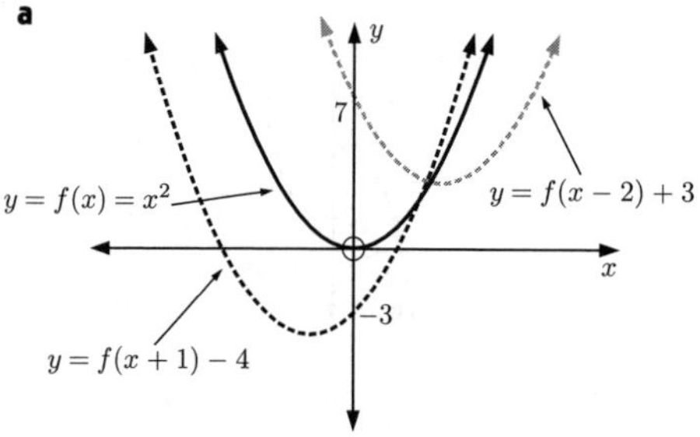

b

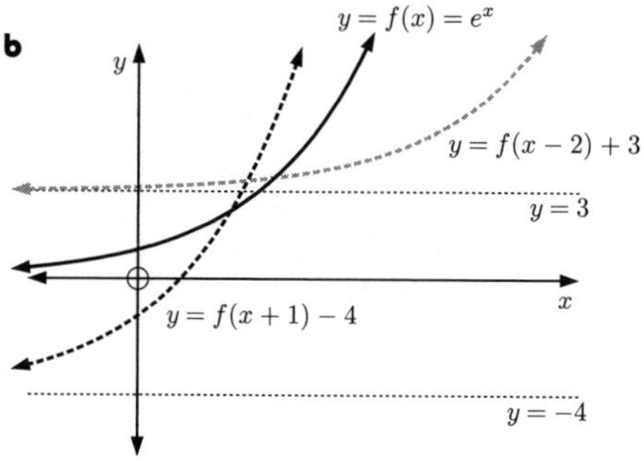

c

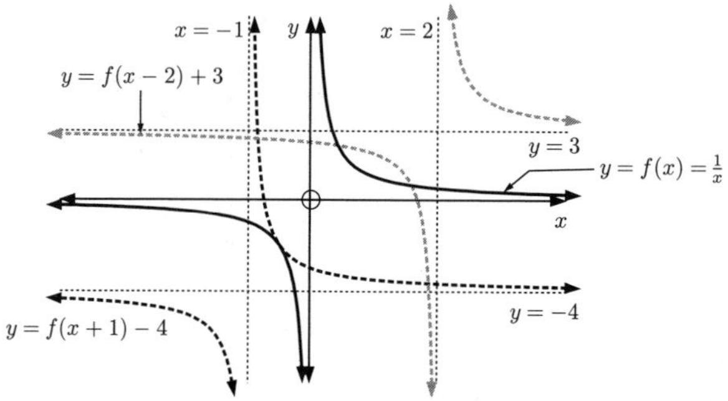

6 A translation of 2 units right and 3 units down.

a

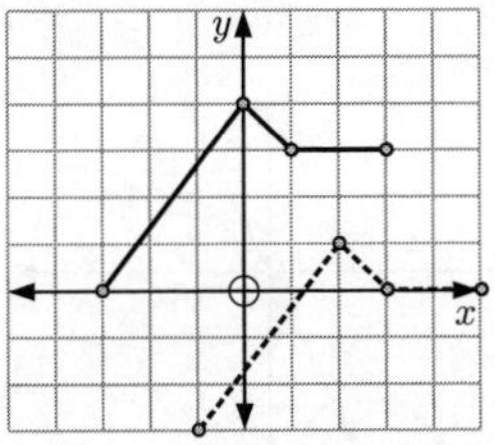

b

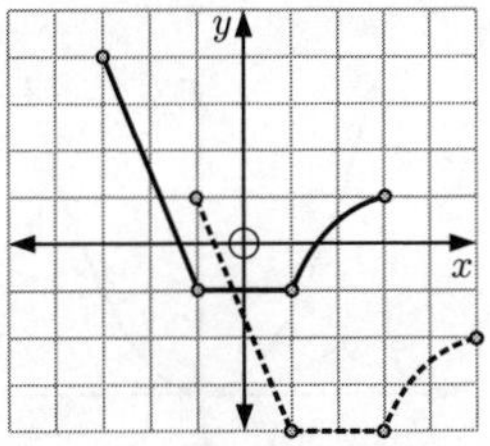

7 **a** The transformation from $f(x) = x^2$ to $g(x) = (x-3)^2 + 2$ is a translation of 3 units right and 2 units up.

i (0, 0) is translated to (3, 2).

ii $(-3, 9)$ is translated to (0, 11).

iii $f(2) = 2^2 = 4$
$\therefore$ (2, 4) is translated to (5, 6).

b The transformation from $g(x)$ back to $f(x)$ is a translation of $\begin{pmatrix} -3 \\ -2 \end{pmatrix}$.

i (1, 6) is translated to $(-2, 4)$.

ii $(-2, 27)$ is translated to $(-5, 25)$.

iii $(1\frac{1}{2}, 4\frac{1}{4})$ is translated to $(-1\frac{1}{2}, 2\frac{1}{4})$.

EXERCISE 5B.2

1 **a**

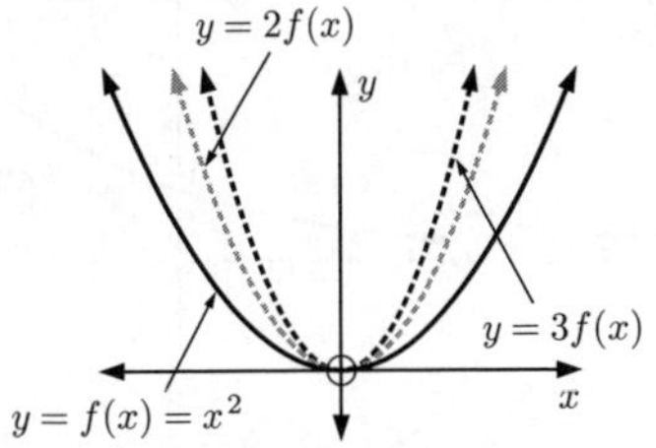

b

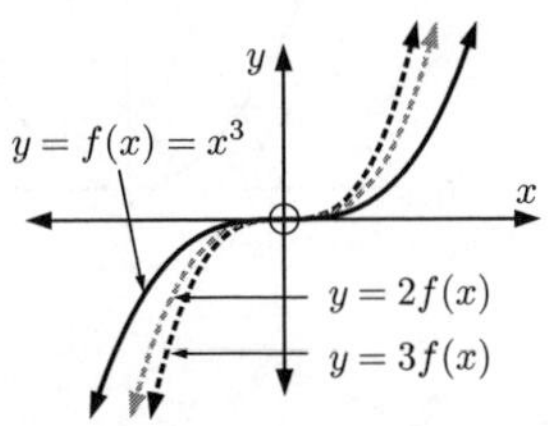

c

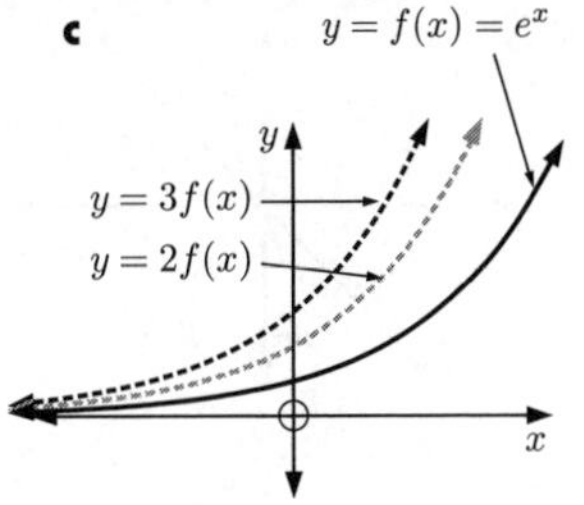

d

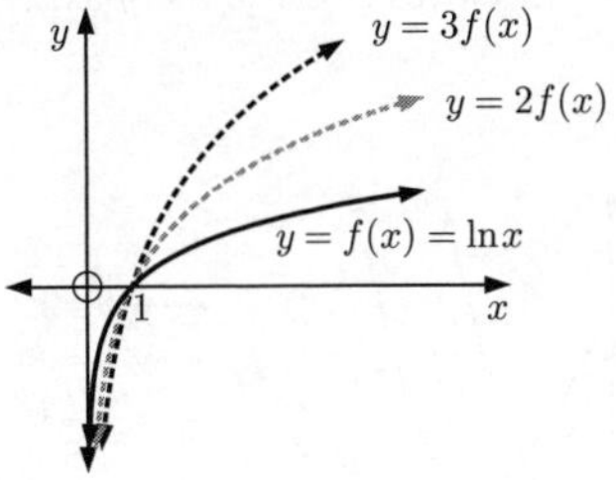

e

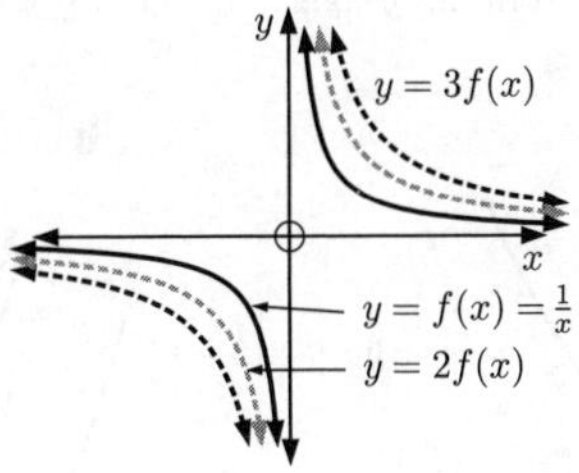

2 **a**

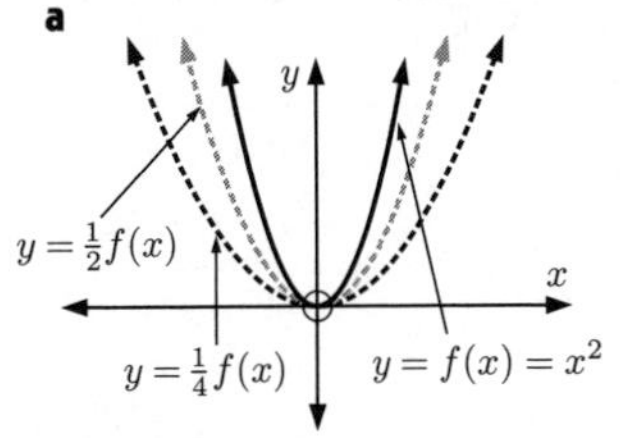

b

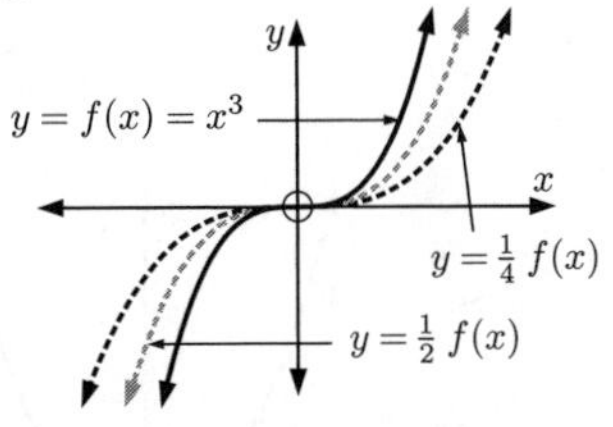

c

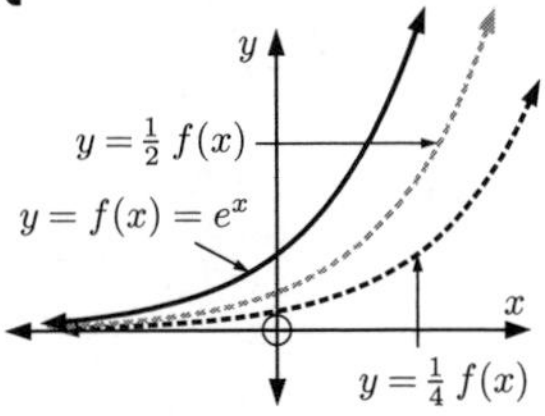

3 p affects the vertical stretching of the graph of $y = f(x)$ by a factor of p.
If $p > 1$, the graph moves further from the x-axis.
If $0 < p < 1$, the graph moves closer to the x-axis.

4 **a**

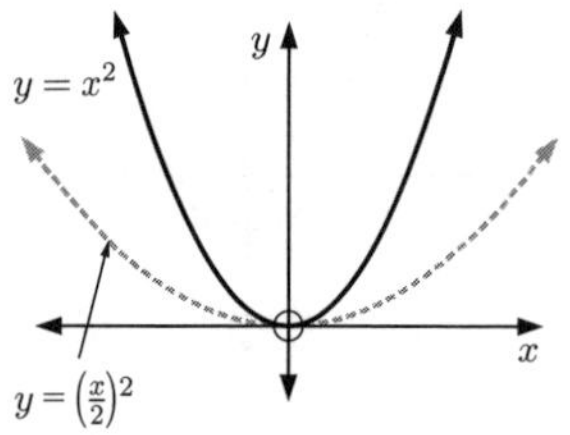

b

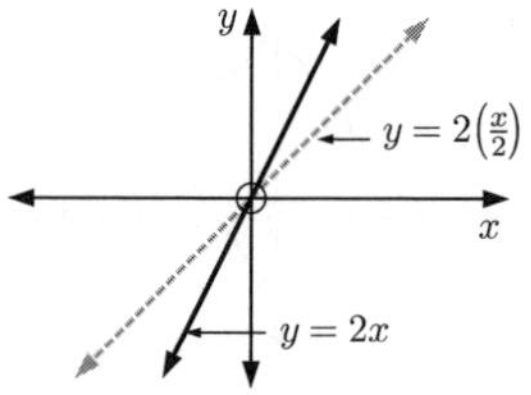

c

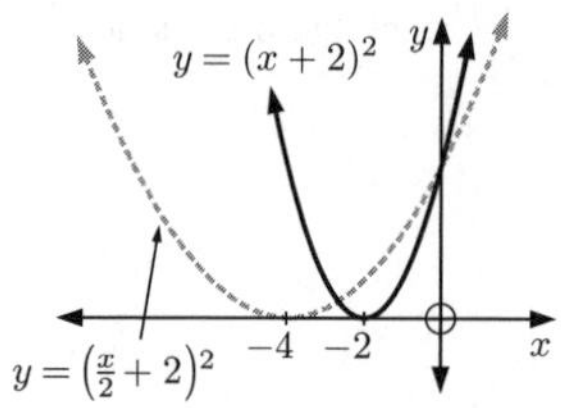

5 **a**

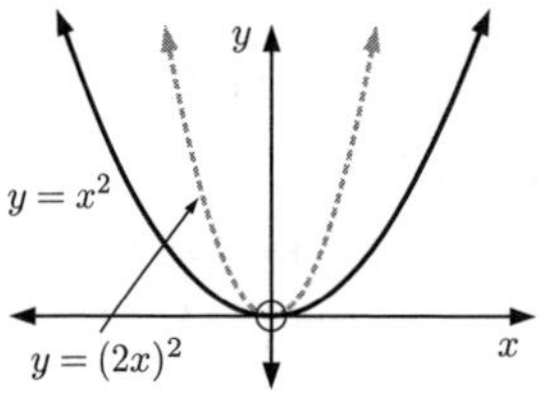

b

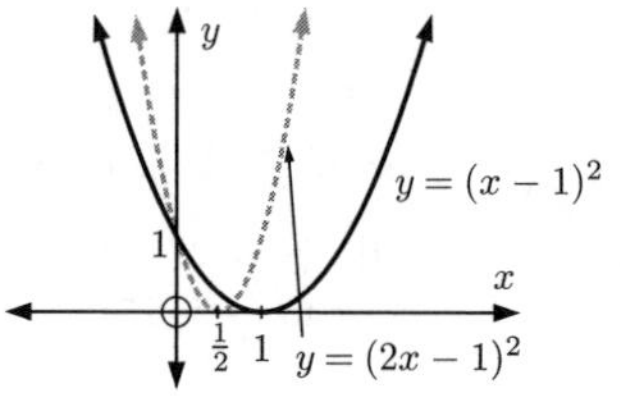

c

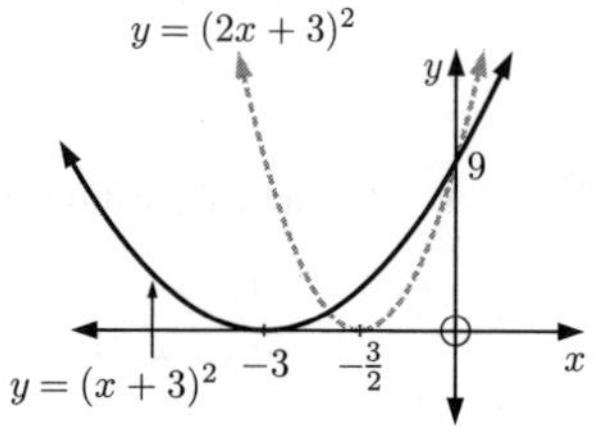

6 **a**

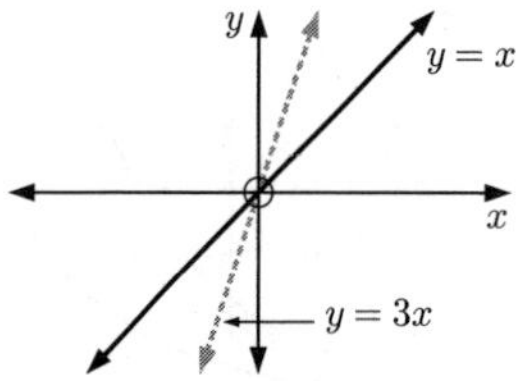

b

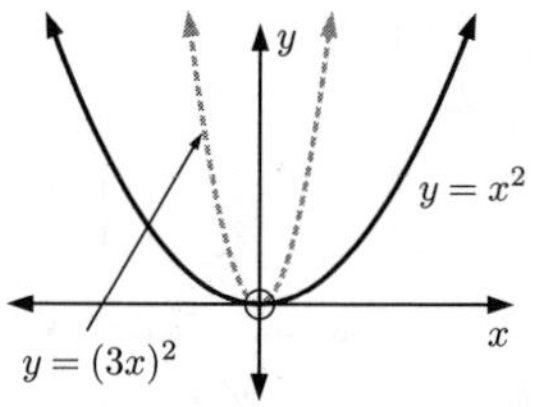

c

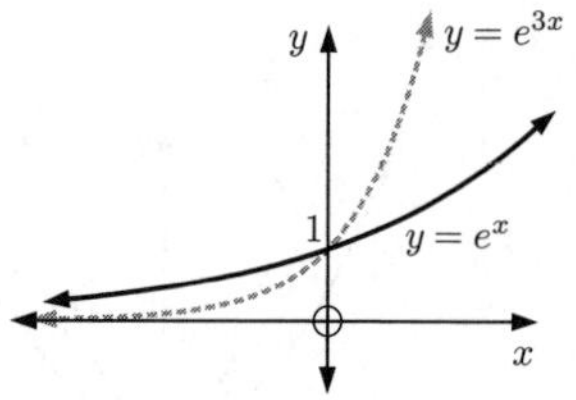

7 q affects the horizontal stretching of $y = f(x)$ by a factor of q.
If $q > 1$, it moves further from the y-axis. If $0 < q < 1$, it moves closer to the y-axis.

8 **a**

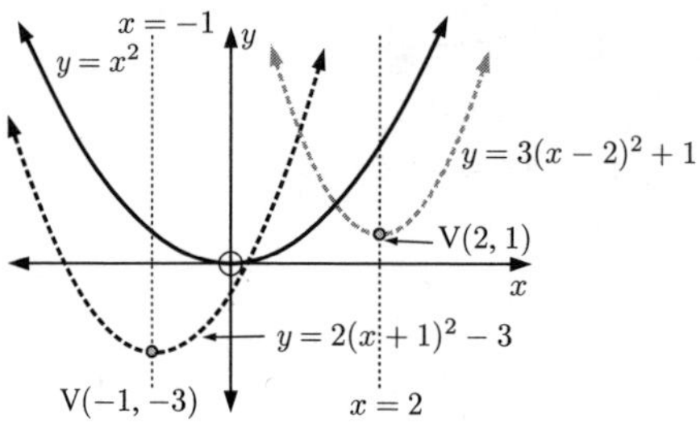

b

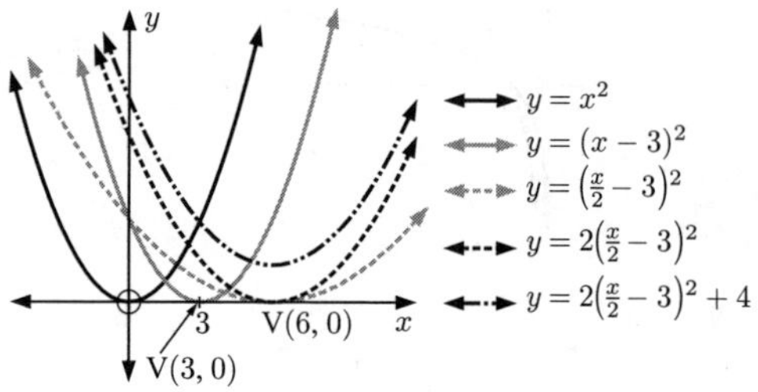

c

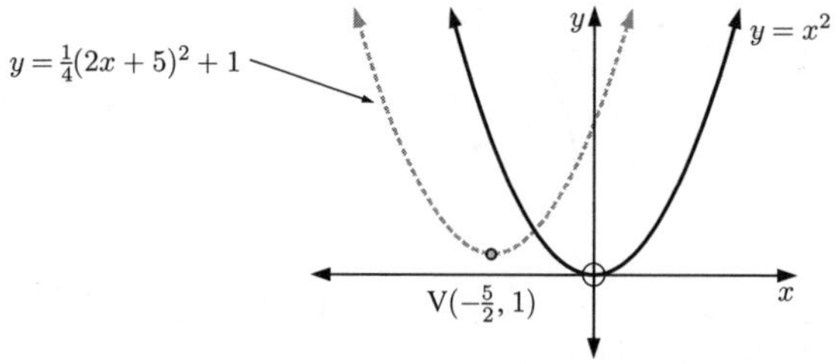

9 **a** The transformation from $y = f(x)$ to $y = 3\,f(2x)$ is a horizontal stretch of factor $\frac{1}{2}$ followed by a vertical stretch of factor 3.

i $(3, -5) \to (\frac{3}{2}, -5) \to (\frac{3}{2}, -15)$ $\quad\therefore$ $(3, -5)$ is translated to $(\frac{3}{2}, -15)$

ii $(1, 2) \to (\frac{1}{2}, 2) \to (\frac{1}{2}, 6)$ $\quad\therefore$ $(1, 2)$ is translated to $(\frac{1}{2}, 6)$

iii $(-2, 1) \to (-1, 1) \to (-1, 3)$ $\quad\therefore$ $(-2, 1)$ is translated to $(-1, 3)$

b The translation from $y = 3\,f(2x)$ back to $y = f(x)$ is a vertical stretch of factor $\frac{1}{3}$ followed by a horizontal stretch of factor 2.

i $(2, 1) \to (2, \frac{1}{3}) \to (4, \frac{1}{3})$ $\quad\therefore$ $(4, \frac{1}{3})$ is the point on $y = f(x)$

ii $(-3, 2) \to (-3, \frac{2}{3}) \to (-6, \frac{2}{3})$ $\quad\therefore$ $(-6, \frac{2}{3})$ is the point on $y = f(x)$

iii $(-7, 3) \to (-7, 1) \to (-14, 1)$ $\quad\therefore$ $(-14, 1)$ is the point on $y = f(x)$

EXERCISE 5B.3

1 **a**

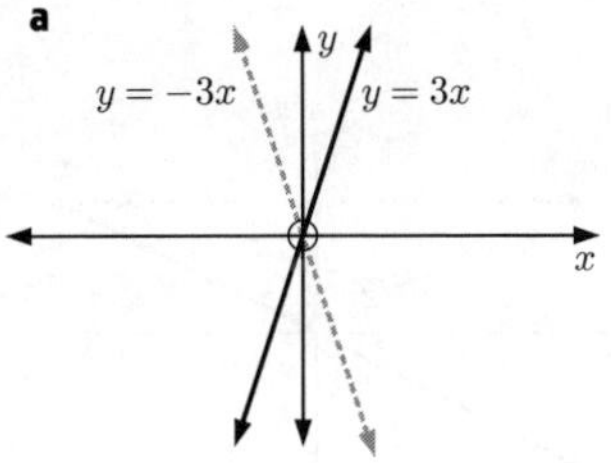

b

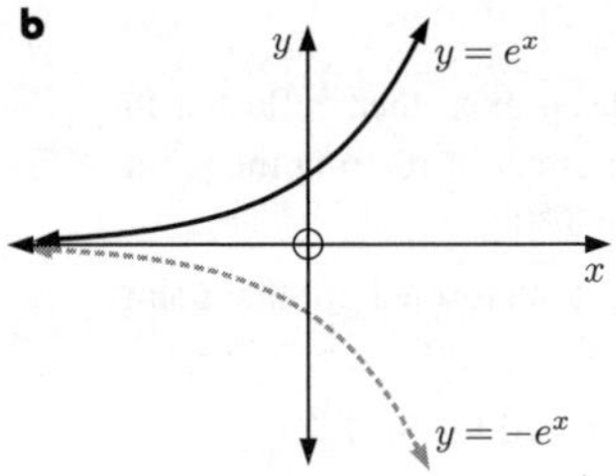

c

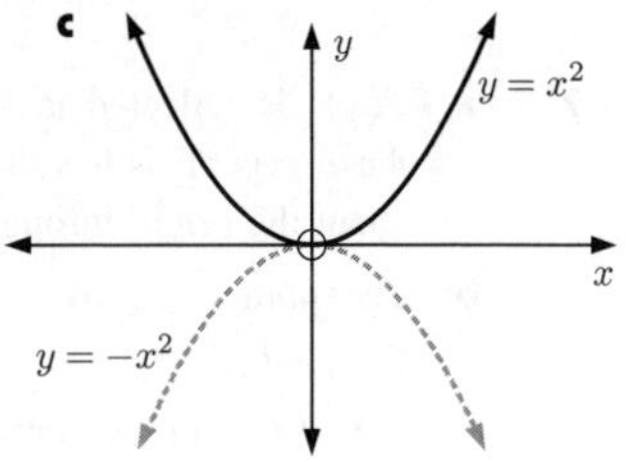

d

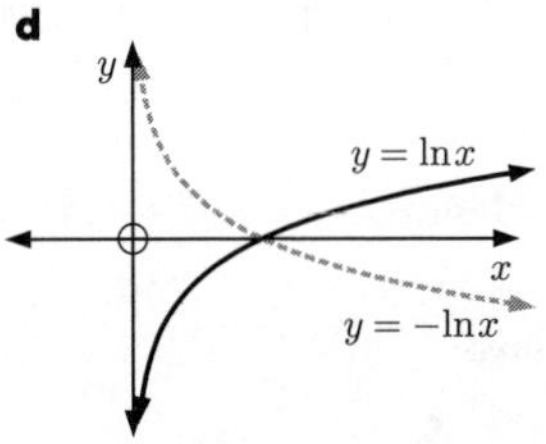

e

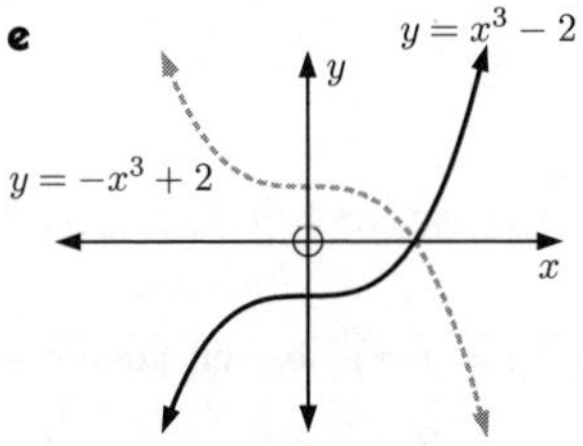

f

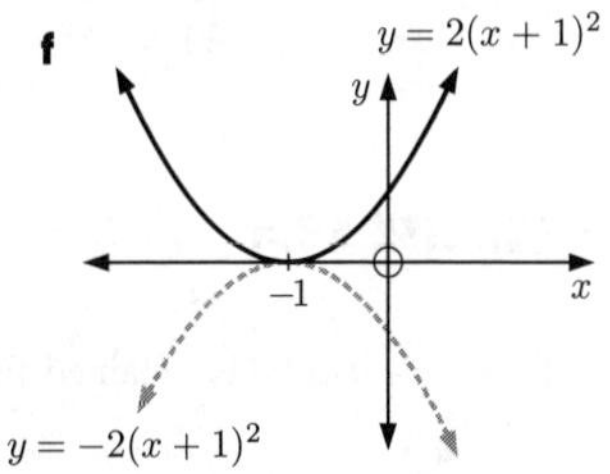

2 $y = -f(x)$ is the reflection of $y = f(x)$ in the x-axis.

3 **a** **i**
$$f(x) = 2x + 1$$
$$\therefore\ f(-x) = 2(-x) + 1 = -2x + 1$$

ii
$$f(x) = x^2 + 2x + 1$$
$$\therefore\ f(-x) = (-x)^2 + 2(-x) + 1 = x^2 - 2x + 1$$

iii
$$f(x) = x^3$$
$$\therefore\ f(-x) = (-x)^3 = -x^3$$

b **i**

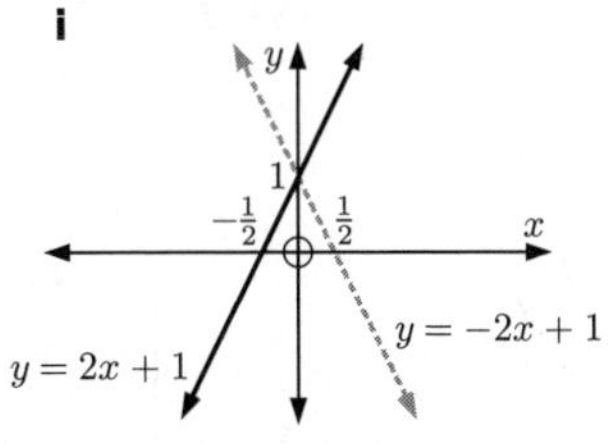

ii

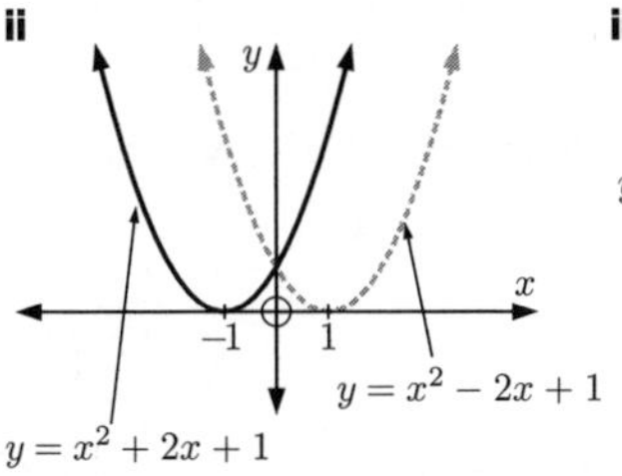

iii

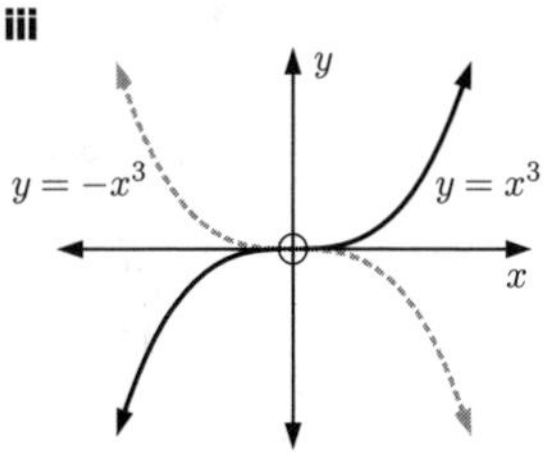

4 $y = f(-x)$ is the reflection of $y = f(x)$ in the y-axis.

5 **a** To transform $y = f(x)$ to $g(x) = -f(x)$, we reflect $y = f(x)$ in the x-axis. To do this we keep the x-coordinates the same and take the negative of the y-coordinates.

i $(3, 0)$ is transformed to $(3, 0)$ $\quad$ **ii** $(2, -1)$ is transformed to $(2, 1)$

iii $(-3, 2)$ is transformed to $(-3, -2)$

b To find the points on $f(x)$ corresponding to $g(x)$, we again take the negative of the y-coordinates.

i The point transformed to $(7, -1)$ is $(7, 1)$.

ii The point transformed to $(-5, 0)$ is $(-5, 0)$.

iii The point transformed to $(-3, -2)$ is $(-3, 2)$.

6 **a** To transform $y = f(x)$ to $h(x) = f(-x)$, we reflect $y = f(x)$ in the y-axis.
To do this we keep the y-coordinates the same and take the negative of the x-coordinates.

i $(2, -1)$ is transformed to $(-2, -1)$. **ii** $(0, 3)$ is transformed to $(0, 3)$.

iii $(-1, 2)$ is transformed to $(1, 2)$.

b To find the points on $f(x)$ corresponding to $h(x)$, we again take the negative of the x-coordinates.

i The point transformed to $(5, -4)$ is $(-5, -4)$.

ii The point transformed to $(0, 3)$ is $(0, 3)$.

iii The point transformed to $(2, 3)$ is $(-2, 3)$.

7 **a** $f(x)$ is reflected in the y-axis, then reflected in the x-axis. This has the effect of rotating the point about the origin through $180°$.

b The point (a, b) is transformed to the point $(-a, -b)$.

i $(3, -7)$ is transformed to $(-3, 7)$

ii The point that transforms to $(-5, -1)$ is $(5, 1)$.

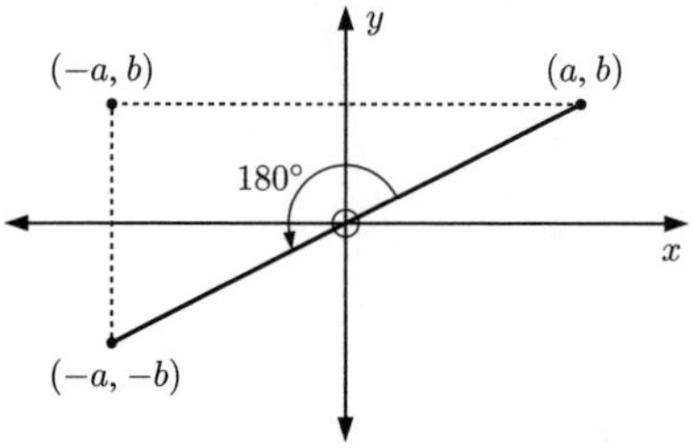

EXERCISE 5B.4

1 $y = -f(x)$ is obtained from $y = f(x)$ by reflecting it in the x-axis.

a

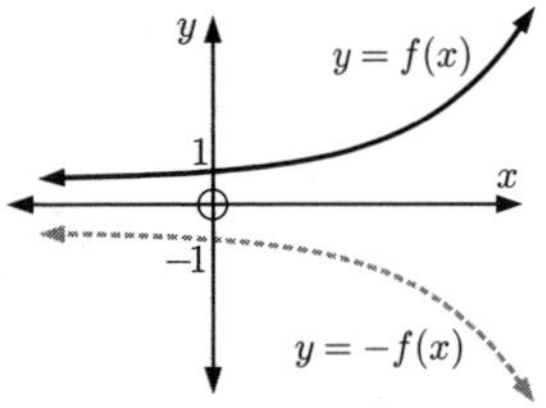

b

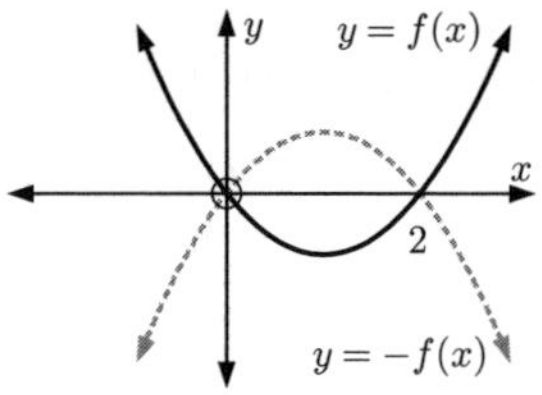

c

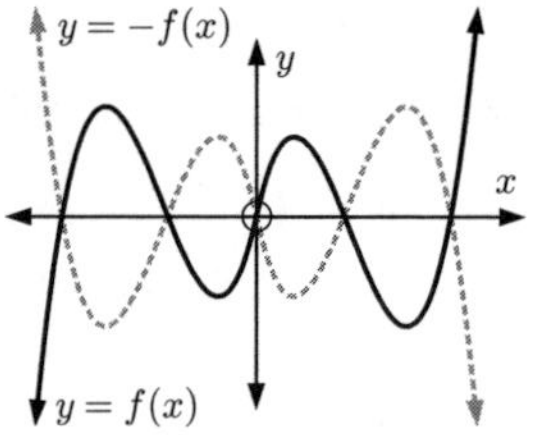

2 $y = f(-x)$ is obtained from $y = f(x)$ by reflecting it in the y-axis.

a

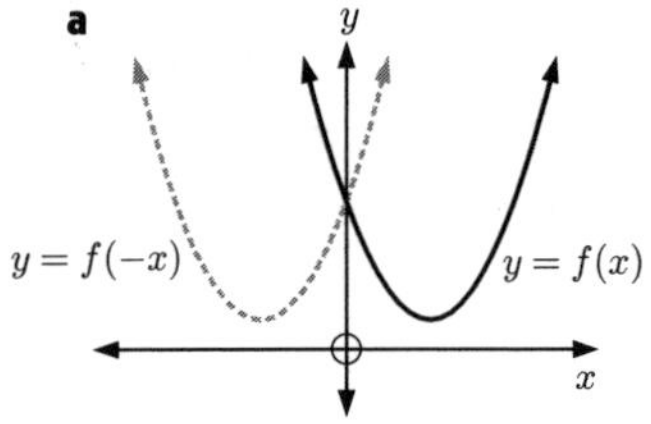

b

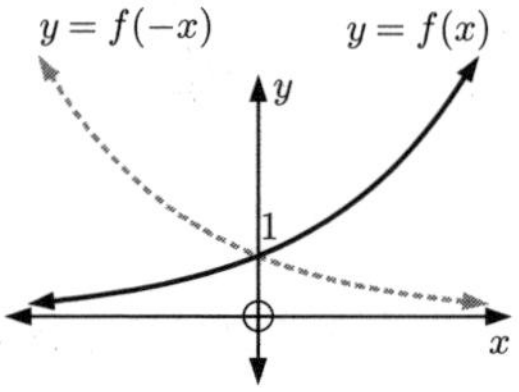

c

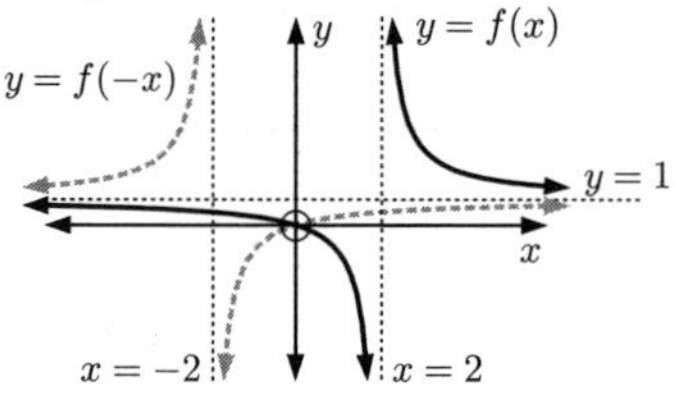

3 $y = 2x^4$ and $y = 6x^4$ are 'thinner' than $y = x^4$ and $y = \frac{1}{2}x^4$ is 'fatter'

$\therefore$ **a** is **A**, **b** is **B**, **c** is **D** and **d** is **C**.

4 **a**

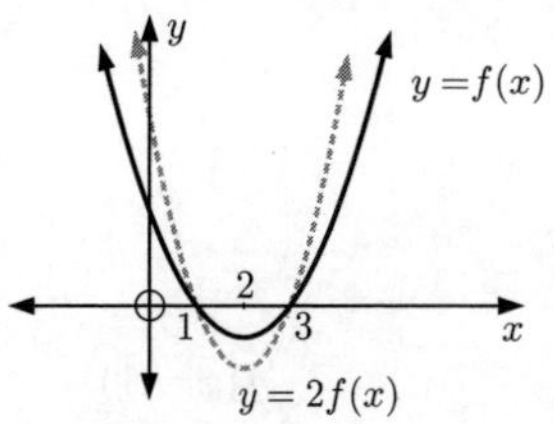

b

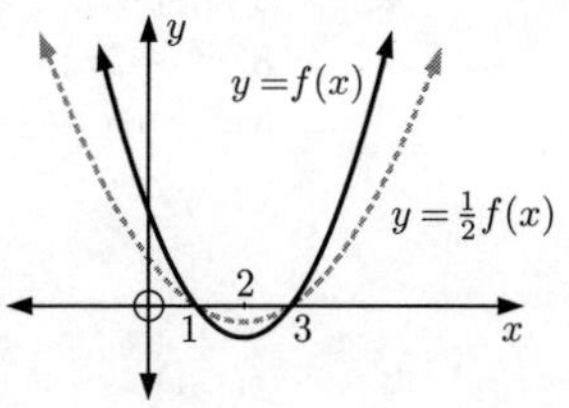

c

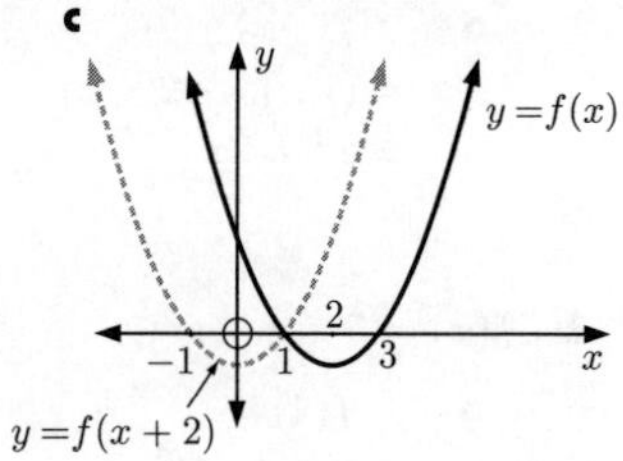

d

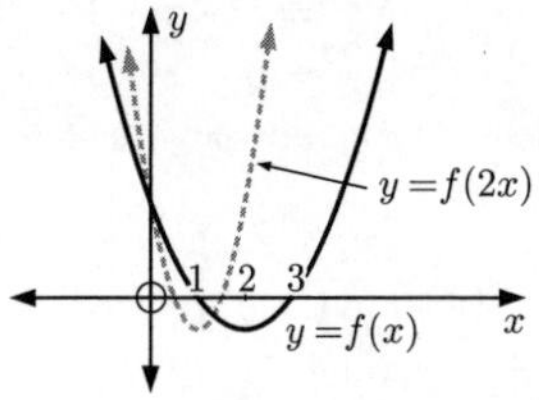

e

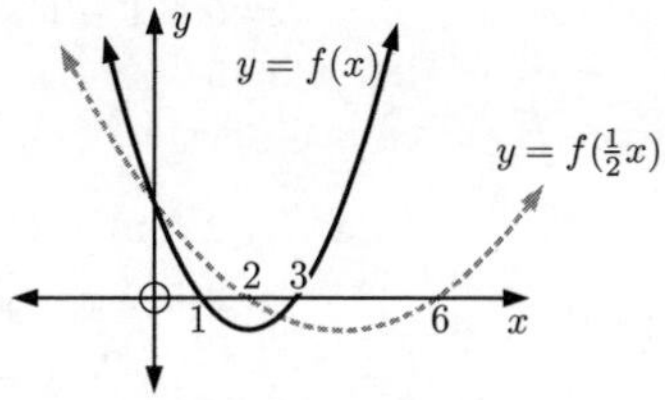

5

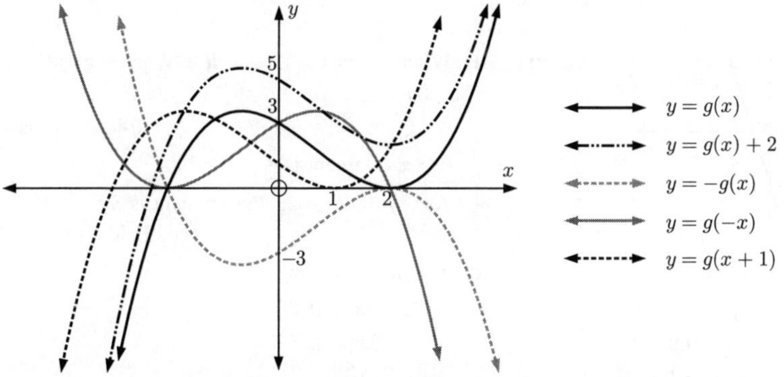

6

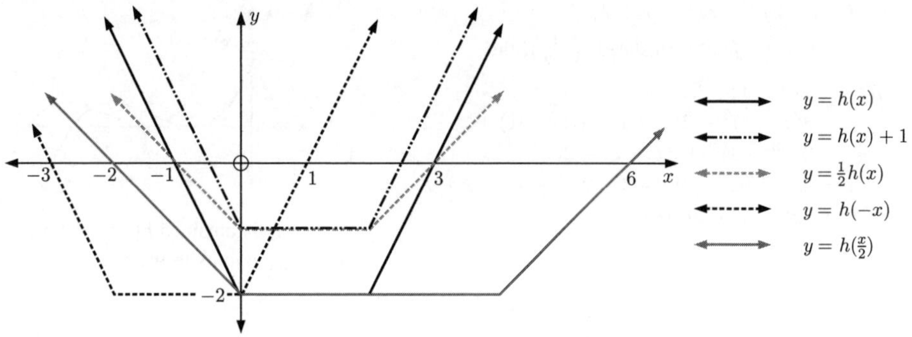

REVIEW SET 5A

1 $f(x) = x^2 - 2x$

a $f(3)$
$= 3^2 - 2(3)$
$= 9 - 6$
$= 3$

b $f(-2)$
$= (-2)^2 - 2(-2)$
$= 4 + 4$
$= 8$

c $f(2x)$
$= (2x)^2 - 2(2x)$
$= 4x^2 - 4x$

d $f(-x)$
$= (-x)^2 - 2(-x)$
$= x^2 + 2x$

e $3\,f(x) - 2$
$= 3(x^2 - 2x) - 2$
$= 3x^2 - 6x - 2$

2 $f(x) = 5 - x - x^2$

a $f(4)$
$= 5 - 4 - 4^2$
$= 1 - 16$
$= -15$

b $f(-1)$
$= 5 - (-1) - (-1)^2$
$= 5 + 1 - 1$
$= 5$

c $f(x-1)$
$= 5 - (x-1) - (x-1)^2$
$= 5 - x + 1 - [x^2 - 2x + 1]$
$= 6 - x - x^2 + 2x - 1$
$= -x^2 + x + 5$

d $f\left(\frac{x}{2}\right)$
$= 5 - \left(\frac{x}{2}\right) - \left(\frac{x}{2}\right)^2$
$= 5 - \frac{1}{2}x - \frac{1}{4}x^2$

e $2\,f(x) - f(-x)$
$= 2(5 - x - x^2) - \left[5 - (-x) - (-x)^2\right]$
$= 10 - 2x - 2x^2 - [5 + x - x^2]$
$= 10 - 2x - 2x^2 - 5 - x + x^2$
$= -x^2 - 3x + 5$

3 **a**

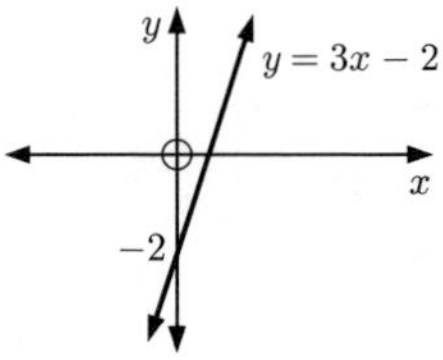

b **i** When $y = 0$,
$3x - 2 = 0$
$\therefore\ x = \frac{2}{3}$
$\therefore$ x-intercept is $\frac{2}{3}$

ii When $x = 0$,
$y = 0 - 2 = -2$
$\therefore$ y-intercept is -2

iii As $y = 3x - 2$, the gradient is 3
{coefficient of x}

c **i** When $x = 0.3$,
$y = 3(0.3) - 2$
$= 0.9 - 2$
$= -1.1$

ii When $y = 0.7$,
$3x - 2 = 0.7$
$\therefore\ 3x = 2.7$
$\therefore\ x = 0.9$

4 $f(x) = 3x^3 - 2x^2 + x + 2$

If $g(x)$ is $f(x)$ translated $\binom{1}{-2}$, then

$g(x) = f(x-1) - 2$
$= 3(x-1)^3 - 2(x-1)^2 + (x-1) + 2 - 2$
$= 3(x^3 - 3x^2 + 3x - 1) - 2(x^2 - 2x + 1) + x - 1$
$= 3x^3 - 9x^2 + 9x - 3 - 2x^2 + 4x - 2 + x - 1$
$= 3x^3 - 11x^2 + 14x - 6$

5

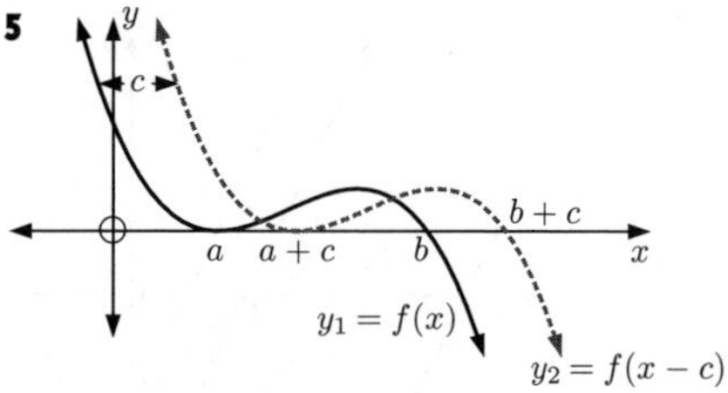

y_2 is obtained by translating y_1 c units to the right.

6 **a, b**

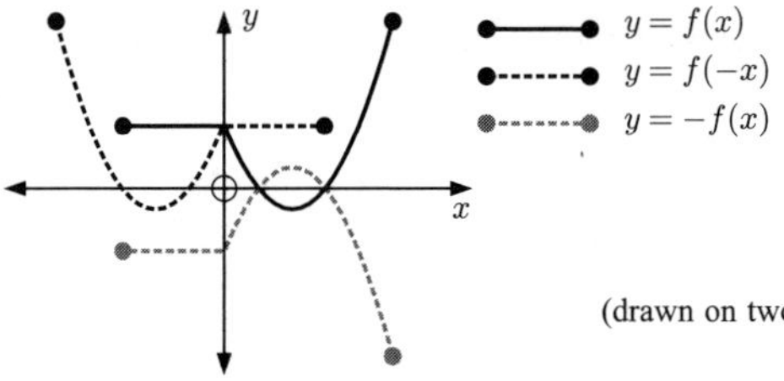

c, d

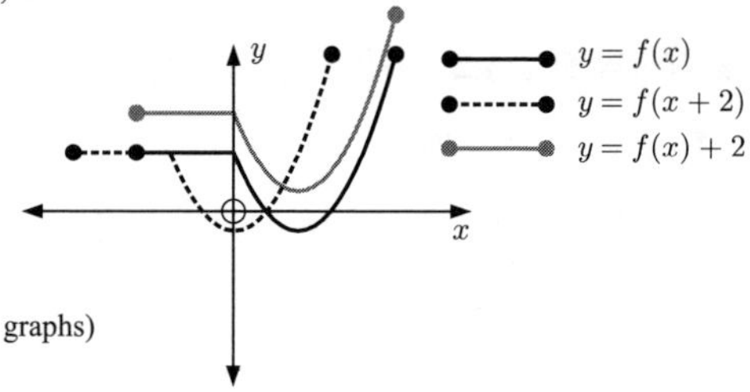

(drawn on two graphs)

7 **a**

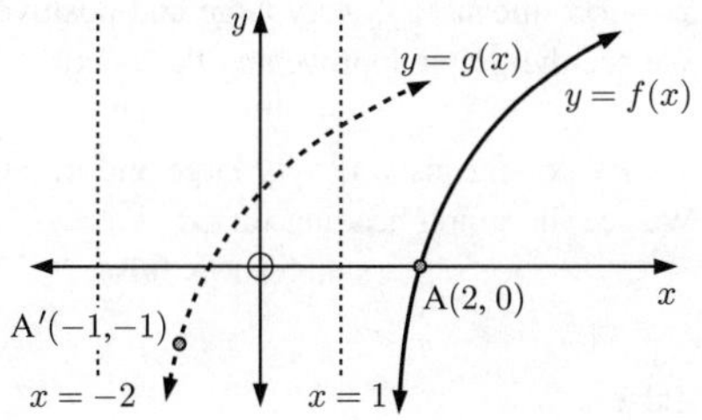

b $f(x+3)-1$ is a translation of $f(x)$ by $\binom{-3}{-1}$.

$\therefore$ vertical asymptote is at $x = 1-3 = -2$.

c A(2, 0) translated by $\binom{-3}{-1}$ gives $(2-3, 0-1)$ which is $A'(-1, -1)$.

8

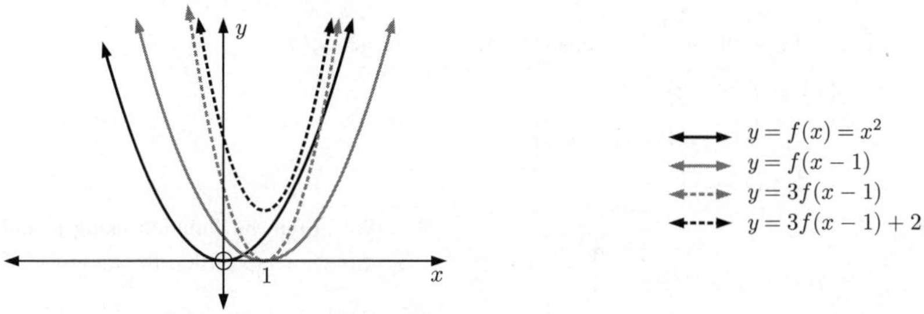

REVIEW SET 5B

1 $f(x) = x^2 - 2$

a $f(-3) = (-3)^2 - 2$
$= 9 - 2$
$= 7$

b $f(x+4) = (x+4)^2 - 2$
$= x^2 + 8x + 16 - 2$
$= x^2 + 8x + 14$

c $3f(x) + 5 = 3(x^2 - 2) + 5$
$= 3x^2 - 6 + 5$
$= 3x^2 - 1$

2 **a**

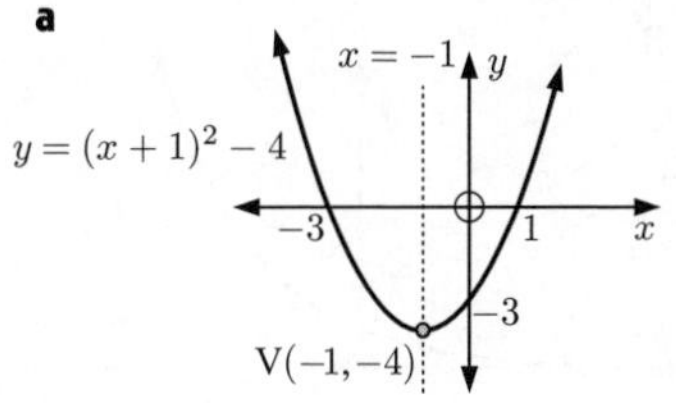

b **i** When $y = 0$,
$(x+1)^2 - 4 = 0$
$\therefore (x+1)^2 = 4$
$\therefore x + 1 = \pm 2$
$\therefore x = 2 - 1$ or $-2 - 1$
$\therefore x = 1$ or -3
$\therefore$ x-intercepts are 1, -3

ii When $x = 0$,
$y = 1^2 - 4$
$= -3$
$\therefore$ y-intercept is -3

c $y = (x+1)^2 - 4$ is obtained from $y = x^2$ under a translation of $\binom{-1}{-4}$. $y = x^2$ has its vertex at (0, 0), so the vertex of $y = (x+1)^2 - 4$ must be $(-1, -4)$.

3

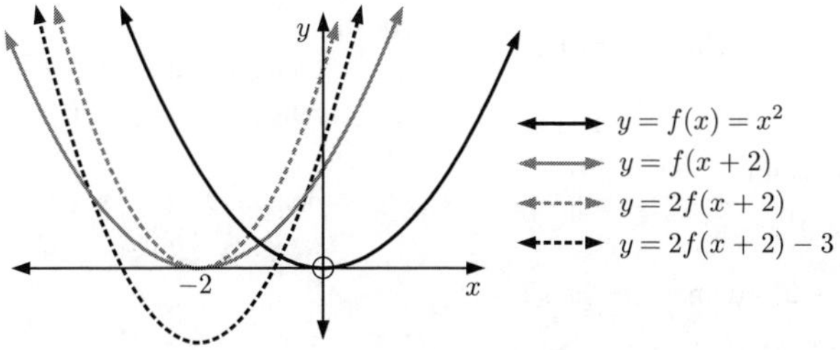

4 **a**

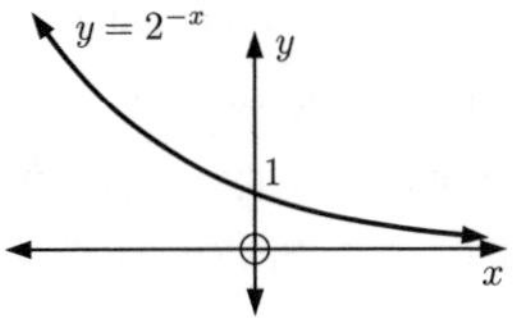

b **i** $x \to \infty$ means x is very large and positive.
We see the graph approaching the x-axis
$\therefore \ y \to 0 \ \therefore$ **true**.

ii $x \to -\infty$ means x is very large and negative.
We see the graph heading for ∞
$\therefore$ statement is **false**.

iii When $x = 0$, $y = 2^0 = 1 \neq \frac{1}{2}$ $\therefore$ **false**.

iv The graph is above the x-axis for all x $\therefore$ $2^{-x} > 0$ for all x $\therefore$ **true**.

5 **a** $f(x) = (x+1)^2 + 4$ is translated by $\binom{2}{4}$ to get $g(x)$.

$$\begin{aligned}\therefore\ g(x) &= f(x-2) + 4 \\ &= \left[((x-2)+1)^2 + 4\right] + 4 \\ &= (x-1)^2 + 8\end{aligned}$$

b

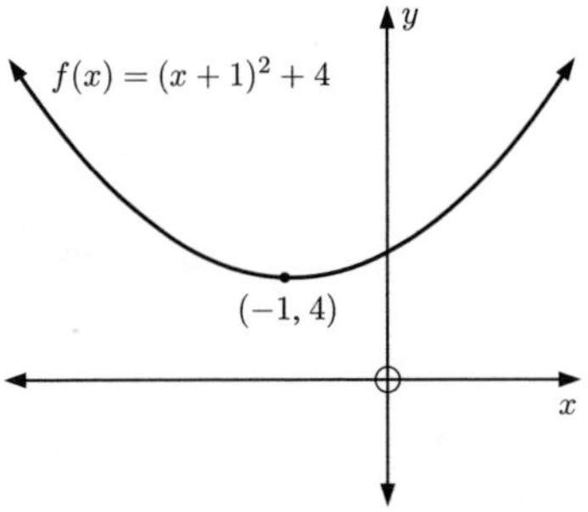

We graph the function using technology, and from this we can see that the range is $\{y \mid y \geqslant 4\}$.

c $g(x)$ is $f(x)$ translated by $\binom{2}{4}$, so the minimum value of $g(x)$ is $4 + 4 = 8$.
$\therefore$ the range of $g(x)$ is $\{y \mid y \geqslant 8\}$.

6 **a** **i** Under translation $\binom{1}{-2}$, $y = \frac{1}{x}$ becomes $y = \frac{1}{x-1} - 2$

ii

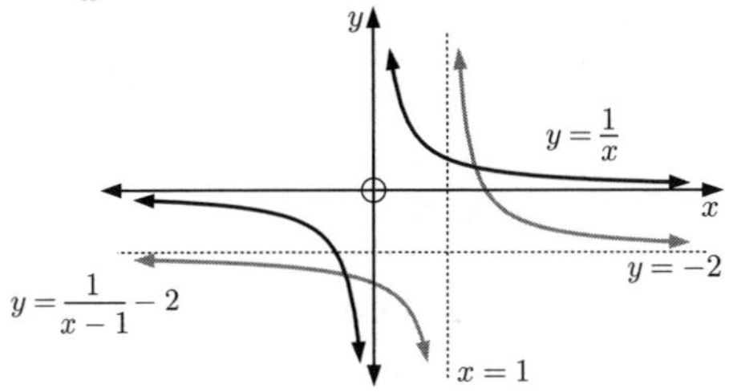

iii For $y = \frac{1}{x}$, V.A. is $x = 0$, H.A. is $y = 0$.
For $y = \frac{1}{x-1} - 2$, V.A. is $x = 1$,
H.A. is $y = -2$.

iv For $y = \frac{1}{x}$, domain is $\{x \mid x \neq 0\}$,
range is $\{y \mid y \neq 0\}$.
For $y = \frac{1}{x-1} - 2$, domain is $\{x \mid x \neq 1\}$,
range is $\{y \mid y \neq -2\}$.

b **i** Under translation $\binom{1}{-2}$, $y = 2^x$ becomes $y = 2^{x-1} - 2$

ii

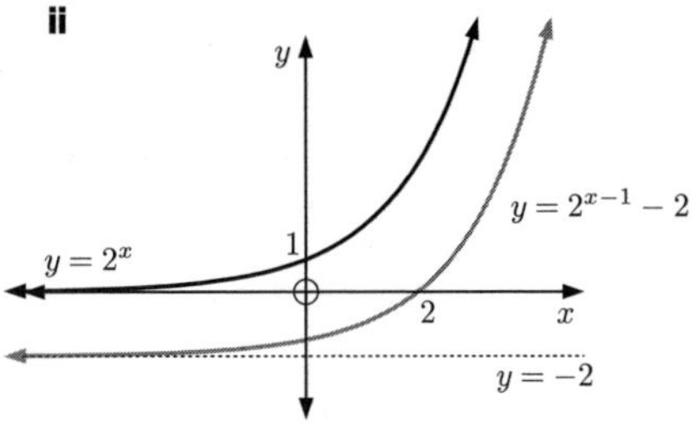

iii For $y = 2^x$, H.A. is $y = 0$, no V.A.
For $y = 2^{x-1} - 2$, H.A. is $y = -2$, no V.A.

iv For $y = 2^x$,
domain is $\{x \mid x \in \mathbb{R}\}$,
range is $\{y \mid y > 0\}$.
For $y = 2^{x-1} - 2$,
domain is $\{x \mid x \in \mathbb{R}\}$,
range is $\{y \mid y > -2\}$.

c **i** Under translation $\binom{1}{-2}$, $y = \log_4 x$ becomes $y = \log_4(x-1) - 2$

ii

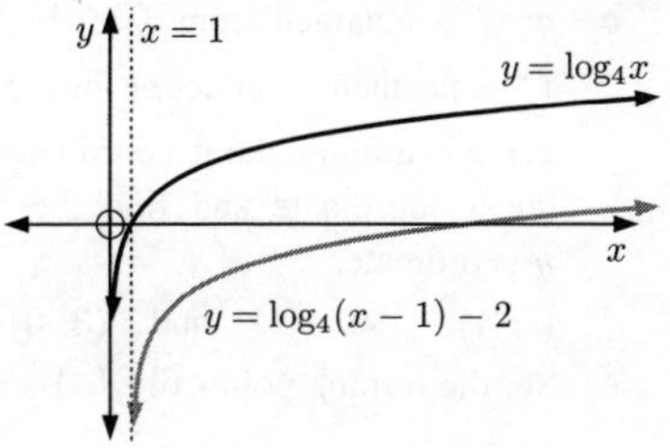

iii For $y = \log_4 x$,
V.A. is $x = 0$, no H.A.
For $y = \log_4(x-1) - 2$,
V.A. is $x = 1$, no H.A.

iv For $y = \log_4 x$, domain is $\{x \mid x > 0\}$,
range is $\{y \mid y \in \mathbb{R}\}$.
For $y = \log_4(x-1) - 2$,
domain is $\{x \mid x > 1\}$,
range is $\{y \mid y \in \mathbb{R}\}$.

7 **a**

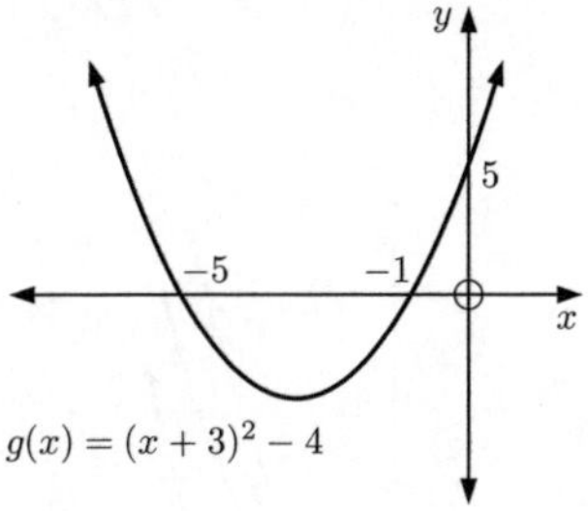

b When $y = 0$, $(x+3)^2 - 4 = 0$
$\therefore\ (x+3)^2 = 4$
$\therefore\ x + 3 = \pm 2$
$\therefore\ x = -5$ or -1
$\therefore$ the x-intercepts are $-5, -1$
When $x = 0$, $y = (0+3)^2 - 4$
$= 9 - 4$
$= 5$
$\therefore$ the y-intercept is 5.

c $g(x)$ is obtained from $y = x^2$ under a transformation of $\binom{-3}{-4}$. $y = x^2$ has $(0, 0)$ as its vertex, so the vertex of $g(x)$ is $(-3, -4)$.

8

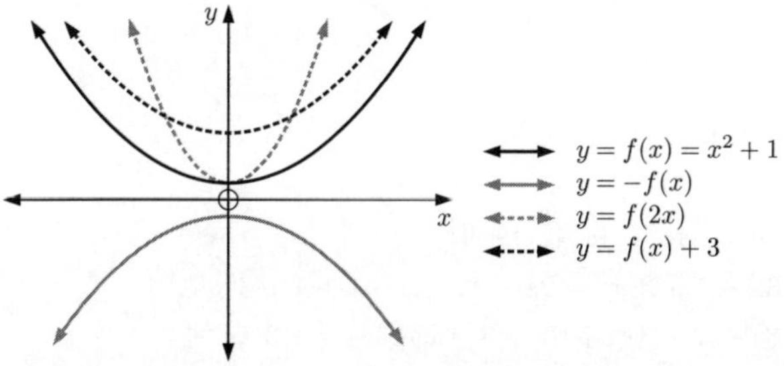

REVIEW SET 5C

1 $f(x) = \dfrac{4}{x}$

a $f(-4) = \dfrac{4}{-4} = -1$

b $f(2x) = \dfrac{4}{2x} = \dfrac{2}{x}$

c $f\left(\dfrac{x}{2}\right) = \dfrac{4}{\frac{x}{2}} = 4 \times \dfrac{2}{x} = \dfrac{8}{x}$

d $4\,f(x+2) - 3 = 4\left(\dfrac{4}{x+2}\right) - 3 = \dfrac{16}{x+2} - 3 \quad \left(\textit{or} \quad \dfrac{16 - 3(x+2)}{x+2} = \dfrac{10 - 3x}{x+2}\right)$

2 So that you can see the answers more easily, they have been drawn on two graphs.

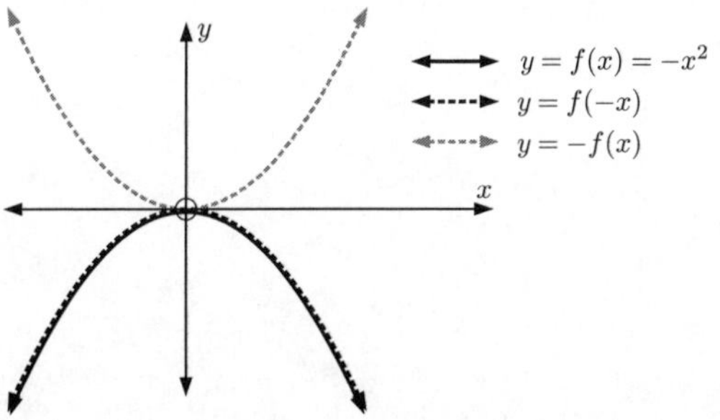

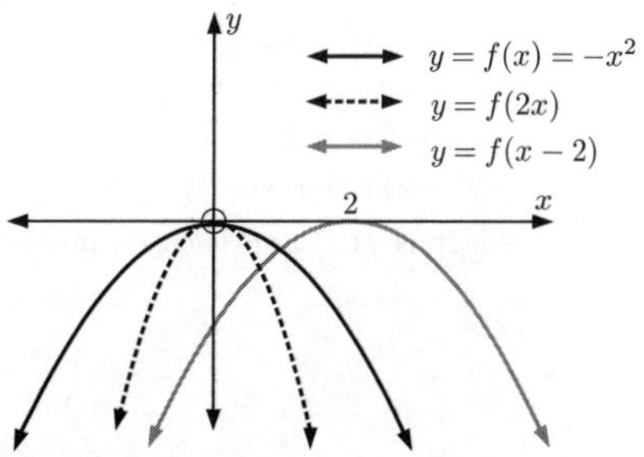

3 **a**

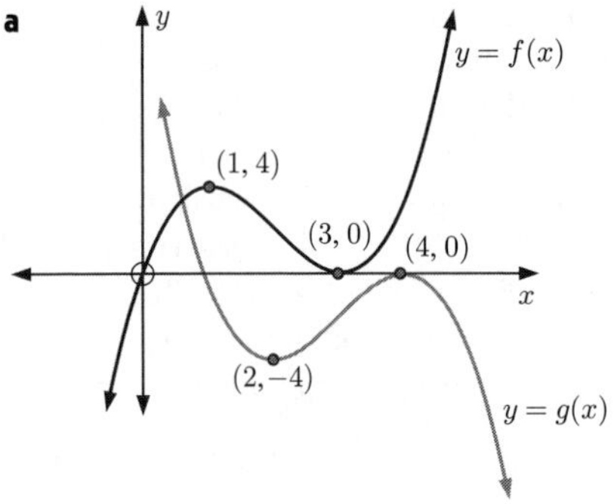

b $g(x)$ is obtained from $f(x)$ by a translation of $\binom{1}{0}$ and then a reflection in the x-axis. So, to get the turning point coordinates we add 1 to the x-coordinate and find the negative of the y-coordinate.

$(1, 4) \mapsto (2, -4)$ and $(3, 0) \mapsto (4, 0)$.

So, the turning points of $g(x)$ are $(2, -4)$ and $(4, 0)$.

4 $f(x) = x^2$ is first reflected in the x-axis to become $-f(x) = -x^2$

The function is then translated by $\binom{-3}{2}$ to become

$$\begin{aligned} -f(x+3)+2 &= -(x+3)^2+2 \\ &= -(x^2+6x+9)+2 \\ \therefore\quad g(x) &= -x^2-6x-7 \end{aligned}$$

5

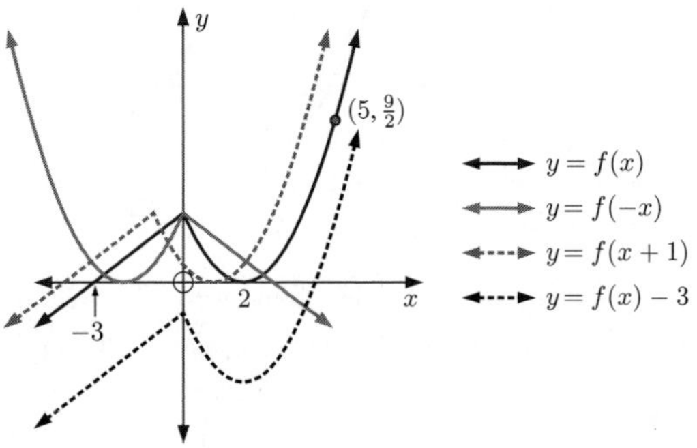

6 $f(x) = x^3 + 3x^2 - x + 4$

$$\begin{aligned} g(x) &= f(x+1)+3 \\ &= \left[(x+1)^3+3(x+1)^2-(x+1)+4\right]+3 \\ &= x^3+3x^2+3x+1+3(x^2+2x+1)-x-1+4+3 \\ &= x^3+3x^2+3x+1+3x^2+6x+3-x-1+4+3 \\ &= x^3+6x^2+8x+10 \end{aligned}$$

7 **a** $f(x) = 3x + 2$

i A translation of 2 units to the left gives
$$\begin{aligned} y &= f(x+2) \\ &= 3(x+2)+2 \\ &= 3x+8 \end{aligned}$$

ii A translation of 6 units upwards gives
$$\begin{aligned} y &= f(x)+6 \\ &= 3x+2+6 \\ &= 3x+8 \end{aligned}$$

b $f(x) = ax + b$ translated k units to the left gives

$$\begin{aligned} y &= f(x+k) \\ &= a(x+k)+b \\ &= ax+ak+b \\ &= (ax+b)+ka \\ &= f(x)+ka \end{aligned}$$

which is $f(x)$ translated ka units upwards.

Chapter 6
QUADRATIC EQUATIONS AND FUNCTIONS

EXERCISE 6A.1

1 **a** $4x^2+7x=0$
$\therefore\ x(4x+7)=0$
$\therefore\ x=0$ or $4x+7=0$ {Null Factor law}
$\therefore\ x=0$ or $-\frac{7}{4}$

b $6x^2+2x=0$
$\therefore\ 2x(3x+1)=0$
$\therefore\ x=0$ or $3x+1=0$ {Null Factor law}
$\therefore\ x=0$ or $-\frac{1}{3}$

c $3x^2-7x=0$
$\therefore\ x(3x-7)=0$
$\therefore\ x=0$ or $3x-7=0$ {Null Factor law}
$\therefore\ x=0$ or $\frac{7}{3}$

d $2x^2-11x=0$
$\therefore\ x(2x-11)=0$
$\therefore\ x=0$ or $2x-11=0$ {Null Factor law}
$\therefore\ x=0$ or $\frac{11}{2}$

e $3x^2=8x$
$\therefore\ 3x^2-8x=0$
$\therefore\ x(3x-8)=0$
$\therefore\ x=0$ or $3x-8=0$ {Null Factor law}
$\therefore\ x=0$ or $\frac{8}{3}$

f $9x=6x^2$
$\therefore\ 6x^2-9x=0$
$\therefore\ 3x(2x-3)=0$
$\therefore\ x=0$ or $2x-3=0$ {Null Factor law}
$\therefore\ x=0$ or $\frac{3}{2}$

g $x^2-5x+6=0$
$\therefore\ (x-2)(x-3)=0$
$\therefore\ x-2=0$ or $x-3=0$ {Null Factor law}
$\therefore\ x=2$ or 3

h $x^2=2x+8$
$\therefore\ x^2-2x-8=0$
$\therefore\ (x-4)(x+2)=0$
$\therefore\ x-4=0$ or $x+2=0$ {Null Factor law}
$\therefore\ x=-2$ or 4

i $x^2+21=10x$
$\therefore\ x^2-10x+21=0$
$\therefore\ (x-3)(x-7)=0$
$\therefore\ x-3=0$ or $x-7=0$ {Null Factor law}
$\therefore\ x=3$ or 7

j $9+x^2=6x$
$\therefore\ x^2-6x+9=0$
$\therefore\ (x-3)^2=0$
$\therefore\ x-3=0$
$\therefore\ x=3$

k $x^2+x=12$
$\therefore\ x^2+x-12=0$
$\therefore\ (x+4)(x-3)=0$
$\therefore\ x+4=0$ or $x-3=0$ {Null Factor law}
$\therefore\ x=-4$ or 3

l $x^2+8x=33$
$\therefore\ x^2+8x-33=0$
$\therefore\ (x+11)(x-3)=0$
$\therefore\ x+11=0$ or $x-3=0$ {Null Factor law}
$\therefore\ x=-11$ or 3

2 **a** $9x^2-12x+4=0$
$\therefore\ (3x-2)^2=0$
$\therefore\ x=\frac{2}{3}$

b $2x^2-13x-7=0$
$\therefore\ (2x+1)(x-7)=0$
$\therefore\ x=-\frac{1}{2}$ or 7

c $3x^2=16x+12$
$\therefore\ 3x^2-16x-12=0$
$\therefore\ (3x+2)(x-6)=0$
$\therefore\ x=-\frac{2}{3}$ or 6

d $3x^2+5x=2$
$\therefore\ 3x^2+5x-2=0$
$\therefore\ (3x-1)(x+2)=0$
$\therefore\ x=\frac{1}{3}$ or -2

e $2x^2+3=5x$
$\therefore\ 2x^2-5x+3=0$
$\therefore\ (2x-3)(x-1)=0$
$\therefore\ x=\frac{3}{2}$ or 1

f $3x^2+8x+4=0$
$\therefore\ (3x+2)(x+2)=0$
$\therefore\ x=-\frac{2}{3}$ or -2

g $3x^2=10x+8$
$\therefore\ 3x^2-10x-8=0$
$\therefore\ (3x+2)(x-4)=0$
$\therefore\ x=-\frac{2}{3}$ or 4

h $4x^2+4x=3$
$\therefore\ 4x^2+4x-3=0$
$\therefore\ (2x+3)(2x-1)=0$
$\therefore\ x=-\frac{3}{2}$ or $\frac{1}{2}$

i $4x^2=11x+3$
$\therefore\ 4x^2-11x-3=0$
$\therefore\ (4x+1)(x-3)=0$
$\therefore\ x=-\frac{1}{4}$ or 3

j $12x^2=11x+15$
$\therefore\ 12x^2-11x-15=0$
$\therefore\ (4x+3)(3x-5)=0$
$\therefore\ x=-\frac{3}{4}$ or $\frac{5}{3}$

k $7x^2+6x=1$
$\therefore\ 7x^2+6x-1=0$
$\therefore\ (7x-1)(x+1)=0$
$\therefore\ x=\frac{1}{7}$ or -1

l $15x^2+2x=56$
$\therefore\ 15x^2+2x-56=0$
$\therefore\ (15x-28)(x+2)=0$
$\therefore\ x=\frac{28}{15}$ or -2

3 **a**
$$\begin{aligned}(x+1)^2 &= 2x^2-5x+11\\ \therefore\ x^2+2x+1 &= 2x^2-5x+11\\ \therefore\ x^2-7x+10 &= 0\\ \therefore\ (x-2)(x-5) &= 0\\ \therefore\ x &= 2 \text{ or } 5\end{aligned}$$

b
$$\begin{aligned}(x+2)(1-x) &= -4\\ \therefore\ x-x^2+2-2x &= -4\\ \therefore\ x^2+x-6 &= 0\\ \therefore\ (x+3)(x-2) &= 0\\ \therefore\ x &= -3 \text{ or } 2\end{aligned}$$

c
$$\begin{aligned}5-4x^2 &= 3(2x+1)+2\\ \therefore\ 5-4x^2 &= 6x+3+2\\ \therefore\ 4x^2+6x &= 0\\ \therefore\ 2x(2x+3) &= 0\\ \therefore\ x &= 0 \text{ or } -\tfrac{3}{2}\end{aligned}$$

d
$$\begin{aligned}x+\frac{2}{x} &= 3\\ \therefore\ x^2+2 &= 3x\\ \therefore\ x^2-3x+2 &= 0\\ \therefore\ (x-1)(x-2) &= 0\\ \therefore\ x &= 1 \text{ or } 2\end{aligned}$$

e
$$\begin{aligned}2x-\frac{1}{x} &= -1\\ \therefore\ 2x^2-1 &= -x\\ \therefore\ 2x^2+x-1 &= 0\\ \therefore\ (2x-1)(x+1) &= 0\\ \therefore\ x &= \tfrac{1}{2} \text{ or } -1\end{aligned}$$

f
$$\begin{aligned}\frac{x+3}{1-x} &= -\frac{9}{x}\\ \therefore\ x(x+3) &= -9(1-x)\\ \therefore\ x^2+3x &= -9+9x\\ \therefore\ x^2-6x+9 &= 0\\ \therefore\ (x-3)^2 &= 0\\ \therefore\ x &= 3\end{aligned}$$

EXERCISE 6A.2

1 **a**
$$\begin{aligned}(x+5)^2 &= 2\\ \therefore\ x+5 &= \pm\sqrt{2}\\ \therefore\ x &= -5\pm\sqrt{2}\end{aligned}$$

b $(x+6)^2 = -11$ has no real solutions as $(x+6)^2$ cannot be negative.

c
$$\begin{aligned}(x-4)^2 &= 8\\ \therefore\ x-4 &= \pm\sqrt{8}\\ \therefore\ x &= 4\pm 2\sqrt{2}\end{aligned}$$

d
$$\begin{aligned}(x-8)^2 &= 7\\ \therefore\ x-8 &= \pm\sqrt{7}\\ \therefore\ x &= 8\pm\sqrt{7}\end{aligned}$$

e
$$\begin{aligned}2(x+3)^2 &= 10\\ \therefore\ (x+3)^2 &= 5\\ \therefore\ x+3 &= \pm\sqrt{5}\\ \therefore\ x &= -3\pm\sqrt{5}\end{aligned}$$

f
$$\begin{aligned}3(x-2)^2 &= 18\\ \therefore\ (x-2)^2 &= 6\\ \therefore\ x-2 &= \pm\sqrt{6}\\ \therefore\ x &= 2\pm\sqrt{6}\end{aligned}$$

g
$$\begin{aligned}(x+1)^2+1 &= 11\\ \therefore\ (x+1)^2 &= 10\\ \therefore\ x+1 &= \pm\sqrt{10}\\ \therefore\ x &= -1\pm\sqrt{10}\end{aligned}$$

h
$$\begin{aligned}(2x+1)^2 &= 3\\ \therefore\ 2x+1 &= \pm\sqrt{3}\\ \therefore\ 2x &= -1\pm\sqrt{3}\\ \therefore\ x &= -\tfrac{1}{2}\pm\tfrac{1}{2}\sqrt{3}\end{aligned}$$

i
$$\begin{aligned}(1-3x)^2-7 &= 0\\ \therefore\ (1-3x)^2 &= 7\\ \therefore\ 1-3x &= \pm\sqrt{7}\\ \therefore\ 3x &= 1\pm\sqrt{7}\\ \therefore\ x &= \tfrac{1}{3}\pm\tfrac{\sqrt{7}}{3}\end{aligned}$$

2 **a**
$$\begin{aligned}x^2-4x+1 &= 0\\ \therefore\ x^2-4x &= -1\\ \therefore\ x^2-4x+(-2)^2 &= -1+(-2)^2\\ \therefore\ (x-2)^2 &= 3\\ \therefore\ x-2 &= \pm\sqrt{3}\\ \therefore\ x &= 2\pm\sqrt{3}\end{aligned}$$

b
$$\begin{aligned}x^2+6x+2 &= 0\\ \therefore\ x^2+6x &= -2\\ \therefore\ x^2+6x+3^2 &= -2+3^2\\ \therefore\ (x+3)^2 &= 7\\ \therefore\ x+3 &= \pm\sqrt{7}\\ \therefore\ x &= -3\pm\sqrt{7}\end{aligned}$$

c
$$\begin{aligned}x^2-14x+46 &= 0\\ \therefore\ x^2-14x &= -46\\ \therefore\ x^2-14x+(-7)^2 &= -46+(-7)^2\\ \therefore\ (x-7)^2 &= 3\\ \therefore\ x-7 &= \pm\sqrt{3}\\ \therefore\ x &= 7\pm\sqrt{3}\end{aligned}$$

d
$$\begin{aligned}x^2 &= 4x+3\\ \therefore\ x^2-4x &= 3\\ \therefore\ x^2-4x+(-2)^2 &= 3+(-2)^2\\ \therefore\ (x-2)^2 &= 7\\ \therefore\ x-2 &= \pm\sqrt{7}\\ \therefore\ x &= 2\pm\sqrt{7}\end{aligned}$$

e
$$x^2+6x+7=0$$
$$\therefore\ x^2+6x=-7$$
$$\therefore\ x^2+6x+3^2=-7+3^2$$
$$\therefore\ (x+3)^2=2$$
$$\therefore\ x+3=\pm\sqrt{2}$$
$$\therefore\ x=-3\pm\sqrt{2}$$

f
$$x^2=2x+6$$
$$\therefore\ x^2-2x=6$$
$$\therefore\ x^2-2x+(-1)^2=6+(-1)^2$$
$$\therefore\ (x-1)^2=7$$
$$\therefore\ x-1=\pm\sqrt{7}$$
$$\therefore\ x=1\pm\sqrt{7}$$

g
$$x^2+6x=2$$
$$\therefore\ x^2+6x+3^2=2+3^2$$
$$\therefore\ (x+3)^2=11$$
$$\therefore\ x+3=\pm\sqrt{11}$$
$$\therefore\ x=-3\pm\sqrt{11}$$

h
$$x^2+10=8x$$
$$\therefore\ x^2-8x=-10$$
$$\therefore\ x^2-8x+(-4)^2=-10+(-4)^2$$
$$\therefore\ (x-4)^2=6$$
$$\therefore\ x-4=\pm\sqrt{6}$$
$$\therefore\ x=4\pm\sqrt{6}$$

i
$$x^2+6x=-11$$
$$\therefore\ x^2+6x+3^2=-11+3^2$$
$$\therefore\ (x+3)^2=-2$$

$\therefore$ x has no real solutions, since the perfect square cannot be negative.

3 **a**
$$2x^2+4x+1=0$$
$$\therefore\ x^2+2x+\tfrac{1}{2}=0$$
$$\therefore\ x^2+2x=-\tfrac{1}{2}$$
$$\therefore\ x^2+2x+1^2=-\tfrac{1}{2}+1^2$$
$$\therefore\ (x+1)^2=\tfrac{1}{2}$$
$$\therefore\ x+1=\pm\tfrac{1}{\sqrt{2}}$$
$$\therefore\ x=-1\pm\tfrac{1}{\sqrt{2}}$$

b
$$2x^2-10x+3=0$$
$$\therefore\ x^2-5x+\tfrac{3}{2}=0$$
$$\therefore\ x^2-5x=-\tfrac{3}{2}$$
$$\therefore\ x^2-5x+(-\tfrac{5}{2})^2=-\tfrac{3}{2}+(-\tfrac{5}{2})^2$$
$$\therefore\ (x-\tfrac{5}{2})^2=-\tfrac{3}{2}+\tfrac{25}{4}$$
$$\therefore\ (x-\tfrac{5}{2})^2=\tfrac{19}{4}$$
$$\therefore\ x-\tfrac{5}{2}=\pm\tfrac{\sqrt{19}}{2}$$
$$\therefore\ x=\tfrac{5}{2}\pm\tfrac{\sqrt{19}}{2}$$

c
$$3x^2+12x+5=0$$
$$\therefore\ x^2+4x+\tfrac{5}{3}=0$$
$$\therefore\ x^2+4x=-\tfrac{5}{3}$$
$$\therefore\ x^2+4x+2^2=-\tfrac{5}{3}+2^2$$
$$\therefore\ (x+2)^2=\tfrac{7}{3}$$
$$\therefore\ x+2=\pm\sqrt{\tfrac{7}{3}}$$
$$\therefore\ x=-2\pm\sqrt{\tfrac{7}{3}}$$

d
$$3x^2=6x+4$$
$$\therefore\ x^2=2x+\tfrac{4}{3}$$
$$\therefore\ x^2-2x=\tfrac{4}{3}$$
$$\therefore\ x^2-2x+(-1)^2=\tfrac{4}{3}+(-1)^2$$
$$\therefore\ (x-1)^2=\tfrac{7}{3}$$
$$\therefore\ x-1=\pm\sqrt{\tfrac{7}{3}}$$
$$\therefore\ x=1\pm\sqrt{\tfrac{7}{3}}$$

e
$$5x^2-15x+2=0$$
$$\therefore\ x^2-3x+\tfrac{2}{5}=0$$
$$\therefore\ x^2-3x=-\tfrac{2}{5}$$
$$\therefore\ x^2-3x+(-\tfrac{3}{2})^2=-\tfrac{2}{5}+(-\tfrac{3}{2})^2$$
$$\therefore\ (x-\tfrac{3}{2})^2=-\tfrac{2}{5}+\tfrac{9}{4}=\tfrac{37}{20}$$
$$\therefore\ x-\tfrac{3}{2}=\pm\sqrt{\tfrac{37}{20}}$$
$$\therefore\ x=\tfrac{3}{2}\pm\sqrt{\tfrac{37}{20}}$$

f
$$4x^2+4x=5$$
$$\therefore\ x^2+x=\tfrac{5}{4}$$
$$\therefore\ x^2+x+(\tfrac{1}{2})^2=\tfrac{5}{4}+(\tfrac{1}{2})^2$$
$$\therefore\ (x+\tfrac{1}{2})^2=\tfrac{6}{4}$$
$$\therefore\ x+\tfrac{1}{2}=\pm\tfrac{\sqrt{6}}{2}$$
$$\therefore\ x=-\tfrac{1}{2}\pm\tfrac{\sqrt{6}}{2}$$

EXERCISE 6A.3

1 **a** $x^2 - 4x - 3 = 0$
has $a = 1,\quad b = -4,\quad c = -3$

$$\therefore\quad x = \frac{-(-4) \pm \sqrt{(-4)^2 - 4(1)(-3)}}{2(1)}$$
$$= \frac{4 \pm \sqrt{28}}{2}$$
$$= \frac{4 \pm 2\sqrt{7}}{2}$$
$$= 2 \pm \sqrt{7}$$

b $x^2 + 6x + 7 = 0$
has $a = 1,\quad b = 6,\quad c = 7$

$$\therefore\quad x = \frac{-6 \pm \sqrt{6^2 - 4(1)(7)}}{2(1)}$$
$$= \frac{-6 \pm \sqrt{8}}{2}$$
$$= \frac{-6 \pm 2\sqrt{2}}{2}$$
$$= -3 \pm \sqrt{2}$$

c $x^2 + 1 = 4x$
$\therefore\quad x^2 - 4x + 1 = 0$
which has $a = 1,\quad b = -4,\quad c = 1$

$$\therefore\quad x = \frac{-(-4) \pm \sqrt{(-4)^2 - 4(1)(1)}}{2(1)}$$
$$= \frac{4 \pm \sqrt{12}}{2}$$
$$= \frac{4 \pm 2\sqrt{3}}{2}$$
$$= 2 \pm \sqrt{3}$$

d $x^2 + 4x = 1$
$\therefore\quad x^2 + 4x - 1 = 0$
which has $a = 1,\quad b = 4,\quad c = -1$

$$\therefore\quad x = \frac{-4 \pm \sqrt{4^2 - 4(1)(-1)}}{2(1)}$$
$$= \frac{-4 \pm \sqrt{20}}{2}$$
$$= \frac{-4 \pm 2\sqrt{5}}{2}$$
$$= -2 \pm \sqrt{5}$$

e $x^2 - 4x + 2 = 0$
has $a = 1,\quad b = -4,\quad c = 2$

$$\therefore\quad x = \frac{-(-4) \pm \sqrt{(-4)^2 - 4(1)(2)}}{2(1)}$$
$$= \frac{4 \pm \sqrt{8}}{2}$$
$$= \frac{4 \pm 2\sqrt{2}}{2}$$
$$= 2 \pm \sqrt{2}$$

f $2x^2 - 2x - 3 = 0$
has $a = 2,\quad b = -2,\quad c = -3$

$$\therefore\quad x = \frac{-(-2) \pm \sqrt{(-2)^2 - 4(2)(-3)}}{2(2)}$$
$$= \frac{2 \pm \sqrt{28}}{4}$$
$$= \frac{2 \pm 2\sqrt{7}}{4}$$
$$= \tfrac{1}{2} \pm \tfrac{\sqrt{7}}{2}$$

g $(3x + 1)^2 = -2x$
$\therefore\quad 9x^2 + 6x + 1 = -2x$
$\therefore\quad 9x^2 + 8x + 1 = 0$
which has $a = 9,\quad b = 8,\quad c = 1$

$$\therefore\quad x = \frac{-8 \pm \sqrt{8^2 - 4(9)(1)}}{2(9)}$$
$$= \frac{-8 \pm \sqrt{28}}{18}$$
$$= \frac{-8 \pm 2\sqrt{7}}{18} \quad \text{or} \quad -\tfrac{4}{9} \pm \tfrac{\sqrt{7}}{9}$$

h $(x + 3)(2x + 1) = 9$
$\therefore\quad 2x^2 + x + 6x + 3 = 9$
$\therefore\quad 2x^2 + 7x - 6 = 0$
which has $a = 2,\quad b = 7,\quad c = -6$

$$\therefore\quad x = \frac{-7 \pm \sqrt{7^2 - 4(2)(-6)}}{2(2)}$$
$$= \frac{-7 \pm \sqrt{49 + 48}}{4}$$
$$= -\tfrac{7}{4} \pm \tfrac{\sqrt{97}}{4}$$

2 **a**
$$(x+2)(x-1) = 2-3x$$
$$\therefore \quad x^2 - x + 2x - 2 = 2 - 3x$$
$$\therefore \quad x^2 + 4x - 4 = 0$$
which has $a = 1,\quad b = 4,\quad c = -4$
$$\therefore \quad x = \frac{-4 \pm \sqrt{4^2 - 4(1)(-4)}}{2(1)} = \frac{-4 \pm \sqrt{32}}{2} = \frac{-4 \pm 4\sqrt{2}}{2} = -2 \pm 2\sqrt{2}$$

b
$$(2x+1)^2 = 3 - x$$
$$\therefore \quad 4x^2 + 4x + 1 = 3 - x$$
$$\therefore \quad 4x^2 + 5x - 2 = 0$$
which has $a = 4,\quad b = 5,\quad c = -2$
$$\therefore \quad x = \frac{-5 \pm \sqrt{5^2 - 4(4)(-2)}}{2(4)} = \frac{-5 \pm \sqrt{25+32}}{8} = -\tfrac{5}{8} \pm \tfrac{\sqrt{57}}{8}$$

c
$$(x-2)^2 = 1 + x$$
$$\therefore \quad x^2 - 4x + 4 = 1 + x$$
$$\therefore \quad x^2 - 5x + 3 = 0$$
which has $a = 1,\quad b = -5,\quad c = 3$
$$\therefore \quad x = \frac{-(-5) \pm \sqrt{(-5)^2 - 4(1)(3)}}{2(1)} = \frac{5 \pm \sqrt{25-12}}{2} = \tfrac{5}{2} \pm \tfrac{\sqrt{13}}{2}$$

d
$$\frac{x-1}{2-x} = 2x + 1$$
$$\therefore \quad x - 1 = (2x+1)(2-x)$$
$$\therefore \quad x - 1 = 4x - 2x^2 + 2 - x$$
$$\therefore \quad 2x^2 - 2x - 3 = 0$$
which has $a = 2,\quad b = -2,\quad c = -3$
$$\therefore \quad x = \frac{-(-2) \pm \sqrt{(-2)^2 - 4(2)(-3)}}{2(2)} = \frac{2 \pm \sqrt{28}}{4} = \frac{2 \pm 2\sqrt{7}}{4} \text{ or } \tfrac{1}{2} \pm \tfrac{\sqrt{7}}{2}$$

e
$$x - \frac{1}{x} = 1$$
$$\therefore \quad x^2 - 1 = x$$
$$\therefore \quad x^2 - x - 1 = 0$$
which has $a = 1,\quad b = -1,\quad c = -1$
$$\therefore \quad x = \frac{-(-1) \pm \sqrt{(-1)^2 - 4(1)(-1)}}{2(1)} = \frac{1 \pm \sqrt{1+4}}{2} = \tfrac{1}{2} \pm \tfrac{\sqrt{5}}{2}$$

f
$$2x - \frac{1}{x} = 3$$
$$\therefore \quad 2x^2 - 1 = 3x$$
$$\therefore \quad 2x^2 - 3x - 1 = 0$$
which has $a = 2,\quad b = -3,\quad c = -1$
$$\therefore \quad x = \frac{-(-3) \pm \sqrt{(-3)^2 - 4(2)(-1)}}{2(2)} = \frac{3 \pm \sqrt{9+8}}{4} = \tfrac{3}{4} \pm \tfrac{\sqrt{17}}{4}$$

EXERCISE 6B

1 **a** $x^2 + 7x - 3 = 0$
has $a = 1,\quad b = 7,\quad c = -3$
$$\therefore \quad \Delta = b^2 - 4ac = 7^2 - 4(1)(-3) = 61$$
Since $\Delta > 0$, there are two distinct real solutions.

b $3x^2 + 2x - 1 = 0$
has $a = 3,\quad b = 2,\quad c = -1$
$$\therefore \quad \Delta = b^2 - 4ac = 2^2 - 4(3)(-1) = 16$$
Since $\Delta > 0$, there are two distinct real solutions.

c $5x^2 + 4x - 3 = 0$
has $a = 5$, $b = 4$, $c = -3$
$\therefore \quad \Delta = b^2 - 4ac$
$= 4^2 - 4(5)(-3)$
$= 76$
Since $\Delta > 0$, there are two distinct real solutions.

d $x^2 + x + 5 = 0$
has $a = 1$, $b = 1$, $c = 5$
$\therefore \quad \Delta = b^2 - 4ac$
$= 1^2 - 4(1)(5)$
$= -19$
Since $\Delta < 0$, there are no real roots.

e $16x^2 - 8x + 1 = 0$
has $a = 16$, $b = -8$, $c = 1$
$\therefore \quad \Delta = b^2 - 4ac$
$= (-8)^2 - 4(16)(1)$
$= 0$
$\therefore$ there is one repeated real root.

2 **a** $6x^2 - 5x - 6 = 0$
has $a = 6$, $b = -5$, $c = -6$
$\therefore \quad \Delta = b^2 - 4ac$
$= (-5)^2 - 4(6)(-6)$
$= 169$
$\therefore \quad \sqrt{\Delta} = 13$, so the equation has rational roots.

b $2x^2 - 7x - 5 = 0$
has $a = 2$, $b = -7$, $c = -5$
$\therefore \quad \Delta = b^2 - 4ac$
$= (-7)^2 - 4(2)(-5)$
$= 89$
$\therefore \quad \sqrt{\Delta} = \sqrt{89}$, so the equation does not have rational roots.

c $3x^2 + 4x + 1 = 0$
has $a = 3$, $b = 4$, $c = 1$
$\therefore \quad \Delta = b^2 - 4ac$
$= 4^2 - 4(3)(1)$
$= 4$
$\therefore \quad \sqrt{\Delta} = 2$, so the equation has rational roots.

d $6x^2 - 47x - 8 = 0$
has $a = 6$, $b = -47$, $c = -8$
$\therefore \quad \Delta = b^2 - 4ac$
$= (-47)^2 - 4(6)(-8)$
$= 2401$
$\therefore \quad \sqrt{\Delta} = 49$,
so the equation has rational roots.

e $4x^2 - 3x + 2 = 0$
has $a = 4$, $b = -3$, $c = 2$
$\therefore \quad \Delta = b^2 - 4ac$
$= (-3)^2 - 4(4)2$
$= -23$
Since $\Delta < 0$, the equation does not have rational roots.

f $8x^2 + 2x - 3 = 0$
has $a = 8$, $b = 2$, $c = -3$
$\therefore \quad \Delta = b^2 - 4ac$
$= 2^2 - 4(8)(-3)$
$= 100$
$\therefore \quad \sqrt{\Delta} = 10$, so the equation has rational roots.

3 **a** For $x^2 + 4x + m = 0$,
$a = 1$, $b = 4$, $c = m$
So, $\Delta = b^2 - 4ac$
$= 4^2 - 4(1)(m)$
$= 16 - 4m$
which has sign diagram

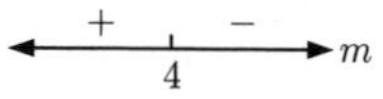

i For a repeated root, $\Delta = 0$
$\therefore \quad m = 4$
ii For two distinct real roots, $\Delta > 0$
$\therefore \quad m < 4$
iii For no real roots, $\Delta < 0$
$\therefore \quad m > 4$

b For $mx^2 + 3x + 2 = 0$,
$a = m$, $b = 3$, $c = 2$
So, $\Delta = b^2 - 4ac$
$= 3^2 - 4(m)(2)$
$= 9 - 8m$
which has sign diagram

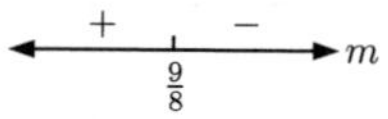

i For a repeated root, $\Delta = 0$
$\therefore \quad m = \frac{9}{8}$
ii For two distinct real roots, $\Delta > 0$
$\therefore \quad m < \frac{9}{8}$
iii For no real roots, $\Delta < 0$
$\therefore \quad m > \frac{9}{8}$

c For $mx^2 - 3x + 1 = 0$,
$a = m, \quad b = -3, \quad c = 1$
So, $\Delta = b^2 - 4ac$
$= (-3)^2 - 4(m)(1)$
$= 9 - 4m$

which has sign diagram

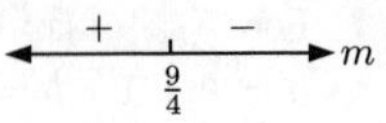

i For a repeated root, $\Delta = 0$ $\quad \therefore \; m = \frac{9}{4}$

ii For two distinct real roots, $\Delta > 0$ $\quad \therefore \; m < \frac{9}{4}$

iii For no real roots, $\Delta < 0$ $\quad \therefore \; m > \frac{9}{4}$

4 **a** For $2x^2 + kx - k = 0$,
$a = 2, \quad b = k, \quad c = -k$
So, $\Delta = b^2 - 4ac$
$= k^2 - 4(2)(-k)$
$= k^2 + 8k$
$= k(k+8)$

which has sign diagram

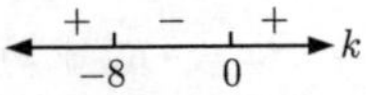

i For two distinct real roots, $\Delta > 0$
$\therefore \; k < -8$ or $k > 0$

ii For two real roots, $\Delta \geqslant 0$
$\therefore \; k \leqslant -8$ or $k \geqslant 0$

iii For a repeated root, $\Delta = 0$
$\therefore \; k = -8$ or 0

iv For no real roots, $\Delta < 0$
$\therefore \; -8 < k < 0$

b For $kx^2 - 2x + k = 0$,
$a = k, \quad b = -2, \quad c = k$
So, $\Delta = b^2 - 4ac$
$= (-2)^2 - 4(k)(k)$
$= 4 - 4k^2$
$= 4(1+k)(1-k)$

which has sign diagram

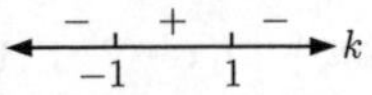

i For two distinct real roots, $\Delta > 0$
$\therefore \; -1 < k < 1$

ii For two real roots, $\Delta \geqslant 0$
$\therefore \; -1 \leqslant k \leqslant 1$

iii For a repeated root, $\Delta = 0$
$\therefore \; k = -1$ or 1

iv For no real roots, $\Delta < 0$
$\therefore \; k < -1$ or $k > 1$

c For $x^2 + (k+2)x + 4 = 0$,
$a = 1, \quad b = k+2, \quad c = 4$
So, $\Delta = b^2 - 4ac$
$= (k+2)^2 - 4(1)(4)$
$= k^2 + 4k + 4 - 16$
$= k^2 + 4k - 12$
$= (k+6)(k-2)$

which has sign diagram

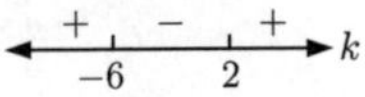

i For two distinct real roots, $\Delta > 0$
$\therefore \; k < -6$ or $k > 2$

ii For two real roots, $\Delta \geqslant 0$
$\therefore \; k \leqslant -6$ or $k \geqslant 2$

iii For a repeated root, $\Delta = 0$
$\therefore \; k = -6$ or 2

iv For no real roots, $\Delta < 0$
$\therefore \; -6 < k < 2$

d For $2x^2 + (k-2)x + 2 = 0$,
$a = 2, \quad b = k-2, \quad c = 2$
So, $\Delta = b^2 - 4ac$
$= (k-2)^2 - 4(2)(2)$
$= k^2 - 4k + 4 - 16$
$= k^2 - 4k - 12$
$= (k-6)(k+2)$

which has sign diagram

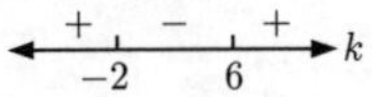

i For two distinct real roots, $\Delta > 0$
$\therefore \; k < -2$ or $k > 6$

ii For two real roots, $\Delta \geqslant 0$
$\therefore \; k \leqslant -2$ or $k \geqslant 6$

iii For a repeated root, $\Delta = 0$
$\therefore \; k = -2$ or 6

iv For no real roots, $\Delta < 0$
$\therefore \; -2 < k < 6$

e For $x^2 + (3k-1)x + (2k+10) = 0$,
$a = 1, \quad b = 3k-1, \quad c = 2k+10$
So, $\Delta = b^2 - 4ac$
$= (3k-1)^2 - 4(1)(2k+10)$
$= 9k^2 - 6k + 1 - 8k - 40$
$= 9k^2 - 14k - 39$
$= (9k+13)(k-3)$

which has sign diagram

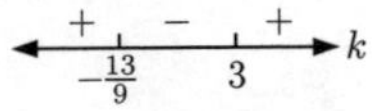

i For two distinct real roots, $\Delta > 0$
$\therefore \; k < -\frac{13}{9}$ or $k > 3$

ii For two real roots, $\Delta \geqslant 0$
$\therefore \; k \leqslant -\frac{13}{9}$ or $k \geqslant 3$

iii For a repeated root, $\Delta = 0$
$\therefore \; k = -\frac{13}{9}$ or 3

iv For no real roots, $\Delta < 0$
$\therefore \; -\frac{13}{9} < k < 3$

f For $(k+1)x^2 + kx + k = 0$,
$a = k+1, \quad b = k, \quad c = k$
So, $\Delta = b^2 - 4ac$
$= k^2 - 4(k+1)(k)$
$= k^2 - 4k^2 - 4k$
$= -3k^2 - 4k$
$= -k(3k+4)$
which has sign diagram

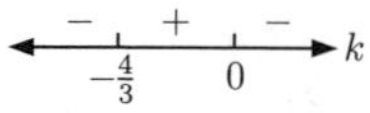

i For two distinct real roots, $\Delta > 0$
$\therefore \quad -\frac{4}{3} < k < 0$

ii For two real roots, $\Delta \geqslant 0$
$\therefore \quad -\frac{4}{3} \leqslant k \leqslant 0$

iii For a repeated root, $\Delta = 0$
$\therefore \quad k = -\frac{4}{3}$ or 0

iv For no real roots, $\Delta < 0$
$\therefore \quad k < -\frac{4}{3}$ or $k > 0$

EXERCISE 6C.1

1 **a** $y = (x-4)(x+2)$ has x-intercepts -2 and 4 and y-intercept -8

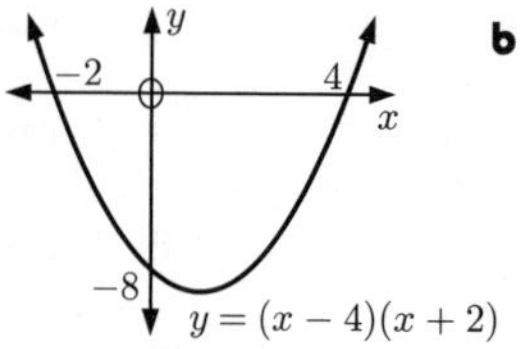

b $y = -(x-4)(x+2)$ has x-intercepts -2 and 4 and y-intercept 8

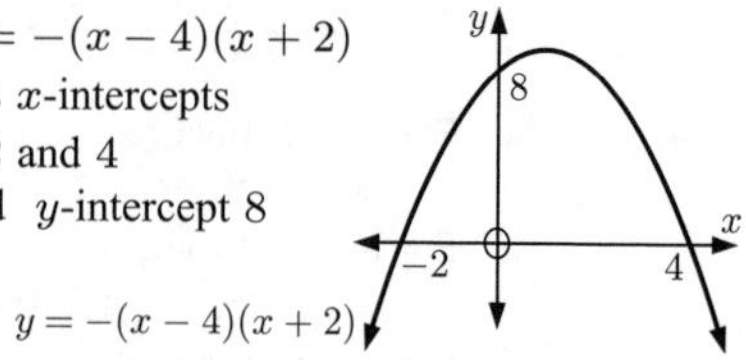

c $y = 2(x+3)(x+5)$ has x-intercepts -5 and -3 and y-intercept 30

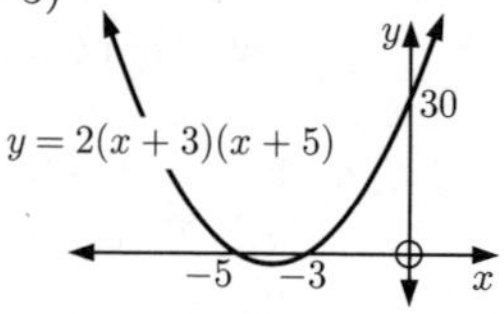

d $y = -3x(x+4)$ has x-intercepts 0 and -4 and y-intercept 0

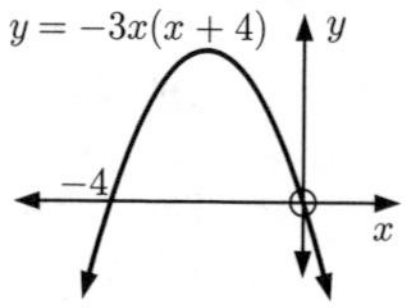

e $y = 2(x+3)^2$ has x-intercept -3 and y-intercept 18

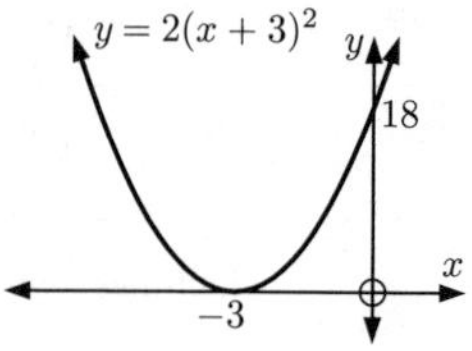

f $y = -\frac{1}{4}(x+2)^2$ has x-intercept -2 and y-intercept -1

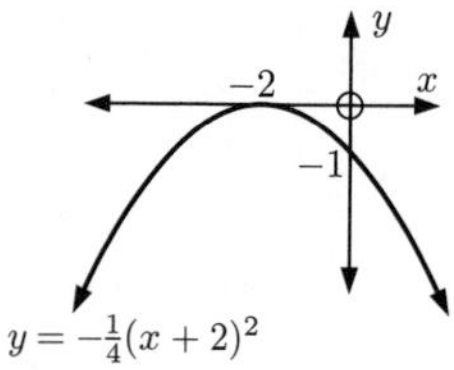

2 **a** The average of the x-intercepts is 1, so the axis of symmetry is $x = 1$.
b The average of the x-intercepts is 1, so the axis of symmetry is $x = 1$.
c The average of the x-intercepts is -4, so the axis of symmetry is $x = -4$.
d The average of the x-intercepts is -2, so the axis of symmetry is $x = -2$.
e The only x-intercept is -3, so the axis of symmetry is $x = -3$.
f The only x-intercept is -2, so the axis of symmetry is $x = -2$.

3 **a** C **b** E **c** B **d** F **e** G **f** H **g** A **h** D

4 **a** The vertex is $(1, 3)$. The axis of symmetry is $x = 1$. The y-intercept is 4.

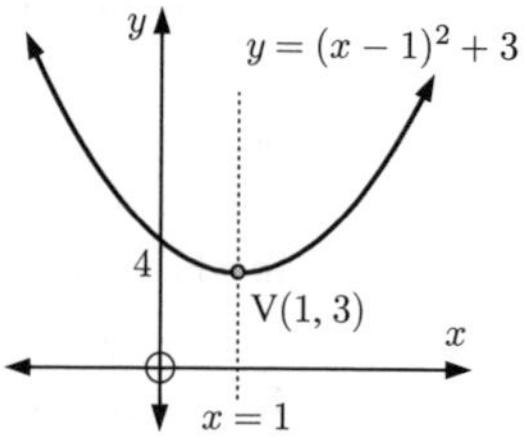

b The vertex is $(-2, 1)$. The axis of symmetry is $x = -2$. The y-intercept is 9.

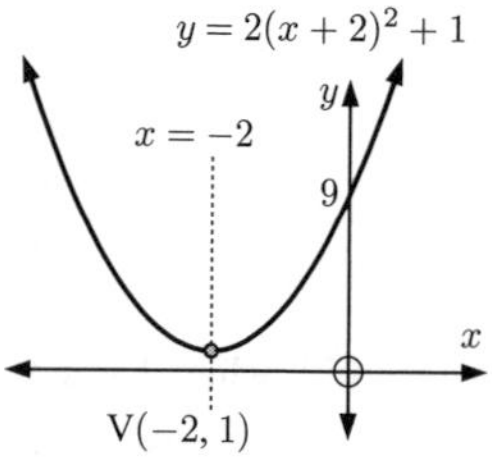

c The vertex is (1, −3).
The axis of symmetry is $x=1$.
The y-intercept is −5.

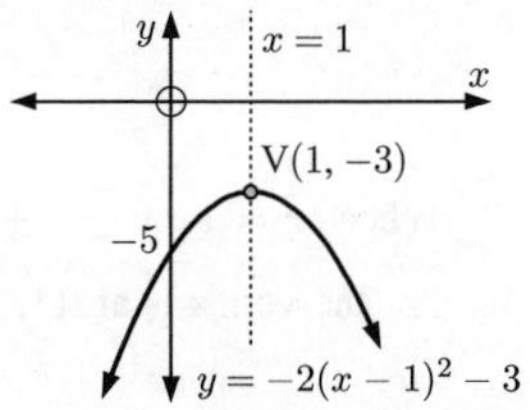

d The vertex is (3, 2).
The axis of symmetry is $x=3$.
The y-intercept is $\frac{13}{2}$.

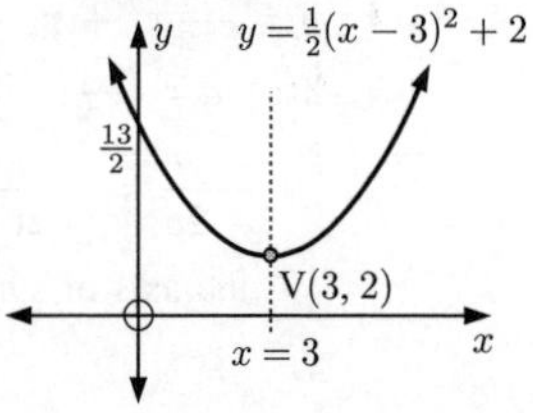

e The vertex is (1, 4).
The axis of symmetry is $x=1$.
The y-intercept is $3\frac{2}{3}$.

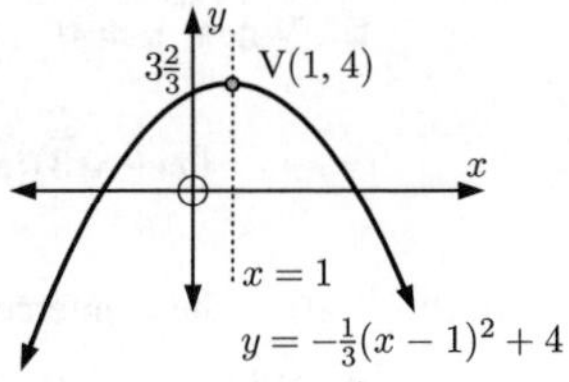

f The vertex is (−2, −3).
The axis of symmetry is $x=-2$.
The y-intercept is $-3\frac{2}{5}$.

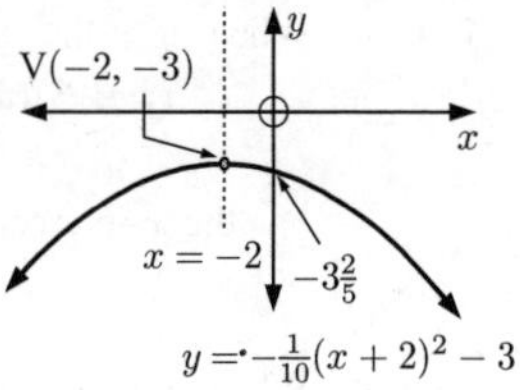

5 **a** G **b** A **c** E **d** B **e** I **f** C **g** D **h** F **i** H

6 **a** $y=x^2-4x+2$
has $a=1,\ b=-4,\ c=2$
$\therefore\ -\dfrac{b}{2a}=-\dfrac{(-4)}{2(1)}=2$
$\therefore$ the axis of symmetry is $x=2$.
When $x=2$,
$y=2^2-4\times 2+2=-2$
$\therefore$ the vertex is at $(2,\ -2)$.

b $y=x^2+2x-3$
has $a=1,\ b=2,\ c=-3$
$\therefore\ -\dfrac{b}{2a}=-\dfrac{2}{2(1)}=-1$
$\therefore$ the axis of symmetry is $x=-1$.
When $x=-1$,
$y=(-1)^2+2(-1)-3=-4$
$\therefore$ the vertex is at $(-1,\ -4)$.

c $y=2x^2+4$
has $a=2,\ b=0,\ c=4$
$\therefore\ -\dfrac{b}{2a}=-\dfrac{0}{2(2)}=0$
$\therefore$ the axis of symmetry is $x=0$.
When $x=0,\ y=4$
$\therefore$ the vertex is at $(0,\ 4)$.

d $y=-3x^2+1$
has $a=-3,\ b=0,\ c=1$
$\therefore\ -\dfrac{b}{2a}=-\dfrac{0}{2(-3)}=0$
$\therefore$ the axis of symmetry is $x=0$.
When $x=0,\ y=1$
$\therefore$ the vertex is at $(0,\ 1)$.

e $y=2x^2+8x-7$
has $a=2,\ b=8,\ c=-7$
$\therefore\ -\dfrac{b}{2a}=-\dfrac{8}{2(2)}=-2$
$\therefore$ the axis of symmetry is $x=-2$.
When $x=-2$,
$y=2(-2)^2+8(-2)-7=-15$
$\therefore$ the vertex is at $(-2,\ -15)$.

f $y=-x^2-4x-9$
has $a=-1,\ b=-4,\ c=-9$
$\therefore\ -\dfrac{b}{2a}=-\dfrac{(-4)}{2(-1)}=-2$
$\therefore$ the axis of symmetry is $x=-2$.
When $x=-2,\ y=-(-2)^2-4(-2)-9$
$=-4+8-9$
$=-5$
$\therefore$ the vertex is at $(-2,\ -5)$.

g $y=2x^2+6x-1$
has $a=2,\ b=6,\ c=-1$
$\therefore\ -\dfrac{b}{2a}=-\dfrac{6}{2(2)}=-\dfrac{3}{2}$
$\therefore$ the axis of symmetry is $x=-\frac{3}{2}$.
When $x=-\frac{3}{2},\ y=2(-\frac{3}{2})^2+6(-\frac{3}{2})-1$
$=\frac{9}{2}-9-1$
$=-\frac{11}{2}$
$\therefore$ the vertex is at $(-\frac{3}{2},\ -\frac{11}{2})$.

h $y=2x^2-10x+3$
has $a=2,\ b=-10,\ c=3$
$\therefore\ -\dfrac{b}{2a}=-\dfrac{(-10)}{2(2)}=\dfrac{5}{2}$
$\therefore$ the axis of symmetry is $x=\frac{5}{2}$.
When $x=\frac{5}{2},\ y=2(\frac{5}{2})^2-10(\frac{5}{2})+3$
$=\frac{25}{2}-\frac{50}{2}+3$
$=-\frac{19}{2}$
$\therefore$ the vertex is at $(\frac{5}{2},\ -\frac{19}{2})$.

i $y = -\frac{1}{2}x^2 + x - 5$

has $a = -\frac{1}{2},\ b = 1,\ c = -5$

$\therefore\ -\dfrac{b}{2a} = -\dfrac{1}{2(-\frac{1}{2})} = 1$

$\therefore$ the axis of symmetry is $x = 1$.

When $x = 1,\ y = -\frac{1}{2}(1)^2 + 1 - 5 = -\frac{9}{2}$

$\therefore$ the vertex is at $(1, -\frac{9}{2})$.

7 **a** When $y = 0,\ x^2 - 9 = 0$

$\therefore\ (x+3)(x-3) = 0$

$\therefore\ x = \pm 3$

$\therefore$ the x-intercepts are ± 3

b When $y = 0,\ 2x^2 - 6 = 0$

$\therefore\ x^2 - 3 = 0$

$\therefore\ (x + \sqrt{3})(x - \sqrt{3}) = 0$

$\therefore\ x = \pm\sqrt{3}$

$\therefore$ the x-intercepts are $\pm\sqrt{3}$

c When $y = 0,\ x^2 + 7x + 10 = 0$

$\therefore\ (x+5)(x+2) = 0$

$\therefore\ x = -5$ or -2

$\therefore$ the x-intercepts are -5 and -2

d When $y = 0,\ x^2 + x - 12 = 0$

$\therefore\ (x+4)(x-3) = 0$

$\therefore\ x = -4$ or 3

$\therefore$ the x-intercepts are -4 and 3

e When $y = 0,\ 4x - x^2 = 0$

$\therefore\ x(4 - x) = 0$

$\therefore\ x = 0$ or 4

$\therefore$ the x-intercepts are 0 and 4

f When $y = 0,\ -x^2 - 6x - 8 = 0$

$\therefore\ x^2 + 6x + 8 = 0$

$\therefore\ (x+4)(x+2) = 0$

$\therefore\ x = -4$ or -2

$\therefore$ the x-intercepts are -4 and -2

g When $y = 0,\ -2x^2 - 4x - 2 = 0$

$\therefore\ x^2 + 2x + 1 = 0$

$\therefore\ (x+1)^2 = 0$

$\therefore\ x = -1$

$\therefore$ the x-intercept is -1 (touching)

h When $y = 0,\ 4x^2 - 24x + 36 = 0$

$\therefore\ x^2 - 6x + 9 = 0$

$\therefore\ (x-3)^2 = 0$

$\therefore\ x = 3$

$\therefore$ the x-intercept is 3 (touching)

i When $y = 0,\ x^2 - 4x + 1 = 0$

$a = 1,\ b = -4$ and $c = 1$

$$\therefore\ x = \frac{-(-4) \pm \sqrt{(-4)^2 - 4(1)(1)}}{2(1)} = \frac{4 \pm \sqrt{12}}{2} = \frac{4 \pm 2\sqrt{3}}{2} = 2 \pm \sqrt{3}$$

$\therefore$ the x-intercepts are $2 \pm \sqrt{3}$

j When $y = 0,\ x^2 + 4x - 3 = 0$

$a = 1,\ b = 4$ and $c = -3$

$$\therefore\ x = \frac{-4 \pm \sqrt{4^2 - 4(1)(-3)}}{2(1)} = \frac{-4 \pm \sqrt{28}}{2} = \frac{-4 \pm 2\sqrt{7}}{2} = -2 \pm \sqrt{7}$$

$\therefore$ the x-intercepts are $-2 \pm \sqrt{7}$

k When $y = 0,\ x^2 - 6x - 2 = 0$

$a = 1,\ b = -6$ and $c = -2$

$$\therefore\ x = \frac{-(-6) \pm \sqrt{(-6)^2 - 4(1)(-2)}}{2(1)} = \frac{6 \pm \sqrt{44}}{2} = \frac{6 \pm 2\sqrt{11}}{2} = 3 \pm \sqrt{11}$$

$\therefore$ the x-intercepts are $3 \pm \sqrt{11}$

l When $y = 0,\ x^2 + 8x + 11 = 0$

$a = 1,\ b = 8$ and $c = 11$

$$\therefore\ x = \frac{-8 \pm \sqrt{8^2 - 4(1)(11)}}{2(1)} = \frac{-8 \pm \sqrt{20}}{2} = \frac{-8 \pm 2\sqrt{5}}{2} = -4 \pm \sqrt{5}$$

$\therefore$ the x-intercepts are $-4 \pm \sqrt{5}$

8 **a** **i** $y = x^2 - 2x + 5$
has $a = 1,\ \ b = -2,\ \ c = 5$

$\therefore\ -\dfrac{b}{2a} = -\dfrac{(-2)}{2(1)} = 1$

$\therefore$ the axis of symmetry is $x = 1$

ii When $x = 1$,

$y = 1^2 - 2(1) + 5$
$= 1 - 2 + 5$
$= 4$

$\therefore$ the vertex is at $(1,\ 4)$

iii When $x = 0,\ \ y = 5$,
so the y-intercept is 5
When $y = 0,\ \ x^2 - 2x + 5 = 0$

$\therefore\ x = \dfrac{-(-2) \pm \sqrt{(-2)^2 - 4(1)(5)}}{2(1)}$

$= \dfrac{2 \pm \sqrt{4 - 20}}{2}$

This has no real solutions,
so there are no x-intercepts.

iv

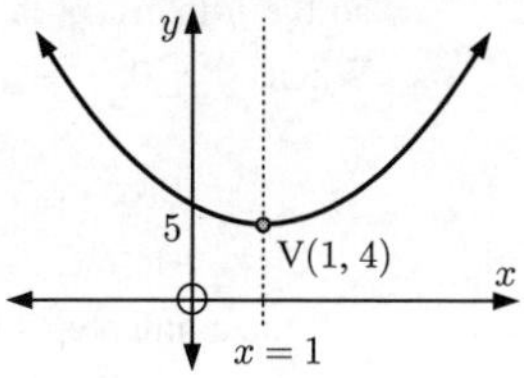

b **i** $y = x^2 + 4x - 1$
has $a = 1,\ \ b = 4,\ \ c = -1$

$\therefore\ -\dfrac{b}{2a} = -\dfrac{4}{2(1)} = -2$

$\therefore$ the axis of symmetry is $x = -2$

ii When $x = -2$,

$y = (-2)^2 + 4(-2) - 1$
$= 4 - 8 - 1$
$= -5$

$\therefore$ the vertex is at $(-2,\ -5)$

iii When $x = 0,\ \ y = -1$,
so the y-intercept is -1.
When $y = 0,\ \ x^2 + 4x - 1 = 0$

$\therefore\ x = \dfrac{-4 \pm \sqrt{4^2 - 4(1)(-1)}}{2(1)}$

$= \dfrac{-4 \pm \sqrt{20}}{2}$

$= \dfrac{-4 \pm 2\sqrt{5}}{2}$

$= -2 \pm \sqrt{5}$

$\therefore$ the x-intercepts are $-2 \pm \sqrt{5}$

iv

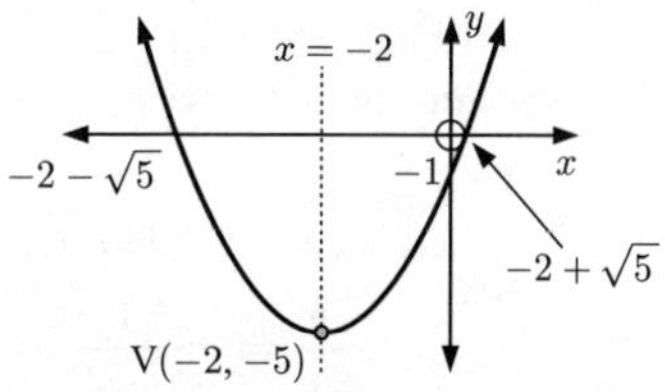

c **i** $y = 2x^2 - 5x + 2$
has $a = 2,\ \ b = -5,\ \ c = 2$

$\therefore\ -\dfrac{b}{2a} = -\dfrac{(-5)}{2(2)} = \dfrac{5}{4}$

$\therefore$ the axis of symmetry is $x = \frac{5}{4}$

ii When $x = \frac{5}{4}$,

$y = 2(\frac{5}{4})^2 - 5(\frac{5}{4}) + 2$
$= \frac{25}{8} - \frac{25}{4} + 2$
$= -\frac{9}{8}$

$\therefore$ the vertex is at $(\frac{5}{4},\ -\frac{9}{8})$

iii When $x = 0,\ y = 2$,
so the y-intercept is 2.
When $y = 0,\ \ 2x^2 - 5x + 2 = 0$

$\therefore\ \ (2x - 1)(x - 2) = 0$

$\therefore\ \ x = \frac{1}{2}$ or 2

$\therefore$ the x-intercepts are $\frac{1}{2}$ and 2

iv

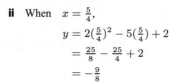

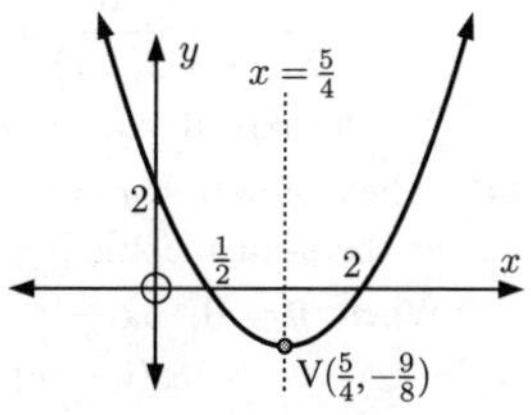

d **i** $y = -x^2 + 3x - 2$
has $a = -1, \quad b = 3, \quad c = -2$

$\therefore \quad -\dfrac{b}{2a} = -\dfrac{3}{2(-1)} = \dfrac{3}{2}$

$\therefore$ the axis of symmetry is $x = \frac{3}{2}$

ii When $x = \frac{3}{2}$, $y = -(\frac{3}{2})^2 + 3(\frac{3}{2}) - 2$
$= -\frac{9}{4} + \frac{9}{2} - 2$
$= \frac{1}{4}$

$\therefore$ the vertex is at $(\frac{3}{2}, \frac{1}{4})$

iii When $x = 0$, $y = -2$,
so the y-intercept is -2.
When $y = 0$, $\quad -x^2 + 3x - 2 = 0$
$\therefore \quad x^2 - 3x + 2 = 0$
$\therefore \quad (x - 1)(x - 2) = 0$
$\therefore \quad x = 1$ or 2

$\therefore$ the x-intercepts are 1 and 2

iv

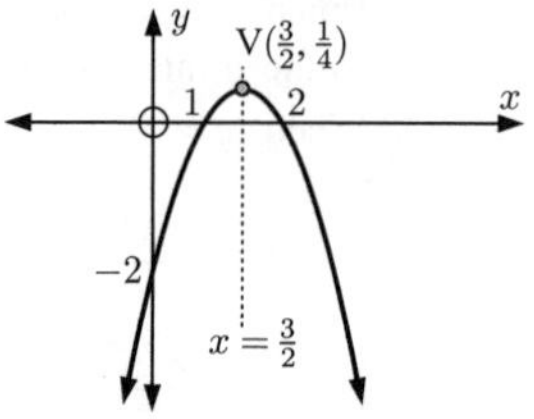

e **i** $y = -3x^2 + 4x - 1$
has $a = -3, \quad b = 4, \quad c = -1$

$\therefore \quad -\dfrac{b}{2a} = -\dfrac{4}{2(-3)} = \dfrac{2}{3}$

$\therefore$ the axis of symmetry is $x = \frac{2}{3}$

ii When $x = \frac{2}{3}$, $y = -3(\frac{2}{3})^2 + 4(\frac{2}{3}) - 1$
$= -\frac{4}{3} + \frac{8}{3} - 1$
$= \frac{1}{3}$

$\therefore$ the vertex is at $(\frac{2}{3}, \frac{1}{3})$

iii When $x = 0$, $y = -1$,
so the y-intercept is -1.
When $y = 0$, $\quad -3x^2 + 4x - 1 = 0$
$\therefore \quad 3x^2 - 4x + 1 = 0$
$\therefore \quad (3x - 1)(x - 1) = 0$
$\therefore \quad x = \frac{1}{3}$ or 1

$\therefore$ the x-intercepts are $\frac{1}{3}$ and 1

iv

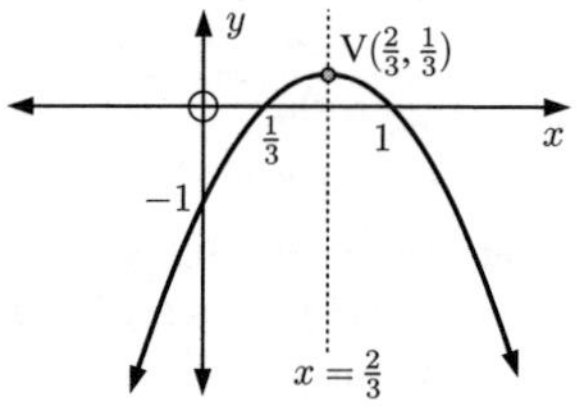

f **i** $y = -2x^2 + x + 1$
has $a = -2, \quad b = 1, \quad c = 1$

$\therefore \quad -\dfrac{b}{2a} = -\dfrac{1}{2(-2)} = \dfrac{1}{4}$

$\therefore$ the axis of symmetry is $x = \frac{1}{4}$

ii When $x = \frac{1}{4}$, $y = -2(\frac{1}{4})^2 + \frac{1}{4} + 1$
$= -\frac{1}{8} + \frac{1}{4} + 1$
$= \frac{9}{8}$

$\therefore$ the vertex is at $(\frac{1}{4}, \frac{9}{8})$

iii When $x = 0$, $y = 1$,
so the y-intercept is 1.
When $y = 0$, $\quad -2x^2 + x + 1 = 0$
$\therefore \quad 2x^2 - x - 1 = 0$
$\therefore \quad (2x + 1)(x - 1) = 0$
$\therefore \quad x = -\frac{1}{2}$ or 1

$\therefore$ the x-intercepts are $-\frac{1}{2}$ and 1

iv

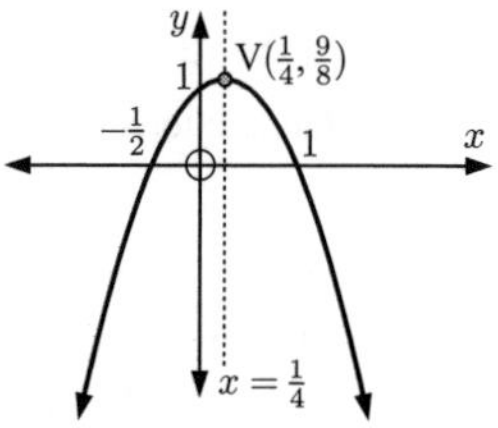

g **i** $y = 6x - x^2$
has $a = -1, \quad b = 6, \quad c = 0$

$\therefore \quad -\dfrac{b}{2a} = -\dfrac{6}{2(-1)} = 3$

$\therefore$ the axis of symmetry is $x = 3$

ii When $x = 3$, $y = 6 \times 3 - 3^2$
$= 9$

$\therefore$ the vertex is at $(3, 9)$

iii When $x = 0$, $y = 0$,
so the y-intercept is 0.
When $y = 0$, $\quad 6x - x^2 = 0$
$\therefore \quad x(6 - x) = 0$
$\therefore \quad x = 0$ or 6

$\therefore$ the x-intercepts are 0 and 6

iv

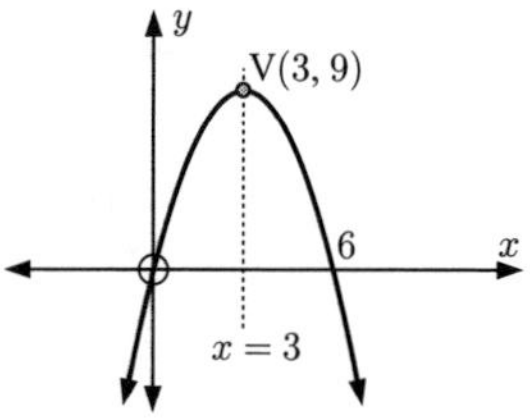

h **i** $y = -x^2 - 6x - 8$
has $a = -1$, $b = -6$, $c = -8$

$\therefore \quad -\dfrac{b}{2a} = -\dfrac{(-6)}{2(-1)} = -3$

$\therefore$ the axis of symmetry is $x = -3$

ii When $x = -3$,

$$\begin{aligned} y &= -(-3)^2 - 6(-3) - 8 \\ &= -9 + 18 - 8 \\ &= 1 \end{aligned}$$

$\therefore$ the vertex is at $(-3, 1)$

iii When $x = 0$, $y = -8$,
so the y-intercept is -8.
When $y = 0$, $-x^2 - 6x - 8 = 0$

$$\begin{aligned} \therefore \quad x^2 + 6x + 8 &= 0 \\ \therefore \quad (x+4)(x+2) &= 0 \\ \therefore \quad x &= -4 \text{ or } -2 \end{aligned}$$

$\therefore$ the x-intercepts are -4 and -2

iv

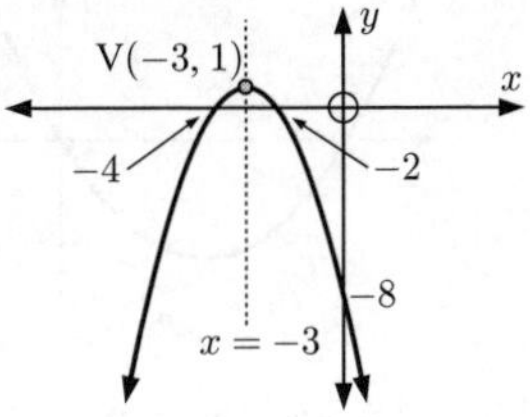

i **i** $y = -\frac{1}{4}x^2 + 2x + 1$
has $a = -\frac{1}{4}$, $b = 2$, $c = 1$

$\therefore \quad -\dfrac{b}{2a} = -\dfrac{2}{2(-\frac{1}{4})} = 4$

$\therefore$ the axis of symmetry is $x = 4$

ii When $x = 4$, $y = -\frac{1}{4}(4)^2 + 2(4) + 1$

$$\begin{aligned} &= -4 + 8 + 1 \\ &= 5 \end{aligned}$$

$\therefore$ the vertex is at $(4, 5)$

iii When $x = 0$, $y = 1$,
so the y-intercept is 1.

When $y = 0$, $-\frac{1}{4}x^2 + 2x + 1 = 0$

$\therefore \quad x^2 - 8x - 4 = 0$

$$\begin{aligned} \therefore \quad x &= \frac{-(-8) \pm \sqrt{(-8)^2 - 4(1)(-4)}}{2(1)} \\ &= \frac{8 \pm \sqrt{80}}{2} \\ &= \frac{8 \pm 4\sqrt{5}}{2} \\ &= 4 \pm 2\sqrt{5} \end{aligned}$$

$\therefore$ the x-intercepts are $4 \pm 2\sqrt{5}$.

iv

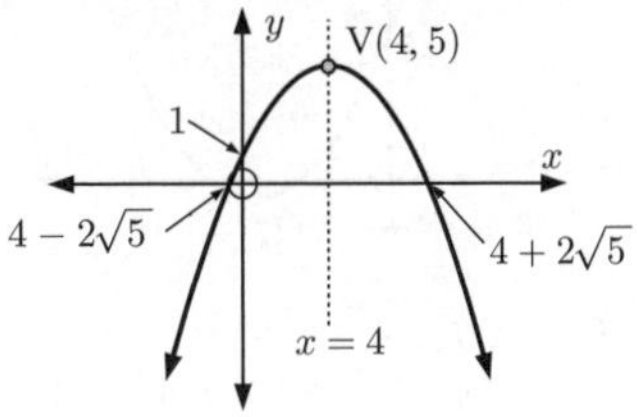

EXERCISE 6C.2

1 **a**

$$\begin{aligned} y &= x^2 - 2x + 3 \\ \therefore \quad y &= x^2 - 2x + 1^2 + 3 - 1^2 \\ \therefore \quad y &= (x-1)^2 + 2 \end{aligned}$$

$\therefore$ vertex is $(1, 2)$, y-intercept is 3

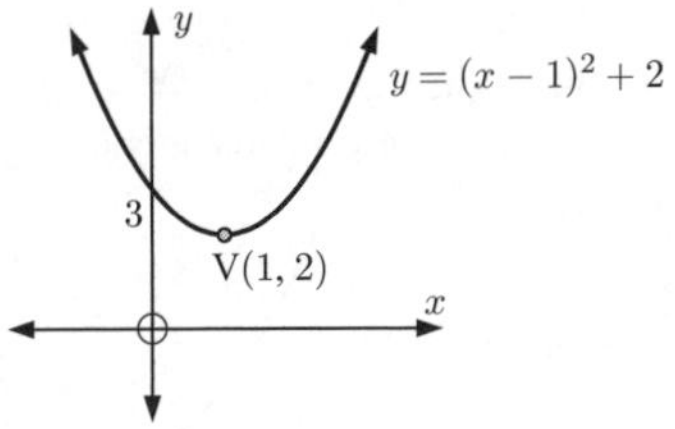

b

$$\begin{aligned} y &= x^2 + 4x - 2 \\ \therefore \quad y &= x^2 + 4x + 2^2 - 2 - 2^2 \\ \therefore \quad y &= (x+2)^2 - 6 \end{aligned}$$

$\therefore$ vertex is $(-2, -6)$, y-intercept is -2

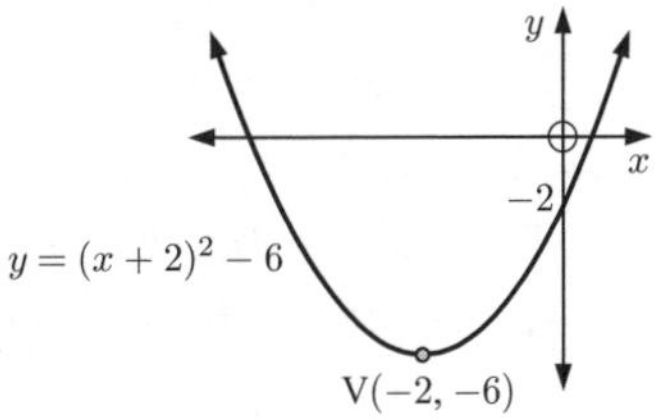

c

$$\begin{aligned} y &= x^2 - 4x \\ \therefore \quad y &= x^2 - 4x + 2^2 - 2^2 \\ \therefore \quad y &= (x-2)^2 - 4 \end{aligned}$$

$\therefore$ vertex is $(2, -4)$, y-intercept is 0

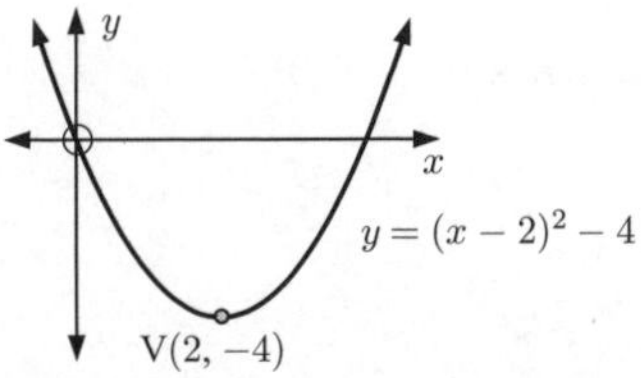

d $y = x^2 + 3x$

$\therefore\ y = x^2 + 3x + (\frac{3}{2})^2 - (\frac{3}{2})^2$

$\therefore\ y = (x + \frac{3}{2})^2 - \frac{9}{4}$

$\therefore$ vertex is $(-\frac{3}{2}, -\frac{9}{4})$, y-intercept is 0

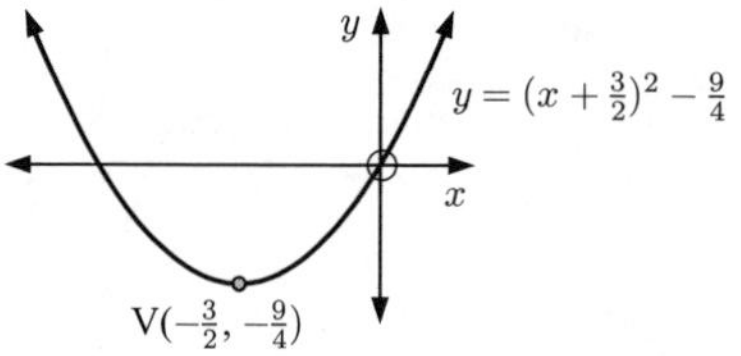

e $y = x^2 + 5x - 2$

$\therefore\ y = x^2 + 5x + (\frac{5}{2})^2 - 2 - (\frac{5}{2})^2$

$\therefore\ y = (x + \frac{5}{2})^2 - \frac{33}{4}$

$\therefore$ vertex is $(-\frac{5}{2}, -\frac{33}{4})$, y-intercept is -2

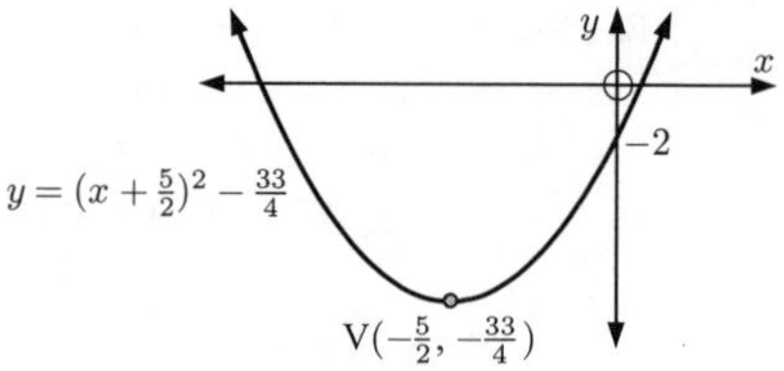

f $y = x^2 - 3x + 2$

$\therefore\ y = x^2 - 3x + (\frac{3}{2})^2 + 2 - (\frac{3}{2})^2$

$\therefore\ y = (x - \frac{3}{2})^2 - \frac{1}{4}$

$\therefore$ vertex is $(\frac{3}{2}, -\frac{1}{4})$, y-intercept is 2

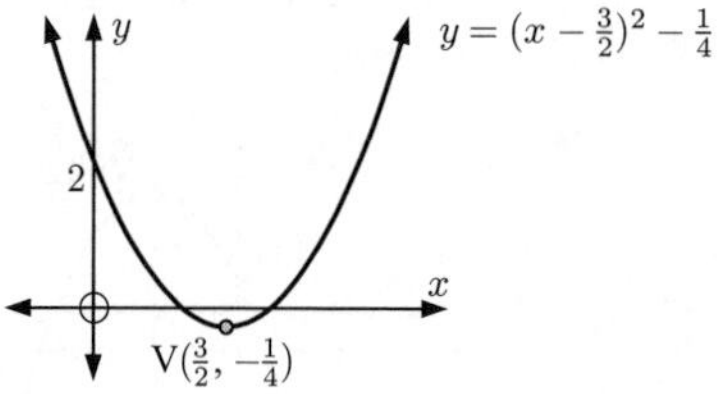

g $y = x^2 - 6x + 5$

$\therefore\ y = x^2 - 6x + 3^2 + 5 - 3^2$

$\therefore\ y = (x - 3)^2 - 4$

$\therefore$ vertex is $(3, -4)$, y-intercept is 5

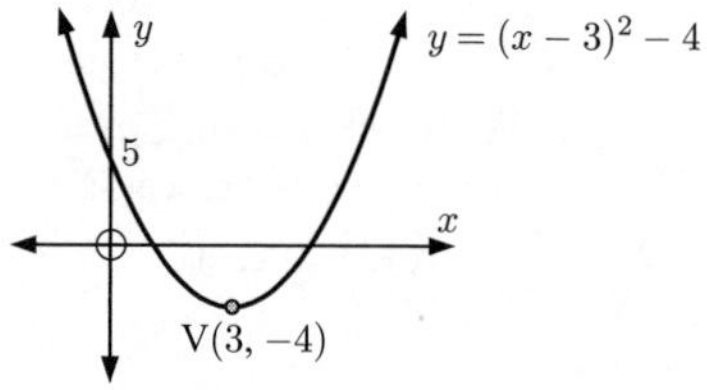

h $y = x^2 + 8x - 2$

$\therefore\ y = x^2 + 8x + 4^2 - 2 - 4^2$

$\therefore\ y = (x + 4)^2 - 18$

$\therefore$ vertex is $(-4, -18)$, y-intercept is -2

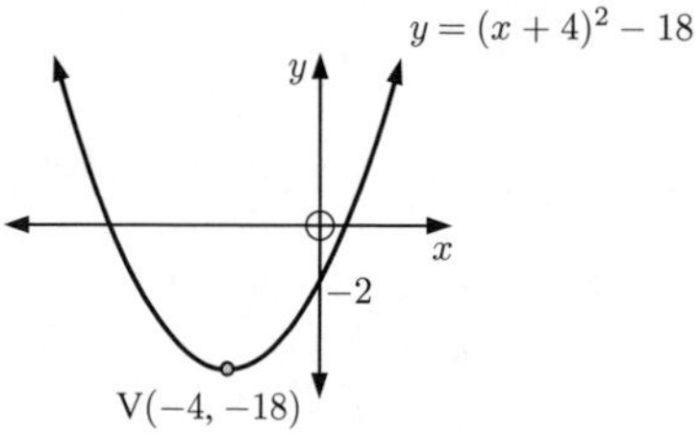

i $y = x^2 - 5x + 1$

$\therefore\ y = x^2 - 5x + (\frac{5}{2})^2 + 1 - (\frac{5}{2})^2$

$\therefore\ y = (x - \frac{5}{2})^2 - \frac{21}{4}$

$\therefore$ vertex is $(\frac{5}{2}, -5\frac{1}{4})$, y-intercept is 1

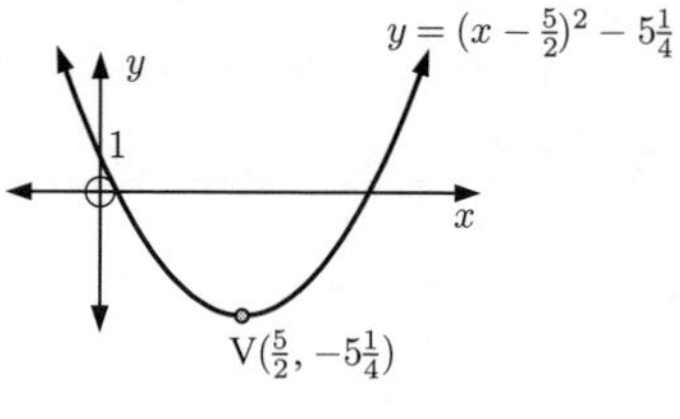

2 **a** **i** $y = 2x^2 + 4x + 5$

$= 2[x^2 + 2x + \frac{5}{2}]$

$= 2[x^2 + 2x + 1^2 - 1^2 + \frac{5}{2}]$

$= 2[(x + 1)^2 + \frac{3}{2}]$

$= 2(x + 1)^2 + 3$

ii The vertex is $(-1, 3)$.

iii When $x = 0$, $y = 5$

$\therefore$ the y-intercept is 5

iv

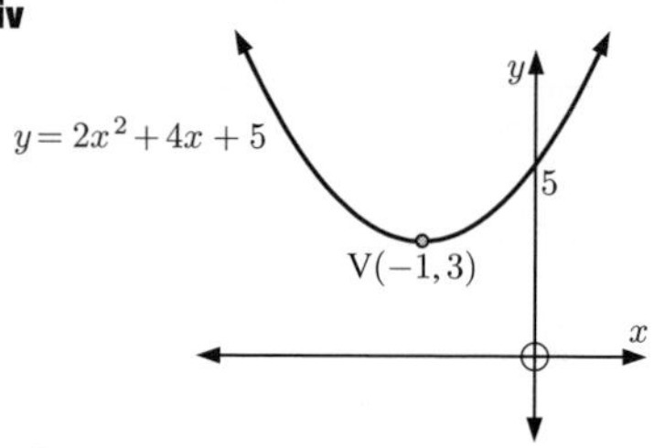

b **i** $y = 2x^2 - 8x + 3$
$= 2[x^2 - 4x + \frac{3}{2}]$
$= 2[x^2 - 4x + 2^2 - 2^2 + \frac{3}{2}]$
$= 2[(x-2)^2 - \frac{5}{2}]$
$= 2(x-2)^2 - 5$

ii The vertex is $(2, -5)$.

iii When $x = 0$, $y = 3$
$\therefore$ the y-intercept is 3

iv

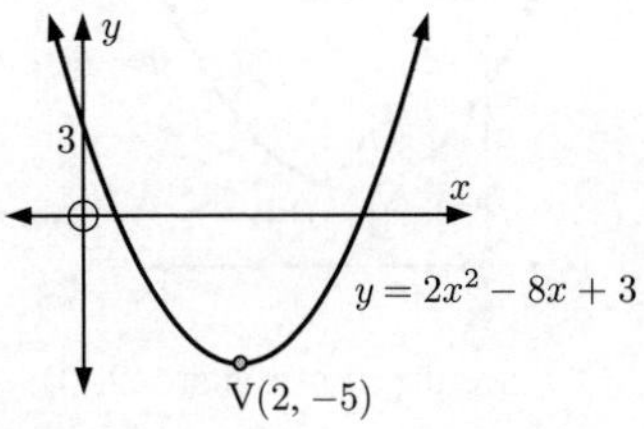

c **i** $y = 2x^2 - 6x + 1$
$= 2[x^2 - 3x + \frac{1}{2}]$
$= 2[x^2 - 3x + (\frac{3}{2})^2 - (\frac{3}{2})^2 + \frac{1}{2}]$
$= 2[(x - \frac{3}{2})^2 - \frac{7}{4}]$
$= 2(x - \frac{3}{2})^2 - \frac{7}{2}$

ii The vertex is $(\frac{3}{2}, -\frac{7}{2})$.

iii When $x = 0$, $y = 1$
$\therefore$ the y-intercept is 1

iv

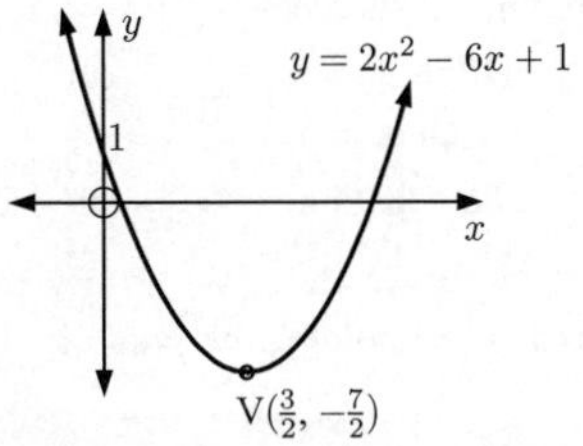

d **i** $y = 3x^2 - 6x + 5$
$= 3[x^2 - 2x + \frac{5}{3}]$
$= 3[x^2 - 2x + 1^2 - 1^2 + \frac{5}{3}]$
$= 3[(x-1)^2 + \frac{2}{3}]$
$= 3(x-1)^2 + 2$

ii The vertex is $(1, 2)$.

iii When $x = 0$, $y = 5$
$\therefore$ the y-intercept is 5

iv

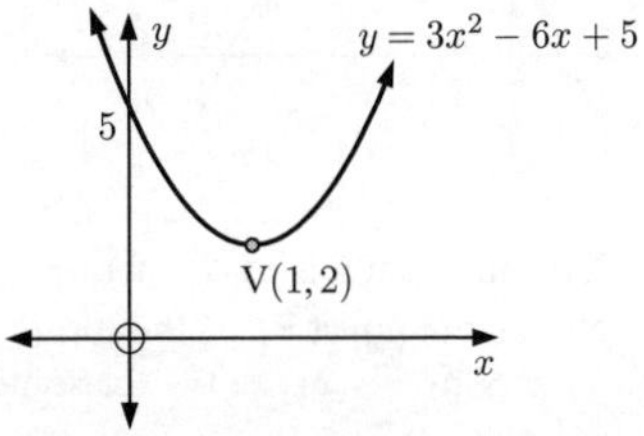

e **i** $y = -x^2 + 4x + 2$
$= -[x^2 - 4x - 2]$
$= -[x^2 - 4x + 2^2 - 2^2 - 2]$
$= -[(x-2)^2 - 6]$
$= -(x-2)^2 + 6$

ii The vertex is $(2, 6)$.

iii When $x = 0$, $y = 2$
$\therefore$ the y-intercept is 2

iv

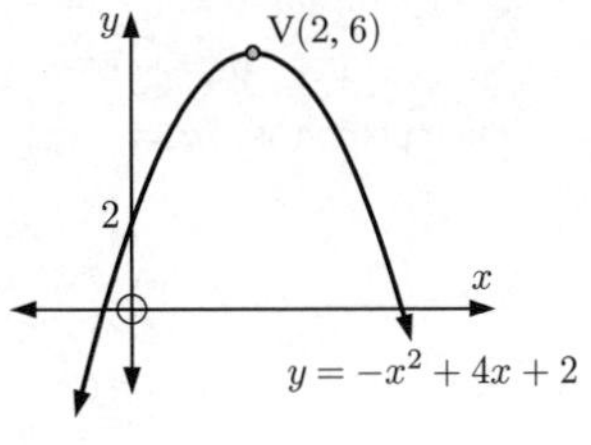

f **i** $y = -2x^2 - 5x + 3$
$= -2[x^2 + \frac{5}{2}x - \frac{3}{2}]$
$= -2[x^2 + \frac{5}{2}x + (\frac{5}{4})^2 - (\frac{5}{4})^2 - \frac{3}{2}]$
$= -2[(x + \frac{5}{4})^2 - \frac{25}{16} - \frac{24}{16}]$
$= -2[(x + \frac{5}{4})^2 - \frac{49}{16}]$
$= -2(x + \frac{5}{4})^2 + \frac{49}{8}$

ii The vertex is $(-\frac{5}{4}, \frac{49}{8})$.

iii When $x = 0$, $y = 3$
$\therefore$ the y-intercept is 3

iv

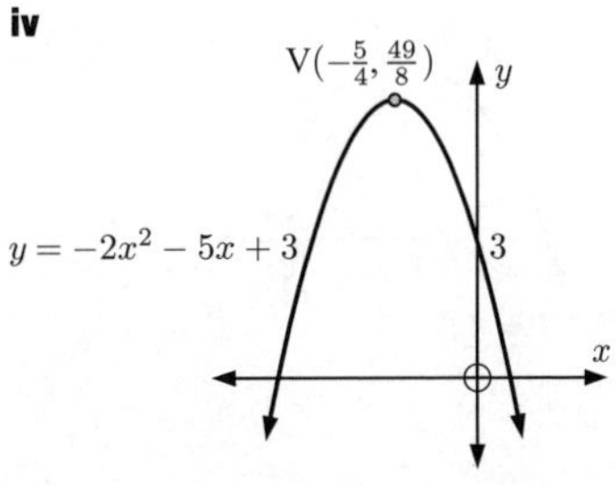

3 **a** Using technology, the graph is

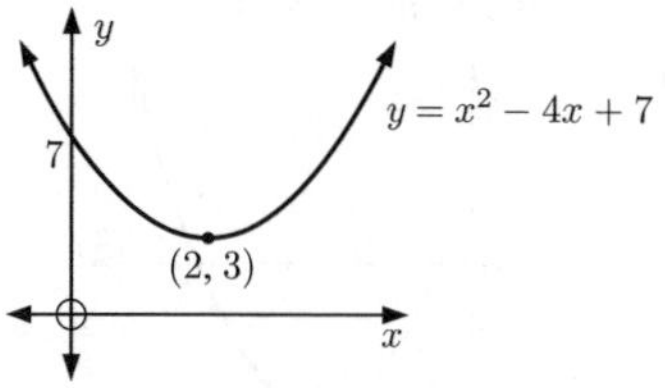

Since the vertex is at $(2, 3)$,
the function must be of the form
$y = a(x-2)^2 + 3$ for some a.
$\therefore$ when $x = 0$,
$y = a(-2)^2 + 3 = 4a + 3$
but the y-intercept is 7
$\therefore$ $4a + 3 = 7$
$\therefore$ $a = 1$
$\therefore$ the equation is $y = (x-2)^2 + 3$

b Using technology, the graph is

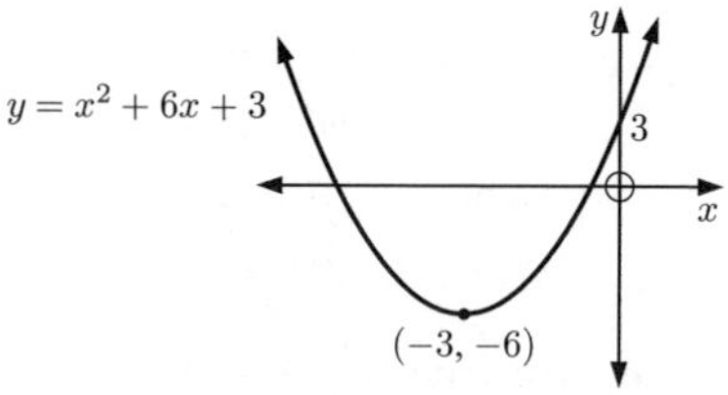

Since the vertex is at $(-3, -6)$,
the function must be of the form
$y = a(x+3)^2 - 6$ for some a.
$\therefore$ when $x = 0$,
$y = a \times 3^2 - 6 = 9a - 6$
but the y-intercept is 3
$\therefore$ $9a - 6 = 3$
$\therefore$ $a = 1$
$\therefore$ the equation is $y = (x+3)^2 - 6$

c Using technology, the graph is

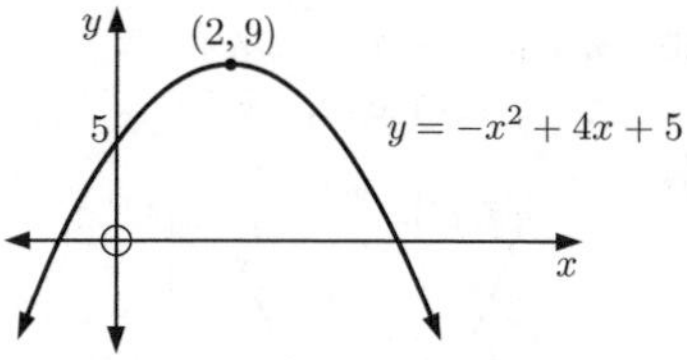

Since the vertex is at $(2, 9)$,
the function must be of the form
$y = a(x-2)^2 + 9$ for some a.
$\therefore$ when $x = 0$,
$y = a(-2)^2 + 9 = 4a + 9$
but the y-intercept is 5
$\therefore$ $4a + 9 = 5$
$\therefore$ $a = -1$
$\therefore$ the equation is $y = -(x-2)^2 + 9$

d Using technology, the graph is

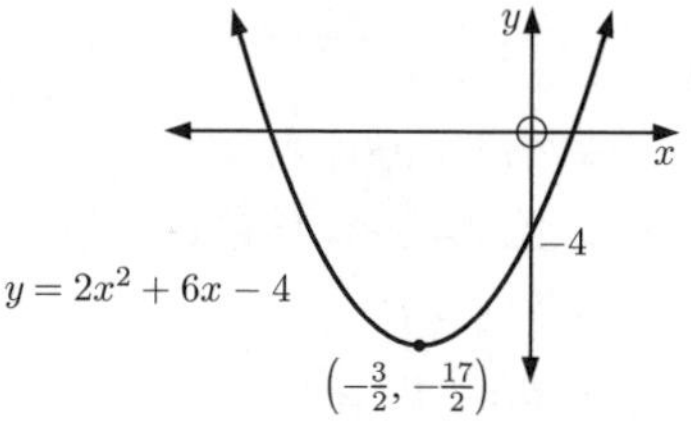

Since the vertex is at $(-\frac{3}{2}, -\frac{17}{2})$,
the function must be of the form
$y = a(x + \frac{3}{2})^2 - \frac{17}{2}$ for some a.
$\therefore$ when $x = 0$,
$y = a(\frac{3}{2})^2 - \frac{17}{2} = \frac{9}{4}a - \frac{17}{2}$
but the y-intercept is -4
$\therefore$ $\frac{9}{4}a - \frac{17}{2} = -4$
$\therefore$ $\frac{9}{4}a = \frac{9}{2}$
$\therefore$ $a = 2$
$\therefore$ the equation is $y = 2(x + \frac{3}{2})^2 - \frac{17}{2}$

e Using technology, the graph is

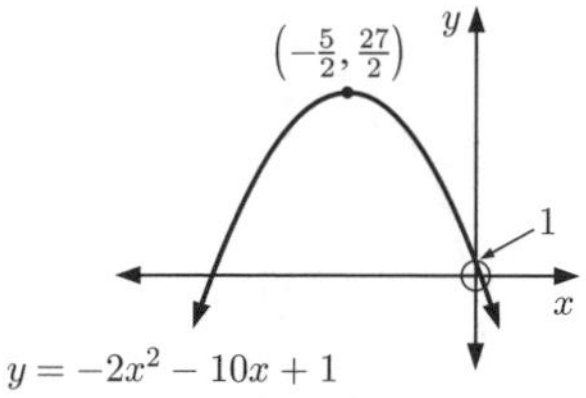

Since the vertex is at $(-\frac{5}{2}, \frac{27}{2})$,
the function must be of the form
$y = a(x + \frac{5}{2})^2 + \frac{27}{2}$ for some a.
$\therefore$ when $x = 0$,
$y = a(\frac{5}{2})^2 + \frac{27}{2} = \frac{25}{4}a + \frac{27}{2}$
but the y-intercept is 1
$\therefore$ $\frac{25}{4}a + \frac{27}{2} = 1$
$\therefore$ $\frac{25}{4}a = -\frac{25}{2}$
$\therefore$ $a = -2$
$\therefore$ the equation is $y = -2(x + \frac{5}{2})^2 + \frac{27}{2}$

f Using technology, the graph is

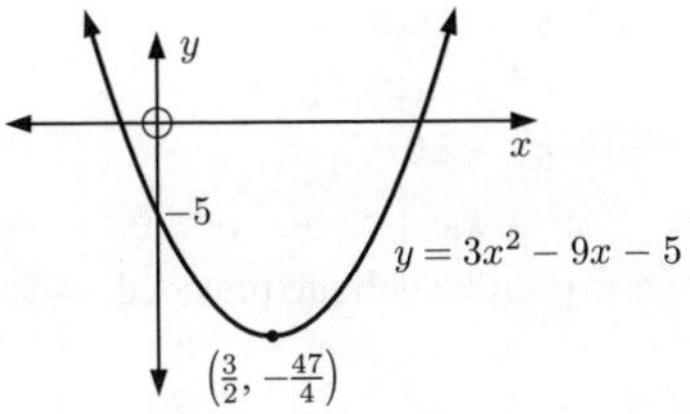

Since the vertex is at $(\frac{3}{2}, -\frac{47}{4})$, the function must be of the form

$y = a(x - \frac{3}{2})^2 - \frac{47}{4}$ for some a.

$\therefore$ when $x = 0$,

$y = a(-\frac{3}{2})^2 - \frac{47}{4} = \frac{9}{4}a - \frac{47}{4}$

but the y-intercept is -5

$\therefore \frac{9}{4}a - \frac{47}{4} = -5$

$\therefore \frac{9}{4}a = \frac{27}{4}$

$\therefore a = 3$

$\therefore$ the equation is $y = 3(x - \frac{3}{2})^2 - \frac{47}{4}$

EXERCISE 6C.3

1 **a** $y = x^2 + 7x - 2$ has $a = 1,\ b = 7,\ c = -2$

$\therefore \Delta = b^2 - 4ac$
$= 7^2 - 4(1)(-2)$
$= 57 > 0$

$\therefore$ the graph cuts the x-axis twice.

b $y = x^2 + 8x + 16$ has $a = 1,\ b = 8,\ c = 16$

$\therefore \Delta = b^2 - 4ac$
$= 8^2 - 4(1)(16)$
$= 0$

$\therefore$ the graph touches the x-axis.

c $y = -2x^2 + 3x + 1$ has $a = -2,\ b = 3,\ c = 1$

$\therefore \Delta = b^2 - 4ac$
$= 3^2 - 4(-2)(1)$
$= 17 > 0$

$\therefore$ the graph cuts the x-axis twice.

d $y = 6x^2 + 5x - 4$ has $a = 6,\ b = 5,\ c = -4$

$\therefore \Delta = b^2 - 4ac$
$= 5^2 - 4(6)(-4)$
$= 121 > 0$

$\therefore$ the graph cuts the x-axis twice.

e $y = -x^2 + x + 6$ has $a = -1,\ b = 1,\ c = 6$

$\therefore \Delta = b^2 - 4ac$
$= 1^2 - 4(-1)(6)$
$= 25 > 0$

$\therefore$ the graph cuts the x-axis twice.

f $y = 9x^2 + 6x + 1$ has $a = 9,\ b = 6,\ c = 1$

$\therefore \Delta = b^2 - 4ac$
$= 6^2 - 4(9)(1)$
$= 0$

$\therefore$ the graph touches the x-axis.

2 **a** $x^2 - 3x + 6$ has $a = 1,\ b = -3,\ c = 6$

$\therefore \Delta = b^2 - 4ac$
$= (-3)^2 - 4(1)(6)$
$= -15$

Since $a > 0$ and $\Delta < 0$, $x^2 - 3x + 6$ is positive definite.

$\therefore x^2 - 3x + 6 > 0$ for all x.

b $4x - x^2 - 6$ has $a = -1,\ b = 4,\ c = -6$

$\therefore \Delta = b^2 - 4ac$
$= 4^2 - 4(-1)(-6)$
$= -8$

Since $a < 0$ and $\Delta < 0$, $4x - x^2 - 6$ is negative definite.

$\therefore 4x - x^2 - 6 < 0$ for all x.

c $2x^2 - 4x + 7$ has $a = 2,\ b = -4,\ c = 7$

$\therefore \Delta = b^2 - 4ac$
$= (-4)^2 - 4(2)(7)$
$= -40$

Since $a > 0$ and $\Delta < 0$, $2x^2 - 4x + 7$ is positive definite.

d $-2x^2 + 3x - 4$ has $a = -2,\ b = 3,\ c = -4$

$\therefore \Delta = b^2 - 4ac$
$= 3^2 - 4(-2)(-4)$
$= -23$

Since $a < 0$ and $\Delta < 0$, $-2x^2 + 3x - 4$ is negative definite.

3 $3x^2 + kx - 1$
has $a = 3, \quad b = k, \quad c = -1$

$\therefore \quad \Delta = b^2 - 4ac$
$= k^2 - 4(3)(-1)$
$= k^2 + 12$

$\therefore \quad \Delta > 0$ for all k
$\{$as $k^2 \geqslant 0$ for all $k\}$

$\therefore \quad 3x^2 + kx - 1$ has two real distinct roots for all k.

$\therefore$ it can never be positive definite.

4 $2x^2 + kx + 2$
has $a = 2, \quad b = k, \quad c = 2$

$\therefore \quad \Delta = b^2 - 4ac$
$= k^2 - 4(2)(2)$
$= k^2 - 16$

Now $2x^2 + kx + 2$ has $a > 0$.

$\therefore$ it is positive definite provided $k^2 - 16 < 0$
$\therefore \quad k^2 < 16$
$\therefore \quad -4 < k < 4$

EXERCISE 6D

1 **a** The x-intercepts are 1 and 2.
$\therefore \quad y = a(x-1)(x-2)$
for some $a \neq 0$.
But the y-intercept is 4.
$\therefore \quad a(-1)(-2) = 4$
$\therefore \quad 2a = 4$
$\therefore \quad a = 2$
$\therefore \quad y = 2(x-1)(x-2)$

b The graph touches the x-axis when $x = 2$.
$\therefore \quad y = a(x-2)^2$
for some $a \neq 0$.
But the y-intercept is 8.
$\therefore \quad a(-2)^2 = 8$
$\therefore \quad 4a = 8$
$\therefore \quad a = 2$
$\therefore \quad y = 2(x-2)^2$

c The x-intercepts are 1 and 3.
$\therefore \quad y = a(x-1)(x-3)$
for some $a \neq 0$.
But the y-intercept is 3.
$\therefore \quad a(-1)(-3) = 3$
$\therefore \quad 3a = 3$
$\therefore \quad a = 1$
$\therefore \quad y = (x-1)(x-3)$

d The x-intercepts are -1 and 3.
$\therefore \quad y = a(x+1)(x-3)$
for some $a \neq 0$.
But the y-intercept is 3.
$\therefore \quad a(1)(-3) = 3$
$\therefore \quad -3a = 3$
$\therefore \quad a = -1$
$\therefore \quad y = -(x+1)(x-3)$

e The graph touches the x-axis when $x = 1$.
$\therefore \quad y = a(x-1)^2$
for some $a \neq 0$.
But the y-intercept is -3.
$\therefore \quad a(-1)^2 = -3$
$\therefore \quad a = -3$
$\therefore \quad y = -3(x-1)^2$

f The x-intercepts are -2 and 3.
$\therefore \quad y = a(x+2)(x-3)$
for some $a \neq 0$.
But the y-intercept is 12.
$\therefore \quad a(2)(-3) = 12$
$\therefore \quad -6a = 12$
$\therefore \quad a = -2$
$\therefore \quad y = -2(x+2)(x-3)$

2 **a** As the axis of symmetry is $x = 3$, the other x-intercept is 4.
$\therefore \quad y = a(x-2)(x-4)$
for some $a \neq 0$.
But the y-intercept $= 12$
$\therefore \quad a(-2)(-4) = 12$
$\therefore \quad 8a = 12$
$\therefore \quad a = \frac{12}{8} = \frac{3}{2}$
$\therefore \quad y = \frac{3}{2}(x-2)(x-4)$

b As the axis of symmetry is $x = -1$, the other x-intercept is 2.
$\therefore \quad y = a(x+4)(x-2)$
for some $a \neq 0$.
But the y-intercept $= 4$
$\therefore \quad a(4)(-2) = 4$
$\therefore \quad -8a = 4$
$\therefore \quad a = -\frac{1}{2}$
$\therefore \quad y = -\frac{1}{2}(x+4)(x-2)$

c The graph touches the x-axis at $x = -3$,
$\therefore \quad y = a(x+3)^2$ for some $a \neq 0$.
But the y-intercept is -12, so $a(3)^2 = -12$
$\therefore \quad 9a = -12$
$\therefore \quad a = -\frac{12}{9} = -\frac{4}{3}$
$\therefore \quad y = -\frac{4}{3}(x+3)^2$

3 **a** Since the x-intercepts are 5 and 1, the equation is $y = a(x-5)(x-1)$ for some $a \neq 0$.

But when $x = 2$, $y = -9$

$\therefore\ -9 = a(2-5)(2-1)$

$\therefore\ -9 = a(-3)(1)$

$\therefore\ -3a = -9$

$\therefore\ a = 3$

$\therefore$ the equation is $y = 3(x-5)(x-1)$

$\therefore\ y = 3(x^2 - 6x + 5)$

$\therefore\ y = 3x^2 - 18x + 15$

b Since the x-intercepts are 2 and $-\frac{1}{2}$, the equation is $y = a(x-2)(x+\frac{1}{2})$ for some $a \neq 0$.

But when $x = 3$, $y = -14$

$\therefore\ -14 = a(3-2)(3+\frac{1}{2})$

$\therefore\ -14 = a(1)(\frac{7}{2})$

$\therefore\ \frac{7}{2}a = -14$

$\therefore\ a = -4$

$\therefore$ the equation is $y = -4(x-2)(x+\frac{1}{2})$

$\therefore\ y = -4(x^2 - \frac{3}{2}x - 1)$

$\therefore\ y = -4x^2 + 6x + 4$

c Since the graph touches the x-axis at 3, its equation is $y = a(x-3)^2$, for some $a \neq 0$.

But when $x = -2$, $y = -25$

$\therefore\ -25 = a(-2-3)^2$

$\therefore\ -25 = 25a$

$\therefore\ a = -1$

$\therefore$ the equation is $y = -(x-3)^2$

$\therefore\ y = -(x^2 - 6x + 9)$

$\therefore\ y = -x^2 + 6x - 9$

d Since the graph touches the x-axis at -2, its equation is $y = a(x+2)^2$, for some $a \neq 0$.

But when $x = -1$, $y = 4$

$\therefore\ 4 = a(-1+2)^2$

$\therefore\ 4 = a$

$\therefore$ the equation is $y = 4(x+2)^2$

$\therefore\ y = 4(x^2 + 4x + 4)$

$\therefore\ y = 4x^2 + 16x + 16$

e Since the graph cuts the x-axis at 3 and has axis of symmetry $x = 2$, it must also cut the x-axis at 1.

$\therefore$ the x-intercepts are 3 and 1, and the equation is $y = a(x-3)(x-1)$ for some $a \neq 0$.

But when $x = 5$, $y = 12$

$\therefore\ 12 = a(5-3)(5-1)$

$\therefore\ 12 = a(2)(4)$

$\therefore\ 8a = 12$

$\therefore\ a = \frac{3}{2}$

$\therefore$ the equation is $y = \frac{3}{2}(x-3)(x-1)$

$\therefore\ y = \frac{3}{2}(x^2 - 4x + 3)$

$\therefore\ y = \frac{3}{2}x^2 - 6x + \frac{9}{2}$

f Since the graph cuts the x-axis at 5 and has axis of symmetry $x = 1$, it must also cut the x-axis at -3.

$\therefore$ the x-intercepts are 5 and -3, and the equation is $y = a(x-5)(x+3)$ for some $a \neq 0$.

But when $x = 2$, $y = 5$

$\therefore\ 5 = a(2-5)(2+3)$

$\therefore\ 5 = a(-3)(5)$

$\therefore\ -15a = 5$

$\therefore\ a = -\frac{1}{3}$

$\therefore$ the equation is $y = -\frac{1}{3}(x-5)(x+3)$

$\therefore\ y = -\frac{1}{3}(x^2 - 2x - 15)$

$\therefore\ y = -\frac{1}{3}x^2 + \frac{2}{3}x + 5$

4 **a** The vertex is $(2, 4)$, so the quadratic has equation $y = a(x-2)^2 + 4$ for some $a \neq 0$.

But the graph passes through the origin

$\therefore\ 0 = a(0-2)^2 + 4$

$\therefore\ 4a + 4 = 0$

$\therefore\ a = -1$

$\therefore$ the equation is $y = -(x-2)^2 + 4$

b The vertex is $(2, -1)$, so the quadratic has equation $y = a(x-2)^2 - 1$ for some $a \neq 0$.

But the graph passes through $(0, 7)$

$\therefore\ 7 = a(0-2)^2 - 1$

$\therefore\ 7 = 4a - 1$

$\therefore\ a = 2$

$\therefore$ the equation is $y = 2(x-2)^2 - 1$

c The vertex is $(3, 8)$,
so the quadratic has equation
$y = a(x-3)^2 + 8$ for some $a \neq 0$.
But the graph passes through $(1, 0)$
$\therefore \ 0 = a(1-3)^2 + 8$
$\therefore \ 0 = 4a + 8$
$\therefore \ a = -2$
$\therefore$ the equation is $y = -2(x-3)^2 + 8$

d The vertex is $(4, -6)$,
so the quadratic has equation
$y = a(x-4)^2 - 6$ for some $a \neq 0$.
But the graph passes through $(7, 0)$
$\therefore \ 0 = a(7-4)^2 - 6$
$\therefore \ 9a - 6 = 0$
$\therefore \ a = \frac{2}{3}$
$\therefore$ the equation is $y = \frac{2}{3}(x-4)^2 - 6$

e The vertex is $(2, 3)$,
so the quadratic has equation
$y = a(x-2)^2 + 3$ for some $a \neq 0$.
But the graph passes through $(3, 1)$
$\therefore \ 1 = a(3-2)^2 + 3$
$\therefore \ 1 = a + 3$
$\therefore \ a = -2$
$\therefore$ the equation is $y = -2(x-2)^2 + 3$

f The vertex is $(\frac{1}{2}, -\frac{3}{2})$,
so the quadratic has equation
$y = a(x-\frac{1}{2})^2 - \frac{3}{2}$ for some $a \neq 0$.
But the graph passes through $(\frac{3}{2}, \frac{1}{2})$
$\therefore \ \frac{1}{2} = a(\frac{3}{2} - \frac{1}{2})^2 - \frac{3}{2}$
$\therefore \ \frac{1}{2} = a - \frac{3}{2}$
$\therefore \ a = 2$
$\therefore$ the equation is $y = 2(x-\frac{1}{2})^2 - \frac{3}{2}$

EXERCISE 6E

1 **a** $y = x^2 - 2x + 8$ meets $y = x + 6$
when $x^2 - 2x + 8 = x + 6$
$\therefore \ x^2 - 3x + 2 = 0$
$\therefore \ (x-1)(x-2) = 0$
$\therefore \ x = 1$ or 2

Substituting into $y = x + 6$,
when $x = 1$, $y = 7$
and when $x = 2$, $y = 8$
$\therefore$ the graphs meet at $(1, 7)$ and $(2, 8)$

b $y = -x^2 + 3x + 9$ meets $y = 2x - 3$
when $-x^2 + 3x + 9 = 2x - 3$
$\therefore \ x^2 - x - 12 = 0$
$\therefore \ (x-4)(x+3) = 0$
$\therefore \ x = 4$ or -3

Substituting into $y = 2x - 3$,
when $x = -3$, $y = 2(-3) - 3 = -9$
and when $x = 4$, $y = 2(4) - 3 = 5$
$\therefore$ the graphs meet at $(-3, -9)$ and $(4, 5)$

c $y = x^2 - 4x + 3$ meets $y = 2x - 6$
when $x^2 - 4x + 3 = 2x - 6$
$\therefore \ x^2 - 6x + 9 = 0$
$\therefore \ (x-3)^2 = 0$
$\therefore \ x = 3$
Substituting into $y = 2x - 6$,
when $x = 3$, $y = 0$
$\therefore$ the graphs touch at $(3, 0)$

d $y = -x^2 + 4x - 7$ meets $y = 5x - 4$
when $-x^2 + 4x - 7 = 5x - 4$
$\therefore \ x^2 + x + 3 = 0$
which has $a = 1$, $b = 1$, $c = 3$
$\therefore \ x = \dfrac{-1 \pm \sqrt{1^2 - 4(1)(3)}}{2}$
$= \dfrac{-1 \pm \sqrt{-11}}{2}$
$\therefore$ there are no real solutions
$\therefore$ the graphs do not meet.

2 **a** $(0.59, 5.59)$ and $(3.41, 8.41)$ **b** $(3, -4)$ touching
c graphs do not meet **d** $(-2.56, -18.81)$ and $(1.56, 1.81)$

3 **a** $y = x^2$ meets $y = x + 2$
when $x^2 = x + 2$
$\therefore \ x^2 - x - 2 = 0$
$\therefore \ (x+1)(x-2) = 0$
$\therefore \ x = -1$ or 2

Substituting into $y = x + 2$,
when $x = -1$, $y = 1$
and when $x = 2$, $y = 4$
$\therefore$ the graphs meet at $(-1, 1)$ and $(2, 4)$.

b $y = x^2 + 2x - 3$ meets $y = x - 1$
when $x^2 + 2x - 3 = x - 1$
$\therefore$ $x^2 + x - 2 = 0$
$\therefore$ $(x-1)(x+2) = 0$
$\therefore$ $x = 1$ or -2

Substituting into $y = x - 1$,
when $x = 1$, $y = 0$
and when $x = -2$, $y = -3$
$\therefore$ the graphs meet at $(1, 0)$ and $(-2, -3)$.

c $y = 2x^2 - x + 3$ meets $y = 2 + x + x^2$
when $2x^2 - x + 3 = 2 + x + x^2$
$\therefore$ $x^2 - 2x + 1 = 0$
$\therefore$ $(x-1)^2 = 0$
$\therefore$ $x = 1$

Substituting into $y = 2 + x + x^2$
when $x = 1$, $y = 2 + 1 + 1 = 4$
$\therefore$ the graphs meet at $(1, 4)$.

d Substituting $y = x + 3$ into $xy = 4$
gives $x(x+3) = 4$
$\therefore$ $x^2 + 3x - 4 = 0$
$\therefore$ $(x+4)(x-1) = 0$
$\therefore$ $x = -4$ or 1

Substituting into $y = x + 3$,
when $x = -4$, $y = -1$
and when $x = 1$, $y = 4$
$\therefore$ the graphs meet at $(-4, -1)$ and $(1, 4)$.

5 $y = 3x + c$ is a tangent to $y = x^2 - 5x + 7$ if they meet at exactly one point (touch).
$y = x^2 - 5x + 7$ meets $y = 3x + c$ when $x^2 - 5x + 7 = 3x + c$
$\therefore$ $x^2 - 8x + 7 - c = 0$

The graphs meet exactly once when this equation has a repeated root $\therefore$ $\Delta = 0$
$\therefore$ $(-8)^2 - 4(1)(7 - c) = 0$
$\therefore$ $64 - 28 + 4c = 0$
$\therefore$ $4c = -36$
$\therefore$ $c = -9$

6 $y = mx - 2$ is a tangent to $y = x^2 - 4x + 2$ if they meet at exactly one point (touch).
$y = x^2 - 4x + 2$ meets $y = mx - 2$ when $x^2 - 4x + 2 = mx - 2$
$\therefore$ $x^2 - (m+4)x + 4 = 0$

The graphs meet exactly once when this equation has a repeated root $\therefore$ $\Delta = 0$
$\therefore$ $(m+4)^2 - 4(1)(4) = 0$
$\therefore$ $m^2 + 8m + 16 - 16 = 0$
$\therefore$ $m(m+8) = 0$
$\therefore$ $m = 0$ or -8

7 Lines with y-intercept 1 have the form $y = mx + 1$.
$y = mx + 1$ is a tangent to $y = 3x^2 + 5x + 4$ if they meet at exactly one point (touch).
$y = 3x^2 + 5x + 4$ meets $y = mx + 1$ when $3x^2 + 5x + 4 = mx + 1$
$\therefore$ $3x^2 + (5 - m)x + 3 = 0$

The graphs meet exactly once when this equation has a repeated root $\therefore$ $\Delta = 0$
$\therefore$ $(5 - m)^2 - 4(3)(3) = 0$
$\therefore$ $25 - 10m + m^2 - 36 = 0$
$\therefore$ $m^2 - 10m - 11 = 0$
$\therefore$ $(m+1)(m-11) = 0$
$\therefore$ $m = -1$ or 11
$\therefore$ the required lines have gradient -1 or 11.

8 **a** $y = x + c$ meets $y = 2x^2 - 3x - 7$
when $2x^2 - 3x - 7 = x + c$
$\therefore\ 2x^2 - 4x - 7 - c = 0$
The graphs will never meet if this equation has no real roots $\therefore\ \Delta < 0$
$\therefore\ (-4)^2 - 4(2)(-7 - c) < 0$
$\therefore\ 16 + 56 + 8c < 0$
$\therefore\ 8c < -72$
$\therefore\ c < -9$

b Choose c such that $c < -9$, for example $c = -10$:

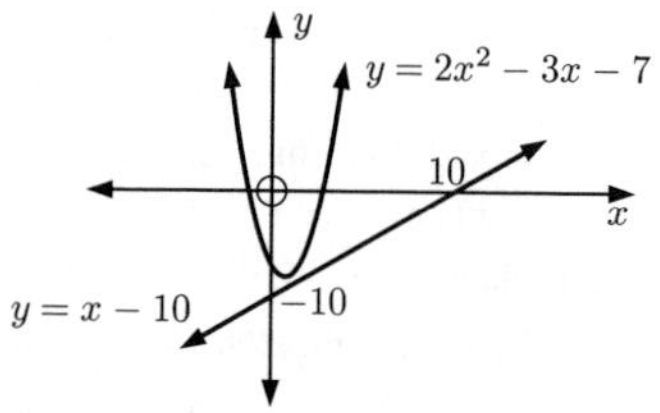

EXERCISE 6F

1 Let the smaller of the integers be x.
The other integer is $(x + 12)$.
$\therefore$ the sum of their squares is
$$x^2 + (x + 12)^2 = 74$$
$\therefore\ x^2 + x^2 + 24x + 144 = 74$
$\therefore\ 2x^2 + 24x + 70 = 0$
$\therefore\ x^2 + 12x + 35 = 0$
$\therefore\ (x + 7)(x + 5) = 0$
$\therefore\ x = -7$ or -5
So, the integers are -7 and 5, or -5 and 7.

2 Let the number be x, so its reciprocal is $\dfrac{1}{x}$.
They have sum $x + \dfrac{1}{x} = 5\frac{1}{5}$
$\therefore\ x^2 + 1 = \frac{26}{5}x$
$\therefore\ x^2 - \frac{26}{5}x + 1 = 0$
$\therefore\ 5x^2 - 26x + 5 = 0$
$\therefore\ (5x - 1)(x - 5) = 0$
$\therefore\ x = \frac{1}{5}$ or 5
So, the number is either $\frac{1}{5}$ or 5.

3 Let the number be x so its square is x^2.
$\therefore$ the sum is $x + x^2 = 210$
$\therefore\ x^2 + x - 210 = 0$
$\therefore\ (x + 15)(x - 14) = 0$
$\therefore\ x = -15$ or 14
But x is a natural number, so $x > 0$,
$\therefore$ the number is 14.

4 Suppose the numbers are x and $(x + 2)$.
Then $x(x + 2) = 360$
$\therefore\ x^2 + 2x - 360 = 0$
$\therefore\ (x + 20)(x - 18) = 0$
$\therefore\ x = -20$ or 18
$\therefore$ the numbers are -20 and -18, or 18 and 20.

5 Suppose the numbers are x and $(x + 2)$.
Then $x(x + 2) = 255$
$\therefore\ x^2 + 2x - 255 = 0$
$\therefore\ (x + 17)(x - 15) = 0$
$\therefore\ x = -17$ or 15
$\therefore$ the numbers are -17 and -15, or 15 and 17.

6 If the polygon has n sides, then
$$\frac{n}{2}(n - 3) = 90$$
$\therefore\ \frac{1}{2}n^2 - \frac{3}{2}n = 90$
$\therefore\ n^2 - 3n - 180 = 0$
$\therefore\ (n - 15)(n + 12) = 0$
$\therefore\ n = -12$ or 15
$\therefore$ the polygon has 15 sides. {as $n > 0$}

7 If the width of the rectangle is w cm, then its length is $(w + 4)$ cm.
$\therefore$ the area is $w(w + 4) = 26$
$\therefore\ w^2 + 4w - 26 = 0$
which has $a = 1$, $b = 4$, $c = -26$
$$\therefore\ w = \frac{-4 \pm \sqrt{4^2 - 4(1)(-26)}}{2(1)}$$
$$= \frac{-4 \pm \sqrt{120}}{2} = -2 \pm \sqrt{30}$$
But $w > 0$, so $w = -2 + \sqrt{30}$
≈ 3.477 cm
So, the width is approximately 3.48 cm.

8 **a** The base has sides of length x cm, so the areas of the top and bottom surfaces are both x^2 cm^2.

The box has height $(x+1)$ cm, so the area of each of the side faces is $x(x+1)$ cm^2.

$\therefore$ the total surface area is

$$A = 2x^2 + 4x(x+1)$$
$$= 2x^2 + 4x^2 + 4x$$
$$= 6x^2 + 4x \text{ cm}^2$$

b $$6x^2 + 4x = 240$$
$$\therefore \quad 3x^2 + 2x - 120 = 0$$
$$\therefore \quad (3x+20)(x-6) = 0$$
$$\therefore \quad x = -\tfrac{20}{3} \text{ or } 6$$

but $x > 0$, so $x = 6$ cm

$\therefore$ the box is $6 \text{ cm} \times 6 \text{ cm} \times 7 \text{ cm}$

9 Suppose the tin plate was $x \text{ cm} \times x \text{ cm}$.
When $3 \text{ cm} \times 3 \text{ cm}$ squares are cut from the corners, the base of the open box formed is $(x-6) \text{ cm} \times (x-6) \text{ cm}$.
The open box has height 3 cm, so its volume is $3 \times (x-6) \times (x-6) = 80$

$$\therefore \quad 3(x^2 - 12x + 36) = 80$$
$$3x^2 - 36x + 108 = 80$$
$$\therefore \quad 3x^2 - 36x + 28 = 0$$

which has $a = 3, \quad b = -36, \quad c = 28$

$$\therefore \quad x = \frac{-(-36) \pm \sqrt{(-36)^2 - 4(3)(28)}}{2(3)}$$
$$= \frac{36 \pm \sqrt{960}}{6} \quad \text{and since } x > 6,$$
$$x = 6 + \frac{\sqrt{960}}{6} \approx 11.16 \text{ cm}$$

$\therefore$ the original piece of tinplate was about 11.2 cm square.

10

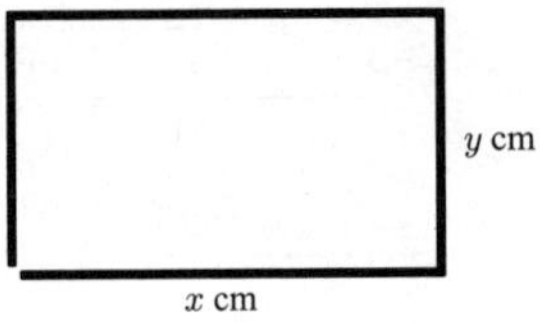

Suppose one side of the rectangle has length x cm and the other has length y cm.
The perimeter is $(2x+2y)$ cm,

so $2x + 2y = 20$

$$\therefore \quad 2y = 20 - 2x$$
$$\therefore \quad y = 10 - x$$

The area of the rectangle is therefore $x(10-x)$ cm^2.

If the area is 30 cm^2, then

$$x(10-x) = 30$$
$$\therefore \quad 10x - x^2 = 30$$
$$\therefore \quad x^2 - 10x + 30 = 0$$

which has $a = 1, \quad b = -10, \quad c = 30$

$$\therefore \quad x = \frac{-(-10) \pm \sqrt{(-10)^2 - 4(1)(30)}}{2(1)}$$
$$= \frac{10 \pm \sqrt{100 - 120}}{2}$$
$$= \frac{10 \pm \sqrt{-20}}{2}$$

$\therefore$ x has no real solutions, so it is not possible.

11 The smaller rectangle is similar to the original rectangle.

$$\therefore \quad \frac{\text{AB}}{\text{AD}} = \frac{\text{BC}}{\text{BY}}$$

Suppose $\text{AB} = x$ units, and $\text{AD} = \text{BC}$
$= 1$ unit

$$\therefore \quad \frac{x}{1} = \frac{1}{x-1}$$
$$\therefore \quad x(x-1) = 1$$
$$\therefore \quad x^2 - x - 1 = 0$$

which has $a = 1, \quad b = -1, \quad c = -1$

$$\therefore \quad x = \frac{-(-1) \pm \sqrt{(-1)^2 - 4(1)(-1)}}{2(1)}$$
$$= \frac{1 \pm \sqrt{1+4}}{2}$$
$$= \frac{1 \pm \sqrt{5}}{2}$$
$$\therefore \quad x = \frac{1+\sqrt{5}}{2}, \quad \text{since } x > 0$$

But $x = \dfrac{\text{AB}}{\text{AD}}$, which is the golden ratio

$\therefore$ the golden ratio is $\dfrac{1+\sqrt{5}}{2}$

12

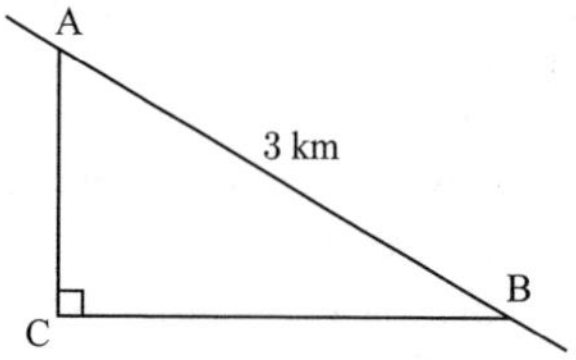

Suppose AC is x hundred metres,
so BC is $(x+4)$ hundred metres.

Now $AC^2 + BC^2 = AB^2$ {Pythagoras}

$\therefore \quad x^2 + (x+4)^2 = 30^2$

$\therefore \quad x^2 + x^2 + 8x + 16 = 900$

$\therefore \quad 2x^2 + 8x - 884 = 0$

$\therefore \quad x^2 + 4x - 442 = 0$

which has $a = 1, \quad b = 4, \quad c = -442$

$$\therefore \quad x = \frac{-4 \pm \sqrt{4^2 - 4(1)(-442)}}{2(1)}$$

$$= \frac{-4 \pm \sqrt{1784}}{2}$$

$$= \frac{-4 + \sqrt{1784}}{2} \approx 19.12 \quad \{\text{as } x > 0\}$$

$\therefore$ AC $\approx$ 19.12 hundred metres
and BC $\approx$ 23.12 hundred metres

$\therefore$ since the paddock is triangular,
its area is $\frac{1}{2} \times$ AC $\times$ BC
$= 221$ hectares.

13

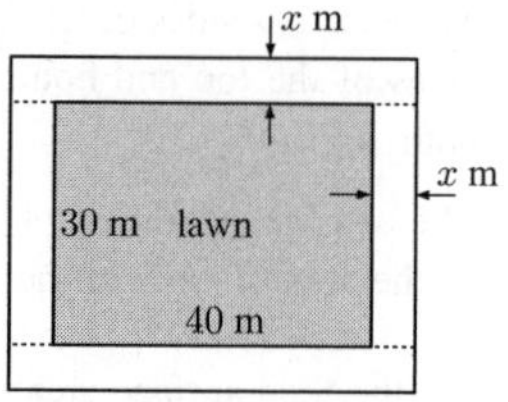

Suppose the concrete has width x m around the lawn. We divide the concrete up into four regions as shown.

The smaller regions have area $30x$ m^2, whilst the larger regions have area $x(40+2x)$ m^2.

Now the total area of concrete is one quarter the area of the lawn.

$\therefore \quad 2 \times 30x + 2 \times x(40+2x) = \frac{1}{4} \times 30 \times 40$

$\therefore \quad 60x + 80x + 4x^2 = 300$

$\therefore \quad 4x^2 + 140x - 300 = 0$

$\therefore \quad x^2 + 35x - 75 = 0$

which has $a = 1, \quad b = 35, \quad c = -75$

$$\therefore \quad x = \frac{-35 \pm \sqrt{35^2 - 4(1)(-75)}}{2(1)}$$

$$= \frac{-35 \pm \sqrt{1525}}{2}$$

But $x > 0$,

so $x = \dfrac{-35 + \sqrt{1525}}{2} \approx 2.026$ m

$\therefore$ the path is about 2.03 m wide.

14 Suppose Hassan's speed is h km h^{-1}. We know that speed $= \dfrac{\text{distance}}{\text{time}}$, so time $= \dfrac{\text{distance}}{\text{speed}}$

$\therefore$ if it takes Hassan t hours, $t = \dfrac{40}{h}$ .... (1)

Now Chuong says he will drive home at speed $(h+40)$ km h^{-1} and arrive in time $(t - \frac{1}{3})$ hrs.

$\therefore \quad t - \frac{1}{3} = \dfrac{40}{h+40}$, and so $t = \dfrac{40}{h+40} + \frac{1}{3}$ .... (2)

Using (1) and (2), $\dfrac{40}{h} = \dfrac{40}{h+40} + \frac{1}{3}$

$\therefore \quad 40(h+40) = 40h + \frac{1}{3}h(h+40)$

$\therefore \quad 40h + 1600 = 40h + \frac{1}{3}h^2 + \frac{40}{3}h$

$\therefore \quad h^2 + 40h - 4800 = 0$

which has $a = 1, \quad b = 40, \quad c = -4800.$

$$\therefore \quad h = \frac{-40 \pm \sqrt{40^2 - 4(1)(-4800)}}{2(1)}$$

$$= \frac{-40 \pm \sqrt{1600 + 19\,200}}{2}$$

$$= \frac{-40 \pm \sqrt{20\,800}}{2}$$

But $h > 0$, so $h = \dfrac{-40 + \sqrt{20\,800}}{2} \approx 52.1$ km h^{-1}.

So, Hassan's speed is approximately 52.1 km h^{-1}.

15 Suppose the speed of the plane is x km h^{-1}. We know speed $= \dfrac{\text{distance}}{\text{time}}$, so time $= \dfrac{\text{distance}}{\text{speed}}$.

Using the information given, $\dfrac{1000}{x} = \dfrac{1000}{x-120} - \frac{1}{2}$

$\therefore \quad 1000(x-120) = 1000x - \frac{1}{2}x(x-120)$

$\therefore \quad 1000x - 120\,000 = 1000x - \frac{1}{2}x^2 + 60x$

$\therefore \quad x^2 - 120x - 240\,000 = 0$ which has $a = 1$, $b = -120$, $c = -240\,000$

$$\therefore \quad x = \frac{-(-120) \pm \sqrt{(-120)^2 - 4(1)(-240\,000)}}{2(1)} = \frac{120 \pm \sqrt{974\,400}}{2}$$

But $x > 0$, so $x = \dfrac{120 + \sqrt{974\,400}}{2} \approx 553.6$ km h^{-1}.

$\therefore$ he plane has speed approximately 554 km h^{-1}.

16 Suppose the express train travels at x km h^{-1}. We know speed $= \dfrac{\text{distance}}{\text{time}}$, so time $= \dfrac{\text{distance}}{\text{speed}}$.

$\therefore$ it takes the express train $\dfrac{160}{x}$ hours and the normal train $\dfrac{160}{x-10}$ hours.

$\therefore \quad \dfrac{160}{x} + \frac{1}{2} = \dfrac{160}{x-10}$

$\therefore \quad 160(x-10) + \frac{1}{2}x(x-10) = 160x$

$\therefore \quad 160x - 1600 + \frac{1}{2}x^2 - 5x = 160x$

$\therefore \quad x^2 - 10x - 3200 = 0$ which has $a = 1$, $b = -10$, $c = -3200$

$$\therefore \quad x = \frac{-(-10) \pm \sqrt{(-10)^2 - 4(1)(-3200)}}{2(1)} = \frac{10 \pm \sqrt{12\,900}}{2}$$

But $x > 0$, so $x = \dfrac{10 + \sqrt{12\,900}}{2} \approx 61.8$ km h^{-1}

$\therefore$ the express train travels on average at about 61.8 km h^{-1}.

17 Suppose n elderly citizens ended up going on the trip, so the cost per person was $\$\dfrac{160}{n}$.
If the original number of elderly citizens had gone, there would have been $(n+8)$,
and the cost per person would have been $\$\dfrac{160}{n+8}$.

Hence $\dfrac{160}{n} = \dfrac{160}{n+8} + 1$

$\therefore \quad 160(n+8) = 160n + n(n+8)$

$\therefore \quad 160n + 1280 = 160n + n^2 + 8n$

$\therefore \quad n^2 + 8n - 1280 = 0$

$\therefore \quad (n-32)(n+40) = 0$

$\therefore$ since $n > 0$, $n = 32$. So, 32 elderly citizens went on the trip.

18 We fit a set of axes to the tunnel, with the origin at ground level at the midpoint of the tunnel.
The parabola has vertex $(0, 8)$, so it has equation

$y = a(x-0)^2 + 8$

$\therefore \quad y = ax^2 + 8$

When $x = 3$, $y = 0$, so

$0 = a(3^2) + 8$

$\therefore \quad 9a = -8$

$\therefore \quad a = -\frac{8}{9}$

$\therefore$ the equation of the parabola is $y = -\frac{8}{9}x^2 + 8$.

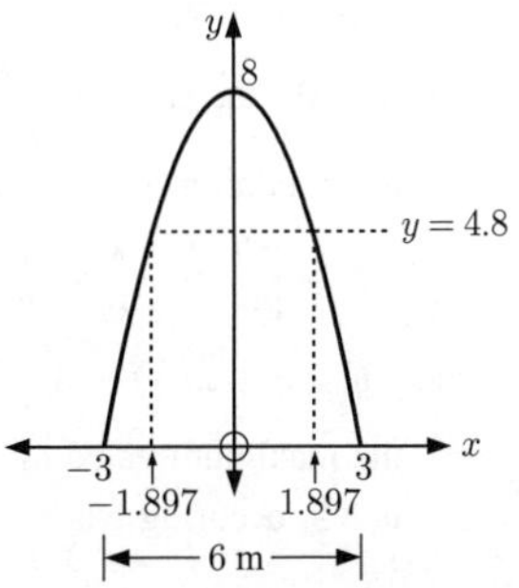

We use this equation to find the width of the tunnel 4.8 m above ground level (at the top of the truck).

When $y = 4.8$, $-\frac{8}{9}x^2 + 8 = 4.8$

$\therefore \quad 3.2 = \frac{8}{9}x^2$

$\therefore \quad x^2 = 3.6$

$\therefore \quad x \approx \pm 1.897$

$\therefore$ width of tunnel $\approx 2 \times 1.897$

≈ 3.79 m

$\therefore$ at 4.8 m above ground level, the tunnel is only 3.79 m wide, and the truck is 3.9 m wide.

$\therefore$ the truck will not fit through the tunnel.

19 **a** With the axes as described, the parabola has vertex $(0, 70)$.

$\therefore$ its equation is $y = ax^2 + 70$ for some $a \neq 0$.

The end of the bridge is 80 m from A,

so the arch meets the vertical end supports at the point $(80, 6)$.

$(80, 6)$ must lie on the curve, so $6 = a(80)^2 + 70$

$\therefore \quad 6400a = -64$ and so $a = -\frac{1}{100}$

$\therefore$ the arch has equation $y = -\frac{1}{100}x^2 + 70$

b The supports occur every 10 m.

When $x = 10$, $y = -\frac{1}{100} \times 10^2 + 70 = 69$ m

When $x = 20$, $y = -\frac{1}{100} \times 20^2 + 70 = 66$ m

When $x = 30$, $y = -\frac{1}{100} \times 30^2 + 70 = 61$ m

When $x = 40$, $y = -\frac{1}{100} \times 40^2 + 70 = 54$ m

When $x = 50$, $y = -\frac{1}{100} \times 50^2 + 70 = 45$ m

When $x = 60$, $y = -\frac{1}{100} \times 60^2 + 70 = 34$ m

When $x = 70$, $y = -\frac{1}{100} \times 70^2 + 70 = 21$ m

and the supports at $x = -10, -20, ..., -70$ m have the same lengths.

$\therefore$ the other supports have lengths 21 m, 34 m, 45 m, 54 m, 61 m, 66 m and 69 m.

EXERCISE 6G

1 **a** For $y = x^2 - 2x$,
$a = 1, \quad b = -2, \quad c = 0$.

As $a > 0$, the shape is 

$\therefore$ the minimum value occurs when

$$x = \frac{-b}{2a} = \frac{2}{2} = 1$$

and $y = 1^2 - 2(1) = -1$

$\therefore$ the minimum value of $y = x^2 - 2x$ is -1, occurring when $x = 1$.

b For $y = 7 - 2x - x^2$,
$a = -1, \quad b = -2, \quad c = 7$.

As $a < 0$, the shape is 

$\therefore$ the maximum value occurs when

$$x = \frac{-b}{2a} = -\frac{-2}{2(-1)} = -1$$

and $y = 7 - 2(-1) - (-1)^2 = 8$

$\therefore$ the maximum value of $y = 7 - 2x - x^2$ is 8, occuring when $x = -1$.

c For $y = 8 + 2x - 3x^2$,
$a = -3, \quad b = 2, \quad c = 8$.

As $a < 0$, the shape is 

$\therefore$ the maximum value occurs when

$$x = \frac{-b}{2a} = \frac{-2}{-6} = \tfrac{1}{3}$$

and $y = 8 + 2(\frac{1}{3}) - 3(\frac{1}{3})^2 = 8\frac{1}{3}$

$\therefore$ the maximum value of $y = 8 + 2x - 3x^2$ is $8\frac{1}{3}$, occuring when $x = \frac{1}{3}$.

d For $y = 2x^2 + x - 1$,
$a = 2, \quad b = 1, \quad c = -1$.

As $a > 0$, the shape is 

$\therefore$ the minimum value occurs when

$$x = \frac{-b}{2a} = -\tfrac{1}{4}$$

and $y = 2(-\frac{1}{4})^2 + (-\frac{1}{4}) - 1$

$= \frac{1}{8} - \frac{1}{4} - 1 = -1\frac{1}{8}$

$\therefore$ the minimum value of $y = 2x^2 + x - 1$ is $-1\frac{1}{8}$, occuring when $x = -\frac{1}{4}$.

e For $y = 4x^2 - x + 5$,
$a = 4, \quad b = -1, \quad c = 5.$

As $a > 0$, the shape is 

$\therefore$ the minimum value occurs when

$$x = \frac{-b}{2a} = \tfrac{1}{8}$$

and $y = 4(\tfrac{1}{8})^2 - \tfrac{1}{8} + 5$

$= \tfrac{1}{16} - \tfrac{1}{8} + 5$

$= 4\tfrac{15}{16}$

$\therefore$ the minimum value of $y = 4x^2 - x + 5$ is $4\tfrac{15}{16}$, occuring when $x = \tfrac{1}{8}$.

f For $y = 7x - 2x^2$,
$a = -2, \quad b = 7, \quad c = 0.$

As $a < 0$, the shape is 

$\therefore$ the maximum value occurs when

$$x = \frac{-b}{2a} = \frac{-7}{-4} = \tfrac{7}{4}$$

and $y = 7(\tfrac{7}{4}) - 2(\tfrac{7}{4})^2$

$= \tfrac{49}{4} - \tfrac{49}{8}$

$= \tfrac{49}{8}$ or $6\tfrac{1}{8}$

$\therefore$ the maximum value of $y = 7x - 2x^2$ is $6\tfrac{1}{8}$, occuring when $x = \tfrac{7}{4}$.

2 For $P = -3x^2 + 240x - 800$,
$a = -3, \quad b = 240, \quad c = -800.$

As $a < 0$, the shape is 

$\therefore$ the maximum profit occurs when

$$x = \frac{-b}{2a} = \frac{-240}{-6} = 40$$

and $P = -3(40)^2 + 240(40) - 800$

$= 4000$

$\therefore$ 40 refrigerators should be made each day, for a maximum profit of \$4000.

3 **a** Let the other side be y m long.
The perimeter is 200 m.

$\therefore \quad 2x + 2y = 200$

$\therefore \quad x + y = 100$

$\therefore \quad y = 100 - x$

$\therefore$ the area $A = xy$

$\therefore \quad A = x(100 - x)$

$\therefore \quad A = 100x - x^2$

b $A = 100x - x^2$ is a quadratic function with
$a = -1, \quad b = 100, \quad c = 0.$

As $a < 0$, the shape is 

$\therefore$ the area is maximised when

$$x = \frac{-b}{2a} = \frac{-100}{-2} = 50$$

and $y = 100 - 50 = 50$

$\therefore$ the area of the rectangle is maximised when $x = y = 50$, which is when the rectangle is a square.

4 Let the dimensions of the paddock be x m $\times$ y m.
If 1000 m of fence is available, then

$2x + y = 1000 \qquad$ {perimeter}

$\therefore \quad y = 1000 - 2x \quad$ ...... (1)

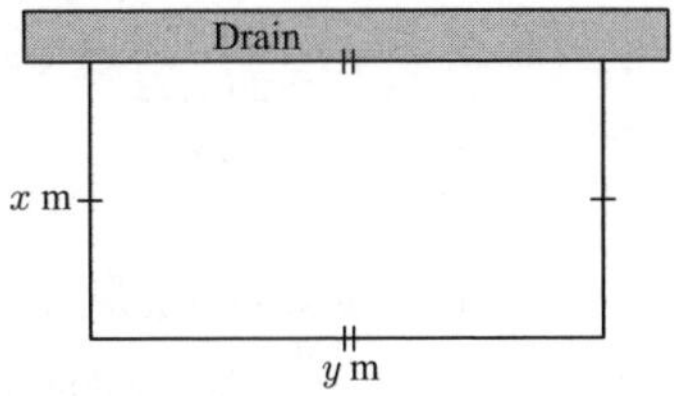

The area of the enclosure $A = xy$

Since $y = 1000 - 2x$, $\quad A = x(1000 - 2x)$

$= 1000x - 2x^2$

$\therefore \quad A = -2x^2 + 1000x$

A is a quadratic and $a < 0$, so its shape is 

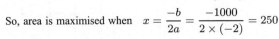

So, area is maximised when $x = \dfrac{-b}{2a} = \dfrac{-1000}{2 \times (-2)} = 250$

and when $x = 250$, $y = 1000 - 2(250) = 500$

$\therefore$ the paddock has a maximum area when the dimensions are 250 m $\times$ 500 m.

5 **a**

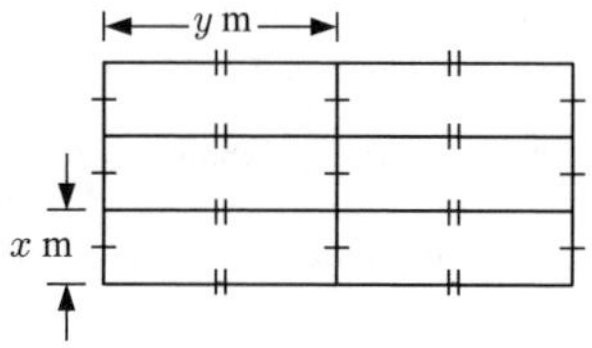

The length of fence required for this enclosure is $9x+8y$. If 1800 m is available for this enclosure, then $9x+8y=1800$.

b If $9x+8y=1800$, then $y=\dfrac{1800-9x}{8}$.

The area of each pen is $A=xy$. Substituting $y=\dfrac{1800-9x}{8}$ into A we get

$$A=x\left(\frac{1800-9x}{8}\right)$$

$$\therefore\quad A=\frac{1800x}{8}-\frac{9x^2}{8}$$

$$\therefore\quad A=-\tfrac{9}{8}x^2+225x$$

c The area is a quadratic function with $a<0$, so its shape is

So, at $x=\dfrac{-b}{2a}$ we have a maximum

$$\therefore\quad x=\frac{-225}{2\times(-\frac{9}{8})}=100, \quad \text{and when } x=100,\quad y=\frac{1800-9(100)}{8}=112.5$$

Hence, the area is maximised when the dimensions are 100 m $\times$ 112.5 m.

6 **a** Let x m $\times$ y m be the dimensions of a single pen as shown below.

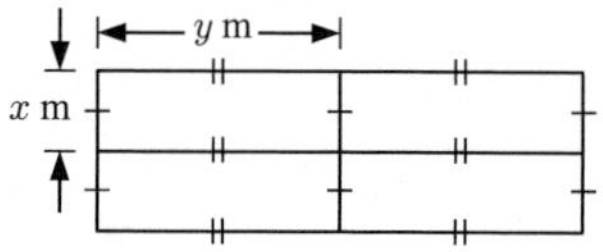

Hence, the total length of fence required is $6x+6y$.

If there is 500 m of fence available, then $6x+6y=500$

$$\therefore\quad x+y=83\tfrac{1}{3}$$

$$\therefore\quad y=83\tfrac{1}{3}-x \quad \text{...... (1)}$$

The area of each pen will be $A=xy$ and substituting equation (1), we have

$$A=x(83\tfrac{1}{3}-x)$$

$$\therefore\quad A=-x^2+83\tfrac{1}{3}x$$

which is a quadratic with $a<0$ $\therefore$ its shape is

Hence, at $x=\dfrac{-b}{2a}$ we have a maximum value of A.

$$\therefore\quad x=\frac{-83\frac{1}{3}}{2(-1)}=41\tfrac{2}{3} \quad \text{and so} \quad y=83\tfrac{1}{3}-x=41\tfrac{2}{3}$$

$\therefore$ the dimensions that maximise the area are $41\frac{2}{3}$ m $\times$ $41\frac{2}{3}$ m.

b Let x m $\times$ y m be the dimensions of a single pen as shown below.

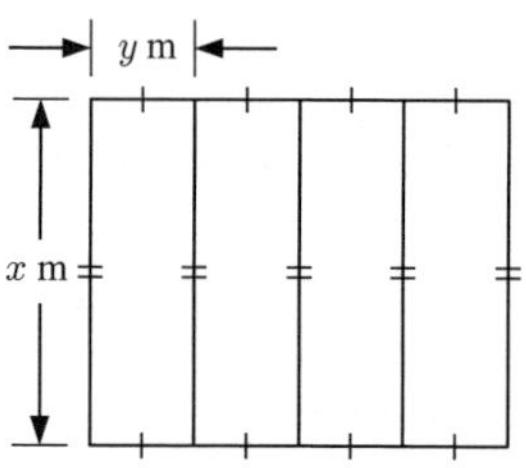

Hence, the total length of fence required is $5x+8y$.

If there is 500 m of fence available, then

$$5x+8y=500$$

$$\therefore\quad 8y=500-5x$$

$$\therefore\quad y=\frac{500-5x}{8}$$

$$\therefore\quad y=62\tfrac{1}{2}-\tfrac{5}{8}x \quad \text{...... (1)}$$

The area of each pen will be $A = xy$ and substituting equation (1), we have

$A = x(62\frac{1}{2} - \frac{5}{8}x)$

$\therefore \quad A = -\frac{5}{8}x^2 + 62\frac{1}{2}x$ which is a quadratic with $a < 0$, $\therefore$ its shape is 

Hence, when $x = \dfrac{-b}{2a}$ we have a maximum value of A.

$\therefore \quad x = \dfrac{-62\frac{1}{2}}{2 \times (-\frac{5}{8})} = 50$ and substituting $x = 50$ into $y = 62\frac{1}{2} - \frac{5}{8}x$, we have

$y = 31\frac{1}{4}$

$\therefore$ the dimensions that maximise the area are $50 \text{ m} \times 31\frac{1}{4}$ m.

7 **a** The graphs of $y = x^2 - 3x$ and $y = 2x - x^2$

meet where $x^2 - 3x = 2x - x^2$

$\therefore \quad 2x^2 - 5x = 0$

$\therefore \quad x(2x - 5) = 0$

$\therefore \quad x = 0$ or $2\frac{1}{2}$

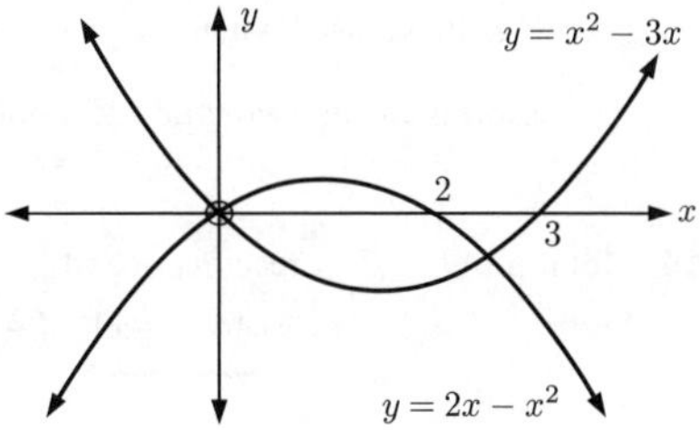

b The vertical separation between the curves is given by

$S = (2x - x^2) - (x^2 - 3x)$ $\quad \{y = 2x - x^2$ is above $y = x^2 - 3x$ for $0 \leqslant x \leqslant 2\frac{1}{2}\}$

$\therefore \quad S = 2x - x^2 - x^2 + 3x$

$\therefore \quad S = -2x^2 + 5x$

Thus S is a quadratic function with $a < 0$ so the shape is 

$\therefore$ the maximum separation occurs when $x = \dfrac{-b}{2a} = \dfrac{-5}{-4} = \frac{5}{4}$

and $S = -2(\frac{5}{4})^2 + 5(\frac{5}{4})$

$= -\frac{25}{8} + \frac{25}{4}$

$= \frac{25}{8}$ or $3\frac{1}{8}$

$\therefore$ the maximum vertical separation between the curves for $0 \leqslant x \leqslant 2\frac{1}{2}$ is $3\frac{1}{8}$ units.

8 **a**

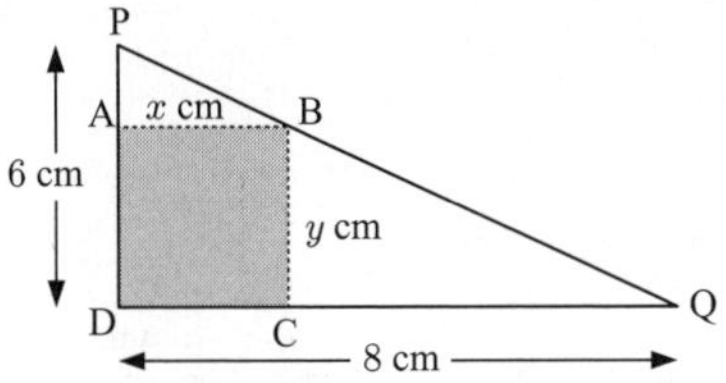

Δs PAB and PDQ are similar

{$\widehat{APB}$ is common,

$\widehat{ABP} = \widehat{DQP}$ as [AB] ∥ [DQ]}

$\therefore \quad \dfrac{PA}{PD} = \dfrac{AB}{DQ}$

$\therefore \quad \dfrac{6 - y}{6} = \dfrac{x}{8}$

$\therefore \quad 6 - y = \frac{3}{4}x$

$\therefore \quad y = 6 - \frac{3}{4}x$

b Rectangle ABCD has area $A = xy$

$= x(6 - \frac{3}{4}x)$

$= -\frac{3}{4}x^2 + 6x$

which is a quadratic with $a < 0$

$\therefore$ the shape is 

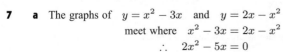

$\therefore$ the area is maximised when

$x = \dfrac{-b}{2a} = \dfrac{-6}{-\frac{3}{2}} = 4$

and when $x = 4$, $y = 6 - \frac{3}{4}(4)$

$= 3$

$\therefore$ the dimensions of rectangle ABCD of maximum area is $4 \text{ cm} \times 3$ cm.

9 Total profit, $P = \text{receipts} - \text{costs}$

$$\therefore \quad P = (550x - 2x^2) - \underbrace{\left(50 + \frac{400}{x}\right)}_{\text{cost for one}} \underset{\uparrow\ x \text{ of them}}{x}$$

$$\therefore \quad P = 550x - 2x^2 - 50x - 400$$

$$\therefore \quad P = -2x^2 + 500x - 400 \text{ dollars}$$

which is a quadratic in x, with $a = -2$, $b = 500$, $c = -400$.

Since $a < 0$, the shape is 

$\therefore$ P is maximised when $x = \dfrac{-b}{2a} = \dfrac{-500}{-4} = 125$

$\therefore$ profit is maximised when 125 pot-belly stoves are produced per week.

10 Total profit, $P = \text{receipts} - \text{cost}$

$$= \underset{\uparrow\ x \text{ of them}}{x}\, \underbrace{(44 - \tfrac{1}{5}x)}_{\text{selling price of each}} - (\tfrac{1}{10}x^2 + 20x + 25)$$

$$\therefore \quad P = 44x - \tfrac{1}{5}x^2 - \tfrac{1}{10}x^2 - 20x - 25$$

$$\therefore \quad P = -\tfrac{3}{10}x^2 + 24x - 25$$

which is a quadratic in x, with $a = -\frac{3}{10}$, $b = 24$, $c = -25$.

Since $a < 0$, the shape is  , so P is maximised when $x = \dfrac{-b}{2a} = \dfrac{-24}{2(-\frac{3}{10})} = 40$

$\therefore$ profit is maximised when 40 toasters are produced each day.

11 Total profit, $P = \text{receipts} - \text{costs}$

$$= (1000x - 3x^2) - \underbrace{\left(60 + \frac{800}{x}\right)}_{\text{cost for one}} \underset{\uparrow\ x \text{ of them}}{x}$$

$$\therefore \quad P = 1000x - 3x^2 - 60x - 800$$

$$\therefore \quad P = -3x^2 + 940x - 800$$

which is a quadratic in x, with $a = -3$, $b = 940$ and $c = -800$.

Since $a < 0$, the shape is  , so P is maximised when $x = \dfrac{-b}{2a} = \dfrac{-940}{2(-3)} = 156\frac{2}{3}$

But only a whole number can be produced, so $x = 157$.

$\therefore$ need to make 157 barbeques per week to maximise profits.

REVIEW SET 6A

1 **a** The x-intercepts are -2 and 1.

b The axis of symmetry lies midway between the x-intercepts, so its equation is $x = -\frac{1}{2}$.

c When $x = 0$, $y = -2(2)(-1) = 4$

$\therefore$ the y-intercept is 4

d When $x = -\frac{1}{2}$, $y = -2(-\frac{1}{2} + 2)(-\frac{1}{2} - 1)$
$= -2(\frac{3}{2})(-\frac{3}{2})$
$= \frac{9}{2}$

$\therefore$ the vertex is $(-\frac{1}{2}, \frac{9}{2})$

e

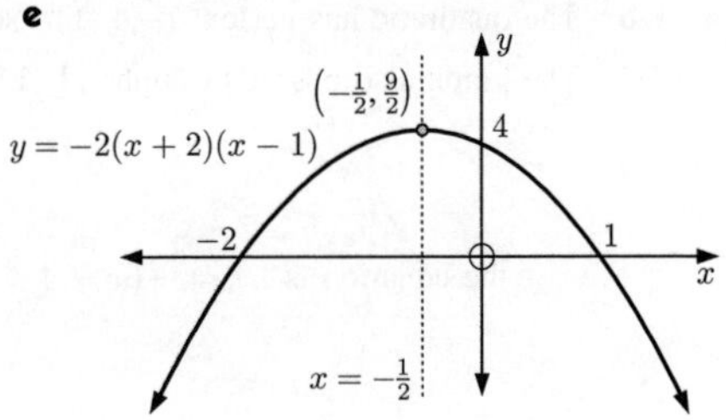

2 **a** $3x^2 - 12x = 0$
$\therefore$ $3x(x-4) = 0$
$\therefore$ $x = 0$ or 4

b $3x^2 - x - 10 = 0$
$\therefore$ $(3x+5)(x-2) = 0$
$\therefore$ $x = -\frac{5}{3}$ or 2

c $x^2 - 11x = 60$
$\therefore$ $x^2 - 11x - 60 = 0$
$\therefore$ $(x+4)(x-15) = 0$
$\therefore$ $x = -4$ or 15

3 **a** $x^2 + 5x + 3 = 0$
has $a = 1$, $b = 5$, $c = 3$

$$\therefore \quad x = \frac{-5 \pm \sqrt{5^2 - 4(1)(3)}}{2(1)}$$
$$= -\frac{5}{2} \pm \frac{\sqrt{13}}{2}$$

b $3x^2 + 11x - 2 = 0$
has $a = 3$, $b = 11$, $c = -2$

$$\therefore \quad x = \frac{-11 \pm \sqrt{11^2 - 4(3)(-2)}}{2(3)}$$
$$= -\frac{11}{6} \pm \frac{\sqrt{145}}{6}$$

4 $x^2 + 7x - 4 = 0$
$\therefore$ $x^2 + 7x + (\frac{7}{2})^2 - (\frac{7}{2})^2 - 4 = 0$
$\therefore$ $(x + \frac{7}{2})^2 - \frac{49}{4} - 4 = 0$
$\therefore$ $(x + \frac{7}{2})^2 = \frac{65}{4}$
$\therefore$ $x + \frac{7}{2} = \pm\frac{\sqrt{65}}{2}$
$\therefore$ $x = -\frac{7}{2} \pm \frac{\sqrt{65}}{2}$

5 **a** $y = (x-2)^2 - 4$ has vertex $(2, -4)$
and axis of symmetry $x = 2$.
When $x = 0$, $y = (-2)^2 - 4 = 0$
so the y-intercept is 0.

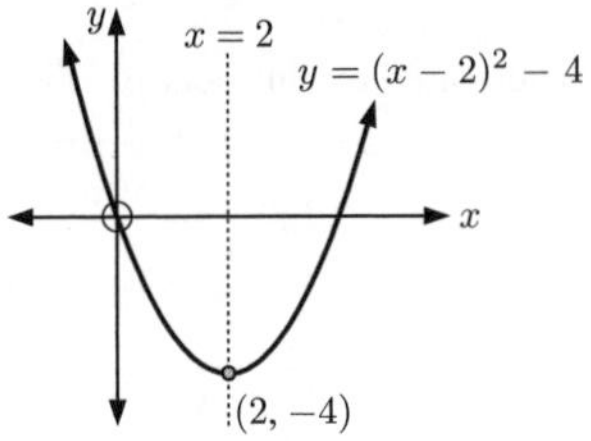

b $y = -\frac{1}{2}(x+4)^2 + 6$ has vertex $(-4, 6)$
and axis of symmetry $x = -4$.
When $x = 0$, $y = -\frac{1}{2}(4)^2 + 6 = -2$
so the y-intercept is -2.

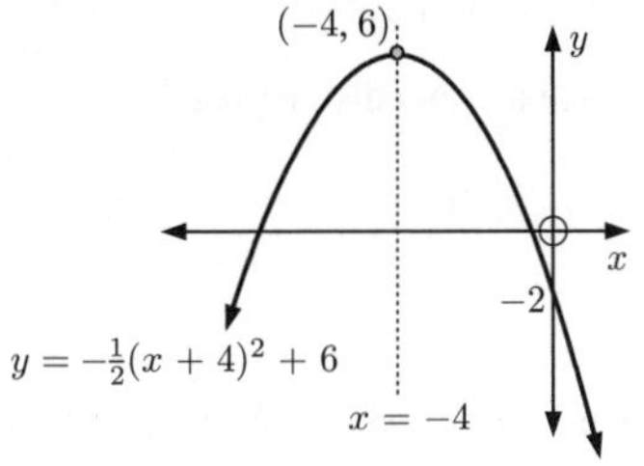

6 **a** The graph touches the x-axis at 4, so its vertex is $(4, 0)$.
$\therefore$ its equation is $y = a(x-4)^2$ for some $a \neq 0$.
The graph also passes through $(2, 12)$ $\therefore$ $a(2-4)^2 = 12$
$\therefore$ $4a = 12$
$\therefore$ $a = 3$

$\therefore$ the equation is $y = 3(x-4)^2$ which is $y = 3(x^2 - 8x + 16)$
or $y = 3x^2 - 24x + 48$

b The quadratic has vertex $(-4, 1)$, so its equation is $y = a(x+4)^2 + 1$ for some $a \neq 0$.

The graph also passes through $(1, 11)$ $\therefore\ 11 = a(1+4)^2 + 1$

$\therefore\ 25a = 10$

$\therefore\ a = \frac{2}{5}$

$\therefore$ the equation is $y = \frac{2}{5}(x+4)^2 + 1$ which is $y = \frac{2}{5}(x^2 + 8x + 16) + 1$

or $y = \frac{2}{5}x^2 + \frac{16}{5}x + \frac{37}{5}$

7 $y = -2x^2 + 4x + 3$ has $a = -2$, $b = 4$ and $c = 3$

Since $a < 0$, the graph has shape  and will have a maximum.

The axis of symmetry is $x = -\dfrac{b}{2a} = -\dfrac{4}{2(-2)} = 1$

When $x = 1$, $y = -2(1)^2 + 4(1) + 3$

$= 5$

$\therefore$ the maximum is 5, and this occurs when $x = 1$.

8 $y = x^2 - 3x$ meets $y = 3x^2 - 5x - 24$

when $x^2 - 3x = 3x^2 - 5x - 24$

$\therefore\ 2x^2 - 2x - 24 = 0$

$\therefore\ x^2 - x - 12 = 0$

$\therefore\ (x-4)(x+3) = 0$

$\therefore\ x = 4$ or -3

Substituting into $y = x^2 - 3x$,

when $x = 4$, $y = 4^2 - 3 \times 4 = 4$

and when $x = -3$, $y = (-3)^2 - 3(-3)$

$= 9 + 9 = 18$

$\therefore$ the graphs meet at $(4, 4)$ and $(-3, 18)$.

9 $y = -2x^2 + 5x + k$

has $a = -2$, $b = 5$ and $c = k$.

$\therefore\ \Delta = b^2 - 4ac$

$= 5^2 - 4(-2)k$

$= 25 + 8k$

The graph does not cut the x-axis if $\Delta < 0$

$\therefore\ 25 + 8k < 0$

$\therefore\ 8k < -25$

$\therefore\ k < -\frac{25}{8}$

So, $k < -3\frac{1}{8}$

10 $2x^2 - 3x + m = 0$

has $a = 2$, $b = -3$ and $c = m$

$\therefore\ \Delta = b^2 - 4ac$

$= (-3)^2 - 4(2)m$

$= 9 - 8m$

a There is a repeated root if $\Delta = 0$

$\therefore\ 9 - 8m = 0$

$\therefore\ m = \frac{9}{8}$

b There are two distinct real roots if $\Delta > 0$

$\therefore\ 9 - 8m > 0$

$\therefore\ 8m < 9$

$\therefore\ m < \frac{9}{8}$

c There are no real roots if $\Delta < 0$

$\therefore\ 9 - 8m < 0$

$\therefore\ 8m > 9$

$\therefore\ m > \frac{9}{8}$

11 Let the number be x, so its reciprocal is $\dfrac{1}{x}$.

$\therefore\ x + \dfrac{1}{x} = 2\frac{1}{30} = \frac{61}{30}$

$\therefore\ x^2 + 1 = \frac{61}{30}x$

$\therefore\ 30x^2 + 30 = 61x$

$\therefore\ 30x^2 - 61x + 30 = 0$

$\therefore\ (6x - 5)(5x - 6) = 0$

$\therefore\ x = \frac{5}{6}$ or $\frac{6}{5}$

$\therefore$ the number is $\frac{5}{6}$ or $\frac{6}{5}$

12 Let the line with y-intercept (0, 10) have equation $y = mx + 10$.

$y = 3x^2 + 7x - 2$ meets this line when $3x^2 + 7x - 2 = mx + 10$

$\therefore\ 3x^2 + (7 - m)x - 12 = 0$

For $y = mx + 10$ to be tangential to $y = 3x^2 + 7x - 2$, this equation must have exactly one solution, so there is a repeated root.

$\therefore\ \Delta = 0$

$\therefore\ (7 - m)^2 - 4(3)(-12) = 0$

$\therefore\ 49 - 14m + m^2 + 144 = 0$

$\therefore\ m^2 - 14m + 193 = 0$

$\therefore\ m = \dfrac{14 \pm \sqrt{(-14)^2 - 4(1)(193)}}{2}$

$\therefore\ m = \dfrac{14 \pm \sqrt{-576}}{2}$ which has no real solutions.

$\therefore$ no line with y-intercept (0, 10) can be tangential to $y = 3x^2 + 7x - 2$.

13 **a** The axis of symmetry for $f(x)$ is $x = 1$.

$\therefore\ -\dfrac{m}{2 \times 1} = 1$ and so $m = -2$.

Now $f(1) = 3$ $\therefore\ 1^2 - 2(1) + n = 3$

$\therefore\ n = 4$

So, $m = -2$ and $n = 4$.

b $f(x) = x^2 - 2x + 4$

$\therefore\ f(3) = 3^2 - 2(3) + 4$

$= 7$

$\therefore\ k = 7$

c $g(x) = f(x - 1) + 2$ is obtained from $f(x)$ by a translation through $\binom{1}{2}$. $f(x)$ has vertex (1, 3), so $g(x)$ has vertex $(1 + 1, 2 + 3) = (2, 5)$.

d The domain of $f(x)$ is $\{x \mid x \in \mathbb{R}\}$, and the range of $f(x)$ is $\{y \mid y \geqslant 3\}$.
The domain of $g(x)$ is $\{x \mid x \in \mathbb{R}\}$, and the range of $g(x)$ is $\{y \mid y \geqslant 5\}$.

REVIEW SET 6B

1 **a** $y = 2x^2 + 6x - 3$

$= 2[x^2 + 3x - \frac{3}{2}]$

$= 2[x^2 + 3x + (\frac{3}{2})^2 - (\frac{3}{2})^2 - \frac{3}{2}]$

$= 2[(x + \frac{3}{2})^2 - \frac{9}{4} - \frac{3}{2}]$

$= 2[(x + \frac{3}{2})^2 - \frac{15}{4}]$

$= 2(x + \frac{3}{2})^2 - \frac{15}{2}$

b The vertex is $(-\frac{3}{2}, -\frac{15}{2})$.

c When $x = 0$, $y = -3$

$\therefore$ the y-intercept is -3.

d

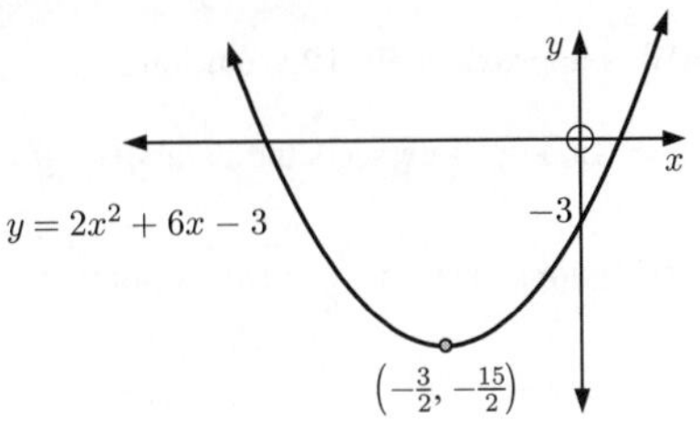

2 **a** $x \approx 0.586$ or 3.414

b $x \approx -0.186$ or 2.686

3 $y = -x^2 + 2x = x(2 - x)$

$\therefore$ the graph has x-intercepts 0 and 2, and y-intercept 0

Its axis of symmetry is midway between the x-intercepts,

at $x = 1$

and when $x = 1$, $y = -1^2 + 2 = 1$

$\therefore$ the vertex is (1, 1).

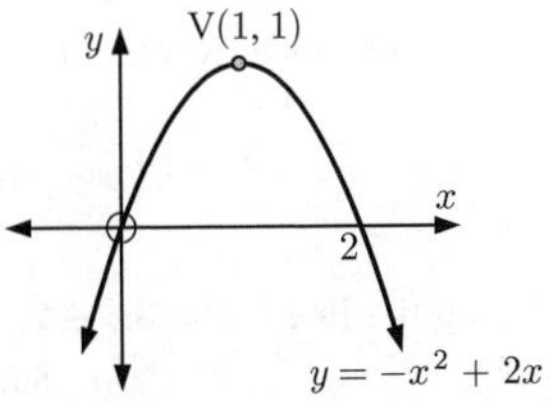

4 $y = -3x^2 + 8x + 7$ has $a = -3$, $b = 8$ and $c = 7$

The axis of symmetry is $x = -\dfrac{b}{2a} = -\dfrac{8}{2(-3)} = \frac{4}{3}$

When $x = \frac{4}{3}$, $y = -3(\frac{4}{3})^2 + 8(\frac{4}{3}) + 7$

$= -\frac{16}{3} + \frac{32}{3} + 7 = \frac{37}{3}$

$\therefore$ the axis of symmetry is $x = \frac{4}{3}$ and the vertex is $(\frac{4}{3}, \frac{37}{3})$ or $(\frac{4}{3}, 12\frac{1}{3})$.

5 **a** $2x^2 - 5x - 7 = 0$

has $a = 2$, $b = -5$, $c = -7$

$\therefore \Delta = b^2 - 4ac$

$= (-5)^2 - 4(2)(-7)$

$= 25 + 56$

$= 81$

$\therefore \Delta > 0$ and $\sqrt{\Delta} = 9$

$\therefore$ there are two distinct real rational roots

b $3x^2 - 24x + 48 = 0$

has $a = 3$, $b = -24$, $c = 48$

$\therefore \Delta = b^2 - 4ac$

$= (-24)^2 - 4(3)(48)$

$= 576 - 576$

$= 0$

$\therefore$ there is a repeated real root

6 Suppose [AB] is x cm in length. Then, using the information given, we can label the diagram:

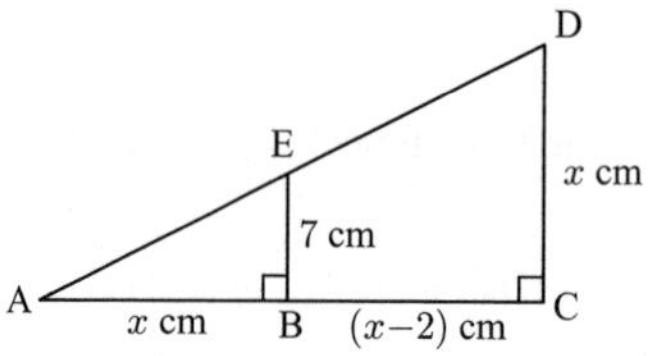

Now by similar triangles, $\dfrac{\text{BE}}{\text{AB}} = \dfrac{\text{CD}}{\text{AC}}$

$\therefore \dfrac{7}{x} = \dfrac{x}{x + (x-2)}$

$\therefore \dfrac{7}{x} = \dfrac{x}{2x-2}$

$\therefore 7(2x - 2) = x^2$

$\therefore 14x - 14 = x^2$

$\therefore x^2 - 14x + 14 = 0$

which has $a = 1$, $b = -14$ and $c = 14$

$$\therefore x = \frac{-(-14) \pm \sqrt{(-14)^2 - 4(1)(14)}}{2(1)} = \frac{14 \pm \sqrt{140}}{2}$$

Now $x - 2 > 0$, so $x = \dfrac{14 + \sqrt{140}}{2} \approx 12.92$ cm

$\therefore$ [AB] is approximately 12.9 cm long.

7 **a** $y = 3x + c$ intersects the parabola $y = x^2 + x - 5$ when $x^2 + x - 5 = 3x + c$

$\therefore x^2 - 2x - 5 - c = 0$

The graphs meet in two distinct points when this equation has two distinct real roots.

$\therefore \Delta > 0$

$\therefore (-2)^2 - 4(1)(-5 - c) > 0$

$\therefore 4 + 20 + 4c > 0$

$\therefore 4c > -24$

$\therefore c > -6$

b Choose c such that $c > -6$, for example, $c = -2$.

The graphs meet where $x^2 + x - 5 = 3x - 2$

$\therefore x^2 - 2x - 3 = 0$

$\therefore (x+1)(x-3) = 0$

$\therefore x = -1$ or 3

Using the line $y = 3x - 2$, when $x = -1$, $y = 3(-1) - 2 = -5$

and when $x = 3$, $y = 3(3) - 2 = 7$

$\therefore$ the points of intersection are $(-1, -5)$ and $(3, 7)$.

8 **a** $y = 2x^2 + 4x - 1$
has $a = 2$, $b = 4$ and $c = -1$

The axis of symmetry is $x = -\dfrac{b}{2a}$

$\therefore \quad x = -\dfrac{4}{2 \times 2}$

$\therefore \quad x = -1$

b When $x = -1$, $y = 2(-1)^2 + 4(-1) - 1$
$= 2 - 4 - 1$
$= -3$

$\therefore$ the vertex is $(-1, -3)$

c When $x = 0$, $y = -1$,
so the y-intercept is -1.
When $y = 0$, $2x^2 + 4x - 1 = 0$

$$\therefore \quad x = \frac{-4 \pm \sqrt{4^2 - 4(2)(-1)}}{2(2)}$$
$$= \frac{-4 \pm \sqrt{24}}{4}$$
$$= \frac{-4 \pm 2\sqrt{6}}{4} = -1 \pm \tfrac{1}{2}\sqrt{6}$$

$\therefore$ the x-intercepts are $-1 \pm \frac{1}{2}\sqrt{6}$.

d

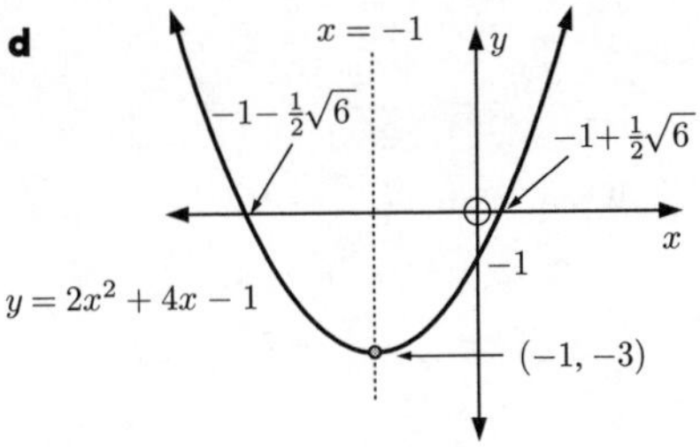

9 Since the container has a square base, the original tinplate must have been square.
Suppose its side was x cm long, so the base of the container is $(x - 8)$ cm by $(x - 8)$ cm.
The height of the container is 4 cm, so its capacity is $4(x - 8)(x - 8)$ cm^3.

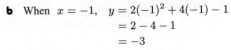

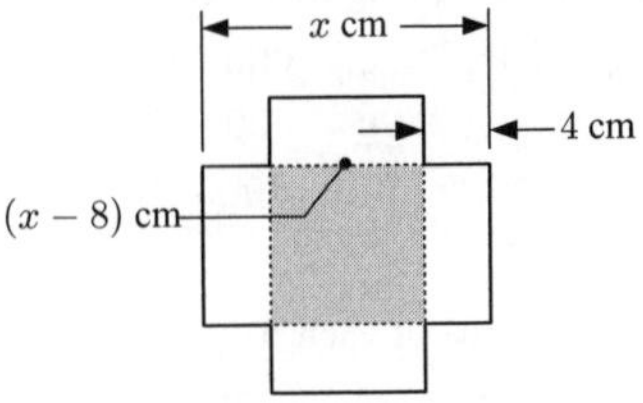

$\therefore \quad 4(x - 8)^2 = 120$
$\therefore \quad (x - 8)^2 = 30$
$\therefore \quad x - 8 = \pm\sqrt{30}$
$\therefore \quad x = 8 \pm \sqrt{30}$

Clearly, $x > 8$, so $x = 8 + \sqrt{30} \approx 13.48$

$\therefore$ the tinplate was about 13.5 cm by 13.5 cm.

10 $y = -x^2 - 5x + 3$ meets $y = x^2 + 3x + 11$ when $-x^2 - 5x + 3 = x^2 + 3x + 11$
$\therefore \quad 2x^2 + 8x + 8 = 0$
$\therefore \quad x^2 + 4x + 4 = 0$
$\therefore \quad (x + 2)^2 = 0$
$\therefore \quad x = -2$

Substituting into $y = x^2 + 3x + 11$, when $x = -2$, $y = (-2)^2 + 3(-2) + 11$
$= 4 - 6 + 11$
$= 9$

$\therefore$ the graphs touch at $(-2, 9)$.

11 **a** $y = 3x^2 + 4x + 7$
has $a = 3$, $b = 4$ and $c = 7$
Since $a > 0$,
the graph
has shape

and so has a minimum.
This occurs on the axis of symmetry

$$x = -\frac{b}{2a}$$

$$\therefore \quad x = -\frac{4}{2(3)} = -\tfrac{2}{3}$$

When $x = -\frac{2}{3}$,

$$\begin{aligned} y &= 3(-\tfrac{2}{3})^2 + 4(-\tfrac{2}{3}) + 7 \\ &= \tfrac{4}{3} - \tfrac{8}{3} + 7 \\ &= \tfrac{17}{3} \end{aligned}$$

$\therefore$ the minimum is $\frac{17}{3}$ when $x = -\frac{2}{3}$

b $y = -2x^2 - 5x + 2$
has $a = -2$, $b = -5$ and $c = 2$
Since $a < 0$,
the graph
has shape

and so has a maximum.
This occurs on the axis of symmetry

$$x = -\frac{b}{2a}$$

$$\therefore \quad x = -\frac{-5}{2(-2)} = -\tfrac{5}{4}$$

When $x = -\frac{5}{4}$,

$$\begin{aligned} y &= -2(-\tfrac{5}{4})^2 - 5(-\tfrac{5}{4}) + 2 \\ &= -\tfrac{25}{8} + \tfrac{25}{4} + 2 \\ &= \frac{-25 + 50 + 16}{8} \\ &= \tfrac{41}{8} \end{aligned}$$

$\therefore$ the maximum is $\frac{41}{8}$ when $x = -\frac{5}{4}$

12 **a** The total length of fencing is $(8x + 9y)$ m

$$\begin{aligned} \therefore \quad 8x + 9y &= 600 \\ \therefore \quad 9y &= 600 - 8x \\ \therefore \quad y &= \frac{600 - 8x}{9} \end{aligned}$$

b The area of each pen is

$$\begin{aligned} A &= xy \\ &= x\left(\frac{600 - 8x}{9}\right) \text{ m}^2 \end{aligned}$$

c

$$\begin{aligned} A &= x\left(\frac{600 - 8x}{9}\right) \\ &= \tfrac{600}{9}x - \tfrac{8}{9}x^2 \end{aligned}$$

which has $a = -\frac{8}{9}$, $b = \frac{600}{9}$
Since $a < 0$, A is maximised at the axis of symmetry, which is $x = -\dfrac{b}{2a}$

$$\therefore \quad x = -\frac{\frac{600}{9}}{2(-\frac{8}{9})} = \frac{600}{16}$$

$$\therefore \quad x = \tfrac{75}{2}$$

When $x = \frac{75}{2}$, $y = \dfrac{600 - 8(\frac{75}{2})}{9} = 33\frac{1}{3}$

$\therefore$ for maximum area, each pen should be $37\frac{1}{2}$ m $\times$ $33\frac{1}{3}$ m.

d The maximum area of each pen is

$$\begin{aligned} & 37\tfrac{1}{2} \times 33\tfrac{1}{3} \\ &= \tfrac{75}{2} \times \tfrac{100}{3} \\ &= 1250 \text{ m}^2 \end{aligned}$$

13 **a** $9x^2 - kx + 4$ touches the x-axis if

$$\begin{aligned} \Delta &= 0 \\ \therefore \quad (-k)^2 - 4(9)(4) &= 0 \\ \therefore \quad k^2 - 144 &= 0 \\ \therefore \quad k &= \pm 12 \end{aligned}$$

b The functions intersect when

$$\begin{aligned} 9x^2 - 12x + 4 &= 9x^2 + 12x + 4 \\ \therefore \quad 24x &= 0 \\ \therefore \quad x &= 0 \end{aligned}$$

$f(0) = 9(0)^2 - 12(0) + 4 = 4$
The two functions intersect at $(0, 4)$.

c

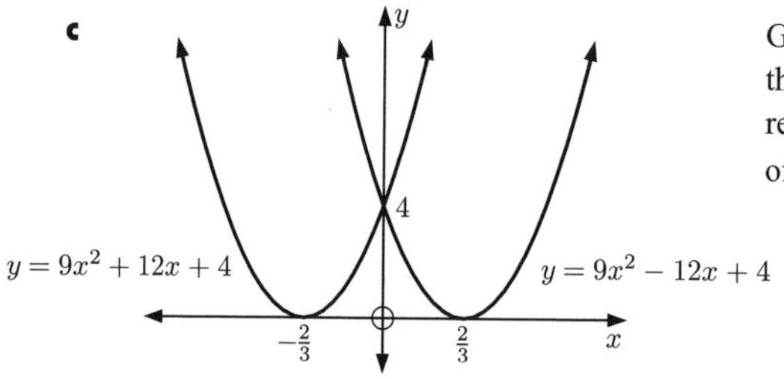

Graphing the functions using technology, we see that one function is transformed to the other by a reflection in the y-axis, *or* by a horizontal translation of $\frac{4}{3}$ units.

REVIEW SET 6C

1 **a** The axis of symmetry is $x = 2$.

b When $x = 2$, $y = \frac{1}{2}(2-2)^2 - 4$
$= -4$

$\therefore$ the vertex is $(2, -4)$

c When $x = 0$, $y = \frac{1}{2}(-2)^2 - 4$
$= -2$

$\therefore$ the y-intercept is -2

d

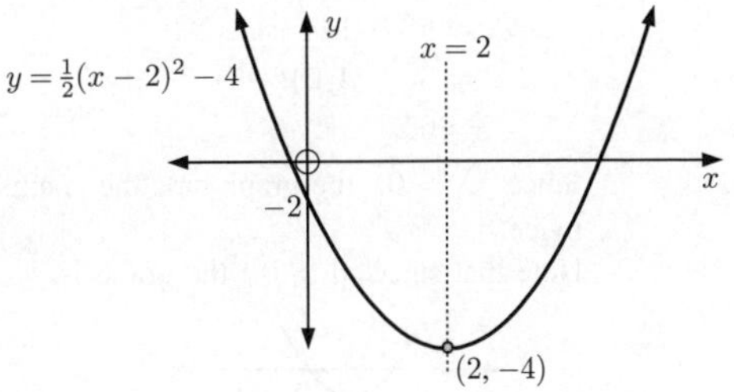

2 **a** $x^2 - 5x - 3 = 0$
has $a = 1$, $b = -5$, $c = -3$

$$\therefore \quad x = \frac{-(-5) \pm \sqrt{(-5)^2 - 4(1)(-3)}}{2(1)}$$
$$= \tfrac{5}{2} \pm \tfrac{\sqrt{37}}{2}$$

b $2x^2 - 7x - 3 = 0$
has $a = 2$, $b = -7$, $c = -3$

$$\therefore \quad x = \frac{-(-7) \pm \sqrt{(-7)^2 - 4(2)(-3)}}{2(2)}$$
$$= \tfrac{7}{4} \pm \tfrac{\sqrt{73}}{4}$$

3 **a** $x^2 - 7x + 3 = 0$
has $a = 1$, $b = -7$ and $c = 3$

$$\therefore \quad x = \frac{-(-7) \pm \sqrt{(-7)^2 - 4(1)(3)}}{2(1)}$$
$$= \frac{7 \pm \sqrt{49 - 12}}{2}$$
$$= \frac{7 \pm \sqrt{37}}{2}$$
$$\therefore \quad x = \tfrac{7}{2} \pm \tfrac{\sqrt{37}}{2}$$

b $2x^2 - 5x + 4 = 0$
has $a = 2$, $b = -5$ and $c = 4$

$$\therefore \quad x = \frac{-(-5) \pm \sqrt{(-5)^2 - 4(2)(4)}}{2(2)}$$
$$= \frac{5 \pm \sqrt{25 - 32}}{4}$$
$$\therefore \quad x = \frac{5 \pm \sqrt{-7}}{4}$$

$\therefore$ x has no real solutions.

4 **a** The graph has vertex $(2, -20)$, so its equation is
$y = a(x-2)^2 - 20$ for some $a \neq 0$.
Now an x-intercept is 5
$\therefore$ $a(5-2)^2 - 20 = 0$
$\therefore$ $9a = 20$ and so $a = \frac{20}{9}$
So the equation is $y = \frac{20}{9}(x-2)^2 - 20$.

b Since one x-intercept is 7 and the axis of symmetry is $x = 4$, the other x-intercept is $x = 1$.
$\therefore$ the graph has equation
$y = a(x-7)(x-1)$ for some $a \neq 0$.
The y-intercept is -2
$\therefore$ $a(-7)(-1) = -2$
$\therefore$ $a = -\frac{2}{7}$
$\therefore$ the equation is $y = -\frac{2}{7}(x-7)(x-1)$.

c The graph has vertex $(-3, 0)$, so its equation is
$y = a(x+3)^2$ for some $a \neq 0$.
The y-intercept is 2
$\therefore$ $a(3)^2 = 2$
$\therefore$ $9a = 2$ and so $a = \frac{2}{9}$
So the equation is $y = \frac{2}{9}(x+3)^2$.

5 **a** $y = 2x^2 + 3x - 7$
has $a = 2$, $b = 3$ and $c = -7$

$\therefore \quad \Delta = b^2 - 4ac$
$= 3^2 - 4(2)(-7)$
$= 65$

Since $\Delta > 0$, the graph cuts the x-axis twice.
Note that since $a > 0$, the graph is

b $y = -3x^2 - 7x + 4$
has $a = -3$, $b = -7$ and $c = 4$

$\therefore \quad \Delta = b^2 - 4ac$
$= (-7)^2 - 4(-3)4$
$= 97$

Since $\Delta > 0$, the graph cuts the x-axis twice.
Note that since $a < 0$, the graph is

6 **a** $y = -2x^2 + 3x + 2$
has $a = -2$, $b = 3$ and $c = 2$

$\therefore \quad \Delta = b^2 - 4ac$
$= 3^2 - 4(-2)(2)$
$= 25$

Since $\Delta > 0$, the function is neither positive definite nor negative definite.

b $y = 3x^2 + x + 11$
has $a = 3$, $b = 1$ and $c = 11$

$\therefore \quad \Delta = b^2 - 4ac$
$= 1^2 - 4(3)(11)$
$= -131$

$\therefore$ $\Delta < 0$, and since $a > 0$, the function is positive definite.

7 **a** The graph has x-intercepts ± 3, so its equation is
$y = a(x + 3)(x - 3)$ for some $a \neq 0$.
Its y-intercept is -27, so
$a(3)(-3) = -27$
$\therefore \quad -9a = -27$
$\therefore \quad a = 3$
$\therefore$ the equation is $y = 3(x + 3)(x - 3)$

b The quadratic has vertex $(2, 25)$
$\therefore$ its equation is $y = a(x - 2)^2 + 25$
The y-intercept is 1, so
$a(-2)^2 + 25 = 1$
$\therefore \quad 4a = -24$
$\therefore \quad a = -6$
$\therefore$ the equation is $y = -6(x - 2)^2 + 25$

8 Let the hypotenuse have length x cm.
$\therefore$ the longer of the remaining sides has length $(x - 2)$ cm, and the third side has length $(x - 9)$ cm.

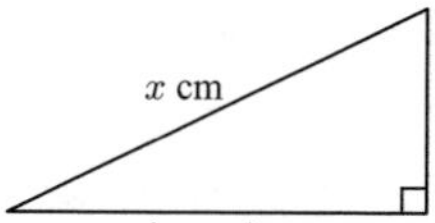

By Pythagoras' theorem, $(x - 2)^2 + (x - 9)^2 = x^2$
$\therefore \quad x^2 - 4x + 4 + x^2 - 18x + 81 = x^2$
$\therefore \quad x^2 - 22x + 85 = 0$
$\therefore \quad (x - 5)(x - 17) = 0$
$\therefore \quad x = 5$ or 17

If x was 5, the shortest side would have negative length. So, the hypotenuse has length 17 cm.

9 Let the line with gradient -3 and y-intercept c have equation $y = -3x + c$.
$y = -3x + c$ is tangential to $y = 2x^2 - 5x + 1$ if they meet at exactly one point.
$y = 2x^2 - 5x + 1$ meets $y = -3x + c$ when $2x^2 - 5x + 1 = -3x + c$
$\therefore \quad 2x^2 - 2x + 1 - c = 0$

The graphs meet exactly once when this equation has a repeated root $\therefore \quad \Delta = 0$
$\therefore \quad (-2)^2 - 4(2)(1 - c) = 0$
$\therefore \quad 4 - 8 + 8c = 0$
$\therefore \quad 8c = 4$
$\therefore \quad c = \frac{1}{2}$

$\therefore$ the y-intercept of the line is $\frac{1}{2}$.

10 $y = x^2 - 2x + k$ has $a = 1$, $b = -2$ and $c = k$

$\therefore \quad \Delta = b^2 - 4ac$
$= (-2)^2 - 4(1)k$
$= 4 - 4k$

The graph cuts the x-axis twice if $\Delta > 0$
$\therefore \quad 4 - 4k > 0$
$\therefore \quad 4k < 4$
$\therefore \quad k < 1$

11 The x-intercepts are 3 and -2, so the equation is $y = a(x-3)(x+2)$ for some $a \neq 0$.
But the y-intercept is 24 $\quad \therefore \quad a(-3)(2) = 24$
$\therefore \quad -6a = 24$
$\therefore \quad a = -4$

$\therefore$ the equation is $y = -4(x-3)(x+2)$ $\quad \therefore \quad y = -4(x^2 - x - 6)$
$\therefore \quad y = -4x^2 + 4x + 24$

12 $y = mx - 10$ is a tangent to $y = 3x^2 + 7x + 2$ if they meet at exactly one point (touch).
$y = 3x^2 + 7x + 2$ meets $y = mx - 10$ when $3x^2 + 7x + 2 = mx - 10$
$\therefore \quad 3x^2 + (7-m)x + 12 = 0$

The graphs meet exactly once when this equation has a repeated root $\quad \therefore \quad \Delta = 0$
$\therefore \quad (7-m)^2 - 4(3)(12) = 0$
$\therefore \quad 49 - 14m + m^2 - 144 = 0$
$\therefore \quad m^2 - 14m - 95 = 0$
$\therefore \quad (m+5)(m-19) = 0$
$\therefore \quad m = -5$ or 19

13 **a** **i** The x-intercepts are $-m$ and $-n$. Since $m > n$, $-m < -n$.
$\therefore$ A is $(-m, 0)$, B is $(-n, 0)$.

ii The axis of symmetry is $x = \dfrac{-m-n}{2}$.

b **i** The graph cuts the x-axis twice, so Δ has positive sign.

ii The graph has shape , so a has negative sign.

Chapter 7
THE BINOMIAL EXPANSION

EXERCISE 7A

1 **a** $(p+q)^3$
$= p^3 + 3p^2q + 3pq^2 + q^3$

b $(x+1)^3$
$= x^3 + 3x^2(1)^1 + 3x(1)^2 + (1)^3$
$= x^3 + 3x^2 + 3x + 1$

c $(x-3)^3$
$= x^3 + 3x^2(-3) + 3x(-3)^2 + (-3)^3$
$= x^3 - 9x^2 + 27x - 27$

d $(2+x)^3$
$= 2^3 + 3(2)^2x + 3(2)x^2 + x^3$
$= 8 + 12x + 6x^2 + x^3$

e $(3x-1)^3$
$= (3x)^3 + 3(3x)^2(-1) + 3(3x)(-1)^2 + (-1)^3$
$= 27x^3 - 27x^2 + 9x - 1$

f $(2x+5)^3$
$= (2x)^3 + 3(2x)^2(5) + 3(2x)(5)^2 + (5)^3$
$= 8x^3 + 60x^2 + 150x + 125$

g $(3x - \frac{1}{3})^3 = (3x)^3 + 3(3x)^2(-\frac{1}{3}) + 3(3x)(-\frac{1}{3})^2 + (-\frac{1}{3})^3$
$= 27x^3 - 9x^2 + x - \frac{1}{27}$

h $\left(2x + \dfrac{1}{x}\right)^3 = (2x)^3 + 3(2x)^2\left(\dfrac{1}{x}\right) + 3(2x)\left(\dfrac{1}{x}\right)^2 + \left(\dfrac{1}{x}\right)^3$
$= 8x^3 + 12x + \dfrac{6}{x} + \dfrac{1}{x^3}$

2 **a** $(1+x)^4 = 1^4 + 4(1)^3x + 6(1)^2x^2 + 4(1)x^3 + x^4$
$= 1 + 4x + 6x^2 + 4x^3 + x^4$

b $(p-q)^4 = p^4 + 4p^3(-q) + 6p^2(-q)^2 + 4p(-q)^3 + (-q)^4$
$= p^4 - 4p^3q + 6p^2q^2 - 4pq^3 + q^4$

c $(x-2)^4 = x^4 + 4x^3(-2)^1 + 6x^2(-2)^2 + 4x(-2)^3 + (-2)^4$
$= x^4 - 8x^3 + 24x^2 - 32x + 16$

d $(3-x)^4 = 3^4 + 4(3)^3(-x) + 6(3)^2(-x)^2 + 4(3)(-x)^3 + (-x)^4$
$= 81 - 108x + 54x^2 - 12x^3 + x^4$

e $(1+2x)^4 = 1^4 + 4(1)^3(2x) + 6(1)^2(2x)^2 + 4(1)(2x)^3 + (2x)^4$
$= 1 + 8x + 24x^2 + 32x^3 + 16x^4$

f $(2x+3)^4 = (2x)^4 + 4(2x)^3(3)^1 + 6(2x)^2(3)^2 + 4(2x)(3)^3 + (3)^4$
$= 16x^4 + 12 \times 8x^3 + 54 \times 4x^2 + 108 \times 2x + 81$
$= 16x^4 + 96x^3 + 216x^2 + 216x + 81$

g $\left(x + \dfrac{1}{x}\right)^4 = x^4 + 4x^3\left(\dfrac{1}{x}\right) + 6x^2\left(\dfrac{1}{x}\right)^2 + 4x\left(\dfrac{1}{x}\right)^3 + \left(\dfrac{1}{x}\right)^4$
$= x^4 + 4x^2 + 6 + \dfrac{4}{x^2} + \dfrac{1}{x^4}$

h $\left(2x - \dfrac{1}{x}\right)^4 = (2x)^4 + 4(2x)^3\left(-\dfrac{1}{x}\right) + 6(2x)^2\left(-\dfrac{1}{x}\right)^2 + 4(2x)\left(-\dfrac{1}{x}\right)^3 + \left(-\dfrac{1}{x}\right)^4$
$= 16x^4 - 32x^2 + 24 - \dfrac{8}{x^2} + \dfrac{1}{x^4}$

3 **a** $(x+2)^5 = x^5 + 5x^4(2) + 10x^3(2)^2 + 10x^2(2)^3 + 5x(2)^4 + 2^5$
$= x^5 + 10x^4 + 40x^3 + 80x^2 + 80x + 32$

b $(x-2y)^5 = x^5 + 5x^4(-2y) + 10x^3(-2y)^2 + 10x^2(-2y)^3 + 5x(-2y)^4 + (-2y)^5$
$= x^5 - 10x^4y + 40x^3y^2 - 80x^2y^3 + 80xy^4 - 32y^5$

c $(1+2x)^5 = 1^5 + 5(1)^4(2x) + 10(1)^3(2x)^2 + 10(1)^2(2x)^3 + 5(1)(2x)^4 + (2x)^5$
$= 1 + 10x + 40x^2 + 80x^3 + 80x^4 + 32x^5$

d $\left(x-\frac{1}{x}\right)^5 = x^5 + 5x^4\left(-\frac{1}{x}\right) + 10x^3\left(-\frac{1}{x}\right)^2 + 10x^2\left(-\frac{1}{x}\right)^3 + 5x\left(-\frac{1}{x}\right)^4 + \left(-\frac{1}{x}\right)^5$
$= x^5 - 5x^3 + 10x - \frac{10}{x} + \frac{5}{x^3} - \frac{1}{x^5}$

4 **a**
1 5 10 10 5 1 ⟵ the 5th row
1 6 15 20 15 6 1 ⟵ the 6th row

b **i** $(x+2)^6 = x^6 + 6x^5(2) + 15x^4(2)^2 + 20x^3(2)^3 + 15x^2(2)^4 + 6x(2)^5 + (2)^6$
$= x^6 + 12x^5 + 60x^4 + 160x^3 + 240x^2 + 192x + 64$

ii $(2x-1)^6 = (2x)^6 + 6(2x)^5(-1) + 15(2x)^4(-1)^2 + 20(2x)^3(-1)^3 + 15(2x)^2(-1)^4$
$+ 6(2x)(-1)^5 + (-1)^6$
$= 64x^6 - 6 \times 32x^5 + 15 \times 16x^4 - 20 \times 8x^3 + 15 \times 4x^2 - 6 \times 2x + 1$
$= 64x^6 - 192x^5 + 240x^4 - 160x^3 + 60x^2 - 12x + 1$

iii $\left(x+\frac{1}{x}\right)^6 = x^6 + 6x^5\left(\frac{1}{x}\right) + 15x^4\left(\frac{1}{x}\right)^2 + 20x^3\left(\frac{1}{x}\right)^3 + 15x^2\left(\frac{1}{x}\right)^4 + 6x\left(\frac{1}{x}\right)^5 + \left(\frac{1}{x}\right)^6$
$= x^6 + 6x^4 + 15x^2 + 20 + \frac{15}{x^2} + \frac{6}{x^4} + \frac{1}{x^6}$

5 **a** $\left(1+\sqrt{2}\right)^3 = (1)^3 + 3(1)^2\left(\sqrt{2}\right) + 3(1)\left(\sqrt{2}\right)^2 + \left(\sqrt{2}\right)^3$
$= 1 + 3\sqrt{2} + 3 \times 2 + 2 \times \sqrt{2}$
$= 1 + 3\sqrt{2} + 6 + 2\sqrt{2}$
$= 7 + 5\sqrt{2}$

b $\left(\sqrt{5}+2\right)^4 = \left(\sqrt{5}\right)^4 + 4\left(\sqrt{5}\right)^3(2) + 6\left(\sqrt{5}\right)^2(2)^2 + 4\left(\sqrt{5}\right)(2)^3 + 2^4$
$= 25 + 8 \times 5\sqrt{5} + 24 \times 5 + 32\sqrt{5} + 16$
$= 25 + 40\sqrt{5} + 120 + 32\sqrt{5} + 16$
$= 161 + 72\sqrt{5}$

c $\left(2-\sqrt{2}\right)^5$
$= (2)^5 + 5(2)^4\left(-\sqrt{2}\right) + 10(2)^3\left(-\sqrt{2}\right)^2 + 10(2)^2\left(-\sqrt{2}\right)^3 + 5(2)^1\left(-\sqrt{2}\right)^4 + \left(-\sqrt{2}\right)^5$
$= 32 - 80\sqrt{2} + 160 - 80\sqrt{2} + 40 - 4\sqrt{2}$
$= 232 - 164\sqrt{2}$

6 **a** $(2+x)^6 = (2)^6 + 6(2)^5x + 15(2)^4x^2 + 20(2)^3x^3 + 15(2)^2x^4 + 6(2)x^5 + x^6$
$= 64 + 192x + 240x^2 + 160x^3 + 60x^4 + 12x^5 + x^6$

b $(2.01)^6$ is obtained by letting $x = 0.01$
$\therefore$ $(2.01)^6 = 64 + 192 \times (0.01) + 240 \times (0.01)^2 + 160 \times (0.01)^3$
$+ 60 \times (0.01)^4 + 12 \times (0.01)^5 + (0.01)^6$
$= 65.944\,160\,601\,201$

	64
	1.92
	0.024
	0.000 16
	0.000 000 6
	0.000 000 001 2
+	0.000 000 000 001
	65.944 160 601 201

7 $(2x+3)(x+1)^4$
$= (2x+3)(x^4+4x^3+6x^2+4x+1)$
$= 2x^5+8x^4+12x^3+8x^2+2x+3x^4+12x^3+18x^2+12x+3$
$= 2x^5+11x^4+24x^3+26x^2+14x+3$

8 **a** $(3a+b)^5 = (3a)^5+5(3a)^4b+10(3a)^3b^2+ ....$
$\therefore$ the coefficient of a^3b^2 is $10\times 3^3 = 270$

b $(2a+3b)^6 = (2a)^6+6(2a)^5(3b)+15(2a)^4(3b)^2+20(2a)^3(3b)^3+ ....$
$\therefore$ the coefficient of a^3b^3 is $20\times 2^3\times 3^3 = 4320$

EXERCISE 7B

1 **a** $(1+2x)^{11} = 1^{11}+\binom{11}{1}1^{10}(2x)^1+\binom{11}{2}1^9(2x)^2+ .... +\binom{11}{10}1^1(2x)^{10}+(2x)^{11}$
$= 1+\binom{11}{1}(2x)+\binom{11}{2}(2x)^2+ .... +\binom{11}{10}(2x)^{10}+(2x)^{11}$

b $\left(3x+\frac{2}{x}\right)^{15} = (3x)^{15}+\binom{15}{1}(3x)^{14}\left(\frac{2}{x}\right)+\binom{15}{2}(3x)^{13}\left(\frac{2}{x}\right)^2+....+\binom{15}{14}(3x)\left(\frac{2}{x}\right)^{14}+\left(\frac{2}{x}\right)^{15}$

c $\left(2x-\frac{3}{x}\right)^{20}$
$= (2x)^{20}+\binom{20}{1}(2x)^{19}\left(-\frac{3}{x}\right)+\binom{20}{2}(2x)^{18}\left(-\frac{3}{x}\right)^2+ .... +\binom{20}{19}(2x)\left(-\frac{3}{x}\right)^{19}+\left(-\frac{3}{x}\right)^{20}$

2 **a** For $(2x+5)^{15}$, $a=(2x)$, $b=5$ and $n=15$
Now $T_{r+1}=\binom{n}{r}a^{n-r}b^r$ and letting $r=5$ gives $T_6=\binom{15}{5}(2x)^{10}5^5$.

b For $(x^2+y)^9$, $a=(x^2)$, $b=y$ and $n=9$
Now $T_{r+1}=\binom{n}{r}a^{n-r}b^r$ and letting $r=3$ gives $T_4=\binom{9}{3}(x^2)^6(y)^3$.

c For $\left(x-\frac{2}{x}\right)^{17}$, $a=x$, $b=\left(-\frac{2}{x}\right)$ and $n=17$
Now $T_{r+1}=\binom{n}{r}a^{n-r}b^r$ and letting $r=9$ gives $T_{10}=\binom{17}{9}x^8\left(-\frac{2}{x}\right)^9$.

d For $\left(2x^2-\frac{1}{x}\right)^{21}$, $a=(2x^2)$, $b=\left(-\frac{1}{x}\right)$ and $n=21$
Now $T_{r+1}=\binom{n}{r}a^{n-r}b^r$ and letting $r=8$ gives $T_9=\binom{21}{8}(2x^2)^{13}\left(-\frac{1}{x}\right)^8$.

3 **a** In $(3+2x^2)^{10}$, $a=3$, $b=(2x^2)$ and $n=10$
Now $T_{r+1}=\binom{n}{r}a^{n-r}b^r$
$= \binom{10}{r}3^{10-r}(2x^2)^r$
$= \binom{10}{r}3^{10-r}2^rx^{2r}$

We now let $2r=10$
$\therefore\ r=5$
So, $T_6=\underbrace{\binom{10}{5}3^52^5}\,x^{10}$
$\therefore$ the coefficient is $\binom{10}{5}3^52^5$.

b In $\left(2x^2-\frac{3}{x}\right)^6$, $a=(2x^2)$, $b=\left(-\frac{3}{x}\right)$ and $n=6$
Now $T_{r+1}=\binom{n}{r}a^{n-r}b^r$
$= \binom{6}{r}\left(2x^2\right)^{6-r}\left(-\frac{3}{x}\right)^r$
$= \binom{6}{r}2^{6-r}x^{12-2r}\,\frac{(-3)^r}{x^r}$
$= \binom{6}{r}2^{6-r}(-3)^rx^{12-3r}$

We now let $12-3r=3$
$\therefore\ 3r=9$
$\therefore\ r=3$
So, $T_4=\underbrace{\binom{6}{3}2^3(-3)^3}\,x^3$
$\therefore$ the coefficient is $\binom{6}{3}2^3(-3)^3$.

c In $(2x^2-3y)^6$, $a=(2x^2)$, $b=(-3y)$ and $n=6$

Now $T_{r+1}=\binom{n}{r}a^{n-r}b^r$

$=\binom{6}{r}\left(2x^2\right)^{6-r}(-3y)^r$

$=\binom{6}{r}2^{6-r}x^{12-2r}(-3)^r y^r$

$=\binom{6}{r}2^{6-r}(-3)^r x^{12-2r}y^r$

We find r such that

$12-2r=6$ and $r=3$

$\therefore\ r=3$ is the solution

So, $T_4=\underbrace{\binom{6}{3}2^3(-3)^3}x^6y^3$

$\therefore$ the coefficient is $\binom{6}{3}2^3(-3)^3$.

d In $\left(2x^2-\frac{1}{x}\right)^{12}$, $a=(2x^2)$, $b=\left(-\frac{1}{x}\right)$ and $n=12$

Now $T_{r+1}=\binom{n}{r}a^{n-r}b^r$

$=\binom{12}{r}\left(2x^2\right)^{12-r}\left(-\frac{1}{x}\right)^r$

$=\binom{12}{r}2^{12-r}x^{24-2r}\frac{(-1)^r}{x^r}$

$=\binom{12}{r}2^{12-r}(-1)^r x^{24-3r}$

We now let $24-3r=12$

$\therefore\ 3r=12$

$\therefore\ r=4$

So, $T_5=\underbrace{\binom{12}{4}2^8(-1)^4}x^{12}$

$\therefore$ the coefficient is $\binom{12}{4}2^8(-1)^4$.

4 **a** For $\left(x+\frac{2}{x^2}\right)^{15}$, $a=x$, $b=\frac{2}{x^2}$ and $n=15$

Now $T_{r+1}=\binom{n}{r}a^{n-r}b^r=\binom{15}{r}x^{15-r}\left(\frac{2}{x^2}\right)^r=\binom{15}{r}x^{15-r}\frac{2^r}{x^{2r}}=\binom{15}{r}2^r x^{15-3r}$

The constant term does not contain x. $\therefore\ 15-3r=0$ $\quad\therefore\ r=5$

so $T_6=\binom{15}{5}2^5x^0$ $\therefore$ the constant term is $\binom{15}{5}2^5$.

b For $\left(x-\frac{3}{x^2}\right)^9$, $a=x$, $b=\left(-\frac{3}{x^2}\right)$ and $n=9$

Now $T_{r+1}=\binom{n}{r}a^{n-r}b^r$

$=\binom{9}{r}x^{9-r}\left(-\frac{3}{x^2}\right)^r$

$=\binom{9}{r}x^{9-r}\frac{(-3)^r}{x^{2r}}$

$=\binom{9}{r}(-3)^r x^{9-3r}$

The constant term does not contain x.

$\therefore\ 9-3r=0$

$\therefore\ r=3$

so $T_4=\binom{9}{3}(-3)^3x^0$

$\therefore$ the constant term is $\binom{9}{3}(-3)^3$.

5 **a, b**

Row 0	1		
Row 1	1 1 ⟵	sum $=1+1$	$=2$ $=2^1$
Row 2	1 2 1 ⟵	sum $=1+2+1$	$=4$ $=2^2$
Row 3	1 3 3 1 ⟵	sum $=1+3+3+1$	$=8$ $=2^3$
Row 4	1 4 6 4 1 ⟵	sum $=1+4+6+4+1$	$=16$ $=2^4$
Row 5	1 5 10 10 5 1 ⟵	sum $=1+5+10+10+5+1$	$=32$ $=2^5$

c The sum of the numbers in row n of Pascal's triangle is 2^n.

d $(1+x)^n$

$=\binom{n}{0}1^n+\binom{n}{1}1^{n-1}x+\binom{n}{2}1^{n-2}x^2+\binom{n}{3}1^{n-3}x^3+\ldots..+\binom{n}{n-1}1^1x^{n-1}+\binom{n}{n}x^n$

$=\binom{n}{0}+\binom{n}{1}x+\binom{n}{2}x^2+\binom{n}{3}x^3+\ldots..+\binom{n}{n-1}x^{n-1}+\binom{n}{n}x^n$ {as all powers of 1 are 1}

Now letting $x=1$ gives LHS $=(1+1)^n=2^n$

and RHS $=\binom{n}{0}+\binom{n}{1}+\binom{n}{2}+\binom{n}{3}+\ldots..+\binom{n}{n-1}+\binom{n}{n}$

$\therefore\ \binom{n}{0}+\binom{n}{1}+\binom{n}{2}+\ldots..+\binom{n}{n-1}+\binom{n}{n}=2^n$

6 **a** $(x+2)(x^2+1)^8$

$= (x+2)\left[(x^2)^8 + \binom{8}{1}(x^2)^7 1 + \binom{8}{2}(x^2)^6 1^2 + \ldots\ldots + \binom{8}{6}(x^2)^2 1^6 + \binom{8}{7}(x^2)^1 1^7 + \binom{8}{8} 1^8\right]$

only terms which when multiplied give an x^5

$\therefore$ coefficient of x^5 is $1 \times \binom{8}{6} = \binom{8}{6} = 28.$

b $(2-x)(3x+1)^9$

$= (2-x)\left[(3x)^9 + \binom{9}{1}(3x)^8 + \binom{9}{2}(3x)^7 + \binom{9}{3}(3x)^6 + \binom{9}{4}(3x)^5 + \ldots\ldots\right]$

$\therefore$ coefficient of x^6 is $2 \times \binom{9}{3} \times 3^6 + (-1) \times \binom{9}{4} \times 3^5 = 2\binom{9}{3}3^6 - \binom{9}{4}3^5 = 91\,854$

7 In $(x^2y - 2y^2)^6$, $a = (x^2y)$, $b = (-2y^2)$ and $n = 6$.

Now $T_{r+1} = \binom{n}{r} a^{n-r} b^r$

$= \binom{6}{r} (x^2y)^{6-r} (-2y^2)^r$

$= \binom{6}{r} x^{12-2r} y^{6-r} (-2)^r y^{2r}$

$= \binom{6}{r} (-2)^r x^{12-2r} y^{6+r}$

Since x and y are raised to the same power, $12 - 2r = 6 + r$

$\therefore\ 3r = 6$

$\therefore\ r = 2$

$T_3 = \binom{6}{2}(-2)^2 x^8 y^8$

$= 60x^8y^8$

8 **a** $(1+x)^n$ has $T_3 = \binom{n}{2} 1^{n-2} x^2 = \binom{n}{2} x^2$ and $n \geqslant 2$

But this term is $36x^2$ $\therefore\ \binom{n}{2} = 36$

$\therefore\ \dfrac{n(n-1)}{2} = 36$

$\therefore\ n(n-1) = 72$

$\therefore\ n^2 - n - 72 = 0$

$\therefore\ (n-9)(n+8) = 0$

$\therefore\ n = 9$ or -8

But $n \geqslant 2$, so $n = 9$

and $T_4 = \binom{n}{3} 1^{n-3} x^3$

$= \binom{9}{3} x^3$

$= 84x^3$

b $(1+kx)^n = 1^n + \binom{n}{1} 1^{n-1} (kx)^1 + \binom{n}{2} 1^{n-2} (kx)^2 + \ldots\ldots$

$= 1 + \binom{n}{1} kx + \binom{n}{2} k^2x^2 + \ldots\ldots$

$\therefore\ \binom{n}{1} k = -12$ and $\binom{n}{2} k^2 = 60$

$\therefore\ nk = -12$ and $\dfrac{n(n-1)}{2} k^2 = 60$

$\therefore\ n(n-1)k^2 = 120$

But $k = -\dfrac{12}{n}$ $\therefore\ n(n-1)\dfrac{144}{n^2} = 120$

$\therefore\ 144(n-1) = 120n \quad \{n \geqslant 2\}$

$\therefore\ 144n - 120n = 144$

$\therefore\ 24n = 144$

$\therefore\ n = 6$ and so $k = -2$

9 $T_{r+1} = \binom{n}{r}a^{n-r}b^r$ where $n = 10,\ a = (x^2),\ b = \left(\frac{1}{ax}\right)$

$= \binom{10}{r}(x^2)^{10-r}\left(\frac{1}{ax}\right)^r$

$= \binom{10}{r}x^{20-2r} \times \frac{1}{a^r x^r}$

$= \binom{10}{r}x^{20-3r} \times \frac{1}{a^r}$

We let $20 - 3r = 11$

$\therefore\ 3r = 9$

$\therefore\ r = 3$

and $T_4 = \binom{10}{3}x^{11} \times \frac{1}{a^3}$

$= \frac{\binom{10}{3}}{a^3}x^{11}$

So, $\frac{\binom{10}{3}}{a^3} = 15$

$\therefore\ \frac{120}{a^3} = 15$

$\therefore\ a^3 = 8$

$\therefore\ a = 2$

REVIEW SET 7

1 **a** $(x-2y)^3 = x^3 + 3x^2(-2y) + 3x(-2y)^2 + (-2y)^3$

$= x^3 - 6x^2y + 12xy^2 - 8y^3$

b $(3x+2)^4 = (3x)^4 + 4(3x)^3(2) + 6(3x)^2(2)^2 + 4(3x)(2)^3 + (2)^4$

$= 81x^4 + 216x^3 + 216x^2 + 96x + 16$

2 In the expansion of $(2x+5)^6$, $a = (2x),\ b = 5,\ n = 6$

$T_{r+1} = \binom{n}{r}a^{n-r}b^r$

$= \binom{6}{r}(2x)^{6-r}5^r$

$= \binom{6}{r}2^{6-r}x^{6-r}5^r$

For the coefficient of x^3 we let $6 - r = 3$

$\therefore\ r = 3$

and $T_4 = \underbrace{\binom{6}{3}2^3 5^3}\,x^3$

$\therefore$ the coefficient is $\binom{6}{3}2^3 5^3 = 20\,000$.

3 In the expansion of $\left(2x^2 - \frac{1}{x}\right)^6$, $a = (2x^2),\ b = \left(-\frac{1}{x}\right),\ n = 6$

$T_{r+1} = \binom{n}{r}a^{n-r}b^r$

$= \binom{6}{r}(2x^2)^{6-r}\left(-\frac{1}{x}\right)^r$

$= \binom{6}{r}2^{6-r}x^{12-2r}(-1)^r x^{-r}$

$= \binom{6}{r}2^{6-r}(-1)^r x^{12-3r}$

For the constant term we let $12 - 3r = 0$

$\therefore\ r = 4$

and $T_5 = \underbrace{\binom{6}{4}2^2(-1)^4}\,x^0$

$\therefore$ the constant term is $\binom{6}{4}2^2(-1)^4 = 60$.

4 The sixth row of Pascal's triangle is 1 6 15 20 15 6 1

$\therefore\ (a+b)^6 = a^6 + 6a^5b + 15a^4b^2 + 20a^3b^3 + 15a^2b^4 + 6ab^5 + b^6$

a $(x-3)^6 = x^6 + 6x^5(-3) + 15x^4(-3)^2 + 20x^3(-3)^3 + 15x^2(-3)^4 + 6x(-3)^5 + (-3)^6$

$= x^6 - 18x^5 + 135x^4 - 540x^3 + 1215x^2 - 1458x + 729$

b $\left(1 + \frac{1}{x}\right)^6$

$= (1)^6 + 6(1)^5\left(\frac{1}{x}\right) + 15(1)^4\left(\frac{1}{x}\right)^2 + 20(1)^3\left(\frac{1}{x}\right)^3 + 15(1)^2\left(\frac{1}{x}\right)^4 + 6(1)\left(\frac{1}{x}\right)^5 + \left(\frac{1}{x}\right)^6$

$= 1 + \frac{6}{x} + \frac{15}{x^2} + \frac{20}{x^3} + \frac{15}{x^4} + \frac{6}{x^5} + \frac{1}{x^6}$

5 $$\begin{aligned}(\sqrt{3}+2)^5 &= (\sqrt{3})^5+5(\sqrt{3})^4(2)+10(\sqrt{3})^3(2)^2+10(\sqrt{3})^2(2)^3+5(\sqrt{3})^1(2)^4+2^5\\ &= 9\sqrt{3}+90+120\sqrt{3}+240+80\sqrt{3}+32\\ &= 362+209\sqrt{3}\end{aligned}$$

6 $$\begin{aligned}(4+x)^3 &= 4^3+3(4)^2x^1+3(4)^1x^2+x^3\\ &= 64+48x+12x^2+x^3\end{aligned}$$

Letting $x=0.02$ gives $$\begin{aligned}(4.02)^3 &= 64+48(0.02)+12(0.02)^2+(0.02)^3\\ &= 64+0.96+0.0048+0.000\,008\\ &= 64.964\,808\end{aligned}$$

7 In $\left(2x-\dfrac{3}{x^2}\right)^{12}$, $a=(2x)$, $b=\left(-\dfrac{3}{x^2}\right)$, $n=12$

$$\begin{aligned}T_{r+1} &= \binom{n}{r}a^{n-r}b^r\\ &= \binom{12}{r}(2x)^{12-r}\left(-\frac{3}{x^2}\right)^r\\ &= \binom{12}{r}2^{12-r}x^{12-r}\,\frac{(-3)^r}{x^{2r}}\\ &= \binom{12}{r}2^{12-r}(-3)^r\,x^{12-3r}\end{aligned}$$

For the coefficient of x^{-6} we let $12-3r=-6$

$\therefore\ 3r=18$

$\therefore\ r=6$

So, $T_7=\underbrace{\binom{12}{6}2^6(-3)^6}\,x^{-6}$

$\therefore$ the coefficient is $\binom{12}{6}2^6(-3)^6=43\,110\,144.$

8 $$\begin{aligned}&(2x+3)(x-2)^6\\ &= (2x+3)\left[x^6+\binom{6}{1}x^5(-2)+\binom{6}{2}x^4(-2)^2+.....\right]\end{aligned}$$

$\therefore$ the coefficient of x^5 is $2\times\binom{6}{2}\times(-2)^2+3\times\binom{6}{1}\times(-2)=8\binom{6}{2}-6\binom{6}{1}=84$

9 $$\begin{aligned}(m-2n)^{10} &= m^{10}+\binom{10}{1}m^9(-2n)+\binom{10}{2}m^8(-2n)^2+....+(-2n)^{10}\\ &= m^{10}-20m^9n+45m^8(4n^2)+....+1024n^{10}\\ &= m^{10}-20m^9n+180m^8n^2+....+1024n^{10}\end{aligned}$$

$\therefore\ k=180$

10 $(1+cx)(1+x)^4=(1+cx)(1^4+\binom{4}{1}1^3x+\binom{4}{2}1^2x^2+\binom{4}{3}1x^3+x^4)$

$\therefore$ coefficient of x^3 is $1\times\binom{4}{3}\times1+c\times\binom{4}{2}\times1^2=4+6c$

But the coefficient of x^3 is 22, so $4+6c=22$

$\therefore\ 6c=18$ $\quad\therefore\ c=3$

11 **a** 7 terms

b In $\left(3x^2+\dfrac{1}{x}\right)^6$, $a=(3x^2)$, $b=\left(\dfrac{1}{x}\right)$ and $n=6$.

Now $$\begin{aligned}T_{r+1} &= \binom{n}{r}a^{n-r}b^r\\ &= \binom{6}{r}(3x^2)^{6-r}\left(\frac{1}{x}\right)^r\\ &= \binom{6}{r}3^{6-r}x^{12-2r}\frac{1}{x^r}\\ &= \binom{6}{r}3^{6-r}x^{12-3r}\end{aligned}$$

The constant term does not contain x

$\therefore\ 12-3r=0$

$\therefore\ r=4$

So, $T_5=\binom{6}{4}3^2x^0=135$

c To find a term involving x^5, we solve

$12-3r=5$

$\therefore\ 3r=7$

$\therefore\ r=\frac{7}{3}$

But r must be an integer.

$\therefore$ no term in the expansion involves x^5.

Chapter 8
THE UNIT CIRCLE AND RADIAN MEASURE

EXERCISE 8A

1 **a** $180^o = \pi$ radians
$\therefore\ 90^o = \frac{\pi}{2}$ radians

b $180^o = \pi$ radians
$\therefore\ 60^o = \frac{\pi}{3}$ radians

c $180^o = \pi$ radians
$\therefore\ 30^o = \frac{\pi}{6}$ radians

d $180^o = \pi$ radians
$\therefore\ 18^o = \frac{\pi}{10}$ radians

e $180^o = \pi$ radians
$\therefore\ 9^o = \frac{\pi}{20}$ radians

f $180^o = \pi$ radians
$\therefore\ 45^o = \frac{\pi}{4}$ radians
$\therefore\ 135^o = \frac{3\pi}{4}$ radians

g $180^o = \pi$ radians
$\therefore\ 45^o = \frac{\pi}{4}$ radians
$\therefore\ 225^o = \frac{5\pi}{4}$ radians

h $180^o = \pi$ radians
$\therefore\ 90^o = \frac{\pi}{2}$ radians
$\therefore\ 270^o = \frac{3\pi}{2}$ radians

i $360^o = 2 \times 180^o$
$= 2\pi$ radians

j $720^o = 4 \times 180^o$
$= 4\pi$ radians

k $180^o = \pi$ radians
$\therefore\ 45^o = \frac{\pi}{4}$ radians
$\therefore\ 315^o = \frac{7\pi}{4}$ radians

l $180^o = \pi$ radians
$\therefore\ 540^o = 3\pi$ radians

m $180^o = \pi$ radians
$\therefore\ 36^o = \frac{\pi}{5}$ radians

n $180^o = \pi$ radians
$\therefore\ 10^o = \frac{\pi}{18}$ radians
$\therefore\ 80^o = \frac{8\pi}{18}$ radians
$= \frac{4\pi}{9}$ radians

o $180^o = \pi$ radians
$\therefore\ 10^o = \frac{\pi}{18}$ radians
$\therefore\ 230^o = \frac{23\pi}{18}$ radians

2 **a** 36.7^o
$= 36.7 \times \frac{\pi}{180}$ radians
≈ 0.641 radians

b 137.2^o
$= 137.2 \times \frac{\pi}{180}$ radians
≈ 2.39 radians

c 317.9^o
$= 317.9 \times \frac{\pi}{180}$ radians
≈ 5.55 radians

d 219.6^o
$= 219.6 \times \frac{\pi}{180}$ radians
≈ 3.83 radians

e 396.7^o
$= 396.7 \times \frac{\pi}{180}$ radians
≈ 6.92 radians

3 **a** $\frac{\pi}{5} = \frac{180^o}{5} = 36^o$

b $\frac{3\pi}{5} = \frac{3\times180^o}{5} = 108^o$

c $\frac{3\pi}{4} = \frac{3\times180^o}{4} = 135^o$

d $\frac{\pi}{18} = \frac{180^o}{18} = 10^o$

e $\frac{\pi}{9} = \frac{180^o}{9} = 20^o$

f $\frac{7\pi}{9} = \frac{7\times180^o}{9} = 140^o$

g $\frac{\pi}{10} = \frac{180^o}{10} = 18^o$

h $\frac{3\pi}{20} = \frac{3\times180^o}{20} = 27^o$

i $\frac{5\pi}{6} = \frac{5\times180^o}{6} = 150^o$

j $\frac{\pi}{8} = \frac{180^o}{8} = 22.5^o$

4 **a** 2^c
$= 2 \times \frac{180}{\pi}$ degrees
$\approx 114.59^o$

b 1.53^c
$= 1.53 \times \frac{180}{\pi}$ degrees
$\approx 87.66^o$

c 0.867^c
$= 0.867 \times \frac{180}{\pi}$ degrees
$\approx 49.68^o$

d 3.179^c
$= 3.179 \times \frac{180}{\pi}$ degrees
$\approx 182.14^o$

e 5.267^c
$= 5.267 \times \frac{180}{\pi}$ degrees
$\approx 301.78^o$

5 **a**

Degrees	0	45	90	135	180	225	270	315	360
Radians	0	$\frac{\pi}{4}$	$\frac{\pi}{2}$	$\frac{3\pi}{4}$	π	$\frac{5\pi}{4}$	$\frac{3\pi}{2}$	$\frac{7\pi}{4}$	2π

b

Degrees	0	30	60	90	120	150	180	210	240	270	300	330	360
Radians	0	$\frac{\pi}{6}$	$\frac{\pi}{3}$	$\frac{\pi}{2}$	$\frac{2\pi}{3}$	$\frac{5\pi}{6}$	π	$\frac{7\pi}{6}$	$\frac{4\pi}{3}$	$\frac{3\pi}{2}$	$\frac{5\pi}{3}$	$\frac{11\pi}{6}$	2π

EXERCISE 8B

1 **a** **i** arc length $= \frac{7\pi}{4} \times 9$
≈ 49.5 cm

ii area $= \frac{1}{2} \times \frac{7\pi}{4} \times 9^2$
≈ 223 cm^2

b **i** arc length $= 4.67 \times 4.93$
≈ 23.0 cm

ii area $= \frac{1}{2}(4.67) \times 4.93^2$
≈ 56.8 cm^2

2 **a** $\theta = 107.9^o$, $\quad l = 5.92$

$\therefore \quad \left(\frac{107.9}{360}\right) \times 2\pi \times r = 5.92$

$\therefore \quad r = \dfrac{5.92 \times 360}{107.9 \times 2 \times \pi}$

$\therefore \quad r \approx 3.14$ m

b area $= \left(\frac{107.9}{360}\right) \times \pi \times (3.1436)^2$
≈ 9.30 m^2

3 **a** area $= \frac{1}{2}\theta r^2$

$\therefore \quad 20.8 = \frac{1}{2}(1.19) \times r^2$

$\therefore \quad \dfrac{20.8 \times 2}{1.19} = r^2$

$\therefore \quad r = \sqrt{\dfrac{20.8 \times 2}{1.19}}$

$\therefore \quad r \approx 5.91$ cm

b perimeter
$= l + 2r$
$\approx 1.19 \times 5.912 + 2 \times 5.912$
≈ 18.9 cm

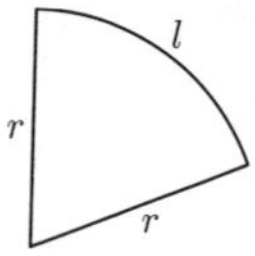

4 **a**

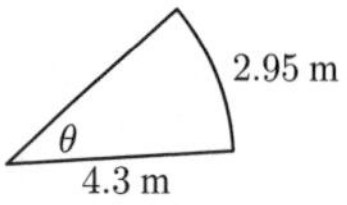

$l = \theta \times r$

$\therefore \quad 2.95 = \theta \times 4.3$

$\therefore \quad \theta \approx 0.686^c$

b area $= \frac{1}{2}\theta r^2$

$\therefore \quad 30 = \frac{1}{2} \times \theta \times 10^2$

$\therefore \quad \dfrac{30 \times 2}{100} = \theta$

$\therefore \quad \theta = 0.6^c$

5 **a**

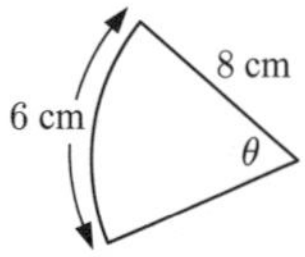

$l = \theta r$

$\therefore \quad 6 = \theta \times 8$

$\therefore \quad \theta = \frac{6}{8}$

$\therefore \quad \theta = 0.75^c$

area $= \frac{1}{2}\theta r^2$
$= \frac{1}{2}(0.75) \times 8^2$
$= 24$ cm^2

b

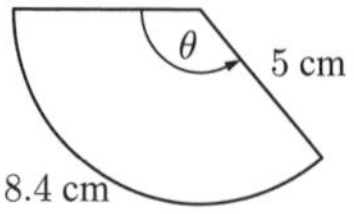

$l = \theta r$

$\therefore \quad 8.4 = \theta \times 5$

$\therefore \quad \theta = \frac{8.4}{5}$

$\therefore \quad \theta = 1.68^c$

area $= \frac{1}{2}\theta r^2$
$= \frac{1}{2}(1.68) \times 5^2$
$= 21$ cm^2

c

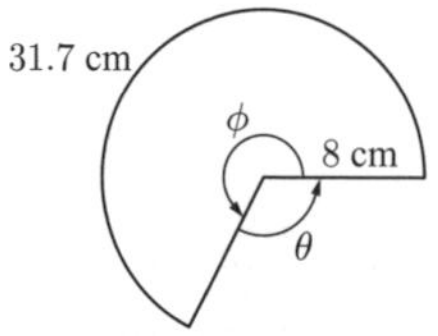

$l = \phi r$

$\therefore \quad 31.7 = \phi \times 8$

$\therefore \quad \phi = \frac{31.7}{8}$

$\therefore \quad \phi \approx 3.96^c$

But $\theta = 2\pi - \phi$

$\therefore \quad \theta \approx 2.32^c$

area $= \frac{1}{2}\phi r^2$
$= \frac{1}{2} \times \frac{31.7}{8} \times 8^2$
$= 126.8$ cm^2

6

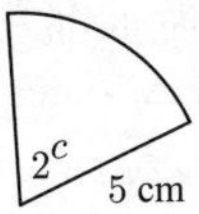

$$\begin{aligned}\text{arc length} &= \theta r\\ &= 2\times 5\\ &= 10 \text{ cm}\end{aligned}$$

$$\begin{aligned}\text{area} &= \tfrac{1}{2}\theta r^2\\ &= \tfrac{1}{2}\times 2\times 5^2\\ &= 25 \text{ cm}^2\end{aligned}$$

7

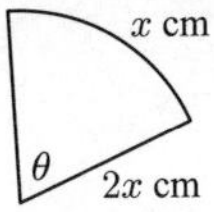

$$\begin{aligned}\text{arc length} &= \theta r\\ \therefore\quad x &= \theta(2x)\\ \therefore\quad \theta &= \tfrac{1}{2}\end{aligned}$$

$$\begin{aligned}\text{area} &= \tfrac{1}{2}\theta r^2\\ &= \tfrac{1}{2}\times(\tfrac{1}{2})\times(2x)^2\\ &= x^2 \text{ cm}^2\end{aligned}$$

8 **a**

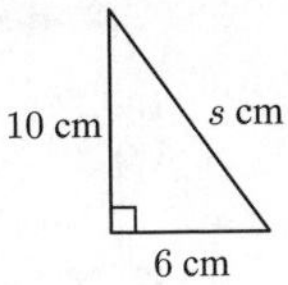

$$\begin{aligned}s^2 &= 6^2+10^2 \quad \{\text{Pythagoras}\}\\ \therefore\quad s &= \sqrt{6^2+10^2}\\ \therefore\quad s &\approx 11.6619\\ \therefore\quad s &\approx 11.7\end{aligned}$$

b $r = s \approx 11.7$

c
$$\begin{aligned}\text{arclength} &= \text{circumference of cone base}\\ &= 2\pi\times 6\\ &\approx 37.6991\\ &\approx 37.7 \text{ cm}\end{aligned}$$

d
$$\begin{aligned}\text{arc length} &= \theta r\\ \therefore\quad 37.6991 &\approx \theta\times 11.6619\\ \therefore\quad \theta &\approx \frac{37.6991}{11.6619}\\ \therefore\quad \theta &\approx 3.23 \text{ radians}\end{aligned}$$

9

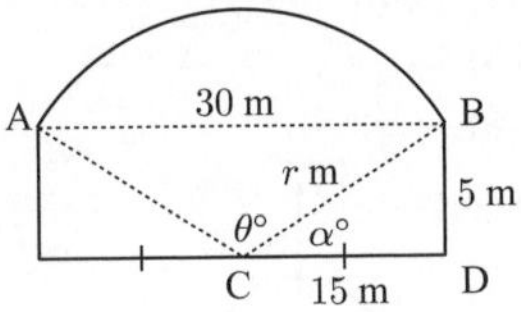

a
$$\begin{aligned}\tan\alpha &= \tfrac{5}{15}\\ \therefore\quad \alpha &= \tan^{-1}\left(\tfrac{1}{3}\right)\\ \therefore\quad \alpha &\approx 18.43\end{aligned}$$

b
$$\begin{aligned}\theta + 2\alpha &= 180 \quad \{\text{angles on a line}\}\\ \therefore\quad \theta &\approx 180 - 2\times 18.43\\ \therefore\quad \theta &\approx 143.1\end{aligned}$$

c
$$\begin{aligned}\text{area} &= 2\times\text{area of }\Delta\text{CDB} + \text{area of sector}\\ &= 2\times\tfrac{1}{2}\times \text{CD}\times\text{BD} + \left(\tfrac{\theta}{360}\right)\times\pi\times r^2\end{aligned}$$

Now $r^2 = 5^2 + 15^2 = 250$

$$\begin{aligned}\therefore\quad \text{area} &\approx 2\times\tfrac{1}{2}\times 15\times 5 + \left(\tfrac{143.1}{360}\right)\times\pi\times 250\\ &\approx 387 \text{ m}^2\end{aligned}$$

10 Since [AT] is a tangent, $\widehat{\text{OTA}}$ is a right angle.

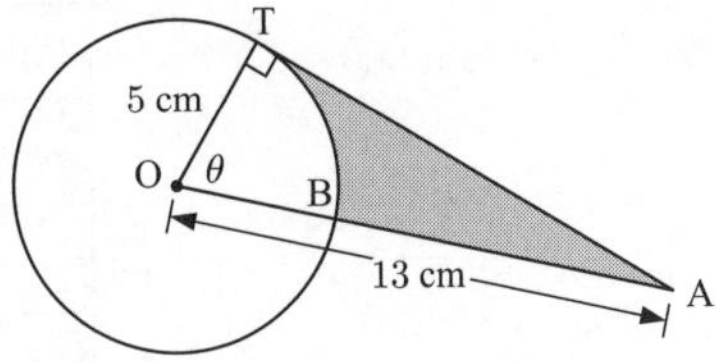

$$\begin{aligned}\therefore\quad \cos\theta &= \tfrac{5}{13}\\ \therefore\quad \theta &\approx 67.38^o\end{aligned}$$

$$\begin{aligned}\text{arc length BT} &= \left(\tfrac{\theta}{360}\right)\times 2\pi r\\ &\approx \tfrac{67.38}{360}\times 2\times\pi\times 5\\ &\approx 5.88 \text{ cm}\end{aligned}$$

$$\begin{aligned}\text{AT}^2 + \text{OT}^2 &= \text{OA}^2 \quad \{\text{Pythagoras}\}\\ \therefore\quad \text{AT}^2 &= 13^2 - 5^2\\ \therefore\quad \text{AT} &= 12 \text{ cm}\end{aligned}$$

$$\begin{aligned}\therefore\quad \text{perimeter} &= \text{AT} + \text{arc length BT} + \text{AB}\\ &\approx 12 + 5.88 + (13-5)\\ &\approx 25.9 \text{ cm}\end{aligned}$$

11 **a**
$$\begin{aligned}l &= \left(\tfrac{\theta}{360}\right)\times 2\pi r\\ &= \frac{\frac{1}{60}}{360}\times 2\times\pi\times 6370 \text{ km}\\ &\approx 1.853 \text{ km}\end{aligned}$$

b
$$\text{speed} = \frac{\text{distance}}{\text{time}} \qquad \therefore\quad \text{time} = \frac{\text{distance}}{\text{speed}}$$

$$\begin{aligned} &= \frac{2130 \text{ km}}{480 \text{ n miles h}^{-1}}\\ &= \frac{2130 \text{ km}}{480\times 1.853 \text{ km h}^{-1}}\\ &\approx 2.395 \text{ hours}\\ &\approx 2 \text{ hours } 24 \text{ min}\end{aligned}$$

12

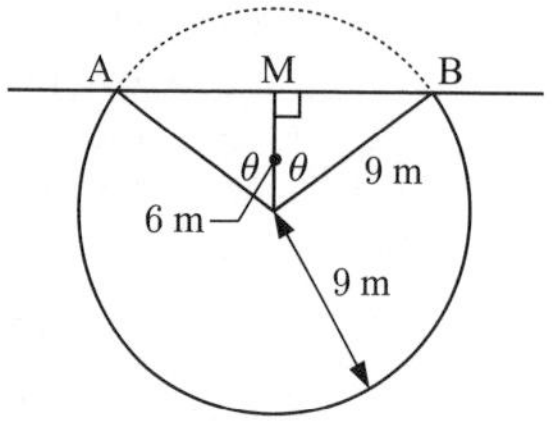

$\cos\theta = \frac{6}{9} = \frac{2}{3}$

$\therefore \quad \theta = \cos^{-1}\left(\frac{2}{3}\right)$

$\therefore \quad \theta \approx 48.19^o$

So, $360 - 2\theta \approx 263.62^o$

Now $\text{MB} = \sqrt{9^2 - 6^2}$

$= \sqrt{45}$

$\therefore$ available feeding area

= area of Δ + area of sector

$\approx \frac{1}{2} \times 2 \times \sqrt{45} \times 6 + \left(\frac{263.62}{360}\right) \times \pi \times 9^2$

$\approx 227\ \text{m}^2$

EXERCISE 8C.1

1 **a**

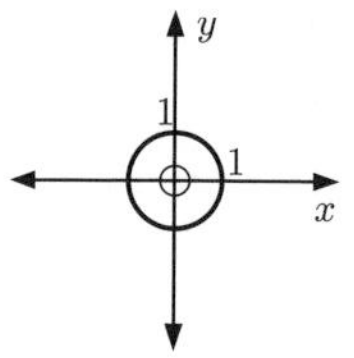

b

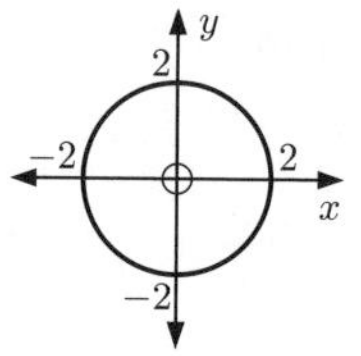

c

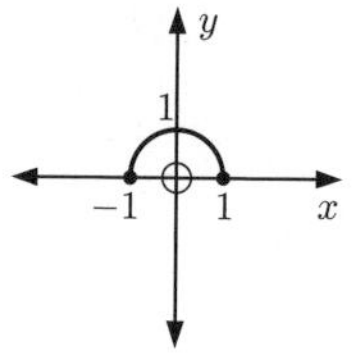

2 **a** **i** A($\cos 26^o$, $\sin 26^o$), B($\cos 146^o$, $\sin 146^o$), C($\cos 199^o$, $\sin 199^o$)

ii A(0.899, 0.438), B(−0.829, 0.559), C(−0.946, −0.326)

b **i** A($\cos 123^o$, $\sin 123^o$), B($\cos 251^o$, $\sin 251^o$), C($\cos(-35^o)$, $\sin(-35^o)$)

ii A(−0.545, 0.839), B(−0.326, −0.946), C(0.819, −0.574)

3

θ (degrees)	0°	90°	180°	270°	360°	450°
θ (radians)	0	$\frac{\pi}{2}$	π	$\frac{3\pi}{2}$	2π	$\frac{5\pi}{2}$
sine	0	1	0	−1	0	1
cosine	1	0	−1	0	1	0
tangent	0	undef.	0	undef.	0	undef.

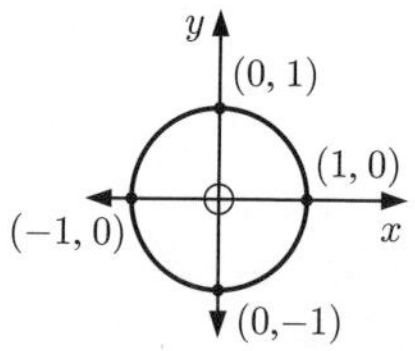

4 **a** **i** $\frac{1}{\sqrt{2}} \approx 0.707$

ii $\frac{\sqrt{3}}{2} \approx 0.866$

b

θ (degrees)	30°	45°	60°	135°	150°	240°	315°
θ (radians)	$\frac{\pi}{6}$	$\frac{\pi}{4}$	$\frac{\pi}{3}$	$\frac{3\pi}{4}$	$\frac{5\pi}{6}$	$\frac{4\pi}{3}$	$\frac{7\pi}{4}$
sine	$\frac{1}{2}$	$\frac{1}{\sqrt{2}}$	$\frac{\sqrt{3}}{2}$	$\frac{1}{\sqrt{2}}$	$\frac{1}{2}$	$-\frac{\sqrt{3}}{2}$	$-\frac{1}{\sqrt{2}}$
cosine	$\frac{\sqrt{3}}{2}$	$\frac{1}{\sqrt{2}}$	$\frac{1}{2}$	$-\frac{1}{\sqrt{2}}$	$-\frac{\sqrt{3}}{2}$	$-\frac{1}{2}$	$\frac{1}{\sqrt{2}}$
tangent	$\frac{1}{\sqrt{3}}$	1	$\sqrt{3}$	−1	$-\frac{1}{\sqrt{3}}$	$\sqrt{3}$	−1

5 **a** **i** 0.985 **ii** 0.985 **iii** 0.866 **iv** 0.866 **v** 0.5 **vi** 0.5 **vii** 0.707 **viii** 0.707

b $\sin(180^o - \theta) = \sin\theta$

c

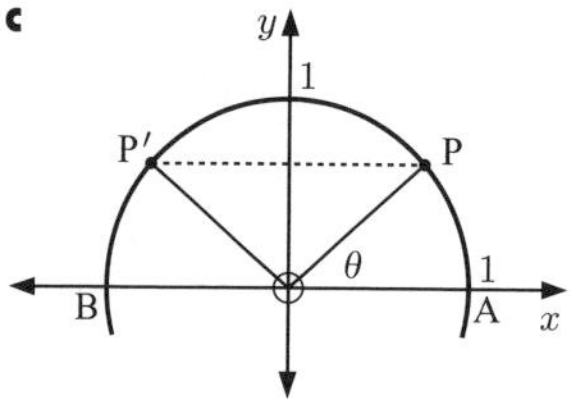

The diagram shows P reflected in the y-axis to P′, so $\text{P}'\widehat{\text{O}}\text{B} = \text{P}\widehat{\text{O}}\text{A} = \theta$, and P′ has coordinates $(-\cos\theta, \sin\theta)$.

But $\text{A}\widehat{\text{O}}\text{P}' = 180^o - \theta$ $\{\text{A}\widehat{\text{O}}\text{P}' + \text{P}'\widehat{\text{O}}\text{B} = 180^o\}$,

so P′ has coordinates $(\cos(180^o - \theta), \sin(180^o - \theta))$.

$\therefore$ $\sin(180^o - \theta) = \sin\theta$ {equating y-coordinates of P′}

d **i** $180^o - 45^o = 135^o$ **ii** $180^o - 51^o = 129^o$ **iii** $\pi - \frac{\pi}{3} = \frac{2\pi}{3}$

iv $\pi - \frac{\pi}{6} = \frac{5\pi}{6}$ {using $\sin(180^o - \theta) = \theta$}

6 **a** **i** 0.342 **ii** −0.342 **iii** 0.5 **iv** −0.5 **v** 0.906 **vi** −0.906
vii 0.174 **viii** −0.174

b $\cos(180^o - \theta) = -\cos\theta$

c

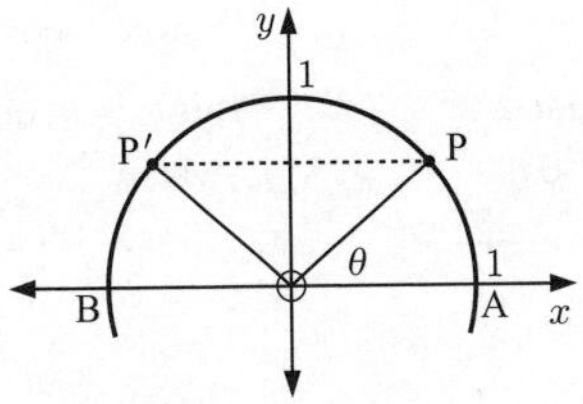

The diagram shows P reflected in the y-axis to P′, so $P'\widehat{O}B = P\widehat{O}A = \theta$, and P′ has coordinates $(-\cos\theta, \sin\theta)$.
But $A\widehat{O}P' = 180^o - \theta$ $\{A\widehat{O}P' + P'\widehat{O}B = 180^o\}$,
so P′ has coordinates $(\cos(180^o - \theta), \sin(180^o - \theta))$.
$\therefore \cos(180^o - \theta) = -\cos\theta$ {equating x-coordinates of P′}

d **i** $180^o - 40^o = 140^o$ **ii** $180^o - 19^o = 161^o$ **iii** $\pi - \frac{\pi}{5} = \frac{4\pi}{5}$
iv $\pi - \frac{2\pi}{5} = \frac{3\pi}{5}$ {using $\cos(180^o - \theta) = -\cos\theta$}

7 **a** $\sin 137^o$
$= \sin(180 - 137)^o$
$= \sin 43^o$
≈ 0.6820

b $\sin 59^o$
$= \sin(180 - 59)^o$
$= \sin 121^o$
≈ 0.8572

c $\cos 143^o$
$= -\cos(180 - 143)^o$
$= -\cos 37^o$
≈ -0.7986

d $\cos 24^o$
$= -\cos(180 - 24)^o$
$= -\cos 156^o$
≈ 0.9135

e $\sin 115^o$
$= \sin(180 - 115)^o$
$= \sin 65^o$
≈ 0.9063

f $\cos 132^o$
$= -\cos(180 - 132)^o$
$= -\cos 48^o$
≈ -0.6691

8 **a**

Quadrant	*Degree measure*	*Radian measure*	$\cos\theta$	$\sin\theta$	$\tan\theta$
1	$0^o < \theta < 90^o$	$0 < \theta < \frac{\pi}{2}$	+ve	+ve	+ve
2	$90^o < \theta < 180^o$	$\frac{\pi}{2} < \theta < \pi$	−ve	+ve	−ve
3	$180^o < \theta < 270^o$	$\pi < \theta < \frac{3\pi}{2}$	−ve	−ve	+ve
4	$270^o < \theta < 360^o$	$\frac{3\pi}{2} < \theta < 2\pi$	+ve	−ve	−ve

b **i** 1 and 4 **ii** 2 and 3 **iii** 3 **iv** 2

9 **a** $A\widehat{O}Q = 180^o - \theta$
b [OQ] is a reflection of [OP] in the y-axis and so Q has coordinates $(-\cos\theta, \sin\theta)$.
c $\cos(180^o - \theta) = -\cos\theta$, $\sin(180^o - \theta) = \sin\theta$

EXERCISE 8C.2

1 **a** $\cos^2\theta + \sin^2\theta = 1$
$\therefore \cos^2\theta + (\frac{1}{2})^2 = 1$
$\therefore \cos^2\theta = \frac{3}{4}$
$\therefore \cos\theta = \pm\frac{\sqrt{3}}{2}$

b $\cos^2\theta + \sin^2\theta = 1$
$\therefore \cos^2\theta + (-\frac{1}{3})^2 = 1$
$\therefore \cos^2\theta = \frac{8}{9}$
$\therefore \cos\theta = \pm\frac{\sqrt{8}}{3}$
$= \pm\frac{2\sqrt{2}}{3}$

c $\cos^2\theta + \sin^2\theta = 1$
$\therefore \cos^2\theta + 0^2 = 1$
$\therefore \cos\theta = \pm 1$

d $\cos^2\theta + \sin^2\theta = 1$
$\therefore \cos^2\theta + (-1)^2 = 1$
$\therefore \cos\theta = 0$

2 **a** $\cos^2\theta + \sin^2\theta = 1$
$\therefore (\frac{4}{5})^2 + \sin^2\theta = 1$
$\therefore \sin^2\theta = \frac{9}{25}$
$\therefore \sin\theta = \pm\frac{3}{5}$

b $\cos^2\theta + \sin^2\theta = 1$
$\therefore (-\frac{3}{4})^2 + \sin^2\theta = 1$
$\therefore \sin^2\theta = \frac{7}{16}$
$\therefore \sin\theta = \pm\frac{\sqrt{7}}{4}$

c $\cos^2\theta + \sin^2\theta = 1$
$\therefore 1^2 + \sin^2\theta = 1$
$\therefore \sin^2\theta = 0$
$\therefore \sin\theta = 0$

d $\cos^2\theta + \sin^2\theta = 1$
$\therefore 0^2 + \sin^2\theta = 1$
$\therefore \sin\theta = \pm 1$

3 **a**
$\cos^2\theta + \sin^2\theta = 1$
$\therefore \quad \frac{4}{9} + \sin^2\theta = 1$
$\therefore \quad \sin^2\theta = \frac{5}{9}$
$\therefore \quad \sin\theta = \pm\frac{\sqrt{5}}{3}$
But θ is in quadrant 1
where $\sin\theta > 0$
$\therefore \quad \sin\theta = \frac{\sqrt{5}}{3}$

b
$\cos^2\theta + \sin^2\theta = 1$
$\therefore \quad \cos^2\theta + \frac{4}{25} = 1$
$\therefore \quad \cos^2\theta = \frac{21}{25}$
$\therefore \quad \cos\theta = \pm\frac{\sqrt{21}}{5}$
But θ is in quadrant 2
where $\cos\theta < 0$
$\therefore \quad \cos\theta = -\frac{\sqrt{21}}{5}$

c
$\cos^2\theta + \sin^2\theta = 1$
$\therefore \quad \cos^2\theta + \frac{9}{25} = 1$
$\therefore \quad \cos^2\theta = \frac{16}{25}$
$\therefore \quad \cos\theta = \pm\frac{4}{5}$
But θ is in quadrant 4
where $\cos\theta > 0$
$\therefore \quad \cos\theta = \frac{4}{5}$

d
$\cos^2\theta + \sin^2\theta = 1$
$\therefore \quad \frac{25}{169} + \sin^2\theta = 1$
$\therefore \quad \sin^2\theta = \frac{144}{169}$
$\therefore \quad \sin\theta = \pm\frac{12}{13}$
But θ is in quadrant 3
where $\sin\theta < 0$
$\therefore \quad \sin\theta = -\frac{12}{13}$

4 **a**
$\cos^2 x + \sin^2 x = 1$
$\therefore \quad \cos^2 x + \frac{1}{9} = 1$
$\therefore \quad \cos^2 x = \frac{8}{9}$
$\therefore \quad \cos x = \pm\frac{2\sqrt{2}}{3}$
But x is in quadrant 2
where $\cos x < 0$
$\therefore \quad \cos x = -\frac{2\sqrt{2}}{3}$
and so $\quad \tan x = \dfrac{\sin x}{\cos x} = \dfrac{\frac{1}{3}}{-\frac{2\sqrt{2}}{3}} = -\frac{1}{2\sqrt{2}}$

b
$\cos^2 x + \sin^2 x = 1$
$\therefore \quad \frac{1}{25} + \sin^2 x = 1$
$\therefore \quad \sin^2 x = \frac{24}{25}$
$\therefore \quad \sin x = \pm\frac{2\sqrt{6}}{5}$
But x is in quadrant 4
where $\sin x < 0$
$\therefore \quad \sin x = -\frac{2\sqrt{6}}{5}$
and so $\quad \tan x = \dfrac{\sin x}{\cos x} = \dfrac{-\frac{2\sqrt{6}}{5}}{\frac{1}{5}} = -2\sqrt{6}$

c
$\cos^2 x + \sin^2 x = 1$
$\therefore \quad \cos^2 x + \frac{1}{3} = 1$
$\therefore \quad \cos^2 x = \frac{2}{3}$
$\therefore \quad \cos x = \pm\frac{\sqrt{2}}{\sqrt{3}}$
But x is in quadrant 3
where $\cos x < 0$
$\therefore \quad \cos x = -\frac{\sqrt{2}}{\sqrt{3}}$
and so $\quad \tan x = \dfrac{\sin x}{\cos x} = \dfrac{-\frac{1}{\sqrt{3}}}{-\frac{\sqrt{2}}{\sqrt{3}}} = \frac{1}{\sqrt{2}}$

d
$\cos^2 x + \sin^2 x = 1$
$\therefore \quad \frac{9}{16} + \sin^2 x = 1$
$\therefore \quad \sin^2 x = \frac{7}{16}$
$\therefore \quad \sin x = \pm\frac{\sqrt{7}}{4}$
But x is in quadrant 2
where $\sin x > 0$
$\therefore \quad \sin x = \frac{\sqrt{7}}{4}$
and so $\quad \tan x = \dfrac{\sin x}{\cos x} = \dfrac{\frac{\sqrt{7}}{4}}{-\frac{3}{4}} = -\frac{\sqrt{7}}{3}$

5 **a**
$\dfrac{\sin x}{\cos x} = \frac{2}{3}$
$\therefore \quad \sin x = \frac{2}{3}\cos x$

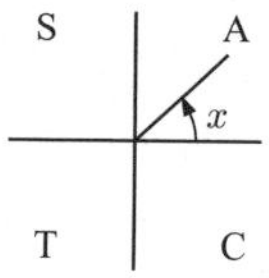

Now $\quad \cos^2 x + \sin^2 x = 1$
$\therefore \quad \cos^2 x + \frac{4}{9}\cos^2 x = 1$
$\therefore \quad \frac{13}{9}\cos^2 x = 1$
$\therefore \quad \cos x = \pm\frac{3}{\sqrt{13}}$
But x is in quadrant 1
$\therefore \quad \cos x$ and $\sin x$ are positive.
$\therefore \quad \cos x = \frac{3}{\sqrt{13}}, \quad \sin x = \frac{2}{\sqrt{13}}$

b
$\dfrac{\sin x}{\cos x} = -\frac{4}{3}$
$\therefore \quad \sin x = -\frac{4}{3}\cos x$

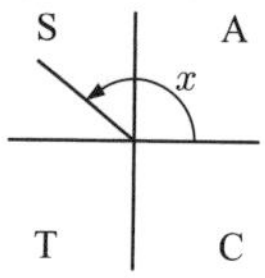

Now $\quad \cos^2 x + \sin^2 x = 1$
$\therefore \quad \cos^2 x + \frac{16}{9}\cos^2 x = 1$
$\therefore \quad \frac{25}{9}\cos^2 x = 1$
$\therefore \quad \cos x = \pm\frac{3}{5}$
But x is in quadrant 2
$\therefore \quad \cos x$ is negative and $\sin x$ is positive.
$\therefore \quad \cos x = -\frac{3}{5}, \quad \sin x = \frac{4}{5}$

c $\dfrac{\sin x}{\cos x} = \frac{\sqrt{5}}{3}$

$\therefore \quad \sin x = \frac{\sqrt{5}}{3}\cos x$

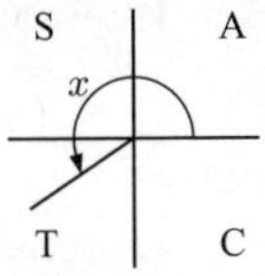

Now $\cos^2 x + \sin^2 x = 1$

$\therefore \quad \cos^2 x + \frac{5}{9}\cos^2 x = 1$

$\therefore \quad \frac{14}{9}\cos^2 x = 1$

$\therefore \quad \cos x = \pm\frac{3}{\sqrt{14}}$

But x is in quadrant 3

$\therefore \quad \cos x$ and $\sin x$ are both negative.

$\therefore \quad \cos x = -\frac{3}{\sqrt{14}}, \quad \sin x = -\frac{\sqrt{5}}{\sqrt{14}}$

d $\dfrac{\sin x}{\cos x} = -\frac{12}{5}$

$\therefore \quad \sin x = -\frac{12}{5}\cos x$

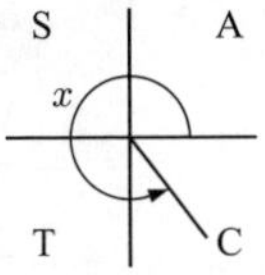

Now $\cos^2 x + \sin^2 x = 1$

$\therefore \quad \cos^2 x + \frac{144}{25}\cos^2 x = 1$

$\therefore \quad \frac{169}{25}\cos^2 x = 1$

$\therefore \quad \cos x = \pm\frac{5}{13}$

But x is in quadrant 4

$\therefore \quad \cos x$ is positive and $\sin x$ is negative.

$\therefore \quad \cos x = \frac{5}{13}, \quad \sin x = -\frac{12}{13}$

EXERCISE 8C.3

1

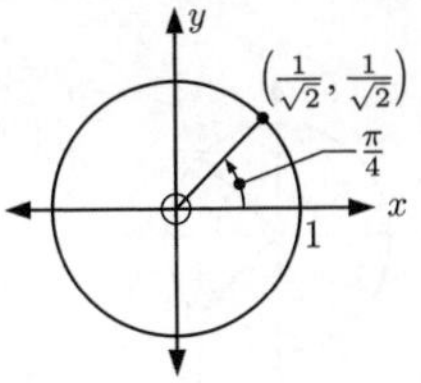

So $\cos(\frac{\pi}{4}) = \frac{1}{\sqrt{2}}$

$\sin(\frac{\pi}{4}) = \frac{1}{\sqrt{2}}$

$\tan(\frac{\pi}{4}) = \dfrac{\frac{1}{\sqrt{2}}}{\frac{1}{\sqrt{2}}} = 1$

You should draw separate unit circle diagrams for each case.

	a	b	c	d	e
$\sin\theta$	$\frac{1}{\sqrt{2}}$	$-\frac{1}{\sqrt{2}}$	$-\frac{1}{\sqrt{2}}$	0	$-\frac{1}{\sqrt{2}}$
$\cos\theta$	$\frac{1}{\sqrt{2}}$	$-\frac{1}{\sqrt{2}}$	$\frac{1}{\sqrt{2}}$	-1	$-\frac{1}{\sqrt{2}}$
$\tan\theta$	1	1	-1	0	1

2

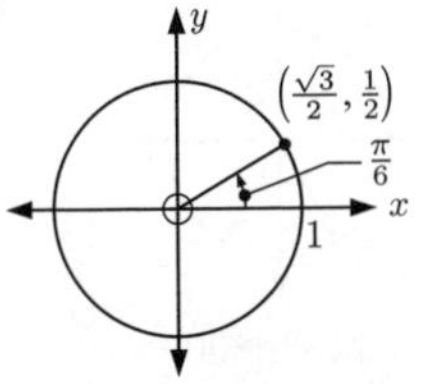

So $\cos(\frac{\pi}{6}) = \frac{\sqrt{3}}{2}$

$\sin(\frac{\pi}{6}) = \frac{1}{2}$

$\tan(\frac{\pi}{6}) = \dfrac{\frac{1}{2}}{\frac{\sqrt{3}}{2}} = \frac{1}{\sqrt{3}}$

You should draw separate unit circle diagrams for each case.

	a	b	c	d	e
$\sin\beta$	$\frac{1}{2}$	$\frac{\sqrt{3}}{2}$	$-\frac{1}{2}$	$-\frac{\sqrt{3}}{2}$	$-\frac{1}{2}$
$\cos\beta$	$\frac{\sqrt{3}}{2}$	$-\frac{1}{2}$	$-\frac{\sqrt{3}}{2}$	$\frac{1}{2}$	$\frac{\sqrt{3}}{2}$
$\tan\beta$	$\frac{1}{\sqrt{3}}$	$-\sqrt{3}$	$\frac{1}{\sqrt{3}}$	$-\sqrt{3}$	$-\frac{1}{\sqrt{3}}$

3 **a** $\sin^2 60^o$

$= \sin 60^o \times \sin 60^o$

$= \frac{\sqrt{3}}{2} \times \frac{\sqrt{3}}{2}$

$= \frac{3}{4}$

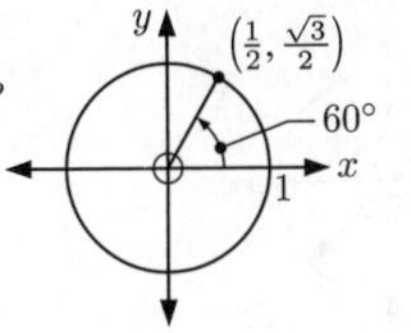

b $\sin 30^o \cos 60^o$

$= \frac{1}{2} \times \frac{1}{2}$

$= \frac{1}{4}$

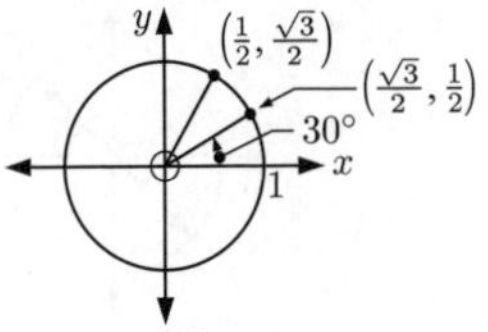

c $4\sin 60^o \cos 30^o$

$= 4\left(\frac{\sqrt{3}}{2}\right)\left(\frac{\sqrt{3}}{2}\right)$

$= 3$

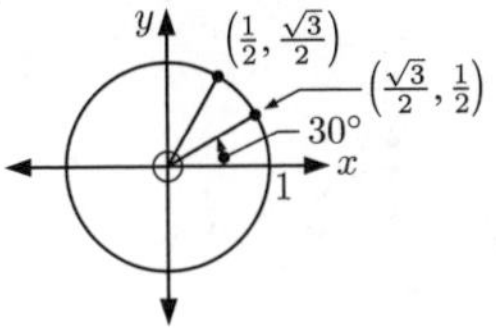

d $1 - \cos^2(\frac{\pi}{6})$

$= 1 - \left(\frac{\sqrt{3}}{2}\right)^2$

$= 1 - \frac{3}{4}$

$= \frac{1}{4}$

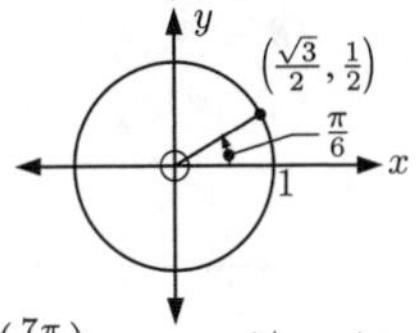

e $\sin^2(\frac{2\pi}{3}) - 1$

$= \left(\frac{\sqrt{3}}{2}\right)^2 - 1$

$= \frac{3}{4} - 1$

$= -\frac{1}{4}$

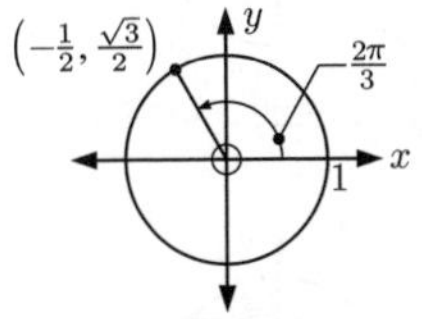

f $\cos^2(\frac{\pi}{4}) - \sin(\frac{7\pi}{6})$

$= \left(\frac{1}{\sqrt{2}}\right)^2 - (-\frac{1}{2})$

$= \frac{1}{2} + \frac{1}{2}$

$= 1$

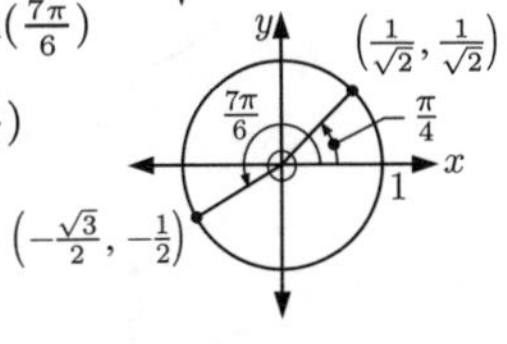

g $\sin(\frac{3\pi}{4}) - \cos(\frac{5\pi}{4})$

$= \frac{1}{\sqrt{2}} - \left(-\frac{1}{\sqrt{2}}\right)$

$= \frac{1}{\sqrt{2}} + \frac{1}{\sqrt{2}}$

$= \frac{2}{\sqrt{2}}$ or $\sqrt{2}$

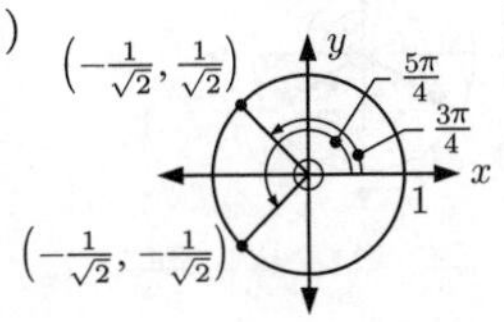

h $1 - 2\sin^2(\frac{7\pi}{6})$

$= 1 - 2(-\frac{1}{2})^2$

$= 1 - 2 \times \frac{1}{4}$

$= \frac{1}{2}$

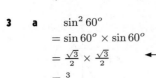

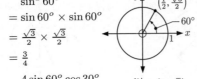

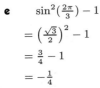

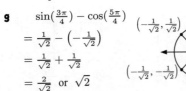

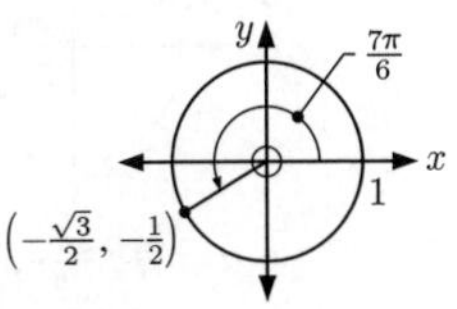

i $\cos^2(\frac{5\pi}{6}) - \sin^2(\frac{5\pi}{6})$
$= \left(-\frac{\sqrt{3}}{2}\right)^2 - (\frac{1}{2})^2$
$= \frac{3}{4} - \frac{1}{4}$
$= \frac{1}{2}$

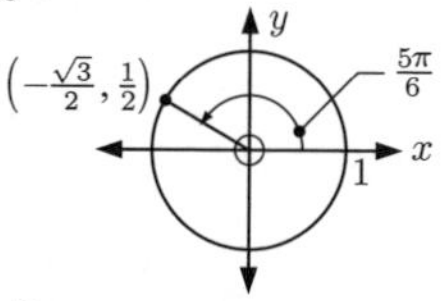

j $\tan^2(\frac{\pi}{3}) - 2\sin^2(\frac{\pi}{4})$
$= \left(\sqrt{3}\right)^2 - 2\left(\frac{1}{\sqrt{2}}\right)^2$
$= 3 - 2(\frac{1}{2})$
$= 2$

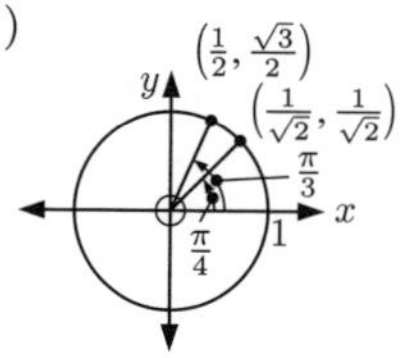

k $2\tan(-\frac{5\pi}{4}) - \sin(\frac{3\pi}{2})$
$= 2(-1) - (-1)$
$= -1$

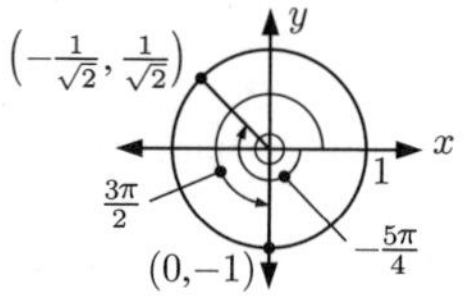

l $\dfrac{2\tan 150^o}{1 - \tan^2 150^o}$

$= \dfrac{2(-\frac{1}{\sqrt{3}})}{1 - (-\frac{1}{\sqrt{3}})^2}$

$= \dfrac{-\frac{2}{\sqrt{3}}}{1 - \frac{1}{3}}$

$= \dfrac{-\frac{2}{\sqrt{3}}}{\frac{2}{3}} = -\frac{3}{\sqrt{3}} = -\sqrt{3}$

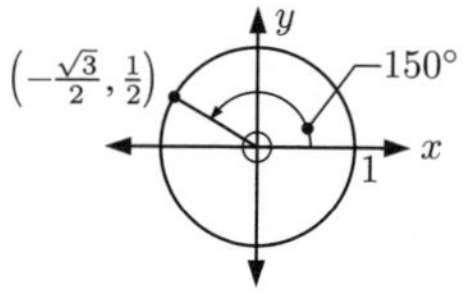

4 **a**

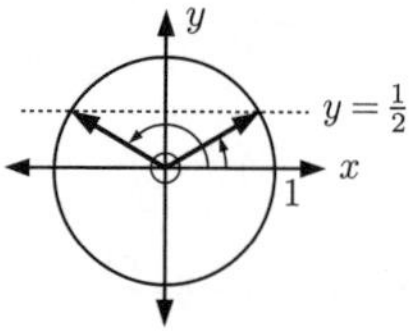

$\theta = 30^o, 150^o$

b

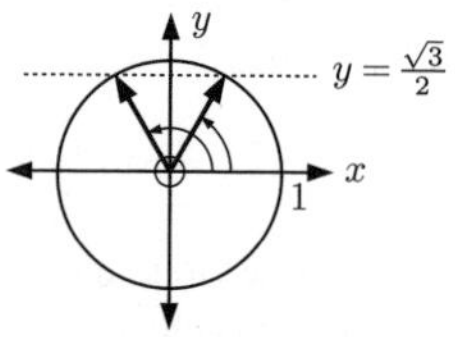

$\theta = 60^o, \ 120^o$

c

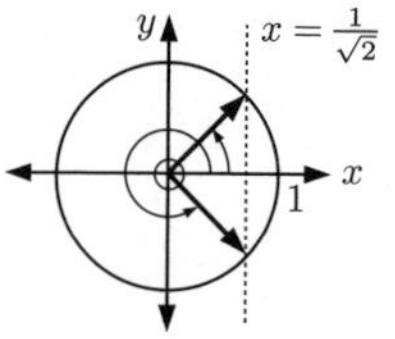

$\theta = 45^o, 315^o$

d

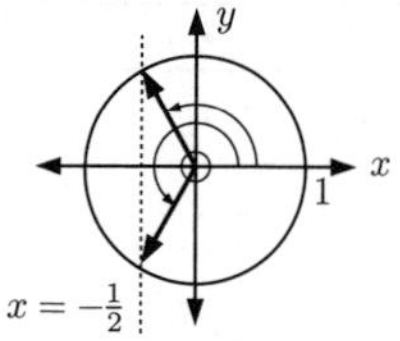

$\theta = 120^o, 240^o$

e

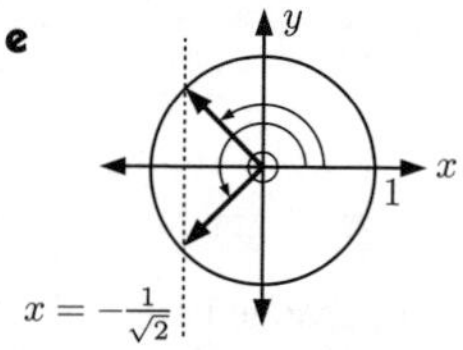

$\theta = 135^o, 225^o$

f

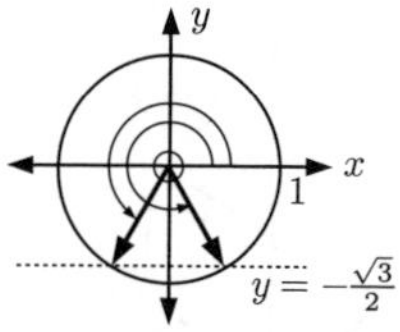

$\theta = 240^o, 300^o$

5 **a**

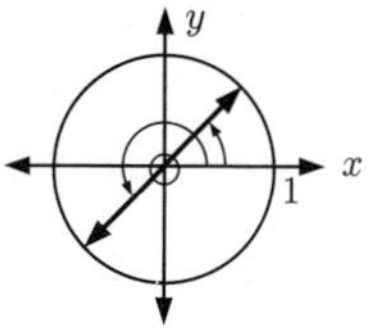

$\theta = \frac{\pi}{4}, \frac{5\pi}{4}$

b

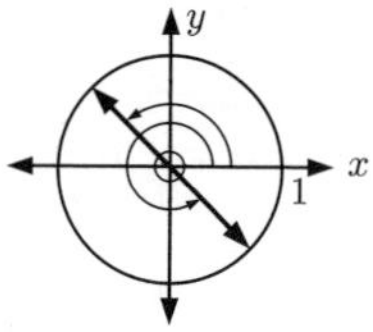

$\theta = \frac{3\pi}{4}, \frac{7\pi}{4}$

c

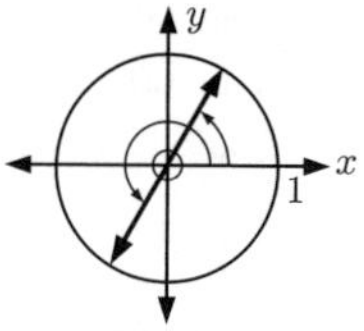

$\theta = \frac{\pi}{3}, \frac{4\pi}{3}$

d

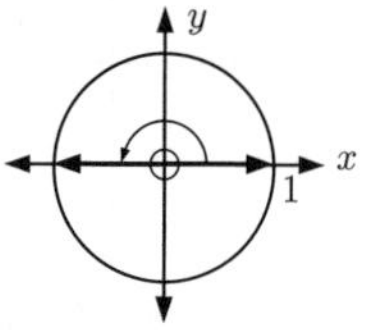

$\theta = 0, \pi, 2\pi$

e

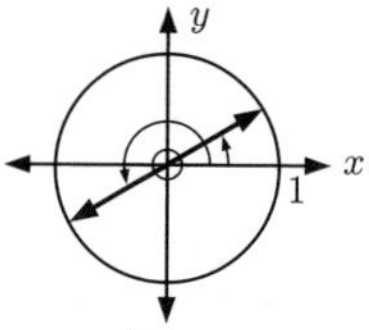

$\theta = \frac{\pi}{6}, \frac{7\pi}{6}$

f

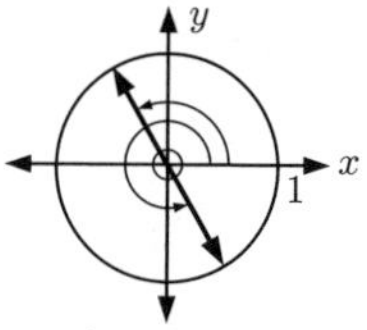

$\theta = \frac{2\pi}{3}, \frac{5\pi}{3}$

6 **a**

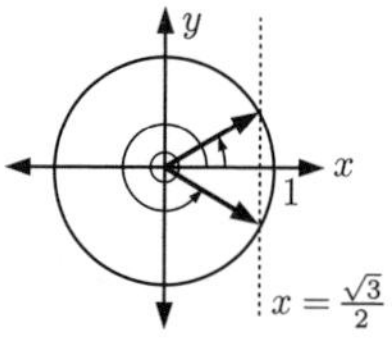

$\theta = \frac{\pi}{6}, \frac{11\pi}{6}, \frac{13\pi}{6}, \frac{23\pi}{6}$

b

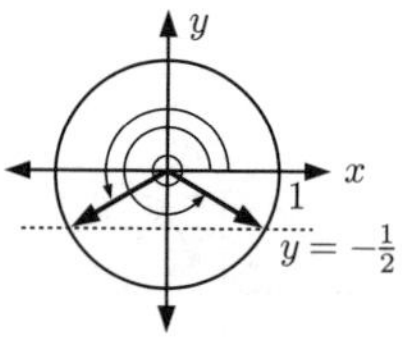

$\theta = \frac{7\pi}{6}, \frac{11\pi}{6}, \frac{19\pi}{6}, \frac{23\pi}{6}$

c

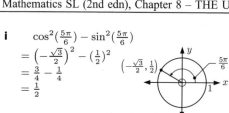

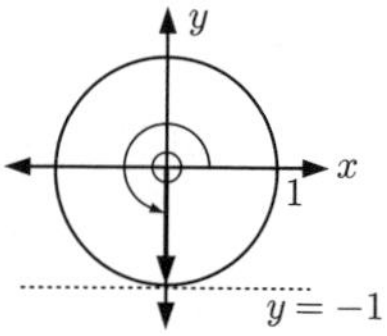

$\theta = \frac{3\pi}{2}, \frac{7\pi}{2}$

7 **a** $\cos\theta = \frac{1}{2}$

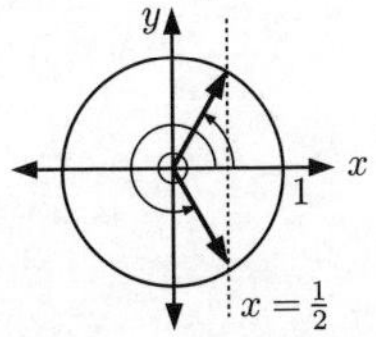

$\therefore \ \theta = \frac{\pi}{3}, \frac{5\pi}{3}$

b $\sin\theta = \frac{\sqrt{3}}{2}$

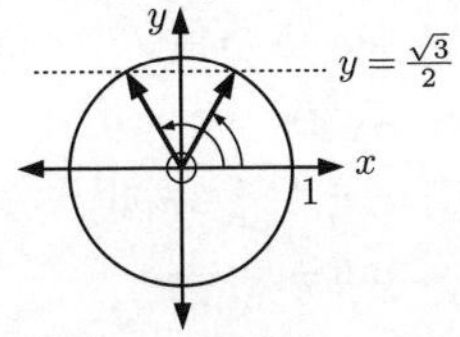

$\therefore \ \theta = \frac{\pi}{3}, \frac{2\pi}{3}$

c $\cos\theta = -1$

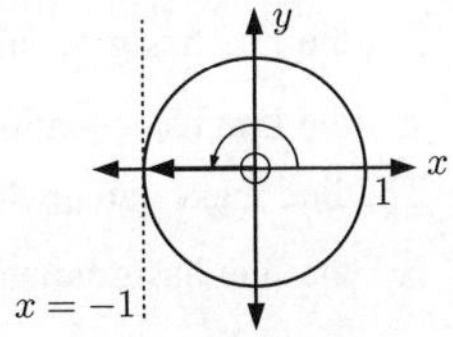

$\therefore \ \theta = \pi$

d $\sin\theta = 1$

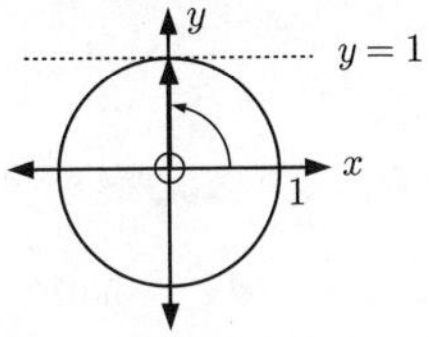

$\therefore \ \theta = \frac{\pi}{2}$

e $\cos\theta = -\frac{1}{\sqrt{2}}$

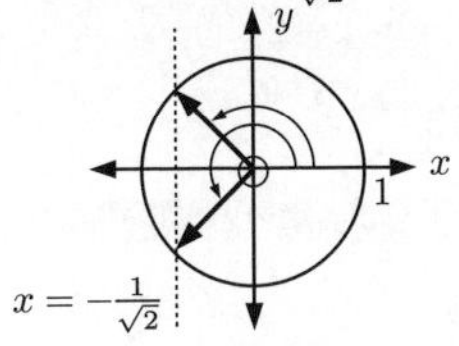

$\therefore \ \theta = \frac{3\pi}{4}, \frac{5\pi}{4}$

f $\sin^2\theta = 1$

$\therefore \ \sin\theta = \pm 1$

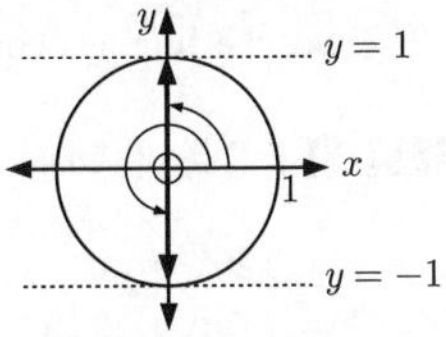

$\therefore \ \theta = \frac{\pi}{2}, \frac{3\pi}{2}$

g $\cos^2\theta = 1$

$\therefore \ \cos\theta = \pm 1$

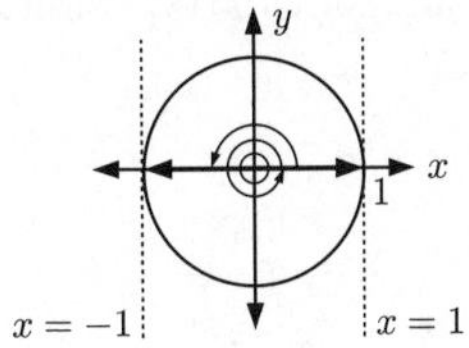

$\therefore \ \theta = 0, \pi, 2\pi$

h $\cos^2\theta = \frac{1}{2}$

$\therefore \ \cos\theta = \pm\frac{1}{\sqrt{2}}$

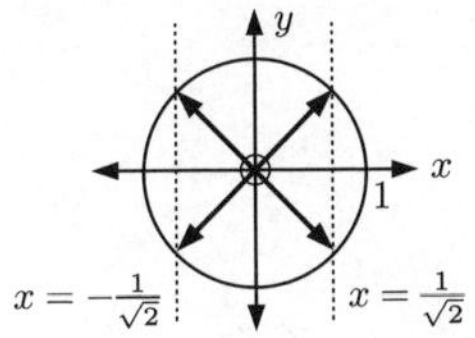

$\therefore \ \theta = \frac{\pi}{4}, \frac{3\pi}{4}, \frac{5\pi}{4}, \frac{7\pi}{4}$

i $\tan\theta = -\frac{1}{\sqrt{3}}$

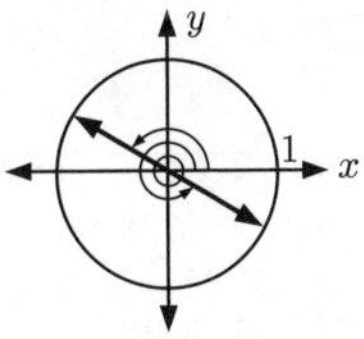

$\therefore \ \theta = \frac{5\pi}{6}, \frac{11\pi}{6}$

j $\tan^2\theta = 3$

$\therefore \ \tan\theta = \pm\sqrt{3}$

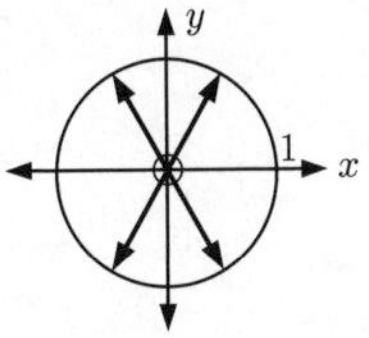

$\therefore \ \theta = \frac{\pi}{3}, \frac{2\pi}{3}, \frac{4\pi}{3}, \frac{5\pi}{3}$

EXERCISE 8D

1 **a** The line has gradient $m = \tan 60^o = \sqrt{3}$ and y-intercept 0.

$\therefore$ the line has equation $y = \sqrt{3}x$.

b The line makes an angle of $\frac{\pi}{2} - \frac{\pi}{4} = \frac{\pi}{4}$ with the positive x-axis.

$\therefore$ the line has gradient $m = \tan\frac{\pi}{4} = 1$ and y-intercept 0.

$\therefore$ the line has equation $y = x$.

c The line makes an angle of $\pi - \frac{\pi}{6} = \frac{5\pi}{6}$ with the positive x-axis.

$\therefore$ the line has gradient $m = \tan\frac{5\pi}{6} = -\frac{1}{\sqrt{3}}$ and y-intercept 0.

$\therefore$ the line has equation $y = -\frac{1}{\sqrt{3}}x$.

2 **a** The line makes an angle of $\frac{\pi}{2} - \frac{\pi}{6} = \frac{\pi}{3}$ with the positive x-axis.

$\therefore$ the line has gradient $m = \tan\frac{\pi}{3} = \sqrt{3}$ and y-intercept 2.

$\therefore$ the line has equation $y = \sqrt{3}x + 2$.

b The line makes an angle of $\pi - \frac{\pi}{3} = \frac{2\pi}{3}$ with the positive x-axis.

$\therefore$ the line has gradient $m = \tan\frac{2\pi}{3} = -\sqrt{3}$ and y-intercept 0.

$\therefore$ the line has equation $y = -\sqrt{3}x$.

c The line makes an angle of $\frac{\pi}{2} - \frac{\pi}{3} = \frac{\pi}{6}$ with the positive x-axis.

$\therefore$ the line has gradient $m = \tan\frac{\pi}{6} = \frac{1}{\sqrt{3}}$.

$\therefore$ $y = \frac{1}{\sqrt{3}}x + c$, where c is a constant.

But when $x = 2\sqrt{3}$, $y = 0$, so $0 = \frac{2\sqrt{3}}{\sqrt{3}} + c$

$\therefore$ $c = -2$

So, the line has equation $y = \frac{1}{\sqrt{3}}x - 2$.

REVIEW SET 8A

1 **a** $120^o = \left(120 \times \frac{\pi}{180}\right)^c = \frac{2\pi}{3}^c$

b $225^o = 5 \times 45^o = 5 \times \frac{\pi}{4}^c = \frac{5\pi}{4}^c$

c $150^o = 5 \times 30^o = 5 \times \frac{\pi}{6}^c = \frac{5\pi}{6}^c$

d $540^o = 3 \times 180^o = 3\pi^c$

2 **a** $\sin\frac{2\pi}{3} = \sin(\pi - \frac{2\pi}{3}) = \sin\frac{\pi}{3}$

$\therefore$ $\theta = \frac{\pi}{3}$

b $\sin 165^o = \sin(180 - 165)^o = \sin 15^o$

$\therefore$ $\theta = 15^o$

c $\cos 276^o = \cos(360 - 276)^o = \cos 84^o$

$\therefore$ $\theta = 84^o$

3 **a** $\sin 159^o = \sin(180 - 159)^o = \sin 21^o \approx 0.358$

b $\cos 92^o = -\cos(180 - 92)^o = -\cos 88^o \approx -0.035$

c $\cos 75^o = -\cos(180 - 75)^o = -\cos 105^o \approx 0.259$

d $\sin 227^o = \sin(-47)^o = -\sin 47^o \approx -0.731$

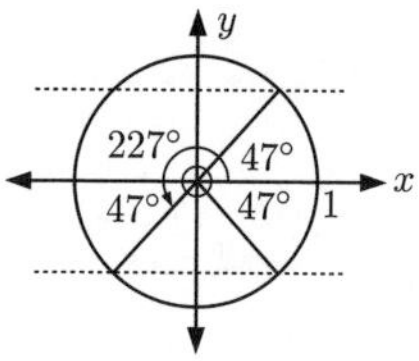

4

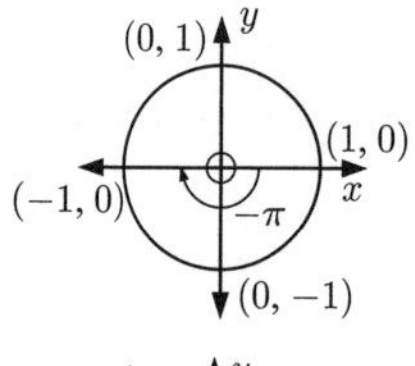

a $\cos 360^o = 1$, $\sin 360^o = 0$

b $\cos(-\pi) = -1$, $\sin(-\pi) = 0$

5

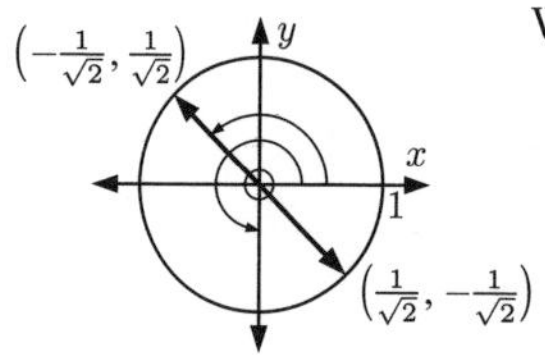

When $\cos\theta = -\sin\theta$,

$\frac{\sin\theta}{\cos\theta} = -1$

$\therefore$ $\tan\theta = -1$

and this only occurs at the two points shown.

So, $\theta = \frac{3\pi}{4}, \frac{7\pi}{4}$

6 **a**

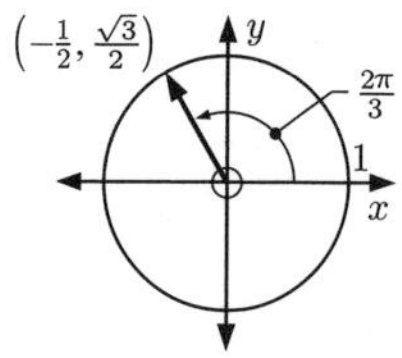

$\sin\left(\frac{2\pi}{3}\right) = \frac{\sqrt{3}}{2}$

$\cos\left(\frac{2\pi}{3}\right) = -\frac{1}{2}$

b

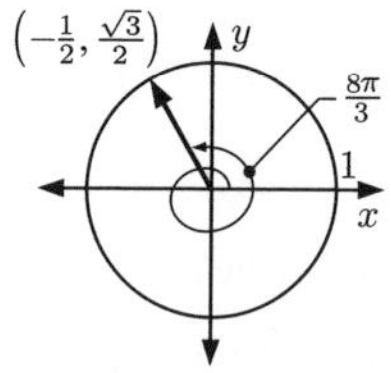

$\sin\left(\frac{8\pi}{3}\right) = \frac{\sqrt{3}}{2}$

$\cos\left(\frac{8\pi}{3}\right) = -\frac{1}{2}$

7 $\cos^2 x + \sin^2 x = 1$

$\therefore\ \cos^2 x + \frac{1}{16} = 1$

$\therefore\ \cos^2 x = \frac{15}{16}$

$\therefore\ \cos x = \pm\frac{\sqrt{15}}{4}$

But x is in quadrant 3 where $\cos x < 0$

$\therefore\ \cos x = -\frac{\sqrt{15}}{4}$

and so $\tan x = \dfrac{\sin x}{\cos x} = \dfrac{-\frac{1}{4}}{-\frac{\sqrt{15}}{4}} = \frac{1}{\sqrt{15}}$

8 **a** $\cos 138^o$

$= -\cos(180 - 138)^o$

$= -\cos 42^o$

≈ -0.743

b

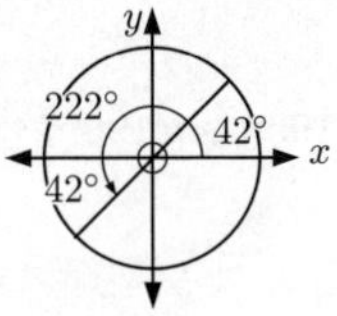

$\cos 222^o$

$= -\cos 42^o$

≈ -0.743

c

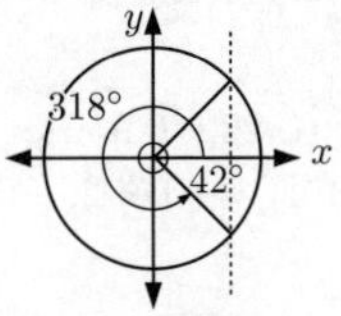

$\cos 318^o$

$= \cos 42^o$

≈ 0.743

d

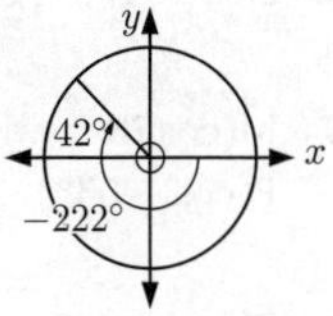

$\cos(-222^o)$

$= -\cos 42^o$

≈ -0.743

9 $\cos^2\theta + \sin^2\theta = 1$

$\therefore\ \frac{9}{16} + \sin^2\theta = 1$

$\therefore\ \sin^2\theta = \frac{7}{16}$

$\therefore\ \sin\theta = \pm\frac{\sqrt{7}}{4}$

10 **a**

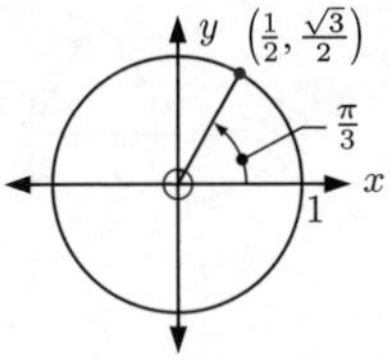

$2\sin(\frac{\pi}{3})\cos(\frac{\pi}{3})$

$= 2(\frac{\sqrt{3}}{2})(\frac{1}{2})$

$= \frac{\sqrt{3}}{2}$

b

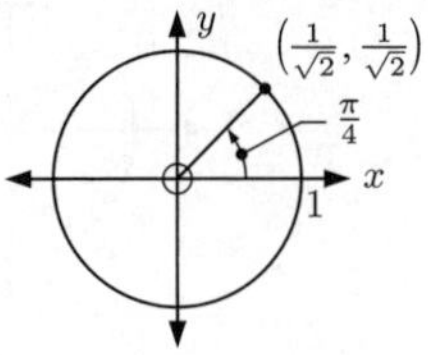

$\tan^2(\frac{\pi}{4}) - 1$

$= 1^2 - 1$

$= 0$

c

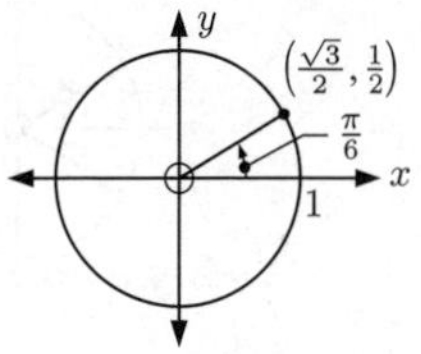

$\cos^2(\frac{\pi}{6}) - \sin^2(\frac{\pi}{6})$

$= \left(\frac{\sqrt{3}}{2}\right)^2 - \left(\frac{1}{2}\right)^2$

$= \frac{3}{4} - \frac{1}{4} = \frac{1}{2}$

11 $\dfrac{\sin x}{\cos x} = -\frac{3}{2}$

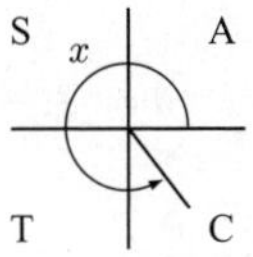

$\therefore\ \sin x = -\frac{3}{2}\cos x$

Now $\cos^2 x + \sin^2 x = 1$

$\therefore\ \cos^2 x + \frac{9}{4}\cos^2 x = 1$

$\therefore\ \frac{13}{4}\cos^2 x = 1$

$\therefore\ \cos x = \pm\frac{2}{\sqrt{13}}$

But x is in quadrant 4, so $\cos x$ is positive and $\sin x$ is negative.

$\therefore\ \cos x = \frac{2}{\sqrt{13}},\ \ \sin x = -\frac{3}{\sqrt{13}}$

So, **a** $\sin x = -\frac{3}{\sqrt{13}}$ **b** $\cos x = \frac{2}{\sqrt{13}}$

REVIEW SET 8B

1 **a** The point is $(\cos 320^o, \sin 320^o) \approx (0.766, -0.643)$.

b The point is $(\cos 163^o, \sin 163^o) \approx (-0.956, 0.292)$.

2 **a** 71°

$= \left(71 \times \frac{\pi}{180}\right)^c$

$\approx 1.239^c$

b 124.6^o

$= \left(124.6 \times \frac{\pi}{180}\right)^c$

$\approx 2.175^c$

c -142°

$= \left(-142 \times \frac{\pi}{180}\right)^c$

$\approx -2.478^c$

3 **a** 3^c
$= \left(3 \times \frac{180}{\pi}\right)^o$
$\approx 171.89^o$

b 1.46^c
$= \left(1.46 \times \frac{180}{\pi}\right)^o$
$\approx 83.65^o$

c 0.435^c
$= \left(0.435 \times \frac{180}{\pi}\right)^o$
$\approx 24.92^o$

d -5.271^c
$= \left(-5.271 \times \frac{180}{\pi}\right)^o$
$\approx -302.01^o$

4 area $= \frac{1}{2} \times \frac{5\pi}{12} \times 13^2 \approx 111 \text{ cm}^2$

5 $\text{M}(\cos 73^o, \sin 73^o) \approx (0.292, 0.956)$, $\text{N}(\cos 190^o, \sin 190^o) \approx (-0.985, -0.174)$,
$\text{P}(\cos(-53^o), \sin(-53^o)) \approx (0.602, -0.799)$

6 The x-coordinate of A $= -0.222$
$\therefore \quad \cos\theta = -0.222$
$\therefore \quad \theta = \cos^{-1}(-0.222)$
$\therefore \quad \theta \approx 102.8^o, 257.2^o$
$\therefore \quad \theta \approx 103^o$ {taking angle to positive x-axis}

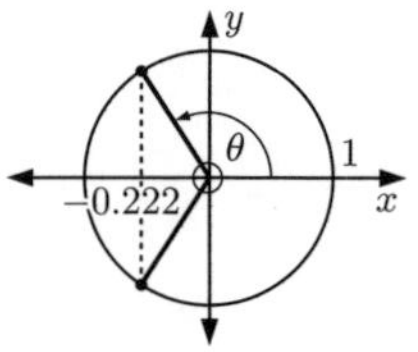

7 **a**

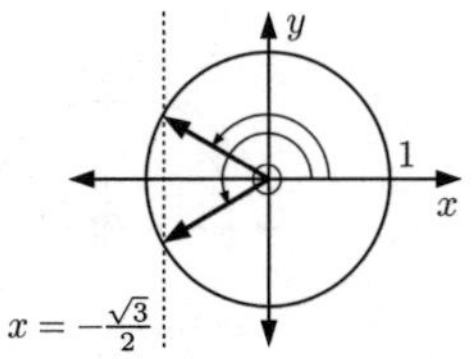

$\therefore \quad \theta = 150^o$ or 210°

b

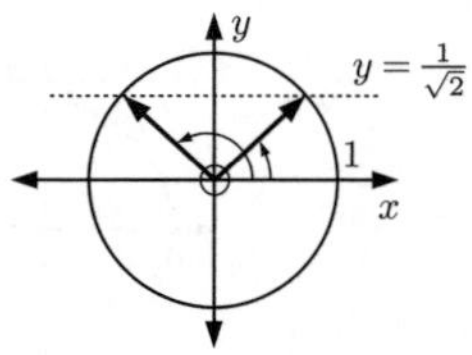

$\therefore \quad \theta = 45^o$ or 135°

c

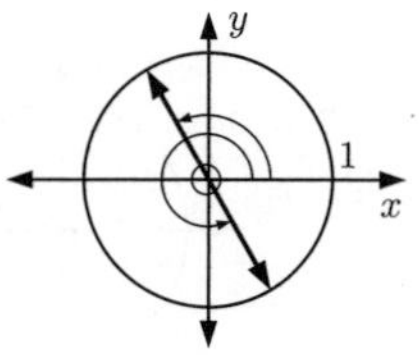

$\therefore \quad \theta = 120^o$ or 300°

8 **a**

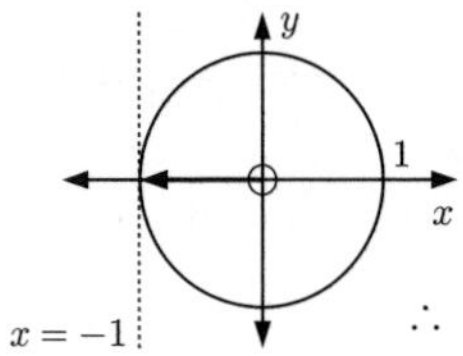

$\therefore \quad \theta = \pi$

b $\sin^2\theta = \frac{3}{4}$
$\therefore \quad \sin\theta = \pm\frac{\sqrt{3}}{2}$

$\therefore \quad \theta = \frac{\pi}{3}, \frac{2\pi}{3}, \frac{4\pi}{3}, \frac{5\pi}{3}$

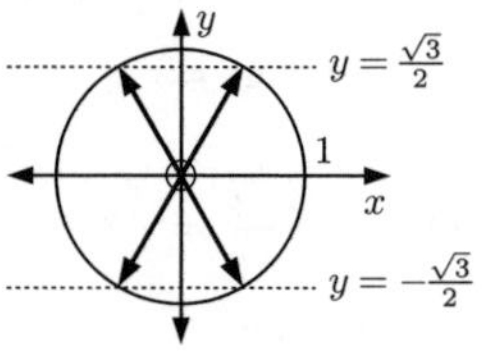

9 **a** $\sin 47^o$
$= \sin(180 - 47)^o$
$= \sin 133^o$
$\therefore \quad \theta = 133^o$

b $\sin\frac{\pi}{15}$
$= \sin(\pi - \frac{\pi}{15})$
$= \sin\frac{14\pi}{15}$
$\therefore \quad \theta = \frac{14\pi}{15}$

c $\cos 186^o$
$= \cos(360 - 186)^o$
$= \cos 174^o$
$\therefore \quad \theta = 174^o$

10

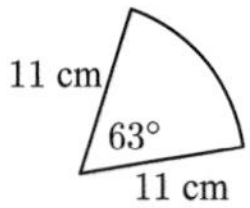

perimeter $= 2 \times 11 + \left(\frac{63}{360}\right) \times 2\pi \times 11$
≈ 34.1 cm

area $= \left(\frac{63}{360}\right) \times \pi \times 11^2$
$\approx 66.5 \text{ cm}^2$

11

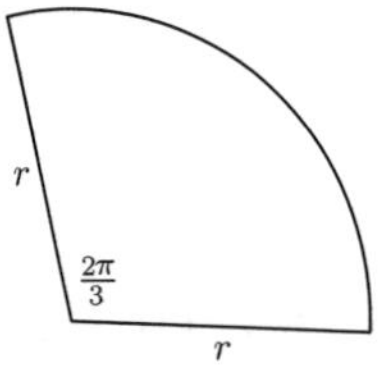

perimeter $= 2r + (\frac{2\pi}{3})r$
$\therefore \quad 36 = r\left(2 + \frac{2\pi}{3}\right)$
$\therefore \quad r = \dfrac{36}{2 + \frac{2\pi}{3}}$ cm
$\therefore \quad r \approx 8.79$ cm

area $\approx \frac{1}{2}(\frac{2\pi}{3}) \times (8.7925)^2$
$\approx 81.0 \text{ cm}^2$

REVIEW SET 8C

1 **a** $\frac{2\pi}{5} = \frac{2\times180^o}{5}$
$= 72^o$

b $\frac{5\pi}{4} = \frac{5\times180^o}{4}$
$= 225^o$

c $\frac{7\pi}{9} = \frac{7\times180^o}{9}$
$= 140^o$

d $\frac{11\pi}{6} = \frac{11\times180^o}{6}$
$= 330^o$

2

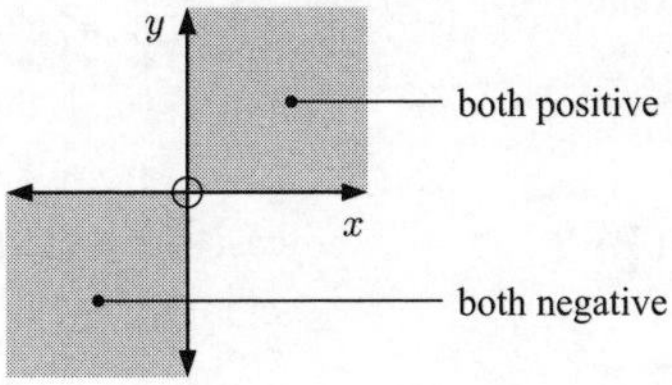

3 **a**

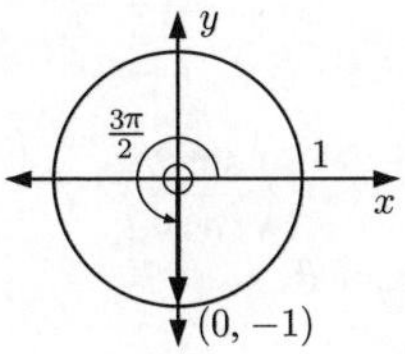

$\therefore \quad \cos\left(\frac{3\pi}{2}\right) = 0$
$\sin\left(\frac{3\pi}{2}\right) = -1$

b

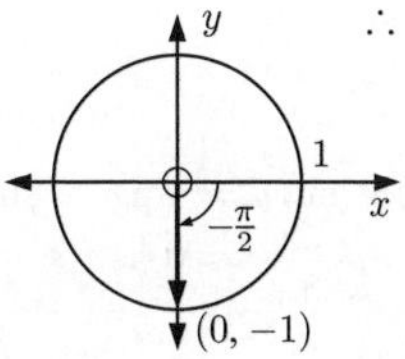

$\therefore \quad \cos\left(-\frac{\pi}{2}\right) = 0$
$\sin\left(-\frac{\pi}{2}\right) = -1$

4 **a**

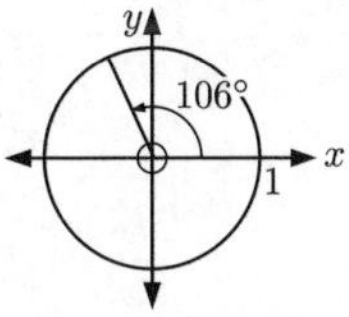

$\sin 106^o$
$= \sin(180 - 106)^o$
$= \sin 74^o$
≈ 0.961

b

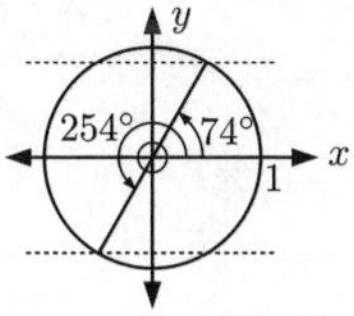

$254^o = 74^o + 180^o$
$\therefore \quad \sin 254^o = -\sin 74^o$
≈ -0.961

c

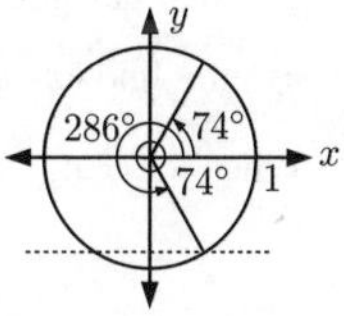

$\sin 286^o = -\sin 74^o$
≈ -0.961

d

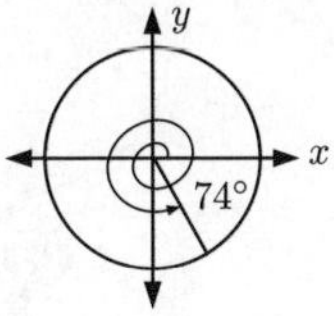

$646^o = 360^o + 286^o$
$\therefore \quad \sin 646^o = \sin 286^o$
≈ -0.961 {from **c**}

5 **a** **i** $\theta = 60^o$ {equilateral triangle} **ii** $\theta = \frac{\pi}{3}$ radians **b** arc length $= \theta r = \frac{\pi}{3}$ units

6

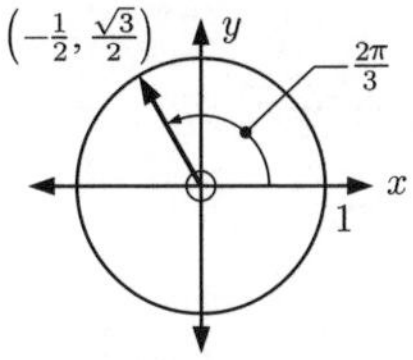

$\tan^2(\frac{2\pi}{3})$
$= (-\sqrt{3})^2$
$= 3$

7

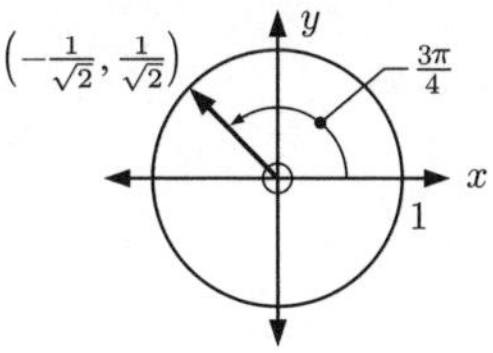

$f(\frac{3\pi}{4}) = \cos(\frac{3\pi}{4}) - \sin(\frac{3\pi}{4})$
$= -\frac{1}{\sqrt{2}} - \frac{1}{\sqrt{2}}$
$= -\frac{2}{\sqrt{2}}$
$= -\sqrt{2}$

8 **a** $\cos^2\theta + \sin^2\theta = 1$
$\therefore \quad \frac{9}{16} + \sin^2\theta = 1$
$\therefore \quad \sin^2\theta = \frac{7}{16}$
$\therefore \quad \sin\theta = \pm\frac{\sqrt{7}}{4}$
But θ is in quadrant 2 where $\sin\theta > 0$
$\therefore \quad \sin\theta = \frac{\sqrt{7}}{4}$

b $\tan\theta = \dfrac{\sin\theta}{\cos\theta} = \dfrac{\frac{\sqrt{7}}{4}}{-\frac{3}{4}} = -\frac{\sqrt{7}}{3}$

c $\sin(\theta + \pi)$
$= -\sin\theta$
$= -\frac{\sqrt{7}}{4}$

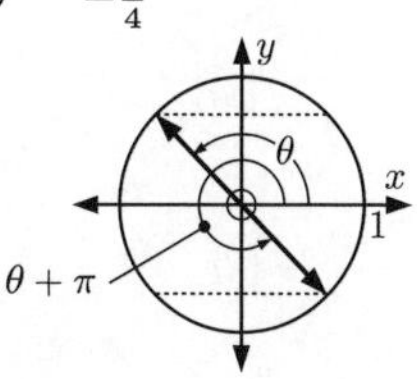

9 **a**

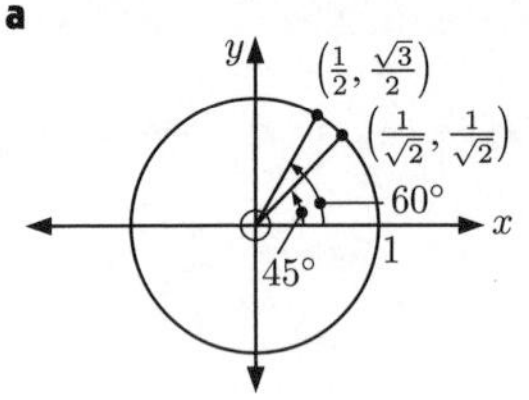

b

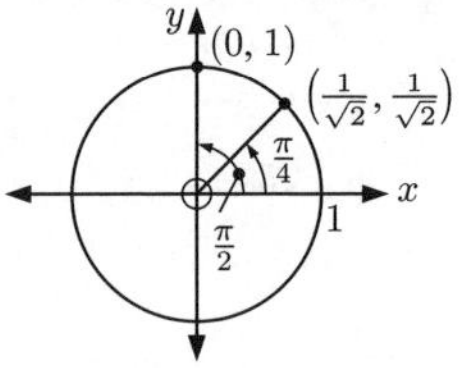

c

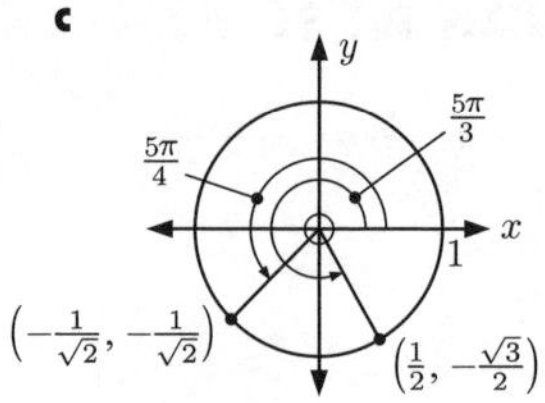

$\tan^2 60^o - \sin^2 45^o$
$= (\sqrt{3})^2 - \left(\frac{1}{\sqrt{2}}\right)^2$
$= 3 - \frac{1}{2}$
$= 2\frac{1}{2}$

$\cos^2(\frac{\pi}{4}) + \sin(\frac{\pi}{2})$
$= \left(\frac{1}{\sqrt{2}}\right)^2 + 1$
$= \frac{1}{2} + 1$
$= 1\frac{1}{2}$

$\cos(\frac{5\pi}{3}) - \tan(\frac{5\pi}{4})$
$= \frac{1}{2} - 1$
$= -\frac{1}{2}$

10 **a** $\sin(\pi - \theta) - \sin\theta = \sin\theta - \sin\theta$
$= 0$

b $\cos\theta\tan\theta = \cos\theta\left(\dfrac{\sin\theta}{\cos\theta}\right)$
$= \sin\theta$

11 **a** The line has gradient $m = \tan(-30^o) = -\frac{1}{\sqrt{3}}$ and y-intercept 0.
$\therefore$ the line has equation $y = -\frac{1}{\sqrt{3}}x$.

b When $x = k,\ y = 2$ $\quad\therefore\ 2 = -\frac{1}{\sqrt{3}}k$
$\therefore\ k = -2\sqrt{3}$

12 [AB], [AC] and [BC] are all radii,
so AB = AC = BC = r.
Hence ΔABC is equilateral
and so $\widehat{CAB} = 60^o$.

$\therefore\ \sin 60^o = \dfrac{CD}{AC}$

$\therefore\ CD = \sin 60^o \times AC = \frac{\sqrt{3}}{2}r$

$\therefore$ area of $\Delta = \frac{1}{2}(r)(\frac{\sqrt{3}}{2}r)$
$= \frac{\sqrt{3}}{4}r^2$

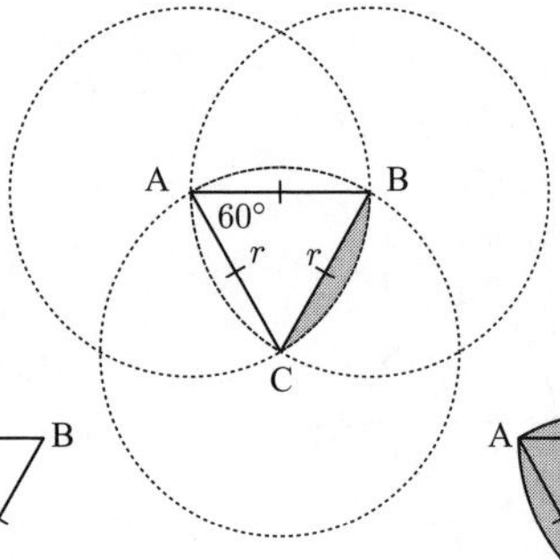

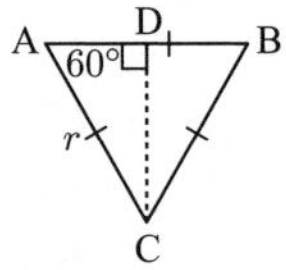

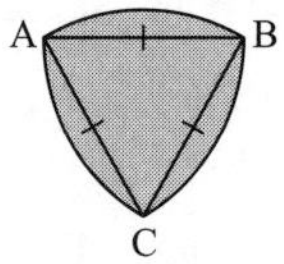

shaded area of sector
$= \text{area of sector} - \text{area of }\Delta$
$= \frac{60}{360}\pi r^2 - \frac{\sqrt{3}}{4}r^2$
$= \frac{\pi}{6}r^2 - \frac{\sqrt{3}}{4}r^2$

$\therefore$ shaded area of figure
$= 3\left[\frac{\pi}{6}r^2 - \frac{\sqrt{3}}{4}r^2\right] + \frac{\sqrt{3}}{4}r^2$
$= \frac{\pi}{2}r^2 - \frac{3\sqrt{3}}{4}r^2 + \frac{\sqrt{3}}{4}r^2$
$= \frac{\pi}{2}r^2 - \frac{2}{4}\sqrt{3}r^2$
$= \dfrac{r^2}{2}\left(\pi - \sqrt{3}\right)$

Chapter 9
NON-RIGHT ANGLED TRIANGLE TRIGONOMETRY

EXERCISE 9A

1 **a** area
$= \frac{1}{2} \times 9 \times 10 \times \sin 40^o$
$\approx 28.9 \text{ cm}^2$

b area
$= \frac{1}{2} \times 25 \times 31 \times \sin 82^o$
$\approx 384 \text{ km}^2$

c area
$= \frac{1}{2} \times 10.2 \times 6.4 \times \sin \frac{2\pi}{3}$
$\approx 28.3 \text{ cm}^2$

2 area $= 150 \text{ cm}^2$
$\therefore \frac{1}{2} \times 17 \times x \times \sin 68^o = 150$
$\therefore x = \dfrac{2 \times 150}{17 \times \sin 68^o}$
$\therefore x \approx 19.0$

3

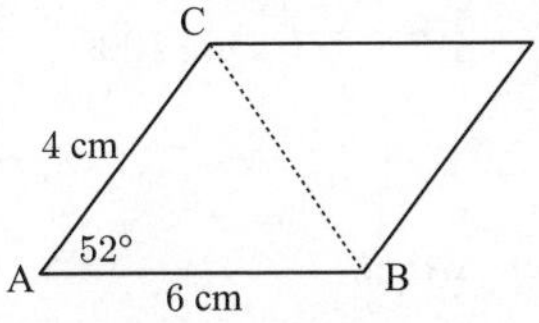

area
$= 2 \times \text{area } \Delta\text{ABC}$
$= 2 \times \frac{1}{2} \times 4 \times 6 \times \sin 52^o$
$\approx 18.9 \text{ cm}^2$

4

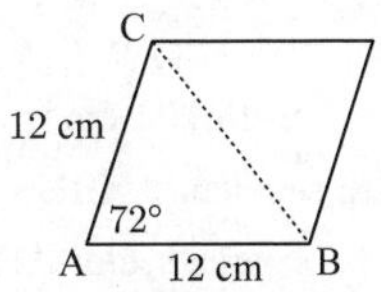

area
$= 2 \times \text{area } \Delta\text{ABC}$
$= 2 \times \frac{1}{2} \times 12^2 \times \sin 72^o$
$\approx 137 \text{ cm}^2$

5

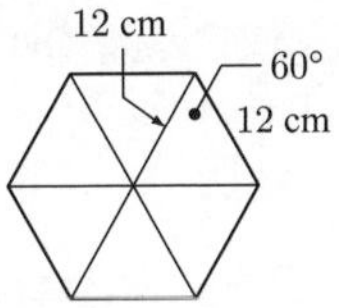

area $= 6 \times$ area of Δ
$= 6 \times \frac{1}{2} \times 12^2 \times \sin 60^o$
$\approx 374 \text{ cm}^2$

6

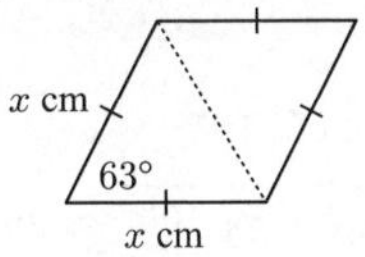

area $= 2 \times \frac{1}{2}x^2 \sin 63^o$
$\therefore x^2 \sin 63^o = 50$
$\therefore x^2 = \dfrac{50}{\sin 63^o}$
$\therefore x = \sqrt{\dfrac{50}{\sin 63^o}}$
$\therefore x \approx 7.49$
So, sides are 7.49 cm long.

7

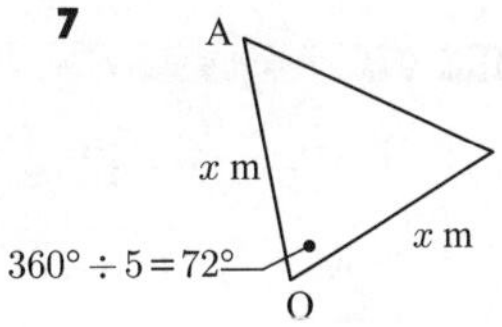

area of $\Delta = \frac{338}{5}$
$\therefore \frac{1}{2}x^2 \sin 72^o = \frac{338}{5}$
$\therefore x^2 = \dfrac{2 \times 338}{5 \times \sin 72^o}$
$\therefore x = \sqrt{\dfrac{2 \times 338}{5 \times \sin 72^o}}$
$\therefore x \approx 11.9$
So, OA ≈ 11.9 m

8 **a** If the included angle is θ
then $\frac{1}{2} \times 5 \times 8 \times \sin\theta = 15$
$\therefore 20 \sin\theta = 15$
$\therefore \sin\theta = \frac{3}{4}$
Now $\arcsin(\frac{3}{4}) \approx 48.6^o$
$\therefore \theta \approx 48.6^o$ or $(180 - 48.6)^o$
$\therefore \theta \approx 48.6^o$ or 131.4^o

b Likewise,
$\frac{1}{2} \times 45 \times 53 \times \sin\theta = 800$
$\therefore \sin\theta = \frac{800 \times 2}{45 \times 53}$
Now $\arcsin\left(\frac{800 \times 2}{45 \times 53}\right) \approx 42.1^o$
$\therefore \theta \approx 42.1^o$ or $(180 - 42.1)^o$
$\therefore \theta \approx 42.1^o$ or 137.9^o

9

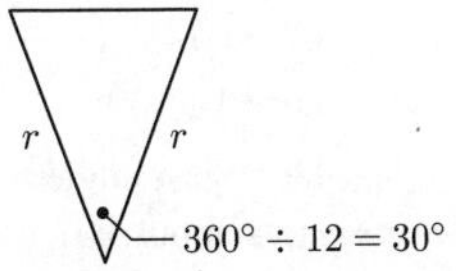

total area of 8 coins
$= 8 \times 12 \times \frac{1}{2}r^2 \sin 30^o$
$= 48r^2(\frac{1}{2})$
$= 24r^2$

area of \$10 note
$= 8r \times 4r$
$= 32r^2$

fraction covered
$= \dfrac{24r^2}{32r^2}$
$= \frac{3}{4} \quad \therefore \frac{1}{4}$ is uncovered

10 **a** shaded area
$= \text{area of sector} - \text{area of triangle}$
$= \frac{1}{2} \times 1.5 \times 12^2 - \frac{1}{2} \times 12^2 \times \sin(1.5)$
$\approx 36.2 \text{ cm}^2$

b shaded area
$= \text{area of triangle} - \text{area of sector}$
$= \frac{1}{2} \times 12 \times 30 \times \sin(0.66) - \frac{1}{2} \times 0.66 \times 12^2$
$\approx 62.8 \text{ cm}^2$

c shaded area
$= \text{area of sector} - \text{area of triangle}$
$= \left(\frac{135}{360}\right) \times \pi \times 7^2 - \frac{1}{2} \times 7^2 \times \sin 135^o$
$\approx 40.4 \text{ mm}^2$

11 area segment AXBD
$= \text{area sector ACBD} - \text{area } \Delta\text{ACB}$
$= \left(\frac{100}{360}\right) \times \pi \times 7.3^2 - \frac{1}{2} \times 7.3 \times 7.3 \times \sin 100^o$
$\approx 20.264 \text{ cm}^2$

area segment AXBE
$= \text{area sector AFBE} - \text{area } \Delta\text{AFB}$
$= \left(\frac{80}{360}\right) \times \pi \times 8.7^2 - \frac{1}{2} \times 8.7 \times 8.7 \times \sin 80^o$
$\approx 15.572 \text{ cm}^2$

$\therefore$ shaded area = area segment AXBD − area segment AXBE
$\approx 20.264 - 15.572$
$\approx 4.69 \text{ cm}^2$

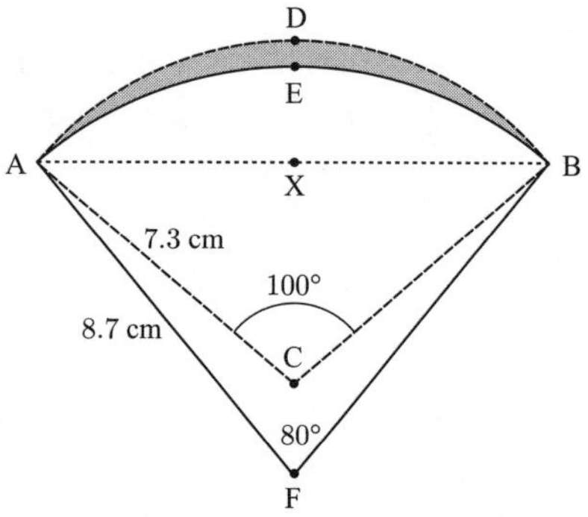

EXERCISE 9B

1 **a** $\text{BC}^2 = 21^2 + 15^2 - 2 \times 21 \times 15 \times \cos 105^o$

$\therefore \quad \text{BC} = \sqrt{21^2 + 15^2 - 2 \times 21 \times 15 \times \cos 105^o} \approx 28.8 \text{ cm}$

b $\text{PQ}^2 = 6.3^2 + 4.8^2 - 2 \times 6.3 \times 4.8 \times \cos 32^o$

$\therefore \quad \text{PQ} = \sqrt{6.3^2 + 4.8^2 - 2 \times 6.3 \times 4.8 \times \cos 32^o} \approx 3.38 \text{ km}$

c $\text{KM}^2 = 6.2^2 + 14.8^2 - 2 \times 6.2 \times 14.8 \times \cos 72^o$

$\therefore \quad \text{KM} = \sqrt{6.2^2 + 14.8^2 - 2 \times 6.2 \times 14.8 \times \cos 72^o} \approx 14.2 \text{ m}$

2 $\cos A = \dfrac{12^2 + 13^2 - 11^2}{2 \times 12 \times 13}$

$\therefore \quad A = \cos^{-1}\left(\frac{192}{312}\right)$
$\therefore \quad A \approx 52.0^o$

$\cos B = \dfrac{13^2 + 11^2 - 12^2}{2 \times 13 \times 11}$

$\therefore \quad B = \cos^{-1}\left(\frac{146}{286}\right)$
$\therefore \quad B \approx 59.3^o$

$C = 180^o - A - B$
$\approx 68.7^o$

3 $\cos Q = \dfrac{5^2 + 7^2 - 10^2}{2 \times 5 \times 7}$

$\therefore \quad Q = \cos^{-1}\left(\frac{-26}{70}\right)$
$\therefore \quad Q \approx 112^o$

4 **a**

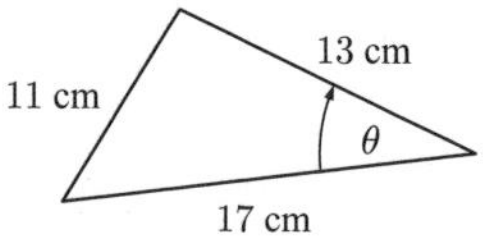

The smallest angle is opposite the shortest side.

$\cos\theta = \dfrac{13^2 + 17^2 - 11^2}{2 \times 13 \times 17}$

$\therefore \quad \theta = \cos^{-1}\left(\frac{337}{442}\right)$
$\therefore \quad \theta \approx 40.3^o$

So, the smallest angle measures 40.3^o.

b

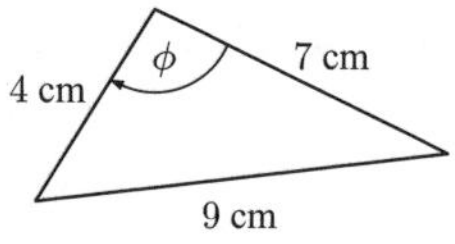

The largest angle is opposite the longest side.

$\cos\phi = \dfrac{4^2 + 7^2 - 9^2}{2 \times 4 \times 7}$

$\therefore \quad \phi = \cos^{-1}\left(-\frac{16}{56}\right)$
$\therefore \quad \phi \approx 106.60^o$

So, the largest angle measures about 107°.

5 **a** $\cos\theta = \dfrac{2^2+5^2-4^2}{2\times2\times5}$

$= \frac{13}{20}$

$= 0.65$

b $x^2 = 5^2+3^2-2\times5\times3\times\cos\theta$

$\therefore\ x = \sqrt{5^2+3^2-2\times5\times3\times0.65}$

$\therefore\ x \approx 3.81$

6 **a** $7^2 = x^2+6^2-2\times x\times6\times\cos 60^o$

$\therefore\ 49 = x^2+36-12x\times(\frac{1}{2})$

$\therefore\ x^2-6x-13=0$

$\therefore\ x = \dfrac{6\pm\sqrt{36-4(1)(-13)}}{2}$

$= \dfrac{6\pm\sqrt{88}}{2}$

$= 3\pm\sqrt{22}$

But $x>0$, so $x = 3+\sqrt{22}$.

b $5^2 = x^2+3^2-2\times x\times3\times\cos 120^o$

$\therefore\ 25 = x^2+9-6x\times(-\frac{1}{2})$

$\therefore\ x^2+3x-16=0$

$\therefore\ x = \dfrac{-3\pm\sqrt{9-4(1)(-16)}}{2}$

$= \dfrac{-3\pm\sqrt{73}}{2}$

But $x>0$, so $x = \dfrac{-3+\sqrt{73}}{2}$.

c $5^2 = (2x)^2+x^2-2\times(2x)\times x\times\cos 60^o$

$\therefore\ 25 = 4x^2+x^2-4x^2(\frac{1}{2})$

$\therefore\ 3x^2 = 25$

$\therefore\ x^2 = \frac{25}{3}$

$\therefore\ x = \pm\frac{5}{\sqrt{3}}$ But $x>0$, so $x = \frac{5}{\sqrt{3}}$.

7 **a** $11^2 = x^2+8^2-2\times x\times8\times\cos 70^o$

$\therefore\ 121 = x^2+64-16x\cos 70^o$

$\therefore\ x^2-(16\cos 70^o)x-57=0$

Using the quadratic formula or technology,
$x\approx -5.29$ or 10.8.
But $x>0$, so $x\approx 10.8$.

b $13^2 = x^2+5^2-2\times x\times5\times\cos 130^o$

$\therefore\ 169 = x^2+25-10x\cos 130^o$

$\therefore\ x^2-(10\cos 130^o)x-144=0$

Using the quadratic formula or technology,
$x\approx -15.6$ or 9.21.
But $x>0$, so $x\approx 9.21$.

c $5^2 = x^2+6^2-2\times x\times6\times\cos 40^o$

$\therefore\ 25 = x^2+36-12x\cos 40^o$

$\therefore\ x^2-(12\cos 40^o)x+11=0$

Using the quadratic formula or technology, $x\approx 1.41$ or 7.78.

EXERCISE 9C.1

1 **a** By the sine rule,

$\dfrac{x}{\sin 48^o} = \dfrac{23}{\sin 37^o}$

$\therefore\ x = \dfrac{23\times\sin 48^o}{\sin 37^o}$

$\therefore\ x\approx 28.4$

b By the sine rule,

$\dfrac{x}{\sin 115^o} = \dfrac{11}{\sin 48^o}$

$\therefore\ x = \dfrac{11\times\sin 115^o}{\sin 48^o}$

$\therefore\ x\approx 13.4$

c By the sine rule,

$\dfrac{x}{\sin 51^o} = \dfrac{4.8}{\sin 80^o}$

$\therefore\ x = \dfrac{4.8\times\sin 51^o}{\sin 80^o}$

$\therefore\ x\approx 3.79$

2 **a**

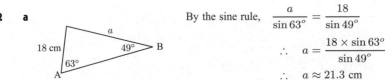

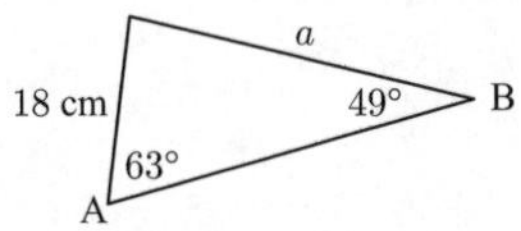

By the sine rule, $\dfrac{a}{\sin 63^o} = \dfrac{18}{\sin 49^o}$

$\therefore\ a = \dfrac{18\times\sin 63^o}{\sin 49^o}$

$\therefore\ a\approx 21.3$ cm

b

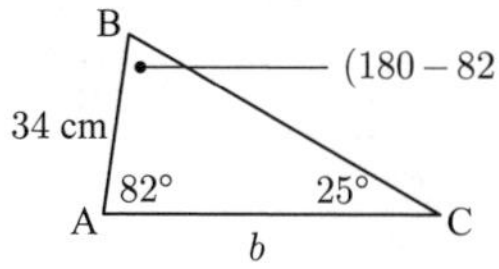

By the sine rule, $\dfrac{b}{\sin 73^o} = \dfrac{34}{\sin 25^o}$

$\therefore \quad b = \dfrac{34 \times \sin 73^o}{\sin 25^o}$

$\therefore \quad b \approx 76.9$ cm

c

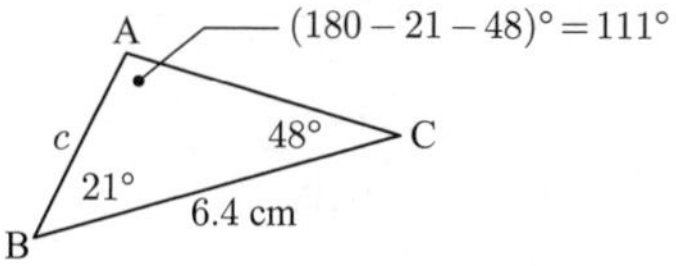

By the sine rule, $\dfrac{c}{\sin 48^o} = \dfrac{6.4}{\sin 111^o}$

$\therefore \quad c = \dfrac{6.4 \times \sin 48^o}{\sin 111^o}$

$\therefore \quad c \approx 5.09$ cm

EXERCISE 9C.2

1 By the sine rule, $\dfrac{\sin C}{11} = \dfrac{\sin 40^o}{8}$

$\therefore \quad \sin C = \dfrac{11 \times \sin 40^o}{8}$

$\therefore \quad C = \sin^{-1}\left(\dfrac{11 \times \sin 40^o}{8}\right)$ or its supplement

$\therefore \quad C \approx 62.1^o$ or $(180 - 62.1)^o$

$\therefore \quad C \approx 62.1^o$ or 117.9^o

2 **a** $\dfrac{\sin A}{a} = \dfrac{\sin B}{b}$

$\therefore \quad \sin A = \dfrac{14.6 \times \sin 65^o}{17.4}$

$\therefore \quad A = \sin^{-1}\left(\dfrac{14.6 \times \sin 65^o}{17.4}\right)$

or its supplement

$\therefore \quad A \approx 49.5^o$ or $180^o - 49.5^o$

$\therefore \quad A \approx 49.5^o$ or 130.5^o

Check: $A = 130.5^o$ is impossible as $A + B = 130.5^o + 65^o$ is already over 180°. $\therefore \quad A \approx 49.5^o$

b $\dfrac{\sin B}{43.8} = \dfrac{\sin 43^o}{31.4}$

$\therefore \quad \sin B = \dfrac{43.8 \times \sin 43^o}{31.4}$

$\therefore \quad B = \sin^{-1}\left(\dfrac{43.8 \times \sin 43^o}{31.4}\right)$

or its supplement

$\therefore \quad B \approx 72.0^o$ or 108°

both of which are possible as $108 + 43 = 151$ which is < 180.

c $\dfrac{\sin C}{4.8} = \dfrac{\sin 71^o}{6.5}$

$\therefore \quad \sin C = \dfrac{4.8 \times \sin 71^o}{6.5}$

$\therefore \quad C = \sin^{-1}\left(\dfrac{4.8 \times \sin 71^o}{6.5}\right)$ or its supplement

$\therefore \quad C \approx 44.3^o$ or 135.7^o

But $135.7 + 71 > 180$, so this case is impossible $\therefore \quad C \approx 44.3^o$

3 The third angle is $180^o - 85^o - 68^o = 27^o$

Now $\dfrac{\sin 85^o}{11.4} \approx 0.087\,38$ and $\dfrac{\sin 27^o}{9.8} \approx 0.046\,32$

This is not possible since $\dfrac{\sin 85^o}{11.4} \neq \dfrac{\sin 27^o}{9.8}$ violates the sine rule.

4

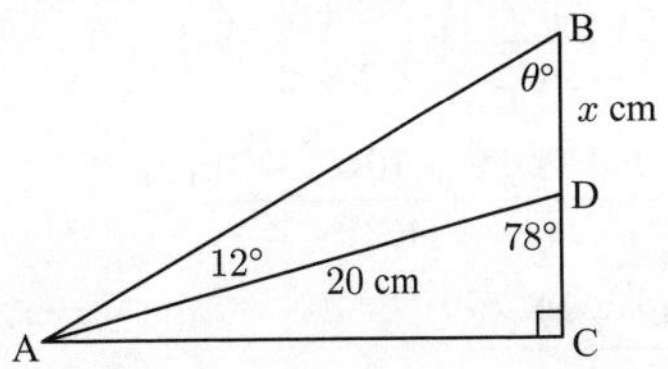

In ΔABD,
$$\theta = 78 - 12$$
$$\therefore \quad \widehat{ABC} = 66^o$$

Now $\dfrac{x}{\sin 12^o} = \dfrac{20}{\sin 66^o}$

$$\therefore \quad x = \frac{20 \times \sin 12^o}{\sin 66^o}$$
$$\therefore \quad x \approx 4.55$$
$$\therefore \quad \text{BD} \approx 4.55 \text{ cm}$$

5 First we find the length of the diagonal, d m.

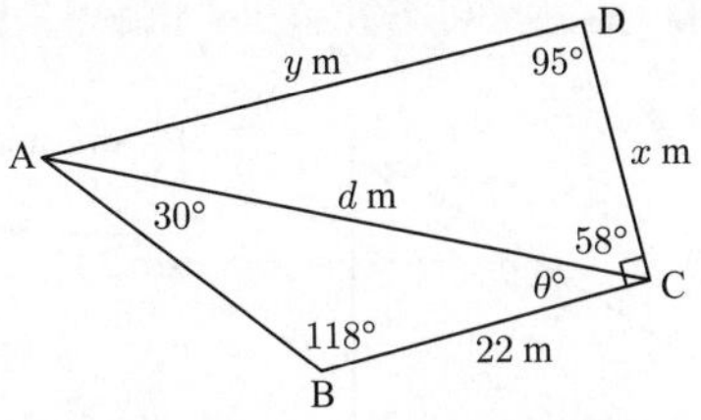

$$\frac{d}{\sin 118^o} = \frac{22}{\sin 30^o}$$
$$\therefore \quad d = \frac{22 \times \sin 118^o}{\sin 30^o}$$
$$\therefore \quad d \approx 38.85$$

Now $\theta = 180 - 30 - 118 = 32$
$$\therefore \quad \widehat{ACD} = 58^o$$

Using the sine rule,

$$\frac{y}{\sin 58^o} = \frac{38.85}{\sin 95^o}$$
$$\therefore \quad y \approx \frac{38.85 \times \sin 58^o}{\sin 95^o}$$
$$\therefore \quad y \approx 33.1$$

and

$$\frac{x}{\sin(180 - 95 - 58)^o} \approx \frac{38.85}{\sin 95^o}$$
$$\therefore \quad x \approx \frac{38.85 \times \sin 27^o}{\sin 95^o}$$
$$\therefore \quad x \approx 17.7$$

6 **a**

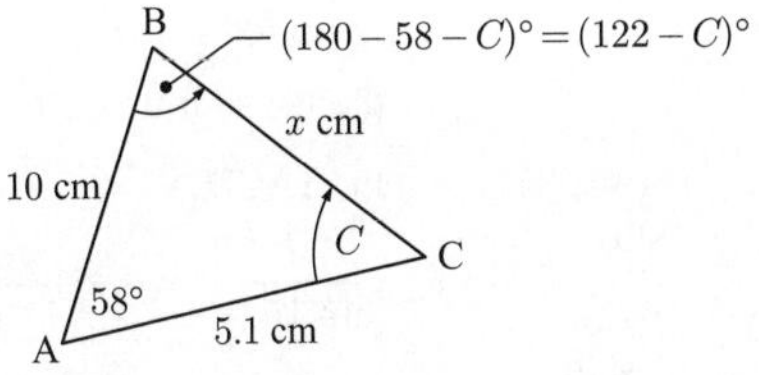

$$\frac{\sin C}{10} = \frac{\sin(122 - C)}{5.1}$$
$$\therefore \quad 5.1 \sin C = 10 \sin(122 - C)$$

Using technology,
$$C \approx 91.3^o$$

b Let BC $= x$ cm $\quad \therefore \quad x^2 = 10^2 + 5.1^2 - 2 \times 10 \times 5.1 \cos 58^o$

$$\therefore \quad x = \sqrt{10^2 + 5.1^2 - 20 \times 5.1 \times \cos 58^o}$$
$$\therefore \quad x \approx 8.4828$$

and $\cos C = \dfrac{5.1^2 + 8.4828^2 - 10^2}{2 \times 5.1 \times 8.4828} \approx -0.023\,09$

$$\therefore \quad C \approx \arccos(-0.02309) \approx 91.3^o$$

c "When faced with using either the sine rule or the cosine rule, it is better to use the *cosine rule* as it avoids the *ambiguous case*."

7

$$9^2 = x^2 + 7^2 - 2 \times x \times 7 \times \cos 30^o$$
$$\therefore \quad 81 = x^2 + 49 - 14x(\tfrac{\sqrt{3}}{2})$$
$$\therefore \quad x^2 - \tfrac{14\sqrt{3}}{2}x - 32 = 0$$

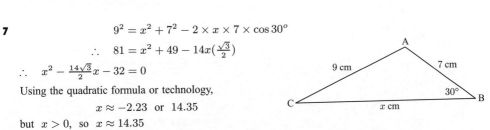

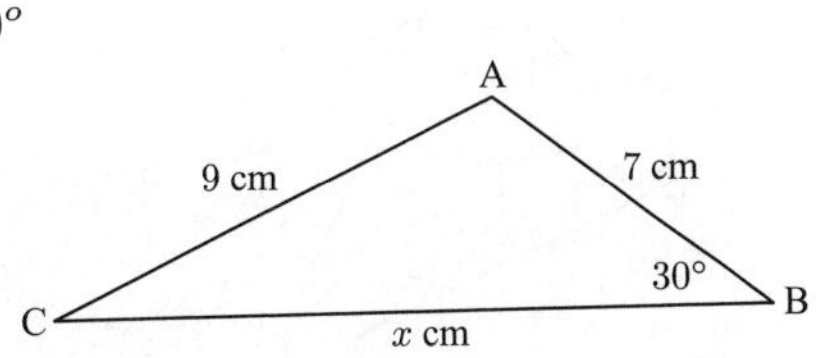

Using the quadratic formula or technology,
$$x \approx -2.23 \quad \text{or} \quad 14.35$$

but $x > 0$, so $x \approx 14.35$

$\therefore$ area of triangle $\approx \frac{1}{2} \times 7 \times 14.35 \times \sin 30^o \approx 25.1 \text{ cm}^2$

8

$$\frac{2x-5}{\sin 45^o} = \frac{x+3}{\sin 30^o}$$

$$\therefore \quad (2x-5)\sin 30^o = (x+3)\sin 45^o$$

$$\therefore \quad \frac{2x-5}{2} = \frac{x+3}{\sqrt{2}}$$

$$\therefore \quad 2\sqrt{2}x - 5\sqrt{2} = 2x + 6$$

$$\therefore \quad -6 - 5\sqrt{2} = x(2 - 2\sqrt{2})$$

$$\therefore \quad x = \left(\frac{-6-5\sqrt{2}}{2-2\sqrt{2}}\right)\left(\frac{2+2\sqrt{2}}{2+2\sqrt{2}}\right)$$

$$= \frac{-12 - 12\sqrt{2} - 10\sqrt{2} - 10(2)}{4 - 4(2)}$$

$$= \frac{-32 - 22\sqrt{2}}{-4}$$

$$= 8 + \tfrac{11}{2}\sqrt{2}$$

EXERCISE 9D

1

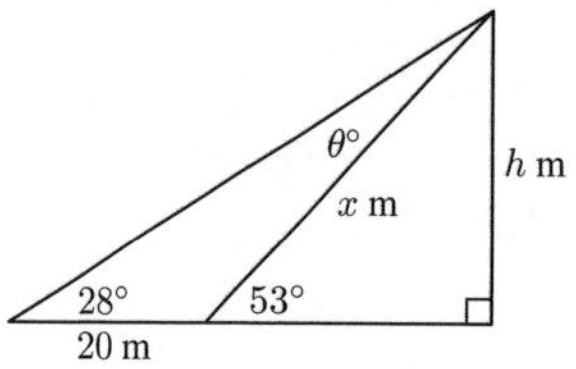

$$\theta^o + 28^o = 53^o$$

{exterior angle of a Δ theorem}

$$\therefore \quad \theta = 25$$

By the sine rule,

$$\frac{x}{\sin 28^o} = \frac{20}{\sin 25^o}$$

$$\therefore \quad x \approx \frac{20 \times \sin 28^o}{\sin 25^o}$$

$$\therefore \quad x \approx 22.22$$

and $\sin 53^o = \dfrac{h}{x}$

$$\therefore \quad h = x \sin 53^o \approx 22.22 \times \sin 53^o \approx 17.7$$

$\therefore$ the pole is 17.7 m high.

2

$$\text{PR}^2 = 63^2 + 175^2 - 2 \times 63 \times 175 \times \cos 112^o$$

$$\therefore \quad \text{PR} = \sqrt{63^2 + 175^2 - 2 \times 63 \times 175 \times \cos 112^o}$$

$$\therefore \quad \text{PR} \approx 207 \text{ m}$$

3

$$\cos T = \frac{220^2 + 340^2 - 165^2}{2 \times 220 \times 340}$$

$$\therefore \quad T = \cos^{-1}\left(\frac{136\,775}{149\,600}\right)$$

$$\therefore \quad T \approx 23.9$$

$\therefore$ the tee shot was 23.9^o off line.

4

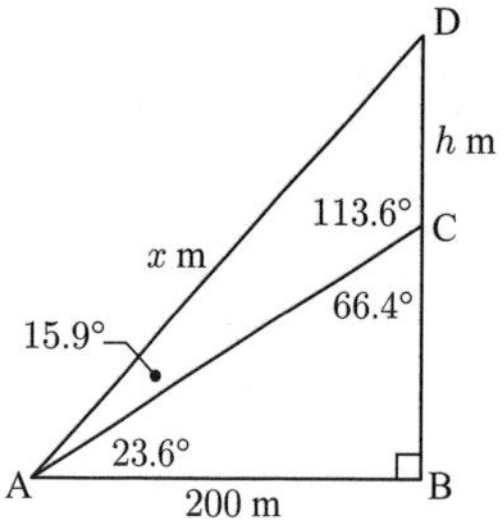

In ΔABD,

$$\cos(23.6 + 15.9)^o = \frac{200}{x}$$

$$\therefore \quad x = \frac{200}{\cos 39.5^o}$$

$$\therefore \quad x \approx 259.2$$

In ΔACD,

$$\frac{h}{\sin 15.9^o} = \frac{x}{\sin 113.6^o}$$

$$\therefore \quad h \approx \frac{259.2 \times \sin 15.9^o}{\sin 113.6^o}$$

$$\therefore \quad h \approx 77.5$$

$\therefore$ the tower is 77.5 m high.

5

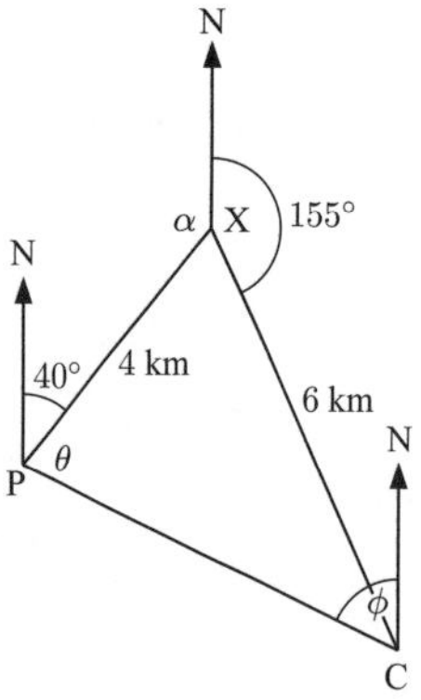

a **i** $\alpha = 140^o$ {co-interior angles}

$\therefore$ $\widehat{\text{PXC}} = 360^o - 140^o - 155^o$ {angles at a point}

$= 65^o$

So, $\text{PC}^2 = 4^2 + 6^2 - 2 \times 4 \times 6 \cos 65^o$

$$\therefore \quad \text{PC} = \sqrt{16 + 36 - 48\cos 65^o} \approx 5.6315 \text{ km}$$

$\therefore$ Esko hikes 5.63 km.

ii

$$\cos\theta \approx \frac{4^2 + 5.6315^2 - 6^2}{2 \times 4 \times 5.6315}$$

$$\therefore \quad \theta \approx 74.9^o$$

$\therefore$ bearing $= 40^o + \theta \approx 114.9^o$

$\therefore$ Esko hikes on a bearing of 115°.

b **i** $\text{speed} = \dfrac{\text{distance}}{\text{time}} \quad \Rightarrow \quad \text{time} = \dfrac{\text{distance}}{\text{speed}}$

$\therefore \quad \text{time}_{\text{Ritva}} = \dfrac{4+6}{10} = 1 \text{ hour} \quad \text{and} \quad \text{time}_{\text{Esko}} \approx \dfrac{5.6315}{6} \approx 0.9386 \text{ hours}$

$\approx 56.32 \text{ min}$

So Esko arrives at the campsite first.

ii $60 - 56.32 = 3.68$

Esko needs to wait about 3.68 minutes before Ritva arrives.

c $\phi \approx 180^o - 114.9^o \approx 65.1^o$ {co-interior angles}

$\therefore \quad 360^o - \phi \approx 295^o$

The return bearing is 295°.

6

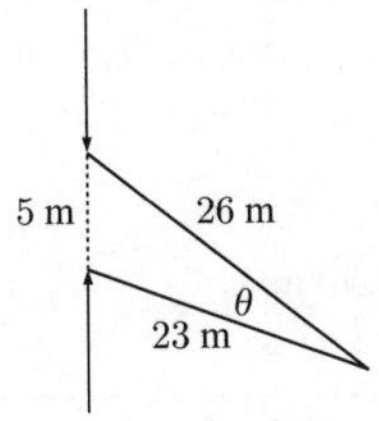

$\cos\theta = \dfrac{23^2 + 26^2 - 5^2}{2 \times 23 \times 26}$

$\therefore \quad \theta = \cos^{-1}\left(\frac{1180}{1196}\right)$

$\therefore \quad \theta \approx 9.38^o$

$\therefore$ the angle of view is 9.38^o.

7

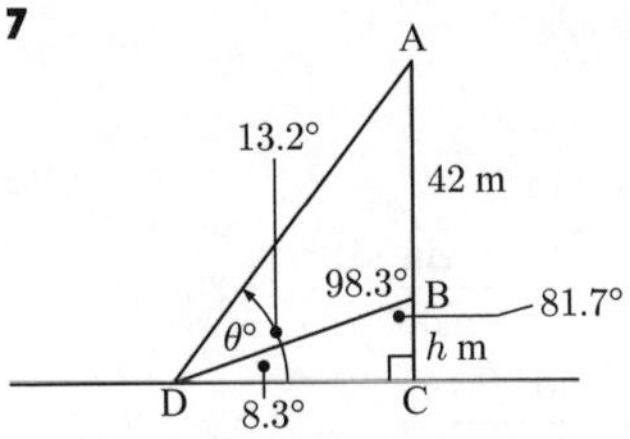

$\theta = 13.2^o - 8.3^o = 4.9^o$

In ΔABD,

$\dfrac{\text{AD}}{\sin 98.3^o} = \dfrac{42}{\sin 4.9^o}$

$\therefore \quad \text{AD} = \dfrac{42 \times \sin 98.3^o}{\sin 4.9^o}$

$\therefore \quad \text{AD} \approx 486.56 \text{ m}$

In ΔADC,

$\sin 13.2^o = \dfrac{h+42}{\text{AD}}$

$\therefore \quad h + 42 \approx 486.56 \times \sin 13.2^o$

$\therefore \quad h + 42 \approx 111.1$

$\therefore \quad h \approx 69.1$

$\therefore$ the hill is 69.1 m high.

8 **a, b**

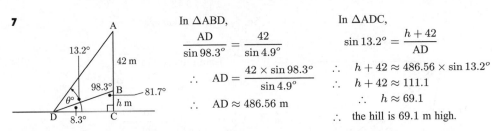

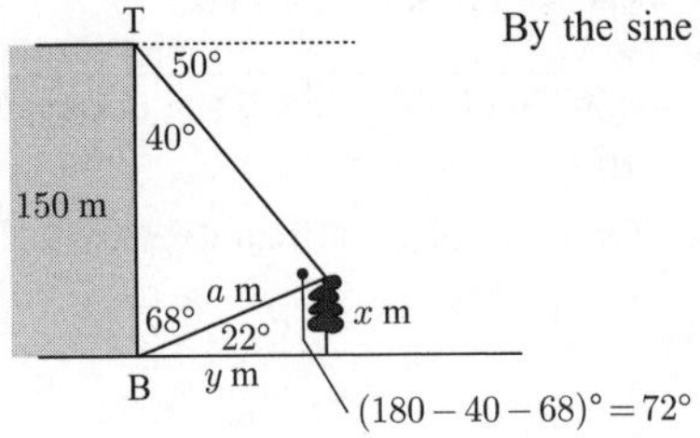

By the sine rule, $\dfrac{a}{\sin 40^o} = \dfrac{150}{\sin 72^o}$

$\therefore \quad a = \dfrac{150 \times \sin 40^o}{\sin 72^o}$

$\therefore \quad a \approx 101.38$

Now $\sin 22^o \approx \dfrac{x}{101.38}$ and $\cos 22^o \approx \dfrac{y}{101.38}$

$\therefore \quad x \approx 101.38 \times \sin 22^o \qquad \therefore \quad y \approx 101.38 \times \cos 22^o$

$\therefore \quad x \approx 38.0 \qquad \therefore \quad y \approx 94.0$

$\therefore$ the tree is 38.0 m high and 94.0 m from the building.

9 Using Pythagoras' theorem

$\text{RQ} = \sqrt{4^2 + 7^2} = \sqrt{65} \text{ m}$

$\text{PQ} = \sqrt{8^2 + 7^2} = \sqrt{113} \text{ cm}$

$\text{PR} = \sqrt{8^2 + 4^2} = \sqrt{80} \text{ cm}$

Now $\cos Q = \dfrac{(\sqrt{113})^2 + (\sqrt{65})^2 - (\sqrt{80})^2}{2 \times \sqrt{113} \times \sqrt{65}}$

$\therefore \quad \cos Q \approx \left(\frac{98}{171.4}\right)$

$\therefore \quad Q \approx \cos^{-1}\left(\frac{98}{171.4}\right)$

$\therefore \quad Q \approx 55.1$ So, $\widehat{\text{PQR}}$ measures 55.1^o

10

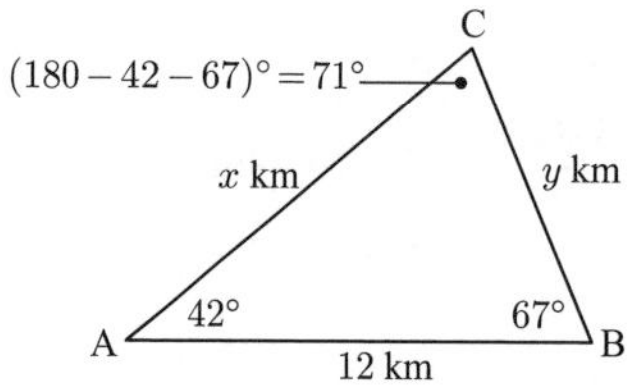

$$\frac{x}{\sin 67^o} = \frac{12}{\sin 71^o} = \frac{y}{\sin 42^o}$$

$\therefore \quad x = \dfrac{12 \times \sin 67^o}{\sin 71^o}$ and $y = \dfrac{12 \times \sin 42^o}{\sin 71^o}$

$\therefore \quad x \approx 11.7$ $\therefore \quad y \approx 8.49$

So, C is 11.7 km from A and 8.49 km from B.

11 **a** $\text{QS} = \sqrt{8^2 + 12^2 - 2 \times 8 \times 12 \times \cos 70^o}$

≈ 11.93

$\therefore \quad \text{area} \approx \frac{1}{2} \times 8 \times 12 \times \sin 70^o + \frac{1}{2} \times 10 \times 11.93 \times \sin 30^o$

$\approx 74.9 \text{ km}^2$

b 1 ha is 100 m $\times$ 100 m

$= 0.1 \text{ km} \times 0.1 \text{ km}$

$= 0.01 \text{ km}^2$

$\therefore \quad 1 \text{ km}^2 = 100 \text{ ha}$

$\therefore \quad \text{area} \approx 7490 \text{ ha}$

12

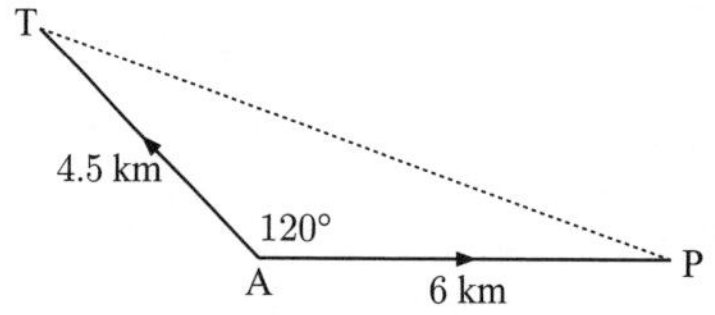

Distance = speed $\times$ time

So, after 45 min = 0.75 h,

$\text{AT} = 6 \times 0.75 = 4.5 \text{ km}$

$\text{AP} = 8 \times 0.75 = 6 \text{ km}$

Now $\text{PT} = \sqrt{4.5^2 + 6^2 - 2 \times 4.5 \times 6 \times \cos 120^o}$

$\therefore \quad \text{PT} \approx 9.12$

So, they are 9.12 km apart.

13

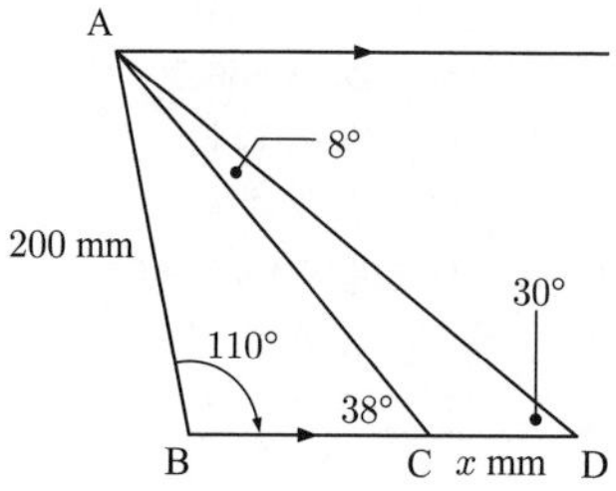

In ΔABC, $\dfrac{\text{AC}}{\sin 110^o} = \dfrac{200}{\sin 38^o}$

$\therefore \quad \text{AC} = \dfrac{200 \times \sin 110^o}{\sin 38^o} \approx 305.26$

and in ΔACD, $\dfrac{x}{\sin 8^o} \approx \dfrac{305.26}{\sin 30^o}$

$\therefore \quad x \approx \dfrac{305.26 \times \sin 8^o}{\sin 30^o} \approx 84.968$

$\therefore$ the metal strip is 85.0 mm wide.

14

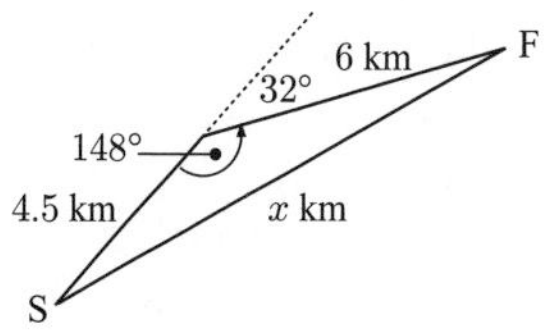

$x = \sqrt{6^2 + (4.5)^2 - 2 \times 6 \times 4.5 \times \cos 148^o}$

$\therefore \quad x \approx 10.1$

$\therefore$ the orienteer is 10.1 km from the start.

15

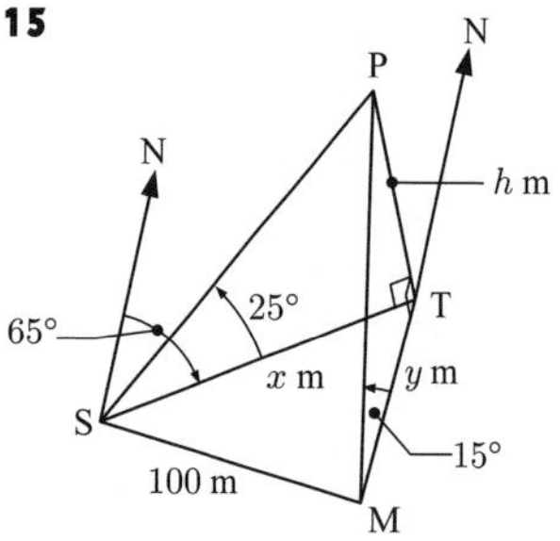

In ΔPST, $\tan 25^o = \dfrac{h}{x}$ In ΔPMT, $\tan 15^o = \dfrac{h}{y}$

$\therefore \quad x = \dfrac{h}{\tan 25^o}$ $\therefore \quad y = \dfrac{h}{\tan 15^o}$

$\approx 2.145h$ $\approx 3.732h$

But $\widehat{\text{STM}} = 65^o$ {equal alternate angles}

and $100^2 = x^2 + y^2 - 2xy \cos 65^o$

$\therefore \quad 10\,000 \approx (2.145h)^2 + (3.732h)^2 - 2 \times (2.145)(3.732)h^2 \cos 65^o$

$\therefore \quad 10\,000 \approx 11.762\, h^2$

$\therefore \quad h^2 \approx 850.17$

$\therefore \quad h \approx 29.2$ So, the tree is 29.2 m high.

16

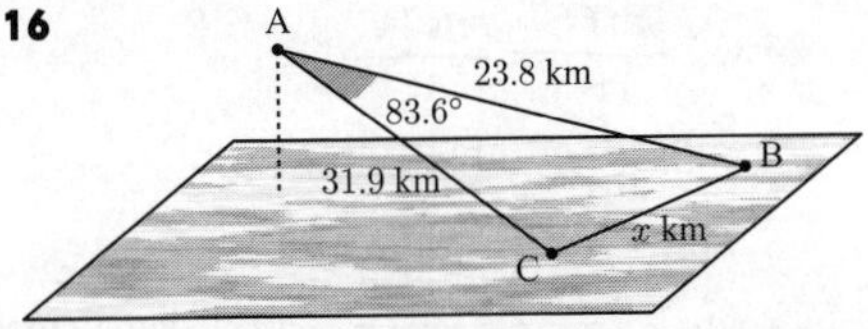

By the cosine rule

$$x^2 = 23.8^2 + 31.9^2 - 2 \times 23.8 \times 31.9 \times \cos 83.6^o$$

$$\therefore \quad x = \sqrt{23.8^2 + 31.9^2 - 2 \times 23.8 \times 31.9 \times \cos 83.6^o}$$

$$\therefore \quad x \approx 37.6$$

$\therefore$ B and C are 37.6 km apart.

REVIEW SET 9A

1
$$\text{area} = \tfrac{1}{2} \times 7 \times 8 \times \sin 30^o = 28 \times \tfrac{1}{2} = 14 \text{ km}^2$$

2 If the unknown is an angle, use the cosine rule to avoid the ambiguous case.

3 **a** By the cosine rule, $7^2 = 8^2 + x^2 - 2 \times 8 \times x \times \cos 60^o$

$$\therefore \quad 49 = 64 + x^2 - 16x\left(\tfrac{1}{2}\right)$$

$$\therefore \quad 49 = 64 + x^2 - 8x$$

$$\therefore \quad x^2 - 8x + 15 = 0$$

$$\therefore \quad (x-3)(x-5) = 0$$

$$\therefore \quad x = 3 \text{ or } 5$$

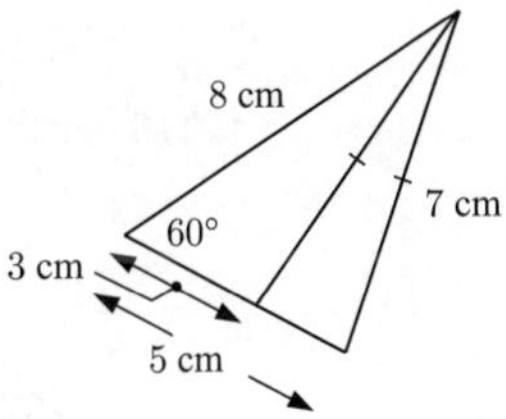

b Kady's response should be "Please supply me with additional information as there are two possibilities. Which one do you want?"

4 area $= 42$ cm^2

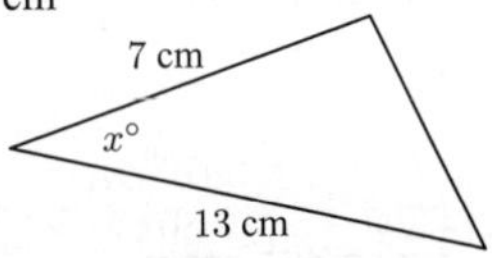

$$\therefore \quad \tfrac{1}{2} \times 7 \times 13 \times \sin x^o = 42$$

$$\therefore \quad \sin x^o = \frac{42 \times 2}{7 \times 13} = \tfrac{12}{13}$$

5 Total distance travelled $= x + 10$ km

$\therefore$ AB $= (x+10) - 4 = x + 6$ km

Now $(x+6)^2 = x^2 + 10^2 - 2 \times x \times 10 \times \cos 120^o$

$$\therefore \quad \cancel{x^2} + 12x + 36 = \cancel{x^2} + 100 - 20x\left(-\tfrac{1}{2}\right)$$

$$\therefore \quad 12x + 36 = 100 + 10x$$

$$\therefore \quad 2x = 64$$

$$\therefore \quad x = 32$$

$\therefore$ the boat travelled $x + 10 = 42$ km.

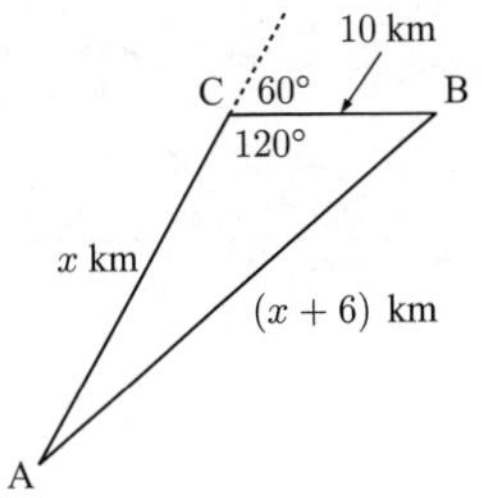

6 shaded area = area of sector $-$ area of Δ

$$= \tfrac{1}{2} \times \tfrac{13\pi}{18} \times 7^2 - \tfrac{1}{2} \times 7 \times 7 \times \sin\left(\tfrac{13\pi}{18}\right)$$

$$= \tfrac{49}{2}\left(\tfrac{13\pi}{18} - \sin\left(\tfrac{13\pi}{18}\right)\right)$$

REVIEW SET 9B

1 **a**
$$\cos x^o = \frac{13^2 + 19^2 - 11^2}{2 \times 13 \times 19}$$

$$\therefore \quad \cos x^o = \tfrac{409}{494}$$

$$\therefore \quad x^o = \cos^{-1}\left(\tfrac{409}{494}\right)$$

$$\therefore \quad x \approx 34.1$$

b
$$x = \sqrt{15^2 + 17^2 - 2 \times 15 \times 17 \times \cos 72^o}$$

$$\therefore \quad x \approx 18.9$$

2 $\text{AC} = \sqrt{11^2 + 9.8^2 - 2 \times 11 \times 9.8 \times \cos 74^o}$

$\therefore \quad \text{AC} \approx 12.554 \text{ cm}$

$\therefore \quad \text{AC} \approx 12.6 \text{ cm}$

Now $\dfrac{\sin C}{11} = \dfrac{\sin 74^o}{\text{AC}}$

$\therefore \quad \sin C \approx \dfrac{11 \times \sin 74^o}{12.554}$

$\therefore \quad C \approx \sin^{-1}\left(\dfrac{11 \times \sin 74^o}{12.554}\right)$ or its supplement

$\therefore \quad C \approx 57.4^o$ or 122.6^o

↑ impossible as $122.6 + 74 > 180$

$\therefore \quad C$ measures 57.4^o

$\therefore \quad A$ measures 48.6^o.

3 $\text{DB} = \sqrt{7^2 + 11^2 - 2 \times 7 \times 11 \times \cos 110^o} \approx 14.922 \text{ cm}$

$\therefore$ total area = area ΔABD + area ΔBCD

$\approx \frac{1}{2} \times 7 \times 11 \times \sin 110^o + \frac{1}{2} \times 16 \times 14.922 \times \sin 40^o$

$\approx 113 \text{ cm}^2$

4

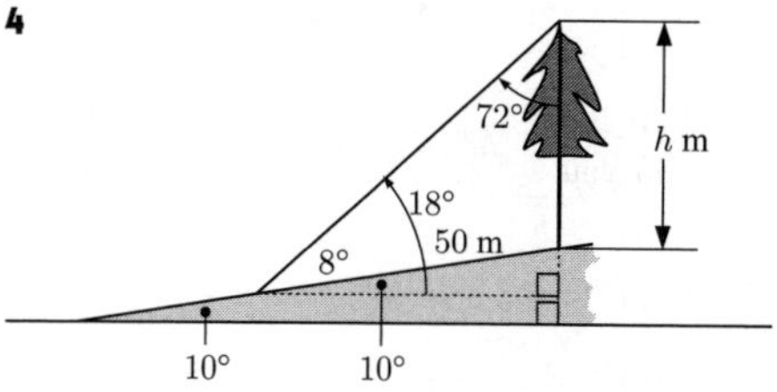

$\dfrac{h}{\sin 8^o} = \dfrac{50}{\sin 72^o}$

$\therefore \quad h = \dfrac{50 \times \sin 8^o}{\sin 72^o}$

$\therefore \quad h \approx 7.32$

So, the tree is 7.32 m high.

5

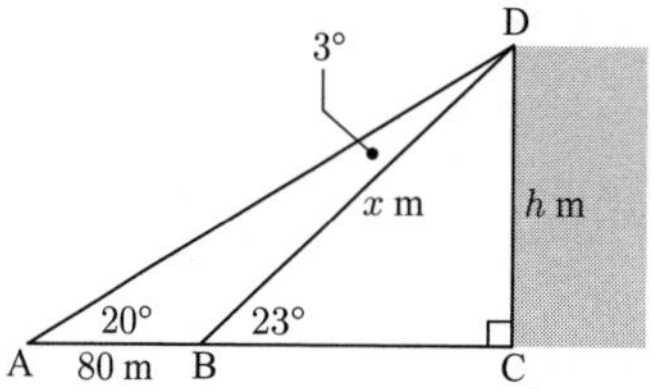

In ΔABD, $\dfrac{x}{\sin 20^o} = \dfrac{80}{\sin 3^o}$

$\therefore \quad x = \dfrac{80 \times \sin 20^o}{\sin 3^o} \approx 522.8$

Now $\sin 23^o = \dfrac{h}{x}$

$\therefore \quad h \approx 522.8 \times \sin 23^o$

$\therefore \quad h \approx 204$

So the building is 204 m tall.

6

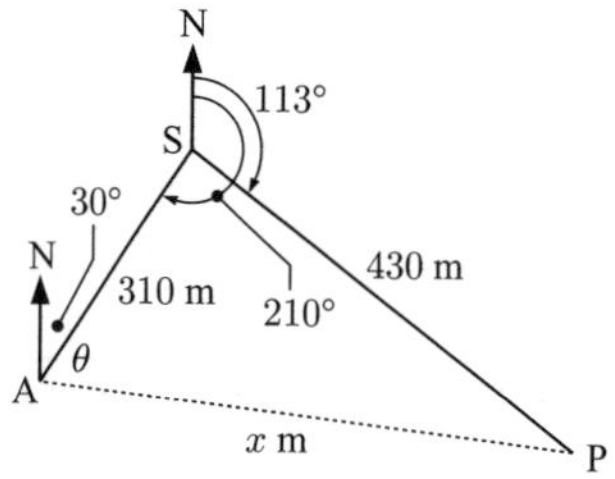

$\text{A}\widehat{\text{S}}\text{P} = 210^o - 113^o = 97^o$

$\therefore \quad x^2 = 310^2 + 430^2 - 2 \times 310 \times 430 \times \cos 97^o$

$\therefore \quad x = \sqrt{310^2 + 430^2 - 2 \times 310 \times 430 \times \cos 97^o}$

$\therefore \quad x \approx 559.9$

$\therefore$ Peter and Alix are 560 m apart.

and $\cos\theta \approx \dfrac{310^2 + 559.9^2 - 430^2}{2 \times 310 \times 559.9}$

$\therefore \quad \theta \approx 49.7$

and $30 + \theta \approx 79.7$

$\therefore$ the bearing of Peter from Alix is 079.7^o.

REVIEW SET 9C

1 **a** $\cos x^o = \dfrac{11^2 + 19^2 - 13^2}{2 \times 11 \times 19}$

$\therefore \cos x^o = \frac{313}{418}$

$\therefore x^o = \cos^{-1}\left(\frac{313}{418}\right)$

$\therefore x \approx 41.5$

b $x = \sqrt{14^2 + 21^2 - 2 \times 14 \times 21 \times \cos 47^o}$

$\therefore x \approx 15.4$

2 area $= 80$ cm^2

$\therefore \frac{1}{2} \times 11.3 \times 19.2 \sin x^o = 80$

$\therefore \sin x^o = \dfrac{160}{11.3 \times 19.2}$

Now $\arcsin\left(\dfrac{160}{11.3 \times 19.2}\right) \approx 47.5^o$

$\therefore x \approx 47.5$ or $180 - 47.5$

$\therefore x \approx 47.5$ or 132.5

3 Using Pythagoras,

$\text{ED} = \sqrt{6^2 + 3^2} = \sqrt{45}$ m

$\text{DG} = \sqrt{4^2 + 3^2} = \sqrt{25} = 5$ m

$\text{EG} = \sqrt{6^2 + 4^2} = \sqrt{52}$ m

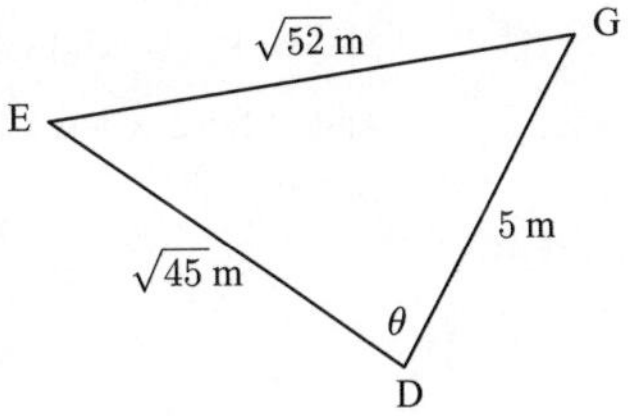

Using the cosine rule, $\cos\theta = \dfrac{(\sqrt{45})^2 + 5^2 - (\sqrt{52})^2}{2 \times \sqrt{45} \times 5}$

$\therefore \theta = \cos^{-1}\left(\frac{18}{10\sqrt{45}}\right)$

$\therefore \theta \approx 74.4^o$ Thus $\widehat{\text{EDG}}$ measures 74.4^o.

4 **a** $\text{BD} = \sqrt{120^2 + 125^2 - 2 \times 120 \times 125 \cos 75^o}$

≈ 149.2 m

The area of the block = area of ΔABD + area of ΔBCD

$\approx \frac{1}{2} \times 120 \times 125 \times \sin 75^o + \frac{1}{2} \times 149.2 \times 90 \times \sin 30^o$

$\approx 10\,600$ m^2

b ≈ 1.06 ha {$10\,000$ m^2 = 1 ha}

5

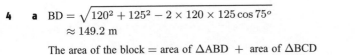

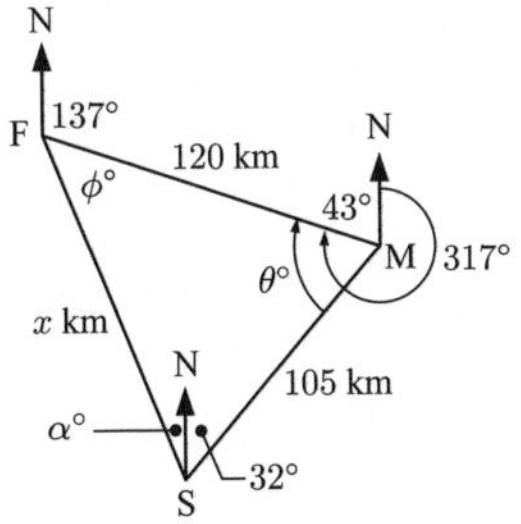

In 45 minutes, $140 \times \frac{3}{4} = 105$ km is travelled.

In 40 minutes, $180 \times \frac{2}{3} = 120$ km is travelled.

We notice that $\theta + 43 + 32 = 180$ {co-interior angles add to 180°}

$\therefore \theta = 105$

Using the cosine rule, $x = \sqrt{120^2 + 105^2 - 2 \times 120 \times 105 \times \cos 105^o}$

$\therefore x \approx 178.74$

So, the car is 179 km from the start.

Now $\dfrac{\sin\phi^o}{105} \approx \dfrac{\sin 105^o}{178.74}$

$\therefore \sin\phi^o \approx \dfrac{105 \times \sin 105^o}{178.74}$

$\therefore \phi \approx 34.6$

$\therefore \alpha \approx 180 - 105 - 34.6 - 32 \approx 8.4 \approx 8$

So, the bearing from its starting point is 352°.

6 **a**

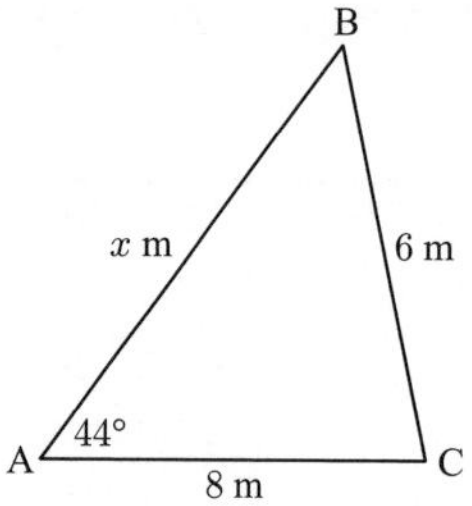

By the cosine rule, $6^2 = x^2 + 8^2 - 2 \times x \times 8 \times \cos 44^o$

$\therefore \quad 36 = x^2 + 64 - 16x \times \cos 44^o$

$\therefore \quad x^2 - 11.51x + 28 \approx 0$

$$\therefore \quad x \approx \frac{11.51 \pm \sqrt{11.51^2 - 4(1)(28)}}{2}$$

$$\therefore \quad x \approx \frac{11.51 \pm 4.524}{2}$$

$\therefore \quad x \approx 8.02 \text{ or } 3.49$

Frank needs additional information as there are two possible cases:

(1) when AB $\approx$ 8.02 m and

(2) when AB $\approx$ 3.49 m

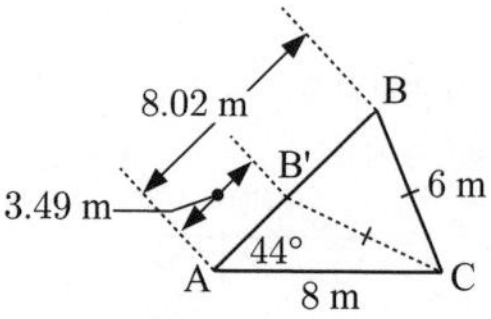

b Volume = area × depth

$= \frac{1}{2} \times 8 \times x \times \sin 44^o \times 0.1$ and is a maximum when $x \approx 8.02$ m

$\approx 4 \times 8.02 \times \sin 44^o \times 0.1$

$\approx 2.23 \text{ m}^3$

Chapter 10
ADVANCED TRIGONOMETRY

EXERCISE 10A

1 **a**

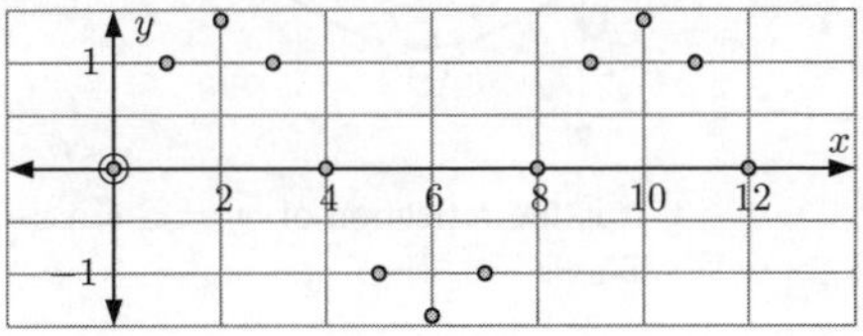

Data exhibits periodic behaviour.

b

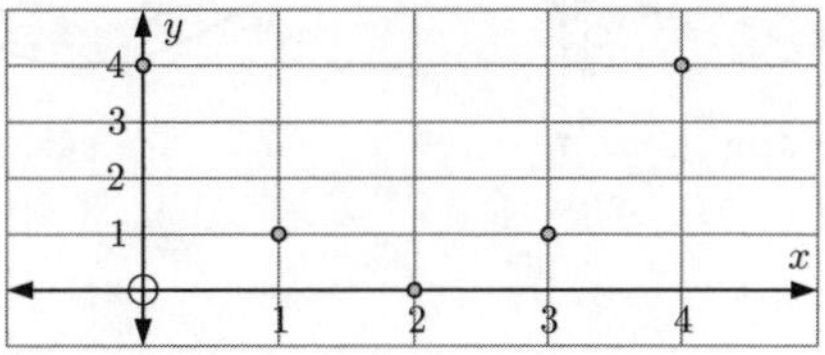

Not enough information to say data is periodic. It may in fact be quadratic.

c

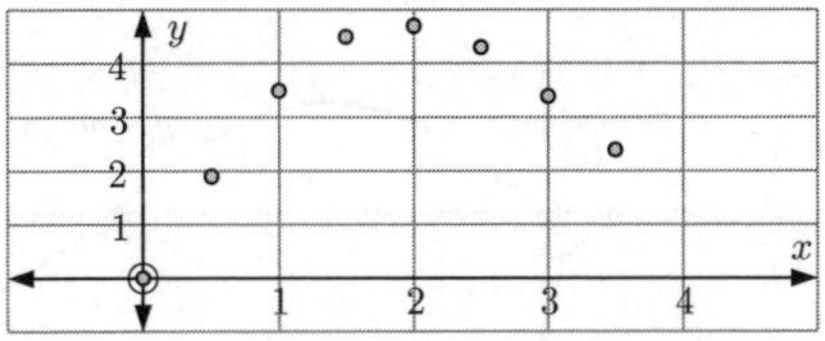

Not enough information to say data is periodic. It may in fact be quadratic.

d

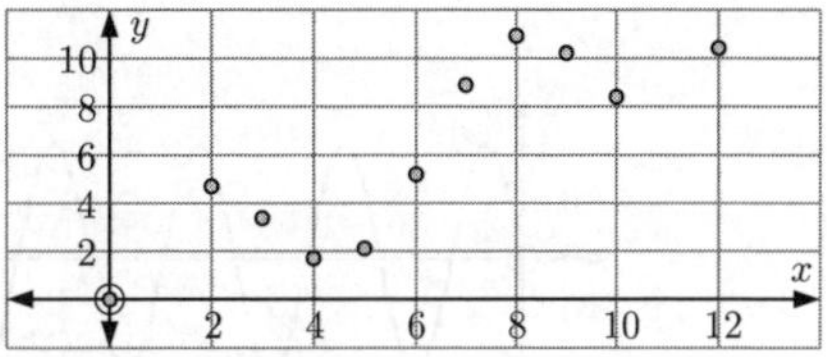

Not enough information to say data is periodic.

2 **a**

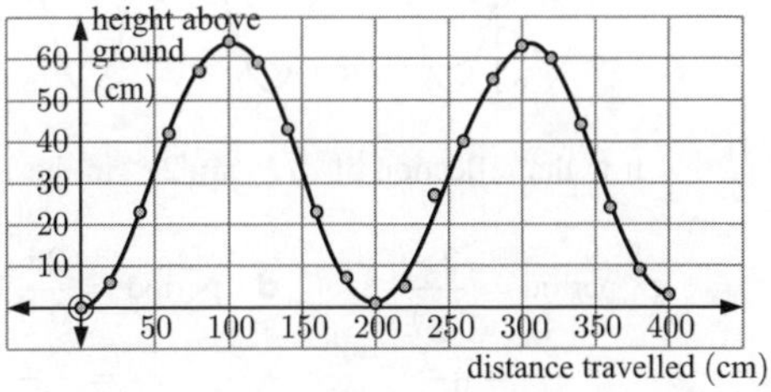

b The data is periodic.

i The minimum value from the table is 0 and the maximum value is 64.
So, the principal axis is $y \approx \frac{0+64}{2}$,
$\therefore\ y \approx 32$.

ii The maximum value is ≈ 64 cm.

iii The period is ≈ 200 cm.

iv The amplitude is ≈ 32 cm.

c A curve can be fitted to the data as the distance travelled is continuous.

3 **a** periodic **b** periodic **c** periodic **d** not periodic **e** periodic **f** periodic

EXERCISE 10B.1

1 **a** $y = 3\sin x$
has amplitude 3 and period $\frac{2\pi}{1} = 2\pi$
When $x = 0$, $y = 0$.

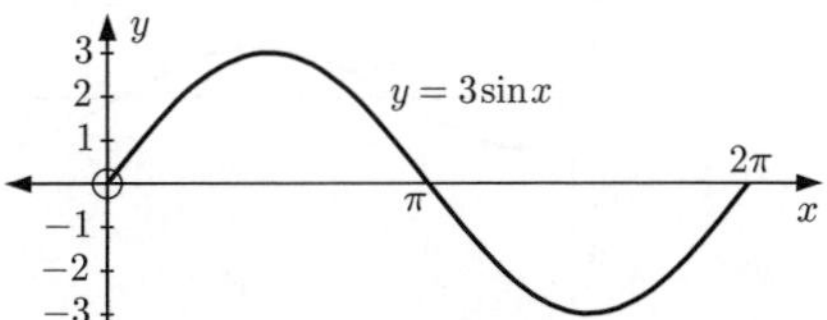

b $y = -3\sin x$
has amplitude $|-3| = 3$
and period $\frac{2\pi}{1} = 2\pi$.
When $x = 0$, $y = 0$.

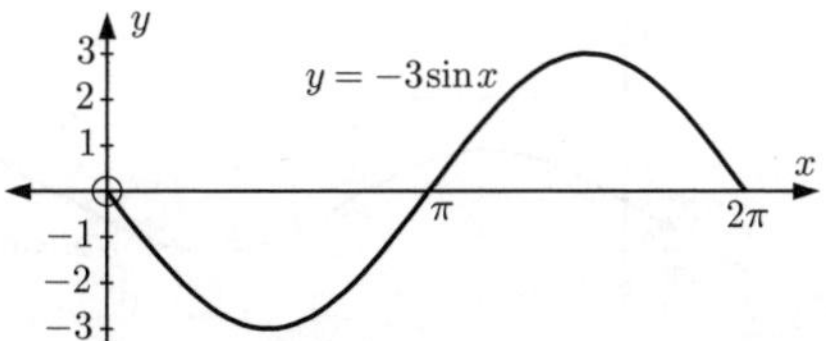

It is the reflection of $y = 3\sin x$ in the x-axis.

c $y = \frac{3}{2}\sin x$
has amplitude $\frac{3}{2}$ and period $\frac{2\pi}{1} = 2\pi$.
When $x = 0$, $y = 0$.

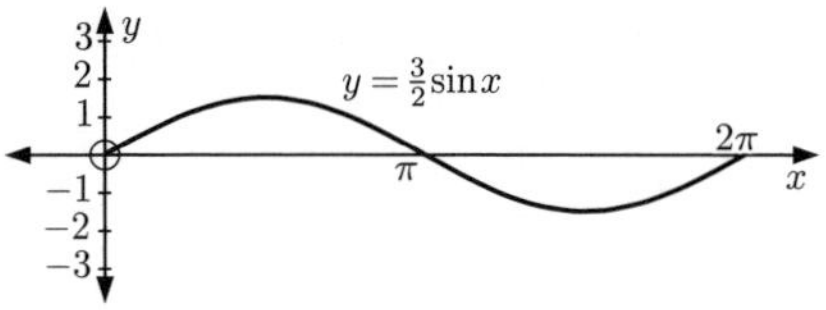

d $y = -\frac{3}{2}\sin x$
has amplitude $\left|-\frac{3}{2}\right| = \frac{3}{2}$
and period $\frac{2\pi}{1} = 2\pi$.

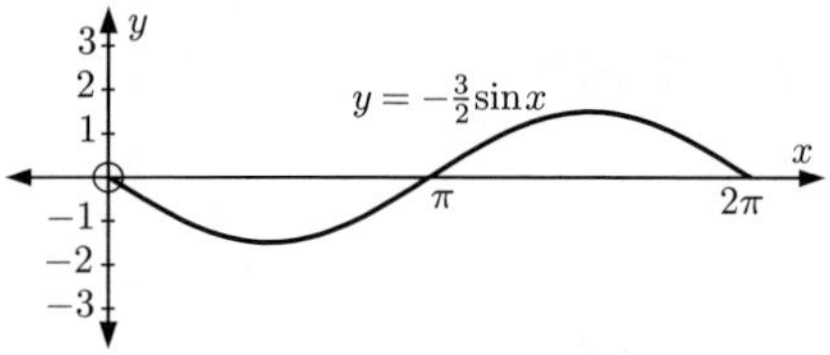

It is the reflection of $y = \frac{3}{2}\sin x$ in the x-axis.

2 **a** $y = \sin 3x$
has amplitude 1 and period $\frac{2\pi}{3}$.
When $x = 0$, $y = 0$.

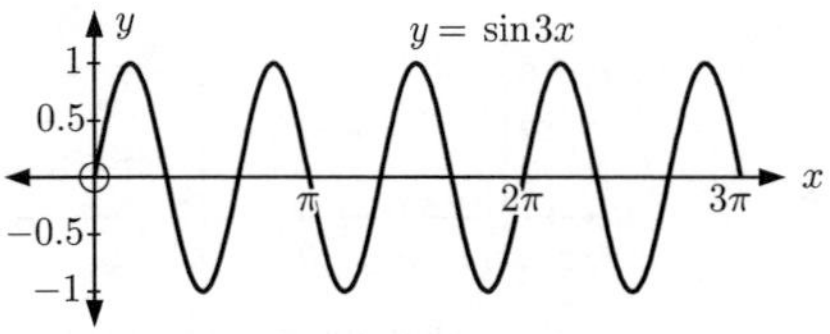

b $y = \sin\left(\frac{x}{2}\right)$
has amplitude 1 and period $\dfrac{2\pi}{\frac{1}{2}} = 4\pi$.
When $x = 0$, $y = 0$.

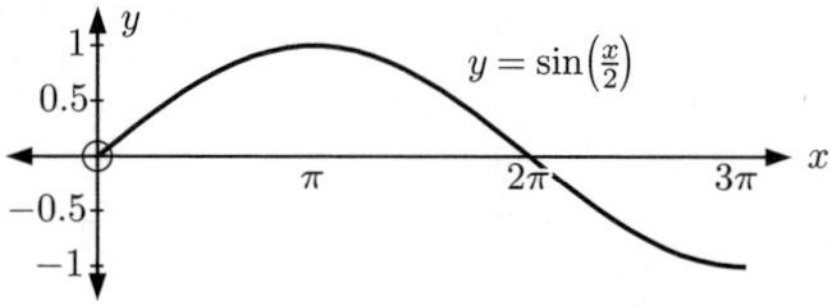

c $y = \sin(-2x)$
has amplitude 1 and period $\dfrac{2\pi}{|-2|} = \pi$.
When $x = 0$, $y = 0$.

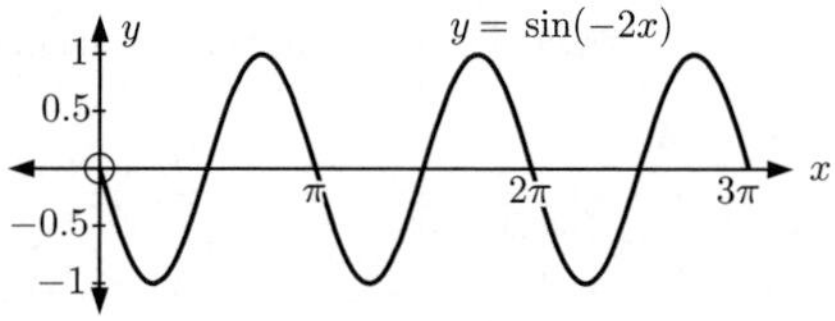

It is the reflection of $y = \sin 2x$ in the y-axis.

3 **a** $\text{period} = \dfrac{2\pi}{4} = \frac{\pi}{2}$
b $\text{period} = \dfrac{2\pi}{|-4|} = \frac{\pi}{2}$
c $\text{period} = \dfrac{2\pi}{\left(\frac{1}{3}\right)} = 6\pi$
d $\text{period} = \dfrac{2\pi}{0.6} = \frac{20\pi}{6} = \frac{10\pi}{3}$

4 **a** $\dfrac{2\pi}{b} = 5\pi$ $\therefore\ b = \frac{2}{5}$
b $\dfrac{2\pi}{b} = \dfrac{2\pi}{3}$ $\therefore\ b = 3$
c $\dfrac{2\pi}{b} = 12\pi$ $\therefore\ b = \frac{1}{6}$
d $\dfrac{2\pi}{b} = 4$ $\therefore\ b = \frac{\pi}{2}$
e $\dfrac{2\pi}{b} = 100$ $\therefore\ b = \frac{2\pi}{100} = \frac{\pi}{50}$

EXERCISE 10B.2

1 **a**

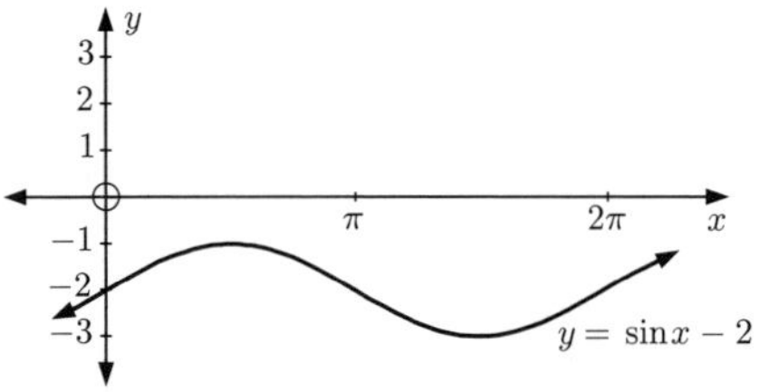

This is the graph of $y = \sin x$ translated by $\binom{0}{-2}$.

b

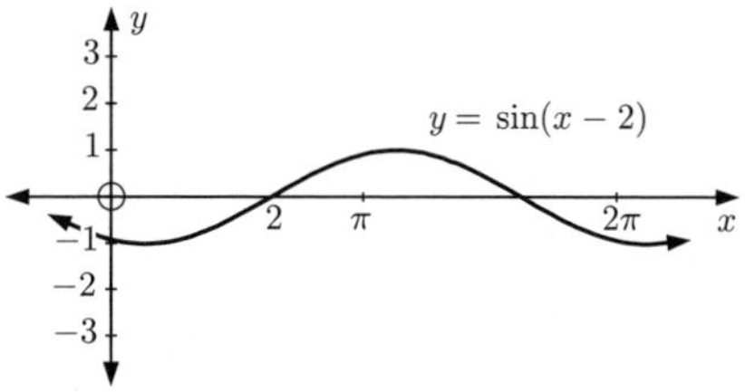

This is the graph of $y = \sin x$ translated by $\binom{2}{0}$.

c

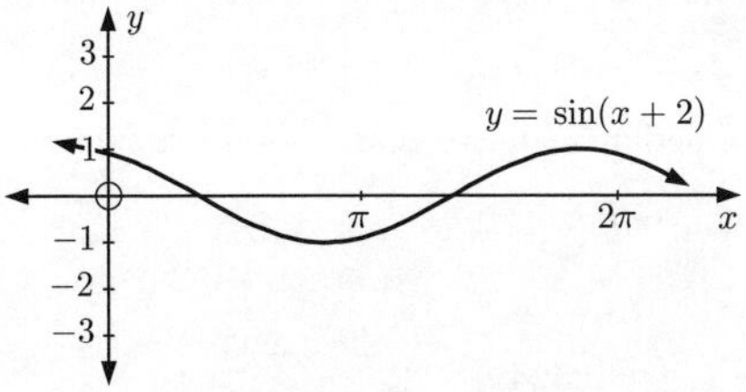

This is the graph of $y = \sin x$ translated by $\binom{-2}{0}$.

d

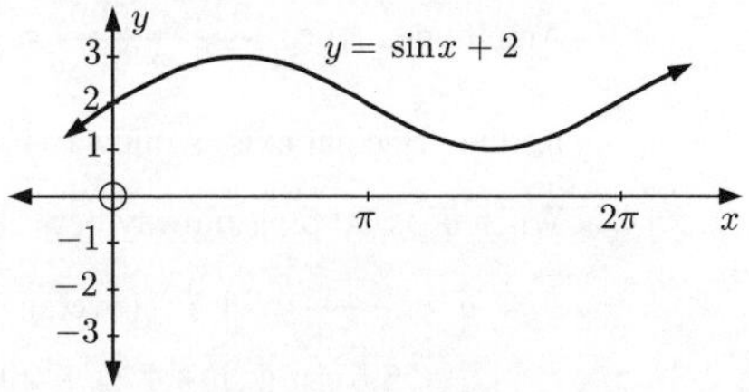

This is the graph of $y = \sin x$ translated by $\binom{0}{2}$.

e

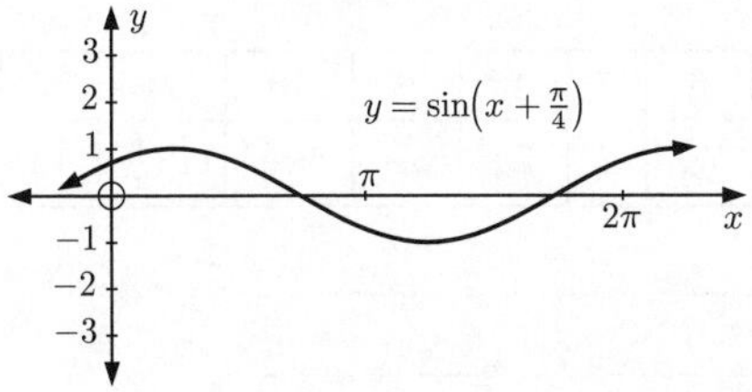

This is the graph of $y = \sin x$ translated by $\begin{pmatrix} -\frac{\pi}{4} \\ 0 \end{pmatrix}$.

f

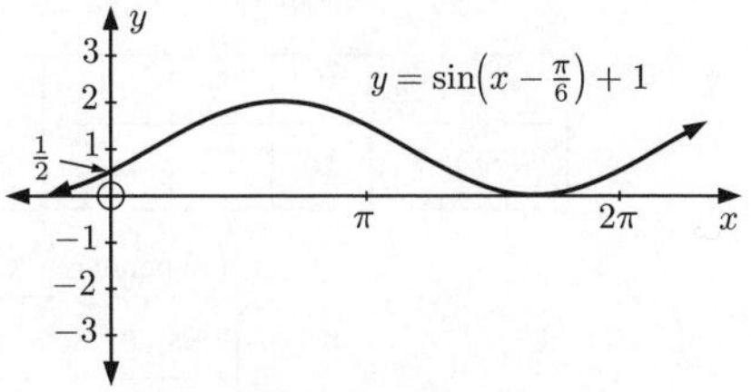

This is the graph of $y = \sin x$ translated by $\begin{pmatrix} \frac{\pi}{6} \\ 1 \end{pmatrix}$.

2 **a** period $= \dfrac{2\pi}{5} = \frac{2\pi}{5}$ **b** period $= \dfrac{2\pi}{\left(\frac{1}{4}\right)} = 8\pi$ **c** period $= \dfrac{2\pi}{|-2|} = \pi$

3 **a** $\dfrac{2\pi}{b} = 3\pi$ $\therefore\ b = \frac{2}{3}$ **b** $\dfrac{2\pi}{b} = \dfrac{\pi}{10}$ $\therefore\ b = 20$ **c** $\dfrac{2\pi}{b} = 100\pi$ $\therefore\ b = \frac{2}{100} = \frac{1}{50}$ **d** $\dfrac{2\pi}{b} = 50$ $\therefore\ b = \frac{2\pi}{50} = \frac{\pi}{25}$

4 **a** A translation of $\binom{0}{-1}$, or vertically down 1 unit.

b A translation of $\begin{pmatrix} \frac{\pi}{4} \\ 0 \end{pmatrix}$, or horizontally $\frac{\pi}{4}$ units right.

c A vertical stretch of factor 2. **d** A horizontal stretch of factor $\frac{1}{4}$.

e A vertical stretch of factor $\frac{1}{2}$. **f** A horizontal stretch of factor 4.

g A reflection in the x-axis. **h** A translation of $\binom{-2}{-3}$.

i A vertical stretch of factor 2 followed by a horizontal stretch of factor $\frac{1}{3}$.

j A translation of $\begin{pmatrix} \frac{\pi}{3} \\ 2 \end{pmatrix}$.

EXERCISE 10C

1 **a**

Month, t	1	2	3	4	5	6	7	8	9	10	11	12
Temp, T	15	14	15	18	21	25	27	26	24	20	18	16

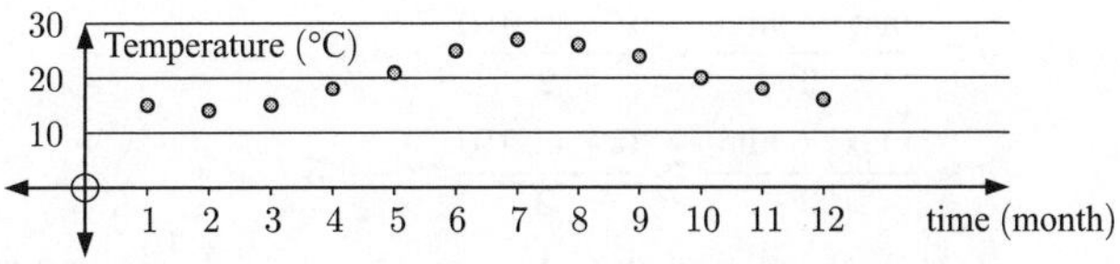

The period is 12 months so $\dfrac{2\pi}{b} = 12$ $\therefore\ b = \frac{\pi}{6}$ {assuming $b > 0$}.

Amplitude, $a \approx \dfrac{\text{max.} - \text{min.}}{2} \approx \dfrac{27-14}{2} \approx 6.5$

As the principal axis is midway between min. and max., then $d \approx \dfrac{27+14}{2} \approx 20.5$

When T is 20.5 (midway between min. and max.)

$c \approx \dfrac{2+7}{2} \approx 4.5$ {average of t values}

$\therefore \quad T \approx 6.5 \sin \frac{\pi}{6}(t-4.5) + 20.5$ where $\frac{\pi}{6} \approx 0.524$.

b Using technology, $T \approx 6.14 \sin(0.575t - 2.70) + 20.4$

$\therefore \quad T \approx 6.14 \sin 0.575(t - 4.70) + 20.4$

2 **a**

Month, t	1	2	3	4	5	6	7	8	9	10	11	12
Temp, T	15	16	$14\frac{1}{2}$	12	10	$7\frac{1}{2}$	7	$7\frac{1}{2}$	$8\frac{1}{2}$	$10\frac{1}{2}$	$12\frac{1}{2}$	14

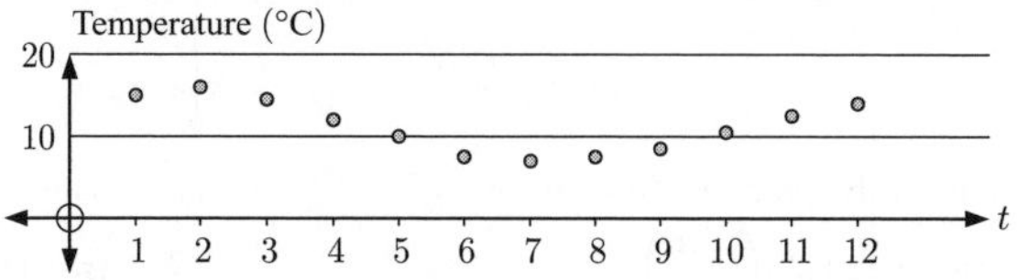

The period is $\dfrac{2\pi}{b} = 12 \quad \therefore \quad b = \frac{\pi}{6} \quad \{b > 0\}$

Amplitude, $a \approx \dfrac{\text{max.} - \text{min.}}{2} \approx \dfrac{16-7}{2} \approx 4.5$

As the principal axis is midway between min. and max. then $d \approx \dfrac{16+7}{2} \approx 11.5$

At min., $t = 7$ and at max., $t = 2 + 12 = 14 \quad \therefore \quad c \approx \dfrac{7+14}{2} \approx 10.5$

So, $T \approx 4.5 \sin \frac{\pi}{6}(t - 10.5) + 11.5$

b Using technology, $T \approx 4.29 \sin(0.533t + 0.769) + 11.2$

$\therefore \quad T \approx 4.29 \sin 0.533(t + 1.44) + 11.2$

Note: (1) $\frac{\pi}{6} \approx 0.524$ ✓

(2) $1.44 - (-10.5) = 11.94$
≈ 12

3

Month, t	1	2	3	4	5	6	7	8	9	10	11	12
Temp, T	0	-4	-10	-15	-16	-17	-18	-19	-17	-13	-6	-1

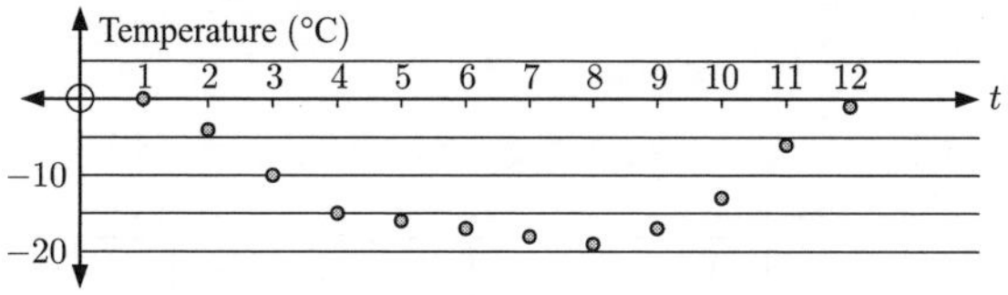

The period is $\dfrac{2\pi}{b} = 12 \quad \therefore \quad b = \frac{\pi}{6} \quad \{b > 0\}$

Amplitude, $a \approx \dfrac{\text{max.} - \text{min.}}{2} \approx \dfrac{0-(-19)}{2} \approx 9.5$

$d \approx \dfrac{\text{max.} + \text{min.}}{2} \approx \dfrac{0+(-19)}{2} \approx -9.5$

At min., $t = 8$ and at max., $t = 1 + 12 = 13 \quad \therefore \quad c \approx \dfrac{8+13}{2} \approx 10.5$

So, $T \approx 9.5 \sin \frac{\pi}{6}(t - 10.5) - 9.5$

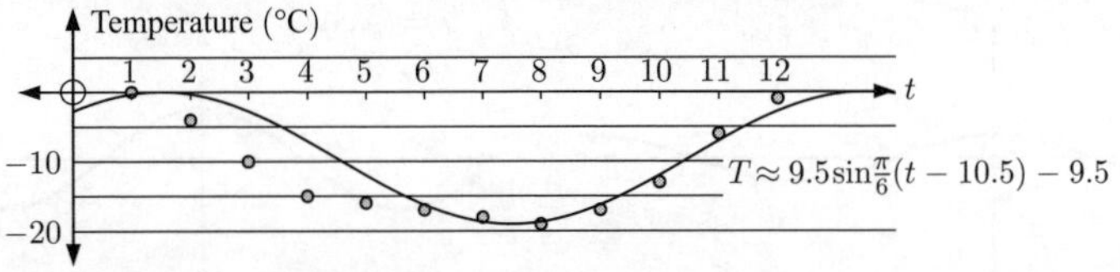

The model is not very appropriate.

4 **a** For the model $H = a\sin b(t - c) + d$

$$\text{period} = \frac{2\pi}{b} = 12.4 \text{ hours} \quad \therefore \quad b = \frac{2\pi}{12.4} \approx 0.507$$

We let the principal axis be 0, so $d = 0$

$\therefore$ the amplitude $a = 7$, so the min. is -7, and the max. is $+7$

Let $t = 0$ correspond to 'low tide' $\therefore$ $t = 6.2$ corresponds to 'high tide'

$\therefore \quad c = \dfrac{0 + 6.2}{2} = 3.1$

So, $H \approx 7\sin 0.507(t - 3.1) + 0$

$\therefore \quad H \approx 7\sin 0.507(t - 3.1)$

b

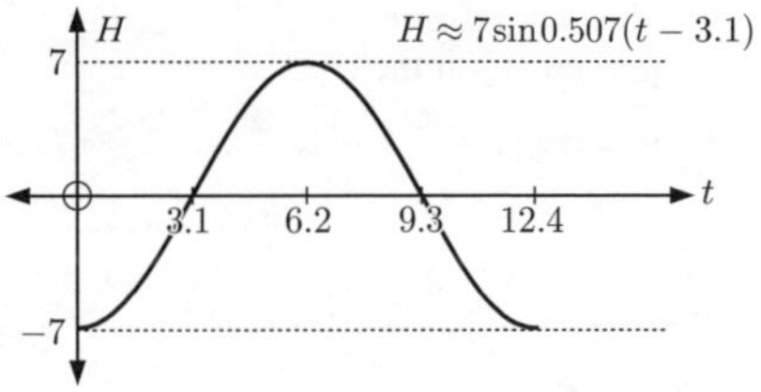

5 Let the model be $H = a\sin b(t - c) + d$ metres

When $t = 0$, $H = 2$ (min.) and when $t = 50$, $H = 22$ (max.)

$$\text{period} = \frac{2\pi}{b} = 100 \quad \therefore \quad b = \frac{2\pi}{100} = \frac{\pi}{50}$$

$a = 10$ {from the diagram} $\qquad d = \dfrac{\text{max.} + \text{min.}}{2} = \dfrac{22 + 2}{2} = 12$

$c = \dfrac{0 + 50}{2} = 25$ {values of t at min. and max.} $\quad \therefore \quad H = 10\sin\dfrac{\pi}{50}(t - 25) + 12$

EXERCISE 10D

1 **a** $y = \cos x + 2$

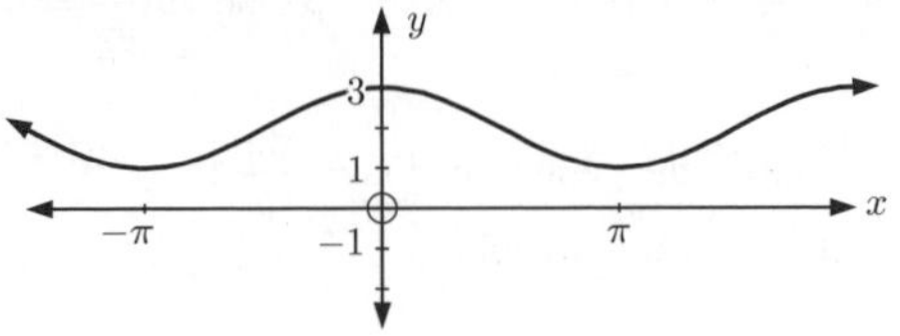

This is a vertical translation of $y = \cos x$ through $\binom{0}{2}$.

b $y = \cos x - 1$

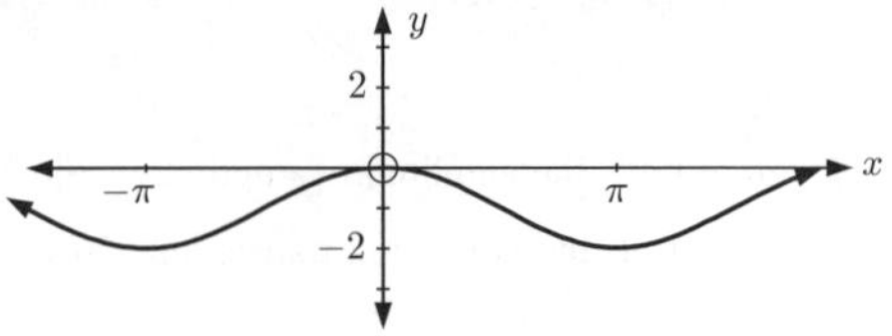

This is a vertical translation of $y = \cos x$ through $\binom{0}{-1}$.

c $y = \cos(x - \frac{\pi}{4})$

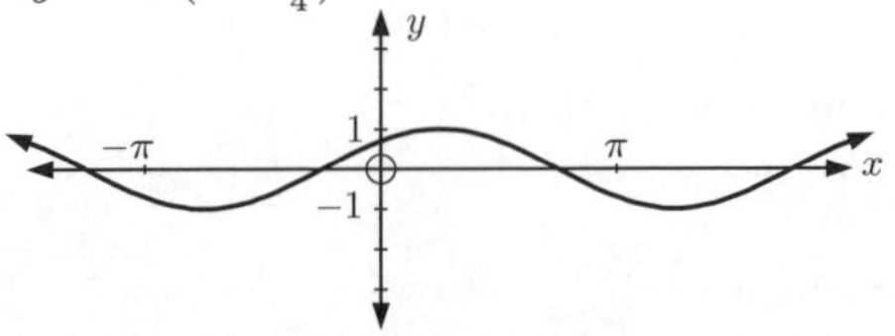

This is a horizontal translation of $y = \cos x$ through $\begin{pmatrix} \frac{\pi}{4} \\ 0 \end{pmatrix}$.

d $y = \cos(x + \frac{\pi}{6})$

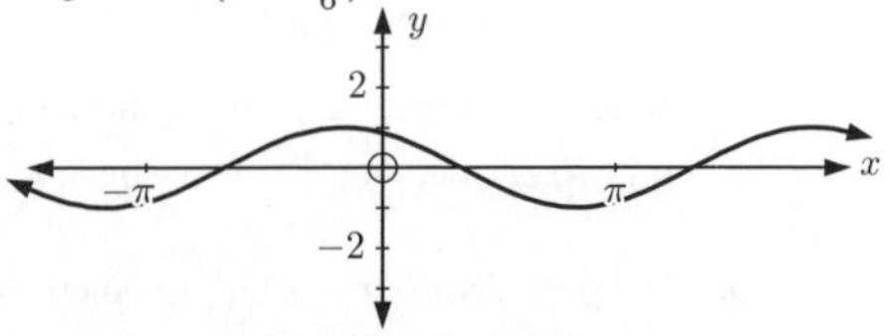

This is a horizontal translation of $y = \cos x$ through $\begin{pmatrix} -\frac{\pi}{6} \\ 0 \end{pmatrix}$.

e $y = \frac{2}{3}\cos x$

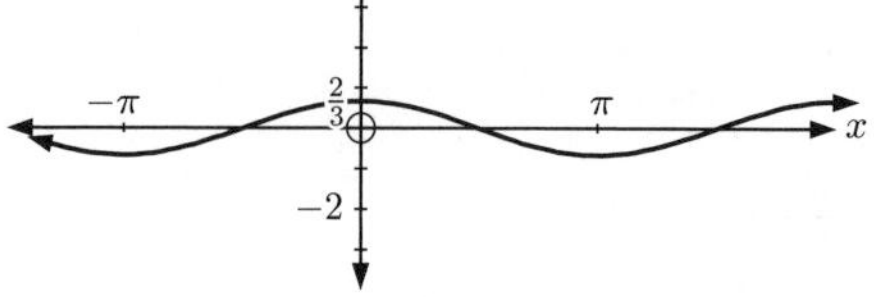

This is a vertical stretch of $y = \cos x$ with factor $\frac{2}{3}$.

f $y = \frac{3}{2}\cos x$

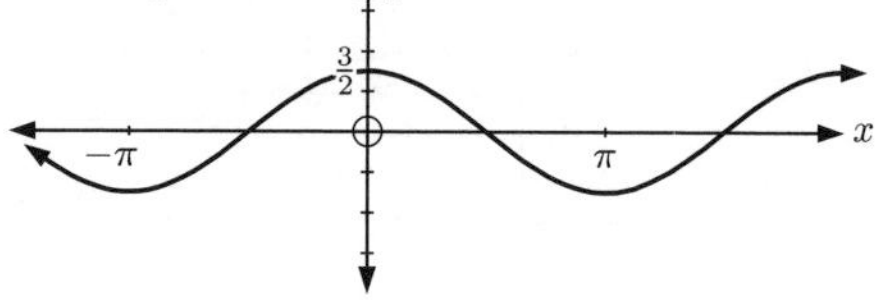

This is a vertical stretch of $y = \cos x$ with factor $\frac{3}{2}$.

g $y = -\cos x$

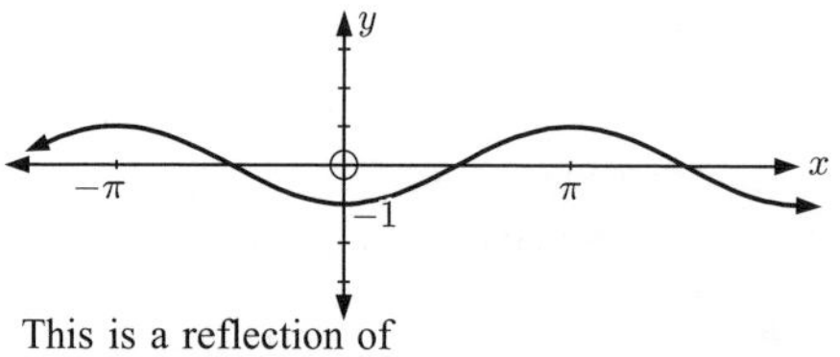

This is a reflection of $y = \cos x$ in the x-axis.

h $y = \cos(x - \frac{\pi}{6}) + 1$

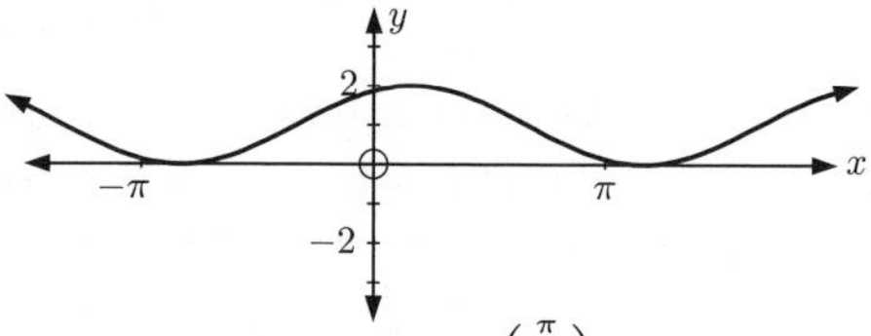

This is a translation of $\begin{pmatrix} \frac{\pi}{6} \\ 1 \end{pmatrix}$.

i $y = \cos(x + \frac{\pi}{4}) - 1$

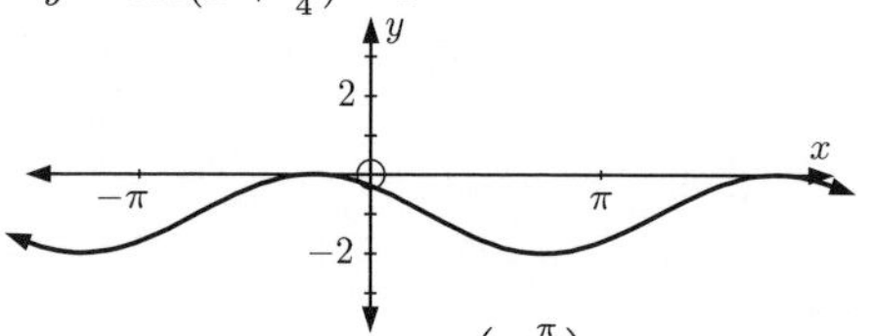

This is a translation of $\begin{pmatrix} -\frac{\pi}{4} \\ -1 \end{pmatrix}$.

j $y = \cos 2x$

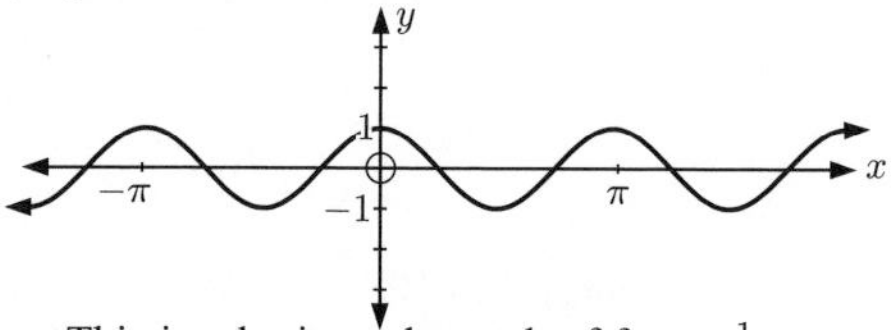

This is a horizontal stretch of factor $\frac{1}{2}$.

k $y = \cos\left(\frac{x}{2}\right)$

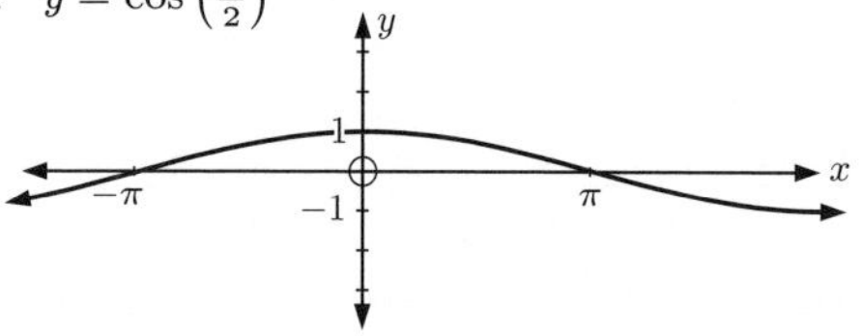

This is a horizontal stretch of factor 2.

l $y = 3\cos 2x$

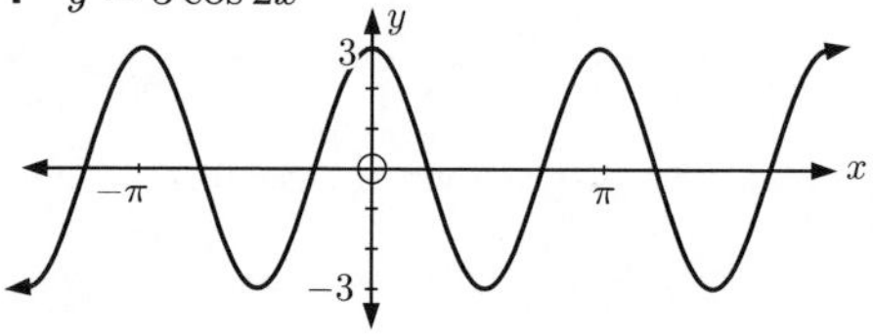

This is a horizontal stretch of factor $\frac{1}{2}$ followed by a vertical stretch of factor 3.

2 **a** period $= \frac{2\pi}{3}$ **b** period $= \dfrac{2\pi}{\frac{1}{3}} = 6\pi$ **c** period $= \dfrac{2\pi}{\frac{\pi}{50}} = 100$

3 a controls the amplitude {amplitude $= |a|$}. b controls the period {period $= \dfrac{2\pi}{|b|}$}.
c controls the horizontal translation. d controls the vertical translation.

4 **a** If $y = a\cos b(x - c) + d$, then $a = 2$, $\pi = \dfrac{2\pi}{b}$ $\therefore$ $b = 2$
c and d are 0 as there is no horizontal or vertical shift. $\therefore$ $y = 2\cos(2x)$

b If $y = a\cos b(x - c) + d$, then $a = 1$, $4\pi = \dfrac{2\pi}{b}$ $\therefore$ $b = \frac{1}{2}$
A vertical shift of 2 units, no horizontal shift $\therefore$ $d = 2$, $c = 0$.
So, $y = \cos(\frac{1}{2}x) + 2$ or $y = \cos\left(\frac{x}{2}\right) + 2$.

c If $y = a\cos b(x - c) + d$, then $a = -5$, $6 = \dfrac{2\pi}{b}$ $\therefore$ $b = \frac{\pi}{3}$
$c = d = 0$ {as there is no translation} $\therefore$ $y = -5\cos\left(\frac{\pi}{3}x\right)$

EXERCISE 10E.1

1 **a** $\tan 0^o = 0$ **b** $\tan 15^o \approx 0.268$ **c** $\tan 20^o \approx 0.364$ **d** $\tan 25^o \approx 0.466$

e $\tan 35^o \approx 0.700$ **f** $\tan 45^o = 1$ **g** $\tan 50^o \approx 1.19$ **h** $\tan 55^o \approx 1.43$

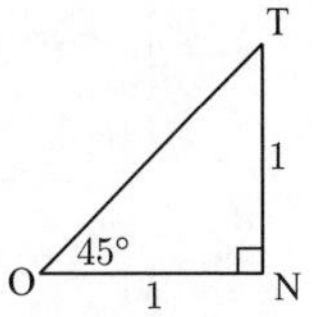

2 In ΔTON, ON = NT = 1 {Δ is isosceles} $\tan 45^o = \dfrac{\text{NT}}{\text{ON}} = \frac{1}{1} = 1$

EXERCISE 10E.2

1 **a** **i** $y = \tan(x - \frac{\pi}{2})$ is $y = \tan x$ translated $\begin{pmatrix} \frac{\pi}{2} \\ 0 \end{pmatrix}$.

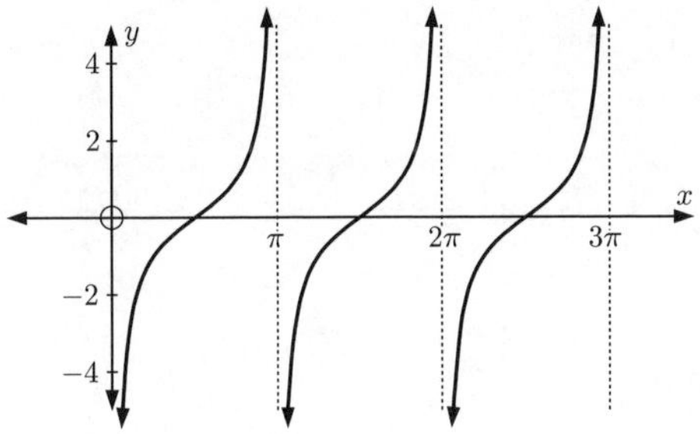

ii $y = -\tan x$ is $y = \tan x$ reflected in the x-axis.

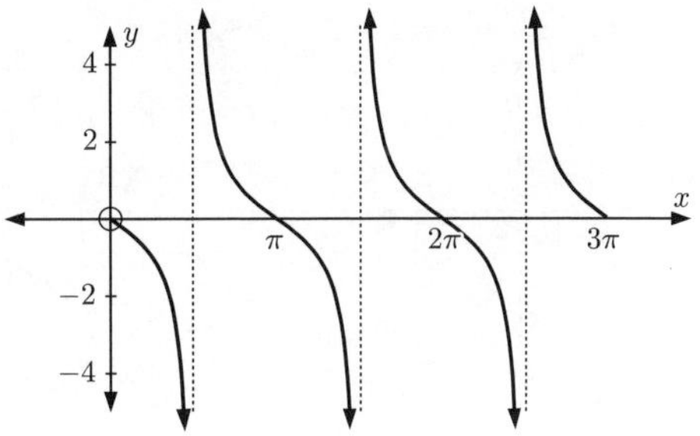

iii $y = \tan 3x$ comes from $y = \tan x$ under a horizontal stretch of factor $\frac{1}{3}$.

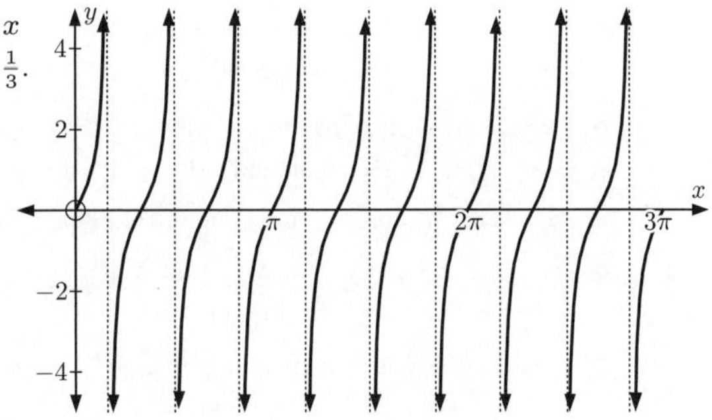

2 **a** translation through $\binom{1}{0}$ **b** reflection in x-axis **c** horizontal stretch, factor 2

3 **a** period $= \frac{\pi}{1} = \pi$ **b** period $= \frac{\pi}{2}$ **c** period $= \frac{\pi}{n}$

EXERCISE 10F

1 **a** amplitude $= |1| = 1$ **b** amplitude undefined **c** amplitude $= |-1| = 1$

2 **a** period $= \dfrac{\pi}{1} = \pi$ **b** period $= \dfrac{2\pi}{\frac{1}{3}} = 6\pi$ **c** period $= \dfrac{2\pi}{2} = \pi$

3 **a** $\dfrac{2\pi}{b} = 2\pi$ $\therefore\ b = 1$ **b** $\dfrac{2\pi}{b} = \dfrac{2\pi}{3}$ $\therefore\ b = 3$ **c** $\dfrac{\pi}{b} = \dfrac{\pi}{2}$ $\therefore\ b = 2$ **d** $\dfrac{2\pi}{b} = 4$ $\therefore\ b = \frac{\pi}{2}$

4 **a**

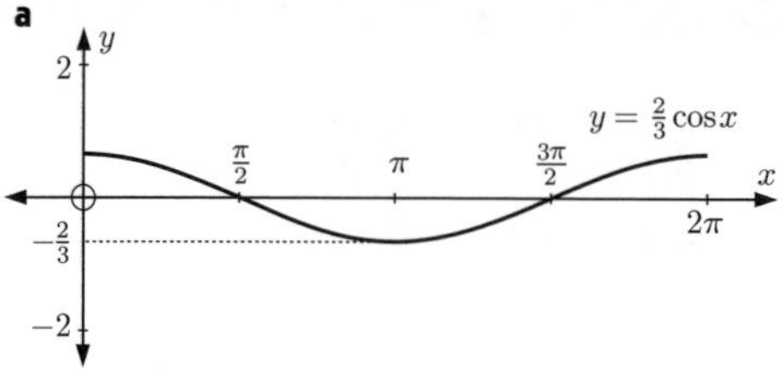

b

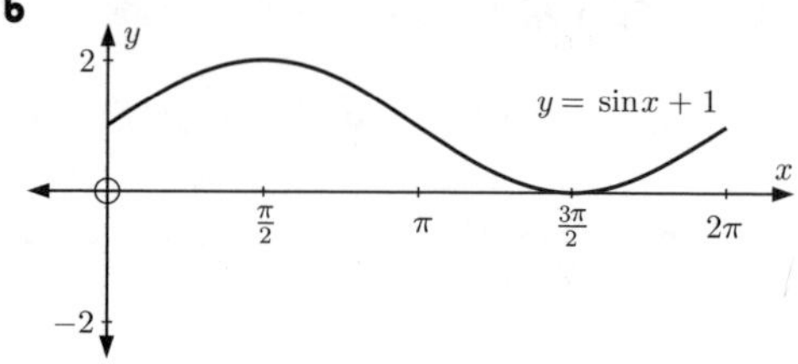

c

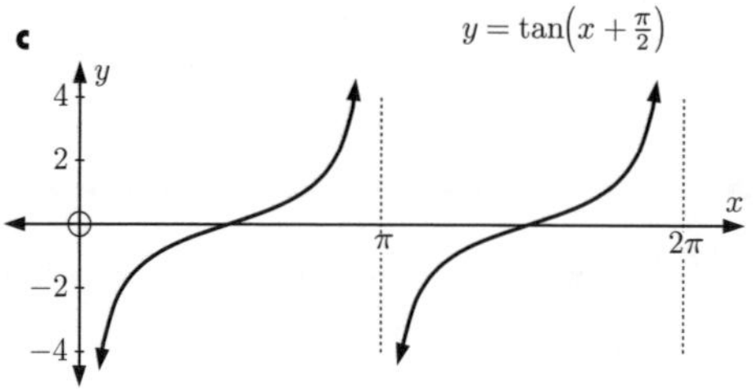

d

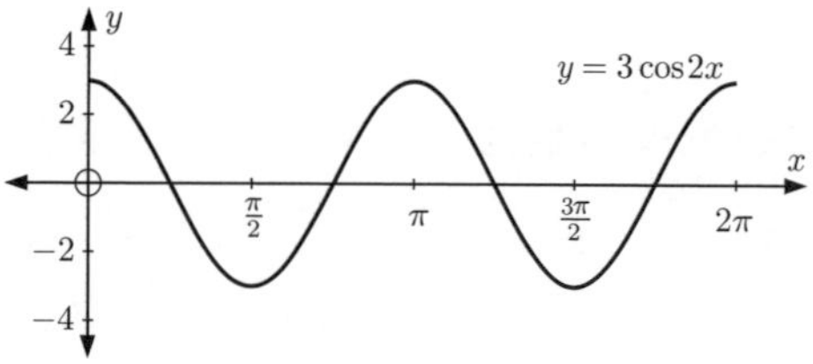

e

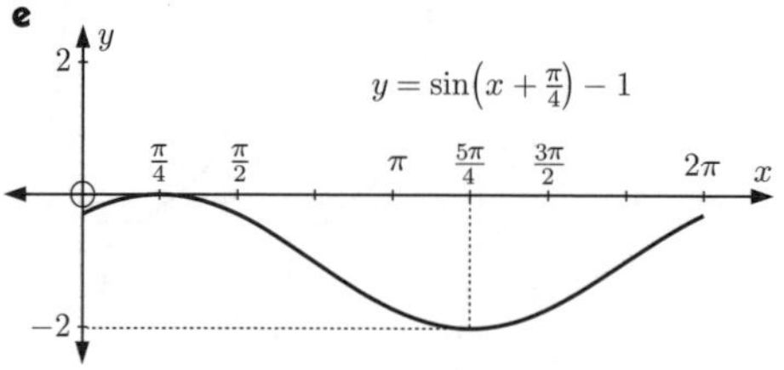

f

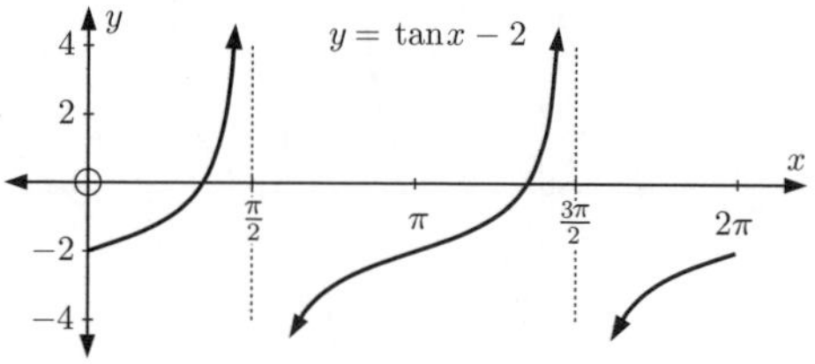

5 **a** $y = -\sin 5x$ has maximum value $-(-1) = 1$ $\{$when $\sin 5x = -1\}$
and minimum value $-(1) = -1$ $\{$when $\sin 5x = 1\}$

b $y = 3\cos x$ has maximum value $3(1) = 3$ $\{$when $\cos x = 1\}$
and minimum value $3(-1) = -3$ $\{$when $\cos x = -1\}$

c $y = 2\tan x$ has no maximum or minimum values.

d $y = -\cos 2x + 3$ has maximum value $-(-1) + 3 = 4$ $\{$when $\cos 2x = -1\}$
and minimum value $-(1) + 3 = 2$ $\{$when $\cos 2x = 1\}$

e $y = 1 + 2\sin x$ has maximum value $1 + 2(1) = 3$ $\{$when $\sin x = 1\}$
and minimum value $1 + 2(-1) = -1$ $\{$when $\sin x = -1\}$

f $y = \sin\left(x - \frac{\pi}{2}\right) - 3$ has maximum value $1 - 3 = -2$ $\{$when $\sin\left(x - \frac{\pi}{2}\right) = 1\}$
and minimum value $-1 - 3 = -4$ $\{$when $\sin\left(x - \frac{\pi}{2}\right) = -1\}$

6 **a** vertical stretch, factor $\frac{1}{2}$
b horizontal stretch, factor 4
c reflection in the x-axis
d vertical translation down 2 units
e horizontal translation $\frac{\pi}{4}$ units to the left
f reflection in the y-axis

7 The amplitude is 2, so $m = 2$.
The principal axis is $y = -3$, so $n = -3$.

8 The period is 2π, so $\frac{\pi}{p} = 2\pi$

$\therefore\ p = \frac{1}{2}$

The graph has undergone a vertical translation of 1 unit, so $q = 1$.

EXERCISE 10G.1

1

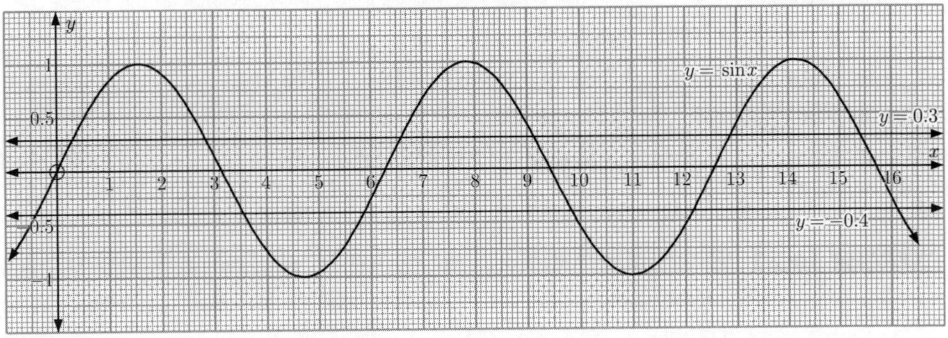

a When $\sin x = 0.3$, $x \approx 0.3, 2.8, 6.6, 9.1, 12.9$

b When $\sin x = -0.4$, $x \approx 5.9, 9.8, 12.2$

2

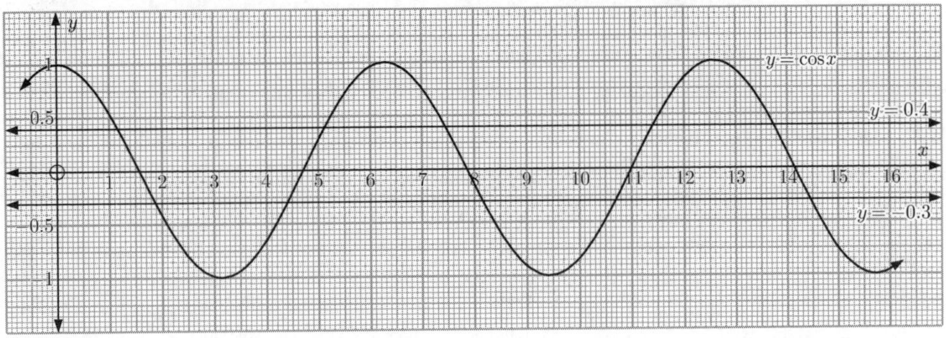

a When $\cos x = 0.4$, $x \approx 1.2, 5.1, 7.4$

b When $\cos x = -0.3$, $x \approx 4.4, 8.2, 10.7$

3

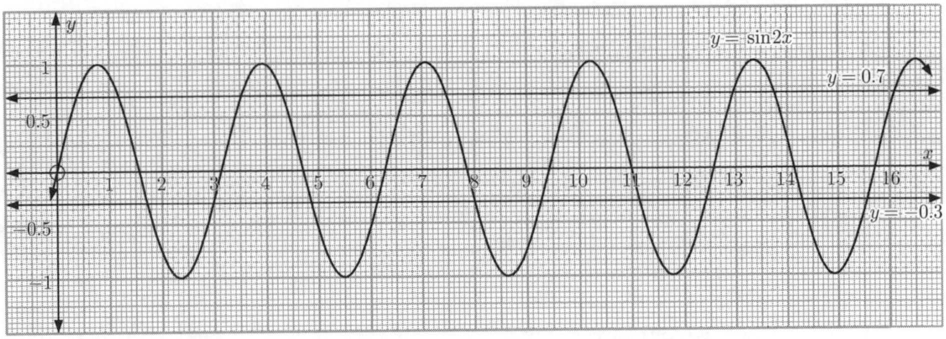

a When $\sin 2x = 0.7$, $x \approx 0.4, 1.2, 3.5, 4.3, 6.7, 7.5, 9.8, 10.6, 13.0, 13.7$

b When $\sin 2x = -0.3$, $x \approx 1.7, 3.0, 4.9, 6.1, 8.0, 9.3, 11.1, 12.4, 14.3, 15.6$

4

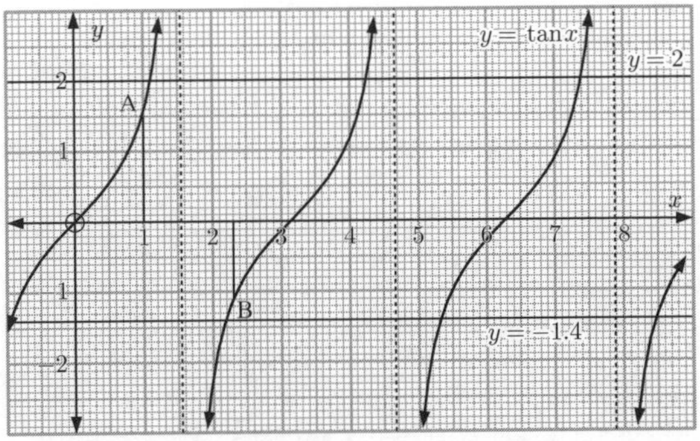

a **i** $\tan 1 \approx 1.6$ {point A}

ii $\tan 2.3 \approx -1.1$ {point B}

Using technology, we see that $\tan 1 \approx 1.557$ and $\tan 2.3 \approx -1.119$.

b **i** When $\tan x = 2$, $x \approx 1.1, 4.2, 7.4$

ii When $\tan x = -1.4$, $x \approx 2.2, 5.3$

EXERCISE 10G.2

1 Using technology:

a $\sin(x+2) = 0.0652$ when $x \approx 1.08,\ 4.35$

b $\sin^2 x + \sin x - 1 = 0$ when $x \approx 0.666,\ 2.48$

c $x\tan(\frac{x^2}{10}) = x^2 - 6x + 1$ when $x \approx 0.171,\ 4.92$

d $2\sin(2x)\cos x = \ln x$ when $x \approx 1.31,\ 2.03,\ 2.85$

2 $\cos(x-1) + \sin(x+1) = 6x + 5x^2 - x^3$ when $x \approx -0.951,\ 0.234,\ 5.98$

EXERCISE 10G.3

1 **a** $x = \frac{\pi}{6} + \frac{k12\pi}{6}$ and $0 \leqslant x \leqslant \frac{36\pi}{6}$

$\therefore\ x = \frac{\pi}{6},\ \frac{13\pi}{6},\ \frac{25\pi}{6}$

b $x = -\frac{\pi}{3} + \frac{k6\pi}{3}$ and $-\frac{6\pi}{3} \leqslant x \leqslant \frac{6\pi}{3}$

$\therefore\ x = -\frac{\pi}{3},\ \frac{5\pi}{3}$

c $x = -\frac{\pi}{2} + \frac{k2\pi}{2}$ and $-\frac{8\pi}{2} \leqslant x \leqslant \frac{8\pi}{2}$

$\therefore\ x = -\frac{7\pi}{2},\ -\frac{5\pi}{2},\ -\frac{3\pi}{2},\ -\frac{\pi}{2},\ \frac{\pi}{2},\ \frac{3\pi}{2},\ \frac{5\pi}{2},\ \frac{7\pi}{2}$

d $x = \frac{5\pi}{6} + \frac{k3\pi}{6}$ and $0 \leqslant x \leqslant \frac{24\pi}{6}$

$\therefore\ x = \frac{2\pi}{6},\ \frac{5\pi}{6},\ \frac{8\pi}{6},\ \frac{11\pi}{6},\ \frac{14\pi}{6},\ \frac{17\pi}{6},\ \frac{20\pi}{6},\ \frac{23\pi}{6}$

$\therefore\ x = \frac{\pi}{3},\ \frac{5\pi}{6},\ \frac{4\pi}{3},\ \frac{11\pi}{6},\ \frac{7\pi}{3},\ \frac{17\pi}{6},\ \frac{10\pi}{3},\ \frac{23\pi}{6}$

2 **a** $\cos x = -\frac{1}{2},\quad 0 \leqslant x \leqslant 5\pi$

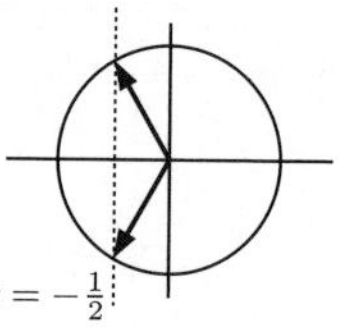

$\therefore\ x = \left.\begin{matrix}\frac{2\pi}{3}\\ \frac{4\pi}{3}\end{matrix}\right\} + k2\pi$

$\therefore\ x = \frac{2\pi}{3},\ \frac{4\pi}{3},\ \frac{8\pi}{3},\ \frac{10\pi}{3},\ \frac{14\pi}{3}$

b $2\sin x - 1 = 0,\quad -360^o \leqslant x \leqslant 360^o$

$\therefore\ \sin x = \frac{1}{2}$

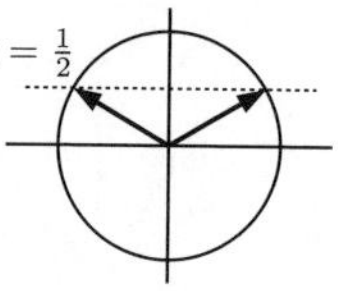

$\therefore\ x = \left.\begin{matrix}30^o\\ 150^o\end{matrix}\right\} + k\,360^o$

$\therefore\ x = -330^o,\ -210^o,\ 30^o,\ 150^o$

c $2\cos x + \sqrt{3} = 0,\quad 0 \leqslant x \leqslant 3\pi$

$\therefore\ \cos x = -\frac{\sqrt{3}}{2}$

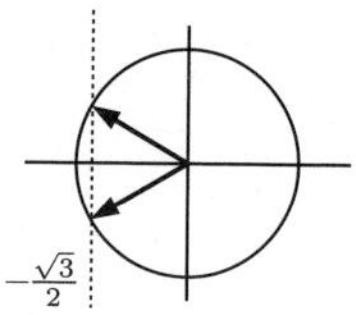

$\therefore\ x = \left.\begin{matrix}\frac{5\pi}{6}\\ \frac{7\pi}{6}\end{matrix}\right\} + k2\pi$

$\therefore\ x = \frac{5\pi}{6},\ \frac{7\pi}{6},\ \frac{17\pi}{6}$

d $\cos(x - \frac{2\pi}{3}) = \frac{1}{2},\quad -2\pi \leqslant x \leqslant 2\pi$

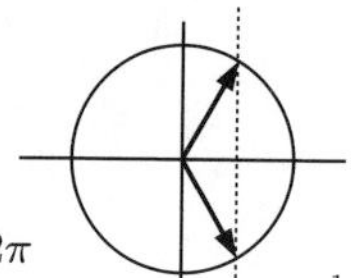

$\therefore\ x - \frac{2\pi}{3} = \left.\begin{matrix}\frac{\pi}{3}\\ \frac{5\pi}{3}\end{matrix}\right\} + k2\pi$

$\therefore\ x = \left.\begin{matrix}\pi\\ \frac{7\pi}{3}\end{matrix}\right\} + k2\pi$

$\therefore\ x = -\frac{5\pi}{3},\ -\pi,\ \frac{\pi}{3},\ \pi$

e $2\sin(x + \frac{\pi}{3}) = 1,\quad -3\pi \leqslant x \leqslant 3\pi$

$\therefore\ \sin(x + \frac{\pi}{3}) = \frac{1}{2}$

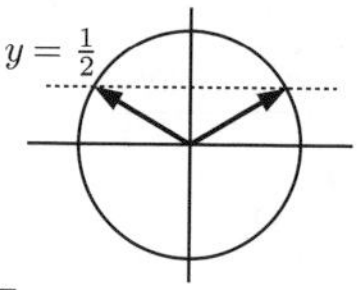

$\therefore\ x + \frac{\pi}{3} = \left.\begin{matrix}\frac{\pi}{6}\\ \frac{5\pi}{6}\end{matrix}\right\} + k2\pi$

$\therefore\ x = \left.\begin{matrix}-\frac{\pi}{6}\\ \frac{\pi}{2}\end{matrix}\right\} + k2\pi$

$\therefore\ x = -\frac{13\pi}{6},\ -\frac{3\pi}{2},\ -\frac{\pi}{6},\ \frac{\pi}{2},\ \frac{11\pi}{6},\ \frac{5\pi}{2}$

f $\sqrt{2}\sin(x - \frac{\pi}{4}) + 1 = 0,\quad 0 \leqslant x \leqslant 3\pi$

$\therefore\ \sin(x - \frac{\pi}{4}) = -\frac{1}{\sqrt{2}}$

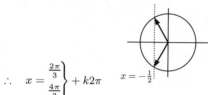

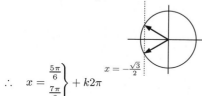

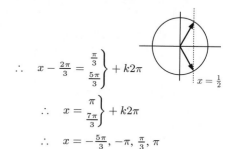

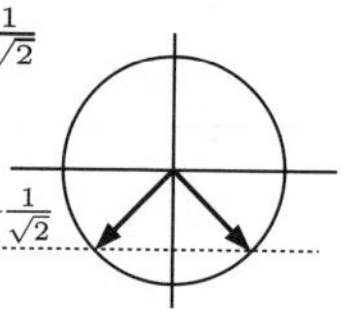

$\therefore\ x - \frac{\pi}{4} = \left.\begin{matrix}\frac{5\pi}{4}\\ \frac{7\pi}{4}\end{matrix}\right\} + k2\pi$

$\therefore\ x = \left.\begin{matrix}\frac{3\pi}{2}\\ 2\pi\end{matrix}\right\} + k2\pi$

$\therefore\ x = 0,\ \frac{3\pi}{2},\ 2\pi$

g $3\cos 2x + 3 = 0, \quad 0 \leqslant x \leqslant 3\pi$

$\therefore \quad \cos 2x = -1$

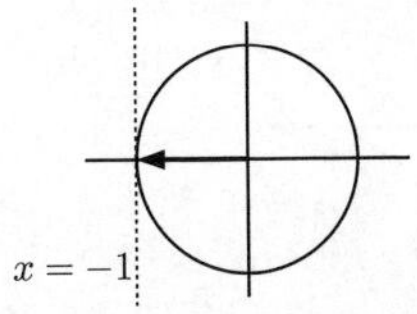

$\therefore \quad 2x = \pi + k2\pi$

$\therefore \quad x = \frac{\pi}{2} + k\pi$

$\therefore \quad x = \frac{\pi}{2}, \frac{3\pi}{2}, \frac{5\pi}{2}$

h $\sin\left(4(x - \frac{\pi}{4})\right) = 0, \quad 0 \leqslant x \leqslant \pi$

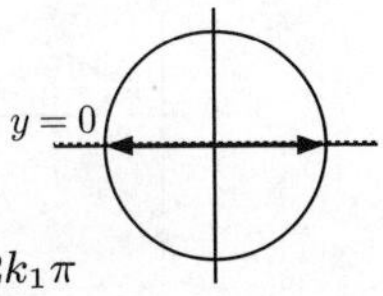

$\therefore \quad 4(x - \frac{\pi}{4}) = \left.\begin{matrix} 0 \\ \pi \end{matrix}\right\} + 2k_1\pi$

$\therefore \quad 4(x - \frac{\pi}{4}) = 0 + k_2\pi$

$\therefore \quad x - \frac{\pi}{4} = 0 + k_2\frac{\pi}{4}$

$\therefore \quad x = \frac{\pi}{4} + k_2\frac{\pi}{4}$

$\therefore \quad x = 0, \frac{\pi}{4}, \frac{\pi}{2}, \frac{3\pi}{4}, \pi$

i $4\cos 3x + 2 = 0, \quad -\pi \leqslant x \leqslant \pi$

$\therefore \quad \cos 3x = -\frac{1}{2}$

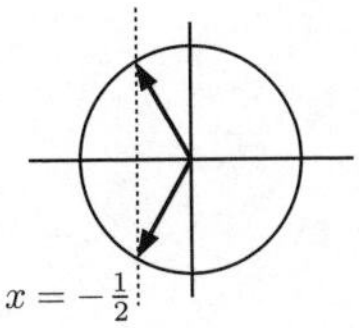

$\therefore \quad 3x = \left.\begin{matrix} \frac{2\pi}{3} \\ \frac{4\pi}{3} \end{matrix}\right\} + k\,2\pi$

$\therefore \quad x = \left.\begin{matrix} \frac{2\pi}{9} \\ \frac{4\pi}{9} \end{matrix}\right\} + k\,\frac{2\pi}{3}$

$\therefore \quad x = -\frac{8\pi}{9}, -\frac{4\pi}{9}, -\frac{2\pi}{9}, \frac{2\pi}{9}, \frac{4\pi}{9}, \frac{8\pi}{9}$

j $2\sin\left(2(x - \frac{\pi}{3})\right) = -\sqrt{3}, \quad 0 \leqslant x \leqslant 2\pi$

$\therefore \quad \sin\left(2(x - \frac{\pi}{3})\right) = -\frac{\sqrt{3}}{2}$

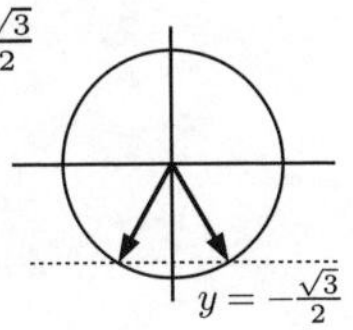

$\therefore \quad 2(x - \frac{\pi}{3}) = \left.\begin{matrix} \frac{4\pi}{3} \\ \frac{5\pi}{3} \end{matrix}\right\} + k\,2\pi$

$\therefore \quad x - \frac{\pi}{3} = \left.\begin{matrix} \frac{2\pi}{3} \\ \frac{5\pi}{6} \end{matrix}\right\} + k\pi$

$\therefore \quad x = \left.\begin{matrix} \pi \\ \frac{7\pi}{6} \end{matrix}\right\} + k\pi$

$\therefore \quad x = 0, \frac{\pi}{6}, \pi, \frac{7\pi}{6}, 2\pi$

3 $X = \tan^{-1}\left(\sqrt{3}\right) = \frac{\pi}{3} + k\pi$

a $\tan(x - \frac{\pi}{6}) = \sqrt{3}, \quad 0 \leqslant x \leqslant 2\pi$

$\therefore \quad x - \frac{\pi}{6} = \frac{\pi}{3} + k\pi$

$\therefore \quad x = \frac{\pi}{2} + k\pi$

$\therefore \quad x = \frac{\pi}{2}, \frac{3\pi}{2}$

b $\tan 4x = \sqrt{3}, \quad 0 \leqslant x \leqslant 2\pi$

$\therefore \quad 4x = \frac{\pi}{3} + k\pi$

$\therefore \quad x = \frac{\pi}{12} + \frac{k\pi}{4}$

$\therefore \quad x = \frac{\pi}{12}, \frac{\pi}{3}, \frac{7\pi}{12}, \frac{5\pi}{6}, \frac{13\pi}{12}, \frac{4\pi}{3}, \frac{19\pi}{12}, \frac{11\pi}{6}$

c $\tan^2 x = 3, \quad 0 \leqslant x \leqslant 2\pi$

$\therefore \quad \tan x = \pm\sqrt{3}$

$\therefore \quad x = \left.\begin{matrix} \frac{\pi}{3} \\ -\frac{\pi}{3} \end{matrix}\right\} + k\pi$

$\therefore \quad x = \frac{\pi}{3}, \frac{2\pi}{3}, \frac{4\pi}{3}, \frac{5\pi}{3}$

4 **a** The zeros of $y = \sin 2x$ are the solutions of $\sin 2x = 0$ $\{0^o \leqslant x \leqslant 180^o\}$

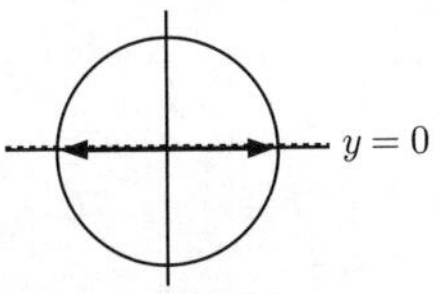

$\therefore \quad 2x = 0^o + k\,180^o$

$\therefore \quad x = 0^o + k\,90^o$

$\therefore \quad x = 0^o, 90^o, 180^o$

b The zeros of $y = \sin(x - \frac{\pi}{4})$ are the solutions of $\sin(x - \frac{\pi}{4}) = 0$ $\{0 \leqslant x \leqslant 3\pi\}$

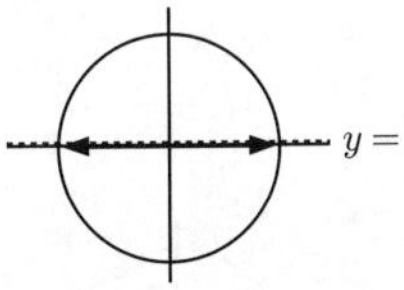

$\therefore \quad x - \frac{\pi}{4} = 0 + k\pi$

$\therefore \quad x = \frac{\pi}{4} + k\pi$

$\therefore \quad x = \frac{\pi}{4}, \frac{5\pi}{4}, \frac{9\pi}{4}$

5 **a**

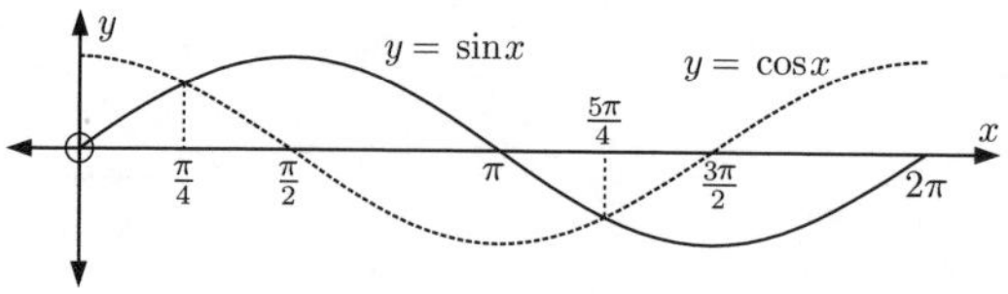

b $x = \frac{\pi}{4}$ or $\frac{5\pi}{4}$

c If $\sin x = \cos x$ then $\frac{\sin x}{\cos x} = 1$

$\therefore \tan x = 1$

$\therefore x = \frac{\pi}{4} + k\pi$

$\therefore x = \frac{\pi}{4}, \frac{5\pi}{4}$

6 **a** $\sin x = -\cos x, \quad 0 \leqslant x \leqslant 2\pi$

$\therefore \frac{\sin x}{\cos x} = -1$

$\therefore \tan x = -1$

$\therefore x = \frac{3\pi}{4} + k\pi$

$\therefore x = \frac{3\pi}{4}, \frac{7\pi}{4}$

b $\sin(3x) = \cos(3x), \quad 0 \leqslant x \leqslant 2\pi$

$\therefore \frac{\sin(3x)}{\cos(3x)} = 1$

$\therefore \tan(3x) = 1$

$\therefore 3x = \frac{\pi}{4} + k\pi$

$\therefore x = \frac{\pi}{12} + \frac{k\pi}{3}$

$\therefore x = \frac{\pi}{12}, \frac{5\pi}{12}, \frac{9\pi}{12}, \frac{13\pi}{12}, \frac{17\pi}{12}, \frac{21\pi}{12}$

$\therefore x = \frac{\pi}{12}, \frac{5\pi}{12}, \frac{3\pi}{4}, \frac{13\pi}{12}, \frac{17\pi}{12}, \frac{7\pi}{4}$

c $\sin(2x) = \sqrt{3}\cos(2x), \quad 0 \leqslant x \leqslant 2\pi$

$\therefore \frac{\sin(2x)}{\cos(2x)} = \sqrt{3}$

$\therefore \tan(2x) = \sqrt{3}$

$\therefore 2x = \frac{\pi}{3} + k\pi$

$\therefore x = \frac{\pi}{6} + \frac{k\pi}{2}$

$\therefore x = \frac{\pi}{6}, \frac{4\pi}{6}, \frac{7\pi}{6}, \frac{10\pi}{6}$

$\therefore x = \frac{\pi}{6}, \frac{2\pi}{3}, \frac{7\pi}{6}, \frac{5\pi}{3}$

EXERCISE 10H

1 **a** $P(t) = 7500 + 3000\sin(\frac{\pi t}{8}), \quad 0 \leqslant t \leqslant 12$

i $P(0) = 7500 + 3000\sin 0$
$= 7500 + 0$
$= 7500$ grasshoppers

ii $P(5) = 7500 + 3000\sin(\frac{5\pi}{8})$
$\approx 10\,271.63$
$\approx 10\,300$ grasshoppers

b The greatest value of $P(t)$ occurs when $\sin\left(\frac{\pi t}{8}\right) = 1$, so the greatest population is $7500 + 3000 = 10\,500$ grasshoppers when $\frac{\pi t}{8} = \frac{\pi}{2} + k2\pi$

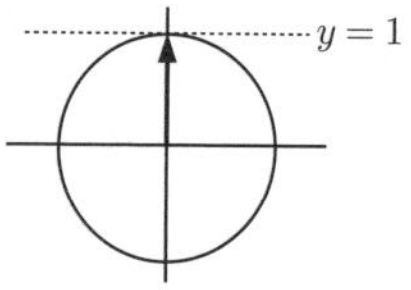

$\therefore \frac{t}{8} = \frac{1}{2} + 2k$

$\therefore t = 4 + 16k$

$\therefore t = 4 \quad \{\text{as } 0 \leqslant t \leqslant 12\}$

So the greatest population occurs after 4 weeks.

c **i** When $P(t) = 9000$,

$7500 + 3000\sin\left(\frac{\pi t}{8}\right) = 9000$

$\therefore 3000\sin\left(\frac{\pi t}{8}\right) = 1500$

$\therefore \sin\left(\frac{\pi t}{8}\right) = \frac{1}{2}$

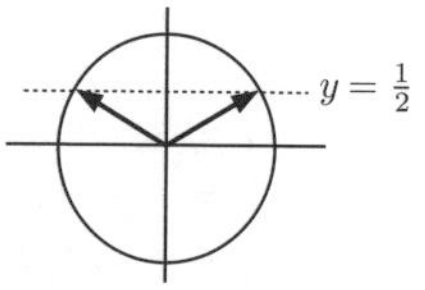

$\therefore \frac{\pi t}{8} = \left.\begin{matrix}\frac{\pi}{6}\\ \frac{5\pi}{6}\end{matrix}\right\} + k2\pi$

$\therefore \frac{t}{8} = \left.\begin{matrix}\frac{1}{6}\\ \frac{5}{6}\end{matrix}\right\} + k\,2$

$\therefore t = \left.\begin{matrix}\frac{4}{3}\\ \frac{20}{3}\end{matrix}\right\} + k\,16$

$\therefore t = 1\frac{1}{3}$ or $6\frac{2}{3}$

So, the population is 9000 at $1\frac{1}{3}$ weeks and $6\frac{2}{3}$ weeks.

ii When $P(t) = 6000$,

$7500 + 3000\sin\left(\frac{\pi t}{8}\right) = 6000$

$\therefore \quad 3000\sin\left(\frac{\pi t}{8}\right) = -1500$

$\therefore \quad \sin\left(\frac{\pi t}{8}\right) = -\frac{1}{2}$

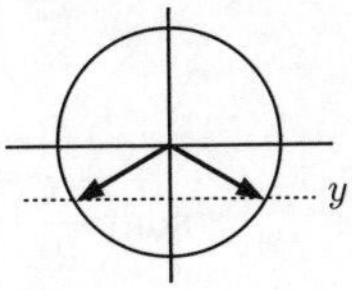

$\therefore \quad \frac{\pi t}{8} = \left.\begin{matrix}\frac{7\pi}{6}\\ \frac{11\pi}{6}\end{matrix}\right\} + k2\pi$

$\therefore \quad \frac{t}{8} = \left.\begin{matrix}\frac{7}{6}\\ \frac{11}{6}\end{matrix}\right\} + k\,2$

$\therefore \quad t = \left.\begin{matrix}\frac{28}{3}\\ \frac{44}{3}\end{matrix}\right\} + k\,16$

$\therefore \quad t = 9\frac{1}{3}$

So, the population is 6000 at $9\frac{1}{3}$ weeks.

d If $P(t) > 10\,000$, then

$7500 + 3000\sin\left(\frac{\pi t}{8}\right) > 10\,000$

$\therefore \quad 3000\sin\left(\frac{\pi t}{8}\right) > 2500$

$\therefore \quad \sin\left(\frac{\pi t}{8}\right) > \frac{5}{6}$

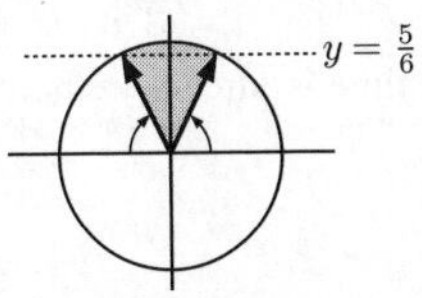

Solving $\sin\left(\frac{\pi t}{8}\right) = \frac{5}{6}$ using technology

$t \approx 2.51$ or 5.49 So, $2.51 \leqslant t \leqslant 5.49$ weeks.

2 $H(t) = 20 - 19\sin\left(\frac{2\pi t}{3}\right)$

a $H(0) = 20 - 19(0)$

$= 20$ m

So, at time $t = 0$, the light is 20 m above the ground.

b H is smallest when $\sin\left(\frac{2\pi t}{3}\right) = 1$

$\therefore \quad \frac{2\pi t}{3} = \frac{\pi}{2} + k2\pi$

$\therefore \quad \frac{2t}{3} = \frac{1}{2} + k\,2$

$\therefore \quad t = \frac{3}{4} + k\,3$

$\therefore \quad t = \frac{3}{4}$ min $\{$as $k = 0\}$

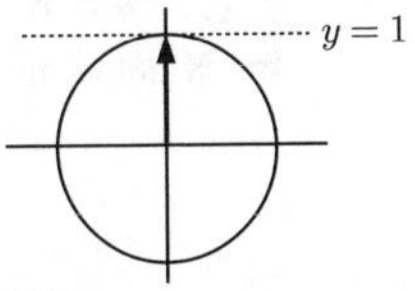

c period $= \dfrac{2\pi}{\frac{2\pi}{3}} = 3$ min

$\therefore$ one revolution takes 3 minutes

d

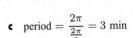

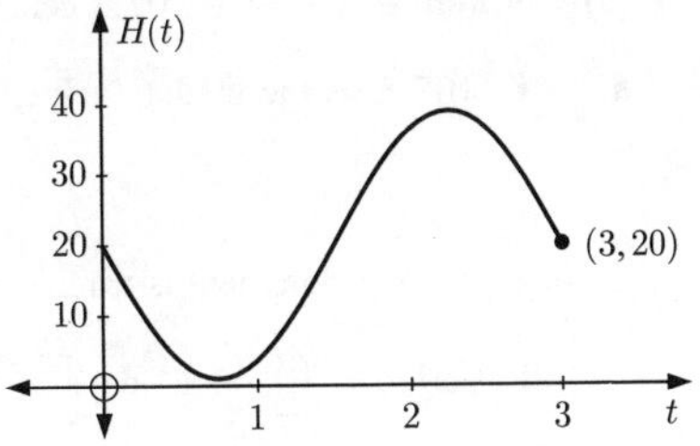

3 $P(t) = 400 + 250\sin\left(\frac{\pi t}{2}\right)$ years

a $P(0) = 400 + 250(0)$

$= 400$ water buffalo

b **i** $P(\frac{1}{2}) = 400 + 250\sin\left(\dfrac{\pi(\frac{1}{2})}{2}\right)$

$= 400 + 250\sin(\frac{\pi}{4})$

$= 400 + 250 \times \frac{1}{\sqrt{2}}$

≈ 577 water buffalo

ii $P(2) = 400 + 250\sin\pi$

$= 400 + 250(0)$

$= 400$ water buffalo

c $P(1) = 400 + 250\sin(\frac{\pi}{2})$

$= 400 + 250 \times 1$

$= 650$ water buffalo

This is the maximum herd size.

d $P(t)$ is smallest when $\sin\left(\frac{\pi t}{2}\right) = -1$ and is $400 - 250 = 150$ water buffalo.

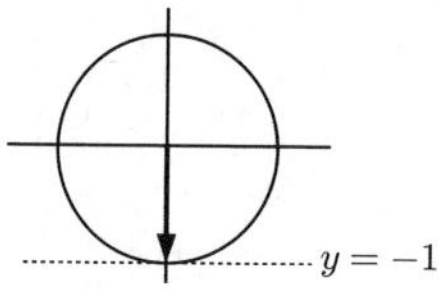

It occurs when $\dfrac{\pi t}{2} = \dfrac{3\pi}{2} + k2\pi$

$\therefore \quad \dfrac{t}{2} = \frac{3}{2} + k\,2$

$\therefore \quad t = 3 + 4k$

So, the first time is after 3 years.

e If $P(t) > 500$ then

$400 + 250\sin\left(\frac{\pi t}{2}\right) > 500$

$\therefore \quad 250\sin\left(\frac{\pi t}{2}\right) > 100$

$\therefore \quad \sin\left(\frac{\pi t}{2}\right) > \frac{2}{5}$

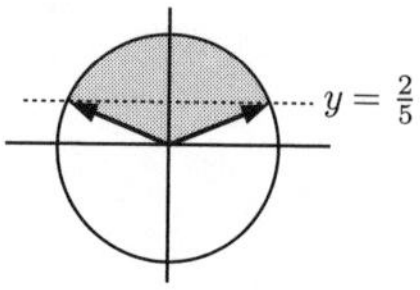

$\sin\left(\dfrac{\pi t}{2}\right) = \frac{2}{5}$ when

$\dfrac{\pi t}{2} = 0.4115$ or $\pi - 0.4115$

$\therefore \quad t \approx 0.262$ or 1.74

So, for $\sin\left(\frac{\pi t}{2}\right) > \frac{2}{5}$, $0.26 < t < 1.74$

$\therefore$ the herd first exceeded 500 when $t \approx 0.262$ years.

4 **a** The period is 4 seconds.

$\therefore \quad \dfrac{2\pi}{b} = 4$

$\therefore \quad b = \frac{\pi}{2}$

Amplitude is 3

$\therefore \quad a = 3$

$d = 1 + 3 = 4$

$c = 0$

$\therefore \quad H(t) = 3\cos\frac{\pi}{2}(t - 0) + 4$ metres

$\therefore \quad H(t) = 3\cos(\frac{\pi}{2}t) + 4$ metres

Check: When $t = 0$, $H(0) = 3\cos 0 + 4 = 7$ ✓

b X enters the water when $H(t) = 2$

$\therefore \quad 3\cos\left(\frac{\pi t}{2}\right) + 4 = 2$

$\therefore \quad \cos\left(\frac{\pi t}{2}\right) = -\frac{2}{3}$

Using technology, $t \approx 1.46$ seconds

5 $C(t) = 9.2\sin\frac{\pi}{7}(t - 4) + 107.8$ cents L^{-1}

a **i** 107.8 is the median value. Values are between $107.8 - 9.2 = 98.6$ cents L^{-1} (min.) and $107.8 + 9.2 = 117.0$ cents L^{-1} (max.)

$\therefore$ the statement is true.

ii period $= \dfrac{2\pi}{\frac{\pi}{7}} = 14$ days $\therefore$ true

b $C(7) = 9.2\sin\frac{\pi}{7}(3) + 107.8 \approx 116.8$ cents L^{-1}

c When $C(t) = \$1.10\ \text{L}^{-1}$ then $9.2\sin\frac{\pi}{7}(t - 4) + 107.8 = 110$

$\therefore \quad \sin\frac{\pi}{7}(t-4) = \dfrac{2.2}{9.2} \approx 0.239\,13$

$\therefore \quad \frac{\pi}{7}(t-4) \approx \left.\begin{matrix} 0.2415 \\ \pi - 0.2415 \end{matrix}\right\} + k\,2\pi$

$\therefore \quad t - 4 \approx \left.\begin{matrix} 0.538 \\ 6.462 \end{matrix}\right\} + 14k$

$\therefore \quad t \approx \left.\begin{matrix} 4.538 \\ 10.462 \end{matrix}\right\} + 14k$

So, the price is \$1.10 per litre on the 5th, 11th, 19th and 25th days.

d The minimum cost per litre is $-9.2 + 107.8 = 98.6$ cents L^{-1}

when $\sin \frac{\pi}{7}(t-4) = -1$

$\therefore \quad \frac{\pi}{7}(t-4) = \frac{3\pi}{2}$

$\therefore \quad \dfrac{t-4}{7} = \frac{3}{2}$

$\therefore \quad 2t - 8 = 21$

$\therefore \quad 2t = 29$

$\therefore \quad t = 14.5 \pm 14k$

{period is 14 days}

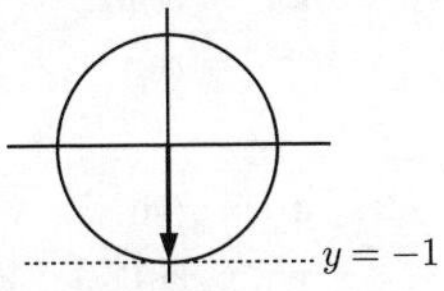

So, the minimum occurred on the 1st day and the 15th day.

EXERCISE 10I.1

1 **a** $\sin\theta + \sin\theta$
$= 2\sin\theta$

b $2\cos\theta + \cos\theta$
$= 3\cos\theta$

c $3\sin\theta - \sin\theta$
$= 2\sin\theta$

d $3\sin\theta - 2\sin\theta$
$= \sin\theta$

e $\cos\theta - 3\cos\theta$
$= -2\cos\theta$

f $2\cos\theta - 5\cos\theta$
$= -3\cos\theta$

2 **a** $3\sin^2\theta + 3\cos^2\theta$
$= 3(\sin^2\theta + \cos^2\theta)$
$= 3(1)$
$= 3$

b $-2\sin^2\theta - 2\cos^2\theta$
$= -2(\sin^2\theta + \cos^2\theta)$
$= -2(1)$
$= -2$

c $-\cos^2\theta - \sin^2\theta$
$= -(\cos^2\theta + \sin^2\theta)$
$= -(1)$
$= -1$

d $3 - 3\sin^2\theta$
$= 3(1 - \sin^2\theta)$
$= 3\cos^2\theta$

e $4 - 4\cos^2\theta$
$= 4(1 - \cos^2\theta)$
$= 4\sin^2\theta$

f $\cos^3\theta + \cos\theta\sin^2\theta$
$= \cos\theta(\cos^2\theta + \sin^2\theta)$
$= \cos\theta(1)$
$= \cos\theta$

g $\cos^2\theta - 1$
$= 1 - \sin^2\theta - 1$
$= -\sin^2\theta$

h $\sin^2\theta - 1$
$= 1 - \cos^2\theta - 1$
$= -\cos^2\theta$

i $2\cos^2\theta - 2$
$= -2(1 - \cos^2\theta)$
$= -2\sin^2\theta$

j $\dfrac{1 - \sin^2\theta}{\cos^2\theta}$
$= \dfrac{\cos^2\theta}{\cos^2\theta}$
$= 1$

k $\dfrac{1 - \cos^2\theta}{\sin\theta}$
$= \dfrac{\sin^2\theta}{\sin\theta}$
$= \sin\theta$

l $\dfrac{\cos^2\theta - 1}{-\sin\theta}$
$= \dfrac{1 - \sin^2\theta - 1}{-\sin\theta}$
$= \dfrac{-\sin^2\theta}{-\sin\theta}$
$= \sin\theta$

3 **a** $3\tan x - \tan x$
$= 2\tan x$

b $\tan x - 4\tan x$
$= -3\tan x$

c $\tan x \cos x$
$= \dfrac{\sin x}{\cos x} \times \cos x$
$= \sin x$

d $\dfrac{\sin x}{\tan x}$
$= \sin x \div \dfrac{\sin x}{\cos x}$
$= \sin x \times \dfrac{\cos x}{\sin x}$
$= \cos x$

e $3\sin x + 2\cos x \tan x$
$= 3\sin x + 2\cos x \dfrac{\sin x}{\cos x}$
$= 3\sin x + 2\sin x$
$= 5\sin x$

f $\dfrac{2\tan x}{\sin x}$
$= 2\left(\dfrac{\sin x}{\cos x}\right) \div \dfrac{\sin x}{1}$
$= \dfrac{2\sin x}{\cos x} \times \dfrac{1}{\sin x}$
$= \dfrac{2}{\cos x}$

4 **a** $(1 + \sin\theta)^2$
$= 1 + 2\sin\theta + \sin^2\theta$

b $(\sin\alpha - 2)^2$
$= \sin^2\alpha - 4\sin\alpha + 4$

c $(\tan\alpha - 1)^2$
$= \tan^2\alpha - 2\tan\alpha + 1$

d $(\sin\alpha + \cos\alpha)^2$
$= \sin^2\alpha + 2\sin\alpha\cos\alpha + \cos^2\alpha$
$= 1 + 2\sin\alpha\cos\alpha$

e $(\sin\beta - \cos\beta)^2$
$= \sin^2\beta - 2\sin\beta\cos\beta + \cos^2\beta$
$= 1 - 2\sin\beta\cos\beta$

f $-(2 - \cos\alpha)^2$
$= -[4 - 4\cos\alpha + \cos^2\alpha]$
$= -4 + 4\cos\alpha - \cos^2\alpha$

EXERCISE 10I.2

1 **a** $1 - \sin^2\theta$
$= (1 + \sin\theta)(1 - \sin\theta)$

b $\sin^2\alpha - \cos^2\alpha$
$= (\sin\alpha + \cos\alpha)(\sin\alpha - \cos\alpha)$

c $\tan^2\alpha - 1$
$= (\tan\alpha + 1)(\tan\alpha - 1)$

d $2\sin^2\beta - \sin\beta$
$= \sin\beta(2\sin\beta - 1)$

e $2\cos\phi + 3\cos^2\phi$
$= \cos\phi(2 + 3\cos\phi)$

f $3\sin^2\theta - 6\sin\theta$
$= 3\sin\theta(\sin\theta - 2)$

g $\tan^2\theta + 5\tan\theta + 6$
$= (\tan\theta + 2)(\tan\theta + 3)$

h $2\cos^2\theta + 7\cos\theta + 3$
$= (2\cos\theta + 1)(\cos\theta + 3)$

i $6\cos^2\alpha - \cos\alpha - 1$
$= (3\cos\alpha + 1)(2\cos\alpha - 1)$

2 **a** $\dfrac{1 - \sin^2\alpha}{1 - \sin\alpha}$
$= \dfrac{(1 + \sin\alpha)\cancel{(1 - \sin\alpha)}}{\cancel{1 - \sin\alpha}_{\,1}}$
$= 1 + \sin\alpha$

b $\dfrac{\tan^2\beta - 1}{\tan\beta + 1}$
$= \dfrac{\cancel{(\tan\beta + 1)}(\tan\beta - 1)}{\cancel{\tan\beta + 1}_{\,1}}$
$= \tan\beta - 1$

c $\dfrac{\cos^2\phi - \sin^2\phi}{\cos\phi + \sin\phi}$
$= \dfrac{\cancel{(\cos\phi + \sin\phi)}(\cos\phi - \sin\phi)}{\cancel{\cos\phi + \sin\phi}_{\,1}}$
$= \cos\phi - \sin\phi$

d $\dfrac{\cos^2\phi - \sin^2\phi}{\cos\phi - \sin\phi}$
$= \dfrac{(\cos\phi + \sin\phi)\cancel{(\cos\phi - \sin\phi)}}{\cancel{\cos\phi - \sin\phi}_{\,1}}$
$= \cos\phi + \sin\phi$

e $\dfrac{\sin\alpha + \cos\alpha}{\sin^2\alpha - \cos^2\alpha}$
$= \dfrac{{}^1\cancel{\sin\alpha + \cos\alpha}}{\cancel{(\sin\alpha + \cos\alpha)}(\sin\alpha - \cos\alpha)}$
$= \dfrac{1}{\sin\alpha - \cos\alpha}$

f $\dfrac{3 - 3\sin^2\theta}{6\cos\theta} = \dfrac{3(1 - \sin^2\theta)}{6\cos\theta}$
$= \dfrac{3\cos^2\theta}{6\cos\theta}$
$= \dfrac{\cos\theta}{2}$

3 **a** $(\cos\theta + \sin\theta)^2 + (\cos\theta - \sin\theta)^2$
$= \cos^2\theta + \cancel{2\cos\theta\sin\theta} + \sin^2\theta$
$\quad + \cos^2\theta - \cancel{2\cos\theta\sin\theta} + \sin^2\theta$
$= 2\cos^2\theta + 2\sin^2\theta$
$= 2(\cos^2\theta + \sin^2\theta)$
$= 2(1)$
$= 2$

b $(2\sin\theta + 3\cos\theta)^2 + (3\sin\theta - 2\cos\theta)^2$
$= 4\sin^2\theta + \cancel{12\sin\theta\cos\theta} + 9\cos^2\theta$
$\quad + 9\sin^2\theta - \cancel{12\sin\theta\cos\theta} + 4\cos^2\theta$
$= 13\sin^2\theta + 13\cos^2\theta$
$= 13(\sin^2\theta + \cos^2\theta)$
$= 13(1)$
$= 13$

c $(1-\cos\theta)\left(1+\dfrac{1}{\cos\theta}\right)$

$= 1 + \dfrac{1}{\cos\theta} - \cos\theta - 1$

$= \dfrac{1}{\cos\theta} - \cos\theta$

$= \dfrac{1}{\cos\theta} - \cos\theta\left(\dfrac{\cos\theta}{\cos\theta}\right)$

$= \dfrac{1-\cos^2\theta}{\cos\theta}$

$= \dfrac{\sin^2\theta}{\cos\theta}$

$= \tan\theta\sin\theta$

d $\left(1+\dfrac{1}{\sin\theta}\right)(\sin\theta - \sin^2\theta)$

$= \sin\theta - \sin^2\theta + 1 - \sin\theta$

$= 1 - \sin^2\theta$

$= \cos^2\theta$

e $\dfrac{\sin\theta}{1+\cos\theta} + \dfrac{1+\cos\theta}{\sin\theta}$

$= \dfrac{\sin^2\theta + (1+\cos\theta)(1+\cos\theta)}{\sin\theta(1+\cos\theta)}$

$= \dfrac{\sin^2\theta + 1 + 2\cos\theta + \cos^2\theta}{\sin\theta(1+\cos\theta)}$

$= \dfrac{1+1+2\cos\theta}{\sin\theta(1+\cos\theta)}$

$= \dfrac{2\cancel{(1+\cos\theta)}}{\sin\theta\cancel{(1+\cos\theta)}}$

$= \dfrac{2}{\sin\theta}$

f $\dfrac{\sin\theta}{1-\cos\theta} - \dfrac{\sin\theta}{1+\cos\theta}$

$= \dfrac{\sin\theta(1+\cos\theta) - \sin\theta(1-\cos\theta)}{(1-\cos\theta)(1+\cos\theta)}$

$= \dfrac{\cancel{\sin\theta} + \sin\theta\cos\theta - \cancel{\sin\theta} + \sin\theta\cos\theta}{1-\cos^2\theta}$

$= \dfrac{2\sin\theta\cos\theta}{\sin^2\theta}$

$= \dfrac{2\cancel{\sin\theta}\cos\theta}{\cancel{\sin\theta}\sin\theta}$

$= \dfrac{2\cos\theta}{\sin\theta}$

$= \dfrac{2}{\tan\theta}$

g $\dfrac{1}{1-\sin\theta} + \dfrac{1}{1+\sin\theta}$

$= \dfrac{1+\sin\theta+1-\sin\theta}{(1-\sin\theta)(1+\sin\theta)}$

$= \dfrac{2}{1-\sin^2\theta}$

$= \dfrac{2}{\cos^2\theta}$

EXERCISE 10J

1 **a** $\sin 2\theta = 2\sin\theta\cos\theta$

$= 2(\frac{4}{5})(\frac{3}{5})$

$= \frac{24}{25}$

b $\cos 2\theta = \cos^2\theta - \sin^2\theta$

$= \frac{9}{25} - \frac{16}{25}$

$= -\frac{7}{25}$

2 **a** $\cos 2A = 2\cos^2 A - 1$

$= 2(\frac{1}{3})^2 - 1$

$= 2 \times \frac{1}{9} - 1$

$= \frac{2}{9} - 1$

$= -\frac{7}{9}$

b $\cos 2\phi = 1 - 2\sin^2\phi$

$= 1 - 2(-\frac{2}{3})^2$

$= 1 - 2(\frac{4}{9})$

$= 1 - \frac{8}{9}$

$= \frac{1}{9}$

3 **a** $\sin\alpha = -\frac{2}{3}$

α is in Q3

$\therefore \quad \cos\alpha < 0$

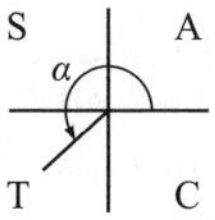

$\cos^2\alpha + \sin^2\alpha = 1$

$\therefore \quad \cos^2\alpha + \frac{4}{9} = 1$

$\therefore \quad \cos^2\alpha = \frac{5}{9}$

$\therefore \quad \cos\alpha = -\frac{\sqrt{5}}{3}$

$\sin 2\alpha = 2\sin\alpha\cos\alpha$

$= 2\left(-\frac{2}{3}\right)\left(-\frac{\sqrt{5}}{3}\right)$

$= \frac{4\sqrt{5}}{9}$

b $\cos\beta = \frac{2}{5}$

β is in Q4

$\therefore \quad \sin\beta < 0$

$\cos^2\beta + \sin^2\beta = 1$

$\therefore \quad \frac{4}{25} + \sin^2\beta = 1$

$\therefore \quad \sin^2\beta = \frac{21}{25}$

$\therefore \quad \sin\beta = -\frac{\sqrt{21}}{5}$

$\sin 2\beta = 2\sin\beta\cos\beta$

$= 2\left(-\frac{\sqrt{21}}{5}\right)\left(\frac{2}{5}\right)$

$= -\frac{4\sqrt{21}}{25}$

4 α is acute $\therefore$ $\cos\alpha$ and $\sin\alpha$ are positive

a $\cos 2\alpha = 2\cos^2\alpha - 1$

$\therefore \quad -\frac{7}{9} = 2\cos^2\alpha - 1$

$\therefore \quad 2\cos^2\alpha = \frac{2}{9}$

$\therefore \quad \cos^2\alpha = \frac{1}{9}$

$\therefore \quad \cos\alpha = \frac{1}{3}$

b $\sin\alpha = \sqrt{1 - \cos^2\alpha}$

$= \sqrt{1 - \frac{1}{9}}$

$= \sqrt{\frac{8}{9}}$

$= \frac{2\sqrt{2}}{3}$

5 $\left[\cos(\frac{\pi}{12}) + \sin(\frac{\pi}{12})\right]^2$

$= \cos^2(\frac{\pi}{12}) + 2\cos(\frac{\pi}{12})\sin(\frac{\pi}{12}) + \sin^2(\frac{\pi}{12})$

$= 1 + 2\cos(\frac{\pi}{12})\sin(\frac{\pi}{12})$

$= 1 + \sin(\frac{\pi}{6}) \qquad \{\sin 2A = 2\cos A\sin A\}$

$= 1 + \frac{1}{2}$

$= \frac{3}{2}$

6 **a** $2\sin\alpha\cos\alpha$

$= \sin 2\alpha$

b $4\cos\alpha\sin\alpha$

$= 2(2\sin\alpha\cos\alpha)$

$= 2\sin 2\alpha$

c $\sin\alpha\cos\alpha$

$= \frac{1}{2}(2\sin\alpha\cos\alpha)$

$= \frac{1}{2}\sin 2\alpha$

d $2\cos^2\beta - 1$

$= \cos 2\beta$

e $1 - 2\cos^2\phi$

$= -(2\cos^2\phi - 1)$

$= -\cos 2\phi$

f $1 - 2\sin^2 N$

$= \cos 2N$

g $2\sin^2 M - 1$

$= -(1 - 2\sin^2 M)$

$= -\cos 2M$

h $\cos^2\alpha - \sin^2\alpha$

$= \cos 2\alpha$

i $\sin^2\alpha - \cos^2\alpha$

$= -(\cos^2\alpha - \sin^2\alpha)$

$= -\cos 2\alpha$

j $2\sin 2A\cos 2A$

$= \sin 2(2A)$

$= \sin 4A$

k $2\cos 3\alpha\sin 3\alpha$

$= \sin 2(3\alpha)$

$= \sin 6\alpha$

l $2\cos^2 4\theta - 1$

$= \cos 2(4\theta)$

$= \cos 8\theta$

m $1 - 2\cos^2 3\beta$

$= -(2\cos^2 3\beta - 1)$

$= -\cos 2(3\beta)$

$= -\cos 6\beta$

n $1 - 2\sin^2 5\alpha$

$= \cos 2(5\alpha)$

$= \cos 10\alpha$

o $2\sin^2 3D - 1$

$= -(1 - 2\sin^2 3D)$

$= -\cos 2(3D)$

$= -\cos 6D$

p $\cos^2 2A - \sin^2 2A$

$= \cos 2(2A)$

$= \cos 4A$

q $\cos^2(\frac{\alpha}{2}) - \sin^2(\frac{\alpha}{2})$

$= \cos 2(\frac{\alpha}{2})$

$= \cos\alpha$

r $2\sin^2 3P - 2\cos^2 3P$

$= -2[\cos^2 3P - \sin^2 3P]$

$= -2\cos 2(3P)$

$= -2\cos 6P$

7 **a** $(\sin\theta + \cos\theta)^2$
$= \sin^2\theta + 2\sin\theta\cos\theta + \cos^2\theta$
$= \sin^2\theta + \cos^2\theta + 2\sin\theta\cos\theta$
$= 1 + \sin 2\theta$

b $\cos^4\theta - \sin^4\theta$
$= (\cos^2\theta + \sin^2\theta)(\cos^2\theta - \sin^2\theta)$
$= 1 \times \cos 2\theta$
$= \cos 2\theta$

8 **a** $\sin 2x + \sin x = 0$
$\therefore\ 2\sin x\cos x + \sin x = 0$
$\therefore\ \sin x(2\cos x + 1) = 0$
$\therefore\ \sin x = 0 \quad \text{or} \quad \cos x = -\frac{1}{2}$

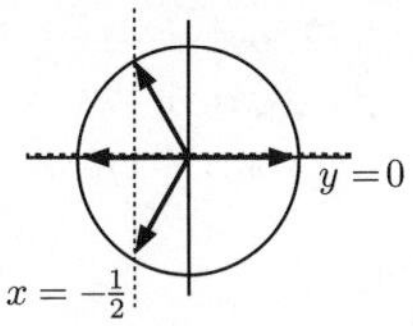

$\therefore\ x = 0, \frac{2\pi}{3}, \pi, \frac{4\pi}{3}, 2\pi$

b $\sin 2x - 2\cos x = 0$
$\therefore\ 2\sin x\cos x - 2\cos x = 0$
$\therefore\ 2\cos x(\sin x - 1) = 0$
$\therefore\ \cos x = 0 \quad \text{or} \quad \sin x = 1$

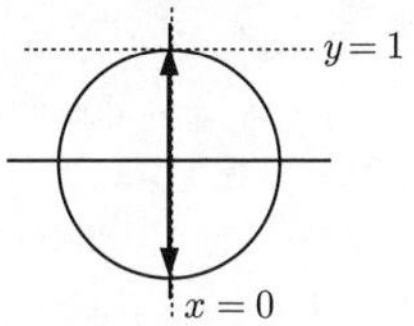

$\therefore\ x = \frac{\pi}{2}, \frac{3\pi}{2}$

c $\sin 2x + 3\sin x = 0$
$\therefore\ 2\sin x\cos x + 3\sin x = 0$
$\therefore\ \sin x(2\cos x + 3) = 0$
$\therefore\ \sin x = 0 \quad \text{or} \quad \cos x = -\frac{3}{2}$
↑
impossible

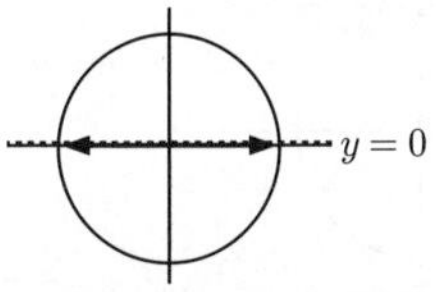

$\therefore\ x = 0, \pi, 2\pi$

9 **a** $\frac{1}{2} - \frac{1}{2}\cos 2\theta$
$= \frac{1}{2} - \frac{1}{2}(1 - 2\sin^2\theta)$
$= \frac{1}{2} - \frac{1}{2} + \sin^2\theta$
$= \sin^2\theta$

b $\frac{1}{2} + \frac{1}{2}\cos 2\theta$
$= \frac{1}{2} + \frac{1}{2}(2\cos^2\theta - 1)$
$= \frac{1}{2} + \cos^2\theta - \frac{1}{2}$
$= \cos^2\theta$

EXERCISE 10K

1 **a** $2\sin^2 x + \sin x = 0$
$\therefore\ \sin x(2\sin x + 1) = 0$
$\therefore\ \sin x = 0 \text{ or } -\frac{1}{2}$

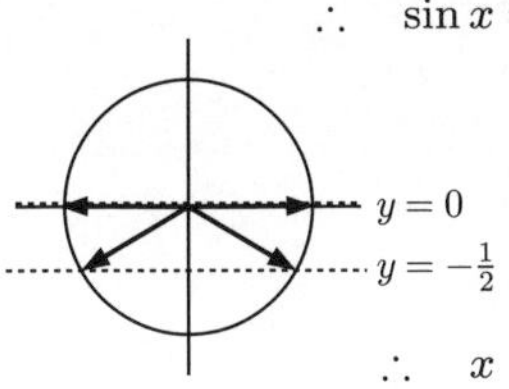

$\therefore\ x = 0, \pi, \frac{7\pi}{6}, \frac{11\pi}{6}, 2\pi$

b $2\cos^2 x = \cos x$
$\therefore\ 2\cos^2 x - \cos x = 0$
$\therefore\ \cos x(2\cos x - 1) = 0$
$\therefore\ \cos x = 0 \text{ or } \frac{1}{2}$

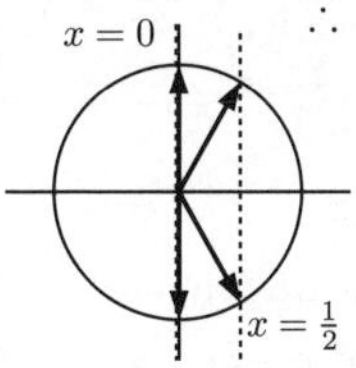

$\therefore\ x = \frac{\pi}{3}, \frac{\pi}{2}, \frac{3\pi}{2}, \frac{5\pi}{3}$

c $2\cos^2 x + \cos x - 1 = 0$
$\therefore\ (2\cos x - 1)(\cos x + 1) = 0$
$\therefore\ \cos x = \frac{1}{2} \text{ or } -1$

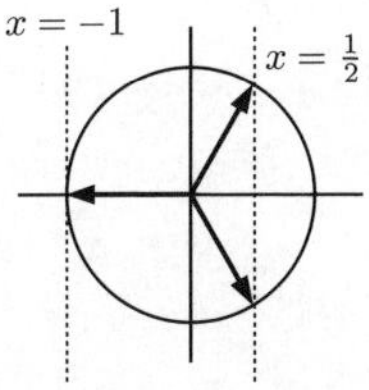

$\therefore\ x = \frac{\pi}{3}, \pi, \frac{5\pi}{3}$

d $2\sin^2 x + 3\sin x + 1 = 0$
$\therefore\ (2\sin x + 1)(\sin x + 1) = 0$
$\therefore\ \sin x = -\frac{1}{2} \text{ or } -1$

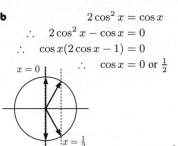

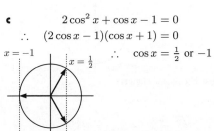

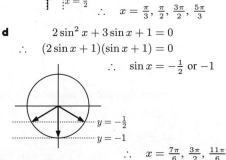

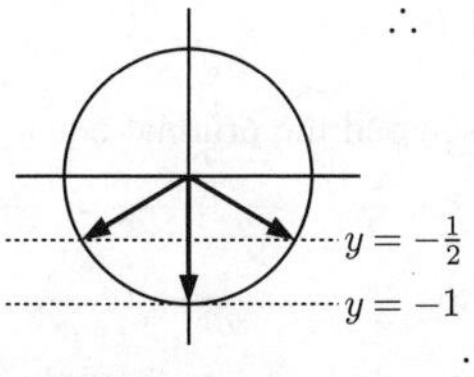

$\therefore\ x = \frac{7\pi}{6}, \frac{3\pi}{2}, \frac{11\pi}{6}$

e

$$\sin^2 x = 2 - \cos x$$
$$\therefore\ 1 - \cos^2 x = 2 - \cos x$$
$$\therefore\ \cos^2 x - \cos x + 1 = 0$$
$$\text{where}\quad \Delta = (-1)^2 - 4(1)(1) = 1 - 4 = -3$$

$\therefore$ no real solutions exist

2 **a**

$$\cos 2x - \cos x = 0$$
$$\therefore\ (2\cos^2 x - 1) - \cos x = 0$$
$$\therefore\ 2\cos^2 x - \cos x - 1 = 0$$
$$\therefore\ (2\cos x + 1)(\cos x - 1) = 0$$
$$\therefore\ \cos x = -\tfrac{1}{2} \text{ or } 1$$

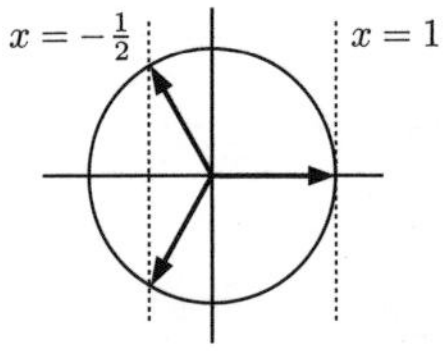

$$\therefore\ x = 0,\ \tfrac{2\pi}{3},\ \tfrac{4\pi}{3},\ 2\pi$$

b

$$\cos 2x + 3\cos x = 1$$
$$\therefore\ (2\cos^2 x - 1) + 3\cos x = 1$$
$$\therefore\ 2\cos^2 x + 3\cos x - 2 = 0$$
$$\therefore\ (2\cos x - 1)(\cos x + 2) = 0$$
$$\therefore\ \cos x = \tfrac{1}{2}$$
$$\{-1 \leqslant \cos x \leqslant 1\}$$

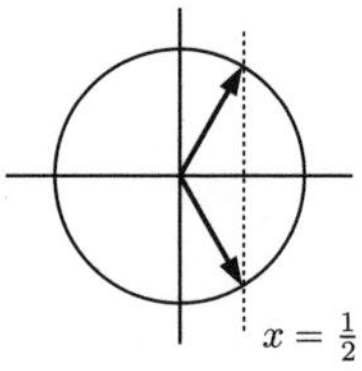

$$\therefore\ x = \tfrac{\pi}{3},\ \tfrac{5\pi}{3}$$

c

$$\cos 2x + \sin x = 0$$
$$\therefore\ (1 - 2\sin^2 x) + \sin x = 0$$
$$\therefore\ -2\sin^2 x + \sin x + 1 = 0$$
$$\therefore\ 2\sin^2 x - \sin x - 1 = 0$$
$$\therefore\ (2\sin x + 1)(\sin x - 1) = 0$$
$$\therefore\ \sin x = -\tfrac{1}{2} \text{ or } 1$$
$$\therefore\ x = \tfrac{\pi}{2},\ \tfrac{7\pi}{6},\ \tfrac{11\pi}{6}$$

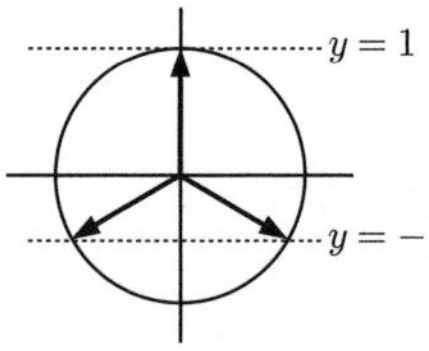

d

$$\sin 4x = \sin 2x$$
$$\therefore\ 2\sin 2x \cos 2x = \sin 2x$$
$$\therefore\ 2\sin 2x \cos 2x - \sin 2x = 0$$
$$\therefore\ \sin 2x(2\cos 2x - 1) = 0$$
$$\therefore\ \sin 2x = 0 \quad\text{or}\quad \cos 2x = \tfrac{1}{2}$$
$$\therefore\ 2x = 0,\ \tfrac{\pi}{3},\ \pi,\ \tfrac{5\pi}{3},\ 2\pi,\ \tfrac{7\pi}{3},\ 3\pi,\ \tfrac{11\pi}{3},\ 4\pi \quad \{0 \leqslant 2x \leqslant 4\pi\}$$
$$\therefore\ x = 0,\ \tfrac{\pi}{6},\ \tfrac{\pi}{2},\ \tfrac{5\pi}{6},\ \pi,\ \tfrac{7\pi}{6},\ \tfrac{3\pi}{2},\ \tfrac{11\pi}{6},\ 2\pi$$

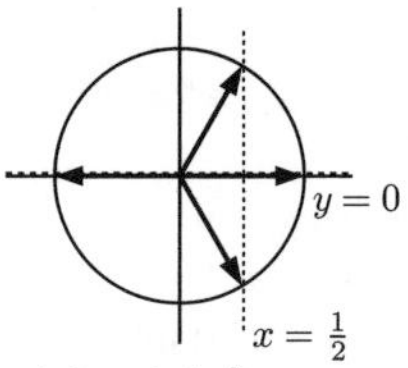

e $\sin x + \cos x = \sqrt{2}$

Squaring both sides we get:

$$\sin^2 x + 2\sin x \cos x + \cos^2 x = 2$$
$$\therefore\ \sin 2x + 1 = 2$$
$$\therefore\ \sin 2x = 1$$
$$\therefore\ 2x = \tfrac{\pi}{2},\ \tfrac{5\pi}{2} \quad \{0 \leqslant 2x \leqslant 4\pi\}$$
$$\therefore\ x = \tfrac{\pi}{4},\ \tfrac{5\pi}{4}$$

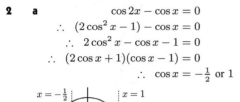

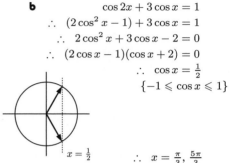

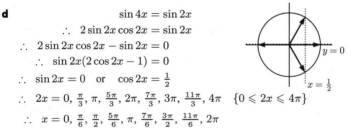

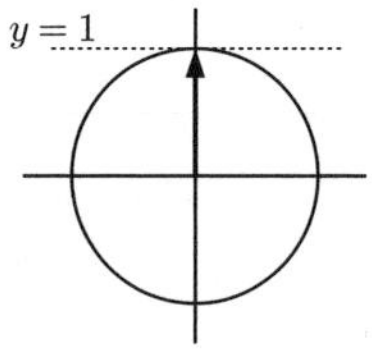

Since we squared the original equation, we must check our answers.

$$\sin\tfrac{\pi}{4} + \cos\tfrac{\pi}{4} = \tfrac{1}{\sqrt{2}} + \tfrac{1}{\sqrt{2}} = \tfrac{2}{\sqrt{2}} = \sqrt{2} \quad ✓$$
$$\sin\tfrac{5\pi}{4} + \cos\tfrac{5\pi}{4} = -\tfrac{1}{\sqrt{2}} + (-\tfrac{1}{\sqrt{2}}) = -\tfrac{2}{\sqrt{2}} = -\sqrt{2} \quad ×$$

$\therefore\ x = \frac{\pi}{4}$ is the only solution

f $$2\cos^2 x = 3\sin x$$
$$\therefore\ 2(1-\sin^2 x) = 3\sin x$$
$$\therefore\ 2\sin^2 x + 3\sin x - 2 = 0$$
$$\therefore\ (2\sin x - 1)(\sin x + 2) = 0$$
$$\therefore\ \sin x = \tfrac{1}{2} \quad \{-1 \leqslant \sin x \leqslant 1\}$$
$$\therefore\ x = \tfrac{\pi}{6},\ \tfrac{5\pi}{6}$$

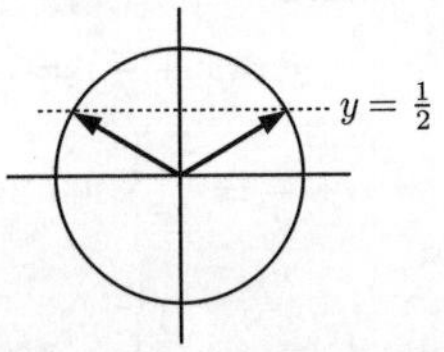

REVIEW SET 10A

1 $y = 4\sin x$ has amplitude 4.

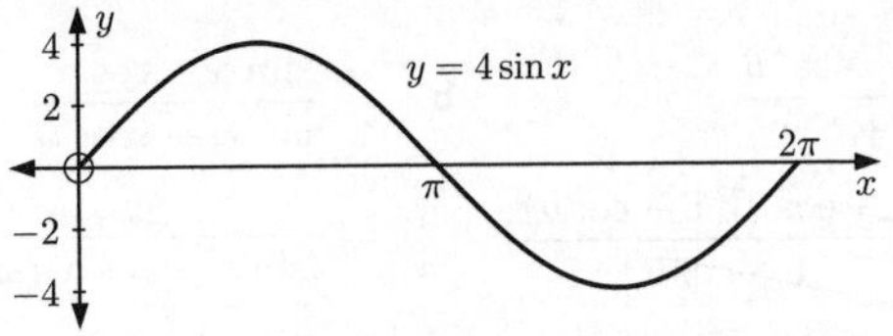

2 **a** $-1 \leqslant \sin x \leqslant 1$

$\therefore$ $1+\sin x$ has minimum $1+(-1)=0$ and maximum $1+1=2$.

b $-1 \leqslant \cos 3x \leqslant 1$

$\therefore$ $-2\cos 3x$ has minimum $-2(1) = -2$ and maximum $-2(-1) = 2$.

3 **a** $2\sin x = -1, \quad 0 \leqslant x \leqslant 4\pi$

$\therefore\ \sin x = -\frac{1}{2}$

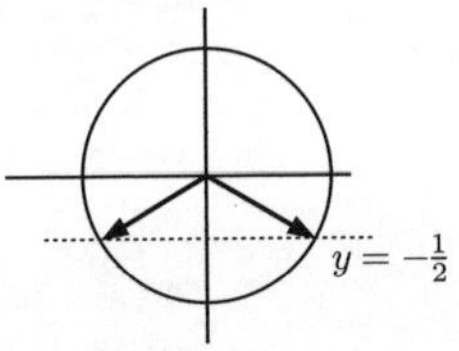

$$\therefore\ x = \left.\begin{matrix}\frac{7\pi}{6}\\ \frac{11\pi}{6}\end{matrix}\right\} + k2\pi$$

$$\therefore\ x = \tfrac{7\pi}{6},\ \tfrac{11\pi}{6},\ \tfrac{19\pi}{6},\ \tfrac{23\pi}{6}$$

b $\sqrt{2}\sin x - 1 = 0, \quad -2\pi \leqslant x \leqslant 2\pi$

$\therefore\ \sin x = \frac{1}{\sqrt{2}}$

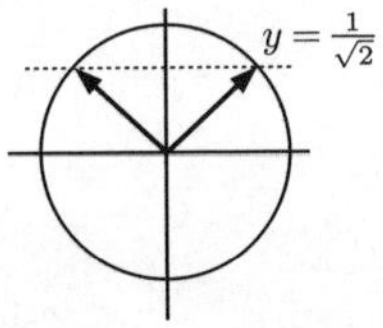

$$\therefore\ x = \left.\begin{matrix}\frac{\pi}{4}\\ \frac{3\pi}{4}\end{matrix}\right\} + k2\pi$$

$$\therefore\ x = -\tfrac{7\pi}{4},\ -\tfrac{5\pi}{4},\ \tfrac{\pi}{4},\ \tfrac{3\pi}{4}$$

4 **a** $2\sin 3x + \sqrt{3} = 0, \quad 0 \leqslant x \leqslant 2\pi$

$\therefore\ \sin 3x = -\frac{\sqrt{3}}{2}$

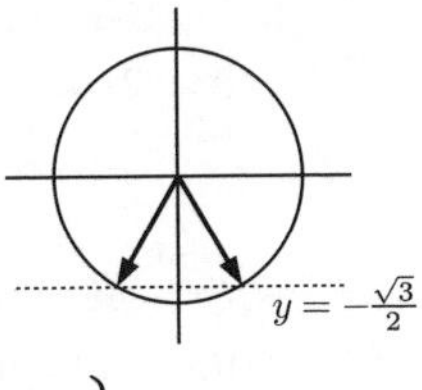

$$\therefore\ 3x = \left.\begin{matrix}\frac{4\pi}{3}\\ \frac{5\pi}{3}\end{matrix}\right\} + k2\pi$$

$$\therefore\ x = \left.\begin{matrix}\frac{4\pi}{9}\\ \frac{5\pi}{9}\end{matrix}\right\} + k\,\frac{2\pi}{3}$$

So, the x-intercepts are $\frac{4\pi}{9},\ \frac{5\pi}{9},\ \frac{10\pi}{9},\ \frac{11\pi}{9},\ \frac{16\pi}{9},\ \frac{17\pi}{9}$.

b $\sqrt{2}\sin(x+\frac{\pi}{4}) = 0, \quad 0 \leqslant x \leqslant 3\pi$

$\therefore\ \sin(x+\frac{\pi}{4}) = 0$

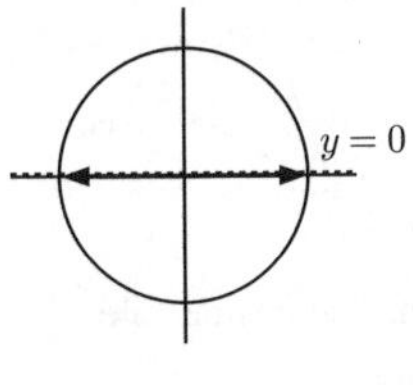

$$\therefore\ x + \tfrac{\pi}{4} = 0 + k\pi$$
$$\therefore\ x = -\tfrac{\pi}{4} + k\pi$$

So, the x-intercepts are $\frac{3\pi}{4},\ \frac{7\pi}{4},\ \frac{11\pi}{4}$.

5 $\sqrt{2}\cos(x+\frac{\pi}{4})-1=0, \quad 0 \leqslant x \leqslant 4\pi$

$\therefore \quad \cos(x+\frac{\pi}{4})=\frac{1}{\sqrt{2}}$

$\therefore \quad x+\frac{\pi}{4}=\left.\begin{matrix}\frac{\pi}{4}\\ \frac{7\pi}{4}\end{matrix}\right\}+k2\pi$

$\therefore \quad x=\left.\begin{matrix}0\\ \frac{3\pi}{2}\end{matrix}\right\}+k2\pi$

$\therefore \quad x=0, \frac{3\pi}{2}, 2\pi, \frac{7\pi}{2}, 4\pi$

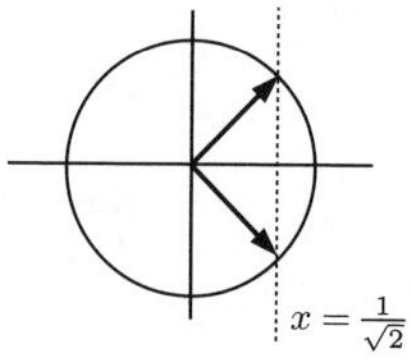

6 **a** $\dfrac{1-\cos^2\theta}{1+\cos\theta}$

$=\dfrac{\cancel{(1+\cos\theta)}(1-\cos\theta)}{\cancel{1+\cos\theta}_{\,1}}$

$=1-\cos\theta$

b $\dfrac{\sin\alpha-\cos\alpha}{\sin^2\alpha-\cos^2\alpha}$

$=\dfrac{^1\,\cancel{\sin\alpha-\cos\alpha}}{(\sin\alpha+\cos\alpha)\cancel{(\sin\alpha-\cos\alpha)}}$

$=\dfrac{1}{\sin\alpha+\cos\alpha}$

c $\dfrac{4\sin^2\alpha-4}{8\cos\alpha}$

$=\dfrac{-4(1-\sin^2\alpha)}{8\cos\alpha}$

$=\dfrac{-4\cos^2\alpha}{8\cos\alpha}$

$=\dfrac{-\cos\alpha}{2}$

7

$\cos^2\alpha+\sin^2\alpha=1$

$\therefore \quad \cos^2\alpha+\frac{9}{16}=1$

$\therefore \quad \cos^2\alpha=\frac{7}{16}$

$\therefore \quad \cos\alpha=\pm\frac{\sqrt{7}}{4}$

But in Q3, $\cos\alpha<0$

$\therefore \quad \cos\alpha=-\frac{\sqrt{7}}{4}$

$\sin 2\alpha=2\sin\alpha\cos\alpha$

$=2(-\frac{3}{4})(-\frac{\sqrt{7}}{4})$

$=\frac{3\sqrt{7}}{8}$

8 $\dfrac{\sin 2\alpha-\sin\alpha}{\cos 2\alpha-\cos\alpha+1}=\dfrac{2\sin\alpha\cos\alpha-\sin\alpha}{2\cos^2\alpha-1-\cos\alpha+1}$

$=\dfrac{\sin\alpha\cancel{(2\cos\alpha-1)}}{\cos\alpha\cancel{(2\cos\alpha-1)}}=\dfrac{\sin\alpha}{\cos\alpha}=\tan\alpha$

9

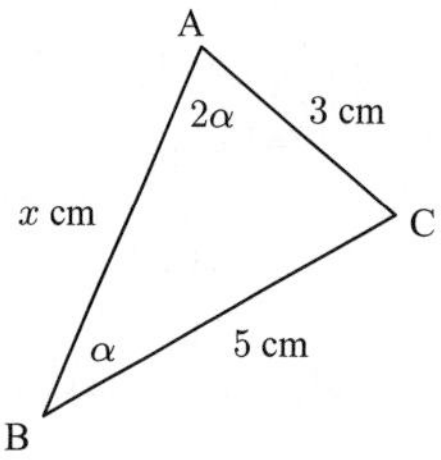

a Using the sine rule, $\dfrac{\sin 2\alpha}{5}=\dfrac{\sin\alpha}{3}$

$\therefore \quad \dfrac{2\cancel{\sin\alpha}\cos\alpha}{\cancel{\sin\alpha}_{\,1}}=\frac{5}{3}$

$\therefore \quad 2\cos\alpha=\frac{5}{3} \quad \{\sin\alpha\neq 0\}$

$\therefore \quad \cos\alpha=\frac{5}{6}$

b Using the cosine rule,

$3^2=x^2+5^2-2\times x\times 5\times\cos\alpha$

$\therefore \quad 9=x^2+25-10x\left(\frac{5}{6}\right)$

$\therefore \quad x^2-\frac{25}{3}x+16=0$

$\therefore \quad 3x^2-25x+48=0$

c $3x^2-25x+48=0$

$\therefore \quad (3x-16)(x-3)=0$

$\therefore \quad x=\frac{16}{3}$ or 3

But when $x=3$, ABC is isosceles,

so $\widehat{BCA}=\alpha$

$\therefore \quad \alpha+\alpha+2\alpha=180$ and so $\alpha=45$

which contradicts $\cos\alpha=\frac{5}{6}$

$\therefore \quad x=\frac{16}{3}$ is the only valid solution.

10 $P(t) = 5 + 2\sin\left(\frac{\pi t}{3}\right)$, $0 \leqslant t \leqslant 8$, where $P(t)$ is in thousands of water beetles.

a $P(0) = 5 + 2\sin 0$
$= 5$

So, 5000 water beetles.

b Smallest $P = 5 + 2(-1) = 3$
Largest $P = 5 + 2(1) = 7$
$\therefore$ smallest is 3000 water beetles
largest is 7000 water beetles

c If population is > 6000,
then $P(t) > 6$

$\therefore\ 5 + 2\sin\left(\frac{\pi t}{3}\right) > 6$

$\therefore\ 2\sin\left(\frac{\pi t}{3}\right) > 1$

$\therefore\ \sin\left(\frac{\pi t}{3}\right) > \frac{1}{2}$

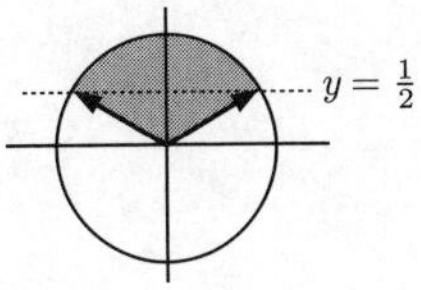

$\therefore\ \frac{\pi}{6} + k2\pi < \frac{\pi t}{3} < \frac{5\pi}{6} + k\,2\pi$

$\therefore\ \frac{1}{2} + 6k < t < \frac{5}{2} + 6k$

$\therefore\ 0.5 < t < 2.5,\quad 6.5 < t \leqslant 8\quad$ {since $0 \leqslant t \leqslant 8$}

REVIEW SET 10B

1 $y = \sin 3x$ has period $\frac{2\pi}{3}$.

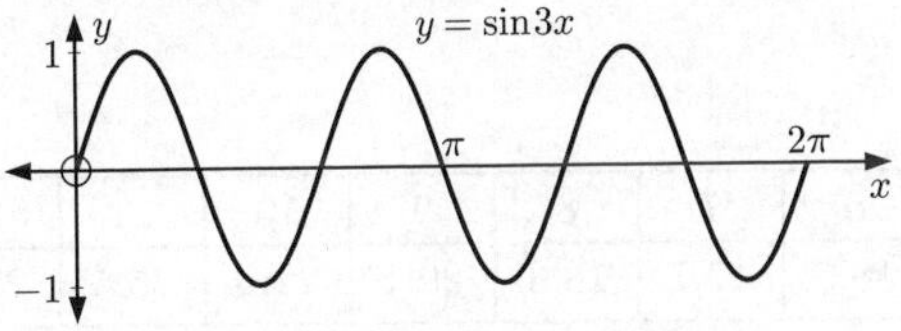

2 **a** period $= \dfrac{2\pi}{\frac{1}{3}} = 6\pi$

b period $= \frac{\pi}{4}$

3 **a** $\sin x = 0.382$, $0 \leqslant x \leqslant 8$
$\therefore\ x \approx 0.392, 2.75, 6.68$

b $\tan\left(\frac{x}{2}\right) = -0.458$, $0 \leqslant x \leqslant 8$
$\therefore\ x \approx 5.42$

4 **a**

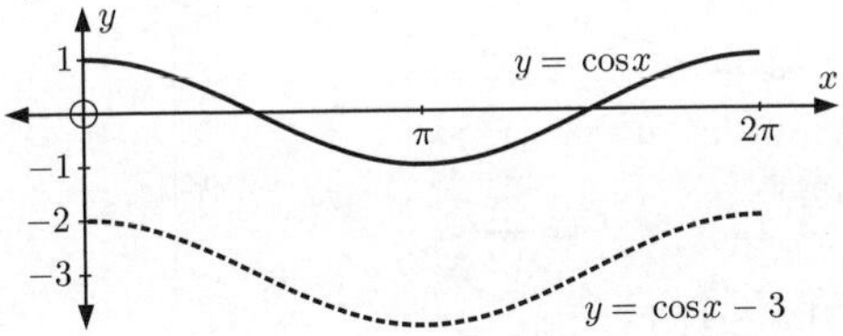

b

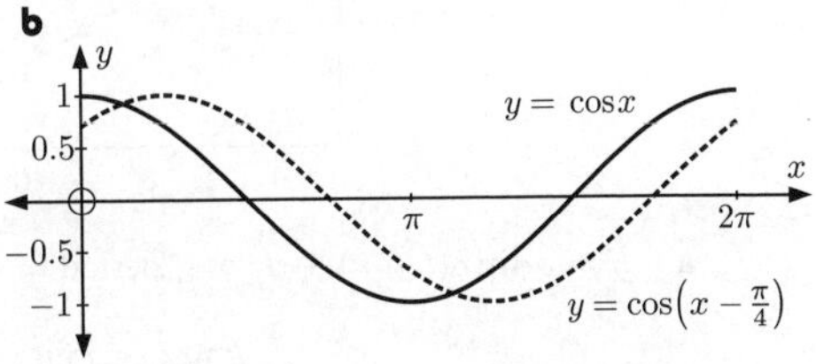

c

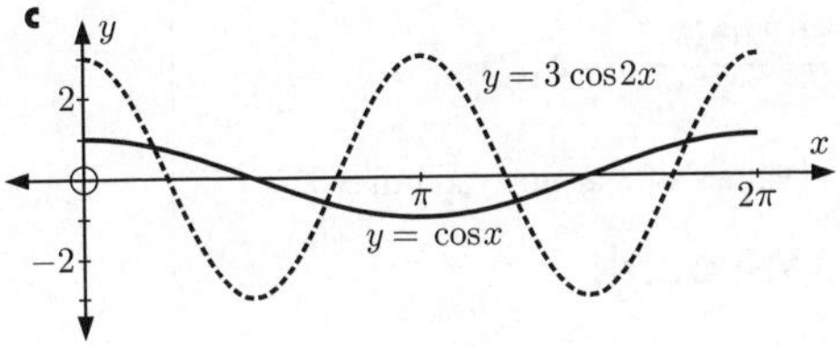

d

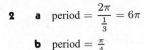

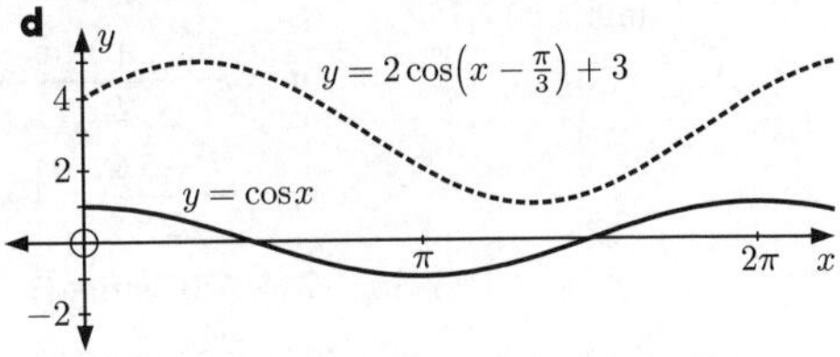

5 **a** $\cos x = 0.4379$, $0 \leqslant x \leqslant 10$
$\therefore\ x \approx 1.12, 5.17, 7.40$

b $\cos(x - 2.4) = -0.6014$, $0 \leqslant x \leqslant 6$
$\therefore\ x \approx 0.184, 4.62$

6 **a** $\sin 2A = 2\sin A\cos A$
$= 2(\frac{5}{13})(\frac{12}{13})$
$= \frac{120}{169}$

b $\cos 2A = \cos^2 A - \sin^2 A$
$= (\frac{12}{13})^2 - (\frac{5}{13})^2$
$= \frac{144-25}{169}$
$= \frac{119}{169}$

7 **a** **i** $\tan x = 4$
$\therefore\ x \approx 1.33, 4.47, 7.61$

ii $\tan\left(\frac{x}{4}\right) = 4$
$\therefore\ x \approx 5.30$

iii $\tan(x - 1.5) = 4$
$\therefore\ x \approx 2.83, 5.97, 9.11$

b **i** $\tan(x + \frac{\pi}{6}) = -\sqrt{3}, \quad -\pi \leqslant x \leqslant \pi$

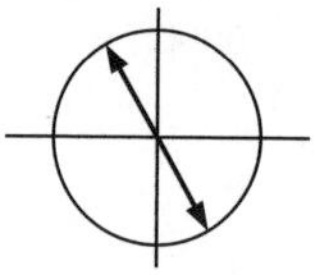

$\therefore \quad x + \frac{\pi}{6} = \frac{2\pi}{3} + k\pi$

$\therefore \quad x = \frac{\pi}{2} + k\pi$

$\therefore \quad x = -\frac{\pi}{2}, \frac{\pi}{2}$

ii $\tan 2x = -\sqrt{3}, \quad -\pi \leqslant x \leqslant \pi$

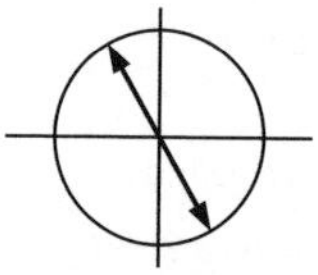

$\therefore \quad 2x = \frac{2\pi}{3} + k\pi$

$\therefore \quad x = \frac{\pi}{3} + \frac{k\pi}{2}$

$\therefore \quad x = -\frac{2\pi}{3}, -\frac{\pi}{6}, \frac{\pi}{3}, \frac{5\pi}{6}$

iii $\tan^2 x - 3 = 0, \quad -\pi \leqslant x \leqslant \pi$

$\therefore \quad \tan x = \pm\sqrt{3}$

$\therefore \quad x = \left.\begin{matrix} \frac{\pi}{3} \\ \frac{2\pi}{3} \end{matrix}\right\} + k\pi$

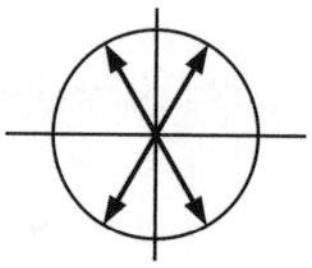

$\therefore \quad x = -\frac{2\pi}{3}, -\frac{\pi}{3}, \frac{\pi}{3}, \frac{2\pi}{3}$

c $3\tan(x - 1.2) = -2$

$\therefore \quad x \approx 0.612, 3.75, 6.90$

8

Month	1	2	3	4	5	6	7	8	9	10	11	12
Temp	31.5	31.8	29.5	25.4	21.5	18.8	17.7	18.3	20.1	22.4	25.5	28.8

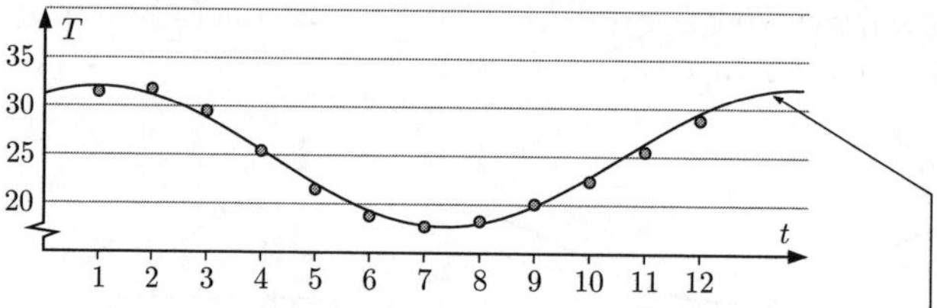

a $T = a \sin b(t - c) + d \qquad$ period $= \dfrac{2\pi}{b} = 12, \quad \therefore \quad b = \dfrac{2\pi}{12} = \dfrac{\pi}{6}$

max. $= 31.8$ $\quad \therefore \quad a = \dfrac{\text{max.} - \text{min.}}{2} \approx \dfrac{31.8 - 17.7}{2} \approx 7.05$

min. $= 17.7$

$d = \dfrac{\text{max.} + \text{min.}}{2} \approx \dfrac{31.8 + 17.7}{2} \approx 24.75$

$c = \dfrac{7 + 14}{2} = 10.5$ {values of t at min. and max.}

So, $T \approx 7.05 \sin \frac{\pi}{6}(t - 10.5) + 24.75$

b From technology, $T \approx 7.21 \sin(0.488t + 1.082) + 24.75$

$\approx 7.21 \sin 0.488(t + 2.22) + 24.75$

REVIEW SET 10C

1 **a** $\sin^2 x - \sin x - 2 = 0, \quad 0 \leqslant x \leqslant 2\pi$

$\therefore \quad (\sin x - 2)(\sin x + 1) = 0$

$\therefore \quad \sin x = 2$ or -1

But $\sin x$ values lie between -1 and 1 inclusive

$\therefore \quad \sin x = -1$

$\therefore \quad x = \frac{3\pi}{2}$

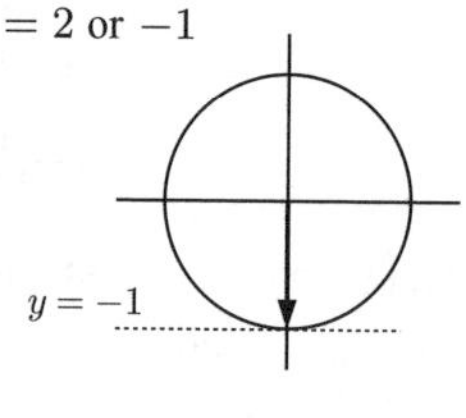

b $4\sin^2 x = 1, \quad 0 \leqslant x \leqslant 2\pi$

$\therefore \quad \sin^2 x = \frac{1}{4}$

$\therefore \quad \sin x = \pm\frac{1}{2}$

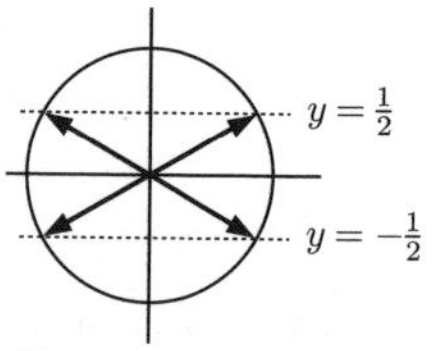

$\therefore \quad x = \frac{\pi}{6}, \frac{5\pi}{6}, \frac{7\pi}{6}, \frac{11\pi}{6}$

2 **a**

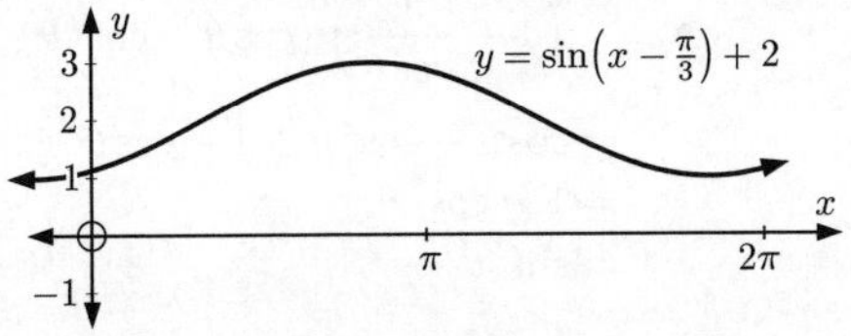

b $f(x)$ has minimum value $-1 + 2 = 1$ and maximum value $1 + 2 = 3$

$\therefore$ $f(x) = k$ will have solutions for $1 \leqslant k \leqslant 3$

3 **a** If $y = a\cos b(t - c) + d$

then $a = -4$, $\dfrac{2\pi}{b} = \pi$

$\therefore$ $b = 2$

$c = d = 0$

$\therefore$ $y = -4\cos 2x$

b If $y = a\cos b(x - c) + d$

then $a = 1$, $\dfrac{2\pi}{b} = 8$ $\therefore$ $b = \frac{\pi}{4}$

$d = \dfrac{\text{max.} + \text{min.}}{2} = \dfrac{3+1}{2} = 2$

$c = 0$

So, $y = \cos\left(\frac{\pi}{4}x\right) + 2$

4 **a** $\tan(x - \frac{\pi}{3}) = \frac{1}{\sqrt{3}}$, $0 \leqslant x \leqslant 4\pi$

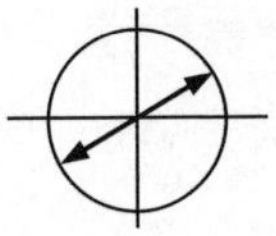

$\therefore$ $x - \frac{\pi}{3} = \frac{\pi}{6} + k\pi$

$\therefore$ $x = \frac{\pi}{2} + k\pi$

$\therefore$ $x = \frac{\pi}{2}, \frac{3\pi}{2}, \frac{5\pi}{2}, \frac{7\pi}{2}$

b $\cos(x + \frac{2\pi}{3}) = \frac{1}{2}$, $-2\pi \leqslant x \leqslant 2\pi$

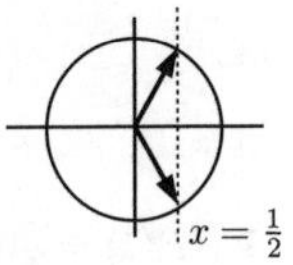

$x + \frac{2\pi}{3} = \left.\begin{matrix}\frac{\pi}{3}\\ \frac{5\pi}{3}\end{matrix}\right\} + k2\pi$

$\therefore$ $x = \left.\begin{matrix}-\frac{\pi}{3}\\ \pi\end{matrix}\right\} + k2\pi$

$\therefore$ $x = -\pi, -\frac{\pi}{3}, \pi, \frac{5\pi}{3}$

5 **a**

$\cos^3\theta + \sin^2\theta\cos\theta$
$= \cos\theta(\cos^2\theta + \sin^2\theta)$
$= \cos\theta(1)$
$= \cos\theta$

b

$\dfrac{\cos^2\theta - 1}{\sin\theta} = \dfrac{-(1 - \cos^2\theta)}{\sin\theta}$
$= -\dfrac{\sin^2\theta}{\sin\theta}$
$= -\sin\theta$

c

$5 - 5\sin^2\theta$
$= 5(1 - \sin^2\theta)$
$= 5\cos^2\theta$

d

$\dfrac{\sin^2\theta - 1}{\cos\theta} = -\dfrac{(1 - \sin^2\theta)}{\cos\theta}$
$= -\dfrac{\cos^2\theta}{\cos\theta}$
$= -\cos\theta$

6 **a**

$(2\sin\alpha - 1)^2$
$= 4\sin^2\alpha - 4\sin\alpha + 1$

b

$(\cos\alpha - \sin\alpha)^2$
$= \cos^2\alpha - 2\sin\alpha\cos\alpha + \sin^2\alpha$
$= \cos^2\alpha + \sin^2\alpha - 2\sin\alpha\cos\alpha$
$= 1 - \sin 2\alpha$

7 **a** $\dfrac{\cos\theta}{1+\sin\theta}+\dfrac{1+\sin\theta}{\cos\theta}$

$=\dfrac{\cos^2\theta+(1+\sin\theta)^2}{(1+\sin\theta)\cos\theta}$

$=\dfrac{\cos^2\theta+1+2\sin\theta+\sin^2\theta}{(1+\sin\theta)\cos\theta}$

$=\dfrac{2+2\sin\theta}{(1+\sin\theta)\cos\theta}$ $\{\cos^2\theta+\sin^2\theta=1\}$

$=\dfrac{2\cancel{(1+\sin\theta)}}{\cancel{(1+\sin\theta)}\cos\theta}$

$=\dfrac{2}{\cos\theta}$

b $\left(1+\dfrac{1}{\cos\theta}\right)\left(\cos\theta-\cos^2\theta\right)$

$=\cancel{\cos\theta}-\cos^2\theta+1-\cancel{\cos\theta}$

$=1-\cos^2\theta$

$=\sin^2\theta$

8 $\tan\theta=-\frac{2}{3}$, $\frac{\pi}{2}<\theta<\pi$

$\therefore\ \dfrac{\sin\theta}{\cos\theta}=-\frac{2}{3}$

$\therefore\ \sin\theta=-2k,\quad \cos\theta=3k$

but $\cos^2\theta+\sin^2\theta=1$

$\therefore\ 9k^2+4k^2=1$

$\therefore\ 13k^2=1$

$\therefore\ k=\pm\frac{1}{\sqrt{13}}$

But in Q2,

$\sin\theta>0,\quad \cos\theta<0$

$\therefore\ k=-\frac{1}{\sqrt{13}}$

$\therefore\ \sin\theta=\frac{2}{\sqrt{13}},\quad \cos\theta=-\frac{3}{\sqrt{13}}$

9 $P(t)=40+12\sin\frac{2\pi}{7}\left(t-\frac{37}{12}\right)$ mg

a $P(t)$ has a minimum of

$40+12(-1)=28$ mg per m^3

b when $\sin\frac{2\pi}{7}(t-\frac{37}{12})=-1$

$\therefore\ \frac{2\pi}{7}(t-\frac{37}{12})=\frac{3\pi}{2}+k2\pi$

$\therefore\ \frac{2}{7}(t-\frac{37}{12})=\frac{3}{2}+k\,2$

So, $t-\frac{37}{12}=\frac{21}{4}+k\,7$

$\therefore\ t=8\frac{1}{3}+k\,7$

$\therefore\ t=1\frac{1}{3},\ 8\frac{1}{3},\ 15\frac{1}{3}$, and so on.

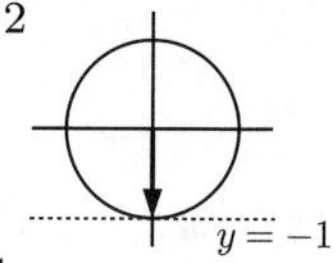

$\therefore$ on Mondays at 8.00 am

$\{1\frac{1}{3}$ days after midnight Saturday$\}$

Chapter 11

MATRICES

EXERCISE 11A

1 **a** 1 row and 4 columns $\therefore$ 1×4 **b** 2 rows and 1 column $\therefore$ 2×1

c 2 rows and 2 columns $\therefore$ 2×2 **d** 3 rows and 3 columns $\therefore$ 3×3

2 **a** $\begin{pmatrix} 2 & 1 & 6 & 1 \end{pmatrix}$ **b** $\begin{pmatrix} 1.95 \\ 2.35 \\ 0.15 \\ 0.95 \end{pmatrix}$ **c** $(2\times 1.95)+(1\times 2.35)+(6\times 0.15)+(1\times 0.95)$ represents the total cost of the groceries.

3

$$\begin{array}{ccc} 200\text{ g} & 300\text{ g} & 500\text{ g} \end{array}$$
$$\begin{pmatrix} 1000 & 1500 & 1250 \\ 1500 & 1000 & 1000 \\ 800 & 2300 & 1300 \\ 1200 & 1200 & 1200 \end{pmatrix} \begin{array}{l} \text{week 1} \\ \text{week 2} \\ \text{week 3} \\ \text{week 4} \end{array}$$

4

$$\begin{array}{cccc} \text{pies} & \text{pasties} & \text{rolls} & \text{buns} \end{array}$$
$$\begin{pmatrix} 40 & 50 & 55 & 40 \\ 25 & 65 & 44 & 30 \\ 35 & 40 & 40 & 35 \\ 35 & 40 & 35 & 50 \end{pmatrix} \begin{array}{l} \text{Friday} \\ \text{Saturday} \\ \text{Sunday} \\ \text{Monday} \end{array} \quad \text{(in dozens)}$$

EXERCISE 11B.1

1 **a** $\mathbf{A}+\mathbf{B}=\begin{pmatrix} 3 & 4 \\ 5 & 2 \end{pmatrix}+\begin{pmatrix} 6 & -3 \\ -2 & 1 \end{pmatrix}=\begin{pmatrix} 9 & 1 \\ 3 & 3 \end{pmatrix}$

b $\mathbf{A}+\mathbf{B}+\mathbf{C}=\begin{pmatrix} 9 & 1 \\ 3 & 3 \end{pmatrix}+\begin{pmatrix} -3 & 7 \\ -4 & -2 \end{pmatrix}=\begin{pmatrix} 6 & 8 \\ -1 & 1 \end{pmatrix}$

c $\mathbf{B}+\mathbf{C}=\begin{pmatrix} 6 & -3 \\ -2 & 1 \end{pmatrix}+\begin{pmatrix} -3 & 7 \\ -4 & -2 \end{pmatrix}=\begin{pmatrix} 3 & 4 \\ -6 & -1 \end{pmatrix}$

d $\mathbf{C}+\mathbf{B}-\mathbf{A}=\begin{pmatrix} -3 & 7 \\ -4 & -2 \end{pmatrix}+\begin{pmatrix} 6 & -3 \\ -2 & 1 \end{pmatrix}-\begin{pmatrix} 3 & 4 \\ 5 & 2 \end{pmatrix}=\begin{pmatrix} 0 & 0 \\ -11 & -3 \end{pmatrix}$

2 **a** $\mathbf{P}+\mathbf{Q}=\begin{pmatrix} 3 & 5 & -11 \\ 10 & 2 & 6 \\ -2 & -1 & 7 \end{pmatrix}+\begin{pmatrix} 17 & -4 & 3 \\ -2 & 8 & -8 \\ 3 & -4 & 11 \end{pmatrix}=\begin{pmatrix} 20 & 1 & -8 \\ 8 & 10 & -2 \\ 1 & -5 & 18 \end{pmatrix}$

b $\mathbf{P}-\mathbf{Q}=\begin{pmatrix} 3 & 5 & -11 \\ 10 & 2 & 6 \\ -2 & -1 & 7 \end{pmatrix}-\begin{pmatrix} 17 & -4 & 3 \\ -2 & 8 & -8 \\ 3 & -4 & 11 \end{pmatrix}=\begin{pmatrix} -14 & 9 & -14 \\ 12 & -6 & 14 \\ -5 & 3 & -4 \end{pmatrix}$

c $\mathbf{Q}-\mathbf{P}=\begin{pmatrix} 17 & -4 & 3 \\ -2 & 8 & -8 \\ 3 & -4 & 11 \end{pmatrix}-\begin{pmatrix} 3 & 5 & -11 \\ 10 & 2 & 6 \\ -2 & -1 & 7 \end{pmatrix}=\begin{pmatrix} 14 & -9 & 14 \\ -12 & 6 & -14 \\ 5 & -3 & 4 \end{pmatrix}$

3 **a**

$$\begin{array}{cc} \text{Friday} & \text{Saturday} \\ \begin{pmatrix} 85 \\ 92 \\ 52 \end{pmatrix} & \begin{pmatrix} 102 \\ 137 \\ 49 \end{pmatrix} \end{array}$$

b Total for Friday and Saturday $=\begin{pmatrix} 85 \\ 92 \\ 52 \end{pmatrix}+\begin{pmatrix} 102 \\ 137 \\ 49 \end{pmatrix}=\begin{pmatrix} 187 \\ 229 \\ 101 \end{pmatrix}$

4 **a** **i** Cost price $\begin{pmatrix} 1.72 \\ 27.85 \\ 0.92 \\ 2.53 \\ 3.56 \end{pmatrix}$ **ii** Selling price $\begin{pmatrix} 1.79 \\ 28.75 \\ 1.33 \\ 2.25 \\ 3.51 \end{pmatrix}$

b In order to find David's profit/loss matrix we subtract the cost price matrix from the selling price matrix.

c Profit/Loss matrix $= \begin{pmatrix} 1.79 \\ 28.75 \\ 1.33 \\ 2.25 \\ 3.51 \end{pmatrix} - \begin{pmatrix} 1.72 \\ 27.85 \\ 0.92 \\ 2.53 \\ 3.56 \end{pmatrix} = \begin{pmatrix} 0.07 \\ 0.90 \\ 0.41 \\ -0.28 \\ -0.05 \end{pmatrix}$

5 **a** $\begin{array}{r} \\ \text{fridges} \\ \text{stoves} \\ \text{microwaves} \end{array}\begin{array}{c} \begin{array}{cc}\text{Lou} & \text{Rose}\end{array} \\ \begin{pmatrix} 23 & 19 \\ 17 & 29 \\ 31 & 24 \end{pmatrix}\end{array}$

b $\begin{array}{r} \\ \text{fr} \\ \text{st} \\ \text{mi} \end{array}\begin{array}{c} \begin{array}{cc}\text{L} & \text{R}\end{array} \\ \begin{pmatrix} 18 & 25 \\ 7 & 13 \\ 36 & 19 \end{pmatrix}\end{array}$

c Total sales for November and December

$= \begin{pmatrix} 23 & 19 \\ 17 & 29 \\ 31 & 24 \end{pmatrix} + \begin{pmatrix} 18 & 25 \\ 7 & 13 \\ 36 & 19 \end{pmatrix} = \begin{pmatrix} 41 & 44 \\ 24 & 42 \\ 67 & 43 \end{pmatrix}$

6 **a** $\begin{pmatrix} x & x^2 \\ 3 & -1 \end{pmatrix} = \begin{pmatrix} y & 4 \\ 3 & y+1 \end{pmatrix}$

Equating corresponding elements:

$x = y,\ x^2 = 4$ and $-1 = y + 1$

$\therefore\ x = \pm 2$ and $y = -2$

But $x = y \quad \therefore\ x = y = -2$

b $\begin{pmatrix} x & y \\ y & x \end{pmatrix} = \begin{pmatrix} -y & x \\ x & -y \end{pmatrix}$

Equating corresponding elements:

$\left.\begin{array}{l} x = -y \\ y = x \end{array}\right\} \quad \therefore\ y = 0,\ x = 0$

7 **a** $\mathbf{A} + \mathbf{B} = \begin{pmatrix} 2 & 1 \\ 3 & -1 \end{pmatrix} + \begin{pmatrix} -1 & 2 \\ 2 & 3 \end{pmatrix}$

$= \begin{pmatrix} 2+(-1) & 1+2 \\ 3+2 & -1+3 \end{pmatrix}$

$= \begin{pmatrix} 1 & 3 \\ 5 & 2 \end{pmatrix}$

$\mathbf{B} + \mathbf{A} = \begin{pmatrix} -1 & 2 \\ 2 & 3 \end{pmatrix} + \begin{pmatrix} 2 & 1 \\ 3 & -1 \end{pmatrix}$

$= \begin{pmatrix} -1+2 & 2+1 \\ 2+3 & 3+(-1) \end{pmatrix}$

$= \begin{pmatrix} 1 & 3 \\ 5 & 2 \end{pmatrix}$

b $\mathbf{A} + \mathbf{B} = \mathbf{B} + \mathbf{A}$ for all 2×2 matrices **A** and **B** because addition of numbers is commutative.

8 **a** $(\mathbf{A} + \mathbf{B}) + \mathbf{C}$

$= \left[\begin{pmatrix} -1 & 0 \\ 1 & 5 \end{pmatrix} + \begin{pmatrix} 3 & 4 \\ -1 & -2 \end{pmatrix}\right] + \begin{pmatrix} 4 & -1 \\ -1 & 3 \end{pmatrix}$

$= \begin{pmatrix} 2 & 4 \\ 0 & 3 \end{pmatrix} + \begin{pmatrix} 4 & -1 \\ -1 & 3 \end{pmatrix}$

$= \begin{pmatrix} 6 & 3 \\ -1 & 6 \end{pmatrix}$

$\mathbf{A} + (\mathbf{B} + \mathbf{C})$

$= \begin{pmatrix} -1 & 0 \\ 1 & 5 \end{pmatrix} + \left[\begin{pmatrix} 3 & 4 \\ -1 & -2 \end{pmatrix} + \begin{pmatrix} 4 & -1 \\ -1 & 3 \end{pmatrix}\right]$

$= \begin{pmatrix} -1 & 0 \\ 1 & 5 \end{pmatrix} + \begin{pmatrix} 7 & 3 \\ -2 & 1 \end{pmatrix}$

$= \begin{pmatrix} 6 & 3 \\ -1 & 6 \end{pmatrix}$

b Let $\mathbf{A} = \begin{pmatrix} a & b \\ c & d \end{pmatrix}$, $\mathbf{B} = \begin{pmatrix} p & q \\ r & s \end{pmatrix}$ and $\mathbf{C} = \begin{pmatrix} w & x \\ y & z \end{pmatrix}$

$\therefore\ (\mathbf{A} + \mathbf{B}) + \mathbf{C}$

$= \left[\begin{pmatrix} a & b \\ c & d \end{pmatrix} + \begin{pmatrix} p & q \\ r & s \end{pmatrix}\right] + \begin{pmatrix} w & x \\ y & z \end{pmatrix}$

$= \begin{pmatrix} a+p & b+q \\ c+r & d+s \end{pmatrix} + \begin{pmatrix} w & x \\ y & z \end{pmatrix}$

$= \begin{pmatrix} a+p+w & b+q+x \\ c+r+y & d+s+z \end{pmatrix}$

$\mathbf{A} + (\mathbf{B} + \mathbf{C})$

$= \begin{pmatrix} a & b \\ c & d \end{pmatrix} + \left[\begin{pmatrix} p & q \\ r & s \end{pmatrix} + \begin{pmatrix} w & x \\ y & z \end{pmatrix}\right]$

$= \begin{pmatrix} a & b \\ c & d \end{pmatrix} + \begin{pmatrix} p+w & q+x \\ r+y & s+z \end{pmatrix}$

$= \begin{pmatrix} a+p+w & b+q+x \\ c+r+y & d+s+z \end{pmatrix}$

$= (\mathbf{A} + \mathbf{B}) + \mathbf{C}$

EXERCISE 11B.2

1 **a** $2\mathbf{B} = 2\begin{pmatrix} 6 & 12 \\ 24 & 6 \end{pmatrix} = \begin{pmatrix} 12 & 24 \\ 48 & 12 \end{pmatrix}$

b $\frac{1}{3}\mathbf{B} = \frac{1}{3}\begin{pmatrix} 6 & 12 \\ 24 & 6 \end{pmatrix} = \begin{pmatrix} 2 & 4 \\ 8 & 2 \end{pmatrix}$

c $\frac{1}{12}\mathbf{B} = \frac{1}{12}\begin{pmatrix} 6 & 12 \\ 24 & 6 \end{pmatrix} = \begin{pmatrix} \frac{1}{2} & 1 \\ 2 & \frac{1}{2} \end{pmatrix}$

d $-\frac{1}{2}\mathbf{B} = -\frac{1}{2}\begin{pmatrix} 6 & 12 \\ 24 & 6 \end{pmatrix} = \begin{pmatrix} -3 & -6 \\ -12 & -3 \end{pmatrix}$

2 **a** $\mathbf{A} + \mathbf{B}$

$= \begin{pmatrix} 2 & 3 & 5 \\ 1 & 6 & 4 \end{pmatrix} + \begin{pmatrix} 1 & 2 & 1 \\ 1 & 2 & 3 \end{pmatrix}$

$= \begin{pmatrix} 2+1 & 3+2 & 5+1 \\ 1+1 & 6+2 & 4+3 \end{pmatrix}$

$= \begin{pmatrix} 3 & 5 & 6 \\ 2 & 8 & 7 \end{pmatrix}$

b $\mathbf{A} - \mathbf{B}$

$= \begin{pmatrix} 2 & 3 & 5 \\ 1 & 6 & 4 \end{pmatrix} - \begin{pmatrix} 1 & 2 & 1 \\ 1 & 2 & 3 \end{pmatrix}$

$= \begin{pmatrix} 2-1 & 3-2 & 5-1 \\ 1-1 & 6-2 & 4-3 \end{pmatrix}$

$= \begin{pmatrix} 1 & 1 & 4 \\ 0 & 4 & 1 \end{pmatrix}$

c $2\mathbf{A} + \mathbf{B}$

$= 2\begin{pmatrix} 2 & 3 & 5 \\ 1 & 6 & 4 \end{pmatrix} + \begin{pmatrix} 1 & 2 & 1 \\ 1 & 2 & 3 \end{pmatrix}$

$= \begin{pmatrix} 4 & 6 & 10 \\ 2 & 12 & 8 \end{pmatrix} + \begin{pmatrix} 1 & 2 & 1 \\ 1 & 2 & 3 \end{pmatrix}$

$= \begin{pmatrix} 4+1 & 6+2 & 10+1 \\ 2+1 & 12+2 & 8+3 \end{pmatrix}$

$= \begin{pmatrix} 5 & 8 & 11 \\ 3 & 14 & 11 \end{pmatrix}$

d $3\mathbf{A} - \mathbf{B}$

$= 3\begin{pmatrix} 2 & 3 & 5 \\ 1 & 6 & 4 \end{pmatrix} - \begin{pmatrix} 1 & 2 & 1 \\ 1 & 2 & 3 \end{pmatrix}$

$= \begin{pmatrix} 6 & 9 & 15 \\ 3 & 18 & 12 \end{pmatrix} - \begin{pmatrix} 1 & 2 & 1 \\ 1 & 2 & 3 \end{pmatrix}$

$= \begin{pmatrix} 6-1 & 9-2 & 15-1 \\ 3-1 & 18-2 & 12-3 \end{pmatrix}$

$= \begin{pmatrix} 5 & 7 & 14 \\ 2 & 16 & 9 \end{pmatrix}$

3 **a** $2\mathbf{H} = \begin{pmatrix} 12 \\ 24 \\ 120 \\ 60 \end{pmatrix}$ **b** $\frac{1}{2}\mathbf{H} = \begin{pmatrix} 3 \\ 6 \\ 30 \\ 15 \end{pmatrix}$ **c** $\mathbf{H} + \frac{1}{2}\mathbf{H} = \frac{3}{2}\mathbf{H} = \begin{pmatrix} 9 \\ 18 \\ 90 \\ 45 \end{pmatrix}$

4 **a** Increase of 15% $= 1.15\begin{pmatrix} 30 & 40 & 40 & 60 \\ 50 & 40 & 30 & 75 \\ 40 & 40 & 50 & 50 \\ 10 & 20 & 20 & 15 \end{pmatrix} = \begin{pmatrix} 35 & 46 & 46 & 69 \\ 58 & 46 & 35 & 86 \\ 46 & 46 & 58 & 58 \\ 12 & 23 & 23 & 17 \end{pmatrix}$ rounded to the nearest whole number.

b Decrease of 15% $= 0.85\begin{pmatrix} 30 & 40 & 40 & 60 \\ 50 & 40 & 30 & 75 \\ 40 & 40 & 50 & 50 \\ 10 & 20 & 20 & 15 \end{pmatrix} = \begin{pmatrix} 26 & 34 & 34 & 51 \\ 43 & 34 & 26 & 64 \\ 34 & 34 & 43 & 43 \\ 9 & 17 & 17 & 13 \end{pmatrix}$ rounded to the nearest whole number.

5 **a**

Weekdays	Weekends	
75	136	DVD
27	43	VHS
102	129	games

b $\begin{pmatrix} 75 \\ 27 \\ 102 \end{pmatrix} + \begin{pmatrix} 136 \\ 43 \\ 129 \end{pmatrix} = \begin{pmatrix} 211 \\ 70 \\ 231 \end{pmatrix}$

c The sum matrix of **b** represents total weekly average hirings.

6 The matrix is $12\mathbf{F} = 12\begin{pmatrix} 1 \\ 4 \\ 2 \\ 1 \end{pmatrix} = \begin{pmatrix} 12 \\ 48 \\ 24 \\ 12 \end{pmatrix}$

EXERCISE 11B.3

1 **a** $\mathbf{A}+2\mathbf{A}=3\mathbf{A}$

b $3\mathbf{B}-3\mathbf{B}=\mathbf{O}$

c $\mathbf{C}-2\mathbf{C}=-\mathbf{C}$

d $-\mathbf{B}+\mathbf{B}=\mathbf{O}$

e $2(\mathbf{A}+\mathbf{B})=2\mathbf{A}+2\mathbf{B}$

f $-(\mathbf{A}+\mathbf{B})=-\mathbf{A}-\mathbf{B}$

g $-(2\mathbf{A}-\mathbf{C})$
$=-2\mathbf{A}+\mathbf{C}$

h $3\mathbf{A}-(\mathbf{B}-\mathbf{A})$
$=3\mathbf{A}-\mathbf{B}+\mathbf{A}$
$=4\mathbf{A}-\mathbf{B}$

i $\mathbf{A}+2\mathbf{B}-(\mathbf{A}-\mathbf{B})$
$=\mathbf{A}+2\mathbf{B}-\mathbf{A}+\mathbf{B}$
$=3\mathbf{B}$

2 **a**
$$\mathbf{X}+\mathbf{B}=\mathbf{A}$$
$$\therefore\ \mathbf{X}+\mathbf{B}+(-\mathbf{B})=\mathbf{A}+(-\mathbf{B})$$
$$\therefore\ \mathbf{X}+\mathbf{O}=\mathbf{A}-\mathbf{B}$$
$$\therefore\ \mathbf{X}=\mathbf{A}-\mathbf{B}$$

b
$$\mathbf{B}+\mathbf{X}=\mathbf{C}$$
$$\therefore\ \mathbf{B}+\mathbf{X}+(-\mathbf{B})=\mathbf{C}+(-\mathbf{B})$$
$$\therefore\ \mathbf{O}+\mathbf{X}=\mathbf{C}-\mathbf{B}$$
$$\therefore\ \mathbf{X}=\mathbf{C}-\mathbf{B}$$

c
$$4\mathbf{B}+\mathbf{X}=2\mathbf{C}$$
$$\therefore\ 4\mathbf{B}+\mathbf{X}+(-4\mathbf{B})=2\mathbf{C}+(-4\mathbf{B})$$
$$\therefore\ \mathbf{O}+\mathbf{X}=2\mathbf{C}-4\mathbf{B}$$
$$\therefore\ \mathbf{X}=2\mathbf{C}-4\mathbf{B}$$

d
$$2\mathbf{X}=\mathbf{A}$$
$$\therefore\ \tfrac{1}{2}(2\mathbf{X})=\tfrac{1}{2}\mathbf{A}$$
$$\therefore\ 1\mathbf{X}=\tfrac{1}{2}\mathbf{A}\quad\therefore\ \mathbf{X}=\tfrac{1}{2}\mathbf{A}$$

e
$$3\mathbf{X}=\mathbf{B}$$
$$\therefore\ \tfrac{1}{3}(3\mathbf{X})=\tfrac{1}{3}\mathbf{B}$$
$$\therefore\ 1\mathbf{X}=\tfrac{1}{3}\mathbf{B}\quad\therefore\ \mathbf{X}=\tfrac{1}{3}\mathbf{B}$$

f
$$\mathbf{A}-\mathbf{X}=\mathbf{B}$$
$$\therefore\ \mathbf{A}-\mathbf{X}+\mathbf{X}=\mathbf{B}+\mathbf{X}$$
$$\therefore\ \mathbf{A}+\mathbf{O}=\mathbf{B}+\mathbf{X}$$
$$\therefore\ \mathbf{A}=\mathbf{B}+\mathbf{X}$$
$$\text{and}\ \ \mathbf{A}+(-\mathbf{B})=\mathbf{B}+\mathbf{X}+(-\mathbf{B})$$
$$\therefore\ \mathbf{A}-\mathbf{B}=\mathbf{X}+\mathbf{O}$$
$$\therefore\ \mathbf{A}-\mathbf{B}=\mathbf{X}$$
$$\therefore\ \mathbf{X}=\mathbf{A}-\mathbf{B}$$

g
$$\tfrac{1}{2}\mathbf{X}=\mathbf{C}$$
$$\therefore\ 2(\tfrac{1}{2}\mathbf{X})=2\mathbf{C}$$
$$\therefore\ 1\mathbf{X}=2\mathbf{C}\quad\therefore\ \mathbf{X}=2\mathbf{C}$$

h
$$2(\mathbf{X}+\mathbf{A})=\mathbf{B}$$
$$\therefore\ \tfrac{1}{2}[2(\mathbf{X}+\mathbf{A})]=\tfrac{1}{2}\mathbf{B}$$
$$\therefore\ 1(\mathbf{X}+\mathbf{A})=\tfrac{1}{2}\mathbf{B}$$
$$\therefore\ \mathbf{X}+\mathbf{A}=\tfrac{1}{2}\mathbf{B}$$
$$\therefore\ \mathbf{X}+\mathbf{A}+(-\mathbf{A})=\tfrac{1}{2}\mathbf{B}+(-\mathbf{A})$$
$$\therefore\ \mathbf{X}+\mathbf{O}=\tfrac{1}{2}\mathbf{B}-\mathbf{A}$$
$$\therefore\ \mathbf{X}=\tfrac{1}{2}\mathbf{B}-\mathbf{A}$$

i
$$\mathbf{A}-4\mathbf{X}=\mathbf{C}$$
$$\therefore\ \mathbf{A}-4\mathbf{X}+4\mathbf{X}=\mathbf{C}+4\mathbf{X}$$
$$\therefore\ \mathbf{A}+\mathbf{O}=\mathbf{C}+4\mathbf{X}$$
$$\therefore\ \mathbf{A}=\mathbf{C}+4\mathbf{X}$$
$$\text{and}\ \ \mathbf{A}-\mathbf{C}=4\mathbf{X}$$
$$\therefore\ \tfrac{1}{4}(\mathbf{A}-\mathbf{C})=\tfrac{1}{4}(4\mathbf{X})$$
$$\therefore\ \mathbf{X}=\tfrac{1}{4}(\mathbf{A}-\mathbf{C})$$

3 **a**
If $\tfrac{1}{3}\mathbf{X}=\mathbf{M}$
then $3(\tfrac{1}{3}\mathbf{X})=3\mathbf{M}$
$$\therefore\ \mathbf{X}=3\mathbf{M}=3\begin{pmatrix}1&2\\3&6\end{pmatrix}=\begin{pmatrix}3&6\\9&18\end{pmatrix}$$

b If $4\mathbf{X}=\mathbf{N}$ then $\tfrac{1}{4}(4\mathbf{X})=\tfrac{1}{4}\mathbf{N}$
$$\therefore\ \mathbf{X}=\tfrac{1}{4}\mathbf{N}$$
$$\therefore\ \mathbf{X}=\tfrac{1}{4}\begin{pmatrix}2&-1\\3&5\end{pmatrix}=\begin{pmatrix}\frac{1}{2}&-\frac{1}{4}\\\frac{3}{4}&\frac{5}{4}\end{pmatrix}$$

c
If $\mathbf{A}-2\mathbf{X}=3\mathbf{B}$
then $\mathbf{A}-2\mathbf{X}+2\mathbf{X}=3\mathbf{B}+2\mathbf{X}$
$$\therefore\ \mathbf{A}=3\mathbf{B}+2\mathbf{X}$$
$$\therefore\ \mathbf{A}+(-3\mathbf{B})=3\mathbf{B}+2\mathbf{X}+(-3\mathbf{B})$$
$$\therefore\ \mathbf{A}-3\mathbf{B}=2\mathbf{X}$$
$$\therefore\ \tfrac{1}{2}(\mathbf{A}-3\mathbf{B})=\tfrac{1}{2}(2\mathbf{X})$$
$$\therefore\ \tfrac{1}{2}(\mathbf{A}-3\mathbf{B})=1\mathbf{X}$$
$$\therefore\ \mathbf{X}=\tfrac{1}{2}(\mathbf{A}-3\mathbf{B})$$
$$=\tfrac{1}{2}\left[\begin{pmatrix}1&0\\-1&2\end{pmatrix}-3\begin{pmatrix}1&4\\-1&1\end{pmatrix}\right]$$
$$=\tfrac{1}{2}\begin{pmatrix}-2&-12\\2&-1\end{pmatrix}=\begin{pmatrix}-1&-6\\1&-\frac{1}{2}\end{pmatrix}$$

EXERCISE 11B.4

1 **a** $\begin{pmatrix} 3 & -1 \end{pmatrix}\begin{pmatrix} 5 \\ 4 \end{pmatrix} = (3 \times 5 + (-1) \times 4)$
$= (15 - 4)$
$= (11)$

b $\begin{pmatrix} 1 & 3 & 2 \end{pmatrix}\begin{pmatrix} 5 \\ 1 \\ 7 \end{pmatrix} = (1 \times 5 + 3 \times 1 + 2 \times 7)$
$= (5 + 3 + 14)$
$= (22)$

c $\begin{pmatrix} 6 & -1 & 2 & 3 \end{pmatrix}\begin{pmatrix} 1 \\ 0 \\ -1 \\ 4 \end{pmatrix} = (6 \times 1 + (-1) \times 0 + 2 \times (-1) + 3 \times 4)$
$= (6 + 0 - 2 + 12)$
$= (16)$

2 $\begin{pmatrix} w & x & y & z \end{pmatrix}\begin{pmatrix} 1 \\ 1 \\ 1 \\ 1 \end{pmatrix} = \begin{pmatrix} w + x + y + z \end{pmatrix}$

$\therefore$ $\frac{1}{4}(w + x + y + z)$, which is the average of w, x, y and z, can be represented as $\begin{pmatrix} w & x & y & z \end{pmatrix}\begin{pmatrix} \frac{1}{4} \\ \frac{1}{4} \\ \frac{1}{4} \\ \frac{1}{4} \end{pmatrix}$.

3 **a** $\mathbf{Q} = \begin{pmatrix} 4 \\ 3 \\ 2 \end{pmatrix}$, $\mathbf{P} = \begin{pmatrix} 27 & 35 & 39 \end{pmatrix}$

b total cost $= \mathbf{PQ} = \begin{pmatrix} 27 & 35 & 39 \end{pmatrix}\begin{pmatrix} 4 \\ 3 \\ 2 \end{pmatrix}$
$= (27 \times 4 + 35 \times 3 + 39 \times 2)$
$= (291)$ $\therefore$ total cost is \$291

4 **a** $\mathbf{P} = \begin{pmatrix} 10 & 6 & 3 & 1 \end{pmatrix}$

$\mathbf{N} = \begin{pmatrix} 3 \\ 2 \\ 4 \\ 2 \end{pmatrix}$

b total points $= \mathbf{PN} = \begin{pmatrix} 10 & 6 & 3 & 1 \end{pmatrix}\begin{pmatrix} 3 \\ 2 \\ 4 \\ 2 \end{pmatrix}$
$= (10 \times 3 + 6 \times 2 + 3 \times 4 + 1 \times 2)$
$= (30 + 12 + 12 + 2)$
$= (56)$ So, the number of points awarded is 56.

EXERCISE 11B.5

1 $\mathbf{A} = \begin{pmatrix} 4 & 2 & 1 \end{pmatrix}$ which is 1 row $\times$ 3 columns, $\mathbf{B} = \begin{pmatrix} 1 & 2 & 1 \\ 0 & 1 & 0 \end{pmatrix}$ which is 2 rows $\times$ 3 columns.

AB cannot be found because the number of columns in **A** does not equal the number of rows in **B**.

2 **A** is $2 \times n$ and **B** is $m \times 3$.

a We can find **AB** if the number of columns in **A** equals the number of rows in **B**, $\therefore$ if $n = m$.

b If **AB** can be found its order is 2×3.

c **BA** cannot be found because the number of columns in **B** does not equal the number of rows in **A**.

3 **a** **i** **A** is 2×2 and **B** is 1×2 $\therefore$ **AB** does not exist.
(not equal)

ii **B** is 1×2 and **A** is 2×2

$\therefore$ **BA** is 1×2

$\mathbf{BA} = \begin{pmatrix} 5 & 6 \end{pmatrix}\begin{pmatrix} 2 & 1 \\ 3 & 4 \end{pmatrix}$
$= \begin{pmatrix} 5 \times 2 + 6 \times 3 & 5 \times 1 + 6 \times 4 \end{pmatrix}$
$= \begin{pmatrix} 10 + 18 & 5 + 24 \end{pmatrix}$
$= \begin{pmatrix} 28 & 29 \end{pmatrix}$

b **i** **A** is 1×3 and **B** is 3×1, $\therefore$ **AB** is 1×1 and

$$\mathbf{AB} = \begin{pmatrix} 2 & 0 & 3 \end{pmatrix} \begin{pmatrix} 1 \\ 4 \\ 2 \end{pmatrix} = (2 \times 1 + 0 \times 4 + 3 \times 2) = (2 + 0 + 6) = (8)$$

ii **B** is 3×1 and **A** is 1×3, $\therefore$ **BA** is 3×3 and

$$\mathbf{BA} = \begin{pmatrix} 1 \\ 4 \\ 2 \end{pmatrix} \begin{pmatrix} 2 & 0 & 3 \end{pmatrix} = \begin{pmatrix} 1 \times 2 & 1 \times 0 & 1 \times 3 \\ 4 \times 2 & 4 \times 0 & 4 \times 3 \\ 2 \times 2 & 2 \times 0 & 2 \times 3 \end{pmatrix} = \begin{pmatrix} 2 & 0 & 3 \\ 8 & 0 & 12 \\ 4 & 0 & 6 \end{pmatrix}$$

4 **a** $\begin{pmatrix} 1 & 2 & 1 \end{pmatrix} \begin{pmatrix} 2 & 3 & 1 \\ 0 & 1 & 0 \\ 1 & 0 & 2 \end{pmatrix}$ is 1×3 by 3×3 $\therefore$ resultant matrix is 1×3

$$= \begin{pmatrix} 1 \times 2 + 2 \times 0 + 1 \times 1 & 1 \times 3 + 2 \times 1 + 1 \times 0 & 1 \times 1 + 2 \times 0 + 1 \times 2 \end{pmatrix}$$
$$= \begin{pmatrix} 2 + 0 + 1 & 3 + 2 + 0 & 1 + 0 + 2 \end{pmatrix}$$
$$= \begin{pmatrix} 3 & 5 & 3 \end{pmatrix}$$

b $\begin{pmatrix} 1 & 0 & -1 \\ -1 & 1 & 0 \\ 0 & -1 & 1 \end{pmatrix} \begin{pmatrix} 2 \\ 3 \\ 4 \end{pmatrix}$ is 3×3 by 3×1 $\therefore$ resultant matrix is 3×1

$$= \begin{pmatrix} 1 \times 2 + 0 \times 3 + (-1) \times 4 \\ (-1) \times 2 + 1 \times 3 + 0 \times 4 \\ 0 \times 2 + (-1) \times 3 + 1 \times 4 \end{pmatrix} = \begin{pmatrix} 2 + 0 - 4 \\ -2 + 3 + 0 \\ 0 - 3 + 4 \end{pmatrix} = \begin{pmatrix} -2 \\ 1 \\ 1 \end{pmatrix}$$

5 **a** $\mathbf{C} = \begin{pmatrix} 12.5 \\ 9.5 \end{pmatrix}$ $\quad \mathbf{N} = \begin{pmatrix} 2375 & 5156 \\ 2502 & 3612 \end{pmatrix}$ (columns: adults, children; rows: first day, second day)

b **N** is 2×2 and **C** is 2×1 $\therefore$ **NC** is 2×1

$$\mathbf{NC} = \begin{pmatrix} 2375 & 5156 \\ 2502 & 3612 \end{pmatrix} \begin{pmatrix} 12.5 \\ 9.5 \end{pmatrix} = \begin{pmatrix} 2375 \times 12.5 + 5156 \times 9.5 \\ 2502 \times 12.5 + 3612 \times 9.5 \end{pmatrix} = \begin{pmatrix} 29\,687.5 + 48\,982 \\ 31\,275 + 34\,314 \end{pmatrix}$$

$$= \begin{pmatrix} 78\,669.5 \\ 65\,589 \end{pmatrix} \begin{matrix} \text{income from day 1} \\ \text{income from day 2} \end{matrix}$$

c Total income $= \$78\,669.50 + \$65\,589 = \$144\,258.50$

6 **a** $\mathbf{R} = \begin{pmatrix} 1 & 1 \\ 1 & 2 \\ 2 & 3 \end{pmatrix}$ (columns: me, friend; rows: hammers, screwdrivers, cans of paint)

b $\mathbf{P} = \begin{pmatrix} 7 & 3 & 19 \\ 6 & 2 & 22 \end{pmatrix} \begin{matrix} \text{store A} \\ \text{store B} \end{matrix}$

c **P** is 2×3 and **R** is 3×2, $\therefore$ **PR** is 2×2

$$\mathbf{PR} = \begin{pmatrix} 7 & 3 & 19 \\ 6 & 2 & 22 \end{pmatrix} \begin{pmatrix} 1 & 1 \\ 1 & 2 \\ 2 & 3 \end{pmatrix} = \begin{pmatrix} 7 \times 1 + 3 \times 1 + 19 \times 2 & 7 \times 1 + 3 \times 2 + 19 \times 3 \\ 6 \times 1 + 2 \times 1 + 22 \times 2 & 6 \times 1 + 2 \times 2 + 22 \times 3 \end{pmatrix}$$

$$\therefore \quad \mathbf{PR} = \begin{pmatrix} 7 + 3 + 38 & 7 + 6 + 57 \\ 6 + 2 + 44 & 6 + 4 + 66 \end{pmatrix} = \begin{pmatrix} 48 & 70 \\ 52 & 76 \end{pmatrix}$$

d My costs at Store A are €48; my friend's costs at Store B are €76.

e My costs at Store B are €52. Therefore I should shop at Store A, which is cheaper.

EXERCISE 11B.6

1 **a** $\begin{pmatrix} 16 & 18 & 15 \\ 13 & 21 & 16 \\ 10 & 22 & 24 \end{pmatrix}$ **b** $\begin{pmatrix} 10 & 6 & -7 \\ 9 & 3 & 0 \\ 4 & -4 & -10 \end{pmatrix}$ **c** $\begin{pmatrix} 22 & 0 & 132 & 176 & 198 \\ 44 & 154 & 88 & 110 & 0 \\ 176 & 44 & 88 & 88 & 132 \end{pmatrix}$ **d** $\begin{pmatrix} 115 \\ 136 \\ 46 \\ 106 \end{pmatrix}$

2 **a** nights, breakfasts, dinners

Numbers matrix $\mathbf{N} = \begin{pmatrix} 3 & 3 & 2 \end{pmatrix}$

b Prices matrix (columns: Bay View, Terrace, Staunton Star; rows: room, breakfast, dinner)

$$\mathbf{P} = \begin{pmatrix} 125 & 150 & 140 \\ 44 & 40 & 40 \\ 75 & 80 & 65 \end{pmatrix} \begin{matrix} \text{room} \\ \text{breakfast} \\ \text{dinner} \end{matrix}$$

c Total prices for each venue = numbers matrix × prices matrix = **NP**

N is 1×3 and **P** is 3×3 $\therefore$ **NP** is 1×3

$$\mathbf{NP} = \begin{pmatrix} 3 & 3 & 2 \end{pmatrix} \begin{pmatrix} 125 & 150 & 140 \\ 44 & 40 & 40 \\ 75 & 80 & 65 \end{pmatrix} = \begin{pmatrix} 657 & 730 & 670 \end{pmatrix} \quad \{\text{using technology}\}$$

$\therefore$ \$657 for Bay View, \$730 for Terrace, \$670 for Staunton Star.

d Total prices $= \begin{pmatrix} 2 & 1 & 1 \end{pmatrix} \begin{pmatrix} 125 & 150 & 140 \\ 44 & 40 & 40 \\ 75 & 80 & 65 \end{pmatrix} = \begin{pmatrix} 369 & 420 & 385 \end{pmatrix}$ {using technology}

$\therefore$ \$369 for Bay View, \$420 for Terrace, \$385 for Staunton Star.

e To include both scenarios we calculate

$$\begin{pmatrix} 3 & 3 & 2 \\ 2 & 1 & 1 \end{pmatrix} \begin{pmatrix} 125 & 150 & 140 \\ 44 & 40 & 40 \\ 75 & 80 & 65 \end{pmatrix} = \begin{pmatrix} 657 & 730 & 670 \\ 369 & 420 & 385 \end{pmatrix}, \text{ using technology}$$

3 Prices matrix $= \begin{pmatrix} 125 \\ 315 \\ 405 \\ 375 \end{pmatrix}$

$$\text{Total income} = \begin{matrix} \text{numbers matrix} \\ \times \text{ prices matrix} \end{matrix} = \begin{pmatrix} 50 & 42 & 18 & 65 \\ 65 & 37 & 25 & 82 \\ 120 & 29 & 23 & 75 \\ 42 & 36 & 19 & 72 \end{pmatrix} \begin{pmatrix} 125 \\ 315 \\ 405 \\ 375 \end{pmatrix} = \begin{pmatrix} 51\,145 \\ 60\,655 \\ 61\,575 \\ 51\,285 \end{pmatrix} \quad \{\text{using technology}\}$$

$\therefore$ total income = \$51 145 + \$60 655 + \$61 575 + \$51 285 = \$224 660

4 **a** Income matrix $\mathbf{I} = \begin{pmatrix} 125 & 195 & 225 \end{pmatrix}$, Cost matrix $\mathbf{C} = \begin{pmatrix} 85 & 120 & 130 \end{pmatrix}$,

Numbers (bookings) matrix $\mathbf{N} = \begin{pmatrix} 15 & 12 & 13 & 11 & 14 & 16 & 8 \\ 4 & 3 & 6 & 2 & 0 & 4 & 7 \\ 3 & 1 & 4 & 4 & 3 & 2 & 0 \end{pmatrix}$

Profit per day

= (income from room) × (bookings per day) − (maintenance cost per room) × (bookings per day)

= **IN** − **CN**

$$= \begin{pmatrix} 125 & 195 & 225 \end{pmatrix} \begin{pmatrix} 15 & 12 & 13 & 11 & 14 & 16 & 8 \\ 4 & 3 & 6 & 2 & 0 & 4 & 7 \\ 3 & 1 & 4 & 4 & 3 & 2 & 0 \end{pmatrix} - \begin{pmatrix} 85 & 120 & 130 \end{pmatrix} \begin{pmatrix} 15 & 12 & 13 & 11 & 14 & 16 & 8 \\ 4 & 3 & 6 & 2 & 0 & 4 & 7 \\ 3 & 1 & 4 & 4 & 3 & 2 & 0 \end{pmatrix}$$

$= \begin{pmatrix} 1185 & 800 & 1350 & 970 & 845 & 1130 & 845 \end{pmatrix}$ {using technology}

$\therefore$ profit for the week = \$1185 + \$800 + \$1350 + \$970 + \$845 + \$1130 + \$845 = \$7125

b If the hotel maintained every room every day we would need to calculate
(income from room) × (bookings per day) − (maintenance costs per room) × (number of rooms)

$$= \begin{pmatrix}125 & 195 & 225\end{pmatrix}\begin{pmatrix}15 & 12 & 13 & 11 & 14 & 16 & 8\\ 4 & 3 & 6 & 2 & 0 & 4 & 7\\ 3 & 1 & 4 & 4 & 3 & 2 & 0\end{pmatrix} - \begin{pmatrix}85 & 120 & 130\end{pmatrix}\begin{pmatrix}20 & 20 & 20 & 20 & 20 & 20 & 20\\ 15 & 15 & 15 & 15 & 15 & 15 & 15\\ 5 & 5 & 5 & 5 & 5 & 5 & 5\end{pmatrix}$$

$\therefore$ using technology, profit per day $= \begin{pmatrix}-820 & -1840 & -455 & -1485 & -1725 & -920 & -1785\end{pmatrix}$

$\therefore$ the profit per week would be $(-\$820) + ... + (-\$1785) = -\$9030$, or a loss of \$9030.

c Profit per room matrix = income per room matrix − cost per room matrix

$$= \mathbf{I} - \mathbf{C} = \begin{pmatrix}125 & 195 & 225\end{pmatrix} - \begin{pmatrix}85 & 120 & 130\end{pmatrix}$$

$$= \begin{pmatrix}40 & 75 & 95\end{pmatrix}$$

$\therefore$ the result in **a** can be calculated using:

$$\begin{pmatrix}40 & 75 & 95\end{pmatrix}\begin{pmatrix}15 & 12 & 13 & 11 & 14 & 16 & 8\\ 4 & 3 & 6 & 2 & 0 & 4 & 7\\ 3 & 1 & 4 & 4 & 3 & 2 & 0\end{pmatrix}$$

This checks using technology.

EXERCISE 11B.7

1 $\mathbf{AB} = \begin{pmatrix}1 & 0\\ 1 & 2\end{pmatrix}\begin{pmatrix}-1 & 1\\ 0 & 3\end{pmatrix} = \begin{pmatrix}-1+0 & 1+0\\ -1+0 & 1+6\end{pmatrix} = \begin{pmatrix}-1 & 1\\ -1 & 7\end{pmatrix}$

$\mathbf{BA} = \begin{pmatrix}-1 & 1\\ 0 & 3\end{pmatrix}\begin{pmatrix}1 & 0\\ 1 & 2\end{pmatrix} = \begin{pmatrix}-1+1 & 0+2\\ 0+3 & 0+6\end{pmatrix} = \begin{pmatrix}0 & 2\\ 3 & 6\end{pmatrix}$

$\mathbf{AB} \neq \mathbf{BA}$ $\therefore$ in the general case **AB** does not necessarily equal **BA**.

2 $\mathbf{AO} = \begin{pmatrix}a & b\\ c & d\end{pmatrix}\begin{pmatrix}0 & 0\\ 0 & 0\end{pmatrix} = \begin{pmatrix}0+0 & 0+0\\ 0+0 & 0+0\end{pmatrix} = \begin{pmatrix}0 & 0\\ 0 & 0\end{pmatrix} = \mathbf{O}$

$\mathbf{OA} = \begin{pmatrix}0 & 0\\ 0 & 0\end{pmatrix}\begin{pmatrix}a & b\\ c & d\end{pmatrix} = \begin{pmatrix}0+0 & 0+0\\ 0+0 & 0+0\end{pmatrix} = \begin{pmatrix}0 & 0\\ 0 & 0\end{pmatrix} = \mathbf{O}$ $\therefore$ $\mathbf{AO} = \mathbf{OA} = \mathbf{O}$

3 **a** Suppose $\mathbf{A} = \begin{pmatrix}1 & 2\\ 3 & 4\end{pmatrix}$, $\mathbf{B} = \begin{pmatrix}1 & 1\\ 1 & 1\end{pmatrix}$ and $\mathbf{C} = \begin{pmatrix}2 & 3\\ 1 & 1\end{pmatrix}$.

$\mathbf{A}(\mathbf{B}+\mathbf{C})$

$= \begin{pmatrix}1 & 2\\ 3 & 4\end{pmatrix}\begin{pmatrix}3 & 4\\ 2 & 2\end{pmatrix}$

$= \begin{pmatrix}3+4 & 4+4\\ 9+8 & 12+8\end{pmatrix}$

$= \begin{pmatrix}7 & 8\\ 17 & 20\end{pmatrix}$

$\mathbf{AB}+\mathbf{AC}$

$= \begin{pmatrix}1 & 2\\ 3 & 4\end{pmatrix}\begin{pmatrix}1 & 1\\ 1 & 1\end{pmatrix} + \begin{pmatrix}1 & 2\\ 3 & 4\end{pmatrix}\begin{pmatrix}2 & 3\\ 1 & 1\end{pmatrix}$

$= \begin{pmatrix}1+2 & 1+2\\ 3+4 & 3+4\end{pmatrix} + \begin{pmatrix}2+2 & 3+2\\ 6+4 & 9+4\end{pmatrix}$

$= \begin{pmatrix}3 & 3\\ 7 & 7\end{pmatrix} + \begin{pmatrix}4 & 5\\ 10 & 13\end{pmatrix}$

$= \begin{pmatrix}7 & 8\\ 17 & 20\end{pmatrix}$

$= \mathbf{A}(\mathbf{B}+\mathbf{C})$

b $\mathbf{A} = \begin{pmatrix}a & b\\ c & d\end{pmatrix}$, $\mathbf{B} = \begin{pmatrix}p & q\\ r & s\end{pmatrix}$ and $\mathbf{C} = \begin{pmatrix}w & x\\ y & z\end{pmatrix}$

$\mathbf{A}(\mathbf{B}+\mathbf{C}) = \begin{pmatrix}a & b\\ c & d\end{pmatrix}\begin{pmatrix}p+w & q+x\\ r+y & s+z\end{pmatrix} = \begin{pmatrix}ap+aw+br+by & aq+ax+bs+bz\\ cp+cw+dr+dy & cq+cx+ds+dz\end{pmatrix}$

$$\mathbf{AB}+\mathbf{AC}=\begin{pmatrix}a & b\\ c & d\end{pmatrix}\begin{pmatrix}p & q\\ r & s\end{pmatrix}+\begin{pmatrix}a & b\\ c & d\end{pmatrix}\begin{pmatrix}w & x\\ y & z\end{pmatrix}$$
$$=\begin{pmatrix}ap+br & aq+bs\\ cp+dr & cq+ds\end{pmatrix}+\begin{pmatrix}aw+by & ax+bz\\ cw+dy & cx+dz\end{pmatrix}$$
$$=\begin{pmatrix}ap+aw+br+by & aq+ax+bs+bz\\ cp+cw+dr+dy & cq+cx+ds+dz\end{pmatrix}=\mathbf{A}(\mathbf{B}+\mathbf{C})$$

c Using the matrices in **a**,

$$(\mathbf{AB})\mathbf{C}=\begin{pmatrix}3 & 3\\ 7 & 7\end{pmatrix}\begin{pmatrix}2 & 3\\ 1 & 1\end{pmatrix}=\begin{pmatrix}6+3 & 9+3\\ 14+7 & 21+7\end{pmatrix}=\begin{pmatrix}9 & 12\\ 21 & 28\end{pmatrix}$$

$$\mathbf{A}(\mathbf{BC})=\begin{pmatrix}1 & 2\\ 3 & 4\end{pmatrix}\left[\begin{pmatrix}1 & 1\\ 1 & 1\end{pmatrix}\begin{pmatrix}2 & 3\\ 1 & 1\end{pmatrix}\right]=\begin{pmatrix}1 & 2\\ 3 & 4\end{pmatrix}\begin{pmatrix}3 & 4\\ 3 & 4\end{pmatrix}=\begin{pmatrix}3+6 & 4+8\\ 9+12 & 12+16\end{pmatrix}=\begin{pmatrix}9 & 12\\ 21 & 28\end{pmatrix}=(\mathbf{AB})\mathbf{C}$$

d Using the matrices in **b**,

$$(\mathbf{AB})\mathbf{C}=\begin{pmatrix}ap+br & aq+bs\\ cp+dr & cq+ds\end{pmatrix}\begin{pmatrix}w & x\\ y & z\end{pmatrix}$$
$$=\begin{pmatrix}apw+brw+aqy+bsy & apx+brx+aqz+bsz\\ cpw+drw+cqy+dsy & cpx+drx+cqz+dsz\end{pmatrix}$$

$$\mathbf{A}(\mathbf{BC})=\begin{pmatrix}a & b\\ c & d\end{pmatrix}\left[\begin{pmatrix}p & q\\ r & s\end{pmatrix}\begin{pmatrix}w & x\\ y & z\end{pmatrix}\right]=\begin{pmatrix}a & b\\ c & d\end{pmatrix}\begin{pmatrix}pw+qy & px+qz\\ rw+sy & rx+sz\end{pmatrix}$$
$$=\begin{pmatrix}apw+brw+aqy+bsy & apx+brx+aqz+bsz\\ cpw+drw+cqy+dsy & cpx+drx+cqz+dsz\end{pmatrix}$$
$$=(\mathbf{AB})\mathbf{C}$$

4 **a** If $\begin{pmatrix}a & b\\ c & d\end{pmatrix}\begin{pmatrix}w & x\\ y & z\end{pmatrix}=\begin{pmatrix}a & b\\ c & d\end{pmatrix}$ then $\begin{pmatrix}aw+by & ax+bz\\ cw+dy & cx+dz\end{pmatrix}=\begin{pmatrix}a & b\\ c & d\end{pmatrix}$

If $w=z=1$ and $x=y=0$, then

$$\text{LHS}=\begin{pmatrix}a(1)+b(0) & a(0)+b(1)\\ c(1)+d(0) & c(0)+d(1)\end{pmatrix}=\begin{pmatrix}a & b\\ c & d\end{pmatrix}=\text{RHS}\ \checkmark$$

b In **a** we showed that if $\mathbf{X}=\begin{pmatrix}1 & 0\\ 0 & 1\end{pmatrix}$ then $\mathbf{AX}=\mathbf{A}$.

$$\mathbf{XA}=\begin{pmatrix}1 & 0\\ 0 & 1\end{pmatrix}\begin{pmatrix}a & b\\ c & d\end{pmatrix}=\begin{pmatrix}a+0 & b+0\\ 0+c & 0+d\end{pmatrix}=\begin{pmatrix}a & b\\ c & d\end{pmatrix}=\mathbf{A}.$$

$\therefore$ $\mathbf{AI}=\mathbf{IA}=\mathbf{A}$ for all 2×2 matrices $\mathbf{A}$ where $\mathbf{I}=\mathbf{X}=\begin{pmatrix}1 & 0\\ 0 & 1\end{pmatrix}$.

5 **a**
$$\mathbf{A}^2=\begin{pmatrix}2 & 1\\ 3 & -2\end{pmatrix}\begin{pmatrix}2 & 1\\ 3 & -2\end{pmatrix}=\begin{pmatrix}4+3 & 2+(-2)\\ 6+(-6) & 3+4\end{pmatrix}=\begin{pmatrix}7 & 0\\ 0 & 7\end{pmatrix}$$

b
$$\mathbf{A}^3=\begin{pmatrix}5 & -1\\ 2 & 4\end{pmatrix}\begin{pmatrix}5 & -1\\ 2 & 4\end{pmatrix}\begin{pmatrix}5 & -1\\ 2 & 4\end{pmatrix}$$
$$=\begin{pmatrix}25+(-2) & -5+(-4)\\ 10+8 & -2+16\end{pmatrix}\begin{pmatrix}5 & -1\\ 2 & 4\end{pmatrix}$$
$$=\begin{pmatrix}23 & -9\\ 18 & 14\end{pmatrix}\begin{pmatrix}5 & -1\\ 2 & 4\end{pmatrix}$$
$$=\begin{pmatrix}115+(-18) & -23+(-36)\\ 90+28 & -18+56\end{pmatrix}=\begin{pmatrix}97 & -59\\ 118 & 38\end{pmatrix}$$

6 **a** $\mathbf{A} = \begin{pmatrix} 1 & 2 \\ 3 & 4 \\ 5 & 6 \end{pmatrix}$ $\therefore$ $\mathbf{A}^2$ is 3×2 by 3×2. $2 \neq 3$ so $\mathbf{A}^2$ does not exist.

b We can square a matrix when the number of columns equals the number of rows, that is, if it is a square matrix.

7 $\mathbf{I}^2 = \begin{pmatrix} 1 & 0 \\ 0 & 1 \end{pmatrix}\begin{pmatrix} 1 & 0 \\ 0 & 1 \end{pmatrix} = \begin{pmatrix} 1+0 & 0+0 \\ 0+0 & 0+1 \end{pmatrix} = \begin{pmatrix} 1 & 0 \\ 0 & 1 \end{pmatrix} = \mathbf{I}$ $\therefore$ $\mathbf{I}^3 = \mathbf{I}\mathbf{I}^2 = \mathbf{I}\mathbf{I} = \mathbf{I}^2 = \mathbf{I}$

EXERCISE 11B.8

1 **a**
$$\begin{aligned} &\mathbf{A}(\mathbf{A}+\mathbf{I}) \\ &= \mathbf{A}^2 + \mathbf{A}\mathbf{I} \\ &= \mathbf{A}^2 + \mathbf{A} \end{aligned}$$

b
$$\begin{aligned} &(\mathbf{B} + 2\mathbf{I})\mathbf{B} \\ &= \mathbf{B}^2 + 2\mathbf{I}\mathbf{B} \\ &= \mathbf{B}^2 + 2\mathbf{B} \end{aligned}$$

c
$$\begin{aligned} &\mathbf{A}(\mathbf{A}^2 - 2\mathbf{A} + \mathbf{I}) \\ &= \mathbf{A}^3 - 2\mathbf{A}^2 + \mathbf{A}\mathbf{I} \\ &= \mathbf{A}^3 - 2\mathbf{A}^2 + \mathbf{A} \end{aligned}$$

d
$$\begin{aligned} &\mathbf{A}(\mathbf{A}^2 + \mathbf{A} - 2\mathbf{I}) \\ &= \mathbf{A}^3 + \mathbf{A}^2 - 2\mathbf{A}\mathbf{I} \\ &= \mathbf{A}^3 + \mathbf{A}^2 - 2\mathbf{A} \end{aligned}$$

e
$$\begin{aligned} &(\mathbf{A}+\mathbf{B})(\mathbf{C}+\mathbf{D}) \\ &= (\mathbf{A}+\mathbf{B})\mathbf{C} + (\mathbf{A}+\mathbf{B})\mathbf{D} \\ &= \mathbf{AC} + \mathbf{BC} + \mathbf{AD} + \mathbf{BD} \end{aligned}$$

f
$$\begin{aligned} &(\mathbf{A}+\mathbf{B})^2 \\ &= (\mathbf{A}+\mathbf{B})(\mathbf{A}+\mathbf{B}) \\ &= (\mathbf{A}+\mathbf{B})\mathbf{A} + (\mathbf{A}+\mathbf{B})\mathbf{B} \\ &= \mathbf{A}^2 + \mathbf{BA} + \mathbf{AB} + \mathbf{B}^2 \end{aligned}$$

g
$$\begin{aligned} &(\mathbf{A}+\mathbf{B})(\mathbf{A}-\mathbf{B}) \\ &= (\mathbf{A}+\mathbf{B})\mathbf{A} - (\mathbf{A}+\mathbf{B})\mathbf{B} \\ &= \mathbf{A}^2 + \mathbf{BA} - \mathbf{AB} - \mathbf{B}^2 \end{aligned}$$

h
$$\begin{aligned} &(\mathbf{A}+\mathbf{I})^2 \\ &= (\mathbf{A}+\mathbf{I})(\mathbf{A}+\mathbf{I}) \\ &= (\mathbf{A}+\mathbf{I})\mathbf{A} + (\mathbf{A}+\mathbf{I})\mathbf{I} \\ &= \mathbf{A}^2 + \mathbf{IA} + \mathbf{AI} + \mathbf{I}^2 \\ &= \mathbf{A}^2 + \mathbf{A} + \mathbf{A} + \mathbf{I} \\ &= \mathbf{A}^2 + 2\mathbf{A} + \mathbf{I} \end{aligned}$$

i
$$\begin{aligned} &(3\mathbf{I}-\mathbf{B})^2 \\ &= (3\mathbf{I}-\mathbf{B})(3\mathbf{I}-\mathbf{B}) \\ &= (3\mathbf{I}-\mathbf{B})3\mathbf{I} - (3\mathbf{I}-\mathbf{B})\mathbf{B} \\ &= 9\mathbf{I}^2 - 3\mathbf{BI} - 3\mathbf{IB} + \mathbf{B}^2 \\ &= 9\mathbf{I} - 3\mathbf{B} - 3\mathbf{B} + \mathbf{B}^2 \\ &= 9\mathbf{I} - 6\mathbf{B} + \mathbf{B}^2 \end{aligned}$$

2 **a** $\mathbf{A}^2 = 2\mathbf{A} - \mathbf{I}$

$$\begin{aligned} \therefore\ \mathbf{A}^3 &= \mathbf{A} \times \mathbf{A}^2 \\ &= \mathbf{A}(2\mathbf{A}-\mathbf{I}) \\ &= 2\mathbf{A}^2 - \mathbf{AI} \\ &= 2(2\mathbf{A}-\mathbf{I}) - \mathbf{A} \\ &= 4\mathbf{A} - 2\mathbf{I} - \mathbf{A} \\ &= 3\mathbf{A} - 2\mathbf{I} \end{aligned}$$

and

$$\begin{aligned} \mathbf{A}^4 &= \mathbf{A} \times \mathbf{A}^3 \\ &= \mathbf{A}(3\mathbf{A}-2\mathbf{I}) \\ &= 3\mathbf{A}^2 - 2\mathbf{AI} \\ &= 3(2\mathbf{A}-\mathbf{I}) - 2\mathbf{A} \\ &= 6\mathbf{A} - 3\mathbf{I} - 2\mathbf{A} \\ &= 4\mathbf{A} - 3\mathbf{I} \end{aligned}$$

b $\mathbf{B}^2 = 2\mathbf{I} - \mathbf{B}$

$$\begin{aligned} \therefore\ \mathbf{B}^3 &= \mathbf{B} \times \mathbf{B}^2 \\ &= \mathbf{B}(2\mathbf{I}-\mathbf{B}) \\ &= 2\mathbf{BI} - \mathbf{B}^2 \\ &= 2\mathbf{B} - (2\mathbf{I}-\mathbf{B}) \\ &= 2\mathbf{B} - 2\mathbf{I} + \mathbf{B} \\ &= 3\mathbf{B} - 2\mathbf{I} \end{aligned}$$

and

$$\begin{aligned} \mathbf{B}^4 &= \mathbf{B} \times \mathbf{B}^3 \\ &= \mathbf{B}(3\mathbf{B}-2\mathbf{I}) \\ &= 3\mathbf{B}^2 - 2\mathbf{BI} \\ &= 3(2\mathbf{I}-\mathbf{B}) - 2\mathbf{B} \\ &= 6\mathbf{I} - 3\mathbf{B} - 2\mathbf{B} \\ &= 6\mathbf{I} - 5\mathbf{B} \end{aligned}$$

and

$$\begin{aligned} \mathbf{B}^5 &= \mathbf{B} \times \mathbf{B}^4 \\ &= \mathbf{B}(6\mathbf{I}-5\mathbf{B}) \\ &= 6\mathbf{BI} - 5\mathbf{B}^2 \\ &= 6\mathbf{B} - 5(2\mathbf{I}-\mathbf{B}) \\ &= 6\mathbf{B} - 10\mathbf{I} + 5\mathbf{B} \\ &= 11\mathbf{B} - 10\mathbf{I} \end{aligned}$$

c $\mathbf{C}^2 = 4\mathbf{C} - 3\mathbf{I}$

$$\begin{aligned} \mathbf{C}^3 &= \mathbf{C} \times \mathbf{C}^2 \\ &= \mathbf{C}(4\mathbf{C}-3\mathbf{I}) \\ &= 4\mathbf{C}^2 - 3\mathbf{CI} \\ &= 4(4\mathbf{C}-3\mathbf{I}) - 3\mathbf{C} \\ &= 16\mathbf{C} - 12\mathbf{I} - 3\mathbf{C} \\ &= 13\mathbf{C} - 12\mathbf{I} \end{aligned}$$

$$\begin{aligned} \mathbf{C}^5 &= \mathbf{C}^2 \times \mathbf{C}^3 \\ &= (4\mathbf{C}-3\mathbf{I})(13\mathbf{C}-12\mathbf{I}) \\ &= (4\mathbf{C}-3\mathbf{I})13\mathbf{C} - (4\mathbf{C}-3\mathbf{I})12\mathbf{I} \\ &= 52\mathbf{C}^2 - 39\mathbf{IC} - 48\mathbf{CI} + 36\mathbf{I}^2 \\ &= 52(4\mathbf{C}-3\mathbf{I}) - 39\mathbf{C} - 48\mathbf{C} + 36\mathbf{I} \\ &= 208\mathbf{C} - 156\mathbf{I} - 87\mathbf{C} + 36\mathbf{I} \\ &= 121\mathbf{C} - 120\mathbf{I} \end{aligned}$$

3 **a** If $\mathbf{A}^2 = \mathbf{I}$:

i $\mathbf{A}(\mathbf{A}+2\mathbf{I}) = \mathbf{A}^2 + 2\mathbf{A}\mathbf{I} = \mathbf{I} + 2\mathbf{A}$

ii $(\mathbf{A}-\mathbf{I})^2 = (\mathbf{A}-\mathbf{I})(\mathbf{A}-\mathbf{I}) = (\mathbf{A}-\mathbf{I})\mathbf{A} - (\mathbf{A}-\mathbf{I})\mathbf{I} = \mathbf{A}^2 - \mathbf{I}\mathbf{A} - \mathbf{A}\mathbf{I} + \mathbf{I}^2 = \mathbf{I} - \mathbf{A} - \mathbf{A} + \mathbf{I} = 2\mathbf{I} - 2\mathbf{A}$

iii $\mathbf{A}(\mathbf{A}+3\mathbf{I})^2 = \mathbf{A}(\mathbf{A}+3\mathbf{I})(\mathbf{A}+3\mathbf{I}) = \mathbf{A}[(\mathbf{A}+3\mathbf{I})\mathbf{A} + (\mathbf{A}+3\mathbf{I})3\mathbf{I}] = \mathbf{A}(\mathbf{A}^2 + 3\mathbf{I}\mathbf{A} + 3\mathbf{A}\mathbf{I} + 9\mathbf{I}^2) = \mathbf{A}(\mathbf{I} + 3\mathbf{A} + 3\mathbf{A} + 9\mathbf{I}) = \mathbf{A}(10\mathbf{I} + 6\mathbf{A}) = 10\mathbf{A}\mathbf{I} + 6\mathbf{A}^2 = 10\mathbf{A} + 6\mathbf{I}$

b If $\mathbf{A}^3 = \mathbf{I}$, $\mathbf{A}^2(\mathbf{A}+\mathbf{I})^2 = \mathbf{A}^2(\mathbf{A}^2 + 2\mathbf{A} + \mathbf{I}) = \mathbf{A}^4 + 2\mathbf{A}^3 + \mathbf{A}^2\mathbf{I} = \mathbf{A}(\mathbf{A}^3) + 2\mathbf{A}^3 + \mathbf{A}^2\mathbf{I} = \mathbf{A}\mathbf{I} + 2\mathbf{I} + \mathbf{A}^2 = \mathbf{A}^2 + \mathbf{A} + 2\mathbf{I}$

c If $\mathbf{A}^2 = \mathbf{O}$:

i $\mathbf{A}(2\mathbf{A}-3\mathbf{I}) = 2\mathbf{A}^2 - 3\mathbf{A}\mathbf{I} = 2\mathbf{O} - 3\mathbf{A} = -3\mathbf{A}$

ii $\mathbf{A}(\mathbf{A}+2\mathbf{I})(\mathbf{A}-\mathbf{I}) = \mathbf{A}[(\mathbf{A}+2\mathbf{I})\mathbf{A} - (\mathbf{A}+2\mathbf{I})\mathbf{I}] = \mathbf{A}(\mathbf{A}^2 + 2\mathbf{I}\mathbf{A} - \mathbf{A}\mathbf{I} - 2\mathbf{I}^2) = \mathbf{A}(\mathbf{O} + \mathbf{A} - 2\mathbf{I}) = \mathbf{A}^2 - 2\mathbf{A}\mathbf{I} = \mathbf{O} - 2\mathbf{A} = -2\mathbf{A}$

iii $\mathbf{A}(\mathbf{A}+\mathbf{I})^3 = \mathbf{A}(\mathbf{A}+\mathbf{I})(\mathbf{A}+\mathbf{I})^2 = \mathbf{A}(\mathbf{A}+\mathbf{I})(\mathbf{A}^2 + 2\mathbf{A} + \mathbf{I}) = (\mathbf{A}^2 + \mathbf{A}\mathbf{I})(\mathbf{O} + 2\mathbf{A} + \mathbf{I}) = (\mathbf{O} + \mathbf{A})(2\mathbf{A} + \mathbf{I}) = 2\mathbf{A}^2 + \mathbf{A}\mathbf{I} = 2\mathbf{O} + \mathbf{A} = \mathbf{A}$

4 **a** $\mathbf{AB} = \begin{pmatrix} 1 & 0 \\ 0 & 0 \end{pmatrix}\begin{pmatrix} 0 & 0 \\ 0 & 1 \end{pmatrix} = \begin{pmatrix} 0+0 & 0+0 \\ 0+0 & 0+0 \end{pmatrix} = \begin{pmatrix} 0 & 0 \\ 0 & 0 \end{pmatrix} = \mathbf{O}$

b $\mathbf{A}^2 = \begin{pmatrix} \frac{1}{2} & \frac{1}{2} \\ \frac{1}{2} & \frac{1}{2} \end{pmatrix}\begin{pmatrix} \frac{1}{2} & \frac{1}{2} \\ \frac{1}{2} & \frac{1}{2} \end{pmatrix} = \begin{pmatrix} \frac{1}{4}+\frac{1}{4} & \frac{1}{4}+\frac{1}{4} \\ \frac{1}{4}+\frac{1}{4} & \frac{1}{4}+\frac{1}{4} \end{pmatrix} = \begin{pmatrix} \frac{1}{2} & \frac{1}{2} \\ \frac{1}{2} & \frac{1}{2} \end{pmatrix} = \mathbf{A}$

c
$$\begin{aligned} \mathbf{A}^2 &= \mathbf{A} \\ \therefore\ \mathbf{A}^2 - \mathbf{A} &= \mathbf{O} \\ \therefore\ \mathbf{A}(\mathbf{A}-\mathbf{I}) &= \mathbf{O} \\ \therefore\ \mathbf{A} = \mathbf{O} &\text{ or } \mathbf{A}-\mathbf{I} = \mathbf{O} \\ \therefore\ \mathbf{A} = \mathbf{O} &\text{ or } \mathbf{I} \end{aligned}$$

The argument contains a false step. As the example in **a** illustrates, $\mathbf{AB} = \mathbf{O}$ does not imply that $\mathbf{A} = \mathbf{O}$ or $\mathbf{B} = \mathbf{O}$. This is a property of real numbers that does not hold for matrices. Therefore it is false to say that if $\mathbf{A}(\mathbf{A}-\mathbf{I}) = \mathbf{O}$, then $\mathbf{A} = \mathbf{O}$ or $\mathbf{A}-\mathbf{I}$ equals $\mathbf{O}$.

d Let $\mathbf{A} = \begin{pmatrix} a & b \\ c & d \end{pmatrix}$. If $\mathbf{A}^2 = \mathbf{A}$, $\begin{pmatrix} a & b \\ c & d \end{pmatrix}\begin{pmatrix} a & b \\ c & d \end{pmatrix} = \begin{pmatrix} a & b \\ c & d \end{pmatrix}$

$$\therefore\ \begin{pmatrix} a^2+bc & ab+bd \\ ac+cd & bc+d^2 \end{pmatrix} = \begin{pmatrix} a & b \\ c & d \end{pmatrix}$$

Equating corresponding elements:

$a^2 + bc = a \quad \therefore \quad bc = a(1-a)$ .... (1)

$ab + bd = b \quad \therefore \quad b(a+d-1) = 0$ .... (2)

$ac + cd = c \quad \therefore \quad c(a+d-1) = 0$ .... (3)

$bc + d^2 = d \quad \therefore \quad bc = d(1-d)$ .... (4)

<u>If $a+d-1 \neq 0$</u> then from (2) and (3), $b = c = 0$.

$\therefore$ from (1) and (4), $a = 0$ or 1 and $d = 0$ or 1

$\therefore$ $a = 0,\ d = 0$ or $a = 1,\ d = 1$ or $a = 0,\ d = 1$ or $a = 1,\ d = 0$

where the last two cases are not possible as $a + d \neq 1$.

So, if $a = 0,\ d = 0$ then $\mathbf{A} = \begin{pmatrix} 0 & 0 \\ 0 & 0 \end{pmatrix}$ and if $a = 1,\ d = 1$ then $\mathbf{A} = \begin{pmatrix} 1 & 0 \\ 0 & 1 \end{pmatrix}$

<u>If $a+d-1=0$</u> then $d=1-a$ and $c=\dfrac{a-a^2}{b}$

So $\mathbf{A}$ is $\begin{pmatrix} a & b \\ \dfrac{a-a^2}{b} & 1-a \end{pmatrix}$ provided $b \neq 0$.

5 Suppose $\mathbf{A} = \begin{pmatrix} 1 & -1 \\ 1 & -1 \end{pmatrix}$.

$\therefore \quad \mathbf{A}^2 = \begin{pmatrix} 1 & -1 \\ 1 & -1 \end{pmatrix}\begin{pmatrix} 1 & -1 \\ 1 & -1 \end{pmatrix} = \begin{pmatrix} 1+(-1) & -1+1 \\ 1+(-1) & -1+1 \end{pmatrix} = \begin{pmatrix} 0 & 0 \\ 0 & 0 \end{pmatrix}$

$\therefore$ $\mathbf{A} \neq \mathbf{O}$ but $\mathbf{A}^2 = \mathbf{O}$. So, "if $\mathbf{A}^2 = \mathbf{O}$ then $\mathbf{A} = \mathbf{O}$" is a false statement.

6 **a** Since $\mathbf{A}^2 = a\mathbf{A} + b\mathbf{I}$, $\begin{pmatrix} 1 & 2 \\ -1 & 2 \end{pmatrix}\begin{pmatrix} 1 & 2 \\ -1 & 2 \end{pmatrix} = a\begin{pmatrix} 1 & 2 \\ -1 & 2 \end{pmatrix} + b\begin{pmatrix} 1 & 0 \\ 0 & 1 \end{pmatrix}$

$\therefore \quad \begin{pmatrix} 1+(-2) & 2+4 \\ -1+(-2) & -2+4 \end{pmatrix} = \begin{pmatrix} a & 2a \\ -a & 2a \end{pmatrix} + \begin{pmatrix} b & 0 \\ 0 & b \end{pmatrix}$

$\therefore \quad \begin{pmatrix} -1 & 6 \\ -3 & 2 \end{pmatrix} = \begin{pmatrix} a+b & 2a \\ -a & 2a+b \end{pmatrix}$

$\therefore \quad a+b=-1$ and $2a=6$

$\therefore \quad a=3$ and $b=-4$

Checking for consistency: $-a=-3$, $2a+b=6+(-4)=2$ ✓

$\therefore \quad \mathbf{A}^2 = 3\mathbf{A} - 4\mathbf{I}$

b Since $\mathbf{A}^2 = a\mathbf{A} + b\mathbf{I}$, $\begin{pmatrix} 3 & 1 \\ 2 & -2 \end{pmatrix}\begin{pmatrix} 3 & 1 \\ 2 & -2 \end{pmatrix} = a\begin{pmatrix} 3 & 1 \\ 2 & -2 \end{pmatrix} + b\begin{pmatrix} 1 & 0 \\ 0 & 1 \end{pmatrix}$

$\therefore \quad \begin{pmatrix} 9+2 & 3+(-2) \\ 6+(-4) & 2+4 \end{pmatrix} = \begin{pmatrix} 3a & a \\ 2a & -2a \end{pmatrix} + \begin{pmatrix} b & 0 \\ 0 & b \end{pmatrix}$

$\therefore \quad \begin{pmatrix} 11 & 1 \\ 2 & 6 \end{pmatrix} = \begin{pmatrix} 3a+b & a \\ 2a & -2a+b \end{pmatrix}$

$\therefore \quad 3a+b=11$ and $a=1$

$\therefore \quad a=1$ and $b=8$

Checking for consistency $2a=2(1)=2$, $-2a+b=-2(1)+8=6$ ✓

$\therefore \quad \mathbf{A}^2 = \mathbf{A} + 8\mathbf{I}$

7 $\mathbf{A}^2 = p\mathbf{A} + q\mathbf{I}$

$\therefore \quad \begin{pmatrix} 1 & 2 \\ -1 & -3 \end{pmatrix}\begin{pmatrix} 1 & 2 \\ -1 & -3 \end{pmatrix} = p\begin{pmatrix} 1 & 2 \\ -1 & -3 \end{pmatrix} + q\begin{pmatrix} 1 & 0 \\ 0 & 1 \end{pmatrix}$

$\therefore \quad \begin{pmatrix} 1+(-2) & 2+(-6) \\ -1+3 & -2+9 \end{pmatrix} = \begin{pmatrix} p & 2p \\ -p & -3p \end{pmatrix} + \begin{pmatrix} q & 0 \\ 0 & q \end{pmatrix}$

$\therefore \quad \begin{pmatrix} -1 & -4 \\ 2 & 7 \end{pmatrix} = \begin{pmatrix} p+q & 2p \\ -p & -3p+q \end{pmatrix}$

$\therefore \quad p+q=-1$ and $2p=-4$

$\therefore \quad p=-2$ and $q=1$

Checking for consistency $-p=-(-2)=2$, $-3p+q=-3(-2)+1=7$ ✓

$\therefore \quad \mathbf{A}^2 = -2\mathbf{A} + \mathbf{I}$

a $\mathbf{A}^3 = \mathbf{A} \times \mathbf{A}^2$
$= \mathbf{A}(-2\mathbf{A} + \mathbf{I})$
$= -2\mathbf{A}^2 + \mathbf{AI}$
$= -2(-2\mathbf{A} + \mathbf{I}) + \mathbf{A}$
$= 4\mathbf{A} - 2\mathbf{I} + \mathbf{A}$
$= 5\mathbf{A} - 2\mathbf{I}$

b $\mathbf{A}^4 = \mathbf{A} \times \mathbf{A}^3$
$= \mathbf{A}(5\mathbf{A} - 2\mathbf{I})$
$= 5\mathbf{A}^2 - 2\mathbf{AI}$
$= 5(-2\mathbf{A} + \mathbf{I}) - 2\mathbf{A}$
$= -10\mathbf{A} + 5\mathbf{I} - 2\mathbf{A}$
$= -12\mathbf{A} + 5\mathbf{I}$

EXERCISE 11C.1

1 **a** $\begin{pmatrix} 5 & 6 \\ 2 & 3 \end{pmatrix}\begin{pmatrix} 3 & -6 \\ -2 & 5 \end{pmatrix} = \begin{pmatrix} 3 & 0 \\ 0 & 3 \end{pmatrix} = 3\mathbf{I}$

$\therefore \quad \begin{pmatrix} 5 & 6 \\ 2 & 3 \end{pmatrix} \times \frac{1}{3}\begin{pmatrix} 3 & -6 \\ -2 & 5 \end{pmatrix} = \mathbf{I}$

$\therefore \quad \begin{pmatrix} 5 & 6 \\ 2 & 3 \end{pmatrix}^{-1} = \frac{1}{3}\begin{pmatrix} 3 & -6 \\ -2 & 5 \end{pmatrix}$

$= \begin{pmatrix} 1 & -2 \\ -\frac{2}{3} & \frac{5}{3} \end{pmatrix}$

b $\begin{pmatrix} 3 & -4 \\ 1 & 2 \end{pmatrix}\begin{pmatrix} 2 & 4 \\ -1 & 3 \end{pmatrix} = \begin{pmatrix} 10 & 0 \\ 0 & 10 \end{pmatrix} = 10\mathbf{I}$

$\therefore \quad \begin{pmatrix} 3 & -4 \\ 1 & 2 \end{pmatrix} \times \frac{1}{10}\begin{pmatrix} 2 & 4 \\ -1 & 3 \end{pmatrix} = \mathbf{I}$

$\therefore \quad \begin{pmatrix} 3 & -4 \\ 1 & 2 \end{pmatrix}^{-1} = \frac{1}{10}\begin{pmatrix} 2 & 4 \\ -1 & 3 \end{pmatrix}$

$= \begin{pmatrix} 0.2 & 0.4 \\ -0.1 & 0.3 \end{pmatrix}$

2 **a** $|\mathbf{A}| = 12 - 14$
$= -2$

b $|\mathbf{A}| = 2 - 3$
$= -1$

c $|\mathbf{A}| = 0 - 0$
$= 0$

d $|\mathbf{A}| = 1 - 0$
$= 1$

3 **a** $\det \mathbf{B} = 12 - -14$
$= 26$

b $\det \mathbf{B} = 6 - 0$
$= 6$

c $\det \mathbf{B} = 0 - 1$
$= -1$

d $\det \mathbf{B} = a^2 - -a$
$= a^2 + a$

4 $\mathbf{A} = \begin{pmatrix} 2 & -1 \\ -1 & -1 \end{pmatrix}$

a $|\mathbf{A}| = 2(-1) - (-1)(-1)$
$= -3$

b $|\mathbf{A}| = -3$
$\therefore \quad |\mathbf{A}|^2 = (-3)^2$
$= 9$

c $2\mathbf{A} = 2\begin{pmatrix} 2 & -1 \\ -1 & -1 \end{pmatrix}$
$= \begin{pmatrix} 4 & -2 \\ -2 & -2 \end{pmatrix}$
$\therefore \quad |2\mathbf{A}| = 4(-2) - (-2)(-2)$
$= -12$

5 Let $\mathbf{A} = \begin{pmatrix} a & b \\ c & d \end{pmatrix}$ then $k\mathbf{A} = \begin{pmatrix} ka & kb \\ kc & kd \end{pmatrix}$ and $|k\mathbf{A}| = ka(kd) - kb(kc)$
$= k^2(ad - bc)$
$= k^2\,|\mathbf{A}|$

6 $\mathbf{A} = \begin{pmatrix} a & b \\ c & d \end{pmatrix}$, $\mathbf{B} = \begin{pmatrix} w & x \\ y & z \end{pmatrix}$

a $|\mathbf{A}| = ad - bc$
and $|\mathbf{B}| = wz - xy$

b $\mathbf{AB} = \begin{pmatrix} a & b \\ c & d \end{pmatrix}\begin{pmatrix} w & x \\ y & z \end{pmatrix}$
$= \begin{pmatrix} aw + by & ax + bz \\ cw + dy & cx + dz \end{pmatrix}$
$\therefore \quad |\mathbf{AB}| = (aw + by)(cx + dz) - (ax + bz)(cw + dy)$

c Expanding brackets,
$|\mathbf{AB}| = awcx + awdz + bycx + bydz - axcw - axdy - bzcw - bzdy$
$= wz(ad - bc) - xy(ad - bc)$
$= (ad - bc)(wz - xy)$
$= |\mathbf{A}|\,|\mathbf{B}|$

7 $\mathbf{A}=\begin{pmatrix}1&2\\3&4\end{pmatrix}$, $\mathbf{B}=\begin{pmatrix}-1&2\\0&1\end{pmatrix}$

a **i** $|\mathbf{A}|=1(4)-2(3)$ $=4-6$ $=-2$

ii $|2\mathbf{A}|=2^2\,|\mathbf{A}|$ $=4(-2)$ $=-8$

iii $|-\mathbf{A}|=(-1)^2\,|\mathbf{A}|$ $=1(-2)$ $=-2$

iv $|\mathbf{B}|=(-1)(1)-2(0)$ $=-1$ $\therefore\ |-3\mathbf{B}|=(-3)^2\,|\mathbf{B}|$ $=9(-1)$ $=-9$

v $|\mathbf{AB}|=|\mathbf{A}|\,|\mathbf{B}|$ $=(-2)(-1)$ $=2$

b *Check:*

ii $2\mathbf{A}=2\begin{pmatrix}1&2\\3&4\end{pmatrix}=\begin{pmatrix}2&4\\6&8\end{pmatrix}$

$|2\mathbf{A}|=2(8)-4(6)=-8$ ✓

iii $-\mathbf{A}=\begin{pmatrix}-1&-2\\-3&-4\end{pmatrix}$

$|-\mathbf{A}|=(-1)(-4)-(-2)(-3)=-2$ ✓

iv $-3\mathbf{B}=-3\begin{pmatrix}-1&2\\0&1\end{pmatrix}$ $=\begin{pmatrix}3&-6\\0&-3\end{pmatrix}$

$|-3\mathbf{B}|=(3)(-3)-(-6)(0)$ $=-9$ ✓

v $\mathbf{AB}=\begin{pmatrix}1&2\\3&4\end{pmatrix}\begin{pmatrix}-1&2\\0&1\end{pmatrix}$ $=\begin{pmatrix}-1+0&2+2\\-3+0&6+4\end{pmatrix}=\begin{pmatrix}-1&4\\-3&10\end{pmatrix}$

$|\mathbf{AB}|=(-1)(10)-4(-3)=2$ ✓

8 **a** $\begin{pmatrix}2&4\\-1&5\end{pmatrix}^{-1}=\dfrac{1}{2(5)-4(-1)}\begin{pmatrix}5&-4\\-(-1)&2\end{pmatrix}=\frac{1}{14}\begin{pmatrix}5&-4\\1&2\end{pmatrix}$

b $\begin{pmatrix}1&0\\1&-1\end{pmatrix}^{-1}=\dfrac{1}{1(-1)-0(1)}\begin{pmatrix}-1&0\\-1&1\end{pmatrix}=-\begin{pmatrix}-1&0\\-1&1\end{pmatrix}=\begin{pmatrix}1&0\\1&-1\end{pmatrix}$

c $\begin{pmatrix}2&4\\1&2\end{pmatrix}^{-1}$ does not exist, since $ad-bc=2(2)-4(1)=0$

d $\begin{pmatrix}1&0\\0&1\end{pmatrix}^{-1}=\dfrac{1}{1(1)-0(0)}\begin{pmatrix}1&0\\0&1\end{pmatrix}=\begin{pmatrix}1&0\\0&1\end{pmatrix}$

e $\begin{pmatrix}3&5\\-6&-10\end{pmatrix}^{-1}$ does not exist, since $ad-bc=3(-10)-5(-6)=0$

f $\begin{pmatrix}-1&2\\4&7\end{pmatrix}^{-1}=\dfrac{1}{(-1)(7)-2(4)}\begin{pmatrix}7&-2\\-4&-1\end{pmatrix}=-\frac{1}{15}\begin{pmatrix}7&-2\\-4&-1\end{pmatrix}$

g $\begin{pmatrix}3&4\\-1&2\end{pmatrix}^{-1}=\dfrac{1}{3(2)-(-1)(4)}\begin{pmatrix}2&-4\\-(-1)&3\end{pmatrix}=\frac{1}{10}\begin{pmatrix}2&-4\\1&3\end{pmatrix}$

h $\begin{pmatrix}-1&-1\\2&3\end{pmatrix}^{-1}=\dfrac{1}{(-1)3-(-1)2}\begin{pmatrix}3&-(-1)\\-2&-1\end{pmatrix}=-\begin{pmatrix}3&1\\-2&-1\end{pmatrix}=\begin{pmatrix}-3&-1\\2&1\end{pmatrix}$

EXERCISE 11C.2

1 **a** $\begin{pmatrix}1&2\\3&4\end{pmatrix}\begin{pmatrix}x\\y\end{pmatrix}=\begin{pmatrix}x+2y\\3x+4y\end{pmatrix}$

b $\begin{pmatrix}2&3\\1&-4\end{pmatrix}\begin{pmatrix}a\\b\end{pmatrix}=\begin{pmatrix}2a+3b\\a-4b\end{pmatrix}$

2 **a** $\left.\begin{array}{r}3x-y=8\\2x+3y=6\end{array}\right\}$ can be written as $\begin{pmatrix}3&-1\\2&3\end{pmatrix}\begin{pmatrix}x\\y\end{pmatrix}=\begin{pmatrix}8\\6\end{pmatrix}$

b $\left.\begin{array}{r}4x-3y=11\\3x+2y=-5\end{array}\right\}$ can be written as $\begin{pmatrix}4&-3\\3&2\end{pmatrix}\begin{pmatrix}x\\y\end{pmatrix}=\begin{pmatrix}11\\-5\end{pmatrix}$

c $\left.\begin{array}{r} 3a-b=6 \\ 2a+7b=-4 \end{array}\right\}$ can be written as $\begin{pmatrix} 3 & -1 \\ 2 & 7 \end{pmatrix}\begin{pmatrix} a \\ b \end{pmatrix}=\begin{pmatrix} 6 \\ -4 \end{pmatrix}$

3 **a** $\left.\begin{array}{r} 2x-y=6 \\ x+3y=14 \end{array}\right\}$ can be written as $\begin{pmatrix} 2 & -1 \\ 1 & 3 \end{pmatrix}\begin{pmatrix} x \\ y \end{pmatrix}=\begin{pmatrix} 6 \\ 14 \end{pmatrix}$

$\therefore \quad \begin{pmatrix} x \\ y \end{pmatrix}=\begin{pmatrix} 2 & -1 \\ 1 & 3 \end{pmatrix}^{-1}\begin{pmatrix} 6 \\ 14 \end{pmatrix}=\frac{1}{7}\begin{pmatrix} 3 & 1 \\ -1 & 2 \end{pmatrix}\begin{pmatrix} 6 \\ 14 \end{pmatrix}$

$\therefore \quad \begin{pmatrix} x \\ y \end{pmatrix}=\frac{1}{7}\begin{pmatrix} 18+14 \\ -6+28 \end{pmatrix}=\begin{pmatrix} \frac{32}{7} \\ \frac{22}{7} \end{pmatrix}$ and so $x=\frac{32}{7},\ y=\frac{22}{7}$

b $\left.\begin{array}{r} 5x-4y=5 \\ 2x+3y=-13 \end{array}\right\}$ can be written as $\begin{pmatrix} 5 & -4 \\ 2 & 3 \end{pmatrix}\begin{pmatrix} x \\ y \end{pmatrix}=\begin{pmatrix} 5 \\ -13 \end{pmatrix}$

$\therefore \quad \begin{pmatrix} x \\ y \end{pmatrix}=\begin{pmatrix} 5 & -4 \\ 2 & 3 \end{pmatrix}^{-1}\begin{pmatrix} 5 \\ -13 \end{pmatrix}=\frac{1}{23}\begin{pmatrix} 3 & 4 \\ -2 & 5 \end{pmatrix}\begin{pmatrix} 5 \\ -13 \end{pmatrix}$

$\therefore \quad \begin{pmatrix} x \\ y \end{pmatrix}=\frac{1}{23}\begin{pmatrix} 15+(-52) \\ -10+(-65) \end{pmatrix}=\begin{pmatrix} -\frac{37}{23} \\ -\frac{75}{23} \end{pmatrix}$ and so $x=-\frac{37}{23},\ y=-\frac{75}{23}$

c $\left.\begin{array}{r} x-2y=7 \\ 5x+3y=-2 \end{array}\right\}$ can be written as $\begin{pmatrix} 1 & -2 \\ 5 & 3 \end{pmatrix}\begin{pmatrix} x \\ y \end{pmatrix}=\begin{pmatrix} 7 \\ -2 \end{pmatrix}$

$\therefore \quad \begin{pmatrix} x \\ y \end{pmatrix}=\begin{pmatrix} 1 & -2 \\ 5 & 3 \end{pmatrix}^{-1}\begin{pmatrix} 7 \\ -2 \end{pmatrix}=\frac{1}{13}\begin{pmatrix} 3 & 2 \\ -5 & 1 \end{pmatrix}\begin{pmatrix} 7 \\ -2 \end{pmatrix}$

$\therefore \quad \begin{pmatrix} x \\ y \end{pmatrix}=\frac{1}{13}\begin{pmatrix} 21+(-4) \\ -35+(-2) \end{pmatrix}=\begin{pmatrix} \frac{17}{13} \\ -\frac{37}{13} \end{pmatrix}$ and so $x=\frac{17}{13},\ y=-\frac{37}{13}$

d $\left.\begin{array}{r} 3x+5y=4 \\ 2x-y=11 \end{array}\right\}$ can be written as $\begin{pmatrix} 3 & 5 \\ 2 & -1 \end{pmatrix}\begin{pmatrix} x \\ y \end{pmatrix}=\begin{pmatrix} 4 \\ 11 \end{pmatrix}$

$\therefore \quad \begin{pmatrix} x \\ y \end{pmatrix}=\begin{pmatrix} 3 & 5 \\ 2 & -1 \end{pmatrix}^{-1}\begin{pmatrix} 4 \\ 11 \end{pmatrix}=\frac{1}{-13}\begin{pmatrix} -1 & -5 \\ -2 & 3 \end{pmatrix}\begin{pmatrix} 4 \\ 11 \end{pmatrix}$

$\therefore \quad \begin{pmatrix} x \\ y \end{pmatrix}=-\frac{1}{13}\begin{pmatrix} -4+(-55) \\ -8+33 \end{pmatrix}=\begin{pmatrix} \frac{59}{13} \\ \frac{-25}{13} \end{pmatrix}$ and so $x=\frac{59}{13},\ y=-\frac{25}{13}$

e $\left.\begin{array}{r} 4x-7y=8 \\ 3x-5y=0 \end{array}\right\}$ can be written as $\begin{pmatrix} 4 & -7 \\ 3 & -5 \end{pmatrix}\begin{pmatrix} x \\ y \end{pmatrix}=\begin{pmatrix} 8 \\ 0 \end{pmatrix}$

$\therefore \quad \begin{pmatrix} x \\ y \end{pmatrix}=\begin{pmatrix} 4 & -7 \\ 3 & -5 \end{pmatrix}^{-1}\begin{pmatrix} 8 \\ 0 \end{pmatrix}=1\begin{pmatrix} -5 & 7 \\ -3 & 4 \end{pmatrix}\begin{pmatrix} 8 \\ 0 \end{pmatrix}$

$\therefore \quad \begin{pmatrix} x \\ y \end{pmatrix}=\begin{pmatrix} -40+0 \\ -24+0 \end{pmatrix}$ and so $x=-40,\ y=-24$

f $\left.\begin{array}{r} 7x+11y=18 \\ 11x-7y=-11 \end{array}\right\}$ can be written as $\begin{pmatrix} 7 & 11 \\ 11 & -7 \end{pmatrix}\begin{pmatrix} x \\ y \end{pmatrix}=\begin{pmatrix} 18 \\ -11 \end{pmatrix}$

$\therefore \quad \begin{pmatrix} x \\ y \end{pmatrix}=\begin{pmatrix} 7 & 11 \\ 11 & -7 \end{pmatrix}^{-1}\begin{pmatrix} 18 \\ -11 \end{pmatrix}$

$\therefore \quad \begin{pmatrix} x \\ y \end{pmatrix}=\frac{1}{-170}\begin{pmatrix} -7 & -11 \\ -11 & 7 \end{pmatrix}\begin{pmatrix} 18 \\ -11 \end{pmatrix}=-\frac{1}{170}\begin{pmatrix} -126+121 \\ -198-77 \end{pmatrix}$

$\therefore \quad \begin{pmatrix} x \\ y \end{pmatrix}=-\frac{1}{170}\begin{pmatrix} -5 \\ -275 \end{pmatrix}=\begin{pmatrix} \frac{1}{34} \\ \frac{55}{34} \end{pmatrix}$ and so $x=\frac{1}{34},\ y=\frac{55}{34}$

4 **a** If $\mathbf{AX}=\mathbf{B}$ then $\mathbf{A}^{-1}\mathbf{AX}=\mathbf{A}^{-1}\mathbf{B}$ {premultiply by $\mathbf{A}^{-1}$}

$\therefore\ \mathbf{IX}=\mathbf{A}^{-1}\mathbf{B}$

$\therefore\ \mathbf{X}=\mathbf{A}^{-1}\mathbf{B}$

If $\mathbf{XA}=\mathbf{B}$ then $\mathbf{XAA}^{-1}=\mathbf{BA}^{-1}$ {postmultiply by $\mathbf{A}^{-1}$}

$\therefore\ \mathbf{XI}=\mathbf{BA}^{-1}$

$\therefore\ \mathbf{X}=\mathbf{BA}^{-1}$

b **i** $\mathbf{X}\begin{pmatrix}1&2\\5&-1\end{pmatrix}=\begin{pmatrix}14&-5\\22&0\end{pmatrix}$

$$\therefore\ \mathbf{X}=\begin{pmatrix}14&-5\\22&0\end{pmatrix}\begin{pmatrix}1&2\\5&-1\end{pmatrix}^{-1}$$

$$=\begin{pmatrix}14&-5\\22&0\end{pmatrix}\tfrac{1}{-11}\begin{pmatrix}-1&-2\\-5&1\end{pmatrix}$$

$$=-\tfrac{1}{11}\begin{pmatrix}-14+25&-28+(-5)\\-22+0&-44+0\end{pmatrix}$$

$$=\begin{pmatrix}-1&3\\2&4\end{pmatrix}$$

ii $\begin{pmatrix}1&3\\2&-1\end{pmatrix}\mathbf{X}=\begin{pmatrix}1&-3\\4&2\end{pmatrix}$

$$\therefore\ \mathbf{X}=\begin{pmatrix}1&3\\2&-1\end{pmatrix}^{-1}\begin{pmatrix}1&-3\\4&2\end{pmatrix}$$

$$=\tfrac{1}{-7}\begin{pmatrix}-1&-3\\-2&1\end{pmatrix}\begin{pmatrix}1&-3\\4&2\end{pmatrix}$$

$$=-\tfrac{1}{7}\begin{pmatrix}-1+(-12)&3+(-6)\\-2+4&6+2\end{pmatrix}$$

$$=\begin{pmatrix}\frac{13}{7}&\frac{3}{7}\\-\frac{2}{7}&-\frac{8}{7}\end{pmatrix}$$

5 **a** **i** $|\mathbf{A}|=0$ when $2k-(-6)=0$ $\therefore\ k=-3$ $\therefore$ $\mathbf{A}$ is singular when $k=-3$

ii $\mathbf{A}=\begin{pmatrix}k&1\\-6&2\end{pmatrix}$ $\therefore\ \mathbf{A}^{-1}=\dfrac{1}{2k+6}\begin{pmatrix}2&-1\\6&k\end{pmatrix}$, provided that $k\neq-3$

b **i** $|\mathbf{A}|=0$ when $3k-0=0$ $\therefore\ k=0$ $\therefore$ $\mathbf{A}$ is singular when $k=0$

ii $\mathbf{A}=\begin{pmatrix}3&-1\\0&k\end{pmatrix}$ $\therefore\ \mathbf{A}^{-1}=\dfrac{1}{3k}\begin{pmatrix}k&1\\0&3\end{pmatrix}$, provided that $k\neq0$

c **i** $|\mathbf{A}|=0$ when

$$k(k+1)-2=0$$
$$k^2+k-2=0$$
$$(k-1)(k+2)=0$$

$\therefore\ k=1$ or -2 $\therefore$ $\mathbf{A}$ is singular when $k=1$ or -2

ii $\mathbf{A}=\begin{pmatrix}k+1&2\\1&k\end{pmatrix}$ $\therefore\ \mathbf{A}^{-1}=\dfrac{1}{k(k+1)-2}\begin{pmatrix}k&-2\\-1&k+1\end{pmatrix}$

$$=\frac{1}{(k+2)(k-1)}\begin{pmatrix}k&-2\\-1&k+1\end{pmatrix},$$ provided that $k\neq-2$ or 1

6 **a** **i** $\begin{pmatrix}2&-3\\4&-1\end{pmatrix}\begin{pmatrix}x\\y\end{pmatrix}=\begin{pmatrix}8\\11\end{pmatrix}$

and $|\mathbf{A}|=-2--12$

$=-2+12$

$=10$

ii As $|\mathbf{A}|\neq0$, the system has a unique solution

$$\begin{pmatrix}x\\y\end{pmatrix}=\begin{pmatrix}2&-3\\4&-1\end{pmatrix}^{-1}\begin{pmatrix}8\\11\end{pmatrix}$$

$$=\tfrac{1}{10}\begin{pmatrix}-1&3\\-4&2\end{pmatrix}\begin{pmatrix}8\\11\end{pmatrix}$$

$$=\tfrac{1}{10}\begin{pmatrix}25\\-10\end{pmatrix}$$

$$=\begin{pmatrix}\frac{5}{2}\\-1\end{pmatrix}$$

$\therefore\ x=\frac{5}{2},\ y=-1$

b **i** $\begin{pmatrix} 2 & k \\ 4 & -1 \end{pmatrix}\begin{pmatrix} x \\ y \end{pmatrix} = \begin{pmatrix} 8 \\ 11 \end{pmatrix}$

and $|\mathbf{A}| = -2 - 4k$

ii The system has a unique solution if $-2 - 4k \neq 0 \quad \therefore \quad k \neq -\frac{1}{2}$

$$\begin{pmatrix} x \\ y \end{pmatrix} = \begin{pmatrix} 2 & k \\ 4 & -1 \end{pmatrix}^{-1}\begin{pmatrix} 8 \\ 11 \end{pmatrix}$$

$$= \frac{1}{-2-4k}\begin{pmatrix} -1 & -k \\ -4 & 2 \end{pmatrix}\begin{pmatrix} 8 \\ 11 \end{pmatrix}$$

$$= \frac{1}{-2-4k}\begin{pmatrix} -8-11k \\ -10 \end{pmatrix}$$

$$\therefore \quad x = \frac{8+11k}{2+4k}, \quad y = \frac{5}{1+2k}, \quad k \neq -\tfrac{1}{2}$$

is the unique solution

iii When $k = -\frac{1}{2}$, the equations are

$\begin{cases} 2x - \frac{1}{2}y = 8 \\ 4x - y = 11 \end{cases}$ or $\begin{cases} 4x - y = 16 \\ 4x - y = 11 \end{cases}$

So, we have no solutions (as the lines are parallel and so do not meet).

EXERCISE 11C.3

1 $\mathbf{A} = \begin{pmatrix} 2 & 1 \\ 0 & 1 \end{pmatrix}$, $\mathbf{B} = \begin{pmatrix} 1 & 2 \\ -1 & 0 \end{pmatrix}$ and $\mathbf{C} = \begin{pmatrix} 0 & 3 \\ 1 & 2 \end{pmatrix}$

Since $\mathbf{AXB} = \mathbf{C}$,

then $\mathbf{A}^{-1}\mathbf{AXB} = \mathbf{A}^{-1}\mathbf{C}$ {premultiply by $\mathbf{A}^{-1}$}

$\therefore \quad \mathbf{IXB} = \mathbf{A}^{-1}\mathbf{C}$

$\therefore \quad \mathbf{XB} = \mathbf{A}^{-1}\mathbf{C}$

and

$\mathbf{XBB}^{-1} = \mathbf{A}^{-1}\mathbf{CB}^{-1}$ {postmultiply by $\mathbf{B}^{-1}$}

$\therefore \quad \mathbf{XI} = \mathbf{A}^{-1}\mathbf{CB}^{-1}$

$\therefore \quad \mathbf{X} = \mathbf{A}^{-1}\mathbf{CB}^{-1}$

$$\therefore \quad \mathbf{X} = \tfrac{1}{2}\begin{pmatrix} 1 & -1 \\ 0 & 2 \end{pmatrix}\begin{pmatrix} 0 & 3 \\ 1 & 2 \end{pmatrix}\tfrac{1}{2}\begin{pmatrix} 0 & -2 \\ 1 & 1 \end{pmatrix} = \tfrac{1}{4}\begin{pmatrix} 0+(-1) & 3+(-2) \\ 0+2 & 0+4 \end{pmatrix}\begin{pmatrix} 0 & -2 \\ 1 & 1 \end{pmatrix}$$

$$\therefore \quad \mathbf{X} = \tfrac{1}{4}\begin{pmatrix} -1 & 1 \\ 2 & 4 \end{pmatrix}\begin{pmatrix} 0 & -2 \\ 1 & 1 \end{pmatrix} = \tfrac{1}{4}\begin{pmatrix} 0+1 & 2+1 \\ 0+4 & -4+4 \end{pmatrix} = \tfrac{1}{4}\begin{pmatrix} 1 & 3 \\ 4 & 0 \end{pmatrix} = \begin{pmatrix} \frac{1}{4} & \frac{3}{4} \\ 1 & 0 \end{pmatrix}$$

2 **a** If $\mathbf{A} = \mathbf{A}^{-1}$, then $\mathbf{A}^2 = \mathbf{AA} = \mathbf{AA}^{-1} = \mathbf{I}$

b If $\begin{pmatrix} a & b \\ b & a \end{pmatrix}$ is its own inverse, then $\begin{pmatrix} a & b \\ b & a \end{pmatrix}\begin{pmatrix} a & b \\ b & a \end{pmatrix} = \begin{pmatrix} 1 & 0 \\ 0 & 1 \end{pmatrix}$

$$\therefore \quad \begin{pmatrix} a^2+b^2 & 2ab \\ 2ab & b^2+a^2 \end{pmatrix} = \begin{pmatrix} 1 & 0 \\ 0 & 1 \end{pmatrix}$$

$\therefore \quad a^2 + b^2 = 1$ and $2ab = 0$

If $2ab = 0$, then $a = 0$ or $b = 0$

and $b^2 = 1$ and $a^2 = 1$

$\therefore \quad a = 0$ and $b = \pm 1$ $\quad\therefore \quad a = \pm 1$ and $b = 0$

This gives four possible combinations: $\begin{pmatrix} 1 & 0 \\ 0 & 1 \end{pmatrix}, \begin{pmatrix} -1 & 0 \\ 0 & -1 \end{pmatrix}, \begin{pmatrix} 0 & 1 \\ 1 & 0 \end{pmatrix}, \begin{pmatrix} 0 & -1 \\ -1 & 0 \end{pmatrix}$

3 **a** $\mathbf{A} = \begin{pmatrix} 1 & 2 \\ -1 & 0 \end{pmatrix} \quad \therefore \quad \mathbf{A}^{-1} = \frac{1}{2}\begin{pmatrix} 0 & -2 \\ 1 & 1 \end{pmatrix} = \begin{pmatrix} 0 & -1 \\ \frac{1}{2} & \frac{1}{2} \end{pmatrix}$

$$\therefore \quad (\mathbf{A}^{-1})^{-1} = \frac{1}{\frac{1}{2}}\begin{pmatrix} \frac{1}{2} & 1 \\ -\frac{1}{2} & 0 \end{pmatrix} = \begin{pmatrix} 1 & 2 \\ -1 & 0 \end{pmatrix} = \mathbf{A}$$

b If $\mathbf{A}^{-1} = \mathbf{B}$

then $(\mathbf{A}^{-1})^{-1}(\mathbf{A}^{-1}) = \mathbf{B}^{-1}\mathbf{B} = \mathbf{I}$

and $(\mathbf{A}^{-1})(\mathbf{A}^{-1})^{-1} = \mathbf{BB}^{-1} = \mathbf{I}$

c We can deduce from **b** that $(\mathbf{A}^{-1})^{-1} = \mathbf{A}$ and $\mathbf{A}$ is the inverse of $\mathbf{A}^{-1}$.

4 **a** $\mathbf{A}=\begin{pmatrix}1 & 1\\ 2 & -1\end{pmatrix}$, $\mathbf{B}=\begin{pmatrix}0 & 1\\ 2 & -3\end{pmatrix}$

i $\mathbf{A}^{-1}=\frac{1}{-3}\begin{pmatrix}-1 & -1\\ -2 & 1\end{pmatrix}=\begin{pmatrix}\frac{1}{3} & \frac{1}{3}\\ \frac{2}{3} & -\frac{1}{3}\end{pmatrix}$

ii $\mathbf{B}^{-1}=\frac{1}{-2}\begin{pmatrix}-3 & -1\\ -2 & 0\end{pmatrix}=\begin{pmatrix}\frac{3}{2} & \frac{1}{2}\\ 1 & 0\end{pmatrix}$

iii
$$\begin{aligned}(\mathbf{AB})^{-1}&=\left[\begin{pmatrix}1 & 1\\ 2 & -1\end{pmatrix}\begin{pmatrix}0 & 1\\ 2 & -3\end{pmatrix}\right]^{-1}\\ &=\begin{pmatrix}0+2 & 1+(-3)\\ 0+(-2) & 2+3\end{pmatrix}^{-1}\\ &=\begin{pmatrix}2 & -2\\ -2 & 5\end{pmatrix}^{-1}\\ &=\frac{1}{6}\begin{pmatrix}5 & 2\\ 2 & 2\end{pmatrix}\text{ or }\begin{pmatrix}\frac{5}{6} & \frac{1}{3}\\ \frac{1}{3} & \frac{1}{3}\end{pmatrix}\end{aligned}$$

iv
$$\begin{aligned}(\mathbf{BA})^{-1}&=\left[\begin{pmatrix}0 & 1\\ 2 & -3\end{pmatrix}\begin{pmatrix}1 & 1\\ 2 & -1\end{pmatrix}\right]^{-1}\\ &=\begin{pmatrix}0+2 & 0+(-1)\\ 2+(-6) & 2+3\end{pmatrix}^{-1}\\ &=\begin{pmatrix}2 & -1\\ -4 & 5\end{pmatrix}^{-1}\\ &=\frac{1}{6}\begin{pmatrix}5 & 1\\ 4 & 2\end{pmatrix}\text{ or }\begin{pmatrix}\frac{5}{6} & \frac{1}{6}\\ \frac{2}{3} & \frac{1}{3}\end{pmatrix}\end{aligned}$$

v
$$\begin{aligned}\mathbf{A}^{-1}\mathbf{B}^{-1}&=\begin{pmatrix}\frac{1}{3} & \frac{1}{3}\\ \frac{2}{3} & -\frac{1}{3}\end{pmatrix}\begin{pmatrix}\frac{3}{2} & \frac{1}{2}\\ 1 & 0\end{pmatrix}\\ &=\begin{pmatrix}\frac{3}{6}+\frac{1}{3} & \frac{1}{6}+0\\ \frac{6}{6}+\left(-\frac{1}{3}\right) & \frac{2}{6}+0\end{pmatrix}\\ &=\begin{pmatrix}\frac{5}{6} & \frac{1}{6}\\ \frac{2}{3} & \frac{1}{3}\end{pmatrix}\end{aligned}$$

vi
$$\begin{aligned}\mathbf{B}^{-1}\mathbf{A}^{-1}&=\begin{pmatrix}\frac{3}{2} & \frac{1}{2}\\ 1 & 0\end{pmatrix}\begin{pmatrix}\frac{1}{3} & \frac{1}{3}\\ \frac{2}{3} & -\frac{1}{3}\end{pmatrix}\\ &=\begin{pmatrix}\frac{3}{6}+\frac{2}{6} & \frac{3}{6}+\left(-\frac{1}{6}\right)\\ \frac{1}{3}+0 & \frac{1}{3}+0\end{pmatrix}\\ &=\begin{pmatrix}\frac{5}{6} & \frac{1}{3}\\ \frac{1}{3} & \frac{1}{3}\end{pmatrix}\end{aligned}$$

b Choose appropriate matrices and repeat question **a**.

c The results of **a** and **b** suggest that $(\mathbf{AB})^{-1}=\mathbf{B}^{-1}\mathbf{A}^{-1}$ and $(\mathbf{BA})^{-1}=\mathbf{A}^{-1}\mathbf{B}^{-1}$.

d
$$\begin{aligned}&(\mathbf{AB})(\mathbf{B}^{-1}\mathbf{A}^{-1})\\ &=\mathbf{A}(\mathbf{BB}^{-1})\mathbf{A}^{-1}\\ &=\mathbf{AIA}^{-1}\\ &=\mathbf{AA}^{-1}=\mathbf{I}\end{aligned}\qquad\text{and}\qquad\begin{aligned}&(\mathbf{B}^{-1}\mathbf{A}^{-1})(\mathbf{AB})\\ &=\mathbf{B}^{-1}(\mathbf{A}^{-1}\mathbf{A})\mathbf{B}\\ &=\mathbf{B}^{-1}\mathbf{IB}\\ &=\mathbf{B}^{-1}\mathbf{B}=\mathbf{I}\end{aligned}$$

$\therefore$ $(\mathbf{AB})(\mathbf{B}^{-1}\mathbf{A}^{-1})=(\mathbf{B}^{-1}\mathbf{A}^{-1})(\mathbf{AB})=\mathbf{I}$

$\therefore$ $\mathbf{AB}$ and $\mathbf{B}^{-1}\mathbf{A}^{-1}$ are inverses.

5 $(k\mathbf{A})\left(\frac{1}{k}\mathbf{A}^{-1}\right)=k\times\frac{1}{k}(\mathbf{AA}^{-1})=\mathbf{I}$ and $\left(\frac{1}{k}\mathbf{A}^{-1}\right)(k\mathbf{A})=\frac{1}{k}\times k(\mathbf{A}^{-1}\mathbf{A})=\mathbf{I}$

$\therefore$ $(k\mathbf{A})\left(\frac{1}{k}\mathbf{A}^{-1}\right)=\left(\frac{1}{k}\mathbf{A}^{-1}\right)(k\mathbf{A})=\mathbf{I}$ $\qquad\therefore$ $k\mathbf{A}$ and $\frac{1}{k}\mathbf{A}^{-1}$ are inverses.

6 $\mathbf{X}=\mathbf{AY}$ and $\mathbf{Y}=\mathbf{BZ}$

a $\mathbf{X}=\mathbf{AY}=\mathbf{A(BZ)}=\mathbf{ABZ}$

b $(\mathbf{AB})^{-1}\mathbf{X}=(\mathbf{AB})^{-1}\mathbf{ABZ}$ $\quad$ {premultiply by $(\mathbf{AB})^{-1}$}

$(\mathbf{AB})^{-1}\mathbf{X}=\mathbf{IZ}$

$\therefore$ $\mathbf{Z}=\mathbf{B}^{-1}\mathbf{A}^{-1}\mathbf{X}$ $\quad$ {as $(\mathbf{AB})^{-1}=\mathbf{B}^{-1}\mathbf{A}^{-1}$}

7 If $\mathbf{A}=\begin{pmatrix}3 & 2\\ -2 & -1\end{pmatrix}$, let $\mathbf{A}^2=p\mathbf{A}+q\mathbf{I}$

$$\therefore\quad\begin{pmatrix}3 & 2\\ -2 & -1\end{pmatrix}\begin{pmatrix}3 & 2\\ -2 & -1\end{pmatrix}=p\begin{pmatrix}3 & 2\\ -2 & -1\end{pmatrix}+q\begin{pmatrix}1 & 0\\ 0 & 1\end{pmatrix}$$

$$\therefore\quad\begin{pmatrix}9+(-4) & 6+(-2)\\ -6+2 & -4+1\end{pmatrix}=\begin{pmatrix}3p & 2p\\ -2p & -p\end{pmatrix}+\begin{pmatrix}q & 0\\ 0 & q\end{pmatrix}$$

$$\therefore\quad\begin{pmatrix}5 & 4\\ -4 & -3\end{pmatrix}=\begin{pmatrix}3p+q & 2p\\ -2p & -p+q\end{pmatrix}$$

$\therefore \quad 5 = 3p + q \quad \text{and} \quad 4 = 2p$

$\therefore \quad p = 2 \quad \text{and} \quad q = -1 \quad \text{and} \quad \mathbf{A}^2 = 2\mathbf{A} - \mathbf{I}$

Checking for consistency: $-2p = -2(2) = -4$ ✓

$-p + q = -2 + (-1) = -3$ ✓

$$\begin{aligned}
\text{Now} \quad \mathbf{A}^2 &= 2\mathbf{A} - \mathbf{I} \\
\therefore \quad \mathbf{A}^{-1}\mathbf{A}^2 &= \mathbf{A}^{-1}2\mathbf{A} - \mathbf{A}^{-1}\mathbf{I} \quad \{\text{premultiplying by } \mathbf{A}^{-1}\} \\
\therefore \quad \mathbf{A}^{-1}\mathbf{A}\mathbf{A} &= 2\mathbf{A}^{-1}\mathbf{A} - \mathbf{A}^{-1} \\
\therefore \quad \mathbf{I}\mathbf{A} &= 2\mathbf{I} - \mathbf{A}^{-1} \\
\therefore \quad \mathbf{A} &= 2\mathbf{I} - \mathbf{A}^{-1} \\
\therefore \quad \mathbf{A}^{-1} &= 2\mathbf{I} - \mathbf{A}
\end{aligned}$$

8 In each example we premultiply by $\mathbf{A}^{-1}$.

a

$$\begin{aligned}
\mathbf{A}^2 &= 4\mathbf{A} - \mathbf{I} \\
\therefore \quad \mathbf{A}^{-1}\mathbf{A}^2 &= \mathbf{A}^{-1}(4\mathbf{A} - \mathbf{I}) \\
\therefore \quad \mathbf{A}^{-1}\mathbf{A}\mathbf{A} &= 4\mathbf{A}^{-1}\mathbf{A} - \mathbf{A}^{-1}\mathbf{I} \\
\therefore \quad \mathbf{I}\mathbf{A} &= 4\mathbf{I} - \mathbf{A}^{-1} \\
\therefore \quad \mathbf{A} - 4\mathbf{I} &= -\mathbf{A}^{-1} \\
\therefore \quad \mathbf{A}^{-1} &= 4\mathbf{I} - \mathbf{A}
\end{aligned}$$

b

$$\begin{aligned}
5\mathbf{A} &= \mathbf{I} - \mathbf{A}^2 \\
\therefore \quad \mathbf{A}^{-1}5\mathbf{A} &= \mathbf{A}^{-1}(\mathbf{I} - \mathbf{A}^2) \\
\therefore \quad 5\mathbf{A}^{-1}\mathbf{A} &= \mathbf{A}^{-1}\mathbf{I} - \mathbf{A}^{-1}\mathbf{A}\mathbf{A} \\
\therefore \quad 5\mathbf{I} &= \mathbf{A}^{-1} - \mathbf{I}\mathbf{A} \\
\therefore \quad 5\mathbf{I} &= \mathbf{A}^{-1} - \mathbf{A} \\
\therefore \quad \mathbf{A}^{-1} &= 5\mathbf{I} + \mathbf{A}
\end{aligned}$$

c

$$\begin{aligned}
2\mathbf{I} &= 3\mathbf{A}^2 - 4\mathbf{A} \\
\therefore \quad \mathbf{A}^{-1}2\mathbf{I} &= \mathbf{A}^{-1}3\mathbf{A}^2 - \mathbf{A}^{-1}4\mathbf{A} \\
\therefore \quad 2\mathbf{A}^{-1} &= 3\mathbf{A}^{-1}\mathbf{A}\mathbf{A} - 4\mathbf{A}^{-1}\mathbf{A} \\
\therefore \quad 2\mathbf{A}^{-1} &= 3\mathbf{I}\mathbf{A} - 4\mathbf{I} \\
\therefore \quad 2\mathbf{A}^{-1} &= 3\mathbf{A} - 4\mathbf{I} \\
\therefore \quad \mathbf{A}^{-1} &= \tfrac{3}{2}\mathbf{A} - 2\mathbf{I}
\end{aligned}$$

9 If $\mathbf{AB} = \mathbf{A}$ and $\mathbf{BA} = \mathbf{B}$,

$$\begin{aligned}
\text{then} \quad \mathbf{A}^2 &= \mathbf{AA} \\
&= (\mathbf{AB})\mathbf{A} \\
&= \mathbf{A}(\mathbf{BA}) \quad \{\text{associative rule}\} \\
&= \mathbf{AB} \\
\therefore \quad \mathbf{A}^2 &= \mathbf{A}
\end{aligned}$$

$ab = ac$ implies that $b = c$ for non-zero real numbers, but this property does not hold for matrices.
Thus from $\mathbf{AB} = \mathbf{AI} = \mathbf{A}$ it does not follow that $\mathbf{B} = \mathbf{I}$.

10

$$\begin{aligned}
\text{If} \quad \mathbf{AB} &= \mathbf{AC} \\
\text{then} \quad \mathbf{A}^{-1}\mathbf{AB} &= \mathbf{A}^{-1}\mathbf{AC} \\
&\{\text{premultiplying by } \mathbf{A}^{-1}\} \\
\therefore \quad \mathbf{IB} &= \mathbf{IC} \\
\therefore \quad \mathbf{B} &= \mathbf{C}
\end{aligned}$$

We need $\mathbf{A}^{-1}$ to exist, so $|\mathbf{A}| \neq 0$.
So, if $\mathbf{AB} = \mathbf{AC}$ and $|\mathbf{A}| \neq 0$,
then $\mathbf{B} = \mathbf{C}$.

11 If $\mathbf{X} = \mathbf{P}^{-1}\mathbf{AP}$ and $\mathbf{A}^3 = \mathbf{I}$

$$\begin{aligned}
\text{then} \quad \mathbf{X}^3 &= (\mathbf{P}^{-1}\mathbf{AP})(\mathbf{P}^{-1}\mathbf{AP})(\mathbf{P}^{-1}\mathbf{AP}) \\
&= (\mathbf{P}^{-1}\mathbf{A})(\mathbf{PP}^{-1})\mathbf{A}(\mathbf{PP}^{-1})\mathbf{AP} \\
&\qquad \{\text{associative rule}\} \\
&= \mathbf{P}^{-1}\mathbf{AIAIAP} \\
&= \mathbf{P}^{-1}\mathbf{AAAP} \\
&= \mathbf{P}^{-1}\mathbf{A}^3\mathbf{P} \\
&= \mathbf{P}^{-1}\mathbf{IP} \\
&= \mathbf{P}^{-1}\mathbf{P} = \mathbf{I}
\end{aligned}$$

12 If $a\mathbf{A}^2 + b\mathbf{A} + c\mathbf{I} = \mathbf{O}$
and $\mathbf{X} = \mathbf{P}^{-1}\mathbf{AP}$ then

$$\begin{aligned}
&a\mathbf{X}^2 + b\mathbf{X} + c\mathbf{I} \\
&= a(\mathbf{P}^{-1}\mathbf{AP})(\mathbf{P}^{-1}\mathbf{AP}) + b\mathbf{P}^{-1}\mathbf{AP} + c\mathbf{I} \\
&= a\mathbf{P}^{-1}\mathbf{A}(\mathbf{PP}^{-1})\mathbf{AP} + b\mathbf{P}^{-1}\mathbf{AP} + c\mathbf{I} \\
&= a\mathbf{P}^{-1}\mathbf{A}^2\mathbf{P} + b\mathbf{P}^{-1}\mathbf{AP} + c\mathbf{I} \\
&= \mathbf{P}^{-1}(a\mathbf{A}^2 + b\mathbf{A} + c\mathbf{I})\mathbf{P} \\
&= \mathbf{P}^{-1}\mathbf{OP} \\
&= \mathbf{O}
\end{aligned}$$

EXERCISE 11D.1

1 **a** $\begin{vmatrix} 2 & 3 & 0 \\ -1 & 2 & 1 \\ 2 & 0 & 5 \end{vmatrix} = 2\begin{vmatrix} 2 & 1 \\ 0 & 5 \end{vmatrix} - 3\begin{vmatrix} -1 & 1 \\ 2 & 5 \end{vmatrix} + 0\begin{vmatrix} -1 & 2 \\ 2 & 0 \end{vmatrix} = 2(10-0) - 3(-5-2) + 0 = 41$

b $\begin{vmatrix} -1 & 2 & -3 \\ 1 & 0 & 0 \\ -1 & 2 & 1 \end{vmatrix} = -1\begin{vmatrix} 0 & 0 \\ 2 & 1 \end{vmatrix} - 2\begin{vmatrix} 1 & 0 \\ -1 & 1 \end{vmatrix} + (-3)\begin{vmatrix} 1 & 0 \\ -1 & 2 \end{vmatrix}$

$= -1(0-0) - 2(1-0) - 3(2-0)$
$= -8$

c $\begin{vmatrix} 2 & 1 & 3 \\ -1 & 1 & 2 \\ 2 & 1 & 3 \end{vmatrix} = 2\begin{vmatrix} 1 & 2 \\ 1 & 3 \end{vmatrix} - 1\begin{vmatrix} -1 & 2 \\ 2 & 3 \end{vmatrix} + 3\begin{vmatrix} -1 & 1 \\ 2 & 1 \end{vmatrix}$

$= 2(3-2) - 1(-3-4) + 3(-1-2)$
$= 0$

d $\begin{vmatrix} 1 & 0 & 0 \\ 0 & 2 & 0 \\ 0 & 0 & 3 \end{vmatrix} = 1\begin{vmatrix} 2 & 0 \\ 0 & 3 \end{vmatrix} - 0\begin{vmatrix} 0 & 0 \\ 0 & 3 \end{vmatrix} + 0\begin{vmatrix} 0 & 2 \\ 0 & 0 \end{vmatrix} = 1(6-0) = 6$

e $\begin{vmatrix} 0 & 0 & 2 \\ 0 & 1 & 0 \\ 3 & 0 & 0 \end{vmatrix} = 0\begin{vmatrix} 1 & 0 \\ 0 & 0 \end{vmatrix} - 0\begin{vmatrix} 0 & 0 \\ 3 & 0 \end{vmatrix} + 2\begin{vmatrix} 0 & 1 \\ 3 & 0 \end{vmatrix} = 2(0-3) = -6$

f $\begin{vmatrix} 4 & 1 & 3 \\ -1 & 0 & 2 \\ -1 & 1 & 1 \end{vmatrix} = 4\begin{vmatrix} 0 & 2 \\ 1 & 1 \end{vmatrix} - 1\begin{vmatrix} -1 & 2 \\ -1 & 1 \end{vmatrix} + 3\begin{vmatrix} -1 & 0 \\ -1 & 1 \end{vmatrix}$

$= 4(0-2) - 1(-1--2) + 3(-1-0)$
$= -12$

2 **a** $\begin{vmatrix} x & 2 & 9 \\ 3 & 1 & 2 \\ -1 & 0 & x \end{vmatrix} = x\begin{vmatrix} 1 & 2 \\ 0 & x \end{vmatrix} - 2\begin{vmatrix} 3 & 2 \\ -1 & x \end{vmatrix} + 9\begin{vmatrix} 3 & 1 \\ -1 & 0 \end{vmatrix}$

$= x(x) - 2(3x - -2) + 9(1)$
$= x^2 - 6x - 4 + 9$
$= x^2 - 6x + 5$
$= (x-5)(x-1)$

The matrix is singular if its determinant is 0, which occurs when $x = 5$ or $x = 1$.

b This means that when $x = 1$ or 5, the matrix does not have an inverse.

3 **a** $\begin{vmatrix} a & 0 & 0 \\ 0 & b & 0 \\ 0 & 0 & c \end{vmatrix} = a\begin{vmatrix} b & 0 \\ 0 & c \end{vmatrix} - 0\begin{vmatrix} 0 & 0 \\ 0 & c \end{vmatrix} + 0\begin{vmatrix} 0 & b \\ 0 & 0 \end{vmatrix} = a(bc-0) = abc$

b $\begin{vmatrix} 0 & x & y \\ -x & 0 & z \\ -y & -z & 0 \end{vmatrix} = 0\begin{vmatrix} 0 & z \\ -z & 0 \end{vmatrix} - x\begin{vmatrix} -x & z \\ -y & 0 \end{vmatrix} + y\begin{vmatrix} -x & 0 \\ -y & -z \end{vmatrix}$

$= -x(0 - -yz) + y(xz)$
$= -xyz + xyz$
$= 0$

c $\begin{vmatrix} a & b & c \\ b & c & a \\ c & a & b \end{vmatrix} = a\begin{vmatrix} c & a \\ a & b \end{vmatrix} - b\begin{vmatrix} b & a \\ c & b \end{vmatrix} + c\begin{vmatrix} b & c \\ c & a \end{vmatrix}$

$= a(cb - a^2) - b(b^2 - ac) + c(ba - c^2)$
$= abc - a^3 - b^3 + abc + abc - c^3$
$= 3abc - a^3 - b^3 - c^3$

4 $\begin{cases} x+2y-3z=5 \\ 2x-y-z=8 \\ kx+y+2z=14 \end{cases}$ has matrix equation $\underbrace{\begin{pmatrix} 1 & 2 & -3 \\ 2 & -1 & -1 \\ k & 1 & 2 \end{pmatrix}}_{\mathbf{A}}\underbrace{\begin{pmatrix} x \\ y \\ z \end{pmatrix}}_{\mathbf{X}} = \underbrace{\begin{pmatrix} 5 \\ 8 \\ 14 \end{pmatrix}}_{\mathbf{B}}$

$\therefore$ $\mathbf{X}=\mathbf{A}^{-1}\mathbf{B}$ has a unique solution if $|\mathbf{A}| \neq 0$.

Now $\begin{vmatrix} 1 & 2 & -3 \\ 2 & -1 & -1 \\ k & 1 & 2 \end{vmatrix} = 1\begin{vmatrix} -1 & -1 \\ 1 & 2 \end{vmatrix} - 2\begin{vmatrix} 2 & -1 \\ k & 2 \end{vmatrix} + (-3)\begin{vmatrix} 2 & -1 \\ k & 1 \end{vmatrix}$

$$\begin{aligned} &= 1(-2--1)-2(4--k)-3(2--k) \\ &= -1-8-2k-6-3k \\ &= -15-5k \end{aligned}$$

$\therefore$ there is a unique solution for all $k \neq -3$.

5 $\begin{cases} 2x-y-4z=8 \\ 3x-ky+z=1 \\ 5x-y+kz=-2 \end{cases}$ has matrix equation $\underbrace{\begin{pmatrix} 2 & -1 & -4 \\ 3 & -k & 1 \\ 5 & -1 & k \end{pmatrix}}_{\mathbf{A}}\underbrace{\begin{pmatrix} x \\ y \\ z \end{pmatrix}}_{\mathbf{X}} = \underbrace{\begin{pmatrix} 8 \\ 1 \\ -2 \end{pmatrix}}_{\mathbf{B}}$

This has a unique solution if $|\mathbf{A}| \neq 0$.

Now $\begin{vmatrix} 2 & -1 & -4 \\ 3 & -k & 1 \\ 5 & -1 & k \end{vmatrix} = 2\begin{vmatrix} -k & 1 \\ -1 & k \end{vmatrix} - (-1)\begin{vmatrix} 3 & 1 \\ 5 & k \end{vmatrix} + (-4)\begin{vmatrix} 3 & -k \\ 5 & -1 \end{vmatrix}$

$$\begin{aligned} &= 2(-k^2--1)+1(3k-5)-4(-3--5k) \\ &= -2k^2+2+3k-5+12-20k \\ &= -2k^2-17k+9 \\ &= -(2k^2+17k-9) \\ &= -(2k-1)(k+9) \end{aligned}$$

$\therefore$ there is a unique solution for all $k \neq \frac{1}{2}$ or -9.

6 **a** $\begin{vmatrix} k & 2 & 1 \\ 2 & k & 2 \\ 1 & 2 & 1 \end{vmatrix} = 0$

$$\begin{aligned} \therefore\ k\begin{vmatrix} k & 2 \\ 2 & 1 \end{vmatrix} - 2\begin{vmatrix} 2 & 2 \\ 1 & 1 \end{vmatrix} + 1\begin{vmatrix} 2 & k \\ 1 & 2 \end{vmatrix} &= 0 \\ \therefore\ k(k-4)-2(2-2)+(4-k) &= 0 \\ \therefore\ k^2-4k+4-k &= 0 \\ \therefore\ k^2-5k+4 &= 0 \\ \therefore\ (k-1)(k-4) &= 0 \end{aligned}$$

and so $k=1$ or 4

b $\begin{vmatrix} 1 & k & 3 \\ k & 1 & -1 \\ 3 & 4 & 2 \end{vmatrix} = 7 \quad \therefore\ 1\begin{vmatrix} 1 & -1 \\ 4 & 2 \end{vmatrix} - k\begin{vmatrix} k & -1 \\ 3 & 2 \end{vmatrix} + 3\begin{vmatrix} k & 1 \\ 3 & 4 \end{vmatrix} = 7$

$$\begin{aligned} \therefore\ 1(2--4)-k(2k--3)+3(4k-3) &= 7 \\ \therefore\ 6-2k^2-3k+12k-9 &= 7 \\ \therefore\ 2k^2-9k+10 &= 0 \\ \therefore\ (2k-5)(k-2) &= 0 \end{aligned}$$

and so $k=\frac{5}{2}$ or 2

EXERCISE 11D.2

1 $\begin{pmatrix} 2 & 0 & 3 \\ 1 & 5 & 2 \\ 1 & -3 & 1 \end{pmatrix}\begin{pmatrix} -11 & 9 & 15 \\ -1 & 1 & 1 \\ 8 & -6 & -10 \end{pmatrix} = \begin{pmatrix} 2 & 0 & 0 \\ 0 & 2 & 0 \\ 0 & 0 & 2 \end{pmatrix} = 2\mathbf{I}$

$$\therefore \begin{pmatrix} 2 & 0 & 3 \\ 1 & 5 & 2 \\ 1 & -3 & 1 \end{pmatrix} \times \tfrac{1}{2}\begin{pmatrix} -11 & 9 & 15 \\ -1 & 1 & 1 \\ 8 & -6 & -10 \end{pmatrix} = \mathbf{I} \quad \text{and so} \quad \begin{pmatrix} 2 & 0 & 3 \\ 1 & 5 & 2 \\ 1 & -3 & 1 \end{pmatrix}^{-1} = \begin{pmatrix} -\frac{11}{2} & \frac{9}{2} & \frac{15}{2} \\ -\frac{1}{2} & \frac{1}{2} & \frac{1}{2} \\ 4 & -3 & -5 \end{pmatrix}$$

2 **a** $\mathbf{A}^{-1} = \begin{pmatrix} \frac{5}{4} & \frac{3}{4} & -\frac{7}{4} \\ -\frac{1}{4} & -\frac{3}{4} & \frac{3}{4} \\ -\frac{3}{4} & -\frac{1}{4} & \frac{5}{4} \end{pmatrix}$ **b** $\mathbf{A}^{-1} = \begin{pmatrix} -\frac{11}{2} & \frac{9}{2} & \frac{15}{2} \\ -\frac{1}{2} & \frac{1}{2} & \frac{1}{2} \\ 4 & -3 & -5 \end{pmatrix}$

3 **a** $\mathbf{B}^{-1} \approx \begin{pmatrix} 0.050\,23 & -0.011\,48 & -0.066\,34 \\ 4.212 \times 10^{-4} & 0.013\,53 & 0.027\,75 \\ -0.029\,90 & 0.039\,33 & 0.030\,06 \end{pmatrix} \approx \begin{pmatrix} 0.050 & -0.011 & -0.066 \\ 0.000 & 0.014 & 0.028 \\ -0.030 & 0.039 & 0.030 \end{pmatrix}$

b $\mathbf{B}^{-1} \approx \begin{pmatrix} 1.596 & -0.9964 & -0.1686 \\ -3.224 & 1.925 & 0.6291 \\ 2.000 & -1.086 & -0.3958 \end{pmatrix} \approx \begin{pmatrix} 1.596 & -0.996 & -0.169 \\ -3.224 & 1.925 & 0.629 \\ 2.000 & -1.086 & -0.396 \end{pmatrix}$

4 Check that $\mathbf{AA}^{-1} = \mathbf{I}$ in **2**, and $\mathbf{BB}^{-1} = \mathbf{I}$ in **3**.

EXERCISE 11E

1 **a** $\left.\begin{array}{r} x - y - z = 2 \\ x + y + 3z = 7 \\ 9x - y - 3z = -1 \end{array}\right\}$ has matrix equation $\begin{pmatrix} 1 & -1 & -1 \\ 1 & 1 & 3 \\ 9 & -1 & -3 \end{pmatrix}\begin{pmatrix} x \\ y \\ z \end{pmatrix} = \begin{pmatrix} 2 \\ 7 \\ -1 \end{pmatrix}$

b $\left.\begin{array}{r} 2x + y - z = 3 \\ y + 2z = 6 \\ x - y + z = 13 \end{array}\right\}$ has matrix equation $\begin{pmatrix} 2 & 1 & -1 \\ 0 & 1 & 2 \\ 1 & -1 & 1 \end{pmatrix}\begin{pmatrix} x \\ y \\ z \end{pmatrix} = \begin{pmatrix} 3 \\ 6 \\ 13 \end{pmatrix}$

c $\left.\begin{array}{r} a + b - c = 7 \\ a - b + c = 6 \\ 2a + b - 3c = -2 \end{array}\right\}$ has matrix equation $\begin{pmatrix} 1 & 1 & -1 \\ 1 & -1 & 1 \\ 2 & 1 & -3 \end{pmatrix}\begin{pmatrix} a \\ b \\ c \end{pmatrix} = \begin{pmatrix} 7 \\ 6 \\ -2 \end{pmatrix}$

2 $\mathbf{AB} = \begin{pmatrix} 2 & 1 & -1 \\ -1 & 2 & 1 \\ 0 & 6 & 1 \end{pmatrix}\begin{pmatrix} 4 & 7 & -3 \\ -1 & -2 & 1 \\ 6 & 12 & -5 \end{pmatrix}$

$$= \begin{pmatrix} 8-1-6 & 14-2-12 & -6+1+5 \\ -4-2+6 & -7-4+12 & 3+2-5 \\ 0-6+6 & 0-12+12 & 0+6-5 \end{pmatrix}$$

$$= \begin{pmatrix} 1 & 0 & 0 \\ 0 & 1 & 0 \\ 0 & 0 & 1 \end{pmatrix} \qquad \therefore \ \mathbf{AB} = \mathbf{I} \text{ and so } \mathbf{A} = \mathbf{B}^{-1}$$

$\left.\begin{array}{r} 4a + 7b - 3c = -8 \\ -a - 2b + c = 3 \\ 6a + 12b - 5c = -15 \end{array}\right\}$ has matrix equation $\begin{pmatrix} 4 & 7 & -3 \\ -1 & -2 & 1 \\ 6 & 12 & -5 \end{pmatrix}\begin{pmatrix} a \\ b \\ c \end{pmatrix} = \begin{pmatrix} -8 \\ 3 \\ -15 \end{pmatrix}$

$$\therefore \begin{pmatrix} a \\ b \\ c \end{pmatrix} = \begin{pmatrix} 4 & 7 & -3 \\ -1 & -2 & 1 \\ 6 & 12 & -5 \end{pmatrix}^{-1}\begin{pmatrix} -8 \\ 3 \\ -15 \end{pmatrix}$$

$$\therefore \begin{pmatrix} a \\ b \\ c \end{pmatrix} = \begin{pmatrix} 2 & 1 & -1 \\ -1 & 2 & 1 \\ 0 & 6 & 1 \end{pmatrix}\begin{pmatrix} -8 \\ 3 \\ -15 \end{pmatrix}$$

$$\therefore \begin{pmatrix} a \\ b \\ c \end{pmatrix} = \begin{pmatrix} -16+3+15 \\ 8+6-15 \\ 0+18-15 \end{pmatrix} = \begin{pmatrix} 2 \\ -1 \\ 3 \end{pmatrix} \qquad \therefore \ a = 2, \ b = -1, \ c = 3$$

3 $\mathbf{MN} = \begin{pmatrix} 5 & 3 & -7 \\ -1 & -3 & 3 \\ -3 & -1 & 5 \end{pmatrix}\begin{pmatrix} 3 & 2 & 3 \\ 1 & -1 & 2 \\ 2 & 1 & 3 \end{pmatrix}$

$= \begin{pmatrix} 15+3-14 & 10-3-7 & 15+6-21 \\ -3-3+6 & -2+3+3 & -3-6+9 \\ -9-1+10 & -6+1+5 & -9-2+15 \end{pmatrix}$

$= \begin{pmatrix} 4 & 0 & 0 \\ 0 & 4 & 0 \\ 0 & 0 & 4 \end{pmatrix} = 4\mathbf{I}$ $\quad \therefore \quad (\frac{1}{4}\mathbf{M})\mathbf{N} = \mathbf{I}$ and so $\frac{1}{4}\mathbf{M} = \mathbf{N}^{-1}$

Now $\left.\begin{array}{r} 3u+2v+3w=18 \\ u-v+2w=6 \\ 2u+v+3w=16 \end{array}\right\}$ has matrix equation $\begin{pmatrix} 3 & 2 & 3 \\ 1 & -1 & 2 \\ 2 & 1 & 3 \end{pmatrix}\begin{pmatrix} u \\ v \\ w \end{pmatrix} = \begin{pmatrix} 18 \\ 6 \\ 16 \end{pmatrix}$

$\therefore \quad \begin{pmatrix} u \\ v \\ w \end{pmatrix} = \begin{pmatrix} 3 & 2 & 3 \\ 1 & -1 & 2 \\ 2 & 1 & 3 \end{pmatrix}^{-1}\begin{pmatrix} 18 \\ 6 \\ 16 \end{pmatrix} = \frac{1}{4}\begin{pmatrix} 5 & 3 & -7 \\ -1 & -3 & 3 \\ -3 & -1 & 5 \end{pmatrix}\begin{pmatrix} 18 \\ 6 \\ 16 \end{pmatrix}$

$\therefore \quad \begin{pmatrix} u \\ v \\ w \end{pmatrix} = \frac{1}{4}\begin{pmatrix} 90+18-112 \\ -18-18+48 \\ -54-6+80 \end{pmatrix} = \frac{1}{4}\begin{pmatrix} -4 \\ 12 \\ 20 \end{pmatrix} = \begin{pmatrix} -1 \\ 3 \\ 5 \end{pmatrix} \quad \therefore \quad u=-1, \quad v=3, \quad w=5$

4 **a** $\begin{pmatrix} 3 & 2 & -1 \\ 1 & -1 & 2 \\ 2 & 3 & -1 \end{pmatrix}\begin{pmatrix} x \\ y \\ z \end{pmatrix} = \begin{pmatrix} 14 \\ -8 \\ 13 \end{pmatrix}$

$\therefore \quad \begin{pmatrix} x \\ y \\ z \end{pmatrix} = \begin{pmatrix} 3 & 2 & -1 \\ 1 & -1 & 2 \\ 2 & 3 & -1 \end{pmatrix}^{-1}\begin{pmatrix} 14 \\ -8 \\ 13 \end{pmatrix}$

Using technology, $x = \frac{23}{10}$, $y = \frac{13}{10}$, $z = -\frac{9}{2}$

b $\begin{pmatrix} 1 & -1 & -2 \\ 5 & 1 & 2 \\ 3 & -4 & -1 \end{pmatrix}\begin{pmatrix} x \\ y \\ z \end{pmatrix} = \begin{pmatrix} 4 \\ -6 \\ 17 \end{pmatrix}$

$\therefore \quad \begin{pmatrix} x \\ y \\ z \end{pmatrix} = \begin{pmatrix} 1 & -1 & -2 \\ 5 & 1 & 2 \\ 3 & -4 & -1 \end{pmatrix}^{-1}\begin{pmatrix} 4 \\ -6 \\ 17 \end{pmatrix}$

Using technology, $x = -\frac{1}{3}$, $y = -\frac{95}{21}$, $z = \frac{2}{21}$

c $\begin{pmatrix} 1 & 3 & -1 \\ 2 & 1 & 1 \\ 1 & -1 & -2 \end{pmatrix}\begin{pmatrix} x \\ y \\ z \end{pmatrix} = \begin{pmatrix} 15 \\ 7 \\ 0 \end{pmatrix}$

$\therefore \quad \begin{pmatrix} x \\ y \\ z \end{pmatrix} = \begin{pmatrix} 1 & 3 & -1 \\ 2 & 1 & 1 \\ 1 & -1 & -2 \end{pmatrix}^{-1}\begin{pmatrix} 15 \\ 7 \\ 0 \end{pmatrix}$

Using technology, $x = 2$, $y = 4$, $z = -1$

5 **a** $x=2, \ y=-1, \ z=5$ **b** $x=4, \ y=-2, \ z=1$

c $x=4, \ y=-3, \ z=2$ **d** $x=4, \ y=6, \ z=-7$

e $x=3, \ y=11, \ z=-7$ **f** $x \approx 0.326, \ y \approx 7.65, \ z \approx 4.16$

6 **a** Let x be the cost of a football in dollars,
y be the cost of baseball in dollars, and
z be the cost of a basketball in dollars.

b Using technology, $x=14, \ y=11, \ z=17$.

Cost of 4 footballs and 5 baseballs is $4x+5y = 4(14)+5(11) = \$111$

$\therefore$ amount left for basketballs is $\$315 - \$111 = \$204$

Number of basketballs bought $= \frac{204}{17} = 12$

7 **a** System of equations is:
$$2x+3y+8z=352$$
$$x+5y+4z=274$$
$$x+2y+11z=351$$

b Using technology, $x=42,\ y=28,\ z=23$.
So, the salaries are: manager €42 000, clerk €28 000 and labourer €23 000.

c Salary bill is $3x+8y+37z$
$=3(42)+8(28)+37(23)$
$=1201$ thousands of euros or €1 201 000

8 **a** Let x be the cost in dollars of 1 kg of cashews,
y be the cost in dollars of 1 kg of macadamias, and
z be the cost in dollars of 1 kg of Brazil nuts.
The cost of 1 kg of mix A is $0.5x+0.3y+0.2z=12.5$,
the cost of 1 kg of mix B is $0.2x+0.4y+0.4z=12.4$,
the cost of 1 kg of mix C is $0.6x+0.1y+0.3z=11.7$.
Using technology, $x=12,\ y=15$ and $z=10$
So, the cost of 1 kg of cashews is $12, the cost of 1 kg of macadamias is $15 and the cost of 1 kg of Brazil nuts is $10.

b Cost per kg of 400 g cashews, 200 g macadamias and 400 g Brazil nuts
$=0.4\times 12+0.2\times 15+0.4\times 10$ dollars
$=\$11.80$

9 **a** Number of students who study Chemistry is $\frac{1}{3}p+\frac{1}{3}q+\frac{2}{5}r=27$ .... (1)
number of students who study Maths is $\frac{1}{2}p+\frac{2}{3}q+\frac{1}{5}r=35$ .... (2)
number of students who study Geography is $\frac{1}{4}p+\frac{1}{3}q+\frac{3}{5}r=30$ .... (3)
The required system of equations is
$$5p+5q+6r=405 \quad \{(1)\times 15\}$$
$$15p+20q+6r=1050 \quad \{(2)\times 30\}$$
$$15p+20q+36r=1800 \quad \{(3)\times 60\}$$

b Using technology, $p=24,\ q=27,\ r=25$.

10 **a** As t is the number of years after 2006, then
profit in year 2006 is $P(0)=b+\dfrac{c}{4}=160\,000$
profit in year 2007 is $P(1)=a+b+\dfrac{c}{5}=198\,000$
profit in year 2008 is $P(2)=2a+b+\dfrac{c}{6}=240\,000$
Using technology, $a=50\,000,\ b=100\,000$ and $c=240\,000$.

b Using the model given, the profit in 2005 would be
$P(-1)=-a+b+\dfrac{c}{3}=-50\,000+100\,000+80\,000=130\,000$
$\therefore$ the profit in 2005 does fit the model.

c Predicted profit in 2009 is $P(3)=3a+b+\dfrac{c}{7}=3(50\,000)+100\,000+\dfrac{240\,000}{7}$
$\approx £284\,000$

Predicted profit in 2011 is $P(5)=5a+b+\dfrac{c}{9}=5(50\,000)+100\,000+\dfrac{240\,000}{9}$
$\approx £377\,000$

REVIEW SET 11A

1 **a** $\mathbf{A}+\mathbf{B} = \begin{pmatrix} 3 & 2 \\ 0 & -1 \end{pmatrix} + \begin{pmatrix} 1 & 0 \\ -2 & 4 \end{pmatrix} = \begin{pmatrix} 4 & 2 \\ -2 & 3 \end{pmatrix}$

b $3\mathbf{A} = 3\begin{pmatrix} 3 & 2 \\ 0 & -1 \end{pmatrix} = \begin{pmatrix} 9 & 6 \\ 0 & -3 \end{pmatrix}$

c $-2\mathbf{B} = -2\begin{pmatrix} 1 & 0 \\ -2 & 4 \end{pmatrix} = \begin{pmatrix} -2 & 0 \\ 4 & -8 \end{pmatrix}$

d $\mathbf{A}-\mathbf{B} = \begin{pmatrix} 3 & 2 \\ 0 & -1 \end{pmatrix} - \begin{pmatrix} 1 & 0 \\ -2 & 4 \end{pmatrix} = \begin{pmatrix} 2 & 2 \\ 2 & -5 \end{pmatrix}$

e $\mathbf{B}-2\mathbf{A} = \begin{pmatrix} 1 & 0 \\ -2 & 4 \end{pmatrix} - \begin{pmatrix} 6 & 4 \\ 0 & -2 \end{pmatrix} = \begin{pmatrix} -5 & -4 \\ -2 & 6 \end{pmatrix}$

f $3\mathbf{A}-2\mathbf{B} = 3\begin{pmatrix} 3 & 2 \\ 0 & -1 \end{pmatrix} - 2\begin{pmatrix} 1 & 0 \\ -2 & 4 \end{pmatrix} = \begin{pmatrix} 7 & 6 \\ 4 & -11 \end{pmatrix}$

g $\mathbf{AB} = \begin{pmatrix} 3 & 2 \\ 0 & -1 \end{pmatrix}\begin{pmatrix} 1 & 0 \\ -2 & 4 \end{pmatrix} = \begin{pmatrix} -1 & 8 \\ 2 & -4 \end{pmatrix}$

h $\mathbf{BA} = \begin{pmatrix} 1 & 0 \\ -2 & 4 \end{pmatrix}\begin{pmatrix} 3 & 2 \\ 0 & -1 \end{pmatrix} = \begin{pmatrix} 3 & 2 \\ -6 & -8 \end{pmatrix}$

i $\mathbf{A}^{-1} = \frac{1}{-3}\begin{pmatrix} -1 & -2 \\ 0 & 3 \end{pmatrix} = \begin{pmatrix} \frac{1}{3} & \frac{2}{3} \\ 0 & -1 \end{pmatrix}$

j $\mathbf{A}^2 = \begin{pmatrix} 3 & 2 \\ 0 & -1 \end{pmatrix}\begin{pmatrix} 3 & 2 \\ 0 & -1 \end{pmatrix} = \begin{pmatrix} 9 & 4 \\ 0 & 1 \end{pmatrix}$

k $\mathbf{ABA} = (\mathbf{AB})\mathbf{A} = \begin{pmatrix} -1 & 8 \\ 2 & -4 \end{pmatrix}\begin{pmatrix} 3 & 2 \\ 0 & -1 \end{pmatrix} = \begin{pmatrix} -3 & -10 \\ 6 & 8 \end{pmatrix}$

l $(\mathbf{AB})^{-1} = \begin{pmatrix} -1 & 8 \\ 2 & -4 \end{pmatrix}^{-1} = \frac{1}{4-16}\begin{pmatrix} -4 & -8 \\ -2 & -1 \end{pmatrix} = \frac{1}{-12}\begin{pmatrix} -4 & -8 \\ -2 & -1 \end{pmatrix} = \begin{pmatrix} \frac{1}{3} & \frac{2}{3} \\ \frac{1}{6} & \frac{1}{12} \end{pmatrix}$

2 **a** Equating corresponding elements,

$a=-a$
$b-2=3$
$c=2-c \quad \therefore \quad a=0, \quad b=5$
$d=-4 \qquad\qquad c=1, \quad d=-4$

b Equating corresponding elements,

$3+b=a$
$2a-a=2$
$b+c=2 \quad \therefore \quad a=2, \quad b=-1$
$-2+d=6 \qquad\qquad c=3, \quad d=8$

3 **a**
$\mathbf{B}-\mathbf{Y}=\mathbf{A}$
$\therefore \ -\mathbf{Y}=\mathbf{A}-\mathbf{B}$
$\therefore \ \mathbf{Y}=-(\mathbf{A}-\mathbf{B})$
$\therefore \ \mathbf{Y}=\mathbf{B}-\mathbf{A}$

b
$2\mathbf{Y}+\mathbf{C}=\mathbf{D}$
$\therefore \ 2\mathbf{Y}=\mathbf{D}-\mathbf{C}$
$\therefore \ \mathbf{Y}=\frac{1}{2}(\mathbf{D}-\mathbf{C})$

c
$\mathbf{AY}=\mathbf{B}$
$\therefore \ \mathbf{A}^{-1}\mathbf{AY}=\mathbf{A}^{-1}\mathbf{B}$
$\therefore \ \mathbf{IY}=\mathbf{A}^{-1}\mathbf{B}$
$\therefore \ \mathbf{Y}=\mathbf{A}^{-1}\mathbf{B}$

d
$\mathbf{YB}=\mathbf{C}$
$\therefore \ \mathbf{YBB}^{-1}=\mathbf{CB}^{-1}$
$\therefore \ \mathbf{YI}=\mathbf{CB}^{-1}$
$\therefore \ \mathbf{Y}=\mathbf{CB}^{-1}$

e
$\mathbf{C}-\mathbf{AY}=\mathbf{B}$
$\therefore \ -\mathbf{AY}=\mathbf{B}-\mathbf{C}$
$\therefore \ \mathbf{AY}=\mathbf{C}-\mathbf{B}$
$\therefore \ \mathbf{A}^{-1}\mathbf{AY}=\mathbf{A}^{-1}(\mathbf{C}-\mathbf{B})$
$\therefore \ \mathbf{Y}=\mathbf{A}^{-1}(\mathbf{C}-\mathbf{B})$

f
$\mathbf{AY}^{-1}=\mathbf{B}$
$\therefore \ \mathbf{A}^{-1}\mathbf{AY}^{-1}=\mathbf{A}^{-1}\mathbf{B}$
$\therefore \ \mathbf{Y}^{-1}=\mathbf{A}^{-1}\mathbf{B}$
$\therefore \ (\mathbf{Y}^{-1})^{-1}=(\mathbf{A}^{-1}\mathbf{B})^{-1}$
$\therefore \ \mathbf{Y}=\mathbf{B}^{-1}(\mathbf{A}^{-1})^{-1}$
$\therefore \ \mathbf{Y}=\mathbf{B}^{-1}\mathbf{A}$

4 If $\mathbf{P}+\mathbf{Q}=\mathbf{O}$, then equating corresponding elements gives:

$$\begin{aligned} a+(-3)&=0\\ 2+a-5&=0\\ 5+(-5)&=0\\ -3+a&=0 \end{aligned}$$

$a=3$ is the only value of a that satisfies all these equations, so $a=3$.

5 $\begin{pmatrix}1&0\\1&1\end{pmatrix}\begin{pmatrix}a&b\\c&d\end{pmatrix}=\begin{pmatrix}1&0\\1&1\end{pmatrix}$

$\therefore \begin{pmatrix}a&b\\a+c&b+d\end{pmatrix}=\begin{pmatrix}1&0\\1&1\end{pmatrix}$

$\therefore$ $a=1$, $b=0$, $a+c=1$, $b+d=1$

So, $a=1$, $b=0$, $c=0$, $d=1$

$\therefore$ matrix is $\begin{pmatrix}1&0\\0&1\end{pmatrix}=\mathbf{I}$

6 **a** $\mathbf{AB}=\begin{pmatrix}4&3&2\end{pmatrix}\begin{pmatrix}1\\2\\0\end{pmatrix}=(10)$

b $\mathbf{BA}=\begin{pmatrix}1\\2\\0\end{pmatrix}\begin{pmatrix}4&3&2\end{pmatrix}=\begin{pmatrix}4&3&2\\8&6&4\\0&0&0\end{pmatrix}$

c $\mathbf{AC}=\begin{pmatrix}4&3&2\end{pmatrix}\begin{pmatrix}1&2&3\\3&2&1\\1&2&3\end{pmatrix}=\begin{pmatrix}15&18&21\end{pmatrix}$

d $\mathbf{CA}$ does not exist as $\mathbf{C}$ is 3×3 and $\mathbf{A}$ is 1×3.

e $\mathbf{CB}=\begin{pmatrix}1&2&3\\3&2&1\\1&2&3\end{pmatrix}\begin{pmatrix}1\\2\\0\end{pmatrix}=\begin{pmatrix}5\\7\\5\end{pmatrix}$

7 $\mathbf{A}=2\mathbf{A}^{-1}$

a $$\begin{aligned}\mathbf{A}^2&=\mathbf{A}\times\mathbf{A}\\&=\mathbf{A}(2\mathbf{A}^{-1})\\&=2\mathbf{A}\mathbf{A}^{-1}\\&=2\mathbf{I}\end{aligned}$$

b $$\begin{aligned}(\mathbf{A}-\mathbf{I})(\mathbf{A}+3\mathbf{I})&=(\mathbf{A}-\mathbf{I})\mathbf{A}+(\mathbf{A}-\mathbf{I})3\mathbf{I}\\&=\mathbf{A}^2-\mathbf{IA}+3\mathbf{AI}-3\mathbf{I}^2\\&=2\mathbf{I}-\mathbf{A}+3\mathbf{A}-3\mathbf{I}\quad\{\text{from } \mathbf{a}\}\\&=2\mathbf{A}-\mathbf{I}\end{aligned}$$

8 $$\mathbf{AB}=\begin{pmatrix}1&2&3\\2&5&7\\-2&-4&-5\end{pmatrix}\begin{pmatrix}3&-2&-1\\-4&1&-1\\2&0&1\end{pmatrix}=\begin{pmatrix}3-8+6&-2+2+0&-1-2+3\\6-20+14&-4+5+0&-2-5+7\\-6+16-10&4-4+0&2+4-5\end{pmatrix}=\begin{pmatrix}1&0&0\\0&1&0\\0&0&1\end{pmatrix}$$

$$\mathbf{BA}=\begin{pmatrix}3&-2&-1\\-4&1&-1\\2&0&1\end{pmatrix}\begin{pmatrix}1&2&3\\2&5&7\\-2&-4&-5\end{pmatrix}=\begin{pmatrix}3-4+2&6-10+4&9-14+5\\-4+2+2&-8+5+4&-12+7+5\\2+0-2&4+0-4&6+0-5\end{pmatrix}=\begin{pmatrix}1&0&0\\0&1&0\\0&0&1\end{pmatrix}$$

$\therefore$ $\mathbf{AB}=\mathbf{BA}=\mathbf{I}$ $\quad\therefore$ $\mathbf{A}^{-1}=\mathbf{B}$

9 In matrix form $\begin{pmatrix}k&3\\1&k+2\end{pmatrix}\begin{pmatrix}x\\y\end{pmatrix}=\begin{pmatrix}-6\\2\end{pmatrix}$ has a unique solution if $\begin{vmatrix}k&3\\1&k+2\end{vmatrix}\neq0$

$\therefore\ k^2+2k-3\neq0$

$\therefore\ (k+3)(k-1)\neq0$

$\therefore\ k\neq-3$ or 1

So, the system of equations has a unique solution for all $k\in\mathbb{R}$, $k\neq1$ or -3.

10 $\begin{pmatrix} 2 & -1 \\ 3 & 4 \end{pmatrix} \mathbf{X} = \begin{pmatrix} 3 & 5 & 1 \\ -1 & 13 & 18 \end{pmatrix}$

$$\therefore \quad \mathbf{X} = \begin{pmatrix} 2 & -1 \\ 3 & 4 \end{pmatrix}^{-1} \begin{pmatrix} 3 & 5 & 1 \\ -1 & 13 & 18 \end{pmatrix}$$
$$= \tfrac{1}{11} \begin{pmatrix} 4 & 1 \\ -3 & 2 \end{pmatrix} \begin{pmatrix} 3 & 5 & 1 \\ -1 & 13 & 18 \end{pmatrix}$$
$$= \tfrac{1}{11} \begin{pmatrix} 11 & 33 & 22 \\ -11 & 11 & 33 \end{pmatrix}$$
$$\therefore \quad \mathbf{X} = \begin{pmatrix} 1 & 3 & 2 \\ -1 & 1 & 3 \end{pmatrix}$$

11 $\begin{vmatrix} m & 3 \\ m & m \end{vmatrix} = 18$

$\therefore \quad m^2 - 3m = 18$

$\therefore \quad m^2 - 3m - 18 = 0$

$\therefore \quad (m+3)(m-6) = 0$

$\therefore \quad m = -3, 6$

12 $\begin{vmatrix} a+b & c & c \\ a & b+c & a \\ b & b & c+a \end{vmatrix} = (a+b)\begin{vmatrix} b+c & a \\ b & c+a \end{vmatrix} - c\begin{vmatrix} a & a \\ b & c+a \end{vmatrix} + c\begin{vmatrix} a & b+c \\ b & b \end{vmatrix}$

$$= (a+b)\,[(b+c)(c+a) - ab] - c\,[a(c+a) - ab] + c\,[ab - (b+c)b]$$
$$= (a+b)(bc + ab + c^2 + ac - ab) - ac^2 - a^2c + abc + abc - b^2c - bc^2$$
$$= abc + b^2c + ac^2 + bc^2 + a^2c + abc - ac^2 - a^2c + abc + abc - b^2c - bc^2$$
$$= 4abc$$

13 $\mathbf{MN} = \begin{pmatrix} -6 & -4 \\ 13 & -3 \end{pmatrix}$

$\therefore \quad \mathbf{M}\begin{pmatrix} 2 & -1 \\ 3 & 2 \end{pmatrix} = \begin{pmatrix} -6 & -4 \\ 13 & -3 \end{pmatrix}$

$$\therefore \quad \mathbf{M} = \begin{pmatrix} -6 & -4 \\ 13 & -3 \end{pmatrix}\begin{pmatrix} 2 & -1 \\ 3 & 2 \end{pmatrix}^{-1}$$
$$= \begin{pmatrix} -6 & -4 \\ 13 & -3 \end{pmatrix} \frac{1}{4-(-3)} \begin{pmatrix} 2 & 1 \\ -3 & 2 \end{pmatrix}$$
$$= \tfrac{1}{7}\begin{pmatrix} 0 & -14 \\ 35 & 7 \end{pmatrix}$$
$$= \begin{pmatrix} 0 & -2 \\ 5 & 1 \end{pmatrix}$$

14 **a** $\begin{pmatrix} 2 & 1 & 1 \\ 1 & 1 & 1 \\ 2 & 2 & 1 \end{pmatrix}\begin{pmatrix} a & b & 0 \\ b & 0 & a \\ 0 & 2 & b \end{pmatrix} = \mathbf{I}$

$$\therefore \quad \begin{pmatrix} 2a+b & 2b+2 & a+b \\ a+b & b+2 & a+b \\ 2a+2b & 2b+2 & 2a+b \end{pmatrix} = \begin{pmatrix} 1 & 0 & 0 \\ 0 & 1 & 0 \\ 0 & 0 & 1 \end{pmatrix}$$

$\therefore \quad 2a+b=1, \quad 2b+2=0, \quad a+b=0, \quad b+2=1$ and $2a+2b=0$.

$b=-1$ and $a=1$ satisfy these equations.

$\therefore \quad a=1, \ b=-1.$

$$\therefore \quad \begin{pmatrix} 2 & 1 & 1 \\ 1 & 1 & 1 \\ 2 & 2 & 1 \end{pmatrix}^{-1} = \begin{pmatrix} 1 & -1 & 0 \\ -1 & 0 & 1 \\ 0 & 2 & -1 \end{pmatrix}$$

b Writing the system of equations in matrix form,

$$\begin{pmatrix} 2 & 1 & 1 \\ 1 & 1 & 1 \\ 2 & 2 & 1 \end{pmatrix}\begin{pmatrix} x \\ y \\ z \end{pmatrix} = \begin{pmatrix} 1 \\ 6 \\ 5 \end{pmatrix}$$

$$\therefore \quad \begin{pmatrix} 1 & -1 & 0 \\ -1 & 0 & 1 \\ 0 & 2 & -1 \end{pmatrix}\begin{pmatrix} 2 & 1 & 1 \\ 1 & 1 & 1 \\ 2 & 2 & 1 \end{pmatrix}\begin{pmatrix} x \\ y \\ z \end{pmatrix} = \begin{pmatrix} 1 & -1 & 0 \\ -1 & 0 & 1 \\ 0 & 2 & -1 \end{pmatrix}\begin{pmatrix} 1 \\ 6 \\ 5 \end{pmatrix}$$

$$\therefore \quad \begin{pmatrix} x \\ y \\ z \end{pmatrix} = \begin{pmatrix} -5 \\ 4 \\ 7 \end{pmatrix}$$ and so $x=-5, \ y=4, \ z=7.$

REVIEW SET 11B

1 This system in matrix form is $\begin{pmatrix} 3 & -1 & 2 \\ 2 & 3 & -1 \\ 1 & -2 & 3 \end{pmatrix}\begin{pmatrix} x \\ y \\ z \end{pmatrix} = \begin{pmatrix} 8 \\ -3 \\ 9 \end{pmatrix}$

$$\therefore \begin{pmatrix} x \\ y \\ z \end{pmatrix} = \begin{pmatrix} 3 & -1 & 2 \\ 2 & 3 & -1 \\ 1 & -2 & 3 \end{pmatrix}^{-1}\begin{pmatrix} 8 \\ -3 \\ 9 \end{pmatrix}$$

$$\therefore \begin{pmatrix} x \\ y \\ z \end{pmatrix} = \begin{pmatrix} 1 \\ -1 \\ 2 \end{pmatrix}$$

So, $x = 1$, $y = -1$, $z = 2$. {using technology}

2 **a** Given $\begin{cases} 3x - 4y = 2 \\ 5x + 2y = -1 \end{cases}$ we can write

$$\begin{pmatrix} 3 & -4 \\ 5 & 2 \end{pmatrix}\begin{pmatrix} x \\ y \end{pmatrix} = \begin{pmatrix} 2 \\ -1 \end{pmatrix}$$

$$\therefore \begin{pmatrix} x \\ y \end{pmatrix} = \begin{pmatrix} 3 & -4 \\ 5 & 2 \end{pmatrix}^{-1}\begin{pmatrix} 2 \\ -1 \end{pmatrix} = \tfrac{1}{26}\begin{pmatrix} 2 & 4 \\ -5 & 3 \end{pmatrix}\begin{pmatrix} 2 \\ -1 \end{pmatrix} = \tfrac{1}{26}\begin{pmatrix} 0 \\ -13 \end{pmatrix} = \begin{pmatrix} 0 \\ -\frac{1}{2} \end{pmatrix}$$

$\therefore$ $x = 0$, $y = -\frac{1}{2}$

b Given $\begin{cases} 4x - y = 5 \\ 2x + 3y = 9 \end{cases}$ we can write

$$\begin{pmatrix} 4 & -1 \\ 2 & 3 \end{pmatrix}\begin{pmatrix} x \\ y \end{pmatrix} = \begin{pmatrix} 5 \\ 9 \end{pmatrix}$$

$$\therefore \begin{pmatrix} x \\ y \end{pmatrix} = \begin{pmatrix} 4 & -1 \\ 2 & 3 \end{pmatrix}^{-1}\begin{pmatrix} 5 \\ 9 \end{pmatrix} = \tfrac{1}{14}\begin{pmatrix} 3 & 1 \\ -2 & 4 \end{pmatrix}\begin{pmatrix} 5 \\ 9 \end{pmatrix} = \tfrac{1}{14}\begin{pmatrix} 24 \\ 26 \end{pmatrix} = \begin{pmatrix} \frac{12}{7} \\ \frac{13}{7} \end{pmatrix}$$

$\therefore$ $x = \frac{12}{7}$, $y = \frac{13}{7}$

c $\mathbf{X}\begin{pmatrix} 3 & 4 \\ 1 & 1 \end{pmatrix} = \begin{pmatrix} 5 & 4 \\ 0 & -2 \end{pmatrix}$

$$\therefore \mathbf{X} = \begin{pmatrix} 5 & 4 \\ 0 & -2 \end{pmatrix}\begin{pmatrix} 3 & 4 \\ 1 & 1 \end{pmatrix}^{-1}$$

$$\therefore \mathbf{X} = \begin{pmatrix} 5 & 4 \\ 0 & -2 \end{pmatrix}\tfrac{1}{-1}\begin{pmatrix} 1 & -4 \\ -1 & 3 \end{pmatrix}$$

$$\therefore \mathbf{X} = \begin{pmatrix} 5 & 4 \\ 0 & -2 \end{pmatrix}\begin{pmatrix} -1 & 4 \\ 1 & -3 \end{pmatrix}$$

$$\therefore \mathbf{X} = \begin{pmatrix} -1 & 8 \\ -2 & 6 \end{pmatrix}$$

d $\begin{pmatrix} 2 & 0 \\ -1 & 1 \end{pmatrix}\mathbf{X} = \begin{pmatrix} -1 \\ 2 \end{pmatrix}$

$$\therefore \mathbf{X} = \begin{pmatrix} 2 & 0 \\ -1 & 1 \end{pmatrix}^{-1}\begin{pmatrix} -1 \\ 2 \end{pmatrix}$$

$$\therefore \mathbf{X} = \tfrac{1}{2}\begin{pmatrix} 1 & 0 \\ 1 & 2 \end{pmatrix}\begin{pmatrix} -1 \\ 2 \end{pmatrix}$$

$$\therefore \mathbf{X} = \tfrac{1}{2}\begin{pmatrix} -1 \\ 3 \end{pmatrix}$$

$$\therefore \mathbf{X} = \begin{pmatrix} -\frac{1}{2} \\ \frac{3}{2} \end{pmatrix}$$

e $\begin{pmatrix} 1 & 1 \\ 1 & -2 \end{pmatrix}\mathbf{X} = \begin{pmatrix} 5 \\ 4 \end{pmatrix}$

$$\therefore \mathbf{X} = \begin{pmatrix} 1 & 1 \\ 1 & -2 \end{pmatrix}^{-1}\begin{pmatrix} 5 \\ 4 \end{pmatrix}$$

$$\therefore \mathbf{X} = \tfrac{1}{-3}\begin{pmatrix} -2 & -1 \\ -1 & 1 \end{pmatrix}\begin{pmatrix} 5 \\ 4 \end{pmatrix}$$

$$\therefore \mathbf{X} = \tfrac{1}{-3}\begin{pmatrix} -14 \\ -1 \end{pmatrix}$$

$$\therefore \mathbf{X} = \begin{pmatrix} \frac{14}{3} \\ \frac{1}{3} \end{pmatrix}$$

f $\begin{pmatrix} 1 & 1 \\ -1 & 1 \end{pmatrix}\mathbf{X}\begin{pmatrix} 2 & 1 \\ 1 & -1 \end{pmatrix} = \begin{pmatrix} 5 & 1 \\ 0 & 3 \end{pmatrix}$

$$\therefore \mathbf{X} = \begin{pmatrix} 1 & 1 \\ -1 & 1 \end{pmatrix}^{-1}\begin{pmatrix} 5 & 1 \\ 0 & 3 \end{pmatrix}\begin{pmatrix} 2 & 1 \\ 1 & -1 \end{pmatrix}^{-1}$$

$$\therefore \mathbf{X} = \tfrac{1}{2}\begin{pmatrix} 1 & -1 \\ 1 & 1 \end{pmatrix}\begin{pmatrix} 5 & 1 \\ 0 & 3 \end{pmatrix}\tfrac{1}{-3}\begin{pmatrix} -1 & -1 \\ -1 & 2 \end{pmatrix}$$

$$\therefore \mathbf{X} = -\tfrac{1}{6}\begin{pmatrix} 5 & -2 \\ 5 & 4 \end{pmatrix}\begin{pmatrix} -1 & -1 \\ -1 & 2 \end{pmatrix}$$

$$\therefore \mathbf{X} = -\tfrac{1}{6}\begin{pmatrix} -3 & -9 \\ -9 & 3 \end{pmatrix} = \begin{pmatrix} \frac{1}{2} & \frac{3}{2} \\ \frac{3}{2} & -\frac{1}{2} \end{pmatrix}$$

3 **a** Using technology, $\mathbf{A}^{-1} = \begin{pmatrix} 2 & 1 & 1 \\ 4 & -7 & 3 \\ 3 & -2 & -1 \end{pmatrix}^{-1} = \begin{pmatrix} \frac{1}{4} & -\frac{1}{52} & \frac{5}{26} \\ \frac{1}{4} & -\frac{5}{52} & -\frac{1}{26} \\ \frac{1}{4} & \frac{7}{52} & -\frac{9}{26} \end{pmatrix}$

b In matrix form, the system is $\begin{pmatrix} 2 & 1 & 1 \\ 4 & -7 & 3 \\ 3 & -2 & -1 \end{pmatrix}\begin{pmatrix} x \\ y \\ z \end{pmatrix} = \begin{pmatrix} 8 \\ 10 \\ 1 \end{pmatrix}$

$$\therefore \quad \begin{pmatrix} x \\ y \\ z \end{pmatrix} = \begin{pmatrix} \frac{1}{4} & -\frac{1}{52} & \frac{5}{26} \\ \frac{1}{4} & -\frac{5}{52} & -\frac{1}{26} \\ \frac{1}{4} & \frac{7}{52} & -\frac{9}{26} \end{pmatrix}\begin{pmatrix} 8 \\ 10 \\ 1 \end{pmatrix} = \begin{pmatrix} 2 \\ 1 \\ 3 \end{pmatrix}$$

So, $x = 2, \ y = 1, \ z = 3$.

4 $\begin{vmatrix} x & 2 & 0 \\ 2 & x+1 & -2 \\ 0 & -2 & x+2 \end{vmatrix} = x\begin{vmatrix} x+1 & -2 \\ -2 & x+2 \end{vmatrix} - 2\begin{vmatrix} 2 & -2 \\ 0 & x+2 \end{vmatrix} + 0\begin{vmatrix} 2 & x+1 \\ 0 & -2 \end{vmatrix}$

$= x\,[(x+1)(x+2) - 4] - 2\,[2(x+2) - 0]$
$= x(x^2 + 3x + 2 - 4) - 4(x+2)$
$= x(x^2 + 3x - 2) - 4x - 8$
$= x^3 + 3x^2 - 2x - 4x - 8$
$= x^3 + 3x^2 - 6x - 8$
$= (x+4)(x+1)(x-2)$ {using technology}

$\therefore \quad (x+4)(x+1)(x-2) = 0$
$\therefore \quad x = -4, -1$ or 2

5 $\mathbf{A} = \begin{pmatrix} -2 & 3 \\ 4 & -1 \end{pmatrix}$, $\mathbf{B} = \begin{pmatrix} -7 & 9 \\ 9 & -3 \end{pmatrix}$, $\mathbf{C} = \begin{pmatrix} -1 & 0 & 3 \\ 0 & 2 & 1 \end{pmatrix}$

a $2\mathbf{A} - 2\mathbf{B} = \begin{pmatrix} -4 & 6 \\ 8 & -2 \end{pmatrix} - \begin{pmatrix} -14 & 18 \\ 18 & -6 \end{pmatrix} = \begin{pmatrix} 10 & -12 \\ -10 & 4 \end{pmatrix}$

b **A** is 2×2 and **C** is 2×3 $\therefore$ **AC** is 2×3

$\mathbf{AC} = \begin{pmatrix} -2 & 3 \\ 4 & -1 \end{pmatrix}\begin{pmatrix} -1 & 0 & 3 \\ 0 & 2 & 1 \end{pmatrix} = \begin{pmatrix} 2+0 & 0+6 & -6+3 \\ -4+0 & 0-2 & 12-1 \end{pmatrix} = \begin{pmatrix} 2 & 6 & -3 \\ -4 & -2 & 11 \end{pmatrix}$

c **C** is 2×3 and **B** is 2×2 $\therefore$ **CB** is not possible.

d $\mathbf{DA} = \mathbf{B}$

$\therefore \quad \mathbf{DAA}^{-1} = \mathbf{BA}^{-1}$ {postmultiplying by $\mathbf{A}^{-1}$}

$\therefore \quad \mathbf{D} = \mathbf{BA}^{-1}$

$\therefore \quad \mathbf{D} = \begin{pmatrix} -7 & 9 \\ 9 & -3 \end{pmatrix} \dfrac{1}{-2(-1) - 3(4)} \begin{pmatrix} -1 & -3 \\ -4 & -2 \end{pmatrix} = -\frac{1}{10}\begin{pmatrix} 7-36 & 21-18 \\ -9+12 & -27+6 \end{pmatrix}$

$\therefore \quad \mathbf{D} = \begin{pmatrix} \frac{29}{10} & -\frac{3}{10} \\ -\frac{3}{10} & \frac{21}{10} \end{pmatrix}$

6 **a** **i** If $\mathbf{AB} = \mathbf{B}$ then $\mathbf{ABB}^{-1} = \mathbf{BB}^{-1}$ provided $\mathbf{B}^{-1}$ exists.

$\therefore \quad \mathbf{A} = \mathbf{I}$ provided $\mathbf{B}^{-1}$ exists, $\therefore$ provided that $|\mathbf{B}| \neq 0$.

ii $(\mathbf{A}+\mathbf{B})^2 = (\mathbf{A}+\mathbf{B})(\mathbf{A}+\mathbf{B})$
$= \mathbf{A}^2 + \mathbf{AB} + \mathbf{BA} + \mathbf{B}^2$
$= \mathbf{A}^2 + 2\mathbf{AB} + \mathbf{B}^2$ provided that $\mathbf{AB} = \mathbf{BA}$.

b $\mathbf{M} = \begin{pmatrix} k & 2 \\ 2 & k \end{pmatrix}\begin{pmatrix} k-1 & -2 \\ -3 & k \end{pmatrix}$

$\therefore \quad |\mathbf{M}| = \begin{vmatrix} k & 2 \\ 2 & k \end{vmatrix}\begin{vmatrix} k-1 & -2 \\ -3 & k \end{vmatrix} = (k^2-4)(k^2-k-6)$
$= (k+2)(k-2)(k+2)(k-3)$

Since $\mathbf{M}^{-1}$ exists, $|\mathbf{M}| \neq 0$.
$\therefore \quad k$ is any real number $\neq 3$ or ± 2.

7 In matrix form: $\begin{pmatrix} 2 & 3 & 1 \\ 5 & 2 & -4 \\ 3 & -1 & -2 \end{pmatrix}\begin{pmatrix} x \\ y \\ z \end{pmatrix} = \begin{pmatrix} 4 \\ 0 \\ -7 \end{pmatrix}$

$\therefore \quad \begin{pmatrix} x \\ y \\ z \end{pmatrix} = \begin{pmatrix} 2 & 3 & 1 \\ 5 & 2 & -4 \\ 3 & -1 & -2 \end{pmatrix}^{-1}\begin{pmatrix} 4 \\ 0 \\ -7 \end{pmatrix}$

Using technology, $x=-2$, $y=3$, $z=-1$.

8 **a** $s = at^2 + bt + c$
At $t=1$, $s(1)=63 \quad \therefore \quad a+b+c=63$
At $t=2$, $s(2)=72 \quad \therefore \quad 4a+2b+c=72$
At $t=7$, $s(7)=27 \quad \therefore \quad 49a+7b+c=27$

In matrix form, $\begin{pmatrix} 1 & 1 & 1 \\ 4 & 2 & 1 \\ 49 & 7 & 1 \end{pmatrix}\begin{pmatrix} a \\ b \\ c \end{pmatrix} = \begin{pmatrix} 63 \\ 72 \\ 27 \end{pmatrix}$

Using technology, $a=-3$, $b=18$, $c=48$ and $\therefore \quad s(t) = -3t^2+18t+48$

b $s(0)=48 \quad \therefore$ the height of the cliff is 48 m.

c The rock reaches sea level when $s(t)=0$
$\therefore \quad -3(t^2-6t-16)=0$
$\therefore \quad -3(t-8)(t+2)=0$
$\therefore \quad t=8$ or -2
but $t \geqslant 0$, so it reaches sea level after 8 seconds.

9
$\mathbf{AXB} = \mathbf{C}$
$\therefore \quad \mathbf{A}^{-1}\mathbf{AXBB}^{-1} = \mathbf{A}^{-1}\mathbf{CB}^{-1}$
$\therefore \quad \mathbf{IXI} = \mathbf{A}^{-1}\mathbf{CB}^{-1}$
$\therefore \quad \mathbf{X} = \mathbf{A}^{-1}\mathbf{CB}^{-1}$

So $\mathbf{X} = \begin{pmatrix} 1 & -3 \\ 2 & 1 \end{pmatrix}^{-1}\begin{pmatrix} -12 & -11 \\ -10 & -1 \end{pmatrix}\begin{pmatrix} -1 & 2 \\ 3 & 1 \end{pmatrix}^{-1}$

$\therefore \quad \mathbf{X} = \begin{pmatrix} 0 & -2 \\ 1 & 1 \end{pmatrix}$ {using technology}

10 **a** $C(0)=80 \quad \therefore \quad a(0)+b(0)+c(0)+d=80 \quad \therefore \quad d=80$

b $C(1)=100 \quad \therefore \quad a+b+c+80=100$
$C(2)=148 \quad \therefore \quad 8a+4b+2c+80=148$
$C(4)=376 \quad \therefore \quad 64a+16b+4c+80=376$

$\begin{pmatrix} 1 & 1 & 1 \\ 8 & 4 & 2 \\ 64 & 16 & 4 \end{pmatrix}\begin{pmatrix} a \\ b \\ c \end{pmatrix} = \begin{pmatrix} 20 \\ 68 \\ 296 \end{pmatrix}$

$\therefore \quad \begin{pmatrix} a \\ b \\ c \end{pmatrix} = \begin{pmatrix} 1 & 1 & 1 \\ 8 & 4 & 2 \\ 64 & 16 & 4 \end{pmatrix}^{-1}\begin{pmatrix} 20 \\ 68 \\ 296 \end{pmatrix} = \begin{pmatrix} 2 \\ 8 \\ 10 \end{pmatrix}$ {using technology}

$\therefore \quad a=2$, $b=8$, $c=10$

Note: In the first print run of the text there was an extra question here. It was deleted in later prints as it copied question **3 b**.

11 $\mathbf{A} = \begin{pmatrix} 1 & 2 & 1 \\ 2 & 4 & 6 \\ 3 & 1 & 2 \end{pmatrix}$ $\mathbf{B} = \begin{pmatrix} -1 & 2 & -3 \\ 2 & -1 & 4 \\ 3 & 4 & 1 \end{pmatrix}$

$$\mathbf{AB} = \begin{pmatrix} 1 & 2 & 1 \\ 2 & 4 & 6 \\ 3 & 1 & 2 \end{pmatrix}\begin{pmatrix} -1 & 2 & -3 \\ 2 & -1 & 4 \\ 3 & 4 & 1 \end{pmatrix} = \begin{pmatrix} -1+4+3 & 2-2+4 & -3+8+1 \\ -2+8+18 & 4-4+24 & -6+16+6 \\ -3+2+6 & 6-1+8 & -9+4+2 \end{pmatrix}$$

$$\therefore \quad \mathbf{AB} = \begin{pmatrix} 6 & 4 & 6 \\ 24 & 24 & 16 \\ 5 & 13 & -3 \end{pmatrix}$$

$$\therefore \quad \det \mathbf{AB} = 6\begin{vmatrix} 24 & 16 \\ 13 & -3 \end{vmatrix} - 4\begin{vmatrix} 24 & 16 \\ 5 & -3 \end{vmatrix} + 6\begin{vmatrix} 24 & 24 \\ 5 & 13 \end{vmatrix}$$
$$= 6(-72-208) - 4(-72-80) + 6(312-120)$$
$$= 80$$

$$\det \mathbf{A} = 1\begin{vmatrix} 4 & 6 \\ 1 & 2 \end{vmatrix} - 2\begin{vmatrix} 2 & 6 \\ 3 & 2 \end{vmatrix} + 1\begin{vmatrix} 2 & 4 \\ 3 & 1 \end{vmatrix}$$
$$= 1(8-6) - 2(4-18) + 1(2-12)$$
$$= 2 + 28 - 10$$
$$= 20$$

$$\text{and} \quad \det \mathbf{B} = -1\begin{vmatrix} -1 & 4 \\ 4 & 1 \end{vmatrix} - 2\begin{vmatrix} 2 & 4 \\ 3 & 1 \end{vmatrix} + (-3)\begin{vmatrix} 2 & -1 \\ 3 & 4 \end{vmatrix}$$
$$= -1(-1-16) - 2(2-12) - 3(8 - -3)$$
$$= 17 + 20 - 33$$
$$= 4$$

$$\therefore \quad \det \mathbf{A} \times \det \mathbf{B} = 20 \times 4 = 80 = \det(\mathbf{AB})$$

12 **a** Let €x be the cost of an opera ticket
€y be the cost of a play ticket
€z be the cost of a concert ticket

$$\therefore \quad 3x + 2y + 5z = 267$$
$$2x + 3y + z = 145$$
$$x + 5y + 4z = 230$$

b So, $\begin{pmatrix} 3 & 2 & 5 \\ 2 & 3 & 1 \\ 1 & 5 & 4 \end{pmatrix}\begin{pmatrix} x \\ y \\ z \end{pmatrix} = \begin{pmatrix} 267 \\ 145 \\ 230 \end{pmatrix}$ $\therefore \begin{pmatrix} x \\ y \\ z \end{pmatrix} = \begin{pmatrix} 3 & 2 & 5 \\ 2 & 3 & 1 \\ 1 & 5 & 4 \end{pmatrix}^{-1}\begin{pmatrix} 267 \\ 145 \\ 230 \end{pmatrix}$

$\therefore \begin{pmatrix} x \\ y \\ z \end{pmatrix} = \begin{pmatrix} 32 \\ 18 \\ 27 \end{pmatrix}$ using technology

$\therefore$ the cost of each ticket is €32 for an opera, €18 for a play, €27 for a concert.

c Total cost = 4 × €32 + 1 × €18 + 2 × €27
= €128 + €18 + €54 = €200

REVIEW SET 11C

1 **a** $2\mathbf{B} = 2\begin{pmatrix} 2 & 4 \\ 0 & 1 \\ 3 & 2 \end{pmatrix} = \begin{pmatrix} 4 & 8 \\ 0 & 2 \\ 6 & 4 \end{pmatrix}$

b $\frac{1}{2}\mathbf{B} = \frac{1}{2}\begin{pmatrix} 2 & 4 \\ 0 & 1 \\ 3 & 2 \end{pmatrix} = \begin{pmatrix} 1 & 2 \\ 0 & \frac{1}{2} \\ \frac{3}{2} & 1 \end{pmatrix}$

c $\mathbf{AB} = \begin{pmatrix} 1 & 2 & 3 \end{pmatrix}\begin{pmatrix} 2 & 4 \\ 0 & 1 \\ 3 & 2 \end{pmatrix} = \begin{pmatrix} 11 & 12 \end{pmatrix}$

$1 \times 3 \longleftrightarrow 3 \times 2$

d $\mathbf{B}$ is 3×2 and $\mathbf{A}$ is 1×3 — not equal $\therefore$ **BA** does not exist.

2 **a** $\mathbf{P} + \mathbf{Q} = \begin{pmatrix} 1 & 2 \\ 1 & 0 \\ 2 & 3 \end{pmatrix} + \begin{pmatrix} 3 & 0 \\ 1 & 4 \\ 1 & 1 \end{pmatrix} = \begin{pmatrix} 4 & 2 \\ 2 & 4 \\ 3 & 4 \end{pmatrix}$

b $\mathbf{Q}-\mathbf{P}=\begin{pmatrix}3&0\\1&4\\1&1\end{pmatrix}-\begin{pmatrix}1&2\\1&0\\2&3\end{pmatrix}=\begin{pmatrix}2&-2\\0&4\\-1&-2\end{pmatrix}$

c $\frac{3}{2}\mathbf{P}-\mathbf{Q}=\begin{pmatrix}\frac{3}{2}&3\\\frac{3}{2}&0\\3&\frac{9}{2}\end{pmatrix}-\begin{pmatrix}3&0\\1&4\\1&1\end{pmatrix}=\begin{pmatrix}-\frac{3}{2}&3\\\frac{1}{2}&-4\\2&\frac{7}{2}\end{pmatrix}$

3 In matrix form: $\begin{pmatrix}1&4\\k&3\end{pmatrix}\begin{pmatrix}x\\y\end{pmatrix}=\begin{pmatrix}2\\-6\end{pmatrix}$

This system has a unique solution when $\begin{vmatrix}1&4\\k&3\end{vmatrix}\neq 0$

$\therefore\ 3-4k\neq 0$

$\therefore\ k\neq\frac{3}{4}$

So, k can take any real value except $\frac{3}{4}$.

4 $\mathbf{A}^2=\begin{pmatrix}-1&2\\3&0\end{pmatrix}\begin{pmatrix}-1&2\\3&0\end{pmatrix}=\begin{pmatrix}7&-2\\-3&6\end{pmatrix}$

Now $\mathbf{A}^2+\mathbf{A}+k\mathbf{I}=\mathbf{O}$

$\therefore\ \begin{pmatrix}7&-2\\-3&6\end{pmatrix}+\begin{pmatrix}-1&2\\3&0\end{pmatrix}+k\begin{pmatrix}1&0\\0&1\end{pmatrix}=\begin{pmatrix}0&0\\0&0\end{pmatrix}$

$\therefore\ \begin{cases}7+(-1)+k=0\\-2+2=0\\-3+3=0\\6+k=0\end{cases}$ and so $k=-6$.

5 **a** $\mathbf{A}^{-1}=\frac{1}{42-40}\begin{pmatrix}7&-8\\-5&6\end{pmatrix}=\begin{pmatrix}\frac{7}{2}&-4\\-\frac{5}{2}&3\end{pmatrix}$

b $\mathbf{A}^{-1}$ does not exist as $|\mathbf{A}|=-24--24=0$

c $\mathbf{A}^{-1}=\frac{1}{-3}\begin{pmatrix}-3&-5\\6&11\end{pmatrix}=\begin{pmatrix}1&\frac{5}{3}\\-2&-\frac{11}{3}\end{pmatrix}$

6 **a** $3\mathbf{A}=3\begin{pmatrix}-3&2&2\\1&-1&0\end{pmatrix}=\begin{pmatrix}-9&6&6\\3&-3&0\end{pmatrix}$

b $\mathbf{AB}=\begin{pmatrix}-3&2&2\\1&-1&0\end{pmatrix}\begin{pmatrix}2&4\\-3&1\\1&2\end{pmatrix}=\begin{pmatrix}-10&-6\\5&3\end{pmatrix}$

c $\mathbf{BA}=\begin{pmatrix}2&4\\-3&1\\1&2\end{pmatrix}\begin{pmatrix}-3&2&2\\1&-1&0\end{pmatrix}=\begin{pmatrix}-2&0&4\\10&-7&-6\\-1&0&2\end{pmatrix}$

d $\mathbf{A}$ is 2×3 and $\mathbf{C}$ is 2×2 $\therefore$ $\mathbf{AC}$ does not exist.

e $\mathbf{BC}=\begin{pmatrix}2&4\\-3&1\\1&2\end{pmatrix}\begin{pmatrix}-2&5\\1&3\end{pmatrix}=\begin{pmatrix}0&22\\7&-12\\0&11\end{pmatrix}$

7 $\mathbf{AB}=\begin{pmatrix}x&1\\4&-3x\end{pmatrix}\begin{pmatrix}2\\-1\end{pmatrix}=\begin{pmatrix}9\\23\end{pmatrix}$

$\therefore\ 2x-1=9$ {first row}

$\therefore\ 2x=10$

$\therefore\ x=5$

Check: $8+3x=23$ {second row}

$3x=15$

$x=5$ ✓

8 **a** $5\mathbf{A}^2 - 6\mathbf{A} = 3\mathbf{I}$

$\therefore\ \mathbf{A}(5\mathbf{A} - 6\mathbf{I}) = 3\mathbf{I}$

$\therefore\ \mathbf{A} \times \frac{1}{3}(5\mathbf{A} - 6\mathbf{I}) = \mathbf{I}$

$\therefore\ \mathbf{A}(\frac{5}{3}\mathbf{A} - 2\mathbf{I}) = \mathbf{I}$

b $\mathbf{A}^{-1} = \frac{5}{3}\mathbf{A} - 2\mathbf{I}$

9 **a** $\mathbf{A}^3$

$= \begin{pmatrix} -1 & -2 \\ 0 & -3 \end{pmatrix}\begin{pmatrix} -1 & -2 \\ 0 & -3 \end{pmatrix}\begin{pmatrix} -1 & -2 \\ 0 & -3 \end{pmatrix}$

$= \begin{pmatrix} 1 & 8 \\ 0 & 9 \end{pmatrix}\begin{pmatrix} -1 & -2 \\ 0 & -3 \end{pmatrix}$

$= \begin{pmatrix} -1 & -26 \\ 0 & -27 \end{pmatrix}$

b $\mathbf{A}^3 + 2\mathbf{X} = \mathbf{B}$

$\therefore\ 2\mathbf{X} = \mathbf{B} - \mathbf{A}^3$

$\therefore\ \mathbf{X} = \frac{1}{2}\mathbf{B} - \frac{1}{2}\mathbf{A}^3$

$= \frac{1}{2}\begin{pmatrix} 1 & 0 \\ -6 & 7 \end{pmatrix} - \frac{1}{2}\begin{pmatrix} -1 & -26 \\ 0 & -27 \end{pmatrix}$

$= \begin{pmatrix} 1 & 13 \\ -3 & 17 \end{pmatrix}$

10 If $\mathbf{A}^2 = 5\mathbf{A} + 2\mathbf{I}$,

$\mathbf{A}^3 = \mathbf{A}(5\mathbf{A} + 2\mathbf{I})$
$= 5\mathbf{A}^2 + 2\mathbf{AI}$
$= 5(5\mathbf{A} + 2\mathbf{I}) + 2\mathbf{A}$
$= 25\mathbf{A} + 10\mathbf{I} + 2\mathbf{A}$
$= 27\mathbf{A} + 10\mathbf{I}$

$\mathbf{A}^4 = \mathbf{A}(27\mathbf{A} + 10\mathbf{I})$
$= 27\mathbf{A}^2 + 10\mathbf{AI}$
$= 27(5\mathbf{A} + 2\mathbf{I}) + 10\mathbf{A}$
$= 135\mathbf{A} + 54\mathbf{I} + 10\mathbf{A}$
$= 145\mathbf{A} + 54\mathbf{I}$

$\mathbf{A}^5 = \mathbf{A}(145\mathbf{A} + 54\mathbf{I})$
$= 145\mathbf{A}^2 + 54\mathbf{AI}$
$= 145(5\mathbf{A} + 2\mathbf{I}) + 54\mathbf{A}$
$= 725\mathbf{A} + 290\mathbf{I} + 54\mathbf{A}$
$= 779\mathbf{A} + 290\mathbf{I}$

$\mathbf{A}^6 = \mathbf{A}(779\mathbf{A} + 290\mathbf{I})$
$= 779\mathbf{A}^2 + 290\mathbf{AI}$
$= 779(5\mathbf{A} + 2\mathbf{I}) + 290\mathbf{AI}$
$= 4185\mathbf{A} + 1558\mathbf{I}$

11 Sales matrix is $\begin{pmatrix} 42-27 & 54-31 \\ 36-28 & 27-15 \\ 34-28 & 30-22 \end{pmatrix} = \begin{pmatrix} 15 & 23 \\ 8 & 12 \\ 6 & 8 \end{pmatrix}$ Totals matrix is $\begin{pmatrix} 38 \\ 20 \\ 14 \end{pmatrix}$

Profit matrix is $\begin{pmatrix} 0.75 & 0.55 & 1.20 \end{pmatrix}\begin{pmatrix} 38 \\ 20 \\ 14 \end{pmatrix} = (56.3)$ $\therefore$ profit = \$56.30

12 **a** $\mathbf{AB} = \begin{pmatrix} 1 & 2 \\ a & b \end{pmatrix}\begin{pmatrix} c & d \\ 0 & -1 \end{pmatrix}$

$= \begin{pmatrix} c & d-2 \\ ac & ad-b \end{pmatrix}$

b $\mathbf{A}^{-1} = \dfrac{1}{b-2a}\begin{pmatrix} b & -2 \\ -a & 1 \end{pmatrix}$

c $\mathbf{B}^{-1} = -\mathbf{I}$

$\therefore\ \dfrac{1}{-c}\begin{pmatrix} -1 & -d \\ 0 & c \end{pmatrix} = -\begin{pmatrix} 1 & 0 \\ 0 & 1 \end{pmatrix}$

$\therefore\ \begin{pmatrix} \frac{1}{c} & \frac{d}{c} \\ 0 & -1 \end{pmatrix} = \begin{pmatrix} -1 & 0 \\ 0 & -1 \end{pmatrix}$

$\therefore\ c = -1,\ d = 0$

13 **a** If $\mathbf{A}^2 = \mathbf{A} - \mathbf{I}$ then $\mathbf{A}^3 = \mathbf{A}(\mathbf{A} - \mathbf{I}) = \mathbf{A}^2 - \mathbf{AI} = \mathbf{A} - \mathbf{I} - \mathbf{A} = -\mathbf{I}$

$\mathbf{A}^4 = \mathbf{AA}^3 = \mathbf{A}(-\mathbf{I}) = -\mathbf{A}$

$\mathbf{A}^5 = \mathbf{AA}^4 = \mathbf{A}(-\mathbf{A}) = -\mathbf{A}^2 = -(\mathbf{A} - \mathbf{I}) = \mathbf{I} - \mathbf{A}$

$\mathbf{A}^6 = \mathbf{AA}^5 = \mathbf{A}(\mathbf{I} - \mathbf{A}) = \mathbf{AI} - \mathbf{A}^2 = \mathbf{A} - (\mathbf{A} - \mathbf{I}) = \mathbf{I}$

$\mathbf{A}^7 = \mathbf{AA}^6 = \mathbf{AI} = \mathbf{A}$

$\mathbf{A}^8 = \mathbf{AA}^7 = \mathbf{AA} = \mathbf{A}^2 = \mathbf{A} - \mathbf{I}$

b $\mathbf{A}^{6n+3} = (\mathbf{A}^6)^n\mathbf{A}^3$
$= -\mathbf{I}$

$\mathbf{A}^{6n+5} = (\mathbf{A}^6)^n\mathbf{A}^5$
$= \mathbf{I} - \mathbf{A}$

c Now $\mathbf{A}^2 = \mathbf{A} - \mathbf{I}$

$\therefore\ \mathbf{A}^{-1}\mathbf{AA} = \mathbf{A}^{-1}\mathbf{A} - \mathbf{A}^{-1}\mathbf{I}$ {premultiplying by $\mathbf{A}^{-1}$}

$\therefore\ \mathbf{IA} = \mathbf{I} - \mathbf{A}^{-1}$

$\therefore\ \mathbf{A}^{-1} = \mathbf{I} - \mathbf{A}$

Chapter 12
VECTORS IN 2 AND 3 DIMENSIONS

EXERCISE 12A.1

1 **a**

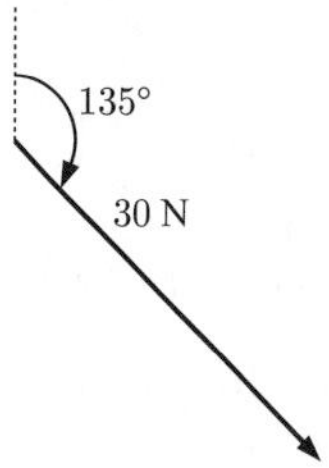

b

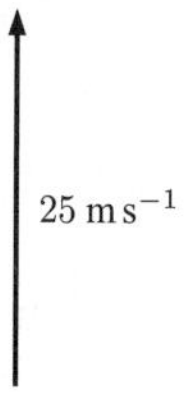

c

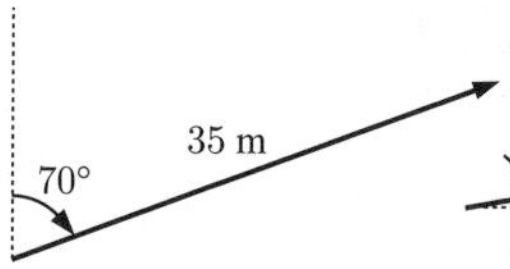

d

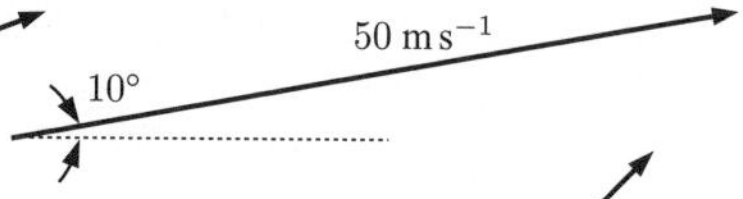

2 **a**

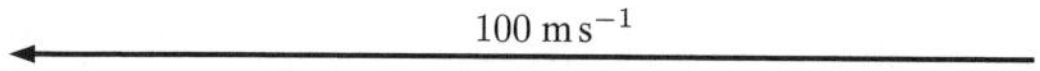

b

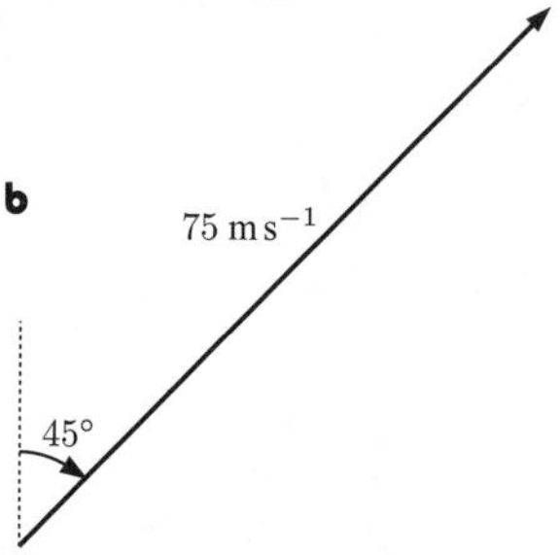

3 **a** scale: 1 cm $\equiv$ 10 N

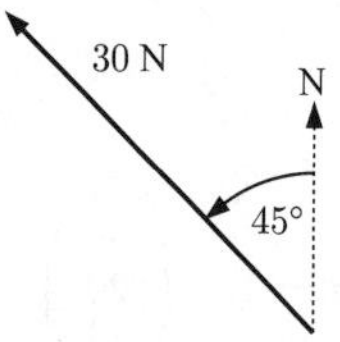

b scale: 1 cm $\equiv$ 10 m s^{-1}

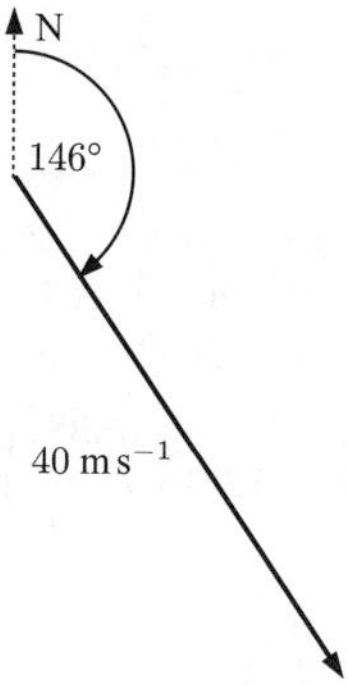

c scale: 1 cm $\equiv$ 10 km

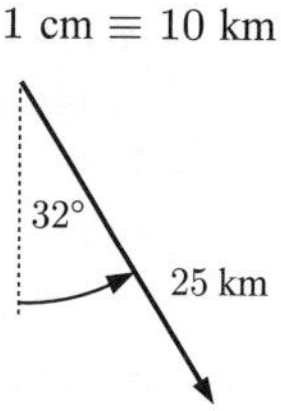

d scale: 1 cm $\equiv$ 30 km h^{-1}

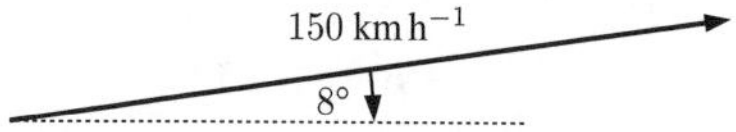

EXERCISE 12A.2

1 **a** If they are equal in magnitude, they have the same length. These are **p**, **q**, **s** and **t**.

b Those parallel are **p**, **q**, **r** and **t**.

c Those in the same direction are: **p** and **r**, **q** and **t**.

d To be equal they must have the same direction and be equal in length $\therefore$ $\mathbf{q} = \mathbf{t}$.

e **p** and **q** are negatives (equal length, but opposite direction). Likewise, **p** and **t** are negatives. We write $\mathbf{p} = -\mathbf{q}$ and $\mathbf{p} = -\mathbf{t}$.

2 **a** True, as they have the same length and are parallel.

b True, as they are sides of an equilateral triangle.

c False, as they do not have the same direction.

d False, as they have opposite directions.

e True, as they have the same length and direction.

f False, as they do not have the same direction.

EXERCISE 12B.1

1 **a**

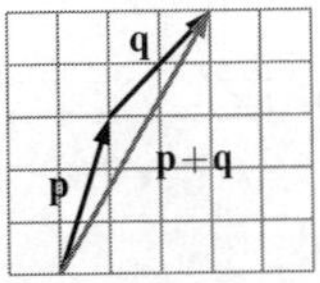

b

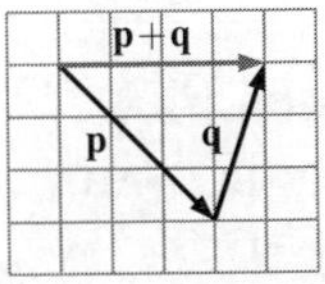

c

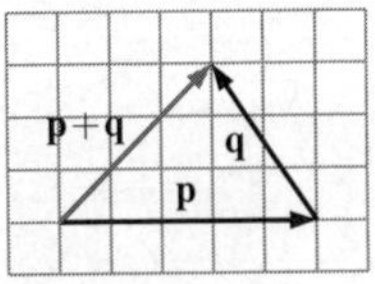

d

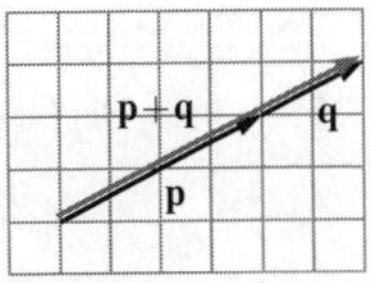

e

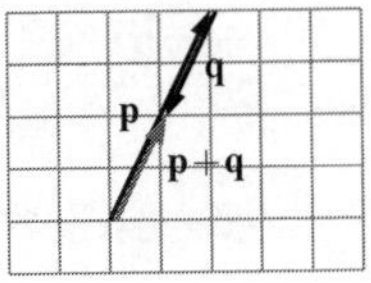

f

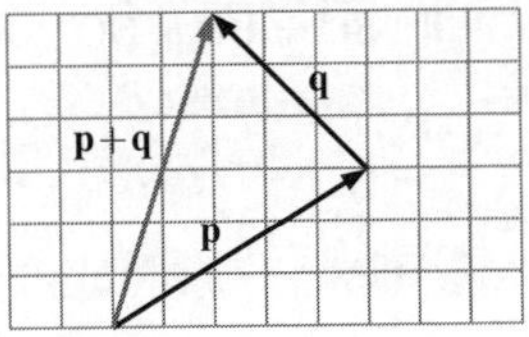

2 **a** $\overrightarrow{AB} + \overrightarrow{BC}$
$= \overrightarrow{AC}$

b $\overrightarrow{BC} + \overrightarrow{CD}$
$= \overrightarrow{BD}$

c $\overrightarrow{AB} + \overrightarrow{BC} + \overrightarrow{CD}$
$= \overrightarrow{AC} + \overrightarrow{CD}$
$= \overrightarrow{AD}$

d $\overrightarrow{AC} + \overrightarrow{CB} + \overrightarrow{BD}$
$= \overrightarrow{AB} + \overrightarrow{BD}$
$= \overrightarrow{AD}$

3 **a** **i**

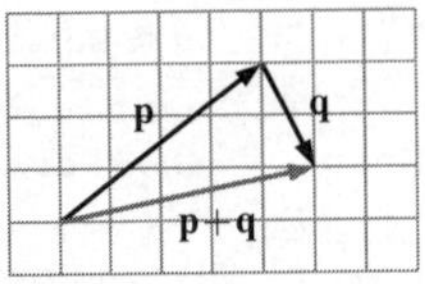

ii

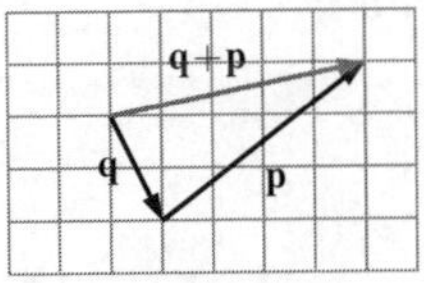

b yes

4 $\overrightarrow{PS} = \overrightarrow{PR} + \overrightarrow{RS}$
$= (\mathbf{a}+\mathbf{b}) + \mathbf{c}$

But $\overrightarrow{PS} = \overrightarrow{PQ} + \overrightarrow{QS}$
$= \mathbf{a} + (\mathbf{b}+\mathbf{c})$

$\therefore \quad (\mathbf{a}+\mathbf{b}) + \mathbf{c} = \mathbf{a} + (\mathbf{b}+\mathbf{c})$ {as both are equal to $\overrightarrow{PS}$}

EXERCISE 12B.2

1 **a**

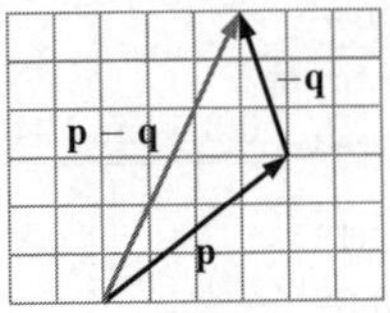

b

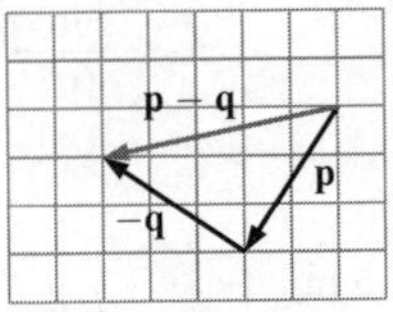

c

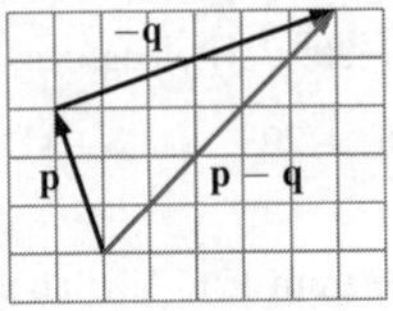

d

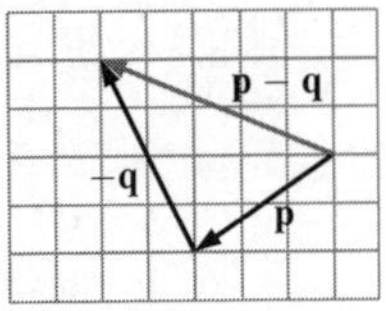

2 **a**

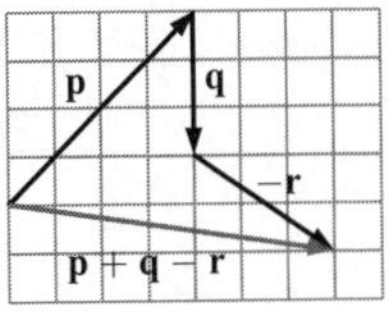

b

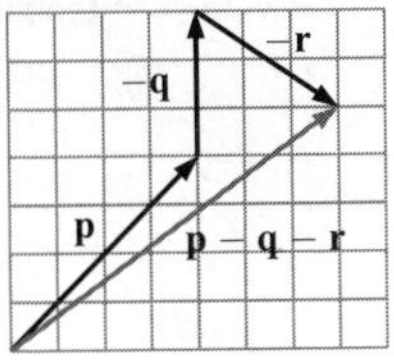

c

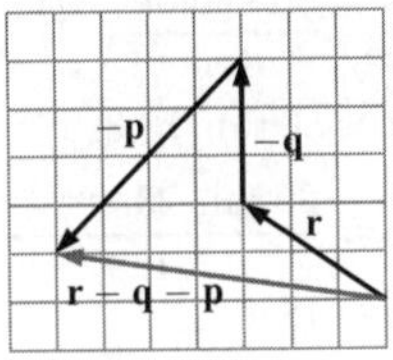

3 **a** $\overrightarrow{AC} + \overrightarrow{CB}$
$= \overrightarrow{AB}$

b $\overrightarrow{AD} - \overrightarrow{BD}$
$= \overrightarrow{AD} + \overrightarrow{DB}$
$= \overrightarrow{AB}$

c $\overrightarrow{AC} + \overrightarrow{CA}$
$= \overrightarrow{AA}$
$= \mathbf{0}$

d $\overrightarrow{AB} + \overrightarrow{BC} + \overrightarrow{CD}$
$= \overrightarrow{AC} + \overrightarrow{CD}$
$= \overrightarrow{AD}$

e $\overrightarrow{BA} - \overrightarrow{CA} + \overrightarrow{CB}$
$= \overrightarrow{BA} + \overrightarrow{AC} + \overrightarrow{CB}$
$= \overrightarrow{BC} + \overrightarrow{CB}$
$= \overrightarrow{BB}$
$= \mathbf{0}$

f $\overrightarrow{AB} - \overrightarrow{CB} - \overrightarrow{DC}$
$= \overrightarrow{AB} + \overrightarrow{BC} + \overrightarrow{CD}$
$= \overrightarrow{AC} + \overrightarrow{CD}$
$= \overrightarrow{AD}$

4 **a** $\mathbf{t} = \mathbf{r} + \mathbf{s}$ **b** $\mathbf{r} = -\mathbf{s} - \mathbf{t}$ **c** $\mathbf{r} = -\mathbf{p} - \mathbf{q} - \mathbf{s}$ **d** $\mathbf{r} = \mathbf{q} - \mathbf{p} + \mathbf{s}$

e $\mathbf{p} = \mathbf{t} + \mathbf{s} + \mathbf{r} - \mathbf{q}$ **f** $\mathbf{p} = -\mathbf{u} + \mathbf{t} + \mathbf{s} - \mathbf{r} - \mathbf{q}$

5 **a** **i** $\overrightarrow{OB} = \overrightarrow{OA} + \overrightarrow{AB} = \mathbf{r} + \mathbf{s}$

ii $\overrightarrow{CA} = \overrightarrow{CB} + \overrightarrow{BA} = -\overrightarrow{BC} - \overrightarrow{AB} = -\mathbf{t} - \mathbf{s}$

iii $\overrightarrow{OC} = \overrightarrow{OA} + \overrightarrow{AB} + \overrightarrow{BC} = \mathbf{r} + \mathbf{s} + \mathbf{t}$

b **i** $\overrightarrow{AD} = \overrightarrow{AB} + \overrightarrow{BD} = \mathbf{p} + \mathbf{q}$

ii $\overrightarrow{BC} = \overrightarrow{BD} + \overrightarrow{DC} = \mathbf{q} + \mathbf{r}$

iii $\overrightarrow{AC} = \overrightarrow{AB} + \overrightarrow{BD} + \overrightarrow{DC} = \mathbf{p} + \mathbf{q} + \mathbf{r}$

EXERCISE 12B.3

1

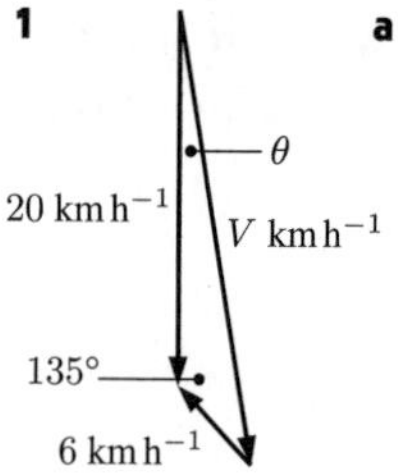

a Using the cosine rule,

$V^2 = 20^2 + 6^2 - 2 \times 20 \times 6 \times \cos 135^o$

$\therefore \quad V \approx 24.6$

$\therefore$ the equivalent speed in still water is 24.6 km h^{-1}.

b Using the sine rule,

$\dfrac{\sin\theta}{6} \approx \dfrac{\sin 135^o}{24.6}$

$\therefore \quad \theta \approx \sin^{-1}\left(\dfrac{6 \times \sin 135^o}{24.6}\right)$

$\therefore \quad \theta \approx 9.93^o$

$\therefore$ the boat should head 9.93^o east of south.

2

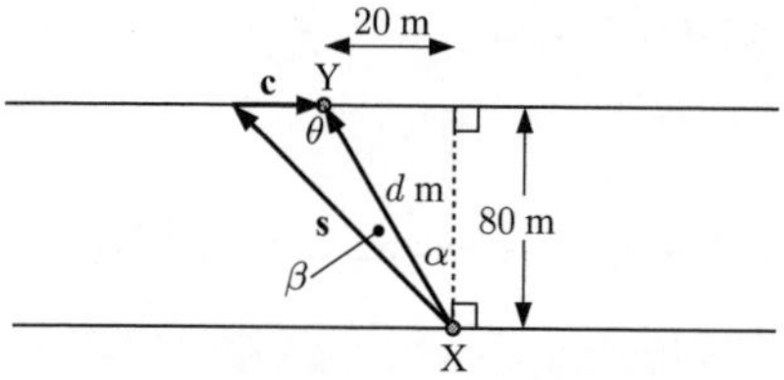

a $d^2 = 80^2 + 20^2$ {Pythagoras}

$\therefore \quad d = \sqrt{80^2 + 20^2}$ $\{d > 0\}$

$\therefore \quad d \approx 82.5$

$\therefore$ the distance from X to Y is about 82.5 m.

b $\alpha = \tan^{-1}(\frac{20}{80}) \approx 14.04^o$

$\therefore \quad \theta \approx 90^o + 14.04^o$ {exterior angle of Δ}

$\therefore \quad \theta \approx 104.04^o$

In t seconds, Stephanie can swim $1.8t$ metres, and the current will move $0.3t$ metres.

$\therefore \quad |\mathbf{s}| = 1.8t$ and $|\mathbf{c}| = 0.3t$

Using the sine rule,

$\dfrac{\sin\beta}{0.3t} = \dfrac{\sin\theta}{1.8t}$

$\beta \approx \sin^{-1}\left(\dfrac{0.3 \times \sin 104.04^o}{1.8}\right)$

$\therefore \quad \beta \approx 9.31^o$

$\therefore \quad \alpha + \beta \approx 23.3^o$

$\therefore$ Stephanie should head 23.3^o west of north.

c $\tan(\alpha + \beta) = \dfrac{20 + 0.3t}{80}$

$\therefore \quad 20 + 0.3t \approx 80\tan(23.3^o)$

$\therefore \quad t \approx \dfrac{80\tan(23.3^o) - 20}{0.3}$

$\therefore \quad t \approx 48.4$

$\therefore$ Stephanie will take 48.4 seconds to cross the river.

EXERCISE 12B.4

1 **a**

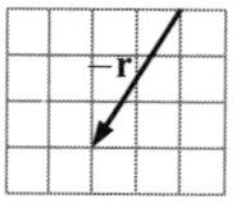

b

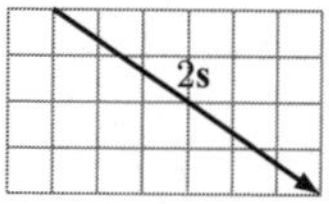

c

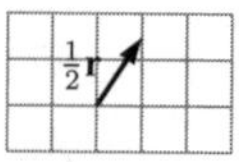

d

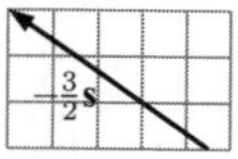

e

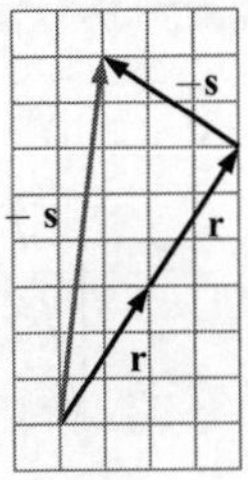

f

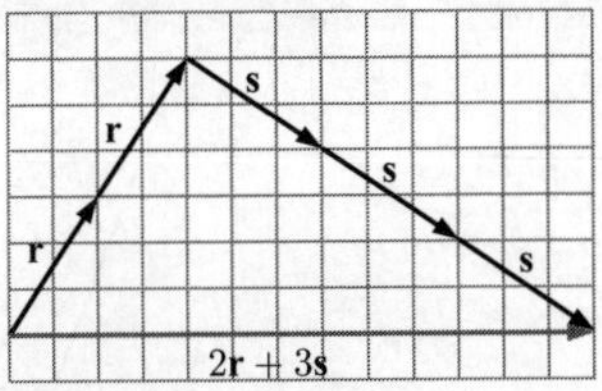

g

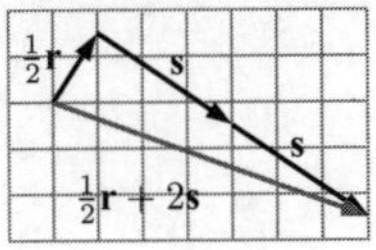

h

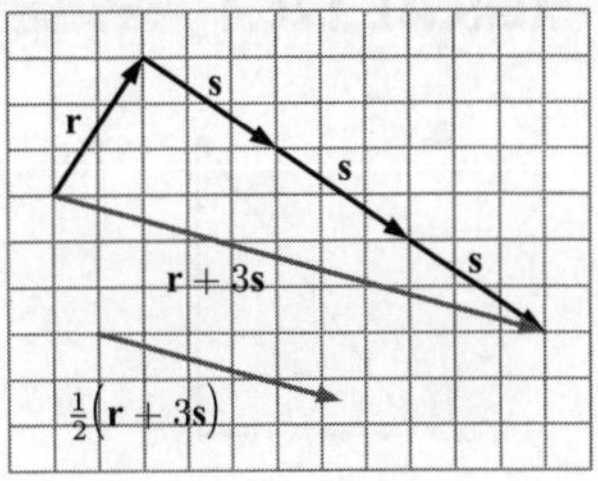

2 **a**

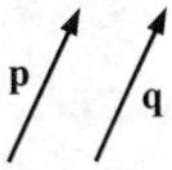

$\mathbf{p} = \mathbf{q}$

b

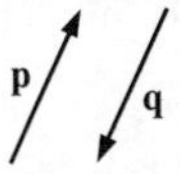

$\mathbf{p} = -\mathbf{q}$

c

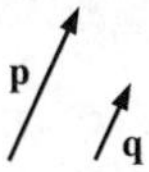

$\mathbf{p} = 2\mathbf{q}$

d

$\mathbf{p} = \frac{1}{3}\mathbf{q}$

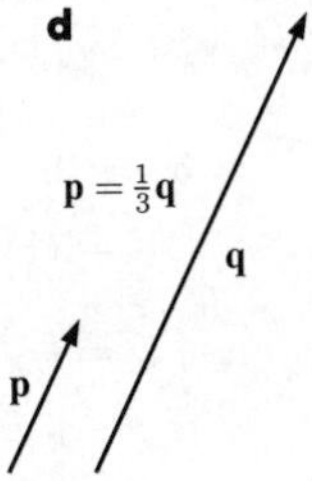

e

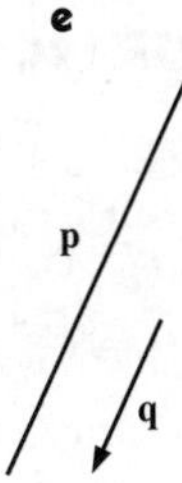

$\mathbf{p} = -3\mathbf{q}$

3 **a**

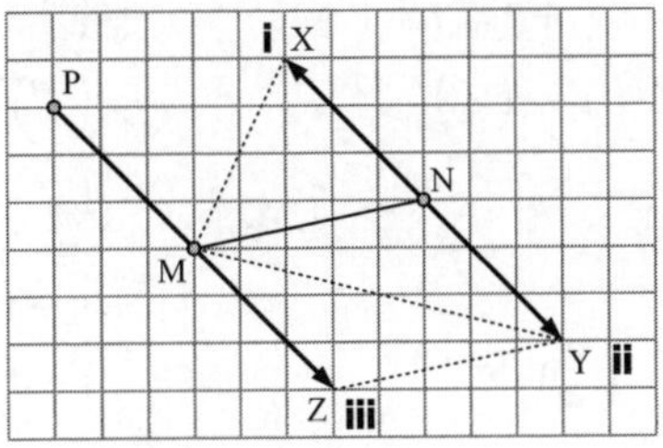

b a parallelogram

4 **a** $\overrightarrow{AB} = -\mathbf{a} + \mathbf{b}$

b

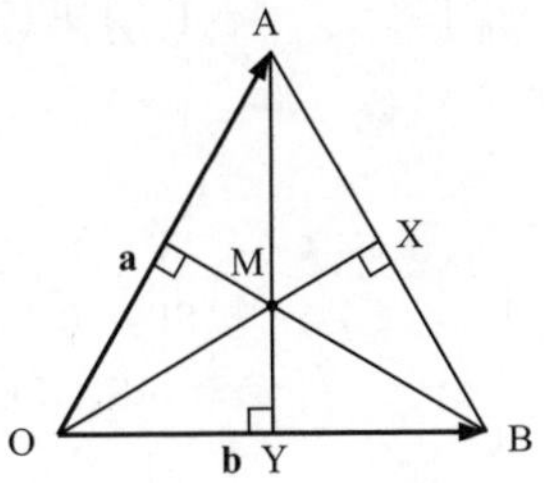

Since OAB is an equilateral triangle, the perpendiculars bisect the sides.

$\therefore \quad \overrightarrow{AX} = \frac{1}{2}\overrightarrow{AB}$

$\therefore \quad \overrightarrow{AX} = \frac{1}{2}(-\mathbf{a} + \mathbf{b})$

Now, $\overrightarrow{OX} = \overrightarrow{OA} + \overrightarrow{AX}$

$\therefore \quad \overrightarrow{OX} = \mathbf{a} + \frac{1}{2}(-\mathbf{a} + \mathbf{b})$

$= \frac{1}{2}\mathbf{a} + \frac{1}{2}\mathbf{b}$

In ΔOXB, $B\widehat{X}O = 90^o$, and $O\widehat{B}X = 60^o$ since ΔOAB is equilateral

$\therefore \quad M\widehat{O}Y = X\widehat{O}B = 30^o$

$\sin M\widehat{O}Y = \dfrac{MY}{OM}$

$\therefore \quad \sin 30^o = \dfrac{MY}{OM}$

$\therefore \quad \frac{1}{2} = \dfrac{MY}{OM}$

$\therefore \quad \frac{1}{2}OM = MY$

By symmetry, $MY = MX$

$\therefore \quad \frac{1}{2}OM = MX$

Now, $OX = OM + MX$

$= OM + \frac{1}{2}OM$

$= \frac{3}{2}OM$

So, $\overrightarrow{OX} = \frac{3}{2}\overrightarrow{OM}$

$\therefore \quad \overrightarrow{OM} = \frac{2}{3}\overrightarrow{OX}$

$= \frac{2}{3}(\frac{1}{2}\mathbf{a} + \frac{1}{2}\mathbf{b})$

$= \frac{1}{3}(\mathbf{a} + \mathbf{b})$

EXERCISE 12C.1

1 **a**

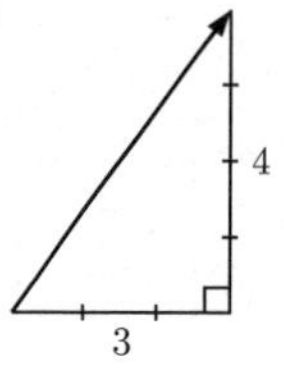

b

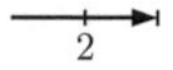

c

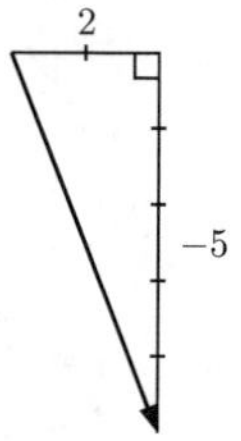

d

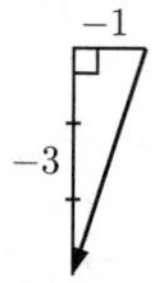

2 **a** $\begin{pmatrix}7\\3\end{pmatrix}$ **b** $\begin{pmatrix}-6\\0\end{pmatrix}$ **c** $\begin{pmatrix}2\\-5\end{pmatrix}$ **d** $\begin{pmatrix}0\\6\end{pmatrix}$ **e** $\begin{pmatrix}-6\\3\end{pmatrix}$ **f** $\begin{pmatrix}-5\\-5\end{pmatrix}$

EXERCISE 12C.2

1 **a** $\mathbf{a}+\mathbf{b} = \begin{pmatrix}-3\\2\end{pmatrix}+\begin{pmatrix}1\\4\end{pmatrix} = \begin{pmatrix}-2\\6\end{pmatrix}$

b $\mathbf{b}+\mathbf{a} = \begin{pmatrix}1\\4\end{pmatrix}+\begin{pmatrix}-3\\2\end{pmatrix} = \begin{pmatrix}-2\\6\end{pmatrix}$

c $\mathbf{b}+\mathbf{c} = \begin{pmatrix}1\\4\end{pmatrix}+\begin{pmatrix}-2\\-5\end{pmatrix} = \begin{pmatrix}-1\\-1\end{pmatrix}$

d $\mathbf{c}+\mathbf{b} = \begin{pmatrix}-2\\-5\end{pmatrix}+\begin{pmatrix}1\\4\end{pmatrix} = \begin{pmatrix}-1\\-1\end{pmatrix}$

e $\mathbf{a}+\mathbf{c} = \begin{pmatrix}-3\\2\end{pmatrix}+\begin{pmatrix}-2\\-5\end{pmatrix} = \begin{pmatrix}-5\\-3\end{pmatrix}$

f $\mathbf{c}+\mathbf{a} = \begin{pmatrix}-2\\-5\end{pmatrix}+\begin{pmatrix}-3\\2\end{pmatrix} = \begin{pmatrix}-5\\-3\end{pmatrix}$

g $\mathbf{a}+\mathbf{a} = \begin{pmatrix}-3\\2\end{pmatrix}+\begin{pmatrix}-3\\2\end{pmatrix} = \begin{pmatrix}-6\\4\end{pmatrix}$

h $\mathbf{b}+\mathbf{a}+\mathbf{c} = \begin{pmatrix}1\\4\end{pmatrix}+\begin{pmatrix}-3\\2\end{pmatrix}+\begin{pmatrix}-2\\-5\end{pmatrix} = \begin{pmatrix}-2\\6\end{pmatrix}+\begin{pmatrix}-2\\-5\end{pmatrix} = \begin{pmatrix}-4\\1\end{pmatrix}$

2 **a** $\mathbf{p}-\mathbf{q} = \begin{pmatrix}-4\\2\end{pmatrix}-\begin{pmatrix}-1\\-5\end{pmatrix} = \begin{pmatrix}-3\\7\end{pmatrix}$

b $\mathbf{q}-\mathbf{r} = \begin{pmatrix}-1\\-5\end{pmatrix}-\begin{pmatrix}3\\-2\end{pmatrix} = \begin{pmatrix}-4\\-3\end{pmatrix}$

c $\mathbf{p}+\mathbf{q}-\mathbf{r} = \begin{pmatrix}-4\\2\end{pmatrix}+\begin{pmatrix}-1\\-5\end{pmatrix}-\begin{pmatrix}3\\-2\end{pmatrix} = \begin{pmatrix}-8\\-1\end{pmatrix}$

d $\mathbf{p}-\mathbf{q}-\mathbf{r} = \begin{pmatrix}-4\\2\end{pmatrix}-\begin{pmatrix}-1\\-5\end{pmatrix}-\begin{pmatrix}3\\-2\end{pmatrix} = \begin{pmatrix}-6\\9\end{pmatrix}$

e $\mathbf{q}-\mathbf{r}-\mathbf{p} = \begin{pmatrix}-1\\-5\end{pmatrix}-\begin{pmatrix}3\\-2\end{pmatrix}-\begin{pmatrix}-4\\2\end{pmatrix} = \begin{pmatrix}0\\-5\end{pmatrix}$

f $\mathbf{r}+\mathbf{q}-\mathbf{p} = \begin{pmatrix}3\\-2\end{pmatrix}+\begin{pmatrix}-1\\-5\end{pmatrix}-\begin{pmatrix}-4\\2\end{pmatrix} = \begin{pmatrix}6\\-9\end{pmatrix}$

3 **a** $\overrightarrow{AC} = \overrightarrow{AB}+\overrightarrow{BC} = -\overrightarrow{BA}+\overrightarrow{BC} = -\begin{pmatrix}2\\-3\end{pmatrix}+\begin{pmatrix}-3\\1\end{pmatrix} = \begin{pmatrix}-5\\4\end{pmatrix}$

b $\overrightarrow{CB} = \overrightarrow{CA}+\overrightarrow{AB} = \begin{pmatrix}2\\-1\end{pmatrix}+\begin{pmatrix}-1\\3\end{pmatrix} = \begin{pmatrix}1\\2\end{pmatrix}$

c $\overrightarrow{SP} = \overrightarrow{SR}+\overrightarrow{RQ}+\overrightarrow{QP} = -\overrightarrow{RS}+\overrightarrow{RQ}-\overrightarrow{PQ} = -\begin{pmatrix}-3\\2\end{pmatrix}+\begin{pmatrix}2\\1\end{pmatrix}-\begin{pmatrix}-1\\4\end{pmatrix} = \begin{pmatrix}6\\-5\end{pmatrix}$

4 **a** $\overrightarrow{AB} = \begin{pmatrix}b_1-a_1\\b_2-a_2\end{pmatrix} = \begin{pmatrix}4-2\\7-3\end{pmatrix} = \begin{pmatrix}2\\4\end{pmatrix}$

b $\overrightarrow{AB} = \begin{pmatrix}b_1-a_1\\b_2-a_2\end{pmatrix} = \begin{pmatrix}1-3\\4--1\end{pmatrix} = \begin{pmatrix}-2\\5\end{pmatrix}$

c $\overrightarrow{AB} = \begin{pmatrix}b_1-a_1\\b_2-a_2\end{pmatrix} = \begin{pmatrix}1--2\\4-7\end{pmatrix} = \begin{pmatrix}3\\-3\end{pmatrix}$

d $\overrightarrow{AB} = \begin{pmatrix}b_1-a_1\\b_2-a_2\end{pmatrix} = \begin{pmatrix}3-2\\0-5\end{pmatrix} = \begin{pmatrix}1\\-5\end{pmatrix}$

e $\overrightarrow{AB} = \begin{pmatrix}b_1-a_1\\b_2-a_2\end{pmatrix} = \begin{pmatrix}6-0\\-1-4\end{pmatrix} = \begin{pmatrix}6\\-5\end{pmatrix}$

f $\overrightarrow{AB} = \begin{pmatrix}b_1-a_1\\b_2-a_2\end{pmatrix} = \begin{pmatrix}0--1\\0--3\end{pmatrix} = \begin{pmatrix}1\\3\end{pmatrix}$

EXERCISE 12C.3

1 **a** $-3\mathbf{p} = -3\begin{pmatrix}1\\5\end{pmatrix} = \begin{pmatrix}-3\\-15\end{pmatrix}$

b $\frac{1}{2}\mathbf{q} = \frac{1}{2}\begin{pmatrix}-2\\4\end{pmatrix} = \begin{pmatrix}-1\\2\end{pmatrix}$

c $\begin{aligned}2\mathbf{p}+\mathbf{q} &= 2\begin{pmatrix}1\\5\end{pmatrix} + \begin{pmatrix}-2\\4\end{pmatrix}\\ &= \begin{pmatrix}2\\10\end{pmatrix} + \begin{pmatrix}-2\\4\end{pmatrix}\\ &= \begin{pmatrix}0\\14\end{pmatrix}\end{aligned}$

d $\begin{aligned}\mathbf{p}-2\mathbf{q} &= \begin{pmatrix}1\\5\end{pmatrix} - 2\begin{pmatrix}-2\\4\end{pmatrix}\\ &= \begin{pmatrix}1\\5\end{pmatrix} - \begin{pmatrix}-4\\8\end{pmatrix}\\ &= \begin{pmatrix}5\\-3\end{pmatrix}\end{aligned}$

e $\begin{aligned}\mathbf{p}-\tfrac{1}{2}\mathbf{r} &= \begin{pmatrix}1\\5\end{pmatrix} - \begin{pmatrix}-\frac{3}{2}\\-\frac{1}{2}\end{pmatrix}\\ &= \begin{pmatrix}\frac{5}{2}\\\frac{11}{2}\end{pmatrix}\end{aligned}$

f $\begin{aligned}2\mathbf{p}+3\mathbf{r} &= 2\begin{pmatrix}1\\5\end{pmatrix} + 3\begin{pmatrix}-3\\-1\end{pmatrix}\\ &= \begin{pmatrix}2\\10\end{pmatrix} + \begin{pmatrix}-9\\-3\end{pmatrix}\\ &= \begin{pmatrix}-7\\7\end{pmatrix}\end{aligned}$

g $\begin{aligned}2\mathbf{q}-3\mathbf{r} &= 2\begin{pmatrix}-2\\4\end{pmatrix} - 3\begin{pmatrix}-3\\-1\end{pmatrix}\\ &= \begin{pmatrix}-4\\8\end{pmatrix} - \begin{pmatrix}-9\\-3\end{pmatrix}\\ &= \begin{pmatrix}5\\11\end{pmatrix}\end{aligned}$

h $\begin{aligned}2\mathbf{p}-\mathbf{q}+\tfrac{1}{3}\mathbf{r} &= \begin{pmatrix}2\\10\end{pmatrix} - \begin{pmatrix}-2\\4\end{pmatrix} + \begin{pmatrix}-1\\-\frac{1}{3}\end{pmatrix}\\ &= \begin{pmatrix}3\\\frac{17}{3}\end{pmatrix}\end{aligned}$

2 **a**

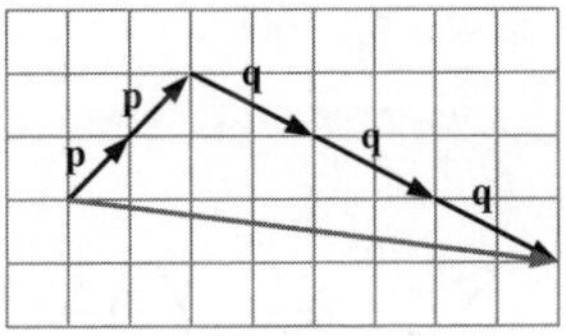

$= \begin{pmatrix}8\\-1\end{pmatrix}$

b

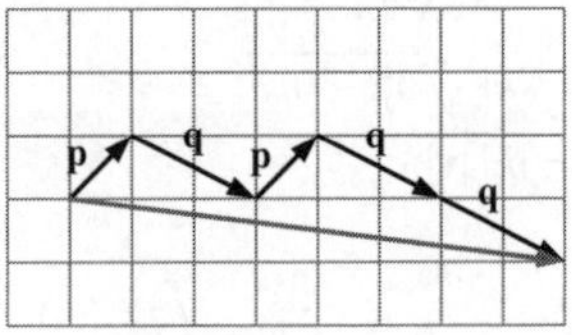

$= \begin{pmatrix}8\\-1\end{pmatrix}$

c

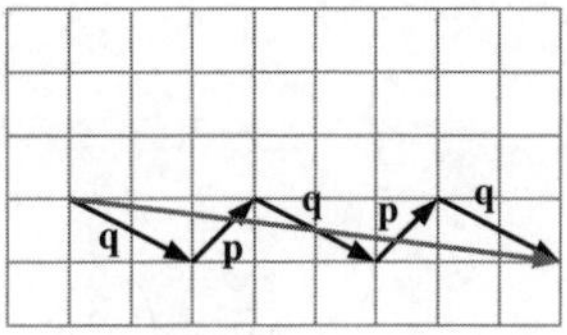

$= \begin{pmatrix}8\\-1\end{pmatrix}$

The vector expressions are equal, as each consists of 2 **p**s and 3 **q**s. Each expression is equal to $2\mathbf{p}+3\mathbf{q}$.

EXERCISE 12C.4

1 **a** $\begin{aligned}|\mathbf{r}| &= \sqrt{2^2+3^2}\\ &= \sqrt{13} \text{ units}\end{aligned}$

b $\begin{aligned}|\mathbf{s}| &= \sqrt{(-1)^2+4^2}\\ &= \sqrt{17} \text{ units}\end{aligned}$

c $\begin{aligned}&\mathbf{r}+\mathbf{s}\\ &= \begin{pmatrix}2\\3\end{pmatrix} + \begin{pmatrix}-1\\4\end{pmatrix}\\ &= \begin{pmatrix}1\\7\end{pmatrix}\\ \therefore\ &|\mathbf{r}+\mathbf{s}|\\ &= \sqrt{1^2+7^2}\\ &= \sqrt{50} \text{ units}\\ &= 5\sqrt{2} \text{ units}\end{aligned}$

d $\begin{aligned}&\mathbf{r}-\mathbf{s}\\ &= \begin{pmatrix}2\\3\end{pmatrix} - \begin{pmatrix}-1\\4\end{pmatrix}\\ &= \begin{pmatrix}3\\-1\end{pmatrix}\\ \therefore\ &|\mathbf{r}-\mathbf{s}|\\ &= \sqrt{3^2+(-1)^2}\\ &= \sqrt{10} \text{ units}\end{aligned}$

e $\begin{aligned}&\mathbf{s}-2\mathbf{r}\\ &= \begin{pmatrix}-1\\4\end{pmatrix} - \begin{pmatrix}4\\6\end{pmatrix}\\ &= \begin{pmatrix}-5\\-2\end{pmatrix}\\ \therefore\ &|\mathbf{s}-2\mathbf{r}|\\ &= \sqrt{(-5)^2+(-2)^2}\\ &= \sqrt{29} \text{ units}\end{aligned}$

2 **a** $\begin{aligned}|\mathbf{p}| &= \sqrt{1^2+3^2}\\ &= \sqrt{10} \text{ units}\end{aligned}$

b $\begin{aligned}2\mathbf{p} &= \begin{pmatrix}2\\6\end{pmatrix}\\ \therefore\ |2\mathbf{p}| &= \sqrt{2^2+6^2}\\ &= \sqrt{4+36}\\ &= \sqrt{40}\\ &= 2\sqrt{10} \text{ units}\end{aligned}$

c $\begin{aligned}-2\mathbf{p} &= \begin{pmatrix}-2\\-6\end{pmatrix}\\ \therefore\ |-2\mathbf{p}| &= \sqrt{(-2)^2+(-6)^2}\\ &= \sqrt{4+36}\\ &= \sqrt{40}\\ &= 2\sqrt{10} \text{ units}\end{aligned}$

d $\begin{aligned}3\mathbf{p} &= \begin{pmatrix}3\\9\end{pmatrix}\\ \therefore\ |3\mathbf{p}| &= \sqrt{3^2+9^2}\\ &= \sqrt{9+81}\\ &= \sqrt{90}\\ &= 3\sqrt{10} \text{ units}\end{aligned}$

e $\begin{aligned}-3\mathbf{p} &= \begin{pmatrix}-3\\-9\end{pmatrix}\\ \therefore\ |-3\mathbf{p}| &= \sqrt{(-3)^2+(-9)^2}\\ &= \sqrt{9+81}\\ &= \sqrt{90}\\ &= 3\sqrt{10} \text{ units}\end{aligned}$

f $\begin{aligned}|\mathbf{q}| &= \sqrt{(-2)^2+4^2}\\ &= \sqrt{4+16}\\ &= \sqrt{20}\\ &= 2\sqrt{5} \text{ units}\end{aligned}$

g $4\mathbf{q} = \binom{-8}{16}$

$\therefore\ |4\mathbf{q}| = \sqrt{(-8)^2 + 16^2}$
$= \sqrt{64 + 256}$
$= \sqrt{320}$
$= 8\sqrt{5}$ units

h $-4\mathbf{q} = \binom{8}{-16}$

$\therefore\ |-4\mathbf{q}| = \sqrt{8^2 + (-16)^2}$
$= \sqrt{64 + 256}$
$= \sqrt{320}$
$= 8\sqrt{5}$ units

i $\frac{1}{2}\mathbf{q} = \binom{-1}{2}$

$\therefore\ \left|\frac{1}{2}\mathbf{q}\right| = \sqrt{(-1)^2 + 2^2}$
$= \sqrt{5}$ units

j $-\frac{1}{2}\mathbf{q} = \binom{1}{-2}$

$\therefore\ \left|-\frac{1}{2}\mathbf{q}\right| = \sqrt{1^2 + (-2)^2}$
$= \sqrt{5}$ units

3 $k\mathbf{v} = \begin{pmatrix} kv_1 \\ kv_2 \end{pmatrix}$ $\therefore\ |k\mathbf{v}| = \sqrt{(kv_1)^2 + (kv_2)^2}$
$= \sqrt{k^2{v_1}^2 + k^2{v_2}^2}$
$= \sqrt{k^2({v_1}^2 + {v_2}^2)}$
$= \sqrt{k^2}\sqrt{{v_1}^2 + {v_2}^2}$
$= |k|\sqrt{{v_1}^2 + {v_2}^2}$
$= |k|\,|\mathbf{v}|$

4 **a** $\overrightarrow{AB} = \begin{pmatrix} 3-2 \\ 5--1 \end{pmatrix} = \binom{1}{6}$

$\therefore\ AB = \sqrt{1^2 + 6^2} = \sqrt{37}$ units

b $\overrightarrow{BA} = \begin{pmatrix} 2-3 \\ -1-5 \end{pmatrix} = \binom{-1}{-6}$

$\therefore\ BA = \sqrt{(-1)^2 + (-6)^2} = \sqrt{37}$ units

c $\overrightarrow{BC} = \begin{pmatrix} -1-3 \\ 4-5 \end{pmatrix} = \binom{-4}{-1}$

$\therefore\ BC = \sqrt{(-4)^2 + (-1)^2} = \sqrt{17}$ units

d $\overrightarrow{DC} = \begin{pmatrix} -1--4 \\ 4--3 \end{pmatrix} = \binom{3}{7}$

$\therefore\ DC = \sqrt{3^2 + 7^2} = \sqrt{58}$ units

e $\overrightarrow{CA} = \begin{pmatrix} 2--1 \\ -1-4 \end{pmatrix} = \binom{3}{-5}$

$\therefore\ CA = \sqrt{3^2 + (-5)^2} = \sqrt{34}$ units

f $\overrightarrow{DA} = \begin{pmatrix} 2--4 \\ -1--3 \end{pmatrix} = \binom{6}{2}$

$\therefore\ DA = \sqrt{6^2 + 2^2} = \sqrt{40} = 2\sqrt{10}$ units

EXERCISE 12D

1 **a**

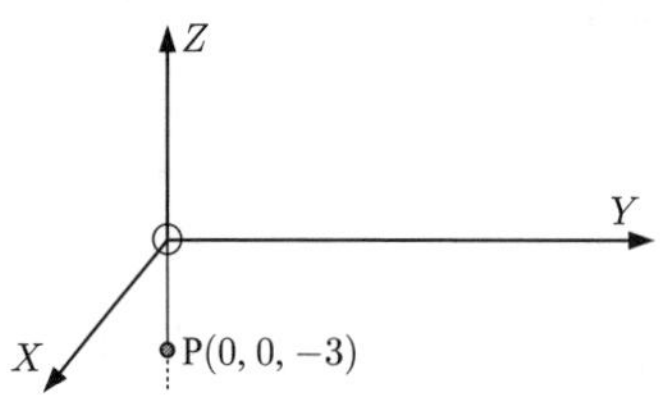

$OP = \sqrt{0^2 + 0^2 + (-3)^2} = 3$ units

b

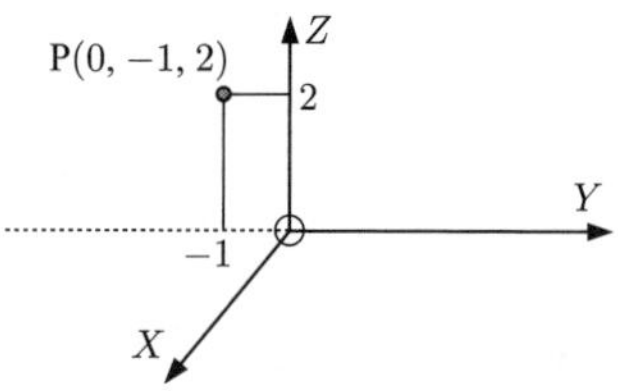

$OP = \sqrt{0^2 + (-1)^2 + 2^2} = \sqrt{5}$ units

c

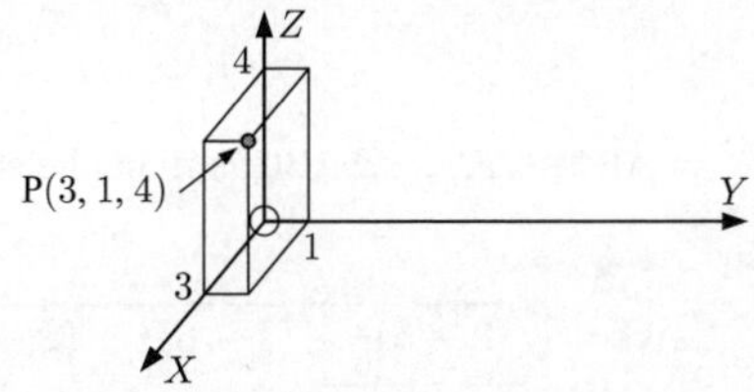

$\text{OP} = \sqrt{3^2 + 1^2 + 4^2}$
$= \sqrt{26}$ units

d

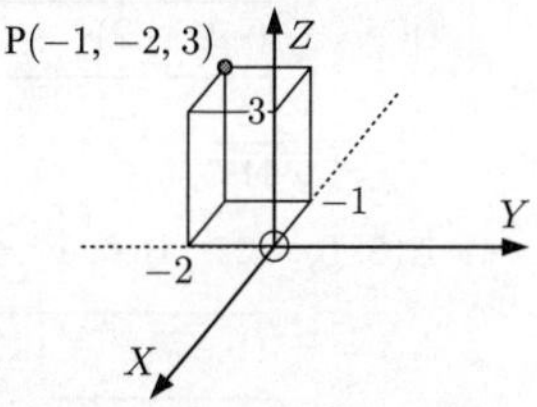

$\text{OP} = \sqrt{(-1)^2 + (-2)^2 + 3^2}$
$= \sqrt{14}$ units

2 **a** **i** AB
$= \sqrt{(0 - -1)^2 + (-1-2)^2 + (1-3)^2}$
$= \sqrt{1+9+4}$
$= \sqrt{14}$ units

ii Midpoint is at $\left(\dfrac{-1+0}{2}, \dfrac{2-1}{2}, \dfrac{3+1}{2}\right)$

which is $\left(-\frac{1}{2}, \frac{1}{2}, 2\right)$.

b **i** AB
$= \sqrt{(2-0)^2 + (-1-0)^2 + (3-0)^2}$
$= \sqrt{4+1+9}$
$= \sqrt{14}$ units

ii Midpoint is at $\left(\dfrac{0+2}{2}, \dfrac{0-1}{2}, \dfrac{0+3}{2}\right)$

which is $\left(1, -\frac{1}{2}, \frac{3}{2}\right)$.

c **i** AB
$= \sqrt{(-1-3)^2 + (0 - -1)^2 + (1 - -1)^2}$
$= \sqrt{16+1+4}$
$= \sqrt{21}$ units

ii Midpoint is at $\left(\dfrac{3-1}{2}, \dfrac{-1+0}{2}, \dfrac{-1+1}{2}\right)$

which is $\left(1, -\frac{1}{2}, 0\right)$.

d **i** AB
$= \sqrt{(0-2)^2 + (1-0)^2 + (0 - -3)^2}$
$= \sqrt{4+1+9}$
$= \sqrt{14}$ units

ii Midpoint is at $\left(\dfrac{2+0}{2}, \dfrac{0+1}{2}, \dfrac{-3+0}{2}\right)$

which is $\left(1, \frac{1}{2}, -\frac{3}{2}\right)$.

3 P(0, 4, 4), Q(2, 6, 5), R(1, 4, 3)

$\text{PQ} = \sqrt{(2-0)^2 + (6-4)^2 + (5-4)^2}$
$= \sqrt{4+4+1}$
$= 3$

$\text{PR} = \sqrt{(1-0)^2 + (4-4)^2 + (3-4)^2}$
$= \sqrt{1+0+1}$
$= \sqrt{2}$

$\text{QR} = \sqrt{(1-2)^2 + (4-6)^2 + (3-5)^2}$
$= \sqrt{1+4+4}$
$= 3$

$\therefore$ PQ = QR and so ΔPQR is isosceles.

4 **a** A(2, −1, 7), B(3, 1, 4), C(5, 4, 5)

$\text{AB} = \sqrt{(3-2)^2 + (1 - -1)^2 + (4-7)^2}$
$= \sqrt{1+4+9}$
$= \sqrt{14}$

$\text{AC} = \sqrt{(5-2)^2 + (4 - -1)^2 + (5-7)^2}$
$= \sqrt{9+25+4}$
$= \sqrt{38}$

$\text{BC} = \sqrt{(5-3)^2 + (4-1)^2 + (5-4)^2}$
$= \sqrt{4+9+1}$
$= \sqrt{14}$

Since AB = BC, ΔABC is isosceles.

b A(0, 0, 3) B(2, 8, 1) C(−9, 6, 18)

$\text{AB} = \sqrt{(2-0)^2 + (8-0)^2 + (1-3)^2}$
$= \sqrt{4+64+4}$
$= \sqrt{72}$

$\text{AC} = \sqrt{(-9-0)^2 + (6-0)^2 + (18-3)^2}$
$= \sqrt{81+36+225}$
$= \sqrt{342}$

$BC = \sqrt{(-9-2)^2 + (6-8)^2 + (18-1)^2}$
$= \sqrt{121 + 4 + 289}$
$= \sqrt{414}$ Since $BC^2 = AB^2 + AC^2$, ΔABC is right angled.

c A(5, 6, −2) B(6, 12, 9) C(2, 4, 2)

$AB = \sqrt{(6-5)^2 + (12-6)^2 + (9--2)^2}$
$= \sqrt{1 + 36 + 121}$
$= \sqrt{158}$

$AC = \sqrt{(2-5)^2 + (4-6)^2 + (2--2)^2}$
$= \sqrt{9 + 4 + 16}$
$= \sqrt{29}$

$BC = \sqrt{(2-6)^2 + (4-12)^2 + (2-9)^2}$
$= \sqrt{16 + 64 + 49}$
$= \sqrt{129}$ Since $AB^2 = AC^2 + BC^2$, ΔABC is right angled.

d A(1, 0, −3) B(2, 2, 0) C(4, 6, 6)

$AB = \sqrt{(2-1)^2 + (2-0)^2 + (0--3)^2}$
$= \sqrt{1^2 + 2^2 + 3^2}$
$= \sqrt{14}$

$AC = \sqrt{(4-1)^2 + (6-0)^2 + (6--3)^2}$
$= \sqrt{3^2 + 6^2 + 9^2}$
$= \sqrt{126}$
$= 3\sqrt{14}$

$BC = \sqrt{(4-2)^2 + (6-2)^2 + (6-0)^2}$
$= \sqrt{2^2 + 4^2 + 6^2}$
$= \sqrt{56} = 2\sqrt{14}$

Since AB + BC = AC, the points A, B and C lie on a straight line, so they do not form a triangle.

5

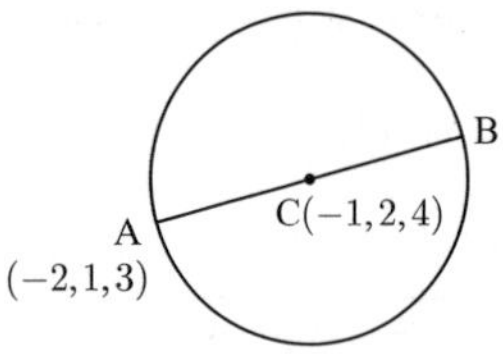

If B is (a, b, c) then $\dfrac{a-2}{2} = -1$, $\dfrac{b+1}{2} = 2$, $\dfrac{c+3}{2} = 4$

$\therefore$ $a = 0$, $b = 3$, $c = 5$

$\therefore$ B is (0, 3, 5)

$r = AC = \sqrt{(-1--2)^2 + (2-1)^2 + (4-3)^2}$
$= \sqrt{1+1+1}$
$= \sqrt{3}$ units

6 **a** $(0, y, 0)$ for any y

b The distance between $(0, y, 0)$ and B(−1, −1, 2) is $\sqrt{(-1)^2 + (-1-y)^2 + 2^2}$.

$\therefore \sqrt{1 + (y+1)^2 + 4} = \sqrt{14}$
$\therefore (y+1)^2 = 9$
$\therefore y + 1 = \pm 3$
$\therefore y = -1 \pm 3$
$\therefore y = -4$ or 2 $\therefore$ the two points are (0, −4, 0) and (0, 2, 0).

EXERCISE 12E

1 **a**

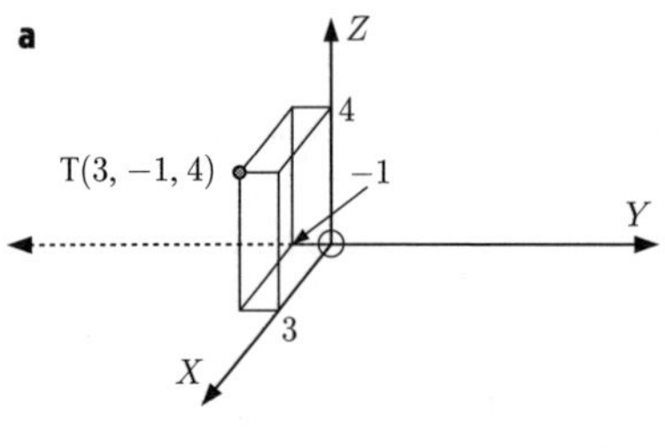

b $\overrightarrow{OT} = \begin{pmatrix} 3 \\ -1 \\ 4 \end{pmatrix}$

c $OT = \sqrt{(3-0)^2 + (-1-0)^2 + (4-0)^2}$
$= \sqrt{9 + 1 + 16}$
$= \sqrt{26}$ units

2 **a** $\overrightarrow{AB} = \begin{pmatrix} 1-(-3) \\ 0-1 \\ -1-2 \end{pmatrix} = \begin{pmatrix} 4 \\ -1 \\ -3 \end{pmatrix}$

$\overrightarrow{BA} = \begin{pmatrix} -3-1 \\ 1-0 \\ 2-(-1) \end{pmatrix} = \begin{pmatrix} -4 \\ 1 \\ 3 \end{pmatrix}$

b $|\overrightarrow{AB}| = \sqrt{4^2+(-1)^2+(-3)^2}$
$= \sqrt{26}$ units

$|\overrightarrow{BA}| = \sqrt{(-4)^2+1^2+3^2}$
$= \sqrt{26}$ units

3 $\overrightarrow{OA} = \begin{pmatrix} 3 \\ 1 \\ 0 \end{pmatrix}$ $\overrightarrow{OB} = \begin{pmatrix} -1 \\ 1 \\ 2 \end{pmatrix}$ $\overrightarrow{AB} = \begin{pmatrix} -1-3 \\ 1-1 \\ 2-0 \end{pmatrix} = \begin{pmatrix} -4 \\ 0 \\ 2 \end{pmatrix}$

4 **a** The position vector of M from N
$= \overrightarrow{NM} = \begin{pmatrix} 4-(-1) \\ -2-2 \\ -1-0 \end{pmatrix} = \begin{pmatrix} 5 \\ -4 \\ -1 \end{pmatrix}.$

b The position vector of N from M
$= \overrightarrow{MN} = \begin{pmatrix} -1-4 \\ 2-(-2) \\ 0-(-1) \end{pmatrix} = \begin{pmatrix} -5 \\ 4 \\ 1 \end{pmatrix}.$

c $MN = \sqrt{(-5)^2+4^2+1^2} = \sqrt{25+16+1} = \sqrt{42}$ units

5 **a** The position vector of A from O
$= \overrightarrow{OA} = \begin{pmatrix} -1 \\ 2 \\ 5 \end{pmatrix}.$

$\therefore\ OA = \sqrt{(-1)^2+2^2+5^2}$
$= \sqrt{1+4+25}$
$= \sqrt{30}$ units

b The position vector of C from A
$= \overrightarrow{AC} = \begin{pmatrix} -3-(-1) \\ 1-2 \\ 0-5 \end{pmatrix} = \begin{pmatrix} -2 \\ -1 \\ -5 \end{pmatrix}.$

$\therefore\ AC = \sqrt{(-2)^2+(-1)^2+(-5)^2}$
$= \sqrt{4+1+25}$
$= \sqrt{30}$ units

c The position vector of B from C
$= \overrightarrow{CB} = \begin{pmatrix} 5 \\ -1 \\ 3 \end{pmatrix}$

and $CB = \sqrt{5^2+(-1)^2+3^2}$
$= \sqrt{25+1+9}$
$= \sqrt{35}$ units

6

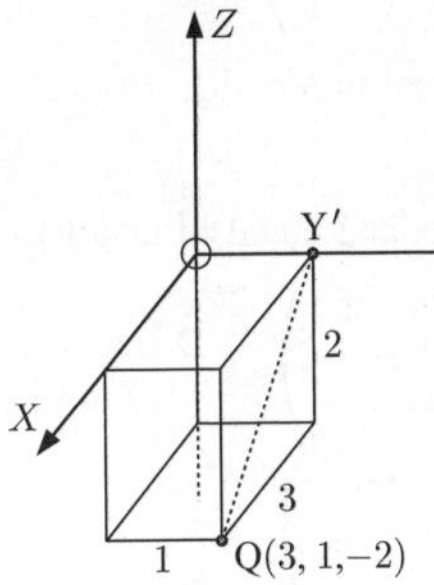

a The distance from Q to the Y-axis is the distance from Q to Y′(0, 1, 0)

$\therefore\ QY' = \sqrt{(3-0)^2+(1-1)^2+(-2-0)^2}$
$= \sqrt{9+4}$
$= \sqrt{13}$ units

b The distance from Q to the origin is
$QO = \sqrt{(3-0)^2+(1-0)^2+(-2-0)^2}$
$= \sqrt{9+1+4}$
$= \sqrt{14}$ units

c The distance from Q to the ZOY plane is the distance from Q to (0, 1, −2), which is 3 units.

7 **a** $\begin{pmatrix} a-4 \\ b-3 \\ c+2 \end{pmatrix} = \begin{pmatrix} 1 \\ 3 \\ -4 \end{pmatrix}$

$\therefore\ \begin{cases} a-4=1 \\ b-3=3 \\ c+2=-4 \end{cases}$

$\therefore\ a=5,\quad b=6,\quad c=-6$

b $\begin{pmatrix} a-5 \\ b-2 \\ c+3 \end{pmatrix} = \begin{pmatrix} 3-a \\ 2-b \\ 5-c \end{pmatrix}$

$\therefore\ \begin{cases} a-5=3-a \\ b-2=2-b \\ c+3=5-c \end{cases}$

$\therefore\ 2a=8,\quad 2b=4,\quad 2c=2$

$\therefore\ a=4,\quad b=2,\quad c=1$

8 **a** $2\begin{pmatrix}1\\0\\3a\end{pmatrix}=\begin{pmatrix}b\\c-1\\2\end{pmatrix}$

$\therefore\ \begin{pmatrix}2\\0\\6a\end{pmatrix}=\begin{pmatrix}b\\c-1\\2\end{pmatrix}$

$\therefore\ 6a=2,\ b=2,\ c-1=0$

$\therefore\ a=\frac{1}{3},\ b=2,\ c=1$

b $\begin{pmatrix}2\\a\\3\end{pmatrix}=\begin{pmatrix}b\\a^2\\a+b\end{pmatrix}$

$\therefore\ b=2,\ a^2=a,\ a+b=3$

$\therefore\ a=1,\ b=2$

c $a\begin{pmatrix}1\\1\\0\end{pmatrix}+b\begin{pmatrix}2\\0\\-1\end{pmatrix}+c\begin{pmatrix}0\\1\\1\end{pmatrix}=\begin{pmatrix}-1\\3\\3\end{pmatrix}$

$\therefore\ a+2b=-1$ .... (1), $a+c=3$ .... (2) and $-b+c=3$ .... (3)

(1) – (2) gives: $2b-c=-4$ .... (4)

Adding (3) and (4) gives: $b=-1$

$\therefore$ using (3), $c=2$

and using (2), $a=1$

$\therefore\ a=1,\ b=-1,\ c=2$

9 A(−1, 3, 4), B(2, 5, −1), C(−1, 2, −2), D(r, s, t)

a If $\overrightarrow{AC}=\overrightarrow{BD}$ then $\begin{pmatrix}-1-(-1)\\2-3\\-2-4\end{pmatrix}=\begin{pmatrix}r-2\\s-5\\t+1\end{pmatrix}$

$\therefore\ r-2=0,\ s-5=-1$ and $t+1=-6$ $\quad\therefore\ r=2,\ s=4$ and $t=-7$

b If $\overrightarrow{AB}=\overrightarrow{DC}$ then $\begin{pmatrix}2-(-1)\\5-3\\-1-4\end{pmatrix}=\begin{pmatrix}-1-r\\2-s\\-2-t\end{pmatrix}$

$\therefore\ -1-r=3,\ 2-s=2$ and $-2-t=-5$ $\quad\therefore\ r=-4,\ s=0$ and $t=3$

10 **a** $\overrightarrow{AB}=\begin{pmatrix}3-1\\-3-2\\2-3\end{pmatrix}=\begin{pmatrix}2\\-5\\-1\end{pmatrix}$ and $\overrightarrow{DC}=\begin{pmatrix}7-5\\-4-1\\5-6\end{pmatrix}=\begin{pmatrix}2\\-5\\-1\end{pmatrix}$.

b ABCD is a parallelogram since its opposite sides are parallel and equal in length.

11 **a** Suppose S is at $(x,\ y,\ z)$. $\quad\overrightarrow{PQ}=\overrightarrow{SR}$ {opposite sides are parallel and equal in length}

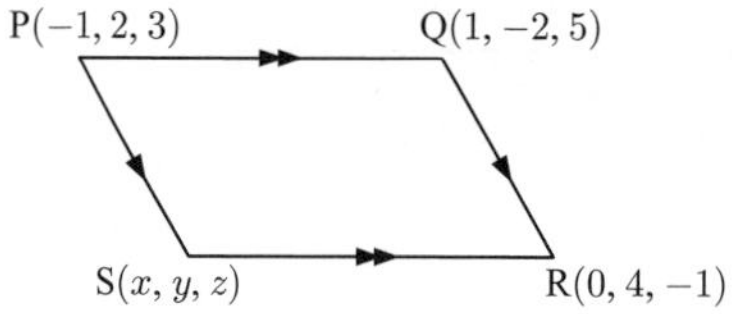

$\therefore\ \begin{pmatrix}1-(-1)\\-2-2\\5-3\end{pmatrix}=\begin{pmatrix}0-x\\4-y\\-1-z\end{pmatrix}$

$\therefore\ \begin{pmatrix}2\\-4\\2\end{pmatrix}=\begin{pmatrix}-x\\4-y\\-1-z\end{pmatrix}$

$\therefore\ -x=2 \qquad 4-y=-4 \qquad -1-z=2$

$\therefore\ x=-2 \qquad y=8 \qquad z=-3$

$\therefore$ S is at (−2, 8, −3).

b The midpoint of [PR] is $\left(\dfrac{-1+0}{2},\ \dfrac{2+4}{2},\ \dfrac{3+(-1)}{2}\right)$ which is $(-\frac{1}{2},\ 3,\ 1)$.

The midpoint of [QS] is $\left(\dfrac{1+(-2)}{2},\ \dfrac{-2+8}{2},\ \dfrac{5+(-3)}{2}\right)$ which is $(-\frac{1}{2},\ 3,\ 1)$.

So, [PR] and [QS] have the same midpoint. ✓

EXERCISE 12F

1 **a** $2\mathbf{x} = \mathbf{q}$
$\therefore \frac{1}{2}(2\mathbf{x}) = \frac{1}{2}\mathbf{q}$
$\therefore \mathbf{x} = \frac{1}{2}\mathbf{q}$

b $\frac{1}{2}\mathbf{x} = \mathbf{n}$
$\therefore 2(\frac{1}{2}\mathbf{x}) = 2\mathbf{n}$
$\therefore \mathbf{x} = 2\mathbf{n}$

c $-3\mathbf{x} = \mathbf{p}$
$\therefore 3\mathbf{x} = -\mathbf{p}$
$\therefore \frac{1}{3}(3\mathbf{x}) = -\frac{1}{3}\mathbf{p}$
$\therefore \mathbf{x} = -\frac{1}{3}\mathbf{p}$

d $\mathbf{q} + 2\mathbf{x} = \mathbf{r}$
$\therefore 2\mathbf{x} = \mathbf{r} - \mathbf{q}$
$\therefore \mathbf{x} = \frac{1}{2}(\mathbf{r} - \mathbf{q})$

e $4\mathbf{s} - 5\mathbf{x} = \mathbf{t}$
$\therefore -5\mathbf{x} = \mathbf{t} - 4\mathbf{s}$
$\therefore 5\mathbf{x} = 4\mathbf{s} - \mathbf{t}$
$\therefore \mathbf{x} = \frac{1}{5}(4\mathbf{s} - \mathbf{t})$

f $4\mathbf{m} - \frac{1}{3}\mathbf{x} = \mathbf{n}$
$\therefore 4\mathbf{m} - \mathbf{n} = \frac{1}{3}\mathbf{x}$
$\therefore \mathbf{x} = 12\mathbf{m} - 3\mathbf{n}$
$= 3(4\mathbf{m} - \mathbf{n})$

2 **a** $2\mathbf{y} = \mathbf{r}$
$\therefore \mathbf{y} = \frac{1}{2}\mathbf{r} = \frac{1}{2}\begin{pmatrix} -2 \\ 3 \end{pmatrix} = \begin{pmatrix} -1 \\ \frac{3}{2} \end{pmatrix}$

b $\frac{1}{2}\mathbf{y} = \mathbf{s}$
$\therefore \mathbf{y} = 2\mathbf{s} = 2\begin{pmatrix} 1 \\ 2 \end{pmatrix} = \begin{pmatrix} 2 \\ 4 \end{pmatrix}$

c $\mathbf{r} + 2\mathbf{y} = \mathbf{s}$
$\therefore 2\mathbf{y} = \mathbf{s} - \mathbf{r}$
$\therefore \mathbf{y} = \frac{1}{2}\mathbf{s} - \frac{1}{2}\mathbf{r}$
$= \begin{pmatrix} \frac{1}{2} \\ 1 \end{pmatrix} - \begin{pmatrix} -1 \\ \frac{3}{2} \end{pmatrix}$
$= \begin{pmatrix} \frac{3}{2} \\ -\frac{1}{2} \end{pmatrix}$

d $3\mathbf{s} - 4\mathbf{y} = \mathbf{r}$
$\therefore 3\mathbf{s} - \mathbf{r} = 4\mathbf{y}$
$\therefore \mathbf{y} = \frac{3}{4}\mathbf{s} - \frac{1}{4}\mathbf{r}$
$= \begin{pmatrix} \frac{3}{4} \\ \frac{3}{2} \end{pmatrix} - \begin{pmatrix} -\frac{1}{2} \\ \frac{3}{4} \end{pmatrix}$
$= \begin{pmatrix} \frac{5}{4} \\ \frac{3}{4} \end{pmatrix}$

3 $k\mathbf{x} = \mathbf{a}$

$\therefore k\begin{pmatrix} x_1 \\ x_2 \end{pmatrix} = \begin{pmatrix} a_1 \\ a_2 \end{pmatrix}$

$\therefore kx_1 = a_1$ and $kx_2 = a_2$

$\therefore x_1 = \frac{1}{k}a_1$ and $x_2 = \frac{1}{k}a_2$

$\therefore \begin{pmatrix} x_1 \\ x_2 \end{pmatrix} = \begin{pmatrix} \frac{1}{k}a_1 \\ \frac{1}{k}a_2 \end{pmatrix} = \frac{1}{k}\begin{pmatrix} a_1 \\ a_2 \end{pmatrix}$ and so $\mathbf{x} = \frac{1}{k}\mathbf{a}$.

4

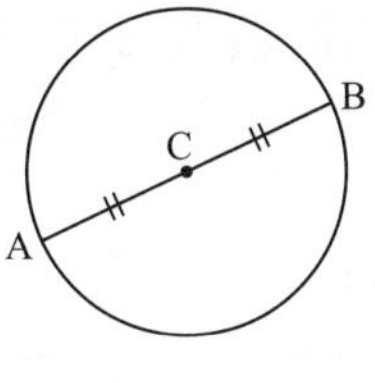

a $\overrightarrow{AC} = \begin{pmatrix} 1-3 \\ 4--2 \end{pmatrix} = \begin{pmatrix} -2 \\ 6 \end{pmatrix}$ $\therefore$ B is $(1-2,\ 4+6)$ or $(-1,\ 10)$.

b $\overrightarrow{AC} = \begin{pmatrix} -1-0 \\ -2-5 \end{pmatrix} = \begin{pmatrix} -1 \\ -7 \end{pmatrix}$ $\therefore$ B is $(-1-1,\ -2-7)$ or $(-2,\ -9)$.

c $\overrightarrow{AC} = \begin{pmatrix} 3--1 \\ 0--4 \end{pmatrix} = \begin{pmatrix} 4 \\ 4 \end{pmatrix}$ $\therefore$ B is $(3+4,\ 0+4)$ or $(7,\ 4)$.

5 **a** M is $\left(\frac{3+-1}{2}, \frac{6+2}{2}\right)$
$\therefore$ M is $(1,\ 4)$

b $\overrightarrow{CA} = \begin{pmatrix} 3--4 \\ 6-1 \end{pmatrix} = \begin{pmatrix} 7 \\ 5 \end{pmatrix}$

$\overrightarrow{CM} = \begin{pmatrix} 1--4 \\ 4-1 \end{pmatrix} = \begin{pmatrix} 5 \\ 3 \end{pmatrix}$

$\overrightarrow{CB} = \begin{pmatrix} -1--4 \\ 2-1 \end{pmatrix} = \begin{pmatrix} 3 \\ 1 \end{pmatrix}$

c $\frac{1}{2}\overrightarrow{CA} + \frac{1}{2}\overrightarrow{CB}$
$= \frac{1}{2}\begin{pmatrix} 7 \\ 5 \end{pmatrix} + \frac{1}{2}\begin{pmatrix} 3 \\ 1 \end{pmatrix}$
$= \begin{pmatrix} 5 \\ 3 \end{pmatrix}$ which is $\overrightarrow{CM}$

6 **a** $2\mathbf{a} + \mathbf{x} = \mathbf{b}$

$\therefore\ \mathbf{x} = \mathbf{b} - 2\mathbf{a}$

$$= \begin{pmatrix} 2 \\ -2 \\ 1 \end{pmatrix} - 2\begin{pmatrix} -1 \\ 2 \\ 3 \end{pmatrix} = \begin{pmatrix} 2 \\ -2 \\ 1 \end{pmatrix} - \begin{pmatrix} -2 \\ 4 \\ 6 \end{pmatrix} = \begin{pmatrix} 4 \\ -6 \\ -5 \end{pmatrix}$$

b $3\mathbf{x} - \mathbf{a} = 2\mathbf{b}$

$\therefore\ 3\mathbf{x} = \mathbf{a} + 2\mathbf{b}$

$\therefore\ \mathbf{x} = \frac{1}{3}(\mathbf{a} + 2\mathbf{b})$

$$= \tfrac{1}{3}\left[\begin{pmatrix} -1 \\ 2 \\ 3 \end{pmatrix} + 2\begin{pmatrix} 2 \\ -2 \\ 1 \end{pmatrix}\right] = \tfrac{1}{3}\left[\begin{pmatrix} -1 \\ 2 \\ 3 \end{pmatrix} + \begin{pmatrix} 4 \\ -4 \\ 2 \end{pmatrix}\right] = \tfrac{1}{3}\begin{pmatrix} 3 \\ -2 \\ 5 \end{pmatrix} = \begin{pmatrix} 1 \\ -\frac{2}{3} \\ \frac{5}{3} \end{pmatrix}$$

c $2\mathbf{b} - 2\mathbf{x} = -\mathbf{a}$

$\therefore\ \mathbf{a} + 2\mathbf{b} = 2\mathbf{x}$

$$\therefore\ \mathbf{x} = \tfrac{1}{2}(\mathbf{a} + 2\mathbf{b}) = \tfrac{1}{2}\begin{pmatrix} 3 \\ -2 \\ 5 \end{pmatrix} \quad \{\text{using } \mathbf{b}\}$$

$$= \begin{pmatrix} \frac{3}{2} \\ -1 \\ \frac{5}{2} \end{pmatrix}$$

7 $$\overrightarrow{AB} = \overrightarrow{AO} + \overrightarrow{OB} = -\overrightarrow{OA} + \overrightarrow{OB} = -\begin{pmatrix} -2 \\ -1 \\ 1 \end{pmatrix} + \begin{pmatrix} 1 \\ 3 \\ -1 \end{pmatrix} = \begin{pmatrix} 2 \\ 1 \\ -1 \end{pmatrix} + \begin{pmatrix} 1 \\ 3 \\ -1 \end{pmatrix} = \begin{pmatrix} 3 \\ 4 \\ -2 \end{pmatrix}$$

$\therefore\ |\overrightarrow{AB}| = \sqrt{3^2 + 4^2 + (-2)^2} = \sqrt{9 + 16 + 4} = \sqrt{29}$ units

8 $$\overrightarrow{OA} = \begin{pmatrix} 2 \\ 1 \\ -2 \end{pmatrix},\ \overrightarrow{OB} = \begin{pmatrix} 0 \\ 3 \\ -4 \end{pmatrix},\ \overrightarrow{OC} = \begin{pmatrix} 1 \\ -2 \\ 1 \end{pmatrix},\ \overrightarrow{OD} = \begin{pmatrix} -2 \\ -3 \\ 2 \end{pmatrix}$$

$$\therefore\ \overrightarrow{BD} = \overrightarrow{BO} + \overrightarrow{OD} = -\overrightarrow{OB} + \overrightarrow{OD} = -\begin{pmatrix} 0 \\ 3 \\ -4 \end{pmatrix} + \begin{pmatrix} -2 \\ -3 \\ 2 \end{pmatrix} = \begin{pmatrix} -2 \\ -6 \\ 6 \end{pmatrix}$$

and $$\overrightarrow{AC} = \overrightarrow{AO} + \overrightarrow{OC} = -\begin{pmatrix} 2 \\ 1 \\ -2 \end{pmatrix} + \begin{pmatrix} 1 \\ -2 \\ 1 \end{pmatrix} = \begin{pmatrix} -1 \\ -3 \\ 3 \end{pmatrix} \quad \therefore\ \overrightarrow{BD} = \begin{pmatrix} -2 \\ -6 \\ 6 \end{pmatrix} = 2\begin{pmatrix} -1 \\ -3 \\ 3 \end{pmatrix} = 2\overrightarrow{AC}$$

9 $$\overrightarrow{AB} = \begin{pmatrix} 2 - -1 \\ 3 - 5 \\ -3 - 2 \end{pmatrix} = \begin{pmatrix} 3 \\ -2 \\ -5 \end{pmatrix}$$

$\therefore$ C is $(2 + 3,\ 3 - 2,\ -3 - 5)$, or $(5,\ 1,\ -8)$,
D is $(5 + 3,\ 1 - 2,\ -8 - 5)$, or $(8,\ -1,\ -13)$,
E is $(8 + 3,\ -1 - 2,\ -13 - 5)$, or $(11,\ -3,\ -18)$.

10

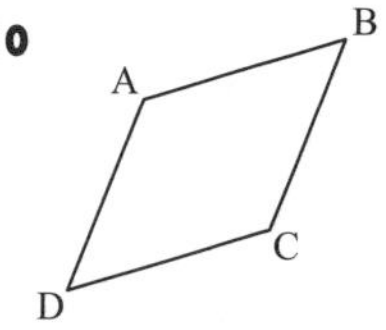

a $$\overrightarrow{AB} = \begin{pmatrix} 4 - 3 \\ 2 - -1 \end{pmatrix} = \begin{pmatrix} 1 \\ 3 \end{pmatrix}$$

$$\overrightarrow{DC} = \begin{pmatrix} -1 - -2 \\ 4 - 1 \end{pmatrix} = \begin{pmatrix} 1 \\ 3 \end{pmatrix}$$

Now $\overrightarrow{AB} = \overrightarrow{DC}$

$\therefore$ sides [AB] and [DC] are equal in length and parallel.

This is sufficient to deduce that ABCD is a parallelogram.

b $$\overrightarrow{AB} = \begin{pmatrix} -1 - 5 \\ 2 - 0 \\ 4 - 3 \end{pmatrix} = \begin{pmatrix} -6 \\ 2 \\ 1 \end{pmatrix}$$

$$\overrightarrow{DC} = \begin{pmatrix} 4 - 10 \\ -3 - -5 \\ 6 - 5 \end{pmatrix} = \begin{pmatrix} -6 \\ 2 \\ 1 \end{pmatrix}$$

So $\overrightarrow{AB} = \overrightarrow{DC}$

$\therefore$ sides [AB] and [DC] are equal in length and parallel.

This is sufficient to deduce that ABCD is a parallelogram.

c $\overrightarrow{AB} = \begin{pmatrix} 1-2 \\ 4--3 \\ -1-2 \end{pmatrix} = \begin{pmatrix} -1 \\ 7 \\ -3 \end{pmatrix}$ So, $\overrightarrow{AB} \neq \overrightarrow{DC}$

$\therefore$ ABCD cannot be a parallelogram.

$\overrightarrow{DC} = \begin{pmatrix} -2--1 \\ 6--1 \\ -2-2 \end{pmatrix} = \begin{pmatrix} -1 \\ 7 \\ -4 \end{pmatrix}$

11 **a** Let D be (a, b).

Now $\overrightarrow{CD} = \overrightarrow{BA}$

$\therefore \begin{pmatrix} a-8 \\ b--2 \end{pmatrix} = \begin{pmatrix} 3-2 \\ 0--1 \end{pmatrix}$

$\therefore \begin{pmatrix} a-8 \\ b+2 \end{pmatrix} = \begin{pmatrix} 1 \\ 1 \end{pmatrix}$

$\therefore$ $a = 9$, $b = -1$

So, D is $(9, -1)$.

b Let R be (a, b, c).

Now $\overrightarrow{SR} = \overrightarrow{PQ}$

$\therefore \begin{pmatrix} a-4 \\ b-0 \\ c-7 \end{pmatrix} = \begin{pmatrix} -2--1 \\ 5-4 \\ 2-3 \end{pmatrix}$

$\therefore \begin{pmatrix} a-4 \\ b \\ c-7 \end{pmatrix} = \begin{pmatrix} -1 \\ 1 \\ -1 \end{pmatrix}$

$\therefore$ $a = 3$, $b = 1$, $c = 6$

So, R is $(3, 1, 6)$.

c Let X be (a, b, c).

Now $\overrightarrow{WX} = \overrightarrow{ZY}$

$\therefore \begin{pmatrix} a--1 \\ b-5 \\ c-8 \end{pmatrix} = \begin{pmatrix} 3-0 \\ -2-4 \\ -2-6 \end{pmatrix}$

$\therefore \begin{pmatrix} a+1 \\ b-5 \\ c-8 \end{pmatrix} = \begin{pmatrix} 3 \\ -6 \\ -8 \end{pmatrix}$

$\therefore$ $a = 2$, $b = -1$, $c = 0$

So, X is $(2, -1, 0)$.

12 **a** $\overrightarrow{BD} = \frac{1}{2}\overrightarrow{OA}$
$= \frac{1}{2}\mathbf{a}$

b $\overrightarrow{AB} = \overrightarrow{AO} + \overrightarrow{OB}$
$= -\mathbf{a} + \mathbf{b}$
$= \mathbf{b} - \mathbf{a}$

c $\overrightarrow{BA} = -\overrightarrow{AB}$
$= -(\mathbf{b} - \mathbf{a})$
$= -\mathbf{b} + \mathbf{a}$ or $\mathbf{a} - \mathbf{b}$

d $\overrightarrow{OD} = \overrightarrow{OB} + \overrightarrow{BD}$
$= \mathbf{b} + \frac{1}{2}\mathbf{a}$

e $\overrightarrow{AD} = \overrightarrow{AO} + \overrightarrow{OD}$
$= -\mathbf{a} + \mathbf{b} + \frac{1}{2}\mathbf{a}$
$= -\frac{1}{2}\mathbf{a} + \mathbf{b}$

f $\overrightarrow{DA} = -\overrightarrow{AD}$
$= \frac{1}{2}\mathbf{a} - \mathbf{b}$

13 **a** $\overrightarrow{AD} = \overrightarrow{AB} + \overrightarrow{BD} = \begin{pmatrix} -1 \\ 3 \\ 2 \end{pmatrix} + \begin{pmatrix} 0 \\ 2 \\ -3 \end{pmatrix} = \begin{pmatrix} -1 \\ 5 \\ -1 \end{pmatrix}$

b $\overrightarrow{CB} = \overrightarrow{CA} + \overrightarrow{AB} = -\overrightarrow{AC} + \overrightarrow{AB} = -\begin{pmatrix} 2 \\ -1 \\ 4 \end{pmatrix} + \begin{pmatrix} -1 \\ 3 \\ 2 \end{pmatrix} = \begin{pmatrix} -2 \\ 1 \\ -4 \end{pmatrix} + \begin{pmatrix} -1 \\ 3 \\ 2 \end{pmatrix} = \begin{pmatrix} -3 \\ 4 \\ -2 \end{pmatrix}$

c $\overrightarrow{CD} = \overrightarrow{CB} + \overrightarrow{BD} = \begin{pmatrix} -3 \\ 4 \\ -2 \end{pmatrix} + \begin{pmatrix} 0 \\ 2 \\ -3 \end{pmatrix}$ {using **b**} $= \begin{pmatrix} -3 \\ 6 \\ -5 \end{pmatrix}$

14 **a** $\mathbf{a} + \mathbf{b} = \begin{pmatrix} 2 \\ -1 \\ 1 \end{pmatrix} + \begin{pmatrix} 1 \\ 2 \\ -3 \end{pmatrix} = \begin{pmatrix} 3 \\ 1 \\ -2 \end{pmatrix}$

b $\mathbf{a} - \mathbf{b} = \begin{pmatrix} 2 \\ -1 \\ 1 \end{pmatrix} - \begin{pmatrix} 1 \\ 2 \\ -3 \end{pmatrix} = \begin{pmatrix} 1 \\ -3 \\ 4 \end{pmatrix}$

c $\mathbf{b} + 2\mathbf{c} = \begin{pmatrix} 1 \\ 2 \\ -3 \end{pmatrix} + 2\begin{pmatrix} 0 \\ 1 \\ -3 \end{pmatrix} = \begin{pmatrix} 1 \\ 2 \\ -3 \end{pmatrix} + \begin{pmatrix} 0 \\ 2 \\ -6 \end{pmatrix} = \begin{pmatrix} 1 \\ 4 \\ -9 \end{pmatrix}$

d $\mathbf{a} - 3\mathbf{c} = \begin{pmatrix} 2 \\ -1 \\ 1 \end{pmatrix} - 3\begin{pmatrix} 0 \\ 1 \\ -3 \end{pmatrix} = \begin{pmatrix} 2 \\ -1 \\ 1 \end{pmatrix} - \begin{pmatrix} 0 \\ 3 \\ -9 \end{pmatrix} = \begin{pmatrix} 2 \\ -4 \\ 10 \end{pmatrix}$

e $\mathbf{a} + \mathbf{b} + \mathbf{c} = \begin{pmatrix} 2 \\ -1 \\ 1 \end{pmatrix} + \begin{pmatrix} 1 \\ 2 \\ -3 \end{pmatrix} + \begin{pmatrix} 0 \\ 1 \\ -3 \end{pmatrix} = \begin{pmatrix} 3 \\ 2 \\ -5 \end{pmatrix}$

f $\mathbf{c} - \frac{1}{2}\mathbf{a} = \begin{pmatrix} 0 \\ 1 \\ -3 \end{pmatrix} - \frac{1}{2}\begin{pmatrix} 2 \\ -1 \\ 1 \end{pmatrix} = \begin{pmatrix} 0 \\ 1 \\ -3 \end{pmatrix} - \begin{pmatrix} 1 \\ -\frac{1}{2} \\ \frac{1}{2} \end{pmatrix} = \begin{pmatrix} -1 \\ \frac{3}{2} \\ -\frac{7}{2} \end{pmatrix}$

g $\mathbf{a}-\mathbf{b}-\mathbf{c}=\begin{pmatrix}2\\-1\\1\end{pmatrix}-\begin{pmatrix}1\\2\\-3\end{pmatrix}-\begin{pmatrix}0\\1\\-3\end{pmatrix}=\begin{pmatrix}1\\-4\\7\end{pmatrix}$

h $2\mathbf{b}-\mathbf{c}+\mathbf{a}=2\begin{pmatrix}1\\2\\-3\end{pmatrix}-\begin{pmatrix}0\\1\\-3\end{pmatrix}+\begin{pmatrix}2\\-1\\1\end{pmatrix}=\begin{pmatrix}2\\4\\-6\end{pmatrix}-\begin{pmatrix}0\\1\\-3\end{pmatrix}+\begin{pmatrix}2\\-1\\1\end{pmatrix}=\begin{pmatrix}4\\2\\-2\end{pmatrix}$

15 **a** $|\mathbf{a}|=\sqrt{(-1)^2+1^2+3^2}$
$=\sqrt{11}$ units

b $|\mathbf{b}|=\sqrt{1^2+(-3)^2+2^2}$
$=\sqrt{14}$ units

c $|\mathbf{b}+\mathbf{c}|=\left|\begin{pmatrix}1\\-3\\2\end{pmatrix}+\begin{pmatrix}-2\\2\\4\end{pmatrix}\right|=\left|\begin{pmatrix}-1\\-1\\6\end{pmatrix}\right|$
$=\sqrt{(-1)^2+(-1)^2+6^2}$
$=\sqrt{1+1+36}$
$=\sqrt{38}$ units

d $|\mathbf{a}-\mathbf{c}|=\left|\begin{pmatrix}-1\\1\\3\end{pmatrix}-\begin{pmatrix}-2\\2\\4\end{pmatrix}\right|=\left|\begin{pmatrix}1\\-1\\-1\end{pmatrix}\right|$
$=\sqrt{1^2+(-1)^2+(-1)^2}$
$=\sqrt{3}$ units

e $|\mathbf{a}|\,\mathbf{b}=\sqrt{11}\begin{pmatrix}1\\-3\\2\end{pmatrix}=\begin{pmatrix}\sqrt{11}\\-3\sqrt{11}\\2\sqrt{11}\end{pmatrix}$

f $\dfrac{1}{|\mathbf{a}|}\,\mathbf{a}=\frac{1}{\sqrt{11}}\begin{pmatrix}-1\\1\\3\end{pmatrix}=\begin{pmatrix}-\frac{1}{\sqrt{11}}\\\frac{1}{\sqrt{11}}\\\frac{3}{\sqrt{11}}\end{pmatrix}$

16 **a** $r\begin{pmatrix}1\\-1\end{pmatrix}+s\begin{pmatrix}2\\5\end{pmatrix}=\begin{pmatrix}-8\\-27\end{pmatrix}$

$\therefore\ \begin{pmatrix}r+2s\\-r+5s\end{pmatrix}=\begin{pmatrix}-8\\-27\end{pmatrix}$

$\therefore\ r+2s=-8$ .... (1)
$-r+5s=-27$
adding $7s=-35$
$\therefore\ s=-5$
and in (1) $r+2(-5)=-8$
$\therefore\ r-10=-8$
$\therefore\ r=2$
So, $r=2,\ s=-5$

b $r\begin{pmatrix}2\\-3\\1\end{pmatrix}+s\begin{pmatrix}1\\7\\2\end{pmatrix}=\begin{pmatrix}7\\-19\\2\end{pmatrix}$

$\therefore\ \begin{cases}2r+s=7 & \text{.... (1)}\\-3r+7s=-19 & \text{.... (2)}\\r+2s=2 & \text{.... (3)}\end{cases}$

$\therefore\ -4r-2s=-14\quad\{-2\times(1)\}$
$r+2s=2$
adding $-3r=-12$
$\therefore\ r=4$
In (1), $2(4)+s=7\quad\therefore\ s=-1$
Checking in (2),
$-3r+7s=-3(4)+7(-1)=-19$ ✓
$\therefore\ r=4,\ s=-1$ satisfies all equations.

EXERCISE 12G

1 Since **a** and **b** are parallel, then $\mathbf{b}=k\mathbf{a}$. $\therefore\ \begin{pmatrix}-6\\r\\s\end{pmatrix}=k\begin{pmatrix}2\\-1\\3\end{pmatrix}=\begin{pmatrix}2k\\-k\\3k\end{pmatrix}$

$\therefore\ 2k=-6,\ r=-k,\ s=3k\quad\therefore\ k=-3,\ r=3,\ s=-9$

2 If $\begin{pmatrix}a\\2\\b\end{pmatrix}$ and $\begin{pmatrix}3\\-1\\2\end{pmatrix}$ are parallel, then $\begin{pmatrix}a\\2\\b\end{pmatrix}=k\begin{pmatrix}3\\-1\\2\end{pmatrix}$.

$\therefore\ 2=-k,\ a=3k,\ b=2k\quad\therefore\ k=-2,\ a=-6$ and $b=-4$

3 **a** Let the vector parallel to **a** be $k\mathbf{a}$.

$$\therefore \quad k\mathbf{a} = k\begin{pmatrix} 2 \\ -1 \\ -2 \end{pmatrix} = \begin{pmatrix} 2k \\ -k \\ -2k \end{pmatrix}$$

Now $k\mathbf{a}$ has length $= 1$,

$$\text{so} \quad \sqrt{(2k)^2 + (-k)^2 + (-2k)^2} = 1$$
$$\therefore \quad 4k^2 + k^2 + 4k^2 = 1$$
$$\therefore \quad 9k^2 = 1$$
$$\therefore \quad k = \pm\tfrac{1}{3}$$

$$\therefore \quad \text{the vector is} \begin{pmatrix} \frac{2}{3} \\ -\frac{1}{3} \\ -\frac{2}{3} \end{pmatrix} \textit{ or } \begin{pmatrix} -\frac{2}{3} \\ \frac{1}{3} \\ \frac{2}{3} \end{pmatrix}$$

b Let the vector parallel to **b** be $k\mathbf{b}$.

$$\therefore \quad k\mathbf{b} = k\begin{pmatrix} -2 \\ -1 \\ 2 \end{pmatrix} = \begin{pmatrix} -2k \\ -k \\ 2k \end{pmatrix}$$

Now $k\mathbf{b}$ has length $= 2$,

$$\text{so} \quad \sqrt{(-2k)^2 + (-k)^2 + (2k)^2} = 2$$
$$\therefore \quad 4k^2 + k^2 + 4k^2 = 4$$
$$\therefore \quad 9k^2 = 4$$
$$\therefore \quad k = \pm\tfrac{2}{3}$$

$$\therefore \quad \text{the vector is} \begin{pmatrix} -\frac{4}{3} \\ -\frac{2}{3} \\ \frac{4}{3} \end{pmatrix} \textit{ or } \begin{pmatrix} \frac{4}{3} \\ \frac{2}{3} \\ -\frac{4}{3} \end{pmatrix}$$

4 **a** $\overrightarrow{AB} = 3\overrightarrow{CD}$ means that $\overrightarrow{AB}$ is parallel to $\overrightarrow{CD}$ and 3 times its length.

b $\overrightarrow{RS} = -\frac{1}{2}\overrightarrow{KL}$ means that $\overrightarrow{RS}$ is parallel to $\overrightarrow{KL}$, half its length, and in the opposite direction.

c

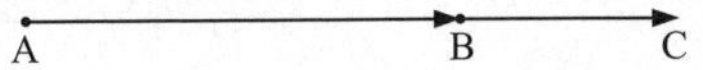

$\overrightarrow{AB} = 2\overrightarrow{BC}$ means that A, B and C are collinear and the length of $\overrightarrow{AB}$ is twice the length of $\overrightarrow{BC}$.

5 $\overrightarrow{OP} = \begin{pmatrix} 3 \\ 2 \\ -1 \end{pmatrix}$, $\overrightarrow{OQ} = \begin{pmatrix} 1 \\ 4 \\ -3 \end{pmatrix}$, $\overrightarrow{OR} = \begin{pmatrix} 2 \\ -1 \\ 2 \end{pmatrix}$, $\overrightarrow{OS} = \begin{pmatrix} -1 \\ -2 \\ 3 \end{pmatrix}$

a $$\overrightarrow{PR} = \overrightarrow{PO} + \overrightarrow{OR} = -\begin{pmatrix} 3 \\ 2 \\ -1 \end{pmatrix} + \begin{pmatrix} 2 \\ -1 \\ 2 \end{pmatrix} = \begin{pmatrix} -1 \\ -3 \\ 3 \end{pmatrix}$$

$$\overrightarrow{QS} = \overrightarrow{QO} + \overrightarrow{OS} = -\begin{pmatrix} 1 \\ 4 \\ -3 \end{pmatrix} + \begin{pmatrix} -1 \\ -2 \\ 3 \end{pmatrix} = \begin{pmatrix} -2 \\ -6 \\ 6 \end{pmatrix} = 2\begin{pmatrix} -1 \\ -3 \\ 3 \end{pmatrix} = 2\overrightarrow{PR} \quad \text{and so [QS]} \parallel \text{[PR]}.$$

b Since $\overrightarrow{QS} = 2\overrightarrow{PR}$, $|\overrightarrow{QS}| = 2|\overrightarrow{PR}|$, and so [QS] is twice as long as [PR].

6
- Consider **a** not parallel to **b**:

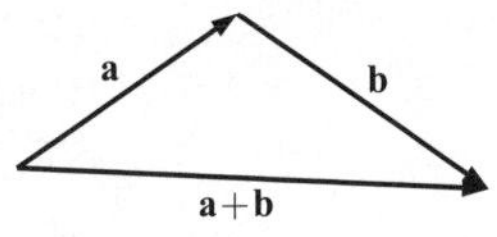

Clearly, $|\mathbf{a}| + |\mathbf{b}| > |\mathbf{a}+\mathbf{b}|$

- Consider **a** parallel to **b**:

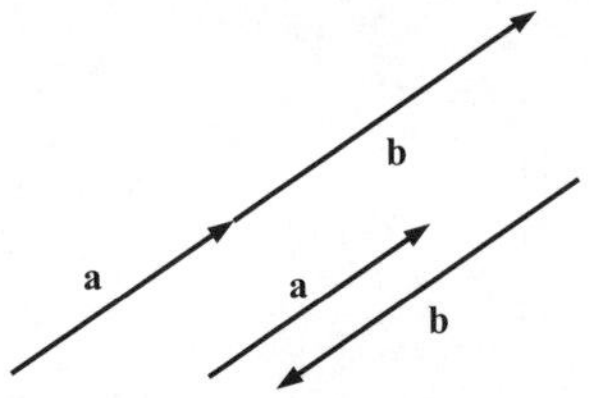

$|\mathbf{a}| + |\mathbf{b}| = |\mathbf{a}+\mathbf{b}|$

or

$|\mathbf{a}| + |\mathbf{b}| > |\mathbf{a}+\mathbf{b}|$

- If $\mathbf{a} = \mathbf{0}$, $\mathbf{b} \neq \mathbf{0}$, then $\mathbf{a}+\mathbf{b} = \mathbf{b}$ $\quad \therefore \quad |\mathbf{a}| + |\mathbf{b}| = \mathbf{0} + |\mathbf{b}| = |\mathbf{b}| = |\mathbf{a}+\mathbf{b}|$

  Similarly if $\mathbf{a} \neq \mathbf{0}$, $\mathbf{b} = \mathbf{0}$, then $\mathbf{a}+\mathbf{b} = \mathbf{a}$

  $$\therefore \quad |\mathbf{a}| + |\mathbf{b}| = |\mathbf{a}| + \mathbf{0} = |\mathbf{a}| = |\mathbf{a}+\mathbf{b}|$$

  If $\mathbf{a} = \mathbf{0}$ and $\mathbf{b} = \mathbf{0}$, then $\mathbf{a}+\mathbf{b} = \mathbf{0}$

  $$\therefore \quad |\mathbf{a}| + |\mathbf{b}| = |\mathbf{a}+\mathbf{b}|$$

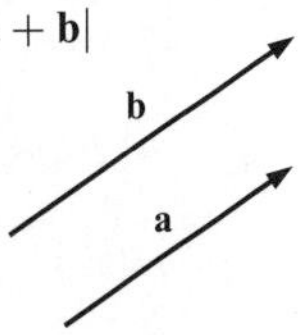

Combining **all** possibilities, $|\mathbf{a}| + |\mathbf{b}| \geqslant |\mathbf{a}+\mathbf{b}|$, or $|\mathbf{a}+\mathbf{b}| \leqslant |\mathbf{a}| + |\mathbf{b}|$

EXERCISE 12H

1 **a** $\sqrt{0^2+(-1)^2}=1$

$\therefore\ \binom{0}{-1}$ is a unit vector.

b $\sqrt{(-\frac{1}{\sqrt{2}})^2+(\frac{1}{\sqrt{2}})^2}=\sqrt{\frac{1}{2}+\frac{1}{2}}=1$

$\therefore\ \begin{pmatrix}-\frac{1}{\sqrt{2}}\\ \frac{1}{\sqrt{2}}\end{pmatrix}$ is a unit vector.

c $\sqrt{(\frac{2}{3})^2+(\frac{1}{3})^2}=\sqrt{\frac{4}{9}+\frac{1}{9}}=\frac{\sqrt{5}}{3}$

$\therefore\ \begin{pmatrix}\frac{2}{3}\\ \frac{1}{3}\end{pmatrix}$ is not a unit vector.

d $\sqrt{(-\frac{3}{5})^2+(-\frac{4}{5})^2}=\sqrt{\frac{9}{25}+\frac{16}{25}}=1$

$\therefore\ \begin{pmatrix}-\frac{3}{5}\\ -\frac{4}{5}\end{pmatrix}$ is a unit vector.

e $\sqrt{(\frac{2}{7})^2+(-\frac{5}{7})^2}=\sqrt{\frac{4}{49}+\frac{25}{49}}=\frac{\sqrt{29}}{7}$

$\therefore\ \begin{pmatrix}\frac{2}{7}\\ -\frac{5}{7}\end{pmatrix}$ is not a unit vector.

2 **a** $2\mathbf{i}-\mathbf{j}$ **b** $-3\mathbf{i}-4\mathbf{j}$ **c** $-3\mathbf{i}$ **d** $7\mathbf{j}$ **e** $\frac{1}{\sqrt{2}}\mathbf{i}-\frac{1}{\sqrt{2}}\mathbf{j}$

3 **a** $\binom{3}{5}$ **b** $\binom{5}{-4}$ **c** $\binom{-4}{0}$ **d** $\binom{0}{3}$ **e** $\begin{pmatrix}\frac{\sqrt{3}}{2}\\ -\frac{1}{2}\end{pmatrix}$

4 **a** $\mathbf{i}-\mathbf{j}+\mathbf{k}=\begin{pmatrix}1\\-1\\1\end{pmatrix}$

$\therefore\ |\mathbf{i}-\mathbf{j}+\mathbf{k}|=\sqrt{3}$ units

b $3\mathbf{i}-\mathbf{j}+\mathbf{k}=\begin{pmatrix}3\\-1\\1\end{pmatrix}$

$\therefore\ |3\mathbf{i}-\mathbf{j}+\mathbf{k}|=\sqrt{9+1+1}=\sqrt{11}$ units

c $\mathbf{i}-5\mathbf{k}=\begin{pmatrix}1\\0\\-5\end{pmatrix}$

$\therefore\ |\mathbf{i}-5\mathbf{k}|=\sqrt{1+25}$
$=\sqrt{26}$ units

d $\frac{1}{2}(\mathbf{j}+\mathbf{k})=\begin{pmatrix}0\\ \frac{1}{2}\\ \frac{1}{2}\end{pmatrix}$

$\therefore\ \left|\frac{1}{2}(\mathbf{j}+\mathbf{k})\right|=\sqrt{(\frac{1}{2})^2+(\frac{1}{2})^2}$
$=\sqrt{\frac{1}{2}}$
$=\frac{1}{\sqrt{2}}$ units

5 **a** length $=1$
$\therefore\ \sqrt{0^2+k^2}=1$
$\therefore\ k^2=1$
$\therefore\ k=\pm1$

b length $=1$
$\therefore\ \sqrt{k^2+0}=1$
$\therefore\ k^2=1$
$\therefore\ k=\pm1$

c length $=1$
$\therefore\ \sqrt{k^2+1}=1$
$\therefore\ k^2+1=1$
$\therefore\ k^2=0$
$\therefore\ k=0$

d length $=1$
$\therefore\ \sqrt{\left(\frac{1}{4}\right)+k^2+\frac{1}{16}}=1$
$\therefore\ \sqrt{k^2+\frac{5}{16}}=1$
$\therefore\ k^2=\frac{11}{16}$
$\therefore\ k=\pm\frac{\sqrt{11}}{4}$

e length $=1$
$\therefore\ \sqrt{k^2+\frac{4}{9}+\frac{1}{9}}=1$
$\therefore\ \sqrt{k^2+\frac{5}{9}}=1$
$\therefore\ k^2=\frac{4}{9}$
$\therefore\ k=\pm\frac{2}{3}$

6 **a** length
$= \sqrt{3^2 + 4^2}$
$= \sqrt{9+16}$
$= \sqrt{25} = 5$ units

b length
$= \sqrt{2^2 + (-1)^2 + 1^2}$
$= \sqrt{4+1+1}$
$= \sqrt{6}$ units

c length
$= \sqrt{1^2 + 2^2 + (-2)^2}$
$= \sqrt{1+4+4}$
$= \sqrt{9} = 3$ units

d length $= \sqrt{(-2.36)^2 + (5.65)^2} \approx 6.12$ units

7 **a** $\mathbf{i} + 2\mathbf{j}$ has length $\sqrt{1^2 + 2^2} = \sqrt{5}$ units $\therefore$ unit vector $= \frac{1}{\sqrt{5}}(\mathbf{i} + 2\mathbf{j})$

b $2\mathbf{i} - 3\mathbf{k}$ has length $\sqrt{2^2 + 0^2 + (-3)^2} = \sqrt{4+9} = \sqrt{13}$ units
$\therefore$ unit vector is $\frac{1}{\sqrt{13}}(2\mathbf{i} - 3\mathbf{k})$

c $-2\mathbf{i} - 5\mathbf{j} - 2\mathbf{k}$ has length $\sqrt{(-2)^2 + (-5)^2 + (-2)^2} = \sqrt{4+25+4} = \sqrt{33}$ units
$\therefore$ unit vector is $\frac{1}{\sqrt{33}}(-2\mathbf{i} - 5\mathbf{j} - 2\mathbf{k})$

8 **a** $\begin{pmatrix} 2 \\ -1 \end{pmatrix}$ has length $\sqrt{2^2 + (-1)^2} = \sqrt{5}$ units
$\therefore$ the unit vector in the same direction is $\frac{1}{\sqrt{5}}\begin{pmatrix} 2 \\ -1 \end{pmatrix}$
$\therefore$ the vector of length 3 units in the same direction is $\frac{3}{\sqrt{5}}\begin{pmatrix} 2 \\ -1 \end{pmatrix} = \begin{pmatrix} \frac{6}{\sqrt{5}} \\ -\frac{3}{\sqrt{5}} \end{pmatrix}$

b $\begin{pmatrix} -1 \\ -4 \end{pmatrix}$ has length $\sqrt{(-1)^2 + (-4)^2} = \sqrt{17}$ units
$\therefore$ the unit vector in the opposite direction is $-\frac{1}{\sqrt{17}}\begin{pmatrix} -1 \\ -4 \end{pmatrix} = \frac{1}{\sqrt{17}}\begin{pmatrix} 1 \\ 4 \end{pmatrix}$
$\therefore$ the vector of length 2 units in the opposite direction is $\frac{2}{\sqrt{17}}\begin{pmatrix} 1 \\ 4 \end{pmatrix} = \begin{pmatrix} \frac{2}{\sqrt{17}} \\ \frac{8}{\sqrt{17}} \end{pmatrix}$

c $\begin{pmatrix} -1 \\ 4 \\ 1 \end{pmatrix}$ has length $\sqrt{(-1)^2 + 4^2 + 1^2} = \sqrt{18} = 3\sqrt{2}$ units
$\therefore$ the unit vector in the same direction is $\frac{1}{3\sqrt{2}}\begin{pmatrix} -1 \\ 4 \\ 1 \end{pmatrix}$
$\therefore$ the vector of length 6 units in the same direction is $\frac{6}{3\sqrt{2}}\begin{pmatrix} -1 \\ 4 \\ 1 \end{pmatrix} = \begin{pmatrix} -\sqrt{2} \\ 4\sqrt{2} \\ \sqrt{2} \end{pmatrix}$

d $\begin{pmatrix} -1 \\ -2 \\ -2 \end{pmatrix}$ has length $\sqrt{(-1)^2 + (-2)^2 + (-2)^2} = \sqrt{9} = 3$ units
$\therefore$ the unit vector in the opposite direction is $-\frac{1}{3}\begin{pmatrix} -1 \\ -2 \\ -2 \end{pmatrix} = \frac{1}{3}\begin{pmatrix} 1 \\ 2 \\ 2 \end{pmatrix}$
$\therefore$ the vector of length 5 units in the opposite direction is $\frac{5}{3}\begin{pmatrix} 1 \\ 2 \\ 2 \end{pmatrix} = \begin{pmatrix} \frac{5}{3} \\ \frac{10}{3} \\ \frac{10}{3} \end{pmatrix}$

EXERCISE 12I

1 **a** $\mathbf{q} \bullet \mathbf{p}$
$= \begin{pmatrix} -1 \\ 5 \end{pmatrix} \bullet \begin{pmatrix} 3 \\ 2 \end{pmatrix}$
$= -3 + 10$
$= 7$

b $\mathbf{q} \bullet \mathbf{r}$
$= \begin{pmatrix} -1 \\ 5 \end{pmatrix} \bullet \begin{pmatrix} -2 \\ 4 \end{pmatrix}$
$= 2 + 20$
$= 22$

c $\mathbf{q} \bullet (\mathbf{p} + \mathbf{r})$
$= \begin{pmatrix} -1 \\ 5 \end{pmatrix} \bullet \left[\begin{pmatrix} 3 \\ 2 \end{pmatrix} + \begin{pmatrix} -2 \\ 4 \end{pmatrix}\right]$
$= \begin{pmatrix} -1 \\ 5 \end{pmatrix} \bullet \begin{pmatrix} 1 \\ 6 \end{pmatrix}$
$= -1 + 30 = 29$

d $3\mathbf{r} \bullet \mathbf{q}$

$$= 3\begin{pmatrix} -2 \\ 4 \end{pmatrix} \bullet \begin{pmatrix} -1 \\ 5 \end{pmatrix}$$
$$= \begin{pmatrix} -6 \\ 12 \end{pmatrix} \bullet \begin{pmatrix} -1 \\ 5 \end{pmatrix}$$
$$= 6 + 60 = 66$$

e $2\mathbf{p} \bullet 2\mathbf{p}$

$$= 2\begin{pmatrix} 3 \\ 2 \end{pmatrix} \bullet 2\begin{pmatrix} 3 \\ 2 \end{pmatrix}$$
$$= \begin{pmatrix} 6 \\ 4 \end{pmatrix} \bullet \begin{pmatrix} 6 \\ 4 \end{pmatrix}$$
$$= 36 + 16 = 52$$

f $\mathbf{i} \bullet \mathbf{p}$

$$= \begin{pmatrix} 1 \\ 0 \end{pmatrix} \bullet \begin{pmatrix} 3 \\ 2 \end{pmatrix}$$
$$= 3 + 0$$
$$= 3$$

g $\mathbf{q} \bullet \mathbf{j} = \begin{pmatrix} -1 \\ 5 \end{pmatrix} \bullet \begin{pmatrix} 0 \\ 1 \end{pmatrix}$

$$= 0 + 5$$
$$= 5$$

h $\mathbf{i} \bullet \mathbf{i} = \begin{pmatrix} 1 \\ 0 \end{pmatrix} \bullet \begin{pmatrix} 1 \\ 0 \end{pmatrix}$

$$= 1 + 0$$
$$= 1$$

2 **a** $\mathbf{a} \bullet \mathbf{b} = \begin{pmatrix} 2 \\ 1 \\ 3 \end{pmatrix} \bullet \begin{pmatrix} -1 \\ 1 \\ 1 \end{pmatrix}$

$$= 2(-1) + 1(1) + 3(1)$$
$$= -2 + 1 + 3$$
$$= 2$$

b $\mathbf{b} \bullet \mathbf{a} = \begin{pmatrix} -1 \\ 1 \\ 1 \end{pmatrix} \bullet \begin{pmatrix} 2 \\ 1 \\ 3 \end{pmatrix}$

$$= (-1)(2) + 1(1) + 1(3)$$
$$= -2 + 1 + 3$$
$$= 2$$

c $|\mathbf{a}|^2 = \left(\sqrt{2^2 + 1^2 + 3^2}\right)^2$

$$= 14$$

d $\mathbf{a} \bullet \mathbf{a} = \begin{pmatrix} 2 \\ 1 \\ 3 \end{pmatrix} \bullet \begin{pmatrix} 2 \\ 1 \\ 3 \end{pmatrix}$

$$= 2(2) + 1(1) + 3(3)$$
$$= 14$$

e $\mathbf{a} \bullet (\mathbf{b} + \mathbf{c})$

$$= \begin{pmatrix} 2 \\ 1 \\ 3 \end{pmatrix} \bullet \left[\begin{pmatrix} -1 \\ 1 \\ 1 \end{pmatrix} + \begin{pmatrix} 0 \\ -1 \\ 1 \end{pmatrix}\right]$$
$$= \begin{pmatrix} 2 \\ 1 \\ 3 \end{pmatrix} \bullet \begin{pmatrix} -1 \\ 0 \\ 2 \end{pmatrix}$$
$$= 2(-1) + 1(0) + 3(2) = 4$$

f $\mathbf{a} \bullet \mathbf{b} + \mathbf{a} \bullet \mathbf{c}$

$$= 2 + \begin{pmatrix} 2 \\ 1 \\ 3 \end{pmatrix} \bullet \begin{pmatrix} 0 \\ -1 \\ 1 \end{pmatrix} \quad \{\text{using } \mathbf{a}\}$$
$$= 2 + 2(0) + 1(-1) + 3(1)$$
$$= 4$$

3 **a** $\mathbf{p} \bullet \mathbf{q}$

$$= \begin{pmatrix} 3 \\ -1 \\ 2 \end{pmatrix} \bullet \begin{pmatrix} -2 \\ 1 \\ 3 \end{pmatrix}$$
$$= -6 - 1 + 6$$
$$= -1$$

b If the angle between $\mathbf{p}$ and $\mathbf{q}$ is θ, then

$$\cos\theta = \frac{\mathbf{p} \bullet \mathbf{q}}{|\mathbf{p}|\,|\mathbf{q}|} = \frac{-1}{\sqrt{3^2 + (-1)^2 + 2^2}\sqrt{(-2)^2 + 1^2 + 3^2}}$$
$$= \frac{-1}{\sqrt{14}\sqrt{14}}$$
$$\therefore \quad \theta = \cos^{-1}(-\tfrac{1}{14}) \approx 94.1^o$$

4 **a** $(\mathbf{i} + \mathbf{j} - \mathbf{k}) \bullet (2\mathbf{j} + \mathbf{k})$

$$= \begin{pmatrix} 1 \\ 1 \\ -1 \end{pmatrix} \bullet \begin{pmatrix} 0 \\ 2 \\ 1 \end{pmatrix}$$
$$= 1(0) + 1(2) - 1(1) = 1$$

b $\mathbf{i} \bullet \mathbf{i} = \begin{pmatrix} 1 \\ 0 \\ 0 \end{pmatrix} \bullet \begin{pmatrix} 1 \\ 0 \\ 0 \end{pmatrix}$

$$= 1$$

c $\mathbf{i} \bullet \mathbf{j} = \begin{pmatrix} 1 \\ 0 \\ 0 \end{pmatrix} \bullet \begin{pmatrix} 0 \\ 1 \\ 0 \end{pmatrix}$

$$= 0$$

5 **a** $\mathbf{p} \bullet \mathbf{q} = |\mathbf{p}|\,|\mathbf{q}|\cos\theta$

$$= 2 \times 5 \times \cos 60^o$$
$$= 5$$

b $\mathbf{p} \bullet \mathbf{q} = |\mathbf{p}|\,|\mathbf{q}|\cos\theta$

$$= 6 \times 3 \times \cos 120^o$$
$$= -9$$

6 $\mathbf{a} \bullet (\mathbf{b}+\mathbf{c})$

$$= \begin{pmatrix} a_1 \\ a_2 \\ a_3 \end{pmatrix} \bullet \left[\begin{pmatrix} b_1 \\ b_2 \\ b_3 \end{pmatrix} + \begin{pmatrix} c_1 \\ c_2 \\ c_3 \end{pmatrix} \right]$$

$$= \begin{pmatrix} a_1 \\ a_2 \\ a_3 \end{pmatrix} \bullet \begin{pmatrix} b_1+c_1 \\ b_2+c_2 \\ b_3+c_3 \end{pmatrix}$$

$= a_1(b_1+c_1)+a_2(b_2+c_2)+a_3(b_3+c_3)$

$= a_1b_1+a_1c_1+a_2b_2+a_2c_2+a_3b_3+a_3c_3$

$= (a_1b_1+a_2b_2+a_3b_3)+(a_1c_1+a_2c_2+a_3c_3)$

$= \mathbf{a} \bullet \mathbf{b} + \mathbf{a} \bullet \mathbf{c}$

$\therefore\ \mathbf{p} \bullet (\mathbf{c}+\mathbf{d}) = \mathbf{p} \bullet \mathbf{c} + \mathbf{p} \bullet \mathbf{d}$

If we let $\mathbf{p} = \mathbf{a}+\mathbf{b}$,

then $(\mathbf{a}+\mathbf{b}) \bullet (\mathbf{c}+\mathbf{d})$

$= \mathbf{p} \bullet (\mathbf{c}+\mathbf{d})$

$= \mathbf{p} \bullet \mathbf{c} + \mathbf{p} \bullet \mathbf{d}$

$= (\mathbf{a}+\mathbf{b}) \bullet \mathbf{c} + (\mathbf{a}+\mathbf{b}) \bullet \mathbf{d}$

$= \mathbf{c} \bullet (\mathbf{a}+\mathbf{b}) + \mathbf{d} \bullet (\mathbf{a}+\mathbf{b})$

$= \mathbf{c} \bullet \mathbf{a} + \mathbf{c} \bullet \mathbf{b} + \mathbf{d} \bullet \mathbf{a} + \mathbf{d} \bullet \mathbf{b}$

$= \mathbf{a} \bullet \mathbf{c} + \mathbf{a} \bullet \mathbf{d} + \mathbf{b} \bullet \mathbf{c} + \mathbf{b} \bullet \mathbf{d}$

7 **a** $\begin{pmatrix} 3 \\ t \end{pmatrix} \bullet \begin{pmatrix} -2 \\ 1 \end{pmatrix} = 0$

$\therefore\ -6+t=0$

$\therefore\ t=6$

b $\begin{pmatrix} t \\ t+2 \end{pmatrix} \bullet \begin{pmatrix} 3 \\ -4 \end{pmatrix} = 0$

$\therefore\ 3t-4(t+2)=0$

$\therefore\ 3t-4t-8=0$

$\therefore\ -t=8$

$\therefore\ t=-8$

c $\begin{pmatrix} t \\ t+2 \end{pmatrix} \bullet \begin{pmatrix} 2-3t \\ t \end{pmatrix} = 0$

$\therefore\ 2t-3t^2+t^2+2t=0$

$\therefore\ -2t^2+4t=0$

$\therefore\ t^2-2t=0$

$\therefore\ t(t-2)=0$

$\therefore\ t=0$ or 2

d $\begin{pmatrix} 3 \\ -1 \\ t \end{pmatrix} \bullet \begin{pmatrix} 2t \\ -3 \\ -4 \end{pmatrix} = 0$

$\therefore\ 3(2t)+(-1)(-3)+t(-4)=0$

$\therefore\ 6t+3-4t=0$

$\therefore\ 2t+3=0$

$\therefore\ t=-\frac{3}{2}$

8 **a** If $\mathbf{p} \parallel \mathbf{q}$ then $\begin{pmatrix} 3 \\ t \end{pmatrix} = k\begin{pmatrix} -2 \\ 1 \end{pmatrix}$ where $k \neq 0$ $\quad\therefore\ 3=-2k$ and $t=k$

$\therefore\ k=-\frac{3}{2}$ and $t=-\frac{3}{2}$

b If $\mathbf{r} \parallel \mathbf{s}$ then $\begin{pmatrix} t \\ t+2 \end{pmatrix} = k\begin{pmatrix} 3 \\ -4 \end{pmatrix}$ where $k \neq 0$

$\therefore\ t=3k$ and $t+2=-4k$

$\therefore\ t+2=-4\left(\frac{t}{3}\right)$

$\therefore\ 3t+6=-4t$

$\therefore\ 7t=-6$

$\therefore\ t=-\frac{6}{7}$

c If $\mathbf{a} \parallel \mathbf{b}$ then $\begin{pmatrix} t \\ t+2 \end{pmatrix} = k\begin{pmatrix} 2-3t \\ t \end{pmatrix}$

$\therefore\ t=k(2-3t)$ and $t+2=kt$

$\therefore\ \dfrac{t}{2-3t}=\dfrac{t+2}{t}$ {equating ks}

$\therefore\ t^2=(t+2)(2-3t)$

$\therefore\ t^2=2t-3t^2+4-6t$

$\therefore\ 4t^2+4t-4=0$

$\therefore\ t^2+t-1=0$ which has $\Delta=1^2-4(1)(-1)=5$

$\therefore\ t=\dfrac{-1\pm\sqrt{5}}{2}$

d If $\mathbf{a} \parallel \mathbf{b}$ then $\begin{pmatrix} 3 \\ -1 \\ t \end{pmatrix} = k \begin{pmatrix} 2t \\ -3 \\ -4 \end{pmatrix}$ where $k \neq 0$

$\therefore\; 3 = 2kt, \quad -1 = -3k \quad \text{and} \quad t = -4k$

$\therefore\; k = \frac{1}{3} \quad \text{and} \quad 3 = \frac{2}{3}t, \quad t = -\frac{4}{3}$

$\therefore\; t = \frac{9}{2}$ and $-\frac{4}{3}$ simultaneously which is impossible.

$\therefore$ the vectors can never be parallel.

9 $\mathbf{a} \bullet \mathbf{b}$

$= \begin{pmatrix} 3 \\ 1 \\ 2 \end{pmatrix} \bullet \begin{pmatrix} -1 \\ 1 \\ 1 \end{pmatrix}$

$= 3(-1) + 1(1) + 2(1)$

$= 0$

$\mathbf{b} \bullet \mathbf{c}$

$= \begin{pmatrix} -1 \\ 1 \\ 1 \end{pmatrix} \bullet \begin{pmatrix} 1 \\ 5 \\ -4 \end{pmatrix}$

$= (-1)(1) + 1(5) + 1(-4)$

$= 0$

$\mathbf{a} \bullet \mathbf{c}$

$= \begin{pmatrix} 3 \\ 1 \\ 2 \end{pmatrix} \bullet \begin{pmatrix} 1 \\ 5 \\ -4 \end{pmatrix}$

$= (3)(1) + 1(5) + 2(-4)$

$= 0$

$\therefore$ **a**, **b** and **c** are mutually perpendicular.

10 **a** $\begin{pmatrix} 1 \\ 1 \\ 5 \end{pmatrix} \bullet \begin{pmatrix} 2 \\ 3 \\ -1 \end{pmatrix} = 1(2) + 1(3) + 5(-1) = 0$

$\therefore \begin{pmatrix} 1 \\ 1 \\ 5 \end{pmatrix}$ and $\begin{pmatrix} 2 \\ 3 \\ -1 \end{pmatrix}$ are perpendicular.

b $\begin{pmatrix} 3 \\ t \\ -2 \end{pmatrix} \bullet \begin{pmatrix} 1-t \\ -3 \\ 4 \end{pmatrix} = 0$

$\therefore\; 3(1-t) + t(-3) + (-2)4 = 0$

$\therefore\; 3 - 3t - 3t - 8 = 0$

$\therefore\; -6t = 5$

$\therefore\; t = -\frac{5}{6}$

11 **a** We have three points: A(−2, 1), B(−2, 5), C(3, 1)

Then $\overrightarrow{AB} = \binom{0}{4}$, $\overrightarrow{AC} = \binom{5}{0}$ and $\overrightarrow{BC} = \binom{5}{-4}$

Now $\overrightarrow{AB} \bullet \overrightarrow{AC} = \binom{0}{4} \bullet \binom{5}{0} = 0 + 0 = 0$

$\therefore$ $\overrightarrow{AB}$ is perpendicular to $\overrightarrow{AC}$ and so ΔABC is right angled at A.

b We have three points: A(4, 7), B(1, 2), C(−1, 6)

Then $\overrightarrow{AB} = \binom{-3}{-5}$, $\overrightarrow{AC} = \binom{-5}{-1}$ and $\overrightarrow{BC} = \binom{-2}{4}$

Now $\overrightarrow{AB} \bullet \overrightarrow{AC} = \binom{-3}{-5} \bullet \binom{-5}{-1} = 15 + 5 = 20$

$\overrightarrow{AB} \bullet \overrightarrow{BC} = \binom{-3}{-5} \bullet \binom{-2}{4} = 6 + (-20) = -14$

$\overrightarrow{AC} \bullet \overrightarrow{BC} = \binom{-5}{-1} \bullet \binom{-2}{4} = 10 + (-4) = 6$

$\therefore$ none of the sides are perpendicular to each other and so ΔABC is not right angled.

c We have three points: A(2, −2), B(5, 7), C(−1, −1)

Then $\overrightarrow{AB} = \binom{3}{9}$, $\overrightarrow{AC} = \binom{-3}{1}$ and $\overrightarrow{BC} = \binom{-6}{-8}$

Now $\overrightarrow{AB} \bullet \overrightarrow{AC} = \binom{3}{9} \bullet \binom{-3}{1} = -9 + 9 = 0$

$\therefore$ $\overrightarrow{AB}$ is perpendicular to $\overrightarrow{AC}$ and so ΔABC is right angled at A.

d We have three points: A(10, 1), B(5, 2), C(7, 4)

Then $\overrightarrow{AB} = \binom{-5}{1}$, $\overrightarrow{AC} = \binom{-3}{3}$ and $\overrightarrow{BC} = \binom{2}{2}$

Now $\overrightarrow{AC} \bullet \overrightarrow{BC} = \binom{-3}{3} \bullet \binom{2}{2} = -6 + 6 = 0$

$\therefore$ $\overrightarrow{AC}$ is perpendicular to $\overrightarrow{BC}$ and so ΔABC is right angled at C.

12 We have three points: A(5, 1, 2), B(6, −1, 0), C(3, 2, 0)

Then $\overrightarrow{AB} = \begin{pmatrix} 1 \\ -2 \\ -2 \end{pmatrix}$, $\overrightarrow{AC} = \begin{pmatrix} -2 \\ 1 \\ -2 \end{pmatrix}$ and $\overrightarrow{BC} = \begin{pmatrix} -3 \\ 3 \\ 0 \end{pmatrix}$

Now $\overrightarrow{AB} \bullet \overrightarrow{AC} = \begin{pmatrix} 1 \\ -2 \\ -2 \end{pmatrix} \bullet \begin{pmatrix} -2 \\ 1 \\ -2 \end{pmatrix} = (-2) + (-2) + 4 = 0$

$\therefore$ $\overrightarrow{AB}$ is perpendicular to $\overrightarrow{AC}$ and so ΔABC is right angled at A.

13 **a**

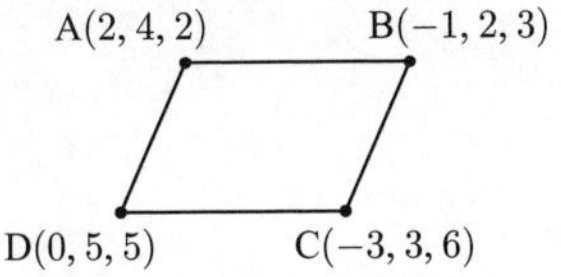

$\overrightarrow{AB} = \begin{pmatrix} -3 \\ -2 \\ 1 \end{pmatrix}$ $\overrightarrow{BC} = \begin{pmatrix} -2 \\ 1 \\ 3 \end{pmatrix}$ $\therefore$ $\overrightarrow{AB}$ is parallel to $\overrightarrow{DC}$ and $\overrightarrow{BC}$ is parallel to $\overrightarrow{AD}$.

$\overrightarrow{DC} = \begin{pmatrix} -3 \\ -2 \\ 1 \end{pmatrix}$ $\overrightarrow{AD} = \begin{pmatrix} -2 \\ 1 \\ 3 \end{pmatrix}$ $\therefore$ ABCD is a parallelogram.

b $|\overrightarrow{AB}| = \sqrt{14}$ units and $|\overrightarrow{BC}| = \sqrt{14}$ units $\therefore$ ABCD is a rhombus.

c $\overrightarrow{AC} \bullet \overrightarrow{BD} = \begin{pmatrix} -5 \\ -1 \\ 4 \end{pmatrix} \bullet \begin{pmatrix} 1 \\ 3 \\ 2 \end{pmatrix} = (-5) \times 1 + (-1) \times 3 + 4(2) = 0$

$\therefore$ $\overrightarrow{AC}$ is perpendicular to $\overrightarrow{BD}$ which illustrates that the diagonals of a rhombus are perpendicular.

14 **a** $x - y = 3$ has gradient $+\frac{1}{1}$ and so has direction vector $\binom{1}{1}$.

$3x + 2y = 11$ has gradient $-\frac{3}{2}$ and so has direction vector $\binom{2}{-3}$.

$\therefore$ $\binom{1}{1} \bullet \binom{2}{-3} = \sqrt{1+1}\sqrt{4+9}\cos\theta$

$\therefore$ $2 - 3 = \sqrt{2}\sqrt{13}\cos\theta$

$\therefore$ $\frac{-1}{\sqrt{26}} = \cos\theta$

$\therefore$ $\theta \approx 101.3^o$ $\therefore$ the angle is $180^o - 101.3^o \approx 78.7^o$

b $y = x + 2$ has gradient $1 = \frac{1}{1}$ and so has direction vector is $\binom{1}{1}$.

$y = 1 - 3x$ has gradient $-3 = \frac{-3}{1}$ and so has direction vector is $\binom{1}{-3}$.

$\therefore$ $\binom{1}{1} \bullet \binom{1}{-3} = \sqrt{1+1}\sqrt{1+9}\cos\theta$

$\therefore$ $1 - 3 = \sqrt{2}\sqrt{10}\cos\theta$

$\therefore$ $\frac{-2}{\sqrt{20}} = \cos\theta$

$\therefore$ $\theta \approx 116.6^o$ $\therefore$ the angle is $180^o - 116.6^o \approx 63.4^o$

c $y + x = 7$ has gradient $-1 = \frac{-1}{1}$ and so has direction vector is $\binom{1}{-1}$.

$x - 3y + 2 = 0$ has gradient $\frac{1}{3}$ and so has direction vector is $\binom{3}{1}$.

$\therefore$ $\binom{1}{-1} \bullet \binom{3}{1} = \sqrt{1+1}\sqrt{9+1}\cos\theta$

$\therefore$ $3 - 1 = \sqrt{2}\sqrt{10}\cos\theta$

$\therefore$ $\frac{2}{\sqrt{20}} = \cos\theta$

$\therefore$ $\theta \approx 63.4^o$ $\therefore$ the angle is 63.4^o

d $y = 2 - x$ has gradient $-1 = \frac{-1}{1}$ and so has has direction vector $\binom{1}{-1}$.

$x - 2y = 7$ has gradient $\frac{1}{2}$ and so has has direction vector $\binom{2}{1}$.

$\therefore$ $\binom{1}{-1} \bullet \binom{2}{1} = \sqrt{1+1}\sqrt{4+1}\cos\theta$

$\therefore$ $2 - 1 = \sqrt{2}\sqrt{5}\cos\theta$

$\therefore$ $\frac{1}{\sqrt{10}} = \cos\theta$

$\therefore$ $\theta \approx 71.6^o$ $\therefore$ the angle is 71.6^o

15 **a** $\binom{5}{2} \bullet \binom{-2}{5} = -10 + 10 = 0$, so $\binom{-2}{5}$ is one such vector.

$\therefore$ required vectors have form $k\binom{-2}{5}$, $k \neq 0$. **Note:** $k\binom{2}{-5}$, $k \neq 0$ is also acceptable.

b $\binom{-1}{-2} \bullet \binom{-2}{1} = 2 - 2 = 0$, so $\binom{-2}{1}$ is one such vector.

$\therefore$ required vectors have form $k\binom{-2}{1}$, $k \neq 0$.

c $\binom{3}{-1} \bullet \binom{1}{3} = 3 - 3 = 0$, so $\binom{1}{3}$ is one such vector.

$\therefore$ required vectors have form $k\binom{1}{3}$, $k \neq 0$.

d $\binom{-4}{3} \bullet \binom{3}{4} = -12 + 12 = 0$, so $\binom{3}{4}$ is one such vector.

$\therefore$ required vectors have form $k\binom{3}{4}$, $k \neq 0$.

e $\binom{2}{0} \bullet \binom{0}{1} = 0 + 0 = 0$, so $\binom{0}{1}$ is one such vector.

$\therefore$ required vectors have form $k\binom{0}{1}$, $k \neq 0$.

16 Given A(3, 0, 1), B(−3, 1, 2) and C(−2, 1, −1),

$$\overrightarrow{BC} = \begin{pmatrix} 1 \\ 0 \\ -3 \end{pmatrix} \text{ and } \overrightarrow{BA} = \begin{pmatrix} 6 \\ -1 \\ -1 \end{pmatrix}$$

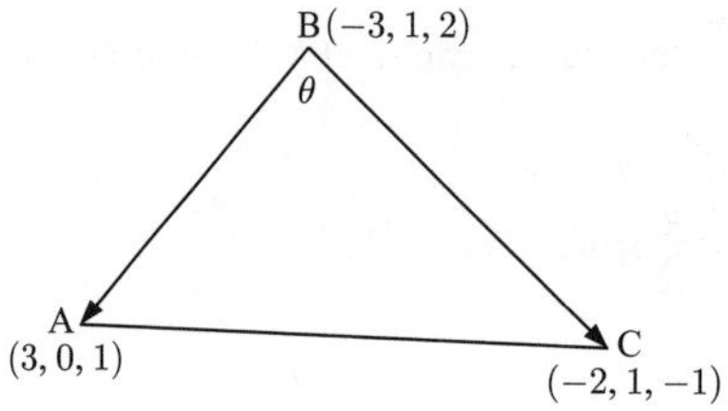

$$\therefore \cos\theta = \frac{\overrightarrow{BC} \bullet \overrightarrow{BA}}{|\overrightarrow{BC}||\overrightarrow{BA}|} = \frac{\begin{pmatrix} 1 \\ 0 \\ -3 \end{pmatrix} \bullet \begin{pmatrix} 6 \\ -1 \\ -1 \end{pmatrix}}{\sqrt{1+9}\sqrt{36+1+1}} = \frac{6+0+3}{\sqrt{10}\sqrt{38}} = \frac{9}{\sqrt{380}}$$

$\therefore \theta \approx 62.5^o$

If $\overrightarrow{BA}$ and $\overrightarrow{CB}$ are used we would find the exterior angle of the triangle at B, which is 117.5^o.

17

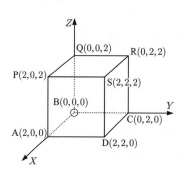

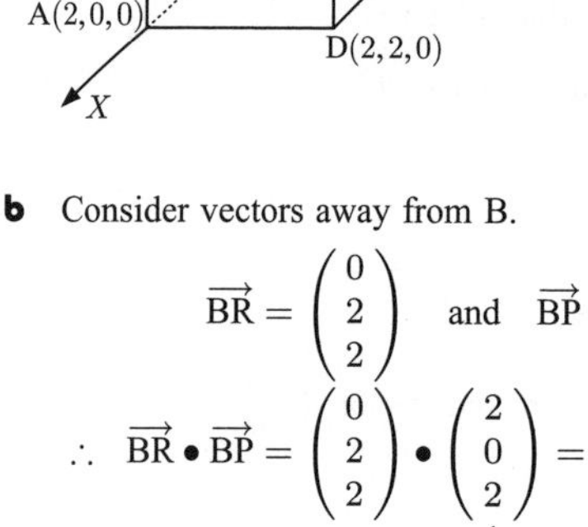

a Suppose the origin is at B.

Now $\overrightarrow{BA} = \begin{pmatrix} 2 \\ 0 \\ 0 \end{pmatrix}$ and $\overrightarrow{BS} = \begin{pmatrix} 2 \\ 2 \\ 2 \end{pmatrix}$

$$\therefore \overrightarrow{BA} \bullet \overrightarrow{BS} = \begin{pmatrix} 2 \\ 0 \\ 0 \end{pmatrix} \bullet \begin{pmatrix} 2 \\ 2 \\ 2 \end{pmatrix} = 4 + 0 + 0 = 4$$

$$\therefore \cos A\widehat{B}S = \frac{4}{\sqrt{4+0+0}\sqrt{4+4+4}} = \frac{4}{2 \times 2\sqrt{3}} = \frac{1}{\sqrt{3}}$$

$\therefore A\widehat{B}S \approx 54.7^o$

b Consider vectors away from B.

$$\overrightarrow{BR} = \begin{pmatrix} 0 \\ 2 \\ 2 \end{pmatrix} \text{ and } \overrightarrow{BP} = \begin{pmatrix} 2 \\ 0 \\ 2 \end{pmatrix}$$

$$\therefore \overrightarrow{BR} \bullet \overrightarrow{BP} = \begin{pmatrix} 0 \\ 2 \\ 2 \end{pmatrix} \bullet \begin{pmatrix} 2 \\ 0 \\ 2 \end{pmatrix} = 0 + 0 + 4 = 4$$

$$\therefore \cos R\widehat{B}P = \frac{4}{\sqrt{0+4+4}\sqrt{4+0+4}} = \frac{4}{\sqrt{8} \times \sqrt{8}} = \tfrac{1}{2} \text{ and so } R\widehat{B}P = 60^o$$

c $\overrightarrow{BP} = \begin{pmatrix} 2 \\ 0 \\ 2 \end{pmatrix}$ and $\overrightarrow{BS} = \begin{pmatrix} 2 \\ 2 \\ 2 \end{pmatrix}$

$$\therefore \overrightarrow{BP} \bullet \overrightarrow{BS} = \begin{pmatrix} 2 \\ 0 \\ 2 \end{pmatrix} \bullet \begin{pmatrix} 2 \\ 2 \\ 2 \end{pmatrix} = 4 + 0 + 4 = 8$$

$$\therefore \cos P\widehat{B}S = \frac{8}{\sqrt{4+4}\sqrt{4+4+4}} = \frac{8}{\sqrt{96}}$$

$\therefore P\widehat{B}S \approx 35.3^o$

18 Suppose the origin is at N.

a

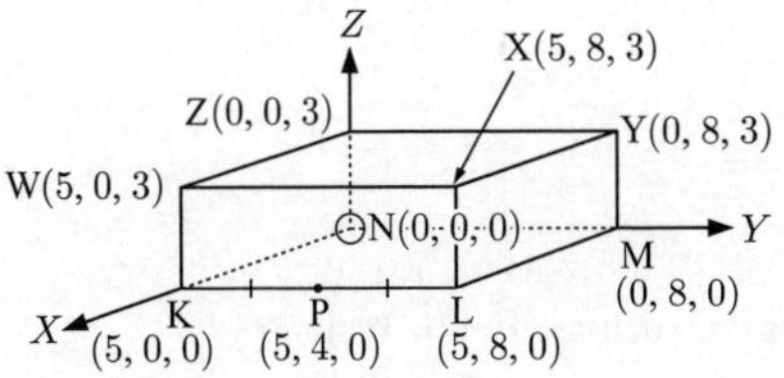

$\overrightarrow{NY} = \begin{pmatrix} 0 \\ 8 \\ 3 \end{pmatrix}$ and $\overrightarrow{NX} = \begin{pmatrix} 5 \\ 8 \\ 3 \end{pmatrix}$

$\overrightarrow{NY} \bullet \overrightarrow{NX} = \begin{pmatrix} 0 \\ 8 \\ 3 \end{pmatrix} \bullet \begin{pmatrix} 5 \\ 8 \\ 3 \end{pmatrix} = 0 + 64 + 9 = 73$

$\therefore \quad \cos \widehat{YNX} = \dfrac{73}{\sqrt{64+9}\sqrt{25+64+9}}$

$= \dfrac{73}{\sqrt{73}\sqrt{98}} = \sqrt{\frac{73}{98}}$

$\therefore \quad \widehat{YNX} \approx 30.3^o$

b $\overrightarrow{NY} = \begin{pmatrix} 0 \\ 8 \\ 3 \end{pmatrix}$ and $\overrightarrow{NP} = \begin{pmatrix} 5 \\ 4 \\ 0 \end{pmatrix}$

$\overrightarrow{NY} \bullet \overrightarrow{NP} = \begin{pmatrix} 0 \\ 8 \\ 3 \end{pmatrix} \bullet \begin{pmatrix} 5 \\ 4 \\ 0 \end{pmatrix}$

$= 0 + 32 + 0$

$= 32$

$\therefore \quad \cos \widehat{YNP} = \dfrac{32}{\sqrt{64+9}\sqrt{25+16}}$

$= \dfrac{32}{\sqrt{73}\sqrt{41}}$

$\therefore \quad \widehat{YNP} \approx 54.2^o$

19 **a** M is the midpoint of [BC]. $\therefore$ M is at $\left(\dfrac{2+1}{2}, \dfrac{2+3}{2}, \dfrac{2+1}{2}\right)$, which is $\left(\frac{3}{2}, \frac{5}{2}, \frac{3}{2}\right)$.

b

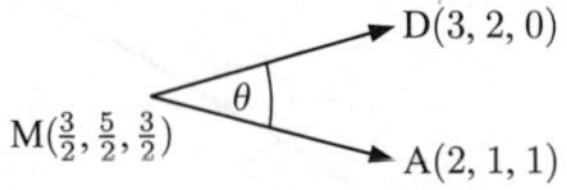

Now $\overrightarrow{MD} = \begin{pmatrix} \frac{3}{2} \\ -\frac{1}{2} \\ -\frac{3}{2} \end{pmatrix}$ and $\overrightarrow{MA} = \begin{pmatrix} \frac{1}{2} \\ -\frac{3}{2} \\ -\frac{1}{2} \end{pmatrix}$

$\therefore \quad \cos\theta = \dfrac{\overrightarrow{MD} \bullet \overrightarrow{MA}}{|\overrightarrow{MD}||\overrightarrow{MA}|} = \dfrac{\begin{pmatrix} \frac{3}{2} \\ -\frac{1}{2} \\ -\frac{3}{2} \end{pmatrix} \bullet \begin{pmatrix} \frac{1}{2} \\ -\frac{3}{2} \\ -\frac{1}{2} \end{pmatrix}}{\sqrt{\frac{9}{4}+\frac{1}{4}+\frac{9}{4}}\sqrt{\frac{1}{4}+\frac{9}{4}+\frac{1}{4}}}$

$\therefore \quad \cos\theta = \dfrac{\frac{3}{4}+\frac{3}{4}+\frac{3}{4}}{\sqrt{\frac{19}{4}}\sqrt{\frac{11}{4}}} = \dfrac{\frac{9}{4}}{\frac{\sqrt{209}}{4}} = \dfrac{9}{\sqrt{209}}$ and so $\theta \approx 51.5^o$

20 **a** $\begin{pmatrix} 2 \\ t \\ t-2 \end{pmatrix} \bullet \begin{pmatrix} t \\ 3 \\ t \end{pmatrix} = 0 \quad \therefore \quad 2t + 3t + t(t-2) = 0$

$\therefore \quad 5t + t^2 - 2t = 0$

$\therefore \quad t^2 + 3t = 0$

$\therefore \quad t(t+3) = 0$ and so $t = 0$ or $t = -3$

b Given that $\mathbf{a} = \begin{pmatrix} 1 \\ 2 \\ 3 \end{pmatrix}$, $\mathbf{b} = \begin{pmatrix} 2 \\ 2 \\ r \end{pmatrix}$ and $\mathbf{c} = \begin{pmatrix} s \\ t \\ 1 \end{pmatrix}$ are mutually perpendicular,

$\mathbf{a} \bullet \mathbf{b} = 0$, $\mathbf{b} \bullet \mathbf{c} = 0$ and $\mathbf{a} \bullet \mathbf{c} = 0$

$\therefore \quad \begin{pmatrix} 1 \\ 2 \\ 3 \end{pmatrix} \bullet \begin{pmatrix} 2 \\ 2 \\ r \end{pmatrix} = 0 \quad \therefore \quad 2 + 4 + 3r = 0$

$\therefore \quad 3r = -6$

$\therefore \quad r = -2$

and $\begin{pmatrix} 2 \\ 2 \\ -2 \end{pmatrix} \bullet \begin{pmatrix} s \\ t \\ 1 \end{pmatrix} = 0 \quad \therefore \quad 2s + 2t - 2 = 0$

$\therefore \quad s + t = 1 \quad .... (1)$

and $\begin{pmatrix}1\\2\\3\end{pmatrix} \bullet \begin{pmatrix}s\\t\\1\end{pmatrix} = 0$ $\quad\therefore\quad s + 2t + 3 = 0$

$\therefore\quad s + 2t = -3 \quad .... (2)$

(2) − (1) gives $t = -4$ and so $s = 5$

$\therefore\quad r = -2,\quad s = 5\quad$ and $\quad t = -4$

21 **a** Choose any vector in the direction of the X-axis, such as $\mathbf{i} = \begin{pmatrix}1\\0\\0\end{pmatrix}$.

Then $\cos\theta = \dfrac{\begin{pmatrix}1\\0\\0\end{pmatrix} \bullet \begin{pmatrix}1\\2\\3\end{pmatrix}}{\left|\begin{pmatrix}1\\0\\0\end{pmatrix}\right|\left|\begin{pmatrix}1\\2\\3\end{pmatrix}\right|} = \dfrac{1}{\sqrt{1}\sqrt{1+4+9}} = \dfrac{1}{\sqrt{14}}$ and so $\theta \approx 74.5^o$.

b A line parallel to the Y-axis has direction vector $\mathbf{j} = \begin{pmatrix}0\\1\\0\end{pmatrix}$.

Then $\cos\theta = \dfrac{\begin{pmatrix}0\\1\\0\end{pmatrix} \bullet \begin{pmatrix}-1\\1\\3\end{pmatrix}}{\sqrt{1}\sqrt{1+1+9}} = \dfrac{1}{\sqrt{11}}$ and so $\theta \approx 72.5^o$.

REVIEW SET 12A

1 **a**

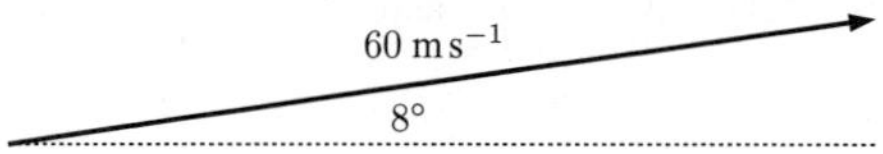

Scale: $1\text{ cm} \equiv 10\text{ m s}^{-1}$

b

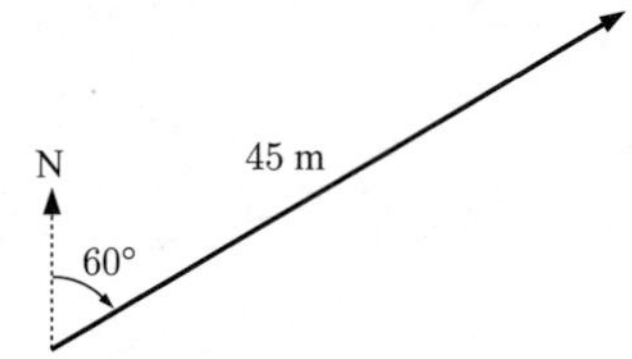

Scale: $1\text{ cm} \equiv 10\text{ m}$

2 **a** $\overrightarrow{AB} - \overrightarrow{CB}$
$= \overrightarrow{AB} + \overrightarrow{BC}$
$= \overrightarrow{AC}$

b $\overrightarrow{AB} + \overrightarrow{BC} - \overrightarrow{DC}$
$= \overrightarrow{AC} + \overrightarrow{CD}$
$= \overrightarrow{AD}$

3 **a** $\mathbf{q} = \mathbf{p} + \mathbf{r}$

b $\mathbf{l} = \mathbf{k} - \mathbf{j} + \mathbf{n} - \mathbf{m}$

4 **a** $\binom{4}{3}$

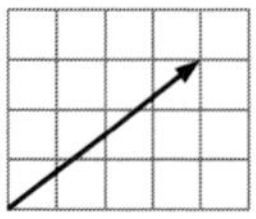

b $\binom{3}{-5}$

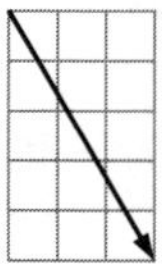

c $\binom{0}{-4}$

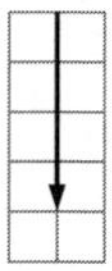

5 $\overrightarrow{SP}$
$= \overrightarrow{SR} + \overrightarrow{RQ} + \overrightarrow{QP}$
$= -\overrightarrow{RS} + \overrightarrow{RQ} - \overrightarrow{PQ}$
$= -\binom{2}{-3} + \binom{-1}{2} - \binom{-4}{1}$
$= \binom{1}{4}$

6 **a** $\overrightarrow{BC} = 2\overrightarrow{OA} = 2\mathbf{p}$

Now $\overrightarrow{AC} = \overrightarrow{AO} + \overrightarrow{OB} + \overrightarrow{BC}$
$= -\mathbf{p} + \mathbf{q} + 2\mathbf{p}$
$= \mathbf{p} + \mathbf{q}$

b $\overrightarrow{OM} = \overrightarrow{OA} + \overrightarrow{AM}$
$= \mathbf{p} + \frac{1}{2}\overrightarrow{AC}$
$= \mathbf{p} + \frac{1}{2}(\mathbf{p} + \mathbf{q})$
$= \frac{3}{2}\mathbf{p} + \frac{1}{2}\mathbf{q}$

7 The vectors are parallel, so $\begin{pmatrix}-12\\-20\\2\end{pmatrix} = k\begin{pmatrix}3\\m\\n\end{pmatrix}$ $\quad\therefore\quad 3k = -12,\quad km = -20,\quad kn = 2$

$\therefore\quad k = -4,\quad m = 5,\quad n = -\frac{1}{2}$

8 $\overrightarrow{CB} = \overrightarrow{CA} + \overrightarrow{AB} = -\overrightarrow{AC} + \overrightarrow{AB} = \begin{pmatrix} 6 \\ -1 \\ 3 \end{pmatrix} + \begin{pmatrix} 2 \\ -7 \\ 4 \end{pmatrix} = \begin{pmatrix} 8 \\ -8 \\ 7 \end{pmatrix}$

9 **a** $\mathbf{p} \bullet \mathbf{q} = \binom{3}{-2} \bullet \binom{-1}{5}$
$= -3 + (-10)$
$= -13$

b $\mathbf{p} - \mathbf{r} = \binom{3}{-2} - \binom{-3}{4}$
$= \binom{6}{-6}$

$\therefore\ \mathbf{q} \bullet (\mathbf{p} - \mathbf{r}) = \binom{-1}{5} \bullet \binom{6}{-6}$
$= -6 - 30$
$= -36$

10

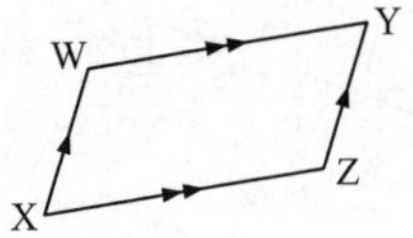

$\overrightarrow{WY} = \begin{pmatrix} 3 - -3 \\ 4 - -1 \end{pmatrix} = \begin{pmatrix} 6 \\ 5 \end{pmatrix}$

$\overrightarrow{XZ} = \begin{pmatrix} 4 - -2 \\ 10 - 5 \end{pmatrix} = \begin{pmatrix} 6 \\ 5 \end{pmatrix}$

So, $\overrightarrow{WY} = \overrightarrow{XZ}$
$\therefore$ [WY] is parallel to [XZ] and they are equal in length. This is sufficient to deduce that WYZX is a parallelogram.

11 **a** $\mathbf{a} - \mathbf{x} = 2\mathbf{b}$
$\therefore\ \mathbf{x} = \mathbf{a} - 2\mathbf{b}$
$= \begin{pmatrix} -1 \\ 3 \\ -2 \end{pmatrix} - 2\begin{pmatrix} 5 \\ -1 \\ 4 \end{pmatrix}$
$= \begin{pmatrix} -11 \\ 5 \\ -10 \end{pmatrix}$

b $\mathbf{b} - 2\mathbf{x} = -\mathbf{a}$
$\therefore\ 2\mathbf{x} = \mathbf{a} + \mathbf{b}$
$\therefore\ \mathbf{x} = \frac{1}{2}(\mathbf{a} + \mathbf{b})$
$= \frac{1}{2}\left[\begin{pmatrix} -1 \\ 3 \\ -2 \end{pmatrix} + \begin{pmatrix} 5 \\ -1 \\ 4 \end{pmatrix}\right]$
$= \begin{pmatrix} 2 \\ 1 \\ 1 \end{pmatrix}$

12 $\overrightarrow{AB} = \begin{pmatrix} -1 - 2 \\ 4 - 3 \end{pmatrix} = \begin{pmatrix} -3 \\ 1 \end{pmatrix}$

$\overrightarrow{AC} = \begin{pmatrix} 3 - 2 \\ k - 3 \end{pmatrix} = \begin{pmatrix} 1 \\ k - 3 \end{pmatrix}$

Now $\overrightarrow{AB} \bullet \overrightarrow{AC} = 0$ {as $\text{B}\widehat{\text{A}}\text{C} = 90^o$}

$\therefore\ \begin{pmatrix} -3 \\ 1 \end{pmatrix} \bullet \begin{pmatrix} 1 \\ k - 3 \end{pmatrix} = 0$
$\therefore\ -3 + k - 3 = 0$
$\therefore\ k = 6$

13 One vector perpendicular to $\binom{-4}{5}$ is $\binom{5}{4}$ as the dot product $= -20 + 20 = 0$
$\therefore$ all vectors have form $k\binom{5}{4}$, $k \neq 0$.

14

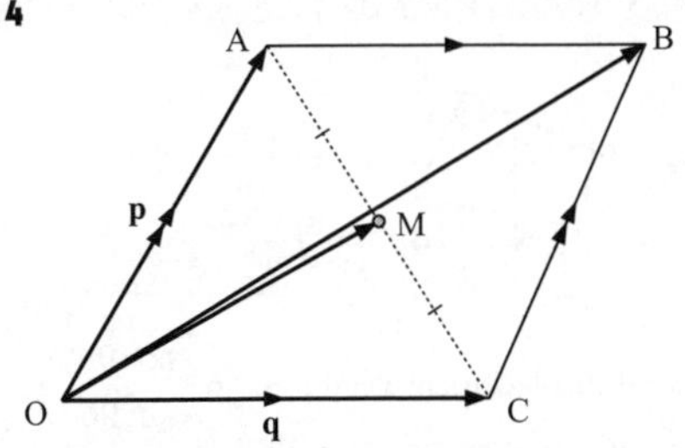

a **i** $\overrightarrow{OB}$
$= \overrightarrow{OA} + \overrightarrow{AB}$
$= \overrightarrow{OA} + \overrightarrow{OC}$
$= \mathbf{p} + \mathbf{q}$

ii $\overrightarrow{OM}$
$= \overrightarrow{OA} + \overrightarrow{AM}$
$= \overrightarrow{OA} + \frac{1}{2}\overrightarrow{AC}$
$= \mathbf{p} + \frac{1}{2}(\overrightarrow{AO} + \overrightarrow{OC})$
$= \mathbf{p} + \frac{1}{2}(-\mathbf{p} + \mathbf{q})$
$= \mathbf{p} - \frac{1}{2}\mathbf{p} + \frac{1}{2}\mathbf{q}$
$= \frac{1}{2}\mathbf{p} + \frac{1}{2}\mathbf{q}$

b We notice that $\overrightarrow{OM} = \frac{1}{2}\overrightarrow{OB}$
$\therefore$ [OM] $\parallel$ [OB] and OM $= \frac{1}{2}$OB
So, O, M and B are collinear (as O is common) and hence M is the midpoint of [OB].

15 $\mathbf{a} \bullet \mathbf{b} = |\mathbf{a}|\,|\mathbf{b}|\cos\theta$
$= 2 \times 4 \times \cos 120^o$
$= 8(-\frac{1}{2})$
$= -4$

$\mathbf{b} \bullet \mathbf{c} = |\mathbf{b}|\,|\mathbf{c}|\cos\theta$
$= 4 \times 5 \times \cos 60^o$
$= 20(\frac{1}{2})$
$= 10$

$\mathbf{a} \bullet \mathbf{c} = |\mathbf{a}|\,|\mathbf{c}|\cos\theta$
$= 2 \times 5 \times \cos 180^o$
$= 10(-1)$
$= -10$

16 $\overrightarrow{JK} = \begin{pmatrix} 6 \\ -3 \\ -3 \end{pmatrix}, \quad \overrightarrow{JL} = \begin{pmatrix} a+4 \\ b-1 \\ -1 \end{pmatrix}$

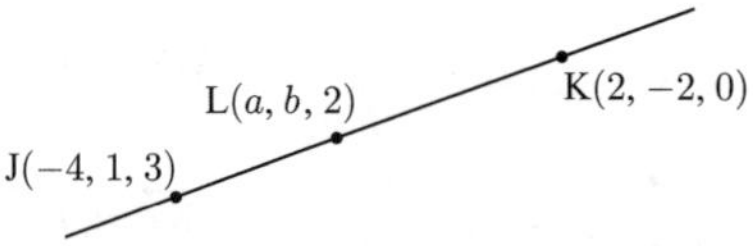

If J, K and L are collinear then $\overrightarrow{JK} \parallel \overrightarrow{JL}$

$\therefore \quad \begin{pmatrix} 6 \\ -3 \\ -3 \end{pmatrix} = k \begin{pmatrix} a+4 \\ b-1 \\ -1 \end{pmatrix}$ for some $k \neq 0$

$\therefore \quad k = 3$

$\therefore \quad a + 4 = 2$ and $b - 1 = -1$

$\therefore \quad a = -2$ and $b = 0$

17

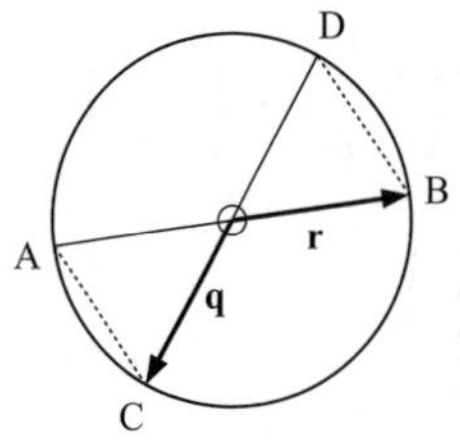

a $\overrightarrow{DB} = \overrightarrow{DO} + \overrightarrow{OB}$
$= \overrightarrow{OC} + \overrightarrow{OB}$
$= \mathbf{q} + \mathbf{r}$

b $\overrightarrow{AC} = \overrightarrow{AO} + \overrightarrow{OC}$
$= \overrightarrow{OB} + \overrightarrow{OC}$
$= \mathbf{r} + \mathbf{q}$

We see that $\overrightarrow{DB} = \overrightarrow{AC}$

$\therefore$ [DB] is parallel to [AC] and equal in length.

18 **a** $\begin{pmatrix} 2-t \\ 3 \\ t \end{pmatrix} \bullet \begin{pmatrix} t \\ 4 \\ t+1 \end{pmatrix} = 0$

$\therefore \quad (2-t)t + 12 + t(t+1) = 0$

$\therefore \quad 2t - t^2 + 12 + t^2 + t = 0$

$\therefore \quad 3t + 12 = 0$

$\therefore \quad t = -4$

b $\overrightarrow{KL} = \begin{pmatrix} -7 \\ 1 \\ 3 \end{pmatrix}, \quad \overrightarrow{KM} = \begin{pmatrix} -2 \\ -2 \\ -1 \end{pmatrix}, \quad \overrightarrow{LM} = \begin{pmatrix} 5 \\ -3 \\ -4 \end{pmatrix}$

Now $\overrightarrow{KM} \bullet \overrightarrow{LM} = \begin{pmatrix} -2 \\ -2 \\ -1 \end{pmatrix} \bullet \begin{pmatrix} 5 \\ -3 \\ -4 \end{pmatrix}$
$= -2(5) - 2(-3) - 1(-4)$
$= 0$

$\therefore$ [KM] and [LM] are perpendicular

$\therefore$ triangle KLM is right angled at M.

REVIEW SET 12B

1 **a**

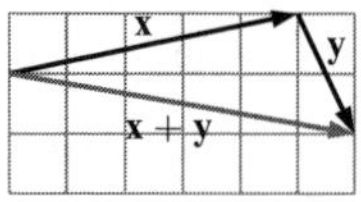

b

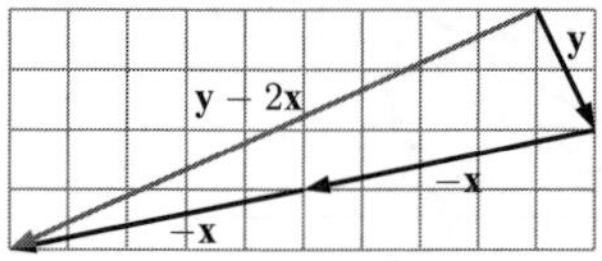

2 Dino's first displacement vector is $9\begin{pmatrix} \cos 246^o \\ \sin 246^o \end{pmatrix}$. His second displacement vector is $6\begin{pmatrix} \cos 96^o \\ \sin 96^o \end{pmatrix}$.

$\therefore$ Dino's resultant displacement vector is $\begin{pmatrix} 9\cos 246^o \\ 9\sin 246^o \end{pmatrix} + \begin{pmatrix} 6\cos 96^o \\ 6\sin 96^o \end{pmatrix} \approx \begin{pmatrix} -4.288 \\ -2.255 \end{pmatrix}$

which has length $\sqrt{(-4.288)^2 + (-2.255)^2} \approx 4.845$.

$\therefore$ Dino's displacement has unit vector $\frac{1}{4.845}\begin{pmatrix} -4.288 \\ -2.255 \end{pmatrix} \approx \begin{pmatrix} -0.8851 \\ -0.4654 \end{pmatrix} \begin{matrix} \leftarrow \cos\theta \\ \leftarrow \sin\theta \end{matrix}$

If $\cos\theta = -0.8851$ and $\sin\theta = -0.4654$, θ is in Quadrant 3

$\therefore$ $\theta = 180^o + \cos^{-1}(0.8851) \approx 207.7^o$

$\therefore$ Dino is 4.84 km from the start at a bearing of 208°.

3 $\overrightarrow{AB} = \begin{pmatrix} 6 \\ 1 \\ -4 \end{pmatrix}$, $\overrightarrow{AC} = \begin{pmatrix} 0 \\ 2 \\ -7 \end{pmatrix}$, $\overrightarrow{BC} = \begin{pmatrix} -6 \\ 1 \\ -3 \end{pmatrix}$

$\therefore$ $AB = \sqrt{6^2 + 1^2 + (-4)^2} = \sqrt{53}$ units $AC = \sqrt{0^2 + 2^2 + (-7)^2} = \sqrt{53}$ units $BC = \sqrt{(-6)^2 + 1^2 + (-3)^2} = \sqrt{46}$ units

$\therefore$ AB = AC, so ABC is an isosceles triangle.

4 **a** $|\mathbf{r}| = \sqrt{4^2 + 1^2} = \sqrt{17}$ units

b $|\mathbf{s}| = \sqrt{(-3)^2 + 2^2} = \sqrt{13}$ units

c $\mathbf{r} + \mathbf{s} = \binom{4}{1} + \binom{-3}{2} = \binom{1}{3}$

$\therefore$ $|\mathbf{r} + \mathbf{s}| = \sqrt{1^2 + 3^2} = \sqrt{10}$ units

d $2\mathbf{s} - \mathbf{r} = 2\binom{-3}{2} - \binom{4}{1} = \binom{-10}{3}$

$\therefore$ $|2\mathbf{s} - \mathbf{r}| = \sqrt{(-10)^2 + 3^2} = \sqrt{109}$ units

5 If θ is the angle then $\begin{pmatrix} 2 \\ -4 \\ 3 \end{pmatrix} \bullet \begin{pmatrix} -1 \\ 1 \\ 3 \end{pmatrix} = \sqrt{4+16+9}\sqrt{1+1+9}\cos\theta$

$\therefore$ $-2 - 4 + 9 = \sqrt{29}\sqrt{11}\cos\theta$

$\therefore$ $\dfrac{3}{\sqrt{29 \times 11}} = \cos\theta$ and so $\theta \approx 80.3^o$

6 $r\begin{pmatrix} -2 \\ 1 \end{pmatrix} + s\begin{pmatrix} 3 \\ -4 \end{pmatrix} = \begin{pmatrix} 13 \\ -24 \end{pmatrix}$

$\therefore$ $\begin{pmatrix} -2r + 3s \\ r - 4s \end{pmatrix} = \begin{pmatrix} 13 \\ -24 \end{pmatrix}$

$\therefore$ $-2r + 3s = 13$

$r - 4s = -24$ .... (1)

$\therefore$ $-2r + 3s = 13$

$2r - 8s = -48$ $\{2 \times (1)\}$

adding $-5s = -35$

$\therefore$ $s = 7$

Now using (1), $r - 4(7) = -24$

$\therefore$ $r = -24 + 28$

$\therefore$ $r = 4$ and $s = 7$

7 **a** $\overrightarrow{PQ} = \begin{pmatrix} -4-2 \\ 4-3 \\ 2--1 \end{pmatrix} = \begin{pmatrix} -6 \\ 1 \\ 3 \end{pmatrix}$

b $PQ = |\overrightarrow{PQ}| = \sqrt{36+1+9} = \sqrt{46}$ units

c The midpoint is at $\left(\dfrac{2+-4}{2}, \dfrac{3+4}{2}, \dfrac{-1+2}{2}\right)$ which is $(-1, \frac{7}{2}, \frac{1}{2})$.

8 **a** $\mathbf{p} \bullet \mathbf{q} = \begin{pmatrix} -1 \\ 2 \\ 1 \end{pmatrix} \bullet \begin{pmatrix} 3 \\ -1 \\ 4 \end{pmatrix} = -3 - 2 + 4 = -1$

b $\mathbf{p} + 2\mathbf{q} - \mathbf{r} = \begin{pmatrix} -1 \\ 2 \\ 1 \end{pmatrix} + \begin{pmatrix} 6 \\ -2 \\ 8 \end{pmatrix} - \begin{pmatrix} 1 \\ 1 \\ 2 \end{pmatrix} = \begin{pmatrix} 4 \\ -1 \\ 7 \end{pmatrix}$

c $\mathbf{p} \bullet \mathbf{r} = |\mathbf{p}|\,|\mathbf{r}|\cos\theta$ $\quad\therefore\quad \begin{pmatrix}-1\\2\\1\end{pmatrix} \bullet \begin{pmatrix}1\\1\\2\end{pmatrix} = \sqrt{1+4+1}\sqrt{1+1+4}\cos\theta$

$\therefore\quad -1+2+2 = \sqrt{6}\sqrt{6}\cos\theta$

$\therefore\quad 3 = 6\cos\theta$

$\therefore\quad \cos\theta = \frac{1}{2}$

$\therefore\quad \theta = 60^o$

9 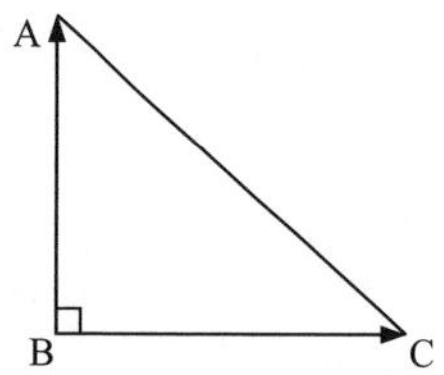

$\overrightarrow{BA} = \begin{pmatrix}4--1\\2-5\\-1-2\end{pmatrix} = \begin{pmatrix}5\\-3\\-3\end{pmatrix}$

$\overrightarrow{BC} = \begin{pmatrix}3--1\\-3-5\\c-2\end{pmatrix} = \begin{pmatrix}4\\-8\\c-2\end{pmatrix}$

But $\overrightarrow{BA} \bullet \overrightarrow{BC} = 0$

$\therefore\quad 20+24-3(c-2) = 0$

$\therefore\quad 44 = 3(c-2)$

$\therefore\quad 3c-6 = 44$

$\therefore\quad 3c = 50$

$\therefore\quad c = \frac{50}{3}$

10 **a** $2\mathbf{a} - 3\mathbf{b} = 2\begin{pmatrix}2\\-3\\1\end{pmatrix} - 3\begin{pmatrix}-1\\2\\3\end{pmatrix} = \begin{pmatrix}4\\-6\\2\end{pmatrix} - \begin{pmatrix}-3\\6\\9\end{pmatrix} = \begin{pmatrix}7\\-12\\-7\end{pmatrix}$

b $\mathbf{a} - 3\mathbf{x} = \mathbf{b}$ $\quad\therefore\quad \mathbf{a} - \mathbf{b} = 3\mathbf{x}$ $\quad\therefore\quad \mathbf{x} = \frac{1}{3}(\mathbf{a}-\mathbf{b}) = \frac{1}{3}\begin{pmatrix}3\\-5\\-2\end{pmatrix} = \begin{pmatrix}1\\-\frac{5}{3}\\-\frac{2}{3}\end{pmatrix}$

11 If the angle is θ then $\begin{pmatrix}3\\1\\-2\end{pmatrix} \bullet \begin{pmatrix}2\\5\\1\end{pmatrix} = \sqrt{9+1+4}\sqrt{4+25+1}\cos\theta$

$\therefore\quad 6+5-2 = \sqrt{14}\sqrt{30}\cos\theta$

$\therefore\quad \dfrac{9}{\sqrt{14\times 30}} = \cos\theta$

$\therefore\quad \theta \approx 64.0^o$

12 Let Q(0, 0, z) be a point on the Z-axis.

$PQ = \sqrt{4^2+(-2)^2+(z-5)^2} = 6$

$\therefore\quad 16+4+(z-5)^2 = 36$

$\therefore\quad (z-5)^2 = 16$

$\therefore\quad z-5 = \pm 4$

$\therefore\quad z = 1$ or 9 $\quad\therefore\quad$ Q is (0, 0, 1) or (0, 0, 9).

13 Since they are perpendicular, $\begin{pmatrix}3\\3-2t\end{pmatrix} \bullet \begin{pmatrix}t^2+t\\-2\end{pmatrix} = 0$

$\therefore\quad 3(t^2+t) - 2(3-2t) = 0$

$\therefore\quad 3t^2+3t-6+4t = 0$

$\therefore\quad 3t^2+7t-6 = 0$

$\therefore\quad (3t-2)(t+3) = 0$

$\therefore\quad t = \frac{2}{3}$ or -3

14 $4x - 5y = 11$ has gradient $\frac{4}{5}$ $\quad\therefore\quad$ it has direction vector $\begin{pmatrix}5\\4\end{pmatrix}$.

$2x + 3y = 7$ has gradient $-\frac{2}{3}$ $\quad\therefore\quad$ it has direction vector $\begin{pmatrix}3\\-2\end{pmatrix}$.

If the angle is θ, $\begin{pmatrix}5\\4\end{pmatrix} \bullet \begin{pmatrix}3\\-2\end{pmatrix} = \sqrt{5^2+4^2}\sqrt{3^2+(-2)^2}\cos\theta$

$$\therefore \quad 15-8 = \sqrt{41}\sqrt{13}\cos\theta$$

$$\therefore \quad \frac{7}{\sqrt{41\times 13}} = \cos\theta$$

$$\therefore \quad \theta \approx 72.3^o \qquad \therefore \quad \text{the angle is } 72.3^o$$

15

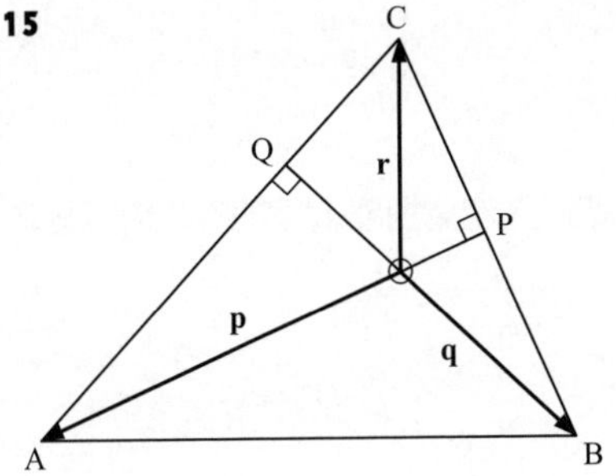

a $\overrightarrow{AC} = \overrightarrow{AO} + \overrightarrow{OC}$
$= -\mathbf{p} + \mathbf{r}$
$= \mathbf{r} - \mathbf{p}$

$\overrightarrow{BC} = \overrightarrow{BO} + \overrightarrow{OC}$
$= -\mathbf{q} + \mathbf{r}$
$= \mathbf{r} - \mathbf{q}$

b [AP] $\perp$ [BC] and [BQ] $\perp$ [AC]

$\therefore \ \mathbf{p} \perp \mathbf{r} - \mathbf{q}$ $\qquad \therefore \ \mathbf{q} \perp (\mathbf{r} - \mathbf{p})$

$\therefore \ \mathbf{p} \bullet (\mathbf{r} - \mathbf{q}) = 0$ $\qquad \therefore \ \mathbf{q} \bullet (\mathbf{r} - \mathbf{p}) = 0$

$\therefore \ \mathbf{p} \bullet \mathbf{r} - \mathbf{p} \bullet \mathbf{q} = 0$ $\qquad \therefore \ \mathbf{q} \bullet \mathbf{r} - \mathbf{q} \bullet \mathbf{p} = 0$

$\therefore \ \mathbf{p} \bullet \mathbf{r} = \mathbf{p} \bullet \mathbf{q}$ $\qquad \therefore \ \mathbf{q} \bullet \mathbf{r} = \mathbf{p} \bullet \mathbf{q}$

$$\therefore \ \mathbf{q} \bullet \mathbf{r} = \mathbf{p} \bullet \mathbf{q} = \mathbf{p} \bullet \mathbf{r}$$

c $\mathbf{r} \bullet \overrightarrow{AB} = \mathbf{r} \bullet (-\mathbf{p} + \mathbf{q})$
$= -\mathbf{r} \bullet \mathbf{p} + \mathbf{r} \bullet \mathbf{q}$
$= -\mathbf{p} \bullet \mathbf{q} + \mathbf{p} \bullet \mathbf{q}$ {from **b**}
$= 0$ and so $\mathbf{r} \perp \overrightarrow{AB}$ $\therefore$ [OC] $\perp$ [AB]

16 **a** $\mathbf{u} \bullet \mathbf{v}$
$= \begin{pmatrix}-4\\2\\1\end{pmatrix} \bullet \begin{pmatrix}-1\\3\\-2\end{pmatrix}$
$= -4(-1) + 2(3) + 1(-2)$
$= 8$

b If θ is the angle between $\mathbf{u}$ and $\mathbf{v}$ then

$$\cos\theta = \frac{\mathbf{u} \bullet \mathbf{v}}{|\mathbf{u}|\,|\mathbf{v}|}$$
$$= \frac{8}{\sqrt{(-4)^2+2^2+1^2}\sqrt{(-1)^2+3^2+(-2)^2}}$$
$$= \frac{8}{\sqrt{21}\sqrt{14}}$$
$$\therefore \quad \theta \approx 62.2^o$$

17 **a** $\begin{pmatrix}r\\4\\3\end{pmatrix} = k\begin{pmatrix}-5\\10\\s\end{pmatrix}$

$$\therefore \begin{cases} r = -5k \\ 4 = 10k \\ 3 = ks \end{cases}$$

$$\therefore \begin{cases} r = -5k & \text{.... (1)} \\ k = \frac{2}{5} \\ s = \dfrac{3}{k} & \text{.... (2)} \end{cases}$$

Substituting $k = \frac{2}{5}$ into (1) and (2) gives

$$r = -5(\tfrac{2}{5}) \quad \text{and} \quad s = \frac{3}{\frac{2}{5}}$$

$$\therefore \quad r = -2 \quad \text{and} \quad s = \tfrac{15}{2}$$

b $|3\mathbf{i} - 2\mathbf{j} + \mathbf{k}| = \sqrt{3^2 + (-2)^2 + 1^2} = \sqrt{14}$

$\therefore$ a unit vector in the direction $3\mathbf{i} - 2\mathbf{j} + \mathbf{k}$ is $\frac{1}{\sqrt{14}}(3\mathbf{i} - 2\mathbf{j} + \mathbf{k})$

$\therefore$ a vector 4 units long and parallel to $3\mathbf{i} - 2\mathbf{j} + \mathbf{k}$ is $\pm\frac{4}{\sqrt{14}}(3\mathbf{i} - 2\mathbf{j} + \mathbf{k})$.

18 M is $\left(\frac{-2+2}{2}, \frac{1+5}{2}, \frac{-3-1}{2}\right)$ or $(0, 3, -2)$.

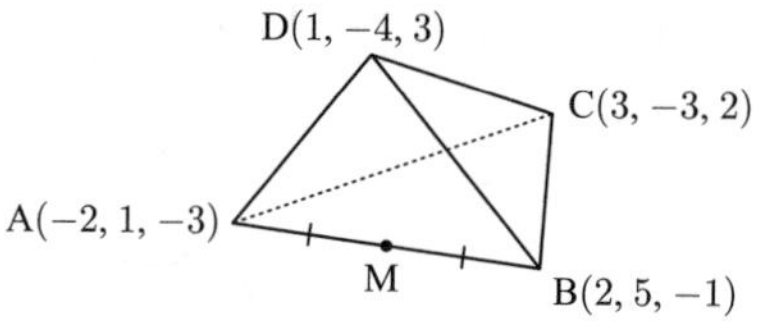

$\therefore \ \overrightarrow{MD} = \begin{pmatrix} 1 \\ -7 \\ 5 \end{pmatrix}, \ \overrightarrow{MC} = \begin{pmatrix} 3 \\ -6 \\ 4 \end{pmatrix}$

$\therefore \ \overrightarrow{MD} \bullet \overrightarrow{MC} = |\overrightarrow{MD}||\overrightarrow{MC}| \cos\theta$

$\therefore \ 3 + 42 + 20 = \sqrt{1+49+25}\sqrt{9+36+16}\cos\theta$

$\therefore \ 65 = \sqrt{75}\sqrt{61}\cos\theta$

$\therefore \ \theta \approx 16.1^o$

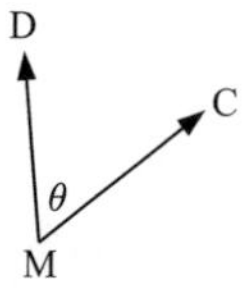

REVIEW SET 12C

1 **a** $\overrightarrow{PR} + \overrightarrow{RQ}$
$= \overrightarrow{PQ}$

b $\overrightarrow{PS} + \overrightarrow{SQ} + \overrightarrow{QR}$
$= \overrightarrow{PQ} + \overrightarrow{QR}$
$= \overrightarrow{PR}$

2 **a** $\mathbf{m} - \mathbf{n} + \mathbf{p} = \begin{pmatrix} 6 \\ -3 \\ 1 \end{pmatrix} - \begin{pmatrix} 2 \\ 3 \\ -4 \end{pmatrix} + \begin{pmatrix} -1 \\ 3 \\ 6 \end{pmatrix} = \begin{pmatrix} 3 \\ -3 \\ 11 \end{pmatrix}$

b $2\mathbf{n} - 3\mathbf{p} = 2\begin{pmatrix} 2 \\ 3 \\ -4 \end{pmatrix} - 3\begin{pmatrix} -1 \\ 3 \\ 6 \end{pmatrix} = \begin{pmatrix} 4 \\ 6 \\ -8 \end{pmatrix} - \begin{pmatrix} -3 \\ 9 \\ 18 \end{pmatrix} = \begin{pmatrix} 7 \\ -3 \\ -26 \end{pmatrix}$

c $\mathbf{m} + \mathbf{p} = \begin{pmatrix} 6 \\ -3 \\ 1 \end{pmatrix} + \begin{pmatrix} -1 \\ 3 \\ 6 \end{pmatrix} = \begin{pmatrix} 5 \\ 0 \\ 7 \end{pmatrix}$ $\therefore \ |\mathbf{m} + \mathbf{p}| = \sqrt{25+0+49}$
$= \sqrt{74}$ units

3 **a** If $\overrightarrow{AB} = \frac{1}{2}\overrightarrow{CD}$ then [AB] ∥ [CD] and [AB] is half the length of [CD].

b If $\overrightarrow{AB} = 2\overrightarrow{AC}$ then [AB] ∥ [AC] and AB = 2AC
$\therefore$ A, B and C are collinear and AB = 2AC.
So, C is the midpoint of [AB].

4 **a** $\overrightarrow{PQ} = \begin{pmatrix} -1-2 \\ 7--5 \\ 9-6 \end{pmatrix} = \begin{pmatrix} -3 \\ 12 \\ 3 \end{pmatrix}$

b $PQ = \sqrt{(-3)^2 + 12^2 + 3^2}$
$= \sqrt{162}$ units

c

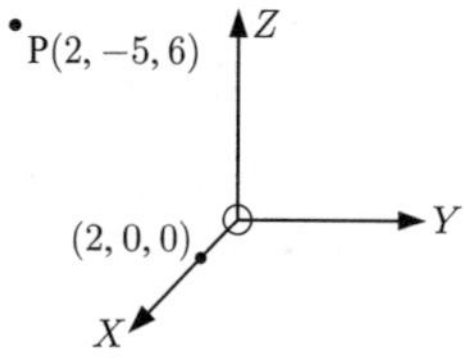

$\therefore$ distance
$= \sqrt{(2-2)^2 + (0--5)^2 + (0-6)^2}$
$= \sqrt{0+25+36}$
$= \sqrt{61}$ units

5 **a** $\overrightarrow{OQ} = \overrightarrow{OR} + \overrightarrow{RQ} = \mathbf{r} + \mathbf{q}$

b $\overrightarrow{PQ} = \overrightarrow{PO} + \overrightarrow{OR} + \overrightarrow{RQ} = -\mathbf{p} + \mathbf{r} + \mathbf{q}$

c $\overrightarrow{ON} = \overrightarrow{OR} + \overrightarrow{RN} = \mathbf{r} + \frac{1}{2}\mathbf{q}$

d $\overrightarrow{MN} = \overrightarrow{MQ} + \overrightarrow{QN}$
$= \frac{1}{2}\overrightarrow{PQ} + \frac{1}{2}\overrightarrow{QR}$
$= \frac{1}{2}(-\mathbf{p} + \mathbf{r} + \mathbf{q}) + \frac{1}{2}(-\mathbf{q})$
$= -\frac{1}{2}\mathbf{p} + \frac{1}{2}\mathbf{r} + \frac{1}{2}\mathbf{q} - \frac{1}{2}\mathbf{q}$
$= \frac{1}{2}\mathbf{r} - \frac{1}{2}\mathbf{p}$

6 **a** $2\mathbf{p}+\mathbf{q}$
$= 2\binom{-3}{1} + \binom{2}{-4}$
$= \binom{-6}{2} + \binom{2}{-4}$
$= \binom{-4}{-2}$

b $\mathbf{q} - 3\mathbf{r}$
$= \binom{2}{-4} - 3\binom{1}{3}$
$= \binom{2}{-4} - \binom{3}{9}$
$= \binom{-1}{-13}$

c $\mathbf{p}-\mathbf{q}+\mathbf{r}$
$= \binom{-3}{1} - \binom{2}{-4} + \binom{1}{3}$
$= \binom{-5}{5} + \binom{1}{3}$
$= \binom{-4}{8}$

7 **a** $\mathbf{p} - 3\mathbf{x} = \mathbf{0}$
$\therefore \quad \mathbf{p} = 3\mathbf{x}$
$\therefore \quad \frac{1}{3}\mathbf{p} = \mathbf{x}$
$\therefore \quad \mathbf{x} = \begin{pmatrix} -1 \\ \frac{1}{3} \end{pmatrix}$

b $2\mathbf{q} - \mathbf{x} = \mathbf{r}$
$\therefore \quad 2\mathbf{q} - \mathbf{r} = \mathbf{x}$
$\therefore \quad \mathbf{x} = 2\binom{2}{-4} - \binom{3}{2}$
$\therefore \quad \mathbf{x} = \binom{1}{-10}$

8 Since $\mathbf{v}$ is parallel to $\mathbf{w}$, the angle θ between $\mathbf{v}$ and $\mathbf{w}$ is either 0° or 180°.

Now, $\mathbf{v} \bullet \mathbf{w} = |\mathbf{v}|\,|\mathbf{w}|\cos\theta$
$= 3 \times 2 \times \cos 0^o \quad \text{or} \quad 3 \times 2 \times \cos 180^o$
$= 6(1) \quad \text{or} \quad 6(-1)$
$= \pm 6$

9 As the vectors are perpendicular,

$$\begin{pmatrix} -4 \\ t+2 \\ t \end{pmatrix} \bullet \begin{pmatrix} t \\ 1+t \\ -3 \end{pmatrix} = 0$$

$\therefore \quad -4t + (t+2)(1+t) - 3t = 0$
$\therefore \quad -4t + t + t^2 + 2 + 2t - 3t = 0$
$\therefore \quad t^2 - 4t + 2 = 0$

$\therefore \quad t = \dfrac{4 \pm \sqrt{16 - 4(1)(2)}}{2}$
$\therefore \quad t = \dfrac{4 \pm \sqrt{8}}{2} = 2 \pm \sqrt{2}$

10 $\overrightarrow{\text{MK}} = \begin{pmatrix} 3-4 \\ 1-1 \\ 4-3 \end{pmatrix} = \begin{pmatrix} -1 \\ 0 \\ 1 \end{pmatrix}$

$\overrightarrow{\text{ML}} = \begin{pmatrix} -2-4 \\ 1-1 \\ 3-3 \end{pmatrix} = \begin{pmatrix} -6 \\ 0 \\ 0 \end{pmatrix}$

$\overrightarrow{\text{MK}} \bullet \overrightarrow{\text{ML}} = |\overrightarrow{\text{MK}}||\overrightarrow{\text{ML}}| \cos M$
$\therefore \quad 6+0+0 = \sqrt{1+0+1}\sqrt{36+0+0}\cos M$
$\therefore \quad 6 = \sqrt{2} \times 6 \cos M$
$\therefore \quad \cos M = \frac{1}{\sqrt{2}}$
$\therefore \quad M = 45^o$
and $\quad K \approx 180^o - 45^o - 11.3^o \approx 123.7^o$

$\overrightarrow{\text{LK}} = \begin{pmatrix} 3--2 \\ 1-1 \\ 4-3 \end{pmatrix} = \begin{pmatrix} 5 \\ 0 \\ 1 \end{pmatrix}$

$\overrightarrow{\text{LM}} = \begin{pmatrix} 4--2 \\ 1-1 \\ 3-3 \end{pmatrix} = \begin{pmatrix} 6 \\ 0 \\ 0 \end{pmatrix}$

$\therefore \quad \overrightarrow{\text{LK}} \bullet \overrightarrow{\text{LM}} = |\overrightarrow{\text{LK}}||\overrightarrow{\text{LM}}| \cos L$
$\therefore \quad 30+0+0 = \sqrt{25+0+1}\sqrt{36+0+0}\cos L$
$\therefore \quad 30 = \sqrt{26} \times 6 \cos L$
$\therefore \quad \frac{5}{\sqrt{26}} = \cos L$
$\therefore \quad L \approx 11.3^o$

11 **a** $\begin{pmatrix} \frac{4}{7} \\ \frac{1}{k} \end{pmatrix}$ is a unit vector if

$\sqrt{\left(\frac{4}{7}\right)^2 + \left(\frac{1}{k}\right)^2} = 1$
$\therefore \quad \frac{16}{49} + \dfrac{1}{k^2} = 1$
$\therefore \quad \dfrac{1}{k^2} = \frac{33}{49}$
$\therefore \quad k = \pm\frac{7}{\sqrt{33}}$

b $\begin{pmatrix} k \\ k \\ k \end{pmatrix}$ is a unit vector if

$\sqrt{k^2 + k^2 + k^2} = 1$
$\therefore \quad 3k^2 = 1$
$\therefore \quad k^2 = \frac{1}{3}$
$\therefore \quad k = \pm\frac{1}{\sqrt{3}}$

12 If D is the origin, (DA) the X-axis, (DC) the Y-axis and (DE) the Z-axis, then A is (4, 0, 0), C is (0, 8, 0) and G is (4, 8, 5).

$$\overrightarrow{AG} = \begin{pmatrix} 4-4 \\ 8-0 \\ 5-0 \end{pmatrix} = \begin{pmatrix} 0 \\ 8 \\ 5 \end{pmatrix} \qquad \overrightarrow{AC} = \begin{pmatrix} 0-4 \\ 8-0 \\ 0-0 \end{pmatrix} = \begin{pmatrix} -4 \\ 8 \\ 0 \end{pmatrix}$$

If the required angle is θ then $\begin{pmatrix} 0 \\ 8 \\ 5 \end{pmatrix} \bullet \begin{pmatrix} -4 \\ 8 \\ 0 \end{pmatrix} = \sqrt{0+64+25}\sqrt{16+64+0}\cos\theta$

$$\therefore \quad 0+64+0 = \sqrt{89}\sqrt{80}\cos\theta$$

$$\therefore \quad \cos\theta = \frac{64}{\sqrt{89\times 80}} \quad \text{and so} \quad \theta \approx 40.7^o$$

13 LHS $= \mathbf{p}\bullet(\mathbf{q}-\mathbf{r})$

$= \binom{3}{-2}\bullet\left[\binom{-2}{5}-\binom{1}{-3}\right]$

$= \binom{3}{-2}\bullet\binom{-3}{8}$

$= -9-16$

$= -25$

RHS $= \mathbf{p}\bullet\mathbf{q}-\mathbf{p}\bullet\mathbf{r}$

$= \binom{3}{-2}\bullet\binom{-2}{5}-\binom{3}{-2}\bullet\binom{1}{-3}$

$= (-6-10)-(3+6)$

$= -16-9$

$= -25$ $\quad\therefore\quad$ LHS = RHS ✓

14 $\overrightarrow{KL} = \begin{pmatrix} 3--2 \\ 2-1 \end{pmatrix} = \begin{pmatrix} 5 \\ 1 \end{pmatrix}$

$\overrightarrow{KM} = \begin{pmatrix} 1--2 \\ -3-1 \end{pmatrix} = \begin{pmatrix} 3 \\ -4 \end{pmatrix}$

Now $\overrightarrow{KL}\bullet\overrightarrow{KM} = |\overrightarrow{KL}||\overrightarrow{KM}|\cos K$

$\therefore \quad \begin{pmatrix} 5 \\ 1 \end{pmatrix}\bullet\begin{pmatrix} 3 \\ -4 \end{pmatrix} = \sqrt{25+1}\sqrt{9+16}\cos K$

$\therefore \quad 15-4 = \sqrt{26}\sqrt{25}\cos K$

$\therefore \quad \cos K = \dfrac{11}{5\sqrt{26}}$

$\therefore \quad K \approx 64.4^o$

$\overrightarrow{LK} = -\overrightarrow{KL} = \begin{pmatrix} -5 \\ -1 \end{pmatrix}$

$\overrightarrow{LM} = \begin{pmatrix} 1-3 \\ -3-2 \end{pmatrix} = \begin{pmatrix} -2 \\ -5 \end{pmatrix}$

Now $\overrightarrow{LK}\bullet\overrightarrow{LM} = |\overrightarrow{LK}||\overrightarrow{LM}|\cos L$

$\therefore \quad \begin{pmatrix} -5 \\ -1 \end{pmatrix}\bullet\begin{pmatrix} -2 \\ -5 \end{pmatrix} = \sqrt{25+1}\sqrt{4+25}\cos L$

$\therefore \quad 10+5 = \sqrt{26}\sqrt{29}\cos L$

$\therefore \quad \cos L = \dfrac{15}{\sqrt{26\times 29}}$

$\therefore \quad L \approx 56.9^o$

$\therefore \quad M \approx 180^o - 56.89^o - 64.44^o \approx 58.7^o$

15 $\overrightarrow{MP}\bullet\overrightarrow{PT} = 0$

Now, $\binom{5}{-1}\bullet\binom{1}{5} = 5+(-5) = 0$

$\therefore$ $\overrightarrow{PT}$ has the form $k\binom{1}{5}$, where k is a scalar.

Also, $|\overrightarrow{MP}| = |\overrightarrow{PT}|$

$\therefore \quad \sqrt{5^2+(-1)^2} = \sqrt{k^2+(5k)^2}$

$\therefore \quad \sqrt{26} = \sqrt{26k^2}$

$= |k|\sqrt{26}$

$\therefore \quad k = \pm 1$

So $\overrightarrow{PT} = \binom{1}{5}$ or $\binom{-1}{-5}$

Finally, $\overrightarrow{OT} = \overrightarrow{OM}+\overrightarrow{MP}+\overrightarrow{PT}$

$= \binom{-2}{4}+\binom{5}{-1}+\binom{1}{5}$ or $\binom{-2}{4}+\binom{5}{-1}+\binom{-1}{-5}$

$= \binom{4}{8}$ or $\binom{2}{-2}$

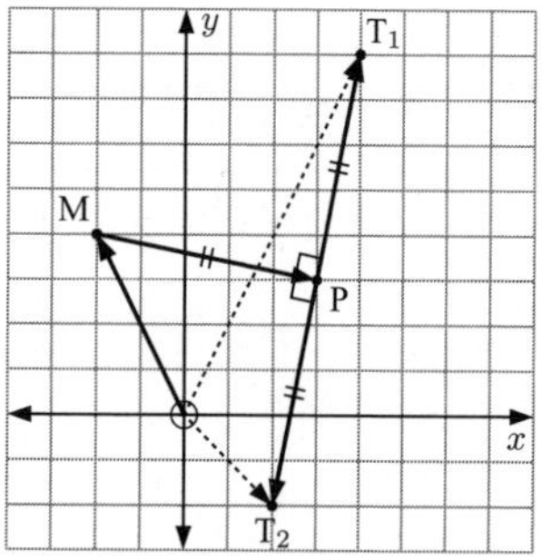

16 **a** $\sqrt{k^2 + (\frac{1}{\sqrt{2}})^2 + (-k)^2} = 1$

$\therefore \quad k^2 + \frac{1}{2} + k^2 = 1$

$\therefore \quad 2k^2 = \frac{1}{2}$

$\therefore \quad k^2 = \frac{1}{4}$

$\therefore \quad k = \pm\frac{1}{2}$

b $\begin{pmatrix} 3 \\ 2 \\ -1 \end{pmatrix}$ has length $\sqrt{3^2 + 2^2 + (-1)^2} = \sqrt{14}$ units

$\therefore$ a unit vector in the opposite direction is $-\frac{1}{\sqrt{14}}\begin{pmatrix} 3 \\ 2 \\ -1 \end{pmatrix}$

$\therefore$ a vector of length 5 units in the opposite direction

is $-\frac{5}{\sqrt{14}}\begin{pmatrix} 3 \\ 2 \\ -1 \end{pmatrix}$.

17 **a** $\mathbf{p} \bullet \mathbf{q}$

$= \begin{pmatrix} 2 \\ -1 \\ 4 \end{pmatrix} \bullet \begin{pmatrix} -1 \\ -4 \\ 2 \end{pmatrix}$

$= 2(-1) - 1(-4) + 4(2)$

$= 10$

b If θ is the angle between $\mathbf{p}$ and $\mathbf{q}$ then

$$\cos\theta = \frac{\mathbf{p} \bullet \mathbf{q}}{|\mathbf{p}|\,|\mathbf{q}|}$$

$$= \frac{10}{\sqrt{2^2 + (-1)^2 + 4^2}\sqrt{(-1)^2 + (-4)^2 + 2^2}}$$

$$= \frac{10}{\sqrt{21}\sqrt{21}}$$

$\therefore \quad \theta \approx 61.6^o$

18 $2\begin{pmatrix} s-1 \\ r+1 \\ t \end{pmatrix} = \begin{pmatrix} 4s \\ 3r \\ r \end{pmatrix} + \begin{pmatrix} r \\ -1 \\ s \end{pmatrix} \quad \therefore \quad \begin{pmatrix} 2s-2 \\ 2r+2 \\ 2t \end{pmatrix} = \begin{pmatrix} 4s+r \\ 3r-1 \\ r+s \end{pmatrix}$

$\therefore \begin{cases} 2s - 2 = 4s + r \\ 2r + 2 = 3r - 1 \\ 2t = r + s \end{cases}$ and so $\begin{cases} r = -2s - 2 & \text{.... (1)} \\ r = 3 & \text{.... (2)} \\ t = \dfrac{r+s}{2} & \text{.... (3)} \end{cases}$

Substituting (2) into (1) gives $3 = -2s - 2$

$\therefore \quad s = -\frac{5}{2}$

Using (3), $t = \dfrac{3 - \frac{5}{2}}{2} = \frac{1}{4}$

$\therefore \quad r = 3, \quad s = -\frac{5}{2}, \quad t = \frac{1}{4}$

19 $\mathbf{u} = \binom{2}{1}, \quad \mathbf{v} = \binom{0}{3}$

$$\cos\theta = \frac{\mathbf{u} \bullet \mathbf{v}}{|\mathbf{u}|\,|\mathbf{v}|}$$

$$= \frac{\binom{2}{1} \bullet \binom{0}{3}}{\sqrt{2^2 + 1^2}\sqrt{3^2}}$$

$$= \frac{3}{\sqrt{5}\sqrt{9}}$$

$$= \frac{1}{\sqrt{5}}$$

Now $\sin^2\theta + \cos^2\theta = 1$

$\therefore \quad \sin^2\theta + \frac{1}{5} = 1$

$\therefore \quad \sin^2\theta = \frac{4}{5}$

$\therefore \quad \sin\theta = \pm\frac{2}{\sqrt{5}}$

But θ is acute, so $\sin\theta = \frac{2}{\sqrt{5}}$.

Chapter 13
LINES AND PLANES IN SPACE

EXERCISE 13A.1

1 **a** **i** $\begin{pmatrix} x \\ y \end{pmatrix} = \begin{pmatrix} 3 \\ -4 \end{pmatrix} + t\begin{pmatrix} 1 \\ 4 \end{pmatrix}, \quad t \in \mathbb{R}$

ii $x = 3 + t, \quad y = -4 + 4t$

$\therefore \; t = x - 3 = \dfrac{y+4}{4}$

$\therefore \; 4x - 12 = y + 4$

$\therefore \; 4x - y = 16$

b **i** If the line has direction vector **b** perpendicular to $\begin{pmatrix} -8 \\ 2 \end{pmatrix}$, then

$\mathbf{b} \bullet \begin{pmatrix} -8 \\ 2 \end{pmatrix} = 0$

$\therefore \; \mathbf{b} = \begin{pmatrix} 2 \\ 8 \end{pmatrix}$ is a reasonable choice

$\therefore \; \begin{pmatrix} x \\ y \end{pmatrix} = \begin{pmatrix} 5 \\ 2 \end{pmatrix} + t\begin{pmatrix} 2 \\ 8 \end{pmatrix}, \quad t \in \mathbb{R}$

ii $x = 5 + 2t, \quad y = 2 + 8t$

$\therefore \; t = \dfrac{x-5}{2} = \dfrac{y-2}{8}$

$\therefore \; 8x - 40 = 2y - 4$

$\therefore \; 8x - 2y = 36$

$\therefore \; 4x - y = 18$

c **i** $\begin{pmatrix} x \\ y \end{pmatrix} = \begin{pmatrix} -6 \\ 0 \end{pmatrix} + t\begin{pmatrix} 3 \\ 7 \end{pmatrix}, \quad t \in \mathbb{R}$

ii $x = -6 + 3t, \quad y = 7t$

$\therefore \; t = \dfrac{x+6}{3} = \dfrac{y}{7}$

$\therefore \; 7x + 42 = 3y$

$\therefore \; 7x - 3y = -42$

d **i** Take $(-1, 11)$ as our fixed point, so $\mathbf{a} = \begin{pmatrix} -1 \\ 11 \end{pmatrix}$.

The direction vector $\mathbf{b} = \begin{pmatrix} -3-(-1) \\ 12-11 \end{pmatrix}$

$= \begin{pmatrix} -2 \\ 1 \end{pmatrix}$.

$\therefore \; \begin{pmatrix} x \\ y \end{pmatrix} = \begin{pmatrix} -1 \\ 11 \end{pmatrix} + t\begin{pmatrix} -2 \\ 1 \end{pmatrix}, \quad t \in \mathbb{R}$

ii $x = -1 - 2t, \quad y = 11 + t$

$\therefore \; t = \dfrac{x+1}{-2} = y - 11$

$\therefore \; x + 1 = -2y + 22$

$\therefore \; x + 2y = 21$

2 $x = -1 + 2t, \; y = 4 - t, \quad t \in \mathbb{R}$

When $t = 0$, $x = -1 + 2(0) = -1$ and $y = 4 - 0 = 4$ $\quad \therefore$ the point is $(-1, 4)$.

When $t = 1$, $x = -1 + 2(1) = 1$ and $y = 4 - 1 = 3$ $\quad \therefore$ the point is $(1, 3)$.

When $t = 3$, $x = -1 + 2(3) = 5$ and $y = 4 - 3 = 1$ $\quad \therefore$ the point is $(5, 1)$.

When $t = -1$, $x = -1 + 2(-1) = -3$ and $y = 4 - -1 = 5$ $\quad \therefore$ the point is $(-3, 5)$.

When $t = -4$, $x = -1 + 2(-4) = -9$ and $y = 4 - -4 = 8$ $\quad \therefore$ the point is $(-9, 8)$.

3 **a** If $t + 2 = 3$ and $1 - 3t = -2$,

then $t = 1$ and $-3t = -3$

$\therefore \; t = 1$

Since $t = 1$ in each case, $(3, -2)$ lies on the line.

b If $(k, 4)$ lies on $x = 1 - 2t, \; y = 1 + t$, then

$k = 1 - 2t$ and $4 = 1 + t$

$\therefore \; t = 3$

and $k = 1 - 6 = -5$

4 **a** $x(0) = 1$ and $y(0) = 2$,

$\therefore$ the initial position is $(1, 2)$

b

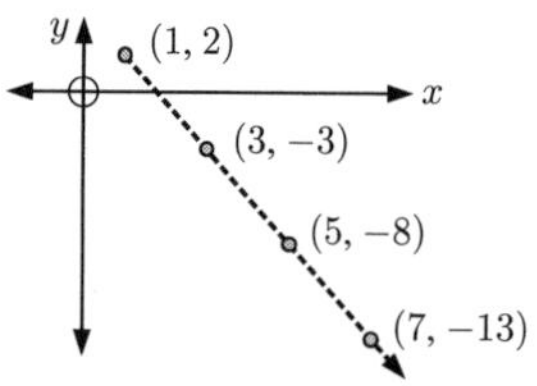

c In 1 second, the x-step is 2 and y-step is -5, which is a distance of $\sqrt{2^2 + (-5)^2} = \sqrt{29}$ cm

$\therefore$ the speed is $\sqrt{29}$ cm s^{-1}.

5 In parametric form: $x = 1 - t, \; y = 5 + 3t, \; t \in \mathbb{R}$

In Cartesian form: $t = \dfrac{x-1}{-1} = \dfrac{y-5}{3}$

$\therefore \; 3x - 3 = -y + 5$

$\therefore \; 3x + y = 8$

EXERCISE 13A.2

1 **a** The vector equation is $\begin{pmatrix} x \\ y \\ z \end{pmatrix} = \begin{pmatrix} 1 \\ 3 \\ -7 \end{pmatrix} + t\begin{pmatrix} 2 \\ 1 \\ 3 \end{pmatrix}, \; t \in \mathbb{R}$

b The vector equation is $\begin{pmatrix} x \\ y \\ z \end{pmatrix} = \begin{pmatrix} 0 \\ 1 \\ 2 \end{pmatrix} + t\begin{pmatrix} 1 \\ 1 \\ -2 \end{pmatrix}, \; t \in \mathbb{R}$

c Since the line is parallel to the X-axis, it has direction vector $\begin{pmatrix} 1 \\ 0 \\ 0 \end{pmatrix}$

$\therefore$ the vector equation is $\begin{pmatrix} x \\ y \\ z \end{pmatrix} = \begin{pmatrix} -2 \\ 2 \\ 1 \end{pmatrix} + t\begin{pmatrix} 1 \\ 0 \\ 0 \end{pmatrix}, \; t \in \mathbb{R}$

2 **a** The parametric equations are:

$x = 5 + (-1)t, \; y = 2 + 2t, \; z = -1 + 6t$

$\therefore \; x = 5 - t, \; y = 2 + 2t, \; z = -1 + 6t, \; t \in \mathbb{R}$

b The parametric equations are:

$x = 0 + 2t, \; y = 2 + (-1)t, \; z = -1 + 3t$

$\therefore \; x = 2t, \; y = 2 - t, \; z = -1 + 3t, \; t \in \mathbb{R}$

c Since the line is perpendicular to the XOY plane, it has direction vector $\begin{pmatrix} 0 \\ 0 \\ 1 \end{pmatrix}$.

$\therefore$ the parametric equations are: $x = 3 + 0t, \; y = 2 + 0t, \; z = -1 + 1t$

$\therefore \; x = 3, \; y = 2, \; z = -1 + t, \; t \in \mathbb{R}$

3 **a** $\overrightarrow{AB} = \begin{pmatrix} -1-1 \\ 3-2 \\ 2-1 \end{pmatrix} = \begin{pmatrix} -2 \\ 1 \\ 1 \end{pmatrix} \quad \therefore \; \begin{pmatrix} x \\ y \\ z \end{pmatrix} = \begin{pmatrix} 1 \\ 2 \\ 1 \end{pmatrix} + t\begin{pmatrix} -2 \\ 1 \\ 1 \end{pmatrix}, \; t \in \mathbb{R}$

b $\overrightarrow{CD} = \begin{pmatrix} 3-0 \\ 1-1 \\ -1-3 \end{pmatrix} = \begin{pmatrix} 3 \\ 0 \\ -4 \end{pmatrix} \quad \therefore \; \begin{pmatrix} x \\ y \\ z \end{pmatrix} = \begin{pmatrix} 0 \\ 1 \\ 3 \end{pmatrix} + t\begin{pmatrix} 3 \\ 0 \\ -4 \end{pmatrix}, \; t \in \mathbb{R}$

c $\overrightarrow{EF} = \begin{pmatrix} 1-1 \\ -1-2 \\ 5-5 \end{pmatrix} = \begin{pmatrix} 0 \\ -3 \\ 0 \end{pmatrix} \quad \therefore \; \begin{pmatrix} x \\ y \\ z \end{pmatrix} = \begin{pmatrix} 1 \\ 2 \\ 5 \end{pmatrix} + t\begin{pmatrix} 0 \\ -3 \\ 0 \end{pmatrix}, \; t \in \mathbb{R}$

d $\overrightarrow{GH} = \begin{pmatrix} 5-0 \\ -1-1 \\ 3--1 \end{pmatrix} = \begin{pmatrix} 5 \\ -2 \\ 4 \end{pmatrix} \quad \therefore \; \begin{pmatrix} x \\ y \\ z \end{pmatrix} = \begin{pmatrix} 0 \\ 1 \\ -1 \end{pmatrix} + t\begin{pmatrix} 5 \\ -2 \\ 4 \end{pmatrix}, \; t \in \mathbb{R}$

4 Given $x = 1 - t, \; y = 3 + t, \; z = 3 - 2t$:

a The line meets the XOY plane when $z = 0$ $\quad \therefore \; 3 - 2t = 0$

$\therefore \; t = \frac{3}{2}$

Then $x = 1 - \frac{3}{2} = -\frac{1}{2}$ and $y = 3 + \frac{3}{2} = \frac{9}{2}$, so the point is $\left(-\frac{1}{2}, \frac{9}{2}, 0\right)$.

b The line meets the YOZ plane when $x = 0$ $\quad \therefore \; 1 - t = 0$

$\therefore \; t = 1$

Then $y = 3 + 1 = 4$ and $z = 3 - 2 = 1$, so the point is $(0, 4, 1)$.

c The line meets the XOZ plane when $y = 0$ $\therefore\ 3 + t = 0$

$\therefore\ t = -3$

Then $x = 1 - (-3) = 4$ and $z = 3 - 2(-3) = 9$, so the point is $(4, 0, 9)$.

5 Given a line with equations $x = 2 - t$, $y = 3 + 2t$ and $z = 1 + t$,

the distance to the point $(1, 0, -2)$ is $\sqrt{(2 - t - 1)^2 + (3 + 2t - 0)^2 + (1 + t + 2)^2}$.

But this distance $= 5\sqrt{3}$ units

$$\therefore\ \sqrt{(1-t)^2 + (3+2t)^2 + (t+3)^2} = 5\sqrt{3}$$
$$\therefore\ (1-t)^2 + (3+2t)^2 + (t+3)^2 = 75$$
$$\therefore\ 1 - 2t + t^2 + 9 + 12t + 4t^2 + t^2 + 6t + 9 = 75$$
$$\therefore\ 6t^2 + 16t - 56 = 0$$
$$\therefore\ 3t^2 + 8t - 28 = 0$$
$$\therefore\ (3t + 14)(t - 2) = 0$$
$$\therefore\ t = -\tfrac{14}{3} \text{ or } t = 2$$

When $t = 2$ the point is $(0, 7, 3)$, and when $t = -\frac{14}{3}$ the point is $\left(\frac{20}{3}, -\frac{19}{3}, -\frac{11}{3}\right)$.

EXERCISE 13A.3

1 L_1 has direction vector $\binom{4}{-3}$ and L_2 has direction vector $\binom{5}{4}$. If θ is the angle between them,

$$\cos\theta = \frac{\left|\binom{4}{-3} \bullet \binom{5}{4}\right|}{\sqrt{16+9}\sqrt{25+16}} = \frac{|20 + (-12)|}{\sqrt{25 \times 41}} = \frac{8}{\sqrt{25 \times 41}}$$

$$\therefore\ \theta = \cos^{-1}\left(\frac{8}{\sqrt{25 \times 41}}\right) \approx 75.5^o$$

$\therefore$ the required angle measures 75.5^o.

2 L_1 has direction vector $\binom{12}{5}$ and L_2 has direction vector $\binom{3}{-4}$. If θ is the angle between them,

$$\cos\theta = \frac{\left|\binom{12}{5} \bullet \binom{3}{-4}\right|}{\sqrt{144+25}\sqrt{9+16}} = \frac{|36 + (-20)|}{13 \times 5} = \frac{16}{65}$$

$$\therefore\ \theta = \cos^{-1}\left(\tfrac{16}{65}\right) \approx 75.7^o$$

3 Line 1 has direction vector $\binom{5}{-2}$ and line 2 has direction vector $\binom{4}{10}$

and $\binom{5}{-2} \bullet \binom{4}{10} = 20 + (-20) = 0$

$\therefore$ the lines are perpendicular.

4 **a** Line 1 has direction vector $\begin{pmatrix} 3 \\ -16 \\ 7 \end{pmatrix}$ and line 2 has direction vector $\begin{pmatrix} 3 \\ 8 \\ -5 \end{pmatrix}$.

If θ is the angle between them,

$$\cos\theta = \frac{\left|\begin{pmatrix} 3 \\ -16 \\ 7 \end{pmatrix} \bullet \begin{pmatrix} 3 \\ 8 \\ -5 \end{pmatrix}\right|}{\sqrt{9+256+49}\sqrt{9+64+25}} = \frac{|9 - 128 - 35|}{\sqrt{314}\sqrt{98}} = \frac{154}{\sqrt{314 \times 98}}$$

$\therefore\ \theta \approx 28.6^o$

b Since $L_1 \perp L_3$, $\begin{pmatrix} 3 \\ -16 \\ 7 \end{pmatrix} \bullet \begin{pmatrix} 0 \\ -3 \\ x \end{pmatrix} = 0$

$$\therefore\ 48 + 7x = 0$$
$$\therefore\ x = -\tfrac{48}{7}$$

EXERCISE 13B.1

1 **a** **i** When $t=0$, $\binom{x}{y}=\binom{-4}{3}$ $\therefore$ the object is at $(-4, 3)$.
ii The velocity vector is $\binom{12}{5}$.
iii The speed is $\sqrt{12^2+5^2} = 13\ \text{m s}^{-1}$

b **i** When $t=0$, $\binom{x}{y}=\binom{0}{-6}$ $\therefore$ the object is at $(0, -6)$.
ii The velocity vector is $\binom{3}{-4}$.
iii The speed is $\sqrt{3^2+(-4)^2} = 5\ \text{m s}^{-1}$

c **i** When $t=0$, $\binom{x}{y}=\binom{-2}{-7}$ $\therefore$ the object is at $(-2, -7)$.
ii The velocity vector is $\binom{-6}{-4}$.
iii The speed is $\sqrt{(-6)^2+(-4)^2} = \sqrt{52}\ \text{m s}^{-1}$

d **i** When $t=0$, $x=5$ and $y=-5$ $\therefore$ the object is at $(5, -5)$.
ii The velocity vector is $\binom{8}{4}$.
iii The speed is $\sqrt{8^2+4^2} = \sqrt{80}\ \text{m s}^{-1}$

2 **a** $\binom{4}{-3}$ has length $\sqrt{4^2+(-3)^2}=5$
$\therefore$ $30\binom{4}{-3}$ has length 150
$\therefore$ the velocity vector is $\binom{120}{-90}$.

b $\binom{24}{7}$ has length $\sqrt{24^2+7^2}=25$
$\therefore$ $\frac{1}{2}\binom{24}{7}$ has length 12.5
$\therefore$ the velocity vector is $\binom{12}{3.5}$.

c $2\mathbf{i}+\mathbf{j}=\binom{2}{1}$ has length $\sqrt{2^2+1^2}=\sqrt{5}$
$\therefore$ $10\sqrt{5}\binom{2}{1}$ has length 50
$\therefore$ the velocity vector is $\binom{20\sqrt{5}}{10\sqrt{5}}$.

3 $-2\mathbf{i}+5\mathbf{j}-14\mathbf{k}$ has length $\sqrt{(-2)^2+5^2+(-14)^2}=\sqrt{4+25+196}=\sqrt{225}=15$

$\therefore$ $6\begin{pmatrix}-2\\5\\-14\end{pmatrix}$ has length 90, so the velocity vector is $\begin{pmatrix}-12\\30\\-84\end{pmatrix}$.

EXERCISE 13B.2

1 **a** $\mathbf{r}=\mathbf{a}+t\mathbf{b}$
$\therefore$ $\binom{x}{y}=\binom{-3}{-2}+t\binom{2}{4}$, $t \geqslant 0$
$\therefore$ $\binom{x}{y}=\binom{-3+2t}{-2+4t}$

b At $t=2.5$, $-3+2t=-3+5=2$
and $-2+4t=-2+10=8$

So, the position vector is $\binom{2}{8}$.

c **i** When the car is due north, $x=0$.
$\therefore$ $-3+2t=0$
$\therefore$ $t=1.5$ seconds

ii When the car is due west, $y=0$.
$\therefore$ $-2+4t=0$
$\therefore$ $t=0.5$ seconds

d

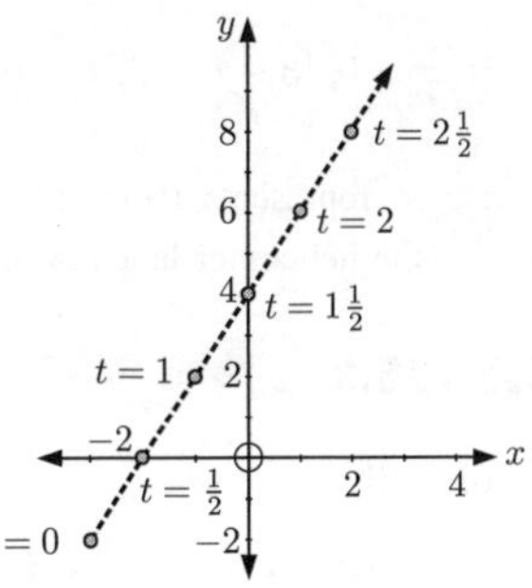

2 Yacht A: $\binom{x_A}{y_B}=\binom{4}{5}+t\binom{1}{-2}$ Yacht B: $\binom{x_B}{y_B}=\binom{1}{-8}+t\binom{2}{1}$, $t \geqslant 0$

a When $t=0$, $\binom{x_A}{y_A}=\binom{4}{5}$ $\therefore$ A is at $(4, 5)$
and $\binom{x_B}{y_B}=\binom{1}{-8}$ $\therefore$ B is at $(1, -8)$.

b For A, the velocity vector is $\binom{1}{-2}$, and for B it is $\binom{2}{1}$.

c Speed of A $=\sqrt{1^2+(-2)^2}=\sqrt{5}\ \text{km h}^{-1}$. Speed of B $=\sqrt{2^2+1^2}=\sqrt{5}\ \text{km h}^{-1}$.

d A has direction vector $\binom{1}{-2}$ and B has direction vector $\binom{2}{1}$.

Since $\binom{1}{-2} \bullet \binom{2}{1} = 2 - 2 = 0$, the paths of the yachts are at right angles to each other.

3 **a** P's torpedo has position $\binom{x_1}{y_1} = \binom{-5}{4} + t\binom{3}{-1}$ and at $t = 0$, the time is 1:34 pm

$\therefore\ x_1(t) = -5 + 3t, \quad y_1(t) = 4 - t.$

b Speed $= \sqrt{3^2 + (-1)^2} = \sqrt{10}$ km min^{-1}

c Q fires its torpedo after a minutes.

$\therefore$ at time t, its torpedo has travelled for $(t - a)$ minutes.

$\therefore\ \binom{x_2}{y_2} = \binom{15}{7} + (t-a)\binom{-4}{-3}, \quad t \geqslant a$

$\therefore\ x_2(t) = 15 - 4(t - a)$ and $y_2(t) = 7 - 3(t - a)$

d They meet when $x_1(t) = x_2(t)$ and $y_1(t) = y_2(t)$

$\therefore\ -5 + 3t = 15 - 4(t - a)$ and $4 - t = 7 - 3(t - a)$

$\therefore\ 7t - 4a = 20$ .... (1) and $2t - 3a = 3$ .... (2)

Solving simultaneously,

$$\begin{aligned} 21t - 12a &= 60 \quad \{3 \times (1)\} \\ -8t + 12a &= -12 \quad \{-4 \times (2)\} \\ \hline \text{adding} \quad 13t &= 48 \end{aligned}$$

$\therefore\ t = \frac{48}{13}$ and $7\left(\frac{48}{13}\right) - 4a = 20$

$\therefore\ t \approx 3.6923$ $\quad\therefore\ 5.8462 = 4a$

$\therefore\ t \approx 3$ min 41.54 sec $\quad\therefore\ a \approx 1.4615 \approx 1$ min 27.7 sec

So, as $a \approx 1.4615$, Q fired at 1:35:28 pm, and the explosion occurred at 1:37:42 pm.

4 **a** $\overrightarrow{AB} = \begin{pmatrix} 3-6 \\ 10-9 \\ 2.5-3 \end{pmatrix} = \begin{pmatrix} -3 \\ 1 \\ -0.5 \end{pmatrix}$

b $|\overrightarrow{AB}| = \sqrt{(-3)^2 + 1^2 + (-0.5)^2}$
$= \sqrt{10.25}$ km

c $\begin{pmatrix} x \\ y \\ z \end{pmatrix} = \begin{pmatrix} 6 \\ 9 \\ 3 \end{pmatrix} + t\begin{pmatrix} -3 \\ 1 \\ -0.5 \end{pmatrix}, \quad t \in \mathbb{R}$

The helicopter travels $\sqrt{10.25}$ km in 10 minutes.

$\therefore$ the helicopter's speed is

$6 \times \sqrt{10.25} \approx 19.2$ km h^{-1}.

d If $z = 0$, $3 + (-0.5)t = 0$

$\therefore\ t = 6$

$t = 1$ represents 10 minutes, so $t = 6$ represents 60 minutes.

$\therefore$ the helicopter lands on the helipad after 1 hour.

EXERCISE 13B.3

1 **a** $6\mathbf{i} - 6\mathbf{j}$

b The length of $\binom{-3}{4} = \sqrt{(-3)^2 + 4^2} = \sqrt{25} = 5$

As the speed is 10 km h^{-1}, the liner has velocity vector $2\binom{-3}{4} = \binom{-6}{8}$.

$\therefore$ the liner has position $\binom{x}{y} = \binom{6}{-6} + t\binom{-6}{8}, \quad t \geqslant 0, \quad t$ in hours.

c The liner is due east when $y = 0$

$\therefore\ -6 + 8t = 0$

$\therefore$ at $t = \frac{3}{4}$ hour

d The liner L is nearest the fishing boat O when $\overrightarrow{OL} \perp \binom{-3}{4}$.

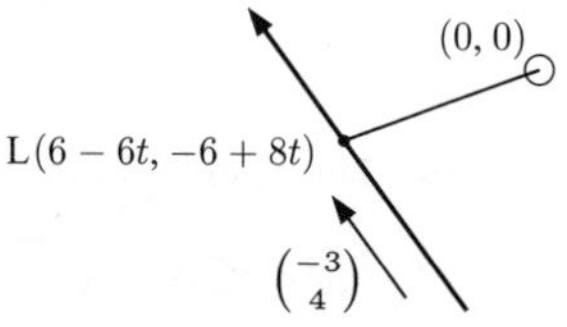

$\therefore\ \overrightarrow{OL} \bullet \binom{-3}{4} = 0$

$\therefore\ \binom{6-6t}{-6+8t} \bullet \binom{-3}{4} = 0$

$\therefore\ (-18+18t)+(-24+32t)=0$

$\therefore\ 50t=42$

$\therefore\ t=0.84 \text{ hours} = 50.4 \text{ minutes}$

and when $t=0.84$, $\text{L}=\begin{pmatrix}6-6(0.84)\\-6+8(0.84)\end{pmatrix}=\begin{pmatrix}0.96\\0.72\end{pmatrix}$

$\therefore$ the liner is closest to the fishing boat after 0.84 hours or 50.4 minutes, when it is at (0.96, 0.72).

2 **a** $|\mathbf{b}|=\sqrt{(-3)^2+(-1)^2}=\sqrt{10}$

As the speed is $40\sqrt{10}$ km h^{-1}, the velocity vector is $40\begin{pmatrix}-3\\-1\end{pmatrix}=\begin{pmatrix}-120\\-40\end{pmatrix}$.

b $\begin{pmatrix}x\\y\end{pmatrix}=\begin{pmatrix}200\\100\end{pmatrix}+t\begin{pmatrix}-120\\-40\end{pmatrix}$, $t\geqslant 0$ {$t=0$ at 12:00 noon}

c At 1:00 pm, $t=1$ and $\begin{pmatrix}x\\y\end{pmatrix}=\begin{pmatrix}200-120\\100-40\end{pmatrix}=\begin{pmatrix}80\\60\end{pmatrix}$

d The distance from O(0, 0) to P_1(80, 60) is $\left|\begin{pmatrix}80\\60\end{pmatrix}\right|=\sqrt{80^2+60^2}=100$ km,

which is when it becomes visible to radar. {within 100 km of O(0, 0)}

e

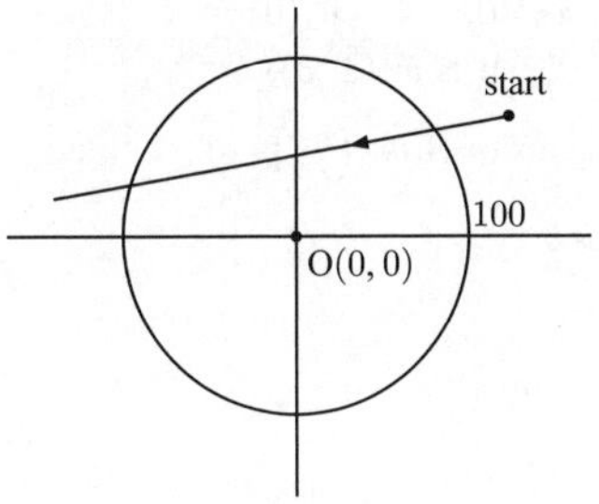

A general point on the path is P$(200-120t,\ 100-40t)$.

Now $\overrightarrow{\text{OP}}=\begin{pmatrix}200-120t\\100-40t\end{pmatrix}$,

and for the closest point $\overrightarrow{\text{OP}}\bullet\begin{pmatrix}-3\\-1\end{pmatrix}=0$

$\therefore\ -3(200-120t)-1(100-40t)=0$

$\therefore\ -700+400t=0$

$\therefore\ t=\frac{7}{4}=1\frac{3}{4}$ hours

$\therefore$ the time when the aircraft is closest is 1:45 pm, and

at this time $\overrightarrow{\text{OP}}=\begin{pmatrix}200-120(\frac{7}{4})\\100-40(\frac{7}{4})\end{pmatrix}=\begin{pmatrix}-10\\30\end{pmatrix}$

$\therefore\ d_{\text{min}}=\sqrt{(-10)^2+30^2}\approx 31.6$ km

f It disappears from radar when $|\overrightarrow{\text{OP}}|=100$ and $t>1\frac{3}{4}$

$\therefore\ \sqrt{(200-120t)^2+(100-40t)^2}=100$

$\therefore\ 40\,000-48\,000t+14\,400t^2+10\,000-8000t+1600t^2=10\,000$

$\therefore\ 16\,000t^2-56\,000t+40\,000=0$

$\therefore\ 16t^2-56t+40=0$ $\{\div 1000\}$

$\therefore\ 2t^2-7t+5=0$ $\{\div 8\}$

$\therefore\ (2t-5)(t-1)=0$

$\therefore\ t=\frac{5}{2}$ {as $t>1\frac{3}{4}$}

So, the aircraft disappears from the radar screen $2\frac{1}{2}$ hours after noon, or at 2:30 pm.

3 **a** At A, $y=0$

$\therefore\ 2x=36$

$\therefore\ x=18$

At B, $x=0$

$\therefore\ 3y=36$

$\therefore\ y=12$

So A is (18, 0) and B is (0, 12).

b $2x+3y=36$

$\therefore\ 3y=36-2x$

$\therefore\ y=\dfrac{36-2x}{3}$

$\therefore$ any point R on the railway track can be written $\text{R}\left(x,\ \dfrac{36-2x}{3}\right)$.

c $\overrightarrow{\text{PR}}=\begin{pmatrix}x-4\\\frac{36-2x}{3}-0\end{pmatrix}=\begin{pmatrix}x-4\\\frac{36-2x}{3}\end{pmatrix}$

$\overrightarrow{\text{AB}}=\begin{pmatrix}0-18\\12-0\end{pmatrix}=\begin{pmatrix}-18\\12\end{pmatrix}$

d The point closest to the railway track is R such that $\overrightarrow{PR} \perp \overrightarrow{AB}$.

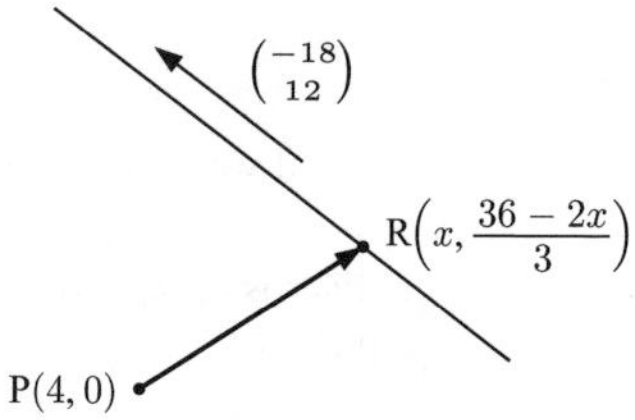

$$\therefore \quad \overrightarrow{PR} \bullet \overrightarrow{AB} = 0$$

$$\therefore \quad \begin{pmatrix} x-4 \\ \frac{36-2x}{3} \end{pmatrix} \bullet \begin{pmatrix} -18 \\ 12 \end{pmatrix} = 0$$

$$\therefore \quad -18(x-4) + 4(36-2x) = 0$$

$$\therefore \quad -18x + 72 + 144 - 8x = 0$$

$$\therefore \quad 26x = 216$$

$$\therefore \quad x = \tfrac{108}{13}$$

Now when $x = \frac{108}{13}$, $\dfrac{36-2x}{3} = 12 - \frac{2}{3}x = 12 - \frac{2}{3}\left(\frac{108}{13}\right) = \frac{84}{13}$

So R is $\left(\frac{108}{13}, \frac{84}{13}\right)$.

$$|\overrightarrow{PR}| = \sqrt{\left(\tfrac{108}{13} - 4\right)^2 + \left(\tfrac{84}{13} - 0\right)^2} = \sqrt{\tfrac{784}{13}} \approx 7.77 \text{ km}$$

The closest point on the track to the camp is $\left(\frac{108}{13}, \frac{84}{13}\right)$, a distance of 7.77 km.

4 For A, $x_A(t) = 3 - t$, $y_A(t) = 2t - 4$ For B, $x_B(t) = 4 - 3t$, $y_B(t) = 3 - 2t$

a When $t = 0$, $x_A(0) = 3$, $y_A(0) = -4$ and $x_B(0) = 4$, $y_B(0) = 3$

$\therefore$ A is at (3, −4). $\therefore$ B is at (4, 3).

b The velocity vector of A is $\binom{-1}{2}$ and the velocity vector of B is $\binom{-3}{-2}$.

c If the angle is θ, $\binom{-1}{2} \bullet \binom{-3}{-2} = \sqrt{1+4}\sqrt{9+4}\cos\theta$

$$\therefore \quad 3 - 4 = \sqrt{5}\sqrt{13}\cos\theta$$

$$\therefore \quad \tfrac{-1}{\sqrt{65}} = \cos\theta \quad \text{and so} \quad \theta \approx 97.1^o$$

d If D is the distance between them, then

$$\begin{aligned} D &= \sqrt{[(4-3t)-(3-t)]^2 + [(3-2t)-(2t-4)]^2} \\ &= \sqrt{[1-2t]^2 + [7-4t]^2} \\ &= \sqrt{1 - 4t + 4t^2 + 49 - 56t + 16t^2} \\ &= \sqrt{20t^2 - 60t + 50} \end{aligned}$$

Under the square root we have a quadratic in t, so D is a minimum when $t = -\dfrac{b}{2a} = \frac{60}{40} = 1\frac{1}{2}$

$\therefore$ $t = 1.5$ hours

5 **a** The direction vector of the line is $\mathbf{b} = \begin{pmatrix} 3 \\ 2 \\ 1 \end{pmatrix}$.

Let the point (3, 0, −1) be P, and $A(2+3t, -1+2t, 4+t)$ be any point on the line.

$$\therefore \quad \overrightarrow{PA} = \begin{pmatrix} 2+3t-3 \\ -1+2t-0 \\ 4+t-(-1) \end{pmatrix} = \begin{pmatrix} -1+3t \\ -1+2t \\ 5+t \end{pmatrix}$$

Now $\overrightarrow{PA}$ and **b** are perpendicular, so $\overrightarrow{PA} \bullet \mathbf{b} = 0$.

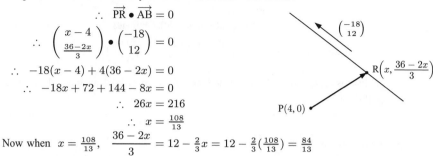

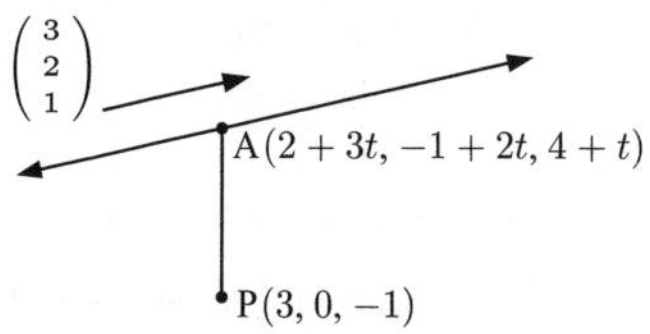

$$\therefore \quad \begin{pmatrix} -1+3t \\ -1+2t \\ 5+t \end{pmatrix} \bullet \begin{pmatrix} 3 \\ 2 \\ 1 \end{pmatrix} = 0$$

$$\therefore \quad -3 + 9t - 2 + 4t + 5 + t = 0$$

$$\therefore \quad 14t = 0$$

$$\therefore \quad t = 0$$

Substituting $t = 0$ into the parametric equations, we obtain the foot of the perpendicular (2, −1, 4).

b When $t=0$, $\overrightarrow{PA}=\begin{pmatrix}-1\\-1\\5\end{pmatrix}$, so $PA=\sqrt{1+1+25}$
$=\sqrt{27}$ units

$\therefore$ the shortest distance from the point to the line is $3\sqrt{3}$ units.

6 **a** The line has direction vector $\mathbf{b}=\begin{pmatrix}2\\3\\1\end{pmatrix}$.

Let the point $(1, 1, 3)$ be P and $A(1+2t, -1+3t, 2+t)$ be any point on the line.

$$\therefore\ \overrightarrow{PA}=\begin{pmatrix}1+2t-1\\-1+3t-1\\2+t-3\end{pmatrix}=\begin{pmatrix}2t\\-2+3t\\-1+t\end{pmatrix}.$$

Now $\overrightarrow{PA}$ and $\mathbf{b}$ are perpendicular, so $\overrightarrow{PA}\bullet\mathbf{b}=0$

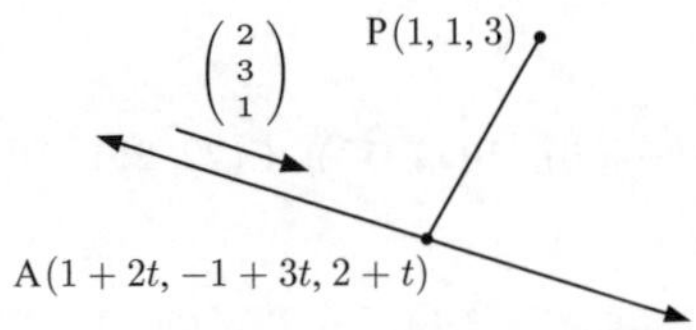

$$\therefore\ \begin{pmatrix}2t\\-2+3t\\-1+t\end{pmatrix}\bullet\begin{pmatrix}2\\3\\1\end{pmatrix}=0$$

$\therefore\ 4t-6+9t-1+t=0$

$\therefore\ 14t=7$

$\therefore\ t=\frac{1}{2}$

Substituting $t=\frac{1}{2}$ into the parametric equations, we obtain the foot of the perpendicular $(2, \frac{1}{2}, \frac{5}{2})$.

b When $t=\frac{1}{2}$, $\overrightarrow{PA}=\begin{pmatrix}1\\-2+\frac{3}{2}\\-1+\frac{1}{2}\end{pmatrix}=\begin{pmatrix}1\\-\frac{1}{2}\\-\frac{1}{2}\end{pmatrix}$

$\therefore\ PA=\sqrt{1+\frac{1}{4}+\frac{1}{4}}=\sqrt{\frac{3}{2}}$ units

$\therefore$ the shortest distance from the point to the line is $\sqrt{\frac{3}{2}}$ units.

EXERCISE 13B.4

1 **a**

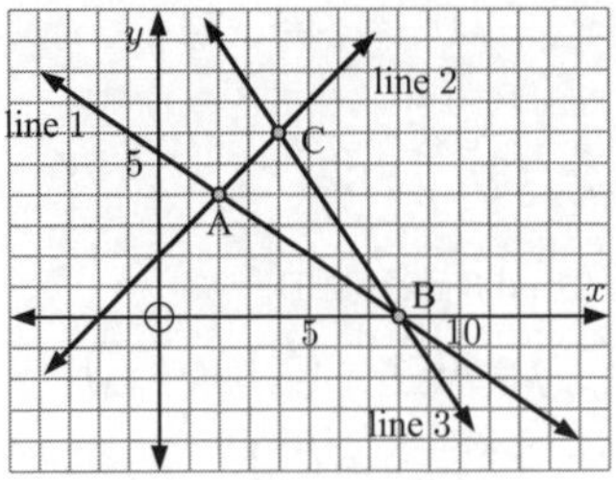

b A is (2, 4), B is (8, 0), C is (4, 6)

c $BC=\sqrt{(4-8)^2+(6-0)^2}=\sqrt{16+36}=\sqrt{52}$ units

$AB=\sqrt{(8-2)^2+(0-4)^2}=\sqrt{36+16}=\sqrt{52}$ units

$\therefore\ BC=AB$ and so ΔABC is isosceles.

d Line 1 and Line 2 meet at A.

$\therefore\ \binom{-1}{6}+\binom{3r}{-2r}=\binom{0}{2}+s\binom{1}{1}$

$\therefore\ \binom{3r-s}{-2r-s}=\binom{1}{-4}$

$\therefore\ 3r-s=1$

and $2r+s=4$

Adding, $5r\ \ =5$ $\therefore\ r=1$

$\therefore\ \binom{x}{y}=\binom{-1}{6}+\binom{3}{-2}=\binom{2}{4}$ ✓

Line 2 and Line 3 meet at C.

$\therefore\ \binom{0}{2}+s\binom{1}{1}=\binom{10}{-3}+t\binom{-2}{3}$

$\therefore\ \binom{s+2t}{s-3t}=\binom{10}{-5}$

$\therefore\ s+2t=10$

$-s+3t=5$

Adding, $5t=15$ $\therefore\ t=3$

$\therefore\ \binom{x}{y}=\binom{10}{-3}+3\binom{-2}{3}=\binom{4}{6}$ ✓

Line 1 and Line 3 meet at B.

$\therefore \quad \begin{pmatrix}-1\\6\end{pmatrix} + r\begin{pmatrix}3\\-2\end{pmatrix} = \begin{pmatrix}10\\-3\end{pmatrix} + t\begin{pmatrix}-2\\3\end{pmatrix}$

$\therefore \quad \begin{pmatrix}3r+2t\\-2r-3t\end{pmatrix} = \begin{pmatrix}11\\-9\end{pmatrix}$

$\therefore \quad 3r + 2t = 11$ .... (1)

$-2r - 3t = -9$ .... (2)

$\therefore \quad 9r + 6t = 33$ $\{3 \times (1)\}$

$-4r - 6t = -18$ $\{2 \times (2)\}$

Adding, $5r = 15$

$\therefore \quad r = 3$

So, $\begin{pmatrix}x\\y\end{pmatrix} = \begin{pmatrix}-1\\6\end{pmatrix} + 3\begin{pmatrix}3\\-2\end{pmatrix} = \begin{pmatrix}8\\0\end{pmatrix}$ ✓

2 **a**

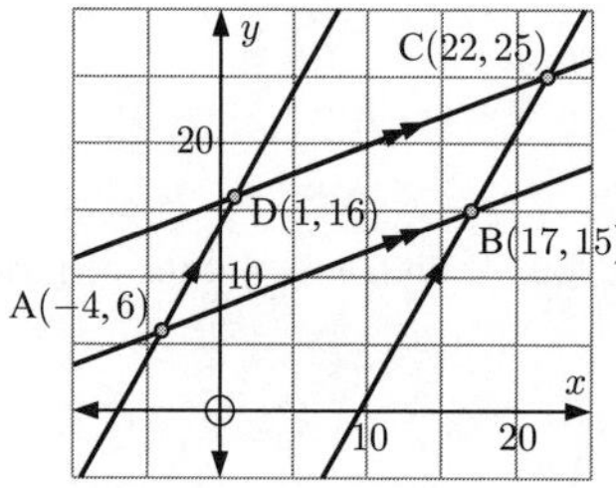

b A(−4, 6), B(17, 15), C(22, 25), D(1, 16)

c Lines 1 and 2 meet at A.

$\therefore \quad \begin{pmatrix}-4\\6\end{pmatrix} + r\begin{pmatrix}7\\3\end{pmatrix} = \begin{pmatrix}-4\\6\end{pmatrix} + s\begin{pmatrix}1\\2\end{pmatrix}$

$\therefore \quad \begin{pmatrix}7r-s\\3r-2s\end{pmatrix} = \begin{pmatrix}0\\0\end{pmatrix}$

$\therefore \quad 7r - s = 0$ .... (1)

and $3r - 2s = 0$

$-14r + 2s = 0$ $\{-2 \times (1)\}$

Adding, $-11r = 0$

$\therefore \quad r = 0$

$\therefore \quad \begin{pmatrix}x\\y\end{pmatrix} = \begin{pmatrix}-4\\6\end{pmatrix}$ ✓

Lines 1 and 4 meet at B.

$\therefore \quad \begin{pmatrix}-4\\6\end{pmatrix} + r\begin{pmatrix}7\\3\end{pmatrix} = \begin{pmatrix}22\\25\end{pmatrix} + u\begin{pmatrix}-1\\-2\end{pmatrix}$

$\therefore \quad \begin{pmatrix}7r+u\\3r+2u\end{pmatrix} = \begin{pmatrix}26\\19\end{pmatrix}$

$\therefore \quad 7r + u = 26$ .... (1)

and $3r + 2u = 19$

$-14r - 2u = -52$ $\{-2 \times (1)\}$

Adding, $-11r = -33$

$\therefore \quad r = 3$

$\therefore \quad \begin{pmatrix}x\\y\end{pmatrix} = \begin{pmatrix}-4\\6\end{pmatrix} + 3\begin{pmatrix}7\\3\end{pmatrix} = \begin{pmatrix}17\\15\end{pmatrix}$ ✓

Lines 3 and 4 meet at C.

$\therefore \quad \begin{pmatrix}22\\25\end{pmatrix} + t\begin{pmatrix}-7\\-3\end{pmatrix} = \begin{pmatrix}22\\25\end{pmatrix} + u\begin{pmatrix}-1\\-2\end{pmatrix}$

$\therefore \quad \begin{pmatrix}-7t+u\\-3t+2u\end{pmatrix} = \begin{pmatrix}0\\0\end{pmatrix}$

$\therefore \quad -7t + u = 0$ .... (1)

and $-3t + 2u = 0$

$14t - 2u = 0$ $\{-2 \times (1)\}$

Adding, $11t = 0$

$\therefore \quad t = 0$

$\therefore \quad \begin{pmatrix}x\\y\end{pmatrix} = \begin{pmatrix}22\\25\end{pmatrix}$ ✓

Lines 2 and 3 meet at D.

$\therefore \quad \begin{pmatrix}-4\\6\end{pmatrix} + s\begin{pmatrix}1\\2\end{pmatrix} = \begin{pmatrix}22\\25\end{pmatrix} + t\begin{pmatrix}-7\\-3\end{pmatrix}$

$\therefore \quad \begin{pmatrix}s+7t\\2s+3t\end{pmatrix} = \begin{pmatrix}26\\19\end{pmatrix}$

$\therefore \quad s + 7t = 26$ .... (1)

and $2s + 3t = 19$

$-2s - 14t = -52$ $\{-2 \times (1)\}$

Adding, $-11t = -33$

$\therefore \quad t = 3$

$\therefore \quad \begin{pmatrix}x\\y\end{pmatrix} = \begin{pmatrix}22\\25\end{pmatrix} + 3\begin{pmatrix}-7\\-3\end{pmatrix} = \begin{pmatrix}1\\16\end{pmatrix}$ ✓

3 **a** Lines 1 and 3 meet at A.

$\therefore \begin{pmatrix}0\\2\end{pmatrix} + r\begin{pmatrix}2\\1\end{pmatrix} = \begin{pmatrix}0\\5\end{pmatrix} + t\begin{pmatrix}1\\-1\end{pmatrix}$

$\therefore \begin{pmatrix}2r-t\\r+t\end{pmatrix} = \begin{pmatrix}0\\3\end{pmatrix}$

$\therefore$ $2r - t = 0$
$r + t = 3$

Adding, $3r = 3$

$\therefore r = 1$

$\therefore \begin{pmatrix}x\\y\end{pmatrix} = \begin{pmatrix}0\\2\end{pmatrix} + \begin{pmatrix}2\\1\end{pmatrix} = \begin{pmatrix}2\\3\end{pmatrix}$

$\therefore$ A is (2, 3)

Lines 1 and 2 meet at B.

$\therefore \begin{pmatrix}0\\2\end{pmatrix} + r\begin{pmatrix}2\\1\end{pmatrix} = \begin{pmatrix}8\\6\end{pmatrix} + s\begin{pmatrix}-1\\-2\end{pmatrix}$

$\therefore \begin{pmatrix}2r+s\\r+2s\end{pmatrix} = \begin{pmatrix}8\\4\end{pmatrix}$

$\therefore$ $-4r - 2s = -16$
$r + 2s = 4$

Adding, $-3r = -12$

$\therefore r = 4$

$\therefore \begin{pmatrix}x\\y\end{pmatrix} = \begin{pmatrix}0\\2\end{pmatrix} + 4\begin{pmatrix}2\\1\end{pmatrix} = \begin{pmatrix}8\\6\end{pmatrix}$

$\therefore$ B is (8, 6)

Lines 2 and 3 meet at C.

$\therefore \begin{pmatrix}8\\6\end{pmatrix} + s\begin{pmatrix}-1\\-2\end{pmatrix} = \begin{pmatrix}0\\5\end{pmatrix} + t\begin{pmatrix}1\\-1\end{pmatrix}$

$\therefore \begin{pmatrix}-s-t\\-2s+t\end{pmatrix} = \begin{pmatrix}-8\\-1\end{pmatrix}$

$\therefore$ $-s - t = -8$
$-2s + t = -1$

Adding, $-3s = -9$

$\therefore s = 3$

$\therefore \begin{pmatrix}x\\y\end{pmatrix} = \begin{pmatrix}8\\6\end{pmatrix} + 3\begin{pmatrix}-1\\-2\end{pmatrix} = \begin{pmatrix}5\\0\end{pmatrix}$

$\therefore$ C is (5, 0)

b A(2, 3), B(8, 6), C(5, 0)

$AB = \sqrt{(8-2)^2 + (6-3)^2}$
$= \sqrt{36+9}$
$= \sqrt{45}$

$BC = \sqrt{(5-8)^2 + (0-6)^2}$
$= \sqrt{9+36}$
$= \sqrt{45}$

The two equal sides are [AB] and [BC] and they have length $\sqrt{45}$ units.

4 **a** Lines (QP) and (PR) meet at P.

$\therefore \begin{pmatrix}3\\-1\end{pmatrix} + r\begin{pmatrix}14\\10\end{pmatrix} = \begin{pmatrix}0\\18\end{pmatrix} + t\begin{pmatrix}5\\-7\end{pmatrix}$

$\therefore \begin{pmatrix}14r-5t\\10r+7t\end{pmatrix} = \begin{pmatrix}-3\\19\end{pmatrix}$

$\therefore$ $14r - 5t = -3$ .... (1)
$10r + 7t = 19$ .... (2)

$\therefore$ $98r - 35t = -21$ {7 × (1)}
$50r + 35t = 95$ {5 × (2)}

Adding, $148r = 74$

$\therefore r = \frac{1}{2}$

$\therefore \begin{pmatrix}x\\y\end{pmatrix} = \begin{pmatrix}3\\-1\end{pmatrix} + \frac{1}{2}\begin{pmatrix}14\\10\end{pmatrix} = \begin{pmatrix}10\\4\end{pmatrix}$

$\therefore$ P is (10, 4)

Lines (QR) and (PR) meet at R.

$\therefore \begin{pmatrix}3\\-1\end{pmatrix} + s\begin{pmatrix}17\\-9\end{pmatrix} = \begin{pmatrix}0\\18\end{pmatrix} + t\begin{pmatrix}5\\-7\end{pmatrix}$

$\therefore \begin{pmatrix}17s-5t\\-9s+7t\end{pmatrix} = \begin{pmatrix}-3\\19\end{pmatrix}$

$\therefore$ $17s - 5t = -3$ .... (1)
$-9s + 7t = 19$ .... (2)

$\therefore$ $119s - 35t = -21$ {7 × (1)}
$-45s + 35t = 95$ {5 × (2)}

Adding, $74s = 74$

$\therefore s = 1$

$\therefore \begin{pmatrix}x\\y\end{pmatrix} = \begin{pmatrix}3\\-1\end{pmatrix} + \begin{pmatrix}17\\-9\end{pmatrix} = \begin{pmatrix}20\\-10\end{pmatrix}$

$\therefore$ R is (20, −10)

Lines (QP) and (PR) meet at Q.

$\begin{pmatrix}3\\-1\end{pmatrix} + r\begin{pmatrix}14\\10\end{pmatrix} = \begin{pmatrix}3\\-1\end{pmatrix} + s\begin{pmatrix}17\\-9\end{pmatrix}$

$\therefore r\begin{pmatrix}14\\10\end{pmatrix} = s\begin{pmatrix}17\\-9\end{pmatrix}$

$\therefore r = s = 0$

So, $\begin{pmatrix}x\\y\end{pmatrix} = \begin{pmatrix}3\\-1\end{pmatrix}$

$\therefore$ Q is (3, −1)

b $\overrightarrow{PQ} = \begin{pmatrix}3-10\\-1-4\end{pmatrix} = \begin{pmatrix}-7\\-5\end{pmatrix}$

$\overrightarrow{PR} = \begin{pmatrix}20-10\\-10-4\end{pmatrix} = \begin{pmatrix}10\\-14\end{pmatrix}$

and $\overrightarrow{PQ} \bullet \overrightarrow{PR} = -70 + 70 = 0$

c $[PQ] \perp [PR]$ $\therefore$ $Q\widehat{P}R = 90^o$

d Area $= \frac{1}{2}|\overrightarrow{PQ}||\overrightarrow{PR}|$
$= \frac{1}{2}\sqrt{49+25}\sqrt{100+196}$
$= 74$ units^2

5 **a** Lines 1 and 4 meet at A.

$\therefore \quad \binom{2}{5} + r\binom{4}{1} = \binom{3}{1} + u\binom{-3}{12}$

$\therefore \quad \binom{4r+3u}{r-12u} = \binom{1}{-4}$

$\therefore \quad 4r + 3u = 1$

$r - 12u = -4 \quad .... (1)$

$\therefore \quad 4r + 3u = 1$

$-4r + 48u = 16 \quad \{-4 \times (1)\}$

Adding, $\quad 51u = 17$

$\therefore \quad u = \frac{1}{3}$

$\therefore \quad \binom{x}{y} = \binom{3}{1} + \frac{1}{3}\binom{-3}{12} = \binom{2}{5}$

$\therefore$ A is (2, 5)

Lines 1 and 2 meet at B.

$\therefore \quad \binom{2}{5} + r\binom{4}{1} = \binom{18}{9} + s\binom{-8}{32}$

$\therefore \quad \binom{4r+8s}{r-32s} = \binom{16}{4}$

$\therefore \quad 4r + 8s = 16 \quad .... (1)$

$r - 32s = 4 \quad .... (2)$

$\therefore \quad r + 2s = 4 \quad \{(1) \div 4\}$

$-r + 32s = -4 \quad \{-1 \times (2)\}$

Adding, $\quad 34s = 0$

$\therefore \quad s = 0$

$\therefore \quad \binom{x}{y} = \binom{18}{9}$

$\therefore$ B is (18, 9)

Lines 2 and 3 meet at C.

$\therefore \quad \binom{18}{9} + s\binom{-8}{32} = \binom{14}{25} + t\binom{-8}{-2}$

$\therefore \quad \binom{-8s+8t}{32s+2t} = \binom{-4}{16}$

$\therefore \quad -8s + 8t = -4 \quad .... (1)$

$32s + 2t = 16$

$\therefore \quad 2s - 2t = 1 \quad \{(1) \div -4\}$

$32s + 2t = 16$

Adding, $\quad 34s = 17$

$\therefore \quad s = \frac{1}{2}$

$\therefore \quad \binom{x}{y} = \binom{18}{9} + \frac{1}{2}\binom{-8}{32} = \binom{14}{25}$

$\therefore$ C is (14, 25)

Lines 3 and 4 meet at D.

$\therefore \quad \binom{14}{25} + t\binom{-8}{-2} = \binom{3}{1} + u\binom{-3}{12}$

$\therefore \quad \binom{-8t+3u}{-2t-12u} = \binom{-11}{-24}$

$\therefore \quad -8t + 3u = -11 \quad .... (1)$

$-2t - 12u = -24 \quad .... (2)$

$\therefore \quad 16t - 6u = 22 \quad \{(-2) \times (1)\}$

$t + 6u = 12 \quad \{(2) \div -2\}$

Adding, $\quad 17t = 34$

$\therefore \quad t = 2$

$\therefore \quad \binom{x}{y} = \binom{14}{25} + 2\binom{-8}{-2} = \binom{-2}{21}$

$\therefore$ D is (−2, 21)

b $\overrightarrow{AC} = \binom{14-2}{25-5} = \binom{12}{20}$

$\overrightarrow{DB} = \binom{18--2}{9-21} = \binom{20}{-12}$

i $|\overrightarrow{AC}| = \sqrt{12^2 + 20^2} = \sqrt{544}$ units

ii $|\overrightarrow{DB}| = \sqrt{20^2 + (-12)^2} = \sqrt{544}$ units

iii $\overrightarrow{AC} \bullet \overrightarrow{DB} = 240 - 240 = 0$

c The diagonals are perpendicular and equal in length, and as their midpoints are the same (at (8, 15)), ABCD is a square.

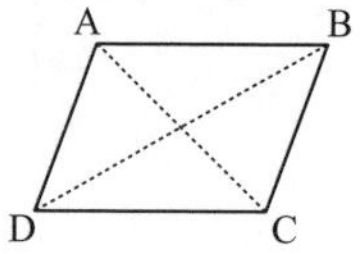

EXERCISE 13C

1 **a** Line 1 has direction vector $\begin{pmatrix} 2 \\ -1 \\ 1 \end{pmatrix}$ and line 2 has direction vector $\begin{pmatrix} 3 \\ -1 \\ 2 \end{pmatrix}$.

As one vector is not a scalar multiple of the other, the lines are not parallel.

Now $\quad 1 + 2t = -2 + 3s \qquad 2 - t = 3 - s \qquad 3 + t = 1 + 2s$

$\therefore \quad 2t - 3s = -3 \quad .... (1) \qquad \therefore \quad -t + s = 1 \quad .... (2) \qquad \therefore \quad t - 2s = -2 \quad .... (3)$

Solving (2) and (3) simultaneously: $\quad -t + s = 1$

$t - 2s = -2$

$-s = -1 \quad \therefore \quad s = 1$ and $t = 0$

and in (1), LHS $= 2t - 3s = 2(0) - 3(1) = -3$ ✓

$\therefore \quad s = 1, \quad t = 0$ satisfies all three equations

$\therefore$ the two lines meet at (1, 2, 3) $\quad$ {using $t = 0$ or $s = 1$}

The acute angle between the lines has $\cos\theta = \dfrac{|6+1+2|}{\sqrt{4+1+1}\,\sqrt{9+1+4}} = \dfrac{9}{\sqrt{84}}$

and so $\theta \approx 10.9^o$

b Line 1 has direction vector $\begin{pmatrix} 2 \\ -12 \\ 12 \end{pmatrix}$ and line 2 has direction vector $\begin{pmatrix} 4 \\ 3 \\ -1 \end{pmatrix}$.

As one vector is not a scalar multiple of the other, the lines are not parallel.

Now $-1+2t = 4s-3$ $\quad$ $2-12t = 3s+2$ $\quad$ $4+12t = -s-1$

$\therefore\ 2t-4s = -2$ $\quad$ $-12t-3s = 0$ $\quad$ $12t+s = -5$ .... (3)

$\therefore\ t-2s = -1$ .... (1) $\quad$ $s = -4t$ .... (2)

Solving (1) and (2) simultaneously: $t-2(-4t) = -1$

$\therefore\ 9t = -1$

$\therefore\ t = -\frac{1}{9}$ and so $s = \frac{4}{9}$

In (3), $12t+s = 12\left(-\frac{1}{9}\right)+\frac{4}{9} = -\frac{12}{9}+\frac{4}{9} = -\frac{8}{9}$, which is not -5.

Since the system is inconsistent, the lines do not intersect, so the lines are skew.

The acute angle between the lines has $\cos\theta = \dfrac{|8-36-12|}{\sqrt{292}\,\sqrt{26}} = \dfrac{40}{\sqrt{7592}}$ and so $\theta \approx 62.7^o$.

c Line 1 has direction vector $\begin{pmatrix} 6 \\ 8 \\ 2 \end{pmatrix}$ and line 2 has direction vector $\begin{pmatrix} 3 \\ 4 \\ 1 \end{pmatrix}$.

As $\begin{pmatrix} 6 \\ 8 \\ 2 \end{pmatrix} = 2\begin{pmatrix} 3 \\ 4 \\ 1 \end{pmatrix}$ the two lines are parallel. Hence, $\theta = 0^o$.

d In line 1 set $x = 2-y = z+2 = t$, so $x = t$, $y = 2-t$ and $z = t-2$.

Line 1 has direction vector $\begin{pmatrix} 1 \\ -1 \\ 1 \end{pmatrix}$ and line 2 has direction vector $\begin{pmatrix} 3 \\ -2 \\ 2 \end{pmatrix}$.

As one vector is not a scalar multiple of the other, the lines are not parallel.

Now $t = 1+3s$ .... (1) $\quad$ $2-t = -2-2s$ $\quad$ $-2+t = 2s+\frac{1}{2}$

$-t+2s = -4$ .... (2) $\quad$ $t-2s = 2\frac{1}{2}$ .... (3)

Solving (1) and (2) simultaneously, $-(1+3s)+2s = -4$

$\therefore\ -s = -3$

$\therefore\ s = 3$ and so $t = 1+3(3) = 10$

Substituting in (3), $t-2s = 10-2(3) = 4 \neq 2\frac{1}{2}$

Since the system is inconsistent, the lines do not meet. $\therefore$ they are skew.

The acute angle between the lines has $\cos\theta = \dfrac{|3+2+2|}{\sqrt{1+1+1}\sqrt{9+4+4}} = \dfrac{7}{\sqrt{3}\sqrt{17}}$

$\therefore\ \theta \approx 11.4^o$

e Line 1 has direction vector $\begin{pmatrix} 1 \\ -1 \\ 2 \end{pmatrix}$ and line 2 has direction vector $\begin{pmatrix} 3 \\ -2 \\ 1 \end{pmatrix}$.

As one vector is not a scalar multiple of the other, the lines are not parallel.

$1+t = 2+3s$ $\quad$ $2-t = 3-2s$ $\quad$ $3+2t = s-5$

$t-3s = 1$ .... (1) $\quad$ $-t+2s = 1$ .... (2) $\quad$ $2t-s = -8$ .... (3)

Solving (1) and (2) simultaneously, $t-3s = 1$

$-t+2s = 1$

Adding, $-s = 2$

$\therefore\ s = -2$ and $t-3(-2) = 1$ $\therefore\ t = -5$

Checking in (3), $2t - s = 2(-5) - (-2) = -10 + 2 = -8$ ✓

Since $s = -2,\ t = -5$ satisfies all three equations, the lines meet.

They meet at $x = 1 + (-5),\ y = 2 - (-5),\ z = 3 + 2(-5)$, or at $(-4, 7, -7)$.

The acute angle between the lines has $\cos\theta = \dfrac{|3+2+2|}{\sqrt{1+1+4}\sqrt{9+4+1}}$

$= \dfrac{7}{\sqrt{84}}$ and so $\theta \approx 40.2^o$

f Line 1 has direction vector $\begin{pmatrix} -2 \\ 1 \\ 0 \end{pmatrix}$ and line 2 has direction vector $\begin{pmatrix} 4 \\ -2 \\ 0 \end{pmatrix}$.

Now $\begin{pmatrix} 4 \\ -2 \\ 0 \end{pmatrix} = -2\begin{pmatrix} -2 \\ 1 \\ 0 \end{pmatrix}$, so the lines are parallel and hence $\theta = 0^o$.

REVIEW SET 13A

1 **a** The vector equation is

$\binom{x}{y} = \binom{-6}{3} + t\binom{4}{-3}$

b The parametric equations are

$x = -6 + 4t, \quad y = 3 - 3t, \quad t \in \mathbb{R}$

2 $(-3, m)$ lies on the line, so

$\binom{-3}{m} = \binom{18}{-2} + \binom{-7t}{4t}$

$\therefore\ -3 = 18 - 7t$ and $m = -2 + 4t$

$\therefore\ 7t = 21$

$\therefore\ t = 3$ and so $m = -2 + 4(3) = 10$

3 $x = 3 + 2t,\ y = -3 + 5t$ {parametric form}

or $t = \dfrac{x-3}{2} = \dfrac{y+3}{5}$

$\therefore\ 5x - 15 = 2y + 6$

$\therefore\ 5x - 2y = 21$ {general Cartesian form}

4 P(2, 0, 1), Q(3, 4, −2), R(−1, 3, 2)

a $\overrightarrow{PQ} = \begin{pmatrix} 1 \\ 4 \\ -3 \end{pmatrix}$

$|\overrightarrow{PQ}| = \sqrt{1+16+9} = \sqrt{26}$ units

and $\overrightarrow{QR} = \begin{pmatrix} -4 \\ -1 \\ 4 \end{pmatrix}$

b Since $\overrightarrow{PQ} = \begin{pmatrix} 1 \\ 4 \\ -3 \end{pmatrix}$ and P is at $(2, 0, 1)$,

the line has parametric equations

$x = 2 + t, \quad y = 0 + 4t, \quad z = 1 - 3t$

$\therefore\ x = 2 + t, \quad y = 4t, \quad z = 1 - 3t, \quad t$ in $\mathbb{R}$

5 Any vector parallel to $\mathbf{i} + r\mathbf{j} + 3\mathbf{k}$ has the form $t\begin{pmatrix} 1 \\ r \\ 3 \end{pmatrix} = \begin{pmatrix} t \\ rt \\ 3t \end{pmatrix}$, t in $\mathbb{R}$.

This is perpendicular to $\begin{pmatrix} 2 \\ -1 \\ 2 \end{pmatrix}$ if $\begin{pmatrix} 2 \\ -1 \\ 2 \end{pmatrix} \bullet \begin{pmatrix} t \\ rt \\ 3t \end{pmatrix} = 0$

$\therefore\ 2t - rt + 6t = 0$

$\therefore\ 8t - rt = 0$

$\therefore\ t(8 - r) = 0$

$\therefore\ t = 0$ or $r = 8$

But if $t = 0$, the vector has zero length.

$\therefore\ r = 8$ and so a vector is $\begin{pmatrix} 1 \\ 8 \\ 3 \end{pmatrix}$.

The length of this vector is $\sqrt{1+64+9} = \sqrt{74}$ units.

$\therefore$ the unit vectors are $\frac{1}{\sqrt{74}}\mathbf{i} + \frac{8}{\sqrt{74}}\mathbf{j} + \frac{3}{\sqrt{74}}\mathbf{k}$ or $-\frac{1}{\sqrt{74}}\mathbf{i} - \frac{8}{\sqrt{74}}\mathbf{j} - \frac{3}{\sqrt{74}}\mathbf{k}$

6 **a** Lines AB and AC meet at A.

$\therefore \begin{pmatrix}4\\-1\end{pmatrix} + t\begin{pmatrix}1\\3\end{pmatrix} = \begin{pmatrix}-1\\0\end{pmatrix} + u\begin{pmatrix}3\\1\end{pmatrix}$

$\therefore \begin{pmatrix}4+t\\-1+3t\end{pmatrix} = \begin{pmatrix}-1+3u\\u\end{pmatrix}$

$\therefore t - 3u = -5$ .... (1)

$3t - u = 1$

$\therefore -3t + 9u = 15 \quad \{-3 \times (1)\}$

$3t - u = 1$

Adding, $8u = 16$

$\therefore u = 2$

So, $\begin{pmatrix}x\\y\end{pmatrix} = \begin{pmatrix}-1\\0\end{pmatrix} + 2\begin{pmatrix}3\\1\end{pmatrix} = \begin{pmatrix}5\\2\end{pmatrix}$

$\therefore$ A is (5, 2).

Lines BC and AC meet at C.

$\therefore \begin{pmatrix}7\\4\end{pmatrix} + s\begin{pmatrix}1\\-1\end{pmatrix} = \begin{pmatrix}-1\\0\end{pmatrix} + u\begin{pmatrix}3\\1\end{pmatrix}$

$\therefore \begin{pmatrix}7+s\\4-s\end{pmatrix} = \begin{pmatrix}-1+3u\\u\end{pmatrix}$

$\therefore s - 3u = -8$

$-s - u = -4$

Adding, $-4u = -12$

$\therefore u = 3$

So, $\begin{pmatrix}x\\y\end{pmatrix} = \begin{pmatrix}-1\\0\end{pmatrix} + 3\begin{pmatrix}3\\1\end{pmatrix} = \begin{pmatrix}8\\3\end{pmatrix}$

$\therefore$ C is (8, 3).

Lines AB and BC meet at B.

$\therefore \begin{pmatrix}4\\-1\end{pmatrix} + t\begin{pmatrix}1\\3\end{pmatrix} = \begin{pmatrix}7\\4\end{pmatrix} + s\begin{pmatrix}1\\-1\end{pmatrix}$

$\therefore \begin{pmatrix}4+t\\-1+3t\end{pmatrix} = \begin{pmatrix}7+s\\4-s\end{pmatrix}$

$\therefore t - s = 3$

$3t + s = 5$

Adding, $4t = 8$

$\therefore t = 2$

So, $\begin{pmatrix}x\\y\end{pmatrix} = \begin{pmatrix}4\\-1\end{pmatrix} + 2\begin{pmatrix}1\\3\end{pmatrix} = \begin{pmatrix}6\\5\end{pmatrix}$

$\therefore$ B is (6, 5).

b $\overrightarrow{AB} = \begin{pmatrix}1\\3\end{pmatrix}$, so $|\overrightarrow{AB}| = \sqrt{1+9} = \sqrt{10}$ units

$\overrightarrow{BC} = \begin{pmatrix}2\\-2\end{pmatrix}$, so $|\overrightarrow{BC}| = \sqrt{4+4} = \sqrt{8}$ units

$\overrightarrow{AC} = \begin{pmatrix}3\\1\end{pmatrix}$, so $|\overrightarrow{AC}| = \sqrt{9+1} = \sqrt{10}$ units

c Triangle ABC is isosceles.

7 **a**

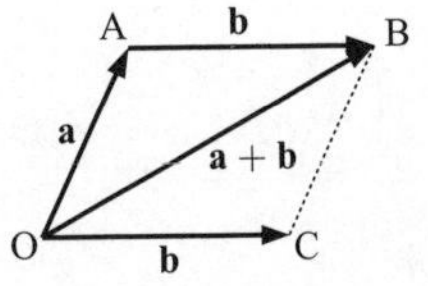

As **a** and **b** are unit vectors, OABC is a rhombus.
But the angles of a rhombus are bisected by its diagonals,
so **a** + **b** bisects the angle between vector **a** and vector **b**.

b

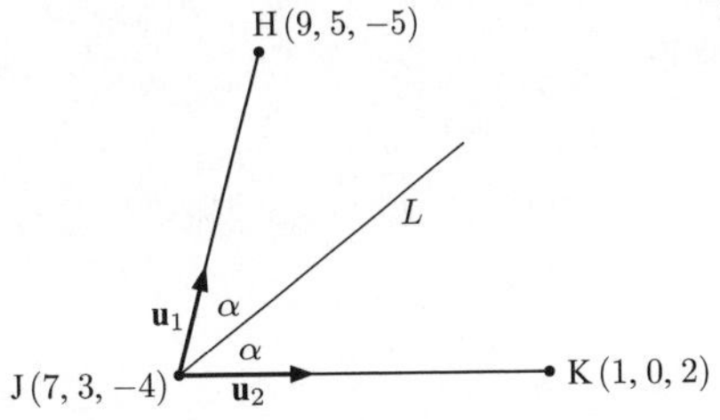

$\overrightarrow{JH} = \begin{pmatrix}9-7\\5-3\\-5--4\end{pmatrix} = \begin{pmatrix}2\\2\\-1\end{pmatrix} \qquad \overrightarrow{JK} = \begin{pmatrix}1-7\\0-3\\2--4\end{pmatrix} = \begin{pmatrix}-6\\-3\\6\end{pmatrix}$

$\therefore \mathbf{u}_1 = \dfrac{1}{\sqrt{4+4+1}}\begin{pmatrix}2\\2\\-1\end{pmatrix} = \begin{pmatrix}\frac{2}{3}\\\frac{2}{3}\\-\frac{1}{3}\end{pmatrix}$

and $\mathbf{u}_2 = \dfrac{1}{\sqrt{36+9+36}}\begin{pmatrix}-6\\-3\\6\end{pmatrix} = \begin{pmatrix}-\frac{2}{3}\\-\frac{1}{3}\\\frac{2}{3}\end{pmatrix}$

and $\mathbf{u}_1 + \mathbf{u}_2 = \begin{pmatrix}0\\\frac{1}{3}\\\frac{1}{3}\end{pmatrix}$, which bisects $\widehat{HJK}$, by **a**

$\therefore$ the line's equation is $\begin{pmatrix}x\\y\\z\end{pmatrix} = \begin{pmatrix}7\\3\\-4\end{pmatrix} + t\begin{pmatrix}0\\\frac{1}{3}\\\frac{1}{3}\end{pmatrix}, \quad t \in \mathbb{R}$

c $\overrightarrow{HK} = \begin{pmatrix} 1-9 \\ 0-5 \\ 2--5 \end{pmatrix} = \begin{pmatrix} -8 \\ -5 \\ 7 \end{pmatrix}$ so (HK) has equation $\begin{pmatrix} x \\ y \\ z \end{pmatrix} = \begin{pmatrix} 1 \\ 0 \\ 2 \end{pmatrix} + s\begin{pmatrix} -8 \\ -5 \\ 7 \end{pmatrix}$.

This line meets L where

$$7 = 1 - 8s, \quad 3 + \frac{t}{3} = -5s \quad \text{and} \quad -4 + \frac{t}{3} = 2 + 7s \quad \text{......} \ (*)$$

$\therefore \ 8s = -6 \quad \therefore \quad s = -\frac{3}{4}$ and so $3 + \frac{t}{3} = \frac{15}{4} \quad \therefore \quad \frac{t}{3} = \frac{3}{4} \quad \therefore \quad t = \frac{9}{4}$

In $(*)$ $\quad \text{LHS} = -4 + \frac{t}{3}$ $\qquad$ $\text{RHS} = 2 + 7s$

$= -4 + \frac{3}{4}$ $\qquad$ $= 2 + 7\left(-\frac{3}{4}\right)$

$= -\frac{13}{4}$ $\qquad$ $= \frac{8}{4} - \frac{21}{4}$

$= -\frac{13}{4}$ ✓

$\therefore \ s = -\frac{3}{4}, \ t = \frac{9}{4}$ satisfy all 3 equations

So, $\begin{pmatrix} x \\ y \\ z \end{pmatrix} = \begin{pmatrix} 1 \\ 0 \\ 2 \end{pmatrix} - \frac{3}{4}\begin{pmatrix} -8 \\ -5 \\ 7 \end{pmatrix} = \begin{pmatrix} 1+6 \\ 0+\frac{15}{4} \\ 2-\frac{21}{4} \end{pmatrix} = \begin{pmatrix} 7 \\ 3\frac{3}{4} \\ -3\frac{1}{4} \end{pmatrix}$

$\therefore \ L$ meets (HK) at $(7, 3\frac{3}{4}, -3\frac{1}{4})$.

8 If A is $(3, -1, -2)$ and $B(5, 3, -4)$ then $\overrightarrow{AB} = \begin{pmatrix} 5-3 \\ 3--1 \\ -4--2 \end{pmatrix} = \begin{pmatrix} 2 \\ 4 \\ -2 \end{pmatrix} = 2\begin{pmatrix} 1 \\ 2 \\ -1 \end{pmatrix}$

$\therefore$ the line has equation $\begin{pmatrix} x \\ y \\ z \end{pmatrix} = \begin{pmatrix} 3 \\ -1 \\ -2 \end{pmatrix} + t\begin{pmatrix} 1 \\ 2 \\ -1 \end{pmatrix}, \quad t \in \mathbb{R}$

and it meets $x^2 + y^2 + z^2 = 26$ where

$$(3+t)^2 + (-1+2t)^2 + (-2-t)^2 = 26$$

$$\therefore \ 9 + 6t + t^2 + 1 - 4t + 4t^2 + 4 + 4t + t^2 - 26 = 0$$

$$\therefore \ 6t^2 + 6t - 12 = 0$$

$$\therefore \ t^2 + t - 2 = 0$$

$$\therefore \ (t+2)(t-1) = 0$$

$$\therefore \ t = -2 \text{ or } 1$$

$\therefore \ \begin{pmatrix} x \\ y \\ z \end{pmatrix} = \begin{pmatrix} 3 \\ -1 \\ -2 \end{pmatrix} - 2\begin{pmatrix} 1 \\ 2 \\ -1 \end{pmatrix}$ or $\begin{pmatrix} 3 \\ -1 \\ -2 \end{pmatrix} + \begin{pmatrix} 1 \\ 2 \\ -1 \end{pmatrix}$

$\therefore$ the line meets the sphere at $(1, -5, 0)$ and $(4, 1, -3)$.

REVIEW SET 13B

1 The vector equation is $\binom{x}{y} = \binom{0}{8} + t\binom{5}{4}, \quad t \in \mathbb{R}$

2 **a** $\overrightarrow{PQ} = \begin{pmatrix} 4--1 \\ 0-2 \\ -1-3 \end{pmatrix}$

$= \begin{pmatrix} 5 \\ -2 \\ -4 \end{pmatrix}$

b For the X-axis, $\mathbf{v} = \begin{pmatrix} 1 \\ 0 \\ 0 \end{pmatrix}$

$$\cos\theta = \frac{\left|\begin{pmatrix} 5 \\ -2 \\ -4 \end{pmatrix} \bullet \begin{pmatrix} 1 \\ 0 \\ 0 \end{pmatrix}\right|}{\sqrt{25+4+16}\sqrt{1+0+0}} = \frac{|5+0+0|}{\sqrt{45}} = \frac{5}{\sqrt{45}}$$

$\therefore \quad \theta \approx 41.8^o$

3 **a** **i** The yacht is initially at $(-6, 10)$, so its initial position vector is $\begin{pmatrix}-6\\10\end{pmatrix}$ or $-6\mathbf{i}+10\mathbf{j}$

ii $-\mathbf{i}-3\mathbf{j}$ has length $\sqrt{(-1)^2+(-3)^2}=\sqrt{10}$

$\therefore\ 5(-\mathbf{i}-3\mathbf{j})$ has length $5\sqrt{10}$

$\therefore$ the direction vector is $-5\mathbf{i}-15\mathbf{j}$

iii $\begin{pmatrix}x\\y\end{pmatrix}=\begin{pmatrix}-6\\10\end{pmatrix}+t\begin{pmatrix}-5\\-15\end{pmatrix}$ $\quad\therefore$ the position vector is $-6\mathbf{i}+10\mathbf{j}+t(-5\mathbf{i}-15\mathbf{j})$

$= (-6-5t)\mathbf{i}+(10-15t)\mathbf{j},\quad t\geqslant 0$

b

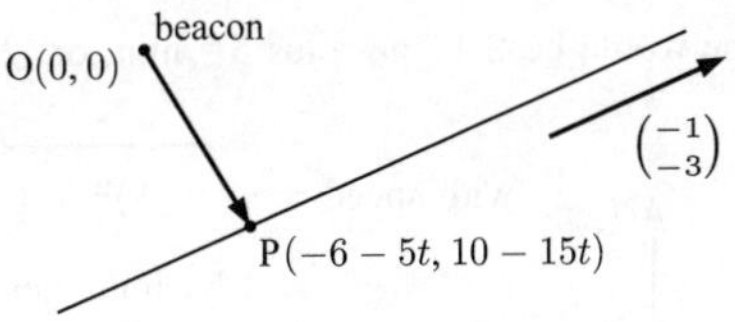

$\overrightarrow{\mathrm{OP}}=\begin{pmatrix}-6-5t\\10-15t\end{pmatrix}$ and $\overrightarrow{\mathrm{OP}}\bullet\begin{pmatrix}-1\\-3\end{pmatrix}=0$

$\therefore\ -1(-6-5t)-3(10-15t)=0$

$\therefore\ 6+5t-30+45t=0$

$\therefore\ 50t=24$

$\therefore\ t=0.48$ h

(or 28.8 min)

c When $t=0.48$, $\overrightarrow{\mathrm{OP}}=\begin{pmatrix}-6-5(0.48)\\10-15(0.48)\end{pmatrix}=\begin{pmatrix}-8.4\\2.8\end{pmatrix}$

and $\mathrm{OP}=\sqrt{(-8.4)^2+(2.8)^2}\approx 8.85$ km

As the closest distance is 8.85 km and the radius is 8 km, the yacht will miss the reef.

4 **a** **i** $\begin{pmatrix}x\\y\end{pmatrix}=\begin{pmatrix}2\\-3\end{pmatrix}+t\begin{pmatrix}4\\-1\end{pmatrix},\quad t\in\mathbb{R}$

ii $x=2+4t,\quad y=-3-t$

$\therefore\ t=\dfrac{x-2}{4}=\dfrac{y+3}{-1}$

$\therefore\ -x+2=4y+12$

$\therefore\ x+4y=-10$

b **i** The line has direction vector $\begin{pmatrix}5-(-1)\\-2-6\end{pmatrix}=\begin{pmatrix}6\\-8\end{pmatrix}$

$\therefore\ \begin{pmatrix}x\\y\end{pmatrix}=\begin{pmatrix}-1\\6\end{pmatrix}+t\begin{pmatrix}6\\-8\end{pmatrix},\quad t\in\mathbb{R}$

ii $x=-1+6t,\quad y=6-8t$

$\therefore\ t=\dfrac{x+1}{6}=\dfrac{y-6}{-8}$

$\therefore\ -8x-8=6y-36$

$\therefore\ 8x+6y=28$

$\therefore\ 4x+3y=14$

5 **a** The XOY plane has $z=0$

$\therefore\ 5-t=0$

$\therefore\ t=5$

So the line meets the XOY plane at $(2+3(5), 1-2(5), 5-5)$ or $(17, -9, 0)$.

b The YOZ plane has $x=0$

$\therefore\ 2+3t=0$

$\therefore\ t=-\frac{2}{3}$

So the line meets the YOZ plane at $(2+3(-\frac{2}{3}), 1-2(-\frac{2}{3}), 5-(-\frac{2}{3}))$ or $(0, \frac{7}{3}, \frac{17}{3})$.

c The XOZ plane has $y=0$

$\therefore\ 1-2t=0$

$\therefore\ t=\frac{1}{2}$

So the line meets the XOZ plane at $(2+3(\frac{1}{2}), 1-2(\frac{1}{2}), 5-\frac{1}{2})$ or $(\frac{7}{2}, 0, \frac{9}{2})$.

6 L_1 has direction vector $\mathbf{b}_1=\begin{pmatrix}5-0\\-2-3\end{pmatrix}=\begin{pmatrix}5\\-5\end{pmatrix}$

L_2 has direction vector $\mathbf{b}_2=\begin{pmatrix}-6-(-2)\\7-4\end{pmatrix}=\begin{pmatrix}-4\\3\end{pmatrix}$

If the angle between the lines is θ,

$$\cos\theta=\frac{|\mathbf{b}_1\bullet\mathbf{b}_2|}{|\mathbf{b}_1|\,|\mathbf{b}_2|}=\frac{|-20-15|}{\sqrt{25+25}\sqrt{16+9}}=\frac{35}{\sqrt{50}\times 5}$$

$\therefore\ \theta\approx 8.13^o$

$\therefore$ the angle between L_1 and L_2 is about 8.13^o.

7 **a** $\begin{pmatrix} x_1(t) \\ y_1(t) \end{pmatrix} = \begin{pmatrix} 2 \\ 4 \end{pmatrix} + t\begin{pmatrix} 1 \\ -3 \end{pmatrix}$ where $t \geqslant 0$. When $t = 0$, the time is 2:17 pm.

$\therefore \quad x_1(t) = 2 + t, \quad y_1(t) = 4 - 3t, \quad t \geqslant 0$

b After time t has passed, submarine Y18's torpedo has been moving for time $(t - 2)$.

$\therefore \quad x_2(t) = 11 - (t - 2), \qquad y_2(t) = 3 + a(t - 2)$

$\therefore \quad x_2(t) = 13 - t \qquad y_2(t) = [3 - 2a] + at, \quad t \geqslant 2$

c They meet where $2 + t = 13 - t$ and $4 - 3t = [3 - 2a] + at$

$\therefore \quad 2t = 11$

$\therefore \quad t = \frac{11}{2}$ $\quad \therefore$ the time would be 2:17 pm plus $5\frac{1}{2}$ min, or 2:22:30 pm

d When $t = \frac{11}{2}$,

$4 - 3\left(\frac{11}{2}\right) = [3 - 2a] + a\left(\frac{11}{2}\right)$

$\therefore \quad -\frac{25}{2} = 3 + \dfrac{7a}{2}$

$\therefore \quad -25 = 6 + 7a$

$\therefore \quad 7a = -31$

$\therefore \quad a = -\frac{31}{7}$

Y18's torpedo has velocity vector $\begin{pmatrix} -1 \\ -\frac{31}{7} \end{pmatrix}$

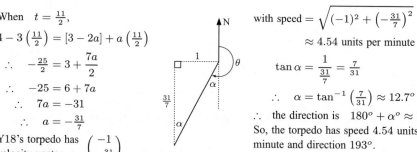

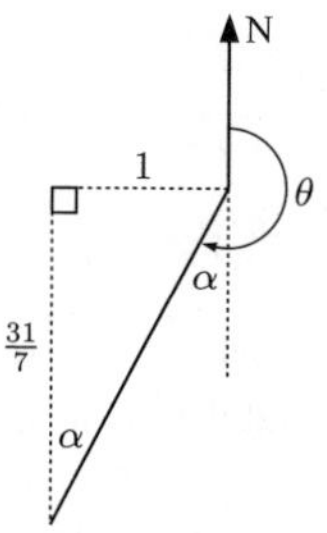

with speed $= \sqrt{(-1)^2 + \left(-\frac{31}{7}\right)^2}$

≈ 4.54 units per minute

$\tan\alpha = \dfrac{1}{\frac{31}{7}} = \frac{7}{31}$

$\therefore \quad \alpha = \tan^{-1}\left(\frac{7}{31}\right) \approx 12.7^o$

$\therefore$ the direction is $180^o + \alpha^o \approx 192.7^o$

So, the torpedo has speed 4.54 units per minute and direction 193°.

8 **a** Line 1 has direction vector $\begin{pmatrix} 1 \\ 2 \\ -1 \end{pmatrix}$ and line 2 has direction vector $\begin{pmatrix} 4 \\ 1 \\ -2 \end{pmatrix}$.

As one vector is not a scalar multiple of the other, the lines are not parallel.

Now, $2 + t = -8 + 4s$ .... (1) $\quad -1 + 2t = s$ .... (2) $\quad 3 - t = 7 - 2s$ .... (3)

Substituting (2) into (1), $2 + t = -8 + 4(-1 + 2t)$

$\therefore \quad 2 + t = -8 - 4 + 8t$

$\therefore \quad 7t = 14$

$\therefore \quad t = 2$

$\therefore \quad s = -1 + 2(2) = 3$

In (3), LHS $= 3 - 2 = 1$ $\qquad$ RHS $= 7 - 2(3) = 1$ ✓

$\therefore \quad s = 3, \ t = 2$ satisfies all three equations.

$\therefore$ the lines meet at (4, 3, 1) {substituting $t = 2$ into line 1}

The angle θ between the lines has

$$\cos\theta = \frac{\left|\begin{pmatrix} 1 \\ 2 \\ -1 \end{pmatrix} \bullet \begin{pmatrix} 4 \\ 1 \\ -2 \end{pmatrix}\right|}{\sqrt{1+4+1}\sqrt{16+1+4}} = \frac{|4+2+2|}{\sqrt{6}\sqrt{21}} = \frac{8}{3\sqrt{14}}$$

$\therefore \quad \theta \approx 44.5^o$

b Line 1 has direction vector $\begin{pmatrix} 1 \\ -2 \\ 3 \end{pmatrix}$ and line 2 has direction vector $\begin{pmatrix} -1 \\ 3 \\ 1 \end{pmatrix}$.

As one vector is not a scalar multiple of the other, the lines are not parallel.

Now, $3 + t = 2 - s \qquad 5 - 2t = 1 + 3s \qquad -1 + 3t = 4 + s$

$\therefore \quad t + s = -1$ .... (1) $\qquad 2t + 3s = 4$ .... (2) $\qquad 3t - s = 5$ .... (3)

Solving (1) and (3) simultaneously:

$$\begin{aligned} t + s &= -1 \\ 3t - s &= 5 \\ \hline \text{Adding,} \quad 4t \quad &= 4 \end{aligned}$$

$\therefore \quad t = 1 \qquad \therefore \quad s = -2$

In (2), LHS $= 2(1) + 3(-2) = -4$ ×

$\therefore$ the system of equations is inconsistent and so the lines are skew.

The angle θ between them has

$$\cos\theta = \frac{\left|\begin{pmatrix}1\\-2\\3\end{pmatrix} \bullet \begin{pmatrix}-1\\3\\1\end{pmatrix}\right|}{\sqrt{1+4+9}\sqrt{1+9+1}} = \frac{|-1-6+3|}{\sqrt{14}\sqrt{11}} = \frac{4}{\sqrt{154}}$$

$\therefore\ \theta \approx 71.2^o$

REVIEW SET 13C

1 The direction vector is $\begin{pmatrix}3\\-1\end{pmatrix}$ which has length $\sqrt{3^2+(-1)^2} = \sqrt{10}$ units

$\therefore\ 2\sqrt{10}\begin{pmatrix}3\\-1\end{pmatrix}$ has length 20. So, the velocity vector is $\begin{pmatrix}6\sqrt{10}\\-2\sqrt{10}\end{pmatrix}$ or $2\sqrt{10}(3\mathbf{i}-\mathbf{j})$

2 **a** $\overrightarrow{AB} = \begin{pmatrix}-4\\0\\5\end{pmatrix}$ $\therefore$ the line is $\begin{pmatrix}x\\y\\z\end{pmatrix} = \begin{pmatrix}3\\2\\-1\end{pmatrix} + t\begin{pmatrix}-4\\0\\5\end{pmatrix}$, $t \in \mathbb{R}$

b The distance from a point on the line to A is $d = \sqrt{(-4t)^2+0^2+(5t)^2} = \sqrt{41t^2}$

$\therefore$ since $d = 2\sqrt{41}$ units, $\sqrt{41t^2} = 2\sqrt{41}$

$\therefore\ t^2 = 4$

$\therefore\ t = \pm 2$ $\therefore$ the points are $(-5, 2, 9)$ and $(11, 2, -11)$.

3 **a** $x(0) = -4$ and $y(0) = 3$, so the initial position is $(-4, 3)$.

b $x(4) = -4 + 8(4) = 28$ and $y(4) = 3 + 6(4) = 27$, so at $t = 4$ the position is $(28, 27)$.

c The velocity vector is $\begin{pmatrix}8\\6\end{pmatrix}$, so the speed is $\sqrt{8^2+6^2} = 10\ \text{m s}^{-1}$ **d** $\begin{pmatrix}8\\6\end{pmatrix}$

4 **a** Line 1 has direction vector $\begin{pmatrix}5\\-2\end{pmatrix}$ and line 4 has direction vector $\begin{pmatrix}-5\\2\end{pmatrix}$.

Now $\begin{pmatrix}5\\-2\end{pmatrix} = -\begin{pmatrix}-5\\2\end{pmatrix}$, so lines 1 and 4 are parallel, $\therefore$ [KL] $\parallel$ [MN].

b $\overrightarrow{KL} = a\begin{pmatrix}5\\-2\end{pmatrix}$, $\overrightarrow{NK} = b\begin{pmatrix}4\\10\end{pmatrix}$, $\overrightarrow{MN} = c\begin{pmatrix}-5\\2\end{pmatrix}$ {for some constants a, b, c}

$\therefore\ \overrightarrow{KL} \bullet \overrightarrow{NK} = ab(20-20) = 0$ and $\overrightarrow{NK} \bullet \overrightarrow{MN} = bc(-20+20) = 0$

$\therefore$ [NK] is perpendicular to both [KL] and [MN].

c Lines 1 and 3 meet at K.

$\therefore\ \begin{pmatrix}2\\19\end{pmatrix} + p\begin{pmatrix}5\\-2\end{pmatrix} = \begin{pmatrix}3\\7\end{pmatrix} + r\begin{pmatrix}4\\10\end{pmatrix}$

$\therefore\ \begin{pmatrix}5p-4r\\-2p-10r\end{pmatrix} = \begin{pmatrix}1\\-12\end{pmatrix}$

$\therefore\ 5p - 4r = 1$ .... (1)

$2p + 10r = 12$ .... (2)

$\therefore\ 25p - 20r = 5$ $\{5 \times (1)\}$

$4p + 20r = 24$ $\{2 \times (2)\}$

Adding, $29p = 29$

$\therefore\ p = 1$ and $\begin{pmatrix}x\\y\end{pmatrix} = \begin{pmatrix}2\\19\end{pmatrix} + \begin{pmatrix}5\\-2\end{pmatrix} = \begin{pmatrix}7\\17\end{pmatrix}$

$\therefore$ K is (7, 17).

Lines 1 and 2 meet at L.

$\therefore\ \begin{pmatrix}2\\19\end{pmatrix} + p\begin{pmatrix}5\\-2\end{pmatrix} = \begin{pmatrix}33\\-5\end{pmatrix} + q\begin{pmatrix}-11\\16\end{pmatrix}$

$\therefore\ \begin{pmatrix}5p+11q\\-2p-16q\end{pmatrix} = \begin{pmatrix}31\\-24\end{pmatrix}$

$\therefore\ 5p + 11q = 31$ .... (1)

$-2p - 16q = -24$ .... (2)

$\therefore\ 10p + 22q = 62$ $\{2 \times (1)\}$

$-10p - 80q = -120$ $\{5 \times (2)\}$

Adding, $-58q = -58$

$\therefore\ q = 1$ and $\begin{pmatrix}x\\y\end{pmatrix} = \begin{pmatrix}33\\-5\end{pmatrix} + \begin{pmatrix}-11\\16\end{pmatrix} = \begin{pmatrix}22\\11\end{pmatrix}$

$\therefore$ L is (22, 11)

Lines 2 and 4 meet at M.

$\therefore \quad \begin{pmatrix} 33 \\ -5 \end{pmatrix} + q\begin{pmatrix} -11 \\ 16 \end{pmatrix} = \begin{pmatrix} 43 \\ -9 \end{pmatrix} + s\begin{pmatrix} -5 \\ 2 \end{pmatrix}$

$\therefore \quad \begin{pmatrix} -11q+5s \\ 16q-2s \end{pmatrix} = \begin{pmatrix} 10 \\ -4 \end{pmatrix}$

$\therefore \quad -11q + 5s = 10 \quad .... (1)$

$16q - 2s = -4 \quad .... (2)$

$\therefore \quad -22q + 10s = 20 \quad \{2 \times (1)\}$

$80q - 10s = -20 \quad \{5 \times (2)\}$

Adding, $58q = 0$

$\therefore \quad q = 0$ and so $\begin{pmatrix} x \\ y \end{pmatrix} = \begin{pmatrix} 33 \\ -5 \end{pmatrix}$

$\therefore$ M is (33, −5)

Lines 3 and 4 meet at N.

$\therefore \quad \begin{pmatrix} 3 \\ 7 \end{pmatrix} + r\begin{pmatrix} 4 \\ 10 \end{pmatrix} = \begin{pmatrix} 43 \\ -9 \end{pmatrix} + s\begin{pmatrix} -5 \\ 2 \end{pmatrix}$

$\therefore \quad \begin{pmatrix} 4r+5s \\ 10r-2s \end{pmatrix} = \begin{pmatrix} 40 \\ -16 \end{pmatrix}$

$\therefore \quad 4r + 5s = 40 \quad .... (1)$

$10r - 2s = -16 \quad .... (2)$

$\therefore \quad 8r + 10s = 80 \quad \{2 \times (1)\}$

$50r - 10s = -80 \quad \{5 \times (2)\}$

Adding, $58r = 0$

$\therefore \quad r = 0$ and so $\begin{pmatrix} x \\ y \end{pmatrix} = \begin{pmatrix} 3 \\ 7 \end{pmatrix}$

$\therefore$ N is (3, 7)

d

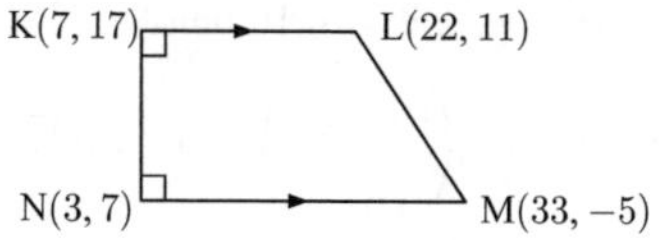

$$\text{KL} = \sqrt{(22-7)^2 + (11-17)^2} = \sqrt{225+36} = \sqrt{261} \text{ units}$$

$$\text{NM} = \sqrt{(33-3)^2 + (-5-7)^2} = \sqrt{900+144} = \sqrt{1044} \text{ units}$$

$$\text{KN} = \sqrt{(7-3)^2 + (17-7)^2} = \sqrt{16+100} = \sqrt{116} \text{ units}$$

$$\therefore \quad \text{area} = \left(\frac{\sqrt{261}+\sqrt{1044}}{2}\right) \times \sqrt{116} = 261 \text{ units}^2$$

5 L_1 has direction vector $\mathbf{b}_1 = \begin{pmatrix} -4 \\ 3 \end{pmatrix}$

L_2 has direction vector $\mathbf{b}_2 = \begin{pmatrix} 5 \\ -12 \end{pmatrix}$

If θ is the angle between them,

$$\cos\theta = \frac{|\mathbf{b}_1 \bullet \mathbf{b}_2|}{|\mathbf{b}_1|\,|\mathbf{b}_2|} = \frac{|-20-36|}{\sqrt{16+9}\sqrt{25+144}} = \frac{56}{5 \times 13}$$

$$\therefore \quad \theta \approx 30.5^o$$

$\therefore$ the angle between L_1 and L_2 is about 30.5^o.

6 **a** $\overrightarrow{\text{AB}} = \begin{pmatrix} 0-3 \\ 2--1 \\ -1-1 \end{pmatrix} = \begin{pmatrix} -3 \\ 3 \\ -2 \end{pmatrix}$

$\therefore \ |\overrightarrow{\text{AB}}| = \sqrt{9+9+4} = \sqrt{22}$ units

b $\begin{pmatrix} x \\ y \\ z \end{pmatrix} = \begin{pmatrix} 3 \\ -1 \\ 1 \end{pmatrix} + t\begin{pmatrix} -3 \\ 3 \\ -2 \end{pmatrix}, \quad t \in \mathbb{R}$

7 If A is the origin, (AB) the X-axis, (AD) the Y-axis, and (AP) the Z-axis, then Q is (4, 0, 7), D is (0, 10, 0) and M is (0, 5, 7).

Now $\overrightarrow{\text{DQ}} = \begin{pmatrix} 4-0 \\ 0-10 \\ 7-0 \end{pmatrix} = \begin{pmatrix} 4 \\ -10 \\ 7 \end{pmatrix}$ and $\overrightarrow{\text{DM}} = \begin{pmatrix} 0-0 \\ 5-10 \\ 7-0 \end{pmatrix} = \begin{pmatrix} 0 \\ -5 \\ 7 \end{pmatrix}$

So, if the required angle is θ, $\cos\theta = \dfrac{\begin{pmatrix} 4 \\ -10 \\ 7 \end{pmatrix} \bullet \begin{pmatrix} 0 \\ -5 \\ 7 \end{pmatrix}}{\sqrt{16+100+49}\sqrt{0+25+49}} = \dfrac{0+50+49}{\sqrt{165}\sqrt{74}} = \dfrac{99}{\sqrt{165 \times 74}}$

and so $\theta \approx 26.4^o$

8 **a** Road A has direction vector $\begin{pmatrix} 15--9 \\ -16-2 \end{pmatrix} = \begin{pmatrix} 24 \\ -18 \end{pmatrix} = 6\begin{pmatrix} 4 \\ -3 \end{pmatrix}$.

So, Road A has equation $\begin{pmatrix} x \\ y \end{pmatrix} = \begin{pmatrix} -9 \\ 2 \end{pmatrix} + t\begin{pmatrix} 4 \\ -3 \end{pmatrix}$, $t \in \mathbb{R}$.

Road B has direction vector $\begin{pmatrix} 21-6 \\ 18--18 \end{pmatrix} = \begin{pmatrix} 15 \\ 36 \end{pmatrix} = 3\begin{pmatrix} 5 \\ 12 \end{pmatrix}$

So, Road B has equation $\begin{pmatrix} x \\ y \end{pmatrix} = \begin{pmatrix} 6 \\ -18 \end{pmatrix} + s\begin{pmatrix} 5 \\ 12 \end{pmatrix}$, $s \in \mathbb{R}$.

b Let A$(-9+4t,\ 2-3t)$ be any point on Road A.

$\therefore\ \overrightarrow{HA} = \begin{pmatrix} -9+4t-4 \\ 2-3t-11 \end{pmatrix} = \begin{pmatrix} -13+4t \\ -9-3t \end{pmatrix}$

The closest point to H(4, 11) on Road A is such that $\overrightarrow{HA} \perp \begin{pmatrix} 4 \\ -3 \end{pmatrix}$.

$\therefore\ \begin{pmatrix} -13+4t \\ -9-3t \end{pmatrix} \bullet \begin{pmatrix} 4 \\ -3 \end{pmatrix} = 0$

$\therefore\ -52+16t+27+9t = 0$

$\therefore\ 25t = 25$

$\therefore\ t = 1$

So A is $(-9+4,\ 2-3)$ or $(-5,\ -1)$.

$\therefore\ \overrightarrow{HA} = \begin{pmatrix} -13+4 \\ -9-3 \end{pmatrix} = \begin{pmatrix} -9 \\ -12 \end{pmatrix}$

$\therefore\ |\overrightarrow{HA}| = \sqrt{81+144} = \sqrt{225} = 15$ km

Now, let B$(6+5s,\ -18+12s)$ be any point on Road B.

$\therefore\ \overrightarrow{HB} = \begin{pmatrix} 6+5s-4 \\ -18+12s-11 \end{pmatrix} = \begin{pmatrix} 2+5s \\ -29+12s \end{pmatrix}$

The closest point to H(4, 11) on Road B is such that $\overrightarrow{HB} \perp \begin{pmatrix} 5 \\ 12 \end{pmatrix}$.

$\therefore\ \begin{pmatrix} 2+5s \\ -29+12s \end{pmatrix} \bullet \begin{pmatrix} 5 \\ 12 \end{pmatrix} = 0$

$\therefore\ 10+25s-348+144s = 0$

$\therefore\ 169s = 338$

$\therefore\ s = 2$

So B is $(6+5(2),\ -18+12(2))$ or $(16,\ 6)$.

$\therefore\ \overrightarrow{HB} = \begin{pmatrix} 2+5(2) \\ -29+12(2) \end{pmatrix} = \begin{pmatrix} 12 \\ -5 \end{pmatrix}$

$\therefore\ |\overrightarrow{HB}| = \sqrt{144+25} = \sqrt{169} = 13$ km

The hiker should head toward Road B, a distance of 13 km.

9 **a** $\overrightarrow{AB} = \begin{pmatrix} 2-4 \\ 1-2 \\ 5--1 \end{pmatrix} = \begin{pmatrix} -2 \\ -1 \\ 6 \end{pmatrix}$ and $\overrightarrow{AC} = \begin{pmatrix} 9-4 \\ 4-2 \\ 1--1 \end{pmatrix} = \begin{pmatrix} 5 \\ 2 \\ 2 \end{pmatrix}$

$\therefore\ \overrightarrow{AB} \bullet \overrightarrow{AC} = (-2)(5)+(-1)(2)+(6)(2) = -10-2+12 = 0$

$\therefore\ \overrightarrow{AB} \perp \overrightarrow{AC}$

b The equation is $\begin{pmatrix} x \\ y \\ z \end{pmatrix} = \begin{pmatrix} 4 \\ 2 \\ -1 \end{pmatrix} + t\begin{pmatrix} -2 \\ -1 \\ 6 \end{pmatrix}$, $t \in \mathbb{R}$

c The equation is $\begin{pmatrix} x \\ y \\ z \end{pmatrix} = \begin{pmatrix} 4 \\ 2 \\ -1 \end{pmatrix} + s\begin{pmatrix} 5 \\ 2 \\ 2 \end{pmatrix}$, $s \in \mathbb{R}$

Chapter 14
DESCRIPTIVE STATISTICS

EXERCISE 14A

1 **a** Heights can take any value from 170 cm to 205 cm, including decimal values such as 181.372 cm. The 'height' variable can take any real number between 170 and 205.

b

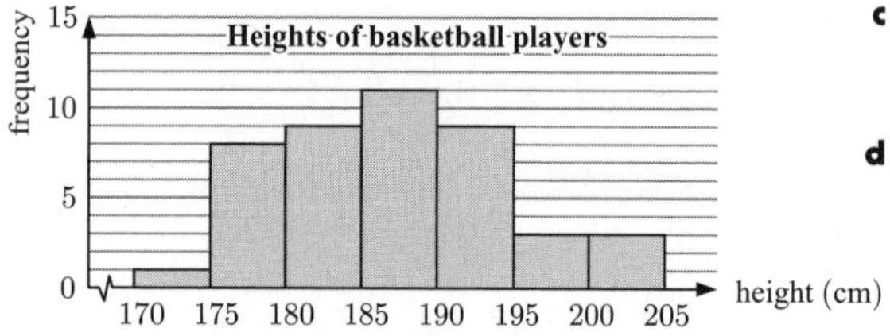

c The modal class is the class occurring most often. This is $185 \leqslant H < 190$ cm.

d The distribution is slightly positively skewed, as there is more of a 'tail' to the right.

2 **a** The data is continuous numerical. Actual time is continuous and could be measured to the nearest second, millisecond and so on. After it has been rounded to the nearest minute it becomes discrete numerical data.

b

Stem	*Leaf*
0	3 6 8 8 8 8
1	0 0 0 0 0 2 2 2 4 4 4 4 5 5 5 5 6 6 6 6 7 8 8 8 8 9
2	0 0 0 1 2 4 5 5 5 6 7 7 8
3	1 2 2 2 3 4 5 7 8
4	0 2 5 5 5 6

1 | 2 means 12 minutes

c The distribution is positively skewed, or skewed to the high end.

d The modal travelling time was between 10 and 20 min, if considering classes. The mode is actually 10.

3 **a** The data is discrete numerical, so a column graph should be used.

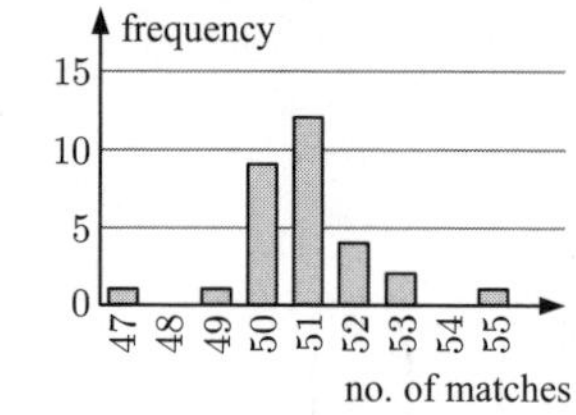

b The data is continuous, so a frequency histogram should be used.

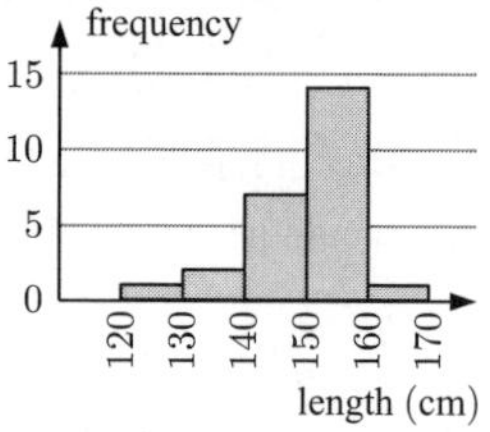

4 **a**

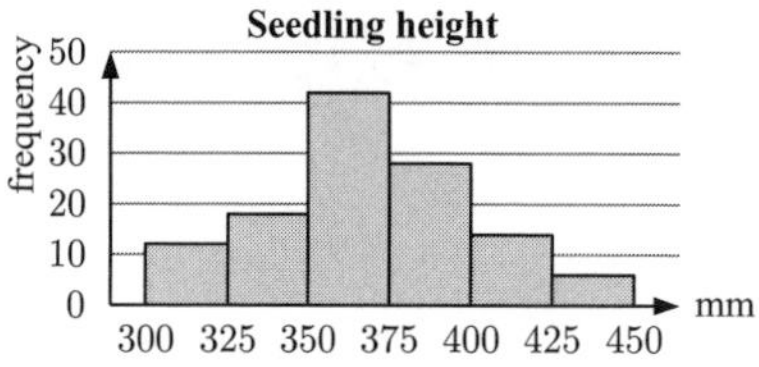

b Number which are $\geqslant 400$ mm is $14 + 6 = 20$ seedlings.

c 120 seedlings have been sampled

$$\therefore \ \% \text{ between 349 and 400} = \frac{42+28}{120} \times 100\%$$
$$= \frac{70}{120} \times 100\%$$
$$\approx 58.3\%$$

d **i** Number

$$= \frac{12+18+42+28}{120} \times 1462$$
$$= \frac{100}{120} \times 1462$$
$$\approx 1218 \text{ seedlings}$$

ii Number

$$= \frac{28+14}{120} \times 1462$$
$$= \frac{42}{120} \times 1462$$
$$\approx 512 \text{ seedlings}$$

EXERCISE 14B.1

1 **a** **i** $\text{mean} = \frac{2+3+3+3+4+....+9+9}{23} = \frac{129}{23} \approx 5.61$

ii median = 12th score (when in order) $= 6$

iii mode = 6 (6 occurs most often)

b **i** $\text{mean} = \frac{10+12+12+15+....+20+21}{15} = \frac{245}{15} \approx 16.3$

ii median = 8th score (when in order) $= 17$

iii mode = 18

c **i** $\text{mean} = \frac{22.4+24.6+21.8+....+23.5}{11} = \frac{273}{11} \approx 24.8$

ii median = 6th score (when in order) $= 24.9$

iii mode = 23.5

2 **a** $\text{mean of set A} = \frac{3+4+4+5+....+10}{13} \approx 6.46$ $\text{mean of set B} = \frac{3+4+4+5+....+15}{13} \approx 6.85$

b median of set A = 7th score = 7 median of set B = 7th score = 7

c The data sets are the same except for the last value, and the last value of set A is less than that of set B. So, the mean of set A is less than that of set B.

d The middle value of both data sets is the same, so the median is the same.

3 **a** $\text{mean} = \frac{23\,000+46\,000+23\,000+....+32\,000}{10} = \$29\,300$

median = middle score when in order of size $= \frac{\$23\,000+\$24\,000}{2} = \$23\,500$

mode = $23 000

b The mode is unsatisfactory because it is the lowest salary. It does not take the higher values into account.

c The median is too close to the lower end of the distribution since the data is positively skewed. So the median is not a satisfactory measure of the middle.

4 **a** $\text{mean} = \frac{3+1+0+0+....+1+0+0}{31} = \frac{99}{31} \approx 3.19$

median = 16th score (when in order) = 0

mode = 0 (most frequently occurring score)

b The median is not in the centre, as the data is very positively skewed.

c The mode is the lowest value. It does not take the higher values into account.

d Yes, 42 and 21. **e** No, as this would ignore actual valid data.

5 **a** $\text{mean} = \frac{43+55+41+37}{4} = \frac{176}{4} = 44$ **b** another 44 points

c $\text{new mean} = \frac{176+25}{5} = 40.2$

d It will increase the new mean to 40.3 as 41 is greater than the old mean of 40.2.
$\left\{\frac{5 \times 40.2 + 41}{6} \approx 40.3\right\}$

6 $\text{mean} = \frac{\text{total}}{12}$ $\therefore\ 15\,467 = \frac{\text{total}}{12}$ $\therefore\ \text{total} = 15\,467 \times 12 = \$185\,604$

7 $\text{mean} = \frac{\text{total}}{12}$ $\therefore\ 262 = \frac{\text{total}}{12}$ $\therefore\ \text{total} = 262 \times 12 = 3144\text{ km}$

8 $\overline{x} = \dfrac{\sum_{i=1}^{n} x_i}{n}$ $\quad\therefore\ 11.6 = \dfrac{\sum_{i=1}^{10} x_i}{10}$ $\quad\therefore\ \sum_{i=1}^{10} x_i = 11.6 \times 10 = 116$

9 Total for first 14 matches $= 14 \times 16.5$ goals $= 231$ goals

$\therefore$ new average $= \dfrac{231 + 21 + 24}{16} = \dfrac{276}{16} = 17.25$ goals per game

10 **a** mean selling price $= \dfrac{146\,400 + 127\,600 + 211\,000 + ... + 162\,500}{10} = \$163\,770$

median selling price $= \dfrac{\text{5th} + \text{6th}}{2} = \dfrac{146\,400 + 148\,000}{2} = \$147\,200$

These figures differ by \$16 570. There are more selling prices at the lower end of the market.

b **i** Use the mean as it tends to inflate the average house value of that district.

ii Use the median as you want to buy at the lowest price possible.

11 $\dfrac{5+9+11+12+13+14+17+x}{8} = 12$

$\therefore\ \dfrac{81+x}{8} = 12$

$\therefore\ 81 + x = 96$

$\therefore\ x = 15$

12 $\dfrac{3+0+a+a+4+a+6+a+3}{9} = 4$

$\therefore\ \dfrac{4a+16}{9} = 4$

$\therefore\ 4a + 16 = 36$

$\therefore\ 4a = 20$

$\therefore\ a = 5$

13 $\dfrac{29+36+32+38+35+34+39+x}{8} = 35$

$\therefore\ \dfrac{243+x}{8} = 35$

$\therefore\ 243 + x = 280$

$\therefore\ x = 37$

So, her 8th result was 37.

14 Total for first 10 measurements $= 10 \times 15.7$
$= 157$

Total for next 20 measurements $= 20 \times 14.3$
$= 286$

$\therefore$ mean $= \dfrac{157 + 286}{30} \approx 14.8$

15 The measurements are $7, 9, 11, 13, 14, 17, 19, a, b$ where $a \leqslant b$

mean $= \dfrac{7+9+11+13+....+a+b}{9} = \dfrac{90+a+b}{9}$ $\quad\therefore\ \dfrac{90+a+b}{9} = 12$

$\therefore\ 90 + a + b = 108$

$\therefore\ a + b = 18$

7 9 11 13 14 17 19

If $b \geqslant 13$, then $a \leqslant 5$ and the median $= 13$ (✗)

If $b = 12$, then $a = 6$ and the median $= 12$ (✓)

The remaining cases are:

a	7	8	9
b	11	10	9
median	11	11	11

So, the other two data values are 6 and 12.

16 Scores were $5\ 7\ 9\ 9\ 10\ a\ b$ where $a \leqslant b$ say.

mean $= \dfrac{5+7+9+9+10+a+b}{7} = 8$

$\therefore\ \dfrac{40+a+b}{7} = 8$

$\therefore\ 40 + a + b = 56$

$\therefore\ a + b = 16 \quad \{a \leqslant 12,\ b \leqslant 12\}$

Possibilities are:

a	5	6	7	8
b	11	10	9	8
	×	×	✓	×

- 8, 8: reject as modes are 8 and 9
- 6, 10: reject as modes are 9 and 10
- 5, 11: reject as modes are 5 and 9

So, the missing results are 7 and 9.

EXERCISE 14B.2

1 **a** The mode is 1, as this is the result which occurs most often.

b The median is the average of the 15th and 16th scores $=\dfrac{1+1}{2}=1$

c

x	f	fx
0	4	0
1	12	12
2	11	22
3	3	9
$\sum$	30	43

$$\text{mean} = \frac{\sum fx}{\sum f} = \frac{43}{30} \approx 1.43$$

2 **a** **i**

x	f	fx
0	5	0
1	8	8
2	13	26
3	8	24
4	6	24
5	3	15
6	3	18
7	2	14
8	1	8
9	0	0
10	0	0
11	1	11
$\sum$	50	148

$$\text{mean} = \frac{\sum fx}{\sum f} = \frac{148}{50} = 2.96$$

ii median
$=$ average of 25th and 26th scores (when in order)
$=\dfrac{2+2}{2}$ {13 scores are 1 or 0; 26 scores are 2, 1 or 0}
$=2$

iii mode $= 2$ {occurs most often}

b

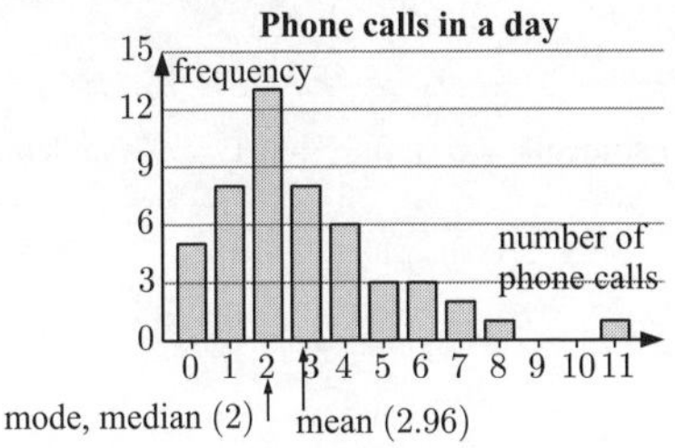

c The distribution is positively skewed. 11 is an outlier.

d The mean takes into account the larger numbers of phone calls.

e The mean, as it best represents all the data.

3 **a** **i** mode $= 49$ {occurs most often}

ii median $=$ average of 15th and 16th values (when in order)

$$= \frac{49+49}{2} = 49 \quad \{9 \text{ are } 47 \text{ or } 48 \text{ and the next } 11 \text{ are } 49\}$$

iii

x	f	fx
47	5	235
48	4	192
49	11	539
50	6	300
51	3	153
52	1	52
$\sum$	30	1471

$$\text{mean} = \frac{\sum fx}{\sum f} = \frac{1471}{30} \approx 49.0$$

b No, as they claim the average is 50 matches per box.

c The sample of only 30 is not large enough. The company could have won its case by arguing that a larger sample would have found an average of 50 matches per box.

4 **a** **i**

x	f	fx
1	5	5
2	28	56
3	15	45
4	8	32
5	2	10
6	1	6
$\sum$	59	154

$$\text{mean} = \frac{\sum fx}{\sum f} = \frac{154}{59} \approx 2.61$$

ii mode $= 2$ {occurs most often}

iii median $=$ 30th score $= 2$

b This school has more children per family (2.61) than the average Australian family (2.2).

c Positive as the higher values are more spread out.

d The mean is higher than the mode and median.

5 **a** **i** mean $= \dfrac{53 + 55 + 56 + 60 + ... + 91}{17} = \dfrac{1175}{17} \approx 69.1$

ii median $=$ 9th score (when in order) $= 67$

iii mode $= 73$ (73 is the only score occurring more than once)

b **i** mean $= \dfrac{3.7 + 4.0 + 4.4 + 4.8 + ... + 8.1}{23} = \dfrac{134.7}{23} \approx 5.86$

ii median $=$ 12th score (when in order) $= 5.8$ **iii** mode $= 6.7$ {occurs most often}

6 **a**

$$\text{mean} = \frac{\sum fx}{\sum f}$$

$$\therefore\ 4.45 = \frac{1 \times 0 + 2 \times 2 + 3 \times 3 + 4 \times 5 + 5 \times x + 6 \times 4 + 7 \times 1}{0 + 2 + 3 + 5 + x + 4 + 1}$$

$$\therefore\ 4.45 = \frac{64 + 5x}{15 + x}$$

$$\therefore\ 4.45(15 + x) = 64 + 5x$$

$$\therefore\ 2.75 = 0.55x$$

$$\therefore\ x = 5$$

b From a total of $2 + 3 + 5 + 5 + 4 + 1 = 20$ students, $5 + 5 + 4 + 1 = 15$ students scored 4 or more.

$\therefore\ \frac{15}{20} = 75\%$ of the students passed.

7 **a** **Without fertiliser**

2	||
3	𝍸 𝍸 |
4	𝍸 𝍸 𝍸 ||||
5	𝍸 𝍸 𝍸 𝍸 𝍸 ||||
6	𝍸 𝍸 𝍸 𝍸 𝍸 𝍸 𝍸 𝍸 𝍸 𝍸 |
7	𝍸 𝍸 𝍸 𝍸 𝍸
8	𝍸 𝍸 ||
9	|

x	f	fx	cf
2	2	4	2
3	11	33	13
4	19	76	32
5	29	145	61
6	51	306	112
7	25	175	137
8	12	96	149
9	1	9	150

i mean $= \dfrac{\sum fx}{\sum f} = \dfrac{844}{150} \approx 5.63$

ii mode $= 6$ {occurs most often}

iii median $=$ average of 75th and 76th scores $= \dfrac{6 + 6}{2} = 6$

b **With fertiliser**

3	||||
4	~~||||~~ ~~||||~~ |||
5	~~||||~~ ~~||||~~ |
6	~~||||~~ ~~||||~~ ~~||||~~ ~~||||~~ ~~||||~~ |||
7	~~||||~~ ~~||||~~ ~~||||~~ ~~||||~~ ~~||||~~ ~~||||~~ ~~||||~~ ~~||||~~ ~~||||~~ ||
8	~~||||~~ ~~||||~~ ~~||||~~ ~~||||~~ ~~||||~~ ||
9	~~||||~~ ~~||||~~ ||||
10	||||
11	|
13	|

x	f	fx	cf
3	4	12	4
4	13	52	17
5	11	55	28
6	28	168	56
7	47	329	103
8	27	216	130
9	14	126	144
10	4	40	148
11	1	11	149
13	1	13	150

i mean $= \dfrac{\sum fx}{\sum f} = \dfrac{1022}{150} \approx 6.81$

ii mode $= 7$ {occurs most often}

iii median = average of 75th and 76th scores $= \dfrac{7+7}{2} = 7$

c The mean best represents the centre for this data.

d Yes, as 6.81 is significantly greater than 5.63.

Note: The total yield of the crop may not have improved as, for example, the number of pods per plant may have decreased when using the fertiliser.

8 **a** mean birth mass $= \dfrac{75+70+80+ \ldots. +83}{8} = \dfrac{567}{8} \approx 70.9$ grams

b mean after 2 weeks $= \dfrac{210+200+200+ \ldots. +230}{8} = \dfrac{1681}{8} \approx 210$ grams

c mean increase $\approx (210.13 - 70.88)$ grams ≈ 139 grams

9 The 31 scores in order are: {15 scores below 10}, 10.1, 10.4, 10.7, 10.9, {12 scores above 11}
Median = 16th score (when in order) = 10.1 cm

10 **a** **Brand A**

x	f	fx
46	1	46
47	2	94
48	3	144
49	7	343
50	10	500
51	20	1020
52	15	780
53	3	159
55	1	55
$\sum$	62	3141

mean $= \dfrac{\sum fx}{\sum f} = \dfrac{3141}{62} \approx 50.7$

Brand B

x	f	fx
48	3	144
49	17	833
50	30	1500
51	7	357
52	2	104
53	1	53
54	1	54
$\sum$	61	3045

mean $= \dfrac{\sum fx}{\sum f} = \dfrac{3045}{61} \approx 49.9$

b Based on average contents, the C.P.S. should not prosecute either manufacturer. To the nearest toothpick, the average contents for A is 51 and for B is 50.

11 **a** **i** median salary $= \dfrac{10\text{th} + 11\text{th}}{2}$ (when in order) $= \dfrac{35\,000 + 28\,000}{2} = €31\,500$

ii modal salary = €28 000 {occurs most often}

iii

x	f	fx
50 000	1	50 000
42 000	3	126 000
35 000	6	210 000
28 000	10	280 000
$\sum$	20	666 000

mean $= \dfrac{\sum fx}{\sum f} = \dfrac{666\,000}{20} = €33\,300$

b The mean, as it is the highest value.

EXERCISE 14B.3

1

midpoint (x)	f	fx
4.5	2	9
14.5	5	72.5
24.5	7	171.5
34.5	27	931.5
44.5	9	400.5
$\sum$	50	1585

$\therefore$ mean result $\approx \frac{1585}{50} \approx 31.7$

2

midpoint (x)	f	fx
2499.5	4	9998
3499.5	4	13 998
4499.5	9	40 495.5
5499.5	14	76 993
6499.5	23	149 488.5
7499.5	16	119 992
$\sum$	70	410 965

a 70

b $\approx$ 411 000 litres

c mean $\approx \frac{\sum fx}{\sum f} \approx \frac{410\,965}{70} \approx$ 5870 litres

3 **a** $5 + 10 + 25 + 40 + 10 + 15 + 10 + 10 = 125$ people

b

midpoint (x)	*frequency* (f)	fx
85	5	425
95	10	950
105	25	2625
115	40	4600
125	10	1250
135	15	2025
145	10	1450
155	10	1550
$\sum$	125	14 875

mean $\approx \frac{\sum fx}{\sum f} \approx \frac{14\,875}{125} \approx$ 119 marks

c $\frac{15}{125} = \frac{3}{25}$ scored < 100

d 20% of 125 people = 25 people and 20 people scored > 140 for the test

$\therefore$ estimate is $140 - \frac{5}{15} \times 10 \approx 137$ marks

EXERCISE 14C.1

1 **a** 2 3 3 3 4 4 4 5 5 5 5 6 6 6 6 6 7 7 8 8 8 9 9 ($n = 23$)

min = 2, Q_1 = 4, median = 6, Q_3 = 7, max = 9

i median = 6 **ii** $Q_1 = 4$, $Q_3 = 7$ **iii** range $= 9 - 2 = 7$ **iv** IQR $= Q_3 - Q_1 = 3$

b 10 12 12 14 15 15 16 16 17 18 18 18 18 19 20 21 22 24 ($n = 18$)

min = 10, Q_1 = 15, median (between 17 and 18), Q_3 = 19, max = 24

i median = 17.5 **ii** $Q_1 = 15$, $Q_3 = 19$ **iii** range $= 24 - 10 = 14$ **iv** IQR $= Q_3 - Q_1 = 4$

c 21.8 22.4 23.5 23.5 24.6 24.9 25.0 25.3 26.1 26.4 29.5 ($n = 11$)

min = 21.8, Q_1 = 23.5, median = 24.9, Q_3 = 26.1, max = 29.5

i median = 24.9 **ii** $Q_1 = 23.5$, $Q_3 = 26.1$ **iii** range $= 29.5 - 21.8 = 7.7$ **iv** IQR $= Q_3 - Q_1 = 2.6$

2 $$\underset{\underset{\text{min}}{\uparrow}}{0}\ \ 0\ \ 0\ \ 0.8\ \ \underbrace{1.4\ \ 1.5}_{\underset{Q_1}{\uparrow}}\ \ 1.6\ \ 1.9\ \ 2.1\ \ \underbrace{2.2\ \ 2.7}_{\underset{\text{median}}{\uparrow}}\ \ 3.0\ \ 3.4\ \ 3.6\ \ \underbrace{3.8\ \ 3.8}_{\underset{Q_3}{\uparrow}}\ \ 4.5\ \ 4.8\ \ 5.2\ \ \underset{\underset{\text{max}}{\uparrow}}{5.2}$$

a median $= \dfrac{2.2+2.7}{2}$ $= 2.45$ $Q_1 = 1.45$ $Q_3 = 3.8$

b range $= 5.2 - 0$ $= 5.2$ IQR $= Q_3 - Q_1$ $= 3.8 - 1.45$ $= 2.35$

c **i** the median $= 2.45$ min **ii** $Q_3 = 3.8$ min

iii The minimum waiting time was 0 minutes and the maximum waiting time was 5.2 minutes. The waiting time was spread over 5.2 minutes.

3 $$\underset{\underset{\text{min}}{\uparrow}}{3}\ \ 4\ \ 7\ \ 9\ \ 10\ \ \underset{\underset{Q_1}{\uparrow}}{13}\ \ 14\ \ 16\ \ 17\ \ 18\ \ 20\ \ \underset{\underset{\text{median}}{\uparrow}}{20}\ \ 23\ \ 25\ \ 26\ \ 29\ \ 29\ \ \underset{\underset{Q_3}{\uparrow}}{29}\ \ 31\ \ 33\ \ 37\ \ 38\ \ \underset{\underset{\text{max}}{\uparrow}}{42}\quad (n=23)$$

a min $= 3$ **b** max $= 42$ **c** median $= 20$ **d** $Q_1 = 13$ **e** $Q_3 = 29$

f range $= 42 - 3 = 39$ **g** IQR $= Q_3 - Q_1 = 29 - 13 = 16$

4 $$\underset{\underset{\text{min}}{\uparrow}}{109}\ \ 111\ \ 113\ \ 114\ \ \underbrace{114\ \ 118}_{\underset{Q_1}{\uparrow}}\ \ 119\ \ 122\ \ 122\ \ \underbrace{124\ \ 124}_{\underset{\text{median}}{\uparrow}}\ \ 126\ \ 128\ \ 129\ \ \underbrace{129\ \ 131}_{\underset{Q_3}{\uparrow}}\ \ 132\ \ 135\ \ 138\ \ \underset{\underset{\text{max}}{\uparrow}}{138}$$

$(n = 20)$

a **i** median $= 124$ cm
ii $Q_3 = 130$ cm, $Q_1 = 116$ cm

b **i** .... 124 cm tall
ii .... 130 cm tall

c **i** range $= 138 - 109$ $= 29$ cm
ii IQR $= Q_3 - Q_1$ $= 14$ cm

d the IQR $= 14$ cm

5 See **Exercise 14B.2** solution to question **7**.

a **Without fertiliser**

i range $= 9 - 2 = 7$ **ii** median $= 6$ **iii** lower quartile $=$ 38th score $= 5$
iv upper quartile $=$ 113th score $= 7$ **v** interquartile range $= 7 - 5 = 2$

b **With fertiliser**

i range $= 13 - 3 = 10$ **ii** median $= 7$ **iii** lower quartile $=$ 38th score $= 6$
iv upper quartile $=$ 113th score $= 8$ **v** interquartile range $= 8 - 6 = 2$

EXERCISE 14C.2

1 **a** **i** median $= 35$ **ii** max. value $= 78$ **iii** min. value $= 13$
iv $Q_3 = 53$ **v** $Q_1 = 26$

b **i** range $= 78 - 13 = 65$ **ii** IQR $= Q_3 - Q_1 = 53 - 26 = 27$

2 **a** highest mark was 98, lowest mark was 25 **b** the median which is 70

c Q_3 which is 85 **d** $Q_1 = 55$ and $Q_3 = 85$ **e** range $= 98 - 25 = 73$

f IQR $= Q_3 - Q_1 = 85 - 55 = 30$

g The data is negatively skewed, so the mean will be slightly lower than the median of 70.
$\therefore$ mean ≈ 67, say.

3 **a** **i** $$\underset{\underset{\text{min}}{\uparrow}}{3}\ \ 4\ \ 5\ \ \underset{\underset{Q_1}{\uparrow}}{5}\ \ 5\ \ 6\ \ \underbrace{6\ \ 6}_{\underset{\text{median}}{\uparrow}}\ \ 7\ \ 7\ \ \underset{\underset{Q_3}{\uparrow}}{8}\ \ 8\ \ 9\ \ \underset{\underset{\text{max}}{\uparrow}}{10}$$

So, min $= 3$, $Q_1 = 5$, median $= 6$, $Q_3 = 8$, max $= 10$

ii

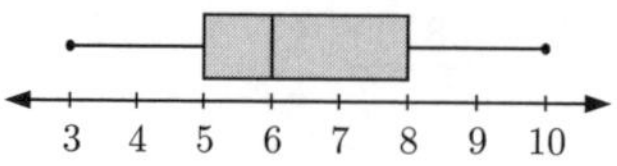

iii range $= 10 - 3$
$= 7$

iv $\text{IQR} = Q_3 - Q_1$
$= 8 - 5$
$= 3$

b **i** 0 1 2 3 4 5 6 6 7 7 7 8 8 8 8 8 8 9 9

↑ min (0), ↑ Q_1 (4), ↑ median (7), ↑ Q_3 (8), ↑ max (9)

So, min $= 0$, $Q_1 = 4$, median $= 7$, $Q_3 = 8$, max $= 9$

ii

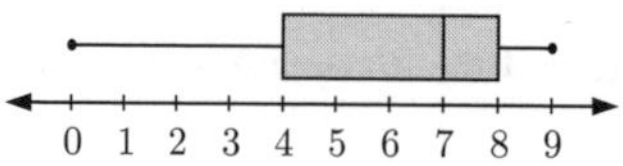

iii range $= 9 - 0$
$= 9$

iv $\text{IQR} = Q_3 - Q_1$
$= 8 - 4$
$= 4$

c **i** min ↓

117 120 123 126 126 128 130 131 131 131 133 135 135 137 144 147 147 149
149 151

↑ max (151), ↑ Q_1 (126 128), ↑ median (131 133), ↑ Q_3 (144 147)

So, min $= 117$, $Q_1 = 127$, median $= 132$, $Q_3 = 145.5$, max $= 151$

ii

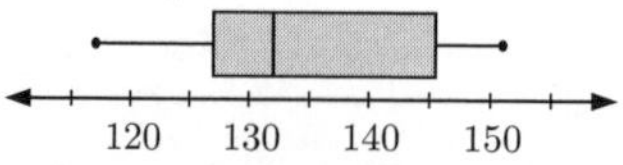

iii range
$= 151 - 117$
$= 34$

iv $\text{IQR} = Q_3 - Q_1$
$= 145.5 - 127$
$= 18.5$

4 **a**

Statistic	*Year 9*	*Year 12*
minimum	1	6
Q_1	5	10
median	7.5	14
Q_3	10	16
maximum	12	17.5

b For the year 9 group

i range $= 12 - 1$
$= 11$

ii $\text{IQR} = 10 - 5$
$= 5$

For the year 12 group

i range $= 17.5 - 6$
$= 11.5$

ii $\text{IQR} = 16 - 10$
$= 6$

c **i** True, as indicated by the median.

ii True, as Q_1 for year 9 $= 5$ and min for year 12 $= 6$.

5 2 3 3 4 4 4 4 5 5 5 5 5 5 5 6 6 6 6 6 6 7 7 7 7 8 8 8 9 9 9 10 12 13

↑ min (2), ↑ Q_1 (5 5), ↑ median (6), ↑ Q_3 (8 8), ↑ max (13)

a median $= 6$, $Q_1 = 5$, $Q_3 = 8$

b $\text{IQR} = 8 - 5 = 3$

c

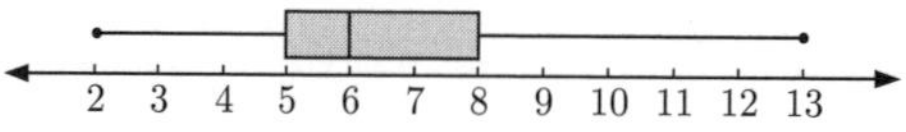

6 **a**

Number of bolts	33	34	35	36	37	38	39	40
Frequency	1	5	7	13	12	8	0	1

↑ min (33), ↑ median is one of these (36), ↑ max (40)

There are 47 scores
$\therefore$ median $=$ 24th $\left(\dfrac{47+1}{2} = 24\right)$

13 scores are 35 or less
26 scores are 36 or less
$\therefore$ median is 36

$Q_1 =$ 12th $\left(\dfrac{23+1}{2} = 12\right)$
$= 35$

$Q_3 =$ 36th
$= 37$

So, min $= 33$, $Q_1 = 35$, median $= 36$, $Q_3 = 37$, max $= 40$

b **i** range $= 40 - 33 = 7$

ii $\text{IQR} = 37 - 35 = 2$

c

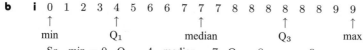

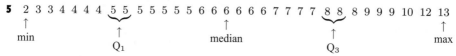

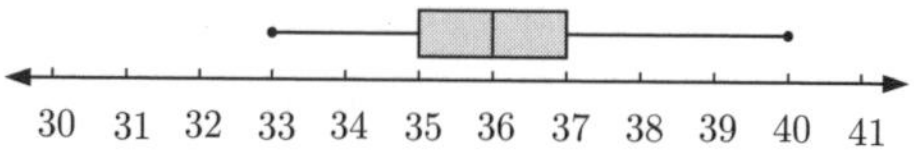

EXERCISE 14D

1

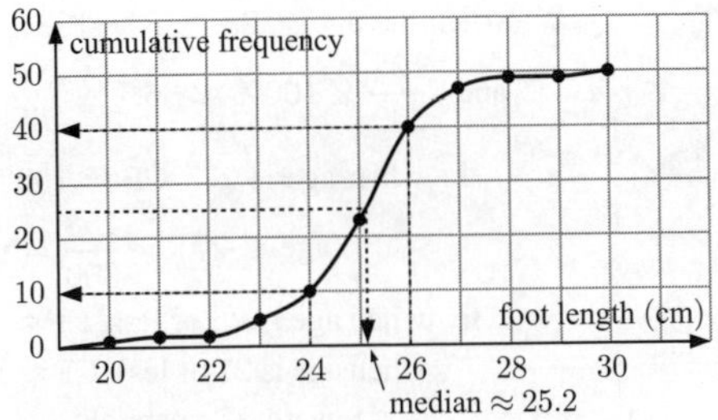

a The median is ≈ 25.2 cm.

b **i** 10 people had a foot length 24 cm or less.
$\therefore$ $50 - 10 = 40$ people had a foot length of 25 cm or more.

ii 40 people

2 **a**

Length (x cm)	*Freq.*	*C. freq.*
$24 \leqslant x < 27$	1	1
$27 \leqslant x < 30$	2	3
$30 \leqslant x < 33$	5	8
$33 \leqslant x < 36$	10	18
$36 \leqslant x < 39$	9	27
$39 \leqslant x < 42$	2	29
$42 \leqslant x < 45$	1	30

b

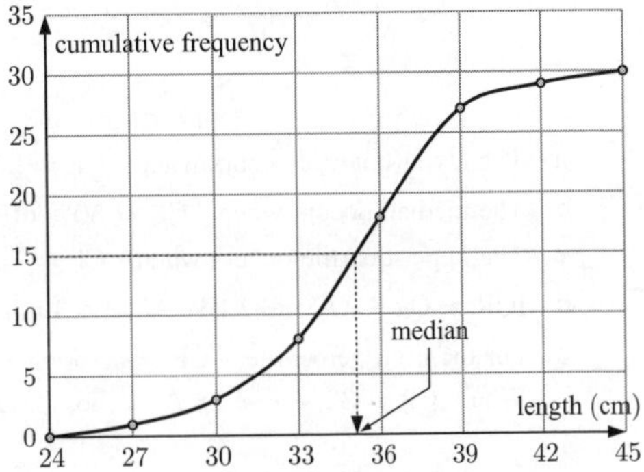

c median ≈ 35

d There are 30 data values. So, the median is the average of the 15th and 16th scores (when in order). In order they are:

24 27 28 30 31 31 32 32 33 33 33 33 34 34 34 35 35 35 36 .... and so on.

$$\text{median} = \frac{34+35}{2} = 34.5$$

So, the median from the graph is a good approximation.

3 **a** For $h = 5$, $\text{CF} \approx 9$ $\therefore$ 9 seedlings have height 5 cm or less

b For $h = 8$, $\text{CF} \approx 43$ $\therefore$ % taller than 8 cm $= \dfrac{60 - 43}{60} \times 100\% \approx 28.3\%$

c The approximate median occurs at $\text{CF} = 30$, $\therefore$ median ≈ 7 cm.

d $\text{IQR} = Q_3 - Q_1 = (h \text{ when } \text{CF} = 45) - (h \text{ when } \text{CF} = 15)$
$\approx 8.3 - 5.9 \approx 2.4$ cm

e 90th percentile occurs when $\text{CF} = 90\% \text{ of } 60 = 54$
$\therefore$ 90th percentile $= 10$
This means that 90% of the seedlings have a height of 10 cm or less.

4

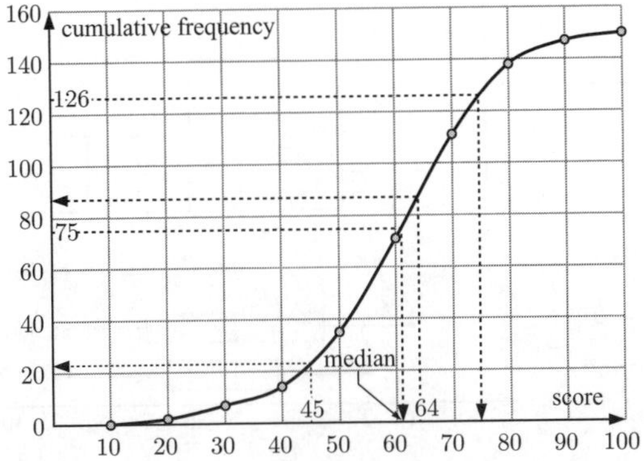

a Total frequency $= 150$
$\therefore$ median ≈ 61 {from the graph}

b When the score $= 64$,
$\text{CF} \approx 87$ {from the graph}
$\therefore$ about 87 students scored less than 65.

c $\approx 36 + 40 = 76$ students scored between 50 and 70.

d For a pass mark of 45, $\text{CF} \approx 24.5$
$\therefore$ 24 or 25 students failed the exam.

e 84% of 150 = 126
So, for a CF of 126, the score value is 76.
So, the minimum credit mark would be 76.

5

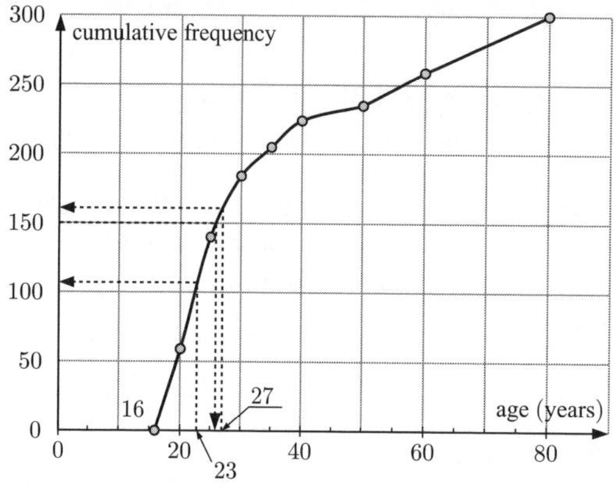

Total frequency $= 300$

a median ≈ 26 years {from the graph}

b when the age $= 23$, CF ≈ 108
and $\dfrac{108}{300} \times 100\% \approx 36\%$

c **i** when age is 27, CF ≈ 158
$\therefore$ $P(\text{age} \leqslant 27) = \dfrac{158}{300} = 0.527$

ii when age is 26 or less, CF ≈ 150 {**a**}
when age is 27 or less, CF ≈ 158 {**c ii**}
$\therefore$ 8 were 27 years old
$\therefore$ $P(\text{aged } 27) \approx \dfrac{8}{300} \approx 0.0267$

6 **a** The lower quartile occurs when CF $= 25\%$ of $80 = 20$ $\therefore$ $Q_1 = 27$ min

b The median occurs when CF $= 50\%$ of $80 = 40$ $\therefore$ median $= 29$ min

c The upper quartile occurs when CF $= 75\%$ of $80 = 60$ $\therefore$ $Q_3 \approx 31.3$ min

d $\text{IQR} = Q_3 - Q_1 \approx 31.3 - 27 \approx 4.3$ min

e For the 40th percentile, CF $= 40\%$ of $80 = 32$
When CF $= 32$, $x \approx 28.2$ So, the 40th percentile is about 28 min 10 sec.

7

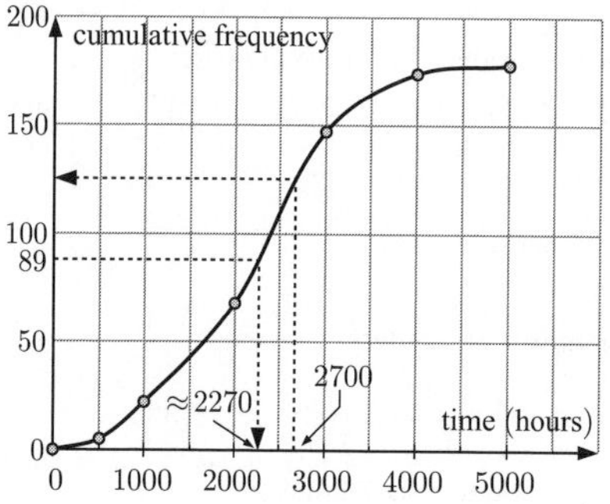

Total frequency $= 178$

a median ≈ 2270 hours

b For a life of 2700 hours CF ≈ 123
and $\dfrac{123}{178} \times 100\% \approx 69.1\%$
So, about 69.1% have a life $\leqslant 2700$ h

c For a life of 1500 h, CF ≈ 45
For a life of 2500 h, CF ≈ 108
and $108 - 45 = 63$
$\therefore$ ≈ 63 had a life between 1500 and 2500 hours.

EXERCISE 14E

1 **a** $\overline{x} \approx 4.87$, $\text{Min}_x = 1$, $Q_1 = 3$, med $= 5$, $Q_3 = 7$, $\text{Max}_x = 9$

b

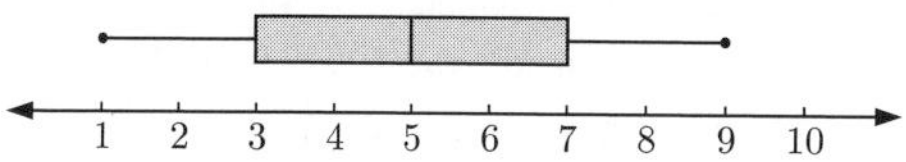

c

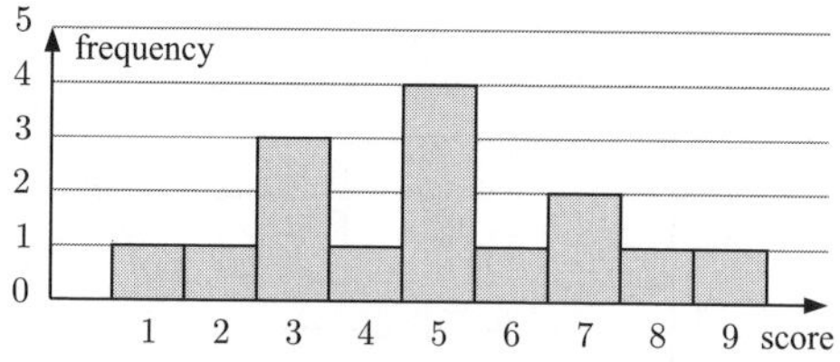

d $\overline{x} \approx 5.24$, $\text{Min}_x = 2$, $Q_1 = 4$, med $= 5$, $Q_3 = 6.5$, $\text{Max}_x = 9$

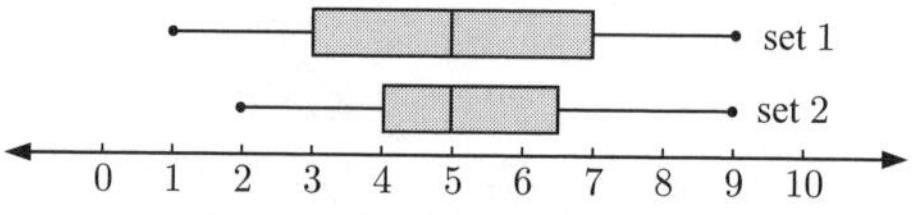

2 **a** discrete

c

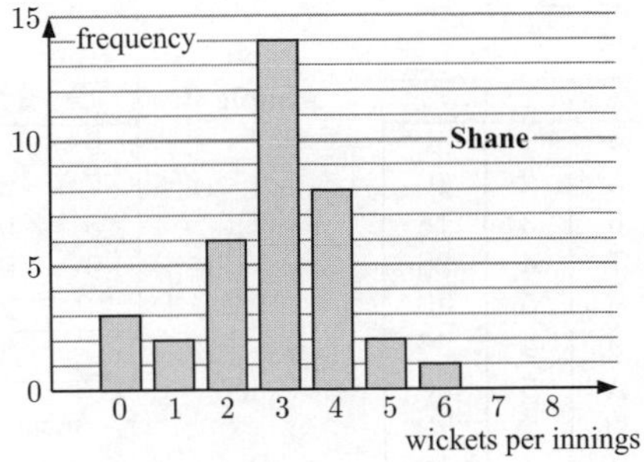

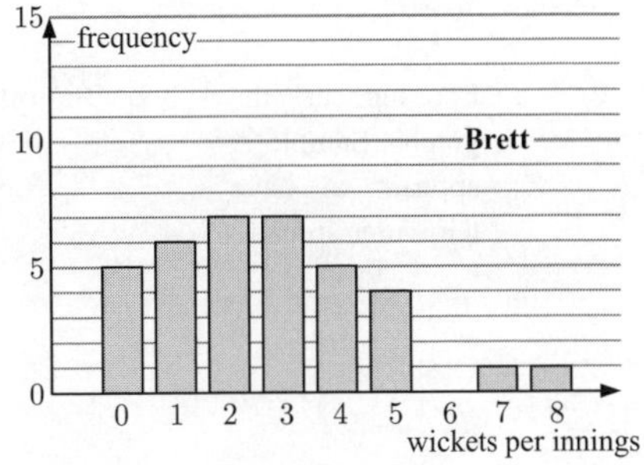

d There are no outliers for Shane. Brett has outliers of 7 and 8 which must not be removed.

e Shane's distribution is reasonably symmetrical. Brett's distribution is positively skewed.

f Shane has a higher mean ($\approx$ 2.89 wickets) compared with Brett ($\approx$ 2.67 wickets). Shane has a higher median (3 wickets) compared with Brett (2.5 wickets). Shane's modal number of wickets is 3 (14 times) compared with Brett, who has two modal values of 2 and 3 (7 times each).

g Shane's range is 6 wickets, compared with Brett's range of 8 wickets. Shane's IQR is 2 wickets, compared with Brett's IQR of 3 wickets. Brett's wicket taking shows greater spread or variability.

h

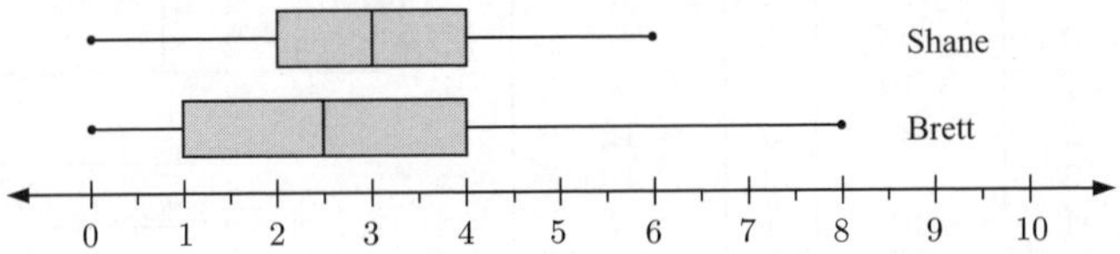

i Generally, Shane takes more wickets than Brett and is a more consistent bowler.

3 **a** continuous

c For the 'New type' globes, 191 hours could be considered an outlier. However, it could be a genuine piece of data, so we will include it in the analysis.

d The mean and median are $\approx 25\%$ and $\approx 19\%$ higher for the 'new type' of globe compared with the 'old type'.
The range is higher for the 'new type' of globe (but has been affected by the 191 hours).
The IQR for each type of globe is almost the same.

	Old type	*New type*
Mean	107	134
Median	110.5	132
Range	56	84
IQR	19	18.5

e

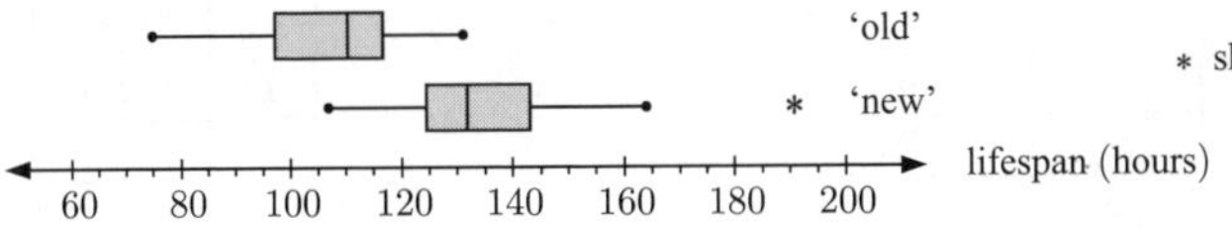

f For the 'old type' of globe, the data is bunched to the right of the median, hence the distribution is negatively skewed. For the 'new type' of globe, the data is bunched to the left of the median, hence the distribution is positively skewed.

g The manufacturer's claim, that the 'new type' of globe has a 20% longer life than the 'old type' seems to be backed up by the 25% higher mean life and 19.5% higher median life.

EXERCISE 14F.1

1 **a** Looking at the graphs, Sample A appears to have the wider spread.

b **Sample A:**

x	f	fx
4	1	4
5	2	10
6	3	18
7	4	28
8	5	40
9	4	36
10	3	30
11	2	22
12	1	12
$\sum$	25	200

$\therefore$ mean $= \frac{200}{25} = 8$

Sample B:

x	f	fx
6	2	12
7	6	42
8	9	72
9	6	54
10	2	20
$\sum$	25	200

$\therefore$ mean $= \frac{200}{25} = 8$

c **Sample A:**

x	$x-\overline{x}$	$(x-\overline{x})^2$	f	$f(x-\overline{x})^2$
4	-4	16	1	16
5	-3	9	2	18
6	-2	4	3	12
7	-1	1	4	4
8	0	0	5	0
9	1	1	4	4
10	2	4	3	12
11	3	9	2	18
12	4	16	1	16
			$\sum$	100

$$\therefore \quad s = \sqrt{\frac{\sum(x-\overline{x})^2}{n}} = \sqrt{\tfrac{100}{25}} = 2$$

Sample B:

x	$x-\overline{x}$	$(x-\overline{x})^2$	f	$f(x-\overline{x})^2$
6	-2	4	2	8
7	-1	1	6	6
8	0	0	9	0
9	1	1	6	6
10	2	4	2	8
			$\sum$	28

$$\therefore \quad s = \sqrt{\frac{\sum(x-\overline{x})^2}{n}} = \sqrt{\tfrac{28}{25}} \approx 1.06$$

The standard deviation is higher for Sample A.

2 **a** **Andrew:** $\overline{x} = \dfrac{23+17+....+28+32}{8} = 25$

x	$x-\overline{x}$	$(x-\overline{x})^2$
23	-2	4
17	-8	64
31	6	36
25	0	0
25	0	0
19	-6	36
28	3	9
32	7	49
	$\sum$	198

$$\therefore \quad s = \sqrt{\frac{\sum(x-\overline{x})^2}{n}} = \sqrt{\tfrac{198}{8}} \approx 4.97$$

Brad: $\overline{x} = \dfrac{9+29+....+38+43}{8} = 30.5$

x	$x-\overline{x}$	$(x-\overline{x})^2$
9	-21.5	462.25
29	-1.5	2.25
41	10.5	110.25
26	-4.5	20.25
14	-16.5	272.25
44	13.5	182.25
38	7.5	56.25
43	12.5	156.25
	$\sum$	1262

$$\therefore \quad s = \sqrt{\frac{\sum(x-\overline{x})^2}{n}} = \sqrt{\tfrac{1262}{8}} \approx 12.6$$

b Andrew, as he has the smaller standard deviation.

3 **a** Rockets have mean $= \dfrac{0+10+1+9+11+0+8+5+6+7}{10} = \dfrac{57}{10} = 5.7$

Bullets have mean $= \dfrac{4+3+4+1+4+11+7+6+12+5}{10} = \dfrac{57}{10} = 5.7$

Rocket's range $= 11 - 0 = 11$ Bullet's range $= 12 - 1 = 11$

b We suspect the Rockets, as they have two zeros.

c **Rockets**

x	$(x-\overline{x})^2$
0	$(5.7)^2$
10	$(4.3)^2$
1	$(4.7)^2$
9	$(3.3)^2$
11	$(5.3)^2$
0	$(5.7)^2$
8	$(2.3)^2$
5	$(0.7)^2$
6	$(0.3)^2$
7	$(1.3)^2$
	152.1

$s=\sqrt{\dfrac{\sum(x-\overline{x})^2}{n}}$

$=\sqrt{\dfrac{152.1}{10}}$

$=3.9$

↑
greater variability

Bullets

x	$(x-\overline{x})^2$
4	$(1.7)^2$
3	$(2.7)^2$
4	$(1.7)^2$
1	$(4.7)^2$
4	$(1.7)^2$
11	$(5.3)^2$
7	$(1.3)^2$
6	$(0.3)^2$
12	$(6.3)^2$
5	$(0.7)^2$
	108.1

$s=\sqrt{\dfrac{\sum(x-\overline{x})^2}{n}}$

$=\sqrt{\dfrac{108.1}{10}}$

≈ 3.29

d The standard deviation, as it takes all values into account, not just the lowest and highest.

4 **a** We suspect variability in standard deviation since the factors may change every day.

b **i** sample mean **ii** sample standard deviation

c less variability in the volume of soft drink per can

5 **a**

x	$(x-\overline{x})^2$
79	10^2
64	5^2
59	10^2
71	2^2
68	1^2
68	1^2
74	5^2
483	256

$\overline{x}=\dfrac{\sum x}{n}$

$=\dfrac{483}{7}=69$

$s=\sqrt{\dfrac{\sum(x-\overline{x})^2}{n}}$

$=\sqrt{\dfrac{256}{7}}$

≈ 6.05 kg

b

x	$(x-\overline{x})^2$
89	10^2
74	5^2
69	10^2
81	2^2
78	1^2
78	1^2
84	5^2
553	256

$\overline{x}=\dfrac{\sum x}{n}$

$=\dfrac{553}{7}$

$=79$

$s\approx 6.05$ kg

c The distribution has simply shifted by 10 kg. The mean increases by 10 kg and the standard deviation remains the same. In general, changing the values by a constant does not affect standard deviation.

6 **a**

x	$(x-\overline{x})^2$
0.8	$(0.21)^2$
1.1	$(0.09)^2$
1.2	$(0.19)^2$
0.9	$(0.11)^2$
1.2	$(0.19)^2$
1.2	$(0.19)^2$
0.9	$(0.11)^2$
0.7	$(0.31)^2$
1.0	$(0.01)^2$
1.1	$(0.09)^2$
10.1	0.289

$\overline{x}=\dfrac{\sum x}{n}$

$=\dfrac{10.1}{10}$

$=1.01$ kg

$=\sqrt{\dfrac{\sum(x-\overline{x})^2}{n}}$

$s=\sqrt{\dfrac{0.289}{10}}$

$=0.17$ kg

b

x	$(x-\overline{x})^2$
1.6	$(0.42)^2$
2.2	$(0.18)^2$
2.4	$(0.38)^2$
1.8	$(0.22)^2$
2.4	$(0.38)^2$
2.4	$(0.38)^2$
1.8	$(0.22)^2$
1.4	$(0.62)^2$
2.0	$(0.02)^2$
2.2	$(0.18)^2$
20.2	1.156

$\overline{x}=\dfrac{\sum x}{n}$

$=\dfrac{20.2}{10}$

$=2.02$ kg

$s=\sqrt{\dfrac{\sum(x-\overline{x})^2}{n}}$

$=\sqrt{\dfrac{1.156}{10}}$

$=0.34$ kg

c Doubling the values doubles the mean and the standard deviation.

7 mean $= \dfrac{1+3+5+7+4+5+p+q}{8} = 5$

$$25+p+q=40$$
$$\therefore\ p+q=15$$
$$\therefore\ q=15-p$$

and $s = \sqrt{\dfrac{(-4)^2+(-2)^2+0^2+2^2+(-1)^2+0^2+(p-5)^2+(q-5)^2}{8}} = \sqrt{5.25}$

$$\therefore\ \frac{16+4+4+1+(p-5)^2+(15-p-5)^2}{8} = 5.25$$
$$\therefore\ 25+p^2-10p+25+100-20p+p^2=42$$
$$\therefore\ 2p^2-30p+108=0$$
$$\therefore\ p^2-15p+54=0$$
$$\therefore\ (p-6)(p-9)=0$$
$$\therefore\ p=6 \text{ or } 9 \text{ and } q=9 \text{ or } 6$$

But $p<q$ $\quad\therefore\ p=6,\ q=9$

8 mean $= \dfrac{3+9+5+5+6+4+a+6+b+8}{10} = 6$

$$\therefore\ \frac{46+a+b}{10}=6$$
$$\therefore\ 46+a+b=60$$
$$\therefore\ a+b=14$$
$$\therefore\ b=14-a$$

and $s = \sqrt{\dfrac{(-3)^2+3^2+(-1)^2+(-1)^2+(-2)^2+(a-6)^2+(b-6)^2+2^2}{10}} = \sqrt{3.2}$

$$\therefore\ 9+9+1+1+4+4+(a-6)^2+(14-a-6)^2=32$$
$$\therefore\ 28+a^2-12a+36+64-16a+a^2=32$$
$$\therefore\ 2a^2-28a+96=0$$
$$\therefore\ a^2-14a+48=0$$
$$\therefore\ (a-6)(a-8)=0$$
$$\therefore\ a=6 \text{ or } 8 \text{ and } b=8 \text{ or } 6$$

But $a>b$ $\quad\therefore\ a=8,\ b=6$

9 **a** $\overline{x} = \dfrac{0.8+0.6+0.7+0.8+0.4+2.8}{6}$

≈ 1.017

x	$(x-\overline{x})^2$
0.8	$(-0.217)^2$
0.6	$(-0.417)^2$
0.7	$(-0.317)^2$
0.8	$(-0.217)^2$
0.4	$(-0.617)^2$
2.8	$(1.783)^2$
$\sum$	3.928

$\therefore\ s = \sqrt{\dfrac{\sum(x-\overline{x})^2}{n}} \approx \sqrt{\dfrac{3.928}{6}}$

≈ 0.809

b $\overline{x} = \dfrac{0.8+0.6+0.7+0.8+0.4}{5}$

$= 0.66$

x	$(x-\overline{x})^2$
0.8	$(0.14)^2$
0.6	$(-0.06)^2$
0.7	$(0.04)^2$
0.8	$(0.14)^2$
0.4	$(-0.26)^2$
$\sum$	0.112

$\therefore\ s = \sqrt{\dfrac{\sum(x-\overline{x})^2}{n}} = \sqrt{\dfrac{0.112}{5}}$

≈ 0.150

c The extreme value greatly increases the standard deviation.

EXERCISE 14F.2

1 **a** $s = \sqrt{\text{variance}}$

$= \sqrt{45.9}$ kg

≈ 6.77 kg

We estimate the standard deviation σ by $s \approx 6.77$ kg.

b Unbiased estimate of μ is $\overline{x} = 93.8$ kg

2 **a** Using technology, $\overline{x} \approx 77.5$ g and $s_n \approx 7.44g$

b Unbiased estimate of μ is $\overline{x} \approx 77.5$ g

Estimate of σ is $s_n \approx 7.44$ g

EXERCISE 14F.3

1 **a**

x	f	fx	$f(x-\overline{x})^2$
0	14	0	41.62
1	18	18	9.44
2	13	26	0.99
3	5	15	8.14
4	3	12	15.54
5	2	10	21.46
6	2	12	36.57
7	1	7	27.83
$\sum$	58	100	161.59

$\overline{x} = \dfrac{\sum fx}{\sum f}$

$= \dfrac{100}{58}$

≈ 1.72 children

$s = \sqrt{\dfrac{\sum f(x-\overline{x})^2}{\sum f}}$

$\approx \sqrt{\dfrac{161.59}{58}}$

≈ 1.67 children

b Estimate of μ is $\overline{x} \approx 1.72$ children

Estimate of σ is $s \approx 1.67$ children

2 **a**

x	f	fx	$f(x-\overline{x})^2$
11	2	22	24.22
12	1	12	6.150
13	4	52	8.762
14	5	70	1.152
15	6	90	1.622
16	4	64	9.242
17	2	34	12.70
18	1	18	12.39
$\sum$	25	362	76.24

$\overline{x} = \dfrac{\sum fx}{\sum f}$

$= \dfrac{362}{25}$

≈ 14.5 years

$s = \sqrt{\dfrac{\sum f(x-\overline{x})^2}{\sum f}}$

$= \sqrt{\dfrac{76.24}{25}}$

≈ 1.75 years

b Estimate of μ is $\overline{x} \approx 14.5$ years

Estimate of σ is $s \approx 1.75$ years

3 **a**

x	f	fx	$f(x-\overline{x})^2$
33	1	33	18.24
35	5	175	25.78
36	7	252	11.31
37	13	481	0.95
38	12	456	6.38
39	8	312	23.92
40	2	80	14.90
$\sum$	48	1789	101.48

$\overline{x} = \dfrac{\sum fx}{\sum f}$

$= \dfrac{1789}{48}$

≈ 37.3 toothpicks

$s = \sqrt{\dfrac{\sum f(x-\overline{x})^2}{\sum f}}$

$= \sqrt{\dfrac{101.48}{48}}$

≈ 1.45 toothpicks

b Estimate of μ is $\overline{x} \approx 37.3$ toothpicks

Estimate of σ is $s \approx 1.45$ toothpicks

4 **a**

Midpoint (x)	f	fx	$f(x-\overline{x})^2$
41	1	41	52.80
43	1	43	27.74
45	3	135	32.01
47	7	329	11.23
49	11	539	5.91
51	5	255	37.35
53	2	106	44.80
$\sum$	30	1448	211.87

$\overline{x} = \dfrac{\sum fx}{\sum f} = \dfrac{1448}{30} \approx 48.3$ cm

$s = \sqrt{\dfrac{\sum f(x-\overline{x})^2}{\sum f}} \approx \sqrt{\dfrac{211.87}{30}} \approx 2.66$ cm

b Estimate of μ is $\overline{x} \approx 48.3$ cm
Estimate of σ is $s \approx 2.66$ cm

5 **a**

Midpoint (x)	f	fx	$f(x-\overline{x})^2$
364.995	17	6204.9	10 881.53
374.995	38	14 249.8	8895.42
384.995	47	18 094.8	1320.23
394.995	57	22 514.7	1259.13
404.995	18	7289.9	3889.62
414.995	10	4150.0	6100.9
424.995	10	4250.0	12 040.9
434.995	3	1305.0	5994.27
$\sum$	200	78 059	50 382

$\overline{x} = \dfrac{\sum fx}{\sum f} = \dfrac{78\,059}{200} \approx \$390.30$

$s = \sqrt{\dfrac{\sum f(x-\overline{x})^2}{\sum f}} = \sqrt{\dfrac{50\,382}{200}} \approx \$15.87$

b Estimate of μ is $\overline{x} \approx \$390.30$
Estimate of σ is $s \approx \$15.87$

EXERCISE 14G

1 **a** 68% of the players are between $184 - 5 = 179$ cm and $184 + 5 = 189$ cm tall.
$\therefore$ 32% of players are shorter than 179 cm or taller than 189 cm.
$\therefore$ by symmetry, $\frac{32}{2} = 16\%$ of players are taller than 189 cm.

b From **a**, 68% of players are between 179 cm and 189 cm tall, and 16% are taller than 189 cm.
$\therefore$ $68 + 16 = 84\%$ of players are taller than 179 cm tall.

c 95% of players are between $184 - 2 \times 5 = 174$ cm and $184 + 2 \times 5 = 194$ cm tall.
Also, 99.7% of players are between $184 - 3 \times 5 = 169$ cm and $184 + 3 \times 5 = 199$ cm tall.
So, $99.7 - 95 = 4.7\%$ of players are between 169 cm and 174 cm, or between 194 cm and 199 cm.
$\therefore$ by symmetry, 2.35% of players are between 194 cm and 199 cm tall.
$\therefore$ $95 + 2.35 \approx 97.4\%$ of players are between 174 cm and 199 cm tall.

d From **c**, 99.7% of players are between 169 cm and 199 cm tall.
$\therefore$ 0.3% of players are either shorter than 169 cm or taller than 199 cm.
$\therefore$ by symmetry, $\frac{0.3}{2} = 0.15\%$ of players are over 199 cm tall.

2 68% of the time, August rainfall will be between $48 - 6 = 42$ and $48 + 6 = 54$ mm.
$\therefore$ $\dfrac{100-68}{2} = 16\%$ of the time, rainfall is less than 42 mm.
16% of $20 = 3.2$
So, over a 20 year period we would expect less than 42 mm of rain to fall in Claudona in August 3 times.

3 **a** 95% of lifesavers took between $10.5 - 2 \times 0.25 = 10$ and $10.5 + 2 \times 0.25 = 11$ minutes.
$\therefore$ 5% of lifesavers took less than 10 minutes or more than 11 minutes.
$\therefore$ by symmetry, $\frac{5}{2} = 2.5\%$ took over 11 minutes.
$\therefore$ 2.5% of $200 = 5$ lifesavers took longer than 11 minutes.

b 68% of competitors took between 10 min 15 s and 10 min 45 s.

$\therefore$ by symmetry, $\dfrac{100-68}{2} = 16\%$ took less than 10 min 15 s.

$\therefore$ 16% of 200 = 32 competitors took less than 10 minutes 15 seconds.

c 68% took between 10 min 15 s and 10 min 45 s.

$\therefore$ 68% of 200 = 136 competitors took between 10 min 15 s and 10 min 45 s.

4 **a** 68% of babies weighed between $3 - 0.2 = 2.8$ kg and $3 + 0.2 = 3.2$ kg.

$\therefore$ by symmetry, $\dfrac{100-68}{2} = 16\%$ weighed more than 3.2 kg.

$\therefore$ $100 - 16 = 84\%$ weighed less than 3.2 kg.

$\therefore$ 84% of 545 $\approx$ 458 babies weighed less than 3.2 kg.

b From **a**, 68% weigh between 2.8 kg and 3.2 kg.

Also, 95% weigh between $3 - 2 \times 0.2 = 2.6$ kg and $3 + 2 \times 0.2 = 3.4$ kg.

$\therefore$ by symmetry, $\dfrac{95-68}{2} = 13.5\%$ weigh between 3.2 kg and 3.4 kg.

$\therefore$ $68 + 13.5 = 81.5\%$ weigh between 2.8 kg and 3.4 kg.

$\therefore$ 81.5% of 545 $\approx$ 444 babies weighed between 2.8 kg and 3.4 kg.

REVIEW SET 14A

1 **a** Diameter of bacteria colonies

Stem	Leaf
0	4 8 9
1	3 5 5 7
2	1 1 5 6 8 8
3	0 1 2 3 4 5 5 6 6 7 7 9
4	0 1 2 7 9

Scale: 0 | 4 means 0.4 cm

b **i** There are 30 colonies.

$\therefore$ median = average of 15th and 16th colonies

$= \dfrac{3.1 + 3.2}{2} = 3.15$ cm

ii range = $4.9 - 0.4 = 4.5$ cm

c The distribution is negatively skewed.

2 Mean $= \dfrac{5+6+8+3+a+b}{6} = 6$

$\therefore \; 22 + a + b = 36$

$\therefore \; a + b = 14$

$\therefore \; b = 14 - a$

and $s = \sqrt{\dfrac{(-1)^2 + 0^2 + 2^2 + (-3)^2 + (a-6)^2 + (b-6)^2}{6}} = \sqrt{3}$

$\therefore \; \dfrac{1 + 4 + 9 + (a-6)^2 + (8-a)^2}{6} = 3$

$\therefore \; 14 + a^2 - 12a + 36 + 64 - 16a + a^2 = 18$

$\therefore \; 2a^2 - 28a + 96 = 0$

$\therefore \; a^2 - 14a + 48 = 0$

$\therefore \; (a-6)(a-8) = 0$

$\therefore \; a = 6$ or 8 and $b = 8$ or 6

$\therefore \; a = 6,\ b = 8$ or $a = 8,\ b = 6$

3 **a** Both boys and girls have 20 member squads, so the median is the average of the 10th and 11th swimmer.

Girls: median $= \dfrac{36.3+36.3}{2} = 36.3$ s *Boys:* median $= \dfrac{34.8+35.0}{2} = 34.9$ s

So:

	Girls	*Boys*
shape	pos. skewed	approx. symm.
median	36.3 s	34.9 s
range	$41.1 - 33.4 = 7.7$ s	$37.0 - 32.1 = 4.9$ s

b The girls' distribution is positively skewed and the boys' distribution is approximately symmetrical. The median swim time for boys is 1.4 seconds lower than for girls but the range of the girls' swim times is 2.8 seconds higher than for boys. The analysis supports the conjecture that boys generally swim faster than girls with less spread of times.

4 **a** 95% of boys are between $179 - 2\times 8 = 163$ cm and $179 + 2\times 8 = 195$ cm.

$\therefore$ 5% are either shorter than 163 cm or taller than 195 cm.

$\therefore$ by symmetry, $\frac{5}{2} = 2.5\%$ of 17 year old boys are taller than 195 cm.

b From **a**, 95% of 17 year old boys have heights between 163 cm and 195 cm.

c 68% of 17 year old boys have heights between $179 - 8 = 171$ cm and $179 + 8 = 187$ cm.

5 11 12 $\underbrace{12\ 13}$ 14 14 15 15 15 $\underbrace{16\ 17}$ 17 18

11 ↑ min; $Q_1 = 12.5$; 15 ↑ median; $Q_3 = 16.5$; 18 ↑ max

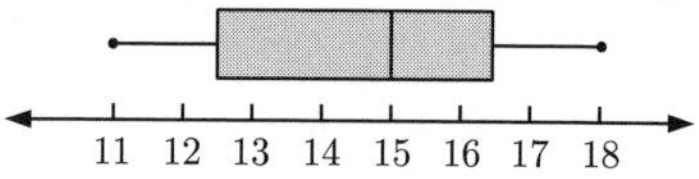

6 **a** median = score for CF of 40
≈ 58.5 seconds

b IQR = (score for CF of 60) – (score for CF of 20)
$\approx 61.5 - 55.5$
≈ 6 seconds

7 **a** 95% of the time the shop sells between $2500 - 2\times 300 = 1900$ and $2500 + 2\times 300 = 3100$ bottles.

$\therefore$ on 5% of days, fewer than 1900 or greater than 3100 bottles are sold.

$\therefore$ by symmetry, $\frac{5}{2} = 2.5\%$ of days see fewer than 1900 bottles sold.

b On 68% of days between $2500 - 300 = 2200$ and $2500 + 300 = 2800$ bottles are sold.

$\therefore$ by symmetry, $\dfrac{100-68}{2} = 16\%$ of days see fewer than 2200 bottles sold.

$\therefore$ on 84% of days, the shop sells more than 2200 bottles.

c From **b**, on 84% of days more than 2200 bottles are sold.
From **a**, on 95% of days between 1900 and 3100 bottles are sold.

$\therefore$ by symmetry, on $\dfrac{100-95}{2} = 2.5\%$ of days more than 3100 bottles are sold.

$\therefore$ on $84 - 2.5 = 81.5\%$ of days between 2200 and 3100 bottles are sold.

8 **a** When $t = 20$, CF ≈ 108 and when $t = 10$, CF ≈ 20. So, approximately $108 - 20 \approx 88$ students spent between 10 and 20 minutes travelling to school.

b If 30% of students spent more than m minutes, 70% of students spent less than m minutes.
70% of 200 students = 140 students.
When CF = 140, $t \approx 24$ minutes $\therefore$ $m \approx 24$.

REVIEW SET 14B

1 **a** highest = 97.5 m, lowest = 64.6 m

b The range = 97.5 − 64.5 = 33

So, if intervals of length 5 are used we need about 7 of them.

We choose $60 \leqslant d < 65$, $65 \leqslant d < 70$, $70 \leqslant d < 75$, and so on.

c

A frequency distribution table for distances thrown by Thabiso

distance (m)	tally	freq. (f)
$60 \leqslant d < 65$	|	1
$65 \leqslant d < 70$	|||	3
$70 \leqslant d < 75$	~~||||~~	5
$75 \leqslant d < 80$	||	2
$80 \leqslant d < 85$	~~||||~~ |||	8
$85 \leqslant d < 90$	~~||||~~ |	6
$90 \leqslant d < 95$	|||	3
$95 \leqslant d < 100$	||	2
	Total	30

d

Frequency histogram displaying the distance Thabiso throws a baseball

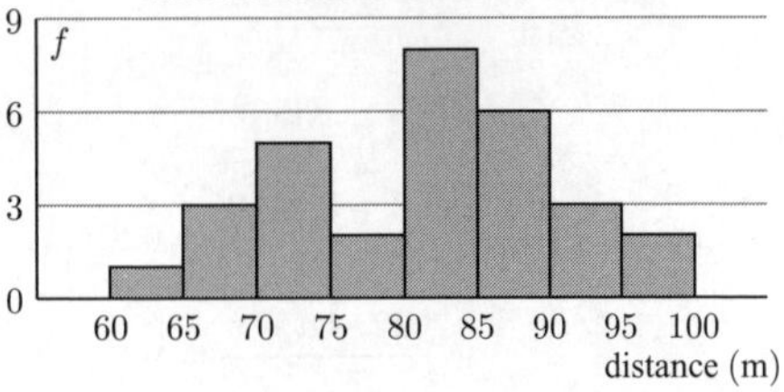

e Using technology:

i $\overline{x} \approx 81.1$ m **ii** median ≈ 83.1 m

2 **a**

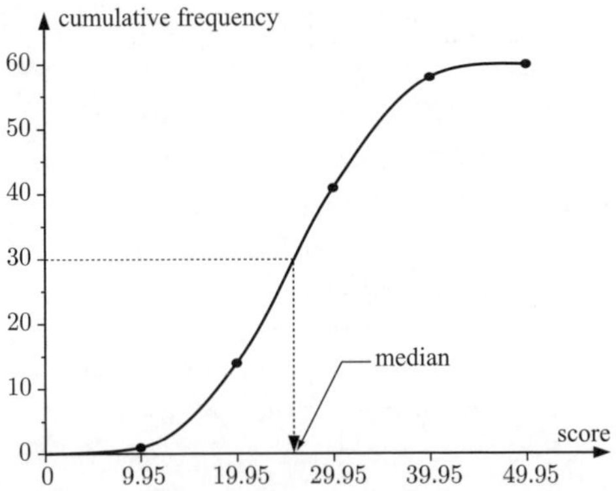

b median ≈ 25.9 (see graph)

c $\text{IQR} = Q_3 - Q_1$

$= (\text{score for CF of } 45)$

$- (\text{score for CF of } 15)$

$\approx 32 - 20 \approx 12$

d

	f	midpt x	fx	$(x-\overline{x})^2$
0 - 9.9	1	4.95	4.95	441
10 - 19.9	13	14.95	194.35	121
20 - 29.9	27	24.95	673.65	1
30 - 39.9	17	34.95	594.15	81
40 - 49.9	2	44.95	89.9	361
	60		1557	

$$\overline{x} = \frac{\sum fx}{\sum x} = \frac{1557}{60} \approx 26.0$$

$$s = \sqrt{\frac{\sum f(x-\overline{x})^2}{\sum f}} = \sqrt{\frac{4140}{60}} \approx 8.31$$

3 Using technology: **a** **i** 101.5 **ii** 98 **iii** 105.5 **b** 7.5 **c** $\overline{x} = 100.2$, $s \approx 7.59$

4 **a** This question could be done using technology or

Litres (x)	f	fx	$f(x-\overline{x})^2$
17.5	5	87.5	1299.38
22.5	13	292.5	1607.71
27.5	17	467.5	636.87
32.5	29	942.5	36.42
37.5	27	1012.5	406.32
42.5	18	765	1419.16
47.5	7	332.5	1348.45
$\sum$	116	3900	6754.31

$$\overline{x} = \frac{\sum fx}{\sum f} = \frac{3900}{116} \approx 33.6 \text{ litres}$$

$$s = \sqrt{\frac{\sum f(x-\overline{x})^2}{\sum f}} \approx \sqrt{\frac{6754.31}{116}} \approx 7.63 \text{ litres}$$

b Estimate of μ is $\overline{x} \approx 33.6$ litres

Estimate of σ is $s \approx 7.63$ litres

5 **a** Using technology, $\overline{x} \approx 49.6$, $s \approx 1.60$.

b This does not justify the claim. A much greater sample is needed.

6 $\underset{\substack{\uparrow \\ \text{min}}}{116}\ \ \underbrace{118\ \ 120}_{\substack{\uparrow \\ Q_1}}\ \ \underbrace{122\ \ 127}_{\substack{\uparrow \\ \text{median}}}\ \ \underbrace{128\ \ 132}_{\substack{\uparrow \\ Q_3}}\ \ \underset{\substack{\uparrow \\ \text{max}}}{135}\qquad (n=8)$

range $= 135 - 116 = 19$

$Q_1 = \dfrac{118+120}{2} = 119$

$Q_3 = \dfrac{128+132}{2} = 130$

$s \approx 6.38$ {technology}

7 $$\text{mean} = \frac{3+a+6+b+13}{5} = 6.8$$

$$\therefore \ \frac{22+a+b}{5} = 6.8$$

$$\therefore \ 22+a+b = 34$$

$$\therefore \ a+b = 12$$

and $$s = \sqrt{\frac{(-3.8)^2+(a-6.8)^2+(-0.8)^2+(b-6.8)^2+6.2^2}{5}} = \sqrt{12.56}$$

$$\therefore \ 14.44+0.64+38.44+(a-6.8)^2+(12-a-6.8)^2 = 5\times 12.56$$

$$\therefore \ 53.52+(a-6.8)^2+(5.2-a)^2 = 62.8$$

$$\therefore \ 53.52+a^2-13.6a+46.24+27.04-10.4a+a^2 = 62.8$$

$$\therefore \ 2a^2-24a+64 = 0$$

$$\therefore \ a^2-12a+32 = 0$$

$$\therefore \ (a-4)(a-8) = 0$$

$$\therefore \ a = 4 \text{ or } 8 \ \text{ and } \ b = 8 \text{ or } 4$$

But $a > b$ $\therefore$ $a = 8$, $b = 4$

REVIEW SET 14C

1

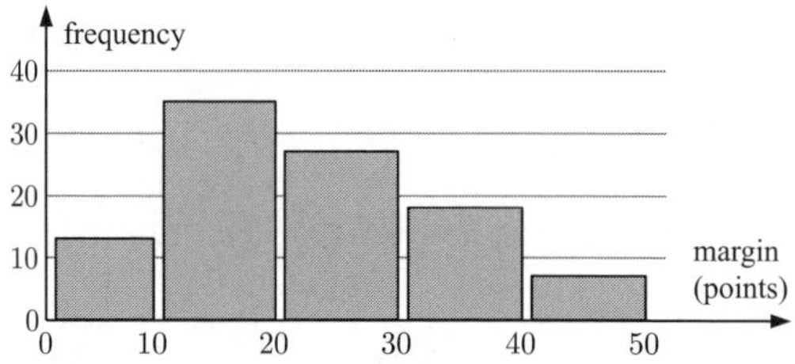

2 **a** $150-138=12$, $162-150=12$, so the values are 1 standard deviation either side of the mean.

$\therefore$ 68% of values lie between 138 and 162.

b 95% of values lie between $150-2\times 12 = 126$ and $150+2\times 12 = 174$.

c From **a**, 68% of values lie between 138 and 162.

From **b**, 95% of values lie between 126 and 174.

$\therefore$ by symmetry, $\dfrac{95-68}{2} = 13.5\%$ of values lie between 162 and 174.

$\therefore$ $95-13.5 = 81.5\%$ of values lie between 126 and 162.

d From **c**, 13.5% of values lie between 162 and 174.

3 Use technology or

Midpoint (x)	f	fx
274.5	14	3843
324.5	34	11 033
374.5	68	25 466
424.5	72	30 564
474.5	54	25 623
524.5	23	12 063.5
574.5	7	4021.5
$\sum$	272	112 614

$$\overline{x} = \frac{\sum fx}{\sum f} = \frac{112\,614}{272}$$

≈ 414 customers

4 **a** Reading from the boxplots

	A	B
Min	11	11.2
Q_1	11.6	12
Median	12	12.6
Q_3	12.6	13.2
Max	13	13.8

b **i** range of A
$= 13 - 11$
$= 2$

range of B
$= 13.8 - 11.2$
$= 2.6$

ii IQR of A
$= 12.6 - 11.6$
$= 1$

IQR of B
$= 13.2 - 12$
$= 1.2$

c **i** The members of squad A generally ran faster because their median time is lower.

ii The times in squad B are more varied because their range and IQR are higher.

5 **a** Using technology with x values 74.995, 84.995, 94.995, and so on,
$\overline{x} \approx$ €103.51 and $s \approx$ €19.40

b Estimate of μ is $\overline{x} \approx$ €103.51
Estimate of σ is $s \approx$ €19.40

6 **a** 68% lies between $\mu - \sigma$ and $\mu + \sigma$, where μ is the mean and σ the standard deviation.

$$\therefore\ \mu = \frac{16.2 + 21.4}{2} = \frac{37.6}{2} = 18.8$$

$\sigma = 21.4 - 18.8 = 2.6$

b The middle 95% of data lies between two standard deviations either side of the mean.
$18.8 + 2 \times 2.6 = 24.0$
$18.8 - 2 \times 2.6 = 13.6$
$\therefore$ the middle 95% of data lies between 13.6 and 24.0

7 Let the number of marks be x.

a When $x = 45$, CF ≈ 120
$\therefore$ about 120 students scored 45 marks or less.

b When CF $= 400$, $x \approx 65$
$\therefore$ the median mark was about 65 marks.

c When CF $= 200$, $x \approx 54$ and when CF $= 600$, $x \approx 75$
$\therefore$ the middle 50% of results lie between 54 and 75 marks.

d IQR $\approx 75 - 54 \approx 21$ marks

e When $x = 55$, CF ≈ 215
$\therefore$ about $\frac{215}{800} \approx 27\%$ of students scored less than 55
$\therefore$ about 73% of students scored 55 or more

f 10% of 800 students = 80 students
When CF $= 800 - 80 = 720$, $x \approx 81$
$\therefore$ a score of 81 marks is required for a 'distinction'.

Chapter 15
PROBABILITY

EXERCISE 15A

1 **a** P(inside a square) $= \frac{113}{145} \approx 0.78$ **b** P(on a line) $= \frac{32}{145} \approx 0.22$

2 Total frequency $= 17 + 38 + 19 + 4 = 78$

a P(20 to 39 seconds) $= \frac{38}{78} \approx 0.487$ **b** P(> 60 seconds) $= \frac{4}{78} \approx 0.051$

c P(between 20 and 59 seconds inclusive) $= \frac{38+19}{78} \approx 0.731$

3

Calls/day	*No. of days*
0	2
1	7
2	11
3	8
4	7
5	4
6	3
7	0
8	1

a Survey lasted $2 + 7 + 11 + 8 + 7 + 4 + 3 + 0 + 1$
$= 43$ days

b **i** P(0 calls) $\approx \frac{2}{43} \approx 0.0465$ **ii** P($\geqslant$ 5 calls) $\approx \frac{4+3+0+1}{43} \approx 0.186$ **iii** P(< 3 calls) $\approx \frac{2+7+11}{43} \approx 0.465$

4 Total frequency
$= 37 + 81 + 48 + 17 + 6 + 1$
$= 190$

a P(4 days gap) $\approx \frac{17}{190} \approx 0.0895$ **b** P(at least 4 days gap) $\approx \frac{17+6+1}{190} \approx 0.126$

EXERCISE 15B

1 **a** {A, B, C, D}

b {BB, BG, GB, GG}

c {ABCD, ABDC, ACBD, ACDB, ADBC, ADCB, BACD, BADC, BCAD, BCDA, BDAC, BDCA, CABD, CADB, CBAD, CBDA, CDAB, CDBA, DABC, DACB, DBAC, DBCA, DCAB, DCBA}

d {GGG, GGB, GBG, BGG, GBB, BGB, BBG, BBB}

2 **a**

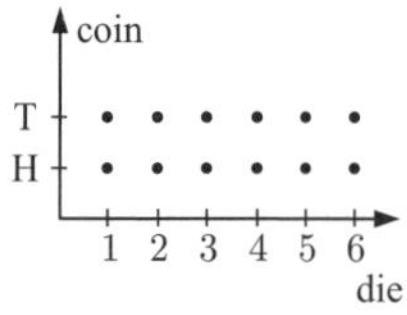

b

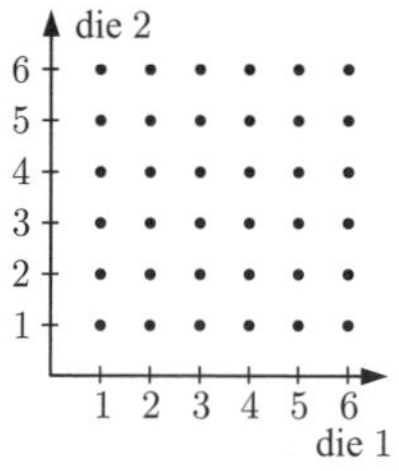

c

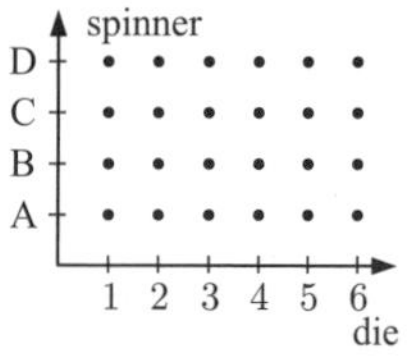

d

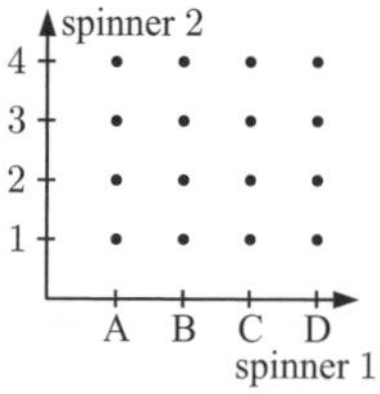

3 **a**

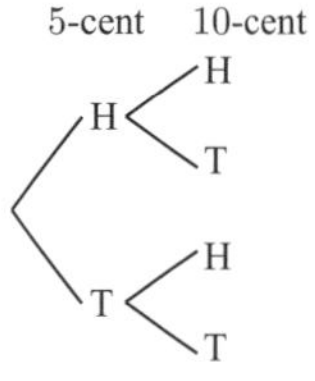

b

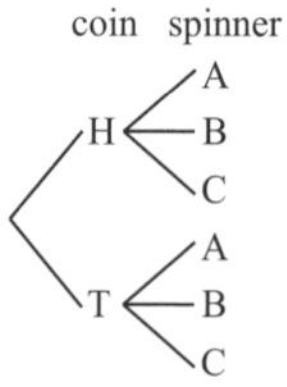

c

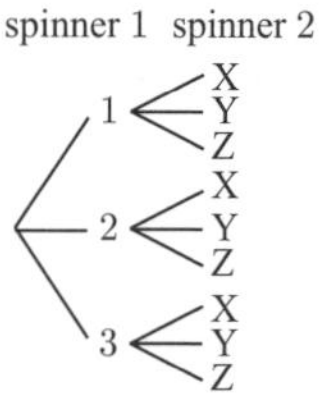

d

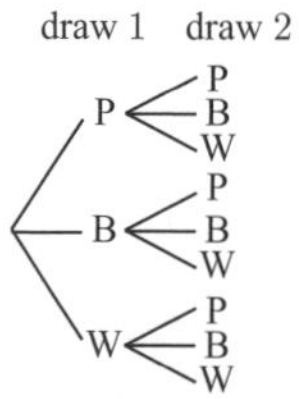

EXERCISE 15C.1

1 Total number of marbles $= 5 + 3 + 7 = 15$

a $P(\text{red}) = \frac{3}{15} = \frac{1}{5}$

b $P(\text{green}) = \frac{5}{15} = \frac{1}{3}$

c $P(\text{blue}) = \frac{7}{15}$

d $P(\text{not red}) = \dfrac{5+7}{15} = \frac{12}{15}$ or $\frac{4}{5}$

e $P(\text{neither green nor blue}) = P(\text{red}) = \frac{1}{5}$

f $P(\text{green or red}) = \dfrac{5+3}{15} = \frac{8}{15}$

2 **a** 8 are brown and so 4 are white.

b **i** $P(\text{brown}) = \frac{8}{12} = \frac{2}{3}$ **ii** $P(\text{white}) = \frac{4}{12} = \frac{1}{3}$

3 **a** $P(\text{multiple of 4})$
$= P(4, 8, 12, 16, 20, 24, 28, 32, 36)$
$= \frac{9}{36}$
$= \frac{1}{4}$

b $P(\text{between 6 and 9 inclusive})$
$= P(6, 7, 8 \text{ or } 9)$
$= \frac{4}{36}$
$= \frac{1}{9}$

c $P(> 20)$
$= P(21, 22, 23, 24, \ldots, 35, 36)$
$= \dfrac{36 - 20}{36}$
$= \frac{16}{36}$
$= \frac{4}{9}$

d $P(9)$
$= \frac{1}{36}$

e $P(\text{multiple of 13})$
$= P(13 \text{ or } 26)$
$= \frac{2}{36}$
$= \frac{1}{18}$

f $P(\text{odd multiple of 3})$
$= P(3, 9, 15, 21, 27, \text{ or } 33)$
$= \frac{6}{36}$
$= \frac{1}{6}$

g $P(\text{multiple of 4 and 6})$
$= P(\text{multiple of 12})$
$= P(12, 24, 36)$
$= \frac{3}{36}$
$= \frac{1}{12}$

h $P(\text{multiple of 4 or 6})$
$= P(4, 6, 8, 12, 16, 18, 20, 24, 28, 30, 32, 36)$
$= \frac{12}{36}$
$= \frac{1}{3}$

4 **a** $P(\text{on Tuesday})$
$= \frac{1}{7}$

b $P(\text{on a weekend})$
$= \frac{2}{7}$

c $P(\text{in July})$
$= \dfrac{4 \times 31}{365 \times 3 + 366}$ {over a 4 year period}
$= \frac{124}{1461}$

d $P(\text{in January or February})$
$= \dfrac{4 \times 31 + 3 \times 28 + 1 \times 29}{3 \times 365 + 1 \times 366}$ {over a 4 year period}
$= \frac{237}{1461} = \frac{79}{487}$

5 Let A denote Antti, K denote Kai and N denote Neda.
Possible orders are: {AKN, ANK, KAN, KNA, NAK, NKA}

a $P(\text{A in middle})$
$= \frac{2}{6}$
$= \frac{1}{3}$

b $P(\text{A at left end})$
$= \frac{2}{6}$
$= \frac{1}{3}$

c $P(\text{A at right end})$
$= \frac{2}{6}$
$= \frac{1}{3}$

d $P(\text{K and N are together}) = \frac{4}{6} = \frac{2}{3}$

6 Let G denote 'a girl' and B denote 'a boy'.

a Possible orders are: {GGG, GGB, GBG, BGG, GBB, BGB, BBG, BBB}

b **i** P(all boys) = P(BBB) = $\frac{1}{8}$

ii P(all girls) = P(GGG) = $\frac{1}{8}$

iii P(boy, then girl, then girl)
= P(BGG)
= $\frac{1}{8}$

iv P(2 girls and a boy)
= P(GGB or GBG or BGG)
= $\frac{3}{8}$

v P(girl is eldest)
= P(GGG or GBG or GBB or GGB)
= $\frac{4}{8} = \frac{1}{2}$

vi P(at least one boy)
= $\frac{7}{8}$ {all except GGG}

7 **a** {ABCD, ABDC, ACBD, ACDB, ADBC, ADCB, BACD, BADC, BCAD, BCDA, BDAC, BDCA, CABD, CADB, CBAD, CBDA, CDAB, CDBA, DABC, DACB, DBAC, DBCA, DCAB, DCBA}

b **i** P(A sits on one end) = $\frac{12}{24} = \frac{1}{2}$

ii P(B sits on one of the two middle seats) = $\frac{12}{24} = \frac{1}{2}$

iii P(A and B are together) = $\frac{12}{24} = \frac{1}{2}$

iv P(A, B and C are together) = $\frac{12}{24} = \frac{1}{2}$

EXERCISE 15C.2

1

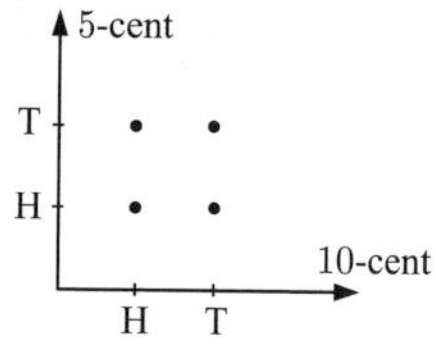

a P(2 heads) = $\frac{1}{4}$

b P(2 tails) = $\frac{1}{4}$

c P(exactly 1 head)
= P(HT or TH)
= $\frac{2}{4}$ or $\frac{1}{2}$

d P(at least one H)
= P(HT or TH or HH)
= $\frac{3}{4}$

2 **a**

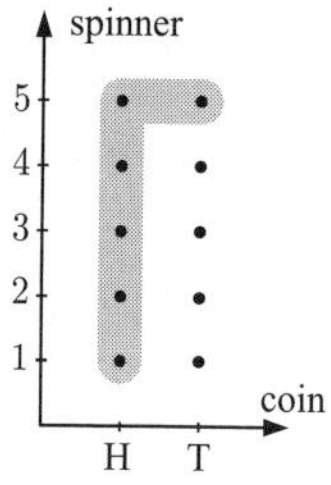

b There are $2 \times 5 = 10$ possible outcomes.

c **i** P(T and 3)
= $\frac{1}{10}$

ii P(H and even)
= P(H2 or H4)
= $\frac{2}{10}$ or $\frac{1}{5}$

iii P(an odd)
= P(H1, T1, H3, T3, H5, T5)
= $\frac{6}{10} = \frac{3}{5}$

iv P(H or 5)
= $\frac{6}{10}$
= $\frac{3}{5}$ {shaded}

3 **a** P(two 3s)
= P((3, 3))
= $\frac{1}{36}$

b P(5 and a 6)
= P((5, 6), (6, 5))
= $\frac{2}{36}$
= $\frac{1}{18}$

c P(5 or a 6)
= $\frac{20}{36}$
= $\frac{5}{9}$

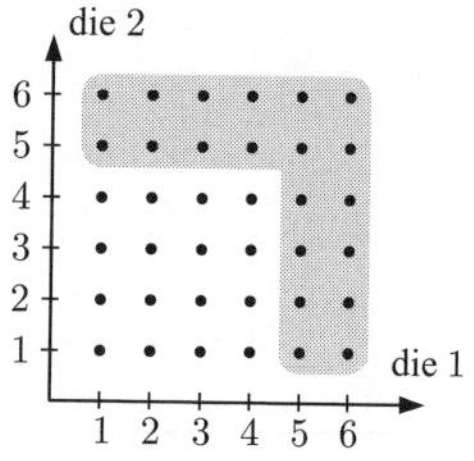

d P(at least one 6)
= $\frac{11}{36}$

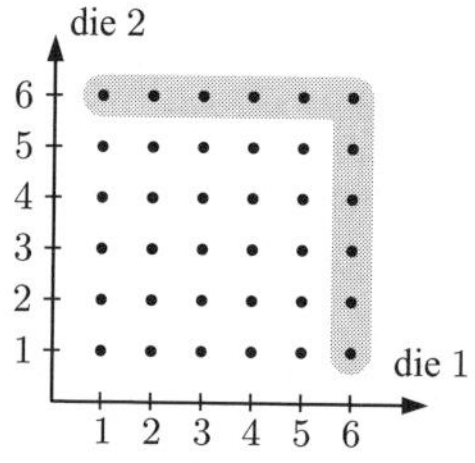

e P(exactly one 6)
= $\frac{10}{36}$
= $\frac{5}{18}$

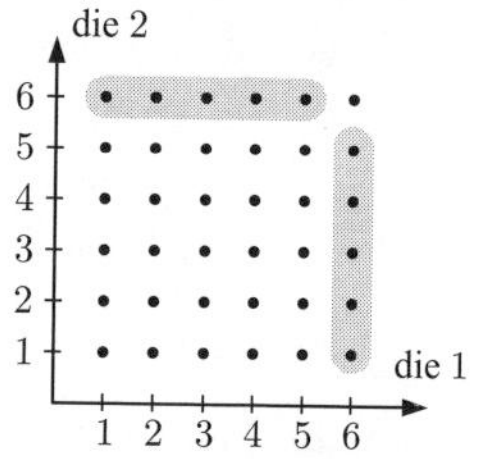

f P(no sixes)

$= \frac{25}{36}$

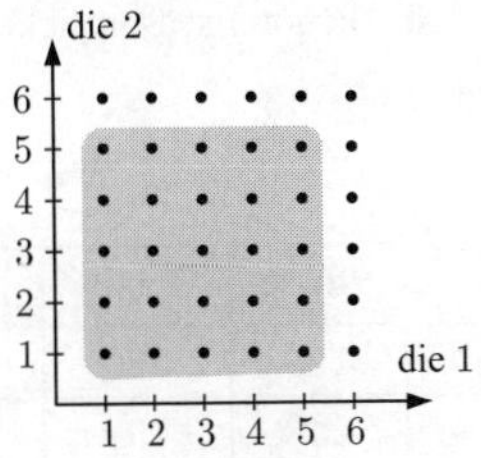

g P(sum of 7)

$= \frac{6}{36}$

$= \frac{1}{6}$

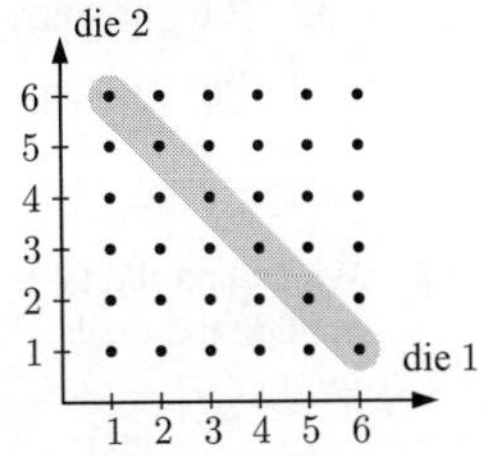

h P(sum > 8)

$= \frac{10}{36}$

$= \frac{5}{18}$

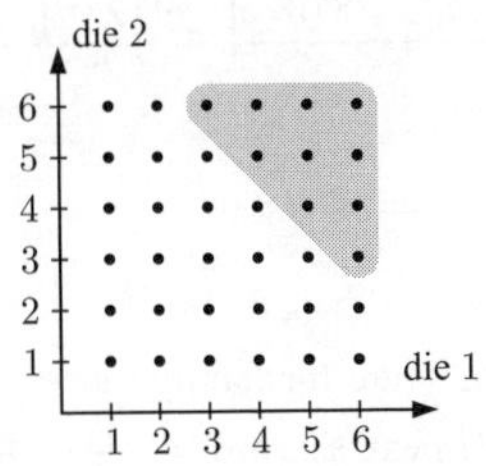

i P(sum of 7 or 11)

$= \dfrac{6+2}{36}$

$= \frac{2}{9}$

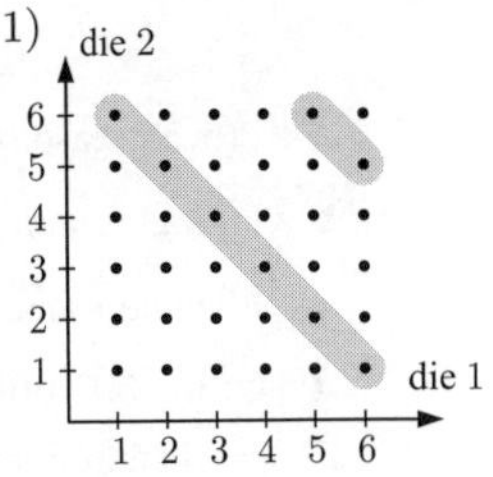

j P(sum no more than 8)

$= \text{P(sum} \leqslant 8)$

$= \frac{26}{36}$

$= \frac{13}{18}$

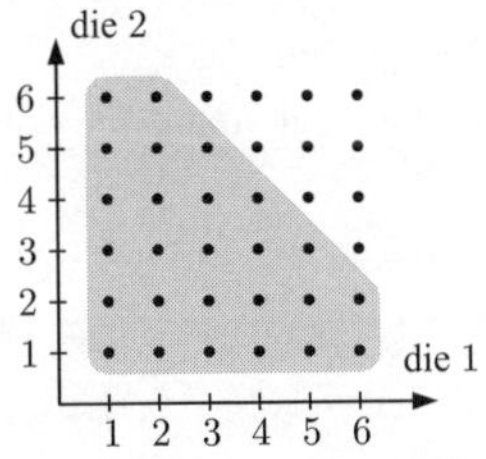

EXERCISE 15D

1 We extend the table to include the totals:

	Employed	Unemployed	Total
Attended university	225	164	389
Did not attend university	197	231	428
Total	422	395	817

a 389 out of the 817 adults surveyed attended university.

$\therefore$ P(attended university) $\approx \frac{389}{817} \approx 0.476$

b 197 out of the 817 adults surveyed did not attend university and are currently employed.

$\therefore$ P(did not attend university and is currently employed) $\approx \frac{197}{817} \approx 0.241$

c 395 out of the 817 adults surveyed were unemployed.

$\therefore$ P(unemployed) $\approx \frac{395}{817} \approx 0.483$

d Of the 389 adults who attended university, 225 are currently employed.

$\therefore$ P(employed given the adult attended university) $\approx \frac{225}{389} \approx 0.578$

e Of the 395 unemployed adults, 164 attended university.

$\therefore$ P(attended university given the adult is currently unemployed) $\approx \frac{164}{395} \approx 0.415$

2 We extend the table to include the totals:

	Adult	Child	Total
Season ticket holder	1824	779	2603
Not a season ticket holder	3247	1660	4907
Total	5071	2439	7510

a Total match attendance was 7510.

b **i** P(child) $= \frac{2439}{7510} \approx 0.325$ **ii** P(not a season ticket holder) $= \frac{4907}{7510} \approx 0.653$

iii P(adult season ticket holder) $= \frac{1824}{7510} \approx 0.243$

3 We extend the table to include the totals:

	Single	Couple	Family	Total
Peak season	125	220	98	443
Off-peak season	248	192	152	592
Total	373	412	250	1035

a P(peak season) $= \frac{443}{1035} \approx 0.428$

b P(single room in the off-peak season) $= \frac{248}{1035} \approx 0.240$

c P(single room or couple room) $= \frac{373+412}{1035} = \frac{785}{1035} \approx 0.758$

d Of the 592 off-peak season bookings, 152 were for family rooms.

$\therefore$ P(family room given it was in the off-peak season) $= \frac{152}{592} \approx 0.257$

e $412 + 250 = 662$ bookings were not for a single room. Of these, $220 + 98 = 318$ were in the peak season.

$\therefore$ P(peak season given it was not a single room) $= \frac{318}{662} \approx 0.480$

EXERCISE 15E.1

1 **a** P(rains on any one day)
$= \frac{6}{7}$

b P(rains on 2 successive days)
$=$ P(R and R)
$= \frac{6}{7} \times \frac{6}{7}$
$= \frac{36}{49}$

c P(rains on 3 successive days)
$=$ P(R and R and R)
$= \frac{6}{7} \times \frac{6}{7} \times \frac{6}{7}$ or $\frac{216}{343}$

2 **a** P(H, then H, then H)
$=$ P(H and H and H)
$= \frac{1}{2} \times \frac{1}{2} \times \frac{1}{2}$
$= \frac{1}{8}$

b P(T, then H, then T)
$=$ P(T and H and T)
$= \frac{1}{2} \times \frac{1}{2} \times \frac{1}{2}$
$= \frac{1}{8}$

3 Let A be the event of photocopier A malfunctioning and B be the event of photocopier B malfunctioning.

a P(both malfunction)
$= \text{P}(A \text{ and } B)$
$= 0.08 \times 0.12$
$= 0.0096$

b P(both work)
$= \text{P}(A' \text{ and } B')$
$= 0.92 \times 0.88$
$= 0.8096$

4 **a** P(they will be happy)
$=$ P(B, then G, then B, then G)
$=$ P(B and G and B and G)
$= \frac{1}{2} \times \frac{1}{2} \times \frac{1}{2} \times \frac{1}{2}$
$= \frac{1}{16}$

b P(they will be unhappy)
$= 1 -$ P(they will be happy)
$= 1 - \frac{1}{16}$
$= \frac{15}{16}$

5 Let J be the event of Jiri hitting the target and B be the event of Benita hitting the target.

a P(both hit) $= \text{P}(JB)$ $= 0.7 \times 0.8$ $= 0.56$

b P(both miss) $= \text{P}(J'B')$ $= 0.3 \times 0.2$ $= 0.06$

c P(J hits and B misses) $= \text{P}(JB')$ $= 0.7 \times 0.2$ $= 0.14$

d P(B hits and J misses) $= \text{P}(BJ')$ $= 0.8 \times 0.3$ $= 0.24$

6 Let H be the event the archer hits the target. $\therefore\ \text{P}(H) = \frac{2}{5},\ \ \text{P}(H') = \frac{3}{5}$

a P(3 hits) $= \text{P}(HHH)$ $= \frac{2}{5} \times \frac{2}{5} \times \frac{2}{5}$ $= \frac{8}{125}$

b P(2 hits then a miss) $= \text{P}(HHH')$ $= \frac{2}{5} \times \frac{2}{5} \times \frac{3}{5}$ $= \frac{12}{125}$

c P(all misses) $= \text{P}(H'H'H')$ $= \frac{3}{5} \times \frac{3}{5} \times \frac{3}{5}$ $= \frac{27}{125}$

EXERCISE 15E.2

1 **a** P(all strawberry creams) $= \text{P(SSS)}$ $= \frac{8}{12} \times \frac{7}{11} \times \frac{6}{10}$ $= \frac{14}{55}$

b P(none is a strawberry cream) $= \text{P(S}'\text{S}'\text{S}')$ $= \frac{4}{12} \times \frac{3}{11} \times \frac{2}{10}$ $= \frac{1}{55}$

2 **a** P(both red) $= \text{P(RR)}$ $= \frac{7}{10} \times \frac{6}{9}$ $= \frac{7}{15}$

b P(GR) $= \frac{3}{10} \times \frac{7}{9}$ $= \frac{7}{30}$

c P(a green and a red) $= \text{P(GR or RG)}$ $= \frac{3}{10} \times \frac{7}{9} + \frac{7}{10} \times \frac{3}{9}$ $= \frac{7}{15}$

3 **a** P(wins first prize) $= \frac{3}{100}$

b P(wins 1st and 2nd) $= \text{P(WW)}$ $= \frac{3}{100} \times \frac{2}{99}$ $\approx 0.000\,606$

c P(wins all 3) $= \text{P(WWW)}$ $= \frac{3}{100} \times \frac{2}{99} \times \frac{1}{98}$ $\approx 0.000\,006\,18$

d P(wins none of them) $= \text{P(W}'\text{W}'\text{W}')$ $= \frac{97}{100} \times \frac{96}{99} \times \frac{95}{98}$ ≈ 0.912

4 **a** P(does not contain captain) $= \text{P(C}'\text{C}'\text{C}')$ $= \frac{6}{7} \times \frac{5}{6} \times \frac{4}{5}$ $= \frac{4}{7}$

b P(does not contain captain or vice captain) $= \text{P(OOO)}$ {O ≡ other} $= \frac{5}{7} \times \frac{4}{6} \times \frac{3}{5}$ $= \frac{2}{7}$

EXERCISE 15F

1 **a**

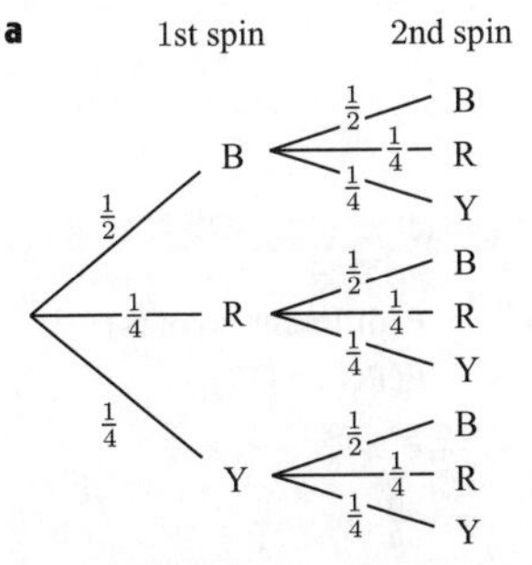

b P(both black) $= \text{P(BB)}$ $= \frac{1}{2} \times \frac{1}{2}$ $= \frac{1}{4}$

c P(both yellow) $= \text{P(YY)}$ $= \frac{1}{4} \times \frac{1}{4}$ $= \frac{1}{16}$

d P(both different)
$= \text{P(BR or BY or RB or RY or YB or YR)}$
$= \frac{1}{2} \times \frac{1}{4} + \frac{1}{2} \times \frac{1}{4} + \frac{1}{4} \times \frac{1}{2} + \frac{1}{4} \times \frac{1}{4} + \frac{1}{4} \times \frac{1}{2} + \frac{1}{4} \times \frac{1}{4}$
$= \frac{4}{8} + \frac{2}{16}$
$= \frac{5}{8}$

e P(B appears on either spin)
$= \text{P(BB or BR or BY or RB or YB)}$
$= \frac{1}{2} \times \frac{1}{2} + \frac{1}{2} \times \frac{1}{4} + \frac{1}{2} \times \frac{1}{4} + \frac{1}{4} \times \frac{1}{2} + \frac{1}{4} \times \frac{1}{2}$
$= 4(\frac{1}{8}) + \frac{1}{4}$
$= \frac{3}{4}$

2

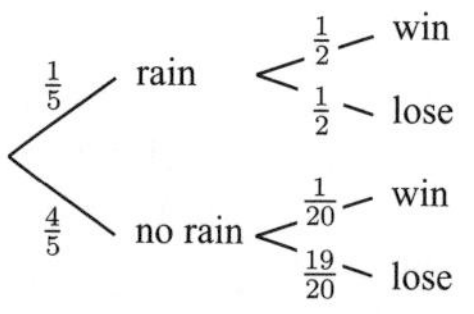

P(Mudlark wins)
$= \text{P(rain and win or no rain and win)}$
$= \frac{1}{5} \times \frac{1}{2} + \frac{4}{5} \times \frac{1}{20}$
$= \frac{1}{10} + \frac{4}{100}$
$= \frac{14}{100}$
$= \frac{7}{50}$

3 P(next is spoiled) $= \text{P(from A and spoiled or from B and spoiled)}$
$= 0.4 \times 0.05 + 0.6 \times 0.02$
$= 0.020 + 0.012$
$= 0.032 \quad (3.2\%)$

4

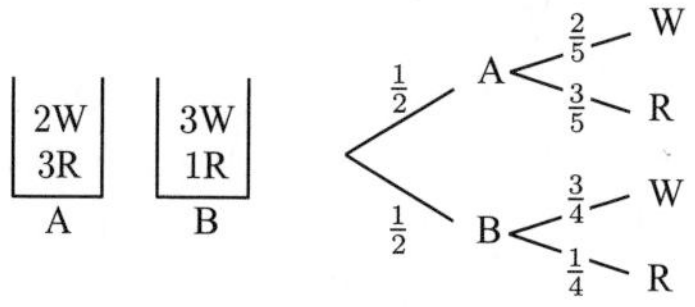

P(red)
$= \text{P(A and red or B and red)}$
$= \frac{1}{2} \times \frac{3}{5} + \frac{1}{2} \times \frac{1}{4}$
$= \frac{3}{10} + \frac{1}{8}$
$= \frac{17}{40}$

5 There are 7 teams above Tottenham and 12 teams below Tottenham.

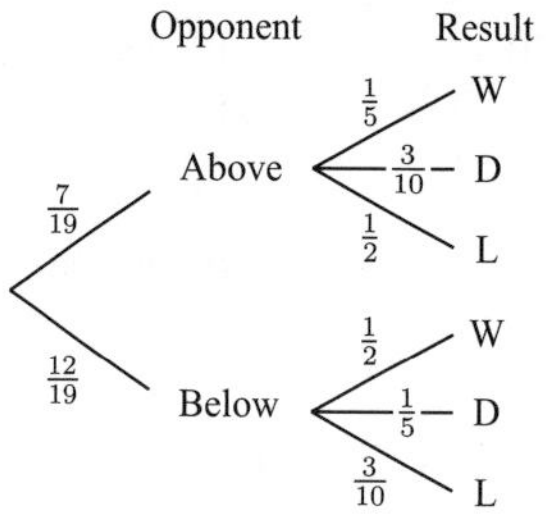

$\therefore$ P(Draw)
$= \frac{7}{19} \times \frac{3}{10} + \frac{12}{19} \times \frac{1}{5}$
$= \frac{21}{190} + \frac{24}{190}$
$= \frac{9}{38}$

6

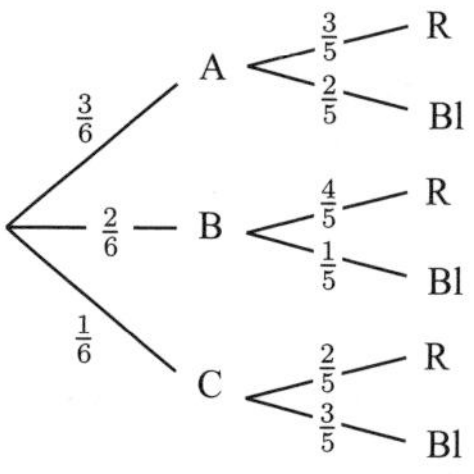

a P(blue) $= \text{P(A and Bl or B and Bl or C and Bl)}$
$= \frac{3}{6} \times \frac{2}{5} + \frac{2}{6} \times \frac{1}{5} + \frac{1}{6} \times \frac{3}{5}$
$= \frac{11}{30}$

b P(red) $= 1 - \text{P(blue)}$
$= 1 - \frac{11}{30}$
$= \frac{19}{30}$

EXERCISE 15G

1

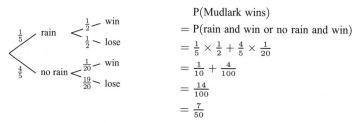

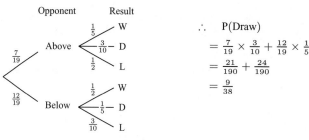

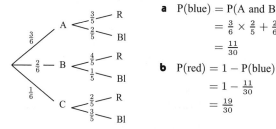

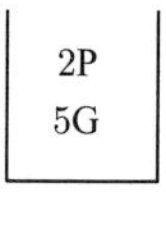

a P(different colours)
$= \text{P(PG or GP)}$
$= \frac{2}{7} \times \frac{5}{7} + \frac{5}{7} \times \frac{2}{7}$
$= \frac{20}{49}$

b P(different colours)
$= \text{P(PG or GP)}$
$= \frac{2}{7} \times \frac{5}{6} + \frac{5}{7} \times \frac{2}{6}$
$= \frac{20}{42}$ or $\frac{10}{21}$

2 **a** P(both odd)
$= \text{P(odd and odd)}$
$= \frac{3}{5} \times \frac{2}{4}$
$= \frac{3}{10}$

b P(both even)
$= \text{P(even and even)}$
$= \frac{2}{5} \times \frac{1}{4}$
$= \frac{1}{10}$

c P(one odd and other even)
$= 1 - \text{P(both odd)} - \text{P(both even)}$
$= 1 - \frac{3}{10} - \frac{1}{10}$
$= \frac{6}{10}$
$= \frac{3}{5}$

3

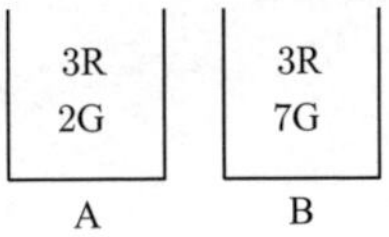

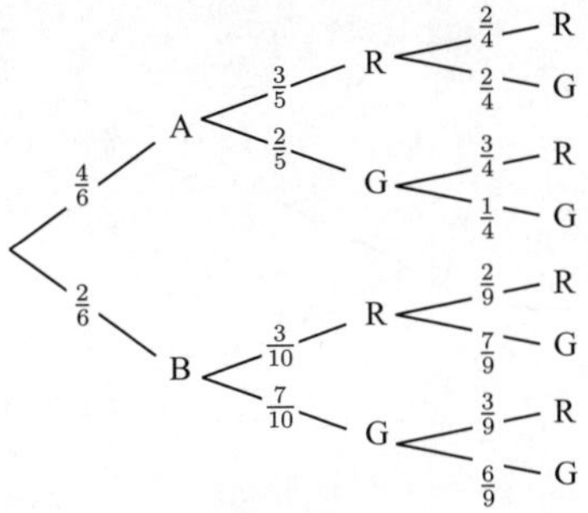

a P(both green)
$= \text{P(AGG or BGG)}$
$= \frac{4}{6} \times \frac{2}{5} \times \frac{1}{4} + \frac{2}{6} \times \frac{7}{10} \times \frac{6}{9}$
$= \frac{1}{15} + \frac{7}{45}$
$= \frac{10}{45}$
$= \frac{2}{9}$

b P(different in colour)
$= 1 - \text{P(both green)} - \text{P(both red)}$
$= 1 - \frac{2}{9} - \text{P(ARR or BRR)}$
$= \frac{7}{9} - (\frac{4}{6} \times \frac{3}{5} \times \frac{2}{4} + \frac{2}{6} \times \frac{3}{10} \times \frac{2}{9})$
$= \frac{7}{9} - (\frac{1}{5} + \frac{1}{45})$
$= \frac{5}{9}$

4

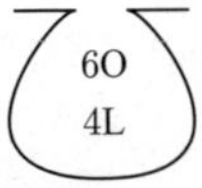

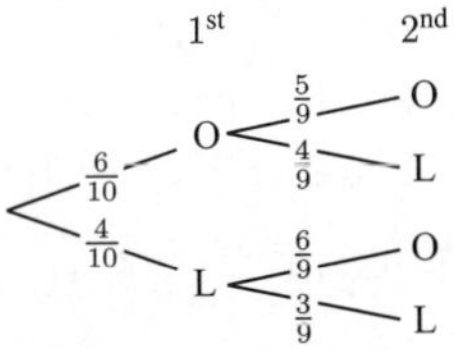

a P(both O)
$= \frac{6}{10} \times \frac{5}{9}$
$= \frac{1}{3}$

b P(both L)
$= \frac{4}{10} \times \frac{3}{9}$
$= \frac{2}{15}$

c P(OL)
$= \frac{6}{10} \times \frac{4}{9}$
$= \frac{4}{15}$

d P(LO)
$= \frac{4}{10} \times \frac{6}{9}$
$= \frac{4}{15}$

$\frac{1}{3} + \frac{2}{15} + \frac{4}{15} + \frac{4}{15}$
$= \frac{5}{15} + \frac{2}{15} + \frac{4}{15} + \frac{4}{15}$
$= \frac{15}{15}$ which is 1

The answer must be 1 as the four categories **a**, **b**, **c**, **d** are all the possibilities that could occur.

5

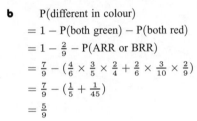

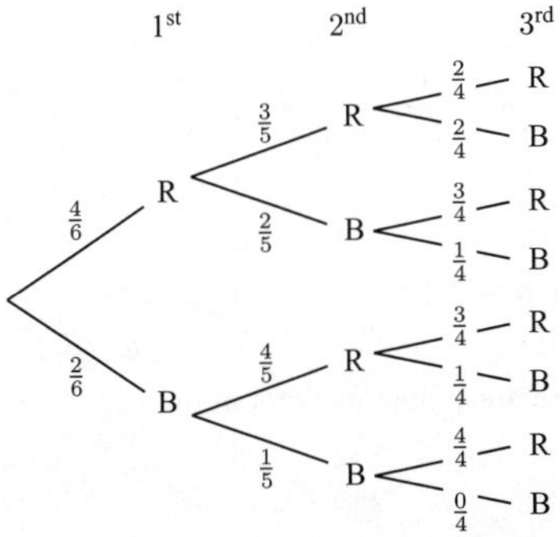

a P(all red)
$= \text{P(RRR)}$
$= \frac{4}{6} \times \frac{3}{5} \times \frac{2}{4}$
$= \frac{1}{5}$

b P(only two are red)
$= \text{P(RRB or RBR or BRR)}$
$= \frac{4}{6} \times \frac{3}{5} \times \frac{2}{4} + \frac{4}{6} \times \frac{2}{5} \times \frac{3}{4} + \frac{2}{6} \times \frac{4}{5} \times \frac{3}{4}$
$= 3 \times \left(\frac{24}{6 \times 5 \times 4}\right)$
$= \frac{3}{5}$

c P(at least two are red)
$= \text{P(all red or only two are red)}$
$= \frac{1}{5} + \frac{3}{5}$ {from **a** and **b**}
$= \frac{4}{5}$

6

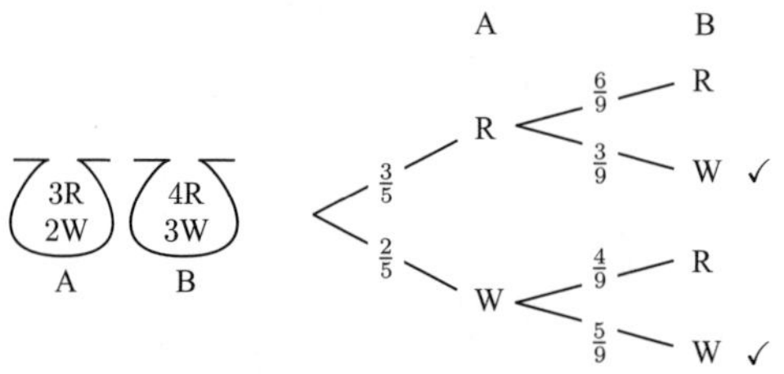

P(marble from B is W)
$= \text{P(RW or WW)}$ {paths ticked}
$= \frac{3}{5} \times \frac{3}{9} + \frac{2}{5} \times \frac{5}{9}$
$= \frac{19}{45}$

7

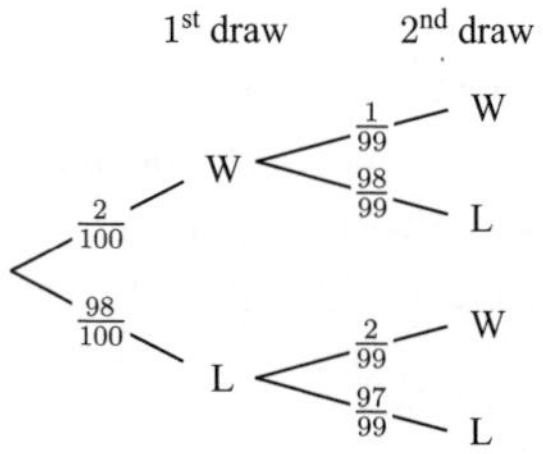

a P(wins both)
$= \text{P(WW)}$
$= \frac{2}{100} \times \frac{1}{99}$
$\approx 0.000\,202$

b P(wins neither)
$= \text{P(LL)}$
$= \frac{98}{100} \times \frac{97}{99}$
≈ 0.960

c P(wins at least one prize) $= 1 - \text{P(wins neither)}$
$= 1 - \frac{98}{100} \times \frac{97}{99}$
≈ 0.0398

8

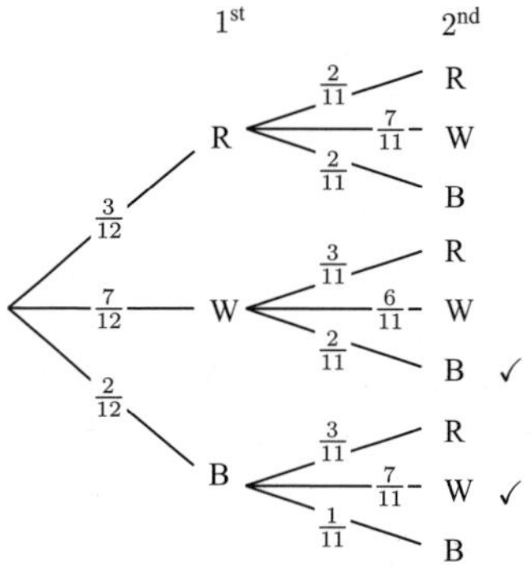

P(one white and one black)
$= \text{P(WB or BW)}$ {paths ticked}
$= \frac{7}{12} \times \frac{2}{11} + \frac{2}{12} \times \frac{7}{11}$
$= \frac{7}{33}$

9 There are $(n + 7)$ markers in total.

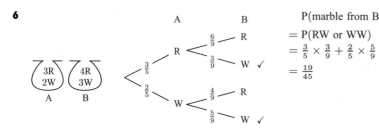

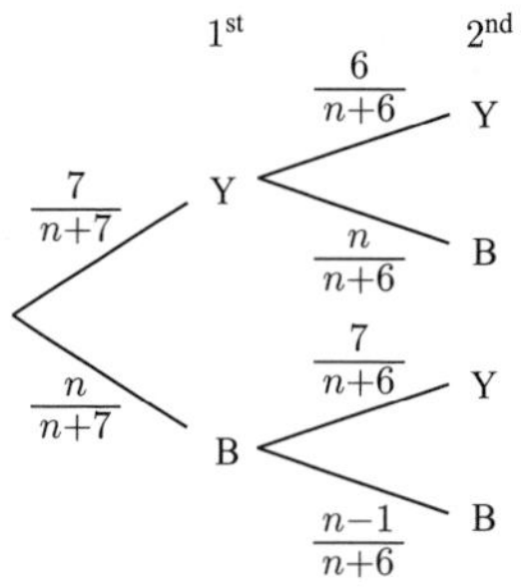

$\text{P(YY)} = \frac{3}{13}$
$\therefore \frac{7}{n+7} \times \frac{6}{n+6} = \frac{3}{13}$
$\therefore \frac{42}{n^2 + 13n + 42} = \frac{3}{13}$
$\therefore 546 = 3n^2 + 39n + 126$
$\therefore 3(n^2 + 13n - 140) = 0$
$\therefore 3(n - 7)(n + 20) = 0$
$\therefore n = 7$ $\{n \geqslant 0\}$
$\therefore$ there are 7 blue markers in the bag.

EXERCISE 15H

1 **a** $(p+q)^4 = p^4 + 4p^3q + 6p^2q^2 + 4pq^3 + q^4$

b $\text{P(3 heads)} = 4p^3q$
$= 4\left(\frac{1}{2}\right)^3\left(\frac{1}{2}\right) \quad \{\text{as } p = q = \frac{1}{2}\}$
$= \frac{1}{4}$

2 **a** $(p+q)^5 = p^5 + 5p^4q + 10p^3q^2 + 10p^2q^3 + 5pq^4 + q^5$

b **i** P(4H and 1T)
$= 5p^4q$
$= 5\left(\frac{1}{2}\right)^4\left(\frac{1}{2}\right)$
$= \frac{5}{32}$

ii P(2H and 3T)
$= 10p^2q^3$
$= 10\left(\frac{1}{2}\right)^2\left(\frac{1}{2}\right)^3$
$= \frac{10}{32}$
$= \frac{5}{16}$

iii P(HHHHT)
$= \left(\frac{1}{2}\right)^4 \times \frac{1}{2}$
$= \frac{1}{32}$

3 **a** $\left(\frac{2}{3}+\frac{1}{3}\right)^4 = \left(\frac{2}{3}\right)^4 + 4\left(\frac{2}{3}\right)^3\left(\frac{1}{3}\right) + 6\left(\frac{2}{3}\right)^2\left(\frac{1}{3}\right)^2 + 4\left(\frac{2}{3}\right)\left(\frac{1}{3}\right)^3 + \left(\frac{1}{3}\right)^4$

b $\text{P(S)} = \frac{2}{3}, \quad \text{P(S}') = \frac{1}{3}$ $\quad$ S′ represents an almond centre

i P(all S)
$= \left(\frac{2}{3}\right)^4$
$= \frac{16}{81}$

ii P(two of each)
$= 6\left(\frac{2}{3}\right)^2\left(\frac{1}{3}\right)^2$
$= \frac{8}{27}$

iii P(at least 2 strawberry creams)
$= $ P(all S or 3S, 1S′ or 2S, 2S′)
$= \left(\frac{2}{3}\right)^4 + 4\left(\frac{2}{3}\right)^3\left(\frac{1}{3}\right) + 6\left(\frac{2}{3}\right)^2\left(\frac{1}{3}\right)^2$
$= \frac{16}{81} + \frac{32}{81} + \frac{24}{81}$
$= \frac{72}{81}$
$= \frac{8}{9}$

4 **a** $\left(\frac{3}{4}+\frac{1}{4}\right)^5 = \left(\frac{3}{4}\right)^5 + 5\left(\frac{3}{4}\right)^4\left(\frac{1}{4}\right)^1 + 10\left(\frac{3}{4}\right)^3\left(\frac{1}{4}\right)^2 + 10\left(\frac{3}{4}\right)^2\left(\frac{1}{4}\right)^3 + 5\left(\frac{3}{4}\right)\left(\frac{1}{4}\right)^4 + \left(\frac{1}{4}\right)^5$

b P('normal' kiwi) $= \frac{3}{4}$, P('flat back') $= \frac{1}{4}$

i P(2 'flat backs')
$= $ P(3F′, 2F)
$= 10 \times \left(\frac{3}{4}\right)^3\left(\frac{1}{4}\right)^2$
$= \frac{135}{512}$

ii P(at least 3 'flat backs')
$= $ P(2F′, 3F or 1F′, 4F or 5F)
$= 10\left(\frac{3}{4}\right)^2\left(\frac{1}{4}\right)^3 + 5\left(\frac{3}{4}\right)\left(\frac{1}{4}\right)^4 + \left(\frac{1}{4}\right)^5$
$= \frac{53}{512}$ on simplifying

iii P(at most 3 'normal' kiwis) $= 1 -$ P(4 or 5 normal kiwis)
$= 1 -$ P(4F′, 1F or 5F′)
$= 1 - \left(5\left(\frac{3}{4}\right)^4\left(\frac{1}{4}\right) + \left(\frac{3}{4}\right)^5\right)$
$= \frac{47}{128}$

5 Let X be the number of Huy's hits.

a Using the binomial expansion,
$P(X=2) = 6\left(\frac{4}{5}\right)^2\left(\frac{1}{5}\right)^2 \approx 0.154$

b $P(X \geqslant 2)$
$= 1 - P(X \leqslant 1)$
$\approx 1 - \left(4\left(\frac{4}{5}\right)\left(\frac{1}{5}\right)^3 + \left(\frac{1}{5}\right)^4\right)$
≈ 0.973

6 Let X be the number of defective light bulbs.

a $P(X=2) \approx 0.0305$ {using technology}

b $P(X \geqslant 1)$
$= 1 - P(X=0)$
$\approx 1 - 0.735$
≈ 0.265

7 If X is the number of questions Raj answers correctly, then X is binomial. There are $n = 10$ independent trials with probability $p = \frac{1}{5}$ of a corrrect answer for each.

$$\begin{aligned} \text{P(Raj passes)} &= \text{P}(X \geqslant 7) \\ &= 1 - \text{P}(X \leqslant 6) \\ &\approx 1 - 0.999\,136 \\ &\approx 0.000\,864 \quad \{\text{or about 9 in } 10\,000\} \end{aligned}$$

8 P(M wins a game against J) $= \frac{2}{3}$ $\quad\therefore$ P(M wins) $= \frac{2}{3}$ $\quad$ P(J wins) $= \frac{1}{3}$

$$\begin{aligned} \text{P(J wins a set 6 games to 4)} &= \text{P}(\underbrace{\text{J wins 5 of the first 9 games}}_{\text{this is binomial with } n = 9 \text{ trials of probability } p = \frac{1}{3}} \textbf{ and } \text{J wins the 10th game}) \\ &\approx 0.1024 \times \tfrac{1}{3} \\ &\approx 0.0341 \end{aligned}$$

9 If there are n dice thrown, $\text{P(no sixes)} = \left(\frac{5}{6}\right)^n$

$\therefore$ $\text{P(at least 1 six)} = 1 - \left(\frac{5}{6}\right)^n$

$\therefore$ need to find the smallest integer n such that

$$\begin{aligned} 1 - \left(\tfrac{5}{6}\right)^n &\geqslant 0.5 \\ \therefore \quad \left(\tfrac{5}{6}\right)^n &\leqslant 0.5 \\ \therefore \quad n \log\left(\tfrac{5}{6}\right) &\leqslant \log(0.5) \\ \therefore \quad n &\geqslant \frac{\log(0.5)}{\log\left(\frac{5}{6}\right)} \qquad \{\log\left(\tfrac{5}{6}\right) < 0\} \\ \therefore \quad n &\geqslant 3.80 \end{aligned}$$

$\therefore$ at least 4 dice are needed.

EXERCISE 15I.1

1 **a** $A = \{1, 2, 3, 6\}$, $B = \{2, 4, 6, 8, 10\}$

b **i** $n(A) = 4$ $\quad$ **ii** $A \cup B = \{1, 2, 3, 4, 6, 8, 10\}$ $\quad$ **iii** $A \cap B = \{2, 6\}$

2 **a**

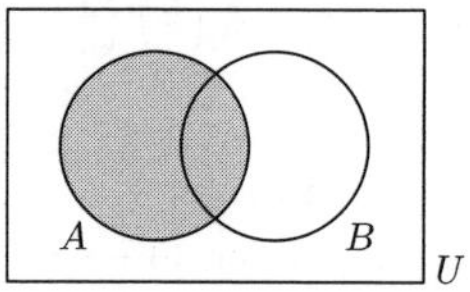

b

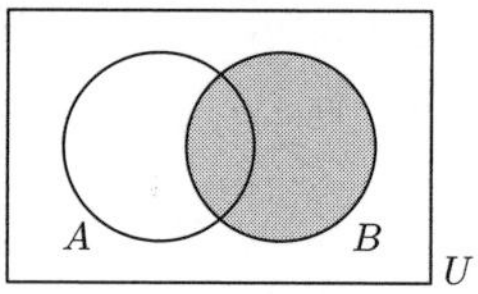

c

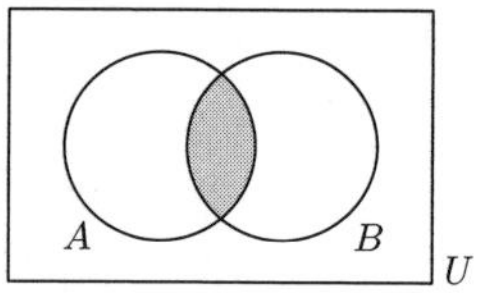

d

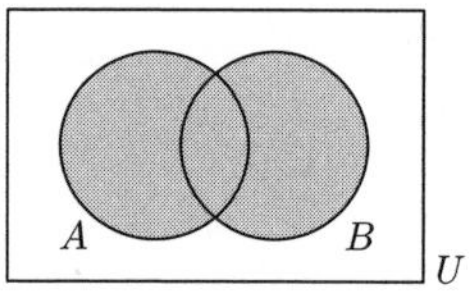

e

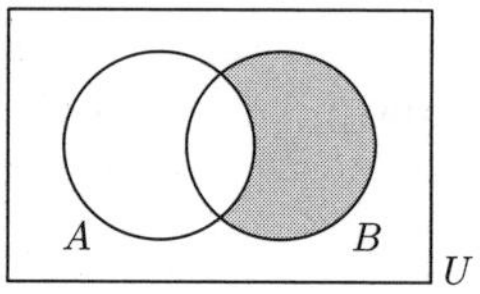

f

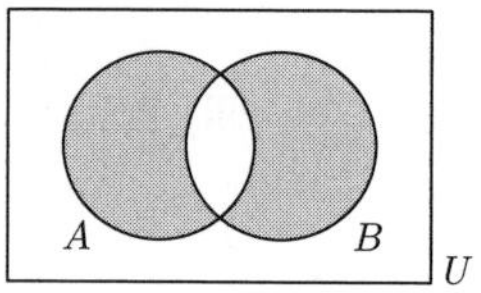

3 **a** Total number in the class $= 3 + 5 + 17 + 4 = 29$

b Number who study both $= 17$ $\quad$ {the intersection}

c Number who study at least one $= 5 + 17 + 4 = 26$ $\quad$ {the union}

d Number who study only Chemistry $= 5$

4 **a** Total number in the survey $= 37 + 9 + 15 + 4 = 65$

b Number who liked both $= 9$ $\quad$ {the intersection}

c Number who liked neither $= 4$

d Number who liked exactly one $= 37 + 15 = 52$

5

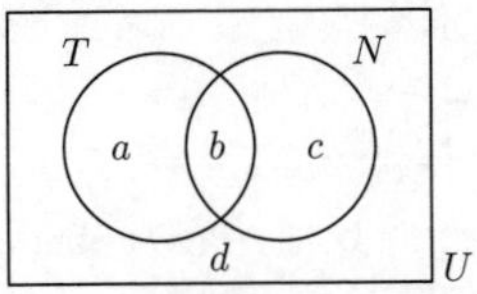

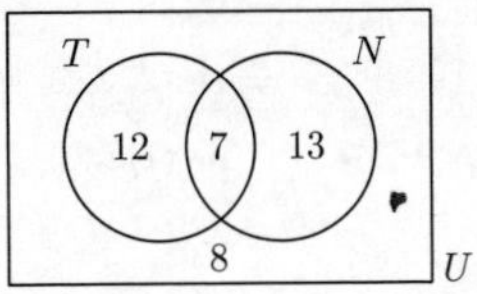

T represents those playing tennis
N represents those playing netball

$$\therefore \begin{cases} a+b+c+d=40 \\ a+b=19 \\ b+c=20 \\ d=8 \end{cases}$$

So, $a+b+c=32$

$\therefore \quad 19+c=32$ and $a+20=32$

$\therefore \quad c=13$ and $a=12$

Hence, $12+b=19$

$\therefore \quad b=7$

a P(plays tennis)
$= \dfrac{12+7}{40}$
$= \frac{19}{40}$

b P(does not play netball)
$= \dfrac{12+8}{40}$
$= \frac{1}{2}$

c P(plays at least one)
$= \dfrac{12+7+13}{40}$
$= \frac{32}{40}$
$= \frac{4}{5}$

d P(plays one and only one)
$= \dfrac{12+13}{40}$
$= \frac{25}{40}$
$= \frac{5}{8}$

e P(plays netball, but not tennis) $= \frac{13}{40}$

f P(plays tennis given plays netball)
$= \dfrac{7}{7+13}$
$= \frac{7}{20}$

6

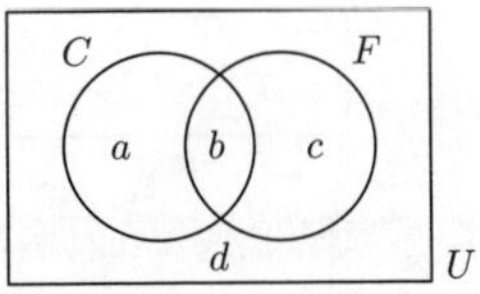

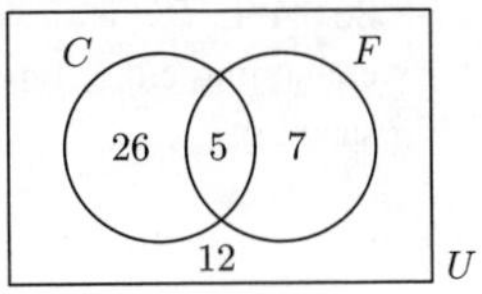

C represents men who gave chocolates.
F represents men who gave flowers.

$$\therefore \begin{cases} a+b+c+d=50 \\ a+b=31 \\ b+c=12 \\ b=5 \end{cases}$$

Thus $c=7$, $a=26$ and $26+5+7+d=50 \quad \therefore \quad d=12$

a P(C or F)
$= \dfrac{26+5+7}{50}$
$= \frac{38}{50}$ or $\frac{19}{25}$

b P(C but not F)
$= \frac{26}{50}$
$= \frac{13}{25}$

c P(neither C nor F)
$= \frac{12}{50}$
$= \frac{6}{25}$

d P(F given that C')
$= \dfrac{7}{7+12}$
$= \frac{7}{19}$

7

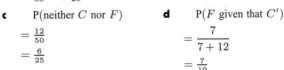

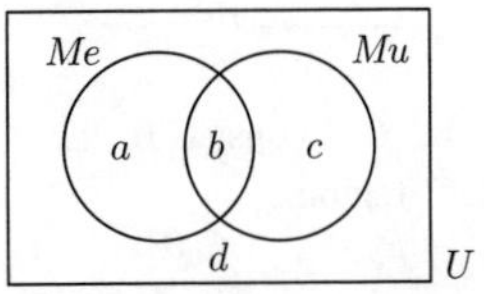

Me represents children who had measles.
Mu represents children who had mumps.

$$\therefore \begin{cases} a+b+c+d=30 \\ a+b=24 \\ b=12 \\ a+b+c=26 \end{cases}$$

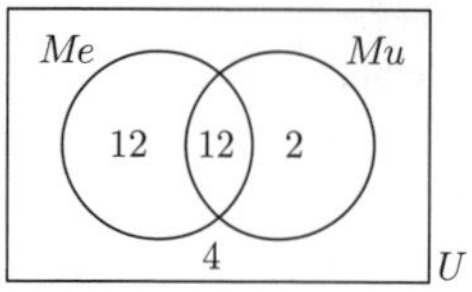

$\therefore \quad 26 + d = 30 \quad \therefore \quad d = 4$

$24 + c = 26 \quad \therefore \quad c = 2$

and $\quad a + 12 = 24 \quad \therefore \quad a = 12$

a $\text{P}(Mu) = \frac{14}{30} = \frac{7}{15}$

b $\text{P}(Mu\text{, but not } Me) = \frac{2}{30} = \frac{1}{15}$

c $\text{P}(\text{neither } Mu \text{ nor } Me) = \frac{4}{30} = \frac{2}{15}$

d $\text{P}(Me \text{ given } Mu) = \frac{12}{14} = \frac{6}{7}$

8 **a**

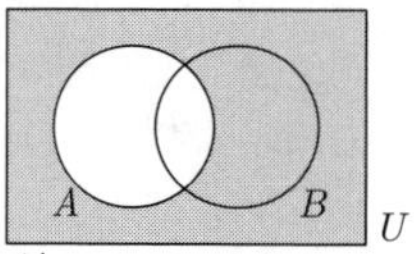

A'

b

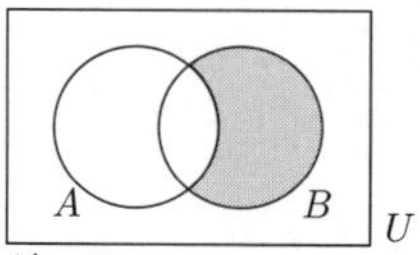

$A' \cap B$

c

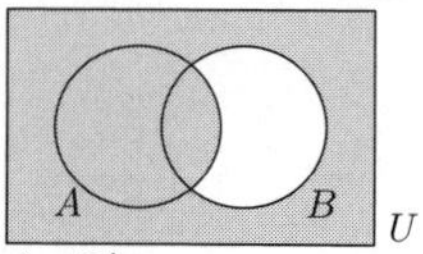

$A \cup B'$

d

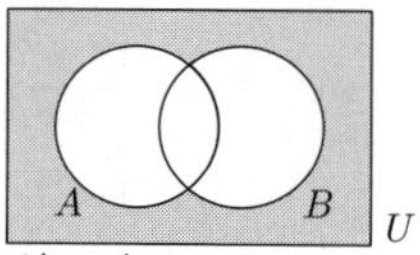

$A' \cap B'$

9 **a**

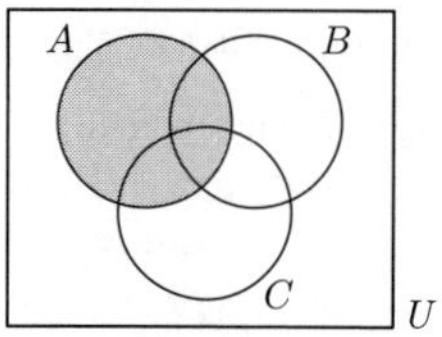

b

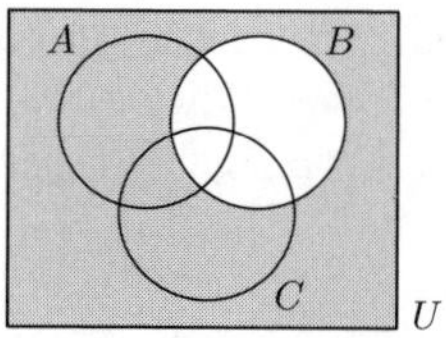

c

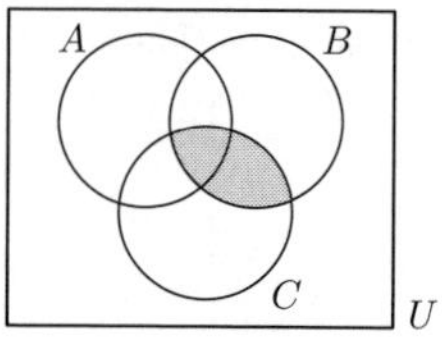

d

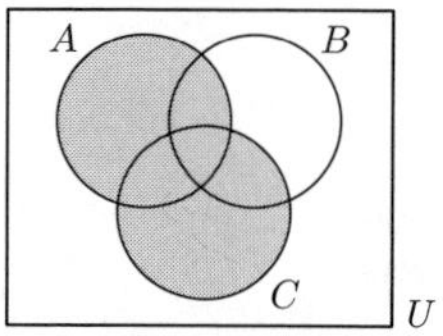

e

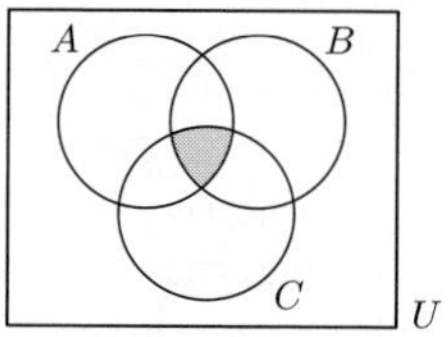

f

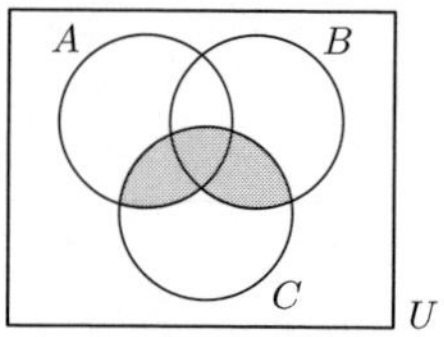

EXERCISE 15I.2

1 **a**

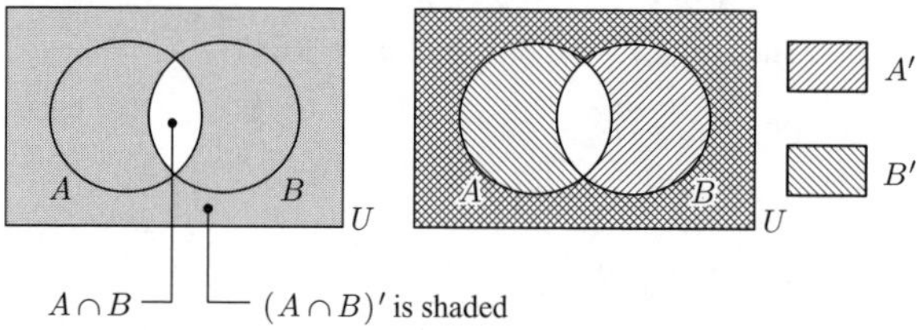

So $A' \cup B'$ is the region containing either type of shading.

Thus, as the regions are the same, $(A \cap B)' = A' \cup B'$ is verified.

b

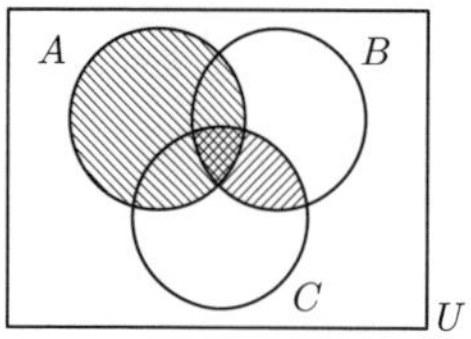

$A \cup (B \cap C)$ consists of the shaded region

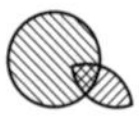

As the two regions are identical

$A \cup (B \cap C) = (A \cup B) \cap (A \cup C)$ is verified.

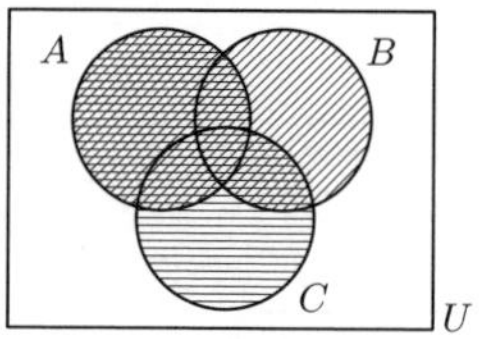

$(A \cup B) \cap (A \cup C)$ consists of the 'double shaded' region.

c

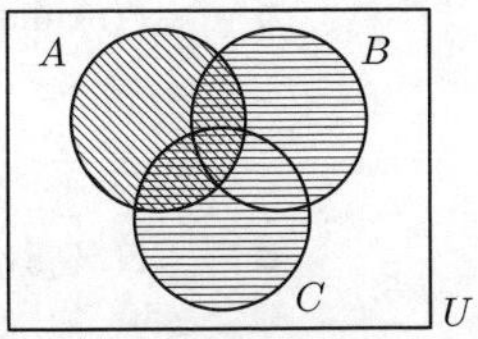

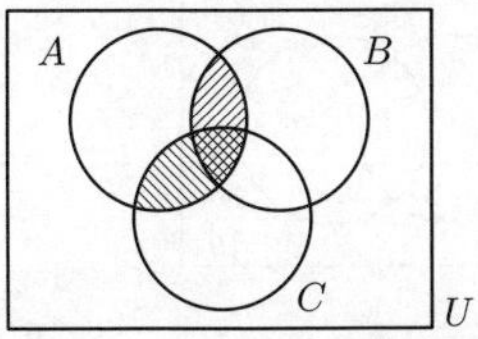

$A \cap (B \cup C)$ consists of the double shaded region 

$(A \cap B) \cup (A \cap C)$ consists of the region shaded. (all forms ▧ and ▨)

As the regions are identical, $A \cap (B \cup C) = (A \cap B) \cup (A \cap C)$ is verified.

2 **a** $A = \{7, 14, 21, 28, 35, ..., 98\}$
$B = \{5, 10, 15, 20, 25, ..., 95\}$

i as $98 = 7 \times 14$, $n(A) = 14$ **ii** as $95 = 5 \times 19$, $n(B) = 19$

iii $A \cap B = \{35, 70\}$ $\therefore$ $n(A \cap B) = 2$

iv $A \cup B = \{5, 7, 10, 14, 15, 20, 21, 25, 28, 30, 35, 40, 42, 45, 49, 50, 55, 56, 60, 63, 65, 70, 75, 77, 80, 84, 85, 90, 91, 95, 98\}$ $\therefore$ $n(A \cup B) = 31$

b $n(A) + n(B) - n(A \cap B)$
$= 14 + 19 - 2$
$= 31$
$= n(A \cup B)$ ✓

c From the diagram, $n(A) + n(B) - n(A \cap B)$
$= (a + b) + (b + c) - b$
$= a + b + c$
$= n(A \cup B)$

3 **a** **i** $\text{P}(B) = \dfrac{n(B)}{n(U)} = \dfrac{b + c}{a + b + c + d}$

ii $\text{P}(A \text{ and } B) = \dfrac{n(A \cap B)}{n(U)} = \dfrac{b}{a + b + c + d}$

iii $\text{P}(A \text{ or } B) = \dfrac{n(A \cup B)}{n(U)} = \dfrac{a + b + c}{a + b + c + d}$

iv $\text{P}(A) + \text{P}(B) - \text{P}(A \text{ and } B) = \dfrac{a + b + b + c - b}{a + b + c + d} = \dfrac{a + b + c}{a + b + c + d}$

b $\text{P}(A \text{ or } B) = \text{P}(A) + \text{P}(B) - \text{P}(A \text{ and } B)$ {using **iii** and **iv**}

EXERCISE 15J

1 **a**

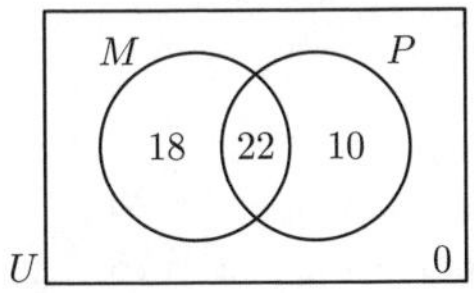

So 22 study both.

b **i** $\text{P}(M \text{ but not } P) = \frac{18}{50} = \frac{9}{25}$

ii $\text{P}(P \text{ given } M) = \dfrac{22}{18 + 22} = \frac{22}{40} = \frac{11}{20}$

2

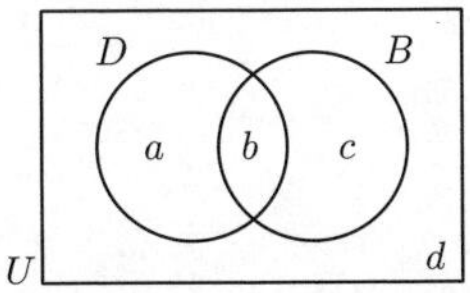

$a + b + c + d = 40$ .... (1)
$a + b = 23$ .... (2)
$b + c = 18$ .... (3)
$a + b + c = 26$ .... (4)

$\therefore$ $d = 14$ {using (1) and (4)}
$23 + c = 26$ and $a + 18 = 26$
$\therefore$ $c = 3$ and $a = 8$
Thus $b = 18 - c = 15$

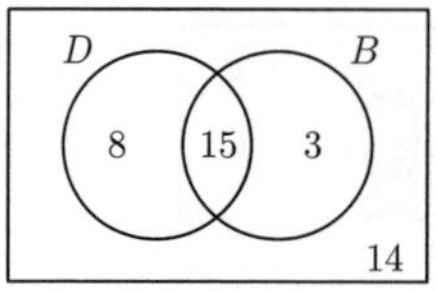

a P(D and B)
$= \frac{15}{40}$
$= \frac{3}{8}$

b P(neither D nor B)
$= \frac{14}{40}$
$= \frac{7}{20}$

c P(D, but not B)
$= \frac{8}{40}$
$= \frac{1}{5}$

d P(B given D)
$= \frac{15}{23}$

3

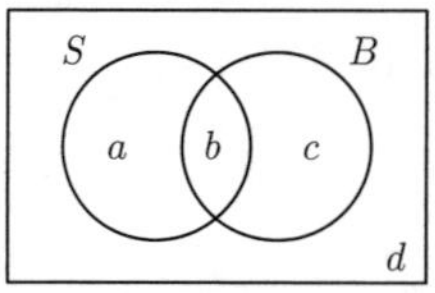

$$a + b + c + d = 50$$
$$a + b = 23$$
$$b + c = 22$$
$$b = 5$$

$\therefore\ c = 17,\ a = 18$

and $18 + 5 + 17 + d = 50$

$\therefore\ d = 10$

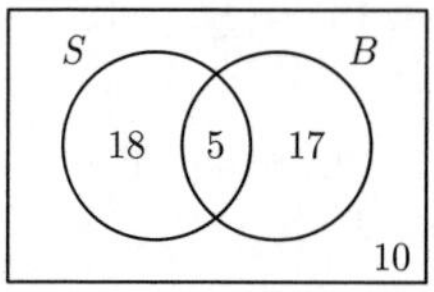

a P(not B)
$= \text{P}(B')$
$= \frac{28}{50}$
$= \frac{14}{25}$

b P(B or S)
$= \dfrac{18 + 5 + 17}{50}$
$= \frac{40}{50}$
$= \frac{4}{5}$

c P(neither B nor S)
$= \frac{10}{50}$
$= \frac{1}{5}$

d P(B, given S)
$= \dfrac{5}{18 + 5}$
$= \frac{5}{23}$

e P(S, given B')
$= \dfrac{18}{18 + 10}$
$= \frac{18}{28}$
$= \frac{9}{14}$

4

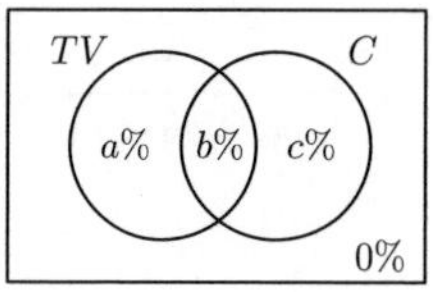

$$\begin{cases} a + b + c = 100 \\ a + b = 90 \quad \therefore\ c = 10 \text{ and } a = 40 \\ b + c = 60 \quad \therefore\ b = 50 \end{cases}$$

$$\text{P(TV, given C)} = \frac{50}{50 + 10} = \frac{5}{6}$$

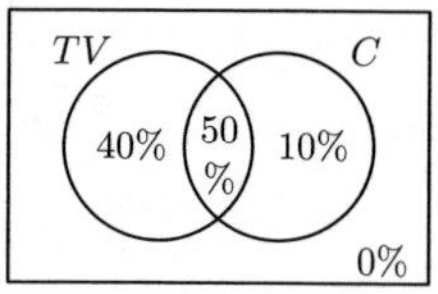

5

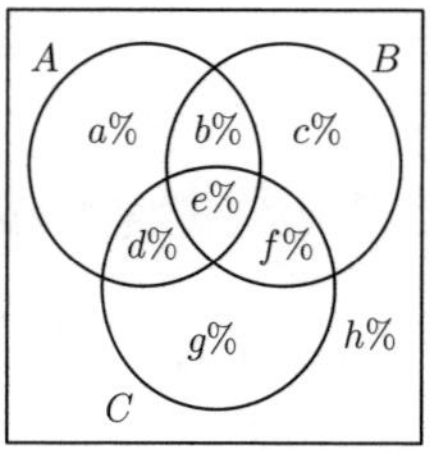

$$a + b + c + d + e + f + g + h = 100$$
$$a + b + d + e = 20$$
$$b + c + e + f = 16$$
$$d + e + f + g = 14$$
$$b + e = 8$$
$$d + e = 5$$
$$e + f = 4$$
$$e = 2$$

$$\therefore\ e = 2,\ f = 2,\ d = 3,\ b = 6,\ \begin{cases} a + 6 + 3 + 2 = 20 \\ 6 + c + 2 + 2 = 16 \\ 3 + 2 + 2 + g = 14 \end{cases} \therefore \begin{cases} a = 9 \\ c = 6 \\ g = 7 \end{cases}$$

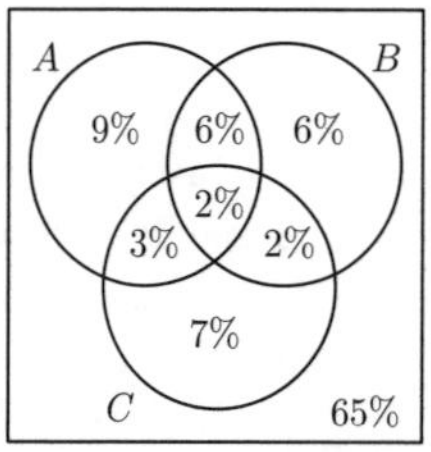

a P(none)
$= \frac{65}{100}$
$= \frac{13}{20}$

b P(at least one)
$= 1 - \text{P(none)}$
$= 1 - \frac{13}{20}$
$= \frac{7}{20}$

c P(exactly one)
$= \dfrac{9 + 6 + 7}{100}$
$= \frac{22}{100}$
$= \frac{11}{50}$

d $\text{P}(A \text{ or } B)$
$= \frac{9+6+6+3+2+2}{100}$
$= \frac{28}{100}$
$= \frac{7}{25}$

e $\text{P}(A, \text{ given at least one})$
$= \frac{9+6+2+3}{35}$
$= \frac{20}{35}$
$= \frac{4}{7}$

f $\text{P}(C, \text{ given } A \text{ or } B \text{ or both})$
$= \frac{3+2+2}{9+6+6+3+2+2}$
$= \frac{7}{28}$
$= \frac{1}{4}$

6

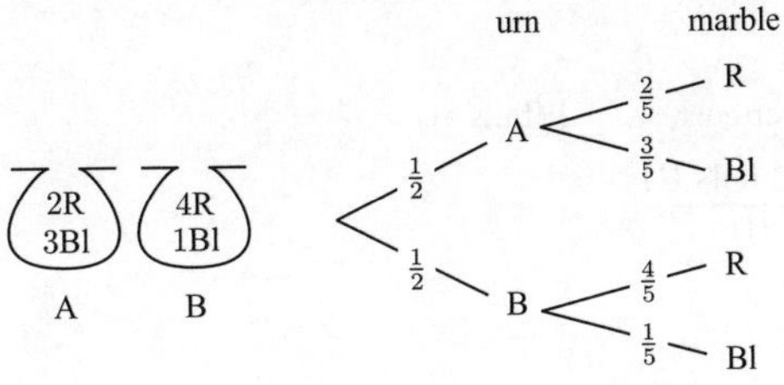

a $\text{P(R)} = \frac{1}{2} \times \frac{2}{5} + \frac{1}{2} \times \frac{4}{5}$
$= \frac{3}{5}$

b $\text{P(B} \mid \text{R)} = \frac{\text{P(B} \cap \text{R)}}{\text{P(R)}}$
$= \frac{\frac{1}{2} \times \frac{4}{5}}{\frac{3}{5}}$
$= \frac{2}{3}$

7

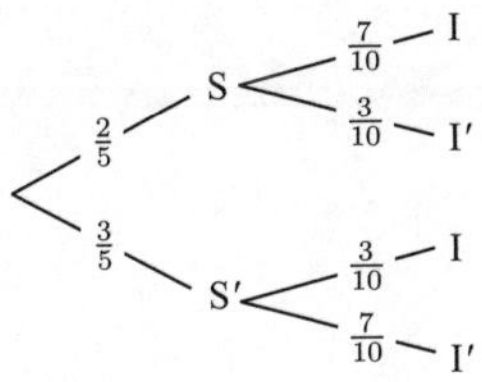

a $\text{P(I)} = \frac{2}{5} \times \frac{7}{10} + \frac{3}{5} \times \frac{3}{10}$
$= \frac{23}{50}$ (or 0.46)

b $\text{P(S} \mid \text{I)} = \frac{\text{P(S} \cap \text{I)}}{\text{P(I)}}$
$= \frac{\frac{2}{5} \times \frac{7}{10}}{\frac{23}{50}}$
$= \frac{14}{23}$

8

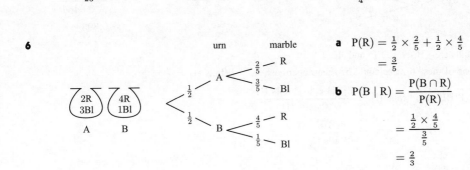

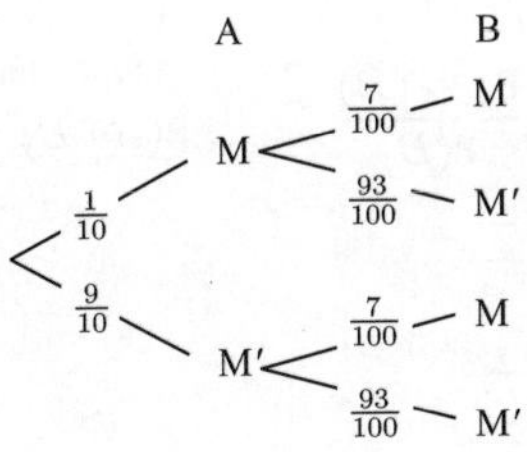

$\text{P(B} \mid \text{at least one malfunctions)}$
$= \frac{\text{P(B} \cap \text{at least one malfunctions)}}{\text{P(at least one malfunctions)}}$
$= \frac{\frac{1}{10} \times \frac{7}{100} + \frac{9}{10} \times \frac{7}{100}}{\frac{1}{10} \times \frac{7}{100} + \frac{1}{10} \times \frac{93}{100} + \frac{9}{10} \times \frac{7}{100}}$
$= \frac{7+63}{7+93+63}$
$= \frac{70}{163}$

9 $\text{P}(B) = 0.5$, $\text{P}(G) = 0.6$, $\text{P}(G \mid B) = 0.9$, where B is "the boy eats his lunch" and G is "the girl eats her lunch"

a P(both eat lunch)
$= \text{P}(B \cap G)$
$= \text{P}(G \mid B) \times \text{P}(B)$ $\quad \left\{ \text{as } \text{P}(G \mid B) = \frac{\text{P}(G \cap B)}{\text{P}(B)} \right\}$
$= 0.9 \times 0.5$
$= 0.45$

b $\text{P}(B \mid G)$
$= \frac{\text{P}(B \cap G)}{\text{P}(G)}$
$= \frac{0.45}{0.6}$
$= 0.75$

c P(at least one eats lunch)
$= \text{P}(B \cup G)$
$= \text{P}(B) + \text{P}(G) - \text{P}(B \cap G)$
$= 0.5 + 0.6 - 0.45$
$= 0.65$

10

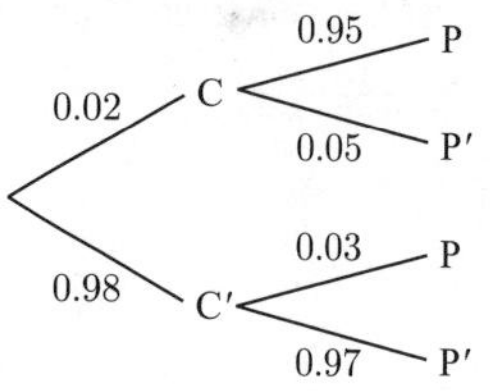

a P(P)
$= 0.02 \times 0.95 + 0.98 \times 0.03$
$= 0.0484$

b $\text{P(C} \mid \text{P)}$
$= \dfrac{\text{P(C} \cap \text{P)}}{\text{P(P)}}$
$= \dfrac{0.02 \times 0.95}{0.0484}$
≈ 0.393

11 The coins are H, H T, T and H, T.
Any one of these 6 faces could be seen uppermost, $\therefore$ P(falls H) $= \frac{3}{6} = \frac{1}{2}$

Now $\text{P(HH coin} \mid \text{falls H)} = \dfrac{\text{P(HH coin} \cap \text{falls H)}}{\text{P(falls H)}}$
$= \dfrac{\text{P(HH)}}{\text{P(falls H)}}$
$= \dfrac{\frac{1}{3}}{\frac{1}{2}}$
$= \frac{2}{3}$

EXERCISE 15K

1 $\text{P}(R \cap S)$
$= \text{P}(R) + \text{P}(S) - \text{P}(R \cup S)$
$= 0.4 + 0.5 - 0.7$
$= 0.2$

Also, $\text{P}(R) \times \text{P}(S)$
$= 0.4 \times 0.5$
$= 0.2$

So, $\text{P}(R \cap S) = \text{P}(R) \times \text{P}(S)$ and hence R and S are independent events.

2 **a** $\text{P}(A \cap B)$
$= \text{P}(A) + \text{P}(B) - \text{P}(A \cup B)$
$= \frac{2}{5} + \frac{1}{3} - \frac{1}{2}$
$= \frac{7}{30}$

b $\text{P}(B \mid A)$
$= \dfrac{\text{P}(B \cap A)}{\text{P}(A)}$
$= \dfrac{\frac{7}{30}}{\frac{2}{5}}$
$= \frac{7}{12}$

c $\text{P}(A \mid B)$
$= \dfrac{\text{P}(A \cap B)}{\text{P}(B)}$
$= \dfrac{\frac{7}{30}}{\frac{1}{3}}$
$= \frac{7}{10}$

A and B are not independent as $\text{P}(A \mid B) \neq \text{P}(A)$.

3 **a** As X and Y are independent
$\text{P}(X \cap Y) = \text{P}(X) \times \text{P}(Y)$
$= 0.5 \times 0.7$
$= 0.35$
$\therefore$ P(both X and Y) $= 0.35$

b P(X or Y)
$= \text{P}(X \cup Y)$
$= \text{P}(X) + \text{P}(Y) - \text{P}(X \cap Y)$
$= 0.5 + 0.7 - 0.35$
$= 0.85$

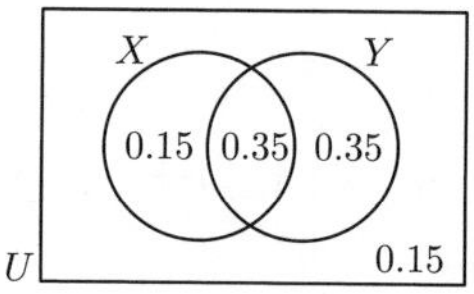

c P(neither X nor Y)
$= 0.15$

d P(X but not Y)
$= 0.15$

e $\text{P}(X \mid Y) = \dfrac{\text{P}(X \cap Y)}{\text{P}(Y)} = \dfrac{0.35}{0.70} = \frac{1}{2}$

4 P(at least one solves it)
$= 1 -$ P(no-one solves it)
$= 1 - \text{P(A}' \text{ and B}' \text{ and C}')$
$= 1 - \frac{2}{5} \times \frac{1}{3} \times \frac{1}{2}$
$= 1 - \frac{1}{15}$
$= \frac{14}{15}$

5 **a** P(at least one 6)

$= 1 - \text{P(no 6s)}$

$= 1 - \text{P}(6' \text{ and } 6' \text{ and } 6')$

$= 1 - \frac{5}{6} \times \frac{5}{6} \times \frac{5}{6}$

$= 1 - \frac{125}{216}$

$= \frac{91}{216}$

b P(at least one 6 in n throws)

$= 1 - (\frac{5}{6})^n$

So we want $1 - (\frac{5}{6})^n > 0.99$

$\therefore \quad -(\frac{5}{6})^n > -0.01$

$\therefore \quad (\frac{5}{6})^n < 0.01$

$\therefore \quad n \log(\frac{5}{6}) < \log(0.01)$

$\therefore \quad n > \dfrac{\log(0.01)}{\log(\frac{5}{6})}$ {as $\log(\frac{5}{6}) < 0$}

$\therefore \quad n > 25.2585$

$\therefore \quad n = 26$

6 A and B are independent, so $\text{P}(A \cap B) = \text{P}(A)\,\text{P}(B)$ .... (1)

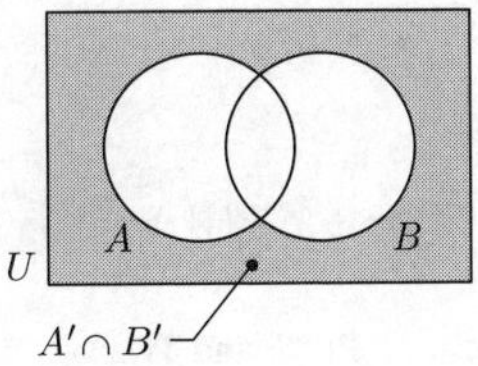

Now $\text{P}(A' \cap B')$

$= 1 - \text{P}(A \cup B)$

$= 1 - [\text{P}(A) + \text{P}(B) - \text{P}(A \cap B)]$

$= 1 - \text{P}(A) - \text{P}(B) + \text{P}(A \cap B)$

$= 1 - \text{P}(A) - \text{P}(B) + \text{P}(A)\,\text{P}(B)$ {using (1)}

$= [1 - \text{P}(A)]\,[1 - \text{P}(B)]$

$= \text{P}(A')\,\text{P}(B')$

$\therefore$ A' and B' are also independent.

7

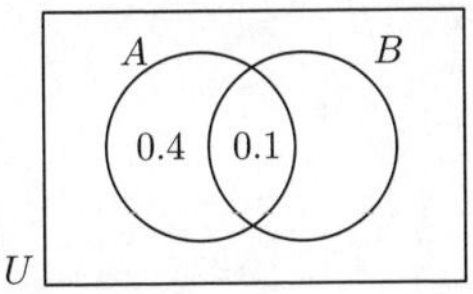

$\therefore \quad \text{P}(A) = 0.5$

and $\text{P}(A \cap B) = \text{P}(A) \times \text{P}(B)$ {A and B are independent}

$\therefore \quad 0.1 = 0.5 \times \text{P}(B)$

$\therefore \quad \text{P}(B) = 0.2$

Now $\text{P}(A \cup B') = \text{P}(A) + \text{P}(B') - \text{P}(A \cap B')$

$= 0.5 + 0.8 - 0.4$

$= 0.9$

8 **a** **i** $\text{P}(C \mid D) = \dfrac{\text{P}(C \cap D)}{\text{P}(D)}$, so $\text{P}(C \cap D) = \text{P}(C \mid D)\,\text{P}(D)$

Similarly, $\text{P}(C \cap D') = \text{P}(C \mid D')\,\text{P}(D')$

Now $\text{P}(C \cap D) + \text{P}(C \cap D') = \text{P}(C)$

$\therefore \quad \text{P}(C \mid D)\,\text{P}(D) + \text{P}(C \mid D')\,\text{P}(D') = \text{P}(C)$

$\therefore \quad \frac{6}{13}\,\text{P}(D) + \frac{3}{7}\,[1 - \text{P}(D)] = \frac{9}{20}$

$\therefore \quad \frac{6}{13}\,\text{P}(D) + \frac{3}{7} - \frac{3}{7}\,\text{P}(D) = \frac{9}{20}$

$\therefore \quad \frac{3}{91}\,\text{P}(D) = \frac{3}{140}$

$\therefore \quad \text{P}(D) = \frac{91}{140}$ or $\frac{13}{20}$

ii $\text{P}(C \cap D) = \text{P}(C \mid D)\,\text{P}(D) = \frac{6}{13} \times \frac{13}{20} = \frac{3}{10}$

Now $\text{P}(C' \cup D') = 1 - \text{P}(C \cap D)$

$= 1 - \frac{3}{10} = \frac{7}{10}$

b $\text{P}(C \cap D) = \frac{3}{10}$ and $\text{P}(C)\,\text{P}(D) = \frac{9}{20} \times \frac{13}{20} = \frac{117}{400}$

$\therefore$ C and D are not independent as $\text{P}(C \cap D) \neq \text{P}(C)\,\text{P}(D)$

REVIEW SET 15A

1 ABCD, ABDC, ACBD, ACDB, ADBC, ADCB, BACD, BADC, BCAD, BCDA, BDAC, BDCA, CABD, CADB, CBAD, CBDA, CDAB, CDBA, DABC, DACB, DBAC, DBCA, DCAB, DCBA

a There are 24 possible orderings.

$\therefore$ P(A is next to C)

$= \frac{12}{24}$ {12 have A next to C}

$= \frac{1}{2}$

b P(exactly one person between A and C)

$= \frac{8}{24}$ {8 have one person between A and C}

$= \frac{1}{3}$

2

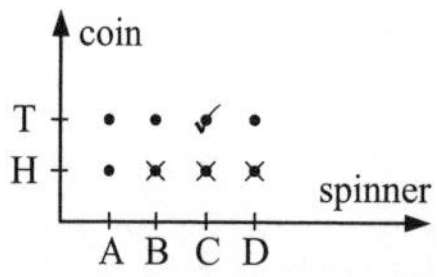

a Consonants are B, C and D

$\therefore$ P(H and a consonant)

$= \frac{3}{8}$ {those with a $\times$}

b P(T and C)

$= \frac{1}{8}$ {those with a $\checkmark$}

c P(T or vowel)

$=$ P(T or A)

$=$ P(T) + P(A) − P(T and A)

$= \frac{4}{8} + \frac{2}{8} - \frac{1}{8}$

$= \frac{5}{8}$

3 $\text{P}(M) = \frac{3}{5}$, $\text{P}(W) = \frac{2}{3}$, where M is the event "the man is alive in 25 years", and W is the event "the woman is alive in 25 years".

a $\text{P}(M \text{ and } W)$

$= \frac{3}{5} \times \frac{2}{3}$

{assuming independence}

$= \frac{2}{5}$

b P(at least one)

$= \text{P}(M \text{ or } W)$

$= \text{P}(M) + \text{P}(W) - \text{P}(M \text{ and } W)$

$= \frac{3}{5} + \frac{2}{3} - \frac{2}{5}$

$= \frac{13}{15}$

c $\text{P}(M' \text{ and } W)$

$= (1 - \frac{3}{5}) \times \frac{2}{3}$

$= \frac{2}{5} \times \frac{2}{3}$

$= \frac{4}{15}$

4 **a** $\text{P}(X \cap Y) = 0$ {X and Y are mutually exclusive events}

b $\text{P}(X \cup Y) = \text{P}(X) + \text{P}(Y) - \text{P}(X \cap Y)$

$\therefore \ 0.8 = \text{P}(X) + 0.35 - 0$

$\therefore \ \text{P}(X) = 0.45$

c $\text{P}(X \text{ or } Y \text{ but not both}) = \text{P}(X \text{ or } Y)$ {X and Y mutually exclusive}

$= \text{P}(X \cup Y)$

$= 0.8$

5 **a** Two events are independent if the occurrence of each event does not influence the occurrence of the other. For A and B independent, $\text{P}(A) \times \text{P}(B) = \text{P}(A \text{ and } B)$.

b Two events, A and B, are disjoint if they have no common outcomes,

$\therefore$ $\text{P}(A \text{ and } B) = 0$ and so $\text{P}(A \text{ or } B) = \text{P}(A) + \text{P}(B)$.

6 **a**

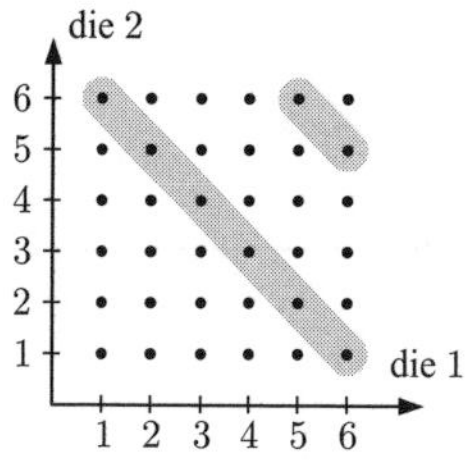

P(sum of 7 or 11)

$= \frac{8}{36}$

$= \frac{2}{9}$

b

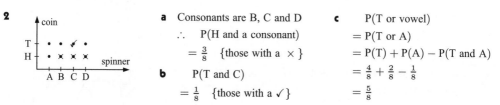

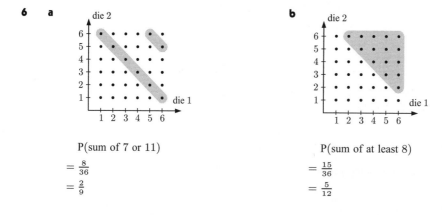

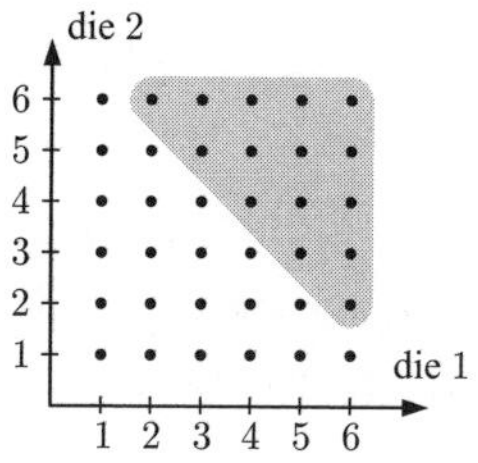

P(sum of at least 8)

$= \frac{15}{36}$

$= \frac{5}{12}$

7

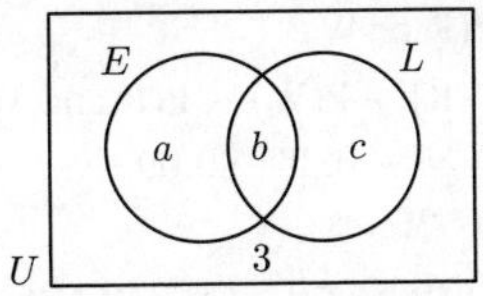

$a+b+c=37 \quad \therefore \quad 22+c=37$ and $a+25=37$

$a+b=22 \quad \therefore \quad c=15$ and $a=12$

$b+c=25$ Hence, $b=22-a=10$

a P(E and L)
$=\frac{10}{40}$
$=\frac{1}{4}$

b P(at least one)
$=\frac{12+10+15}{40}$
$=\frac{37}{40}$

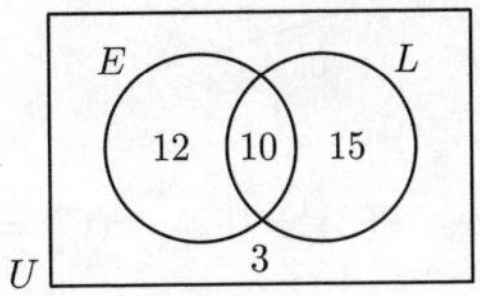

c $\mathrm{P}(E \mid L)=\frac{10}{15+10}=\frac{10}{25}=\frac{2}{5}$

8 Let L_i be the event that the salesman leaves his sunglasses in store i.

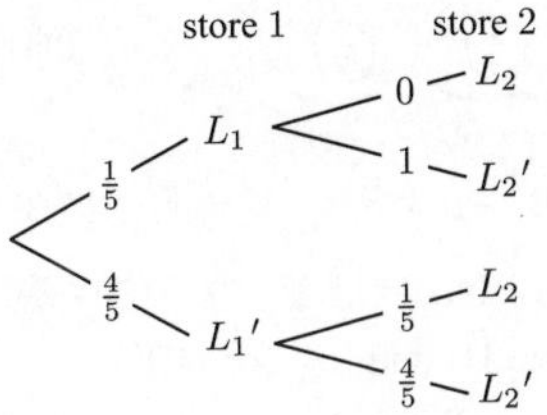

$$\begin{aligned}\mathrm{P}(L_1 \mid L_1 \text{ or } L_2) &= \frac{\mathrm{P}(L_1 \cap (L_1 \text{ or } L_2))}{\mathrm{P}(L_1 \text{ or } L_2)} \\ &= \frac{\mathrm{P}(L_1)}{\mathrm{P}(L_1 L_2' \text{ or } L_1' L_2)} \\ &= \frac{\frac{1}{5}}{\frac{1}{5}+\frac{4}{5}\times\frac{1}{5}} \\ &= \frac{5}{9}\end{aligned}$$

REVIEW SET 15B

1

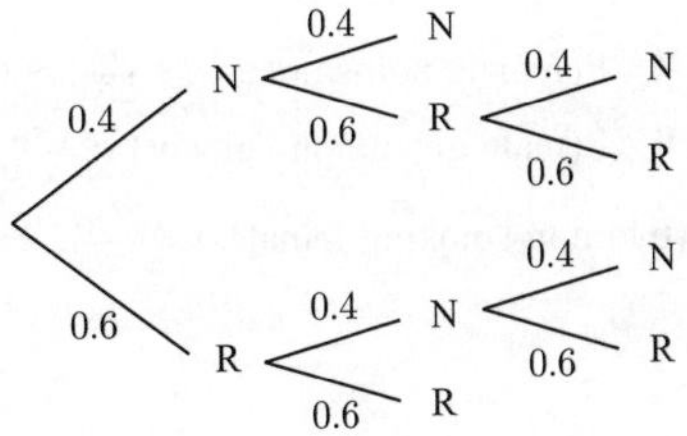

P(Niklas wins)
$=(0.4)(0.4)+(0.4)(0.6)(0.4)+(0.6)(0.4)(0.4)$
$=0.352$

2 **a** P(win first 3 prizes)
$=\mathrm{P(WWW)}$
$=\frac{4}{500}\times\frac{3}{499}\times\frac{2}{498}$
$\approx 1.93\times 10^{-7}$

b P(win at least one of the 3 prizes)
$=1-\mathrm{P}$(wins none of them)
$=1-\mathrm{P(W'W'W')}$
$=1-\frac{496}{500}\times\frac{495}{499}\times\frac{494}{498}$
≈ 0.0239

3

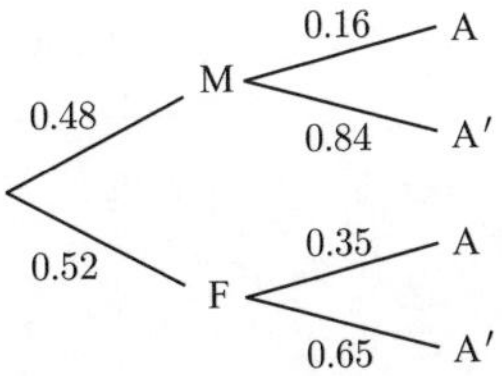

a $\mathrm{P(A)}=\mathrm{P(M\cap A \text{ or } F\cap A)}$
$=0.48\times 0.16+0.52\times 0.35$
$=0.2588 \quad (\approx 0.259)$

b $\mathrm{P(F \mid A)}=\frac{\mathrm{P(F\cap A)}}{\mathrm{P(A)}}$
$=\frac{0.52\times 0.35}{0.2588}$
≈ 0.703

4

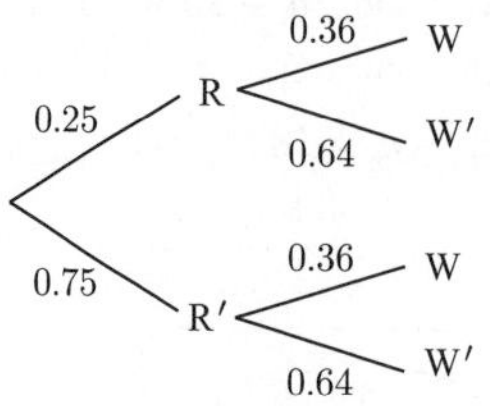

a $P(R \text{ and } W)$
$= 0.25 \times 0.36$
$= 0.09$

b $P(R \text{ or } W)$
$= P(R) + P(W) - P(R \text{ and } W)$
$= 0.25 + 0.36 - 0.09$
$= 0.52$

or $P(R \text{ or } W) = 1 - P(R'W')$
$= 1 - 0.75 \times 0.64$
$= 0.52$

5 $P(A) = 0.1, \quad P(B) = 0.2, \quad P(C) = 0.3$

$\therefore$ P(group solves it) = P(at least one solves it)
= 1 − P(no-one solves it)
$= 1 - P(A' \text{ and } B' \text{ and } C')$
$= 1 - (0.9 \times 0.8 \times 0.7)$
$= 0.496$

6 **a** $\left(\frac{3}{5}+\frac{2}{5}\right)^4 = \underbrace{\left(\frac{3}{5}\right)^4}_{4B} + \underbrace{4\left(\frac{3}{5}\right)^3\left(\frac{2}{5}\right)}_{3B\ 1B'} + \underbrace{6\left(\frac{3}{5}\right)^2\left(\frac{2}{5}\right)^2}_{2B\ 2B'} + \underbrace{4\left(\frac{3}{5}\right)\left(\frac{2}{5}\right)^3}_{1B\ 3B'} + \underbrace{\left(\frac{2}{5}\right)^4}_{4B'}$

$P(B) = \frac{12}{20}$
$= \frac{3}{5}$
$\therefore \quad P(B') = \frac{2}{5}$

b **i** P(2 Blue inks)
$= P(2B \text{ and } 2B')$
$= 6\left(\frac{3}{5}\right)^2\left(\frac{2}{5}\right)^2$
$= \dfrac{6 \times 9 \times 4}{5^4}$
$= \frac{216}{625}$

ii P(at most 2 Blue inks)
$= P(2B \text{ and } 2B' \text{ or } 1B \text{ and } 3B' \text{ or } 4B')$
$= 6\left(\frac{3}{5}\right)^2\left(\frac{2}{5}\right)^2 + 4\left(\frac{3}{5}\right)\left(\frac{2}{5}\right)^3 + \left(\frac{2}{5}\right)^4$
$= \dfrac{6 \times 9 \times 4 + 4 \times 3 \times 8 + 16}{625}$
$= \frac{328}{625}$

7 **a**

	Female	*Male*	*Total*
smoker	20	40	60
non-smoker	70	70	140
Total	90	110	200

b **i** P(female non-smoker) $\approx \frac{70}{200} \approx 0.35$

ii P(male given non-smoker) $\approx \frac{70}{140} \approx 0.5$

c P(two non-smoking females) $= \frac{70}{200} \times \frac{69}{199}$
≈ 0.121

REVIEW SET 15C

1 BBBB, BBBG, BBGB, BGBB, GBBB, BBGG, BGBG, BGGB, GBBG, GBGB, GGBB, BGGG, GBGG, GGBG, GGGB, GGGG

P(2B and 2G)
$= \frac{6}{16}$ ← 6 have 2B and 2G
$= \frac{3}{8}$

2 **a** P(both blue)
$= P(BB)$
$= \frac{5}{12} \times \frac{4}{11}$
$= \frac{5}{33}$

b P(both same colour)
$= P(BB \text{ or } RR \text{ or } YY)$
$= \frac{5}{12} \times \frac{4}{11} + \frac{3}{12} \times \frac{2}{11} + \frac{4}{12} \times \frac{3}{11}$
$= \frac{19}{66}$

c P(at least one R)
= 1 − P(no reds)
$= 1 - P(R'R')$
$= 1 - \frac{9}{12} \times \frac{8}{11}$
$= 1 - \frac{6}{11}$
$= \frac{5}{11}$

d P(exactly one Y)
$= P(YY' \text{ or } Y'Y)$
$= \frac{4}{12} \times \frac{8}{11} + \frac{8}{12} \times \frac{4}{11}$
$= \frac{16}{33}$

3

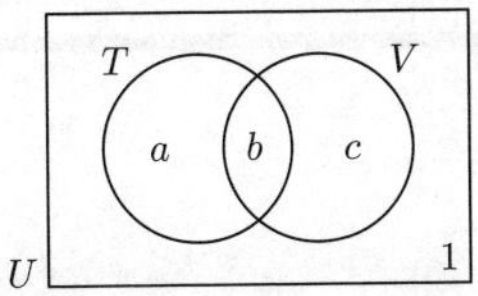

$a+b+c=24$ $\quad\therefore\quad 13+c=24$ and $a+14=24$

$a+b=13$ $\quad\therefore\quad c=11$ and $a=10$

$b+c=14$

Also $b=13-a$

$=3$

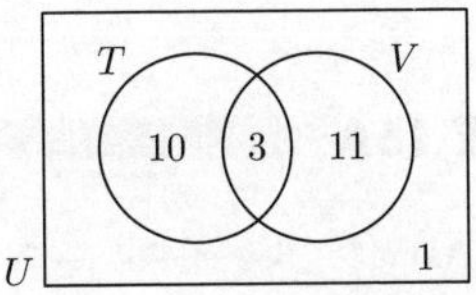

a $\text{P}(T \text{ and } V)$
$=\frac{3}{25}$

b P(at least one)
$=1-\text{P(neither)}$
$=1-\frac{1}{25}$
$=\frac{24}{25}$

c $\text{P}(V \mid T')$
$=\dfrac{11}{11+1}$
$=\frac{11}{12}$

4 **a** There are now 3 red and 5 blue balls remaining.

$\therefore$ P(blue) $=\frac{5}{8}$

b

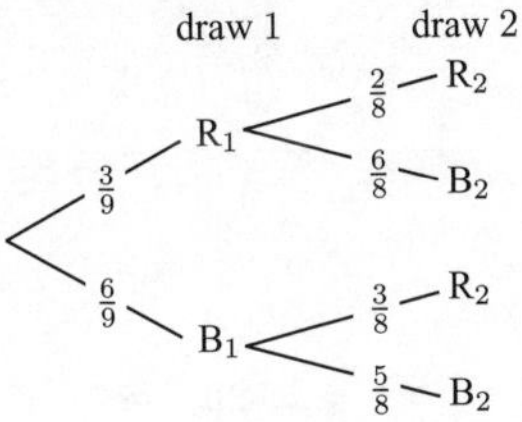

$\text{P}(\text{R}_1 \mid \text{R}_2)=\dfrac{\text{P}(\text{R}_1 \cap \text{R}_2)}{\text{P}(\text{R}_2)}$

$=\dfrac{\frac{3}{9}\times\frac{2}{8}}{\frac{3}{9}\times\frac{2}{8}+\frac{6}{9}\times\frac{3}{8}}$

$=\frac{1}{4}$

5

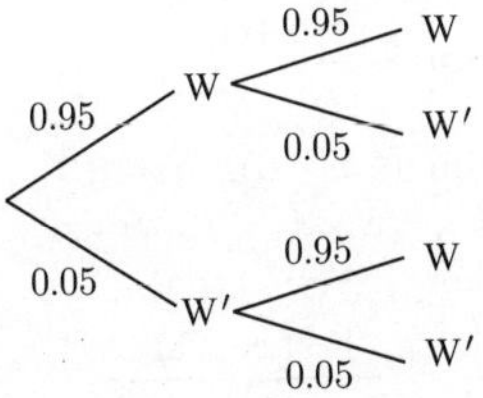

P(works on at least one day)
$=0.95\times0.95+0.95\times0.05+0.05\times0.95$
$=0.9975$

6

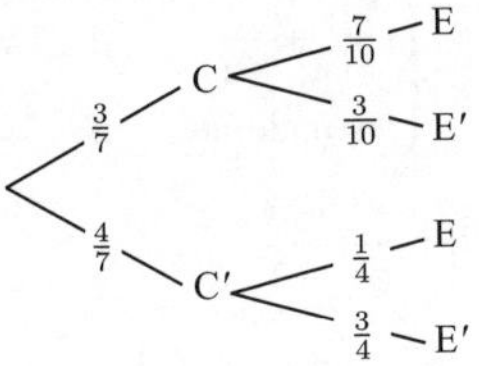

a $\text{P(E)}=\frac{3}{7}\times\frac{7}{10}+\frac{4}{7}\times\frac{1}{4}$
$=\frac{3}{10}+\frac{1}{7}$
$=\frac{31}{70}$

b $\text{P(C} \mid \text{E)}=\dfrac{\text{P(C and E)}}{\text{P(E)}}$
$=\dfrac{\frac{3}{7}\times\frac{7}{10}}{\frac{31}{70}}$
$=\frac{21}{31}$

7 **a** $\left(\frac{4}{5}+\frac{1}{5}\right)^5=\left(\frac{4}{5}\right)^5+5\left(\frac{4}{5}\right)^4\left(\frac{1}{5}\right)^1+10\left(\frac{4}{5}\right)^3\left(\frac{1}{5}\right)^2+10\left(\frac{4}{5}\right)^2\left(\frac{1}{5}\right)^3+5\left(\frac{4}{5}\right)^1\left(\frac{1}{5}\right)^4+\left(\frac{1}{5}\right)^5$

b Let $X=$ the number of goals scored

i P(3 goals **then** 2 misses)
$=\text{P(GGGG}'\text{G}')$
$=(0.8)^3\times(0.2)^2$
≈ 0.0205

ii P(3 goals and 2 misses)
$=\text{P}(X=3)$
$=10\left(\frac{4}{5}\right)^3\left(\frac{1}{5}\right)^2$
≈ 0.205

Chapter 16
INTRODUCTION TO CALCULUS

EXERCISE 16A

1 **a** As $x \to 3$, $x+4 \to 7$

$\therefore \lim_{x \to 3} (x+4) = 7$

b As $x \to -1$, $5-2x \to 7$

$\therefore \lim_{x \to -1} (5-2x) = 7$

c As $x \to 4$, $3x-1 \to 11$

$\therefore \lim_{x \to 4} (3x-1) = 11$

d As $x \to 2$, $5x^2-3x+2 \to 5(4)-3(2)+2 = 16$

$\therefore \lim_{x \to 2} (5x^2-3x+2) = 16$

e As $h \to 0$, $h^2 \to 0$ and $1-h \to 1$

$\therefore \lim_{h \to 0} h^2(1-h) = 0 \times 1 = 0$

f As $x \to -1$, $1-2x \to 3$ and $x^2+1 \to 2$

$\therefore \lim_{x \to -1} \dfrac{1-2x}{x^2+1} = \frac{3}{2}$

g As $x \to 0$, $x^2+5 \to 5$

$\therefore \lim_{x \to 0} (x^2+5) = 5$

h As $x \to -2$, $\dfrac{4}{x} \to -2$

$\therefore \lim_{x \to -2} \dfrac{4}{x} = -2$

2 **a** $\dfrac{x^2-3x}{x} \begin{cases} = x-3 & \text{if } x \neq 0 \\ \text{is undefined} & \text{if } x = 0 \end{cases}$

$\therefore \lim_{x \to 0} \dfrac{x^2-3x}{x}$

$= \lim_{x \to 0} (x-3),\ x \neq 0$

$= -3$

b $\dfrac{2h^2+6h}{h} \begin{cases} = 2h+6 & \text{if } h \neq 0 \\ \text{is undefined} & \text{if } h = 0 \end{cases}$

$\therefore \lim_{h \to 0} \dfrac{2h^2+6h}{h}$

$= \lim_{h \to 0} (2h+6),\ h \neq 0$

$= 6$

c $\dfrac{h^3-8h}{h} \begin{cases} = h^2-8 & \text{if } h \neq 0 \\ \text{is undefined} & \text{if } h = 0 \end{cases}$

$\therefore \lim_{h \to 0} \dfrac{h^3-8h}{h}$

$= \lim_{h \to 0} (h^2-8),\ h \neq 0$

$= -8$

d $\dfrac{x^2-x}{x^2-1} = \dfrac{x(x-1)}{(x+1)(x-1)}$

$\begin{cases} = \dfrac{x}{x+1} & \text{if } x \neq 1 \\ \text{is undefined} & \text{if } x = 1 \end{cases}$

$\therefore \lim_{x \to 1} \dfrac{x^2-x}{x^2-1}$

$= \lim_{x \to 1} \dfrac{x}{x+1},\ x \neq 1$

$= \frac{1}{2}$

e $\dfrac{x^2-2x}{x^2-4} = \dfrac{x(x-2)}{(x+2)(x-2)}$

$\begin{cases} = \dfrac{x}{x+2} & \text{if } x \neq 2 \\ \text{is undefined} & \text{if } x = 2 \end{cases}$

$\therefore \lim_{x \to 2} \dfrac{x^2-2x}{x^2-4}$

$= \lim_{x \to 2} \dfrac{x}{x+2},\ x \neq 2$

$= \frac{2}{4}$ or $\frac{1}{2}$

f $\dfrac{x^2-x-6}{x^2-5x+6} = \dfrac{(x-3)(x+2)}{(x-3)(x-2)}$

$\begin{cases} = \dfrac{x+2}{x-2} & \text{if } x \neq 3 \\ \text{is undefined} & \text{if } x = 3 \end{cases}$

$\therefore \lim_{x \to 3} \dfrac{x^2-x-6}{x^2-5x+6}$

$= \lim_{x \to 3} \dfrac{x+2}{x-2},\ x \neq 3$

$= 5$

3 **a** As x gets larger and positive, $\dfrac{1}{x}$ gets smaller and closer to 0. $\therefore \lim\limits_{x\to\infty} \dfrac{1}{x} = 0$

b $\lim\limits_{x\to\infty} \dfrac{3x-2}{x+1}$

$= \lim\limits_{x\to\infty} \dfrac{3-\frac{2}{x}}{1+\frac{1}{x}}$

$= \frac{3}{1}$

$= 3$

c $\lim\limits_{x\to\infty} \dfrac{1-2x}{3x+2}$

$= \lim\limits_{x\to\infty} \dfrac{\frac{1}{x}-2}{3+\frac{2}{x}}$

$= -\frac{2}{3}$

d $\lim\limits_{x\to\infty} \dfrac{x}{1-x}$

$= \lim\limits_{x\to\infty} \dfrac{1}{\frac{1}{x}-1}$

$= \frac{1}{-1}$

$= -1$

e $\lim\limits_{x\to\infty} \dfrac{x^2+3}{x^2-1}$

$= \lim\limits_{x\to\infty} \dfrac{1+\frac{3}{x^2}}{1-\frac{1}{x^2}}$

$= \frac{1}{1}$

$= 1$

f $\lim\limits_{x\to\infty} \dfrac{x^2-2x+4}{x^2+x-1}$

$= \lim\limits_{x\to\infty} \dfrac{1-\frac{2}{x}+\frac{4}{x^2}}{1+\frac{1}{x}-\frac{1}{x^2}}$

$= \frac{1}{1}$

$= 1$

EXERCISE 16B

1 **a** $f(x) = \dfrac{3x-2}{x+3}$

$= \dfrac{3(x+3)-9-2}{x+3}$

$= 3 - \dfrac{11}{x+3}$

$f(x)$ is undefined when $x=-3$

$\therefore$ $x=-3$ is a vertical asymptote

As $|x| \to \infty$, $f(x) \to 3$

$\therefore$ $y=3$ is a horizontal asymptote

The sign diagram for $f(x)$ is:

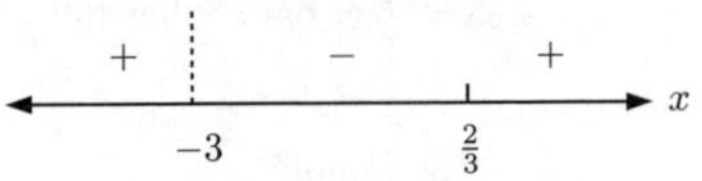

As $x \to -3$ (left), $f(x) \to \infty$

As $x \to -3$ (right), $f(x) \to -\infty$

As $x \to \infty$, $f(x) \to 3$ (below)

As $x \to -\infty$, $f(x) \to 3$ (above)

b $y = \dfrac{x^2-1}{x^2+1} = \dfrac{x^2+1-2}{x^2+1}$

$= 1 - \dfrac{2}{x^2+1}$

y is defined for all $x \in \mathbb{R}$

$\therefore$ no vertical asymptotes exist

As $|x| \to \infty$, $y \to 1$

$\therefore$ $y=1$ is a horizontal asymptote

As $x \to \infty$, $y \to 1$ (below)

As $x \to -\infty$, $y \to 1$ (below)

c $f(x) = \dfrac{x}{x^2+1}$

$f(x)$ is defined for all $x \in \mathbb{R}$

$\therefore$ no vertical asymptotes exist

Now $f(x) = \dfrac{\frac{1}{x}}{1+\frac{1}{x^2}}$

$\therefore$ as $|x| \to \infty$, $f(x) \to \frac{0}{1} = 0$

$\therefore$ $y=0$ is a horizontal asymptote

As $x \to \infty$, $f(x) \to 0$ (above)

As $x \to -\infty$, $f(x) \to 0$ (below)

d $f(x) = \dfrac{x-2}{x^2+x-2} = \dfrac{x-2}{(x+2)(x-1)}$

$f(x)$ is undefined when $x=-2$ or 1

$\therefore$ $x=-2$ and $x=1$ are vertical asymptotes

Now $f(x) = \dfrac{\frac{1}{x}-\frac{2}{x^2}}{1+\frac{1}{x}-\frac{2}{x^2}}$

$\therefore$ as $|x| \to \infty$, $f(x) \to \frac{0}{1} = 0$

$\therefore$ $y=0$ is a horizontal asymptote

The sign diagram for $f(x)$ is:

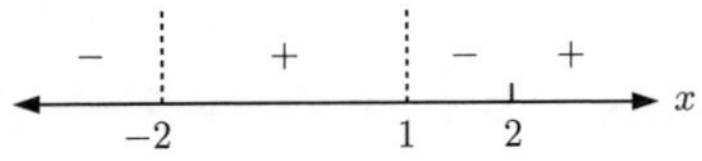

As $x \to -2$ (left), $f(x) \to -\infty$

As $x \to -2$ (right), $f(x) \to \infty$

As $x \to 1$ (left), $f(x) \to \infty$

As $x \to 1$ (right), $f(x) \to -\infty$

As $x \to \infty$, $f(x) \to 0$ (above)

As $x \to -\infty$, $f(x) \to 0$ (below)

EXERCISE 16D.1

1 **a** **i**

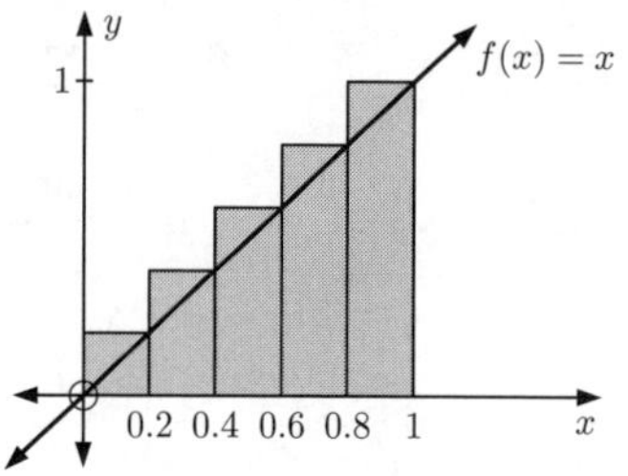

The rectangles are $\frac{1}{5} = 0.2$ units wide.

$$\begin{aligned}A_U &= 0.2 \times f(0.2) + 0.2 \times f(0.4) + 0.2 \times f(0.6) \\ &\quad + 0.2 \times f(0.8) + 0.2 \times f(1) \\ &= 0.2 \times 0.2 + 0.2 \times 0.4 + 0.2 \times 0.6 \\ &\quad + 0.2 \times 0.8 + 0.2 \times 1 \\ &= 0.6 \text{ units}^2\end{aligned}$$

ii

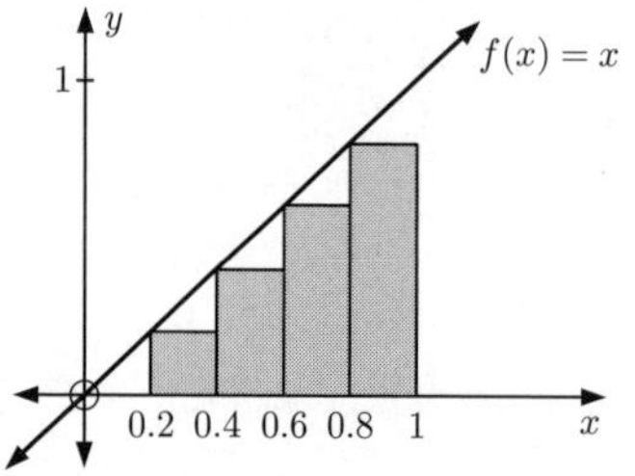

$$\begin{aligned}A_L &= 0.2 \times f(0) + 0.2 \times f(0.2) + 0.2 \times f(0.4) \\ &\quad + 0.2 \times f(0.6) + 0.2 \times f(0.8) \\ &= 0.2 \times 0 + 0.2 \times 0.2 + 0.2 \times 0.4 \\ &\quad + 0.2 \times 0.6 + 0.2 \times 0.8 \\ &= 0.4 \text{ units}^2\end{aligned}$$

b The area between $y = x$ and the x-axis from $x = 0$ to $x = 1$ is a triangle.

$$\begin{aligned}\therefore \quad \text{area} &= \tfrac{1}{2} \times \text{base} \times \text{height} \\ &= \tfrac{1}{2} \times 1 \times 1 \\ &= \tfrac{1}{2} \text{ unit}^2\end{aligned}$$

$\therefore \quad A_L < \text{area} < A_U$, and both A_L and A_U are within 0.1 unit^2, or 20%, of the actual area.

2 The rectangles are $\frac{2}{6} = \frac{1}{3}$ units wide.

a

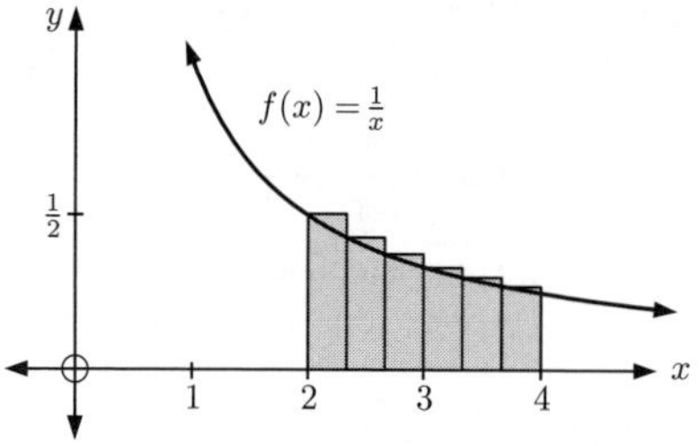

$$\begin{aligned}A_U &= \tfrac{1}{3}f(2) + \tfrac{1}{3}f(\tfrac{7}{3}) + \tfrac{1}{3}f(\tfrac{8}{3}) + \tfrac{1}{3}f(3) \\ &\quad + \tfrac{1}{3}f(\tfrac{10}{3}) + \tfrac{1}{3}f(\tfrac{11}{3}) \\ &= \tfrac{1}{3} \times \tfrac{1}{2} + \tfrac{1}{3} \times \tfrac{3}{7} + \tfrac{1}{3} \times \tfrac{3}{8} + \tfrac{1}{3} \times \tfrac{1}{3} \\ &\quad + \tfrac{1}{3} \times \tfrac{3}{10} + \tfrac{1}{3} \times \tfrac{3}{11} \\ &\approx 0.737 \text{ units}^2\end{aligned}$$

b

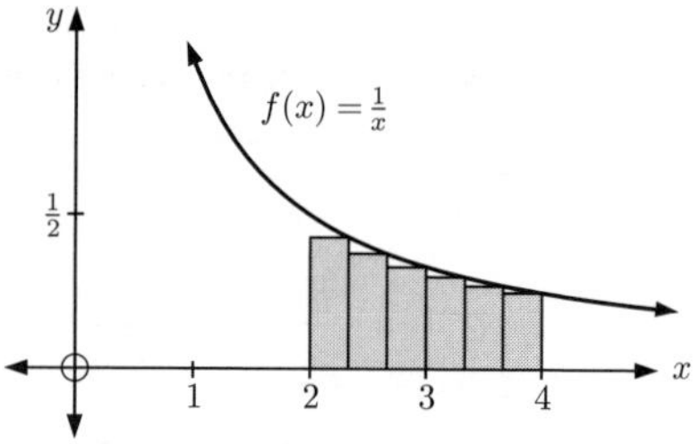

$$\begin{aligned}A_L &= \tfrac{1}{3}f(\tfrac{7}{3}) + \tfrac{1}{3}f(\tfrac{8}{3}) + \tfrac{1}{3}f(3) + \tfrac{1}{3}f(\tfrac{10}{3}) \\ &\quad + \tfrac{1}{3}f(\tfrac{11}{3}) + \tfrac{1}{3}f(4) \\ &= \tfrac{1}{3} \times \tfrac{3}{7} + \tfrac{1}{3} \times \tfrac{3}{8} + \tfrac{1}{3} \times \tfrac{1}{3} + \tfrac{1}{3} \times \tfrac{3}{10} \\ &\quad + \tfrac{1}{3} \times \tfrac{3}{11} + \tfrac{1}{3} \times \tfrac{1}{4} \\ &\approx 0.653 \text{ units}^2\end{aligned}$$

EXERCISE 16D.2

1 Using provided software,

n	A_L	A_U
10	2.1850	2.4850
25	2.2736	2.3936
50	2.3034	2.3634
100	2.3184	2.3484
500	2.3303	2.3363

A_L and A_U converge to $\frac{7}{3}$

2 **a** **i**

n	A_L	A_U
5	0.160 00	0.360 00
10	0.202 50	0.302 50
50	0.240 10	0.260 10
100	0.245 03	0.255 03
500	0.249 00	0.251 00
1000	0.249 50	0.250 50
10 000	0.249 95	0.250 05

ii

n	A_L	A_U
5	0.400 00	0.600 00
10	0.450 00	0.550 00
50	0.490 00	0.510 00
100	0.495 00	0.505 00
500	0.499 00	0.501 00
1000	0.499 50	0.500 50
10 000	0.499 95	0.500 05

iii

n	A_L	A_U
5	0.549 74	0.749 74
10	0.610 51	0.710 51
50	0.656 10	0.676 10
100	0.661 46	0.671 46
500	0.665 65	0.667 65
1000	0.666 16	0.667 16
10 000	0.666 62	0.666 72

iv

n	A_L	A_U
5	0.618 67	0.818 67
10	0.687 40	0.787 40
50	0.738 51	0.758 51
100	0.744 41	0.754 41
500	0.748 93	0.750 93
1000	0.749 47	0.750 47
10 000	0.749 95	0.750 05

b **i** A_L and A_U converge to $0.25 = \frac{1}{4} = \frac{1}{3+1}$

ii A_L and A_U converge to $0.5 = \frac{1}{2} = \frac{1}{1+1}$

iii A_L and A_U converge to $0.6\overline{6} = \frac{2}{3} = \frac{1}{\frac{1}{2}+1}$

iv A_L and A_U converge to $0.75 = \frac{3}{4} = \frac{1}{\frac{1}{3}+1}$

c From **b**, it appears that the area between the graph of $y = x^a$ and the x-axis for $0 \leqslant x \leqslant 1$ is $\dfrac{1}{a+1}$.

3 **a**

n	Rational bounds for π
10	$2.9045 < \pi < 3.3045$
50	$3.0983 < \pi < 3.1783$
100	$3.1204 < \pi < 3.1604$
200	$3.1312 < \pi < 3.1512$
1000	$3.1396 < \pi < 3.1436$
10 000	$3.1414 < \pi < 3.1418$

b $3\frac{10}{71} < \pi < 3\frac{1}{7}$ is approximately
$3.1408 < \pi < 3.1429$
From **a**, this is a better approximation than our estimate using $n = 10, 50, 100, 200, 1000$.
Only $n = 10\,000$ gives us a better estimate.

EXERCISE 16D.3

1 **a**

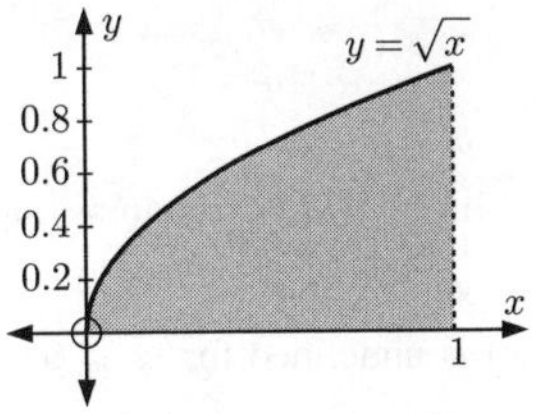

b

n	A_L	A_U
5	0.5497	0.7497
10	0.6105	0.7105
50	0.6561	0.6761
100	0.6615	0.6715
500	0.6656	0.6676

c $\int_0^1 \sqrt{x}\, dx \approx 0.67$

2 **a**

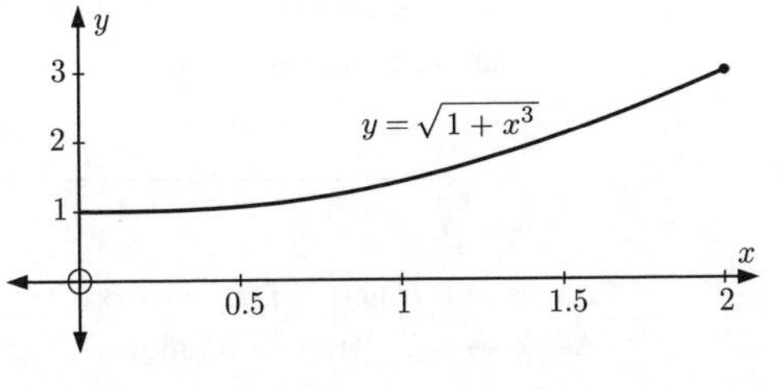

b

n	A_L	A_U
50	3.2016	3.2816
100	3.2214	3.2614
500	3.2373	3.2453

c $\int_0^2 \sqrt{1+x^3}\, dx \approx 3.24$

3 **a**

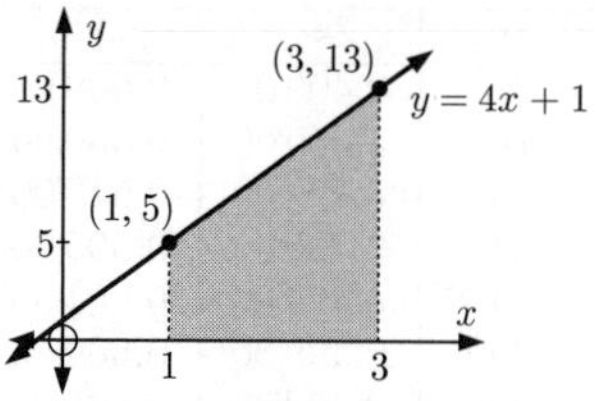

$\int_1^3 (1+4x)\, dx$

= area of the shaded trap.

$= \left(\frac{5+13}{2}\right) \times 2$

$= 18$

b

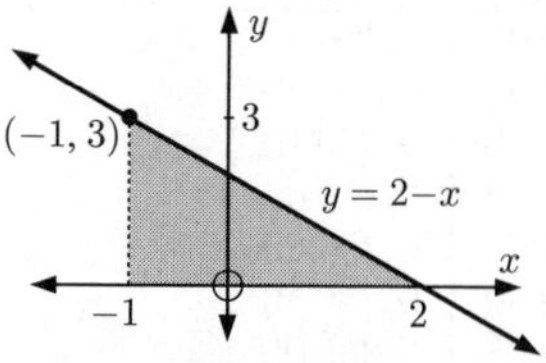

$\int_{-1}^2 (2-x)\, dx$

= area of shaded triangle

$= \frac{1}{2}(3 \times 3)$

$= 4.5$

c

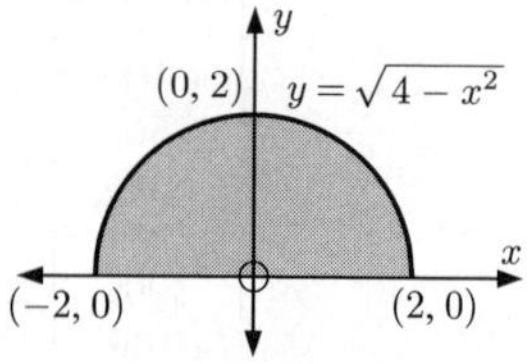

$\int_{-2}^2 \sqrt{4-x^2}\, dx$

= area of semi-circle, radius 2

$= \frac{1}{2}\left(\pi \times 2^2\right)$

$= 2\pi$

REVIEW SET 16

1 **a** $\frac{x^2-4}{2-x} = \frac{(x+2)(x-2)}{(2-x)}$

$\begin{cases} = -(x+2) & \text{if } x \neq 2 \\ \text{is undefined} & \text{if } x = 2 \end{cases}$

$\therefore \lim_{x\to 2} \frac{x^2-4}{2-x}$

$= \lim_{x\to 2} -(x+2), \quad x \neq 2$

$= -4$

b $\frac{\sqrt{x}-2}{x-4} = \frac{\sqrt{x}-2}{(\sqrt{x}+2)(\sqrt{x}-2)}$

$\begin{cases} = \frac{1}{\sqrt{x}+2} & \text{if } x \neq 4 \\ \text{is undefined} & \text{if } x = 4 \end{cases}$

$\therefore \lim_{x\to 4} \frac{\sqrt{x}-2}{x-4}$

$= \lim_{x\to 4} \frac{1}{\sqrt{x}+2}, \quad x \neq 4$

$= \frac{1}{4}$

c $\frac{x^2-16}{x-4} = \frac{(x+4)(x-4)}{x-4}$

$\begin{cases} = x+4 & \text{if } x \neq 4 \\ \text{is undefined} & \text{if } x = 4 \end{cases}$

$\therefore \lim_{x\to 4} \frac{x^2-16}{x-4}$

$= \lim_{x\to 4} x+4, \quad x \neq 4$

$= 8$

d $\lim_{x\to\infty} \frac{1-2x-x^2}{2x^2-4}$

$= \lim_{x\to\infty} \frac{\frac{1}{x^2}-\frac{2}{x}-1}{2-\frac{4}{x^2}}$

$= \frac{-1}{2}$

2 **a** $y = e^{x-2} - 3$

y is defined for all $x \in \mathbb{R}$

$\therefore$ no vertical asymptotes exist

As $x \to -\infty$, $y \to -3$

$\therefore$ $y = -3$ is a horizontal asymptote

As $x \to \infty$, $y \to \infty$

As $x \to -\infty$, $y \to -3$ (above)

b $y = \ln(x^2+3)$ has no asymptotes

c $f(x) = e^{-x} \ln x$

$f(x)$ is undefined for $x \leqslant 0$

$\therefore$ $x = 0$ is a vertical asymptote

As $x \to \infty$, $f(x) \to 0$

$\therefore$ $y = 0$ is a horizontal asymptote

$f(x)$ has sign diagram:

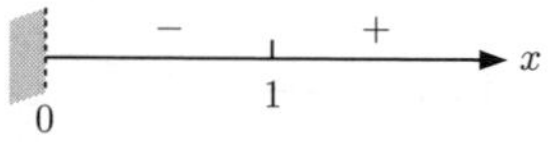

As $x \to 0$ (right), $f(x) \to -\infty$

As $x \to \infty$, $f(x) \to 0$ (above)

d $y = x + \ln(2x-3)$

y is undefined for $x \leqslant \frac{3}{2}$

$\therefore$ $x = \frac{3}{2}$ is a vertical asymptote

As $x \to \frac{3}{2}$ (right), $y \to -\infty$

3 **a**

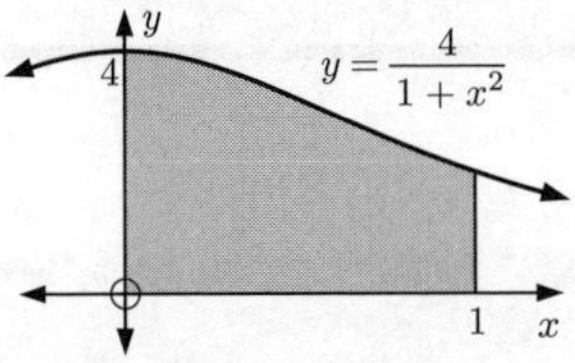

lower rectangles

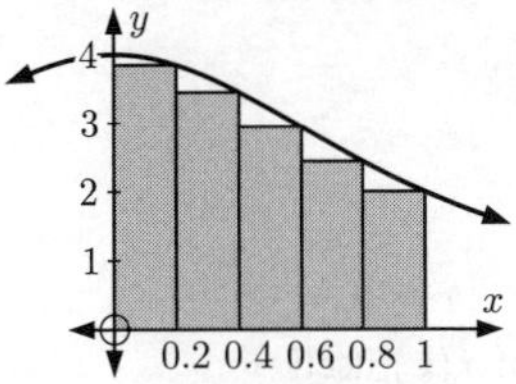

upper rectangles

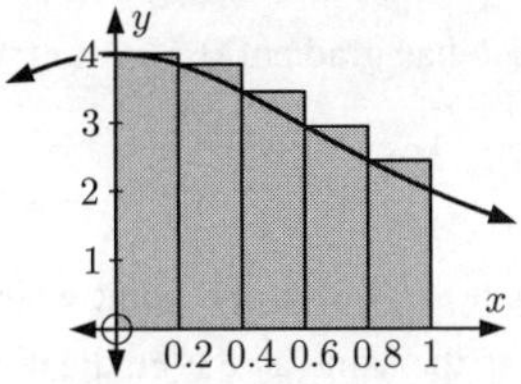

b

n	A_L	A_U
5	2.9349	3.3349
50	3.1215	3.1615
100	3.1316	3.1516
500	3.1396	3.1436

c $\displaystyle\int_0^1 \frac{4}{1+x^2}\,dx \approx 3.1416$

(the average of A_L and A_U for $n = 500$). This value agrees with π to 4 decimal places.

4 **a**

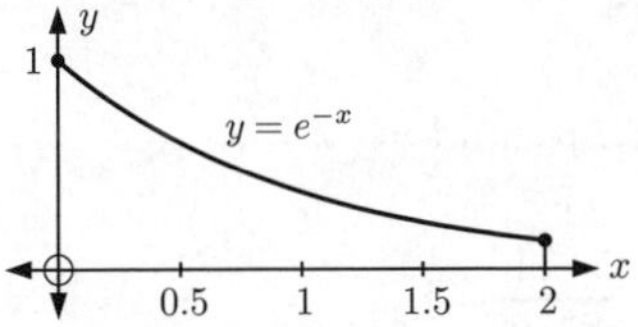

Each rectangle has width $w = \dfrac{2-0}{8} = 0.25$.

For $f(x) = e^{-x}$,

$$f(0) = 1, \quad f(0.25) \approx 0.779, \quad f(0.5) \approx 0.607,$$
$$f(0.75) \approx 0.472, \quad f(1) \approx 0.368, \quad f(1.25) \approx 0.287,$$
$$f(1.5) \approx 0.223, \quad f(1.75) \approx 0.174, \quad f(2) \approx 0.135.$$

b Upper bound $= 0.25\,[f(0) + f(0.25) + f(0.5) + f(0.75) + f(1) + f(1.25) + f(1.5) + f(1.75)]$
$\approx 0.25(3.91)$
≈ 0.977 units^2

Lower bound $= 0.25\,[f(0.25) + f(0.5) + f(0.75) + f(1) + f(1.25) + f(1.5) + f(1.75) + f(2)]$
$\approx 0.25(3.044)$
≈ 0.761 units^2

c Using technology, $A_U \approx 0.8733$ units^2, $A_L \approx 0.8560$ units^2.

5 **a** $\int_0^4 f(x)\,dx =$ area of semi-circle with radius 2
$= \frac{1}{2} \times \pi \times 2^2$
$= 2\pi$

b $\int_4^6 f(x)\,dx =$ area of square
$= 2 \times 2$
$= 4$

6 **a**

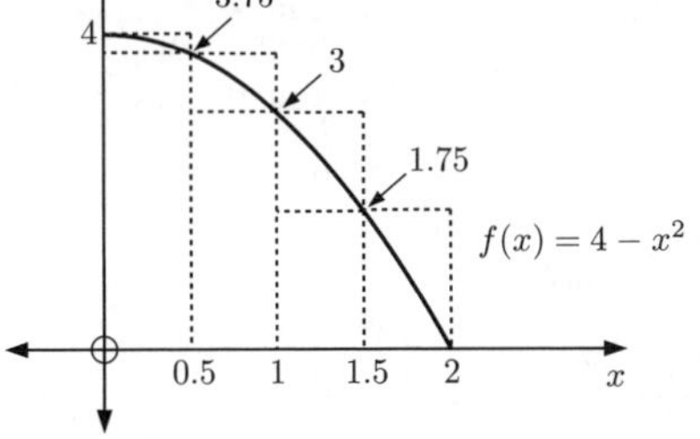

$$A_U = 0.5\,[f(0) + f(0.5) + f(1) + f(1.5)]$$
$$= 0.5(4 + 3.75 + 3 + 1.75)$$
$$= 6.25$$

$$A_L = 0.5\,[f(0.5) + f(1) + f(1.5) + f(2)]$$
$$= 0.5(3.75 + 3 + 1.75 + 0)$$
$$= 4.25$$

$\therefore\ 4.25 < \int_0^2 (4 - x^2)\,dx < 6.25$

$\therefore\ A = 4.25 = \frac{17}{4}, \quad B = 6.25 = \frac{25}{4}$

b an estimate of $\int_0^2 (4 - x^2)\,dx = \dfrac{A+B}{2} = \frac{42}{8} = \frac{21}{4}$

Chapter 17

DIFFERENTIAL CALCULUS

EXERCISE 17A.1

1 **a** $f(2) = 3$

b $f'(2)$ is the gradient of the tangent to $f(x)$ at the point where $x = 2$.
Since $f(x)$ is a straight line, this is the same as the gradient of $f(x)$ itself.
$f(x)$ is a horizontal line, and hence has gradient 0.
$\therefore\ f'(2) = 0$

2 **a** $f(0) = 4$

b $f'(0)$ is the gradient of the tangent to $f(x)$ at the point where $x = 0$.
Since $f(x)$ is a straight line, this is the same as the gradient of $f(x)$ itself.
$f(x)$ goes through (0, 4) and (4, 0), so it has gradient $= \dfrac{0-4}{4-0} = -1$
$\therefore\ f'(0) = -1$

3 The graph shows the tangent to the curve $y = f(x)$ at the point where $x = 2$.
The gradient of this tangent is $f'(2)$.
The tangent passes through (0, 1) and (4, 5), so $f'(2) = \dfrac{5-1}{4-0} = 1$.

The equation of the tangent is $\dfrac{y-1}{x-0} = 1$

$\therefore\ y = x + 1$

When $x = 2$, $y = 3$, so the point of contact is (2, 3).
$\therefore\ f(2) = 3$ and $f'(2) = 1$.

EXERCISE 17A.2

1 **a** $f(x) = x$

$$\begin{aligned}\therefore\ f'(x) &= \lim_{h\to 0} \frac{f(x+h)-f(x)}{h}\\ &= \lim_{h\to 0} \frac{(x+h)-x}{h}\\ &= \lim_{h\to 0} \frac{h}{h}\\ &= \lim_{h\to 0} 1 \quad \{\text{as } h \neq 0\}\\ &= 1\end{aligned}$$

b $f(x) = 5$

$$\begin{aligned}\therefore\ f'(x) &= \lim_{h\to 0} \frac{f(x+h)-f(x)}{h}\\ &= \lim_{h\to 0} \frac{5-5}{h}\\ &= \lim_{h\to 0} \frac{0}{h}\\ &= \lim_{h\to 0} 0 \quad \{\text{as } h \neq 0\}\\ &= 0\end{aligned}$$

c $f(x) = x^3$

$$\begin{aligned}\therefore\ f'(x) &= \lim_{h\to 0} \frac{f(x+h)-f(x)}{h}\\ &= \lim_{h\to 0} \frac{(x+h)^3-x^3}{h}\\ &= \lim_{h\to 0} \frac{x^3+3x^2h+3xh^2+h^3-x^3}{h}\\ &= \lim_{h\to 0} \frac{3x^2h+3xh^2+h^3}{h}\\ &= \lim_{h\to 0} 3x^2+3xh+h^2 \quad \{\text{as } h \neq 0\}\\ &= 3x^2\end{aligned}$$

d $f(x) = x^4 \quad \therefore \quad f'(x) = \lim_{h \to 0} \frac{f(x+h) - f(x)}{h}$

$$= \lim_{h \to 0} \frac{(x+h)^4 - x^4}{h}$$

$$= \lim_{h \to 0} \frac{x^4 + 4x^3h + 6x^2h^2 + 4xh^3 + h^4 - x^4}{h}$$

$$= \lim_{h \to 0} \frac{4x^3h + 6x^2h^2 + 4xh^3 + h^4}{h}$$

$$= \lim_{h \to 0} 4x^3 + 6x^2h + 4xh^2 + h^3 \quad \{\text{as } h \neq 0\}$$

$$= 4x^3$$

2 From **1**, we predict that if $f(x) = x^n$, $f'(x) = nx^{n-1}$, $n \in \mathbb{N}$.

3 **a** $f(x) = 2x + 5$

$\therefore \quad f'(x)$

$$= \lim_{h \to 0} \frac{f(x+h) - f(x)}{h}$$

$$= \lim_{h \to 0} \frac{(2(x+h) + 5) - (2x + 5)}{h}$$

$$= \lim_{h \to 0} \frac{2x + 2h + 5 - 2x - 5}{h}$$

$$= \lim_{h \to 0} \frac{2h}{h}$$

$$= \lim_{h \to 0} 2 \quad \{\text{as } h \neq 0\}$$

$$= 2$$

b $f(x) = -x + 4$

$\therefore \quad f'(x)$

$$= \lim_{h \to 0} \frac{f(x+h) - f(x)}{h}$$

$$= \lim_{h \to 0} \frac{(-(x+h) + 4) - (-x + 4)}{h}$$

$$= \lim_{h \to 0} \frac{-x - h + 4 + x - 4}{h}$$

$$= \lim_{h \to 0} \frac{-h}{h}$$

$$= \lim_{h \to 0} -1 \quad \{\text{as } h \neq 0\}$$

$$= -1$$

c $f(x) = x^2 - 3x$

$$\therefore \quad f'(x) = \lim_{h \to 0} \frac{f(x+h) - f(x)}{h}$$

$$= \lim_{h \to 0} \frac{[(x+h)^2 - 3(x+h)] - [x^2 - 3x]}{h}$$

$$= \lim_{h \to 0} \frac{x^2 + 2xh + h^2 - 3x - 3h - x^2 + 3x}{h}$$

$$= \lim_{h \to 0} \frac{2xh + h^2 - 3h}{h}$$

$$= \lim_{h \to 0} 2x + h - 3 \quad \{\text{as } h \neq 0\}$$

$$= 2x - 3$$

d $f(x) = 2x^2 + x - 1$

$$\therefore \quad f'(x) = \lim_{h \to 0} \frac{f(x+h) - f(x)}{h}$$

$$= \lim_{h \to 0} \frac{[2(x+h)^2 + (x+h) - 1] - [2x^2 + x - 1]}{h}$$

$$= \lim_{h \to 0} \frac{2x^2 + 4xh + 2h^2 + x + h - 1 - 2x^2 - x + 1}{h}$$

$$= \lim_{h \to 0} \frac{4xh + 2h^2 + h}{h}$$
$$= \lim_{h \to 0} 4x + 1 + 2h \quad \{\text{as } h \neq 0\}$$
$$= 4x + 1$$

e $$f(x) = -x^2 + 5x - 3$$
$$\therefore \quad f'(x) = \lim_{h \to 0} \frac{f(x+h) - f(x)}{h}$$
$$= \lim_{h \to 0} \frac{[-(x+h)^2 + 5(x+h) - 3] - [-x^2 + 5x - 3]}{h}$$
$$= \lim_{h \to 0} \frac{-x^2 - 2xh - h^2 + 5x + 5h - 3 + x^2 - 5x + 3}{h}$$
$$= \lim_{h \to 0} \frac{-2xh - h^2 + 5h}{h}$$
$$= \lim_{h \to 0} -2x + 5 - h \quad \{\text{as } h \neq 0\}$$
$$= -2x + 5$$

f $$f(x) = x^3 - 2x^2 + 3$$
$$\therefore \quad f'(x) = \lim_{h \to 0} \frac{f(x+h) - f(x)}{h}$$
$$= \lim_{h \to 0} \frac{[(x+h)^3 - 2(x+h)^2 + 3] - [x^3 - 2x^2 + 3]}{h}$$
$$= \lim_{h \to 0} \frac{(x^3 + 3x^2h + 3xh^2 + h^3 - 2x^2 - 4xh - 2h^2 + 3) - (x^3 - 2x^2 + 3)}{h}$$
$$= \lim_{h \to 0} \frac{3x^2h + 3xh^2 + h^3 - 4xh - 2h^2}{h}$$
$$= \lim_{h \to 0} 3x^2 + 3xh + h^2 - 4x - 2h \quad \{\text{as } h \neq 0\}$$
$$= 3x^2 - 4x$$

EXERCISE 17B

1 a $$f(x) = 3x + 5 \text{ at } x = -2$$
$$\therefore \quad f(-2) = 3(-2) + 5 = -1$$
$$f'(-2) = \lim_{x \to -2} \frac{f(x) - f(-2)}{x + 2}$$
$$= \lim_{x \to -2} \frac{(3x + 5) - (-1)}{x + 2}$$
$$= \lim_{x \to -2} \frac{3x + 6}{x + 2}$$
$$= \lim_{x \to -2} \frac{3\cancel{(x+2)}}{\cancel{x+2}_1}$$
$$= \lim_{x \to -2} 3 \quad \{\text{as } x \neq -2\}$$
$$= 3$$

b $$f(x) = 5 - 2x^2 \text{ at } x = 3$$
$$\therefore \quad f(3) = 5 - 2(3)^2 = -13$$
$$f'(3) = \lim_{x \to 3} \frac{f(x) - f(3)}{x - 3}$$
$$= \lim_{x \to 3} \frac{(5 - 2x^2) - (-13)}{x - 3}$$
$$= \lim_{x \to 3} \frac{18 - 2x^2}{x - 3}$$
$$= \lim_{x \to 3} \frac{-2(x^2 - 9)}{x - 3}$$
$$= \lim_{x \to 3} \frac{-2(x+3)\cancel{(x-3)}}{\cancel{x-3}_1}$$
$$= \lim_{x \to 3} -2(x+3) \quad \{\text{as } x \neq 3\}$$
$$= -2(6)$$
$$= -12$$

c
$$\begin{aligned} f(x) &= x^2+3x-4 \text{ at } x=3 \\ \therefore\ f(3) &= 3^2+3(3)-4 = 14 \\ f'(3) &= \lim_{x\to 3} \frac{f(x)-f(3)}{x-3} \\ &= \lim_{x\to 3} \frac{(x^2+3x-4)-14}{x-3} \\ &= \lim_{x\to 3} \frac{x^2+3x-18}{x-3} \\ &= \lim_{x\to 3} \frac{(x+6)\cancel{(x-3)}}{\cancel{x-3}_1} \\ &= \lim_{x\to 3} x+6 \quad \{\text{as } x\neq 3\} \\ &= 9 \end{aligned}$$

d
$$\begin{aligned} f(x) &= 5-2x-3x^2 \text{ at } x=-2 \\ \therefore\ f(-2) &= 5-2(-2)-3(-2)^2 = -3 \\ f'(-2) &= \lim_{x\to -2} \frac{f(x)-f(-2)}{x-(-2)} \\ &= \lim_{x\to -2} \frac{5-2x-3x^2-(-3)}{x+2} \\ &= \lim_{x\to -2} \frac{-(3x^2+2x-8)}{x+2} \\ &= \lim_{x\to -2} \frac{-(3x-4)\cancel{(x+2)}}{\cancel{x+2}_1} \\ &= \lim_{x\to -2} -(3x-4) \quad \{\text{as } x\neq -2\} \\ &= -(3(-2)-4) \\ &= 10 \end{aligned}$$

2 **a** $f(x)=x^3$
$$\begin{aligned} \therefore\ f'(2) &= \lim_{h\to 0} \frac{f(2+h)-f(2)}{h} \quad \text{where } f(2)=2^3=8 \\ &= \lim_{h\to 0} \frac{(2+h)^3-8}{h} \\ &= \lim_{h\to 0} \frac{8+12h+6h^2+h^3-8}{h} \\ &= \lim_{h\to 0} \frac{12h+6h^2+h^3}{h} \\ &= \lim_{h\to 0} 12+6h+h^2 \quad \{\text{as } h\neq 0\} \\ &= 12 \end{aligned}$$

b $f(x)=x^4$
$$\begin{aligned} \therefore\ f'(3) &= \lim_{h\to 0} \frac{f(3+h)-f(3)}{h} \quad \text{where } f(3)=3^4=81 \\ &= \lim_{h\to 0} \frac{(3+h)^4-81}{h} \\ &= \lim_{h\to 0} \frac{81+108h+54h^2+12h^3+h^4-81}{h} \\ &= \lim_{h\to 0} \frac{108h+54h^2+12h^3+h^4}{h} \\ &= \lim_{h\to 0} 108+54h+12h^2+h^3 \quad \{\text{as } h\neq 0\} \\ &= 108 \end{aligned}$$

EXERCISE 17C

1 **a**
$$\begin{aligned} f(x) &= x^3 \\ \therefore\ f'(x) &= 3x^2 \end{aligned}$$

b
$$\begin{aligned} f(x) &= 2x^3 \\ \therefore\ f'(x) &= 2(3x^2) \\ &= 6x^2 \end{aligned}$$

c
$$\begin{aligned} f(x) &= 7x^2 \\ \therefore\ f'(x) &= 7(2x) \\ &= 14x \end{aligned}$$

d
$$\begin{aligned} f(x) &= 6\sqrt{x} = 6x^{\frac{1}{2}} \\ \therefore\ f'(x) &= 6\left(\tfrac{1}{2}x^{-\frac{1}{2}}\right) \\ &= \frac{3}{\sqrt{x}} \end{aligned}$$

e
$$\begin{aligned} f(x) &= 3\sqrt[3]{x} = 3x^{\frac{1}{3}} \\ \therefore\ f'(x) &= 3\left(\tfrac{1}{3}x^{-\frac{2}{3}}\right) \\ &= \frac{1}{\sqrt[3]{x^2}} \end{aligned}$$

f
$$\begin{aligned} f(x) &= x^2+x \\ \therefore\ f'(x) &= 2x+1 \end{aligned}$$

g $f(x) = 4 - 2x^2$
$\therefore\ f'(x) = 0 - 2 \times 2x$
$= -4x$

h $f(x) = x^2 + 3x - 5$
$\therefore\ f'(x) = 2x + 3 - 0$
$= 2x + 3$

i $f(x) = \frac{1}{2}x^4 - 6x^2$
$\therefore\ f'(x) = \frac{1}{2}(4x^3) - 6(2x)$
$= 2x^3 - 12x$

j $f(x) = \dfrac{3x-6}{x} = 3 - 6x^{-1}$
$\therefore\ f'(x) = 0 - 6(-1x^{-2})$
$= \dfrac{6}{x^2}$

k $f(x) = \dfrac{2x-3}{x^2} = \dfrac{2x}{x^2} - \dfrac{3}{x^2}$
$= 2x^{-1} - 3x^{-2}$
$\therefore\ f'(x) = -2x^{-2} + 6x^{-3} = \dfrac{-2}{x^2} + \dfrac{6}{x^3}$

l $f(x) = \dfrac{x^3+5}{x} = x^2 + 5x^{-1}$
$\therefore\ f'(x) = 2x - 5x^{-2}$
$= 2x - \dfrac{5}{x^2}$

m $f(x) = \dfrac{x^3+x-3}{x}$
$= x^2 + 1 - 3x^{-1}$
$\therefore\ f'(x) = 2x + 0 + 3x^{-2}$
$= 2x + \dfrac{3}{x^2}$

n $f(x) = \dfrac{1}{\sqrt{x}} = x^{-\frac{1}{2}}$
$\therefore\ f'(x) = -\frac{1}{2}x^{-\frac{3}{2}} = -\dfrac{1}{2x\sqrt{x}}$

o $f(x) = (2x-1)^2 = 4x^2 - 4x + 1$
$\therefore\ f'(x) = 8x - 4$

p $f(x) = (x+2)^3$
$= x^3 + 3x^2(2) + 3x(2^2) + 2^3$
$= x^3 + 6x^2 + 12x + 8$
$\therefore\ f'(x) = 3x^2 + 12x + 12$

2 **a** $y = 2.5x^3 - 1.4x^2 - 1.3$
$\therefore\ \dfrac{dy}{dx} = 7.5x^2 - 2.8x$

b $y = \pi x^2$
$\therefore\ \dfrac{dy}{dx} = 2\pi x$

c $y = \dfrac{1}{5x^2} = \frac{1}{5}x^{-2}$
$\therefore\ \dfrac{dy}{dx} = -\frac{2}{5}x^{-3} = -\dfrac{2}{5x^3}$

d $y = 100x$
$\therefore\ \dfrac{dy}{dx} = 100$

e $y = 10(x+1)$
$= 10x + 10$
$\therefore\ \dfrac{dy}{dx} = 10$

f $y = 4\pi x^3$
$\therefore\ \dfrac{dy}{dx} = 12\pi x^2$

3 **a** $\dfrac{d}{dx}(6x+2)$
$= 6$

b $\dfrac{d}{dx}(x\sqrt{x})$
$= \dfrac{d}{dx}(x^{\frac{3}{2}})$
$= \frac{3}{2}x^{\frac{1}{2}}$
$= \frac{3}{2}\sqrt{x}$

c $\dfrac{d}{dx}(5-x)^2$
$= \dfrac{d}{dx}(25 - 10x + x^2)$
$= -10 + 2x$
$= 2x - 10$

d $\dfrac{d}{dx}\left(\dfrac{6x^2-9x^4}{3x}\right)$
$= \dfrac{d}{dx}(2x - 3x^3)$
$= 2 - 9x^2$

e $\dfrac{d}{dx}((x+1)(x-2))$
$= \dfrac{d}{dx}(x^2 - x - 2)$
$= 2x - 1$

f $\dfrac{d}{dx}\left(\dfrac{1}{x^2} + 6\sqrt{x}\right)$
$= \dfrac{d}{dx}\left(x^{-2} + 6x^{\frac{1}{2}}\right)$
$= -2x^{-3} + 3x^{-\frac{1}{2}}$
$= -\dfrac{2}{x^3} + \dfrac{3}{\sqrt{x}}$

g $\dfrac{d}{dx}\left(4x - \dfrac{1}{4x}\right)$

$= \dfrac{d}{dx}\left(4x - \frac{1}{4}x^{-1}\right)$

$= 4 + \frac{1}{4}x^{-2}$

$= 4 + \dfrac{1}{4x^2}$

h $\dfrac{d}{dx}\left(x(x+1)(2x-5)\right)$

$= \dfrac{d}{dx}\left(x(2x^2 - 3x - 5)\right)$

$= \dfrac{d}{dx}\left(2x^3 - 3x^2 - 5x\right)$

$= 6x^2 - 6x - 5$

4 **a** Consider $y = x^2$ when $x = 2$

Now $\dfrac{dy}{dx} = 2x$

$\therefore$ when $x = 2$,

$\dfrac{dy}{dx} = 2(2) = 4$

$\therefore$ the tangent has gradient 4.

b Consider $y = \dfrac{8}{x^2}$ at the point $(9, \frac{8}{81})$

Now $y = 8x^{-2}$

$\therefore \dfrac{dy}{dx} = -16x^{-3} = -\dfrac{16}{x^3}$

$\therefore$ at $(9, \frac{8}{81})$, $x = 9$ and so $\dfrac{dy}{dx} = -\frac{16}{729}$

$\therefore$ the tangent has gradient $-\frac{16}{729}$.

c Consider $y = 2x^2 - 3x + 7$ when $x = -1$

Now $\dfrac{dy}{dx} = 4x - 3$

$\therefore$ when $x = -1$,

$\dfrac{dy}{dx} = 4(-1) - 3 = -7$

$\therefore$ the tangent has gradient -7.

d Consider $y = \dfrac{2x^2 - 5}{x}$ at the point $(2, \frac{3}{2})$

Now $y = 2x - 5x^{-1}$

$\therefore \dfrac{dy}{dx} = 2 + 5x^{-2} = 2 + \dfrac{5}{x^2}$

$\therefore$ at $(2, \frac{3}{2})$, $x = 2$ and so $\dfrac{dy}{dx} = 2 + \frac{5}{4} = \frac{13}{4}$

$\therefore$ the tangent has gradient $\frac{13}{4}$.

e Consider $y = \dfrac{x^2 - 4}{x^2}$ at the point $(4, \frac{3}{4})$

Now $y = 1 - 4x^{-2}$

$\therefore \dfrac{dy}{dx} = 0 + 8x^{-3} = \dfrac{8}{x^3}$

$\therefore$ at $(4, \frac{3}{4})$, $x = 4$ and so

$\dfrac{dy}{dx} = \dfrac{8}{4^3} = \frac{1}{8}$

$\therefore$ the tangent has gradient $\frac{1}{8}$.

f Consider $y = \dfrac{x^3 - 4x - 8}{x^2}$ when $x = -1$

Now $y = x - 4x^{-1} - 8x^{-2}$

$\therefore \dfrac{dy}{dx} = 1 + 4x^{-2} + 16x^{-3}$

$= 1 + \dfrac{4}{x^2} + \dfrac{16}{x^3}$

$\therefore$ when $x = -1$,

$\dfrac{dy}{dx} = 1 + 4 - 16 = -11$

$\therefore$ the tangent has gradient -11.

5 $f(x) = x^2 + (b+1)x + 2c$, $f(2) = 4$, and $f'(-1) = 2$

$\therefore f'(x) = 2x + (b+1)$

But $f'(-1) = 2$, so $2(-1) + b + 1 = 2$

$\therefore -1 + b = 2$

$\therefore b = 3$

So, $f(x) = x^2 + (3+1)x + 2c$

$= x^2 + 4x + 2c$

But $f(2) = 4$, so $2^2 + 4(2) + 2c = 4$

$\therefore 2c = -8$

$\therefore c = -4$

6 **a** $f(x) = 4\sqrt{x} + x = 4x^{\frac{1}{2}} + x$

$\therefore f'(x) = 4\left(\frac{1}{2}\right)x^{-\frac{1}{2}} + 1$

$= \dfrac{2}{\sqrt{x}} + 1$

b $f(x) = \sqrt[3]{x} = x^{\frac{1}{3}}$

$\therefore f'(x) = \frac{1}{3}x^{-\frac{2}{3}}$

$= \dfrac{1}{3\sqrt[3]{x^2}}$

c $f(x) = -\dfrac{2}{\sqrt{x}} = -2x^{-\frac{1}{2}}$

$\therefore\ f'(x) = -2(-\tfrac{1}{2})\,x^{-\frac{3}{2}}$

$= x^{-\frac{3}{2}}$

$= \dfrac{1}{x\sqrt{x}}$

d $f(x) = 2x - \sqrt{x} = 2x - x^{\frac{1}{2}}$

$\therefore\ f'(x) = 2 - \tfrac{1}{2}x^{-\frac{1}{2}}$

$= 2 - \dfrac{1}{2\sqrt{x}}$

e $f(x) = \dfrac{4}{\sqrt{x}} - 5 = 4x^{-\frac{1}{2}} - 5$

$\therefore\ f'(x) = 4(-\tfrac{1}{2})\,x^{-\frac{3}{2}}$

$= -2x^{-\frac{3}{2}}$ or $-\dfrac{2}{x\sqrt{x}}$

f $f(x) = 3x^2 - x\sqrt{x} = 3x^2 - x^{\frac{3}{2}}$

$\therefore\ f'(x) = 6x - \tfrac{3}{2}x^{\frac{1}{2}}$

$= 6x - \tfrac{3}{2}\sqrt{x}$

g $f(x) = \dfrac{5}{x^2\sqrt{x}} = 5x^{-\frac{5}{2}}$

$\therefore\ f'(x) = 5(-\tfrac{5}{2})x^{-\frac{7}{2}}$

$= -\tfrac{25}{2}x^{-\frac{7}{2}}$

$= \dfrac{-25}{2x^3\sqrt{x}}$

h $f(x) = 2x - \dfrac{3}{x\sqrt{x}} = 2x - 3x^{-\frac{3}{2}}$

$\therefore\ f'(x) = 2 - 3(-\tfrac{3}{2})x^{-\frac{5}{2}}$

$= 2 + \tfrac{9}{2}x^{-\frac{5}{2}}$

$= 2 + \dfrac{9}{2x^2\sqrt{x}}$

7 **a** $y = 4x - \dfrac{3}{x} = 4x - 3x^{-1}$ $\therefore$ $\dfrac{dy}{dx} = 4 + 3x^{-2} = 4 + \dfrac{3}{x^2}$

$\dfrac{dy}{dx}$ is the gradient function of $y = 4x - \dfrac{3}{x}$ from which the gradient at any point can be found.

b $S = 2t^2 + 4t$ m $\therefore$ $\dfrac{dS}{dt} = 4t + 4$ m s^{-1}

$\dfrac{dS}{dt}$ is the instantaneous rate of change in position at time t. It is the velocity function.

c $C = 1785 + 3x + 0.002x^2$ dollars.

$\dfrac{dC}{dx} = 3 + 0.002(2x) = 3 + 0.004x$ dollars per toaster

$\dfrac{dC}{dx}$ is the instantaneous rate of change in cost as the number of toasters changes.

EXERCISE 17D.1

1 **a** $g(x) = x^2,\ \ f(x) = 2x + 7$

$\therefore\ g(f(x)) = g(2x + 7) = (2x + 7)^2$

b $g(x) = 2x + 7,\ \ f(x) = x^2$

$g(f(x)) = g(x^2) = 2x^2 + 7$

c $g(x) = \sqrt{x},\ \ f(x) = 3 - 4x$

$g(f(x)) = g(3 - 4x) = \sqrt{3 - 4x}$

d $g(x) = 3 - 4x,\ \ f(x) = \sqrt{x}$

$g(f(x)) = g(\sqrt{x}) = 3 - 4\sqrt{x}$

e $g(x) = \dfrac{2}{x},\ \ f(x) = x^2 + 3$

$g(f(x)) = g(x^2 + 3) = \dfrac{2}{x^2 + 3}$

f $g(x) = x^2 + 3,\ \ f(x) = \dfrac{2}{x}$

$g(f(x)) = g\left(\dfrac{2}{x}\right) = \left(\dfrac{2}{x}\right)^2 + 3 = \dfrac{4}{x^2} + 3$

2 **a** $g(f(x)) = (3x + 10)^3$ $\therefore$ $g(x) = x^3,\ \ f(x) = 3x + 10$

b $g(f(x)) = \dfrac{1}{2x + 4}$ $\therefore$ $g(x) = \dfrac{1}{x},\ \ f(x) = 2x + 4$

c $g(f(x)) = \sqrt{x^2 - 3x}$ $\therefore$ $g(x) = \sqrt{x},\ \ f(x) = x^2 - 3x$

d $g(f(x)) = \dfrac{10}{(3x - x^2)^3}$ $\therefore$ $g(x) = \dfrac{10}{x^3},\ \ f(x) = 3x - x^2$

{other answers are possible for **2**}

EXERCISE 17D.2

1 **a** $\dfrac{1}{(2x-1)^2}$
$= (2x-1)^{-2}$
$= u^{-2}$
where $u = 2x - 1$

b $\sqrt{x^2 - 3x}$
$= (x^2 - 3x)^{\frac{1}{2}}$
$= u^{\frac{1}{2}}$
where $u = x^2 - 3x$

c $\dfrac{2}{\sqrt{2 - x^2}}$
$= 2(2 - x^2)^{-\frac{1}{2}}$
$= 2u^{-\frac{1}{2}}$
where $u = 2 - x^2$

d $\sqrt[3]{x^3 - x^2}$
$= (x^3 - x^2)^{\frac{1}{3}}$
$= u^{\frac{1}{3}}$
where $u = x^3 - x^2$

e $\dfrac{4}{(3-x)^3}$
$= 4(3-x)^{-3}$
$= 4u^{-3}$
where $u = 3 - x$

f $\dfrac{10}{x^2 - 3}$
$= 10(x^2 - 3)^{-1}$
$= 10u^{-1}$
where $u = x^2 - 3$

2 **a** $y = (4x - 5)^2$
$\therefore\ y = u^2$ where $u = 4x - 5$

Now $\dfrac{dy}{dx} = \dfrac{dy}{du}\dfrac{du}{dx}$
$= 2u(4)$
$= 8u$
$= 8(4x - 5)$

b $y = \dfrac{1}{5 - 2x}$
$\therefore\ y = u^{-1}$ where $u = 5 - 2x$
Now $\dfrac{dy}{dx} = \dfrac{dy}{du}\dfrac{du}{dx}$
$= -u^{-2}(-2)$
$= \dfrac{2}{u^2}$
$= \dfrac{2}{(5 - 2x)^2}$

c $y = \sqrt{3x - x^2}$
$\therefore\ y = u^{\frac{1}{2}}$ where $u = 3x - x^2$
Now $\dfrac{dy}{dx} = \dfrac{dy}{du}\dfrac{du}{dx}$
$= \frac{1}{2}u^{-\frac{1}{2}}(3 - 2x)$
$= \dfrac{(3 - 2x)}{2\sqrt{u}}$
$= \dfrac{3 - 2x}{2\sqrt{3x - x^2}}$

d $y = (1 - 3x)^4$
$\therefore\ y = u^4$ where $u = 1 - 3x$
Now $\dfrac{dy}{dx} = \dfrac{dy}{du}\dfrac{du}{dx}$
$= 4u^3(-3)$
$= -12u^3$
$= -12(1 - 3x)^3$

e $y = 6(5 - x)^3$
$\therefore\ y = 6u^3$ where $u = 5 - x$
Now $\dfrac{dy}{dx} = \dfrac{dy}{du}\dfrac{du}{dx}$
$= 18u^2(-1)$
$= -18u^2$
$= -18(5 - x)^2$

f $y = \sqrt[3]{2x^3 - x^2}$
$\therefore\ y = u^{\frac{1}{3}}$ where $u = 2x^3 - x^2$
Now $\dfrac{dy}{dx} = \dfrac{dy}{du}\dfrac{du}{dx}$
$= \frac{1}{3}u^{-\frac{2}{3}}(6x^2 - 2x)$
$= \dfrac{6x^2 - 2x}{3\sqrt[3]{(2x^3 - x^2)^2}}$

g $y = \dfrac{6}{(5x - 4)^2}$
$\therefore\ y = 6u^{-2}$ where $u = 5x - 4$
Now $\dfrac{dy}{dx} = \dfrac{dy}{du}\dfrac{du}{dx} = -12u^{-3}(5)$
$= -\dfrac{60}{u^3}$
$= -\dfrac{60}{(5x - 4)^3}$

h $y = \dfrac{4}{3x - x^2}$

$\therefore\ y = 4u^{-1}$ where $u = 3x - x^2$

Now $\dfrac{dy}{dx} = \dfrac{dy}{du}\dfrac{du}{dx} = -4u^{-2}(3 - 2x)$

$= \dfrac{-4(3-2x)}{u^2}$

$= \dfrac{-4(3-2x)}{(3x - x^2)^2}$

i $y = 2\left(x^2 - \dfrac{2}{x}\right)^3$

$\therefore\ y = 2u^3$ where $u = x^2 - 2x^{-1}$

Now $\dfrac{dy}{dx} = \dfrac{dy}{du}\dfrac{du}{dx}$

$= 6u^2(2x + 2x^{-2})$

$= 6\left(x^2 - \dfrac{2}{x}\right)^2\left(2x + \dfrac{2}{x^2}\right)$

3 **a** $y = \sqrt{1 - x^2}$ at $x = \frac{1}{2}$

$\therefore\ y = \sqrt{u}$ where $u = 1 - x^2$

Now $\dfrac{dy}{dx} = \dfrac{dy}{du}\dfrac{du}{dx} = \frac{1}{2}u^{-\frac{1}{2}}(-2x)$

$= \dfrac{-x}{\sqrt{u}}$

$= \dfrac{-x}{\sqrt{1 - x^2}}$

At $x = \frac{1}{2}$, $\dfrac{dy}{dx} = \dfrac{-\frac{1}{2}}{\sqrt{1 - \frac{1}{4}}} = -\frac{1}{2}\left(\frac{2}{\sqrt{3}}\right)$

$\therefore$ gradient of tangent $= -\frac{1}{\sqrt{3}}$

b $y = (3x + 2)^6$ at $x = -1$

$\therefore\ y = u^6$ where $u = 3x + 2$

Now $\dfrac{dy}{dx} = \dfrac{dy}{du}\dfrac{du}{dx}$

$= 6u^5(3)$

$= 18u^5$

$= 18(3x + 2)^5$

At $x = -1$, $\dfrac{dy}{dx} = 18(-1)^5$

$\therefore$ gradient of tangent $= -18$

c $y = \dfrac{1}{(2x - 1)^4}$ at $x = 1$

$\therefore\ y = u^{-4}$ where $u = 2x - 1$

Now $\dfrac{dy}{dx} = \dfrac{dy}{du}\dfrac{du}{dx} = -4u^{-5}(2)$

$= \dfrac{-8}{u^5}$

$= \dfrac{-8}{(2x - 1)^5}$

At $x = 1$, $\dfrac{dy}{dx} = \dfrac{-8}{1^5}$

$\therefore$ gradient of tangent $= -8$

d $y = 6 \times \sqrt[3]{1 - 2x}$ at $x = 0$

$\therefore\ y = 6u^{\frac{1}{3}}$ where $u = 1 - 2x$

Now $\dfrac{dy}{dx} = \dfrac{dy}{du}\dfrac{du}{dx} = 6(\frac{1}{3})u^{-\frac{2}{3}}(-2)$

$= 2u^{-\frac{2}{3}}(-2)$

$= \dfrac{-4}{\sqrt[3]{u^2}}$

$= \dfrac{-4}{\sqrt[3]{(1 - 2x)^2}}$

At $x = 0$, $\dfrac{dy}{dx} = \dfrac{-4}{\sqrt[3]{1^2}}$

$\therefore$ gradient of tangent $= -4$

e $y = \dfrac{4}{x + 2\sqrt{x}}$ at $x = 4$

$\therefore\ y = 4u^{-1}$ where $u = x + 2x^{\frac{1}{2}}$

Now $\dfrac{dy}{dx} = \dfrac{dy}{du}\dfrac{du}{dx}$

$= -4u^{-2}(1 + x^{-\frac{1}{2}})$

$= -\dfrac{4}{u^2}\left(1 + \dfrac{1}{\sqrt{x}}\right)$

$= \dfrac{-4}{(x + 2\sqrt{x})^2}\left(1 + \dfrac{1}{\sqrt{x}}\right)$

At $x = 4$, $\dfrac{dy}{dx} = \dfrac{-4}{(4 + 4)^2}(1 + \frac{1}{2}) = -\frac{6}{64}$

$\therefore$ gradient of tangent $= -\frac{3}{32}$

f $y = \left(x + \dfrac{1}{x}\right)^3$ at $x = 1$

$\therefore\ y = u^3$ where $u = x + x^{-1}$

Now $\dfrac{dy}{dx} = \dfrac{dy}{du}\dfrac{du}{dx}$

$= 3u^2(1 - x^{-2})$

$= 3\left(x + \dfrac{1}{x}\right)^2\left(1 - \dfrac{1}{x^2}\right)$

At $x = 1$, $\dfrac{dy}{dx} = 3(1 + 1)^2(1 - 1)$

$\therefore$ gradient of tangent $= 0$

4 **a** $y = x^3 \quad \therefore \quad \dfrac{dy}{dx} = 3x^2$

$x = y^{\frac{1}{3}} \quad \therefore \quad \dfrac{dx}{dy} = \frac{1}{3}y^{-\frac{2}{3}}$

$$\begin{aligned}\frac{dy}{dx}\frac{dx}{dy} &= 3x^2\left(\tfrac{1}{3}\right)y^{-\frac{2}{3}}\\ &= x^2(y)^{-\frac{2}{3}}\\ &= x^2(x^3)^{-\frac{2}{3}} \quad \{\text{substituting } y = x^3\}\\ &= x^2(x^{-2})\\ &= x^0\\ &= 1 \quad \text{as required}\end{aligned}$$

b We know that $\dfrac{dy}{du}\dfrac{du}{dx} = \dfrac{dy}{dx}$ {chain rule}

Letting $x = y$, $\dfrac{dy}{du}\dfrac{du}{dy} = \dfrac{dy}{dy}$

$\therefore \quad \dfrac{dy}{du}\dfrac{du}{dy} = 1$

Letting $u = x$, $\dfrac{dy}{dx}\dfrac{dx}{dy} = 1$

EXERCISE 17E

1 **a** $y = x^2(2x-1)$ is the product of $u = x^2$ and $v = 2x - 1$

$\therefore \quad u' = 2x$ and $v' = 2$

Now $\dfrac{dy}{dx} = u'v + uv'$ {product rule}

$$\begin{aligned}\therefore \quad \frac{dy}{dx} &= 2x(2x-1) + x^2(2)\\ &= 2x(2x-1) + 2x^2\end{aligned}$$

b $y = 4x(2x+1)^3$ is the product of $u = 4x$ and $v = (2x+1)^3$

$\therefore \quad u' = 4$ and $v' = 3(2x+1)^2 \times 2 = 6(2x+1)^2$

Now $\dfrac{dy}{dx} = u'v + uv'$ {product rule}

$\therefore \quad \dfrac{dy}{dx} = 4(2x+1)^3 + 24x(2x+1)^2$

c $y = x^2\sqrt{3-x}$ is the product of $u = x^2$ and $v = (3-x)^{\frac{1}{2}}$

$\therefore \quad u' = 2x$ and $v' = \frac{1}{2}(3-x)^{-\frac{1}{2}}(-1) = -\frac{1}{2}(3-x)^{-\frac{1}{2}}$

Now $\dfrac{dy}{dx} = u'v + uv'$ {product rule}

$$\begin{aligned}\therefore \quad \frac{dy}{dx} &= 2x(3-x)^{\frac{1}{2}} + x^2\left[-\tfrac{1}{2}(3-x)^{-\frac{1}{2}}\right]\\ &= 2x\sqrt{3-x} - \frac{x^2}{2\sqrt{3-x}}\end{aligned}$$

d $y = \sqrt{x}(x-3)^2$ is the product of $u = x^{\frac{1}{2}}$ and $v = (x-3)^2$

$\therefore \quad u' = \frac{1}{2}x^{-\frac{1}{2}}$ and $v' = 2(x-3)^1$

Now $\dfrac{dy}{dx} = u'v + uv'$ {product rule}

$\therefore \quad \dfrac{dy}{dx} = \frac{1}{2}x^{-\frac{1}{2}}(x-3)^2 + 2\sqrt{x}(x-3)$

e $y = 5x^2(3x^2-1)^2$ is the product of $u = 5x^2$ and $v = (3x^2-1)^2$

$\therefore \quad u' = 10x$ and $v' = 2(3x^2-1)^1(6x) = 12x(3x^2-1)$

Now $\dfrac{dy}{dx} = u'v + uv'$ {product rule}

$$\begin{aligned}\therefore \quad \frac{dy}{dx} &= 10x(3x^2-1)^2 + 5x^2(12x)(3x^2-1)\\ &= 10x(3x^2-1)^2 + 60x^3(3x^2-1)\end{aligned}$$

f $y = \sqrt{x}(x - x^2)^3$ is the product of $u = x^{\frac{1}{2}}$ and $v = (x - x^2)^3$

$\therefore$ $u' = \frac{1}{2}x^{-\frac{1}{2}}$ and $v' = 3(x - x^2)^2(1 - 2x)$

Now $\dfrac{dy}{dx} = u'v + uv'$ {product rule}

$\therefore$ $\dfrac{dy}{dx} = \frac{1}{2}x^{-\frac{1}{2}}(x - x^2)^3 + 3\sqrt{x}(x - x^2)^2(1 - 2x)$

2 **a** $y = x^4(1 - 2x)^2$ is the product of $u = x^4$ and $v = (1 - 2x)^2$

$\therefore$ $u' = 4x^3$ and $v' = 2(1 - 2x)^1(-2)$

$= -4(1 - 2x)$

Now $\dfrac{dy}{dx} = u'v + uv'$ {product rule}

$\therefore$ $\dfrac{dy}{dx} = 4x^3(1 - 2x)^2 - 4x^4(1 - 2x)$

At $x = -1$, $\dfrac{dy}{dx} = 4(-1)^3(3)^2 - 4(-1)^4(3) = -48$

$\therefore$ gradient of tangent $= -48$

b $y = \sqrt{x}(x^2 - x + 1)^2$ is the product of $u = x^{\frac{1}{2}}$ and $v = (x^2 - x + 1)^2$

$\therefore$ $u' = \frac{1}{2}x^{-\frac{1}{2}}$ and $v' = 2(x^2 - x + 1)(2x - 1)$

Now $\dfrac{dy}{dx} = u'v + uv'$ {product rule}

$\therefore$ $\dfrac{dy}{dx} = \frac{1}{2}x^{-\frac{1}{2}}(x^2 - x + 1)^2 + 2\sqrt{x}(x^2 - x + 1)(2x - 1)$

At $x = 4$, $\dfrac{dy}{dx} = \frac{1}{2}(4)^{-\frac{1}{2}}(13)^2 + 2\sqrt{4}(13)(7) = 406\frac{1}{4}$

$\therefore$ gradient of tangent $= 406\frac{1}{4}$

c $y = x\sqrt{1 - 2x}$ is the product of $u = x$ and $v = (1 - 2x)^{\frac{1}{2}}$

$\therefore$ $u' = 1$ and $v' = \frac{1}{2}(1 - 2x)^{-\frac{1}{2}}(-2)$

$= -(1 - 2x)^{-\frac{1}{2}}$

Now $\dfrac{dy}{dx} = u'v + uv'$ {product rule}

$\therefore$ $\dfrac{dy}{dx} = \sqrt{1 - 2x} - \dfrac{x}{\sqrt{1 - 2x}}$

At $x = -4$, $\dfrac{dy}{dx} = \sqrt{9} - \dfrac{(-4)}{\sqrt{9}} = 3 + \frac{4}{3} = \frac{13}{3}$

$\therefore$ gradient of tangent $= \frac{13}{3}$

d $y = x^3\sqrt{5 - x^2}$ is the product of $u = x^3$ and $v = (5 - x^2)^{\frac{1}{2}}$

$\therefore$ $u' = 3x^2$ and $v' = \frac{1}{2}(5 - x^2)^{-\frac{1}{2}}(-2x)$

$= -x(5 - x^2)^{-\frac{1}{2}}$

Now $\dfrac{dy}{dx} = u'v + uv'$ {product rule}

$\therefore$ $\dfrac{dy}{dx} = 3x^2\sqrt{5 - x^2} - \dfrac{x^4}{\sqrt{5 - x^2}}$

At $x = 1$, $\dfrac{dy}{dx} = 3(1)^2\sqrt{4} - \dfrac{1}{\sqrt{4}} = 6 - \frac{1}{2} = \frac{11}{2}$

$\therefore$ gradient of tangent $= \frac{11}{2}$

3 $y = \sqrt{x}(3-x)^2$ is the product of $u = x^{\frac{1}{2}}$ and $v = (3-x)^2$

$\therefore \quad u' = \frac{1}{2}x^{-\frac{1}{2}}$ and $v' = 2(3-x)^1(-1) = -2(3-x)$

Now $\dfrac{dy}{dx} = u'v + uv'$ {product rule}

$$\therefore \quad \frac{dy}{dx} = \frac{1}{2\sqrt{x}}(3-x)^2 - 2\sqrt{x}(3-x)$$

$$= \frac{(3-x)^2 - (2\sqrt{x})(2\sqrt{x})(3-x)}{2\sqrt{x}}$$

$$= \frac{(3-x)\left[(3-x)-4x\right]}{2\sqrt{x}}$$

$$= \frac{(3-x)(3-5x)}{2\sqrt{x}} \quad \text{as required}$$

Tangents are horizontal when their gradients are 0.

$\dfrac{dy}{dx} = 0$ when $(3-x)(3-5x) = 0$

$\therefore \quad 3 - x = 0$ or $3 - 5x = 0$

$\therefore \quad x = 3$ or $x = \frac{3}{5}$

EXERCISE 17F

1 **a** $y = \dfrac{1+3x}{2-x}$ is a quotient where $u = 1+3x$ and $v = 2-x$

$\therefore \quad u' = 3$ and $v' = -1$

Now $\dfrac{dy}{dx} = \dfrac{u'v - uv'}{v^2}$ {quotient rule}

$$\therefore \quad \frac{dy}{dx} = \frac{3(2-x) - (1+3x)(-1)}{(2-x)^2} = \frac{7}{(2-x)^2}$$

b $y = \dfrac{x^2}{2x+1}$ is a quotient where $u = x^2$ and $v = 2x+1$

$\therefore \quad u' = 2x$ and $v' = 2$

Now $\dfrac{dy}{dx} = \dfrac{u'v - uv'}{v^2}$ {quotient rule}

$$\therefore \quad \frac{dy}{dx} = \frac{2x(2x+1) - x^2(2)}{(2x+1)^2} = \frac{2x^2+2x}{(2x+1)^2}$$

c $y = \dfrac{x}{x^2-3}$ is a quotient where $u = x$ and $v = x^2 - 3$

$\therefore \quad u' = 1$ and $v' = 2x$

Now $\dfrac{dy}{dx} = \dfrac{u'v - uv'}{v^2}$ {quotient rule}

$$\therefore \quad \frac{dy}{dx} = \frac{1(x^2-3) - x(2x)}{(x^2-3)^2} = \frac{-3-x^2}{(x^2-3)^2}$$

d $y = \dfrac{\sqrt{x}}{1-2x}$ is a quotient where $u = x^{\frac{1}{2}}$ and $v = 1 - 2x$

$\therefore \quad u' = \frac{1}{2}x^{-\frac{1}{2}}$ and $v' = -2$

Now $\dfrac{dy}{dx} = \dfrac{u'v - uv'}{v^2}$ {quotient rule}

$$\therefore \quad \frac{dy}{dx} = \frac{\frac{1}{2}x^{-\frac{1}{2}}(1-2x) - \sqrt{x}(-2)}{(1-2x)^2} = \frac{\frac{1}{2}x^{-\frac{1}{2}}(1-2x) + 2\sqrt{x}}{(1-2x)^2}$$

e $y = \dfrac{x^2-3}{3x-x^2}$ is a quotient where $u = x^2 - 3$ and $v = 3x - x^2$

$\therefore \quad u' = 2x$ and $v' = 3 - 2x$

Now $\frac{dy}{dx} = \frac{u'v - uv'}{v^2}$ {quotient rule}

$\therefore \frac{dy}{dx} = \frac{2x(3x - x^2) - (x^2 - 3)(3 - 2x)}{(3x - x^2)^2}$

$= \frac{6x^2 - 2x^3 - 3x^2 + 2x^3 + 9 - 6x}{(3x - x^2)^2} = \frac{3x^2 - 6x + 9}{(3x - x^2)^2}$

f $y = \frac{x}{\sqrt{1 - 3x}}$ is a quotient where $u = x$ and $v = (1 - 3x)^{\frac{1}{2}}$

$\therefore u' = 1$ and $v' = -\frac{3}{2}(1 - 3x)^{-\frac{1}{2}}$

Now $\frac{dy}{dx} = \frac{u'v - uv'}{v^2}$ {quotient rule}

$\therefore \frac{dy}{dx} = \frac{(1 - 3x)^{\frac{1}{2}} - x\left(-\frac{3}{2}(1 - 3x)^{-\frac{1}{2}}\right)}{1 - 3x} = \frac{(1 - 3x)^{\frac{1}{2}} + \frac{3}{2}x(1 - 3x)^{-\frac{1}{2}}}{1 - 3x}$

2 **a** $y = \frac{x}{1 - 2x}$ is a quotient where $u = x$ and $v = 1 - 2x$

$\therefore u' = 1$ and $v' = -2$

Now $\frac{dy}{dx} = \frac{u'v - uv'}{v^2}$ {quotient rule}

$\therefore \frac{dy}{dx} = \frac{1(1 - 2x) - x(-2)}{(1 - 2x)^2} = \frac{1}{(1 - 2x)^2}$

At $x = 1$, $\frac{dy}{dx} = \frac{1}{(1 - 2)^2} = \frac{1}{(-1)^2} = 1$

$\therefore$ the gradient of the tangent $= 1$

b $y = \frac{x^3}{x^2 + 1}$ is a quotient where $u = x^3$ and $v = x^2 + 1$

$\therefore u' = 3x^2$ and $v' = 2x$

Now $\frac{dy}{dx} = \frac{u'v - uv'}{v^2} = \frac{3x^2(x^2 + 1) - x^3(2x)}{(x^2 + 1)^2} = \frac{x^4 + 3x^2}{(x^2 + 1)^2}$

At $x = -1$, $\frac{dy}{dx} = \frac{1 + 3}{(1 + 1)^2} = \frac{4}{4} = 1$

$\therefore$ the gradient of the tangent $= 1$

c $y = \frac{\sqrt{x}}{2x + 1}$ is a quotient where $u = x^{\frac{1}{2}}$ and $v = 2x + 1$

$\therefore u' = \frac{1}{2}x^{-\frac{1}{2}}$ and $v' = 2$

Now $\frac{dy}{dx} = \frac{u'v - uv'}{v^2} = \frac{\frac{1}{2\sqrt{x}}(2x + 1) - \sqrt{x}(2)}{(2x + 1)^2}$

At $x = 4$, $\frac{dy}{dx} = \frac{\frac{9}{4} - 4}{81} = \frac{\left(\frac{9}{4} - 4\right)}{81} \times \frac{4}{4} = \frac{9 - 16}{324}$

$\therefore$ the gradient of the tangent $= -\frac{7}{324}$

d $y = \frac{x^2}{\sqrt{x^2 + 5}}$ is a quotient where $u = x^2$ and $v = (x^2 + 5)^{\frac{1}{2}}$

$\therefore u' = 2x$ and $v' = \frac{1}{2}(x^2 + 5)^{-\frac{1}{2}}(2x)$

$= x(x^2 + 5)^{-\frac{1}{2}}$

Now $\dfrac{dy}{dx} = \dfrac{u'v - uv'}{v^2} = \dfrac{2x\sqrt{x^2+5} - x^2\left(\dfrac{x}{\sqrt{x^2+5}}\right)}{(x^2+5)}$

At $x = -2$, $\dfrac{dy}{dx} = \dfrac{-4(3) - 4\left(\frac{-2}{3}\right)}{9} = \dfrac{\left(-12 + \frac{8}{3}\right)}{9} \times \dfrac{3}{3} = \dfrac{-36+8}{27}$

$\therefore$ the gradient of the tangent $= -\frac{28}{27}$

3 **a** $y = \dfrac{2\sqrt{x}}{1-x}$ is a quotient where $u = 2x^{\frac{1}{2}}$ and $v = 1 - x$

$\therefore$ $u' = x^{-\frac{1}{2}}$ and $v' = -1$

Now $\dfrac{dy}{dx} = \dfrac{u'v - uv'}{v^2}$ {quotient rule}

$\therefore$ $\dfrac{dy}{dx} = \dfrac{\frac{1}{\sqrt{x}}(1-x) - 2\sqrt{x}(-1)}{(1-x)^2} \times \left(\dfrac{\sqrt{x}}{\sqrt{x}}\right) = \dfrac{(1-x)+2x}{\sqrt{x}(1-x)^2} = \dfrac{x+1}{\sqrt{x}(1-x)^2}$ as required

b **i** $\dfrac{dy}{dx} = 0$ when $x + 1 = 0$ $\therefore$ $x = -1$.

However $\dfrac{dy}{dx}$ is not defined for $x \leqslant 0$ because of the $\sqrt{x}$ term. Hence $\dfrac{dy}{dx}$ never equals 0.

ii $\dfrac{dy}{dx}$ is undefined when $x \leqslant 0$ and when $x = 1$.

4 **a** $y = \dfrac{x^2 - 3x + 1}{x+2}$ is a quotient where $u = x^2 - 3x + 1$ and $v = x + 2$

$\therefore$ $u' = 2x - 3$ and $v' = 1$

Now $\dfrac{dy}{dx} = \dfrac{u'v - uv'}{v^2}$ {quotient rule}

$$\therefore \quad \frac{dy}{dx} = \frac{(2x-3)(x+2) - (x^2-3x+1)(1)}{(x+2)^2}$$

$$= \frac{2x^2 + 4x - 3x - 6 - x^2 + 3x - 1}{(x+2)^2}$$

$$= \frac{x^2+4x-7}{(x+2)^2} \quad \text{as required}$$

b **i** $\dfrac{dy}{dx} = 0$ when $x^2 + 4x - 7 = 0$ $\therefore$ $x = \dfrac{-4 \pm \sqrt{44}}{2} = -2 \pm \sqrt{11}$

ii $\dfrac{dy}{dx}$ is undefined when $(x+2)^2 = 0$ $\therefore$ $x = -2$

c $\dfrac{dy}{dx}$ is zero when the tangent to the function is horizontal. This occurs at the function's turning points or points of horizontal inflection.

$\dfrac{dy}{dx}$ is undefined at vertical asymptotes of the function.

EXERCISE 17G

1 **a** We seek the tangent to $y = x - 2x^2 + 3$ at $x = 2$.

When $x = 2$, $y = 2 - 2(2)^2 + 3 = -3$ $\therefore$ the point of contact is $(2, -3)$.

Now $\dfrac{dy}{dx} = 1 - 4x$, so at $x = 2$, $\dfrac{dy}{dx} = 1 - 8 = -7$

$\therefore$ the tangent has equation $\dfrac{y-(-3)}{x-2} = -7$ $\quad\therefore\ y+3 = -7(x-2)$

$\therefore\ y = -7x + 14 - 3$

$\therefore\ y = -7x + 11$

b We seek the tangent to $y = \sqrt{x} + 1 = x^{\frac{1}{2}} + 1$ at $x = 4$.

When $x = 4$, $y = \sqrt{4} + 1 = 3$ $\quad\therefore$ the point of contact is $(4, 3)$.

Now $\dfrac{dy}{dx} = \dfrac{1}{2\sqrt{x}}$, so at $x = 4$, $\dfrac{dy}{dx} = \dfrac{1}{2\sqrt{4}} = \frac{1}{4}$

$\therefore$ the tangent has equation $\dfrac{y-3}{x-4} = \frac{1}{4}$ $\quad\therefore\ 4y - 12 = x - 4$

$\therefore\ 4y = x + 8$

c We seek the tangent to $y = x^3 - 5x$ at $x = 1$.

When $x = 1$, $y = 1^3 - 5(1) = -4$ $\quad\therefore$ the point of contact is $(1, -4)$.

Now $\dfrac{dy}{dx} = 3x^2 - 5$, so at $x = 1$, $\dfrac{dy}{dx} = 3 - 5 = -2$

$\therefore$ the tangent has equation $\dfrac{y-(-4)}{x-1} = -2$ $\quad\therefore\ y + 4 = -2x + 2$

$\therefore\ y = -2x - 2$

d We seek the tangent to $y = \dfrac{4}{\sqrt{x}}$ at $(1, 4)$.

Now $y = \dfrac{4}{\sqrt{x}} = 4x^{-\frac{1}{2}}$, so $\dfrac{dy}{dx} = -2x^{-\frac{3}{2}}$

At $x = 1$, $\dfrac{dy}{dx} = -2\left(1^{-\frac{3}{2}}\right) = -2$

$\therefore$ the tangent has equation $\dfrac{y-4}{x-1} = -2$ $\quad\therefore\ y - 4 = -2x + 2$

$\therefore\ y = -2x + 6$

e We seek the tangent to $y = \dfrac{3}{x} - \dfrac{1}{x^2} = 3x^{-1} - x^{-2}$ at $(-1, -4)$.

$$\text{Now } \frac{dy}{dx} = -3x^{-2} + 2x^{-3}$$
$$= -\frac{3}{x^2} + \frac{2}{x^3}$$

$$\therefore \text{ at } (-1, -4),\ \frac{dy}{dx} = -\frac{3}{(-1)^2} + \frac{2}{(-1)^3}$$
$$= -3 - 2 = -5$$

$\therefore$ the tangent has equation $\dfrac{y-(-4)}{x-(-1)} = -5$ $\quad\therefore\ y + 4 = -5x - 5$

$\therefore\ y = -5x - 9$

f We seek the tangent to $y = 3x^2 - \dfrac{1}{x} = 3x^2 - x^{-1}$ at $x = -1$.

When $x = -1$, $y = 3(-1)^2 - \frac{1}{(-1)} = 4$ $\quad\therefore$ the point of contact is $(-1, 4)$.

$$\text{Now } \frac{dy}{dx} = 6x + x^{-2}$$
$$= 6x + \frac{1}{x^2}$$

$$\therefore \text{ at } x = -1,\ \frac{dy}{dx} = 6(-1) + \frac{1}{(-1)^2} = -5$$

$\therefore$ the tangent has equation $\dfrac{y-4}{x-(-1)} = -5$ $\quad\therefore\ y - 4 = -5x - 5$

$\therefore\ y = -5x - 1$

2 **a** We seek the normal to $y = x^2$ at (3, 9).

Now $\dfrac{dy}{dx} = 2x$ so at $x = 3$, $\dfrac{dy}{dx} = 2(3) = 6 = \frac{6}{1}$

$\therefore$ the normal at (3, 9) has gradient $-\frac{1}{6}$, so the equation of the normal is $\dfrac{y-9}{x-3} = -\frac{1}{6}$

$$\therefore \quad 6y - 54 = -x + 3$$
$$\therefore \quad 6y = -x + 57$$

b We seek the normal to $y = x^3 - 5x + 2$ at $x = -2$.

When $x = -2$, $y = (-2)^3 - 5(-2) + 2 = 4$ and so the point of contact is $(-2, 4)$.

Now $\dfrac{dy}{dx} = 3x^2 - 5$ so at $x = -2$, $\dfrac{dy}{dx} = 3(-2)^2 - 5 = 7$

$\therefore$ the normal at $(-2, 4)$ has gradient $-\frac{1}{7}$, so the equation of the normal is $\dfrac{y-4}{x-(-2)} = -\frac{1}{7}$

$$\therefore \quad 7y - 28 = -(x + 2)$$
$$\therefore \quad 7y = -x + 26$$

c We seek the normal to $y = \dfrac{5}{\sqrt{x}} - \sqrt{x}$ at (1, 4).

Now $y = 5x^{-\frac{1}{2}} - x^{\frac{1}{2}}$ $\quad\therefore\quad \dfrac{dy}{dx} = -\frac{5}{2}x^{-\frac{3}{2}} - \frac{1}{2}x^{-\frac{1}{2}}$

$\therefore$ at $x = 1$, $\dfrac{dy}{dx} = -\frac{5}{2}\left(1^{-\frac{3}{2}}\right) - \frac{1}{2}\left(1^{-\frac{1}{2}}\right) = -\frac{5}{2} - \frac{1}{2} = -3$

$\therefore$ the normal at (1, 4) has gradient $\frac{1}{3}$, so the equation of the normal is $\dfrac{y-4}{x-1} = \frac{1}{3}$

$$\therefore \quad 3y - 12 = x - 1$$
$$\therefore \quad 3y = x + 11$$

d We seek the normal to $y = 8\sqrt{x} - \dfrac{1}{x^2}$ at $x = 1$.

When $x = 1$, $y = 8\sqrt{1} - \dfrac{1}{1^2} = 7$ $\therefore$ the point of contact is (1, 7).

Now $y = 8\sqrt{x} - \dfrac{1}{x^2} = 8x^{\frac{1}{2}} - x^{-2}$ $\quad\therefore\quad \dfrac{dy}{dx} = 4x^{-\frac{1}{2}} + 2x^{-3}$

$\therefore$ at $x = 1$, $\dfrac{dy}{dx} = 4 + 2 = 6$

$\therefore$ the normal at (1, 7) has gradient $-\frac{1}{6}$, so the equation of the normal is $\dfrac{y-7}{x-1} = -\frac{1}{6}$

$$\therefore \quad 6y - 42 = -x + 1$$
$$\therefore \quad 6y = -x + 43$$

3 **a** $y = 2x^3 + 3x^2 - 12x + 1$ $\quad\therefore\quad \dfrac{dy}{dx} = 6x^2 + 6x - 12$

Horizontal tangents have gradient $= 0$ so $6x^2 + 6x - 12 = 0$

$$\therefore \quad x^2 + x - 2 = 0$$
$$\therefore \quad (x + 2)(x - 1) = 0$$
$$\therefore \quad x = -2 \text{ or } x = 1$$

Now at $x = -2$, $y = 2(-2)^3 + 3(-2)^2 - 12(-2) + 1 = 21$

and at $x = 1$, $y = 2(1)^3 + 3(1)^2 - 12(1) + 1 = -6$

$\therefore$ the points of contact are $(-2, 21)$ and $(1, -6)$

$\therefore$ the tangents are $y = -6$ and $y = 21$.

b Now $y = 2\sqrt{x} + \dfrac{1}{\sqrt{x}} = 2x^{\frac{1}{2}} + x^{-\frac{1}{2}}$

$\therefore \dfrac{dy}{dx} = x^{-\frac{1}{2}} - \frac{1}{2}x^{-\frac{3}{2}}$

$= \dfrac{1}{\sqrt{x}} - \dfrac{1}{2x\sqrt{x}}$

Horizontal tangents have gradient $= 0$

$\therefore \dfrac{1}{\sqrt{x}} - \dfrac{1}{2x\sqrt{x}} = 0$

$\therefore \dfrac{2x-1}{2x\sqrt{x}} = 0$

$\therefore 2x - 1 = 0$

$\therefore x = \frac{1}{2}$

Now at $x = \frac{1}{2}$, $y = \dfrac{2\left(\frac{1}{2}\right)+1}{\sqrt{\frac{1}{2}}} = \dfrac{2}{\frac{1}{2}\sqrt{2}} = 2\sqrt{2}$

$\therefore$ the only horizontal tangent touches at the curve at $(\frac{1}{2}, 2\sqrt{2})$.

c Now $y = 2x^3 + kx^2 - 3$

$\therefore \dfrac{dy}{dx} = 6x^2 + 2kx$

When $x = 2$, $\dfrac{dy}{dx} = 4$

$\therefore 6(2)^2 + 2k(2) = 4$

$\therefore 24 + 4k = 4$

$\therefore 4k = -20$

$\therefore k = -5$

d Now $y = 1 - 3x + 12x^2 - 8x^3$ $\therefore \dfrac{dy}{dx} = -3 + 24x - 24x^2$

When $x = 1$, $\dfrac{dy}{dx} = -3 + 24 - 24 = -3$

$\therefore$ the tangent at $(1, 2)$ has gradient -3

The tangents to the curve have gradient -3 when $-3 + 24x - 24x^2 = -3$

$\therefore 24x^2 - 24x = 0$

$\therefore 24x(x-1) = 0$

$\therefore$ when $x = 0$ or $x = 1$

So the other x-value for which the tangent to the curve has gradient -3 is $x = 0$, and when $x = 0$, $y = 1 - 0 + 0 - 0 = 1$

$\therefore$ the tangent to the curve at $(0, 1)$ is parallel to the tangent at $(1, 2)$.

This tangent has equation $\dfrac{y-1}{x-0} = -3$ or $y = -3x + 1$.

4 **a** Now $y = x^2 + ax + b$ $\therefore \dfrac{dy}{dx} = 2x + a$

At $x = 1$, $\dfrac{dy}{dx} = 2 + a$

$\therefore$ the gradient of the tangent to the curve at $x = 1$ will be $2 + a$

However the equation of the tangent is $2x + y = 6$ or $y = -2x + 6$

and so the gradient of the tangent is -2. $\therefore 2 + a = -2$

$\therefore a = -4$

So, the curve is $y = x^2 - 4x + b$.

We also know that the tangent contacts the curve when $x = 1$.

$\therefore 1^2 - 4(1) + b = -2(1) + 6$

$\therefore 1 - 4 + b = 4$

$\therefore b = 7$ $\therefore a = -4, b = 7$

b Now $y = a\sqrt{x} + \dfrac{b}{\sqrt{x}} = ax^{\frac{1}{2}} + bx^{-\frac{1}{2}}$

$\therefore \dfrac{dy}{dx} = \dfrac{a}{2}x^{-\frac{1}{2}} - \dfrac{b}{2}x^{-\frac{3}{2}}$

$\therefore$ at $x = 4$, $\dfrac{dy}{dx} = \dfrac{a}{2}\left(4^{-\frac{1}{2}}\right) - \dfrac{b}{2}\left(4^{-\frac{3}{2}}\right)$

$= \dfrac{a}{2}\left(\dfrac{1}{2}\right) - \dfrac{b}{2}\left(\dfrac{1}{8}\right)$

$= \dfrac{a}{4} - \dfrac{b}{16}$

$\therefore$ the gradient of the tangent to the curve at $x = 4$ will be $\dfrac{a}{4} - \dfrac{b}{16} = \dfrac{4a-b}{16}$

However the equation of the *normal* is $4x + y = 22$, or $y = -4x + 22$

$\therefore$ the normal has gradient -4

$\therefore$ the tangent has gradient $\frac{1}{4}$, and so $\dfrac{4a-b}{16} = \frac{1}{4}$

$$\therefore\ 4a - b = 4$$
$$\therefore\ b = 4a - 4 \quad .... \ (1)$$

Also, at $x = 4$ the normal line intersects the curve.

$$\therefore\ a\sqrt{4} + \frac{b}{\sqrt{4}} = -4(4) + 22$$
$$\therefore\ 2a + \frac{b}{2} = 6$$

Consequently, $2a + \dfrac{4a-4}{2} = 6$ {using (1)}

$$\therefore\ 2a + 2a - 2 = 6$$
$$\therefore\ 4a = 8$$
$$\therefore\ a = 2 \text{ and so } b = 4(2) - 4 = 4 \quad \{\text{from (1)}\}$$

c $y = 2x^2 - 1$

$\therefore\ \dfrac{dy}{dx} = 4x$

$\therefore$ at the point where $x = a$, $\dfrac{dy}{dx} = 4a$

$\therefore$ the gradient of the tangent at the point where $x = a$ is $4a$.

Also, at $x = a$, $y = 2a^2 - 1$.

$\therefore$ the tangent has equation $\dfrac{y - (2a^2 - 1)}{x - a} = 4a$

$$\therefore\ y - 2a^2 + 1 = 4a(x - a)$$
$$\therefore\ y - 2a^2 + 1 = 4ax - 4a^2$$
$$\therefore\ 4ax - y = 2a^2 + 1$$

5 **a** $y = \sqrt{2x+1}$

When $x = 4$, $y = \sqrt{2(4)+1} = 3$, so the point of contact is $(4, 3)$

$$\text{Now } \frac{dy}{dx} = \tfrac{1}{2}(2x+1)^{-\frac{1}{2}}(2) = \frac{1}{\sqrt{2x+1}}$$

$$\therefore \text{ at } x = 4,\ \frac{dy}{dx} = \frac{1}{\sqrt{2(4)+1}} = \tfrac{1}{3}$$

$\therefore$ the tangent has equation $\dfrac{y-3}{x-4} = \frac{1}{3}$ or $3y = x + 5$

b $y = \dfrac{1}{2-x} = (2-x)^{-1}$ $\therefore$ at $x = -1$, $y = \dfrac{1}{2-(-1)} = \frac{1}{3}$

So the point of contact is $(-1, \frac{1}{3})$

$$\text{Now } \frac{dy}{dx} = -1(2-x)^{-2}(-1) = \frac{1}{(2-x)^2}$$

$$\therefore \text{ at } x = -1,\ \frac{dy}{dx} = \frac{1}{(2-(-1))^2} = \tfrac{1}{9}$$

$\therefore$ the tangent has equation $\dfrac{y - \frac{1}{3}}{x - (-1)} = \frac{1}{9}$ $\quad\therefore\ 9y - 3 = x + 1$

$$\therefore\ 9y = x + 4$$

c We seek the tangent to $f(x)=\dfrac{x}{1-3x}$ at $(-1, -\frac{1}{4})$.

$f(x)$ is a quotient where $u=x$ and $v=1-3x$ $\therefore$ $u'=1$ and $v'=-3$

Now $f'(x)=\dfrac{u'v-uv'}{v^2}$ {quotient rule}

$\therefore$ $f'(x)=\dfrac{1(1-3x)-x(-3)}{(1-3x)^2}=\dfrac{1}{(1-3x)^2}$

$\therefore$ $f'(-1)=\dfrac{1}{(1-3(-1))^2}=\frac{1}{16}$

$\therefore$ the tangent has equation $\dfrac{y-\left(-\frac{1}{4}\right)}{x-(-1)}=\frac{1}{16}$ $\therefore$ $16y+4=x+1$

$\therefore$ $16y=x-3$

d We seek the tangent to $f(x)=\dfrac{x^2}{1-x}$ at $(2, -4)$.

$f(x)$ is a quotient where $u=x^2$ and $v=1-x$ $\therefore$ $u'=2x$ and $v'=-1$

Now $f'(x)=\dfrac{u'v-uv'}{v^2}$ {quotient rule}

$\therefore$ $f'(x)=\dfrac{2x(1-x)-x^2(-1)}{(1-x)^2}=\dfrac{2x-2x^2+x^2}{(1-x)^2}=\dfrac{2x-x^2}{(1-x)^2}$

$\therefore$ $f'(2)=\dfrac{2(2)-2^2}{(1-2)^2}=\dfrac{4-4}{1}=0$

As the tangent has gradient 0, it is horizontal.

$\therefore$ its equation is $y=c$

Since the contact point is $(2, -4)$, the tangent has equation $y=-4$.

6 **a** We seek the normal to $y=\dfrac{1}{(x^2+1)^2}$ at $(1, \frac{1}{4})$

As $y=(x^2+1)^{-2}$, $\dfrac{dy}{dx}=-2(x^2+1)^{-3}(2x)=\dfrac{-4x}{(x^2+1)^3}$

$\therefore$ at $x=1$, $\dfrac{dy}{dx}=\dfrac{-4}{(1+1)^3}=\frac{-4}{8}=-\frac{1}{2}$

$\therefore$ the normal at $(1, \frac{1}{4})$ has gradient 2.

So the equation of the normal is $\dfrac{y-\frac{1}{4}}{x-1}=2$

$\therefore$ $y-\frac{1}{4}=2x-2$

$\therefore$ $y=2x-\frac{7}{4}$

b $y=\dfrac{1}{\sqrt{3-2x}}$ $\therefore$ at $x=-3$, $y=\dfrac{1}{\sqrt{3-2(-3)}}=\frac{1}{3}$

$\therefore$ the point of contact is $(-3, \frac{1}{3})$

Now $y=(3-2x)^{-\frac{1}{2}}$

$\therefore$ $\dfrac{dy}{dx}=-\frac{1}{2}(3-2x)^{-\frac{3}{2}}(-2)=(3-2x)^{-\frac{3}{2}}$

$\therefore$ at $x=-3$, $\dfrac{dy}{dx}=(3-2(-3))^{-\frac{3}{2}}=9^{-\frac{3}{2}}=3^{-3}=\frac{1}{27}$

$\therefore$ the normal at $(-3, \frac{1}{3})$ has gradient -27.

So the equation of the normal is $\dfrac{y-\frac{1}{3}}{x-(-3)}=-27$ $\therefore$ $y-\frac{1}{3}=-27(x+3)$

$\therefore$ $y=-27x-\frac{242}{3}$

c $f(x) = \sqrt{x}(1-x)^2$

Since $f(4) = \sqrt{4}(1-4)^2 = 18$, the point of contact is $(4, 18)$

Now $f(x)$ is a product where $u = x^{\frac{1}{2}}$ and $v = (1-x)^2$

$\therefore$ $u' = \frac{1}{2}x^{-\frac{1}{2}}$ and $v' = 2(1-x)(-1) = -2(1-x)$

Now $f'(x) = u'v + uv'$ {product rule}

$\therefore$ $f'(x) = \frac{1}{2}x^{-\frac{1}{2}}(1-x)^2 - x^{\frac{1}{2}}2(1-x)$

$\therefore$ $f'(4) = \frac{1}{2\sqrt{4}}(1-4)^2 - \sqrt{4}(2)(1-4) = \frac{1}{4}(9) - 2(2)(-3) = \frac{57}{4}$

$\therefore$ the normal at (4, 18) has gradient $-\frac{4}{57}$.

So, the equation of the normal is $\dfrac{y-18}{x-4} = -\frac{4}{57}$

$$\therefore \quad 57(y-18) = -4(x-4)$$

$$\therefore \quad 57y = -4x + 1042$$

d $f(x) = \dfrac{x^2-1}{2x+3}$

Since $f(-1) = \dfrac{(-1)^2-1}{2(-1)+3} = \dfrac{0}{1} = 0$ the point of contact is $(-1, 0)$.

Now $f(x)$ is a quotient where $u = x^2 - 1$ and $v = 2x + 3$

$\therefore$ $u' = 2x$ and $v' = 2$

Now $f'(x) = \dfrac{u'v - uv'}{v^2} = \dfrac{2x(2x+3) - (x^2-1)(2)}{(2x+3)^2}$

$$\therefore \quad f'(-1) = \frac{2(-1)(-2+3) - ((-1)^2-1)(2)}{(2(-1)+3)^2} = \frac{-2(1) - (0)(2)}{(1)^2} = -2$$

$\therefore$ the normal at $(-1, 0)$ has gradient $\frac{1}{2}$.

So, the equation of the normal is $\dfrac{y-0}{x-(-1)} = \frac{1}{2}$

or $2y = x + 1$

7 The tangent has equation $3x + y = 5$ or $y = -3x + 5$

$\therefore$ the tangent has gradient -3 .... (1)

Also, at $x = -1$, $y = -3(-1) + 5 = 8$

$\therefore$ the tangent contacts the curve at $(-1, 8)$ .... (2)

Now $y = a(1-bx)^{\frac{1}{2}}$, so $\dfrac{dy}{dx} = \frac{1}{2}a(1-bx)^{-\frac{1}{2}}(-b)$

$$\therefore \quad -3 = \tfrac{1}{2}a(1+b)^{-\frac{1}{2}}(-b) \quad \text{\{using (1)\}}$$

$$\therefore \quad 6 = \frac{ab}{\sqrt{1+b}} \quad .... (3)$$

Using (2), $(-1, 8)$ must lie on the curve $y = a\sqrt{1-bx}$

$$\therefore \quad 8 = a\sqrt{1+b} \quad .... (4)$$

$$\therefore \quad \frac{6\sqrt{1+b}}{b} = \frac{8}{\sqrt{1+b}} \quad \text{\{equating } a\text{s in (3) and (4)\}}$$

$$\therefore \quad 6(1+b) = 8b$$

$$\therefore \quad 6 + 6b = 8b$$

$$\therefore \quad 6 = 2b$$

$$\therefore \quad b = 3 \text{ and } a = \tfrac{8}{\sqrt{4}} = 4$$

8 **a** Consider the tangent to $y = x^3$ at $x = 2$.

When $x = 2$, $y = 2^3 = 8$ so the point of contact is $(2, 8)$

Now $\dfrac{dy}{dx} = 3x^2$ and so at $x = 2$, $\dfrac{dy}{dx} = 3(2)^2 = 12$

$\therefore$ the tangent at $(2, 8)$ has gradient 12 and its equation is
$$\frac{y-8}{x-2} = 12$$
$$\therefore \quad y - 8 = 12x - 24$$
$$\therefore \quad y = 12x - 16$$

$\therefore$ the tangent meets the curve where $12x - 16 = x^3$

$$\therefore \quad x^3 - 12x + 16 = 0$$

Because the tangent touches the curve at $x = 2$, there must be a repeated solution at this point.

$\therefore$ $(x-2)^2$ must be a factor of this cubic

$\therefore$ $(x-2)^2(x+4) = 0$

$\therefore$ the tangent meets the curve again when $x = -4$

When $x = -4$, $y = (-4)^3 = -64$

$\therefore$ the tangent meets the curve again at $(-4, -64)$.

b Consider the tangent to $y = -x^3 + 2x^2 + 1$ at $x = -1$.

When $x = -1$, $y = -(-1)^3 + 2(-1)^2 + 1 = 4$ and so the point of contact is $(-1, 4)$

Now $\dfrac{dy}{dx} = -3x^2 + 4x$ and so at $x = -1$, $\dfrac{dy}{dx} = -3(-1)^2 + 4(-1) = -7$

$\therefore$ the tangent at $(-1, 4)$ has gradient -7 and its equation is
$$\frac{y-4}{x-(-1)} = -7$$
$$\therefore \quad y - 4 = -7(x+1)$$
$$\therefore \quad y = -7x - 3$$

$\therefore$ the tangent meets the curve where $-7x - 3 = -x^3 + 2x^2 + 1$

$$\therefore \quad x^3 - 2x^2 - 7x - 4 = 0$$

Because the tangent touches the curve at $x = -1$, there must be a repeated solution at this point.

$\therefore$ $(x+1)^2$ must be a factor of this cubic

$\therefore$ $(x+1)^2(x-4) = 0$

$\therefore$ the tangent meets the curve again when $x = 4$

When $x = 4$, $y = -(4)^3 + 2(4)^2 + 1 = -64 + 32 + 1 = -31$

$\therefore$ the tangent meets the curve again at $(4, -31)$.

c Consider the tangent to $y = x^3 + \dfrac{4}{x}$ at $x = 1$.

When $x = 1$, $y = 1^3 + \frac{4}{1} = 5$ and so the point of contact is $(1, 5)$

Now $\dfrac{dy}{dx} = 3x^2 - \dfrac{4}{x^2}$ and so at $x = 1$, $\dfrac{dy}{dx} = 3 - 4 = -1$

$\therefore$ the tangent at $(1, 5)$ has gradient -1 and its equation is
$$\frac{y-5}{x-1} = -1$$
$$\therefore \quad y - 5 = -x + 1$$
$$\therefore \quad y = -x + 6$$

$\therefore$ the tangent meets the curve where $-x + 6 = x^3 + \dfrac{4}{x}$

$$\therefore \quad x^3 + x - 6 + \frac{4}{x} = 0$$
$$\therefore \quad x^4 + x^2 - 6x + 4 = 0$$

Using a graphics calculator, this quartic has a graph which touches the x-axis at $x = 1$, and has no other x-intercepts. So, the tangent *never* meets the curve again.

9 **a** Consider the tangent to $y = x^2 - x + 9$ at $x = a$.

When $x = a$, $y = a^2 - a + 9$, so the point of contact is $(a, a^2 - a + 9)$.

Now $\dfrac{dy}{dx} = 2x - 1$ and so at $x = a$, $\dfrac{dy}{dx} = 2a - 1$

$\therefore$ the gradient of the tangent at $(a, a^2 - a + 9)$ is $2a - 1$

$\therefore$ the equation of the tangent is $\dfrac{y - (a^2 - a + 9)}{x - a} = 2a - 1$

$$\therefore \quad y - (a^2 - a + 9) = (2a - 1)(x - a)$$
$$\therefore \quad y = (2a - 1)x - 2a^2 + a + a^2 - a + 9$$
$$\therefore \quad y = (2a - 1)x - a^2 + 9 \quad \text{.... (1)}$$

But this tangent passes through $(0, 0)$, so $0 = a^2 - 9$

$$\therefore \quad (a + 3)(a - 3) = 0$$
$$\therefore \quad a = \pm 3$$

$\therefore$ the tangents are: At $a = 3$: $y = (2(3) - 1)x - 3^2 + 9$ {from (1)}

$\therefore$ $y = 5x$ with contact at $(3, 15)$.

At $a = -3$: $y = (2(-3) - 1)x - (-3)^2 + 9$ {from (1)}

$\therefore$ $y = -7x$ with contact at $(-3, 21)$.

b Let (a, a^3) lie on $y = x^3$.

Now $\dfrac{dy}{dx} = 3x^2$, so at $x = a$, $\dfrac{dy}{dx} = 3a^2$

$\therefore$ the gradient of the tangent at (a, a^3) is $3a^2$

$\therefore$ the equation of the tangent is $\dfrac{y - a^3}{x - a} = 3a^2$ or $y - a^3 = (3a^2)(x - a)$

But this tangent passes through $(-2, 0)$, so $0 - a^3 = 3a^2(-2 - a)$

$$\therefore \quad -a^3 = -6a^2 - 3a^3$$
$$\therefore \quad 2a^3 + 6a^2 = 0$$
$$\therefore \quad 2a^2(a + 3) = 0$$
$$\therefore \quad a = 0 \text{ or } -3$$

If $a = 0$, the tangent equation is $y = 0$ with contact point $(0, 0)$.

If $a = -3$, the tangent equation is $y - (-27) = 27(x + 3)$

$\therefore$ $y = 27x + 54$ with contact point $(-3, -27)$.

c Let $(a, \sqrt{a})$ lie on $y = \sqrt{x}$.

Now $\dfrac{dy}{dx} = \frac{1}{2}x^{-\frac{1}{2}} = \dfrac{1}{2\sqrt{x}}$, so at $x = a$, $\dfrac{dy}{dx} = \dfrac{1}{2\sqrt{a}}$

$\therefore$ the gradient of the tangent at $(a, \sqrt{a})$ is $\dfrac{1}{2\sqrt{a}}$

and the gradient of the normal at this point is $-2\sqrt{a}$.

$\therefore$ the normal has equation $\dfrac{y - \sqrt{a}}{x - a} = -2\sqrt{a}$

or $y - \sqrt{a} = -2\sqrt{a}(x - a)$.

But this normal passes through $(4, 0)$, so $0 - \sqrt{a} = -2\sqrt{a}(4 - a)$

$$\therefore \quad 2\sqrt{a}(4 - a) - \sqrt{a} = 0$$
$$\therefore \quad \sqrt{a}(8 - 2a - 1) = 0$$
$$\therefore \quad \sqrt{a}(7 - 2a) = 0$$
$$\therefore \quad a = 0 \text{ or } \tfrac{7}{2}$$

When $a = 0$, the normal has equation $y = 0$ with contact point $(0, 0)$.

When $a = \frac{7}{2}$, $y - \sqrt{\frac{7}{2}} = -2\sqrt{\frac{7}{2}}\left(x - \frac{7}{2}\right)$

$\therefore \quad \sqrt{2}y - \sqrt{7} = -2\sqrt{7}\left(x - \frac{7}{2}\right)$

$\therefore \quad \sqrt{2}y + 2\sqrt{7}x = 7\sqrt{7} + \sqrt{7}$

$\therefore \quad \sqrt{2}y + 2\sqrt{7}x = 8\sqrt{7}$

$\therefore \quad y = -\sqrt{14}x + 4\sqrt{14}$ with contact point $\left(\frac{7}{2}, \sqrt{\frac{7}{2}}\right)$.

10 **a**

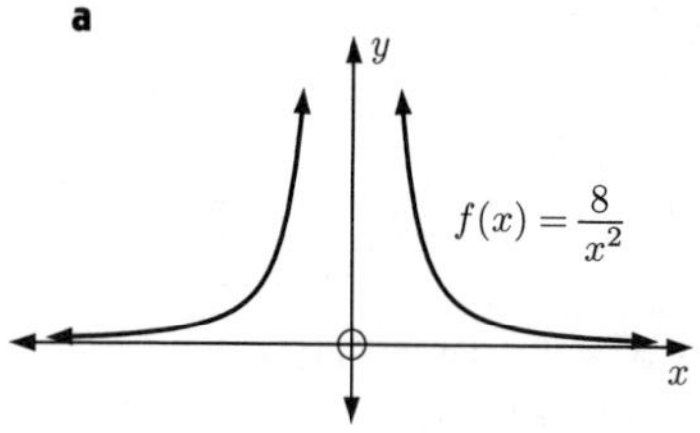

b Let $\left(a, \dfrac{8}{a^2}\right)$ lie on $f(x) = \dfrac{8}{x^2} = 8x^{-2}$

Now $f'(x) = -16x^{-3} = -\dfrac{16}{x^3}$

$\therefore \quad f'(a) = -\dfrac{16}{a^3}$

$\therefore$ the gradient of the tangent at $\left(a, \dfrac{8}{a^2}\right)$ is $-\dfrac{16}{a^3}$

$\therefore$ the equation of the tangent is $\dfrac{y - \frac{8}{a^2}}{x - a} = -\dfrac{16}{a^3}$

$\therefore \quad a^3y - 8a = -16x + 16a$

$\therefore \quad 16x + a^3y = 24a$

c The tangent cuts the x-axis when $y = 0$

$\therefore \quad 16x = 24a$

$\therefore \quad x = \frac{3}{2}a$

$\therefore$ A is $\left(\frac{3}{2}a, 0\right)$.

The tangent cuts the y-axis when $x = 0$

$\therefore \quad a^3y = 24a$

$\therefore \quad y = \dfrac{24}{a^2}$

$\therefore$ B is $\left(0, \dfrac{24}{a^2}\right)$.

d Area of triangle OAB

$= \left|\frac{1}{2} \times \left(\frac{3}{2}a\right) \times \left(\dfrac{24}{a^2}\right)\right|$

$= \dfrac{18}{|a|}$ units^2

As $a \to \infty$, $\dfrac{18}{a} \to 0$

$\therefore$ area $\to 0$

EXERCISE 17H

1 **a** $f(x) = 3x^2 - 6x + 2$

$\therefore \quad f'(x) = 6x - 6$

$\therefore \quad f''(x) = 6$

b $f(x) = 2x^3 - 3x^2 - x + 5$

$\therefore \quad f'(x) = 6x^2 - 6x - 1$

$\therefore \quad f''(x) = 12x - 6$

c $f(x) = \dfrac{2}{\sqrt{x}} - 1 = 2x^{-\frac{1}{2}} - 1$

$\therefore \quad f'(x) = -x^{-\frac{3}{2}}$

$f''(x) = \frac{3}{2}x^{-\frac{5}{2}}$

$= \dfrac{3}{2\sqrt{x^5}}$

d $f(x) = \dfrac{2 - 3x}{x^2} = 2x^{-2} - 3x^{-1}$

$\therefore \quad f'(x) = -4x^{-3} + 3x^{-2}$

$\therefore \quad f''(x) = 12x^{-4} - 6x^{-3}$

$= \dfrac{12 - 6x}{x^4}$

e $f(x) = (1 - 2x)^3$

$\therefore \quad f'(x) = 3(1 - 2x)^2(-2)$

$= -6(1 - 2x)^2$

$\therefore \quad f''(x) = -12(1 - 2x)^1(-2)$

$= 24(1 - 2x) = 24 - 48x$

f $f(x) = \dfrac{x+2}{2x-1}$ is a quotient with $u = x+2$ and $v = 2x-1$

$\therefore\ u' = 1$ and $v' = 2$

$$\therefore\ f'(x) = \frac{1(2x-1) - 2(x+2)}{(2x-1)^2} \quad \{\text{quotient rule}\}$$

$$= \frac{-5}{(2x-1)^2}$$

$$= -5(2x-1)^{-2}$$

$$\therefore\ f''(x) = 10(2x-1)^{-3}(2) = \frac{20}{(2x-1)^3}$$

2 **a** $y = x - x^3$

$$\therefore\ \frac{dy}{dx} = 1 - 3x^2$$

$$\therefore\ \frac{d^2y}{dx^2} = -6x$$

b $y = x^2 - \dfrac{5}{x^2}$

$$= x^2 - 5x^{-2}$$

$$\therefore\ \frac{dy}{dx} = 2x + 10x^{-3}$$

$$\therefore\ \frac{d^2y}{dx^2} = 2 - 30x^{-4} = 2 - \frac{30}{x^4}$$

c $y = 2 - \dfrac{3}{\sqrt{x}}$

$$= 2 - 3x^{-\frac{1}{2}}$$

$$\therefore\ \frac{dy}{dx} = \tfrac{3}{2}x^{-\frac{3}{2}}$$

$$\therefore\ \frac{d^2y}{dx^2} = -\tfrac{9}{4}x^{-\frac{5}{2}}$$

d $y = \dfrac{4-x}{x} = 4x^{-1} - 1$

$$\therefore\ \frac{dy}{dx} = -4x^{-2}$$

$$\therefore\ \frac{d^2y}{dx^2} = 8x^{-3} = \frac{8}{x^3}$$

e $y = (x^2 - 3x)^3$

$$\therefore\ \frac{dy}{dx} = 3(x^2-3x)^2(2x-3)$$

$$= (6x-9)(x^2-3x)^2$$

which is a product where $u = 6x - 9$
and $v = (x^2 - 3x)^2$

$\therefore\ u' = 6$ and $v' = 2(x^2-3x)^1(2x-3)$

$$\therefore\ \frac{d^2y}{dx^2} = 6(x^2-3x)^2 + (6x-9)(2)(x^2-3x)(2x-3)$$

$$= 6(x^2-3x)\left[(x^2-3x) + (2x-3)^2\right]$$

$$= 6(x^2-3x)(x^2-3x+4x^2-12x+9)$$

$$= 6(x^2-3x)(5x^2-15x+9)$$

f $y = x^2 - x + \dfrac{1}{1-x}$

$$= x^2 - x + (1-x)^{-1}$$

$$\therefore\ \frac{dy}{dx} = 2x - 1 + (-1)(1-x)^{-2}(-1)$$

$$= 2x - 1 + (1-x)^{-2}$$

$$\therefore\ \frac{d^2y}{dx^2} = 2 - 2(1-x)^{-3}(-1)$$

$$= 2 + \frac{2}{(1-x)^3}$$

3 **a** $f(x) = 2x^3 - 6x^2 + 5x + 1$

$\therefore\ f'(x) = 6x^2 - 12x + 5$

$\therefore\ f''(x) = 12x - 12$

So, $f''(x) = 0$ when $12x - 12 = 0$

$\therefore\ 12x = 12$

$\therefore\ x = 1$

b $f(x) = \dfrac{x}{x^2+2}$ is a quotient where $u = x$ and $v = x^2+2$

$\therefore\ u' = 1,\ v' = 2x$

$\therefore\ f'(x) = \dfrac{1(x^2+2) - 2x^2}{(x^2+2)^2}$ {quotient rule}

$= \dfrac{2-x^2}{(x^2+2)^2}$

This is another quotient, this time with $u = 2 - x^2$ and $v = (x^2+2)^2$

$\therefore\ u' = -2x,\quad v' = 2(x^2+2)(2x)$

$\therefore\ f''(x) = \dfrac{-2x(x^2+2)^2 - 4x(x^2+2)(2-x^2)}{(x^2+2)^4}$

$= \dfrac{-2x(x^2+2)[x^2+2+2(2-x^2)]}{(x^2+2)^4}$

$= \dfrac{-2x[-x^2+6]}{(x^2+2)^3} = \dfrac{2x[x^2-6]}{(x^2+2)^3}$

So, $f''(x) = 0$ when $2x[x^2-6] = 0$

$\therefore\ x = 0$ or $x^2 - 6 = 0$

$\therefore\ x = 0$ or $x = \pm\sqrt{6}$

4 $f(x) = 2x^3 - x$

$\therefore\ f'(x) = 6x^2 - 1$

$\therefore\ f''(x) = 12x$

By substituting the various values of x into these three functions, we can fill in the table as follows:

x	-1	0	1
$f(x)$	$-$	0	$+$
$f'(x)$	$+$	$-$	$+$
$f''(x)$	$-$	0	$+$

REVIEW SET 17A

1 **a** $f(x) = 7 + x - 3x^2$

$\therefore\ f(3) = 7 + 3 - 3(3)^2 = -17$

b $f'(x) = 1 - 6x$

$\therefore\ f'(3) = 1 - 6(3) = -17$

c $f''(x) = -6$

$\therefore\ f''(3) = -6$

2 Consider $y = -2x^2$. When $x = -1$, $y = -2(-1)^2 = -2$, so the point of contact is $(-1, -2)$.

Now $\dfrac{dy}{dx} = -4x$

$\therefore$ at $x = -1$, $\dfrac{dy}{dx} = -4(-1) = 4$

$\therefore$ the tangent has equation $\dfrac{y-(-2)}{x-(-1)} = 4$ or $y = 4x + 2$.

3 **a** $y = 3x^2 - x^4$

$\therefore\ \dfrac{dy}{dx} = 6x - 4x^3$

b $y = \dfrac{x^3 - x}{x^2} = x - x^{-1}$

$\therefore\ \dfrac{dy}{dx} = 1 + x^{-2} = 1 + \dfrac{1}{x^2}$

4 $f(x) = x^2 + 2x \qquad \therefore\ f'(x) = \lim\limits_{h\to 0} \dfrac{f(x+h) - f(x)}{h}$

$= \lim\limits_{h\to 0} \dfrac{[(x+h)^2 + 2(x+h)] - [x^2+2x]}{h}$

$= \lim\limits_{h\to 0} \dfrac{2xh + h^2 + 2h}{h}$

$= \lim\limits_{h\to 0} 2x + 2 + h$ {as $h \neq 0$}

$= 2x + 2$ *Check:* $f(x) = x^2 + 2x \quad \therefore\ f'(x) = 2x + 2$ ✓

5 Consider $y = \dfrac{1-2x}{x^2}$. When $x = 1$, $y = \dfrac{1-2(1)}{1^2} = -1$, so the point of contact is $(1, -1)$.

Since $y = \dfrac{1}{x^2} - \dfrac{2}{x}$, $\dfrac{dy}{dx} = -2x^{-3} + 2x^{-2} = -\dfrac{2}{x^3} + \dfrac{2}{x^2}$

$\therefore$ at $x = 1$, $\dfrac{dy}{dx} = -2 + 2 = 0$

So, the tangent is a horizontal line, and the normal must be a vertical line of the form $x = k$.
As the normal passes through $(1, -1)$, its equation must be $x = 1$.

6 $y = \dfrac{ax+b}{\sqrt{x}} = a\sqrt{x} + \dfrac{b}{\sqrt{x}} = ax^{\frac{1}{2}} + bx^{-\frac{1}{2}}$

$\therefore \dfrac{dy}{dx} = \dfrac{a}{2}x^{-\frac{1}{2}} - \dfrac{b}{2}x^{-\frac{3}{2}} = \dfrac{a}{2\sqrt{x}} - \dfrac{b}{2x\sqrt{x}}$

The equation of the tangent at $x = 1$ is $2x - y = 1$
or $y = 2x - 1$ so the gradient of the tangent is 2

$\therefore$ at $x = 1$, $\dfrac{dy}{dx} = \dfrac{a}{2} - \dfrac{b}{2} = 2$ $\therefore$ $a - b = 4$
$\therefore$ $a = b + 4$ ..... (1)

Also at $x = 1$, the tangent touches the curve $\therefore$ $\dfrac{a(1)+b}{\sqrt{1}} = 2(1) - 1$

$\therefore$ $a + b = 1$

$\therefore$ $b + 4 + b = 1$ {using (1)}

$\therefore$ $2b = -3$

$\therefore$ $b = -\frac{3}{2}$

and $a = 4 - \frac{3}{2} = \frac{5}{2}$

7 **a** $f(x) = (x^2+3)^4$

$\therefore f'(x) = 4(x^2+3)^3(2x)$

$= 8x(x^2+3)^3$

b $g(x) = \dfrac{\sqrt{x+5}}{x^2}$ is a quotient with

$u = (x+5)^{\frac{1}{2}}$ and $v = x^2$

$\therefore$ $u' = \frac{1}{2}(x+5)^{-\frac{1}{2}}$, $v' = 2x$

$\therefore$ $g'(x) = \dfrac{\frac{1}{2}(x+5)^{-\frac{1}{2}}(x^2) - (x+5)^{\frac{1}{2}}(2x)}{x^4}$

$= \dfrac{\frac{1}{2}x(x+5)^{-\frac{1}{2}} - 2(x+5)^{\frac{1}{2}}}{x^3}$

8 **a** $f(x) = 3x^2 - \dfrac{1}{x} = 3x^2 - x^{-1}$

$\therefore$ $f'(x) = 6x + x^{-2}$

$\therefore$ $f''(x) = 6 - 2x^{-3} = 6 - \dfrac{2}{x^3}$

$\therefore$ $f''(2) = 6 - \dfrac{2}{2^3} = \dfrac{23}{4}$

b $f(x) = \sqrt{x} = x^{\frac{1}{2}}$

$\therefore$ $f'(x) = \frac{1}{2}x^{-\frac{1}{2}}$

$\therefore$ $f''(x) = -\frac{1}{4}x^{-\frac{3}{2}}$

$\therefore$ $f''(2) = -\frac{1}{4}(2^{-\frac{3}{2}})$

$= -\dfrac{1}{4\sqrt{2^3}} = -\dfrac{1}{8\sqrt{2}}$

9 $f(x) = 2x^3 + ax + b$ $\therefore$ $f'(x) = 6x^2 + a$

Now as the gradient at $(-2, 33)$ is 10,

$\therefore$ $f'(-2) = 10$

$\therefore$ $10 = 6(-2)^2 + a$

$\therefore$ $a = -14$

$\therefore$ $f(x) = 2x^3 - 14x + b$

Then, since $(-2, 33)$ lies on the curve,

$f(-2) = 33$

$\therefore$ $2(-2)^3 - 14(-2) + b = 33$

$\therefore$ $-16 + 28 + b = 33$

$\therefore$ $b = 21$

10 $f(x) = 2x^4 - 4x^3 - 9x^2 + 4x + 7$

$\therefore \quad f'(x) = 8x^3 - 12x^2 - 18x + 4$

$\therefore \quad f''(x) = 24x^2 - 24x - 18$

So, $f''(x) = 0$ where $24x^2 - 24x - 18 = 0$

$\therefore \quad 4x^2 - 4x - 3 = 0$

$\therefore \quad (2x+1)(2x-3) = 0$

$\therefore \quad x = -\frac{1}{2}$ or $x = \frac{3}{2}$

11 $y = \dfrac{a}{(x+2)^2} = a(x+2)^{-2}$

The gradient of the line (AB) is $\dfrac{y_2 - y_1}{x_2 - x_1} = \dfrac{8-4}{0-2} = \dfrac{4}{-2} = -2$

$\therefore$ the equation of the tangent is $\dfrac{y-8}{x-0} = -2$ or $y = -2x + 8$

Now $\dfrac{dy}{dx} = -2a(x+2)^{-3}$, so for the given tangent, $-2a(x+2)^{-3} = -2$

$\therefore \quad \dfrac{a}{(x+2)^3} = 1$

$\therefore \quad a = (x+2)^3 \quad .... (1)$

The line (AB) meets the curve where $-2x + 8 = \dfrac{a}{(x+2)^2}$

$\therefore \quad -2x + 8 = \dfrac{(x+2)^3}{(x+2)^2}$ {using (1)}

$\therefore \quad -2x + 8 = x + 2$

$\therefore \quad -3x = -6$

$\therefore \quad x = 2$

and so $a = (2+2)^3 = 64$

12 $y = \dfrac{5}{\sqrt{x}} = 5x^{-\frac{1}{2}}$

$\therefore \quad \dfrac{dy}{dx} = -\frac{5}{2}x^{-\frac{3}{2}}$

$\therefore$ the gradient of the tangent at the point (1, 5) is $-\frac{5}{2}(1)^{-\frac{3}{2}} = -\frac{5}{2}$

$\therefore$ the equation of the tangent is $\dfrac{y-5}{x-1} = -\dfrac{5}{2}$

$\therefore \quad y - 5 = -\frac{5}{2}x + \frac{5}{2}$

$\therefore \quad y = -\frac{5}{2}x + \frac{15}{2}$

Now, P and Q are the y- and x-intercepts, so:

P: $y = -\frac{5}{2}(0) + \frac{15}{2}$
$= \frac{15}{2}$

Q: $0 = -\frac{5}{2}x + \frac{15}{2}$
$\therefore \quad \frac{5}{2}x = \frac{15}{2}$
$\therefore \quad x = 3$

So P is (0, 7.5) and Q is (3, 0).

REVIEW SET 17B

1 **a** $y = 5x - 3x^{-1}$

$\therefore \quad \dfrac{dy}{dx} = 5 + 3x^{-2} = 5 + \dfrac{3}{x^2}$

b $y = (3x^2 + x)^4$

$\therefore \quad \dfrac{dy}{dx} = 4(3x^2 + x)^3(6x + 1)$

c $y = (x^2+1)(1-x^2)^3$ is a product with $u = x^2+1$ and $v = (1-x^2)^3$

$\therefore\ u' = 2x$ and $v' = 3(1-x^2)^2(-2x)$

$\therefore\ \dfrac{dy}{dx} = 2x(1-x^2)^3 - 6x(x^2+1)(1-x^2)^2$ {product rule}

2 $y = x^3 - 3x^2 - 9x + 2 \quad \therefore\ \dfrac{dy}{dx} = 3x^2 - 6x - 9$

Horizontal tangents occur when $\dfrac{dy}{dx} = 0 \quad \therefore\ 3x^2 - 6x - 9 = 0$

$\therefore\ x^2 - 2x - 3 = 0$

$\therefore\ (x-3)(x+1) = 0$

$\therefore\ x = 3$ or $x = -1$

When $x = 3$, the horizontal tangent has equation $y = -25$.
When $x = -1$, the horizontal tangent has equation $y = 7$.

3 Consider the tangent to $y = x^2\sqrt{1-x}$ at $x = -3$.

When $x = -3$, $y = (-3)^2\sqrt{1-(-3)} = 9\sqrt{4} = 18$,

$\therefore$ the point of contact is $(-3, 18)$.

Also, $y = x^2\sqrt{1-x}$ is a product with $u = x^2$ and $v = (1-x)^{\frac{1}{2}}$

$\therefore\ u' = 2x$ and $v' = \frac{1}{2}(1-x)^{-\frac{1}{2}}(-1)$

$\therefore\ \dfrac{dy}{dx} = 2x(1-x)^{\frac{1}{2}} - x^2(\tfrac{1}{2})(1-x)^{-\frac{1}{2}}$

$\therefore$ at $x = -3$, $\dfrac{dy}{dx} = 2(-3)(1-(-3))^{\frac{1}{2}} - (-3)^2(\tfrac{1}{2})(1-(-3))^{-\frac{1}{2}}$

$= -6(2) - 9(\tfrac{1}{2})(\tfrac{1}{2})$

$= -\frac{57}{4}$

$\therefore$ the tangent at $(-3, 18)$ has equation $\dfrac{y-18}{x-(-3)} = -\frac{57}{4}$

$\therefore\ 4y - 72 = -57x - 171$

$\therefore\ 4y = -57x - 99$

Now when $x = 0$, $y = -\frac{99}{4}$ and when $y = 0$, $x = -\frac{99}{57}$

$\therefore$ the area of ΔOAB $= \frac{1}{2}\left(\frac{99}{4}\right)\left(\frac{99}{57}\right) = \frac{3267}{152} \approx 21.5$ units^2

4 Consider the tangent to $y = 2x^3 + 4x - 1$ at $(1, 5)$.

$\dfrac{dy}{dx} = 6x^2 + 4 \quad \therefore$ at $x = 1$, $\dfrac{dy}{dx} = 6(1)^2 + 4 = 10$

$\therefore$ the tangent has equation $\dfrac{y-5}{x-1} = 10$ or $y = 10x - 5$

Now the tangent meets the curve again where $10x - 5 = 2x^3 + 4x - 1$

$\therefore\ 2x^3 - 6x + 4 = 0$

$\therefore\ x^3 - 3x + 2 = 0$

We know that $(x-1)^2$ is a factor since the line is tangent to the curve at $x = 1$.
Consequently, $x^3 - 3x + 2 = (x-1)^2(x+2) = 0$ {since the constant term is 2}
Thus $x = -2$ is the other solution and when $x = -2$, $y = 2(-2)^3 + 4(-2) - 1 = -25$
$\therefore$ the tangent meets the curve again at $(-2, -25)$.

5 Consider $y = 4(ax+1)^{-2}$.

When $x = 0$, $y = 4(0+1)^{-2} = 4$, so the point of contact is $(0, 4)$.

Now $\dfrac{dy}{dx} = -8(ax+1)^{-3}(a) = \dfrac{-8a}{(ax+1)^3} \quad \therefore$ at $x = 0$, $\dfrac{dy}{dx} = -8a$

$\therefore$ the tangent has equation $\dfrac{y-4}{x-0} = -8a$ or $y - 4 = -8ax$

This tangent passes through $(1, 0)$, so $0 - 4 = -8a(1)$ $\therefore$ $a = \frac{1}{2}$

6 $y = 2x^3 + 3x^2 - 10x + 3$

$\therefore$ $\dfrac{dy}{dx} = 6x^2 + 6x - 10$

The gradient of the tangent is 2 where $6x^2 + 6x - 10 = 2$

$\therefore$ $6x^2 + 6x - 12 = 0$

$\therefore$ $6(x^2 + x - 2) = 0$

$\therefore$ $6(x + 2)(x - 1) = 0$

$\therefore$ $x = -2$ or 1

$\therefore$ $y = 2(-2)^3 + 3(-2)^2 - 10(-2) + 3$ or $y = 2(1)^3 + 3(1)^2 - 10(1) + 3$

$= -16 + 12 + 20 + 3$ $\quad = 2 + 3 - 10 + 3$

$= 19$ $\quad = -2$

So, the gradient of the tangent is 2 at $(-2, 19)$ and $(1, -2)$.

7 $y = \sqrt{5 - 4x} = (5 - 4x)^{\frac{1}{2}}$

a $\dfrac{dy}{dx} = \frac{1}{2}(5 - 4x)^{-\frac{1}{2}}(-4)$

$= -2(5 - 4x)^{-\frac{1}{2}}$

b $\dfrac{d^2y}{dx^2} = -2(-\frac{1}{2})(5 - 4x)^{-\frac{3}{2}}(-4)$

$= -4(5 - 4x)^{-\frac{3}{2}}$

8 Consider $y = \dfrac{x+1}{x^2 - 2}$.

When $x = 1$, $y = \dfrac{1+1}{1^2 - 2} = -2$ $\therefore$ the point of contact is $(1, -2)$.

$y = \dfrac{x+1}{x^2-2}$ is a quotient with $u = x + 1$ and $v = x^2 - 2$

$\therefore$ $u' = 1$ and $v' = 2x$

$$\frac{dy}{dx} = \frac{1(x^2 - 2) - (x + 1)(2x)}{(x^2 - 2)^2} \quad \text{\{quotient rule\}}$$

$\therefore$ at $x = 1$, $\dfrac{dy}{dx} = \dfrac{(1 - 2) - 2(1 + 1)}{(1 - 2)^2}$

$= \dfrac{-1 - 4}{1} = -5$

$\therefore$ the normal at $(1, -2)$ has gradient $\frac{1}{5}$.

So the normal has equation $\dfrac{y - (-2)}{x - 1} = \frac{1}{5}$ $\therefore$ $5y + 10 = x - 1$

$\therefore$ $5y = x - 11$

9 $y = 2 - \dfrac{7}{1 + 2x} = 2 - 7(1 + 2x)^{-1}$

$\therefore$ $\dfrac{dy}{dx} = 7(1 + 2x)^{-2} \times 2$ {chain rule}

$= \dfrac{14}{(1 + 2x)^2}$

The tangent is horizontal when the gradient $\dfrac{dy}{dx} = 0$.

But $\dfrac{14}{(1 + 2x)^2}$ is never 0, so $y = 2 - \dfrac{7}{1 + 2x}$ has no horizontal tangents.

10 Let $g(x) = ax^2 + bx + c$.

$g(x)$ has y-intercept (0, 3), so $g(0) = a(0)^2 + b(0) + c = 3$

$\therefore\ c = 3$

$\therefore\ g(x) = ax^2 + bx + 3$

The point (2, 7) lies on $g(x)$, so $g(2) = a(2)^2 + b(2) + 3 = 7$

$\therefore\ 4a + 2b = 4$ .... (1)

Also, $g'(x) = 2ax + b$

$\therefore\ g'(2) = 2a(2) + b = 4a + b$

$\therefore$ the gradient of the tangent to $g(x)$ at (2, 7) is $4a + b$.

But, the tangent at (2, 7) passes through (0, 11), so the gradient $= \dfrac{7-11}{2-0} = -2$

$\therefore\ 4a + b = -2$ .... (2)

Solving (1) and (2) simultaneously,

$$4a + 2b = 4$$
$$4a + b = -2$$

subtracting: $b = 6$

Using (2), $4a + 6 = -2$, so $a = -2$

So, $g(x) = -2x^2 + 6x + 3$

11 **a**

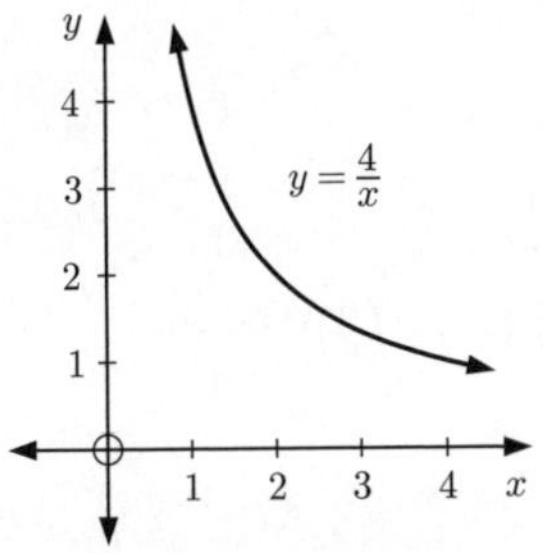

b For $f(x) = \dfrac{4}{x} = 4x^{-1}$,

$f'(x) = -4x^{-2} = -\dfrac{4}{x^2}$ and $f'(k) = -\dfrac{4}{k^2}$, $k > 0$

$\therefore$ the gradient of the tangent to $f(x)$ at $\left(k, \dfrac{4}{k}\right)$ is $-\dfrac{4}{k^2}$

$\therefore$ the equation of the tangent is $\dfrac{y - \frac{4}{k}}{x - k} = -\dfrac{4}{k^2}$

$\therefore\ yk^2 - 4k = -4x + 4k$

$\therefore\ k^2 y = -4x + 8k$

$\therefore\ y = -\dfrac{4}{k^2}x + \dfrac{8}{k}$

c $y = -\dfrac{4}{k^2}x + \dfrac{8}{k}$ cuts the x-axis when $y = 0$

$\therefore\ -\dfrac{4}{k^2}x + \dfrac{8}{k} = 0$

$\therefore\ \dfrac{4}{k^2}x = \dfrac{8}{k}$

$\therefore\ x = 2k$ $\therefore$ A is at $(2k, 0)$

$y = -\dfrac{4}{k^2}x + \dfrac{8}{k}$ cuts the y-axis when $x = 0$

$\therefore\ y = \dfrac{8}{k}$ $\therefore$ B is at $\left(0, \dfrac{8}{k}\right)$

d Area of triangle OAB $= \frac{1}{2}(2k)\left(\dfrac{8}{k}\right) = 8$ units^2

e The gradient of the tangent to $f(x)$ at $\left(k, \dfrac{4}{k}\right)$ is $-\dfrac{4}{k^2}$

$\therefore$ the gradient of the normal to $f(x)$ at $\left(k, \dfrac{4}{k}\right)$ is $\dfrac{k^2}{4}$

$\therefore$ the equation of the normal is $\dfrac{y - \frac{4}{k}}{x - k} = \dfrac{k^2}{4}$

$\therefore\ 4y - \dfrac{16}{k} = k^2 x - k^3$

$$\therefore \quad 4ky - k^3x = 16 - k^4$$

This normal passes through $(1, 1)$, so $4k - k^3 = 16 - k^4$

$$\therefore \quad k^4 - k^3 + 4k - 16 = 0$$

$$\therefore \quad (k-2)(k+2)(k^2 - k + 4) = 0 \quad \{\text{using technology}\}$$

$$\therefore \quad k = \pm 2$$

But $k > 0$, so $k = 2$

REVIEW SET 17C

1 **a** $y = x^3\sqrt{1-x^2}$ is a product where $u = x^3$ and $v = (1-x^2)^{\frac{1}{2}}$

$\therefore \quad u' = 3x^2$ and $v' = \frac{1}{2}(1-x^2)^{-\frac{1}{2}}(-2x) = -x(1-x^2)^{-\frac{1}{2}}$

$$\therefore \quad \frac{dy}{dx} = 3x^2\sqrt{1-x^2} - \frac{x^4}{\sqrt{1-x^2}} \quad \{\text{product rule}\}$$

b $y = \dfrac{x^2 - 3x}{\sqrt{x+1}}$ is a quotient where $u = x^2 - 3x$ and $v = (x+1)^{\frac{1}{2}}$

$\therefore \quad u' = 2x - 3$ and $v' = \frac{1}{2}(x+1)^{-\frac{1}{2}}$

$$\therefore \quad \frac{dy}{dx} = \frac{(2x-3)(x+1)^{\frac{1}{2}} - \frac{1}{2}(x^2-3x)(x+1)^{-\frac{1}{2}}}{x+1} \quad \{\text{quotient rule}\}$$

2 Consider the normal to the curve $y = \dfrac{1}{\sqrt{x}}$ at $x = 4$.

When $x = 4$, $y = \frac{1}{\sqrt{4}} = \frac{1}{2}$, so the point of contact is $(4, \frac{1}{2})$.

Now $\dfrac{dy}{dx} = -\frac{1}{2}x^{-\frac{3}{2}}$ $\quad\therefore$ at $x = 4$, $\dfrac{dy}{dx} = -\frac{1}{2}\left(4^{-\frac{3}{2}}\right) = -\frac{1}{2}\left(\frac{1}{8}\right) = -\frac{1}{16}$

$\therefore$ the normal at $(4, \frac{1}{2})$ has gradient 16.

So the equation is
$$\frac{y - \frac{1}{2}}{x - 4} = 16$$
$$\therefore \quad y - \tfrac{1}{2} = 16x - 64$$
$$\therefore \quad y = 16x - \tfrac{127}{2}$$

3 **a**
$$y = \frac{4}{\sqrt{x}} - 3x = 4x^{-\frac{1}{2}} - 3x$$
$$\therefore \quad \frac{dy}{dx} = -2x^{-\frac{3}{2}} - 3$$
$$= -\frac{2}{x\sqrt{x}} - 3$$

b
$$y = \left(x - \frac{1}{x}\right)^4 = (x - x^{-1})^4$$
$$\therefore \quad \frac{dy}{dx} = 4(x - x^{-1})^3(1 + x^{-2})$$
$$= 4\left(x - \frac{1}{x}\right)^3\left(1 + \frac{1}{x^2}\right)$$

c
$$y = \sqrt{x^2 - 3x} = (x^2 - 3x)^{\frac{1}{2}}$$
$$\therefore \quad \frac{dy}{dx} = \tfrac{1}{2}(x^2 - 3x)^{-\frac{1}{2}}(2x - 3)$$
$$= \frac{2x - 3}{2\sqrt{x^2 - 3x}}$$

4 The tangent shown on the graph passes through $(0, 5)$ and $(5, 0)$.

$\therefore$ the gradient of the tangent is $\dfrac{0-5}{5-0} = -1$, so $f'(3) = -1$.

Also, since the tangent passes through $(0, 5)$, it has equation $\dfrac{y-5}{x-0} = -1$

$$\therefore \quad y - 5 = -x$$
$$\therefore \quad y = -x + 5$$

So when $x = 3$, $y = -3 + 5 = 2$

$\therefore$ the point of contact is (3, 2), and hence $f(3) = 2$.

5 **a** $f(x) = \dfrac{(x+3)^3}{\sqrt{x}}$ is a quotient with $u = (x+3)^3$ and $v = x^{\frac{1}{2}}$

$\therefore$ $u' = 3(x+3)^2$ and $v' = \frac{1}{2}x^{-\frac{1}{2}}$

$\therefore$ $f'(x) = \dfrac{3(x+3)^2\sqrt{x} - \frac{1}{2}x^{-\frac{1}{2}}(x+3)^3}{x}$ {quotient rule}

b $f(x) = x^4\sqrt{x^2+3}$ is a product with $u = x^4$ and $v = (x^2+3)^{\frac{1}{2}}$

$\therefore$ $u' = 4x^3$ and $v' = \frac{1}{2}(x^2+3)^{-\frac{1}{2}}(2x) = x(x^2+3)^{-\frac{1}{2}}$

$\therefore$ $f'(x) = 4x^3\sqrt{x^2+3} + \dfrac{x^5}{\sqrt{x^2+3}}$ {product rule}

6 $y = x^3 + ax + b$ $\therefore$ $\dfrac{dy}{dx} = 3x^2 + a$

$\therefore$ at $x = 1$, $\dfrac{dy}{dx} = 3 + a$

The equation of the tangent at $x = 1$ is $y = 2x$, so the gradient is 2.

$\therefore$ $3 + a = 2$ and so $a = -1$

Also at $x = 1$, the tangent touches the curve.

$\therefore$ $x^3 + ax + b = 2x$ when $x = 1$

$\therefore$ $(1)^3 + (-1)(1) + b = 2(1)$

$\therefore$ $1 - 1 + b = 2$

$\therefore$ $b = 2$

7 $y = x^3 + ax^2 - 4x + 3$ $\therefore$ $\dfrac{dy}{dx} = 3x^2 + 2ax - 4$

The tangent at $x = 1$ is parallel to $y = 3x$, so when $x = 1$, $\dfrac{dy}{dx} = 3$

$\therefore$ $3 = 3(1)^2 + 2a(1) - 4$

$\therefore$ $2a = 4$

$\therefore$ $a = 2$

When $x = 1$, $y = 1^3 + 2(1)^2 - 4(1) + 3 = 2$

The contact point is (1, 2) and since the gradient is 3, the tangent at (1, 2) has equation

$\dfrac{y-2}{x-1} = 3$ $\therefore$ $y - 2 = 3x - 3$

$\therefore$ $y = 3x - 1$

The tangent meets the curve where $x^3 + 2x^2 - 4x + 3 = 3x - 1$

$\therefore$ $x^3 + 2x^2 - 7x + 4 = 0$

Since the line touches the curve at $x = 1$, $(x-1)^2$ must be a factor.

Consequently, $x^3 + 2x^2 - 7x + 4 = (x-1)^2(x+4) = 0$ {since the constant term is 4}

$\therefore$ the curve cuts the tangent when $x = -4$.

When $x = -4$, $y = (-4)^3 + 2(-4)^2 - 4(-4) + 3 = -13$

$\therefore$ the curve cuts the tangent at $(-4, -13)$.

8 Consider the normal to $f(x) = \dfrac{3x}{1+x}$ at (2, 2).

$f(x)$ is a quotient with $u = 3x$ and $v = 1 + x$

$\therefore$ $u' = 3$ and $v' = 1$

$\therefore$ $f'(x) = \dfrac{3(1+x) - 1(3x)}{(1+x)^2} = \dfrac{3}{(1+x)^2}$ {quotient rule}

$\therefore \quad f'(2) = \frac{3}{9} = \frac{1}{3}$

$\therefore$ the normal at (2, 2) has gradient -3

So, the equation of the normal is $\dfrac{y-2}{x-2} = -3$

$\therefore \quad y - 2 = -3(x-2)$

$\therefore \quad y = -3x + 8$

When $x = 0$, $y = 8$ and when $y = 0$, $x = \frac{8}{3}$

$\therefore$ B and C are at (0, 8) and $(\frac{8}{3}, 0)$,

and the distance BC $= \sqrt{\left(0 - \frac{8}{3}\right)^2 + (8-0)^2} = \sqrt{\frac{64}{9} + 64} = \sqrt{\frac{640}{9}} = \frac{8\sqrt{10}}{3}$ units

9 **a** $y = 3x^4 - \dfrac{2}{x} = 3x^4 - 2x^{-1}$

$\therefore \quad \dfrac{dy}{dx} = 12x^3 + 2x^{-2}$

$\therefore \quad \dfrac{d^2y}{dx^2} = 36x^2 - 4x^{-3}$

$= 36x^2 - \dfrac{4}{x^3}$

b $y = x^3 - x + \dfrac{1}{\sqrt{x}} = x^3 - x + x^{-\frac{1}{2}}$

$\therefore \quad \dfrac{dy}{dx} = 3x^2 - 1 - \frac{1}{2}x^{-\frac{3}{2}}$

$\therefore \quad \dfrac{d^2y}{dx^2} = 6x + \frac{3}{4}x^{-\frac{5}{2}}$

10 $f(x) = 3x^3 + ax^2 + b$

$\therefore \quad f'(x) = 9x^2 + 2ax$

Since the tangent at $(-2, 14)$ has gradient 0, $f'(-2) = 0$

$\therefore \quad 36 - 4a = 0$

$\therefore \quad a = 9$

As the point $(-2, 14)$ lies on the curve, $14 = 3(-2)^3 + 9(-2)^2 + b$

$\therefore \quad b = 14 + 24 - 36$

$\therefore \quad b = 2$

$\therefore \quad f'(x) = 9x^2 + 18x$

$\therefore \quad f''(x) = 18x + 18$ and so $f''(-2) = -36 + 18 = -18$

11 The curves $y = \sqrt{3x+1}$ and $y = \sqrt{5x - x^2}$ meet when $\sqrt{3x+1} = \sqrt{5x-x^2}$

Squaring both sides, $3x + 1 = 5x - x^2$

$\therefore \quad x^2 - 2x + 1 = 0$

$\therefore \quad (x-1)^2 = 0$

$\therefore \quad x = 1$

When $x = 1$, $y = \sqrt{3+1} = 2$, so the curves meet at (1, 2).

Now for $y = \sqrt{3x+1} = (3x+1)^{\frac{1}{2}}$

$\dfrac{dy}{dx} = \frac{1}{2}(3x+1)^{-\frac{1}{2}}(3)$

$\therefore$ at (1, 2), $\dfrac{dy}{dx} = \dfrac{3}{2(3+1)^{\frac{1}{2}}} = \frac{3}{4}$

Check: $y = \sqrt{5x - x^2} = (5x - x^2)^{\frac{1}{2}}$

$\dfrac{dy}{dx} = \frac{1}{2}(5x - x^2)^{-\frac{1}{2}}(5 - 2x) = \dfrac{5-2x}{2\sqrt{5x-x^2}}$

$\therefore$ at (1, 2), $\dfrac{dy}{dx} = \dfrac{5-2}{2\sqrt{5-1}} = \frac{3}{4}$ ✓

$\therefore$ the curves have the same gradient of $\frac{3}{4}$ at their point of intersection.

The equation of the common tangent at (1, 2) is $\dfrac{y-2}{x-1} = \frac{3}{4}$

$\therefore \quad 4(y-2) = 3(x-1)$

$\therefore \quad 4y = 3x + 5$

Chapter 18

APPLICATIONS OF DIFFERENTIAL CALCULUS

EXERCISE 18A

1 $P(t) = 2t^2 - 12t + 118$ thousand dollars, $t \geqslant 0$

a $P(0) = \$118\,000$ is the current annual profit

b $\dfrac{dP}{dt} = 4t - 12$ thousand dollars per year

c $\dfrac{dP}{dt}$ is the rate of change in profit with time.

d **i** The profit decreases when $\dfrac{dP}{dt} \leqslant 0$, which occurs when $4t - 12 \leqslant 0$

$\therefore \quad 4t \leqslant 12$

$\therefore \quad t \leqslant 3$

But $t \geqslant 0$, so $0 \leqslant t \leqslant 3$ years.

ii The profit increases on the previous year when $\dfrac{dP}{dt} \geqslant 0$, which is for $t > 3$ years.

e The profit function is a quadratic with $a > 0$ $\therefore$ the shape is 

So, a minimum profit occurs when $\dfrac{dP}{dt} = 0$, which is when $t = 3$ years

and $P(3) = 18 - 36 + 118 = 100$ thousand dollars or $\$100\,000$.

f When $t = 4$, $\dfrac{dP}{dt} = 4$ thousand dollars per year.

So, the profit is increasing at \$4000 per year after 4 years.

When $t = 10$, $\dfrac{dP}{dt} = 28$ thousand dollars per year.

So, the profit is increasing at \$28 000 per year after 10 years.

When $t = 25$, $\dfrac{dP}{dt} = 88$ thousand dollars per year.

So, the profit is increasing at \$88 000 per year after 25 years.

2 $V = 200(50 - t)^2$ m^3

a average rate on $0 \leqslant t \leqslant 5$

$= \dfrac{V(5) - V(0)}{5 - 0}$

$= \dfrac{200(45)^2 - 200(50)^2}{5}$

$= -19\,000$ m^3 per minute

$\therefore$ leaving at $19\,000$ m^3 per minute

b $V'(t) = 400(50 - t)^1 \times (-1)$

$\therefore \quad V'(5) = 400 \times 45 \times -1$

$= -18\,000$ m^3 per minute

$\therefore$ leaving at $18\,000$ m^3 per minute

3 $s(t) = 1.2 + 28.1t - 4.9t^2$ metres

a When released, $t = 0$ and $s(0) = 1.2$ m $\therefore$ it is released 1.2 m above the ground.

b $s'(t) = 28.1 - 9.8t$ m s^{-1} is the instantaneous velocity of the ball at the time t seconds after release.

c When $s'(t) = 0$, $28.1 - 9.8t = 0$ $\therefore$ $t = \dfrac{28.1}{9.8} \approx 2.87$ seconds

So, after 2.87 seconds the ball has stopped and reached its maximum height.

d $s(2.867) = 1.2 + 28.1 \times 2.867 - 4.9 \times 2.867^2 \approx 41.5$ m

So, the maximum height reached is about 41.5 m.

e **i** $s'(0) = 28.1$ m s^{-1}

ii $s'(2) = 28.1 - 19.6$

$= 8.5$ m s^{-1}

iii $s'(5) = 28.1 - 49$

$= -20.9$ m s^{-1}

$\therefore$ speed $= 20.9$ m s^{-1}

If $s'(t) \geqslant 0$, the ball is travelling upwards. If $s'(t) \leqslant 0$, the ball is travelling downwards.

f $s(t) = 0$ when $1.2 + 28.1t - 4.9t^2 = 0$

$\therefore \quad 4.9t^2 - 28.1t - 1.2 = 0$

$$\therefore \quad t = \frac{28.1 \pm \sqrt{28.1^2 - 4(4.9)(-1.2)}}{9.8} \approx -0.0424 \text{ or } 5.777$$

But $t > 0$, so the ball hits the ground after 5.78 seconds.

g $\dfrac{d^2s}{dt^2} = -9.8 \text{ m s}^{-2}$ and is the rate of change in $\dfrac{ds}{dt}$

$\therefore$ the instantaneous acceleration is constant at -9.8 m s^{-2} for the entire motion.

4 **a** $s(t) = bt - 4.9t^2$

$s'(t) = b - 9.8t$

$\therefore \quad s'(0) = b$

$\therefore$ the initial velocity is $b \text{ m s}^{-1}$

b Since $s(14.2) = 0$,

$b(14.2) - 4.9(14.2)^2 = 0$

$\therefore \quad 14.2\,[b - 4.9 \times 14.2] = 0$

$\therefore \quad b = 4.9 \times 14.2$

$\therefore \quad b = 69.58$

$\therefore$ the initial velocity is 69.6 m s^{-1}

EXERCISE 18B

1 $Q = 100 - 10\sqrt{t}, \quad t \geqslant 0$

a **i** At $t = 0$, $Q = 100$ units

ii At $t = 25$, $Q = 50$ units

iii At $t = 100$, $Q = 0$ units

b $\dfrac{dQ}{dt} = -5t^{-\frac{1}{2}} = -\dfrac{5}{\sqrt{t}}$

i At $t = 25$, $\dfrac{dQ}{dt} = -1$ unit per year

$\therefore$ decreasing at 1 unit per year

ii At $t = 50$, $\dfrac{dQ}{dt} = -\frac{5}{\sqrt{50}}$

$= -\frac{1}{\sqrt{2}}$ units per year

$\therefore$ decreasing at $\frac{1}{\sqrt{2}}$ units per year

c $\dfrac{dQ}{dt} = -\dfrac{5}{\sqrt{t}}$ $\quad\therefore$ the skin *loses* the chemical at the rate $R = \dfrac{5}{\sqrt{t}} = 5t^{-\frac{1}{2}}$ units per year.

Now $\dfrac{dR}{dt} = -\frac{5}{2}t^{-\frac{3}{2}} = -\dfrac{5}{2t\sqrt{t}}$

Since $2t\sqrt{t} > 0$ for all $t > 0$, $\dfrac{dR}{dt} < 0$ for all $t > 0$.

$\therefore$ the rate at which the skin loses the chemical is decreasing for all $t > 0$.

2 $H = 20 - \dfrac{97.5}{t+5}$ m, $\quad t \geqslant 0$

a At planting, $t = 0$ $\quad \therefore \quad H(0) = 20 - \dfrac{97.5}{0+5} = 0.5$ m

b $H(4) = 20 - \dfrac{97.5}{4+5} \approx 9.17$ m

$H(8) = 20 - \dfrac{97.5}{8+5} = 12.5$ m

$H(12) = 20 - \dfrac{97.5}{12+5} \approx 14.3$ m

c Now $\dfrac{dH}{dt} = 97.5(t+5)^{-2} = \dfrac{97.5}{(t+5)^2}$

When $t = 0$, $\dfrac{dH}{dt} = \frac{97.5}{25} = 3.9 \text{ m year}^{-1}$

When $t = 5$, $\dfrac{dH}{dt} = \frac{97.5}{100} = 0.975 \text{ m year}^{-1}$

When $t = 10$, $\dfrac{dH}{dt} = \frac{97.5}{225} \approx 0.433 \text{ m year}^{-1}$

d Now $\dfrac{dH}{dt} = \dfrac{97.5}{(t+5)^2}$

Since $(t+5)^2 > 0$ for all $t \geqslant 0$, $\dfrac{dH}{dt} > 0$ for all $t \geqslant 0$

$\therefore$ the height of the tree is always increasing, which means that the tree is always growing.

3 **a** $C(v) = \frac{1}{5}v^2 + 200\,000v^{-1}$ euros

i At $v = 50$ km h^{-1}, $C = €4500$

ii At $v = 100$ km h^{-1}, $C = €4000$

b $\dfrac{dC}{dv} = \frac{2}{5}v - 200\,000v^{-2} = \frac{2}{5}v - \dfrac{200\,000}{v^2}$

i At $v = 30$ km h^{-1}, $\dfrac{dC}{dv} \approx -€210$ per km h^{-1}

ii At $v = 90$ km h^{-1}, $\dfrac{dC}{dv} \approx €11.3$ per km h^{-1}

c The cost is a minimum when $\dfrac{dC}{dv} = 0$, which occurs when $\frac{2}{5}v - \dfrac{200\,000}{v^2} = 0$

$\therefore \quad \frac{2}{5}v = \dfrac{200\,000}{v^2} \qquad \therefore \quad v^3 = 500\,000 \qquad \therefore \quad v \approx 79.4$ km h^{-1}

4 $y = \frac{1}{10}x(x-2)(x-3) = \frac{1}{10}(x^3 - 5x^2 + 6x)$

a When $y = 0$, $x = 0$, 2 or 3

$\therefore$ the lake is between 2 and 3 km from the shoreline.

b $\dfrac{dy}{dx} = \frac{1}{10}(3x^2 - 10x + 6)$

$= \frac{3}{10}x^2 - x + \frac{3}{5}$

When $x = \frac{1}{2}$, $\dfrac{dy}{dx} = \frac{7}{40}$ $\quad\therefore$ land is sloping upwards.

When $x = 1\frac{1}{2}$, $\dfrac{dy}{dx} = -\frac{9}{40}$ $\quad\therefore$ land is sloping downwards.

c The deepest point of the lake occurs when the slope of the land is 0, which is when $\dfrac{dy}{dx} = 0$

$\therefore \quad \frac{1}{10}(3x^2 - 10x + 6) = 0$

$\therefore \quad 3x^2 - 10x + 6 = 0$

$\therefore \quad x = \dfrac{10 \pm \sqrt{100 - 72}}{6} = \dfrac{5 \pm \sqrt{7}}{3}$

but it must be the value between 2 and 3 km, so $x = \dfrac{5 + \sqrt{7}}{3} \approx 2.55$ km from the sea

The depth at this point is $y(2.549) \approx \frac{1}{10}(2.549)(0.549)(-0.451)$

≈ -0.06311 km

≈ 63.1 m below sea level.

5 **a** $V = 50\,000\left(1 - \dfrac{t}{80}\right)^2, \quad 0 \leqslant t \leqslant 80$

$\therefore \quad \dfrac{dV}{dt} = 2 \times 50\,000\left(1 - \dfrac{t}{80}\right)^1 \times \left(-\frac{1}{80}\right) = -1250\left(1 - \dfrac{t}{80}\right)$

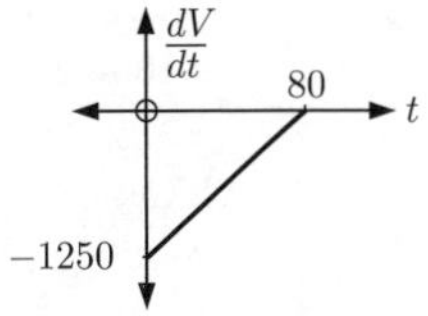

b The outflow was fastest when $t = 0$, when the tap was first opened.

c $\dfrac{dV}{dt} = -1250 + \frac{1250}{80}t \qquad \therefore \quad \dfrac{d^2V}{dt^2} = \frac{1250}{80} = \frac{125}{8}$

Since $\dfrac{d^2V}{dt^2}$ is constant and positive, $\dfrac{dV}{dt}$ is constantly increasing

$\therefore$ the outflow is decreasing at a constant rate.

6 **a** $\dfrac{dP}{dt} = aP\left(1 - \dfrac{P}{b}\right) - cP$ and when $\dfrac{dP}{dt} = 0$, the rate of change of population is zero, so the population is not changing and is stable.

b If $a = 0.06$, $b = 24\,000$, $c = 0.05$ then

$\dfrac{dP}{dt} = 0.06P\left(1 - \dfrac{P}{24\,000}\right) - 0.05P$

$= 0.06P - 0.05P - \dfrac{0.06P^2}{24\,000}$

$= P\left(0.01 - \dfrac{P}{400\,000}\right)$

Now for a stable population, $\dfrac{dP}{dt} = 0$

$\therefore \quad P = 0$ or $\dfrac{P}{400\,000} = 0.01$

$\therefore \quad P = 0$ or 4000

$\therefore$ the stable population is 4000 fish.

c If the harvest rate is 4%, then $\frac{dP}{dt} = 0.06P\left(1 - \frac{P}{24\,000}\right) - 0.04P = P\left(0.02 - \frac{0.06P}{24\,000}\right)$

For a stable population, $\frac{dP}{dt} = 0$ and so $0 = P\left(0.02 - \frac{0.06P}{24\,000}\right)$

$$\therefore \quad P = 0 \text{ or } \frac{0.06P}{24\,000} = 0.02$$

$$\therefore \quad P = 0 \text{ or } \frac{0.02 \times 24\,000}{0.06}$$

$$\therefore \quad P = 0 \text{ or } 8000$$

$\therefore$ the stable population is 8000 fish

7 **a** $C(x) = 0.0003x^3 + 0.02x^2 + 4x + 2250$

$\therefore \quad C'(x) = 0.0009x^2 + 0.04x + 4$ dollars per pair

b $C'(220) = 0.0009(220)^2 + 0.04(220) + 4 = \$56.36$ per pair

This estimates the cost of making the 221st pair of jeans if 220 pairs are currently being made.

c $C(221) - C(220) \approx \$7348.98 - \$7292.40 \approx \$56.58$

This is the actual cost to make the extra pair of jeans (221 instead of 220).

d $C''(x) = 0.0018x + 0.04$

$C''(x) = 0$ when $0.0018x + 0.04 = 0 \quad \therefore \quad x = -\frac{0.04}{0.0018} \approx -22.2$

This is the point when the rate of change is a minimum. However, it is out of the bounds of our model, as we cannot make a negative quantity of jeans.

EXERCISE 18C.1

1 **a** $s(t) = t^2 + 3t - 2, \quad t \geqslant 0$

Average velocity

$$= \frac{s(t_2) - s(t_1)}{t_2 - t_1} = \frac{s(3) - s(1)}{3 - 1} = \frac{16 - 2}{2} = 7 \text{ m s}^{-1}$$

b Average velocity

$$= \frac{s(t_2) - s(t_1)}{t_2 - t_1} = \frac{s(1+h) - s(1)}{(1+h) - 1} = \frac{(1+h)^2 + 3(1+h) - 2 - 2}{h} = \frac{2h + h^2 + 3h}{h} = (5 + h) \text{ m s}^{-1}, \quad h \neq 0$$

c $$\lim_{h \to 0} \frac{s(1+h) - s(1)}{h} = \lim_{h \to 0} 5 + h = 5 \text{ m s}^{-1}$$

This is the instantaneous velocity at $t = 1$ second, or $s'(1)$.

d Average velocity

$$= \frac{s(t_2) - s(t_1)}{t_2 - t_1} = \frac{s(t+h) - s(t)}{(t+h) - t} = \frac{[(t+h)^2 + 3(t+h) - 2] - [t^2 + 3t - 2]}{h} = \frac{2ht + h^2 + 3h}{h} = (2t + 3 + h) \text{ m s}^{-1}, \quad h \neq 0$$

Now $\lim_{h \to 0} \frac{s(t+h) - s(t)}{h} = \lim_{h \to 0} (2t + 3 + h) = (2t + 3) \text{ m s}^{-1}$

This is the instantaneous velocity at t seconds.

2 **a** $s(t) = 5 - 2t^2$ cm

$$\text{Average velocity} = \frac{s(t_2) - s(t_1)}{t_2 - t_1} = \frac{s(5) - s(2)}{5 - 2} = \frac{(-45) - (-3)}{3} = -14 \text{ cm s}^{-1}$$

b
$$\text{Average velocity} = \frac{s(t_2) - s(t_1)}{t_2 - t_1} = \frac{s(2+h) - s(2)}{(2+h) - 2} = \frac{5 - 2(2+h)^2 + 3}{h} = \frac{-8h - 2h^2}{h} = (-8 - 2h) \text{ cm s}^{-1},\ h \neq 0$$

c
$$\lim_{h \to 0} \frac{s(2+h) - s(2)}{h} = \lim_{h \to 0} (-8 - 2h) = -8 \text{ cm s}^{-1}$$

This is the instantaneous velocity when $t = 2$ seconds, or $s'(2)$.

d
$$\lim_{h \to 0} \frac{s(t+h) - s(t)}{h} = \lim_{h \to 0} \frac{[5 - 2(t+h)^2] - [5 - 2t^2]}{h} = \lim_{h \to 0} \frac{-4th - 2h^2}{h} = \lim_{h \to 0} (-4t - 2h) = -4t \text{ cm s}^{-1}$$

This is the instantaneous velocity at t seconds.

3 $v(t) = 2\sqrt{t} + 3 \text{ cm s}^{-1},\ t \geqslant 0$

a Average acceleration
$$= \frac{v(t_2) - v(t_1)}{t_2 - t_1} = \frac{v(4) - v(1)}{4 - 1} = \frac{7 - 5}{3} = \tfrac{2}{3} \text{ cm s}^{-2}$$

b Average acceleration
$$= \frac{v(t_2) - v(t_1)}{t_2 - t_1} = \frac{v(1+h) - v(1)}{(1+h) - 1} = \frac{[2\sqrt{1+h} + 3] - [2\sqrt{1} + 3]}{h} = \frac{2\sqrt{1+h} - 2}{h}$$
$$= \frac{2\left(\sqrt{1+h} - 1\right)}{h} \times \frac{\sqrt{1+h} + 1}{\sqrt{1+h} + 1} = \frac{2(1 + h - 1)}{h\left(\sqrt{1+h} + 1\right)} = \frac{2h}{h\left(\sqrt{1+h} + 1\right)} = \frac{2}{\sqrt{1+h} + 1}$$

c
$$\lim_{h \to 0} \frac{v(1+h) - v(1)}{(1+h) - 1} = \lim_{h \to 0} \frac{2}{\sqrt{1+h} + 1} \quad \{\text{using } \mathbf{b}\}$$
$$= \tfrac{2}{2} = 1 \text{ cm s}^{-2}$$

This is the instantaneous acceleration when $t = 1$ second.

d
$$\lim_{h \to 0} \frac{v(t+h) - v(t)}{h} = \lim_{h \to 0} \frac{2\sqrt{t+h} - 2\sqrt{t}}{h} = \lim_{h \to 0} \frac{2(\sqrt{t+h} - \sqrt{t})}{h} \times \frac{\sqrt{t+h} + \sqrt{t}}{\sqrt{t+h} + t}$$
$$= \lim_{h \to 0} \frac{2h}{h(\sqrt{t+h} + \sqrt{t})} = \frac{2}{2\sqrt{t}} = \frac{1}{\sqrt{t}} \text{ cm s}^{-2}$$

This is the instantaneous acceleration at t seconds.

4 **a** This is the instantaneous velocity at $t = 4$ seconds.

b This is the instantaneous acceleration at $t = 4$ seconds.

EXERCISE 18C.2

1 **a** $s(t) = t^2 - 4t + 3$ cm, $t \geqslant 0$ $\quad \therefore \quad v(t) = 2t - 4$ cm s^{-1} and $a(t) = 2$ cm s^{-2}.

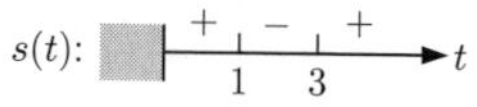

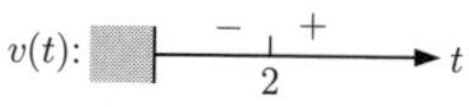

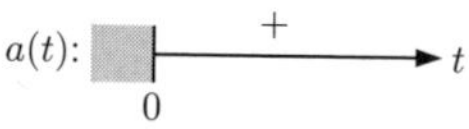

b When $t = 0$, $s(0) = 3$ cm, $v(0) = -4$ cm s^{-1}, $a(0) = 2$ cm s^{-2}

$\therefore$ the object is 3 cm right of O and is moving to the left with a velocity of 4 cm s^{-1} and slowing down, its acceleration being 2 cm s^{-2} to the right.

c When $t = 2$, $s(2) = -1$ cm, $v(2) = 0$ cm s^{-1}, $a(2) = 2$ cm s^{-2}

$\therefore$ the object is 1 cm left of O, momentarily at rest, but with acceleration 2 cm s^{-2} to the right.

d The object reverses direction when $v(t) = 0$, which occurs at $t = 2$ seconds.
At $t = 2$, the particle is 1 cm left of O.

e

f Speed decreases when $v(t)$ and $a(t)$ have opposite signs, which is when $0 \leqslant t \leqslant 2$.

2 $s(t) = 98t - 4.9t^2$ m, $t \geqslant 0$

a $v(t) = 98 - 9.8t$ m s^{-1}
$a(t) = -9.8$ m s^{-2}

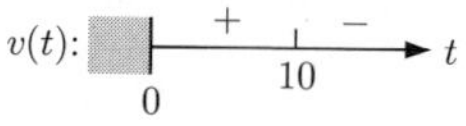

b When $t = 0$, $s(0) = 0$ m, $v(0) = 98$ m s^{-1}

c When $t = 5$, $s(5) = 367.5$ m, $v(5) = 49$ m s^{-1}, $a(5) = -9.8$ m s^{-2}

The stone is 367.5 m above the ground, travelling upwards at 49 m s^{-1}, and slowing down.

When $t = 12$, $s(12) = 470.4$ m, $v(12) = -19.6$ m s^{-1}, $a(12) = -9.8$ m s^{-2}

The stone is 470.4 m above the ground and travelling downwards at 19.6 m s^{-1}, and increasing in speed.

d The maximum height is reached when $v(t) = 0$ m s^{-1}

$\therefore \quad 98 - 9.8t = 0$
$\therefore \quad 9.8t = 98$
$\therefore \quad t = 10$ seconds

$\therefore$ the maximum height is
$s(10) = 98(10) - 4.9(100)$
$= 980 - 490$
$= 490$ m

e The stone is at ground level when $s(t) = 0$ which is when $98t - 4.9t^2 = 0$
$\therefore \quad 4.9t(20 - t) = 0$
$\therefore \quad t = 0$ or 20 seconds
$\therefore$ it hits the ground after 20 seconds.

3 **a** $s(t) = 12t - 2t^3 - 1$ cm, $t \geqslant 0$
$\therefore \quad v(t) = 12 - 6t^2$ cm s^{-1}
and $a(t) = -12t$ cm s^{-2}

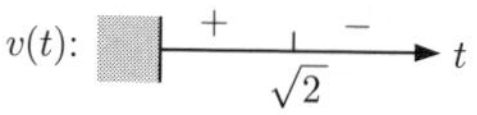

b When $t = 0$, $s(0) = -1$ cm, $v(0) = 12$ cm s^{-1}, $a(0) = 0$ cm s^{-2}

The particle is 1 cm left of O, moving right at 12 cm s^{-1} with constant speed.

c The particle reverses direction when $v(t)=0$ which is when $12-6t^2=0$

$$\therefore\ t^2=2$$
$$\therefore\ t=\sqrt{2}\quad \{t>0\}$$

When $t=\sqrt{2}$, $s(\sqrt{2})=12\sqrt{2}-2(2\sqrt{2})-1$
$=8\sqrt{2}-1$

$\therefore$ the particle is $(8\sqrt{2}-1)$ cm to the right of O.

d **i** From the sign diagrams in **a**, the speed increases for $t\geqslant\sqrt{2}$ seconds.

ii The velocity of the particle never increases $\{a(t)\leqslant 0\}$.

4 **a** $x(t)=t^3-9t^2+24t$ m, $t\geqslant 0$

$v(t)=3t^2-18t+24$ and $a(t)=6t-18$
$=3(t^2-6t+8)$ $\qquad=6(t-3)\ \text{m s}^{-2}$
$=3(t-4)(t-2)\ \text{m s}^{-1}$

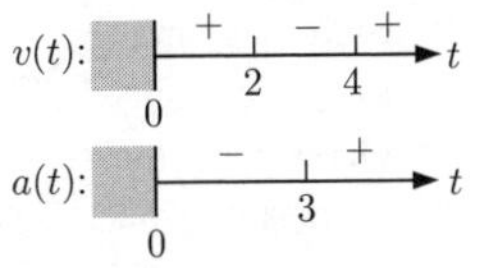

b The particle reverses direction when $v(t)=0$, which occurs at $t=2$ and $t=4$ seconds.

$x(2)=8-36+48$ m and $x(4)=64-144+96$
$=20$ m $\qquad=16$ m

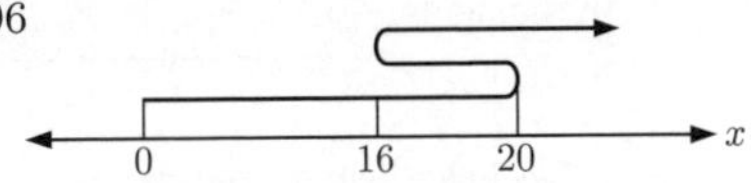

c **i** The speed decreases when $v(t)$ and $a(t)$ have the opposite sign, which is when $0\leqslant t\leqslant 2$ and $3\leqslant t\leqslant 4$.

ii The velocity decreases when $a(t)\leqslant 0$, which is when $0\leqslant t\leqslant 3$.

d When $t=5$, $x(5)=5^3-9(5)^2+24(5)$
$=125-225+120$
$=20$ m

$\therefore$ distance travelled $=20+4+4$ m
$=28$ m

5 **a** Let the equation be $s(t)=at^2+bt+c$

$\therefore\ v(t)=2at+b$

and $a(t)=2a=g$ {gravitational acceleration}

$\therefore\ a=\frac{1}{2}g$ and $v(t)=gt+b$

But when $t=0$, $v(0)=g\times 0+b=b$

$\therefore$ the initial velocity is b

$\therefore\ v(t)=v(0)+gt$ as required

b Now when $t=0$, $s(0)=0$

$\therefore\ a\times 0^2+b\times 0+c=0$

$\therefore\ c=0$

and so $s(t)=\left(\frac{1}{2}g\right)t^2+v(0)t$

$\therefore\ s(t)=v(0)\times t+\frac{1}{2}gt^2$ as required

EXERCISE 18D.1

1 **a**

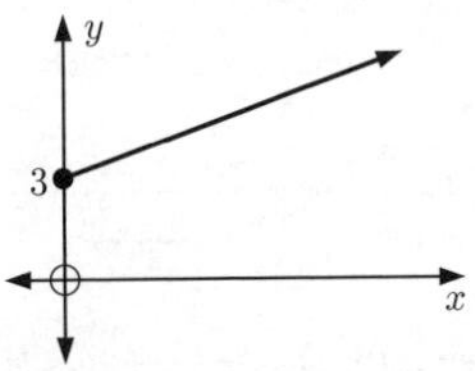

i $x\geqslant 0$ **ii** never

b

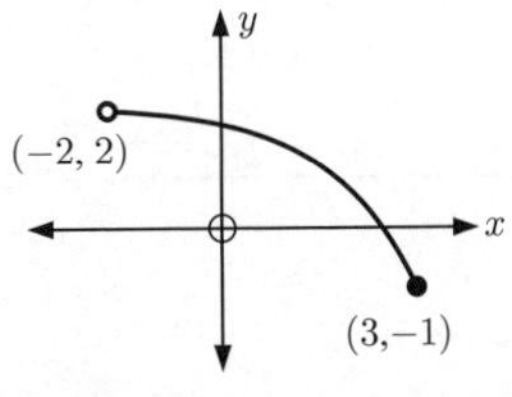

i never **ii** $-2<x\leqslant 3$

c

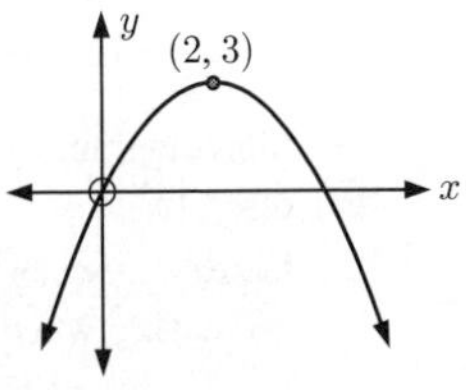

i $x\leqslant 2$ **ii** $x\geqslant 2$

d

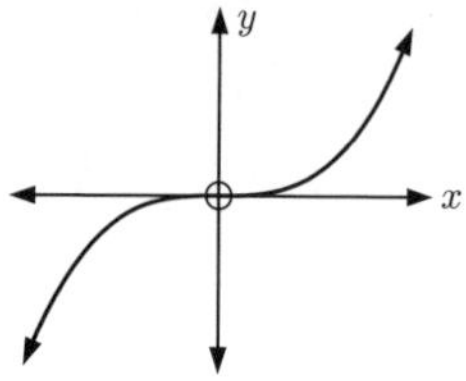

i all real x **ii** never

e

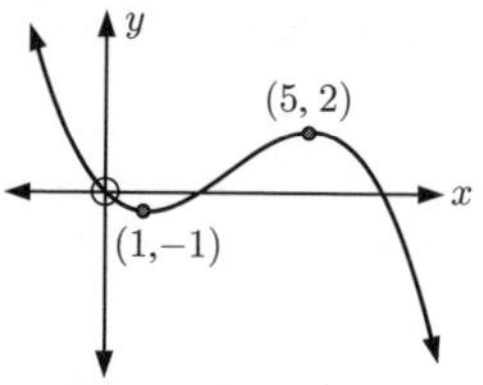

i $1 \leqslant x \leqslant 5$
ii $x \leqslant 1,\ x \geqslant 5$

f

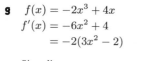

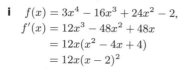

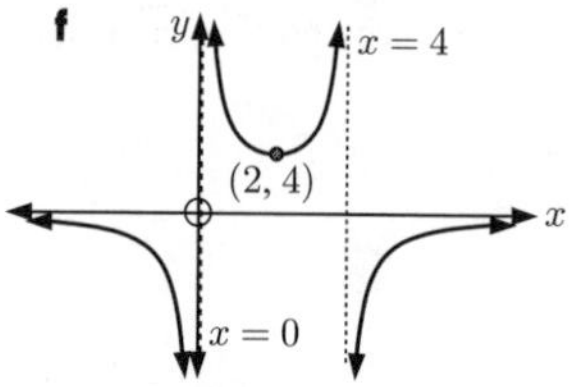

i $2 \leqslant x < 4,\quad x > 4$
ii $x < 0,\quad 0 < x \leqslant 2$

2 **a** $f(x) = x^2,\quad f'(x) = 2x$

Sign diagram of $f'(x)$: $-$ | $+$ (at 0)

increasing when $x \geqslant 0$,
decreasing when $x \leqslant 0$

b $f(x) = -x^3,\quad f'(x) = -3x^2$

Sign diagram of $f'(x)$: $-$ | $-$ (at 0)

decreasing for all x

c $f(x) = 2x^2 + 3x - 4,\quad f'(x) = 4x + 3$

Sign diagram of $f'(x)$: $-$ | $+$ (at $-\frac{3}{4}$)

increasing when $x \geqslant -\frac{3}{4}$,
decreasing when $x \leqslant -\frac{3}{4}$

d $f(x) = \sqrt{x} = x^{\frac{1}{2}}$,

$f'(x) = \frac{1}{2}x^{-\frac{1}{2}} = \dfrac{1}{2\sqrt{x}}$

Sign diagram of $f'(x)$: (undefined for $x < 0$) 0 | $+$

$f(x)$ is only defined when $x \geqslant 0$
increasing when $x \geqslant 0$, never decreasing

e $f(x) = \dfrac{2}{\sqrt{x}} = 2x^{-\frac{1}{2}}$

$f'(x) = -x^{-\frac{3}{2}} = \dfrac{-1}{x\sqrt{x}}$

Sign diagram of $f'(x)$: (undefined for $x \leqslant 0$) 0 | $-$

$f(x)$ is only defined for $x > 0$
never increasing, decreasing when $x > 0$

f $f(x) = x^3 - 6x^2,\quad f'(x) = 3x^2 - 12x$
$= 3x(x - 4)$

Sign diagram of $f'(x)$: $+$ | $-$ | $+$ (at 0, 4)

increasing when $x \leqslant 0$ or $x \geqslant 4$,
decreasing when $0 \leqslant x \leqslant 4$

g $f(x) = -2x^3 + 4x$
$f'(x) = -6x^2 + 4$
$= -2(3x^2 - 2)$

Sign diagram of $f'(x)$: $-$ | $+$ | $-$ (at $-\sqrt{\frac{2}{3}}$, $\sqrt{\frac{2}{3}}$)

increasing for $-\sqrt{\frac{2}{3}} \leqslant x \leqslant \sqrt{\frac{2}{3}}$,
decreasing for $x \leqslant -\sqrt{\frac{2}{3}}$ or $x \geqslant \sqrt{\frac{2}{3}}$

h $f(x) = -4x^3 + 15x^2 + 18x + 3$
$f'(x) = -12x^2 + 30x + 18$
$= -6(2x^2 - 5x - 3)$
$= -6(2x + 1)(x - 3)$

Sign diagram of $f'(x)$: $-$ | $+$ | $-$ (at $-\frac{1}{2}$, 3)

increasing when $-\frac{1}{2} \leqslant x \leqslant 3$,
decreasing when $x \leqslant -\frac{1}{2}$ or $x \geqslant 3$

i $f(x) = 3x^4 - 16x^3 + 24x^2 - 2$,
$f'(x) = 12x^3 - 48x^2 + 48x$
$= 12x(x^2 - 4x + 4)$
$= 12x(x - 2)^2$

Sign diagram of $f'(x)$: $-$ | $+$ | $+$ (at 0, 2)

increasing when $x \geqslant 0$,
decreasing when $x \leqslant 0$

j $f(x) = 2x^3 + 9x^2 + 6x - 7$,
$f'(x) = 6x^2 + 18x + 6$
$= 6(x^2 + 3x + 1)$

$f'(x) = 0$ when $x = \frac{-3\pm\sqrt{9-4}}{2} = \frac{-3\pm\sqrt{5}}{2}$

Sign diagram of $f'(x)$: $+$ | $-$ | $+$ (at $\frac{-3-\sqrt{5}}{2}$, $\frac{-3+\sqrt{5}}{2}$)

increasing for $x \leqslant \frac{-3-\sqrt{5}}{2}$ or $x \geqslant \frac{-3+\sqrt{5}}{2}$,
decreasing for $\frac{-3-\sqrt{5}}{2} \leqslant x \leqslant \frac{-3+\sqrt{5}}{2}$

k $f(x) = x^3 - 6x^2 + 3x - 1,$

$f'(x) = 3x^2 - 12x + 3$
$\quad = 3(x^2 - 4x + 1)$

$f'(x) = 0$ when $x = \frac{4 \pm \sqrt{16-4}}{2} = 2 \pm \sqrt{3}$

Sign diagram of $f'(x)$:

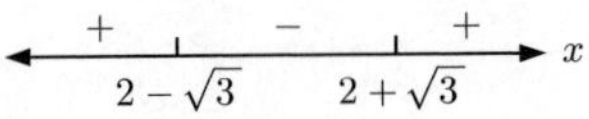

increasing when $x \leqslant 2 - \sqrt{3}$
or $x \geqslant 2 + \sqrt{3}$,
decreasing when $2 - \sqrt{3} \leqslant x \leqslant 2 + \sqrt{3}$

l $f(x) = x - 2\sqrt{x} = x - 2x^{\frac{1}{2}}$

$f'(x) = 1 - x^{-\frac{1}{2}} = 1 - \frac{1}{\sqrt{x}} = \frac{\sqrt{x} - 1}{\sqrt{x}}$

Sign diagram of $f'(x)$:

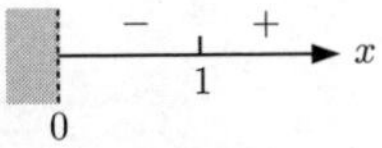

increasing when $x \geqslant 1$,
decreasing when $0 \leqslant x \leqslant 1$

m $f(x) = 3x^4 - 8x^3 - 6x^2 + 24x + 11$

$f'(x) = 12x^3 - 24x^2 - 12x + 24$
$\quad = 12(x^3 - 2x^2 - x + 2)$

Using technology, a root is -1,

$\therefore\ f'(x) = 12(x + 1)(x^2 - 3x + 2)$
$\quad = 12(x + 1)(x - 1)(x - 2)$

Sign diagram of $f'(x)$:

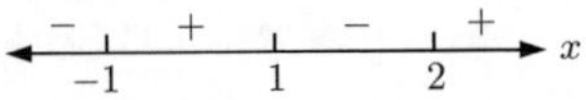

So, $f(x)$ is increasing for $-1 \leqslant x \leqslant 1$ and $x \geqslant 2$, and decreasing for $x \leqslant -1$ and $1 \leqslant x \leqslant 2$.

n $f(x) = x^4 - 4x^3 + 2x^2 + 4x + 1,$

$f'(x) = 4x^3 - 12x^2 + 4x + 4$
$\quad = 4(x^3 - 3x^2 + x + 1)$

Using technology, a root is 1,

$\therefore\ f'(x) = 4(x - 1)(x^2 - 2x - 1)$

$f'(x) = 0$ when $x = 1$ or $x = \frac{2 \pm \sqrt{8}}{2}$

$\therefore\ x = 1$ or $x = 1 \pm \sqrt{2}$

Sign diagram of $f'(x)$:

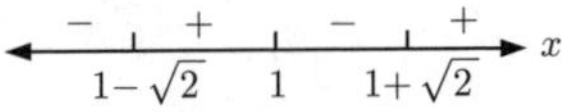

So, $f(x)$ is increasing for $1 - \sqrt{2} \leqslant x \leqslant 1$ and $x \geqslant 1 + \sqrt{2}$, decreasing for $x \leqslant 1 - \sqrt{2}$ and $1 \leqslant x \leqslant 1 + \sqrt{2}$.

3 **a** $f(x) = \frac{4x}{x^2 + 1}$ is a quotient with

$u = 4x$ and $v = x^2 + 1$

$\therefore\ u' = 4$ and $v' = 2x$

$$\therefore\ f'(x) = \frac{4(x^2 + 1) - 4x \times 2x}{(x^2 + 1)^2}$$
$$= \frac{4x^2 + 4 - 8x^2}{(x^2 + 1)^2}$$
$$= \frac{4 - 4x^2}{(x^2 + 1)^2}$$
$$= \frac{-4(x^2 - 1)}{(x^2 + 1)^2}$$
$$= \frac{-4(x + 1)(x - 1)}{(x^2 + 1)^2}$$

Sign diagram of $f'(x)$:

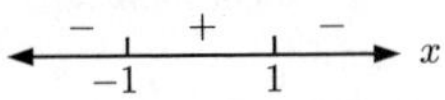

b $f(x)$ is increasing for $-1 \leqslant x \leqslant 1$,
decreasing for $x \leqslant -1$ and $x \geqslant 1$

4 **a** $f(x) = \frac{4x}{(x - 1)^2}$ is a quotient with

$u = 4x$ and $v = (x - 1)^2$

$\therefore\ u' = 4$ and $v' = 2(x - 1)$

$$\therefore\ f'(x) = \frac{4(x - 1)^2 - 8x(x - 1)}{(x - 1)^4}$$
$$= \frac{4(x - 1)\left((x - 1) - 2x\right)}{(x - 1)^4}$$
$$= \frac{4(-1 - x)}{(x - 1)^3}$$
$$= \frac{-4(x + 1)}{(x - 1)^3}$$

Sign diagram of $f'(x)$:

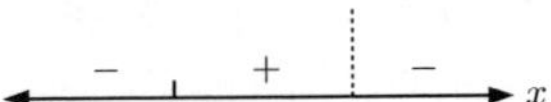

b $f(x)$ is increasing for $-1 \leqslant x < 1$,
decreasing for $x \leqslant -1$ and $x > 1$

5 **a** $f(x) = \dfrac{-x^2+4x-7}{x-1}$ is a quotient with $u = -x^2+4x-7$ and $v = x-1$

$\therefore$ $u' = -2x+4$ and $v' = 1$

$$\therefore\ f'(x) = \frac{(-2x+4)(x-1)-(-x^2+4x-7)(1)}{(x-1)^2}$$

$$= \frac{-2x^2+6x-4+x^2-4x+7}{(x-1)^2}$$

$$= \frac{-x^2+2x+3}{(x-1)^2}$$

$$= \frac{-(x^2-2x-3)}{(x-1)^2}$$

$$= \frac{-(x+1)(x-3)}{(x-1)^2}$$

Sign diagram of $f'(x)$:

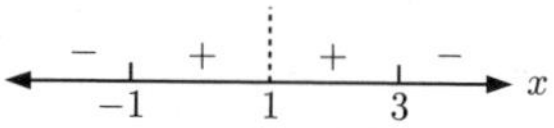

b $f(x)$ is increasing for $-1 \leqslant x < 1$ and $1 < x \leqslant 3$, and decreasing for $x \leqslant -1$ and $x \geqslant 3$.

6 **a** $f(x) = \dfrac{x^3}{x^2-1}$ is a quotient with $u = x^3$ and $v = x^2-1$

$\therefore$ $u' = 3x^2$ and $v' = 2x$

$$\therefore\ f'(x) = \frac{3x^2(x^2-1)-x^3\times 2x}{(x^2-1)^2}$$

$$= \frac{3x^4-3x^2-2x^4}{(x^2-1)^2}$$

$$= \frac{x^2(x^2-3)}{(x^2-1)^2}$$

$$= \frac{x^2(x+\sqrt{3})(x-\sqrt{3})}{(x^2-1)^2}$$

Sign diagram of $f'(x)$:

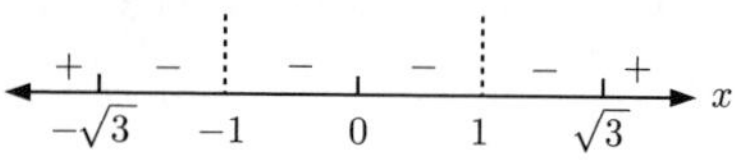

$\therefore$ $f(x)$ is increasing for $x \leqslant -\sqrt{3}$ and $x \geqslant \sqrt{3}$, and decreasing for $-\sqrt{3} \leqslant x < -1$, $-1 < x < 1$ and $1 < x \leqslant \sqrt{3}$.

b

$$f(x) = x^2 + \frac{4}{x-1} = x^2 + 4(x-1)^{-1}$$

$$\therefore\ f'(x) = 2x - 4(x-1)^{-2}\times 1$$

$$= 2x - \frac{4}{(x-1)^2}$$

$$= \frac{2x(x-1)^2-4}{(x-1)^2}$$

$$= \frac{2x(x^2-2x+1)-4}{(x-1)^2}$$

$$= \frac{2x^3-4x^2+2x-4}{(x-1)^2}$$

$$= \frac{(x-2)(2x^2+2)}{(x-1)^2}$$

Sign diagram of $f'(x)$:

$\therefore$ $f(x)$ is increasing for $x \geqslant 2$, and decreasing for $x < 1$ and $1 < x \leqslant 2$.

EXERCISE 18D.2

1 **a** A is a *local maximum*, B is a *horizontal inflection*, C is a *local minimum*.

b $f'(x)$ has sign diagram:

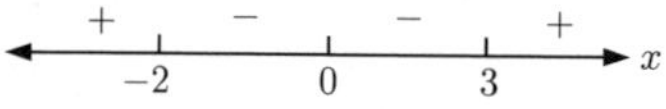

c **i** $f(x)$ is increasing for $x \leqslant -2$ and $x \geqslant 3$ **ii** $f(x)$ is decreasing for $-2 \leqslant x \leqslant 3$.

d $f(x)$ has sign diagram:

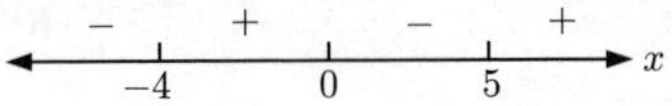

e For **b** we have intervals where the function is increasing (+) or decreasing (−). For **d** we have intervals where the function is above (+) or below (−) the x-axis.

2 **a** $f(x) = x^2 - 2 \quad \therefore \quad f'(x) = 2x$

with sign diagram:

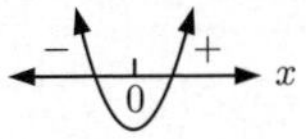

Now $f(0) = -2$,
so there is a local minimum at $(0, -2)$.

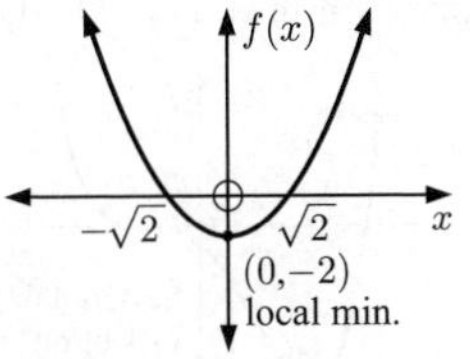

b $f(x) = x^3 + 1 \quad \therefore \quad f'(x) = 3x^2$

with sign diagram:

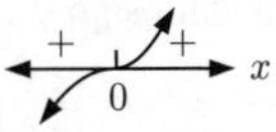

Now $f(0) = 1$,
so there is a horizontal inflection at $(0, 1)$.

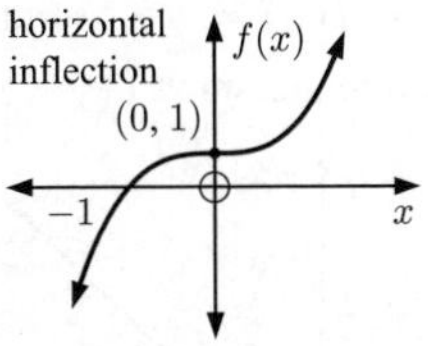

c $f(x) = x^3 - 3x + 2$

$\therefore \quad f'(x) = 3x^2 - 3$
$= 3(x^2 - 1)$
$= 3(x + 1)(x - 1)$

with sign diagram:

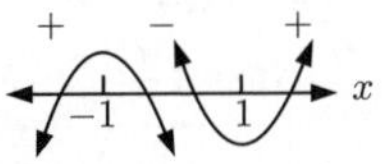

Now $f(-1) = 4$, $f(1) = 0$,
so there is a local maximum at $(-1, 4)$,
and a local minimum at $(1, 0)$.

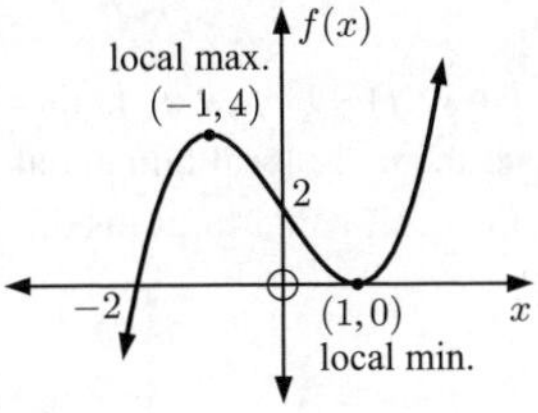

d $f(x) = x^4 - 2x^2$

$\therefore \quad f'(x) = 4x^3 - 4x$
$= 4x(x^2 - 1)$
$= 4x(x + 1)(x - 1)$

with sign diagram:

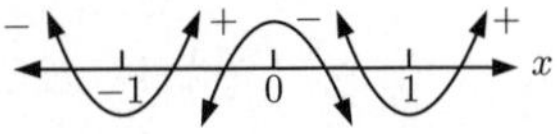

Now $f(-1) = -1$, $f(1) = -1$, $f(0) = 0$,
so there are local minima at $(-1, -1)$ and $(1, -1)$, and a local maximum at $(0, 0)$.

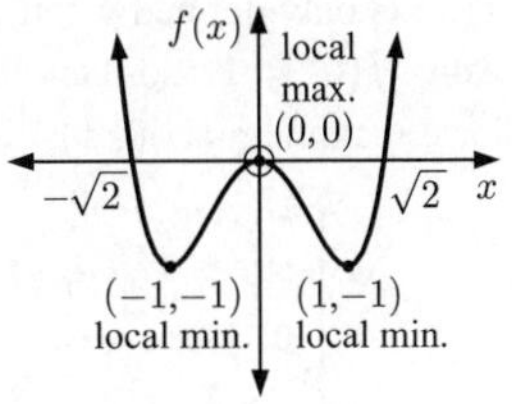

e $f(x) = x^3 - 6x^2 + 12x + 1$

$\therefore \quad f'(x) = 3x^2 - 12x + 12$
$= 3(x^2 - 4x + 4)$
$= 3(x - 2)^2$

with sign diagram:

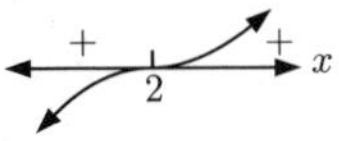

Now $f(2) = 9$, so there is a horizontal inflection at $(2, 9)$.

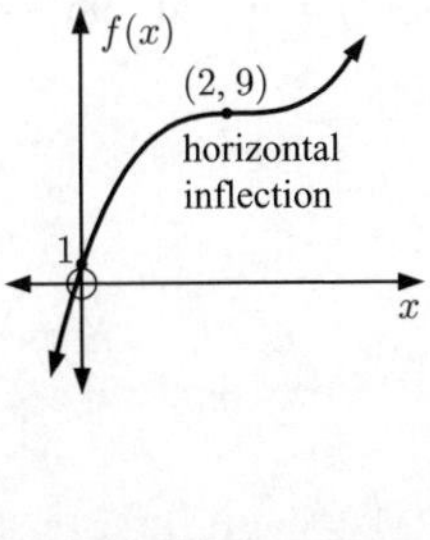

f $f(x) = \sqrt{x} + 2$

$\therefore \quad f'(x) = \frac{1}{2}x^{-\frac{1}{2}}$
$= \dfrac{1}{2\sqrt{x}} \neq 0$

with sign diagram:

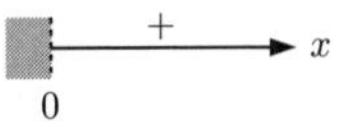

$\therefore$ no stationary points.

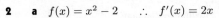

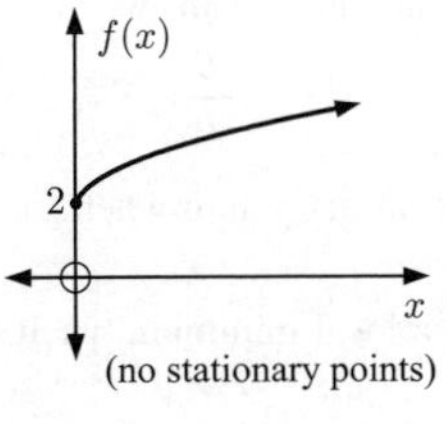

(no stationary points)

g $f(x) = x - \sqrt{x}$

$\therefore \quad f'(x) = 1 - \frac{1}{2}x^{-\frac{1}{2}}$

$= 1 - \dfrac{1}{2\sqrt{x}}$

$= \dfrac{2\sqrt{x} - 1}{2\sqrt{x}}$

with sign diagram:

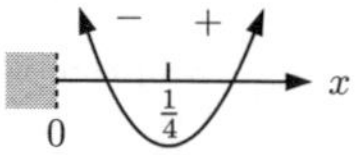

$f(x)$ is defined for all $x \geqslant 0$

Now $f(\frac{1}{4}) = -\frac{1}{4}$, so there is a local minimum at $(\frac{1}{4}, -\frac{1}{4})$.

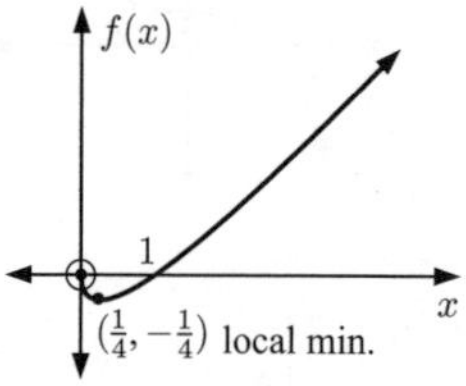

h $f(x) = x^4 - 6x^2 + 8x - 3$

$\therefore \quad f'(x) = 4x^3 - 12x + 8$

$= 4(x^3 - 3x + 2)$

$= 4(x - 1)(x^2 + x - 2)$

$= 4(x - 1)(x + 2)(x - 1)$

with sign diagram:

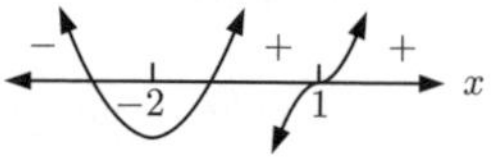

Now $f(-2) = -27$, $f(1) = 0$, so there is a local minimum at $(-2, -27)$, and a horizontal inflection at $(1, 0)$.

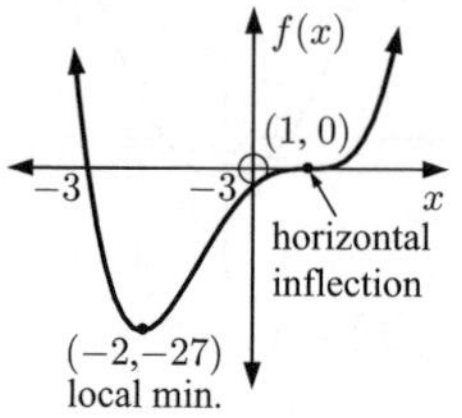

i $f(x) = 1 - x\sqrt{x} = 1 - x^{\frac{3}{2}}$

$\therefore \quad f'(x) = -\frac{3}{2}x^{\frac{1}{2}} = \dfrac{-3\sqrt{x}}{2}$

with sign diagram:

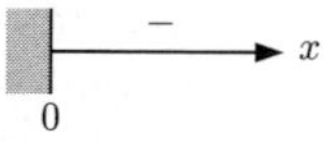

$f(x)$ is only defined when $x \geqslant 0$

Now $f(0) = 1$, so there is a local maximum at $(0, 1)$.

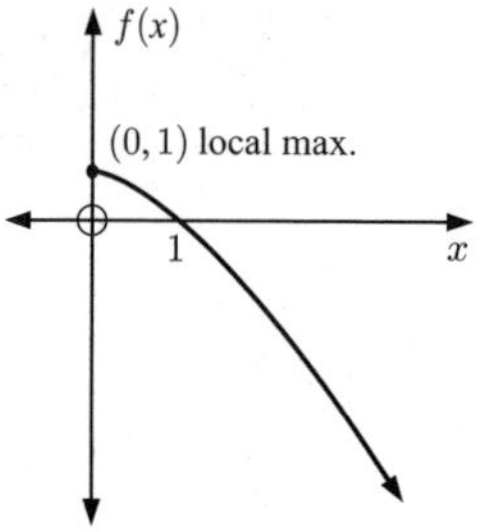

j $f(x) = x^4 - 2x^2 - 8$

$\therefore \quad f'(x) = 4x^3 - 4x$

$= 4x(x^2 - 1)$

$= 4x(x + 1)(x - 1)$

with sign diagram:

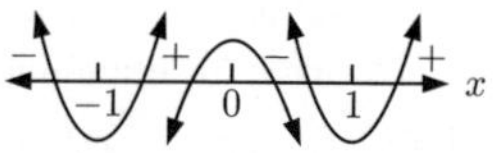

Now $f(-1) = -9$, $f(1) = -9$, $f(0) = -8$, so there are local minima at $(-1, -9)$ and $(1, -9)$, and a local maximum at $(0, -8)$.

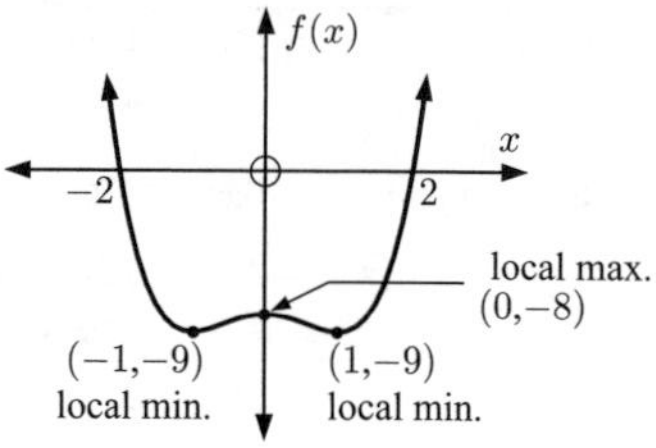

3 $f(x) = ax^2 + bx + c, \quad a \neq 0$

$\therefore \quad f'(x) = 2ax + b$

$f(x)$ has a stationary point when $f'(x) = 0$

$\therefore \quad x = -\dfrac{b}{2a}$

There is a local maximum when $a < 0$ 

and there is a local minimum when $a > 0$ 

4 $f(x) = 2x^3 + ax^2 - 24x + 1$

$\therefore \quad f'(x) = 6x^2 + 2ax - 24$

But $f'(-4) = 0$, so $96 - 8a - 24 = 0$

$\therefore \quad 72 = 8a$

$\therefore \quad a = 9$

5 **a** $f(x) = x^3 + ax + b$

$\therefore\ f'(x) = 3x^2 + a$

But $f'(-2) = 0$

$\therefore\ 3(-2)^2 + a = 0$

$\therefore\ 12 + a = 0$

$\therefore\ a = -12$

Also, $f(-2) = 3$

$\therefore\ (-2)^3 - 12(-2) + b = 3$

$\therefore\ -8 + 24 + b = 3$

$\therefore\ b = -13$

b Now $f(x) = x^3 - 12x - 13$

$\therefore\ f'(x) = 3x^2 - 12$

$= 3(x^2 - 4)$

$= 3(x+2)(x-2)$ with sign diagram:

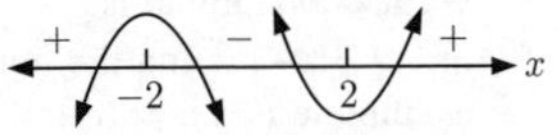

Now $f(2) = -29$, so there is a local maximum at $(-2, 3)$ and a local minimum at $(2, -29)$.

6 Let the cubic polynomial be

$P(x) = ax^3 + bx^2 + cx + d$

$\therefore\ P'(x) = 3ax^2 + 2bx + c$ .... (1)

Now $(0, 2)$ lies on $P(x)$, so $P(0) = 2$

$\therefore\ a(0) + b(0) + c(0) + d = 2$

$\therefore\ d = 2$

The tangent at $(0, 2)$ is $y = 9x + 2$, so

$P'(0) = 9$

$\therefore\ 3a(0) + 2b(0) + c = 9$

$\therefore\ c = 9$ .... (2)

There is a stationary point at $(-1, -7)$, so

$P'(-1) = 0$

$\therefore\ 3a(-1)^2 + 2b(-1) + c = 0$ {using (1)}

$\therefore\ 3a - 2b + c = 0$

So, using (2), $3a - 2b = -9$ .... (3)

Finally, $(-1, -7)$ lies on $P(x)$

$\therefore\ a(-1)^3 + b(-1)^2 + c(-1) + d = -7$

$\therefore\ -a + b - 9 + 2 = -7$

$\therefore\ b - a = 0$

$\therefore\ a = b$

So, using (3), $3a - 2a = -9$

$\therefore\ a = -9$

$\therefore\ a = b = -9$

$\therefore\ P(x) = -9x^3 - 9x^2 + 9x + 2$

7 **a** $f(x) = x^3 - 12x - 2$, for $-3 \leqslant x \leqslant 5$

$\therefore\ f'(x) = 3x^2 - 12$

$= 3(x+2)(x-2)$

which is 0 when $x = -2$ or 2

x	-3	-2	2	5
$f(x)$	7	14	-18	63

$\therefore$ the maximum value is 63 when $x = 5$, and the minimum value is -18 when $x = 2$.

b $f(x) = 4 - 3x^2 + x^3$, for $-2 \leqslant x \leqslant 3$

$\therefore\ f'(x) = -6x + 3x^2$

$= 3x(x-2)$

which is 0 when $x = 0$ or 2

x	-2	0	2	3
$f(x)$	-16	4	0	4

$\therefore$ maximum value is 4 when $x = 0$ or $x = 3$, minimum value is -16 when $x = -2$.

8 $C(x) = 0.0007x^3 - 0.1796x^2 + 14.663x + 160$ for $50 \leqslant x \leqslant 150$

$C'(x) = 0.0021x^2 - 0.3592x + 14.663$

$C'(x) = 0$ when $0.0021x^2 - 0.3592x + 14.663 = 0$

Using technology, $x \approx 103.74$ or $x \approx 67.30$

x	50	67.30	103.74	150
$C(x)$	531.65	546.73	529.80	680.95

$\therefore$ the maximum hourly cost is \$680.95 when 150 hinges are made. The minimum hourly cost is \$529.80 when 104 hinges are made.

EXERCISE 18E.1

1 **a** $f(x) = \dfrac{3x-2}{x+1}$

$= \dfrac{3(x+1)-5}{x+1}$

$= 3 - \dfrac{5}{x+1}$

$\therefore$ vertical asymptote is $x+1=0$ or $x=-1$ and the horizontal asymptote is $y=3$
$\{$as $x \to \pm\infty,\ f(x) \to 3\}$

b $f(x) = \dfrac{x-4}{2x-1}$

$= \dfrac{\frac{1}{2}(2x-1)-3\frac{1}{2}}{2x-1}$

$= \frac{1}{2} + \dfrac{-3\frac{1}{2}}{2x-1}$

$= \frac{1}{2} - \dfrac{7}{2(2x-1)}$

$\therefore$ vertical asymptote is $2x-1=0$ or $x=\frac{1}{2}$ and the horizontal asymptote is $y=\frac{1}{2}$
$\{$as $x \to \pm\infty,\ f(x) \to \frac{1}{2}\}$

c $f(x) = \dfrac{4-2x}{x-1}$

$= \dfrac{-2(x-1)+2}{x-1}$

$= -2 + \dfrac{2}{x-1}$

$\therefore$ vertical asymptote is $x-1=0$ or $x=1$ and the horizontal asymptote is $y=-2$
$\{$as $x \to \pm\infty,\ f(x) \to -2\}$

2 **a** **i** $f(x) = -3 + \dfrac{1}{4-x}$

$\therefore$ V.A. is $4-x=0$ or $x=4$
and H.A. is $y=-3$
$\{$as $x \to \pm\infty,\ y \to -3\}$

ii $f(x) = -3 + (4-x)^{-1}$

$\therefore f'(x) = -(4-x)^{-2} \times (-1)$

$= \dfrac{1}{(4-x)^2}$

Sign diagram of $f'(x)$ is:

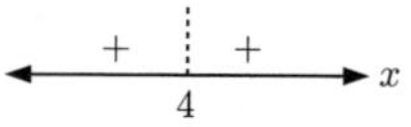

iii When $x=0$, $y = -3 + \frac{1}{4} = -2\frac{3}{4}$

$\therefore$ the y-intercept is $-2\frac{3}{4}$

When $y=0$, $-3 + \dfrac{1}{4-x} = 0$

$\therefore \dfrac{1}{4-x} = 3$

$\therefore 4-x = \frac{1}{3}$

$\therefore x = 3\frac{2}{3}$

$\therefore$ the x-intercept is $3\frac{2}{3}$

iv

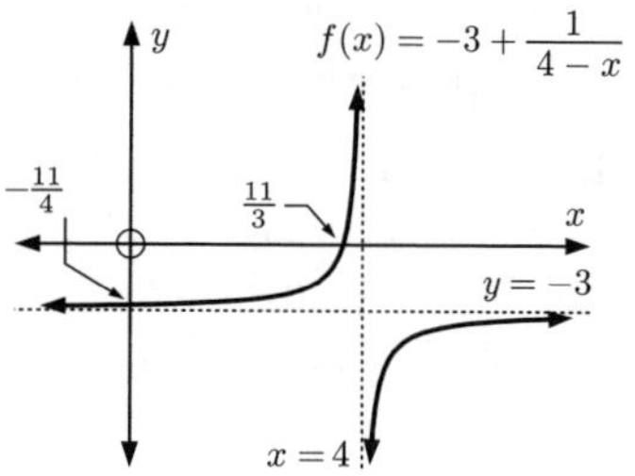

b **i** $f(x) = \dfrac{x}{x+2}$

$= \dfrac{(x+2)-2}{x+2}$

$= 1 - \dfrac{2}{x+2}$

$\therefore$ V.A. is $x+2=0$ or $x=-2$
and H.A. is $y=1$.
$\{$as $x \to \pm\infty,\ y \to 1\}$

ii $f'(x) = \dfrac{1(x+2)-x(1)}{(x+2)^2}$

$= \dfrac{2}{(x+2)^2}$

and has sign diagram:

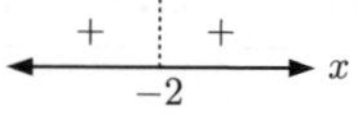

iii When $x = 0$, $f(0) = 0$

$\therefore$ the y-intercept is 0

When $y = 0$, $\dfrac{x}{x+2} = 0$

$\therefore x = 0$

$\therefore$ the x-intercept is 0

iv

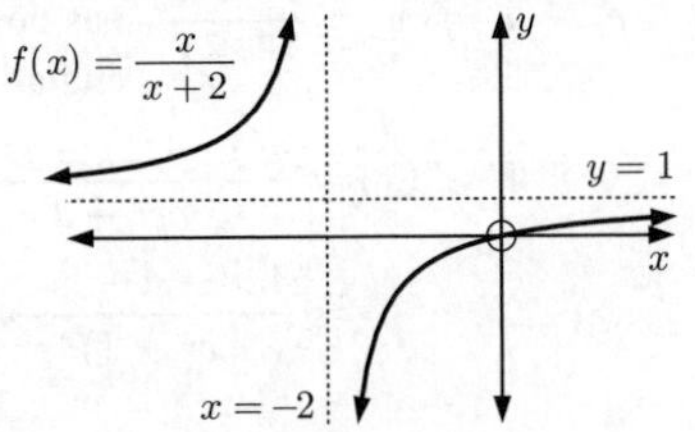

c **i** $f(x) = \dfrac{4x+3}{x-2}$

$= \dfrac{4(x-2)+11}{x-2}$

$= 4 + \dfrac{11}{x-2}$

$\therefore$ V.A. is $x - 2 = 0$ or $x = 2$
and H.A. is $y = 4$
$\{$as $x \to \pm\infty$, $y \to 4\}$

ii $f'(x) = \dfrac{4(x-2) - (4x+3)1}{(x-2)^2}$

$= \dfrac{-11}{(x-2)^2}$

and has sign diagram:

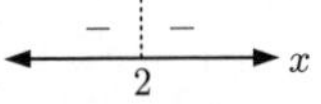

iii when $x = 0$, $f(0) = \frac{3}{-2}$

$\therefore$ the y-intercept is $-1\frac{1}{2}$

when $y = 0$, $4x + 3 = 0$

$\therefore x = -\frac{3}{4}$

$\therefore$ the x-intercept is $-\frac{3}{4}$

iv

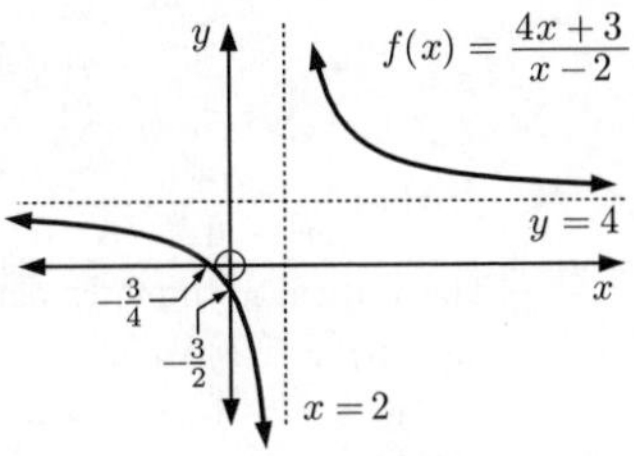

d **i** $f(x) = \dfrac{1-x}{x+2}$

$= \dfrac{-(x+2)+3}{x+2}$

$= -1 + \dfrac{3}{x+2}$

$\therefore$ V.A. is $x + 2 = 0$ or $x = -2$
and H.A. is $y = -1$
$\{$as $x \to \pm\infty$, $y \to -1\}$

ii $f'(x) = \dfrac{-1(x+2) - (1-x)1}{(x+2)^2}$

$= \dfrac{-3}{(x+2)^2}$

and has sign diagram:

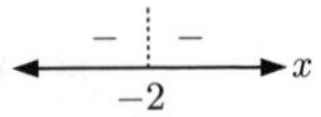

iii when $x = 0$, $f(0) = \frac{1}{2}$

$\therefore$ the y-intercept is $\frac{1}{2}$

when $y = 0$, $\dfrac{1-x}{x+2} = 0$

$\therefore 1 - x = 0$

$\therefore x = 1$

$\therefore$ the x-intercept is 1

iv

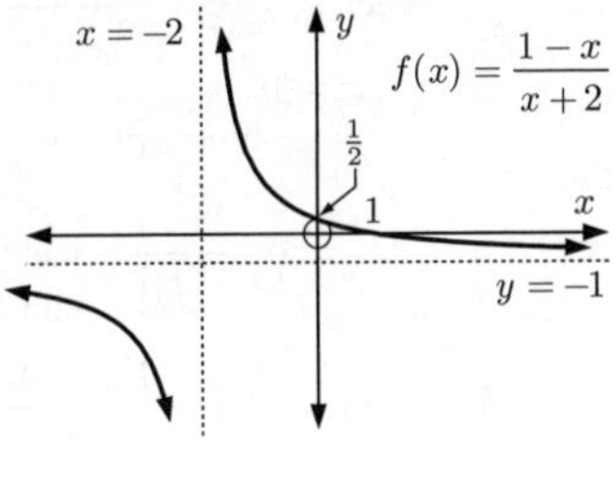

EXERCISE 18E.2

1 **a** $y = \dfrac{2x}{x^2-4} = \dfrac{2x}{(x+2)(x-2)}$

has V.A.s when $x + 2 = 0$ and $x - 2 = 0$

$\therefore$ the V.A.s are $x = -2$ and $x = 2$.

As $x \to \pm\infty$, $y \to 0$, so the H.A. is $y = 0$.

b $y = \dfrac{1-x}{(x+2)^2}$

has V.A. $x + 2 = 0$ or $x = -2$.

As $x \to \pm\infty$, $y \to 0$, so the H.A. is $y = 0$.

c $y = \dfrac{3x+2}{x^2+1}$

$x^2 + 1 = 0$ has no real solutions, so there are no V.A.s.

As $x \to \pm\infty$, $y \to 0$, so the H.A. is $y = 0$.

2 **a** **i** $f(x) = \dfrac{4x}{x^2+1}$ has no vertical asymptotes {as $x^2+1=0$ has no real solutions} and a horizontal asymptote of $y=0$ {as $x \to \pm\infty,\ y \to 0$}

ii $f'(x) = \dfrac{4(x^2+1) - 4x(2x)}{(x^2+1)^2}$

$= \dfrac{4x^2+4-8x^2}{(x^2+1)^2}$

$= \dfrac{4-4x^2}{(x^2+1)^2}$

$= \dfrac{-4(x+1)(x-1)}{(x^2+1)^2}$

and has sign diagram:

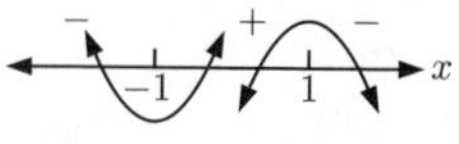

$\therefore$ local minimum is $\left(-1, \dfrac{4(-1)}{1+1}\right)$ or $(-1, -2)$

and local maximum is $\left(1, \dfrac{4(1)}{1+1}\right)$ or $(1, 2)$.

iii When $x=0$, $f(0)=0$

$\therefore$ the y-intercept is 0

when $y=0$, $\dfrac{4x}{x^2+1}=0$

$\therefore$ $x=0$ and so the x-intercept is 0

iv

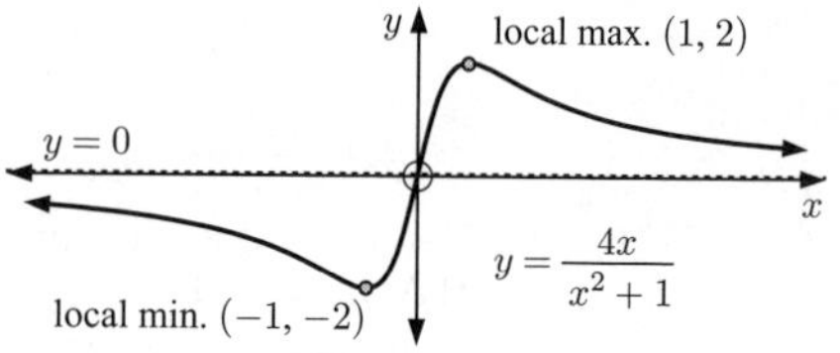

b **i** $f(x) = \dfrac{4x}{x^2-4x-5} = \dfrac{4x}{(x-5)(x+1)}$

The vertical asymptotes occur when $(x-5)(x+1)=0$,

$\therefore$ at $x=-1$ and $x=5$.

The horizontal asymptote is $y=0$

{as $x \to \pm\infty,\ y \to 0$}

ii $f'(x) = \dfrac{4(x^2-4x-5) - 4x(2x-4)}{(x^2-4x-5)^2}$

$= \dfrac{4x^2-16x-20-8x^2+16x}{(x^2-4x-5)^2}$

$= \dfrac{-4(x^2+5)}{(x-5)^2(x+1)^2}$

Since x^2+5 is always positive we have no turning points,

and sign diagram is:

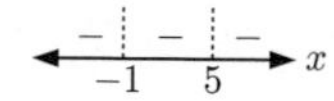

iii When $x=0$, $y=\frac{0}{-5}=0$

$\therefore$ the y-intercept is 0

When $y=0$, $\dfrac{4x}{x^2-4x-5}=0$

$\therefore$ $x=0$

$\therefore$ the x-intercept is 0

iv

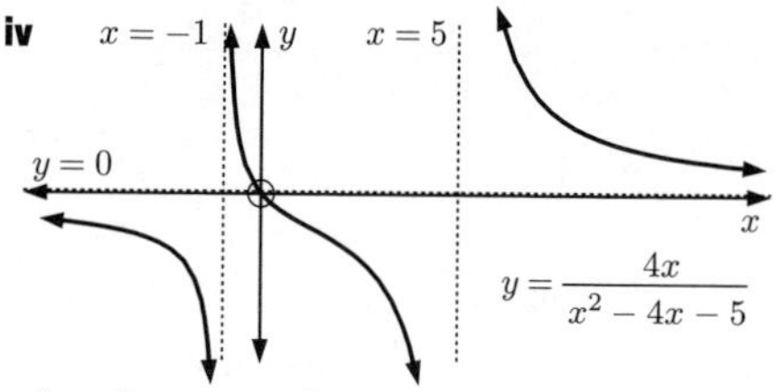

c **i** $f(x) = \dfrac{4x}{(x-1)^2}$ has vertical asymptote $x-1=0$ or $x=1$ and horizontal asymptote $y=0$ {as $x \to \pm\infty,\ y \to 0$}

ii $f'(x) = \dfrac{4(x-1)^2 - 4x \times 2(x-1)(1)}{(x-1)^4}$

$= \dfrac{4(x-1)[x-1-2x]}{(x-1)^4}$

$= \dfrac{4(-x-1)}{(x-1)^3} = \dfrac{-4(x+1)}{(x-1)^3}$

which has sign diagram:

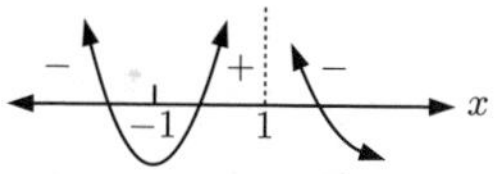

local min. at $\left(-1, \frac{4(-1)}{(-2)^2}\right)$ or $(-1, -1)$

iii When $x=0$, $y=\frac{0}{1}=0$

$\therefore$ the y-intercept is 0

When $y=0$, $\dfrac{4x}{(x-1)^2}=0$

$\therefore$ the x-intercept is 0

iv

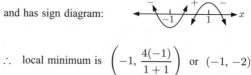

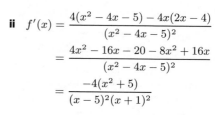

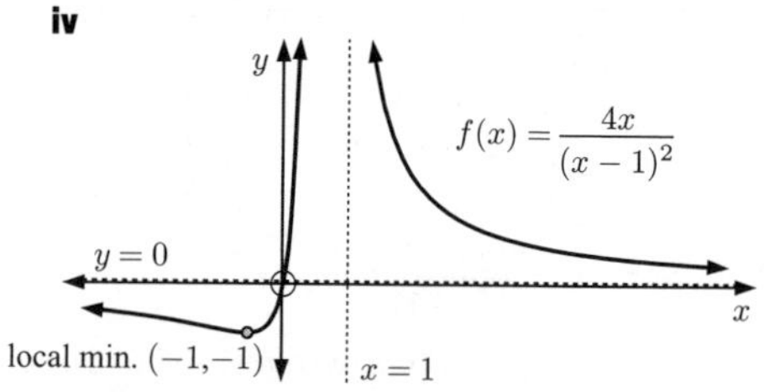

d **i** $f(x) = \dfrac{3x-3}{(x+2)^2}$ has a vertical asymptote when $(x+2)^2 = 0$, so the vertical asymptote is $x = -2$, and horizontal asymptote $y = 0$ $\{$as $x \to \pm\infty,\ y \to 0\}$

ii $f'(x) = \dfrac{3(x+2)^2 - 2(x+2)(3x-3)}{(x+2)^4}$

$= \dfrac{3(x+2) - 2(3x-3)}{(x+2)^3}$

$= \dfrac{3x+6-6x+6}{(x+2)^3}$

$= \dfrac{12-3x}{(x+2)^3} = \dfrac{-3(x-4)}{(x+2)^3}$

Sign diagram of $f'(x)$ is:

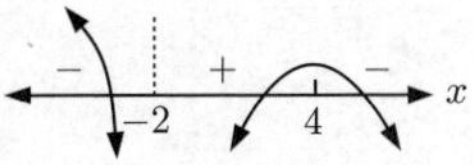

Now $f(4) = \dfrac{12-3}{6^2} = \frac{1}{4}$, so there is a local maximum at $(4, \frac{1}{4})$.

iii When $x = 0$, $y = \dfrac{-3}{2^2} = -\frac{3}{4}$

$\therefore$ the y-intercept is $-\frac{3}{4}$

When $y = 0$, $3x - 3 = 0$

$\therefore\ x = 1$

$\therefore$ the x-intercept is 1

iv

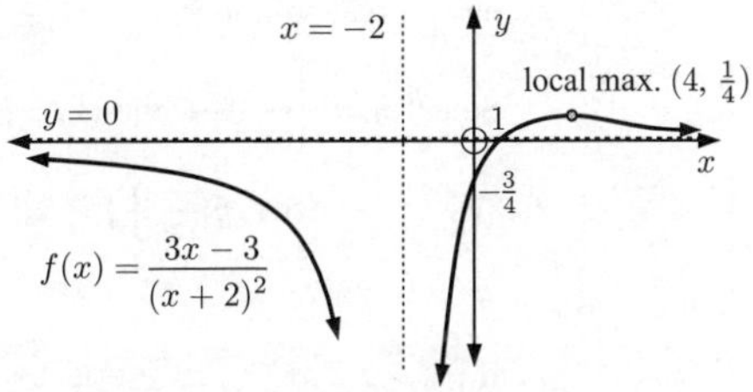

EXERCISE 18E.3

1 **a** $y = \dfrac{2x^2 - x + 2}{x^2 - 1} = \dfrac{2x^2 - x + 2}{(x+1)(x-1)}$

The vertical asymptotes are $x = -1$ and $x = 1$.

We can also write $y = \dfrac{2 - \frac{1}{x} + \frac{2}{x^2}}{1 - \frac{1}{x^2}}$

As $x \to \pm\infty$, $y \to 2$,
so the horizontal asymptote is $y = 2$.

b $y = \dfrac{-x^2 + 2x - 1}{x^2 + x + 1} = \dfrac{-1 + \frac{2}{x} - \frac{1}{x^2}}{1 + \frac{1}{x} + \frac{1}{x^2}}$

Since $x^2 + x + 1 > 0$ for all x, there are no vertical asymptotes.
As $x \to \pm\infty$, $y \to -1$,
so the horizontal asymptote is $y = -1$.

c $y = \dfrac{3x^2 - x + 2}{(x+2)^2}$ has V.A. $x = -2$.

Now $y = \dfrac{3x^2 - x + 2}{x^2 + 4x + 4} = \dfrac{3 - \frac{1}{x} + \frac{2}{x^2}}{1 + \frac{4}{x} + \frac{4}{x^2}}$

As $x \to \pm\infty$, $y \to 3$, so the H.A. is $y = 3$.

2 **a** **i** $y = \dfrac{x^2 - x}{x^2 - x - 6} = \dfrac{x(x-1)}{(x-3)(x+2)}$

has V.A.s when $x - 3 = 0$, $x + 2 = 0$

$\therefore$ the V.A.s are $x = 3$, $x = -2$

Also, since $y = \dfrac{1 - \frac{1}{x}}{1 - \frac{1}{x} - \frac{6}{x^2}}$

the H.A. is $y = 1$
$\{$as $x \to \pm\infty$, $y \to \frac{1}{1}\}$

ii $\dfrac{dy}{dx} = \dfrac{(2x-1)(x^2-x-6) - (x^2-x)(2x-1)}{(x^2-x-6)^2}$

$= \dfrac{(2x-1)(x^2-x-6-x^2+x)}{(x^2-x-6)^2}$

$= \dfrac{-6(2x-1)}{(x^2-x-6)^2}$

Turning points are when $f'(x) = 0$, which occurs when $x = \frac{1}{2}$.
Sign diagram of $f'(x)$ is:

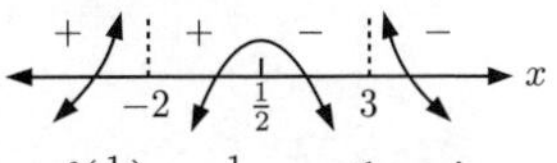

Now $f(\frac{1}{2}) = \frac{1}{25}$, so there is a local maximum at $(\frac{1}{2}, \frac{1}{25})$.

iii When $x = 0$, $y = \frac{0}{-6} = 0$

$\therefore$ the y-intercept is 0

When $y = 0$, $x(x-1) = 0$

$\therefore$ $x = 0$ or 1

$\therefore$ the x-intercepts are 0 and 1

iv

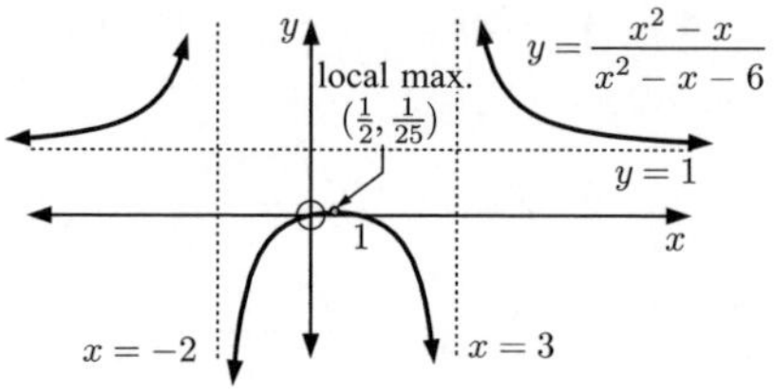

b **i** $y = \dfrac{x^2-1}{x^2+1} = \dfrac{(x+1)(x-1)}{x^2+1}$

has no vertical asymptotes as $x^2 + 1 = 0$ has no real solutions.

Since $y = \dfrac{1 - \frac{1}{x^2}}{1 + \frac{1}{x^2}}$ the H.A. is $y = 1$

$\{$as $x \to \pm\infty$, $y \to \frac{1}{1}\}$

ii $\dfrac{dy}{dx} = \dfrac{2x(x^2+1) - (x^2-1)2x}{(x^2+1)^2}$

$= \dfrac{2x[x^2 + 1 - x^2 + 1]}{(x^2+1)^2}$

$= \dfrac{4x}{(x^2+1)^2}$

and has sign diagram:

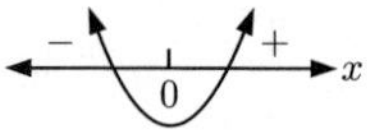

$\therefore$ there is a local minimum at $(0, -1)$.

iii When $x = 0$, $y = \frac{-1}{1} = -1$

$\therefore$ the y-intercept is -1

When $y = 0$, $(x+1)(x-1) = 0$

$\therefore$ $x = \pm 1$

$\therefore$ the x-intercepts are ± 1

iv

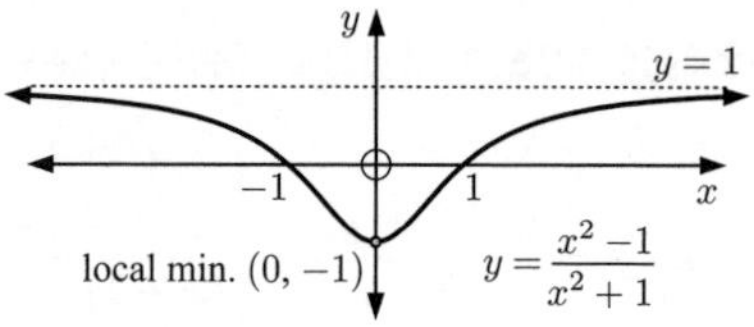

c **i** $y = \dfrac{x^2 - 5x + 4}{x^2 + 5x + 4} = \dfrac{(x-1)(x-4)}{(x+1)(x+4)}$ has V.A.s when $x + 1 = 0$ and $x + 4 = 0$

$\therefore$ the V.A.s are $x = -1$, $x = -4$

Since $y = \dfrac{1 - \frac{5}{x} + \frac{4}{x^2}}{1 + \frac{5}{x} + \frac{4}{x^2}}$ the horizontal asymptote is $y = 1$ $\{$as $x \to \pm\infty$, $y \to \frac{1}{1}\}$

ii $\dfrac{dy}{dx} = \dfrac{(2x-5)(x^2+5x+4) - (x^2-5x+4)(2x+5)}{(x+1)^2(x+4)^2}$

$= \dfrac{[2x^3 + 10x^2 + 8x - 5x^2 - 25x - 20] - [2x^3 + 5x^2 - 10x^2 - 25x + 8x + 20]}{(x+1)^2(x+4)^2}$

$= \dfrac{10x^2 - 40}{(x+1)^2(x+4)^2}$

$= \dfrac{10(x+2)(x-2)}{(x+1)^2(x+4)^2}$ which has sign diagram:

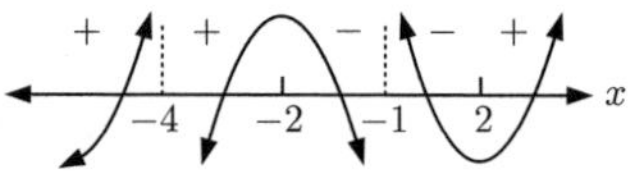

$\therefore$ there is a local maximum at $\left(-2, \dfrac{4 + 10 + 4}{4 - 10 + 4}\right)$ or $(-2, -9)$

and a local minimum at $\left(2, \dfrac{4 - 10 + 4}{4 + 10 + 4}\right)$ or $(2, -\frac{1}{9})$.

iii When $x = 0$, $y = \frac{4}{4} = 1$

$\therefore$ the y-intercept is 1

When $y = 0$, $(x-1)(x-4) = 0$

$\therefore$ $x = 1$ or 4

$\therefore$ the x-intercepts are 1, 4

iv

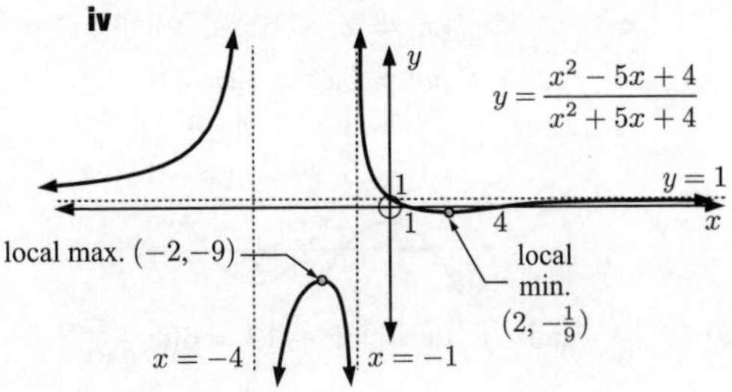

d **i** $y = \dfrac{x^2 - 6x + 5}{(x+1)^2} = \dfrac{(x-1)(x-5)}{(x+1)^2}$ has vertical asymptote $x + 1 = 0$ or $x = -1$,

and horizontal asymptote $y = 1$ $\left\{\text{as } y = \dfrac{1 - \frac{6}{x} + \frac{5}{x^2}}{1 + \frac{2}{x} + \frac{1}{x^2}} \to \frac{1}{1} \text{ as } x \to \pm\infty\right\}$

ii $\dfrac{dy}{dx} = \dfrac{(2x-6)(x+1)^2 - (x^2 - 6x + 5)2(x+1)^1(1)}{(x+1)^4}$

$= \dfrac{(x+1)[2x^2 - 4x - 6 - 2x^2 + 12x - 10]}{(x+1)^4}$

$= \dfrac{8x - 16}{(x+1)^3}$

$= \dfrac{8(x-2)}{(x+1)^3}$ which has sign diagram:

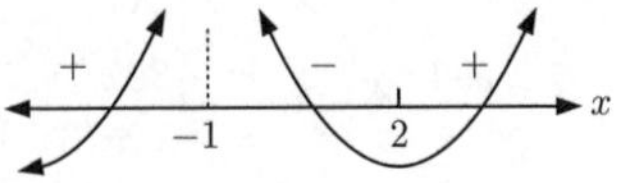

$\therefore$ there is a local minimum at $\left(2, \frac{4-12+5}{3^2}\right)$ or $(2, -\frac{1}{3})$.

iii When $x = 0$, $y = \frac{5}{1} = 5$

$\therefore$ the y-intercept is 5

When $y = 0$, $(x-1)(x-5) = 0$

$\therefore$ $x = 1$ or 5

$\therefore$ the x-intercepts are 1, 5

iv

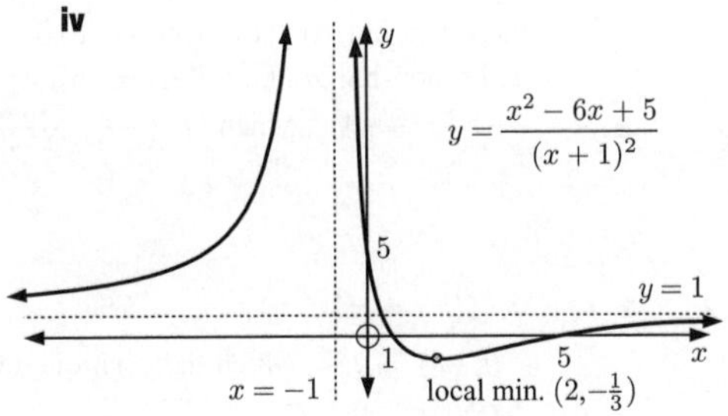

EXERCISE 18F.1

1 **a** $f(x) = x^2 + 3$

$\therefore$ $f'(x) = 2x$

$\therefore$ $f''(x) = 2$

Since $f''(x) \neq 0$,
no points of inflection exist.

b $f(x) = 2 - x^3$

$\therefore$ $f'(x) = -3x^2$

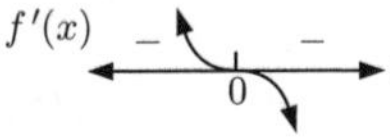

$\therefore$ $f''(x) = -6x$

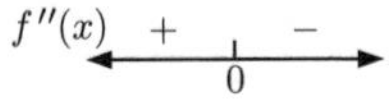

Now $f''(x) = 0$ when $x = 0$,

and $f'(0) = 0$

$\therefore$ there is a horizontal inflection at (0, 2).

c $f(x) = x^3 - 6x^2 + 9x + 1$

$\therefore \quad f'(x) = 3x^2 - 12x + 9$
$= 3(x^2 - 4x + 3)$
$= 3(x-3)(x-1)$

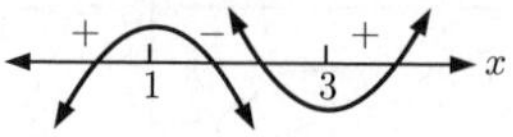

and $f''(x) = 6x - 12 = 6(x-2)$

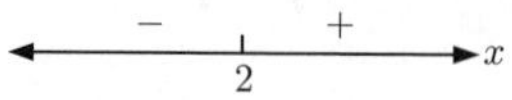

Now $f''(x) = 0$ when $x = 2$
and $f'(2) \neq 0$

$\therefore$ there is a non-horizontal inflection at $(2, f(2))$ which is $(2, 3)$.

d $f(x) = x^3 + 6x^2 + 12x + 5$

$\therefore \quad f'(x) = 3x^2 + 12x + 12$
$= 3(x^2 + 4x + 4)$
$= 3(x+2)^2$

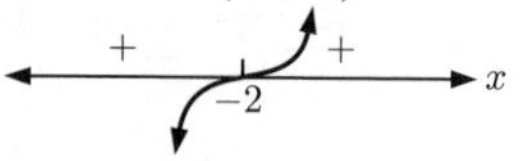

and $f''(x) = 6x + 12 = 6(x+2)$

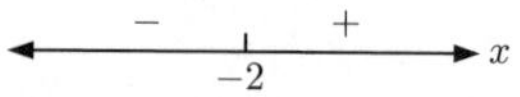

Now $f''(x) = 0$ when $x = -2$
and $f'(-2) = 0$

$\therefore$ there is a horizontal inflection at $(-2, f(-2))$ which is $(-2, -3)$.

e $f(x) = -3x^4 - 8x^3 + 2$

$\therefore \quad f'(x) = -12x^3 - 24x^2$
$= -12x^2(x+2)$

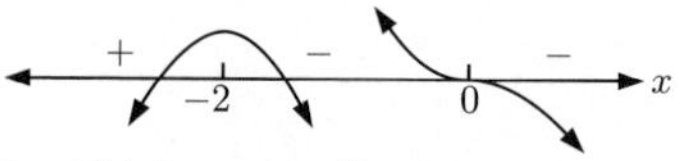

and $f''(x) = -36x^2 - 48x$
$= -12x(3x+4)$

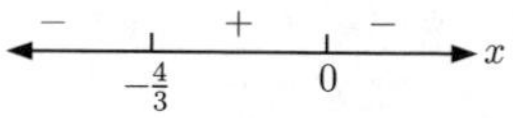

$\therefore$ there is a horizontal inflection at $(0, 2)$, and a non-horizontal inflection at $(-\frac{4}{3}, f(-\frac{4}{3}))$, which is $\left(-\frac{4}{3}, \frac{310}{27}\right)$.

f $f(x) = 3 - \dfrac{1}{\sqrt{x}} = 3 - x^{-\frac{1}{2}}$

$\therefore \quad f'(x) = \frac{1}{2}x^{-\frac{3}{2}}$

and $f''(x) = -\frac{3}{4}x^{-\frac{5}{2}} = \dfrac{-3}{4x^2\sqrt{x}}$

Now $f''(x) \neq 0$ for all x

$\therefore$ there are no points of inflection.

2 **a** $f(x) = x^2$

$\therefore \quad f'(x) = 2x$ which has sign diagram:

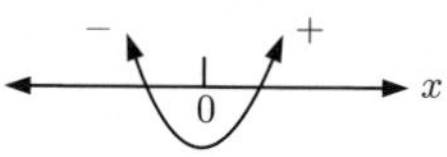

and $f''(x) = 2$

i There is a local minimum at $(0, 0)$.

ii There are no points of inflection as $f''(x) \neq 0$.

iii $f(x)$ is increasing when $x \geqslant 0$, and decreasing when $x \leqslant 0$.

iv $f(x)$ is concave up for all x as $f''(x) > 0$ for all x.

v

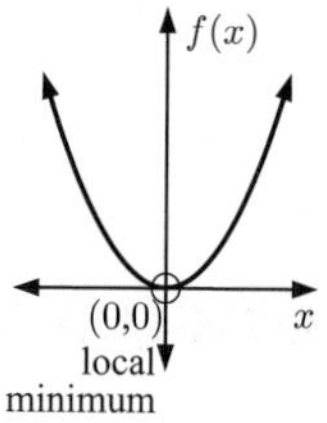

b $f(x) = x^3$

$\therefore \quad f'(x) = 3x^2$ which has sign diagram:

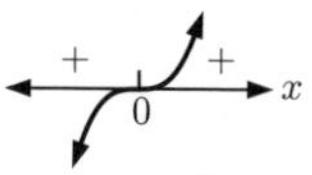

and $f''(x) = 6x$ which has sign diagram:

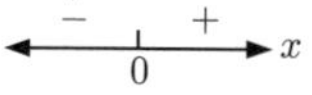

i A horizontal inflection at $(0, 0)$.

ii A horizontal inflection at $(0, 0)$.

iii $f(x)$ is increasing for all real x.

iv $f(x)$ is concave up when $x \geqslant 0$, and concave down when $x \leqslant 0$.

v

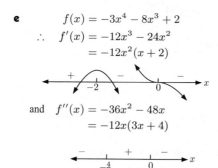

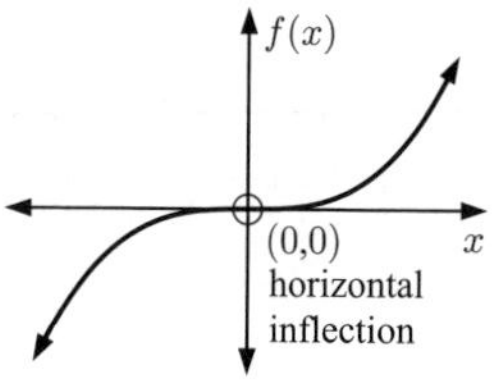

c $f(x) = \sqrt{x}$

$\therefore\ f'(x) = \frac{1}{2}x^{-\frac{1}{2}} = \dfrac{1}{2\sqrt{x}}$ which has sign diagram:

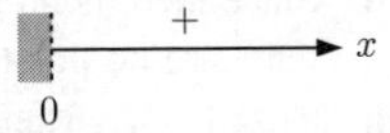

and $f''(x) = -\frac{1}{4}x^{-\frac{3}{2}} = \dfrac{-1}{4x\sqrt{x}}$ which has sign diagram:

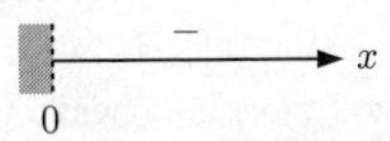

i There are no stationary points as $f'(x) \neq 0$.

ii There are no points of inflection as $f''(x) \neq 0$.

iii $f(x)$ is increasing for all $x \geqslant 0$.

iv $f(x)$ is concave down for all $x \geqslant 0$ as $f''(x) < 0$ for all $x > 0$.

v

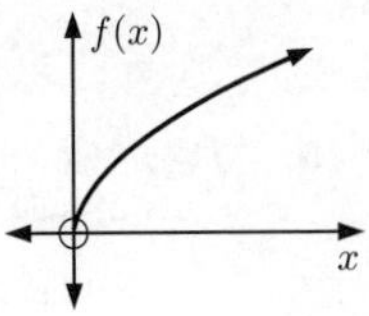

d $f(x) = x^3 - 3x^2 - 24x + 1$

$\therefore\ f'(x) = 3x^2 - 6x - 24$

$= 3(x^2 - 2x - 8)$

$= 3(x-4)(x+2)$ which has sign diagram:

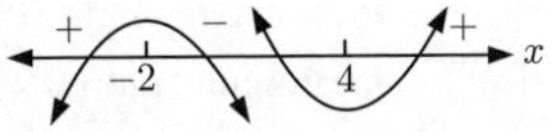

and $f''(x) = 6x - 6$

$= 6(x-1)$ which has sign diagram:

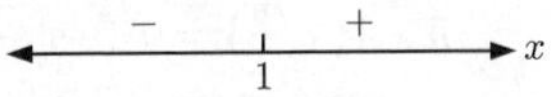

i $f(-2) = 29$, $f(4) = -79$, so there is a local maximum at $(-2, 29)$, and a local minimum at $(4, -79)$.

ii $f(1) = -25$, so there is a non-horizontal inflection at $(1, -25)$.

iii $f(x)$ is increasing for $x \leqslant -2$ and $x \geqslant 4$, and decreasing for $-2 \leqslant x \leqslant 4$.

iv $f(x)$ is concave down for $x \leqslant 1$, and concave up for $x \geqslant 1$.

v

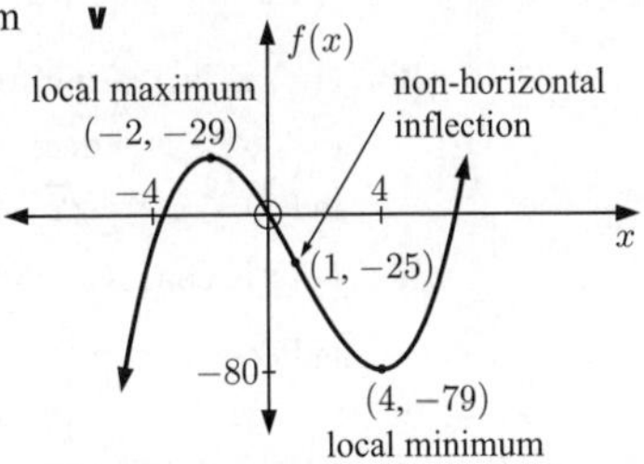

e $f(x) = 3x^4 + 4x^3 - 2$

$\therefore\ f'(x) = 12x^3 + 12x^2$

$= 12x^2(x+1)$ which has sign diagram:

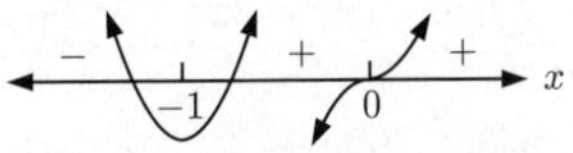

and $f''(x) = 36x^2 + 24x$

$= 12x(3x+2)$ which has sign diagram:

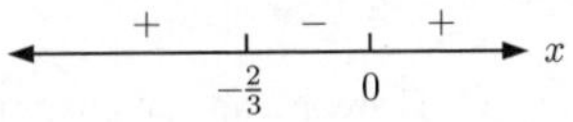

i There is a local minimum at $(-1, f(-1))$ which is $(-1, -3)$, and a horizontal inflection at $(0, -2)$.

ii There is a non-horizontal inflection at $(-\frac{2}{3}, f(-\frac{2}{3}))$ which is $(-\frac{2}{3}, -\frac{70}{27})$, and a horizontal inflection at $(0, -2)$.

iii $f(x)$ is increasing for $x \geqslant -1$, and decreasing for $x \leqslant -1$.

iv $f(x)$ is concave down for $-\frac{2}{3} \leqslant x \leqslant 0$, and concave up for $x \leqslant -\frac{2}{3}$ and $x \geqslant 0$.

v

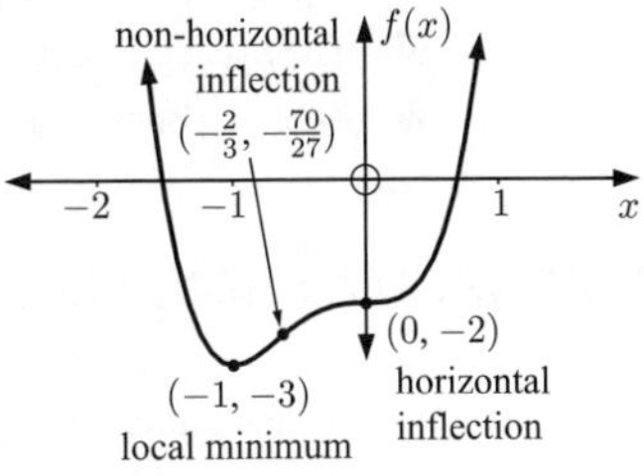

f $f(x) = (x-1)^4$

$\therefore\ f'(x) = 4(x-1)^3$ which has sign diagram:

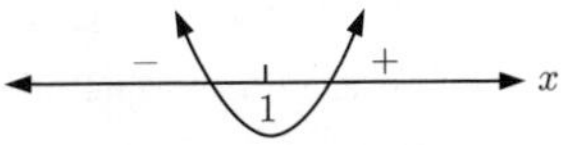

and $f''(x) = 12(x-1)^2$ which has sign diagram:

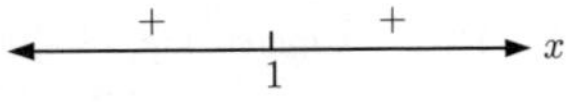

i There is a local minimum at $(1, 0)$.

ii Since there is no sign change in $f''(x)$ at $x = 1$, there are no points of inflection.

iii $f(x)$ is increasing for $x \geqslant 1$, and decreasing for $x \leqslant 1$.

iv $f(x)$ is concave up for all x.

v

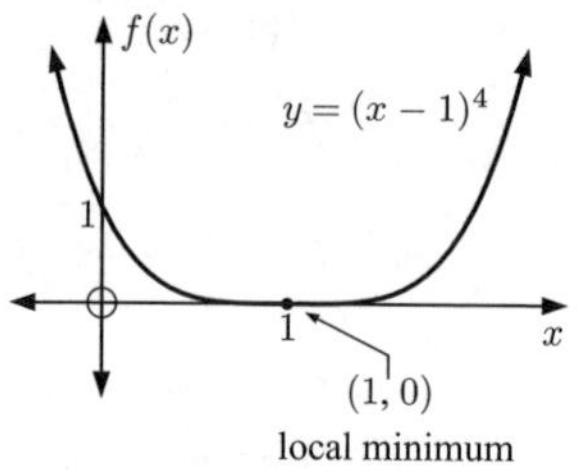

g $f(x) = x^4 - 4x^2 + 3$

$f'(x) = 4x^3 - 8x = 4x(x^2 - 2)$

$= 4x(x + \sqrt{2})(x - \sqrt{2})$ which has sign diagram:

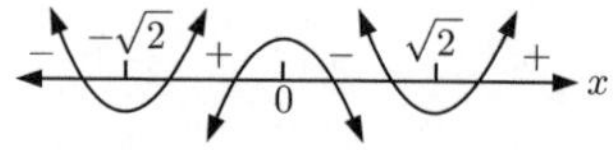

$f''(x) = 12x^2 - 8 = 4(3x^2 - 2)$

$= 4(\sqrt{3}x + \sqrt{2})(\sqrt{3}x - \sqrt{2})$ which has sign diagram:

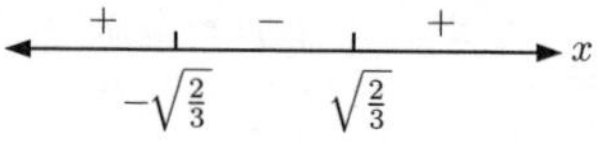

i $f(-\sqrt{2}) = f(\sqrt{2}) = -1$, so there is a local maximum at $(0, 3)$, and local minima at $(\sqrt{2}, -1)$ and $(-\sqrt{2}, -1)$.

ii $f\left(\sqrt{\frac{2}{3}}\right) = f\left(-\sqrt{\frac{2}{3}}\right) = \frac{7}{9}$, so there are non-horizontal inflections at $\left(\sqrt{\frac{2}{3}}, \frac{7}{9}\right)$ and $\left(-\sqrt{\frac{2}{3}}, \frac{7}{9}\right)$.

iii $f(x)$ is increasing for $-\sqrt{2} \leqslant x \leqslant 0$ and $x \geqslant \sqrt{2}$, and decreasing for $x \leqslant -\sqrt{2}$ and $0 \leqslant x \leqslant \sqrt{2}$.

iv $f(x)$ is concave down for $-\sqrt{\frac{2}{3}} \leqslant x \leqslant \sqrt{\frac{2}{3}}$, and concave up for $x \leqslant -\sqrt{\frac{2}{3}}$ and $x \geqslant \sqrt{\frac{2}{3}}$.

v

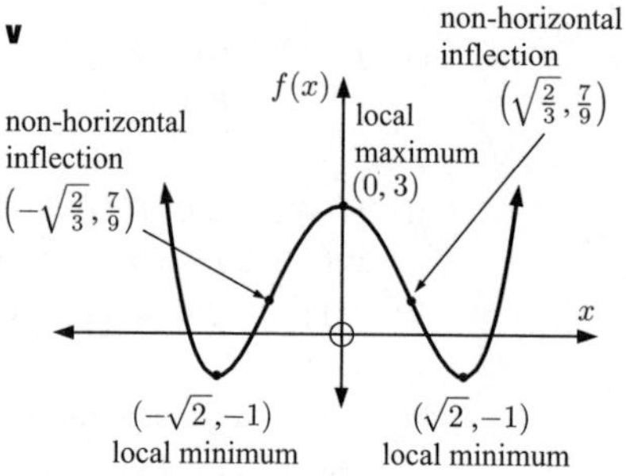

h $f(x) = 3 - \dfrac{4}{\sqrt{x}} = 3 - 4x^{-\frac{1}{2}}, \quad x > 0$

$\therefore \quad f'(x) = 2x^{-\frac{3}{2}} = \dfrac{2}{x\sqrt{x}}$ with sign diagram:

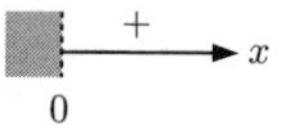

and $f''(x) = -3x^{-\frac{5}{2}} = -\dfrac{3}{x^2\sqrt{x}}$ with sign diagram:

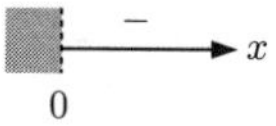

i There are no stationary points as $f'(x) \neq 0$.

ii There are no points of inflection as $f''(x) \neq 0$.

iii $f(x)$ is increasing for all $x > 0$ and never decreasing.

iv $f(x)$ is concave down for all $x > 0$ and never concave up.

v

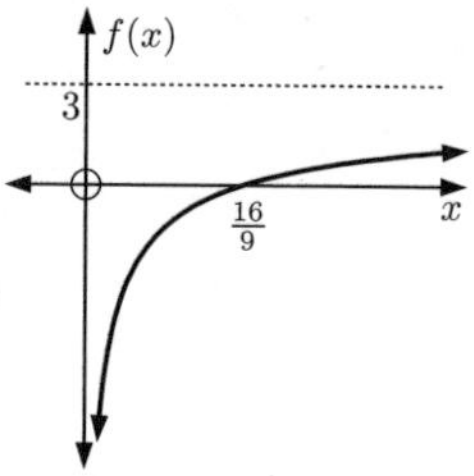

EXERCISE 18F.2

1 **a** $f(x)$ is quadratic, so $f'(x)$ will be linear and $f''(x)$ will be constant.

$f(x)$ is decreasing for $x \leqslant 1$ and increasing for $x \geqslant 1$

$\therefore \quad f'(x) \leqslant 0$ for $x \leqslant 1$ and $f'(x) \geqslant 0$ for $x \geqslant 1$

$\therefore \quad f'(x)$ is an increasing linear function which cuts the x-axis at 1.

As $f'(x)$ is increasing, $f''(x) > 0$.

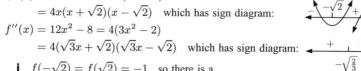

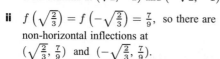

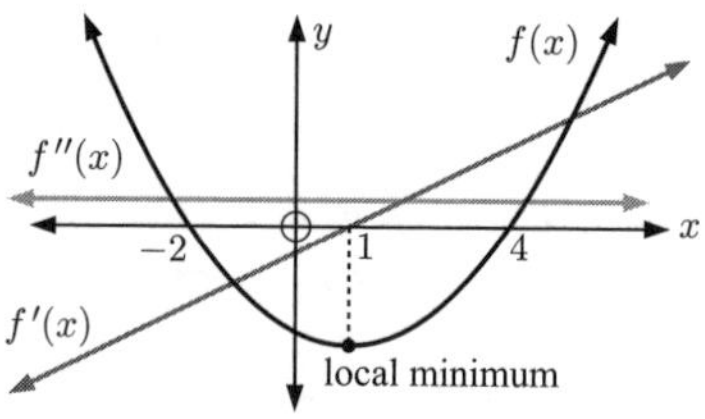

b $f(x)$ is cubic, so $f'(x)$ will be quadratic and $f''(x)$ will be linear.
$f(x)$ has turning points at $x \approx \pm 1$
$\therefore$ $f'(x)$ cuts the x-axis at these points.
$f(x)$ has a non-stationary inflection point at $x = 0$
$\therefore$ $f'(x)$ has a turning point at $x = 0$, and $f''(0) = 0$.
$f(x)$ is concave down for $x \leqslant 0$ and concave up for $x \geqslant 0$
$\therefore$ $f'(x)$ is decreasing for $x \leqslant 0$ and increasing for $x \geqslant 0$
and $f''(x) \leqslant 0$ for $x \leqslant 0$ and $\geqslant 0$ for $x \geqslant 0$.

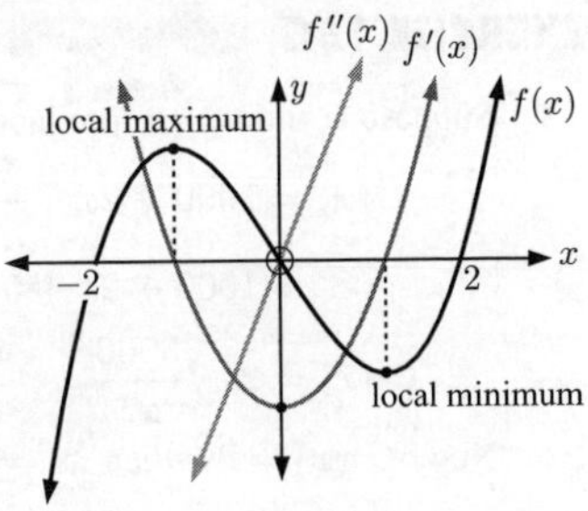

c $f(x)$ is cubic, so $f'(x)$ will be quadratic and $f''(x)$ will be linear.
$f(x)$ has turning points at $x \approx 1$ and $x = 3$
$\therefore$ $f'(x)$ cuts the x-axis at these points.
$f(x)$ has a non-stationary inflection point at $x \approx 2$
$\therefore$ $f'(x)$ has a turning point at $x \approx 2$, and $f''(2) = 0$
$f(x)$ is concave down for $x \leqslant 2$ and concave up for $x \geqslant 2$
$\therefore$ $f'(x)$ is decreasing for $x \leqslant 2$ and increasing for $x \geqslant 2$
and $f''(x) \leqslant 0$ for $x \leqslant 2$ and $\geqslant 0$ for $x \geqslant 2$.

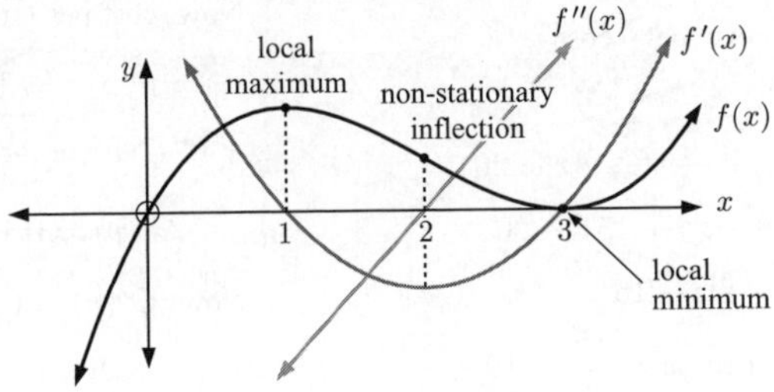

2 **a** $f'(x)$ has sign diagram:

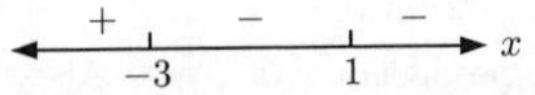

$\therefore$ $f(x)$ is increasing for $x \leqslant -3$, and decreasing for $x \geqslant -3$
$\therefore$ $f(x)$ has a local maximum at $x = -3$
$f'(x)$ has a turning point at $x \approx -1.7$.
At this point, $f''(x) = 0$, but $f'(x) \neq 0$
$\therefore$ $f(x)$ has a non-stationary inflection point here.
$f'(x)$ has another turning point at $x = 1$.
At this point, $f''(x) = 0$ and $f'(x) = 0$
$\therefore$ $f(x)$ has a stationary inflection point at $x = 1$.
A possible graph of $f(x)$ is shown alongside:

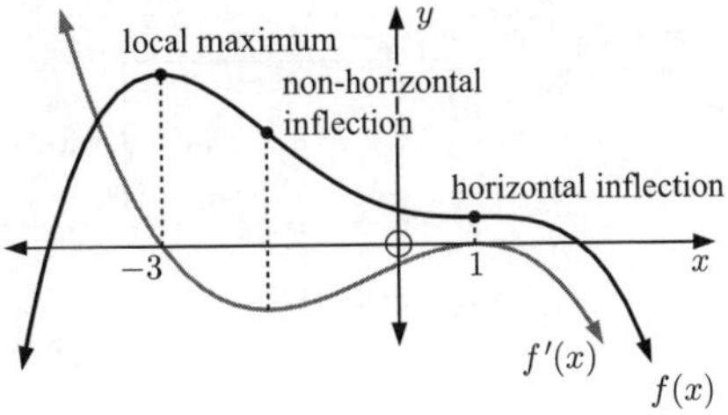

b $f'(x)$ has sign diagram:

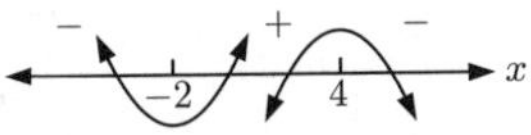

$\therefore$ $f(x)$ has a local minimum at $x = -2$ and a local maximum at $x = 4$
$f'(x)$ has a turning point at $x \approx 1$.
At this point, $f''(x) = 0$, but $f'(x) \neq 0$
$\therefore$ $f(x)$ has a non-stationary inflection point at $x \approx 1$.
A possible graph of $f(x)$ is shown alongside:

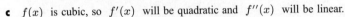

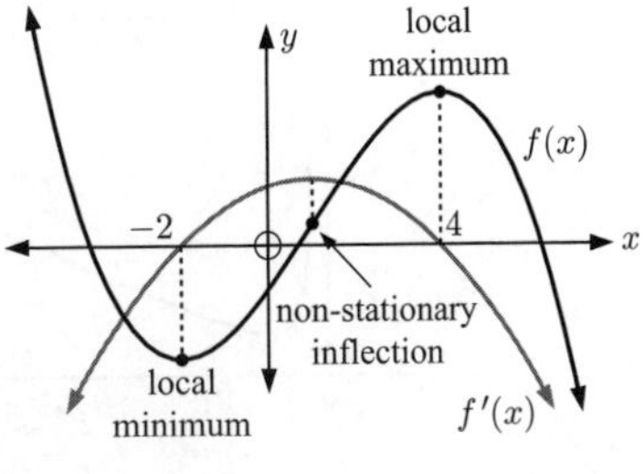

EXERCISE 18G

1 Suppose x fittings are produced daily.

$\therefore \quad C(x) = 1000 + 2x + \dfrac{5000}{x}$

$= 1000 + 2x + 5000x^{-1}$ euros

$\therefore \quad C'(x) = 2 - \dfrac{5000}{x^2}$

Now $C'(x) = 0$ when $x^2 = 2500$

$\therefore \quad x = 50 \quad \{\text{as } x > 0\}$

Also, $C''(x) = 10\,000x^{-3} = \dfrac{10\,000}{x^3}$

which > 0 when $x > 0$.

$\therefore$ the cost is minimised when 50 fittings are produced.

2 $C(x) = 720 + 4x + 0.02x^2$

$p(x) = 15 - 0.002x$

Revenue $R(x) = xp(x) = 15x - 0.002x^2$

Profit $P(x) = \text{revenue} - \text{cost}$

$= (15x - 0.002x^2)$

$\quad - (720 + 4x + 0.02x^2)$

$= -0.022x^2 + 11x - 720$

$\therefore \quad P'(x) = -0.044x + 11$

Now $P'(x) = 0$ when $x = \dfrac{11}{0.044} = 250$

$\therefore$ as $P''(x) = -0.044 < 0$, P is maximised when 250 items are produced.

3 $C(x) = \frac{1}{4}x^2 + 8x + 20$

$p(x) = 23 - \frac{1}{2}x$

Revenue $R(x) = xp(x) = 23x - \frac{1}{2}x^2$

Profit $P(x) = \text{revenue} - \text{cost}$

$= (23x - \frac{1}{2}x^2)$

$\quad - (\frac{1}{4}x^2 + 8x + 20)$

$= -\frac{3}{4}x^2 + 15x - 20$

$\therefore \quad P'(x) = -\frac{3}{2}x + 15$

Now $P'(x) = 0$ when $x = \dfrac{15}{\frac{3}{2}} = 10$

$\therefore$ as $P''(x) = -\frac{3}{2} < 0$, P is maximised when 10 blankets per day are produced.

4 Cost per hour $=$ running costs $+$ other costs

$= \dfrac{v^2}{10} + 62.5$

Now cost per km $= \dfrac{\text{cost per hour}}{\text{km per hour}}$

$\therefore \quad C(v) = \dfrac{\frac{v^2}{10} + 62.5}{v} = 0.1v + 62.5v^{-1}$

$\therefore \quad C'(v) = 0.1 - 62.5v^{-2}$

Now $C'(v) = 0$ when $0.1 = \dfrac{62.5}{v^2}$

$\therefore \quad v^2 = 625$

$\therefore \quad v = 25 \quad \{\text{as } v > 0\}$

Also, $C''(v) = 62.5 \times 2v^{-3} = \dfrac{125}{v^3}$

which is > 0 when $v > 0$

$\therefore$ the minimum cost per km occurs when $v = 25$ km h^{-1}.

5 **a**

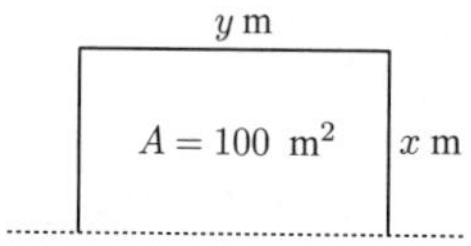

Now $xy = 100$

$\therefore \quad y = \dfrac{100}{x}$

$\therefore \quad L = 2x + y$

$\therefore \quad L = 2x + \dfrac{100}{x}$

b

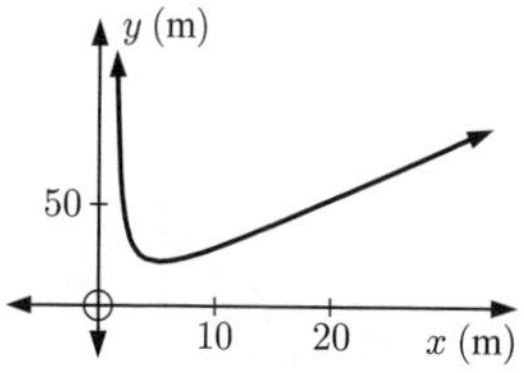

c $\dfrac{dL}{dx} = 2 - 100x^{-2} = 2 - \dfrac{100}{x^2}$

which is 0 when $\dfrac{100}{x^2} = 2$

$\therefore \quad x^2 = 50$

$\therefore \quad x = \sqrt{50} \quad \{x > 0\}$

$\dfrac{d^2L}{dt^2} = 200x^{-3} = \dfrac{200}{x^3} > 0$ for $x > 0$

$\therefore \quad L_{\text{min}} = 2\sqrt{50} + \dfrac{100}{\sqrt{50}}$

$= 2\sqrt{50} + 2\sqrt{50}$

$= 4\sqrt{50}$

$= 20\sqrt{2}$ m when $x = 5\sqrt{2}$ m

$\therefore$ min $L \approx 28.3$ m when $x \approx 7.07$ m

d

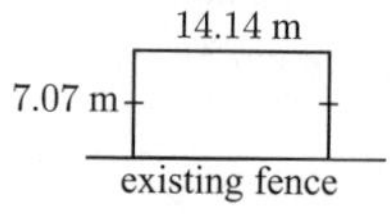

6 **a** Inner length of box $= 2x$ cm

b $$\text{Volume} = 200 \text{ cm}^3$$
$$\therefore \quad x \times 2x \times h = 200$$
$$2x^2h = 200$$
$$\therefore \quad x^2h = 100$$

c From **b**, $h = \dfrac{100}{x^2}$.

Area of inner surface is
$$A(x) = 2(2x \times x) + 2(2x \times h) + 2(x \times h)$$
$$= 4x^2 + 4xh + 2xh$$
$$= 4x^2 + 6xh$$
$$= 4x^2 + \frac{600}{x} \text{ cm}^2$$

d

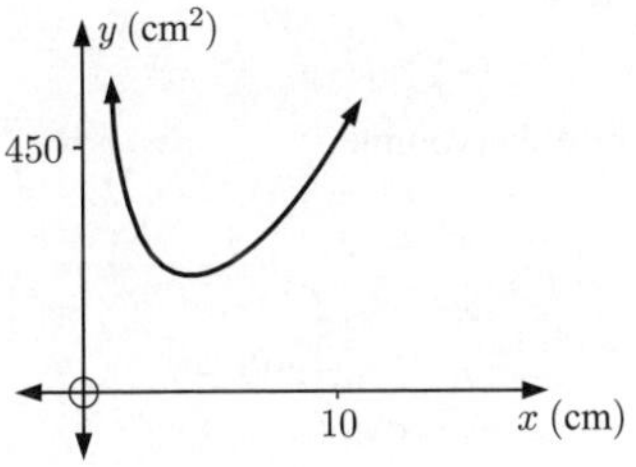

e $$A(x) = 4x^2 + 600x^{-1}$$
$$\therefore \quad A'(x) = 8x - 600x^{-2}$$
$$= 8x - \frac{600}{x^2}$$
$$\therefore \quad A'(x) = 0 \quad \text{when} \quad 8x = \frac{600}{x^2}$$
$$\therefore \quad 8x^3 = 600$$
$$\therefore \quad x^3 = 75$$
$$\therefore \quad x \approx 4.217 \text{ cm}$$
$$A''(x) = 8 + 1200x^{-3}$$
$$= 8 + \frac{1200}{x^3}$$
$$\therefore \quad A''(x) > 0 \qquad \{\text{as } x > 0\}$$
$\therefore$ area is minimised when $x \approx 4.22$ cm
$$\therefore \quad A_{\text{min}} \approx 4(4.217)^2 + \frac{600}{(4.217)}$$
$$\approx 213 \text{ cm}^2$$

f Height $h \approx \dfrac{100}{(4.217)^2}$
$$\approx 5.62 \text{ cm}$$

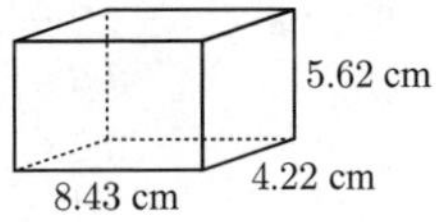

7 **a** Volume of can $= \pi r^2 h$
$$\therefore \quad 1000 = \pi r^2 h \quad \text{(in cm)}$$
$$\therefore \quad h = \frac{1000}{\pi r^2} \text{ cm}$$

b Opening the can up we get:

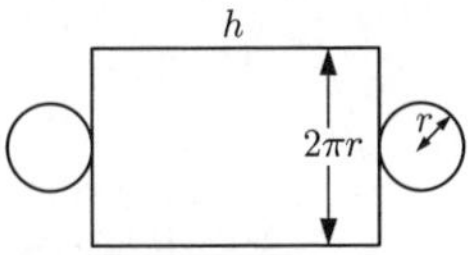

$$\therefore \quad A(r) = \pi r^2 + \pi r^2 + 2\pi r h$$
$$= 2\pi r^2 + 2\pi r h$$
$$= 2\pi r^2 + \frac{2000}{r} \text{ cm}^2$$

c

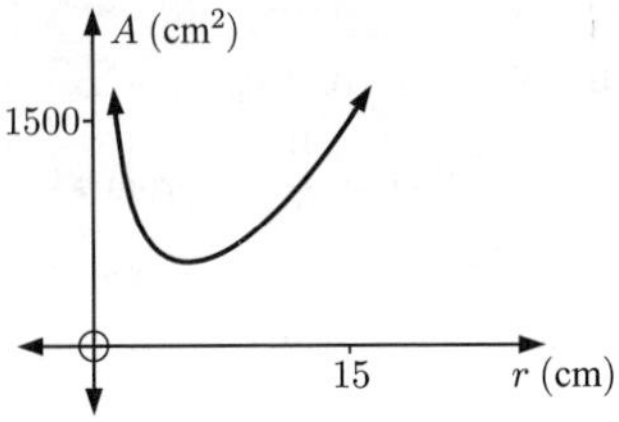

d $$A(r) = 2\pi r^2 + 2000r^{-1}$$
$$A'(r) = 4\pi r - 2000r^{-2} = 4\pi r - \frac{2000}{r^2}$$
So, $A'(r) = 0$ when $4\pi r = \dfrac{2000}{r^2}$
$$r^3 = \frac{2000}{4\pi}$$
$$r = \sqrt[3]{\frac{500}{\pi}}$$
$$\therefore \quad r \approx 5.419 \text{ cm}$$
$$A''(r) = 4\pi + 4000r^{-3} = 4\pi + \frac{4000}{r^3}$$
and as $r > 0$, $A''(r) > 0$

$\therefore$ area is a minimum when $r \approx 5.42$ cm

and $h = \dfrac{1000}{\pi r^2} \approx 10.8$ cm
$$A_{\min} = 2\pi r^2 + 2\pi r h \approx 554 \text{ cm}^2$$

e

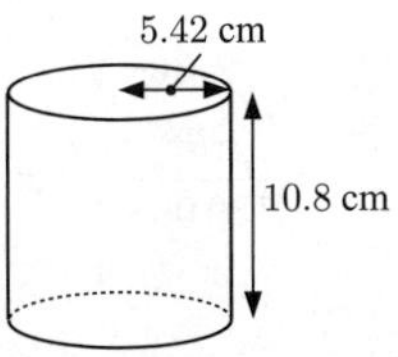

8 **a**

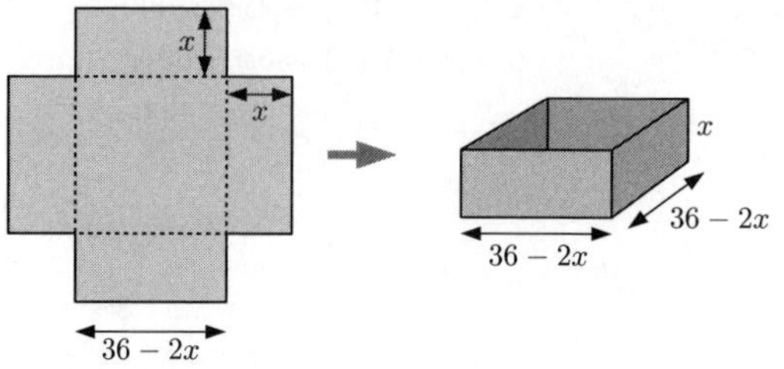

The volume of the container is

$$V = lbd$$
$$= x(36-2x)(36-2x)$$
$$\therefore \quad V = x(36-2x)^2 \text{ cm}^3$$

b Using the product rule,

$$V'(x) = (36-2x)^2 - 4x(36-2x)$$
$$= (36-2x)[(36-2x)-4x]$$
$$= (36-2x)(36-6x)$$
$$\therefore \quad V'(x) = 0 \quad \text{when } x = 6 \text{ or } x = 18$$

Sign diagram of $V'(x)$ is:

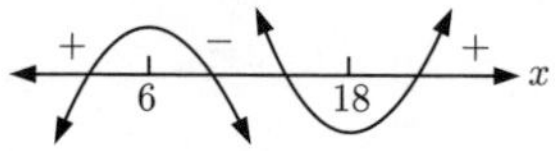

$\therefore$ the volume is maximised when $x = 6$ cm $\{0 \leqslant x \leqslant 18\}$

So, 6 cm × 6 cm squares should be cut out to maximise the volume.

9 **a**

$$P = 2\pi r + 2l$$
$$\therefore \quad 400 = 2\pi(x) + 2l$$
$$\therefore \quad 200 = \pi x + l$$
$$\therefore \quad l = 200 - \pi x$$

Now clearly $x \geqslant 0$ and $l \geqslant 0$

$$\therefore \quad \pi x \leqslant 200$$
$$\therefore \quad x \leqslant \frac{200}{\pi}$$

So, $0 \leqslant x \leqslant \dfrac{200}{\pi}$

b Area

$$A = \pi r^2 + (2x) \times l$$
$$= \pi x^2 + 2xl$$
$$= \pi x^2 + 2x(200 - \pi x)$$
$$= \pi x^2 + 400x - 2\pi x^2$$
$$= 400x - \pi x^2$$

c Now $\dfrac{dA}{dx} = 400 - 2\pi x$

which is 0 when $2\pi x = 400$

$$\therefore \quad x = \frac{200}{\pi} \approx 63.7 \quad \text{and} \quad l = 0$$

Sign diagram of $A'(x)$:

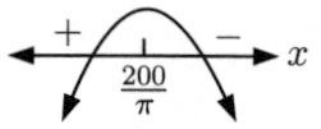

$\therefore$ the area will be a maximum when the track is a circle.

10 **a**

$$\text{Arc AC} = \frac{\theta}{360} \times (2\pi r_{\text{sector}})$$
$$= \frac{\theta}{360}(2 \times \pi \times 10)$$
$$= \frac{\theta\pi}{18}$$

b Now arc AC forms the base of the cone.

$$\therefore \quad 2\pi r = \frac{\theta\pi}{18} \quad \{\text{from } \mathbf{a}\}$$
$$\therefore \quad r = \frac{\theta}{36}$$

c Height of cone $= \sqrt{10^2 - r^2}$ {Pythagoras}

$$\therefore \quad h = \sqrt{100 - \left(\frac{\theta}{36}\right)^2}$$

d

$$V = \tfrac{1}{3}\pi r^2 h$$
$$= \tfrac{1}{3}\pi\left(\frac{\theta}{36}\right)^2 \sqrt{100 - \left(\frac{\theta}{36}\right)^2}$$
$$= \frac{\pi\theta^2}{3 \times 36^2}\sqrt{\frac{129\,600 - \theta^2}{36^2}}$$
$$= \frac{\pi\theta^2}{139\,968}\sqrt{129\,600 - \theta^2}$$

e

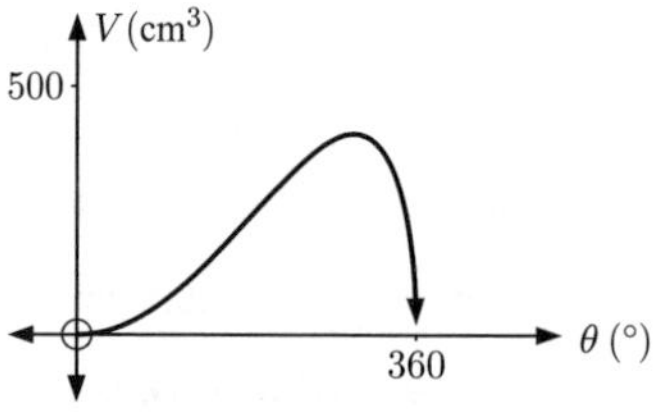

f Now $V'(\theta) = \frac{2\pi\theta}{139\,968}\left(129\,600 - \theta^2\right)^{\frac{1}{2}} + \frac{\pi\theta^2}{139\,968}\left(\frac{1}{2}\right)\left(129\,600 - \theta^2\right)^{-\frac{1}{2}}(-2\theta)$

$$= \frac{\pi\theta}{139\,968}\left(\frac{2\sqrt{129\,600 - \theta^2}}{1} - \frac{\theta^2}{\sqrt{129\,600 - \theta^2}}\right)$$

$$= \frac{\pi\theta}{139\,968}\left(\frac{2(129\,600 - \theta^2) - \theta^2}{\sqrt{129\,600 - \theta^2}}\right)$$

and $V'(\theta) = 0$ when $\theta = 0$ or $2(129\,600 - \theta^2) = \theta^2$

$$259\,200 - 2\theta^2 = \theta^2$$

$$\therefore\ 3\theta^2 = 259\,200$$

$$\therefore\ \theta = \sqrt{86\,400} \quad \{\text{as } \theta > 0\}$$

$$\therefore\ \theta \approx 293.9$$

Sign diagram of $V'(\theta)$ is:

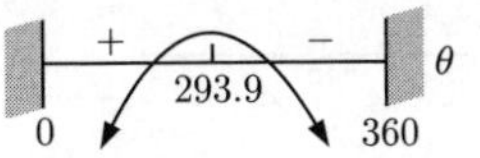

$\therefore$ maximum V occurs when $\theta \approx 294^o$

11 **a** X must lie either between A and C or else at one of the two points. If $x = 0$, then he rows straight to the shore and runs to C. If $x = 6$, then he rows straight to C. $\therefore\ 0 \leqslant x \leqslant 6$

b Now $\text{XC} = 6 - x$

$\therefore$ the time to row from B to X $= \frac{\text{BX}}{8} = \frac{\sqrt{x^2 + 5^2}}{8}$

and the time to run from X to C $= \frac{\text{XC}}{17} = \frac{6 - x}{17}$

$\therefore$ the total time $T(x) = \frac{\sqrt{x^2 + 25}}{8} + \frac{6 - x}{17}$ hours

$$= \tfrac{1}{8}(x^2 + 25)^{\frac{1}{2}} + \tfrac{6}{17} - \tfrac{x}{17}$$

c $\frac{dT}{dx} = \frac{1}{16}(x^2 + 25)^{-\frac{1}{2}}(2x) - \frac{1}{17}$

$$= \frac{x}{8\sqrt{x^2 + 25}} - \frac{1}{17}$$

So, $\frac{dT}{dx} = 0$ when $\frac{x}{8\sqrt{x^2 + 25}} = \frac{1}{17}$

$$17x = 8\sqrt{x^2 + 25}$$

$$\therefore\ 289x^2 = 64(x^2 + 25)$$

$$\therefore\ 225x^2 = 1600$$

$$\therefore\ x^2 = \tfrac{1600}{225}$$

$$\therefore\ x = \tfrac{40}{15} = \tfrac{8}{3}\text{ km}$$

Sign diagram of $\frac{dT}{dx}$:

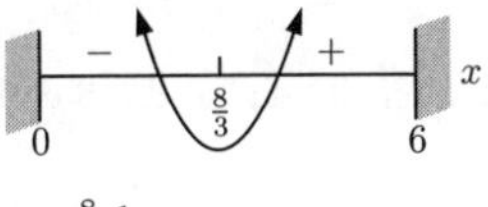

The time taken is minimised if Peter aims for X such that $x = \frac{8}{3}$ km.

12 Let $\text{MX} = x$ km, so $\text{XN} = 5 - x$ km

$\therefore\ \text{AX} = \sqrt{4 + x^2}$ km and $\text{XB} = \sqrt{1 + (5 - x)^2}$ km {Pythagoras}

Now let $P = \text{AX} + \text{XB}$

$$= (4 + x^2)^{\frac{1}{2}} + (26 - 10x + x^2)^{\frac{1}{2}}$$

$$\therefore\ \frac{dP}{dx} = \tfrac{1}{2}(4 + x^2)^{-\frac{1}{2}}(2x) + \tfrac{1}{2}(26 - 10x + x^2)^{-\frac{1}{2}}(2x - 10)$$

$$= \frac{x}{\sqrt{4 + x^2}} + \frac{x - 5}{\sqrt{x^2 - 10x + 26}}$$

Now $\dfrac{dP}{dx} = 0$ when $\dfrac{x}{\sqrt{4+x^2}} = \dfrac{5-x}{\sqrt{x^2-10x+26}}$

$\therefore \quad \dfrac{x^2}{4+x^2} = \dfrac{(5-x)^2}{x^2-10x+26}$ {squaring both sides}

$\therefore \quad x^2(x^2-10x+26) = (4+x^2)(25-10x+x^2)$

$\therefore \quad x^4 - 10x^3 + 26x^2 = 100 - 40x + 4x^2 + 25x^2 - 10x^3 + x^4$

$\therefore \quad 3x^2 - 40x + 100 = 0$

$\therefore \quad (3x-10)(x-10) = 0$

$\therefore \quad x = \frac{10}{3}$ {as $0 \leqslant x \leqslant 5$}

Sign diagram of $\dfrac{dP}{dx}$ is:

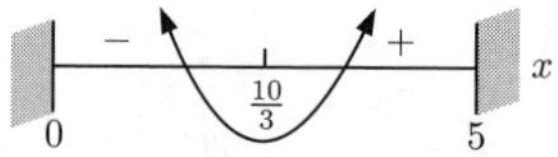

$\therefore$ the minimum length pipeline occurs when $x = \frac{10}{3}$ km

13 $V = \pi r^2 h$

$\therefore \quad 0.1 = \pi r^2 h$ {as 100 L = 0.1 m^3}

$\therefore \quad h = \dfrac{0.1}{\pi r^2}$

Now $A = \pi r^2 + (2\pi r)h = \pi r^2 + 2\pi r\left(\dfrac{0.1}{\pi r^2}\right)$

$\therefore \quad A(r) = \pi r^2 + 0.2r^{-1}$

$\therefore \quad A'(r) = 2\pi r - 0.2r^{-2}$

$= 2\pi r - \dfrac{0.2}{r^2}$

So, $A'(r) = 0$ when $2\pi r = \dfrac{0.2}{r^2}$

$\therefore \quad r^3 = \dfrac{0.2}{2\pi}$

$\therefore \quad r = \sqrt[3]{\dfrac{0.2}{2\pi}} \approx 0.3169$ m

$\therefore \quad r \approx 31.7$ cm

Now $A''(r) = 2\pi + 0.4r^{-3} = 2\pi + \dfrac{0.4}{r^3}$ which is > 0 as $r > 0$

$\therefore$ the minimum area occurs when $r \approx 31.7$ cm and $h \approx \dfrac{1}{10\pi(31.69)} \approx 31.7$ cm

$\therefore \quad r = h \approx 31.7$ cm

14 Now $I \propto \dfrac{s}{d^2}$ where s is the power of the source and d is the distance from it

$\therefore \quad I = \dfrac{ks}{d^2}$ where k is a constant

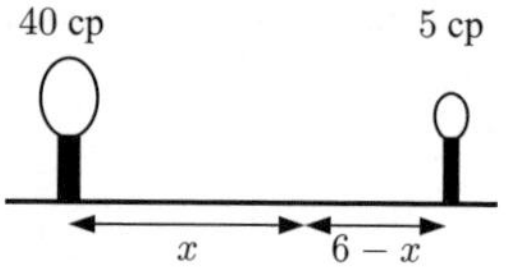

So, the intensity due to the 40 cp bulb $= \dfrac{40k}{x^2}$

and the intensity due to the 5 cp bulb $= \dfrac{5k}{(6-x)^2}$

The total intensity $I = \dfrac{40k}{x^2} + \dfrac{5k}{(6-x)^2}$

$= k[40x^{-2} + 5(6-x)^{-2}]$

$\therefore \quad \dfrac{dI}{dx} = k[-80x^{-3} - 10(6-x)^{-3}(-1)]$

$= k\left[\dfrac{-80}{x^3} + \dfrac{10}{(6-x)^3}\right]$

$\therefore \quad \dfrac{dI}{dx} = 0$ when $\dfrac{80}{x^3} = \dfrac{10}{(6-x)^3}$

$\therefore \quad 8(6-x)^3 = x^3$

$\therefore \quad 2(6-x) = x$ {finding cube roots}

$\therefore \quad 12 - 2x = x$

$\therefore \quad x = 4$

Sign diagram of $\dfrac{dI}{dx}$ is:

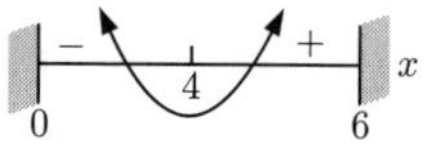

$\therefore$ the darkest point occurs 4 m from the 40 cp lamp.

15 **a** AB $= x$ m

$\therefore$ BC $= (24 - x)$ m $\qquad \therefore \quad D(x) = \sqrt{x^2 + (24-x)^2}$ {Pythagoras}

b
$$[D(x)]^2 = x^2 + (24-x)^2$$
$$= x^2 + 576 - 48x + x^2$$
$$= 2x^2 - 48x + 576$$
$$\therefore \quad \frac{d[D(x)]^2}{dx} = 4x - 48$$
$$\therefore \quad \frac{d[D(x)]^2}{dx} = 0 \text{ when } x = 12$$

Sign diagram for $\dfrac{d[D(x)]^2}{dx}$:

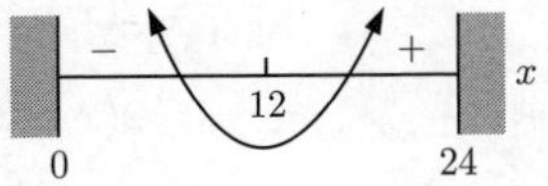

c When AB = BC = 12 m, $D(x)$ is a minimum, and the minimum $D(x) = 12\sqrt{2}$ m ≈ 17.0 m.

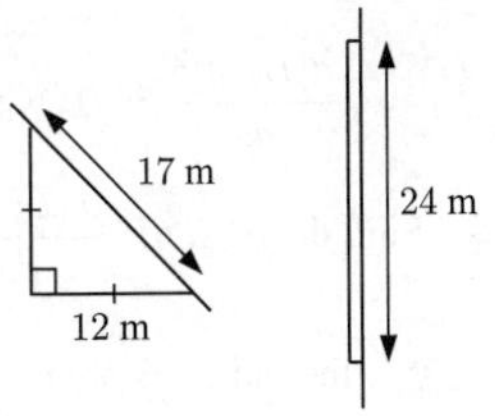

$D(x)$ is a maximum when either $x = 0$ or $x = 24$, when the pen ceases to exist and $D(x) = 24$ m.

16 **a** Consider each boat's position t hours after 1.00 pm.

PA $= 12t$ and QB $= 8t$

$\therefore$ PB $= 100 - 8t$

Using the cosine rule in ΔPAB

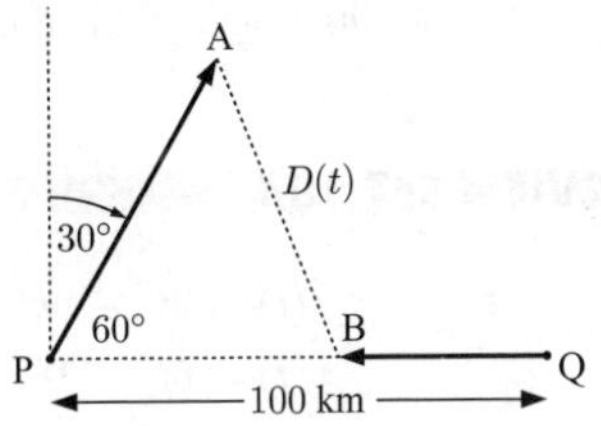

$$D(t)^2 = \text{AP}^2 + \text{BP}^2 - 2\text{AP} \times \text{BP} \cos 60^o$$
$$= (12t)^2 + (100-8t)^2 - 2(12t)(100-8t)\tfrac{1}{2}$$
$$= 144t^2 + (100-8t)^2 - 12t(100-8t)$$
$$= 144t^2 + 10\,000 - 1600t + 64t^2 - 1200t + 96t^2$$
$$= 304t^2 - 2800t + 10\,000$$
$$\therefore \quad D(t) = \sqrt{304t^2 - 2800t + 10\,000}$$

b Now $\dfrac{d[D(t)]^2}{dt} = 608t - 2800$

$$\therefore \quad \frac{d[D(t)]^2}{dt} = 0 \text{ when } t = \frac{2800}{608} \approx 4.605\,26$$

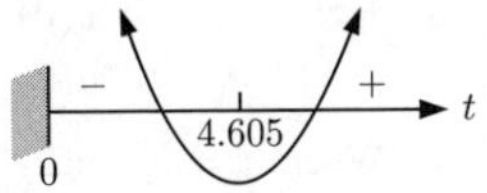

$\therefore$ $D(t)$ is a minimum when $t \approx 4.60526$ hours after 1.00 pm

and $[D(t)]^2_{\text{min}} \approx 304\,(4.6053)^2 - 2800\,(4.6053) + 10\,000$

$\therefore$ $[D(t)]^2_{\text{min}} \approx 3550$ km^2

c The ships are closest when $t = 4.605\,26$ hours which occurs when the time is 4 hours 36 minutes after 1.00 pm.
So, the ships are closest at approximately 5.36 pm.

17 **a** Δs PAB and PRQ are similar.

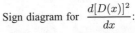

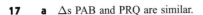

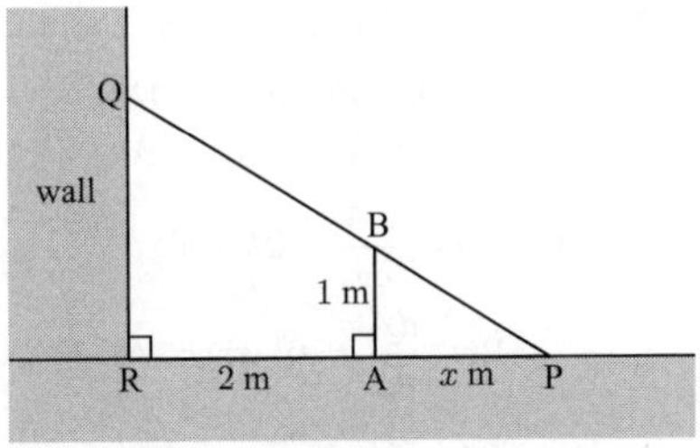

$$\therefore \quad \frac{\text{PA}}{\text{PR}} = \frac{\text{AB}}{\text{RQ}}$$
$$\therefore \quad \frac{x}{x+2} = \frac{1}{\text{QR}} \quad \text{and} \quad \therefore \quad \text{QR} = \frac{x+2}{x}$$

b Now $[L(x)]^2 = \text{RP}^2 + \text{QR}^2$ {Pythagoras}
$$= (x+2)^2 + \left(\frac{x+2}{x}\right)^2$$
$$= (x+2)^2 \times 1 + (x+2)^2 \times \frac{1}{x^2}$$
$$\therefore \quad [L(x)]^2 = (x+2)^2\left(1 + \frac{1}{x^2}\right) \quad \text{as required}$$

c $$[L(x)]^2 = (x+2)^2\left(1+x^{-2}\right)$$

$$\therefore \quad \frac{d[L(x)]^2}{dx} = 2(x+2)\left(1+x^{-2}\right) + (x+2)^2\left(-2x^{-3}\right) \quad \{\text{product rule}\}$$
$$= 2(x+2)\left(1+x^{-2}-(x+2)x^{-3}\right)$$
$$= 2(x+2)\left(1+x^{-2}-x^{-2}-2x^{-3}\right)$$
$$= 2(x+2)\left(1-\frac{2}{x^3}\right)$$
$$= \frac{2(x+2)(x^3-2)}{x^3}$$

$$\therefore \quad \frac{d[L(x)]^2}{dx} = 0 \text{ when } x = \sqrt[3]{2} \qquad \{\text{as } x>0 \text{ and } L(x)>0\}$$

d Sign diagram of $\dfrac{d\,[L(x)]^2}{dx}$ is:

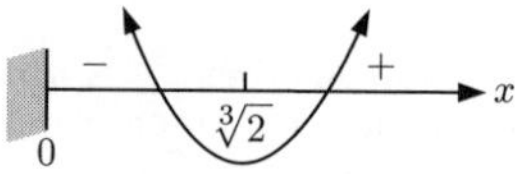

$\therefore$ the ladder is shortest when $x = \sqrt[3]{2}$ m and

its length at this time is $L = \sqrt{(x+2)^2\left(1+\dfrac{1}{x^2}\right)} = \sqrt{(\sqrt[3]{2}+2)^2\left(1+\dfrac{1}{2^{\frac{2}{3}}}\right)} \approx 4.16$ m or 416 cm

REVIEW SET 18A

1 **a** $s(t) = 2t^3 - 9t^2 + 12t - 5$ cm, $t \geqslant 0$

$v(t) = 6t^2 - 18t + 12$
$= 6(t^2 - 3t + 2)$
$= 6(t-2)(t-1)$ cm s^{-1}

and $a(t) = 12t - 18$
$= 6(2t-3)$ cm s^{-2}

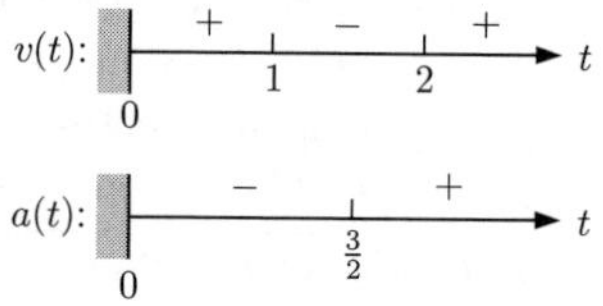

b When $t = 0$, $s(0) = -5$ cm
$v(0) = 12$ cm s^{-1}
$a(0) = -18$ cm s^{-2}

Initially, the particle is 5 cm to the left of O, moving at 12 cm s^{-1} towards the origin and decreasing in speed.

c When $t = 2$, $s(2) = -1$ cm
$v(2) = 0$ cm s^{-1}
$a(2) = 6$ cm s^{-2}

When $t = 2$, the particle is 1 cm to the left of O, instantaneously at rest and increasing in speed towards O.

d The particle changes direction when $t = 1$ and $t = 2$, at $s(1) = 0$ cm, $s(2) = -1$ cm.

e

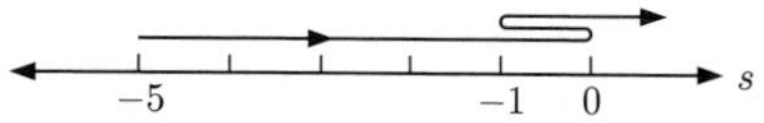

f The speed is increasing when $1 \leqslant t \leqslant \frac{3}{2}$ and $t \geqslant 2$
$\{v(t)$ and $a(t)$ have the same sign$\}$

2 **a** Now if OD $= x$, the coordinates of C are $(x, k - x^2)$.

$\therefore$ the area of ABCD $= 2x \times (k - x^2)$
$\therefore \quad A = 2kx - 2x^3, \quad x > 0$

b Now $\dfrac{dA}{dx} = 2k - 6x^2$

But $\dfrac{dA}{dx} = 0$ when AD $= 2\sqrt{3}$, and this occurs when $x = \sqrt{3}$

$\therefore \quad 2k - 6(\sqrt{3})^2 = 0$
$\therefore \quad 2k - 18 = 0$
$\therefore \quad 2k = 18$
$\therefore \quad k = 9$

Check: $\dfrac{dA}{dx} = 18 - 6x^2$
$= 6(3 - x^2)$
$= 6(\sqrt{3}+x)(\sqrt{3}-x)$

$\dfrac{dA}{dx}$ has sign diagram:

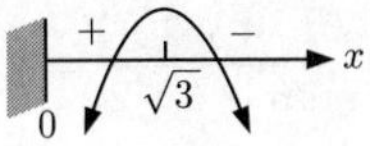

$\therefore$ the maximum occurs when $x = \sqrt{3}$ and AD $= 2\sqrt{3}$.

3 Suppose the sheet is bent x cm from each end. To maximise the water carried we need to maximise the area of cross-section.

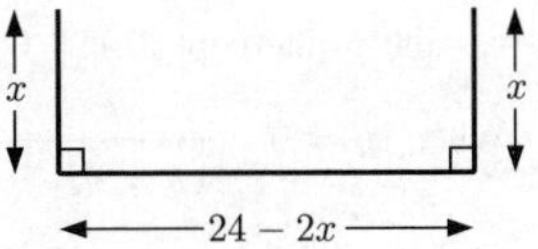

$$A = x(24 - 2x), \quad 0 \leqslant x \leqslant 12$$
$$= 24x - 2x^2$$
$$\therefore \frac{dA}{dx} = 24 - 4x$$

So, $\dfrac{dA}{dx} = 0$ when $x = 6$, and $\dfrac{dA}{dx}$ has sign diagram:

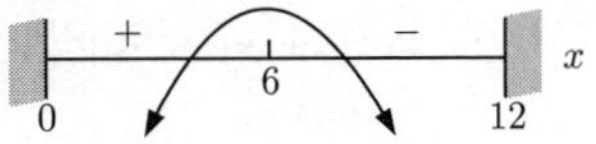

The maximum water is held when $x = 6$ cm

$\therefore$ the bends must be made 6 cm from each end.

4 **a** The vertical asymptote is $x + 3 = 0$ or $x = -3$.

b When $y = 0$, $\dfrac{3x-2}{x+3} = 0$

$$\therefore 3x - 2 = 0$$
$$\therefore x = \tfrac{2}{3}$$

$\therefore$ the x-intercept is $\frac{2}{3}$.

When $x = 0$, $f(0) = \frac{-2}{3}$

$\therefore$ the y-intercept is $-\frac{2}{3}$.

c $$f'(x) = \frac{3(x+3) - (3x-2)(1)}{(x+3)^2}$$
$$= \frac{3x + 9 - 3x + 2}{(x+3)^2}$$
$$= \frac{11}{(x+3)^2}$$

$f'(x)$ has sign diagram:

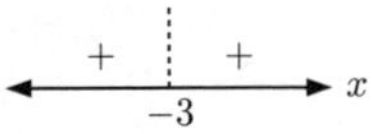

d $f'(x) \neq 0$ for any x, so $f(x)$ has no stationary points.

5 At $x = A$, $f'(x) = 0$ and $f''(x) = 0$

$\therefore$ $f(x)$ has a stationary inflection point at $x = A$.

At $x = B$, $f''(x) = 0$ but $f'(x) \neq 0$

$\therefore$ $f(x)$ has a non-stationary inflection point at $x = B$.

$f'(x)$ is above the x-axis for $x \leqslant C$, and below the x-axis for $x \geqslant C$

$\therefore$ $f(x)$ is increasing for $x \leqslant C$ and decreasing for $x \geqslant C$, so $f(x)$ has a local maximum at $x = C$.

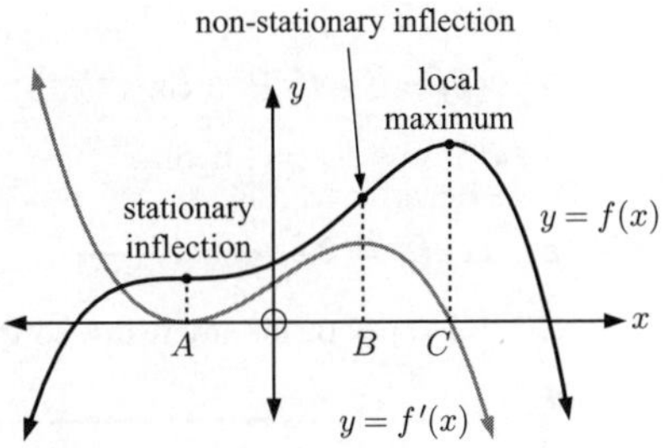

6 **a** $$f(x) = \frac{x-2}{x^2+x-2} = \frac{x-2}{(x+2)(x-1)}$$

$\therefore$ the vertical asymptotes are $x + 2 = 0$ and $x - 1 = 0$, or $x = -2$ and $x = 1$.

Since $f(x) = \dfrac{x-2}{x^2+x-2} = \dfrac{\frac{1}{x} - \frac{2}{x^2}}{1 + \frac{1}{x} - \frac{2}{x^2}}$, the horizontal asymptote is $y = 0$

{as $x \to \pm\infty$, $f(x) \to 0$}

b $$f'(x) = \frac{1(x^2+x-2) - (x-2)(2x+1)}{(x^2+x-2)^2}$$
$$= \frac{x^2 + x - 2 - (2x^2 - 3x - 2)}{(x+2)^2(x-1)^2}$$
$$= \frac{-x^2 + 4x}{(x+2)^2(x-1)^2}$$
$$= \frac{-x(x-4)}{(x+2)^2(x-1)^2}$$

which has sign diagram:

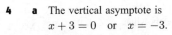

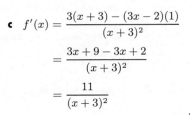

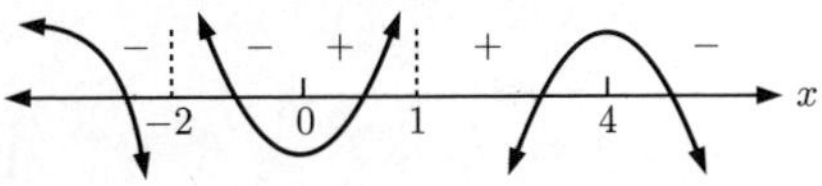

Now $f(0) = \frac{-2}{-2} = 1$ and $f(4) = \dfrac{4-2}{16+4-2} = \frac{1}{9}$

$\therefore$ $f(x)$ has a local minimum at $(0, 1)$ and a local maximum at $(4, \frac{1}{9})$.

c When $x = 0$, $f(0) = 1$

$\therefore$ the y-intercept is 1.

When $y = 0$, $\dfrac{x-2}{(x+2)(x-1)} = 0$

$\therefore$ $x - 2 = 0$

$\therefore$ $x = 2$

$\therefore$ the x-intercept is 2.

d

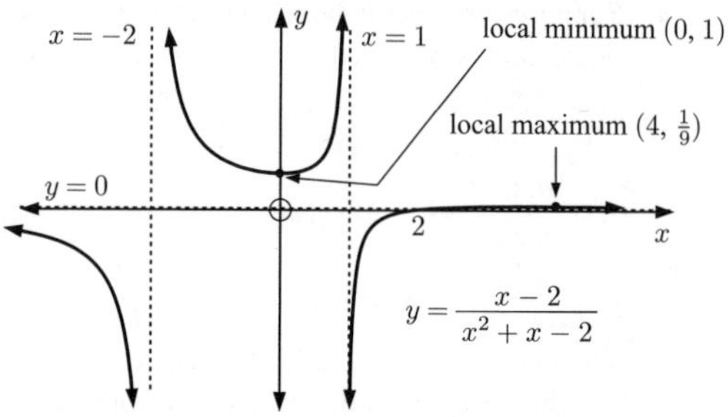

e $\dfrac{x-2}{x^2+x-2} = p$ has two real distinct solutions if the horizontal line $y = p$ meets the graph of $y = \dfrac{x-2}{x^2+x-2}$ in two places. From the graph in **d**, this occurs when $p < 0$, $0 < p < \frac{1}{9}$, and $p > 1$.

7 **a** $s(t) = 2t - \dfrac{4}{t} = 2t - 4t^{-1}$

$\therefore$ $v(t) = 2 + 4t^{-2}$

$= 2 + \dfrac{4}{t^2}$

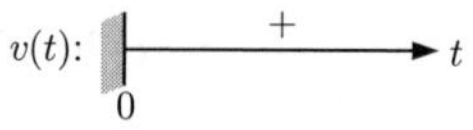

$\therefore$ $a(t) = -8t^{-3}$

$= -\dfrac{8}{t^3}$

b $s(1) = 2(1) - \frac{4}{1} = -2$ cm

$v(1) = 2 + \frac{4}{1^2} = 6$ cm s^{-1}

$a(1) = -\frac{8}{1^3} = -8$ cm s^{-2}

$\therefore$ the particle is 2 cm left of the origin and is moving to the right with velocity 6 cm s^{-1} and slowing down, its acceleration being 8 cm s^{-2} to the left.

c $v(t) = 2 + \dfrac{4}{t^2} = \dfrac{2t^2+4}{t^2}$

$\therefore$ $v(t) \neq 0$ for any real t, so the particle never changes direction.

d

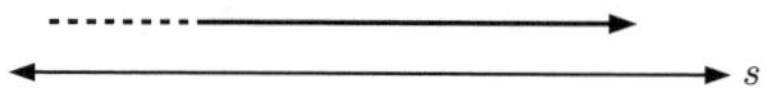

e **i** The velocity is never increasing {acceleration is negative for all $t > 0$}.

ii The speed is never increasing, as $v(t)$ and $a(t)$ have different signs for all $t > 0$.

8 When the box is manufactured its base is $(2k - 2x)$ by $(k - 2x)$ and its height is x cm.

$\therefore$ $V = x(2k - 2x)(k - 2x)$

$\therefore$ $V = x(2k^2 - 4kx - 2xk + 4x^2)$

$= 2k^2x - 6kx^2 + 4x^3$

$\therefore$ $\dfrac{dV}{dx} = 2k^2 - 12kx + 12x^2$

$= 2(6x^2 - 6kx + k^2)$

So, $\dfrac{dV}{dx} = 0$ when $x = \dfrac{6k \pm \sqrt{36k^2 - 4(6)k^2}}{12}$

$= \dfrac{6k \pm k\sqrt{12}}{12}$

$= \dfrac{k}{2} \pm \dfrac{k}{\sqrt{12}}$

$= \dfrac{k}{2} - \dfrac{k}{2\sqrt{3}}$ {as $x \leqslant \dfrac{k}{2}$}

$= \dfrac{k}{2}\left(1 - \frac{1}{\sqrt{3}}\right)$

The sign diagram of $\dfrac{dV}{dx}$ is:

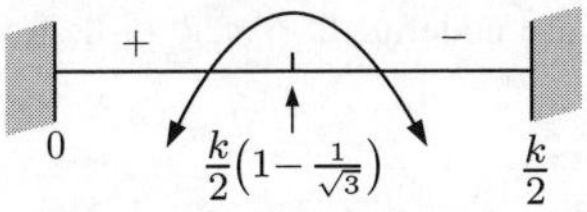

$\therefore$ the maximum capacity occurs when $x = \dfrac{k}{2}\left(1 - \frac{1}{\sqrt{3}}\right)$.

REVIEW SET 18B

1 **a** AC $= 2x$ m

Now ABC is an isosceles triangle.

$\therefore$ XC $= x$

But $\text{BC}^2 = \text{BX}^2 + \text{XC}^2$ {Pythagoras}

$\therefore$ $2500 = \text{BX}^2 + x^2$

$\therefore$ $\text{BX} = \sqrt{2500 - x^2}$

$\therefore$ $A(x) = \frac{1}{2}(2x)\sqrt{2500 - x^2} = x\sqrt{2500 - x^2}$

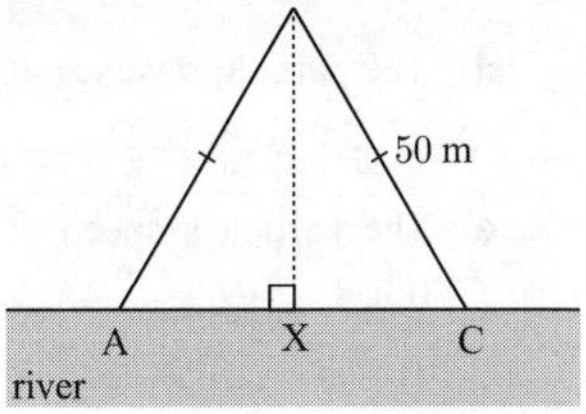

b Now $[A(x)]^2 = x^2(2500 - x^2)$

$\therefore$ $A^2 = 2500x^2 - x^4$

$\therefore$ $\dfrac{d(A^2)}{dx} = 5000x - 4x^3$

$= 4x(1250 - x^2)$

$= 4x(\sqrt{1250} + x)(\sqrt{1250} - x)$

Sign diagram for $\dfrac{d\,[A(x)]^2}{dx}$ is:

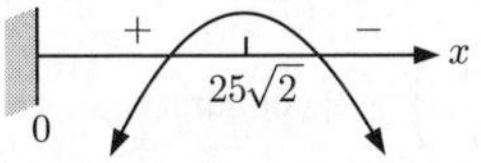

$\therefore$ the maximum area occurs when $x = 25\sqrt{2}$ m ≈ 35.4 m

The corresponding maximum area $= \sqrt{1250} \times \sqrt{1250} = 1250$ m^2.

2 **a** $f(x) = x^3 + ax, \quad a < 0$

$\therefore$ $f'(x) = 3x^2 + a$

$f(x)$ has a turning point at $x = \sqrt{2}$, so $f'(\sqrt{2}) = 0$

$\therefore$ $3\left(\sqrt{2}\right)^2 + a = 0$

$\therefore$ $a = -6$

b $f'(x) = 3x^2 - 6 = 3(x^2 - 2) = 3(x + \sqrt{2})(x - \sqrt{2})$

$\therefore$ $f'(x)$ has sign diagram:

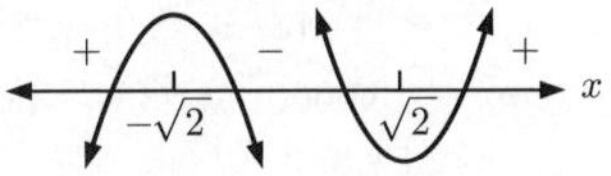

$\therefore$ $f(x)$ has a local maximum at $\left(-\sqrt{2}, (-\sqrt{2})^3 - 6(-\sqrt{2})\right)$ or $\left(-\sqrt{2}, 4\sqrt{2}\right)$, and a local minimum at $\left(\sqrt{2}, \left(\sqrt{2}\right)^3 - 6\sqrt{2}\right)$ or $\left(\sqrt{2}, -4\sqrt{2}\right)$.

c

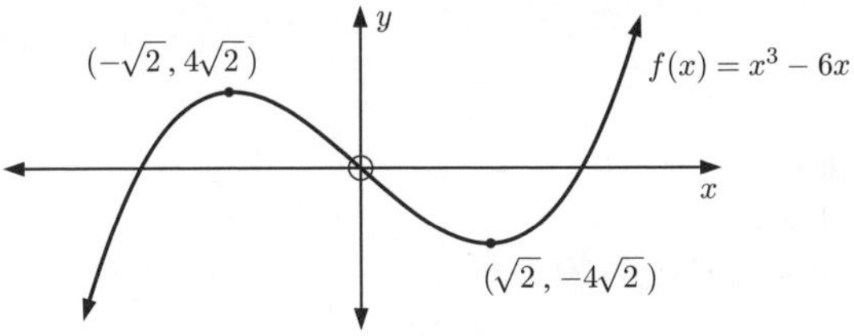

3 **a** $x(t) = 3t - \sqrt{t} = 3t - t^{\frac{1}{2}}$ cm, $t \geqslant 0$

$\therefore$ $v(t) = 3 - \frac{1}{2}t^{-\frac{1}{2}} = 3 - \dfrac{1}{2\sqrt{t}} = \dfrac{6\sqrt{t} - 1}{2\sqrt{t}}$

$\therefore$ $v(t) = 0$ when $\sqrt{t} = \frac{1}{6}$ or $t = \frac{1}{36}$

and $a(t) = \frac{1}{4}t^{-\frac{3}{2}} = \dfrac{1}{4t\sqrt{t}}$ which is always positive

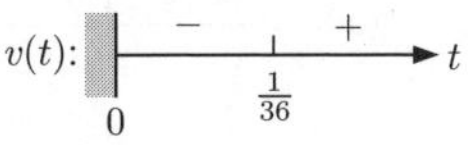

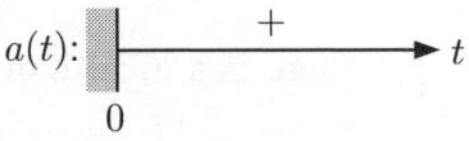

b The velocity and acceleration are undefined. As $t \to 0$ from above, $x \to 0$ cm
$v \to -\infty$ cm s^{-1}
$a \to \infty$ cm s^{-2}

$\therefore$ the particle is at O, moving left and slowing down.

c When $t = 9$, $x(9) = 24$ cm, $v(9) = \frac{17}{6}$ cm s^{-1}, $a(9) = \frac{1}{108}$ cm s^{-2}

the particle is 24 cm right of O, moving right at $\frac{17}{6}$ cm s^{-1} and increasing its speed.

d The particle reverses direction when $t = \frac{1}{36}$ seconds.

$x(\frac{1}{36}) = \frac{3}{36} - \frac{1}{6} = -\frac{3}{36} = -\frac{1}{12}$ $\quad\therefore$ it is $\frac{1}{12} \approx 0.0833$ cm to the left of O.

e The particle's speed decreases when $v(t)$ and $a(t)$ have different signs, which occurs when $0 < t \leqslant \frac{1}{36}$.

4 $C(v) = \dfrac{v^2}{30} + \dfrac{9000}{v}$ dollars per hour

a **i** For $t = 2$ hours at $v = 45$ km h^{-1},

$$\text{cost} = \left(\frac{45^2}{30} + \frac{9000}{45}\right) \times 2 \text{ dollars}$$
$$= \$535.00$$

ii For $t = 5$ hours at $v = 64$ km h^{-1},

$$\text{cost} = \left(\frac{64^2}{30} + \frac{9000}{64}\right) \times 5 \text{ dollars}$$
$$\approx \$1385.79$$
$$\approx \$1390$$

b $C'(v) = \dfrac{2v}{30} - 9000v^{-2} = \dfrac{v}{15} - \dfrac{9000}{v^2}$

i For $v = 50$ km h^{-1}

$$\therefore \quad C'(50) = \frac{50}{15} - \frac{9000}{50^2}$$
$$\approx -\$0.267 \text{ per km h}^{-1}$$

ii For $v = 66$ km h^{-1}

$$\therefore \quad C'(66) = \frac{66}{15} - \frac{9000}{66^2}$$
$$= \$2.33 \text{ per km h}^{-1}$$

c Now $C'(v) = \dfrac{v}{15} - \dfrac{9000}{v^2} = \dfrac{v^3 - 135\,000}{15v^2}$

$\therefore$ $C'(v) = 0$ when $v^3 = 135\,000$

$\therefore$ $v \approx 51.3$

$C'(v)$:

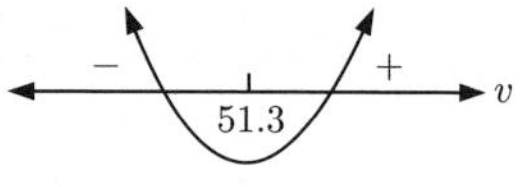

$\therefore$ the minimum cost occurs when $v \approx 51.3$ km h^{-1}

5 $f(x) = \dfrac{x^2 - 1}{x^2 + 1}$

a $f(x)$ cuts the x-axis when $x^2 - 1 = 0$ $\quad\therefore$ $x = \pm 1$

So, the x-intercepts are 1 and -1.

$f(0) = \frac{-1}{1} = -1$, so the y-intercept is -1.

b As $x^2 \geqslant 0$, $x^2 + 1$ can never be 0.

$\therefore$ there are no vertical asymptotes.

c $f'(x) = \dfrac{2x(x^2+1) - (x^2-1)2x}{(x^2+1)^2} = \dfrac{4x}{(x^2+1)^2}$

Sign diagram of $f'(x)$ is:

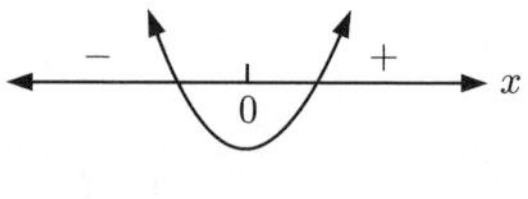

$\therefore$ there is a local minimum at $(0, -1)$.

d $f''(x) = \dfrac{4(x^2+1)^2 - 4x \times 2(x^2+1)(2x)}{(x^2+1)^4}$

$= \dfrac{4(x^2+1) - 16x^2}{(x^2+1)^3}$

$= \dfrac{4 - 12x^2}{(x^2+1)^3}$

$= \dfrac{4(1-3x^2)}{(x^2+1)^3}$

$\therefore \quad f''(x) = 0$ when $3x^2 = 1$

$\therefore \quad x = \pm\sqrt{\frac{1}{3}}$

Sign diagram of $f''(x)$ is:

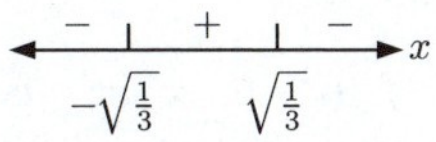

So, $f(x)$ has non-stationary inflections at $x = \pm\sqrt{\frac{1}{3}}$.

e

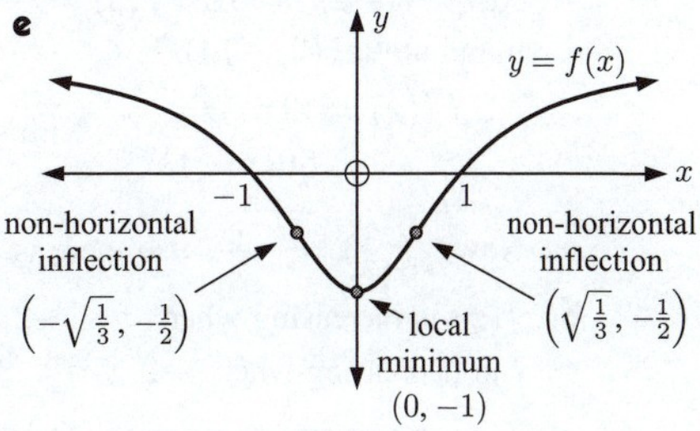

6 $f(x)$ has a turning point at $x = 0$
$\therefore \quad f'(0) = 0$
$f(x)$ is increasing for $x \geqslant 0$, except at the asymptote, so $f'(x)$ is positive for $x \geqslant 0$.
$f(x)$ is decreasing for $x \leqslant 0$, except at the asymptote, so $f'(x)$ is negative for $x \leqslant 0$.
As $x \to \pm\infty$, $f(x)$ becomes closer to horizontal so $f'(x) \to 0$.

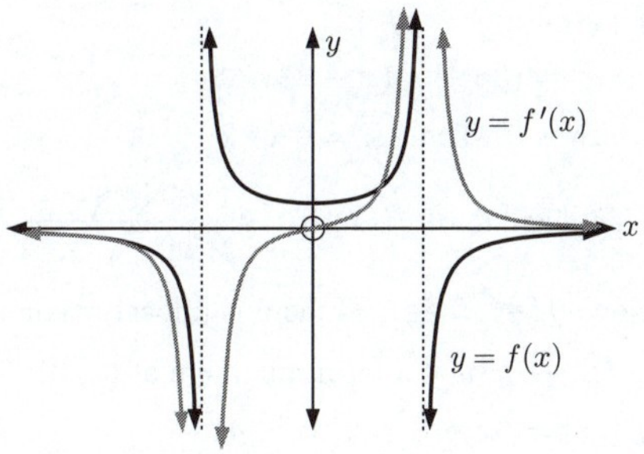

7 **a** $P = 200$ m
But $P = 2x + 2y + \pi x$
$\therefore \quad 200 = 2x + 2y + \pi x$
$\therefore \quad 2y = 200 - 2x - \pi x$
$\therefore \quad y = 100 - x - \frac{\pi}{2}x$

b Area of lawn $= 2x \times y + \frac{1}{2}\pi x^2$
$= 2x\left[100 - x - \frac{\pi}{2}x\right] + \frac{1}{2}\pi x^2$
$= 200x - 2x^2 - \pi x^2 + \frac{1}{2}\pi x^2$
$= 200x - 2x^2 - \frac{1}{2}\pi x^2$
$\therefore \quad A = 200x - \left(2 + \frac{\pi}{2}\right)x^2$ m^2

c $\dfrac{dA}{dx} = 200 - 2\left(2 + \frac{\pi}{2}\right)x = 200 - (4+\pi)x$

$\therefore \quad \dfrac{dA}{dx} = 0$ when $(4+\pi)x = 200 \qquad \therefore \quad x = \dfrac{200}{4+\pi}$

and the sign diagram for $\dfrac{dA}{dx}$ is:

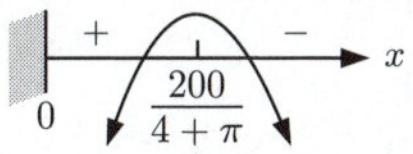

$\therefore$ the maximum area occurs when $x = \dfrac{200}{4+\pi} \approx 28.0$ m

and $y = 100 - x - \frac{\pi}{2}x \approx 28.0$ m

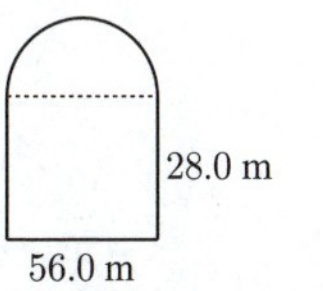

REVIEW SET 18C

1 **a** $f(x) = 2x^3 - 3x^2 - 36x + 7$
$\therefore \quad f'(x) = 6x^2 - 6x - 36$
$= 6(x^2 - x - 6)$
$= 6(x-3)(x+2)$

with sign diagram:

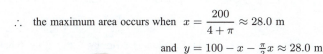

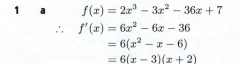

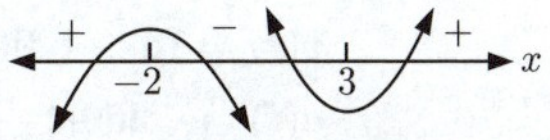

Now $f(-2) = 51$, $f(3) = -74$, so there is a local maximum at $(-2,\ 51)$, and a local minimum at $(3, -74)$.

$$f''(x) = 12x - 6 = 6(2x - 1)$$ with sign diagram:

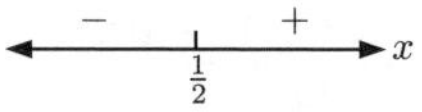

Now $f(\frac{1}{2}) = -\frac{23}{2}$, so there is a non-horizontal inflection at $(\frac{1}{2}, -\frac{23}{2})$.

b $f(x)$ is increasing when $x \leqslant -2$ or $x \geqslant 3$, and decreasing when $-2 \leqslant x \leqslant 3$.

c $f(x)$ is concave up when $x \geqslant \frac{1}{2}$, and concave down when $x \leqslant \frac{1}{2}$.

d

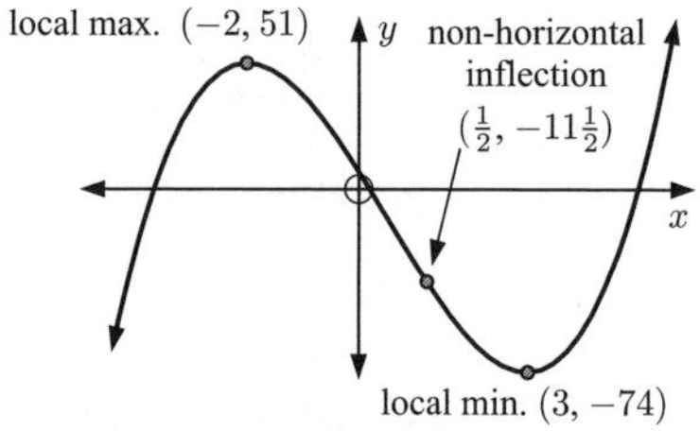

2 $f(x) = x^3 - 4x^2 + 4x = x(x^2 - 4x + 4) = x(x-2)^2$

a $f(0) = 0$, so the y-intercept is 0.
$f(x)$ cuts the x-axis when $y = 0$ $\therefore\ x(x-2)^2 = 0$ $\therefore\ x = 0$ or 2

$\therefore$ the x-intercepts are 0 and 2.

b $f'(x) = 3x^2 - 8x + 4 = (3x - 2)(x - 2)$

which is 0 when $x = \frac{2}{3}$ or 2

Sign diagram of $f'(x)$ is:

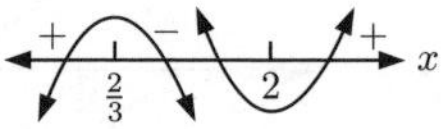

Now $f(\frac{2}{3}) = \frac{32}{27}$, so there is a local maximum at $(\frac{2}{3}, \frac{32}{27})$, and a local minimum at (2, 0).

$f''(x) = 6x - 8 = 2(3x - 4)$

Sign diagram of $f''(x)$ is:

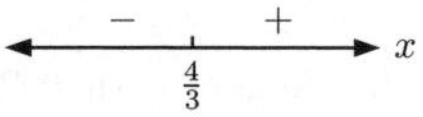

Now $f(\frac{4}{3}) = \frac{16}{27}$, so there is a non-horizontal inflection at $(\frac{4}{3}, \frac{16}{27})$.

c

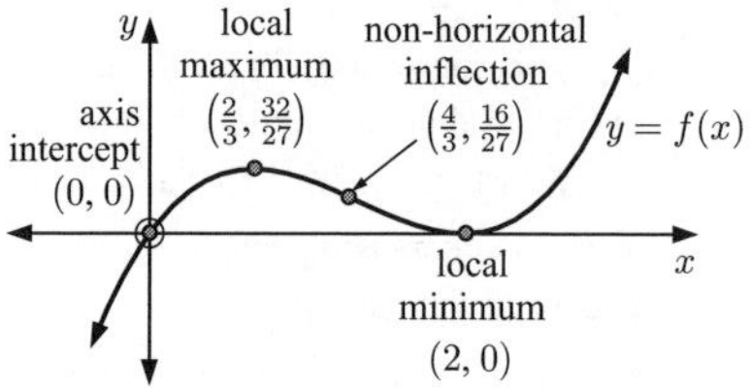

3 **a** Volume $= lbd$

$\therefore\ x^2 y = 1$

$\therefore\ y = \dfrac{1}{x^2}$, $x > 0$

b area $= x^2 + 4xy$

$\therefore$ cost $= (x^2 + 4xy) \times 2$

$\therefore\ C = 2x^2 + 8xy = 2x^2 + \dfrac{8}{x}$ dollars {using **a**}

c $\dfrac{dC}{dx} = 4x - 8x^{-2} = 4x - \dfrac{8}{x^2} = \dfrac{4(x^3 - 2)}{x^2}$

So, $\dfrac{dC}{dx} = 0$ when $x = \sqrt[3]{2}$ m

$\dfrac{dC}{dx}$ has sign diagram:

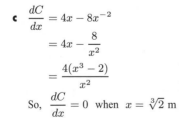

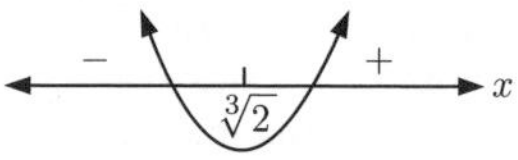

The minimum cost is when $x = \sqrt[3]{2} \approx 1.26$ m

$\therefore\ y = \dfrac{1}{x^2} \approx 0.630$

and the box is 1.26 m by 1.26 m by 0.630 m.

4 **a** $s(t) = 15t - \dfrac{60}{(t-1)^2}$ cm, $t \geqslant 0$

$= 15t - 60(t-1)^{-2}$ cm

$\therefore\ v(t) = 15 + 120(t-1)^{-3}$ cm s^{-1}

$\therefore\ a(t) = -360(t-1)^{-4}$ cm s^{-2}

b When $t = 3$, $s(t) = 30$ cm

$v(t) = 30$ cm s^{-1}

$a(t) = -22.5$ cm s^{-2}

The particle is 30 cm right of O, travelling right at 30 cm s^{-1}, and is slowing down at 22.5 cm s^{-2}.

c $v(t) = 15 + \dfrac{120}{(t-1)^3}$ cm s^{-1}

$v(t) = 0$ when $15 + \dfrac{120}{(t-1)^3} = 0$

$\therefore\ 15(t-1)^3 + 120 = 0$

$\therefore\ (t-1)^3 = -8$

$\therefore\ t = -1$

$a(t) = -360(t-1)^{-4} = \dfrac{-360}{(t-1)^4}$ cm s^{-2}

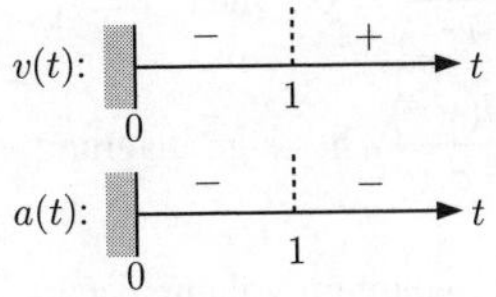

where $(t-1)^4$ is always positive. $\therefore\ a(t) < 0$ for all $t > 0$

The speed increases for $0 \leqslant t < 1$ when $v(t)$ and $a(t)$ have the same sign.

5 a The tree was $H(0) = 6\left(1 - \frac{2}{3}\right) = 2$ metres tall when first planted.

b $t = 3$: $H(3) = 6\left(1 - \frac{2}{3+3}\right) = 4$ metres

$t = 6$: $H(6) = 6\left(1 - \frac{2}{6+3}\right) = 4\frac{2}{3}$ metres

$t = 9$: $H(9) = 6\left(1 - \frac{2}{9+3}\right) = 5$ metres

c $H(t) = 6\left(1 - \dfrac{2}{t+3}\right)$

$= 6 - 12(t+3)^{-1}$

$\therefore\ H'(t) = 12(t+3)^{-2}$

$= \dfrac{12}{(t+3)^2}$

So, $t = 0$: $H'(0) = \frac{12}{3^2} = \frac{4}{3}$ m year^{-1}

$t = 3$: $H'(3) = \frac{12}{6^2} = \frac{1}{3}$ m year^{-1}

$t = 6$: $H'(6) = \frac{12}{9^2} = \frac{4}{27}$ m year^{-1}

$t = 9$: $H'(9) = \frac{12}{12^2} = \frac{1}{12}$ m year^{-1}

d $H'(t) = \dfrac{12}{(t+3)^2}$, and $(t+3)^2 > 0$ for all $t \geqslant 0$

$\therefore\ \dfrac{12}{(t+3)^2} > 0$

$\therefore\ H'(t) > 0$ for all $t \geqslant 0$

This means that the height of the tree is always increasing.

e

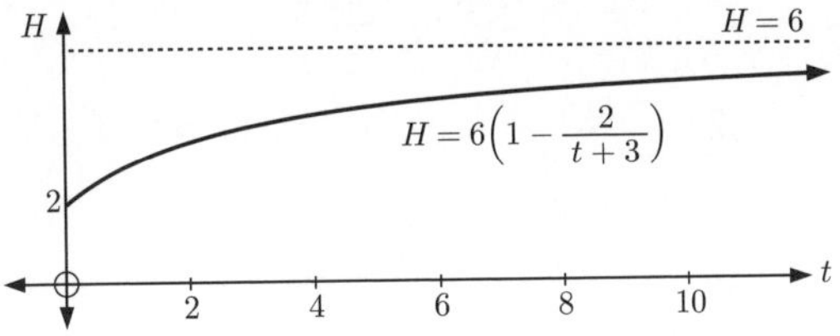

6 a

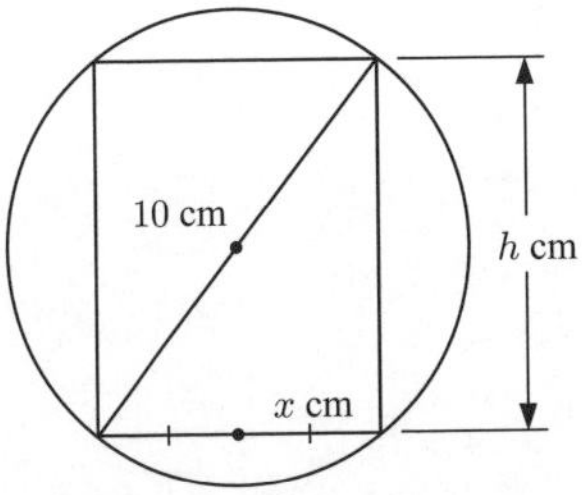

Let the height of the cylinder be h cm.

$\therefore\ (2x)^2 + h^2 = 10^2$ {Pythagoras}

$\therefore\ h = \sqrt{100 - 4x^2}$

$\therefore\ V(x) = $ area of base $\times$ height

$= \pi x^2 \times \sqrt{100 - 4x^2}$

So, $V(x) = \pi x^2 \sqrt{100 - 4x^2}$ cm^3

b Now $V^2 = \pi^2 x^4(100 - 4x^2)$
$= \pi^2(100x^4 - 4x^6)$

$\therefore \dfrac{d(V^2)}{dx} = \pi^2(400x^3 - 24x^5)$
$= 8\pi^2 x^3(50 - 3x^2)$
$= 8\pi^2 x^3(\sqrt{50} + \sqrt{3}x)(\sqrt{50} - \sqrt{3}x)$

$\therefore \dfrac{d(V^2)}{dx} = 0$ when $x = \sqrt{\frac{50}{3}}$ {as $x > 0$}

and $\dfrac{d(V^2)}{dx}$ has sign diagram:

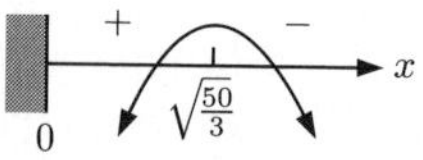

$\therefore$ the maximum volume occurs when $x = \sqrt{\frac{50}{3}} \approx 4.08$

$\therefore$ radius ≈ 4.08 cm, height $= \sqrt{100 - 4\left(\frac{50}{3}\right)} \approx 5.77$ cm

7 At $x = B$, $f''(x) = 0$ but $f'(x) \neq 0$

$\therefore$ $f(x)$ has a non-stationary inflection point at $x = B$.

$f'(x)$ is above the x-axis for $x \leqslant A$ and $x \geqslant C$, and below the x-axis for $A \leqslant x \leqslant C$

$\therefore$ $f(x)$ is increasing for $x \leqslant A$, decreasing for $A \leqslant x \leqslant C$, then increasing for $x \geqslant C$

$\therefore$ $f(x)$ has a local maximum at $x = A$ and a local minimum at $x = C$.

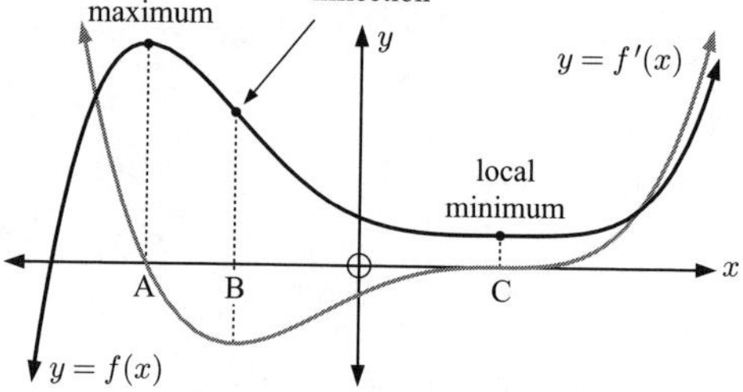

Chapter 19
DERIVATIVES OF EXPONENTIAL AND LOGARITHMIC FUNCTIONS

EXERCISE 19A

1 **a** $f(x) = e^{4x}$
$\therefore\ f'(x) = 4e^{4x}$

b $f(x) = e^x + 3$
$\therefore\ f'(x) = e^x + 0$
$= e^x$

c $f(x) = \exp(-2x)$
$= e^{-2x}$
$\therefore\ f'(x) = -2e^{-2x}$

d $f(x) = e^{\frac{x}{2}}$
$\therefore\ f'(x) = \frac{1}{2}e^{\frac{x}{2}}$

e $f(x) = 2e^{-\frac{x}{2}}$
$\therefore\ f'(x) = 2e^{-\frac{x}{2}}\left(-\frac{1}{2}\right)$
$= -e^{-\frac{x}{2}}$

f $f(x) = 1 - 2e^{-x}$
$\therefore\ f'(x) = 0 - 2e^{-x}(-1)$
$= 2e^{-x}$

g $f(x) = 4e^{\frac{x}{2}} - 3e^{-x}$
$\therefore\ f'(x) = 4e^{\frac{x}{2}}\left(\frac{1}{2}\right) - 3e^{-x}(-1)$
$= 2e^{\frac{x}{2}} + 3e^{-x}$

h $f(x) = \dfrac{e^x + e^{-x}}{2} = \frac{1}{2}(e^x + e^{-x})$
$\therefore\ f'(x) = \frac{1}{2}(e^x + e^{-x}(-1))$
$= \frac{1}{2}(e^x - e^{-x})$

i $f(x) = e^{-x^2}$
$\therefore\ f'(x) = e^{-x^2}(-2x)$
$= -2xe^{-x^2}$

j $f(x) = e^{\frac{1}{x}}$
$\therefore\ f'(x) = e^{\frac{1}{x}}\left(-\dfrac{1}{x^2}\right)$
$= -\dfrac{e^{\frac{1}{x}}}{x^2}$

k $f(x) = 10\left(1 + e^{2x}\right)$
$= 10 + 10e^{2x}$
$\therefore\ f'(x) = 0 + 10e^{2x}\,(2)$
$= 20e^{2x}$

l $f(x) = 20\left(1 - e^{-2x}\right)$
$= 20 - 20e^{-2x}$
$\therefore\ f'(x) = 0 - 20e^{-2x}\,(-2)$
$= 40e^{-2x}$

m $f(x) = e^{2x+1}$
$\therefore\ f'(x) = e^{2x+1}(2)$
$= 2e^{2x+1}$

n $f(x) = e^{\frac{x}{4}}$
$\therefore\ f'(x) = e^{\frac{x}{4}}\left(\frac{1}{4}\right)$
$= \frac{1}{4}e^{\frac{x}{4}}$

o $f(x) = e^{1-2x^2}$
$\therefore\ f'(x) = e^{1-2x^2}(-4x)$
$= -4xe^{1-2x^2}$

p $f(x) = e^{-0.02x}$
$\therefore\ f'(x) = e^{-0.02x} \times (-0.02)$
$= -0.02e^{-0.02x}$

2 **a** $f(x) = xe^x$
$\therefore\ f'(x) = 1e^x + xe^x$ {product rule}
$= e^x + xe^x$

b $f(x) = x^3e^{-x}$
$\therefore\ f'(x) = 3x^2e^{-x} + x^3(-e^{-x})$
{product rule}
$= 3x^2e^{-x} - x^3e^{-x}$

c $f(x) = \dfrac{e^x}{x}$
$\therefore\ f'(x) = \dfrac{e^x x - e^x\,(1)}{x^2}$ {quotient rule}
$= \dfrac{xe^x - e^x}{x^2}$

d $f(x) = \dfrac{x}{e^x}$
$\therefore\ f'(x) = \dfrac{1e^x - xe^x}{(e^x)^2}$ {quotient rule}
$= \dfrac{e^x(1-x)}{(e^x)^2} = \dfrac{1-x}{e^x}$

e $f(x) = x^2e^{3x}$
$\therefore\ f'(x) = 2xe^{3x} + 3x^2e^{3x}$ {product rule}

f $f(x) = \dfrac{e^x}{\sqrt{x}}$
$\therefore\ f'(x) = \dfrac{e^x\sqrt{x} - \dfrac{e^x}{2\sqrt{x}}}{(\sqrt{x})^2}$ {quotient rule}
$= \dfrac{xe^x - \frac{1}{2}e^x}{x\sqrt{x}}$

g $f(x) = \sqrt{x}e^{-x}$

$\therefore\ f'(x) = \frac{1}{2}x^{-\frac{1}{2}}e^{-x} - x^{\frac{1}{2}}e^{-x}$ {product rule}

$= \dfrac{1}{2\sqrt{x}}e^{-x} - \sqrt{x}e^{-x}$

h $f(x) = \dfrac{e^x + 2}{e^{-x} + 1}$

$\therefore\ f'(x) = \dfrac{e^x(e^{-x}+1) - (e^x+2)(-e^{-x})}{(e^{-x}+1)^2}$ {quotient rule}

$= \dfrac{1 + e^x + 1 + 2e^{-x}}{(e^{-x}+1)^2}$

$= \dfrac{e^x + 2 + 2e^{-x}}{(e^{-x}+1)^2}$

3 **a** $f(x) = (e^x + 2)^4$

$= u^4$ where $u = e^x + 2$

$\dfrac{dy}{dx} = \dfrac{dy}{du}\dfrac{du}{dx}$ {chain rule}

$= 4u^3(e^x)$

$\therefore\ f'(x) = 4(e^x+2)^3(e^x)$

$= 4e^x(e^x+2)^3$

b $f(x) = \dfrac{1}{1 - e^{-x}}$

$= u^{-1}$ where $u = 1 - e^{-x}$

$\dfrac{dy}{dx} = \dfrac{dy}{du}\dfrac{du}{dx}$ {chain rule}

$= -u^{-2}(e^{-x})$

$\therefore\ f'(x) = -\dfrac{e^{-x}}{(1-e^{-x})^2}$

c $f(x) = \sqrt{e^{2x} + 10}$

$= u^{\frac{1}{2}}$ where $u = e^{2x} + 10$

$\dfrac{dy}{dx} = \dfrac{dy}{du}\dfrac{du}{dx}$ {chain rule}

$= \frac{1}{2}u^{-\frac{1}{2}}(2e^{2x})$

$\therefore\ f'(x) = \dfrac{e^{2x}}{\sqrt{e^{2x}+10}}$

d $f(x) = \dfrac{1}{(1 - e^{3x})^2}$

$= u^{-2}$ where $u = 1 - e^{3x}$

$\dfrac{dy}{dx} = \dfrac{dy}{du}\dfrac{du}{dx}$ {chain rule}

$= -2u^{-3}(-3e^{3x}) = \dfrac{6e^{3x}}{u^3}$

$\therefore\ f'(x) = \dfrac{6e^{3x}}{(1-e^{3x})^3}$

e $f(x) = \dfrac{1}{\sqrt{1 - e^{-x}}}$

$= u^{-\frac{1}{2}}$ where $u = 1 - e^{-x}$

$\dfrac{dy}{dx} = \dfrac{dy}{du}\dfrac{du}{dx}$ {chain rule}

$= -\frac{1}{2}u^{-\frac{3}{2}}(e^{-x})$

$= \dfrac{-e^{-x}}{2u^{\frac{3}{2}}}$

$\therefore\ f'(x) = \dfrac{-e^{-x}}{2(1-e^{-x})^{\frac{3}{2}}}$

f $f(x) = x\sqrt{1 - 2e^{-x}}$

$= xu^{\frac{1}{2}}$ where $u = 1 - 2e^{-x}$

$\therefore\ f'(x) = 1u^{\frac{1}{2}} + x \times \frac{1}{2}u^{-\frac{1}{2}}\dfrac{du}{dx}$ {product rule and chain rule}

$= 1\sqrt{u} + x\,\frac{1}{2}u^{-\frac{1}{2}}2e^{-x}$

$= \dfrac{\sqrt{1-2e^{-x}}}{1} + \dfrac{xe^{-x}}{\sqrt{1-2e^{-x}}}$

$\therefore\ f'(x) = \dfrac{1 - 2e^{-x} + xe^{-x}}{\sqrt{1-2e^{-x}}}$

4 $y = Ae^{kx}$

a $\dfrac{dy}{dx} = Ae^{kx}(k)$

$= k(Ae^{kx})$

$= ky$

b $\dfrac{d^2y}{dx^2} = \dfrac{d}{dx}kAe^{kx}$ {from **a**}

$= k^2Ae^{kx}$

$= k^2y$

5 $y = 2e^{3x} + 5e^{4x}$ $\quad \therefore \quad \dfrac{dy}{dx} = 6e^{3x} + 20e^{4x}$ and $\dfrac{d^2y}{dx^2} = 18e^{3x} + 80e^{4x}$

$$\begin{aligned}\text{Now } \frac{d^2y}{dx^2} - 7\frac{dy}{dx} + 12y &= \left(18e^{3x} + 80e^{4x}\right) - 7\left(6e^{3x} + 20e^{4x}\right) + 12\left(2e^{3x} + 5e^{4x}\right) \\ &= 18e^{3x} + 80e^{4x} - 42e^{3x} - 140e^{4x} + 24e^{3x} + 60e^{4x} \\ &= e^{3x}\left[18 - 42 + 24\right] + e^{4x}\left[80 - 140 + 60\right] \\ &= e^{3x}(0) + e^{4x}(0) \\ &= 0\end{aligned}$$

$$\therefore \quad \frac{d^2y}{dx^2} - 7\frac{dy}{dx} + 12y = 0$$

6 $f(x) = e^{kx} + x \quad \therefore \quad f'(x) = ke^{kx} + 1$

Now $f'(0) = -8$, so $ke^0 + 1 = -8$

$\therefore \quad k \times 1 = -9$

$\therefore \quad k = -9$

7 **a**

$$\begin{aligned} y &= xe^{-x} \\ \therefore \quad \frac{dy}{dx} &= 1e^{-x} - xe^{-x} \qquad \text{\{product rule\}} \\ &= e^{-x}(1-x) \\ &= \frac{1-x}{e^x} \end{aligned}$$

which has sign diagram:

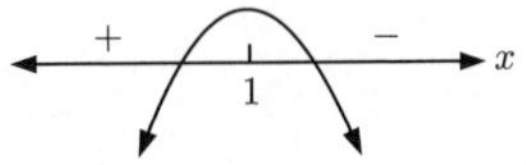

When $x = 1$, $y = 1e^{-1} = \dfrac{1}{e}$, so we have a local maximum at $\left(1, \dfrac{1}{e}\right)$.

b

$$\begin{aligned} y &= x^2e^x \\ \therefore \quad \frac{dy}{dx} &= 2xe^x + x^2e^x \qquad \text{\{product rule\}} \\ &= xe^x(2+x) \end{aligned}$$

which has sign diagram:

When $x = -2$, $y = 4e^{-2}$, and when $x = 0$, $y = 0$.

So, we have a local maximum at $\left(-2, \dfrac{4}{e^2}\right)$, and a local minimum at $(0, 0)$.

c

$$\begin{aligned} y &= \frac{e^x}{x} \\ \therefore \quad \frac{dy}{dx} &= \frac{e^x x - e^x(1)}{x^2} \qquad \text{\{quotient rule\}} \\ &= \frac{e^x(x-1)}{x^2} \end{aligned}$$

which has sign diagram:

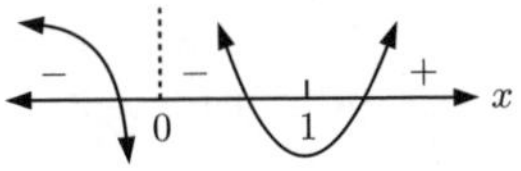

When $x = 1$, $y = \dfrac{e^1}{1} = e$, so we have a local minimum at $(1, e)$.

d

$$\begin{aligned} y &= e^{-x}(x+2) \\ \therefore \quad \frac{dy}{dx} &= -e^{-x}(x+2) + e^{-x} \qquad \text{\{product rule\}} \\ &= e^{-x}(-x-2+1) \\ &= e^{-x}(-x-1) \end{aligned}$$

which has sign diagram:

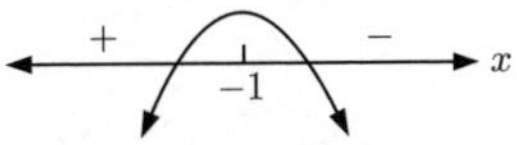

When $x = -1$, $y = e(-1+2) = e$, so we have a local maximum at $(-1, e)$.

EXERCISE 19B

1 **a** $\ln e^2 = 2\ln e = 2(1) = 2$

b $\ln\sqrt{e} = \ln e^{\frac{1}{2}} = \frac{1}{2}\ln e = \frac{1}{2}$

c $\ln\left(\frac{1}{e}\right) = \ln e^{-1} = -1\ln e = -1$

d $\ln\left(\frac{1}{\sqrt{e}}\right) = \ln e^{-\frac{1}{2}} = -\frac{1}{2}\ln e = -\frac{1}{2}$

e $e^{\ln 3} = 3$

f $e^{2\ln 3} = e^{\ln 3^2} = e^{\ln 9} = 9$

g $e^{-\ln 5} = e^{\ln 5^{-1}} = e^{\ln\frac{1}{5}} = \frac{1}{5}$

h $e^{-2\ln 2} = e^{\ln 2^{-2}} = e^{\ln\frac{1}{4}} = \frac{1}{4}$

2 **a** Let $2 = e^x$ $\therefore \ln 2 = x$ $\therefore 2 = e^{\ln 2}$

b Let $10 = e^x$ $\therefore \ln 10 = x$ $\therefore 10 = e^{\ln 10}$

c Let $a = e^x$ $\therefore \ln a = x$ $\therefore a = e^{\ln a}$

d Let $a^x = e^k$ $\therefore \ln a^x = k$ $\therefore x\ln a = k$ $\therefore a^x = e^{x\ln a}$

3 **a** $e^x = 2$ $\therefore \ln e^x = \ln 2$ $\therefore x = \ln 2$

b $e^x = -2$ has no solutions as $e^x > 0$ for all x

c $e^x = 0$ has no solutions as $e^x > 0$ for all x

d $e^{2x} = 2e^x$ $\therefore e^x(e^x - 2) = 0$ $\therefore e^x = 2$ {as $e^x > 0$} $\therefore x = \ln 2$

e $e^x = e^{-x}$ $\therefore x = -x$ $\therefore 2x = 0$ $\therefore x = 0$

f $e^{2x} - 5e^x + 6 = 0$ $\therefore (e^x - 3)(e^x - 2) = 0$ $\therefore e^x = 3$ or 2 $\therefore x = \ln 3$ or $\ln 2$

g $e^x + 2 = 3e^{-x}$ $\therefore e^{2x} + 2e^x = 3$ {$\times e^x$} $\therefore e^{2x} + 2e^x - 3 = 0$ $\therefore (e^x + 3)(e^x - 1) = 0$ $\therefore e^x = -3$ or 1 $\therefore e^x = 1$ {as $e^x > 0$} $\therefore x = \ln 1$ $\therefore x = 0$

h $1 + 12e^{-x} = e^x$ $\therefore e^x + 12 = e^{2x}$ {$\times e^x$} $\therefore e^{2x} - e^x - 12 = 0$ $\therefore (e^x - 4)(e^x + 3) = 0$ $\therefore e^x = 4$ or -3 $\therefore e^x = 4$ {as $e^x > 0$} $\therefore x = \ln 4$

i $e^x + e^{-x} = 3$ $\therefore e^{2x} + 1 = 3e^x$ {$\times e^x$} $\therefore e^{2x} - 3e^x + 1 = 0$

$\therefore e^x = \dfrac{3 \pm \sqrt{9-4}}{2}$

$\therefore e^x = \dfrac{3 \pm \sqrt{5}}{2}$

$\therefore x = \ln\left(\dfrac{3+\sqrt{5}}{2}\right)$ or $\ln\left(\dfrac{3-\sqrt{5}}{2}\right)$

≈ 0.962 or -0.962

4 **a** $y = e^x$ and $y = e^{2x} - 6$ meet when $e^x = e^{2x} - 6$

$\therefore\ e^{2x} - e^x - 6 = 0$

$\therefore\ (e^x - 3)(e^x + 2) = 0$

$\therefore\ e^x = 3$ or -2

$\therefore\ e^x = 3 \quad \{\text{as } e^x > 0\}$

$\therefore\ x = \ln 3$ and $y = e^x = 3$

$\therefore$ they meet at $(\ln 3,\ 3)$.

b $y = 2e^x + 1$ and $y = 7 - e^x$ meet when $2e^x + 1 = 7 - e^x$

$\therefore\ 3e^x = 6$

$\therefore\ e^x = 2$

$\therefore\ x = \ln 2$ and $y = 7 - e^x = 5$

$\therefore$ they meet at $(\ln 2,\ 5)$.

c $y = 3 - e^x$ and $y = 5e^{-x} - 3$ meet when $3 - e^x = 5e^{-x} - 3$

$\therefore\ 3e^x - e^{2x} = 5 - 3e^x \quad \{\times e^x\}$

$\therefore\ e^{2x} - 6e^x + 5 = 0$

$\therefore\ (e^x - 5)(e^x - 1) = 0$

$\therefore\ e^x = 1$ or 5

$\therefore\ x = 0$ or $\ln 5$

When $x = 0$, $y = 3 - e^0 = 3 - 1 = 2$

When $x = \ln 5$, $y = 3 - e^{\ln 5} = 3 - 5 = -2$

$\therefore$ they meet at $(0,\ 2)$ and $(\ln 5,\ -2)$.

5 **a** Consider $f(x) = e^{2x} - 3$

$f(x)$ cuts the x-axis at A when $f(x) = 0$

$\therefore\ e^{2x} - 3 = 0$

$\therefore\ e^{2x} = 3$

$\therefore\ 2x = \ln 3$

$\therefore\ x = \dfrac{\ln 3}{2}$

$\therefore$ A is $\left(\frac{\ln 3}{2},\ 0\right)$

$f(x)$ cuts the y-axis at B when $x = 0$

$\therefore\ f(0) = e^{2\times 0} - 3$
$= e^0 - 3$
$= -2$

$\therefore$ B is $(0,\ -2)$

b $f(x) = e^{2x} - 3$

$\therefore\ f'(x) = 2e^{2x}$

Since $e^{2x} > 0$ for all x, $f'(x) > 0$ for all x, and hence $f(x)$ is increasing for all x.

c $f''(x) = 4e^{2x}$, which is always > 0.

$\therefore\ f(x)$ is concave up for all x.

d

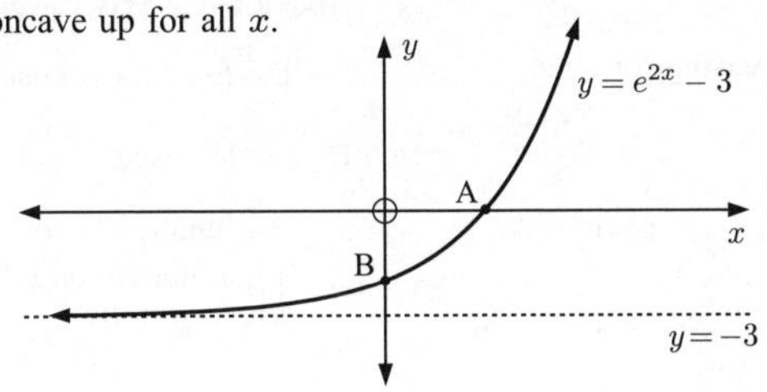

e As $x \to -\infty$, $e^{2x} \to 0$ (from above), so $e^{2x} - 3 \to -3$ (from above)

$\therefore\ y = -3$ is a horizontal asymptote.

6 **a** The x-intercepts occur when $y = 0$

For $f(x) = e^x - 3$, $e^x - 3 = 0$

$\therefore\ e^x = 3$

$\therefore\ x = \ln 3$

and for $g(x) = 3 - \dfrac{5}{e^x}$, $3 - \dfrac{5}{e^x} = 0$

$\therefore\ \dfrac{3e^x - 5}{e^x} = 0$

$\therefore\ 3e^x - 5 = 0$

$\therefore\ e^x = \frac{5}{3}$

$\therefore\ x = \ln \frac{5}{3}$

$\therefore\ f(x)$ has x-intercept $\ln 3$
and $g(x)$ has x-intercept $\ln \frac{5}{3}$.

The y-intercepts occur when $x = 0$

Now $f(0) = e^0 - 3 = -2$ and $g(0) = 3 - \dfrac{5}{e^0} = 3 - 5 = -2$

$\therefore$ both $f(x)$ and $g(x)$ have y-intercept -2.

b As $x \to \infty$, $f(x) \to \infty$
$x \to -\infty$, $f(x) \to -3$ (above)

As $x \to \infty$, $g(x) \to 3$ (below)
$x \to -\infty$, $g(x) \to -\infty$

c $f(x)$ and $g(x)$ meet when

$$\begin{aligned} e^x - 3 &= 3 - 5e^{-x} \\ \therefore\ e^{2x} - 3e^x &= 3e^x - 5 \quad \{\times e^x\} \\ \therefore\ e^{2x} - 6e^x + 5 &= 0 \\ \therefore\ (e^x - 5)(e^x - 1) &= 0 \\ \therefore\ e^x &= 5 \text{ or } 1 \\ \therefore\ x &= \ln 5 \text{ or } 0 \end{aligned}$$

Now $f(\ln 5) = e^{\ln 5} - 3 = 5 - 3 = 2$

and $f(0) = -2$

$\therefore$ $f(x)$ and $g(x)$ meet at $(0, -2)$ and $(\ln 5, 2)$.

d

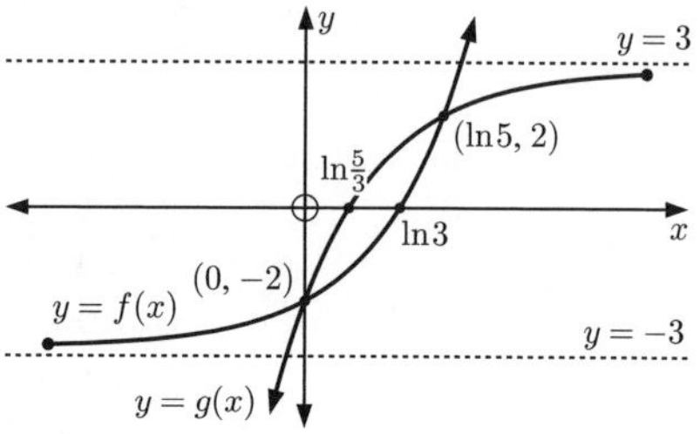

7 **a** Consider $y = e^x - 3e^{-x}$

It cuts the x-axis at P when $y = 0$

$$\begin{aligned} \therefore\ e^x - 3e^{-x} &= 0 \\ \therefore\ e^{2x} - 3 &= 0 \quad \{\times e^x\} \\ \therefore\ e^{2x} &= 3 \\ \therefore\ 2x &= \ln 3 \\ \therefore\ x &= \tfrac{1}{2}\ln 3 \end{aligned}$$

It cuts the y-axis at Q when $x = 0$

$$\begin{aligned} \therefore\ y &= e^0 - 3e^0 \\ &= 1 - 3 \\ &= -2 \end{aligned}$$

$\therefore$ P is $(\frac{1}{2}\ln 3, 0)$ and Q is $(0, -2)$.

b
$$\begin{aligned} \frac{dy}{dx} &= e^x + 3e^{-x} \\ &= e^x + \frac{3}{e^x} \end{aligned}$$

Since $e^x > 0$ for all x,

$\dfrac{dy}{dx} > 0$ for all x

$\therefore$ the function is increasing for all x

c
$$\begin{aligned} \frac{dy}{dx} &= e^x + 3e^{-x} \\ \therefore\ \frac{d^2y}{dx^2} &= e^x - 3e^{-x} \\ &= y \end{aligned}$$

Above the x-axis $y > 0$ $\therefore$ $\dfrac{d^2y}{dx^2} > 0$

$\therefore$ the function is concave up

Below the x-axis $y < 0$ $\therefore$ $\dfrac{d^2y}{dx^2} < 0$

$\therefore$ the function is concave down

$\therefore$ a non-horizontal inflection occurs when $y = 0$

d

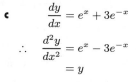

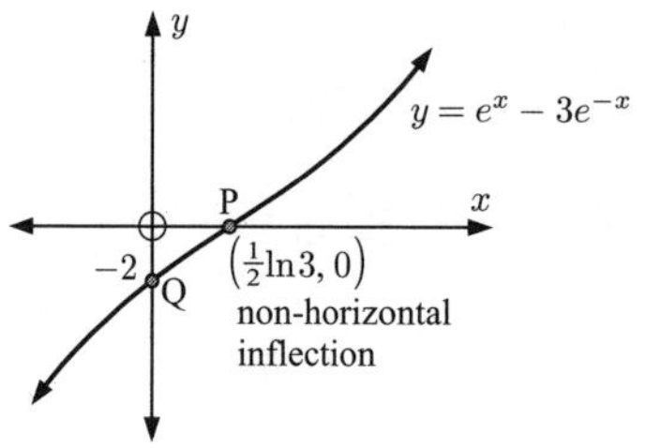

EXERCISE 19C

1 **a**
$$\begin{aligned} y &= \ln(7x) \\ \therefore\ y &= \ln 7 + \ln x \\ \therefore\ \frac{dy}{dx} &= 0 + \frac{1}{x} = \frac{1}{x} \end{aligned}$$

or

$$\begin{aligned} y &= \ln(7x) \\ \therefore\ \frac{dy}{dx} &= \frac{7}{7x} \begin{matrix} \leftarrow f'(x) \\ \leftarrow f(x) \end{matrix} \\ &= \frac{1}{x} \end{aligned}$$

b $y = \ln(2x+1)$

$\therefore \dfrac{dy}{dx} = \dfrac{2}{2x+1} \begin{matrix} \leftarrow f'(x) \\ \leftarrow f(x) \end{matrix}$

c $y = \ln(x - x^2)$

$\therefore \dfrac{dy}{dx} = \dfrac{1-2x}{x-x^2} \begin{matrix} \leftarrow f'(x) \\ \leftarrow f(x) \end{matrix}$

d $y = 3 - 2\ln x$

$\therefore \dfrac{dy}{dx} = 0 - 2\left(\dfrac{1}{x}\right)$

$= -\dfrac{2}{x}$

e $y = x^2 \ln x$

$\therefore \dfrac{dy}{dx} = 2x\ln x + x^2\left(\dfrac{1}{x}\right)$

$= 2x\ln x + x$

f $y = \dfrac{\ln x}{2x}$

$\therefore \dfrac{dy}{dx} = \dfrac{\left(\dfrac{1}{x}\right)2x - \ln x \times 2}{(2x)^2}$

$= \dfrac{2 - 2\ln x}{4x^2}$

$= \dfrac{1 - \ln x}{2x^2}$

g $y = e^x \ln x$

$\therefore \dfrac{dy}{dx} = e^x \ln x + \dfrac{e^x}{x}$

h $y = (\ln x)^2$

$\therefore \dfrac{dy}{dx} = 2(\ln x)^1\left(\dfrac{1}{x}\right)$

$= \dfrac{2\ln x}{x}$

i $y = \sqrt{\ln x} = (\ln x)^{\frac{1}{2}}$

$\therefore \dfrac{dy}{dx} = \frac{1}{2}(\ln x)^{-\frac{1}{2}}\left(\dfrac{1}{x}\right)$

$= \dfrac{1}{2x\sqrt{\ln x}}$

j $y = e^{-x}\ln x$

$\therefore \dfrac{dy}{dx} = -e^{-x}\ln x + e^{-x}\left(\dfrac{1}{x}\right)$

$= \dfrac{e^{-x}}{x} - e^{-x}\ln x$

k $y = \sqrt{x}\ln(2x)$

$\therefore \dfrac{dy}{dx} = \dfrac{1}{2\sqrt{x}}\ln(2x) + \sqrt{x}\left(\dfrac{1}{x}\right)$

$= \dfrac{\ln(2x)}{2\sqrt{x}} + \dfrac{1}{\sqrt{x}}$

l $y = \dfrac{2\sqrt{x}}{\ln x}$

$\therefore \dfrac{dy}{dx} = \dfrac{\dfrac{1}{\sqrt{x}}\ln x - 2\sqrt{x}\left(\dfrac{1}{x}\right)}{(\ln x)^2}$

$= \dfrac{\dfrac{1}{\sqrt{x}}\ln x - \dfrac{2}{\sqrt{x}}}{(\ln x)^2}$

$= \dfrac{\ln x - 2}{\sqrt{x}(\ln x)^2}$

m $y = 3 - 4\ln(1-x)$

$\therefore \dfrac{dy}{dx} = -\dfrac{4}{1-x} \times -1$

$= \dfrac{4}{1-x}$

n $y = x\ln(x^2+1)$

$\therefore \dfrac{dy}{dx} = \ln(x^2+1) + x\dfrac{2x}{x^2+1}$

$= \ln(x^2+1) + \dfrac{2x^2}{x^2+1}$

2 **a** $y = x\ln 5$

$\therefore \dfrac{dy}{dx} = \ln 5$

b $y = \ln(x^3) = 3\ln x$

$\therefore \dfrac{dy}{dx} = 3\left(\dfrac{1}{x}\right) = \dfrac{3}{x}$

c $y = \ln(x^4 + x)$

$\therefore \dfrac{dy}{dx} = \dfrac{4x^3+1}{x^4+x}$

d $y = \ln(10 - 5x)$

$\therefore \dfrac{dy}{dx} = \dfrac{-5}{10-5x} = \dfrac{1}{x-2}$

e

$$y = [\ln(2x+1)]^3$$
$$\therefore \frac{dy}{dx} = 3[\ln(2x+1)]^2 \times \frac{2}{2x+1} = \frac{6[\ln(2x+1)]^2}{2x+1}$$

f

$$y = \frac{\ln(4x)}{x}$$
$$\therefore \frac{dy}{dx} = \frac{\left(\frac{4}{4x}\right)x - \ln(4x) \times 1}{x^2} = \frac{1-\ln(4x)}{x^2}$$

g

$$y = \ln\left(\frac{1}{x}\right) = -\ln x$$
$$\therefore \frac{dy}{dx} = -\frac{1}{x}$$

h

$$y = \ln(\ln x)$$
$$\therefore \frac{dy}{dx} = \frac{\frac{1}{x}}{\ln x} = \frac{1}{x\ln x}$$

i

$$y = \frac{1}{\ln x} = (\ln x)^{-1}$$
$$\therefore \frac{dy}{dx} = -1(\ln x)^{-2} \times \frac{1}{x} = \frac{-1}{x(\ln x)^2}$$

3 **a**

$$y = \ln\sqrt{1-2x} = \ln(1-2x)^{\frac{1}{2}} = \tfrac{1}{2}\ln(1-2x)$$
$$\therefore \frac{dy}{dx} = \tfrac{1}{2} \times \frac{-2}{1-2x} = \frac{-1}{1-2x} = \frac{1}{2x-1}$$

b

$$y = \ln\left(\frac{1}{2x+3}\right) = -\ln(2x+3)$$
$$\therefore \frac{dy}{dx} = -\frac{2}{2x+3}$$

c

$$y = \ln(e^x\sqrt{x}) = \ln e^x + \ln x^{\frac{1}{2}} = \ln e^x + \tfrac{1}{2}\ln x = x + \tfrac{1}{2}\ln x$$
$$\therefore \frac{dy}{dx} = 1 + \tfrac{1}{2}\left(\frac{1}{x}\right) = 1 + \frac{1}{2x}$$

d

$$y = \ln(x\sqrt{2-x}) = \ln x + \ln(2-x)^{\frac{1}{2}} = \ln x + \tfrac{1}{2}\ln(2-x)$$
$$\therefore \frac{dy}{dx} = \frac{1}{x} + \tfrac{1}{2}\left(\frac{-1}{2-x}\right) = \frac{1}{x} - \frac{1}{2(2-x)}$$

e

$$y = \ln\left(\frac{x+3}{x-1}\right) = \ln(x+3) - \ln(x-1)$$
$$\therefore \frac{dy}{dx} = \frac{1}{x+3} - \frac{1}{x-1}$$

f

$$y = \ln\left(\frac{x^2}{3-x}\right) = \ln x^2 - \ln(3-x) = 2\ln x - \ln(3-x)$$
$$\therefore \frac{dy}{dx} = \frac{2}{x} - \frac{-1}{3-x} = \frac{2}{x} + \frac{1}{3-x}$$

g

$$f(x) = \ln\left((3x-4)^3\right) = 3\ln(3x-4)$$
$$\therefore f'(x) = 3 \times \frac{3}{3x-4} = \frac{9}{3x-4}$$

h

$$f(x) = \ln\left(x(x^2+1)\right) = \ln x + \ln(x^2+1)$$
$$\therefore f'(x) = \frac{1}{x} + \frac{2x}{x^2+1}$$

i

$$f(x) = \ln\left(\frac{x^2+2x}{x-5}\right) = \ln(x^2+2x) - \ln(x-5)$$
$$\therefore f'(x) = \frac{2x+2}{x^2+2x} - \frac{1}{x-5}$$

4 **a**

$$y = 2^x = (e^{\ln 2})^x = e^{x\ln 2}$$
$$\therefore \frac{dy}{dx} = e^{x\ln 2} \times \ln 2 = 2^x \ln 2$$

b

$$y = a^x = (e^{\ln a})^x = e^{x\ln a}$$
$$\therefore \frac{dy}{dx} = e^{x\ln a} \times \ln a = a^x \ln a$$

5 $f(x) = \ln(2x - 1) - 3$

a $f(x) = 0$ when $\ln(2x-1) = 3$

$\therefore \quad 2x - 1 = e^3$

$\therefore \quad 2x = e^3 + 1$

$\therefore \quad x = \dfrac{e^3+1}{2} \approx 10.5 \qquad \therefore$ the x-intercept is $\dfrac{e^3+1}{2}$

b $f(0)$ cannot be found as $\ln(-1)$ is not defined. $\quad \therefore$ there is no y-intercept.

c $f'(x) = \dfrac{2}{2x-1} \quad \therefore \quad f'(1) = \dfrac{2}{2-1} = 2 \quad \therefore$ gradient of tangent $= 2$

d $\ln(2x-1)$ has meaning provided $2x - 1 > 0 \quad \therefore \quad 2x > 1$ and so $x > \frac{1}{2}$

$\therefore \quad f(x)$ has meaning provided $x > \frac{1}{2}$

e $f'(x) = 2(2x-1)^{-1}$

$\therefore \quad f''(x) = -2(2x-1)^{-2}(2)$

$= \dfrac{-4}{(2x-1)^2}, \quad x > \frac{1}{2}$

$\therefore$ provided $x > \frac{1}{2}, \quad f''(x) < 0$

$\therefore \quad f(x)$ is concave down when $f(x)$ has meaning.

f

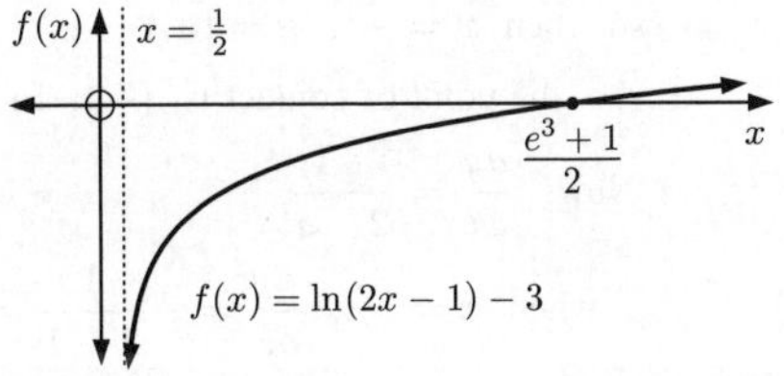

6 **a** $f(x)$ is defined when $\ln x$ is defined $\quad \therefore \quad f(x)$ is defined for $x > 0$

b $f'(x) = \ln x + \dfrac{x}{x}$ {product rule}

$= \ln x + 1$

which is 0 when $\ln x = -1$

$\therefore \quad x = e^{-1}$

Sign diagram of $f'(x)$ is:

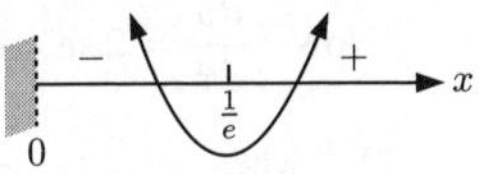

So, there is a local minimum at $\left(\frac{1}{e}, \frac{1}{e}\ln\frac{1}{e}\right)$

$\therefore$ the minimum value of $f(x)$ is $\frac{1}{e}\ln e^{-1} = -\frac{1}{e}$

7 Consider $f(x) = \dfrac{\ln x}{x}$

$\therefore \quad f'(x) = \dfrac{\left(\frac{1}{x}\right)x - \ln x(1)}{x^2} = \dfrac{1 - \ln x}{x^2}$

$\therefore \quad f'(x) = 0$ when $1 - \ln x = 0$

$\therefore \quad \ln x = 1$

$\therefore \quad x = e$

Sign diagram of $f'(x)$ is:

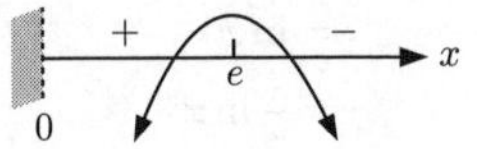

Now $f(e) = \dfrac{\ln e}{e} = \dfrac{1}{e}$

$\therefore$ there is a local maximum at $\left(e, \dfrac{1}{e}\right)$

$\therefore \quad f(x) \leqslant \dfrac{1}{e}$ for all x, and so $\dfrac{\ln x}{x} \leqslant \dfrac{1}{e}$ for all $x > 0$

8 **a** $f(x) = x - \ln x$

$\therefore \quad f'(x) = 1 - \dfrac{1}{x} = \dfrac{x-1}{x}$ and the sign diagram of $f'(x)$ is:

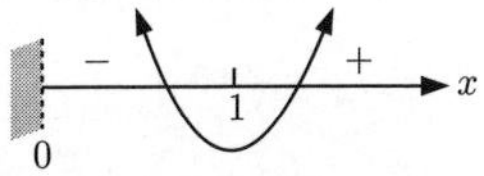

$\therefore \quad f(x)$ has a local minimum at $(1, 1 - \ln 1)$ or $(1, 1)$. This is the only turning point.

b $f(x) \geqslant 1$ for all $x > 0$

$\therefore \quad x - \ln x \geqslant 1$

$\therefore \quad \ln x \leqslant x - 1$ for all $x > 0$

EXERCISE 19D

1 $f(x) = e^{-x}$

$\therefore\ f(1) = e^{-1}$

$\therefore$ the point of contact is $\left(1, \dfrac{1}{e}\right)$.

Now $f'(x) = -e^{-x}$

$\therefore\ f'(1) = -e^{-1} = -\dfrac{1}{e}$

So, the gradient of the tangent is $-\dfrac{1}{e}$

$\therefore$ the tangent has equation $\dfrac{y - \frac{1}{e}}{x - 1} = -\dfrac{1}{e}$

$\therefore\ e\left(y - \dfrac{1}{e}\right) = -(x-1)$

$\therefore\ ey - 1 = -x + 1$

$\therefore\ x + ey = 2$

or $y = -\dfrac{1}{e}x + \dfrac{2}{e}$

2 $y = \ln(2 - x)$

so when $x = -1,\ y = \ln 3$

$\therefore$ the point of contact is $(-1,\ \ln 3)$.

Now $\dfrac{dy}{dx} = \dfrac{-1}{2 - x}$

$\therefore$ when $x = -1,\ \dfrac{dy}{dx} = -\dfrac{1}{2+1} = -\frac{1}{3}$

So, the gradient of the tangent is $-\frac{1}{3}$.

$\therefore$ the tangent has equation $\dfrac{y - \ln 3}{x + 1} = -\frac{1}{3}$

$\therefore\ 3(y - \ln 3) = -(x + 1)$

$\therefore\ 3y - 3\ln 3 = -x - 1$

$\therefore\ x + 3y = 3\ln 3 - 1$

3 $y = x^2e^x$ so when $x = 1,\ y = e$

$\therefore$ the point of contact is $(1, e)$.

Now $\dfrac{dy}{dx} = 2xe^x + x^2e^x$

$\therefore$ when $x = 1,\ \dfrac{dy}{dx} = 2e + e = 3e$

$\therefore$ the tangent has equation $\dfrac{y - e}{x - 1} = 3e$

$\therefore\ y - e = 3ex - 3e$

$\therefore\ y - 3ex = -2e$

$\therefore\ 3ex - y = 2e$

The tangent cuts the x-axis when

$y = 0$

$\therefore\ 3ex = 2e$

$\therefore\ x = \frac{2}{3}$

and the y-axis when

$x = 0$

$\therefore\ -y = 2e$

$\therefore\ y = -2e$

So, A is $(\frac{2}{3}, 0)$ and B is $(0, -2e)$.

4 $y = \ln\sqrt{x}$

$= \ln x^{\frac{1}{2}}$

$= \frac{1}{2}\ln x$

$\therefore$ when $y = -1,\ -1 = \frac{1}{2}\ln x$

$\therefore\ \ln x = -2$

$\therefore\ x = e^{-2}$

$\therefore\ x = \dfrac{1}{e^2}$

$\therefore$ the point of contact is $\left(\dfrac{1}{e^2}, -1\right)$

Now $\dfrac{dy}{dx} = \frac{1}{2}\dfrac{1}{x} = \dfrac{1}{2x}$, so at the point of contact, $\dfrac{dy}{dx} = \dfrac{1}{2e^{-2}} = \dfrac{e^2}{2}$

$\therefore$ the tangent has gradient $\dfrac{e^2}{2}$ and the normal has gradient $-\dfrac{2}{e^2}$

$\therefore$ the normal has equation $\dfrac{y+1}{x - \frac{1}{e^2}} = -\dfrac{2}{e^2}$

$\therefore\ e^2(y+1) = -2\left(x - \dfrac{1}{e^2}\right)$

$\therefore\ e^2y + e^2 = -2x + \dfrac{2}{e^2}$

$\therefore\ 2x + e^2y = -e^2 + \dfrac{2}{e^2}$ or $y = -\dfrac{2}{e^2}x + \dfrac{2}{e^4} - 1$

5 $y = e^x$ so when $x = a$, $y = e^a$

$\therefore$ the point of contact is (a, e^a).

Now $\dfrac{dy}{dx} = e^x$

$\therefore$ at the point (a, e^a), $\dfrac{dy}{dx} = e^a$

$\therefore$ the tangent has equation $\dfrac{y - e^a}{x - a} = e^a$

or $y - e^a = e^a(x - a)$ .... $(*)$

Since the tangent passes through the origin, $(0, 0)$ must satisfy $(*)$

$\therefore$ $0 - e^a = e^a(0 - a)$

$\therefore$ $-e^a = -ae^a$

$\therefore$ $e^a(a - 1) = 0$

$\therefore$ $a = 1$ $\{$as $e^a > 0\}$

So the equation of the tangent is

$y - e = ex - e$ or $y = ex$.

6 **a** $f(x) = \ln x$ is defined for all $x > 0$.

b $f'(x) = \dfrac{1}{x}$ which is > 0 for all $x > 0$

$\therefore$ $f(x)$ is increasing on $x > 0$; its gradient is always positive.

$f''(x) = -x^{-2} = \dfrac{-1}{x^2}$ which is < 0 for all $x > 0$ $\therefore$ $f(x)$ is concave down on $x > 0$.

c

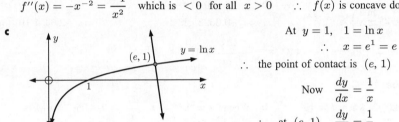

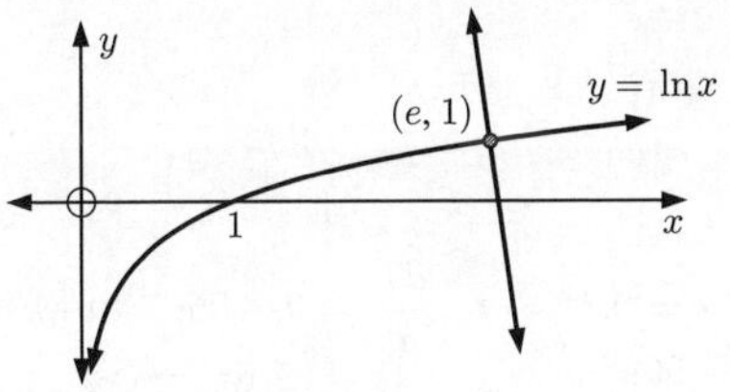

At $y = 1$, $1 = \ln x$

$\therefore$ $x = e^1 = e$

$\therefore$ the point of contact is $(e, 1)$

Now $\dfrac{dy}{dx} = \dfrac{1}{x}$

$\therefore$ at $(e, 1)$, $\dfrac{dy}{dx} = \dfrac{1}{e}$

$\therefore$ the gradient of the tangent is $\dfrac{1}{e}$, and the gradient of the normal is $-e$

$\therefore$ the equation of the normal is $\dfrac{y - 1}{x - e} = -e$

$\therefore$ $y - 1 = -e(x - e)$

$\therefore$ $y - 1 = -ex + e^2$

$\therefore$ $ex + y = 1 + e^2$

7 $y = 3e^{-x}$ and $y = 2 + e^x$ meet when $3e^{-x} = 2 + e^x$

$\therefore$ $3 = 2e^x + e^{2x}$ $\{\times\, e^x\}$

$\therefore$ $e^{2x} + 2e^x - 3 = 0$

$\therefore$ $(e^x + 3)(e^x - 1) = 0$

$\therefore$ $e^x = -3$ or 1

$\therefore$ $e^x = 1$ and so $x = 0$ $\{$as $e^x > 0\}$

Now when $x = 0$, $y = 3e^0 = 3$, so the graphs meet at $(0, 3)$.

For $y = 2 + e^x$, $\dfrac{dy}{dx} = e^x$,

so at the point $(0, 3)$, $\dfrac{dy}{dx} = e^0 = 1$

$\therefore$ the gradient of the tangent at this point is 1

$\therefore$ the tangent has direction vector $\binom{1}{1}$

For $y = 3e^{-x}$, $\dfrac{dy}{dx} = -3e^{-x}$,

so at the point $(0, 3)$, $\dfrac{dy}{dx} = -3$

$\therefore$ the gradient of the tangent at this point is -3

$\therefore$ the tangent has direction vector $\binom{1}{-3}$

If θ is the acute angle between the tangents, then $\cos\theta = \dfrac{|1(1) + 1(-3)|}{\sqrt{1^2 + 1^2}\sqrt{1^2 + (-3)^2}} = \dfrac{|-2|}{\sqrt{2}\sqrt{10}} = \dfrac{2}{\sqrt{20}}$

$\therefore$ $\theta \approx 63.43^o$

8 **a** $W = 20e^{-kt}$ so when $t = 50$ hours, $W = 10$ g

$\therefore$ $20e^{-50k} = 10$

$\therefore$ $e^{-50k} = \frac{1}{2}$

$\therefore$ $-50k = \ln\frac{1}{2} = -\ln 2$ $\therefore$ $k = \frac{1}{50}\ln 2 \approx 0.0139$

b **i** When $t = 0$, $W = 20e^0$ $= 20$ g

ii When $t = 24$, $W = 20e^{-24k}$ $= 20e^{-24\frac{\ln 2}{50}}$ ≈ 14.3 g

iii When $t = 1$ week $= 7 \times 24$ hours $= 168$ hours
$W = 20e^{-168\frac{\ln 2}{50}}$ ≈ 1.95 g

c When $W = 1$ g, $20e^{-\frac{\ln 2}{50}\times t} = 1$

$\therefore \quad e^{-\frac{\ln 2}{50}\times t} = 0.05$

$\therefore \quad -\frac{\ln 2}{50} \times t = \ln 0.05$

$\therefore \quad t = \dfrac{-50\ln 0.05}{\ln 2} \approx 216$ hours or 9 days and 6 minutes

d $\dfrac{dW}{dt}$
$= 20e^{-kt}(-k)$
$= \left(-20\dfrac{\ln 2}{50}\right) \times e^{-\frac{\ln 2}{50}t}$

i When $t = 100$ hours, $\dfrac{dW}{dt} = \left(\dfrac{-20\ln 2}{50}\right)e^{-2\ln 2}$ $\approx -0.0693 \text{ g h}^{-1}$

ii When $t = 1000$ hours, $\dfrac{dW}{dt} = \left(\dfrac{-20\ln 2}{50}\right)e^{-20\ln 2}$ $\approx -2.64 \times 10^{-7} \text{ g h}^{-1}$

e $\dfrac{dW}{dt} = -k(20e^{-kt}) = -kW \quad \therefore \quad \dfrac{dW}{dt} \propto W$

9 $T = 5 + 95e^{-kt}$ ^{o}C

a $T = 20^o$C when $t = 15$

$\therefore \quad 20 = 5 + 95e^{-15k}$

$\therefore \quad 15 = 95e^{-15k}$

$\therefore \quad e^{15k} = \frac{95}{15}$

$\therefore \quad 15k = \ln\left(\frac{19}{3}\right)$

$\therefore \quad k = \frac{1}{15}\ln\left(\frac{19}{3}\right) \approx 0.123$

b When $t = 0$,
$T = 5 + 95e^0$
$= 5 + 95$
$= 100^o$C

c $\dfrac{dT}{dt} = 0 + 95e^{-kt}(-k)$
$= -(95e^{-kt})k$
$= c(T - 5)$ where $c = -k$

d $\dfrac{dT}{dt} = -95e^{-kt} \times k \approx -11.6902e^{-0.1231t}$

i When $t = 0$, $\dfrac{dT}{dt} \approx -11.69$, so the temperature is decreasing at 11.7^oC min^{-1}.

ii When $t = 10$, $\dfrac{dT}{dt} \approx -11.6902e^{-1.231} \approx -3.415$,
so the temperature is decreasing at 3.42^oC min^{-1}.

iii When $t = 20$, $\dfrac{dT}{dt} \approx -11.6902e^{-2.461} \approx -0.998$,
so the temperature is decreasing at 0.998^oC min^{-1}.

10 $H(t) = 20\ln(3t + 2) + 30$ cm, $t \geqslant 0$

a The shrubs were planted when $t = 0$. $H(0) = 20\ln(2) + 30 \approx 43.9$ cm

b When $H = 1$ m $= 100$ cm,
$20\ln(3t + 2) + 30 = 100$

$\therefore \quad 20\ln(3t + 2) = 70$

$\therefore \quad \ln(3t + 2) = 3.5$

$\therefore \quad 3t + 2 = e^{3.5}$

$\therefore \quad 3t = e^{3.5} - 2$

$\therefore \quad t = \dfrac{e^{3.5} - 2}{3}$ years

$\therefore \quad t \approx 10.4$ years

c $\dfrac{dH}{dt} = 20 \times \dfrac{3}{(3t + 2)} = \dfrac{60}{3t + 2}$ cm year^{-1}

i When $t = 3$, $\dfrac{dH}{dt} = \frac{60}{11} \approx 5.4545$
$\therefore$ it is growing at 5.45 cm year^{-1}

ii When $t = 10$, $\dfrac{dH}{dt} = \frac{60}{32} = 1.875$
$\therefore$ it is growing at 1.88 cm year^{-1}

11 **a** $A = s(1 - e^{-kt}), \quad t \geqslant 0$

When $t = 0, \quad A = s(1 - e^0)$
$= s(1 - 1)$
$= 0$

b When $t = 3$, $A = 5$ and $s = 10$

$\therefore \quad 5 = 10(1 - e^{-3k})$
$\therefore \quad 0.5 = 1 - e^{-3k}$
$\therefore \quad e^{-3k} = 0.5$
$\therefore \quad e^{3k} = 2$
$\therefore \quad 3k = \ln 2$
$\therefore \quad k = \frac{1}{3} \ln 2 \approx 0.231$

c $\dfrac{dA}{dt} = ske^{-kt}$

$\therefore$ when $t = 5$ and $s = 10$,

$\dfrac{dA}{dt} = 10\left(\frac{1}{3}\ln 2\right)\left(e^{-\frac{5}{3}\ln 2}\right)$
≈ 0.728 litres per hour

d $\dfrac{dA}{dt} = ske^{-kt}$
$= k\left(se^{-kt}\right)$
$= -k\left(-se^{-kt}\right)$
$= -k(A - s)$
$\therefore \quad \dfrac{dA}{dt} \propto (A - s)$

12 Consider $f(x) = \dfrac{e^x}{x}$.

a $e^x \neq 0$ for all x, so $f(x) \neq 0$ and there is no x-intercept.

$f(0) = \dfrac{e^0}{0}$ is undefined, so there is also no y-intercept.

b As $x \to +\infty$ $f(x) \to \infty$, and as $x \to -\infty$, $f(x) \to 0$ (below)

(As $x \to 0$ (above), $y \to +\infty$, and as $x \to 0$ (below), $y \to -\infty$
$\therefore$ $x = 0$ is a vertical asymptote.)

c Using the quotient rule, $f'(x) = \dfrac{e^x x - e^x(1)}{x^2} = \dfrac{e^x(x-1)}{x^2}$

with sign diagram:

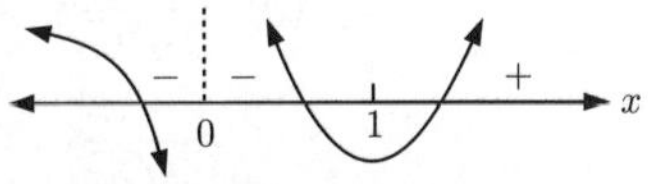

$f(1) = \dfrac{e^1}{1} = e$, so there is a local minimum at $(1, e)$.

d

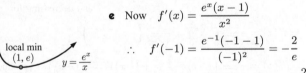

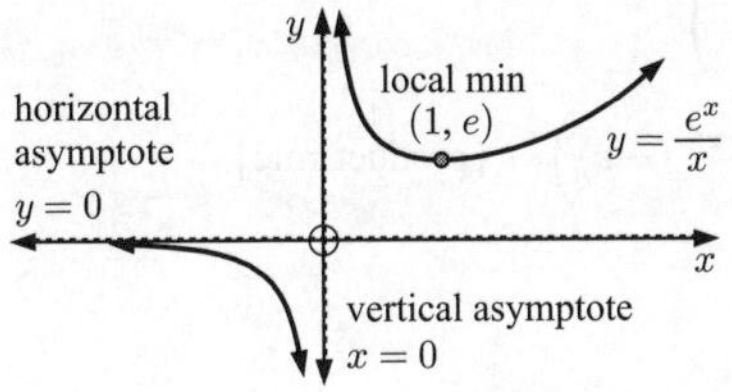

e Now $f'(x) = \dfrac{e^x(x-1)}{x^2}$

$\therefore \quad f'(-1) = \dfrac{e^{-1}(-1-1)}{(-1)^2} = -\dfrac{2}{e}$

$\therefore$ the gradient of the tangent is $= -\dfrac{2}{e}$

When $x = -1$, $y = \dfrac{e^{-1}}{-1} = -\dfrac{1}{e}$

$\therefore$ the equation of tangent is $\dfrac{y - \left(-\frac{1}{e}\right)}{x - (-1)} = -\dfrac{2}{e}$

$\therefore \quad \dfrac{y + \frac{1}{e}}{x + 1} = -\dfrac{2}{e}$

$\therefore \quad e\left(y + \frac{1}{e}\right) = -2(x + 1)$
$ey + 1 = -2x - 2$
$\therefore \quad ey = -2x - 3$

13 **a** $s(t) = 100t + 200e^{-\frac{t}{5}}$ cm, $t \geqslant 0$

$v(t) = 100 - 40e^{-\frac{t}{5}}$ cm s^{-1}

$a(t) = 8e^{-\frac{t}{5}}$ cm s^{-2}

b When $t = 0$, $s(0) = 200$ cm (right of the origin)
$v(0) = 60$ cm s^{-1}
$a(0) = 8$ cm s^{-2}

c As $t \to \infty$, $e^{-\frac{t}{5}} \to 0$,
$\therefore$ $v(t) \to 100$ cm s^{-1} (below)

d

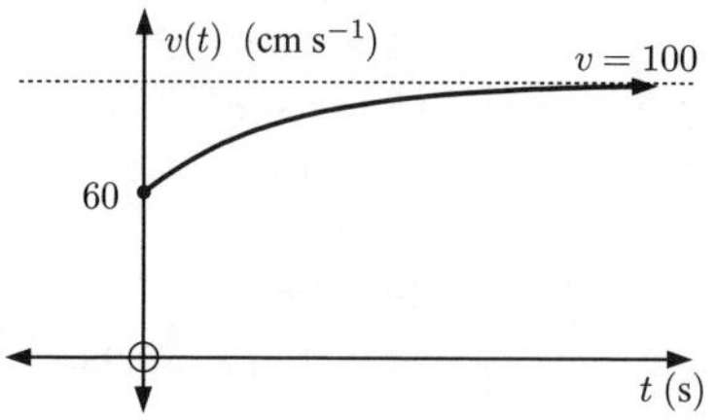

e When $v(t) = 80 \text{ cm s}^{-1}$,

$$100 - 40e^{-\frac{t}{5}} = 80$$
$$\therefore \quad -40e^{-\frac{t}{5}} = -20$$
$$\therefore \quad e^{-\frac{t}{5}} = 0.5$$
$$\therefore \quad -\frac{t}{5} = \ln 0.5$$
$$\therefore \quad t = -5\ln 0.5 \approx 3.47 \text{ s}$$

14 **a** $A(t) = t\ln t + 1, \quad 0 < t \leqslant 5$

$$\therefore \quad A'(t) = \ln t + t \times \frac{1}{t} + 0 \quad \{\text{product rule}\}$$
$$= \ln t + 1$$
$$\therefore \quad A'(t) = 0 \text{ when } \ln t = -1$$
$$\therefore \quad t = e^{-1}$$

and the sign diagram of $A'(t)$ is:

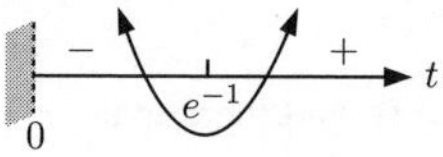

$$\therefore \quad A(t) \text{ is a minimum when } t = \frac{1}{e} \approx 0.3679 \text{ years}$$

$\therefore$ the child's memorising ability is a minimum at 4.41 months old.

b

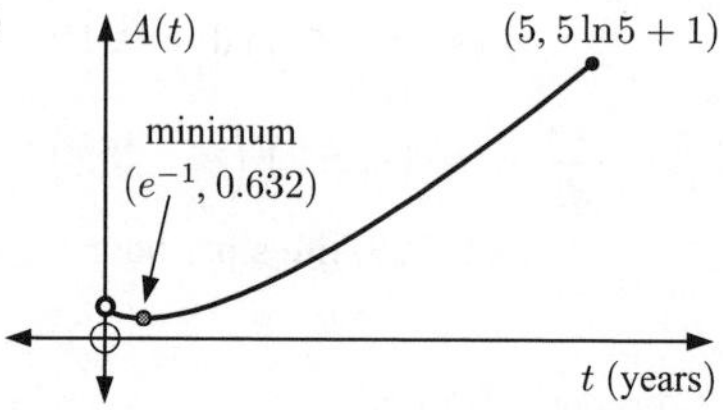

15 **a** $f(x) = \frac{1}{\sqrt{2\pi}} e^{-\frac{1}{2}x^2}$

$$\therefore \quad f'(x) = \frac{1}{\sqrt{2\pi}} e^{-\frac{1}{2}x^2}(-x)$$
$$= \frac{-x}{\sqrt{2\pi}} e^{-\frac{1}{2}x^2}$$
$$\therefore \quad f'(x) = 0 \quad \text{when} \quad x = 0$$

$f'(x)$ has sign diagram:

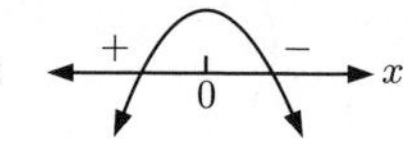

Now $f(0) = \frac{1}{\sqrt{2\pi}}$

so there is a local maximum at $\left(0, \frac{1}{\sqrt{2\pi}}\right)$.

The function is increasing for $x \leqslant 0$ and decreasing for $x \geqslant 0$

b $f'(x) = \frac{-x}{\sqrt{2\pi}} e^{-\frac{1}{2}x^2} = \frac{1}{\sqrt{2\pi}}\left(-xe^{-\frac{1}{2}x^2}\right)$

$$\therefore \quad f''(x) = \frac{1}{\sqrt{2\pi}}\left((-1)\,e^{-\frac{1}{2}x^2} + (-x)\,e^{-\frac{1}{2}x^2}\,(-x)\right) \quad \{\text{product rule}\}$$
$$= \frac{1}{\sqrt{2\pi}} e^{-\frac{1}{2}x^2}\left(x^2 - 1\right)$$
$$= \frac{1}{\sqrt{2\pi}} e^{-\frac{1}{2}x^2}(x+1)(x-1) \quad \text{which has sign diagram:}$$

Now $f(1) = \frac{1}{\sqrt{2\pi}} e^{-\frac{1}{2}} = \frac{1}{\sqrt{2e\pi}}$ and $f(-1) = \frac{1}{\sqrt{2e\pi}}$

$\therefore$ there are points of inflection at $\left(1, \frac{1}{\sqrt{2e\pi}}\right)$ and $\left(-1, \frac{1}{\sqrt{2e\pi}}\right)$.

c As $x \to \infty$, $e^{-\frac{1}{2}x^2} \to 0$ (above),

$\therefore \quad f(x) \to 0$ (above)

As $x \to -\infty$, $e^{-\frac{1}{2}x^2} \to 0$ (above),

$\therefore \quad f(x) \to 0$ (above)

d

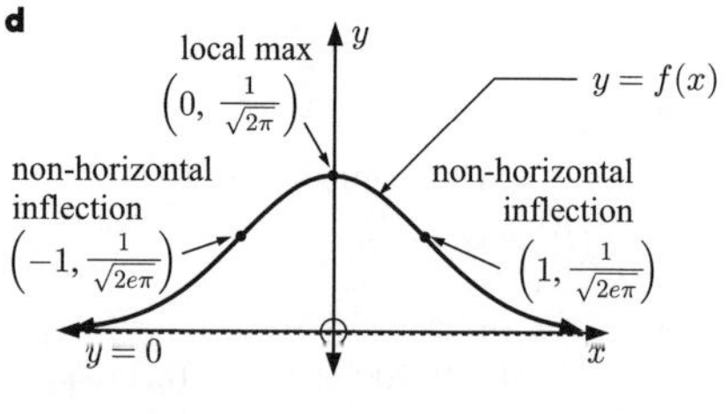

16 $C(x) = 4\ln x + \left(\frac{30-x}{10}\right)^2, \quad x \geqslant 10$

$\therefore \quad C'(x) = \frac{4}{x} + 2\left(\frac{30-x}{10}\right)\left(-\frac{1}{10}\right)$

$= \frac{4}{x} - \frac{30-x}{50}$

$= \frac{200 - x(30-x)}{50x}$

$= \frac{200 - 30x + x^2}{50x}$

$= \frac{(x-10)(x-20)}{50x}$

$C'(x)$ has sign diagram:

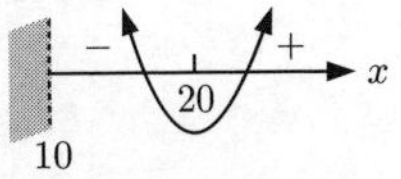

$\therefore$ the minimum cost occurs when $x = 20$ or when 20 kettles per day are produced.

17 Let coordinates of D be $(x, 0)$ where $x > 0$.

$\therefore$ the coordinates of C are (x, e^{-x^2}).

$\therefore$ area ABCD $= 2xe^{-x^2}$

$\therefore \quad \frac{dA}{dx} = 2e^{-x^2} + 2xe^{-x^2}(-2x)$ {product rule}

$= 2e^{-x^2}\left(1 - 2x^2\right)$

$= 2e^{-x^2}(1+\sqrt{2}x)(1-\sqrt{2}x)$

$\frac{dA}{dx}$ has sign diagram:

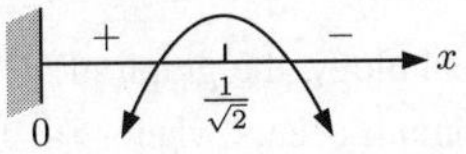

$\therefore$ the area is a maximum when $x = \frac{1}{\sqrt{2}}$

and so C is $\left(\frac{1}{\sqrt{2}}, e^{-\frac{1}{2}}\right)$.

18 $P(x) = R(x) - C(x)$

$\therefore \quad P(x) = \left[1000\ln\left(1 + \frac{x}{400}\right) + 600\right] - [x(1.5) + 300]$

$= 1000\ln(1 + 0.0025x) - 1.5x + 300$

$\therefore \quad P'(x) = 1000\left(\frac{0.0025}{1 + 0.0025x}\right) - 1.5 = \frac{2.5}{1+0.0025x} - 1.5$

$\therefore \quad P'(x) = 0$ when $\frac{2.5}{1+0.0025x} = \frac{3}{2}$

$\therefore \quad 3 + 0.0075x = 5$

$\therefore \quad 0.0075x = 2$

$\therefore \quad x = \frac{2}{0.0075} \approx 266.7$

Sign diagram of $P'(x)$:

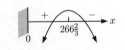

Now $P(266) \approx 410.83$ and $P(267) \approx 410.83$

$\therefore$ to maximise the profit, 266 or 267 torches per day should be produced.

19 **a** $y = ax^2, a > 0$ touches $y = \ln x$ when $ax^2 = \ln x$

If the curves touch when $x = b$ then $ab^2 = \ln b$ .... (1)

Now for $y = ax^2$, $\frac{dy}{dx} = 2ax$ and for $y = \ln x$, $\frac{dy}{dx} = \frac{1}{x}$

$\therefore$ when $x = b$, $\frac{dy}{dx} = 2ab$ $\therefore$ when $x = b$, $\frac{dy}{dx} = \frac{1}{b}$

Since the curves touch each other, they share a common tangent. $\therefore \quad \frac{1}{b} = 2ab$ .... (2)

b Now $ab^2 = \frac{1}{2}$ {from (2)}

and $ab^2 = \ln b$ {from (1)}

$\therefore \quad \ln b = \frac{1}{2}$

$\therefore \quad b = e^{\frac{1}{2}} = \sqrt{e}$

When $x = b = \sqrt{e}$, $y = \ln x = \ln e^{\frac{1}{2}} = \frac{1}{2}$

$\therefore$ the point of contact is $(\sqrt{e}, \frac{1}{2})$.

c $a = \frac{1}{2b^2}$ {from (2)}

$\therefore \quad a = \frac{1}{2(\sqrt{e})^2} = \frac{1}{2e}$

d The tangent has gradient $\dfrac{1}{b} = \dfrac{1}{\sqrt{e}}$ and passes through $(\sqrt{e}, \frac{1}{2})$

$\therefore$ the tangent is $\dfrac{y - \frac{1}{2}}{x - \sqrt{e}} = \dfrac{1}{\sqrt{e}}$ $\quad\therefore\quad y - \frac{1}{2} = \dfrac{1}{\sqrt{e}}\left(x - \sqrt{e}\right)$

$$\therefore\quad y - \tfrac{1}{2} = \frac{1}{\sqrt{e}}x - 1$$

$$\therefore\quad y = e^{-\frac{1}{2}}x - \tfrac{1}{2}$$

20 $$P(t) = \frac{50\,000}{1 + 1000e^{-0.5t}}, \quad 0 \leqslant t \leqslant 25$$

$$= 50\,000(1 + 1000e^{-0.5t})^{-1}$$

$$\therefore\quad P'(t) = -50\,000(1 + 1000e^{-0.5t})^{-2}\left(-500e^{-0.5t}\right)$$

$$= 2.5 \times 10^7 e^{-0.5t}(1 + 1000e^{-0.5t})^{-2}$$

The wasp population is growing the fastest when $\dfrac{dP}{dt}$ is a maximum.

Using technology, the graph of $P'(t)$ can be drawn and the maximum obtained.

The maximum occurs when $t \approx 13.8$ weeks.

21 Consider $y = Ate^{-bt}$, $t \geqslant 0$, $A > 0$, $b > 0$

a This function has a y-intercept when $t = 0$, so $y = A(0)e^{-b\times 0} = 0$ $\therefore$ the y-intercept is 0.

It has a t-intercept when $y = 0$, so $Ate^{-bt} = 0$

$\therefore$ $t = 0$, since $e^{-bt} > 0$ for all t, and $A > 0$

$\therefore$ the only t-intercept is 0.

b $$y = Ate^{-bt}$$

$$\therefore\quad \frac{dy}{dt} = Ae^{-bt} + At(-b)e^{-bt} \quad \{\text{product rule}\}$$

$$= (1 - bt)Ae^{-bt}$$

The function has stationary points when $\dfrac{dy}{dt} = 0$.

Since $Ae^{-bt} > 0$ for all t, this occurs when $1 - bt = 0$

$$\therefore\quad t = \frac{1}{b}$$

The sign diagram of $\dfrac{dy}{dt}$ is:

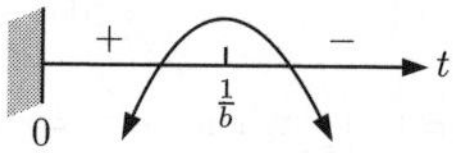

$\therefore$ there is a local maximum at $\left(\dfrac{1}{b}, A\dfrac{1}{b}e^{-b\times\frac{1}{b}}\right)$, or $\left(\dfrac{1}{b}, \dfrac{A}{be}\right)$.

c $$\frac{dy}{dt} = (1 - bt)Ae^{-bt}$$

$$\therefore\quad \frac{d^2y}{dt^2} = -bAe^{-bt} + (1 - bt)(-b)Ae^{-bt} \quad \{\text{product rule}\}$$

$$= Abe^{-bt}(tb - 2)$$

which is 0 when $tb - 2 = 0$ $\{Abe^{-bt} > 0\}$

$$\therefore\quad t = \frac{2}{b}$$

At $t = \dfrac{2}{b}$, $y = A\,\dfrac{2}{b}e^{-b\times\frac{2}{b}} = \dfrac{2A}{be^2}$.

$\therefore$ the point of inflection is $\left(\dfrac{2}{b}, \dfrac{2A}{be^2}\right)$. It is non-stationary, since $\dfrac{dy}{dt} \neq 0$ at $t = \dfrac{2}{b}$.

d

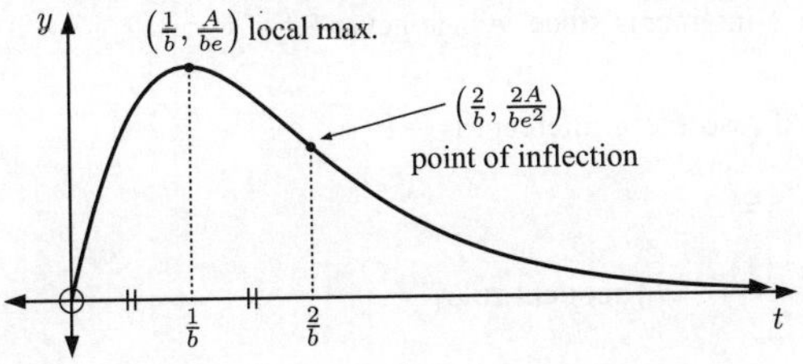

e $E(t)$ has the form Ate^{-bt}, with $A = 750$ and $b = 1.5$.

$\therefore$ the maximum occurs when $t = \frac{1}{b} = \frac{1}{1.5} = \frac{2}{3}$ hour or 40 minutes

REVIEW SET 19A

1 **a**
$$\begin{aligned} y &= e^{x^3+2} \\ &= e^u \quad \text{where } u = x^3+2 \\ \therefore \frac{dy}{dx} &= \frac{dy}{du}\frac{du}{dx} \quad \{\text{chain rule}\} \\ &= e^u \times 3x^2 \\ &= 3x^2e^u \\ &= 3x^2e^{x^3+2} \end{aligned}$$

b
$$\begin{aligned} y &= \ln\left(\frac{x+3}{x^2}\right) \\ &= \ln(x+3) - \ln(x^2) \\ \therefore \frac{dy}{dx} &= \frac{1}{x+3} - \frac{2x}{x^2} \\ &= \frac{1}{x+3} - \frac{2}{x} \end{aligned}$$

2 $y = e^{-x^2}$ so when $x = 1$,

$$y = e^{-1} = \frac{1}{e}$$

$\therefore$ the point of contact is $\left(1, \frac{1}{e}\right)$

Now $\dfrac{dy}{dx} = -2xe^{-x^2}$

$\therefore$ when $x = 1$, $\dfrac{dy}{dx} = -2e^{-1}$

$\therefore$ the gradient of the tangent is $-\dfrac{2}{e}$

and the gradient of the normal is $\dfrac{e}{2}$

$\therefore$ the equation of the normal is $\dfrac{y - \frac{1}{e}}{x-1} = \dfrac{e}{2}$

$$\begin{aligned} \therefore\ 2\left(y - \frac{1}{e}\right) &= e(x-1) \\ \therefore\ 2y - \frac{2}{e} &= ex - e \\ \therefore\ 2y &= ex + \frac{2}{e} - e \\ \therefore\ y &= \frac{e}{2}x + \frac{1}{e} - \frac{e}{2} \end{aligned}$$

3 **a** $f(x)$ and $g(x)$ intersect when $e^{2x} = -e^x + 6$

$$\begin{aligned} \therefore\ (e^x)^2 + e^x - 6 &= 0 \\ \therefore\ (e^x+3)(e^x-2) &= 0 \\ \therefore\ e^x &= -3 \text{ or } 2 \end{aligned}$$

But $e^x > 0$ for all x, so $e^x = 2$

$$\therefore\ x = \ln 2$$

$f(\ln 2) = e^{2\ln 2} = (e^{\ln 2})^2 = 4$

$\therefore$ the point of intersection P is $(\ln 2, 4)$.

b $f'(x) = 2e^{2x}$, so the gradient of the tangent at P is $f'(\ln 2) = 2 \times e^{2\ln 2} = 8$

$\therefore$ the tangent has equation $\dfrac{y-4}{x-\ln 2} = 8$

$$\begin{aligned} \therefore\ y - 4 &= 8x - 8\ln 2 \\ \therefore\ y &= 8x + 4 - 8\ln 2 \end{aligned}$$

4 **a** $f(x) = \dfrac{e^x}{x-1}$ has no x-intercepts since e^x is never 0.

Now $f(0) = \dfrac{e^0}{-1} = -1$ so the y-intercept is -1.

b $f(x)$ is defined for all $x \neq 1$.

c $f'(x) = \dfrac{e^x(x-1) - e^x(1)}{(x-1)^2}$ {quotient rule}

$= \dfrac{e^x(x-2)}{(x-1)^2}$ and has sign diagram:

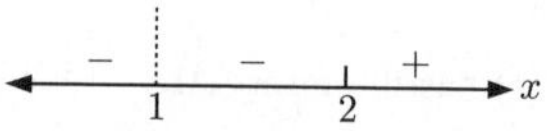

$\therefore$ $f(x)$ is decreasing for $x < 1$ and $1 < x \leqslant 2$, and increasing for $x \geqslant 2$.

$$f''(x) = \frac{[e^x(x-2) + e^x(1)](x-1)^2 - e^x(x-2)[2(x-1)^1(1)]}{(x-1)^4}$$ {product and quotient rules}

$$= \frac{[e^x(x-2+1)(x-1)^2] - 2e^x(x-2)(x-1)}{(x-1)^4}$$

$$= \frac{e^x(x-1)(x-1)^2 - 2e^x(x-2)(x-1)}{(x-1)^4}$$

$$= \frac{e^x(x-1)\,[(x-1)^2 - 2(x-2)]}{(x-1)^4}$$

$$= \frac{e^x(x-1)\,[x^2 - 2x + 1 - 2x + 4]}{(x-1)^4}$$

$$= \frac{e^x(x^2 - 4x + 5)}{(x-1)^3}$$ where the quadratic term has $\Delta < 0$

The sign diagram of $f''(x)$ is:

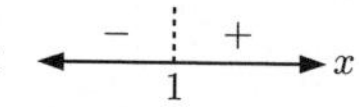

$\therefore$ $f(x)$ is concave down for all $x < 1$ and concave up for all $x > 1$.

d

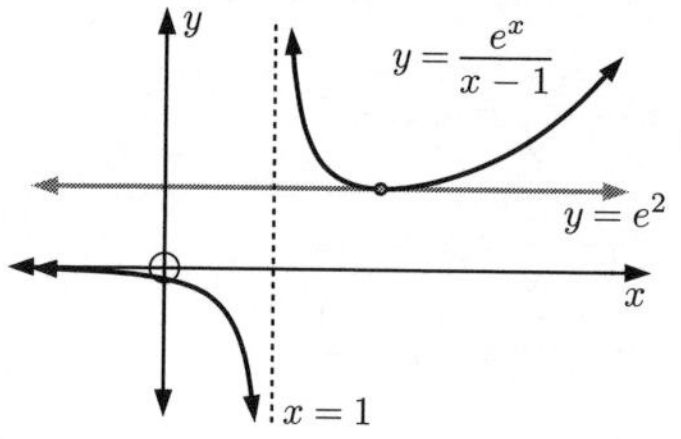

e Now $f(2) = \dfrac{e^2}{2-1} = e^2$

Using **c** we have a local minimum at $(2,\ e^2)$

$\therefore$ the tangent at $x = 2$ is horizontal and is $y = e^2$.

5 $y = \ln(x^2 + 3)$ $\quad\therefore\quad \dfrac{dy}{dx} = \dfrac{2x}{x^2+3}$

When $x = 0$, $\dfrac{dy}{dx} = 0$ so the gradient of the tangent at this point is 0.

But when $x = 0$, $y = \ln(0+3) = \ln 3$

$\therefore$ the tangent is $y = \ln 3$ which does not cut the x-axis.

6 $f(x) = e^{4x} + px + q$

$\therefore$ $f'(x) = 4e^{4x} + p$

At the point where $x = 0$, the tangent to $f(x)$ has equation $y = 5x - 7$, so $f'(0) = 5$

$\therefore$ $4e^0 + p = 5$

$\therefore$ $p = 1$

The tangent meets $f(x)$ when $x = 0$ and $y = 5(0) - 7 = -7$, so $(0, -7)$ must lie on $f(x)$ too.

$\therefore$ $e^{4(0)} + p(0) + q = -7$

$\therefore$ $1 + q = -7$

$\therefore$ $q = -8$

7 **a**
$$3e^x - 5 = -2e^{-x}$$
$$\therefore\ 3e^{2x} - 5e^x = -2 \quad \{\times e^x\}$$
$$\therefore\ 3e^{2x} - 5e^x + 2 = 0$$
$$\therefore\ (3e^x - 2)(e^x - 1) = 0$$
$$\therefore\ e^x = \tfrac{2}{3} \text{ or } 1$$
$$\therefore\ x = \ln\tfrac{2}{3} \text{ or } 0$$

b
$$2\ln x - 3\ln\left(\frac{1}{x}\right) = 10$$
$$\therefore\ 2\ln x - 3\ln(x^{-1}) = 10$$
$$\therefore\ 2\ln x + 3\ln x = 10$$
$$\therefore\ 5\ln x = 10$$
$$\therefore\ \ln x = 2$$
$$\therefore\ x = e^2$$

REVIEW SET 19B

1 $H(t) = 60 + 40\ln(2t+1)$ cm, $t \geqslant 0$

a When first planted, $t = 0$ $\quad \therefore\ H(0) = 60 + 40\ln(1) = 60 + 40(0) = 60$ cm.

b **i** When $H(t) = 150$ cm,
$$\therefore\ 60 + 40\ln(2t+1) = 150$$
$$\therefore\ 40\ln(2t+1) = 90$$
$$\therefore\ \ln(2t+1) = \tfrac{90}{40} = 2.25$$
$$\therefore\ 2t + 1 = e^{2.25}$$
$$\therefore\ 2t = e^{2.25} - 1$$
$$\therefore\ t = \tfrac{1}{2}(e^{2.25} - 1)$$
$$\therefore\ t \approx 4.24 \text{ years}$$

ii When $H(t) = 300$ cm,
$$\therefore\ 60 + 40\ln(2t+1) = 300$$
$$\therefore\ 40\ln(2t+1) = 240$$
$$\therefore\ \ln(2t+1) = 6$$
$$\therefore\ 2t + 1 = e^6$$
$$\therefore\ 2t = e^6 - 1$$
$$\therefore\ t = \tfrac{1}{2}(e^6 - 1)$$
$$\therefore\ t \approx 201 \text{ years}$$

c $H'(t) = 40\left(\dfrac{2}{2t+1}\right) = \dfrac{80}{2t+1}$ cm per year

i When $t = 2$, $H'(2) = \frac{80}{5} = 16$ cm per year

ii When $t = 20$, $H'(20) = \frac{80}{41} \approx 1.95$ cm per year

2 $s(t) = 80e^{-\frac{t}{10}} - 40t$ metres, $t \geqslant 0$

a $v(t) = s'(t) = -8e^{-\frac{t}{10}} - 40$ m s^{-1}

$a(t) = v'(t) = 0.8e^{-\frac{t}{10}}$ m s^{-2}

b When $t = 0$, $s(0) = 80$ m
$v(0) = -48$ m s^{-1}
$a(0) = 0.8$ m s^{-2}

c As $t \to \infty$, $e^{-\frac{t}{10}} \to 0$ $\quad \therefore\ v(t) \to -40$ m s^{-1} (below)

d

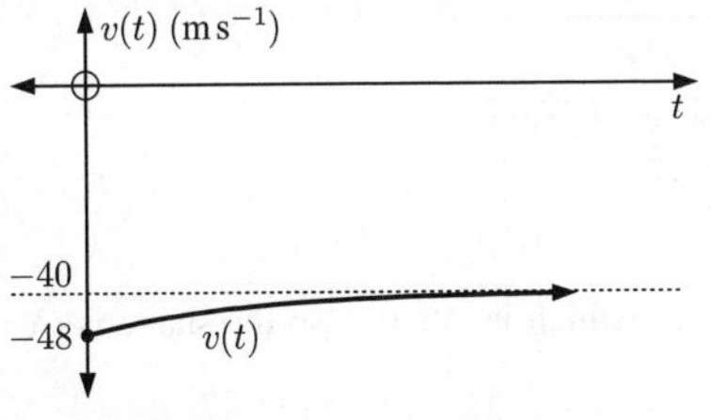

e When $v(t) = -44$ m s^{-1}
$$\therefore\ -8e^{-\frac{t}{10}} - 40 = -44$$
$$\therefore\ -8e^{-\frac{t}{10}} = -4$$
$$\therefore\ e^{-\frac{t}{10}} = 0.5$$
$$\therefore\ -\tfrac{t}{10} = \ln 0.5$$
$$\therefore\ t = -10\ln 0.5$$
$$\therefore\ t \approx 6.93 \text{ seconds}$$

3 $P(x) = R(x) - C(x)$
$$= \left[200\ln\left(1 + \frac{x}{100}\right) + 1000\right] - \left[(x-100)^2 + 200\right]$$
$$= 200\ln(1 + 0.01x) - (x-100)^2 + 800$$

$$\therefore \quad \frac{dP}{dx} = 200\left(\frac{0.01}{1+0.01x}\right) - 2(x-100)^1$$
$$= \frac{2}{1+0.01x} - \frac{2(x-100)}{1}$$
$$= \frac{2-2(x-100)(1+0.01x)}{1+0.01x}$$
$$= \frac{2-2(x+0.01x^2-100-x)}{1+0.01x}$$
$$= \frac{2-0.02x^2+200}{1+0.01x}$$
$$= \frac{202-0.02x^2}{1+0.01x}$$

$\therefore \quad \frac{dP}{dx} = 0$ when $0.02x^2 = 202$

$\therefore \quad x^2 = 10\,100$

$\therefore \quad x = \sqrt{10\,100} \quad \{x > 0\}$

$\therefore \quad x \approx 100.49$

and the sign diagram of $\frac{dP}{dx}$ is: 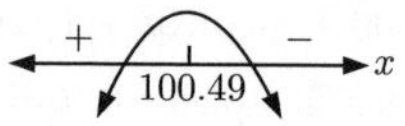

$\therefore$ the maximum profit occurs when $x \approx 100.49$

Now $P(100) \approx \$938.63$ and $P(101) \approx \$938.63$

$\therefore$ the maximum daily profit is \$938.63 when 100 or 101 shirts are made.

4 **a** At time $t = 0$, $V = 20\,000e^{-0.4\times 0}$
$= 20\,000$ dollars

$\therefore$ the purchase price of the car was \$20 000.

b $V' = -0.4(20\,000)e^{-0.4t}$
$= -8000e^{-0.4t}$

At time $t = 10$, $V' = -8000e^{-0.4\times 10}$
≈ -146.53 dollars year^{-1}

$\therefore$ after 10 years, the car is decreasing in value at \$147 per year.

5 $$C(x) = 10\ln x + \left(20 - \frac{x}{10}\right)^2 = 10\ln x + 400 - 4x + \frac{x^2}{100}$$

$$\therefore \quad C'(x) = \frac{10}{x} - 4 + \frac{x}{50} = \frac{500-200x+x^2}{50x}$$

$\therefore \quad C'(x) = 0$ when $x^2 - 200x + 500 = 0$

$$\therefore \quad x = \frac{200 \pm \sqrt{38\,000}}{2}$$
$$\approx 2.53 \text{ or } 197.47$$

But $x \geqslant 50$, so $x \approx 197.47$

Now $C''(x) = -10x^{-2} + \frac{1}{50}$

$\therefore \quad C''(197.47) = -10(197.47)^{-2} + 0.02 \approx 0.02$ which is > 0, so the shape is 

$\therefore$ the minimum cost is when $x \approx 197.47$

Sign diagram for $C'(x)$: 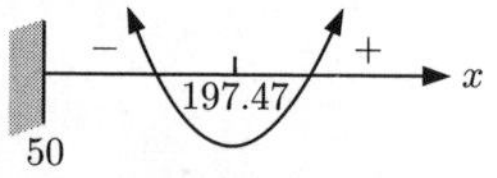

Now $C(197) \approx 52.92$ and $C(198) \approx 52.92$

$\therefore$ the manufacturer needs to produce 197 or 198 clocks per day to minimise costs.

6 **a** $s(t) = 25t - 10\ln t$ cm, $t \geqslant 1$

$\therefore \quad v(t) = 25 - \dfrac{10}{t}$ cm min^{-1}

$\therefore \quad a(t) = 10t^{-2}$

$= \dfrac{10}{t^2}$ cm min^{-2}

b When $t = e$,

$s(e) = 25e - 10\ln e = 25e - 10$ cm

≈ 58.0 cm

$v(e) = 25 - \dfrac{10}{e}$ cm min^{-1}

≈ 21.3 cm min^{-1}

$a(e) = \dfrac{10}{e^2}$ cm min^{-2} ≈ 1.35 cm min^{-2}

c As $t \to \infty$, $\dfrac{10}{t} \to 0$ $\quad \therefore \quad v(t) \to 25$ cm min^{-1} (below)

d

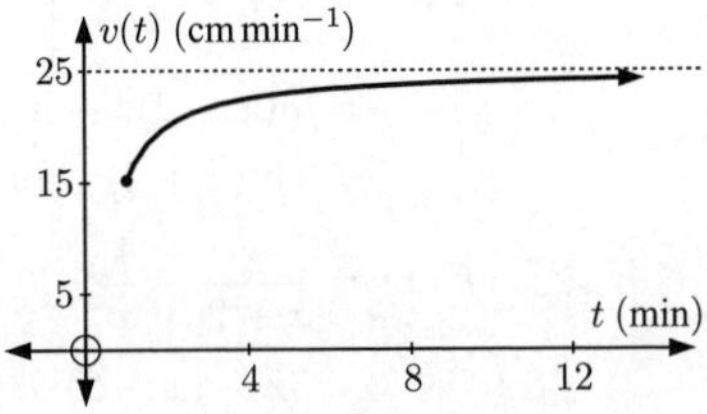

e When $v(t) = 20$ cm min^{-1},

$25 - \dfrac{10}{t} = 20$

$\therefore \quad \dfrac{10}{t} = 5$

$\therefore \quad t = 2$ minutes

REVIEW SET 19C

1 **a** $y = \ln(x^3 - 3x)$

$\therefore \quad \dfrac{dy}{dx} = \dfrac{3x^2 - 3}{x^3 - 3x}$

b $y = \dfrac{e^x}{x^2}$

$\therefore \quad \dfrac{dy}{dx} = \dfrac{e^x(x^2) - e^x(2x)}{x^4}$ {quotient rule}

$= \dfrac{e^x(x-2)}{x^3}$

2 $y = \ln(x^4 + 3)$

$\therefore \quad \dfrac{dy}{dx} = \dfrac{4x^3}{x^4 + 3}$

$\therefore$ when $x = 1$, $\dfrac{dy}{dx} = \dfrac{4(1)^3}{1^4 + 3} = 1$ and $y = \ln(1^4 + 3) = \ln 4$

$\therefore$ the tangent has equation $\dfrac{y - \ln 4}{x - 1} = 1$ or $y = x - 1 + \ln 4$

Now when $x = 0$, $y = \ln 4 - 1$, so the tangent cuts the y-axis at $(0, \ln 4 - 1)$.

3 **a** $e^{2x} = 3e^x$

$\therefore \quad e^{2x} - 3e^x = 0$

$\therefore \quad e^x(e^x - 3) = 0$

$\therefore \quad e^x = 0$ or 3

$\therefore \quad e^x = 3$ {as $e^x > 0$}

$\therefore \quad x = \ln 3$

b $e^{2x} - 7e^x + 12 = 0$

$\therefore \quad (e^x - 3)(e^x - 4) = 0$

$\therefore \quad e^x = 3$ or 4

$\therefore \quad x = \ln 3$ or $\ln 4$

4 $f(x) = e^x - x$

a $f'(x) = e^x - 1$

so $f'(x) = 0$ when $e^x = 1$

$\therefore \quad x = 0$

Sign diagram of $f'(x)$ is:

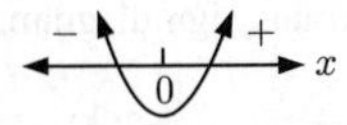

Now $f(0) = e^0 - 0 = 1$

$\therefore$ there is a local minimum at $(0, 1)$.

b As $x \to \infty$, $e^x \to \infty$ faster than x

$\therefore \quad f(x) \to \infty$

c $f''(x) = e^x$

$\therefore \quad f''(x) > 0$ for all x

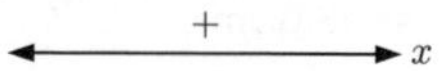

$\therefore \quad f(x)$ is concave up for all x

d

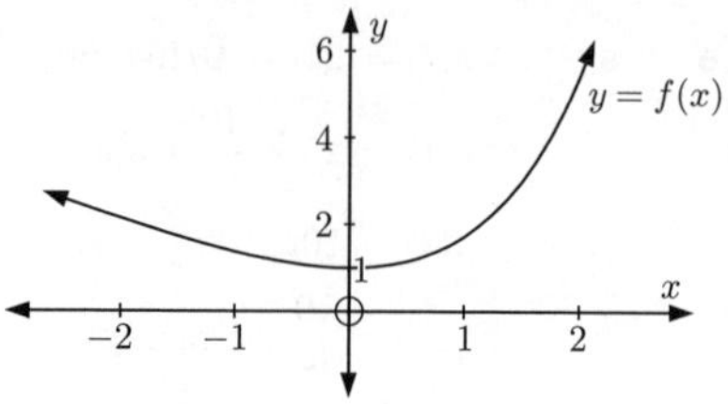

e Since a local minimum exists at $(0, 1)$,

$f(x) \geqslant 1$ for all x

$\therefore \quad e^x - x \geqslant 1$

$\therefore \quad e^x \geqslant x + 1$ for all x

5 **a** $f(x) = \ln(e^x + 3)$

$\therefore \quad f'(x) = \dfrac{e^x}{e^x + 3}$

b
$$\begin{aligned} f(x) &= \ln\left[\frac{(x+2)^3}{x}\right] \\ &= \ln(x+2)^3 - \ln x \\ &= 3\ln(x+2) - \ln x \\ \therefore \quad f'(x) &= \frac{3}{x+2} - \frac{1}{x} \\ &= \frac{3x - (x+2)}{x(x+2)} \\ &= \frac{2x-2}{x(x+2)} \end{aligned}$$

6 **a** $f(x) = x + \ln x$ is defined when $x > 0$

b $f'(x) = 1 + \dfrac{1}{x} = \dfrac{x+1}{x}$ which has sign diagram:

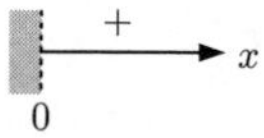

$\therefore \quad f(x)$ is increasing for all $x > 0$.

$f''(x) = -\dfrac{1}{x^2}$ which has sign diagram:

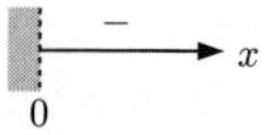

$\therefore \quad f(x)$ is concave down for all $x > 0$.

c

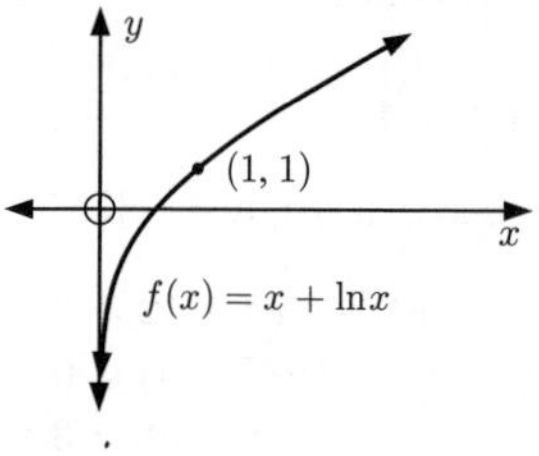

d $f(1) = 1 + \ln(1) = 1$

$\therefore \quad (1, 1)$ is the point of contact.

$f'(1) = \dfrac{1+1}{1} = 2$

$\therefore$ the tangent at $x = 1$ has gradient 2, so the normal has gradient $-\frac{1}{2}$

$\therefore$ the normal has equation $\dfrac{y-1}{x-1} = -\frac{1}{2}$

$\therefore \quad 2y - 2 = -x + 1$

$\therefore \quad x + 2y = 3$

7 Let the coordinates of B be $(x, 0)$, so the coordinates of A are (x, e^{-2x}).

$\therefore$ the area OBAC is $A = xe^{-2x}$

$$\begin{aligned} \therefore \quad \frac{dA}{dx} &= (1)e^{-2x} + x(-2e^{-2x}) \quad \{\text{product rule}\} \\ &= e^{-2x}(1-2x) \\ &= \frac{1-2x}{e^{2x}} \end{aligned}$$

and has sign diagram:

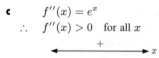

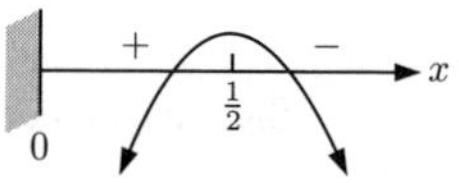

So, the maximum area occurs when $x = \frac{1}{2}$ and $y = e^{-2(\frac{1}{2})} = e^{-1} = \frac{1}{e}$

$\therefore$ the coordinates of A are $\left(\frac{1}{2}, \frac{1}{e}\right)$.

Chapter 20

DERIVATIVES OF TRIGONOMETRIC FUNCTIONS

EXERCISE 20A

1 **a** $y = \sin(2x)$

$\therefore \quad \dfrac{dy}{dx} = \cos(2x)\,\dfrac{d}{dx}(2x)$

$= 2\cos(2x)$

b $y = \sin x + \cos x$

$\therefore \quad \dfrac{dy}{dx} = \cos x - \sin x$

c $y = \cos(3x) - \sin x$

$\therefore \quad \dfrac{dy}{dx} = -\sin(3x) \times 3 - \cos x$

$= -3\sin(3x) - \cos x$

d $y = \sin(x+1)$

$\therefore \quad \dfrac{dy}{dx} = \cos(x+1)\,\dfrac{d}{dx}(x+1)$

$= 1\cos(x+1)$

$= \cos(x+1)$

e $y = \cos(3-2x)$

$\therefore \quad \dfrac{dy}{dx} = -\sin(3-2x) \times -2$

$= 2\sin(3-2x)$

f $y = \tan(5x)$

$\therefore \quad \dfrac{dy}{dx} = \dfrac{1}{\cos^2(5x)} \times 5$

$= \dfrac{5}{\cos^2(5x)}$

g $y = \sin\left(\frac{x}{2}\right) - 3\cos x$

$\therefore \quad \dfrac{dy}{dx} = \frac{1}{2}\cos\left(\frac{x}{2}\right) + 3\sin x$

h $y = 3\tan(\pi x)$

$\therefore \quad \dfrac{dy}{dx} = 3 \times \dfrac{1}{\cos^2(\pi x)} \times \pi$

$= \dfrac{3\pi}{\cos^2(\pi x)}$

i $y = 4\sin x - \cos(2x)$

$\therefore \quad \dfrac{dy}{dx} = 4\cos x + \sin(2x) \times 2$

$= 4\cos x + 2\sin(2x)$

2 **a** $y = x^2 + \cos x$

$\therefore \quad \dfrac{dy}{dx} = 2x - \sin x$

b $y = \tan x - 3\sin x$

$\therefore \quad \dfrac{dy}{dx} = \dfrac{1}{\cos^2 x} - 3\cos x$

c $y = e^x \cos x$

$\therefore \quad \dfrac{dy}{dx} = e^x \cos x + e^x(-\sin x)$

$= e^x \cos x - e^x \sin x$

d $y = e^{-x}\sin x$

$\therefore \quad \dfrac{dy}{dx} = -e^{-x}\sin x + e^{-x}\cos x$

e $y = \ln(\sin x)$

$\therefore \quad \dfrac{dy}{dx} = \dfrac{\cos x}{\sin x}$

f $y = e^{2x}\tan x$

$\therefore \quad \dfrac{dy}{dx} = 2e^{2x}\tan x + \dfrac{e^{2x}}{\cos^2 x}$

g $y = \sin(3x)$

$\therefore \quad \dfrac{dy}{dx} = 3\cos(3x)$

h $y = \cos(\frac{x}{2})$

$\therefore \quad \dfrac{dy}{dx} = -\frac{1}{2}\sin(\frac{x}{2})$

i $y = 3\tan(2x)$

$\therefore \frac{dy}{dx} = \frac{3}{\cos^2(2x)} \times 2$

$= \frac{6}{\cos^2(2x)}$

j $y = x\cos x$

$\therefore \frac{dy}{dx} = 1 \times \cos x + x(-\sin x)$

$= \cos x - x\sin x$

k $y = \frac{\sin x}{x}$

$\therefore \frac{dy}{dx} = \frac{(\cos x)(x) - \sin x \times 1}{x^2}$

$= \frac{x\cos x - \sin x}{x^2}$

l $y = x\tan x$

$\therefore \frac{dy}{dx} = 1 \times \tan x + x \times \frac{1}{\cos^2 x}$

$= \tan x + \frac{x}{\cos^2 x}$

3 **a** $y = \sin(x^2)$

$\therefore \frac{dy}{dx} = 2x\cos(x^2)$

b $y = \cos\left(\sqrt{x}\right) = \cos(x^{\frac{1}{2}})$

$\therefore \frac{dy}{dx} = -\sin(x^{\frac{1}{2}}) \times \frac{1}{2}x^{-\frac{1}{2}}$

$= -\frac{1}{2\sqrt{x}}\sin(\sqrt{x})$

c $y = \sqrt{\cos x} = (\cos x)^{\frac{1}{2}}$

$\therefore \frac{dy}{dx} = \frac{1}{2}(\cos x)^{-\frac{1}{2}} \times (-\sin x)$

$= -\frac{\sin x}{2\sqrt{\cos x}}$

d $y = \sin^2 x = (\sin x)^2$

$\therefore \frac{dy}{dx} = 2\sin x\cos x$

e $y = \cos^3 x = (\cos x)^3$

$\therefore \frac{dy}{dx} = 3\cos^2 x \times (-\sin x)$

$= -3\sin x\cos^2 x$

f $y = \cos x\sin(2x)$

$\therefore \frac{dy}{dx} = (-\sin x)\sin(2x) + \cos x(2\cos(2x))$

$= -\sin x\sin(2x) + 2\cos x\cos(2x)$

g $y = \cos(\cos x)$

$\therefore \frac{dy}{dx} = -\sin(\cos x) \times (-\sin x)$

$= \sin x\sin(\cos x)$

h $y = \cos^3(4x) = (\cos(4x))^3$

$\therefore \frac{dy}{dx} = 3(\cos(4x))^2 \times (-4\sin(4x))$

$= -12\sin(4x)\cos^2(4x)$

i $y = \frac{1}{\sin x} = (\sin x)^{-1}$

$\therefore \frac{dy}{dx} = -1(\sin x)^{-2} \times \cos x$

$= -\frac{\cos x}{\sin^2 x}$

j $y = \frac{1}{\cos(2x)} = (\cos(2x))^{-1}$

$\therefore \frac{dy}{dx} = -1(\cos(2x))^{-2} \times (-2\sin(2x))$

$= \frac{2\sin(2x)}{\cos^2(2x)}$

k $y = \frac{2}{\sin^2(2x)} = 2(\sin(2x))^{-2}$

$\therefore \frac{dy}{dx} = -4(\sin(2x))^{-3} \times 2\cos(2x)$

$= -\frac{8\cos(2x)}{\sin^3(2x)}$

l $y = \frac{8}{\tan^3\left(\frac{x}{2}\right)} = 8\left[\tan\left(\frac{x}{2}\right)\right]^{-3}$

$\therefore \frac{dy}{dx} = -24\left[\tan\left(\frac{x}{2}\right)\right]^{-4} \times \frac{1}{2} \times \frac{1}{\cos^2\left(\frac{x}{2}\right)}$

$= \frac{-12}{\cos^2\left(\frac{x}{2}\right)\tan^4\left(\frac{x}{2}\right)}$

4 **a** $f(x) = 2\sin^3 x - 3\sin x$

$= 2(\sin x)^3 - 3\sin x$

$\therefore f'(x) = 2 \times 3(\sin x)^2 \times (\cos x) - 3\cos x$

$= -3\cos x(1 - 2\sin^2 x)$

$= -3\cos x\cos 2x$

b $f''(x) = -3(-\sin x \times \cos 2x + \cos x \times (-2)\sin 2x)$

$= 3\sin x\cos 2x + 6\cos x\sin 2x$

5 **a** If $y = \sin(2x+3)$, then $\dfrac{dy}{dx} = 2\cos(2x+3)$ and $\dfrac{d^2y}{dx^2} = -4\sin(2x+3)$

$\therefore \quad \dfrac{d^2y}{dx^2} + 4y = -4\sin(2x+3) + 4\sin(2x+3) = 0$

b If $y = 2\sin x + 3\cos x$, then $y' = 2\cos x - 3\sin x$ and $y'' = -2\sin x - 3\cos x$

$\therefore \quad y'' + y = -2\sin x - 3\cos x + 2\sin x + 3\cos x = 0$

c $y = \dfrac{\cos x}{1+\sin x} \quad \therefore \quad \dfrac{dy}{dx} = \dfrac{(-\sin x)(1+\sin x) - \cos x(\cos x)}{(1+\sin x)^2}$

$$= \frac{-\sin x - \sin^2 x - \cos^2 x}{(1+\sin x)^2}$$

$$= \frac{-1-\sin x}{(1+\sin x)^2} \qquad \{\sin^2 x + \cos^2 x = 1\}$$

$$= -\frac{(1+\sin x)}{(1+\sin x)^2}$$

$$= \frac{-1}{1+\sin x}$$

Since $\dfrac{-1}{1+\sin x}$ never equals 0, there are no horizontal tangents.

6 **a** $y = \sin x \quad \therefore \quad \dfrac{dy}{dx} = \cos x$

When $x = 0$, $\dfrac{dy}{dx} = \cos 0 = 1$

$\therefore$ the tangent has equation $\dfrac{y-0}{x-0} = 1$

or $y = x$

b $y = \tan x \quad \therefore \quad \dfrac{dy}{dx} = \dfrac{1}{\cos^2 x}$

When $x = 0$, $\dfrac{dy}{dx} = \dfrac{1}{\cos^2 0} = 1$

$\therefore$ the tangent has equation $\dfrac{y-0}{x-0} = 1$

or $y = x$

c $y = \cos x \quad \therefore \quad \dfrac{dy}{dx} = -\sin x$

When $x = \frac{\pi}{6}$, $y = \frac{\sqrt{3}}{2}$

and $\dfrac{dy}{dx} = -\sin\frac{\pi}{6} = -\frac{1}{2}$

So, the normal has gradient 2,

and its equation is $\dfrac{y - \frac{\sqrt{3}}{2}}{x - \frac{\pi}{6}} = 2$

$\therefore \quad y - \frac{\sqrt{3}}{2} = 2x - \frac{\pi}{3}$

$\therefore \quad 2x - y = \frac{\pi}{3} - \frac{\sqrt{3}}{2}$

d $y = \dfrac{1}{\sin(2x)} = (\sin(2x))^{-1}$

$\therefore \quad \dfrac{dy}{dx} = -1(\sin(2x))^{-2} \times 2\cos(2x)$

$$= -\frac{2\cos(2x)}{(\sin(2x))^2}$$

When $x = \frac{\pi}{4}$, $y = 1$

and $\dfrac{dy}{dx} = -\dfrac{2\cos\frac{\pi}{2}}{\left(\sin\frac{\pi}{2}\right)^2} = 0$

$\therefore$ the gradient of the normal is undefined, so the normal is $x = \frac{\pi}{4}$.

7 $d = 9.3 + 6.8\cos(0.507t)$ m

$\therefore \quad \dfrac{dd}{dt} = -6.8\sin(0.507t) \times 0.507$

$= -3.4476\sin(0.507t)$

a When $t = 8$, $\dfrac{dd}{dt} \approx 2.731 > 0$

$\therefore$ the tide is rising.

b When $t = 8$, the tide is rising at the rate of 2.73 m per hour.

8 **a** $V(t) = 340\sin(100\pi t)$

$\therefore \quad V'(t) = 340\cos(100\pi t) \times 100\pi$

$= 34\,000\pi\cos(100\pi t)$

When $t = 0.01$,

$V'(0.01) = 34\,000\pi \times \cos\pi$

$= -34\,000\pi$ units per second

b When $V(t)$ is a maximum, $V'(t)$ must be 0 units per second.

9 **a** The distance from $A(-x, 0)$ to $P(\cos t, \sin t)$ is fixed at 2 m.

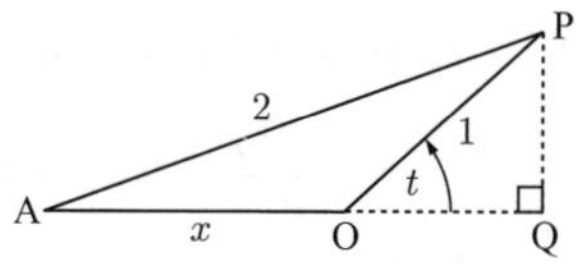

$\cos t = \dfrac{OQ}{1} = OQ$

$\therefore\ (\cos t + x)^2 + \sin^2 t = 2^2$ {Pythagoras in triangle APQ}

$\therefore\ (\cos t + x)^2 = 4 - \sin^2 t$

$\therefore\ x + \cos t = \pm\sqrt{4 - \sin^2 t}$

$\therefore$ since $x > 0$, $x = \sqrt{4 - \sin^2 t} - \cos t$

b Now $\dfrac{dx}{dt} = \frac{1}{2}(4 - \sin^2 t)^{-\frac{1}{2}}(-2\sin t\cos t) + \sin t$

$= \dfrac{-\sin t\cos t}{\sqrt{4 - \sin^2 t}} + \sin t$

i When $t = 0$,
$\sin t = 0$ and $\cos t = 1$

$\therefore\ \dfrac{dx}{dt} = 0 + 0$

$= 0\ \text{m s}^{-1}$

ii When $t = \frac{\pi}{2}$,
$\sin t = 1$ and $\cos t = 0$

$\therefore\ \dfrac{dx}{dt} = 0 + \sin\frac{\pi}{2}$

$= 1\ \text{m s}^{-1}$

iii When $t = \frac{2\pi}{3}$,
$\sin t = \frac{\sqrt{3}}{2}$ and $\cos t = -\frac{1}{2}$

$\therefore\ \dfrac{dx}{dt} = \dfrac{-\frac{\sqrt{3}}{2}(-\frac{1}{2})}{\sqrt{4 - \frac{3}{4}}} + \dfrac{\sqrt{3}}{2}$

$\approx 1.11\ \text{m s}^{-1}$

10 **a** If $f(x) = \sin x$ then $f'(x) = \cos x$

Stationary points occur when $f'(x) = 0$,
which is when $x = \frac{\pi}{2}, \frac{3\pi}{2}$

Sign diagram for $f'(x)$ is:

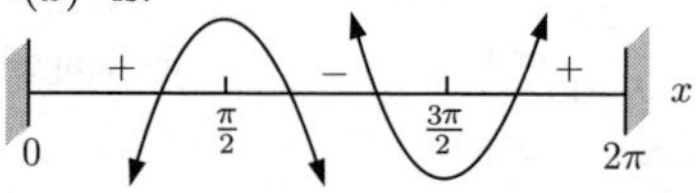

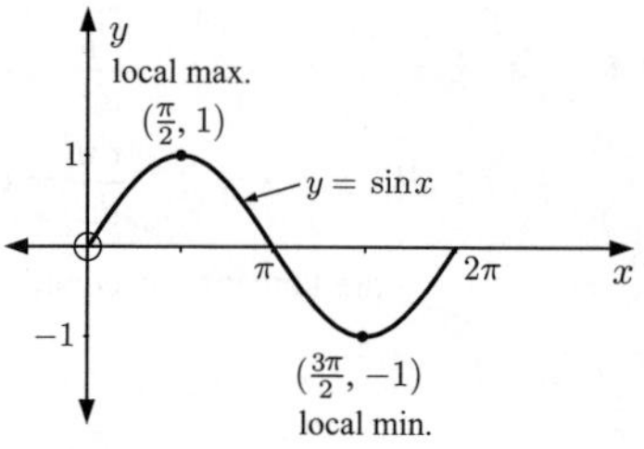

There is a local maximum at $(\frac{\pi}{2}, 1)$
and a local minimum at $(\frac{3\pi}{2}, -1)$.

b If $f(x) = \cos(2x)$ then $f'(x) = -2\sin(2x)$

$\therefore\ f'(x) = 0$ when $-2\sin(2x) = 0$

$\therefore\ \sin(2x) = 0$

$\therefore\ 2x = k\pi$ for any integer k

$\therefore\ x = \dfrac{k\pi}{2}$

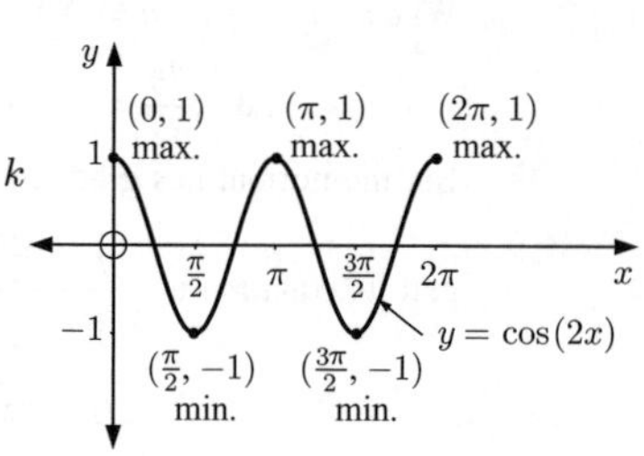

On the domain $0 \leqslant x \leqslant 2\pi$, $f'(x) = 0$
when $x = 0, \frac{\pi}{2}, \pi, \frac{3\pi}{2}$ and 2π.

Sign diagram for $f'(x)$ is:

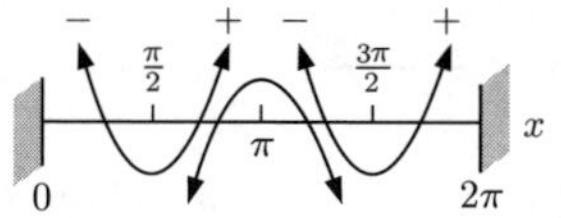

There are local maxima at $(0, 1)$, $(\pi, 1)$, $(2\pi, 1)$ and local minima at $(\frac{\pi}{2}, -1)$, $(\frac{3\pi}{2}, -1)$.

c If $f(x) = \sin^2 x$ then $f'(x) = 2\sin x\cos x = \sin(2x)$

$\therefore\ f'(x) = 0$ when $\sin(2x) = 0$

Using **b**, we know that on the domain $0 \leqslant x \leqslant 2\pi$
$f'(x) = 0$ when $x = 0, \frac{\pi}{2}, \pi, \frac{3\pi}{2}$ and 2π.

Sign diagram for $f'(x)$ is:

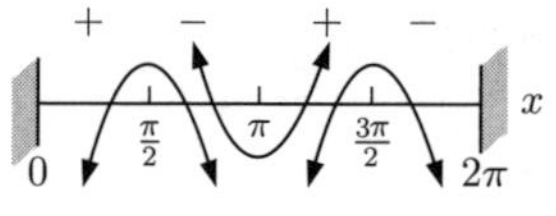

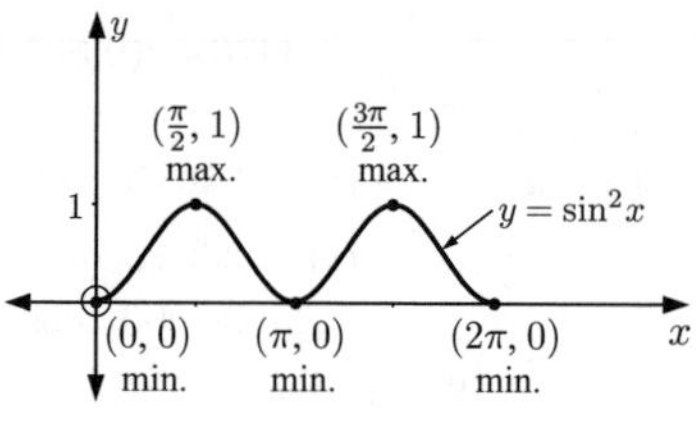

There are local mimima at $(0, 0)$, $(\pi, 0)$, $(2\pi, 0)$ and local maxima at $(\frac{\pi}{2}, 1)$, $(\frac{3\pi}{2}, 1)$.

d If $f(x) = e^{\sin x}$ then $f'(x) = e^{\sin x} \times \cos x$

$\therefore\ f'(x) = 0$ when $\cos x\, e^{\sin x} = 0$

$\therefore\ \cos x = 0 \quad \{e^{\sin x} > 0 \text{ for all } x\}$

$\therefore\ x = \frac{\pi}{2} + k\pi,\quad k$ an integer

On the domain $0 \leqslant x \leqslant 2\pi$, $f'(x) = 0$ when $x = \frac{\pi}{2}, \frac{3\pi}{2}$.

Sign diagram for $f'(x)$ is:

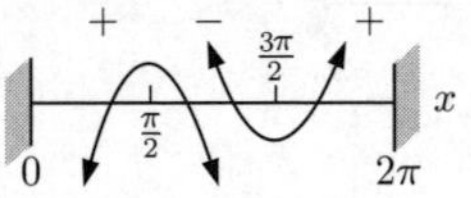

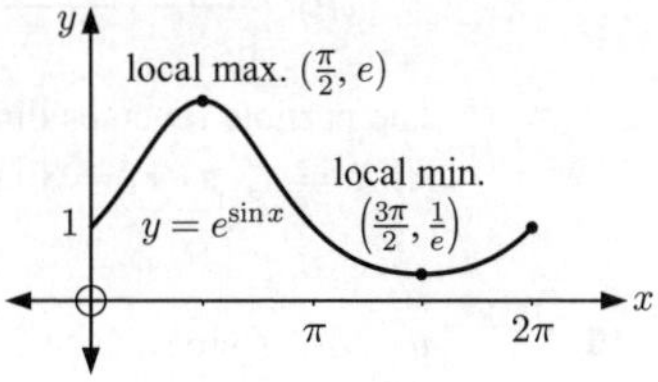

There is a local maximum at $(\frac{\pi}{2}, e)$ and a local minimum at $(\frac{3\pi}{2}, \frac{1}{e})$.

e If $f(x) = \sin(2x) + 2\cos x$ then $f'(x) = 2\cos(2x) - 2\sin x$

$\therefore\ f'(x) = 0$ when $2\cos(2x) - 2\sin x = 0$

$\therefore\ 2(1 - 2\sin^2 x) - 2\sin x = 0$

$\therefore\ -2(2\sin^2 x + \sin x - 1) = 0$

$\therefore\ -2(2\sin x - 1)(\sin x + 1) = 0$

$\therefore$ when $\sin x = \frac{1}{2}$ or $\sin x = -1$

On the domain $0 \leqslant x \leqslant 2\pi$, when $x = \frac{\pi}{6}, \frac{5\pi}{6}, \frac{3\pi}{2}$.

Sign diagram of $f'(x)$:

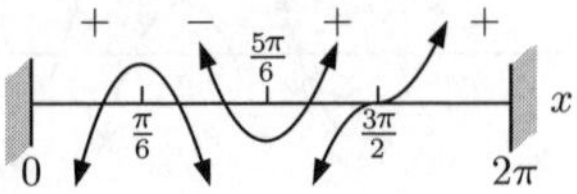

$f(\frac{\pi}{6}) = \sin(\frac{2\pi}{6}) + 2\cos(\frac{\pi}{6})$

$= \frac{\sqrt{3}}{2} + 2 \times \frac{\sqrt{3}}{2} = \frac{3\sqrt{3}}{2}$

$f(\frac{5\pi}{6}) = \sin(\frac{10\pi}{6}) + 2\cos(\frac{5\pi}{6})$

$= -\frac{\sqrt{3}}{2} + 2(-\frac{\sqrt{3}}{2}) = -\frac{3\sqrt{3}}{2}$

$f(\frac{3\pi}{2}) = \sin(3\pi) + 2\cos(\frac{3\pi}{2})$

$= 0 + 2 \times 0 = 0$

$\therefore$ there is a local maximum at $(\frac{\pi}{6}, \frac{3\sqrt{3}}{2})$, a local minimum at $(\frac{5\pi}{6}, -\frac{3\sqrt{3}}{2})$ and a stationary point of inflection at $(\frac{3\pi}{2}, 0)$.

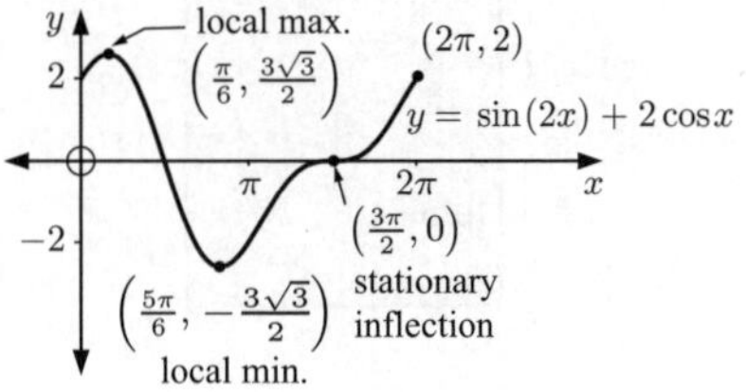

11 $x(t) = 1 - 2\cos t$ cm

$\therefore\ v(t) = x'(t) = 2\sin t$

$\therefore\ a(t) = v'(t) = 2\cos t$

a When $t = 0$,

$x(0) = 1 - 2\cos 0$

$= -1$ cm

$v(0) = 2\sin 0$

$= 0$ cm s^{-1}

$a(0) = 2\cos 0$

$= 2$ cm s^{-2}

b When $t = \frac{\pi}{4}$,

$x(\frac{\pi}{4}) = 1 - \frac{2}{\sqrt{2}}$

$= 1 - \sqrt{2}$ cm

$v(\frac{\pi}{4}) = \frac{2}{\sqrt{2}} = \sqrt{2}$ cm s^{-1}

$a(\frac{\pi}{4}) = \frac{2}{\sqrt{2}} = \sqrt{2}$ cm s^{-2}

The particle is $(\sqrt{2} - 1)$ cm left of the origin, moving right at $\sqrt{2}$ cm s^{-1} with increasing speed.

c We need to look for the points where the velocity equals zero

If $v(t) = 2\sin t = 0$

then $\sin t = 0$

$\therefore \quad t = \pi \quad (0 < t < 2\pi)$

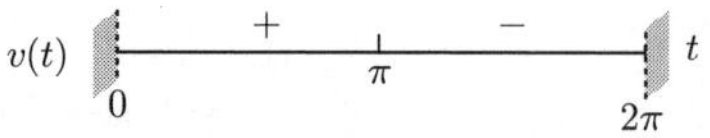

The particle reverses direction when $t = \pi$.

At $t = \pi$, $x(\pi) = 3$ cm.

d The particle's speed is increasing when $v(t) = 2\sin t$ and $a(t) = 2\cos t$ have the same sign.

If $a(t) = 2\cos t = 0$

then $\cos t = 0$

$t = \frac{\pi}{2}, \frac{3\pi}{2} \quad (0 \leqslant t \leqslant 2\pi)$

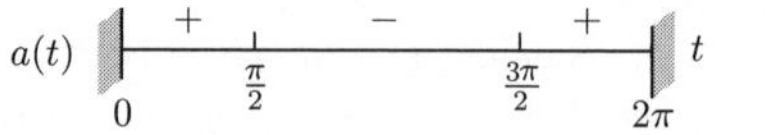

$\therefore$ the particle's speed is increasing when $0 \leqslant t \leqslant \frac{\pi}{2}$ and $\pi \leqslant t \leqslant \frac{3\pi}{2}$.

12 $y = 4e^{-x}\sin x$

$\therefore \quad \dfrac{dy}{dx} = -4e^{-x}\sin x + 4e^{-x}\cos x$

$\therefore$ stationary points occur when $-4e^{-x}\sin x + 4e^{-x}\cos x = 0$

$\therefore \quad 4e^{-x}(\cos x - \sin x) = 0$

$\therefore \quad \cos x - \sin x = 0 \quad \{e^{-x} > 0 \text{ for all } x\}$

$\therefore \quad \sin x = \cos x$

$\therefore \quad \tan x = 1$

$\therefore \quad x = \frac{\pi}{4} + k\pi, \quad k$ an integer

Sign diagram of $\dfrac{dy}{dx}$ is:

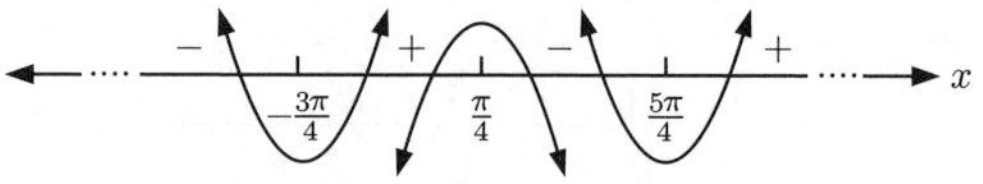

$\therefore \quad y = 4e^{-x}\sin x$ has a maximum when $x = \frac{\pi}{4}$.

EXERCISE 20B

1

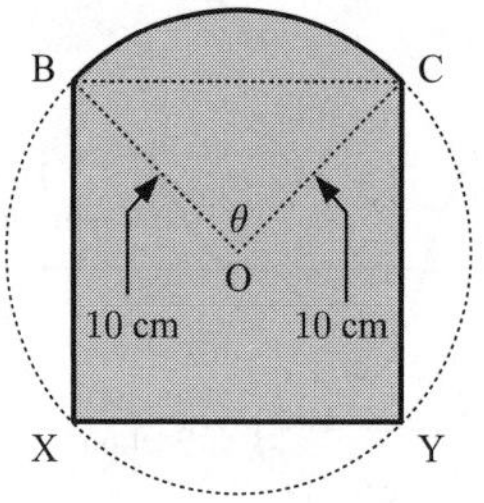

Using the cosine rule in ΔBCO,

$BC^2 = 10^2 + 10^2 - 2 \times 10 \times 10\cos\theta$

$\therefore \quad BC = \sqrt{200 - 200\cos\theta}$

$\therefore \quad XY = \sqrt{200 - 200\cos\theta}$ also

Now $BY^2 = BX^2 + XY^2$ {Pythagoras}

$\therefore \quad 400 = BX^2 + (200 - 200\cos\theta)$

$\therefore \quad BX^2 = 200 + 200\cos\theta$

$\therefore \quad BX = \sqrt{200 + 200\cos\theta}$

The shaded area is equal to the area of the sector plus $\frac{3}{4}$ of the area of BCYX.

$\therefore \quad A = \frac{1}{2}(10)^2\theta + \frac{3}{4}[BX \times BC]$

$= 50\theta + \frac{3}{4}\sqrt{200 + 200\cos\theta}\sqrt{200 - 200\cos\theta}$

$= 50\theta + \frac{3}{4} \times 200\sqrt{1 + \cos\theta}\sqrt{1 - \cos\theta}$

$= 50\theta + 150\sqrt{1 - \cos^2\theta}$

$= 50\theta + 150\sin\theta$

$= 50(\theta + 3\sin\theta)$ as required

$\therefore \quad \dfrac{dA}{d\theta} = 50 + 150\cos\theta = 50(1 + 3\cos\theta)$,

which is zero when $\cos\theta = -\frac{1}{3}$

$\therefore \quad \theta \approx 109.5^o$

The sign diagram of $\dfrac{dA}{d\theta}$ is:

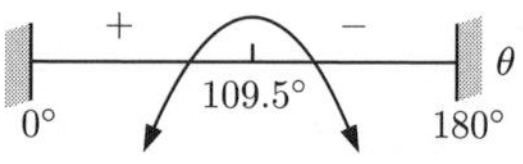

Since $0^o < \theta < 180^o$, A is maximised when $\theta \approx 109.5^o$.

2 **a**

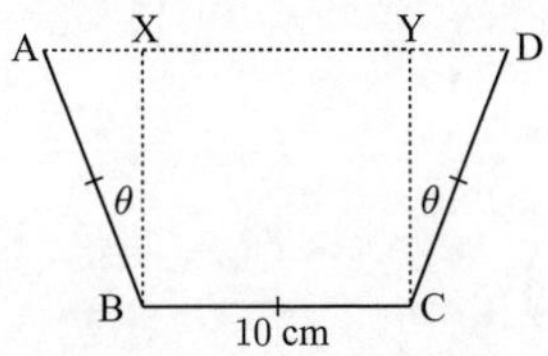

The triangles have height $10\cos\theta$ and width $10\sin\theta$.

$\therefore$ area A
$= \text{area of } \Delta\text{s} + \text{area of rectangle}$
$= 2 \times \frac{1}{2} \times 10\cos\theta \times 10\sin\theta + 10 \times 10\cos\theta$
$= 100\sin\theta\cos\theta + 100\cos\theta$
$= 100\cos\theta(1+\sin\theta)$

b
$$\frac{dA}{d\theta} = 100(-\sin\theta(1+\sin\theta) + \cos\theta \times \cos\theta)$$
$$= 100(-\sin\theta - \sin^2\theta + \cos^2\theta)$$
$$= 100(-\sin\theta - \sin^2\theta + 1 - \sin^2\theta)$$
$$= -100(2\sin^2\theta + \sin\theta - 1)$$
$$= -100(2\sin\theta - 1)(\sin\theta + 1)$$

$\therefore \quad \frac{dA}{d\theta} = 0$ when $2\sin\theta - 1 = 0$ or $\sin\theta + 1 = 0$

$\therefore \quad \sin\theta = \frac{1}{2}$ or $\sin\theta = -1$

c Using **b**, $\frac{dA}{d\theta} = 0$ when $\theta = \frac{\pi}{6}, \frac{5\pi}{6}$ or $\frac{3\pi}{2}$.

But $0 \leqslant \theta \leqslant \frac{\pi}{2}$, so the sign diagram for $\frac{dA}{d\theta}$ is:

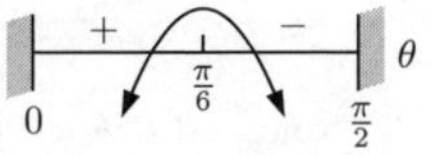

So, the maximum area occurs when $\theta = \frac{\pi}{6} = 30^o$

3

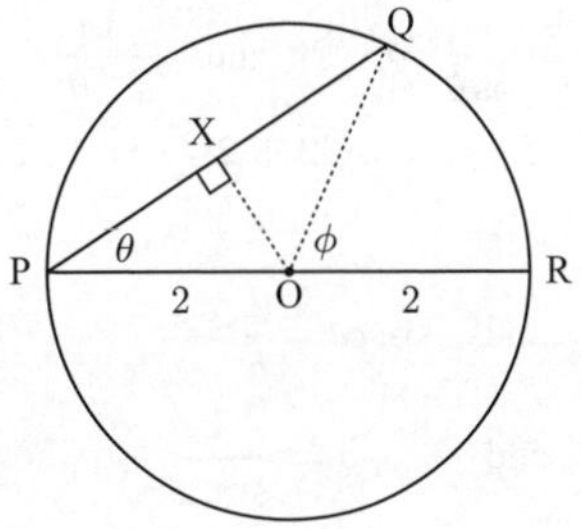

$\frac{\text{PX}}{2} = \cos\theta \quad \therefore \quad \text{PQ} = 2\text{PX} = 4\cos\theta$

$\therefore$ the time taken to row from P to Q is $\frac{4\cos\theta}{3}$ hours

Now $\phi = 2\theta$ {angle at the centre}

But, arc length $\text{QR}_{\text{arc}} = 2\phi$

$\therefore \quad \text{QR}_{\text{arc}} = 4\theta$

and the time taken to walk from Q to R is $\frac{4\theta}{5}$

$\therefore$ the total time from P to R, $T = \frac{4}{3}\cos\theta + \frac{4\theta}{5}$

$\therefore \quad \frac{dT}{d\theta} = -\frac{4}{3}\sin\theta + \frac{4}{5}$

$\therefore \quad \frac{dT}{d\theta} = 0$ when $-\frac{4}{3}\sin\theta = -\frac{4}{5}$

$\therefore \quad \sin\theta = \frac{3}{5}$

$\therefore \quad \theta \approx 0.6435$ radians

$\therefore \quad \theta \approx 36.87^o$

and the sign diagram of $\frac{dT}{d\theta}$ is:

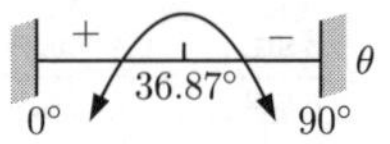

So the maximum time occurs when $\theta \approx 36.9^o$

and the maximum time is $\frac{4}{3}\cos 0.6435 + \frac{4}{5} \times 0.6435$

≈ 1.581 hours

≈ 1 hour 34 min 53 s

4 **a** $\tan\theta = \frac{2}{\text{AX}}$ and $\sin\theta = \frac{2}{\text{BX}}$

$\therefore \quad \text{AX} = \frac{2}{\tan\theta} = \frac{2\cos\theta}{\sin\theta}$ and $\text{BX} = \frac{2}{\sin\theta}$

From the similar Δs, $\frac{L}{\text{BX}} = \frac{\text{AX}+2}{\text{AX}} = 1 + \frac{2}{\text{AX}}$

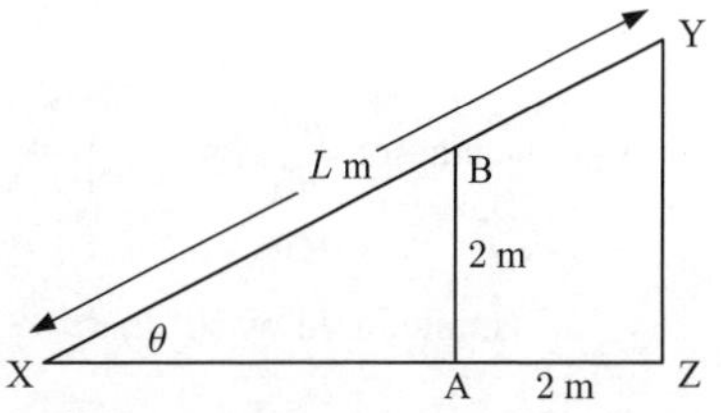

$$\therefore \quad L = \text{BX} + \frac{2\text{BX}}{\text{AX}} = \frac{2}{\sin\theta} + \frac{2\left(\dfrac{2}{\sin\theta}\right)}{\left(\dfrac{2\cos\theta}{\sin\theta}\right)}$$

$$= \frac{2}{\sin\theta} + 2\left(\frac{2}{\sin\theta}\right)\left(\frac{\sin\theta}{2\cos\theta}\right)$$

$$= \frac{2}{\cos\theta} + \frac{2}{\sin\theta} \quad \text{as required}$$

b Now $L = \dfrac{2}{\cos\theta} + \dfrac{2}{\sin\theta} = 2(\cos\theta)^{-1} + 2(\sin\theta)^{-1}$

$$\therefore \quad \frac{dL}{d\theta} = -2(\cos\theta)^{-2} \times (-\sin\theta) - 2(\sin\theta)^{-2} \times \cos\theta$$

$$= \frac{2\sin\theta}{\cos^2\theta} - \frac{2\cos\theta}{\sin^2\theta}$$

$$= \frac{2\sin^3\theta - 2\cos^3\theta}{\sin^2\theta\cos^2\theta}$$

c Now $\dfrac{dL}{d\theta} = 0$ when $2\sin^3\theta - 2\cos^3\theta = 0$

$$\therefore \quad 2\sin^3\theta = 2\cos^3\theta$$

$$\therefore \quad \tan^3\theta = 1$$

$$\therefore \quad \tan\theta = 1$$

$$\therefore \quad \text{since } 0 < \theta < 90^o, \quad \theta = 45^o$$

Sign diagram of $\dfrac{dL}{d\theta}$ is:

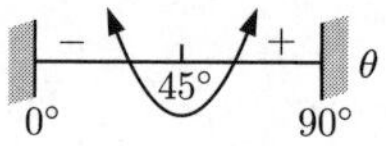

$\therefore$ the ladder is shortest when $\theta = 45^o$

$\therefore \quad \dfrac{1}{\cos\theta} = \sqrt{2}$ and $\dfrac{1}{\sin\theta} = \sqrt{2}$

$\therefore \quad L_{\min} = 2\sqrt{2} + 2\sqrt{2} = 4\sqrt{2}$ m

5

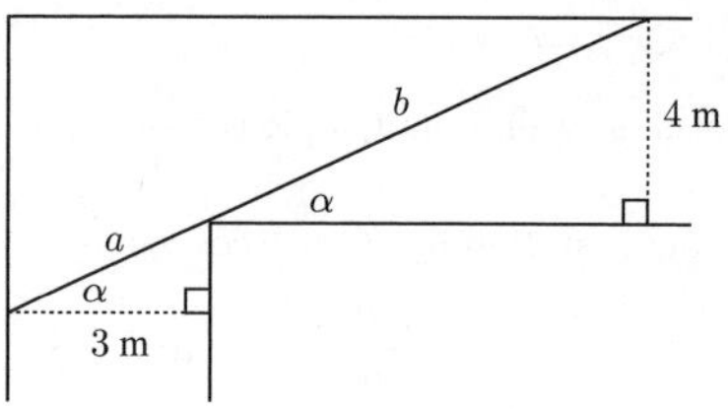

$\cos\alpha = \dfrac{3}{a}$ and $\sin\alpha = \dfrac{4}{b}$

$\therefore \quad a = \dfrac{3}{\cos\alpha}$ and $b = \dfrac{4}{\sin\alpha}$

Now $L = a + b$

$$\therefore \quad L = \frac{3}{\cos\alpha} + \frac{4}{\sin\alpha}$$

$$= 3(\cos\alpha)^{-1} + 4(\sin\alpha)^{-1}$$

$$\therefore \quad \frac{dL}{d\alpha} = -3(\cos\alpha)^{-2} \times (-\sin\alpha) - 4(\sin\alpha)^{-2} \times \cos\alpha$$

$$= \frac{3\sin\alpha}{\cos^2\alpha} - \frac{4\cos\alpha}{\sin^2\alpha}$$

$$= \frac{3\sin^3\alpha - 4\cos^3\alpha}{\cos^2\alpha\sin^2\alpha}$$

$\therefore \quad \dfrac{dL}{d\alpha} = 0$

when $3\sin^3\alpha - 4\cos^3\alpha = 0$

$$\therefore \quad 3\sin^3\alpha = 4\cos^3\alpha$$

$$\therefore \quad \tan^3\alpha = \tfrac{4}{3}$$

$$\therefore \quad \tan\alpha = \sqrt[3]{\tfrac{4}{3}}$$

$$\therefore \quad \alpha \approx 47.74^o$$

Sign diagram of $\dfrac{dL}{d\alpha}$ is:

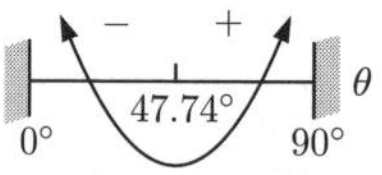

$\therefore \quad L$ is minimised when $\alpha \approx 47.74^o$ and $L = \dfrac{3}{\cos\alpha} + \dfrac{4}{\sin\alpha} \approx 9.87$ m

REVIEW SET 20

1 **a** $\dfrac{d}{dx}(\sin(5x)\ln x) = \dfrac{d}{dx}(\sin(5x))\ln x + \sin(5x)\dfrac{d}{dx}(\ln x)$ {product rule}

$$= 5\cos(5x)\ln x + \frac{\sin(5x)}{x}$$

b $\dfrac{d}{dx}(\sin x\cos(2x)) = \dfrac{d}{dx}(\sin x)\cos(2x) + \sin x\dfrac{d}{dx}(\cos(2x))$ {product rule}

$$= \cos x\cos(2x) + \sin x(-2\sin(2x))$$
$$= \cos x\cos(2x) - 2\sin x\sin(2x)$$

c $\dfrac{d}{dx}\left(e^{-2x}\tan x\right) = \dfrac{d}{dx}\left(e^{-2x}\right)\tan x + e^{-2x}\dfrac{d}{dx}(\tan x)$ {product rule}

$$= -2e^{-2x}\tan x + \frac{e^{-2x}}{\cos^2 x}$$

d $\dfrac{d}{dx}(10x - \sin(10x)) = 10 - 10\cos(10x)$

e $\dfrac{d}{dx}\left(\ln\left(\dfrac{1}{\cos x}\right)\right) = \dfrac{1}{\left(\frac{1}{\cos x}\right)} \times \dfrac{d}{dx}\left(\dfrac{1}{\cos x}\right)$ {chain rule}

$$= \cos x \times \frac{d}{dx}\left((\cos x)^{-1}\right)$$
$$= \cos x \times \left(-(\cos x)^{-2} \times (-\sin x)\right)$$
$$= \frac{\cos x\sin x}{\cos^2 x} = \tan x$$

f $\dfrac{d}{dx}(\sin(5x)\ln(2x)) = \dfrac{d}{dx}(\sin(5x))\ln(2x) + \sin(5x)\dfrac{d}{dx}(\ln(2x))$ {product rule}

$$= 5\cos(5x)\ln(2x) + \sin(5x) \times \frac{2}{2x}$$
$$= 5\cos(5x)\ln(2x) + \frac{\sin(5x)}{x}$$

2 $y = x\tan x$ $\quad\therefore\quad \dfrac{dy}{dx} = 1 \times \tan x + x \times \left(\dfrac{1}{\cos^2 x}\right) = \tan x + \dfrac{x}{\cos^2 x}$

Now $\cos\frac{\pi}{4} = \frac{1}{\sqrt{2}}$ and $\tan\frac{\pi}{4} = 1$

$\therefore$ at $x = \frac{\pi}{4}$, $y = \frac{\pi}{4}$ and $\dfrac{dy}{dx} = 1 + \dfrac{\frac{\pi}{4}}{\left(\frac{1}{\sqrt{2}}\right)^2} = 1 + \frac{\pi}{2}$

$\therefore$ the equation of the tangent is $\dfrac{y - \frac{\pi}{4}}{x - \frac{\pi}{4}} = 1 + \frac{\pi}{2}$

$$\therefore\quad y - \tfrac{\pi}{4} = (1 + \tfrac{\pi}{2})(x - \tfrac{\pi}{4})$$
$$= x - \tfrac{\pi}{4} + \tfrac{\pi}{2}x - \tfrac{\pi^2}{8}$$
$$\therefore\quad y = (1 + \tfrac{\pi}{2})x - \tfrac{\pi^2}{8}$$
$$\therefore\quad 2y = (2 + \pi)x - \tfrac{\pi^2}{4}$$
$$\therefore\quad (2 + \pi)x - 2y = \tfrac{\pi^2}{4} \quad \text{as required}$$

3 **a** $f(x) = 3\sin x - 4\cos(2x)$

$$\therefore\quad f'(x) = 3\cos x + 8\sin(2x)$$
$$\therefore\quad f''(x) = -3\sin x + 16\cos(2x)$$

b $f(x) = x^{\frac{1}{2}}\cos(4x)$

$\therefore\ f'(x) = \frac{1}{2}x^{-\frac{1}{2}}\cos(4x) + x^{\frac{1}{2}}(-4\sin(4x))$ {product rule}

$= \frac{1}{2}x^{-\frac{1}{2}}\cos(4x) - 4x^{\frac{1}{2}}\sin(4x)$

and $f''(x) = -\frac{1}{4}x^{-\frac{3}{2}}\cos(4x) + \frac{1}{2}x^{-\frac{1}{2}}(-4\sin(4x)) - \left[2x^{-\frac{1}{2}}\sin(4x) + 4x^{\frac{1}{2}} \times 4\cos(4x)\right]$

$= -\frac{1}{4}x^{-\frac{3}{2}}\cos(4x) - 4x^{-\frac{1}{2}}\sin(4x) - 16x^{\frac{1}{2}}\cos(4x)$

4 **a** $x(t) = 3 + \sin(2t)$ cm, $t \geqslant 0$ s $\qquad \therefore\ x(0) = 3$ cm

$v(t) = x'(t) = 0 + 2\cos(2t)$ cm s^{-1} $\qquad v(0) = 2$ cm s^{-1}

$a(t) = v'(t) = -4\sin(2t)$ cm s^{-2} $\qquad a(0) = 0$ cm s^{-2}

$\therefore$ initially the particle is 3 cm right of O, moving right at a speed of 2 cm s^{-1}.

b $x'(t) = 0$ when $2\cos(2t) = 0$

$\therefore\ \cos(2t) = 0$

$\therefore\ 2t = \frac{\pi}{2} + k\pi$

For the interval $0 \leqslant t \leqslant \pi$, $t = \frac{\pi}{4}$ or $\frac{3\pi}{4}$

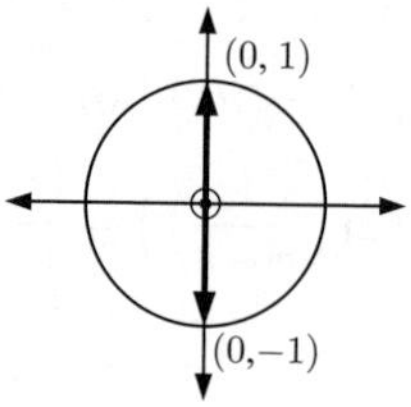

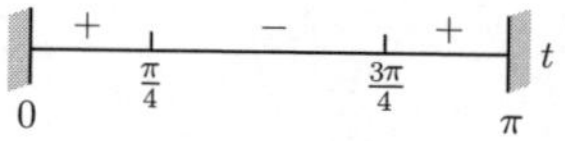

$\therefore$ the particle reverses direction at $t = \frac{\pi}{4}, \frac{3\pi}{4}$

c $x(0) = 3,\quad x\left(\frac{\pi}{4}\right) = 3 + \sin\left(\frac{\pi}{2}\right) = 4,$

$x\left(\frac{3\pi}{4}\right) = 3 + \sin\left(\frac{3\pi}{2}\right) = 3 - 1 = 2,$

$x(\pi) = 3 + \sin(2\pi) = 3$

$\therefore$ the total distance travelled $= 1 + 2 + 1 = 4$ cm.

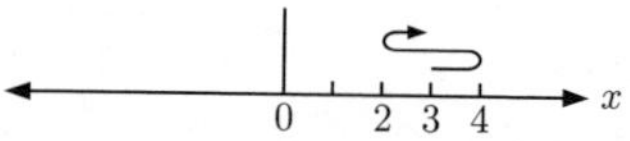

5 **a** $f(x) = \sqrt{\cos x},\ 0 \leqslant x \leqslant 2\pi$

$f(x)$ is meaningful when $\cos x \geqslant 0$,

which is when $0 \leqslant x \leqslant \frac{\pi}{2}$

and $\frac{3\pi}{2} \leqslant x \leqslant 2\pi$.

b $f(x) = (\cos x)^{\frac{1}{2}}$

$\therefore\ f'(x) = \frac{1}{2}(\cos x)^{-\frac{1}{2}}(-\sin x)$

$= \dfrac{-\sin x}{2\sqrt{\cos x}}$

$\therefore\ f'(x) = 0$ when $-\sin x = 0$

For $0 \leqslant x \leqslant 2\pi$, this is when $x = 0, \pi, 2\pi$.

Sign diagram for $f'(x)$ is:

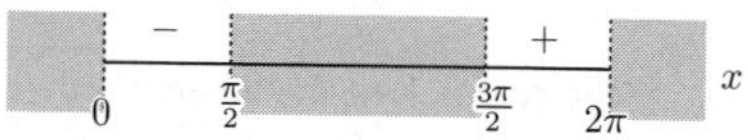

$f(x)$ is increasing for $\frac{3\pi}{2} \leqslant x \leqslant 2\pi$

and decreasing for $0 \leqslant x \leqslant \frac{\pi}{2}$.

c

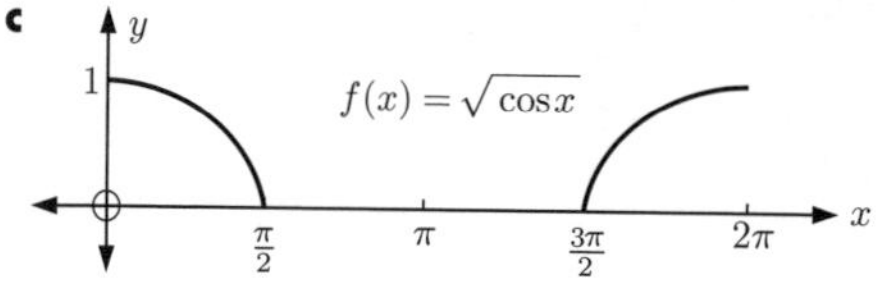

6 **a** $s(t) = 30 + \cos(\pi t)$ cm, $t \geqslant 0$

$\therefore\ v(t) = s'(t) = -\pi\sin(\pi t)$

So, $v(0) = 0$ cm s^{-1}, $v(\frac{1}{2}) = -\pi$ cm s^{-1},

$v(1) = 0$ cm s^{-1}, $v(\frac{3}{2}) = \pi$ cm s^{-1},

$v(2) = 0$ cm s^{-1}

Sign diagram of $v(t)$ is:

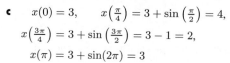

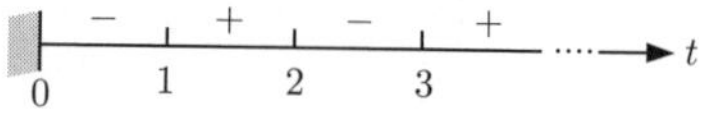

b The cork is falling when $v(t) \leqslant 0$, which is for $0 \leqslant t \leqslant 1,\ 2 \leqslant t \leqslant 3, \ldots$

$\therefore$ the cork is falling for $2n \leqslant t \leqslant 2n + 1,\quad n \in \{0, 1, 2, 3, \ldots\}$

7 **a** $\sin\theta = \dfrac{\text{NA}}{x} = \dfrac{1}{x}$

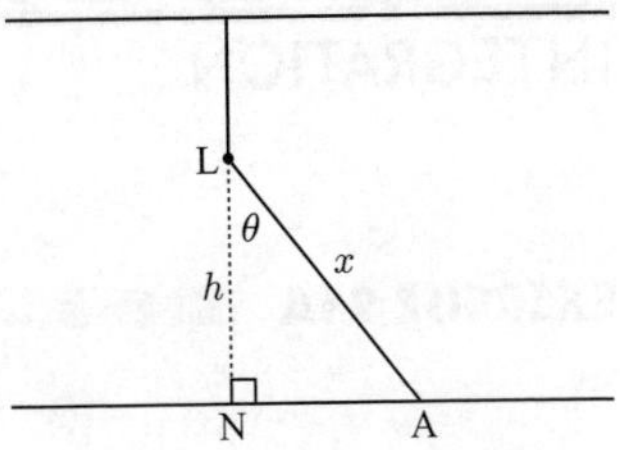

$\therefore \quad \dfrac{1}{x^2} = \sin^2\theta$

$\therefore \quad$ at A, $\quad I = \dfrac{\sqrt{8}\cos\theta}{x^2} = \sqrt{8}\cos\theta\sin^2\theta$

b $\dfrac{dI}{d\theta} = \sqrt{8}(-\sin\theta)\sin^2\theta + \sqrt{8}\cos\theta(2\sin\theta\cos\theta)$

$= \sqrt{8}\sin\theta[2\cos^2\theta - \sin^2\theta]$

$= \sqrt{8}\sin\theta[2(1-\sin^2\theta) - \sin^2\theta]$

$= \sqrt{8}\sin\theta[2 - 3\sin^2\theta]$

$\dfrac{dI}{d\theta} = 0$ when $\sin\theta = \sqrt{\frac{2}{3}}, \quad 0 < \theta < \frac{\pi}{2}$

and the sign diagram of $\dfrac{dI}{d\theta}$ is:

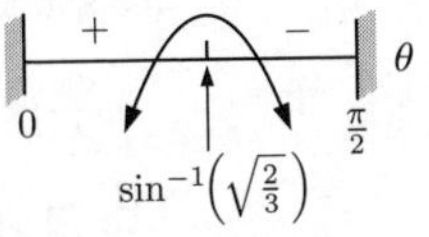

$\therefore \quad$ the maximum illumination at A is obtained when $\sin\theta = \sqrt{\frac{2}{3}}$.

$\therefore \quad x = \dfrac{1}{\sin\theta} = \sqrt{\frac{3}{2}}$

$\therefore \quad h = \sqrt{x^2 - \text{NA}^2} = \sqrt{\frac{3}{2} - 1} = \frac{1}{\sqrt{2}}$

$\therefore \quad$ the bulb is $\frac{1}{\sqrt{2}}$ m above the floor.

8 **a** $y = \dfrac{1}{\sin x} = (\sin x)^{-1}$

$\therefore \quad \dfrac{dy}{dx} = -(\sin x)^{-2}(\cos x)$

$= -\dfrac{\cos x}{\sin^2 x}$

When $x = \frac{\pi}{3}, \quad y = \dfrac{1}{\sin(\frac{\pi}{3})} = \dfrac{2}{\sqrt{3}}$

and $\dfrac{dy}{dx} = -\dfrac{\cos(\frac{\pi}{3})}{\sin^2(\frac{\pi}{3})} = -\dfrac{\frac{1}{2}}{(\frac{\sqrt{3}}{2})^2} = -\frac{2}{3}$

$\therefore \quad$ the tangent has equation $\dfrac{y - \frac{2}{\sqrt{3}}}{x - \frac{\pi}{3}} = -\frac{2}{3}$ which is $3y - 2\sqrt{3} = -2x + \frac{2\pi}{3}$

or $2x + 3y = 2\sqrt{3} + \frac{2\pi}{3}$

b $y = \cos(\frac{x}{2})$

$\therefore \quad \dfrac{dy}{dx} = -\frac{1}{2}\sin(\frac{x}{2})$

When $x = \frac{\pi}{2}, \quad y = \cos(\frac{\pi}{4}) = \frac{1}{\sqrt{2}}$

and $\dfrac{dy}{dx} = -\frac{1}{2}\sin(\frac{\pi}{4}) = -\frac{1}{2\sqrt{2}}$

$\therefore \quad$ the normal has gradient $2\sqrt{2}$, and its equation is $\dfrac{y - \frac{1}{\sqrt{2}}}{x - \frac{\pi}{2}} = 2\sqrt{2}$

$\therefore \quad y - \frac{1}{\sqrt{2}} = 2\sqrt{2}x - \pi\sqrt{2}$

$\therefore \quad y - 2\sqrt{2}x = \frac{1}{\sqrt{2}} - \pi\sqrt{2}$

or $\sqrt{2}y - 4x = 1 - 2\pi$

9 **a** $f(x) = -10\sin 2x\cos 2x, \quad 0 \leqslant x \leqslant \pi$

$\therefore \quad f(x) = -5\sin 4x \quad \{2\sin A\cos A = \sin 2A\}$

b $f'(x) = -20\cos 4x$

If $f'(x) = 0, \quad -20\cos 4x = 0$

$\therefore \quad \cos 4x = 0$

$\therefore \quad 4x = \frac{\pi}{2} + n\pi, \quad n$ any integer

$\therefore \quad x = \frac{\pi}{8} + \frac{n\pi}{4}$

So, for the domain $0 \leqslant x \leqslant \pi, \quad x = \frac{\pi}{8}, \frac{3\pi}{8}, \frac{5\pi}{8}, \frac{7\pi}{8}$

Chapter 21
INTEGRATION

EXERCISE 21A

1 **a** **i** $\dfrac{d}{dx}(x^2) = 2x$

$\therefore \dfrac{d}{dx}\left(\frac{1}{2}x^2\right) = x$

$\therefore$ the antiderivative of x is $\frac{1}{2}x^2$

ii $\dfrac{d}{dx}(x^3) = 3x^2$

$\therefore \dfrac{d}{dx}\left(\frac{1}{3}x^3\right) = x^2$

$\therefore$ the antiderivative of x^2 is $\frac{1}{3}x^3$

iii $\dfrac{d}{dx}(x^6) = 6x^5$

$\therefore \dfrac{d}{dx}\left(\frac{1}{6}x^6\right) = x^5$

$\therefore$ the antiderivative of x^5 is $\frac{1}{6}x^6$

iv $\dfrac{d}{dx}(x^{-1}) = -x^{-2}$

$\therefore \dfrac{d}{dx}(-x^{-1}) = x^{-2}$

$\therefore$ the antiderivative of x^{-2} is $-x^{-1}$ or $-\dfrac{1}{x}$

v $\dfrac{d}{dx}(x^{-3}) = -3x^{-4}$

$\therefore \dfrac{d}{dx}\left(-\frac{1}{3}x^{-3}\right) = x^{-4}$

$\therefore$ the antiderivative of x^{-4} is $-\frac{1}{3}x^{-3}$

vi $\dfrac{d}{dx}\left(x^{\frac{4}{3}}\right) = \frac{4}{3}x^{\frac{1}{3}}$

$\therefore \dfrac{d}{dx}\left(\frac{3}{4}x^{\frac{4}{3}}\right) = x^{\frac{1}{3}}$

$\therefore$ the antiderivative of $x^{\frac{1}{3}}$ is $\frac{3}{4}x^{\frac{4}{3}}$

vii $\dfrac{d}{dx}(x^{\frac{1}{2}}) = \frac{1}{2}x^{-\frac{1}{2}}$

$\therefore \dfrac{d}{dx}(2x^{\frac{1}{2}}) = x^{-\frac{1}{2}}$

$\therefore$ the antiderivative of of $x^{-\frac{1}{2}}$ is $2x^{\frac{1}{2}} = 2\sqrt{x}$

b the antiderivative of x^n is $\dfrac{x^{n+1}}{n+1}$

2 **a** **i** $\dfrac{d}{dx}(e^{2x}) = 2e^{2x}$

$\therefore \dfrac{d}{dx}\left(\frac{1}{2}e^{2x}\right) = e^{2x}$

$\therefore$ the antiderivative of e^{2x} is $\frac{1}{2}e^{2x}$

ii $\dfrac{d}{dx}(e^{5x}) = 5e^{5x}$

$\therefore \dfrac{d}{dx}\left(\frac{1}{5}e^{5x}\right) = e^{5x}$

$\therefore$ the antiderivative of e^{5x} is $\frac{1}{5}e^{5x}$

iii $\dfrac{d}{dx}(e^{\frac{1}{2}x}) = \frac{1}{2}e^{\frac{1}{2}x}$

$\therefore \dfrac{d}{dx}\left(2e^{\frac{1}{2}x}\right) = e^{\frac{1}{2}x}$

$\therefore$ the antiderivative of $e^{\frac{1}{2}x}$ is $2e^{\frac{1}{2}x}$

iv $\dfrac{d}{dx}(e^{0.01x}) = 0.01e^{0.01x}$

$\therefore \dfrac{d}{dx}(100e^{0.01x}) = e^{0.01x}$

$\therefore$ the antiderivative of $e^{0.01x}$ is $100e^{0.01x}$

v $\dfrac{d}{dx}(e^{\pi x}) = \pi e^{\pi x}$

$\therefore \dfrac{d}{dx}\left(\frac{1}{\pi}e^{\pi x}\right) = e^{\pi x}$

$\therefore$ the antiderivative of $e^{\pi x}$ is $\frac{1}{\pi}e^{\pi x}$

vi $\dfrac{d}{dx}\left(e^{\frac{x}{3}}\right) = \frac{1}{3}e^{\frac{x}{3}}$

$\therefore \dfrac{d}{dx}\left(3e^{\frac{x}{3}}\right) = e^{\frac{x}{3}}$

$\therefore$ the antiderivative of $e^{\frac{x}{3}}$ is $3e^{\frac{x}{3}}$

b the antiderivative of e^{kx} is $\dfrac{1}{k}e^{kx}$

3 **a** $\frac{d}{dx}(x^3 + x^2) = 3x^2 + 2x$

$\therefore \quad \frac{d}{dx}\left(2x^3 + 2x^2\right) = 6x^2 + 4x$

$\therefore$ the antiderivative of $6x^2 + 4x$ is $2x^3 + 2x^2$

b $\frac{d}{dx}(e^{3x+1}) = 3e^{3x+1}$

$\therefore \quad \frac{d}{dx}\left(\frac{1}{3}e^{3x+1}\right) = e^{3x+1}$

$\therefore$ the antiderivative of e^{3x+1} is $\frac{1}{3}e^{3x+1}$

c $\frac{d}{dx}(x\sqrt{x}) = \frac{d}{dx}(x^{\frac{3}{2}}) = \frac{3}{2}x^{\frac{1}{2}}$

$= \frac{3}{2}\sqrt{x}$

$\therefore \quad \frac{d}{dx}\left(\frac{2}{3}x\sqrt{x}\right) = \sqrt{x}$

$\therefore$ the antiderivative of $\sqrt{x}$ is $\frac{2}{3}x\sqrt{x}$

d $\frac{d}{dx}((2x+1)^4) = 4(2x+1)^3 \times 2$

$= 8(2x+1)^3$

$\therefore \quad \frac{d}{dx}\left(\frac{1}{8}(2x+1)^4\right) = (2x+1)^3$

$\therefore$ the antiderivative of $(2x+1)^3$ is $\frac{1}{8}(2x+1)^4$

EXERCISE 21B

1 **a** $\int_a^a f(x)\,dx = F(a) - F(a) = 0$

$\int_a^a f(x)\,dx =$ area of the strip between $x = a$ and $x = a$.

This strip has 0 width, so its area $= 0$.

b The antiderivative of c is cx.

$\therefore \quad \int_a^b c\,dx = F(b) - F(a)$

$= cb - ca$

$= c(b - a)$

c $\int_b^a f(x)\,dx = F(a) - F(b)$

$= -[F(b) - F(a)]$

$= -\int_a^b f(x)\,dx$

d If $\frac{d}{dx}F(x) = f(x)$ then

$\frac{d}{dx}c\,F(x) = c\,f(x)$

$\therefore \quad \int_a^b c\,f(x)\,dx = c\,F(b) - c\,F(a)$

$= c[F(b) - F(a)]$

$= c\int_a^b f(x)\,dx$

e $\int_a^b (f(x) + g(x))\,dx = [F(b) + G(b)] - [F(a) + G(a)]$

$= [F(b) - F(a)] + [G(b) - G(a)]$

$= \int_a^b f(x)\,dx + \int_a^b g(x)\,dx$

2 **a** $f(x) = x^3$ has antiderivative $F(x) = \frac{x^4}{4}$

$\therefore \quad$ area $= \int_0^1 x^3\,dx$

$= F(1) - F(0)$

$= \frac{1}{4} - 0$

$= \frac{1}{4}$ units^2

b $f(x) = x^3$ has antiderivative $F(x) = \frac{x^4}{4}$

$\therefore \quad$ area $= \int_1^2 x^3\,dx$

$= F(2) - F(1)$

$= \frac{16}{4} - \frac{1}{4}$

$= 3\frac{3}{4}$ units^2

c $f(x) = x^2 + 3x + 2$ has antiderivative

$F(x) = \frac{x^3}{3} + \frac{3x^2}{2} + 2x$

$\therefore \quad$ area $= \int_1^3 (x^2 + 3x + 2)\,dx$

$= F(3) - F(1)$

$= (\frac{27}{3} + \frac{27}{2} + 6) - (\frac{1}{3} + \frac{3}{2} + 2)$

$= 24\frac{2}{3}$ units^2

d $f(x) = \sqrt{x} = x^{\frac{1}{2}}$ has antiderivative

$F(x) = \frac{x^{\frac{3}{2}}}{\frac{3}{2}} = \frac{2}{3}x\sqrt{x}$

$\therefore \quad$ area $= \int_0^2 \sqrt{x}\,dx$

$= F(2) - F(0)$

$= \frac{2}{3} \times 2\sqrt{2} - 0$

$= \frac{4\sqrt{2}}{3}$ units^2

e $f(x) = e^x$ has antiderivative $F(x) = e^x$

$\therefore$ area $= \int_0^{1.5} e^x \, dx$
$= F(1.5) - F(0)$
$= e^{1.5} - e^0$
$= e^{1.5} - 1$
≈ 3.48 units^2

f $f(x) = \dfrac{1}{\sqrt{x}} = x^{-\frac{1}{2}}$ has antiderivative

$F(x) = \dfrac{x^{\frac{1}{2}}}{\frac{1}{2}} = 2\sqrt{x}$

$\therefore$ area $= \int_1^4 \dfrac{1}{\sqrt{x}} \, dx$
$= F(4) - F(1)$
$= 2\sqrt{4} - 2\sqrt{1}$
$= 2$ units^2

g $f(x) = x^3 + 2x^2 + 7x + 4$ has antiderivative

$F(x) = \dfrac{x^4}{4} + \dfrac{2x^3}{3} + \dfrac{7x^2}{2} + 4x$

$\therefore$ area $= \int_1^{1.25} (x^3 + 2x^2 + 7x + 4) \, dx$
$= F(1.25) - F(1)$
$= [12.381\,18 - 8.416\,67]$
≈ 3.96 units^2

3 Using technology:

a area $= \int_0^{1.5} e^{x^2} \, dx \approx 4.06$ units^2

b area $= \int_2^4 (\ln x)^2 \, dx \approx 2.41$ units^2

c area $= \int_1^2 \sqrt{9 - x^2} \, dx \approx 2.58$ units^2

4 **a** If $\dfrac{d}{dx} F(x) = f(x)$ then $\dfrac{d}{dx}(-F(x)) = -f(x)$

$\therefore \int_a^b (-f(x)) \, dx = -F(b) - (-F(a))$
$= -(F(b) - F(a))$
$= -\int_a^b f(x) \, dx$

b Since $y = -f(x)$ is a reflection of $y = f(x)$ in the x-axis,

shaded area = area between the x-axis and $y = -f(x)$ from $x = a$ to $x = b$
$= \int_a^b (-f(x)) \, dx$
$= -\int_a^b f(x) \, dx$ {using **a**}

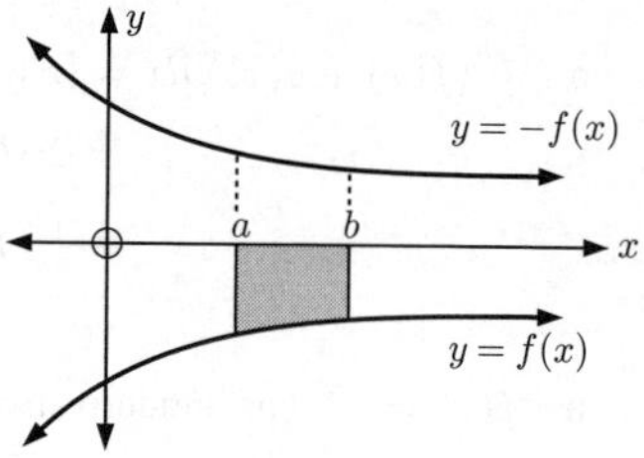

c **i** $\int_0^1 (-x^2) \, dx = -\int_0^1 x^2 \, dx$

Now $f(x) = x^2$ has antiderivative
$F(x) = \frac{1}{3}x^3$

$\therefore \int_0^1 (-x^2) \, dx = -(F(1) - F(0))$
$= -\left(\frac{1}{3} - 0\right)$
$= -\frac{1}{3}$

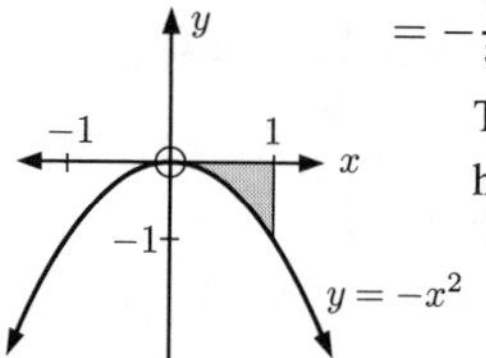

The shaded region has area $\frac{1}{3}$ units^2.

ii $\int_0^1 (x^2 - x) \, dx = -\int_0^1 (x - x^2) \, dx$
$\{x^2 - x \leqslant 0$ for all $x \in [0, 1]\}$

Now $f(x) = x - x^2$ has antiderivative
$F(x) = \frac{1}{2}x^2 - \frac{1}{3}x^3$

$\therefore \int_0^1 (x^2 - x) \, dx = -(F(1) - F(0))$
$= -\left(\frac{1}{2} - \frac{1}{3} - (0 - 0)\right)$
$= -\frac{1}{6}$

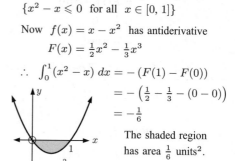

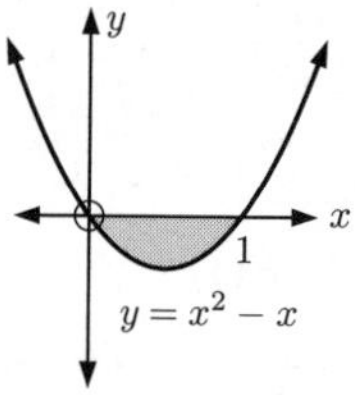

The shaded region has area $\frac{1}{6}$ units^2.

iii $\int_{-2}^{0} 3x\,dx = -\int_{-2}^{0} -3x\,dx$

Now $f(x) = -3x$ has antiderivative

$F(x) = -\frac{3}{2}x^2$

$\therefore \int_{-2}^{0} 3x\,dx = -(F(0) - F(-2))$
$= -(0 - (-6))$
$= -6$

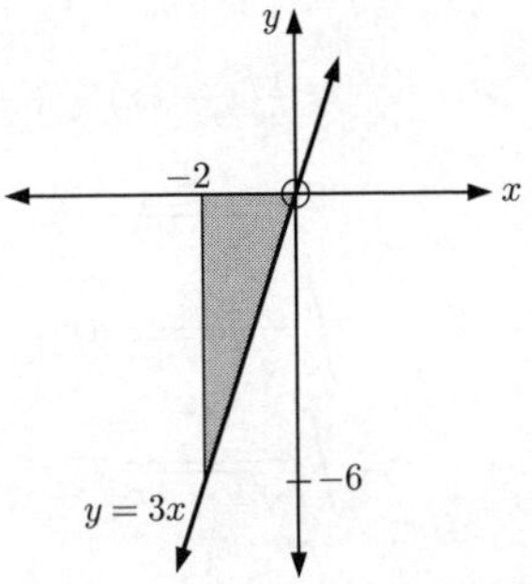

The shaded region has area 6 units^2.

d $\int_0^2 \left(-\sqrt{4-x^2}\right) dx = -\int_0^2 \sqrt{4-x^2}\,dx$

Now $f(x) = \sqrt{4-x^2}$ is the top half of a circle with radius 2 units and centre (0, 0).

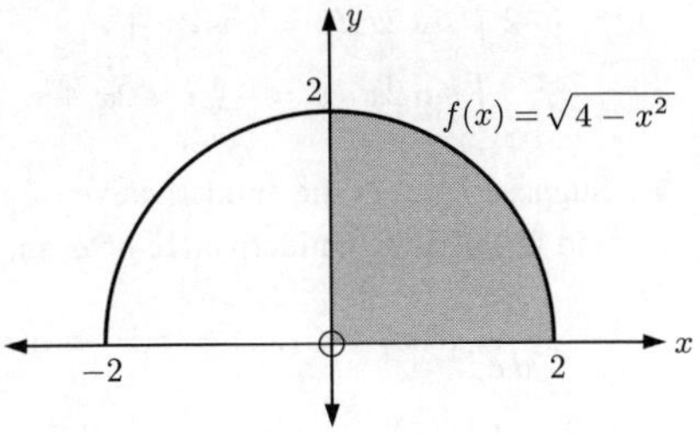

$\therefore \int_0^2 \left(-\sqrt{4-x^2}\right) dx = -\int_0^2 \sqrt{4-x^2}\,dx$
$= -(\text{shaded area})$
$= -\frac{1}{4} \times \pi \times 2^2$
$= -\pi$

EXERCISE 21C.1

1 If $y = x^7$ then $\dfrac{dy}{dx} = 7x^6$

$\therefore \int 7x^6\,dx = x^7 + c_1$

$\therefore 7\int x^6\,dx = x^7 + c_1$

$\therefore \int x^6\,dx = \frac{1}{7}x^7 + c$

2 If $y = x^3 + x^2$ then $\dfrac{dy}{dx} = 3x^2 + 2x$

$\therefore \int (3x^2 + 2x)\,dx = x^3 + x^2 + c$

3 If $y = e^{2x+1}$ then $\dfrac{dy}{dx} = 2e^{2x+1}$

$\therefore \int 2e^{2x+1}\,dx = e^{2x+1} + c_1$

$\therefore 2\int e^{2x+1}\,dx = e^{2x+1} + c_1$

$\therefore \int e^{2x+1}\,dx = \frac{1}{2}e^{2x+1} + c$

4 If $y = (2x+1)^4$

then $\dfrac{dy}{dx} = 4(2x+1)^3 \times 2 = 8(2x+1)^3$

$\therefore \int 8(2x+1)^3\,dx = (2x+1)^4 + c_1$

$\therefore 8\int (2x+1)^3\,dx = (2x+1)^4 + c_1$

$\therefore \int (2x+1)^3\,dx = \frac{1}{8}(2x+1)^4 + c$

5 If $y = x\sqrt{x} = x^{\frac{3}{2}}$

then $\dfrac{dy}{dx} = \frac{3}{2}x^{\frac{1}{2}} = \frac{3}{2}\sqrt{x}$

$\therefore \int \frac{3}{2}\sqrt{x}\,dx = x\sqrt{x} + c_1$

$\therefore \frac{3}{2}\int \sqrt{x}\,dx = x\sqrt{x} + c_1$

$\therefore \int \sqrt{x}\,dx = \frac{2}{3}x\sqrt{x} + c$

6 If $y = \dfrac{1}{\sqrt{x}} = x^{-\frac{1}{2}}$

then $\dfrac{dy}{dx} = -\frac{1}{2}x^{-\frac{3}{2}} = -\dfrac{1}{2x\sqrt{x}}$

$\therefore \displaystyle\int -\frac{1}{2}\left(\frac{1}{x\sqrt{x}}\right) dx = \frac{1}{\sqrt{x}} + c_1$

$\therefore -\frac{1}{2}\displaystyle\int \frac{1}{x\sqrt{x}}\,dx = \frac{1}{\sqrt{x}} + c_1$

$\therefore \displaystyle\int \frac{1}{x\sqrt{x}}\,dx = -\frac{2}{\sqrt{x}} + c$

7 If $y = \cos 2x$

then $\dfrac{dy}{dx} = -2\sin 2x$

$\therefore \int -2\sin 2x \, dx = \cos 2x + c_1$

$\therefore -2\int \sin 2x \, dx = \cos 2x + c_1$

$\therefore \int \sin 2x \, dx = -\frac{1}{2}\cos 2x + c$

8 If $y = \sin(1-5x)$

then $\dfrac{dy}{dx} = -5\cos(1-5x)$

$\therefore \int -5\cos(1-5x)\, dx = \sin(1-5x) + c_1$

$\therefore -5\int \cos(1-5x)\, dx = \sin(1-5x) + c_1$

$\therefore \int \cos(1-5x)\, dx = -\frac{1}{5}\sin(1-5x) + c$

9 Suppose $F(x)$ is the antiderivative of $f(x)$ and $G(x)$ is the antiderivative of $g(x)$.

$\therefore \dfrac{d}{dx}(F(x) + G(x)) = f(x) + g(x)$

$\therefore \int [f(x) + g(x)] \, dx$

$= F(x) + G(x) + c$

$= (F(x) + c_1) + (G(x) + c_2)$

$= \int f(x)\, dx + \int g(x)\, dx$

10 $y = \sqrt{1-4x} = (1-4x)^{\frac{1}{2}}$

$\therefore \dfrac{dy}{dx} = \frac{1}{2}(1-4x)^{-\frac{1}{2}}(-4)$

$= \dfrac{-2}{\sqrt{1-4x}}$

$\therefore \displaystyle\int \frac{-2}{\sqrt{1-4x}}\, dx = \sqrt{1-4x} + c_1$

$\therefore -2\displaystyle\int \frac{1}{\sqrt{1-4x}}\, dx = \sqrt{1-4x} + c_1$

$\therefore \displaystyle\int \frac{1}{\sqrt{1-4x}}\, dx = -\frac{1}{2}\sqrt{1-4x} + c$

11 $\dfrac{d}{dx}(\ln(5-3x+x^2)) = \dfrac{2x-3}{5-3x+x^2}$

Now $5 - 3x + x^2 > 0$ for all x,

as $a > 0$ and $\Delta = -11 < 0$.

$\therefore \displaystyle\int \frac{2x-3}{5-3x+x^2}\, dx = \ln(5-3x+x^2) + c_1$

$\therefore \displaystyle\int \frac{4x-6}{5-3x+x^2}\, dx = 2\ln(5-3x+x^2) + c$

EXERCISE 21C.2

1 **a** $\int (x^4 - x^2 - x + 2)\, dx$

$= \frac{1}{5}x^5 - \frac{1}{3}x^3 - \frac{1}{2}x^2 + 2x + c$

b $\int (\sqrt{x} + e^x)\, dx$

$= \int (x^{\frac{1}{2}} + e^x)\, dx$

$= \dfrac{x^{\frac{3}{2}}}{\frac{3}{2}} + e^x + c$

$= \frac{2}{3}x^{\frac{3}{2}} + e^x + c$

c $\displaystyle\int \left(3e^x - \frac{1}{x}\right) dx$

$= 3e^x - \ln x + c, \ x > 0$

d $\displaystyle\int \left(x\sqrt{x} - \frac{2}{x}\right) dx$

$= \displaystyle\int \left(x^{\frac{3}{2}} - \frac{2}{x}\right) dx$

$= \dfrac{x^{\frac{5}{2}}}{\frac{5}{2}} - 2\ln x + c, \ x > 0$

$= \frac{2}{5}x^{\frac{5}{2}} - 2\ln x + c, \ x > 0$

e $\displaystyle\int \left(\frac{1}{x\sqrt{x}} + \frac{4}{x}\right) dx$

$= \displaystyle\int \left(x^{-\frac{3}{2}} + \frac{4}{x}\right) dx$

$= \dfrac{x^{-\frac{1}{2}}}{-\frac{1}{2}} + 4\ln x + c, \ x > 0$

$= -\dfrac{2}{\sqrt{x}} + 4\ln x + c, \ x > 0$

f $\int (\frac{1}{2}x^3 - x^4 + x^{\frac{1}{3}})\, dx$

$= \frac{1}{2}\dfrac{x^4}{4} - \dfrac{x^5}{5} + \dfrac{x^{\frac{4}{3}}}{\frac{4}{3}} + c$

$= \frac{1}{8}x^4 - \frac{1}{5}x^5 + \frac{3}{4}x^{\frac{4}{3}} + c$

g $\displaystyle\int \left(x^2 + \frac{3}{x}\right) dx$

$= \frac{1}{3}x^3 + 3\ln x + c, \ x > 0$

h $\displaystyle\int \left(\frac{1}{2x} + x^2 - e^x\right) dx$

$= \frac{1}{2}\ln x + \frac{1}{3}x^3 - e^x + c, \ x > 0$

i $\int \left(5e^x + \frac{1}{3}x^3 - \frac{4}{x}\right) dx$

$= 5e^x + \frac{1}{3}\frac{x^4}{4} - 4\ln x + c, \ \ x > 0$

$= 5e^x + \frac{1}{12}x^4 - 4\ln x + c, \ \ x > 0$

2 **a** $\int (3\sin x - 2)\, dx$

$= -3\cos x - 2x + c$

b $\int (4x - 2\cos x)\, dx$

$= 2x^2 - 2\sin x + c$

c $\int (\sin x - 2\cos x + e^x)\, dx$

$= -\cos x - 2\sin x + e^x + c$

d $\int (x^2\sqrt{x} - 10\sin x)\, dx$

$= \int (x^{\frac{5}{2}} - 10\sin x)\, dx$

$= \frac{2}{7}x^{\frac{7}{2}} + 10\cos x + c$

$= \frac{2}{7}x^3\sqrt{x} + 10\cos x + c$

e $\int \left(\frac{x(x-1)}{3} + \cos x\right) dx$

$= \int \left(\frac{x^2}{3} - \frac{x}{3} + \cos x\right) dx$

$= \frac{x^3}{9} - \frac{x^2}{6} + \sin x + c$

f $\int (-\sin x + 2\sqrt{x})\, dx$

$= \int (-\sin x + 2x^{\frac{1}{2}})\, dx$

$= \cos x + \frac{4}{3}x^{\frac{3}{2}} + c$

$= \cos x + \frac{4}{3}x\sqrt{x} + c$

3 **a** $\int (x^2 + 3x - 2)\, dx$

$= \frac{x^3}{3} + \frac{3x^2}{2} - 2x + c$

b $\int \left(\sqrt{x} - \frac{1}{\sqrt{x}}\right) dx$

$= \int \left(x^{\frac{1}{2}} - x^{-\frac{1}{2}}\right) dx$

$= \frac{x^{\frac{3}{2}}}{\frac{3}{2}} - \frac{x^{\frac{1}{2}}}{\frac{1}{2}} + c$

$= \frac{2}{3}x^{\frac{3}{2}} - 2x^{\frac{1}{2}} + c$

c $\int \left(2e^x - \frac{1}{x^2}\right) dx$

$= \int (2e^x - x^{-2})\, dx$

$= 2e^x - \frac{x^{-1}}{-1} + c$

$= 2e^x + \frac{1}{x} + c$

d $\int \left(\frac{1-4x}{x\sqrt{x}}\right) dx$

$= \int \left(\frac{1}{x\sqrt{x}} - \frac{4}{\sqrt{x}}\right) dx$

$= \int (x^{-\frac{3}{2}} - 4x^{-\frac{1}{2}})\, dx$

$= \frac{x^{-\frac{1}{2}}}{-\frac{1}{2}} - \frac{4x^{\frac{1}{2}}}{\frac{1}{2}} + c$

$= -2x^{-\frac{1}{2}} - 8x^{\frac{1}{2}} + c$

e $\int (2x+1)^2\, dx$

$= \int (4x^2 + 4x + 1)\, dx$

$= \frac{4x^3}{3} + \frac{4x^2}{2} + x + c$

$= \frac{4}{3}x^3 + 2x^2 + x + c$

f $\int \frac{x^2 + x - 3}{x}\, dx$

$= \int \left(x + 1 - \frac{3}{x}\right) dx$

$= \frac{x^2}{2} + x - 3\ln x + c, \ \ x > 0$

g $\int \frac{2x-1}{\sqrt{x}}\, dx$

$= \int \left(2x^{\frac{1}{2}} - x^{-\frac{1}{2}}\right) dx$

$= \frac{2x^{\frac{3}{2}}}{\frac{3}{2}} - \frac{x^{\frac{1}{2}}}{\frac{1}{2}} + c$

$= \frac{4}{3}x^{\frac{3}{2}} - 2x^{\frac{1}{2}} + c$

h $\int \frac{x^2 - 4x + 10}{x^2\sqrt{x}}\, dx$

$= \int \left(\frac{x^2}{x^2\sqrt{x}} - \frac{4x}{x^2\sqrt{x}} + \frac{10}{x^2\sqrt{x}}\right) dx$

$= \int \left(x^{-\frac{1}{2}} - 4x^{-\frac{3}{2}} + 10x^{-\frac{5}{2}}\right) dx$

$= \frac{x^{\frac{1}{2}}}{\frac{1}{2}} - \frac{4x^{-\frac{1}{2}}}{-\frac{1}{2}} + \frac{10x^{-\frac{3}{2}}}{-\frac{3}{2}} + c$

$= 2x^{\frac{1}{2}} + 8x^{-\frac{1}{2}} - \frac{20}{3}x^{-\frac{3}{2}} + c$

i $\int (x+1)^3\, dx$

$= \int (x^3 + 3x^2 + 3x + 1)\, dx$

$= \frac{1}{4}x^4 + x^3 + \frac{3}{2}x^2 + x + c$

4 **a** $\int(\sqrt{x}+\frac{1}{2}\cos x)\,dx$
$=\int(x^{\frac{1}{2}}+\frac{1}{2}\cos x)\,dx$
$=\frac{2}{3}x^{\frac{3}{2}}+\frac{1}{2}\sin x+c$

b $\int(2e^t-4\sin t)\,dt$
$=2e^t+4\cos t+c$

c $\displaystyle\int\left(3\cos t-\frac{1}{t}\right)dt$
$=3\sin t-\ln t+c,\ t>0$

5 **a** $\dfrac{dy}{dx}=6$
$\therefore\ y=\int 6\,dx$
$\therefore\ y=6x+c$

b $\dfrac{dy}{dx}=4x^2$
$\therefore\ y=\int 4x^2\,dx$
$\therefore\ y=\frac{4}{3}x^3+c$

c $\dfrac{dy}{dx}=5\sqrt{x}-x^2=5x^{\frac{1}{2}}-x^2$
$\therefore\ y=\int(5x^{\frac{1}{2}}-x^2)\,dx$
$\therefore\ y=\frac{10}{3}x^{\frac{3}{2}}-\frac{1}{3}x^3+c$

d $\dfrac{dy}{dx}=\dfrac{1}{x^2}=x^{-2}$
$\therefore\ y=\int x^{-2}\,dx$
$\therefore\ y=\dfrac{x^{-1}}{-1}+c$
$\therefore\ y=-\dfrac{1}{x}+c$

e $\dfrac{dy}{dx}=2e^x-5$
$\therefore\ y=\int(2e^x-5)\,dx$
$\therefore\ y=2e^x-5x+c$

f $\dfrac{dy}{dx}=4x^3+3x^2$
$\therefore\ y=\int(4x^3+3x^2)\,dx$
$=\dfrac{4x^4}{4}+\dfrac{3x^3}{3}+c$
$\therefore\ y=x^4+x^3+c$

6 **a** $\dfrac{dy}{dx}=(1-2x)^2$
$\therefore\ y=\int(1-2x)^2\,dx$
$=\int(1-4x+4x^2)\,dx$
$=x-\dfrac{4x^2}{2}+\dfrac{4x^3}{3}+c$
$=x-2x^2+\frac{4}{3}x^3+c$

b $\dfrac{dy}{dx}=\sqrt{x}-\dfrac{2}{\sqrt{x}}$
$=x^{\frac{1}{2}}-2x^{-\frac{1}{2}}$
$\therefore\ y=\int(x^{\frac{1}{2}}-2x^{-\frac{1}{2}})\,dx$
$=\dfrac{x^{\frac{3}{2}}}{\frac{3}{2}}-\dfrac{2x^{\frac{1}{2}}}{\frac{1}{2}}+c$
$=\frac{2}{3}x^{\frac{3}{2}}-4x^{\frac{1}{2}}+c$

c $\dfrac{dy}{dx}=\dfrac{x^2+2x-5}{x^2}$
$=1+2x^{-1}-5x^{-2}$
$\therefore\ y=\int(1+2x^{-1}-5x^{-2})\,dx$
$=x+2\ln x-\dfrac{5x^{-1}}{-1}+c,\ x>0$
$=x+2\ln x+\dfrac{5}{x}+c,\ x>0$

7 **a** $f'(x)=x^3-5\sqrt{x}+3$
$=x^3-5x^{\frac{1}{2}}+3$
$\therefore\ f(x)=\int(x^3-5x^{\frac{1}{2}}+3)\,dx$
$=\frac{1}{4}x^4-\frac{10}{3}x^{\frac{3}{2}}+3x+c$

b $f'(x)=2\sqrt{x}(1-3x)$
$=2x^{\frac{1}{2}}-6x^{\frac{3}{2}}$
$\therefore\ f(x)=\int(2x^{\frac{1}{2}}-6x^{\frac{3}{2}})\,dx$
$=\dfrac{2x^{\frac{3}{2}}}{\frac{3}{2}}-\dfrac{6x^{\frac{5}{2}}}{\frac{5}{2}}+c$
$=\frac{4}{3}x^{\frac{3}{2}}-\frac{12}{5}x^{\frac{5}{2}}+c$

c $f'(x)=3e^x-\dfrac{4}{x}$
$\therefore\ f(x)=\displaystyle\int\left(3e^x-\frac{4}{x}\right)dx$
$=3e^x-4\ln x+c,\ x>0$

8 **a**
$$f'(x) = 2x - 1$$
$$\therefore\ f(x) = \int (2x-1)\,dx = \frac{2x^2}{2} - x + c = x^2 - x + c$$
But $f(0) = 3$, so $0 - 0 + c = 3$
$$\therefore\ c = 3$$
$$\therefore\ f(x) = x^2 - x + 3$$

b
$$f'(x) = 3x^2 + 2x$$
$$\therefore\ f(x) = \int (3x^2 + 2x)\,dx = \frac{3x^3}{3} + \frac{2x^2}{2} + c = x^3 + x^2 + c$$
But $f(2) = 5$, so $8 + 4 + c = 5$
$$\therefore\ c = -7$$
$$\therefore\ f(x) = x^3 + x^2 - 7$$

c
$$f'(x) = e^x + \frac{1}{\sqrt{x}} = e^x + x^{-\frac{1}{2}}$$
$$\therefore\ f(x) = \int (e^x + x^{-\frac{1}{2}})\,dx = e^x + 2x^{\frac{1}{2}} + c$$
But $f(1) = 1$, so $e^1 + 2 + c = 1$
$$\therefore\ c = -1 - e$$
$$\therefore\ f(x) = e^x + 2\sqrt{x} - 1 - e$$

d
$$f'(x) = x - \frac{2}{\sqrt{x}} = x - 2x^{-\frac{1}{2}}$$
$$\therefore\ f(x) = \int (x - 2x^{-\frac{1}{2}})\,dx = \frac{x^2}{2} - \frac{2x^{\frac{1}{2}}}{\frac{1}{2}} + c = \tfrac{1}{2}x^2 - 4\sqrt{x} + c$$
But $f(1) = 2$, so $\frac{1}{2} - 4 + c = 2$
$$\therefore\ c = \tfrac{11}{2}$$
$$\therefore\ f(x) = \tfrac{1}{2}x^2 - 4\sqrt{x} + \tfrac{11}{2}$$

9 **a**
$$f'(x) = x^2 - 4\cos x$$
$$\therefore\ f(x) = \int (x^2 - 4\cos x)\,dx = \frac{x^3}{3} - 4\sin x + c$$
But $f(0) = 3$
$$\therefore\ 0 - 4\sin(0) + c = 3$$
$$\therefore\ c = 3$$
$$\therefore\ f(x) = \frac{x^3}{3} - 4\sin x + 3$$

b
$$f'(x) = 2\cos x - 3\sin x$$
$$\therefore\ f(x) = \int (2\cos x - 3\sin x)\,dx = 2\sin x + 3\cos x + c$$
But $f(\frac{\pi}{4}) = \frac{1}{\sqrt{2}}$
$$\therefore\ 2\sin\tfrac{\pi}{4} + 3\cos\tfrac{\pi}{4} + c = \tfrac{1}{\sqrt{2}}$$
$$\therefore\ 2(\tfrac{1}{\sqrt{2}}) + 3(\tfrac{1}{\sqrt{2}}) + c = \tfrac{1}{\sqrt{2}}$$
$$\therefore\ c = -\tfrac{4}{\sqrt{2}}$$
$$\therefore\ c = -2\sqrt{2}$$
$$\therefore\ f(x) = 2\sin x + 3\cos x - 2\sqrt{2}$$

10 **a** Given: $f''(x) = 2x + 1$, $f'(1) = 3$, $f(2) = 7$
$$\therefore\ f'(x) = \int (2x+1)\,dx = \frac{2x^2}{2} + x + c = x^2 + x + c$$
But $f'(1) = 3$ so $1 + 1 + c = 3$
$$\therefore\ c = 1$$
$$\therefore\ f'(x) = x^2 + x + 1$$

Then $f(x) = \int (x^2 + x + 1)\,dx$
$$= \frac{x^3}{3} + \frac{x^2}{2} + x + k$$
But $f(2) = 7$ so $\frac{8}{3} + 2 + 2 + k = 7$
$$\therefore\ k = 7 - 4 - \tfrac{8}{3}$$
$$\therefore\ k = \tfrac{1}{3}$$
$$\therefore\ f(x) = \tfrac{1}{3}x^3 + \tfrac{1}{2}x^2 + x + \tfrac{1}{3}$$

b Given: $f''(x) = 15\sqrt{x} + \dfrac{3}{\sqrt{x}}$, $f'(1) = 12$, $f(0) = 5$

Now $f''(x) = 15x^{\frac{1}{2}} + 3x^{-\frac{1}{2}}$

$\therefore\ f'(x) = \dfrac{15x^{\frac{3}{2}}}{\frac{3}{2}} + \dfrac{3x^{\frac{1}{2}}}{\frac{1}{2}} + c$

$= 10x^{\frac{3}{2}} + 6x^{\frac{1}{2}} + c$

But $f'(1) = 12$ so $10 + 6 + c = 12$

$\therefore\ c = -4$

$\therefore\ f'(x) = 10x^{\frac{3}{2}} + 6x^{\frac{1}{2}} - 4$

$\therefore\ f(x) = \int (10x^{\frac{3}{2}} + 6x^{\frac{1}{2}} - 4)\, dx$

$= \dfrac{10x^{\frac{5}{2}}}{\frac{5}{2}} + \dfrac{6x^{\frac{3}{2}}}{\frac{3}{2}} - 4x + k$

$= 4x^{\frac{5}{2}} + 4x^{\frac{3}{2}} - 4x + k$

But $f(0) = 5$ so $k = 5$

$\therefore\ f(x) = 4x^{\frac{5}{2}} + 4x^{\frac{3}{2}} - 4x + 5$

c Given: $f''(x) = \cos x$, $f'(\frac{\pi}{2}) = 0$ and $f(0) = 3$

Now $f'(x) = \int \cos x\, dx = \sin x + c$

But $f'(\frac{\pi}{2}) = 0$ so $\sin(\frac{\pi}{2}) + c = 0$

$\therefore\ c = -1$

$\therefore\ f'(x) = \sin x - 1$

So, $f(x) = \int (\sin x - 1)\, dx$

$= -\cos x - x + k$

But $f(0) = 3$ so $-\cos 0 - 0 + k = 3$

$\therefore\ -1 + k = 3$

$\therefore\ k = 4$

So, $f(x) = -\cos x - x + 4$

d Given: $f''(x) = 2x$ and that $(1, 0)$ and $(0, 5)$ lie on the curve

Now $f'(x) = \int 2x\, dx = \dfrac{2x^2}{2} + c = x^2 + c$

$\therefore\ f(x) = \int (x^2 + c)\, dx = \dfrac{x^3}{3} + cx + k$

But $f(0) = 5$ so $0 + 0 + k = 5$ and so $k = 5$

and $f(1) = 0$ so $\frac{1}{3} + c + 5 = 0$ and so $c = -5\frac{1}{3}$

$\therefore\ f(x) = \frac{1}{3}x^3 - \frac{16}{3}x + 5$

EXERCISE 21D

1 **a** $\int (2x+5)^3\, dx$

$= \frac{1}{2} \times \dfrac{(2x+5)^4}{4} + c$

$= \frac{1}{8}(2x+5)^4 + c$

b $\displaystyle\int \frac{1}{(3-2x)^2}\, dx$

$= \int (3-2x)^{-2}\, dx$

$= \frac{1}{-2} \times \dfrac{(3-2x)^{-1}}{-1} + c$

$= \dfrac{1}{2(3-2x)} + c$

c $\displaystyle\int \frac{4}{(2x-1)^4}\, dx$

$= \int 4(2x-1)^{-4}\, dx$

$= 4(\frac{1}{2}) \times \dfrac{(2x-1)^{-3}}{-3} + c$

$= \dfrac{-2}{3(2x-1)^3} + c$

d $\int (4x-3)^7\, dx$

$= \frac{1}{4} \times \dfrac{(4x-3)^8}{8} + c$

$= \frac{1}{32}(4x-3)^8 + c$

e $\int \sqrt{3x-4}\, dx$

$= \int (3x-4)^{\frac{1}{2}}\, dx$

$= \frac{1}{3} \times \dfrac{(3x-4)^{\frac{3}{2}}}{\frac{3}{2}} + c$

$= \frac{2}{9}(3x-4)^{\frac{3}{2}} + c$

f $\displaystyle\int \frac{10}{\sqrt{1-5x}}\, dx$

$= \int 10(1-5x)^{-\frac{1}{2}}\, dx$

$= 10(\frac{1}{-5}) \times \dfrac{(1-5x)^{\frac{1}{2}}}{\frac{1}{2}} + c$

$= -4\sqrt{1-5x} + c$

g
$$\begin{aligned}&\int 3(1-x)^4\,dx\\&=3\int(1-x)^4\,dx\\&=3(\tfrac{1}{-1})\times\frac{(1-x)^5}{5}+c\\&=-\tfrac{3}{5}(1-x)^5+c\end{aligned}$$

h
$$\begin{aligned}&\int\frac{4}{\sqrt{3-4x}}\,dx\\&=\int 4(3-4x)^{-\frac{1}{2}}\,dx\\&=4(\tfrac{1}{-4})\times\frac{(3-4x)^{\frac{1}{2}}}{\frac{1}{2}}+c\\&=-2\sqrt{3-4x}+c\end{aligned}$$

2 **a**
$$\begin{aligned}&\int\sin(3x)\;dx\\&=-\tfrac{1}{3}\cos(3x)+c\end{aligned}$$

b
$$\begin{aligned}&\int 2\cos(-4x)+1\;dx\\&=2\times(\tfrac{1}{-4})\sin(-4x)+x+c\\&=-\tfrac{1}{2}\sin(-4x)+x+c\end{aligned}$$

c
$$\begin{aligned}&\int 3\cos\left(\tfrac{x}{2}\right)\,dx\\&=6\sin\left(\tfrac{x}{2}\right)+c\end{aligned}$$

d
$$\begin{aligned}&\int(3\sin(2x)-e^{-x})\;dx\\&=-\tfrac{3}{2}\cos(2x)+e^{-x}+c\end{aligned}$$

e
$$\begin{aligned}&\int 2\sin(2x+\tfrac{\pi}{6})\;dx\\&=-\tfrac{2}{2}\cos(2x+\tfrac{\pi}{6})+c\\&=-\cos(2x+\tfrac{\pi}{6})+c\end{aligned}$$

f
$$\begin{aligned}&\int -3\cos(\tfrac{\pi}{4}-x)\;dx\\&=-3\times(-1)\sin(\tfrac{\pi}{4}-x)+c\\&=3\sin(\tfrac{\pi}{4}-x)+c\end{aligned}$$

g
$$\begin{aligned}&\int\cos(2x)+\sin(2x)\;dx\\&=\tfrac{1}{2}\sin(2x)-\tfrac{1}{2}\cos(2x)+c\end{aligned}$$

h
$$\begin{aligned}&\int 2\sin(3x)+5\cos(4x)\;dx\\&=-\tfrac{2}{3}\cos(3x)+\tfrac{5}{4}\sin(4x)+c\end{aligned}$$

i
$$\begin{aligned}&\int\left(\tfrac{1}{2}\cos(8x)-3\sin x\right)\;dx\\&=\tfrac{1}{2}(\tfrac{1}{8})\sin(8x)+3\cos x+c\\&=\tfrac{1}{16}\sin(8x)+3\cos x+c\end{aligned}$$

3 **a**
$$\begin{aligned}&\frac{dy}{dx}=\sqrt{2x-7}=(2x-7)^{\frac{1}{2}}\\&\therefore\quad y=\tfrac{1}{2}\times\frac{(2x-7)^{\frac{3}{2}}}{\frac{3}{2}}+c\\&\qquad=\tfrac{1}{3}(2x-7)^{\frac{3}{2}}+c\end{aligned}$$

But $y=11$ when $x=8$

$$\begin{aligned}\therefore\quad&\tfrac{1}{3}(16-7)^{\frac{3}{2}}+c=11\\\therefore\quad&\tfrac{1}{3}(27)+c=11\\\therefore\quad&9+c=11\quad\text{and so}\quad c=2\\\therefore\quad&y=\tfrac{1}{3}(2x-7)^{\frac{3}{2}}+2\end{aligned}$$

b $f(x)$ has gradient function $f'(x)=\dfrac{4}{\sqrt{1-x}}=4(1-x)^{-\frac{1}{2}}$

$$\begin{aligned}\therefore\quad f(x)&=4(\tfrac{1}{-1})\times\frac{(1-x)^{\frac{1}{2}}}{\frac{1}{2}}+c\\&=-8\sqrt{1-x}+c\end{aligned}$$

But $y=-11$ when $x=-3$

$$\begin{aligned}\therefore\quad&-8\sqrt{1-(-3)}+c=-11\\\therefore\quad&-8\sqrt{4}+c=-11\\\therefore\quad&-16+c=-11\quad\text{and so}\quad c=5\\\therefore\quad&f(x)=5-8\sqrt{1-x}\end{aligned}$$

Now $f(-8)=5-8\sqrt{1-(-8)}=5-8(3)=-19$, so the point is $(-8,\,-19)$.

4 **a**
$$\begin{aligned}&\int\cos^2 x\;dx\\&=\int(\tfrac{1}{2}+\tfrac{1}{2}\cos(2x))\;dx\\&=\tfrac{1}{2}x+\tfrac{1}{4}\sin(2x)+c\end{aligned}$$

b
$$\begin{aligned}&\int\sin^2 x\;dx\\&=\int(\tfrac{1}{2}-\tfrac{1}{2}\cos(2x))\;dx\\&=\tfrac{1}{2}x-\tfrac{1}{4}\sin(2x)+c\end{aligned}$$

c $\int (1+\cos^2(2x))\,dx$
$= \int (1+\frac{1}{2}+\frac{1}{2}\cos(4x))\,dx$
$= \int (\frac{3}{2}+\frac{1}{2}\cos(4x))\,dx$
$= \frac{3}{2}x+\frac{1}{8}\sin(4x)+c$

d $\int (3-\sin^2(3x))\,dx$
$= \int (3-(\frac{1}{2}-\frac{1}{2}\cos(6x)))\,dx$
$= \int (\frac{5}{2}+\frac{1}{2}\cos(6x))\,dx$
$= \frac{5}{2}x+\frac{1}{12}\sin(6x)+c$

e $\int \frac{1}{2}\cos^2(4x)\,dx$
$= \int \frac{1}{2}(\frac{1}{2}+\frac{1}{2}\cos(8x))\,dx$
$= \int (\frac{1}{4}+\frac{1}{4}\cos(8x))\,dx$
$= \frac{1}{4}x+\frac{1}{32}\sin(8x)+c$

f $\int (1+\cos x)^2\,dx$
$= \int (1+2\cos x+\cos^2 x)\,dx$
$= \int (1+2\cos x+\frac{1}{2}+\frac{1}{2}\cos(2x))\,dx$
$= \int \left(\frac{3}{2}+2\cos x+\frac{1}{2}\cos(2x)\right)dx$
$= \frac{3}{2}x+2\sin x+\frac{1}{4}\sin(2x)+c$

5 **a** $\int 3(2x-1)^2\,dx$
$= 3\int (2x-1)^2\,dx$
$= 3(\frac{1}{2})\,\dfrac{(2x-1)^3}{3}+c$
$= \frac{1}{2}(2x-1)^3+c$

b $\int (x^2-x)^2\,dx$
$= \int (x^4-2x^3+x^2)\,dx$
$= \dfrac{x^5}{5}-\dfrac{2x^4}{4}+\dfrac{x^3}{3}+c$
$= \frac{1}{5}x^5-\frac{1}{2}x^4+\frac{1}{3}x^3+c$

c $\int (1-3x)^3\,dx$
$= (\frac{1}{-3})\,\dfrac{(1-3x)^4}{4}+c$
$= -\frac{1}{12}(1-3x)^4+c$

d $\int (1-x^2)^2\,dx$
$= \int (1-2x^2+x^4)\,dx$
$= x-\frac{2}{3}x^3+\frac{1}{5}x^5+c$

e $\int 4\sqrt{5-x}\,dx$
$= 4\int (5-x)^{\frac{1}{2}}\,dx$
$= 4(\frac{1}{-1})\,\dfrac{(5-x)^{\frac{3}{2}}}{\frac{3}{2}}+c$
$= -\frac{8}{3}(5-x)^{\frac{3}{2}}+c$

f $\int (x^2+1)^3\,dx$
$= \int (x^6+3x^4+3x^2+1)\,dx$
$= \dfrac{x^7}{7}+\dfrac{3x^5}{5}+\dfrac{3x^3}{3}+x+c$
$= \frac{1}{7}x^7+\frac{3}{5}x^5+x^3+x+c$

6 **a** $\int (2e^x+5e^{2x})\,dx$
$= 2e^x+5(\frac{1}{2})e^{2x}+c$
$= 2e^x+\frac{5}{2}e^{2x}+c$

b $\int (3e^{5x-2})\,dx$
$= 3(\frac{1}{5})e^{5x-2}+c$
$= \frac{3}{5}e^{5x-2}+c$

c $\int (e^{7-3x})\,dx$
$= \frac{1}{-3}e^{7-3x}+c$
$= -\frac{1}{3}e^{7-3x}+c$

d $\displaystyle\int \frac{1}{2x-1}\,dx$
$= \frac{1}{2}\ln(2x-1)+c,\ \ 2x-1>0$
$= \frac{1}{2}\ln(2x-1)+c,\ \ x>\frac{1}{2}$

e $\displaystyle\int \frac{5}{1-3x}\,dx$
$= 5\displaystyle\int \frac{1}{1-3x}\,dx$
$= 5(\frac{1}{-3})\ln(1-3x)+c,\ \ 1-3x>0$
$= -\frac{5}{3}\ln(1-3x)+c,\ \ x<\frac{1}{3}$

f $\displaystyle\int \left(e^{-x}-\frac{4}{2x+1}\right)dx$
$= \frac{1}{-1}e^{-x}-4(\frac{1}{2})\ln(2x+1)+c,\ \ 2x+1>0$
$= -e^{-x}-2\ln(2x+1)+c,\ \ x>-\frac{1}{2}$

g $\int (e^x+e^{-x})^2\,dx$
$= \int (e^{2x}+2+e^{-2x})\,dx$
$= \frac{1}{2}e^{2x}+2x+(\frac{1}{-2})e^{-2x}+c$
$= \frac{1}{2}e^{2x}+2x-\frac{1}{2}e^{-2x}+c$

h $\int (e^{-x}+2)^2\,dx$
$= \int (e^{-2x}+4e^{-x}+4)\,dx$
$= \frac{1}{-2}e^{-2x}+4(\frac{1}{-1})e^{-x}+4x+c$
$= -\frac{1}{2}e^{-2x}-4e^{-x}+4x+c$

i $\displaystyle\int \left(x-\frac{5}{1-x}\right)dx$
$= \dfrac{x^2}{2}-5(\frac{1}{-1})\ln(1-x)+c,\ \ 1-x>0$
$= \frac{1}{2}x^2+5\ln(1-x)+c,\ \ x<1$

7 **a** $\frac{dy}{dx} = (1-e^x)^2$

$= 1 - 2e^x + e^{2x}$

$\therefore\ y = x - 2e^x + \frac{1}{2}e^{2x} + c$

b $\frac{dy}{dx} = 1 - 2x + \frac{3}{x+2}$

$\therefore\ y = x - \frac{2x^2}{2} + 3\ln(x+2) + c,\ x+2>0$

$= x - x^2 + 3\ln(x+2) + c,\ x > -2$

c $\frac{dy}{dx} = e^{-2x} + \frac{4}{2x-1}$

$\therefore\ y = \frac{1}{-2}e^{-2x} + 4(\frac{1}{2})\ln(2x-1) + c,\ 2x-1>0$

$= -\frac{1}{2}e^{-2x} + 2\ln(2x-1) + c,\ x > \frac{1}{2}$

8 Differentiating Tracy's answer gives

$\frac{d}{dx}\left(\frac{1}{4}\ln(4x)+c\right) = \frac{1}{4}\left(\frac{1}{4x}\right) \times 4 + 0,\ x>0$

$= \frac{1}{4x},\ x>0$

Differentiating Nadine's answer gives

$\frac{d}{dx}\left(\frac{1}{4}\ln(x)+c\right) = \frac{1}{4}\left(\frac{1}{x}\right) + 0,\ x>0$

$= \frac{1}{4x},\ x>0$

Both answers give the correct derivative and both are correct. This result occurs because $\ln(4x) = \ln 4 + \ln x$. Their answers differ by a constant which is accounted for by c.

9 Given: $f'(x) = p\sin(\frac{1}{2}x)$, $f(0)=1$ and $f(2\pi)=0$

$\therefore\ f(x) = -2p\cos(\frac{1}{2}x) + c$

But $f(0)=1$, so $-2p\cos(0)+c=1$

$\therefore\ -2p + c = 1$

$\therefore\ c = 1 + 2p$ .... (1)

Also, $f(2\pi)=0$, so $-2p\cos(\pi)+c=0$

$\therefore\ 2p + c = 0$

$\therefore\ 2p + 1 + 2p = 0$ {using (1)}

$\therefore\ p = -\frac{1}{4}$

$\therefore\ c = \frac{1}{2}$ {from (1)}

$\therefore\ f(x) = \frac{1}{2}\cos(\frac{1}{2}x) + \frac{1}{2}$

10 $g''(x) = -\sin 2x$

Integrating both sides with respect to x, we get $g'(x) = \frac{1}{2}\cos 2x + c$, c some constant.

So, $g'(\pi) = \frac{1}{2}\cos(2\pi) + c$ and $g'(-\pi) = \frac{1}{2}\cos(-2\pi) + c$

$= \frac{1}{2} + c$ $\qquad\qquad = \frac{1}{2} + c$

$= g'(\pi)$

$\therefore$ the gradients of the tangents to $y = g(x)$ at $x=\pi$ and $x=-\pi$ are equal.

11 **a** $f'(x) = 2e^{-2x}$

$\therefore\ f(x) = 2(\frac{1}{-2})e^{-2x} + c$

$= -e^{-2x} + c$

But $f(0)=3$ so $-e^0 + c = 3$

$\therefore\ c = 4$

$\therefore\ f(x) = -e^{-2x} + 4$

b $f'(x) = 2x - \frac{2}{1-x}$

$\therefore\ f(x) = \frac{2x^2}{2} - \frac{2}{-1}\ln(1-x) + c,\ 1-x>0$

$= x^2 + 2\ln(1-x) + c,\ x<1$

But $f(-1)=3$ so $1 + 2\ln(2) + c = 3$

$\therefore\ c = 2 - 2\ln 2$

$\therefore\ f(x) = x^2 + 2\ln(1-x) + 2 - 2\ln 2,\ x<1$

c $f'(x) = \sqrt{x} + \frac{1}{2}e^{-4x}$

$= x^{\frac{1}{2}} + \frac{1}{2}e^{-4x}$

$\therefore \quad f(x) = \dfrac{x^{\frac{3}{2}}}{\frac{3}{2}} + \frac{1}{2}(\frac{1}{-4})e^{-4x} + c$

$= \frac{2}{3}x^{\frac{3}{2}} - \frac{1}{8}e^{-4x} + c$

But $f(1) = 0$

$\therefore \quad \frac{2}{3} - \frac{1}{8}e^{-4} + c = 0$

$\therefore \quad c = \frac{1}{8}e^{-4} - \frac{2}{3}$

$\therefore \quad f(x) = \frac{2}{3}x^{\frac{3}{2}} - \frac{1}{8}e^{-4x} + \frac{1}{8}e^{-4} - \frac{2}{3}$

12
$$\begin{aligned}(\sin x + \cos x)^2 &= \sin^2 x + 2\sin x\cos x + \cos^2 x\\ &= 1 + \sin 2x\\ \therefore \int(\sin x + \cos x)^2\,dx &= \int(1 + \sin 2x)\,dx\\ &= x - \tfrac{1}{2}\cos 2x + c\end{aligned}$$

13
$$\begin{aligned}(\cos x + 1)^2 &= \cos^2 x + 2\cos x + 1\\ &= (\tfrac{1}{2} + \tfrac{1}{2}\cos 2x) + 2\cos x + 1\\ &= \tfrac{1}{2}\cos 2x + 2\cos x + \tfrac{3}{2}\\ \therefore \int(\cos x + 1)^2\,dx &= \int(\tfrac{1}{2}\cos 2x + 2\cos x + \tfrac{3}{2})\,dx\\ &= \tfrac{1}{4}\sin 2x + 2\sin x + \tfrac{3}{2}x + c\end{aligned}$$

EXERCISE 21E.1

1 **a** $\int_0^1 x^3\,dx = \left[\frac{x^4}{4}\right]_0^1$

$= \frac{1}{4} - 0$

$= \frac{1}{4}$

b $\int_0^2 (x^2 - x)\,dx = \left[\frac{x^3}{3} - \frac{x^2}{2}\right]_0^2$

$= (\frac{8}{3} - 2) - (0 - 0)$

$= \frac{2}{3}$

c $\int_0^1 e^x\,dx = [e^x]_0^1$

$= e^1 - e^0$

$= e - 1$

≈ 1.72

d $\int_0^{\frac{\pi}{6}} \cos x\,dx$

$= [\sin x]_0^{\frac{\pi}{6}}$

$= \sin\frac{\pi}{6} - \sin 0$

$= \frac{1}{2}$

e $\displaystyle\int_1^4 \left(x - \frac{3}{\sqrt{x}}\right) dx$

$= \int_1^4 (x - 3x^{-\frac{1}{2}})\,dx$

$= \left[\dfrac{x^2}{2} - \dfrac{3x^{\frac{1}{2}}}{\frac{1}{2}}\right]_1^4$

$= \left[\dfrac{x^2}{2} - 6\sqrt{x}\right]_1^4$

$= \left[\frac{16}{2} - 12\right] - (\frac{1}{2} - 6)$

$= 1\frac{1}{2}$

f $\displaystyle\int_4^9 \frac{x-3}{\sqrt{x}}\,dx$

$= \int_4^9 (x^{\frac{1}{2}} - 3x^{-\frac{1}{2}})\,dx$

$= \left[\dfrac{x^{\frac{3}{2}}}{\frac{3}{2}} - \dfrac{3x^{\frac{1}{2}}}{\frac{1}{2}}\right]_4^9$

$= \left[\frac{2}{3}x^{\frac{3}{2}} - 6x^{\frac{1}{2}}\right]_4^9$

$= \left[\frac{2}{3}(27) - 6(3)\right] - \left[\frac{2}{3}(8) - 6(2)\right]$

$= (18 - 18) - (\frac{16}{3} - 12)$

$= 6\frac{2}{3}$

g $\displaystyle\int_1^3 \frac{1}{x}\,dx = [\ln(x)]_1^3$

$= \ln 3 - \ln 1$

$= \ln 3 - 0$

$= \ln 3$

≈ 1.10

h $\int_{\frac{\pi}{3}}^{\frac{\pi}{2}} \sin x\,dx$

$= [-\cos x]_{\frac{\pi}{3}}^{\frac{\pi}{2}}$

$= -\cos\frac{\pi}{2} + \cos\frac{\pi}{3}$

$= \frac{1}{2}$

i $\int_1^2 (e^{-x}+1)^2\,dx$
$= \int_1^2 (e^{-2x}+2e^{-x}+1)\,dx$
$= \left[(\frac{1}{-2})e^{-2x} + 2(\frac{1}{-1})e^{-x} + x\right]_1^2$
$= \left[-\dfrac{e^{-2x}}{2} - 2e^{-x} + x\right]_1^2$
$= \left(-\dfrac{e^{-4}}{2} - 2e^{-2} + 2\right) - \left(-\dfrac{e^{-2}}{2} - 2e^{-1} + 1\right)$
≈ 1.52

j $\displaystyle\int_2^6 \frac{1}{\sqrt{2x-3}}\,dx$
$= \int_2^6 (2x-3)^{-\frac{1}{2}}\,dx$
$= \left[\frac{1}{2}\dfrac{(2x-3)^{\frac{1}{2}}}{\frac{1}{2}}\right]_2^6$
$= \left[\sqrt{2x-3}\right]_2^6$
$= \sqrt{9}-\sqrt{1}$
$= 2$

k $\int_0^1 e^{1-x}\,dx = \left[(\frac{1}{-1})e^{1-x}\right]_0^1$
$= \left(\dfrac{e^0}{-1}\right) - \left(\dfrac{e^1}{-1}\right)$
$= -1 + e$
≈ 1.72

l $\int_0^{\frac{\pi}{6}} \sin(3x)\,dx = \left[-\frac{1}{3}\cos(3x)\right]_0^{\frac{\pi}{6}}$
$= -\frac{1}{3}[\cos\frac{\pi}{2} - \cos 0]$
$= -\frac{1}{3}[0-1]$
$= \frac{1}{3}$

m $\int_0^{\frac{\pi}{4}} \cos^2 x\,dx = \int_0^{\frac{\pi}{4}} (\frac{1}{2} + \frac{1}{2}\cos(2x))\,dx$
$= \left[\dfrac{x}{2} + \frac{1}{4}\sin(2x)\right]_0^{\frac{\pi}{4}}$
$= [\frac{\pi}{8} + \frac{1}{4}\sin\frac{\pi}{2}] - 0$
$= \frac{\pi}{8} + \frac{1}{4}$

n $\int_0^{\frac{\pi}{2}} \sin^2 x\,dx = \int_0^{\frac{\pi}{2}} (\frac{1}{2} - \frac{1}{2}\cos(2x))\,dx$
$= \left[\dfrac{x}{2} - \frac{1}{4}\sin(2x)\right]_0^{\frac{\pi}{2}}$
$= [\frac{\pi}{4} - \frac{1}{4}\sin\pi] - 0$
$= \frac{\pi}{4}$

2 Using technology:

a $\int_1^3 \ln x\,dx \approx 1.30$ **b** $\int_{-1}^1 e^{-x^2}\,dx \approx 1.49$ **c** $\displaystyle\int_{\frac{\pi}{4}}^{\frac{\pi}{6}} \sin(\sqrt{x})\,dx \approx -0.189$

EXERCISE 21E.2

1 **a** $\int_1^4 \sqrt{x}\,dx = \int_1^4 x^{\frac{1}{2}}\,dx$
$= \left[\frac{2}{3}x^{\frac{3}{2}}\right]_1^4$
$= \frac{2}{3}(8) - \frac{2}{3}(1)$
$= \frac{14}{3}$

$\int_1^4 (-\sqrt{x})\,dx = \int_1^4 -x^{\frac{1}{2}}\,dx$
$= \left[-\frac{2}{3}x^{\frac{3}{2}}\right]_1^4$
$= -\frac{2}{3}(8) - (-\frac{2}{3}(1))$
$= -\frac{14}{3}$

b $\int_0^1 x^7\,dx = \left[\frac{1}{8}x^8\right]_0^1$
$= \frac{1}{8} - 0 = \frac{1}{8}$

$\int_0^1 (-x^7)\,dx = \left[-\frac{1}{8}x^8\right]_0^1$
$= -\frac{1}{8} - 0 = -\frac{1}{8}$

Property: $\int_a^b [-f(x)]\,dx = -\int_a^b f(x)\,dx$

2 **a** $\int_0^1 x^2\,dx = \left[\frac{1}{3}x^3\right]_0^1$
$= \frac{1}{3} - 0$
$= \frac{1}{3}$

b $\int_1^2 x^2\,dx = \left[\frac{1}{3}x^3\right]_1^2$
$= \frac{1}{3}(8) - \frac{1}{3}(1)$
$= \frac{7}{3}$

c $\int_0^2 x^2\,dx = \left[\frac{1}{3}x^3\right]_0^2$
$= \frac{1}{3}(8) - 0$
$= \frac{8}{3}$

d $\int_0^1 3x^2\,dx = \left[x^3\right]_0^1$
$= 1 - 0$
$= 1$

Properties: $\int_a^b f(x)\,dx + \int_b^c f(x)\,dx = \int_a^c f(x)\,dx$

$\int_a^b c\,f(x)\,dx = c\int_a^b f(x)\,dx,\quad c$ a constant

3 **a** $\int_0^2 (x^3-4x)\,dx = \left[\frac{1}{4}x^4 - 2x^2\right]_0^2$
$= \left[\frac{1}{4}(16) - 2(4)\right] - [0-0]$
$= -4$

b $\int_2^3 (x^3-4x)\,dx = \left[\frac{1}{4}x^4 - 2x^2\right]_2^3$
$= \left[\frac{1}{4}(81) - 2(9)\right] - \left[\frac{1}{4}(16) - 2(4)\right]$
$= \frac{25}{4}$

c $\int_0^3 (x^3-4x)\,dx = \left[\frac{1}{4}x^4 - 2x^2\right]_0^3$
$= \left[\frac{1}{4}(81) - 2(9)\right] - [0-0]$
$= \frac{9}{4}$

4 **a** $\int_0^1 x^2\,dx = \left[\frac{1}{3}x^3\right]_0^1$
$= \frac{1}{3}(1) - 0$
$= \frac{1}{3}$

b $\int_0^1 \sqrt{x}\,dx = \int_0^1 x^{\frac{1}{2}}\,dx$
$= \left[\frac{2}{3}x^{\frac{3}{2}}\right]_0^1$
$= \frac{2}{3}(1) - 0$
$= \frac{2}{3}$

c $\int_0^1 (x^2+\sqrt{x})\,dx$
$= \int_0^1 (x^2 + x^{\frac{1}{2}})\,dx$
$= \left[\frac{1}{3}x^3 + \frac{2}{3}x^{\frac{3}{2}}\right]_0^1$
$= \left[\frac{1}{3}(1) + \frac{2}{3}(1)\right] - [0+0]$
$= 1$

Property: $\int_a^b f(x)\,dx + \int_a^b g(x)\,dx = \int_a^b [f(x)+g(x)]\,dx$

5 **a** $\int_0^3 f(x)\,dx =$ area between $f(x)$ and the x-axis from $x=0$ to $x=3$
$= 2+3+1.5 = 6.5$

b $\int_3^7 f(x)\,dx = -$(area between $f(x)$ and the x-axis from $x=3$ to $x=7$)
$= -\left(\frac{3}{2}+3+\frac{5}{2}+2\right) = -9$

c $\int_2^4 f(x)\,dx =$ (area between $f(x)$ and the x-axis from $x=2$ to $x=3$)
$-$(area between $f(x)$ and the x-axis from $x=3$ to $x=4$)
$= 1.5 - 1.5 = 0$

d $\int_0^7 f(x)\,dx =$ (area between $f(x)$ and the x-axis from $x=0$ to $x=3$)
$-$(area between $f(x)$ and the x-axis from $x=3$ to $x=7$)
$= 6.5 - 9 = -2.5$

6 **a** $\int_0^4 f(x)\,dx =$ area of semi-circle with radius 2
$= \frac{1}{2}\pi(2)^2 = 2\pi$

b $\int_4^6 f(x)\,dx = -$(area of 2 by 2 rectangle)
$= -(2\times 2) = -4$

c $\int_6^8 f(x)\,dx =$ area of semi-circle with radius 1
$= \frac{1}{2}\pi(1)^2 = \frac{\pi}{2}$

d $\int_0^8 f(x)\,dx = \int_0^4 f(x)\,dx + \int_4^6 f(x)\,dx + \int_6^8 f(x)\,dx$
$= 2\pi + (-4) + \frac{\pi}{2} = \frac{5\pi}{2} - 4$

7 **a** $\int_2^4 f(x)\,dx + \int_4^7 f(x)\,dx$
$= \int_2^7 f(x)\,dx$

b $\int_1^3 g(x)\,dx + \int_3^8 g(x)\,dx + \int_8^9 g(x)\,dx$
$= \int_1^9 g(x)\,dx$

8 **a** $\int_1^3 f(x)\,dx + \int_3^6 f(x)\,dx = \int_1^6 f(x)\,dx$
$\therefore \int_3^6 f(x)\,dx = \int_1^6 f(x)\,dx - \int_1^3 f(x)\,dx$
$= (-3) - 2$
$= -5$

b $\int_0^2 f(x)\,dx + \int_2^4 f(x)\,dx + \int_4^6 f(x)\,dx = \int_0^6 f(x)\,dx$
$\therefore \int_2^4 f(x)\,dx = \int_0^6 f(x)\,dx - \int_4^6 f(x)\,dx - \int_0^2 f(x)\,dx$
$= (7) - (-2) - (5)$
$= 4$

9 **a** $\int_1^{-1} f(x)\,dx = -\int_{-1}^1 f(x)\,dx$
$= -(-4)$
$= 4$

b $\int_{-1}^1 (2 + f(x))\,dx = \int_{-1}^1 2\,dx + \int_{-1}^1 f(x)\,dx$
$= [2x]_{-1}^1 + (-4)$
$= (2 - (-2)) - 4$
$= 0$

c $\int_{-1}^1 2\,f(x)\,dx = 2\int_{-1}^1 f(x)\,dx$
$= 2(-4)$
$= -8$

d $\int_{-1}^1 k\,f(x)\,dx = 7$
$\therefore k\int_{-1}^1 f(x)\,dx = 7$
$\therefore k(-4) = 7$
$\therefore k = -\frac{7}{4}$

10 $\int_2^3 (g'(x) - 1)\,dx = \int_2^3 g'(x)\,dx + \int_2^3 -1\,dx$
$= [g(x)]_2^3 + [-x]_2^3$
$= (g(3) - g(2)) + (-3 - (-2))$
$= 5 - 4 - 1$
$= 0$

REVIEW SET 21A

1 **a** $\displaystyle\int \frac{4}{\sqrt{x}}\,dx = 4\int x^{-\frac{1}{2}}\,dx$
$\displaystyle = 4\,\frac{x^{\frac{1}{2}}}{\frac{1}{2}} + c = 8\sqrt{x} + c$

b $\displaystyle\int \frac{3}{1-2x}\,dx = 3\int \frac{1}{1-2x}\,dx$
$= 3(\frac{1}{-2})\ln(1-2x) + c,\ \ 1 - 2x > 0$
$= -\frac{3}{2}\ln(1-2x) + c,\ \ x < \frac{1}{2}$

c $\int \sin(4x - 5)\,dx = -\frac{1}{4}\cos(4x - 5) + c$

d $\int e^{4-3x}\,dx = \frac{1}{-3}e^{4-3x} + c$
$= -\frac{1}{3}e^{4-3x} + c$

2 **a** $\int_{-5}^{-1} \sqrt{1-3x}\,dx = \int_{-5}^{-1} (1-3x)^{\frac{1}{2}}\,dx$
$\displaystyle = \left[\frac{1}{-3} \times \frac{(1-3x)^{\frac{3}{2}}}{\frac{3}{2}}\right]_{-5}^{-1}$
$= -\frac{2}{9}\left[(1-3x)^{\frac{3}{2}}\right]_{-5}^{-1}$
$= -\frac{2}{9}\left(4^{\frac{3}{2}} - 16^{\frac{3}{2}}\right)$
$= -\frac{2}{9}(8 - 64) = 12\frac{4}{9}$

b $\int_0^{\frac{\pi}{2}} \cos\left(\frac{x}{2}\right)\,dx = \left[2\sin\left(\frac{x}{2}\right)\right]_0^{\frac{\pi}{2}}$
$= 2\sin(\frac{\pi}{4}) - 2\sin(0)$
$= 2\left(\frac{1}{\sqrt{2}}\right) - 2(0)$
$= \sqrt{2}$

3
$$y = \sqrt{x^2-4} = (x^2-4)^{\frac{1}{2}}$$
$$\therefore \quad \frac{dy}{dx} = \tfrac{1}{2}\left(x^2-4\right)^{-\frac{1}{2}} \times 2x$$
$$= \frac{x}{(x^2-4)^{\frac{1}{2}}}$$
$$= \frac{x}{\sqrt{x^2-4}}$$
$$\therefore \quad \int \frac{x}{\sqrt{x^2-4}}\, dx = \sqrt{x^2-4} + c$$

4
$$\int_0^b \cos x \, dx = \tfrac{1}{\sqrt{2}}, \quad 0 < b < \pi$$
$$\therefore \quad [\sin x]_0^b = \tfrac{1}{\sqrt{2}}$$
$$\therefore \quad \sin b - \sin 0 = \tfrac{1}{\sqrt{2}}$$
$$\therefore \quad \sin b = \tfrac{1}{\sqrt{2}}$$
$$\therefore \quad b = \tfrac{\pi}{4}, \tfrac{3\pi}{4} \quad \{0 < b < \pi\}$$

5 **a**
$$\int 4\sin^2\left(\tfrac{x}{2}\right) dx$$
$$= \int 4\left(\tfrac{1}{2} - \tfrac{1}{2}\cos x\right) dx$$
$$= \int (2 - 2\cos x)\, dx$$
$$= 2x - 2\sin x + c$$

b
$$\int (2-\cos x)^2\, dx$$
$$= \int (4 - 4\cos x + \cos^2 x)\, dx$$
$$= \int (4 - 4\cos x + \tfrac{1}{2} + \tfrac{1}{2}\cos 2x)\, dx$$
$$= \tfrac{9}{2}x - 4\sin x + \tfrac{1}{4}\sin 2x + c$$

6
$$\frac{d}{dx}\left((3x^2+x)^3\right) = 3(3x^2+x)^2 \times (6x+1)$$
$$\therefore \quad \int 3(3x^2+x)^2(6x+1)\, dx = (3x^2+x)^3 + c_1$$
$$\therefore \quad 3\int (3x^2+x)^2(6x+1)\, dx = (3x^2+x)^3 + c_1$$
$$\therefore \quad \int (3x^2+x)^2(6x+1)\, dx = \tfrac{1}{3}(3x^2+x)^3 + c$$

7 **a**
$$\int_1^4 (f(x)+1)\, dx = \int_1^4 f(x)\, dx + \int_1^4 1\, dx$$
$$= 3 + [x]_1^4$$
$$= 3 + (4-1) = 6$$

b
$$\int_1^2 f(x)\, dx - \int_4^2 f(x)\, dx = \int_1^2 f(x)\, dx + \int_2^4 f(x)\, dx$$
$$= \int_1^4 f(x)\, dx$$
$$= 3$$

8
$$\int_0^a e^{1-2x}\, dx = \frac{e}{4}$$
$$\therefore \quad \left[\tfrac{1}{-2}e^{1-2x}\right]_0^a = \frac{e}{4}$$
$$\therefore \quad (-\tfrac{1}{2}e^{1-2a}) - (-\tfrac{1}{2}e^1) = \frac{e}{4}$$
$$\therefore \quad -\tfrac{1}{2}e^{1-2a} + \frac{e}{2} = \frac{e}{4}$$
$$\therefore \quad \tfrac{1}{2}e^{1-2a} = \frac{e}{4}$$
$$\therefore \quad e^{1-2a} = \frac{e}{2}$$
$$\therefore \quad 1 - 2a = \ln\left(\frac{e}{2}\right) = \ln e - \ln 2$$
$$\therefore \quad 1 - 2a = 1 - \ln 2$$
$$\therefore \quad 2a = \ln 2$$
$$\therefore \quad a = \tfrac{1}{2}\ln 2$$
$$\therefore \quad a = \ln 2^{\frac{1}{2}}$$
$$\therefore \quad a = \ln\sqrt{2}$$

9 Given: $f''(x) = 2\sin(2x)$, $f'(\frac{\pi}{2}) = 0$ and $f(0) = 3$

Now $f'(x) = \int 2\sin(2x)\, dx$
$$= -\cos(2x) + c$$
But $f'(\frac{\pi}{2}) = 0$, so $-\cos(\pi) + c = 0$
$$\therefore \quad -(-1) + c = 0$$
$$\therefore \quad c = -1$$
$$\therefore \quad f'(x) = -\cos(2x) - 1$$

$$\therefore \quad f(x) = \int (-\cos(2x) - 1)\, dx$$
$$= -\tfrac{1}{2}\sin(2x) - x + k$$
But $f(0) = 3$, so $-\frac{1}{2}\sin(0) - 0 + k = 3$
$$\therefore \quad k = 3$$
so $f(x) = -\frac{1}{2}\sin(2x) - x + 3$
$$\therefore \quad f(\tfrac{\pi}{2}) = -\tfrac{1}{2}\sin(\pi) - \tfrac{\pi}{2} + 3$$
$$= 3 - \tfrac{\pi}{2}$$

10 $\int_0^{\frac{\pi}{6}} \sin^2\left(\frac{x}{2}\right) dx$

$= \int_0^{\frac{\pi}{6}} \left(\frac{1}{2} - \frac{1}{2}\cos x\right) dx$

$= \left[\frac{1}{2}x - \frac{1}{2}\sin x\right]_0^{\frac{\pi}{6}}$

$= \frac{\pi}{12} - \frac{1}{2}\left(\frac{1}{2}\right) - 0 + 0$

$= \frac{\pi}{12} - \frac{1}{4}$

REVIEW SET 21B

1 **a** $\frac{dy}{dx} = (x^2-1)^2$

$\therefore \quad y = \int (x^2-1)^2\, dx$

$= \int (x^4 - 2x^2 + 1)\, dx$

$= \frac{1}{5}x^5 - \frac{2}{3}x^3 + x + c$

b $\frac{dy}{dx} = 400 - 20e^{-\frac{x}{2}}$

$\therefore \quad y = \int (400 - 20e^{-\frac{x}{2}})\, dx$

$= 400x - \dfrac{20e^{-\frac{x}{2}}}{-\frac{1}{2}} + c$

$= 400x + 40e^{-\frac{x}{2}} + c$

2 Using technology: **a** $\int_{-2}^{0} 4e^{-x^2}\, dx \approx 3.528$ **b** $\displaystyle\int_0^1 \frac{10x}{\sqrt{3x+1}}\, dx \approx 2.963$

3 $\frac{d}{dx}(\ln x)^2 = 2(\ln x)^1\left(\frac{1}{x}\right)$

$= \dfrac{2\ln x}{x}$

$\therefore \quad \displaystyle\int \frac{2\ln x}{x}\, dx = (\ln x)^2 + c_1$

$\therefore \quad \displaystyle\int \frac{\ln x}{x}\, dx = \frac{1}{2}(\ln x)^2 + c$

4 Given: $f''(x) = 18x + 10, \quad f(0) = -1, \quad f(1) = 13$

$f'(x) = \int (18x+10)\, dx$

$= 9x^2 + 10x + c$

$\therefore \quad f(x) = 3x^3 + 5x^2 + cx + d$

But $f(0) = -1$ so $d = -1$

$\therefore \quad f(x) = 3x^3 + 5x^2 + cx - 1$

And $f(1) = 13$ so $3 + 5 + c - 1 = 13$

$\therefore \quad c + 7 = 13$

$\therefore \quad c = 6$

$\therefore \quad f(x) = 3x^3 + 5x^2 + 6x - 1$

5 Using technology: **a** $\displaystyle\int_3^4 \frac{x}{\sqrt{2x+1}}\, dx \approx 1.236\,17$ **b** $\int_0^1 x^2 e^{x+1}\, dx \approx 1.952\,49$

6 **a** $f''(x) = 3x^2 + 2x$

$\therefore \quad f'(x) = \dfrac{3x^3}{3} + \dfrac{2x^2}{2} + c$

$= x^3 + x^2 + c$

$\therefore \quad f(x) = \dfrac{x^4}{4} + \dfrac{x^3}{3} + cx + d$

But $f(0) = 3$ so $d = 3$

$\therefore \quad f(x) = \dfrac{x^4}{4} + \dfrac{x^3}{3} + cx + 3$

Also, $f(2) = 3$ so $4 + \frac{8}{3} + 2c + 3 = 3$

$\therefore \quad \frac{20}{3} = -2c$

$\therefore \quad c = -\frac{10}{3}$

$\therefore \quad f(x) = \frac{1}{4}x^4 + \frac{1}{3}x^3 - \frac{10}{3}x + 3$

b Now $f'(2) = 2^3 + 2^2 - \frac{10}{3}$

$= 12 - \frac{10}{3}$

$= \frac{26}{3}$

$\therefore$ the normal has gradient $-\frac{3}{26}$

$\therefore$ equation is $\dfrac{y-3}{x-2} = -\frac{3}{26}$

$\therefore \quad y - 3 = -\frac{3}{26}(x-2)$

$\therefore \quad y = -\frac{3}{26}x + \frac{6}{26} + 3$

or $3x + 26y = 84$

7 **a**
$$\begin{aligned}&(e^x+2)^3\\&=(e^x)^3+3(e^x)^2(2)+3(e^x)(2)^2+(2)^3\\&=e^{3x}+6e^{2x}+12e^x+8\end{aligned}$$

b
$$\begin{aligned}&\int_0^1 (e^x+2)^3\,dx\\&=\left[\tfrac{1}{3}e^{3x}+3e^{2x}+12e^x+8x\right]_0^1\\&=\left(\tfrac{1}{3}e^3+3e^2+12e+8\right)-\left(\tfrac{1}{3}+3+12\right)\\&=\tfrac{1}{3}e^3+3e^2+12e-7\tfrac{1}{3}\qquad(\approx 54.1)\end{aligned}$$

c Using technology, $\int_0^1 (e^x+2)^3\,dx\approx 54.1$

REVIEW SET 21C

1 **a**
$$\begin{aligned}&\int\left(2e^{-x}-\frac{1}{x}+3\right)dx\\&=-2e^{-x}-\ln x+3x+c,\quad x>0\end{aligned}$$

b
$$\begin{aligned}&\int\left(\sqrt{x}-\frac{1}{\sqrt{x}}\right)^2dx\\&=\int\left(x-2+\frac{1}{x}\right)dx\\&=\tfrac{1}{2}x^2-2x+\ln x+c,\quad x>0\end{aligned}$$

c
$$\begin{aligned}&\int(3+e^{2x-1})^2\,dx\\&=\int(9+6e^{2x-1}+e^{4x-2})\,dx\\&=9x+3e^{2x-1}+\tfrac{1}{4}e^{4x-2}+c\end{aligned}$$

2
$$\begin{aligned}&f'(x)=x^2-3x+2\\\therefore\quad&f(x)=\frac{x^3}{3}-\frac{3x^2}{2}+2x+c\\\text{But}\quad&f(1)=3\\&\text{so}\quad\tfrac{1}{3}-\tfrac{3}{2}+2+c=3\\\therefore\quad&c=1-\tfrac{1}{3}+1\tfrac{1}{2}\\\therefore\quad&c=2\tfrac{1}{6}\\\therefore\quad&f(x)=\tfrac{1}{3}x^3-\tfrac{3}{2}x^2+2x+2\tfrac{1}{6}\end{aligned}$$

3
$$\begin{aligned}\int_2^3\frac{1}{\sqrt{3x-4}}\,dx&=\int_2^3(3x-4)^{-\frac{1}{2}}\,dx\\&=\left[\tfrac{1}{3}\frac{(3x-4)^{\frac{1}{2}}}{\frac{1}{2}}\right]_2^3\\&=\left[\tfrac{2}{3}\sqrt{3x-4}\right]_2^3\\&=\tfrac{2}{3}\sqrt{5}-\tfrac{2}{3}\sqrt{2}\\&=\tfrac{2}{3}(\sqrt{5}-\sqrt{2})\end{aligned}$$

4
$$\begin{aligned}&\int_0^{\frac{\pi}{3}}\cos^2\left(\tfrac{x}{2}\right)dx\\&=\int_0^{\frac{\pi}{3}}\left(\tfrac{1}{2}+\tfrac{1}{2}\cos x\right)dx\\&=\left[\tfrac{1}{2}x+\tfrac{1}{2}\sin x\right]_0^{\frac{\pi}{3}}\\&=\tfrac{\pi}{6}+\tfrac{1}{2}\left(\tfrac{\sqrt{3}}{2}\right)-0-0\\&=\tfrac{\pi}{6}+\tfrac{\sqrt{3}}{4}\end{aligned}$$

5
$$\begin{aligned}\frac{d}{dx}(e^{-2x}\sin x)&=-2e^{-2x}\sin x+e^{-2x}\cos x\quad\{\text{product rule}\}\\&=e^{-2x}(\cos x-2\sin x)\end{aligned}$$
$$\begin{aligned}\therefore\quad\int_0^{\frac{\pi}{2}}e^{-2x}(\cos x-2\sin x)\,dx&=\left[e^{-2x}\sin x\right]_0^{\frac{\pi}{2}}\\&=e^{-\pi}(1)-e^0(0)=e^{-\pi}\end{aligned}$$

6 If $n\neq-1$, $\int(2x+3)^n\,dx=\tfrac{1}{2}\dfrac{(2x+3)^{n+1}}{n+1}+c=\dfrac{(2x+3)^{n+1}}{2(n+1)}+c$

If $n=-1$, $\int(2x+3)^{-1}\,dx=\displaystyle\int\frac{1}{2x+3}\,dx=\tfrac{1}{2}\ln(2x+3)+c,\quad 2x+3>0$

So,
$$\int(2x+3)^n\,dx=\begin{cases}\dfrac{(2x+3)^{n+1}}{2(n+1)}+c & \text{if } n\neq-1\\[2mm] \tfrac{1}{2}\ln(2x+3)+c & \text{if } n=-1,\quad x>-\tfrac{3}{2}\end{cases}$$

7 $f'(x) = 2\sqrt{x} + \dfrac{a}{\sqrt{x}}$

$= 2x^{\frac{1}{2}} + ax^{-\frac{1}{2}}$

$\therefore \quad f(x) = \frac{4}{3}x^{\frac{3}{2}} + 2ax^{\frac{1}{2}} + c$

$= \dfrac{4x\sqrt{x}}{3} + 2a\sqrt{x} + c$

Now $f(0) = 2$ so $c = 2$

$\therefore \quad f(x) = \dfrac{4x\sqrt{x}}{3} + 2a\sqrt{x} + 2$

Also, $f(1) = 4$ so $\frac{4}{3} + 2a + 2 = 4$

$\therefore \quad 2a = \frac{2}{3}$

$\therefore \quad a = \frac{1}{3}$

$\therefore \quad f'(x) = 2\sqrt{x} + \dfrac{1}{3\sqrt{x}} = \dfrac{6x+1}{3\sqrt{x}}$

Now $f(x)$ is only defined for $x > 0$,

so $f'(x) > 0$ for all x in the domain.

$\therefore$ the function has no stationary points.

8 $$\int_a^{2a} (x^2 + ax + 2)\, dx = \frac{73a}{2}$$

$$\therefore \quad \left[\frac{x^3}{3} + \frac{ax^2}{2} + 2x\right]_a^{2a} = \frac{73a}{2}$$

$$\therefore \quad \left(\frac{8a^3}{3} + \frac{a}{2}(4a^2) + 4a\right) - \left(\frac{a^3}{3} + \frac{a^3}{2} + 2a\right) = \frac{73a}{2}$$

$$\frac{8a^3}{3} + 2a^3 + 4a - \frac{a^3}{3} - \frac{a^3}{2} - 2a = \frac{73a}{2}$$

$$\therefore \quad 16a^3 + 12a^3 + 24a - 2a^3 - 3a^3 - 12a = 219a$$

$$\therefore \quad 23a^3 - 207a = 0$$

$$\therefore \quad 23a(a^2 - 9) = 0$$

$$\therefore \quad 23a(a+3)(a-3) = 0$$

$$\therefore \quad a = 0 \text{ or } a = \pm 3$$

Chapter 22
APPLICATIONS OF INTEGRATION

EXERCISE 22A.1

1 **a**

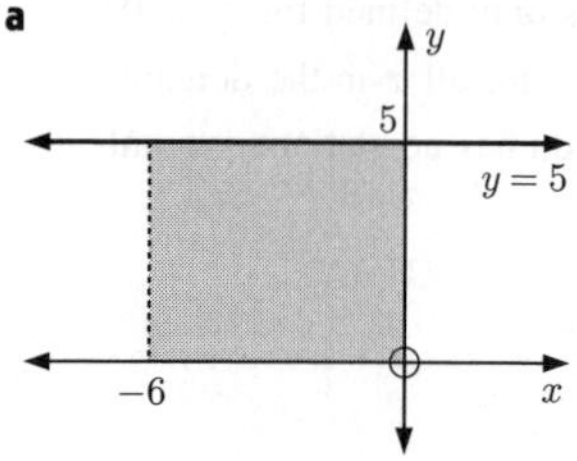

i Area $= 5 \times 6$
$= 30$ units^2

ii Area $= \int_{-6}^{0} 5\, dx$
$= [5x]_{-6}^{0}$
$= 5(0) - 5(-6)$
$= 30$ units^2

b

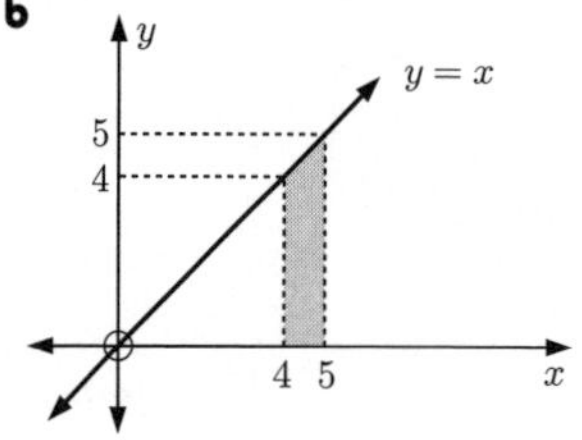

i Area $= \left(\dfrac{4+5}{2}\right) \times 1$
$= \frac{9}{2}$ units^2

ii Area $= \int_{4}^{5} x\, dx$
$= \left[\frac{1}{2}x^2\right]_4^5$
$= \frac{1}{2}(25) - \frac{1}{2}(16)$
$= \frac{9}{2}$ units^2

c

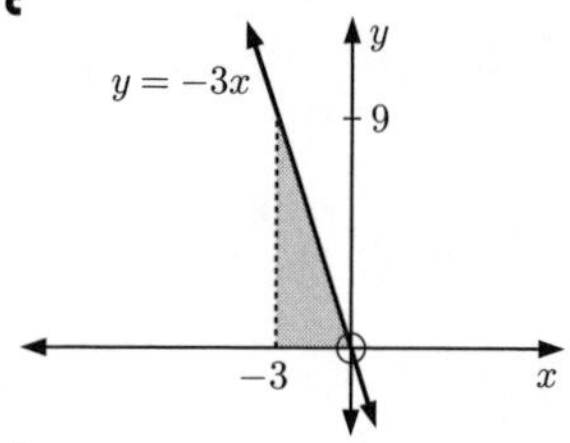

i Area $= \frac{1}{2} \times 3 \times 9$
$= \frac{27}{2}$ units^2

ii Area $= \int_{-3}^{0} -3x\, dx$
$= \left[-\frac{3}{2}x^2\right]_{-3}^{0}$
$= -\frac{3}{2}(0) - (-\frac{3}{2})(9)$
$= \frac{27}{2}$ units^2

d

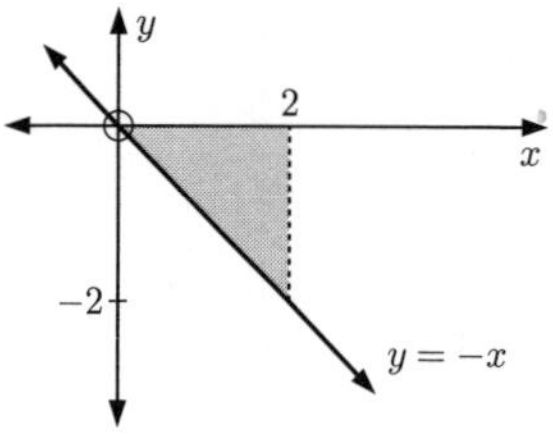

i Area $= \frac{1}{2} \times 2 \times 2$
$= 2$ units^2

ii Area $= -\int_{0}^{2} -x\, dx$
$= -\left[-\frac{1}{2}x^2\right]_0^2$
$= -\left(-\frac{1}{2}(4) - (-\frac{1}{2})(0)\right)$
$= 2$ units^2

2 **a**

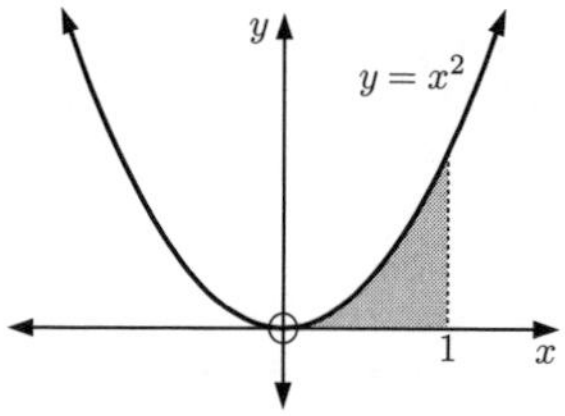

Area $= \int_0^1 x^2\, dx$
$= \left[\dfrac{x^3}{3}\right]_0^1$
$= \frac{1}{3} - 0$
$= \frac{1}{3}$ units^2

b

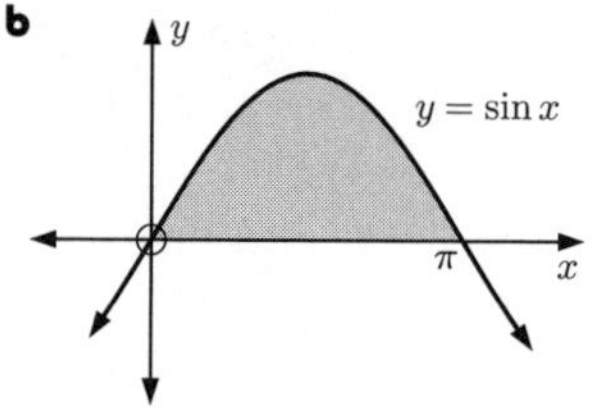

Area $= \int_0^{\pi} \sin x\, dx$
$= [-\cos x]_0^{\pi}$
$= -\cos\pi - (-\cos 0)$
$= 2$ units^2

c

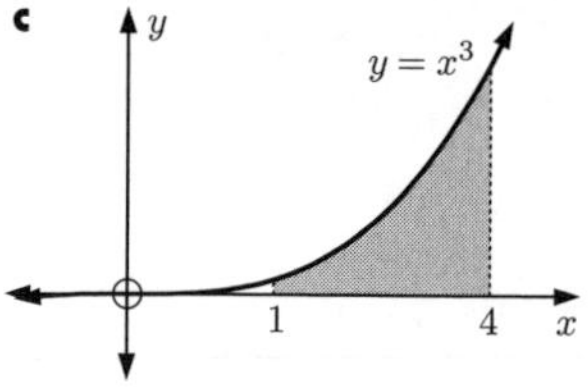

Area $= \int_1^4 x^3\, dx$
$= \left[\dfrac{x^4}{4}\right]_1^4$
$= \frac{256}{4} - \frac{1}{4}$
$= 63\frac{3}{4}$ units^2

d

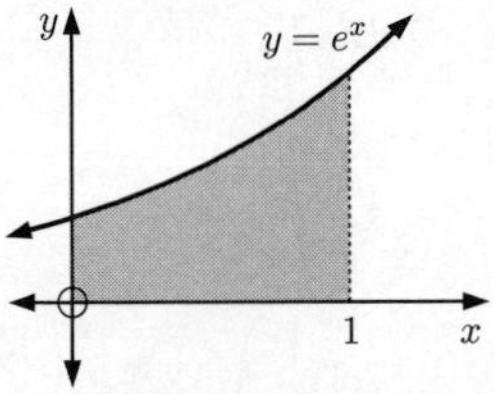

Area $=\int_0^1 e^x\,dx$

$=[e^x]_0^1$

$=e-1$

≈ 1.72 units^2

e The graph cuts the x-axis at $y=0$.

$\therefore\ 6+x-x^2=0$

$\therefore\ (3-x)(2+x)=0$

$\therefore\ x=3$ or -2

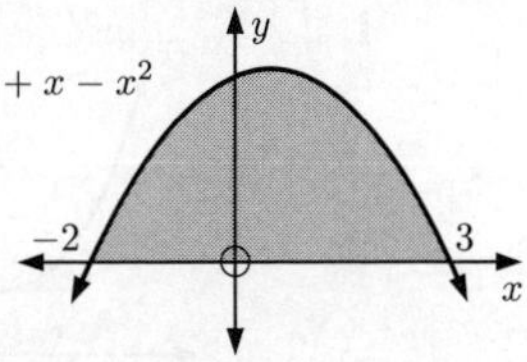

The x-intercepts are 3 and -2.

Area $=\int_{-2}^{3}(6+x-x^2)\,dx$

$=\left[6x+\dfrac{x^2}{2}-\dfrac{x^3}{3}\right]_{-2}^{3}$

$=(18+\frac{9}{2}-9)-(-12+2+\frac{8}{3})$

$=20\frac{5}{6}$ units^2

f

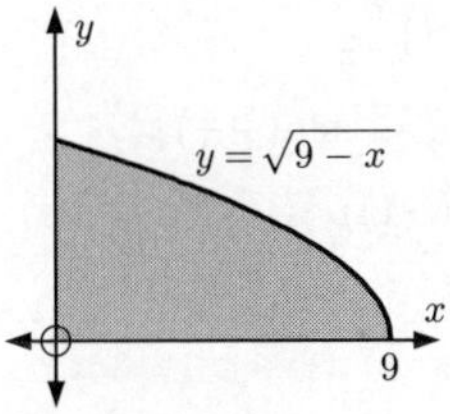

Area $=\int_0^9 (9-x)^{\frac{1}{2}}\,dx$

$=\left[\left(\frac{1}{-1}\right)\dfrac{(9-x)^{\frac{3}{2}}}{\frac{3}{2}}\right]_0^9$

$=-\frac{2}{3}\left[(9-x)^{\frac{3}{2}}\right]_0^9$

$=-\frac{2}{3}(0-27)$

$=18$ units^2

g

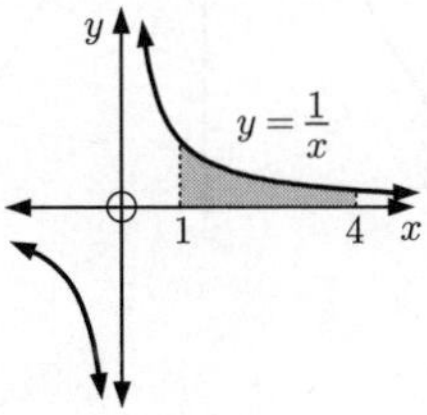

Area $=\displaystyle\int_1^4 \frac{1}{x}\,dx$

$=[\ln x]_1^4\quad \{x>0\}$

$=\ln 4-\ln 1$

$=\ln 4-0$

≈ 1.39 units^2

h

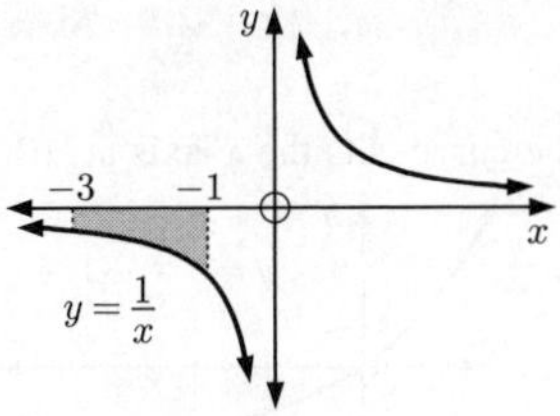

Area $=-\displaystyle\int_{-3}^{-1}\frac{1}{x}\,dx$

$=\displaystyle\int_1^3 \frac{1}{x}\,dx\quad$ {by symmetry}

$=[\ln x]_1^3\quad \{x>0\}$

$=\ln 3-\ln 1$

$=\ln 3-0$

≈ 1.10 units^2

i

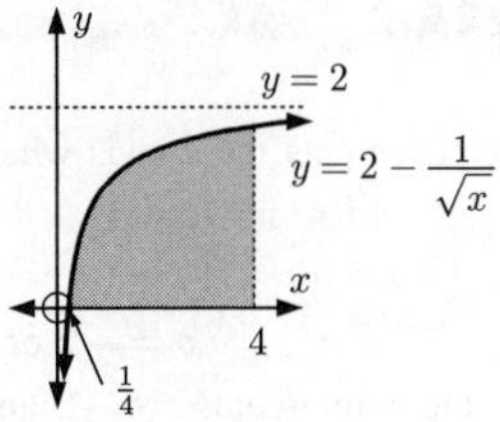

Area $=\displaystyle\int_{\frac{1}{4}}^{4}\left(2-\frac{1}{\sqrt{x}}\right)dx$

$=\displaystyle\int_{\frac{1}{4}}^{4}(2-x^{-\frac{1}{2}})\,dx$

$=\left[2x-\dfrac{x^{\frac{1}{2}}}{\frac{1}{2}}\right]_{\frac{1}{4}}^{4}$

$=\left[2x-2\sqrt{x}\right]_{\frac{1}{4}}^{4}$

$=(8-4)-(\frac{1}{2}-1)$

$=4\frac{1}{2}$ units^2

j

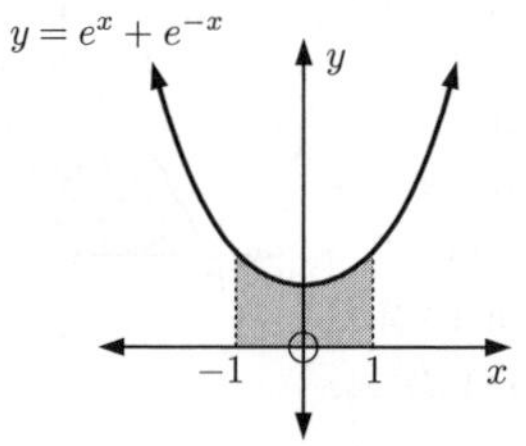

$$\begin{aligned}\text{Area} &= \int_{-1}^{1} (e^x + e^{-x})\,dx \\ &= \left[e^x - e^{-x}\right]_{-1}^{1} \\ &= (e - e^{-1}) - (e^{-1} - e) \\ &= 2e - \frac{2}{e} \\ &\approx 4.70 \text{ units}^2\end{aligned}$$

3

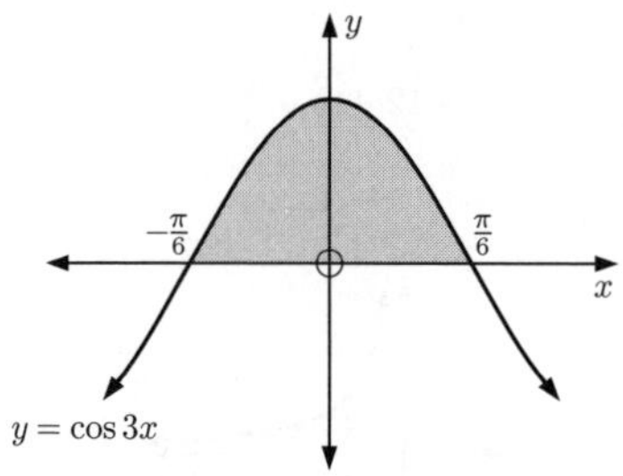

$y = \cos 3x$ has zeros at $\begin{cases} \frac{\pi}{6} \\ -\frac{\pi}{6} \end{cases} + \frac{2k\pi}{3}$, k an integer

$$\begin{aligned}\therefore \quad \text{area} &= \int_{-\frac{\pi}{6}}^{\frac{\pi}{6}} \cos 3x\,dx \\ &= \left[\tfrac{1}{3}\sin 3x\right]_{-\frac{\pi}{6}}^{\frac{\pi}{6}} \\ &= \tfrac{1}{3}\left(\sin(\tfrac{\pi}{2}) - \sin(-\tfrac{\pi}{2})\right) \\ &= \tfrac{1}{3}(1 - (-1)) \\ &= \tfrac{2}{3} \text{ units}^2\end{aligned}$$

4 **a**

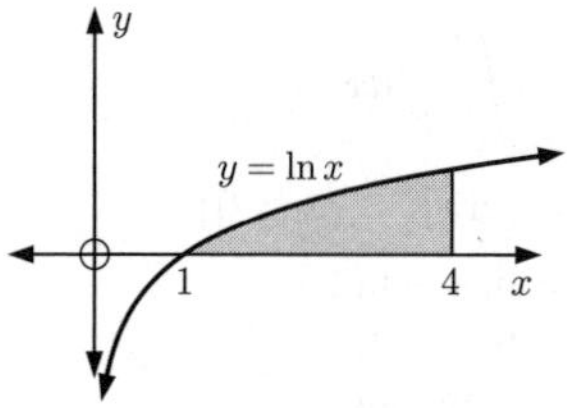

Area ≈ 2.55 units^2

b

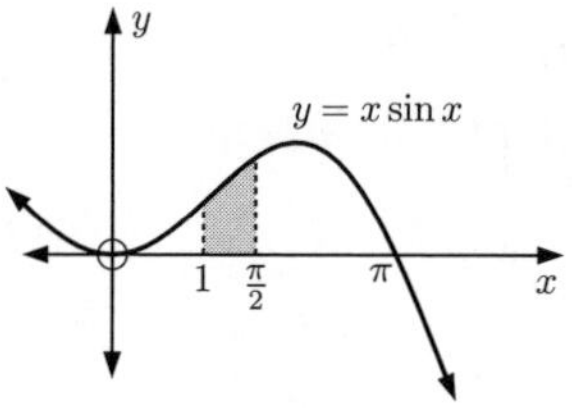

Area ≈ 0.699 units^2

c

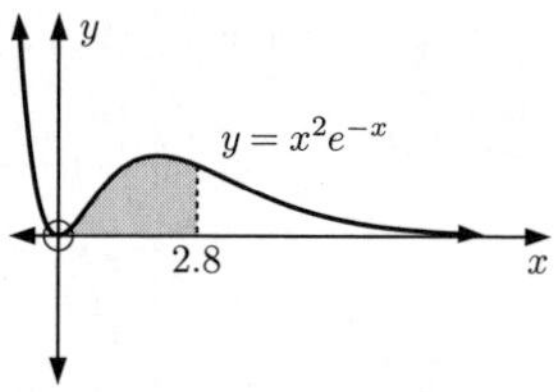

Area ≈ 1.06 units^2

EXERCISE 22A.2

1 **a** The curve cuts the x-axis when $y = 0$.

$\therefore \quad x^2 + x - 2 = 0$

$\therefore \quad (x + 2)(x - 1) = 0$

$\therefore \quad x = -2$ or 1

$\therefore$ the x-intercepts are -2 and 1

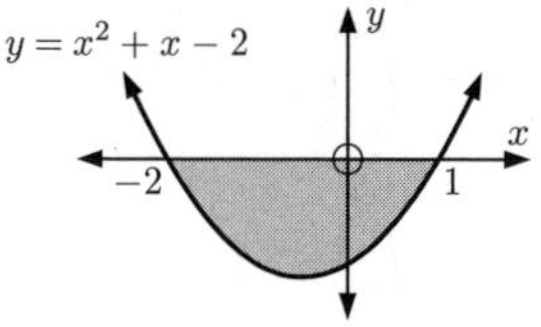

$$\begin{aligned}\text{Area} &= \int_{-2}^{1} [0 - (x^2 + x - 2)]\,dx \\ &= \int_{-2}^{1} (-x^2 - x + 2)\,dx \\ &= \left[-\frac{x^3}{3} - \frac{x^2}{2} + 2x\right]_{-2}^{1} \\ &= (-\tfrac{1}{3} - \tfrac{1}{2} + 2) - (\tfrac{8}{3} - 2 - 4) \\ &= 4\tfrac{1}{2} \text{ units}^2\end{aligned}$$

b The curve cuts the x-axis at $(0, 0)$.

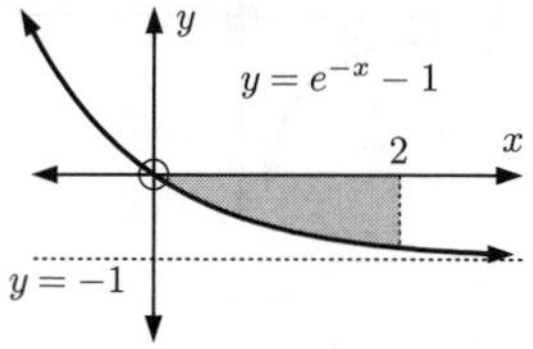

$$\begin{aligned}\text{Area} &= \int_{0}^{2} [0 - (e^{-x} - 1)]\,dx \\ &= \int_{0}^{2} (1 - e^{-x})\,dx \\ &= \left[x + e^{-x}\right]_{0}^{2} \\ &= \left(2 + e^{-2}\right) - \left(0 + e^{0}\right) \\ &= 1 + e^{-2} \text{ units}^2\end{aligned}$$

c The curve cuts the x-axis when $y = 0$.

$\therefore \quad 3x^2 - 8x + 4 = 0$

$\therefore \quad (3x-2)(x-2) = 0$

$\therefore \quad x = 2$ or $\frac{2}{3}$

$\therefore$ the x-intercepts are 2 and $\frac{2}{3}$.

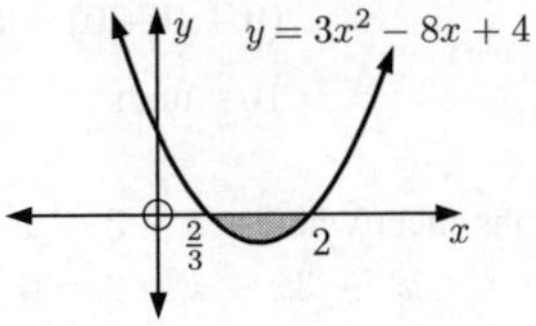

$$\text{Area} = \int_{\frac{2}{3}}^{2} [0 - (3x^2 - 8x + 4)]\, dx$$
$$= \int_{\frac{2}{3}}^{2} (-3x^2 + 8x - 4)\, dx$$
$$= \left[-x^3 + 4x^2 - 4x\right]_{\frac{2}{3}}^{2}$$
$$= (-8 + 16 - 8) - (-\tfrac{8}{27} + \tfrac{16}{9} - \tfrac{8}{3})$$
$$= 1\tfrac{5}{27} \text{ units}^2$$

d

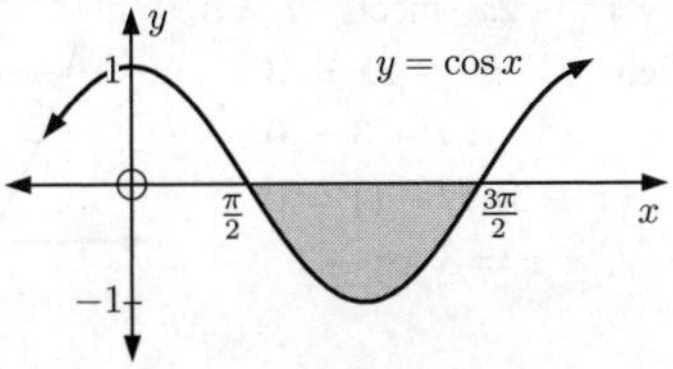

$$\text{Area} = \int_{\frac{\pi}{2}}^{\frac{3\pi}{2}} [0 - \cos x]\, dx$$
$$= \int_{\frac{\pi}{2}}^{\frac{3\pi}{2}} -\cos x\, dx$$
$$= [-\sin x]_{\frac{\pi}{2}}^{\frac{3\pi}{2}}$$
$$= -\sin(\tfrac{3\pi}{2}) - \left(-\sin(\tfrac{\pi}{2})\right)$$
$$= -(-1) - (-1)$$
$$= 2 \text{ units}^2$$

e The curve cuts the x-axis when $y = 0$.

$\therefore \quad x^3 - 4x = 0$

$\therefore \quad x(x^2 - 4) = 0$

$\therefore \quad x(x+2)(x-2) = 0$

$\therefore$ the x-intercepts are 0 and ± 2

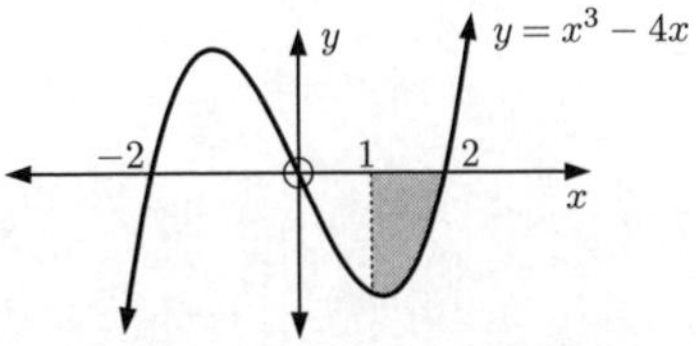

$$\text{Area} = \int_1^2 [0 - (x^3 - 4x)]\, dx$$
$$= \int_1^2 (-x^3 + 4x)\, dx$$
$$= \left[-\frac{x^4}{4} + 2x^2\right]_1^2$$
$$= (-4 + 8) - (-\tfrac{1}{4} + 2)$$
$$= 2\tfrac{1}{4} \text{ units}^2$$

f $y = \sin x - 1$ is the graph of $\sin x$ translated vertically by -1.

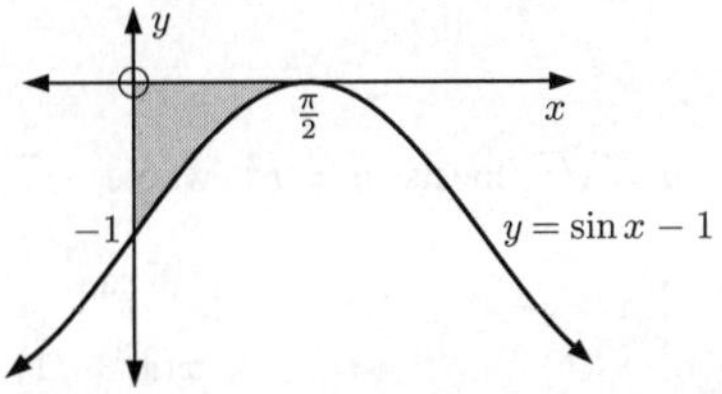

$$\text{Area} = \int_0^{\frac{\pi}{2}} [0 - (\sin x - 1)]\, dx$$
$$= \int_0^{\frac{\pi}{2}} (1 - \sin x)\, dx$$
$$= [x + \cos x]_0^{\frac{\pi}{2}}$$
$$= (\tfrac{\pi}{2} + \cos\tfrac{\pi}{2}) - (0 + \cos 0)$$
$$= \tfrac{\pi}{2} - 1 \text{ units}^2$$

g The curve cuts the x-axis when $y = 0$.

$\therefore \quad \sin^2 x = 0$

$\therefore \quad \sin x = 0$

$\therefore \quad x = 0 + k\pi, \quad k$ an integer

So, the first two non-negative x-intercepts are 0, π.

$$\text{Area} = \int_0^{\pi} [\sin^2 x - 0]\, dx$$
$$= \int_0^{\pi} \left(\tfrac{1}{2} - \tfrac{1}{2}\cos 2x\right) dx$$
$$= \left[\tfrac{1}{2}x - \tfrac{1}{4}\sin 2x\right]_0^{\pi}$$
$$= (\tfrac{1}{2}(\pi) - \tfrac{1}{4}\sin(2\pi)) - (\tfrac{1}{2}(0) - \tfrac{1}{4}\sin(0))$$
$$= \tfrac{\pi}{2} \text{ units}^2$$

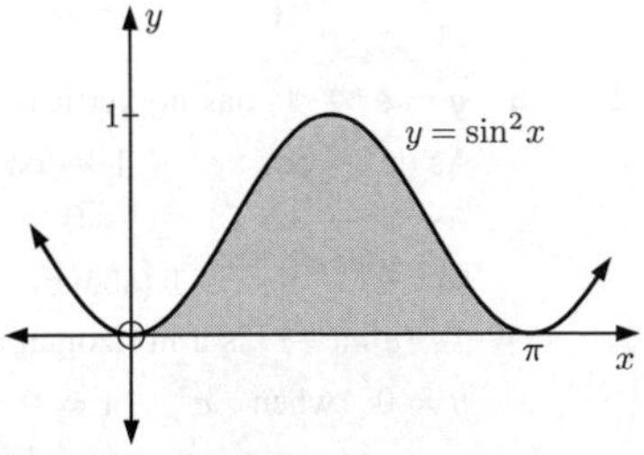

2 $y = x^2 - 2x$ meets $y = 3$
when $x^2 - 2x = 3$
$\therefore \quad x^2 - 2x - 3 = 0$
$\therefore \quad (x-3)(x+1) = 0$
$\therefore \quad x = 3$ or -1

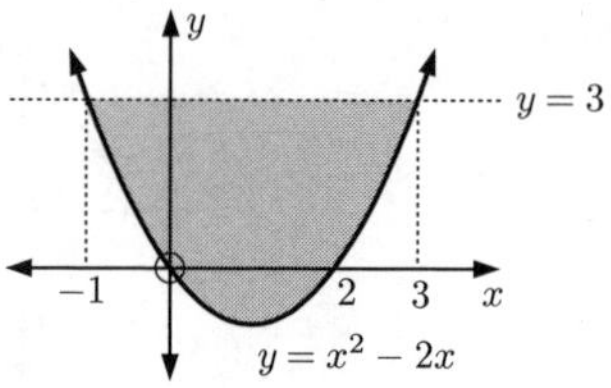

$A = \int_{-1}^{3} [3 - (x^2 - 2x)]\, dx$
$= \int_{-1}^{3} (3 + 2x - x^2)\, dx$
$= \left[3x + x^2 - \frac{x^3}{3}\right]_{-1}^{3}$
$= (9 + 9 - 9) - (-3 + 1 + \frac{1}{3})$
$= 10\frac{2}{3}$ units^2

3 **a**

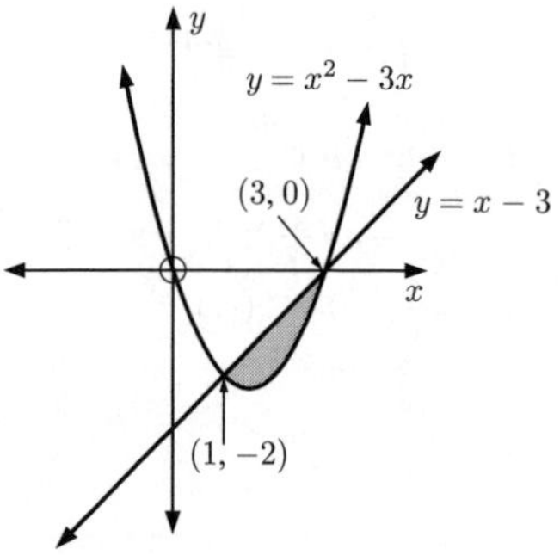

b The graphs meet where $x - 3 = x^2 - 3x$
$\therefore \quad x^2 - 3x - x + 3 = 0$
$\therefore \quad x^2 - 4x + 3 = 0$
$\therefore \quad (x-1)(x-3) = 0$
$\therefore \quad x = 1$ or 3
$\therefore$ the graphs meet at $(1, -2)$ and $(3, 0)$

c Area $= \int_{1}^{3} [(x-3) - (x^2 - 3x)]\, dx$
$= \int_{1}^{3} (-3 + 4x - x^2)\, dx$
$= \left[-3x + 2x^2 - \frac{x^3}{3}\right]_{1}^{3}$
$= (-9 + 18 - 9) - (-3 + 2 - \frac{1}{3})$
$= 1\frac{1}{3}$ units^2

4 $y = \sqrt{x}$ meets $y = x^2$ where $\sqrt{x} = x^2$
$\therefore \quad x = x^4$
$\therefore \quad x^4 - x = 0$
$\therefore \quad x(x^3 - 1) = 0$
$\therefore \quad x(x-1)(x^2 + x + 1) = 0$
$\therefore \quad x = 0$ or 1

The factor $(x^2 + x + 1)$ has no real root since $\Delta = -3$ which is < 0.

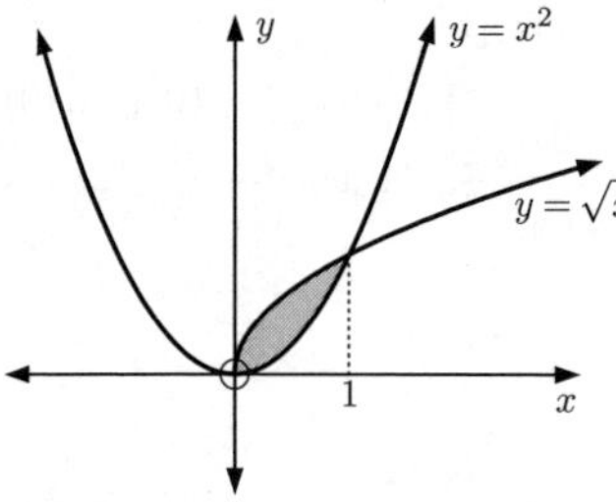

Area $= \int_{0}^{1} \left(\sqrt{x} - x^2\right)\, dx$
$= \int_{0}^{1} (x^{\frac{1}{2}} - x^2)\, dx$
$= \left[\frac{2}{3}x^{\frac{3}{2}} - \frac{x^3}{3}\right]_{0}^{1}$
$= \frac{2}{3} - \frac{1}{3}$
$= \frac{1}{3}$ unit^2

5 **a** $y = e^x - 1$ has no vertical asymptotes.
As $x \to \infty$, $e^x - 1 \to \infty$
As $x \to -\infty$, $e^x \to 0$
so $e^x - 1 \to -1$ (above)
$\therefore \quad y = -1$ is a horizontal asymptote.
$y = 0$ when $e^x - 1 = 0$
$\therefore \quad e^x = 1$
$\therefore \quad x = 0$
$\therefore \quad x$-intercept is $(0, 0)$.
This is also the y-intercept.

$y = 2 - 2e^{-x}$ has no vertical asymptotes.
As $x \to \infty$, $e^{-x} \to 0$
so $2 - 2e^{-x} \to 2$ (below)
$\therefore \quad y = 2$ is a horizontal asymptote.
$y = 0$ when $2 - 2e^{-x} = 0$
$\therefore \quad e^{-x} = 1$
$\therefore \quad x = 0$
$\therefore \quad x$-intercept is $(0, 0)$.
This is also the y-intercept.

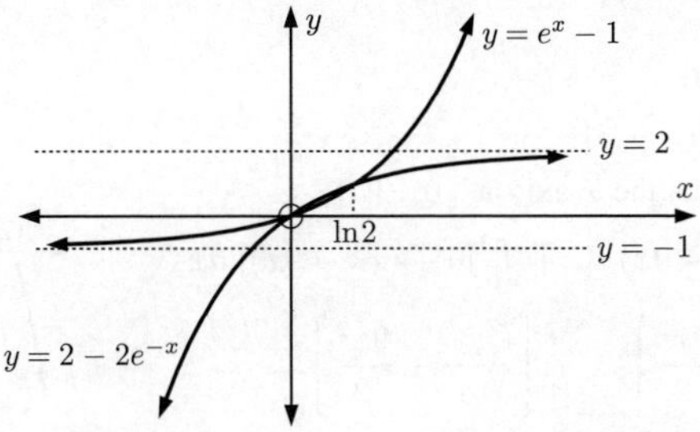

b $y = e^x - 1$ meets $y = 2 - 2e^{-x}$

where $e^x - 1 = 2 - 2e^{-x}$

$\therefore \quad e^{2x} - e^x = 2e^x - 2 \quad \{\times\ e^x\}$

$\therefore \quad e^{2x} - 3e^x + 2 = 0$

$\therefore \quad (e^x - 1)(e^x - 2) = 0$

$\therefore \quad e^x = 1$ or 2

$\therefore \quad x = 0$ or $\ln 2$

$\therefore$ the graphs meet at $(0, 0)$ and $(\ln 2, 1)$.

c $A = \int_0^{\ln 2} [(2 - 2e^{-x}) - (e^x - 1)]\, dx$

$= \int_0^{\ln 2} (3 - e^x - 2e^{-x})\, dx$

$= \left[3x - e^x + 2e^{-x}\right]_0^{\ln 2}$

$= (3\ln 2 - 2 + 1) - (0 - 1 + 2)$

$= 3\ln 2 - 2$

≈ 0.0794 units^2

6 $y = 2e^x$ meets $y = e^{2x}$ where

$2e^x = e^{2x}$

$\therefore \quad e^{2x} - 2e^x = 0$

$\therefore \quad e^x(e^x - 2) = 0$

$\therefore \quad e^x = 2 \quad \{e^x > 0$ for all $x\}$

$\therefore \quad x = \ln 2$

Area $= \int_0^{\ln 2} (2e^x - e^{2x})\, dx$

$= \left[2e^x - \frac{1}{2}e^{2x}\right]_0^{\ln 2}$

$= (4 - 2) - (2 - \frac{1}{2})$

$= \frac{1}{2}$ unit^2

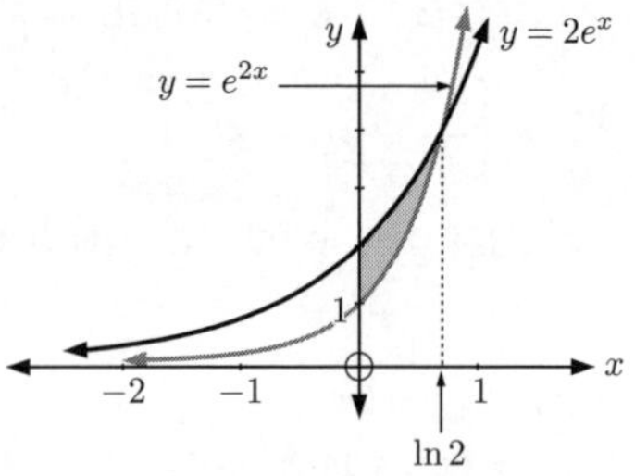

7 $y = 2x$ meets $y = 4x^2$ where

$2x = 4x^2$

$\therefore \quad 4x^2 - 2x = 0$

$\therefore \quad 2x(2x - 1) = 0$

$\therefore \quad x = 0$ or $\frac{1}{2}$

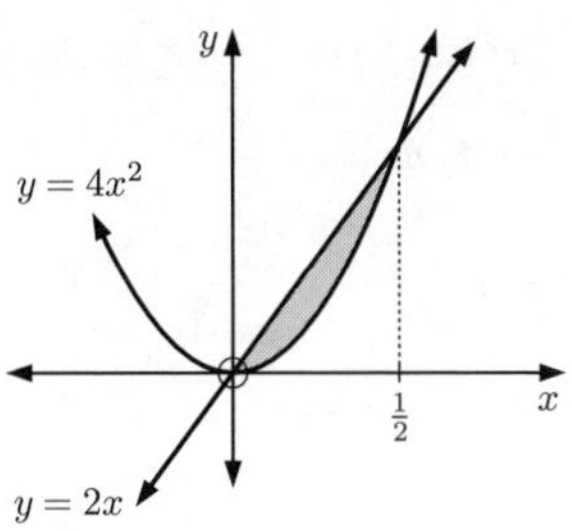

Area

$= \int_0^{\frac{1}{2}} (2x - 4x^2)\, dx$

$= \left[x^2 - \frac{4}{3}x^3\right]_0^{\frac{1}{2}}$

$= \left(\frac{1}{4} - \frac{4}{3}(\frac{1}{8})\right) - (0 - 0)$

$= \frac{1}{12}$ unit^2

8 **a** Now $x^2 + y^2 = 9 \quad \therefore \quad y^2 = 9 - x^2$

$\therefore \quad y = \pm\sqrt{9 - x^2}$

In the upper half of the circle all y-values are $\geqslant 0$

$\therefore \quad y = +\sqrt{9 - x^2}$ is the required equation.

b The shaded area is A where $A = \int_0^3 \sqrt{9 - x^2}\, dx$

This is a quarter of the area of a circle with radius 3 units.

$\therefore \quad A = \frac{1}{4}(\pi \times 3^2) = \frac{9}{4}\pi \approx 7.07$ units^2

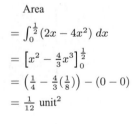

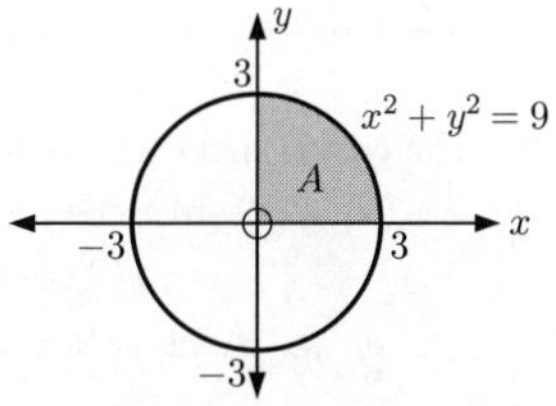

9 **a** $f(x) = x^3 - 9x$
$= x(x^2 - 9)$
$= x(x+3)(x-3)$
$\therefore\ y = f(x)$ cuts the x-axis at $0,\ \pm 3$

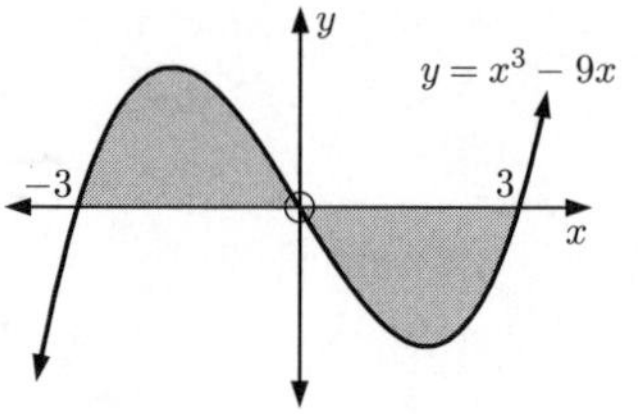

$$\text{Area} = \int_{-3}^{0}(x^3 - 9x)\,dx + \int_0^3 [0 - (x^3 - 9x)]\,dx$$
$$= \left[\frac{x^4}{4} - \frac{9x^2}{2}\right]_{-3}^{0} + \left[-\frac{x^4}{4} + \frac{9x^2}{2}\right]_0^3$$
$$= 0 - (\tfrac{81}{4} - \tfrac{81}{2}) + (-\tfrac{81}{4} + \tfrac{81}{2}) - 0$$
$$= 40\tfrac{1}{2}\ \text{units}^2$$

b $f(x) = -x(x-2)(x-4)$
$= -x^3 + 6x^2 - 8x$
$\therefore\ y = f(x)$ cuts the x-axis at 0, 2 and 4

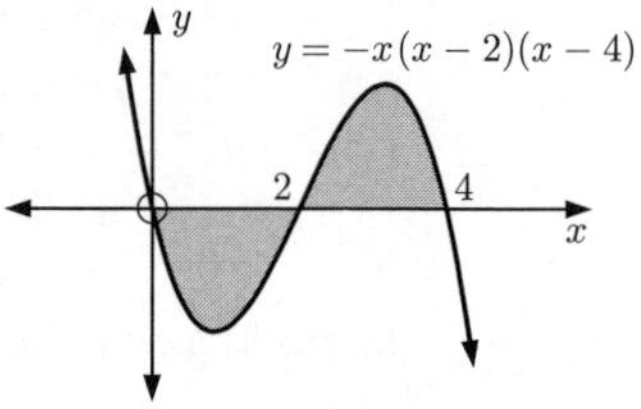

$$\text{Area} = \int_0^2 [0 - (-x^3 + 6x^2 - 8x)]\,dx + \int_2^4 (-x^3 + 6x^2 - 8x)\,dx$$
$$= \int_0^2 (x^3 - 6x^2 + 8x)\,dx + \int_2^4 (-x^3 + 6x^2 - 8x)\,dx$$
$$= \left[\frac{x^4}{4} - 2x^3 + 4x^2\right]_0^2 + \left[-\frac{x^4}{4} + 2x^3 - 4x^2\right]_2^4$$
$$= ([4 - 16 + 16] - 0) + ([-64 + 128 - 64] - [-4 + 16 - 16])$$
$$= 8\ \text{units}^2$$

c $f(x) = x^4 - 5x^2 + 4$
$= (x^2 - 1)(x^2 - 4)$
$= (x+1)(x-1)(x+2)(x-2)$
$\therefore\ y = f(x)$ cuts the x-axis at $\pm 1,\ \pm 2$

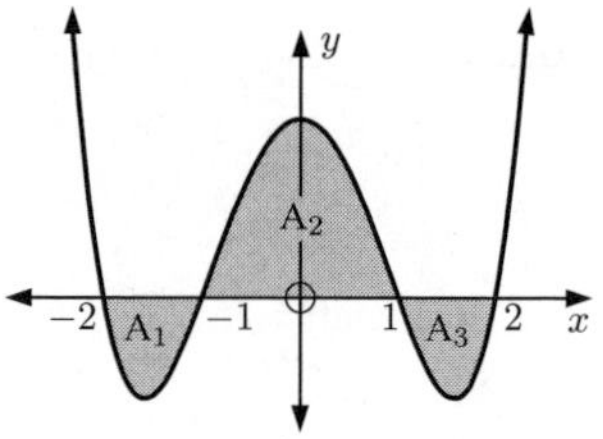

$$A_1 = \int_{-2}^{-1} [0 - (x^4 - 5x^2 + 4)]\,dx$$
$$= \int_{-2}^{-1} (-x^4 + 5x^2 - 4)\,dx$$
$$= \left[-\frac{x^5}{5} + \frac{5x^3}{3} - 4x\right]_{-2}^{-1}$$
$$= (\tfrac{1}{5} - \tfrac{5}{3} + 4) - (\tfrac{32}{5} - \tfrac{40}{3} + 8)$$
$$= \tfrac{22}{15}\ \text{units}^2$$

$$A_2 = \int_{-1}^{1} (x^4 - 5x^2 + 4)\,dx$$
$$= \left[\frac{x^5}{5} - \frac{5x^3}{3} + 4x\right]_{-1}^{1}$$
$$= (\tfrac{1}{5} - \tfrac{5}{3} + 4) - (-\tfrac{1}{5} + \tfrac{5}{3} - 4)$$
$$= \tfrac{76}{15}\ \text{units}^2$$

By symmetry, $A_3 = A_1$ $\quad\therefore$ area $= \tfrac{22}{15} + \tfrac{76}{15} + \tfrac{22}{15} = \tfrac{120}{15} = 8\ \text{units}^2$

10 **a** $y = \sin(2x)$ is the curve C_1 and $y = \sin x$ is the curve C_2.

b The curves meet when $\sin(2x) = \sin x$
$\therefore\ 2\sin x\cos x - \sin x = 0$
$\therefore\ \sin x(2\cos x - 1) = 0$
$\therefore\ \sin x = 0$ or $\cos x = \tfrac{1}{2}$

$\therefore\ x = 0 + k\pi$ or $x = \begin{cases}\frac{\pi}{3}\\ \frac{5\pi}{3}\end{cases} + 2k\pi$, k an integer

$\therefore$ the x-coordinate of A $= \frac{\pi}{3}$ {smallest positive solution}

$\therefore$ A is at $\left(\frac{\pi}{3}, \frac{\sqrt{3}}{2}\right)$

c $\text{Area} = \int_0^{\frac{\pi}{3}} (\sin(2x) - \sin x)\, dx + \int_{\frac{\pi}{3}}^{\pi} (\sin x - \sin(2x))\, dx$

$= \left[-\frac{1}{2}\cos(2x) + \cos x\right]_0^{\frac{\pi}{3}} + \left[-\cos x + \frac{1}{2}\cos(2x)\right]_{\frac{\pi}{3}}^{\pi}$

$= \left(-\frac{1}{2}\cos\frac{2\pi}{3} + \cos\frac{\pi}{3}\right) - \left(-\frac{1}{2}\cos 0 + \cos 0\right) + \left(-\cos\pi + \frac{1}{2}\cos 2\pi\right)$
$\quad - \left(-\cos\frac{\pi}{3} + \frac{1}{2}\cos\frac{2\pi}{3}\right)$

$= (\frac{1}{4} + \frac{1}{2}) - (-\frac{1}{2} + 1) + (1 + \frac{1}{2}) - (-\frac{1}{2} - \frac{1}{4})$

$= 2\frac{1}{2}$ units^2

11

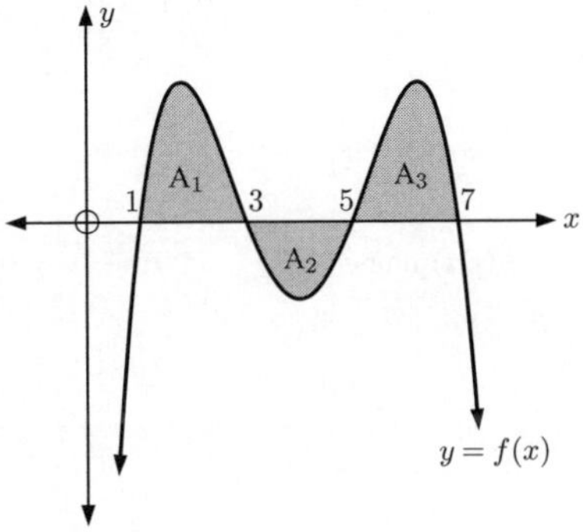

a $\int_1^7 f(x)\, dx$ only gives us the correct area provided that $f(x)$ is positive on the interval $1 \leqslant x \leqslant 7$.
But $f(x)$ is not positive for $3 \leqslant x \leqslant 5$,
so $\int_1^7 f(x)\, dx = A_1 - A_2 + A_3$
which is *not* the shaded area.

b shaded area

$= \int_1^3 f(x)\, dx + \int_3^5 [0 - f(x)]\, dx + \int_5^7 f(x)\, dx$

$= \int_1^3 f(x)\, dx - \int_3^5 f(x)\, dx + \int_5^7 f(x)\, dx$

12 **a** $y = \cos(2x)$ is the curve C_2 and $y = \cos^2 x$ is the curve C_1.

b Point A lies on $y = \cos(2x)$. When $x = 0$, $y = \cos 0 = 1$. $\therefore$ A is at $(0, 1)$.
Point B lies on $y = \cos(2x)$. When $x = \frac{\pi}{4}$, $y = \cos\frac{\pi}{2} = 0$. $\therefore$ B is at $(\frac{\pi}{4}, 0)$.
Point C lies on $y = \cos^2 x$. When $x = \frac{\pi}{2}$, $y = \cos^2\frac{\pi}{2} = 0$. $\therefore$ C is at $(\frac{\pi}{2}, 0)$.
Point D lies on $y = \cos(2x)$. When $x = \frac{3\pi}{4}$, $y = \cos\frac{3\pi}{2} = 0$. $\therefore$ D is at $(\frac{3\pi}{4}, 0)$.
Point E lies where the curves meet. Now $\cos(2\pi) = \cos^2\pi = 1$. $\therefore$ E is at $(\pi, 1)$.

c $A = \int_0^{\pi} (\cos^2 x - \cos(2x))\, dx$

$= \int_0^{\pi} \left(\frac{1}{2} + \frac{1}{2}\cos(2x) - \cos(2x)\right) dx$

$= \int_0^{\pi} \left(\frac{1}{2} - \frac{1}{2}\cos(2x)\right) dx$

$= \left[\frac{x}{2} - \frac{1}{4}\sin(2x)\right]_0^{\pi} = (\frac{\pi}{2} - 0) - (0 - 0) = \frac{\pi}{2}$ units^2

13 **a** The graphs meet when $e^{-x^2} = x^2 - 1$
$\therefore$ $x = \pm 1.1307$ {technology}

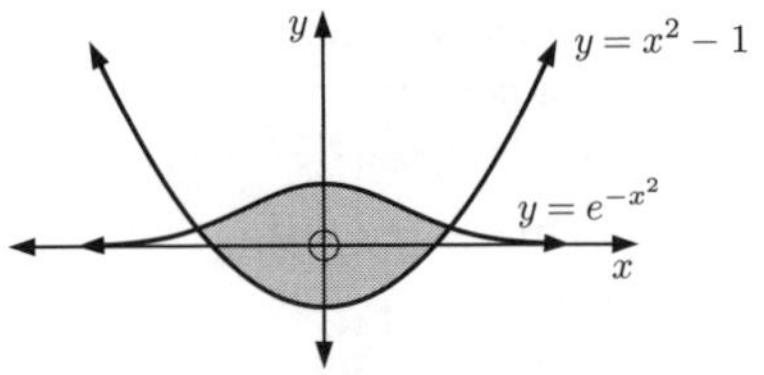

$\therefore$ $\text{area} = \int_{-1.1307}^{1.1307} [e^{-x^2} - (x^2 - 1)]\, dx$
≈ 2.88 units^2 {technology}

b The graphs meet when $x^x = 4x - \frac{1}{10}x^4$
$\therefore$ $x \approx 0.1832$ or 2.2696 {technology}

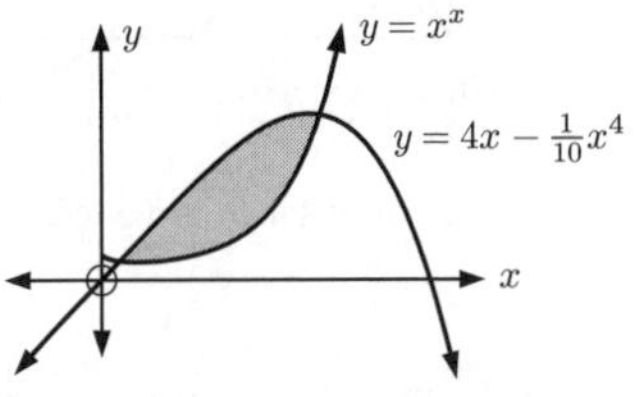

$\therefore$ $\text{area} = \int_{0.1832}^{2.2696} (4x - \frac{1}{10}x^4 - x^x)\, dx$
≈ 4.97 units^2 {technology}

14 Area $= \displaystyle\int_1^k \frac{1}{1+2x}\,dx = 0.2$ units^2

$\therefore \quad \left[\tfrac{1}{2}\ln(1+2x)\right]_1^k = 0.2, \quad 1+2x>0$

$\therefore \quad [\ln(1+2x)]_1^k = 0.4$

$\therefore \quad \ln(1+2k) - \ln 3 = 0.4$

{since $k \geqslant 1$, $1+2x>0$ for all x in the shaded region}

$\therefore \quad \ln\left(\dfrac{1+2k}{3}\right) = 0.4$

$\therefore \quad \dfrac{1+2k}{3} = e^{0.4}$

$\therefore \quad 1+2k = 3e^{0.4}$

$\therefore \quad k = \dfrac{3e^{0.4}-1}{2} \approx 1.7377$

15 Area $= \int_0^b \sqrt{x}\,dx$

$\therefore \quad \int_0^b x^{\frac{1}{2}}\,dx = 1$

$\therefore \quad \left[\tfrac{2}{3}x^{\frac{3}{2}}\right]_0^b = 1$

$\therefore \quad \tfrac{2}{3}b\sqrt{b} - 0 = 1$

$\therefore \quad b\sqrt{b} = \tfrac{3}{2}$

$\therefore \quad b^{\frac{3}{2}} = 1.5$

$\therefore \quad b = (1.5)^{\frac{2}{3}} \approx 1.3104$

16 By symmetry, the area bounded by $x=0$ and $x=a$ is $\frac{1}{2}(6a)$ units^2.

$\therefore \quad \int_0^a (x^2+2)\,dx = 3a$

$\therefore \quad \left[\dfrac{x^3}{3} + 2x\right]_0^a = 3a$

$\therefore \quad \dfrac{a^3}{3} + 2a - 0 = 3a$

$\therefore \quad a^3 + 6a = 9a$

$\therefore \quad a^3 - 3a = 0$

$\therefore \quad a(a^2-3) = 0$

$\therefore \quad a = 0 \text{ or } \pm\sqrt{3} \qquad \therefore \quad a = \sqrt{3} \quad \{\text{as } a>0\}$

EXERCISE 22B.1

1

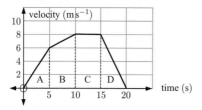

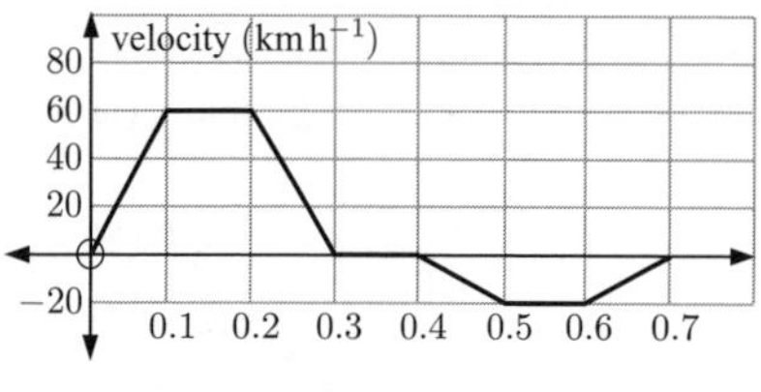

Total distance travelled

$= \text{area A} + \text{area B} + \text{area C} + \text{area D}$

$= \tfrac{1}{2}(5\times 6) + \left(\dfrac{6+8}{2}\right)5 + 5\times 8 + \tfrac{1}{2}(5\times 8)$

$= 15 + 35 + 40 + 20$

$= 110$ m

2

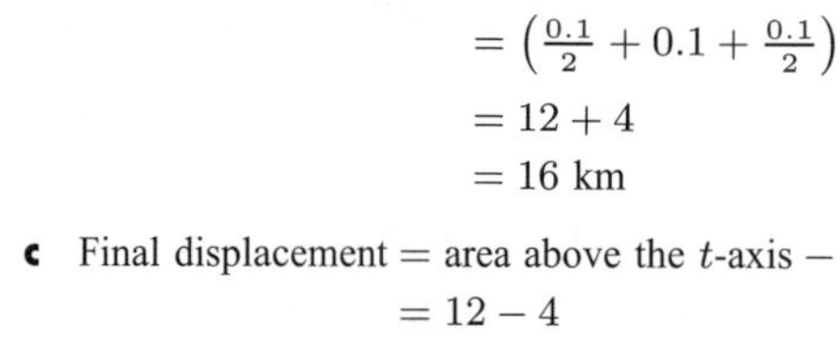

a **i** The graph above the t-axis indicates that the velocity is positive and the car is travelling forwards.

ii The graph below the t-axis indicates that the velocity is negative and the car is travelling backwards.

b Total distance travelled $=$ area above the t-axis $+$ area below the t-axis

$= \left(\frac{0.1}{2} + 0.1 + \frac{0.1}{2}\right)60 + \left(\frac{0.1}{2} + 0.1 + \frac{0.1}{2}\right)20$

$= 12 + 4$

$= 16$ km

c Final displacement $=$ area above the t-axis $-$ area below the t-axis

$= 12 - 4$

$= 8$ km from the starting point in the positive direction

3 **a**

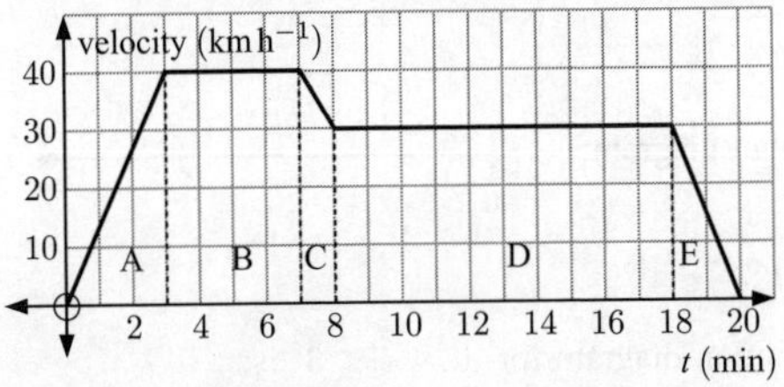

b Total distance travelled

$= \text{area A} + \text{area B} + \text{area C} + \text{area D} + \text{area E}$

$= \frac{1}{60}\left[\frac{1}{2}(3 \times 40) + (40 \times 4) + \left(\frac{40+30}{2}\right)1 + (10 \times 30) + \frac{1}{2}(2 \times 30)\right]$

{the factor $\frac{1}{60}$ accounts for the fact that the times are in minutes while the speeds are in km h^{-1}}

$= \frac{1}{60}[60 + 160 + 35 + 300 + 30]$

$= 9.75$ km

EXERCISE 22B.2

1 **a** $v(t) = s'(t) = 1 - 2t$ cm s^{-1}, $t \geqslant 0$

which has sign diagram:

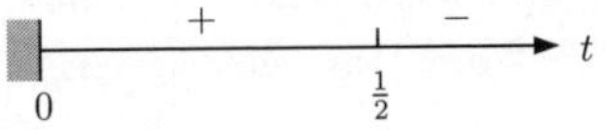

$\therefore$ a direction reversal occurs at $t = \frac{1}{2}$.

Now $s(t) = \int (1 - 2t)\, dt = t - \dfrac{2t^2}{2} + c = t - t^2 + c$

$\therefore$ $s(0) = c$ $\therefore$ motion diagram is:

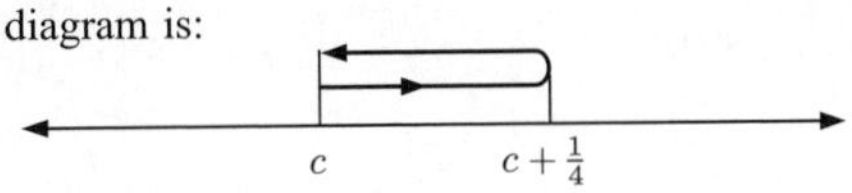

and $s(\frac{1}{2}) = \frac{1}{4} + c$

and $s(1) = c$

$\therefore$ total distance travelled $= (c + \frac{1}{4} - c) + (c + \frac{1}{4} - c)$

$= \frac{1}{2}$ cm

b Displacement $= s(1) - s(0)$

$= c - c$

$= 0$ cm

2 **a** $v(t) = s'(t) = t^2 - t - 2$ cm s^{-1}, $t \geqslant 0$

$= (t-2)(t+1)$

which has sign diagram:

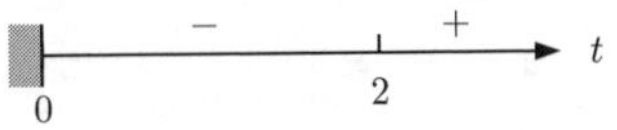

$\therefore$ a direction reversal occurs at $t = 2$.

Now $s(t) = \int (t^2 - t - 2)\, dt = \dfrac{t^3}{3} - \dfrac{t^2}{2} - 2t + c$

$s(0) = c$

$s(2) = c - \frac{10}{3}$ $\therefore$ motion diagram is:

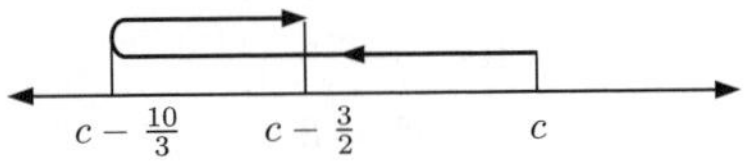

$s(3) = c - \frac{3}{2}$

$\therefore$ total distance travelled $= \left(c - [c - \frac{10}{3}]\right) + \left(c - \frac{3}{2} - [c - \frac{10}{3}]\right)$

$= \frac{10}{3} - \frac{3}{2} + \frac{10}{3}$

$= \frac{31}{6} = 5\frac{1}{6}$ cm

b Displacement $= s(3) - s(0)$

$= c - \frac{3}{2} - c$

$= -\frac{3}{2}$ cm which is $1\frac{1}{2}$ cm left of its starting point.

3 $x'(t) = 16t - 4t^3$ units s^{-1}, $t \geqslant 0$

$= 4t(4 - t^2)$

$= 4t(2+t)(2-t)$ which has sign diagram:

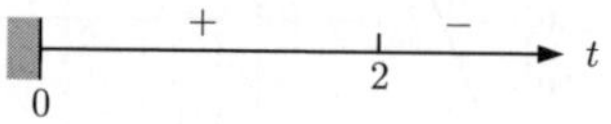

$\therefore$ a direction reversal occurs at $t = 2$.

Now $x(t) = \int (16t - 4t^3)\,dt = 8t^2 - t^4 + c$

a $x(0) = c$ $\therefore$ motion diagram for $0 \leqslant t \leqslant 3$ is:

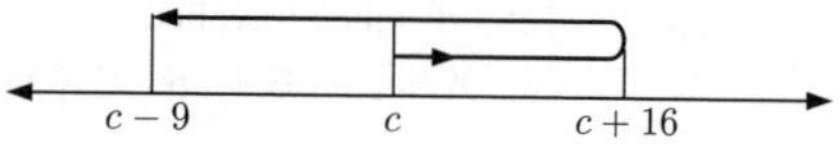

$x(2) = 32 - 16 + c = c + 16$

$x(3) = 72 - 81 + c = c - 9$

$\therefore$ total distance travelled $= (c + 16 - c) + (c + 16 - [c - 9])$

$= 41$ units

b $x(1) = 7 + c = c + 7$ $\therefore$ motion diagram for $1 \leqslant t \leqslant 3$ is:

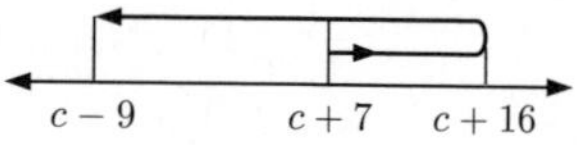

$\therefore$ total distance travelled $= (c + 16 - [c + 7]) + (c + 16 - [c - 9])$

$= 34$ units

4 **a** $v(t) = \cos t \text{ m s}^{-1}$, $t \geqslant 0$

$\therefore$ $v(t)$ has sign diagram:

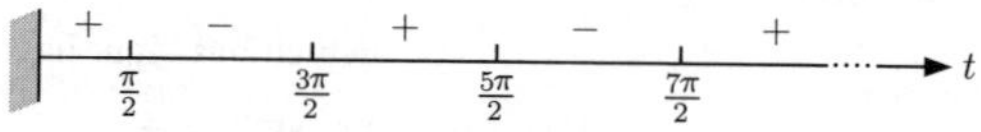

$\therefore$ a direction reversal occurs at $t = \frac{\pi}{2}, \frac{3\pi}{2}, \frac{5\pi}{2}, \frac{7\pi}{2}, \ldots$

$s(t) = \int \cos t\,dt = \sin t + c$

$\therefore$ $s(0) = c$

$s\left(\frac{\pi}{2}\right) = c + 1$

$s\left(\frac{3\pi}{2}\right) = c - 1$

$s\left(\frac{5\pi}{2}\right) = c + 1$

$s\left(\frac{7\pi}{2}\right) = c - 1$

The motion diagram is:

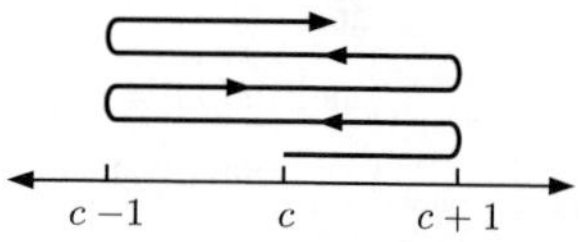

$\therefore$ the particle oscillates between the points $(c - 1)$ and $(c + 1)$.

b distance $= (c + 1) - (c - 1)$

$= 2$ units

5 $v(t) = 50 - 10e^{-0.5t} \text{ m s}^{-1}$, $t \geqslant 0$

a $v(0) = 50 - \dfrac{10}{e^0} = 50 - 10 = 40 \text{ m s}^{-1}$

b $v(3) = 50 - \dfrac{10}{e^{1.5}} \approx 47.8 \text{ m s}^{-1}$

c The velocity reaches 45 m s^{-1}

when $45 = 50 - 10e^{-0.5t}$

$\therefore$ $10e^{-\frac{t}{2}} = 5$

$\therefore$ $e^{\frac{t}{2}} = 2$

$\therefore$ $\dfrac{t}{2} = \ln 2$

$\therefore$ $t = 2\ln 2 \approx 1.39$ seconds

d $v(t) = 50 - \dfrac{10}{e^{\frac{t}{2}}}$

As $t \to \infty$, $\dfrac{10}{e^{\frac{t}{2}}} \to 0$ (from above)

$\therefore$ $v(t) \to 50 \text{ m s}^{-1}$ (below)

e $a(t) = v'(t)$

$= -10e^{-0.5t}(-0.5)$

$= 5e^{-0.5t} \text{ m s}^{-2}$

$= \dfrac{5}{e^{0.5t}} \text{ m s}^{-2}$

$\therefore$ $a(t) > 0$ for all t $\{e^x > 0$ for all $x\}$

$\therefore$ the acceleration is always positive

f

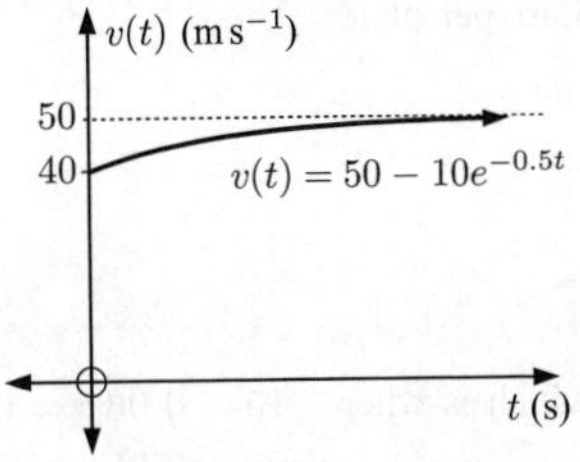

g total distance travelled

$= \int_0^3 (50 - 10e^{-0.5t})\,dt$

$= \left[50t + 20e^{-0.5t}\right]_0^3$

$= 150 + 20e^{-1.5} - 20$

≈ 134 m

6 $a(t) = \dfrac{t}{10} - 3 \ \ \text{m s}^{-2}$

$\therefore \ v(t) = \displaystyle\int \left(\frac{t}{10} - 3\right) dt$

$= \dfrac{t^2}{20} - 3t + c$

But $v(0) = 45 \quad \therefore \quad c = 45$

Now $v(t) = \dfrac{t^2}{20} - 3t + 45$

$= \dfrac{t^2 - 60t + 900}{20}$

$= \dfrac{(t-30)^2}{20}$

Sign diagram of $v(t)$:

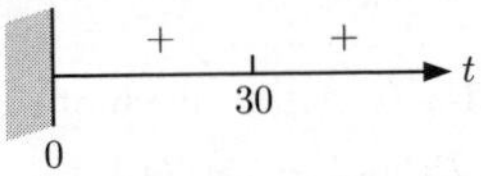

The total distance travelled in the first minute

$= \displaystyle\int_0^{60} \left(\frac{(t-30)^2}{20}\right) dt \quad \{\text{as } (t-30)^2 \geqslant 0\}$

$= \frac{1}{20}\int_0^{60} (t-30)^2\,dt$

$= \dfrac{1}{20}\left[\dfrac{(t-30)^3}{3}\right]_0^{60}$

$= \frac{1}{60}\left((30)^3 - (-30)^3\right)$

$= \frac{1}{60}(30^3 + 30^3)$

$= 900$ m

7 $a(t) = 4e^{-\frac{t}{20}}\,\text{m s}^{-2}$

$\therefore \ v(t) = \int 4e^{-\frac{t}{20}}\,dt$

$= 4\dfrac{1}{-\frac{1}{20}}e^{-\frac{t}{20}} + c$

$= -80e^{-\frac{t}{20}} + c$

Now $v(0) = 20 \text{ m s}^{-1}$

$\therefore \quad c = 100$

$\therefore \quad v(t) = 100 - 80e^{-\frac{t}{20}}$

a As $t \to \infty$, $e^{-\frac{t}{20}} \to 0$ (above) $\quad \therefore \quad v(t) \to 100$ (below)

$\therefore$ the object approaches a limiting velocity of 100 m s^{-1}

b The total distance travelled $= \int_0^{10} (100 - 80e^{-\frac{t}{20}})\,dt \qquad \{v(t) > 0 \text{ for } 0 \leqslant t \leqslant 10\}$

$= [100t + 1600e^{-\frac{t}{20}}]_0^{10}$

$= \left(1000 + 1600e^{-\frac{1}{2}}\right) - (0 + 1600)$

≈ 370 m

EXERCISE 22C

1 The marginal cost is $C'(x) = 3.15 + 0.004x$ € per gadget

$\therefore \quad C(x) = \int (3.15 + 0.004x)\,dx$

$= 3.15x + 0.002x^2 + c$

But $C(0) = 450$ so $c = 450$

$\therefore \quad C(x) = 3.15x + 0.002x^2 + 450$ euros

$\therefore \quad C(800) = 3.15(800) + 0.002(800)^2 + 450$

$=$ €4250

So, the total cost is €4250.

2 **a** The marginal profit is $P'(x) = 15 - 0.03x$ dollars per plate

$$\therefore \quad P(x) = \int (15 - 0.03x)\, dx$$
$$= 15x - 0.015x^2 + c$$

But $P(0) = -650$ so $c = -650$

$\therefore \quad P(x) = 15x - 0.015x^2 - 650$ dollars

b The maximum profit occurs when $P'(x) = 0$, which is when $15 - 0.03x = 0$

$$\therefore \quad 0.03x = 15$$
$$\therefore \quad x = \frac{15}{0.03}$$
$$\therefore \quad x = 500$$

Now $P''(x) = -0.03 < 0$ $\quad\therefore$ the profit is at a maximum when $x = 500$.

The maximum profit $= P(500) = 15(500) - 0.015(500)^2 - 650$
$= \$3100$

c In order for a profit to be made, $P(x)$ must be greater than 0

$\therefore \quad 15x - 0.015x^2 - 650 > 0$

Using technology, the x-intercepts of $P(x)$ are $x_1 = 45.39$ and $x_2 = 954.6$

Since we cannot produce part plates, a profit is made for $46 \leqslant x \leqslant 954$.

3 $E'(t) = 350(80 + 0.15t)^{0.8} - 120(80 + 0.15t)$ calories per day

Total energy needs over the first week $= \int_0^7 E'(t)\, dt$

$$= \int_0^7 [350(80 + 0.15t)^{0.8} - 120(80 + 0.15t)]\, dt$$
$$= \left[\frac{1}{0.15} \times \frac{350(80 + 0.15t)^{1.8}}{1.8} - 9600t - 9t^2 \right]_0^7$$
$$\approx 14\,400 \text{ calories}$$

4 $$\frac{dT}{dx} = \frac{-20}{x^{0.63}} = -20x^{-0.63}$$

$$\therefore \quad T = \int -20x^{-0.63}\, dx$$
$$= \frac{-20x^{0.37}}{0.37} + c$$

Now when $x = 3$, $T = 100$

$$\therefore \quad \frac{-20(3^{0.37})}{0.37} + c = 100$$
$$\therefore \quad c = 100 + \frac{20(3^{0.37})}{0.37} \approx 181.1639$$
$$\therefore \quad T \approx \frac{-20x^{0.37}}{0.37} + 181.1639$$

So, when $x = 6$, $T \approx -104.8925 + 181.1639 \approx 76.27$

$\therefore$ the outer surface temperature is about 76.3°C

EXERCISE 22D.1

1 **a**

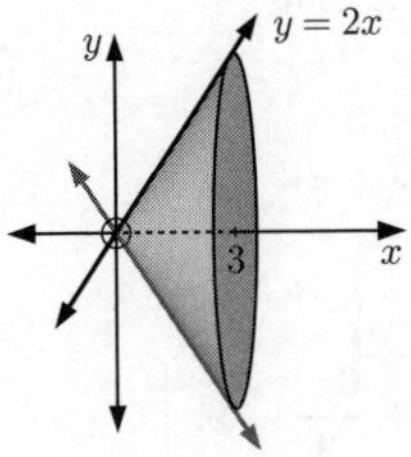

$$\begin{aligned}\text{Volume} &= \pi\int_0^3 (2x)^2\,dx\\ &= 4\pi\int_0^3 x^2\,dx\\ &= 4\pi\left[\tfrac{1}{3}x^3\right]_0^3\\ &= 4\pi(9-0)\\ &= 36\pi \text{ units}^3\end{aligned}$$

b

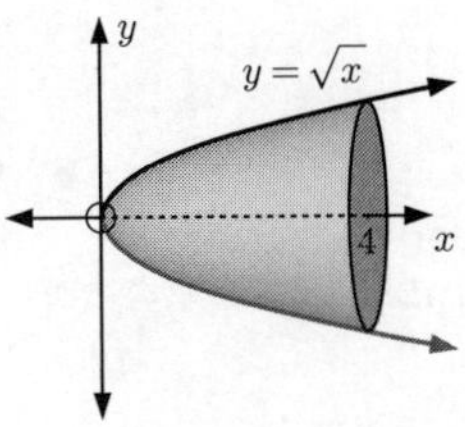

$$\begin{aligned}\text{Volume} &= \pi\int_0^4 (\sqrt{x})^2\,dx\\ &= \pi\int_0^4 x\,dx\\ &= \pi\left[\tfrac{1}{2}x^2\right]_0^4\\ &= \pi(8-0)\\ &= 8\pi \text{ units}^3\end{aligned}$$

c

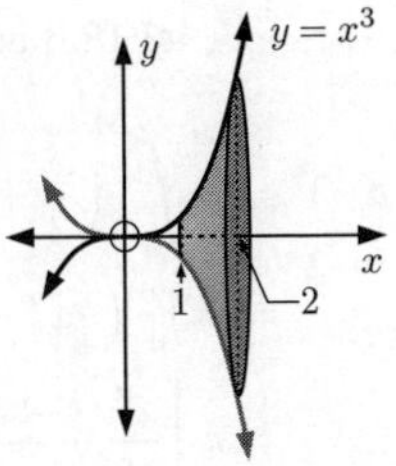

$$\begin{aligned}\text{Volume} &= \pi\int_1^2 (x^3)^2\,dx\\ &= \pi\int_1^2 x^6\,dx\\ &= \pi\left[\tfrac{1}{7}x^7\right]_1^2\\ &= \pi\left(\tfrac{128}{7}-\tfrac{1}{7}\right)\\ &= \tfrac{127\pi}{7} \text{ units}^3\end{aligned}$$

d

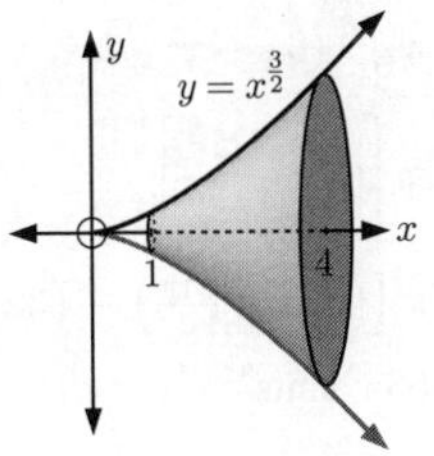

$$\begin{aligned}\text{Volume} &= \pi\int_1^4 (x^{\frac{3}{2}})^2\,dx\\ &= \pi\int_1^4 x^3\,dx\\ &= \pi\left[\tfrac{1}{4}x^4\right]_1^4\\ &= \pi\left(\tfrac{256}{4}-\tfrac{1}{4}\right)\\ &= \tfrac{255\pi}{4} \text{ units}^3\end{aligned}$$

e

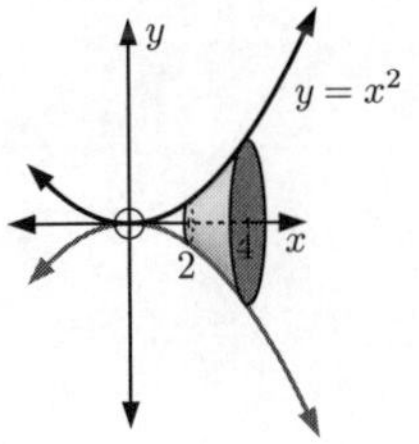

$$\begin{aligned}\text{Volume} &= \pi\int_2^4 (x^2)^2\,dx\\ &= \pi\int_2^4 x^4\,dx\\ &= \pi\left[\tfrac{1}{5}x^5\right]_2^4\\ &= \pi\left(\tfrac{1024}{5}-\tfrac{32}{5}\right)\\ &= \tfrac{992\pi}{5} \text{ units}^3\end{aligned}$$

f

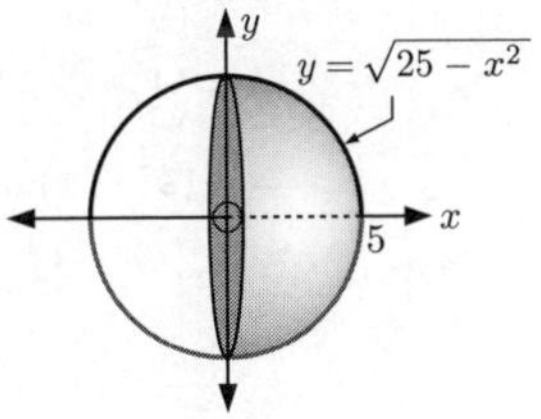

$$\begin{aligned}\text{Volume} &= \pi\int_0^5 (25-x^2)\,dx\\ &= \pi\left[25x-\frac{x^3}{3}\right]_0^5\\ &= \pi\left(125-\tfrac{125}{3}\right)\\ &= \pi\left(\tfrac{2}{3}\right)125\\ &= \tfrac{250\pi}{3} \text{ units}^3\end{aligned}$$

g

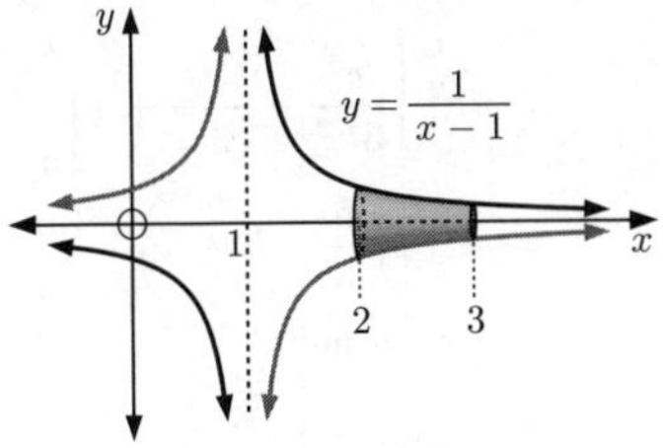

$$\begin{aligned}\text{Volume} &= \pi\int_2^3 \left(\frac{1}{x-1}\right)^2 dx\\ &= \pi\int_2^3 (x-1)^{-2}\,dx\\ &= \pi\left[-\frac{1}{x-1}\right]_2^3\\ &= \pi\left(-\tfrac{1}{2}+1\right)\\ &= \tfrac{\pi}{2} \text{ units}^3\end{aligned}$$

h

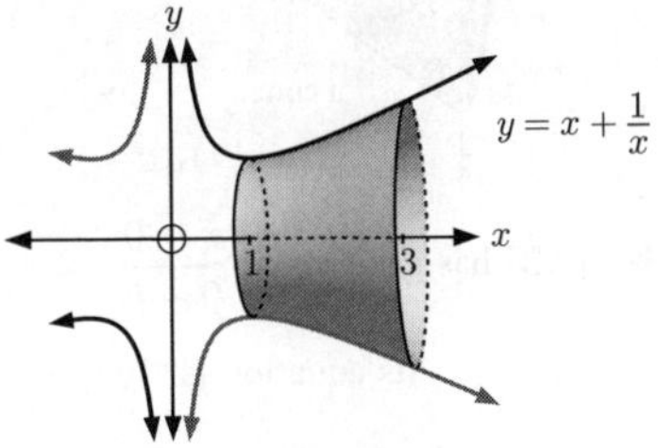

$$\begin{aligned}\text{Volume} &= \pi\int_1^3 \left(x+\frac{1}{x}\right)^2 dx\\ &= \pi\int_1^3 (x^2+2+x^{-2})\,dx\\ &= \pi\left[\frac{x^3}{3}+2x-\frac{1}{x}\right]_1^3\\ &= \pi\left[9+6-\tfrac{1}{3}-\left(\tfrac{1}{3}+2-1\right)\right]\\ &= \tfrac{40\pi}{3} \text{ units}^3\end{aligned}$$

2 **a** Volume $= \pi \int_1^3 \left(\dfrac{x^3}{x^2+1}\right)^2 dx$

$\approx 5.926\pi$ {using technology}

≈ 18.6 units^3

b Volume $= \pi \int_0^2 (e^{\sin x})^2\, dx$

$\approx 9.613\pi$ {using technology}

≈ 30.2 units^3

3 **a** $V = \pi \int_0^6 \left(\dfrac{x}{2}+4\right)^2 dx$

$= \pi \int_0^6 \left(\frac{1}{4}x^2 + 4x + 16\right) dx$

$= \pi \left[\dfrac{x^3}{12} + \dfrac{4x^2}{2} + 16x\right]_0^6$

$= \pi(18 + 72 + 96) - 0$

$= 186\pi$ units^3

b $V = \pi \int_1^2 (x^2+3)^2\, dx$

$= \pi \int_1^2 (x^4 + 6x^2 + 9)\, dx$

$= \pi \left[\dfrac{x^5}{5} + \dfrac{6x^3}{3} + 9x\right]_1^2$

$= \pi \left[\left(\frac{32}{5} + 16 + 18\right) - \left(\frac{1}{5} + 2 + 9\right)\right]$

$= \pi \left(\frac{146}{5}\right)$

$= \frac{146\pi}{5}$ units^3

c $V = \pi \int_0^4 (e^x)^2\, dx$

$= \pi \int_0^4 e^{2x}\, dx$

$= \pi \left[\frac{1}{2}e^{2x}\right]_0^4$

$= \pi \left(\frac{1}{2}e^8 - \frac{1}{2}\right)$

$= \frac{\pi}{2}(e^8 - 1)$ units^3

4 **a** Volume $= \pi \int_5^8 y^2\, dx$

$= \pi \int_5^8 (64 - x^2)\, dx$

$= \pi \left[64x - \dfrac{x^3}{3}\right]_5^8$

$= \pi \left[\left(512 - \frac{512}{3}\right) - \left(320 - \frac{125}{3}\right)\right]$

$= 63\pi$ units^3

b 63π cm^3 ≈ 198 cm^3

5 **a** a cone of base radius r and height h

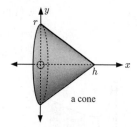

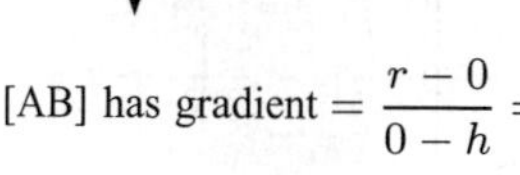

b [AB] has gradient $= \dfrac{r-0}{0-h} = -\dfrac{r}{h}$

$\therefore$ its equation is $y = -\dfrac{r}{h}x + r$

c $V = \pi \int_0^h \left(\dfrac{-r}{h}x + r\right)^2 dx$

$= \pi r^2 \int_0^h \left(-\dfrac{x}{h} + 1\right)^2 dx$

$= \pi r^2 \int_0^h \left(\dfrac{x^2}{h^2} - \dfrac{2x}{h} + 1\right) dx$

$= \pi r^2 \left[\dfrac{x^3}{3h^2} - \dfrac{2x^2}{2h} + x\right]_0^h$

$= \pi r^2 \left[\left(\dfrac{h}{3} - h + h\right) - 0\right]$

$= \frac{1}{3}\pi r^2 h$ units^3

6 **a** a sphere of radius r

b $V = \pi \int_{-r}^r y^2\, dx = 2\pi \int_0^r (r^2 - x^2)\, dx$

$= 2\pi \left[r^2 x - \dfrac{x^3}{3}\right]_0^r$

$= 2\pi\left(r^3 - \dfrac{r^3}{3} - 0\right)$

$= 2\pi \times \frac{2}{3}r^3$

$= \frac{4}{3}\pi r^3$ units^3

7 **a**

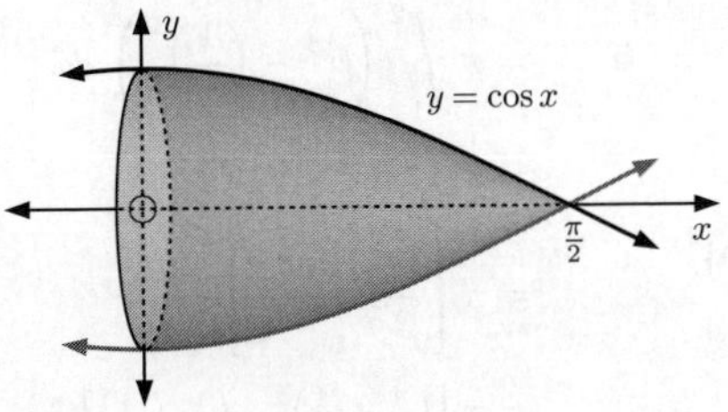

$$\text{Volume} = \pi \int_0^{\frac{\pi}{2}} (\cos x)^2 \, dx$$
$$= \pi \int_0^{\frac{\pi}{2}} \cos^2 x \, dx$$
$$= \pi \int_0^{\frac{\pi}{2}} \left(\tfrac{1}{2} + \tfrac{1}{2}\cos(2x)\right) dx$$
$$= \pi \left[\tfrac{1}{2}x + \tfrac{1}{2}\left(\tfrac{1}{2}\right)\sin(2x)\right]_0^{\frac{\pi}{2}}$$
$$= \pi \left[\tfrac{\pi}{4} + \tfrac{1}{4}\sin\pi - 0\right]$$
$$= \tfrac{\pi^2}{4} \text{ units}^3$$

b

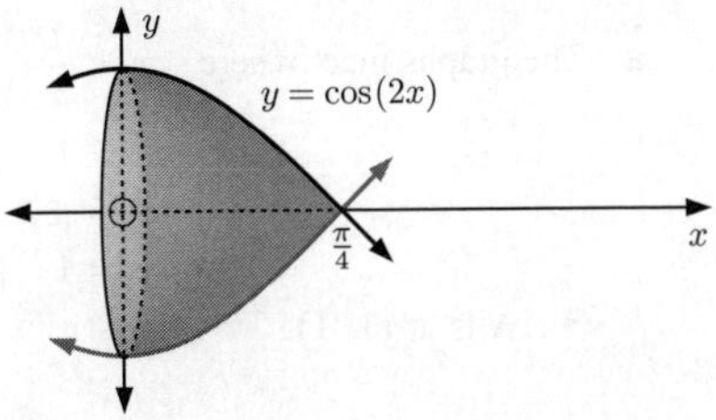

$$\text{Volume} = \pi \int_0^{\frac{\pi}{4}} \cos^2(2x) \, dx$$
$$= \pi \int_0^{\frac{\pi}{4}} \left(\tfrac{1}{2} + \tfrac{1}{2}\cos(4x)\right) dx$$
$$= \pi \left[\tfrac{1}{2}x + \tfrac{1}{2}\left(\tfrac{1}{4}\right)\sin(4x)\right]_0^{\frac{\pi}{4}}$$
$$= \pi \left[\tfrac{\pi}{8} + \tfrac{1}{8}\sin\pi - 0\right]$$
$$= \tfrac{\pi^2}{8} \text{ units}^3$$

8 **a**

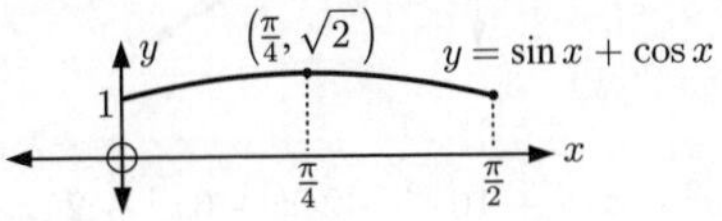

b Volume

$$= \pi \int_0^{\frac{\pi}{4}} (\sin x + \cos x)^2 \, dx$$
$$= \pi \int_0^{\frac{\pi}{4}} (\sin^2 x + 2\sin x\cos x + \cos^2 x) \, dx$$
$$= \pi \int_0^{\frac{\pi}{4}} (1 + \sin(2x)) \, dx$$
$$= \pi \left[x - \tfrac{1}{2}\cos(2x)\right]_0^{\frac{\pi}{4}}$$
$$= \pi \left[\left(\tfrac{\pi}{4} - \tfrac{1}{2}\cos\left(\tfrac{\pi}{2}\right)\right) - \left(0 - \tfrac{1}{2}\cos 0\right)\right]$$
$$= \pi \left(\tfrac{\pi}{4} + \tfrac{1}{2}\right) \text{ units}^3$$

9 **a**

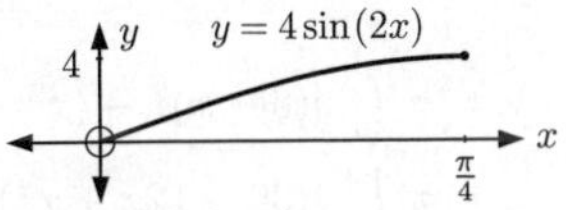

b Volume

$$= \pi \int_0^{\frac{\pi}{4}} (4\sin(2x))^2 \, dx$$
$$= 16\pi \int_0^{\frac{\pi}{4}} \sin^2(2x) \, dx$$
$$= 16\pi \int_0^{\frac{\pi}{4}} \left(\tfrac{1}{2} - \tfrac{1}{2}\cos(4x)\right) dx$$
$$= 16\pi \left[\frac{x}{2} - \tfrac{1}{2}\left(\tfrac{1}{4}\right)\sin(4x)\right]_0^{\frac{\pi}{4}}$$
$$= 16\pi \left[\left(\tfrac{\pi}{8} - \tfrac{1}{8}\sin\pi\right) - \left(0 - \tfrac{1}{8}\sin 0\right)\right]$$
$$= 2\pi^2 \text{ units}^3$$

EXERCISE 22D.2

1 **a** The graphs meet where $4 - x^2 = 3$

$$\therefore \quad x^2 = 1$$
$$\therefore \quad x = \pm 1$$

$\therefore$ A is at $(-1, 3)$ and B is at $(1, 3)$.

b
$$V = \pi \int_{-1}^{1} ((4 - x^2)^2 - 3^2) \, dx$$
$$= \pi \int_{-1}^{1} (16 - 8x^2 + x^4 - 9) \, dx$$
$$= \pi \int_{-1}^{1} (x^4 - 8x^2 + 7) \, dx$$
$$= \pi \left[\frac{x^5}{5} - \frac{8x^3}{3} + 7x\right]_{-1}^{1}$$
$$= \pi \left(\tfrac{1}{5} - \tfrac{8}{3} + 7 - \left(\tfrac{-1}{5} - \tfrac{-8}{3} - 7\right)\right)$$
$$= \tfrac{136\pi}{15} \text{ units}^3$$

2 **a** The graphs meet where $e^{\frac{x}{2}} = e$

$$\therefore \quad e^{\frac{x}{2}} = e^1$$
$$\therefore \quad \frac{x}{2} = 1$$
$$\therefore \quad x = 2$$

$\therefore$ A is at $(2, e)$.

b
$$V = \pi \int_0^2 \left(e^2 - \left(e^{\frac{x}{2}}\right)^2\right) dx$$
$$= \pi \int_0^2 (e^2 - e^x) \, dx$$
$$= \pi \left[e^2 x - e^x\right]_0^2$$
$$= \pi \left[(2e^2 - e^2) - (0 - 1)\right]$$
$$= \pi(e^2 + 1) \text{ units}^3$$

3 **a** The graphs meet where $x = \dfrac{1}{x}$

$\therefore \quad x^2 = 1$

$\therefore \quad x = \pm 1$

$\therefore \quad x = 1 \quad \{\text{as } x > 0\}$

$\therefore$ A is at (1, 1).

b $V = \pi \displaystyle\int_1^2 \left(x^2 - \left(\frac{1}{x}\right)^2\right) dx$

$= \pi \int_1^2 (x^2 - x^{-2})\, dx$

$= \pi \left[\dfrac{x^3}{3} - \dfrac{x^{-1}}{-1}\right]_1^2$

$= \pi \left[\left(\frac{8}{3} + \frac{1}{2}\right) - \left(\frac{1}{3} + 1\right)\right]$

$= \frac{11\pi}{6}$ units^3

4 The graphs meet where $x^2 - 4x + 6 = 6 - x$

$\therefore \quad x^2 - 3x = 0$

$\therefore \quad x(x-3) = 0$

$\therefore \quad x = 0$ or 3

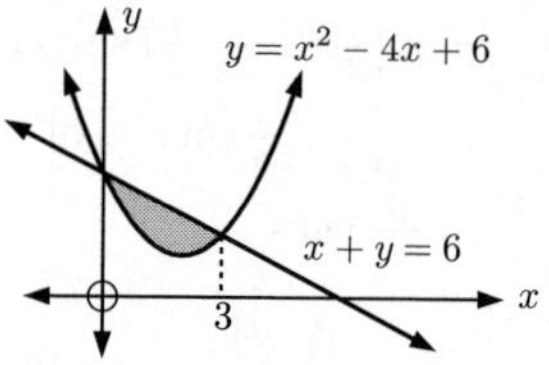

$\therefore \quad V = \pi \int_0^3 [(6-x)^2 - (x^2 - 4x + 6)^2]\, dx$

$= \pi \int_0^3 [(36 - 12x + x^2) - (x^4 - 4x^3 + 6x^2 - 4x^3 + 16x^2 - 24x + 6x^2 - 24x + 36)]\, dx$

$= \pi \int_0^3 (-x^4 + 8x^3 - 27x^2 + 36x)\, dx$

$= \pi \left[-\dfrac{x^5}{5} + 2x^4 - 9x^3 + 18x^2\right]_0^3$

$= \pi \left(-\frac{3^5}{5} + 2(3^4) - 9(27) + 18(9) - 0\right)$

$= \frac{162}{5}\pi$ units^3

5 **a** The curves meet where $\sqrt{x-4} = 1$

$\therefore \quad x - 4 = 1$

$\therefore \quad x = 5$

$\therefore$ A is at (5, 1).

b $V = \pi \int_5^8 \left(\left(\sqrt{x-4}\right)^2 - 1^2\right) dx$

$= \pi \int_5^8 (x - 4 - 1)\, dx$

$= \pi \int_5^8 (x - 5)\, dx$

$= \pi \left[\dfrac{x^2}{2} - 5x\right]_5^8$

$= \pi \left[(32 - 40) - \left(\frac{25}{2} - 25\right)\right]$

$= \frac{9\pi}{2}$ units^3

6 The shaded area $= \displaystyle\int_1^\infty \frac{1}{x}\, dx$

$= \displaystyle\lim_{t \to \infty} \int_1^t \frac{1}{x}\, dx$

$= \displaystyle\lim_{t \to \infty} [\ln(x)]_1^t, \ x > 0$

$= \displaystyle\lim_{t \to \infty} \ln t,$ which is infinite.

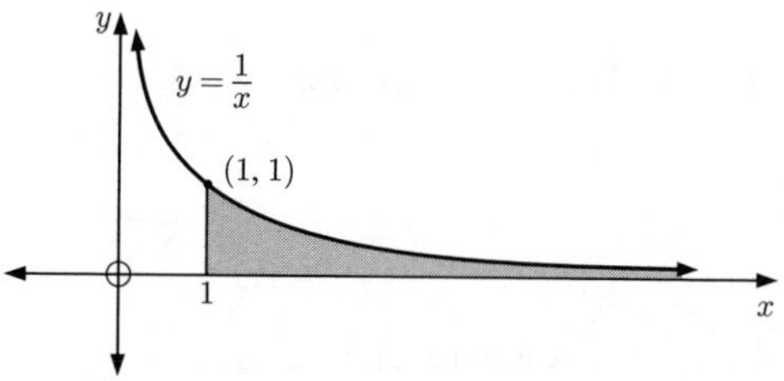

$$\begin{aligned}\text{The volume of revolution} &= \pi \int_1^\infty \left(\frac{1}{x}\right)^2 dx \\ &= \pi \lim_{t\to\infty} \int_1^t x^{-2}\, dx \\ &= \pi \lim_{t\to\infty} \left[-\frac{1}{x}\right]_1^t \\ &= \pi \lim_{t\to\infty} \left(-\frac{1}{t}+1\right) \\ &= \pi, \quad \text{which is finite.}\end{aligned}$$

REVIEW SET 22A

1

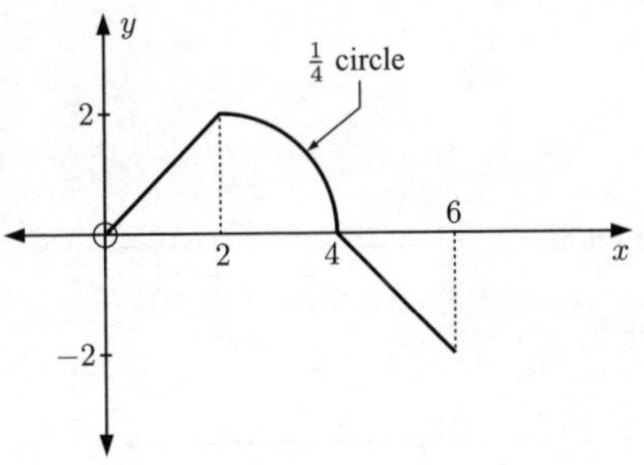

a $\int_0^4 f(x)\,dx = \text{area of triangle} + \text{area of } \frac{1}{4} \text{ circle}$
$= \frac{1}{2}(2\times 2) + \frac{1}{4}\pi(2)^2$
$= 2+\pi$

b $\int_4^6 f(x)\,dx = -\text{area of triangle below } x\text{-axis}$
$= -\frac{1}{2}(2\times 2)$
$= -2$

c $\int_0^6 f(x)\,dx = \int_0^4 f(x)\,dx + \int_4^6 f(x)\,dx$
$= (2+\pi)+(-2)$
$= \pi$

2 shaded area $= \int_a^b [f(x)-g(x)]\,dx + \int_b^c [g(x)-f(x)]\,dx + \int_c^d [f(x)-g(x)]\,dx$

3 $\int_{-1}^3 f(x)\,dx$ gives us the correct area only if $f(x)$ is non-negative on the interval $-1 \leqslant x \leqslant 3$.

In this case $f(x)$ is negative for $1 < x < 3$, so $\int_{-1}^3 f(x)\,dx$ does not provide the correct answer.

(The shaded area which is below the x-axis is given by $\int_1^3 [0-f(x)]\,dx = -\int_1^3 f(x)\,dx$.)

4 $y=k$ meets $y=x^2$ where $x^2=k$ $\quad\therefore\ x=\pm\sqrt{k}$

By symmetry, $\int_0^{\sqrt{k}} (k-x^2)\,dx = \frac{1}{2}\times 5\frac{1}{3} = \frac{1}{2}\times\frac{16}{3}$

$\therefore\ \left[kx - \frac{x^3}{3}\right]_0^{\sqrt{k}} = \frac{8}{3}$

$\therefore\ k\sqrt{k} - \frac{k\sqrt{k}}{3} = \frac{8}{3}$

$\therefore\ \frac{2}{3}k\sqrt{k} = \frac{8}{3}$

$\therefore\ k\sqrt{k} = 4$

$\therefore\ k^{\frac{3}{2}} = 4$

$\therefore\ k = 4^{\frac{2}{3}} = \sqrt[3]{16}$

5

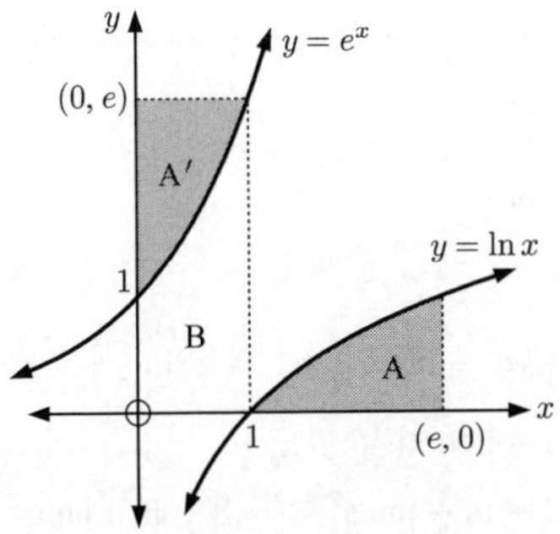

$y=e^x$ and $y=\ln x$ are inverse functions, so they are symmetrical about $y=x$

$\therefore$ area A = area A′

But area A′ + area B = area of rectangle

$\therefore$ area A + area B $= e\times 1 = e$

Since area A $= \int_1^e \ln x\,dx$

and area B $= \int_0^1 e^x\,dx$,

$\int_1^e \ln x\,dx + \int_0^1 e^x\,dx = e$

6 $y = x^2 + 4x + 1$ meets $y = 3x + 3$ where

$$x^2 + 4x + 1 = 3x + 3$$
$$\therefore \quad x^2 + x - 2 = 0$$
$$\therefore \quad (x+2)(x-1) = 0$$
$$\therefore \quad x = -2 \text{ or } 1$$

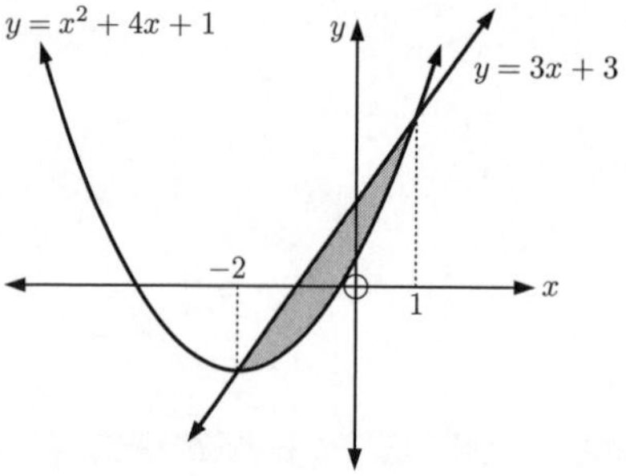

$$\therefore \quad \text{area} = \int_{-2}^{1} [(3x+3) - (x^2 + 4x + 1)]\, dx$$
$$= \int_{-2}^{1} (-x^2 - x + 2)\, dx$$
$$= \left[-\frac{x^3}{3} - \frac{x^2}{2} + 2x \right]_{-2}^{1}$$
$$= \left(-\tfrac{1}{3} - \tfrac{1}{2} + 2\right) - \left(\tfrac{8}{3} - 2 - 4\right)$$
$$= -\tfrac{1}{3} - \tfrac{1}{2} + 2 - \tfrac{8}{3} + 2 + 4$$
$$= 4\tfrac{1}{2} \text{ units}^2$$

7 **a** $v(t) = t^2 - 6t + 8 \text{ m s}^{-1}, \quad t \geqslant 0$

$= (t-4)(t-2)$ which has sign diagram:

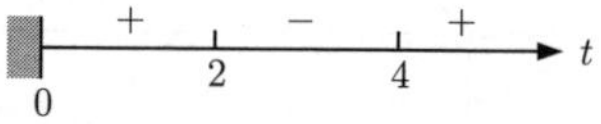

b Now $s(t) = \int (t^2 - 6t + 8)\, dt = \dfrac{t^3}{3} - 3t^2 + 8t + c$

$\therefore \quad s(0) = c$

$s(2) = c + 6\frac{2}{3}$

$s(4) = c + 5\frac{1}{3}$ the motion diagram is:

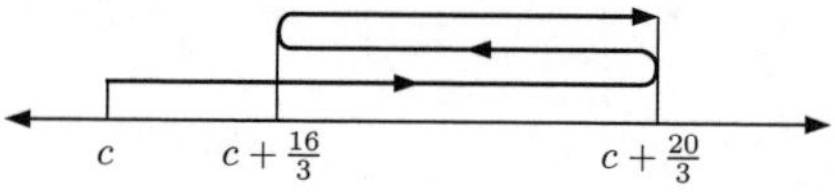

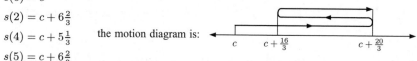

$s(5) = c + 6\frac{2}{3}$

The particle moves in the positive direction initially. When $t = 2$, $6\frac{2}{3}$ m from its starting point, it changes direction. It changes direction again when $t = 4$, $5\frac{1}{3}$ m from its starting point. When $t = 5$ it is $6\frac{2}{3}$ m from its starting point.

c After 5 seconds, the particle is $6\frac{2}{3}$ m to the right of its starting point.

d The total distance travelled $= (c + \frac{20}{3} - c) + [(c + \frac{20}{3}) - (c + \frac{16}{3})] + [(c + \frac{20}{3}) - (c + \frac{16}{3})]$

$= 9\frac{1}{3}$ m

8 Consider $y = 4e^x - 1$.

The x-intercept occurs when $y = 0$

$$\therefore \quad 4e^x - 1 = 0$$
$$\therefore \quad e^x = \tfrac{1}{4}$$
$$\therefore \quad x = \ln \tfrac{1}{4} < 0$$

$y = 4e^x - 1$ is the graph of $y = e^x$ with a vertical stretch of factor 4 and a vertical translation of -1.

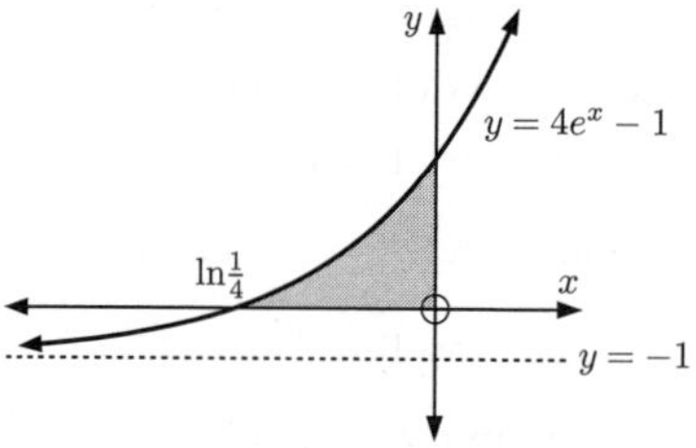

$$\text{Area} = \int_{\ln \frac{1}{4}}^{0} (4e^x - 1)\, dx$$
$$= [4e^x - x]_{\ln \frac{1}{4}}^{0}$$
$$= \left(4e^0 - 0\right) - \left(4e^{\ln \frac{1}{4}} - \ln \tfrac{1}{4}\right)$$
$$= 4 - 0 - 4(\tfrac{1}{4}) + \ln \tfrac{1}{4}$$
$$= 3 + \ln \tfrac{1}{4} \text{ units}^2 \quad (= 3 - \ln 4 \text{ units}^2)$$

9 $y = x^2$ and $y = 4$ meet where $x^2 = 4$

$\therefore \quad x = \pm 2$

But $x > 0$, so $x = 2$

Hence $V = \pi \int_0^2 (4^2 - (x^2)^2)\ dx$

$= \pi \int_0^2 (16 - x^4)\ dx$

$= \pi \left[16x - \dfrac{x^5}{5}\right]_0^2$

$= \pi \left(32 - \frac{32}{5} - 0\right)$

$= \frac{128\pi}{5}$ units^3

REVIEW SET 22B

1 $v(t) = 2t - 3t^2 = t(2 - 3t)$ which has sign diagram:

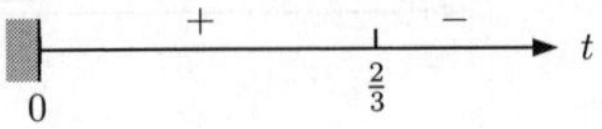

Now $s(t) = \int (2t - 3t^2)\ dt$

$= t^2 - t^3 + c$ metres

and so $s(0) = c$

$s(\frac{2}{3}) = \frac{4}{9} - \frac{8}{27} + c = c + \frac{4}{27}$

$s(1) = 1 - 1 + c = c$

with motion diagram:

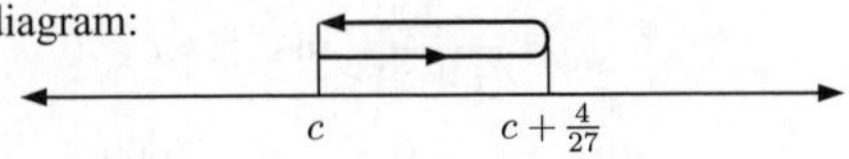

$\therefore$ total distance travelled $= (c + \frac{4}{27} - c) + (c + \frac{4}{27} - c) = \frac{8}{27}$ m ≈ 29.6 cm

2 **a** $f(x) = \dfrac{x}{1 + x^2}$ $\quad \therefore \quad f'(x) = \dfrac{1(1 + x^2) - x(2x)}{(1 + x^2)^2}$ {quotient rule}

$= \dfrac{1 + x^2 - 2x^2}{(1 + x^2)^2}$

$= \dfrac{1 - x^2}{(1 + x^2)^2}$

$= \dfrac{(1 + x)(1 - x)}{(1 + x^2)^2}$

which has sign diagram:

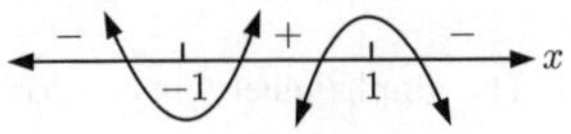

$\therefore$ there is a local minimum at $(-1, -\frac{1}{2})$ and a local maximum at $(1, \frac{1}{2})$.

b As $x \to \infty$, $f(x) \to 0$ (above).
As $x \to -\infty$, $f(x) \to 0$ (below).

c

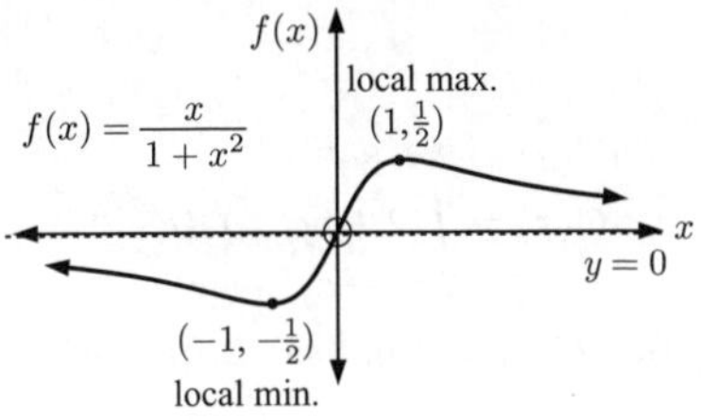

d Area $= \displaystyle\int_{-2}^{0} \left[0 - \frac{x}{1 + x^2}\right] dx$

$= \displaystyle\int_{-2}^{0} \frac{-x}{1 + x^2}\ dx$

≈ 0.805 units^2

3 $v(t) = \sin t$ which has sign diagram:

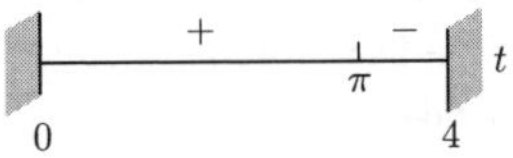

Now $s(t) = \int \sin t\ dt$

$= -\cos t + c$ metres

$\therefore \quad s(0) = -1 + c$

$s(\pi) = 1 + c$

$s(4) = -\cos 4 + c \approx c + 0.654$

motion diagram:

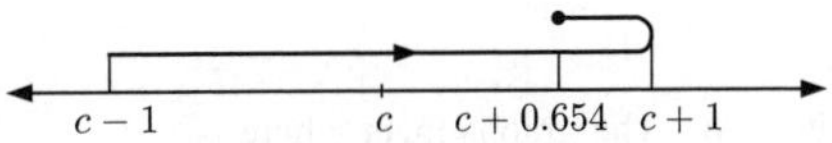

$\therefore$ total distance travelled $= [(c + 1) - (c - 1)] + [(c + 1) - (c + 0.654)]$

≈ 2.35 m

4 $v(t) = \dfrac{100}{(t+2)^2} = 100(t+2)^{-2}\ \text{m s}^{-1}$

a At $t = 0$, $v(0) = \dfrac{100}{2^2} = 25\ \text{m s}^{-1}$. At $t = 3$, $v(3) = \dfrac{100}{5^2} = 4\ \text{m s}^{-1}$.

b As $t \to \infty$, $v(t) \to 0\ \text{m s}^{-1}$ (above)

c

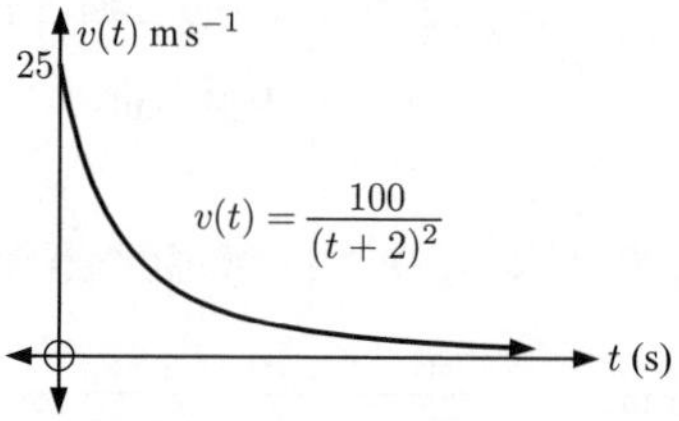

d As $v(t)$ is always positive, the boat is always travelling forwards.

$$\begin{aligned} s(t) &= \int v(t)\, dt \\ &= \int 100(t+2)^{-2}\, dt \\ &= -100(t+2)^{-1} + c \\ &= \frac{-100}{t+2} + c \end{aligned}$$

$\therefore$ $s(0) = c - 50$ m

$\therefore$ when the boat has travelled 30 m, $s(t) = c - 20$

$\therefore$ $c - 20 = \dfrac{-100}{t+2} + c$

$\therefore$ $\dfrac{-100}{t+2} = -20$

$\therefore$ $t + 2 = 5$

$\therefore$ $t = 3$ seconds

e
$$\begin{aligned} a(t) &= v'(t) \\ &= -200(t+2)^{-3} \\ &= \frac{-200}{(t+2)^3}\ \text{m s}^{-2},\quad t \geqslant 0 \end{aligned}$$

f
$$\begin{aligned} \frac{dv}{dt} = \frac{-200}{(t+2)^3} &= -\tfrac{1}{5}\,\frac{1000}{(t+2)^3} \\ &= -\tfrac{1}{5}\left(\frac{100}{(t+2)^2}\right)^{\frac{3}{2}} \\ &= -\tfrac{1}{5}\, v^{\frac{3}{2}} \end{aligned}$$

$\therefore$ $\dfrac{dv}{dt} = -kv^{\frac{3}{2}}$ where $k = \frac{1}{5}$

5 **a** The graphs meet when $\cos 2x = e^{3x}$

Using technology, $x = 0$ and $x \approx -0.7292$

b Shaded area $\approx \int_{-0.7292}^{0} (\cos 2x - e^{3x})\, dx \approx 0.2009$ units^2 {using technology}

6 $C'(x) = 2 + 8e^{-x}$

$\therefore$ $C(x) = \int (2 + 8e^{-x})\, dx$

$= 2x - 8e^{-x} + c$

But $C(0) = 240$, so $2(0) - 8e^0 + c = 240$

$\therefore$ $c = 248$

So $C(x) = 2x - 8e^{-x} + 248$

$\therefore$ $C(80) = 2(80) - 8e^{-80} + 248 \approx 408$

$\therefore$ the total cost is £408 per day.

7 $\int_0^m \sin x\, dx = \frac{1}{2}$

$\therefore$ $[-\cos x]_0^m = \frac{1}{2}$

$\therefore$ $-\cos m + \cos 0 = \frac{1}{2}$

$\therefore$ $\cos m = \frac{1}{2}$

$\therefore$ $m = \frac{\pi}{3}$ $\{0 < m < \frac{\pi}{2}\}$

8 **a** The graphs meet where

$x^2 = \sin x$

$\therefore$ $x = 0$ or ≈ 0.8767 {using technology}

$\therefore$ $a \approx 0.8767$

b area $\approx \int_0^{0.8767} (\sin x - x^2)\, dx$

≈ 0.1357 units^2 {using technology}

9 **a** $y = \cos(2x)$ meets the x-axis where $2x = \frac{\pi}{2}$, or $x = \frac{\pi}{4}$.

$$\therefore \quad V = \pi \int_{\frac{\pi}{16}}^{\frac{\pi}{4}} \cos^2(2x)\,dx = \pi \int_{\frac{\pi}{16}}^{\frac{\pi}{4}} \left(\tfrac{1}{2} + \tfrac{1}{2}\cos(4x)\right)\,dx$$

$$= \pi \left[\tfrac{1}{2}x + \tfrac{1}{8}\sin(4x)\right]_{\frac{\pi}{16}}^{\frac{\pi}{4}}$$

$$= \pi \left[\left(\tfrac{\pi}{8} + \tfrac{1}{8}\sin\pi\right) - \left(\tfrac{\pi}{32} + \tfrac{1}{8}\sin\left(\tfrac{\pi}{4}\right)\right)\right]$$

$$= \pi\left(\tfrac{\pi}{8} - \tfrac{\pi}{32} - \tfrac{1}{8}\left(\tfrac{1}{\sqrt{2}}\right)\right)$$

$$= \pi\left(\tfrac{3\pi}{32} - \tfrac{1}{8\sqrt{2}}\right) \text{ units}^3$$

b $V = \pi \int_0^2 (e^{-x} + 4)^2\,dx$

$= \pi \int_0^2 (e^{-2x} + 8e^{-x} + 16)\,dx$

$= \pi \left[\frac{1}{-2}e^{-2x} + \frac{8}{-1}e^{-x} + 16x\right]_0^2$

$= \pi\left[\left(-\frac{1}{2}e^{-4} - 8e^{-2} + 32\right) - \left(-\frac{1}{2} - 8\right)\right]$

$= \pi\left(\dfrac{81}{2} - \dfrac{1}{2e^4} - \dfrac{8}{e^2}\right)$ units^3

≈ 124 units^3

REVIEW SET 22C

1 $a(t) = 6t - 30$ cm s^{-2}

$\therefore \quad v(t) = \int (6t - 30)\,dt$

$= 3t^2 - 30t + c$

But $v(0) = 27$ so $c = 27$

$\therefore \quad v(t) = 3t^2 - 30t + 27$ cm s^{-1}

$\therefore \quad s(t) = \int (3t^2 - 30t + 27)\,dt$

$= t^3 - 15t^2 + 27t + d$

But $s(0) = 0$ so $d = 0$

$\therefore \quad s(t) = t^3 - 15t^2 + 27t$

Also, $v(t) = 3t^2 - 30t + 27$

$= 3(t^2 - 10t + 9)$

$= 3(t - 1)(t - 9)$

which has sign diagram:

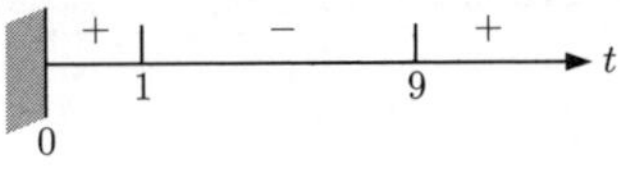

The particle comes to rest for the second time at $t = 9$ seconds.

$s(0) = 0$ cm

$s(1) = 1^3 - 15(1)^2 + 27(1) = 13$ cm

$s(9) = 9^3 - 15(9)^2 + 27(9) = -243$ cm

Motion diagram:

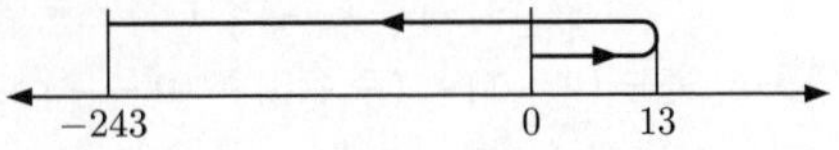

$\therefore$ the total distance travelled is $13 + (13 - (-243)) = 269$ cm

2 **a**

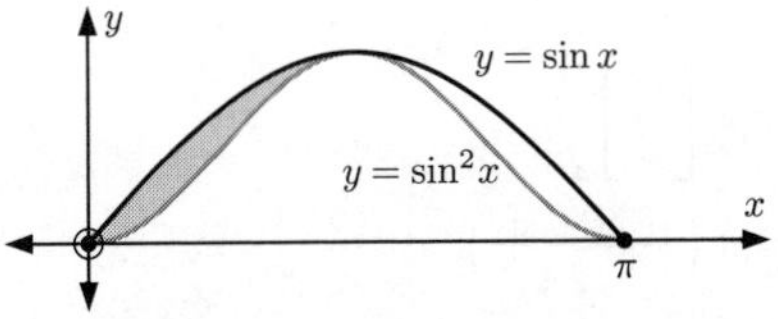

b Area $= \int_0^{\frac{\pi}{2}} (\sin x - \sin^2 x)\,dx$

$= \int_0^{\frac{\pi}{2}} \left(\sin x - \left(\frac{1}{2} - \frac{1}{2}\cos 2x\right)\right)dx$

$= \int_0^{\frac{\pi}{2}} \left(\sin x + \frac{1}{2}\cos 2x - \frac{1}{2}\right)dx$

$= \left[-\cos x + \frac{1}{4}\sin 2x - \frac{1}{2}x\right]_0^{\frac{\pi}{2}}$

$= \left(0 + \frac{1}{4}(0) - \frac{\pi}{4}\right) - (-1 + 0 - 0)$

$= \left(1 - \frac{\pi}{4}\right)$ units^2

3 The area between $x = 0$ and $x = a$ is 2 units^2.

$\therefore \int_0^a e^x \, dx = 2$

$\therefore [e^x]_0^a = 2$

$\therefore e^a - e^0 = 2$

$\therefore e^a = 3$

$\therefore a = \ln 3$

The area between $x = a = \ln 3$ and $x = b$ is 2 units^2.

$\therefore \int_{\ln 3}^b e^x \, dx = 2$

$\therefore [e^x]_{\ln 3}^b = 2$

$\therefore e^b - e^{\ln 3} = 2$

$\therefore e^b - 3 = 2$

$\therefore e^b = 5$

$\therefore b = \ln 5$

4

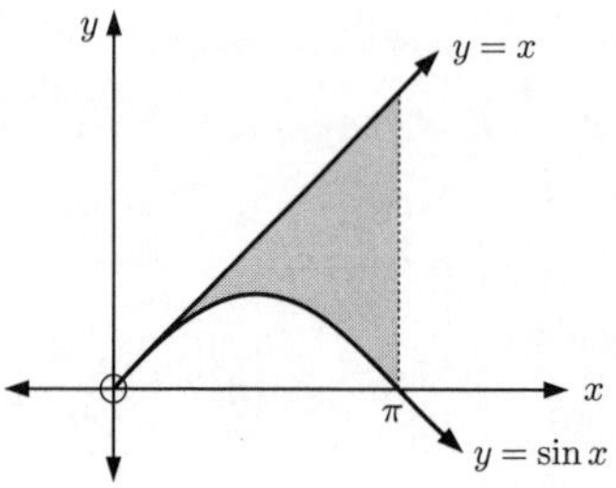

Required area = area of Δ − area under sine curve

$= \frac{1}{2}\pi \times \pi - \int_0^\pi \sin x \, dx$

$= \frac{\pi^2}{2} - [-\cos x]_0^\pi$

$= \frac{\pi^2}{2} - [-\cos \pi + \cos 0]$

$= \left(\frac{\pi^2}{2} - 2\right)$ units^2

5 The graphs meet when $\frac{2}{\pi}x = \sin x$

$\therefore x = -\frac{\pi}{2}, 0, \frac{\pi}{2}$ {using technology}

$\therefore$ area $= \int_{-\frac{\pi}{2}}^0 (\frac{2}{\pi}x - \sin x) \, dx + \int_0^{\frac{\pi}{2}} (\sin x - \frac{2}{\pi}x) \, dx$

$= \left[\frac{x^2}{\pi} + \cos x\right]_{-\frac{\pi}{2}}^0 + \left[-\cos x - \frac{x^2}{\pi}\right]_0^{\frac{\pi}{2}}$

$= (0 + 1) - (\frac{\pi}{4} + 0) + (0 - \frac{\pi}{4}) - (-1 - 0)$

$= (2 - \frac{\pi}{2})$ units^2

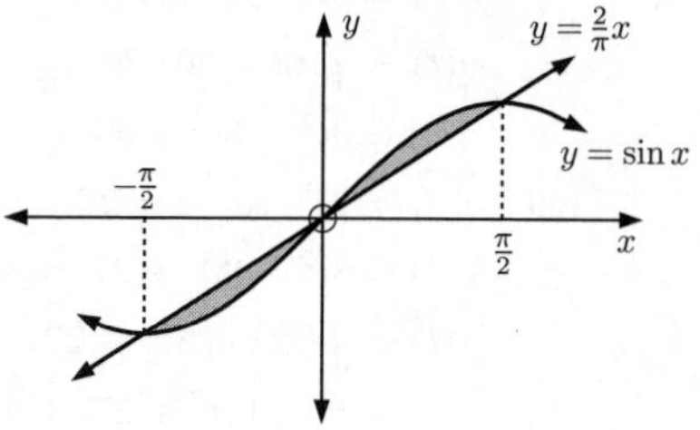

6 The coordinates of B are $(2, 4 + k)$

$\therefore$ area rectangle OABC $= 2 \times (4 + k)$

$= 8 + 2k$

$\therefore$ since the two shaded regions are equal in area, each area is $4 + k$ units^2.

$\therefore \int_0^2 (x^2 + k) \, dx = 4 + k$

$\therefore \left[\frac{x^3}{3} + kx\right]_0^2 = 4 + k$

$\therefore \frac{8}{3} + 2k = 4 + k$

$\therefore k = 4 - \frac{8}{3}$

$\therefore k = \frac{4}{3}$

7 **a**

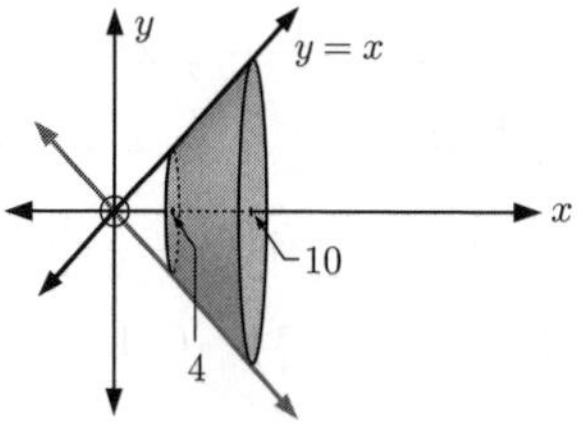

$V = \pi \int_4^{10} x^2 \, dx$

$= \pi \left[\frac{x^3}{3}\right]_4^{10}$

$= \pi \left(\frac{1000}{3} - \frac{64}{3}\right)$

$= \frac{936\pi}{3}$

$= 312\pi$ units^3

b

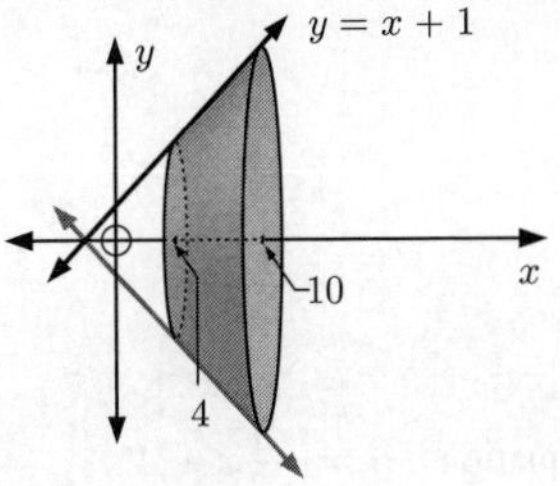

$$V = \pi \int_4^{10} (x+1)^2\, dx$$
$$= \pi \left[\frac{(x+1)^3}{3} \right]_4^{10}$$
$$= \pi \left(\frac{11^3}{3} - \frac{5^3}{3} \right)$$
$$= \frac{1206\pi}{3} = 402\pi \text{ units}^3$$

c

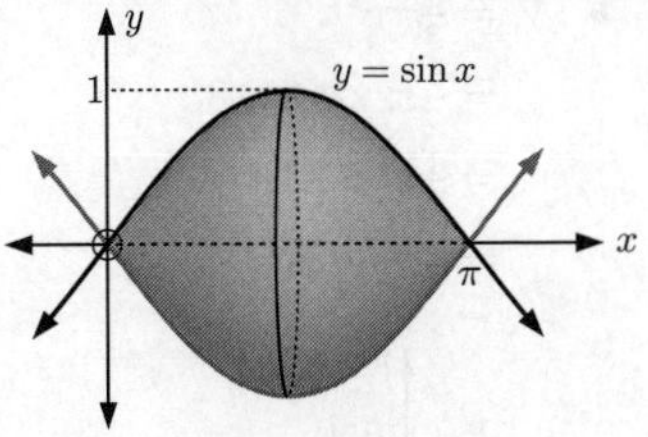

$$V = \pi \int_0^{\pi} \sin^2 x\, dx$$
$$= \pi \int_0^{\pi} \left(\tfrac{1}{2} - \tfrac{1}{2}\cos(2x) \right) dx$$
$$= \pi \left[\tfrac{1}{2}x - \tfrac{1}{2}\left(\tfrac{1}{2}\right)\sin(2x) \right]_0^{\pi}$$
$$= \pi \left(\tfrac{1}{2}\pi - \tfrac{1}{4}\sin 2\pi - 0 \right)$$
$$= \frac{\pi^2}{2} \text{ units}^3$$

d

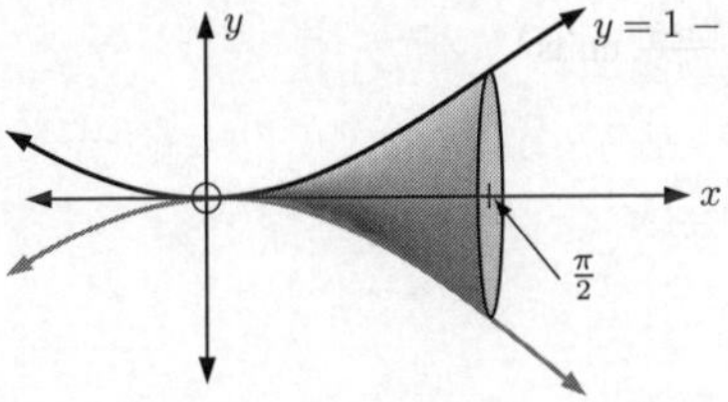

$$V = \pi \int_0^{\frac{\pi}{2}} (1 - \cos x)^2\, dx$$
$$= \pi \int_0^{\frac{\pi}{2}} (1 - 2\cos x + \cos^2 x)\, dx$$
$$= \pi \int_0^{\frac{\pi}{2}} (1 - 2\cos x + \tfrac{1}{2} + \tfrac{1}{2}\cos 2x)\, dx$$
$$= \pi \left[\tfrac{3}{2}x - 2\sin x + \tfrac{1}{4}\sin 2x \right]_0^{\frac{\pi}{2}}$$
$$= \pi \left[\left(\tfrac{3}{2}(\tfrac{\pi}{2}) - 2\sin(\tfrac{\pi}{2}) + \tfrac{1}{4}\sin\pi \right) - \left(\tfrac{3}{2}(0) - 2\sin 0 + \tfrac{1}{4}\sin 0 \right) \right]$$
$$= \pi \left(\tfrac{3\pi}{4} - 2 \right) \text{ units}^3$$

8 $y = \sin x$ and $y = \cos x$ meet where $\sin x = \cos x$

$$\therefore \quad \frac{\sin x}{\cos x} = 1$$
$$\therefore \quad \tan x = 1$$
$$\therefore \quad x = \tfrac{\pi}{4}$$

Hence $V = \pi \int_0^{\frac{\pi}{4}} (\cos^2 x - \sin^2 x)\, dx$

$$= \pi \int_0^{\frac{\pi}{4}} \cos(2x)\, dx$$
$$= \pi \left[\tfrac{1}{2}\sin(2x) \right]_0^{\frac{\pi}{4}}$$
$$= \pi \left(\tfrac{1}{2}\sin\left(\tfrac{\pi}{2}\right) - \tfrac{1}{2}\sin 0 \right)$$
$$= \pi \left(\tfrac{1}{2}(1) - 0 \right)$$
$$= \tfrac{\pi}{2} \text{ units}^3$$

9 $\dfrac{dT}{dx} = \dfrac{k}{x} = kx^{-1}$

$\therefore \quad T = k\ln x + c \quad \{x > 0\}$

When $x = r_1$, $T = T_0$

$\therefore \quad k\ln r_1 + c = T_0$

$\therefore \quad c = T_0 - k\ln r_1$

$\therefore \quad T = k\ln x + T_0 - k\ln r_1$

$$= T_0 + k\ln\left(\frac{x}{r_1}\right)$$

So, when $x = r_2$,

$$T = T_0 + k\ln\left(\frac{r_2}{r_1}\right)$$

$\therefore$ the outer surface has temperature $T_0 + k\ln\left(\dfrac{r_2}{r_1}\right)$

10 **a** $V = \frac{1}{3}\pi r^2 h$

$= \frac{1}{3}\pi \times 4^2 \times 8$

$= \frac{1}{3}\pi \times 128$

$= \frac{128\pi}{3}$ units^3

b

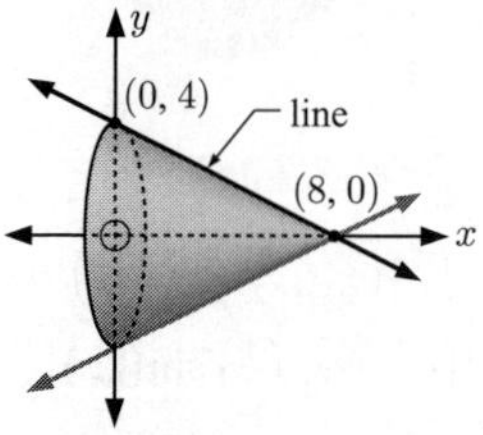

gradient $= \frac{0-4}{8-0} = -\frac{1}{2}$

$\therefore$ the line has equation $y = -\frac{1}{2}x + 4$

$\therefore$ $V = \pi \int_0^8 \left(-\frac{1}{2}x + 4\right)^2 dx$

$= \pi \displaystyle\int_0^8 \left(\frac{x^2}{4} - 4x + 16\right) dx$

$= \pi \left[\frac{x^3}{12} - \frac{4x^2}{2} + 16x\right]_0^8$

$= \pi \left(\frac{128}{3} - 128 + 128 - 0\right)$

$= \frac{128\pi}{3}$ units^3 ✓

Chapter 23

STATISTICAL DISTRIBUTIONS OF DISCRETE RANDOM VARIABLES

EXERCISE 23A

1 **a** The quantity of fat in a sausage is a continuous random variable.

b The mark out of 50 for a geography test is a discrete random variable.

c The weight of a seventeen year old student is a continuous random variable.

d The volume of water in a cup of coffee is a continuous random variable.

e The number of trout in a lake is a discrete random variable.

f The number of hairs on a cat is a discrete random variable.

g The length of hairs on a horse is a continuous random variable.

h The height of a sky-scraper is a continuous random variable.

2 **a** **i** The random variable X is the height of water in the rain gauge.

ii $0 \leqslant X \leqslant 200$ mm **iii** The variable is a continuous random variable.

b **i** The random variable X is the stopping distance.

ii $0 \leqslant X \leqslant 50$ m **iii** The variable is a continuous random variable.

c **i** The random variable X is the number of times that the switch is turned on or off before it fails.

ii X any integer $\geqslant 1$ **iii** The variable is a discrete random variable.

3 **a** Since X is the number of weighing devices that are accurate, $X = 0, 1, 2, 3$ or 4.

b

		YYNN		
		YNYN		
	YYYN	YNNY	NNNY	
	YYNY	NYYN	NNYN	
	YNYY	NYNY	NYNN	
YYYY	NYYY	NNYY	YNNN	NNNN
$(X = 4)$	$(X = 3)$	$(X = 2)$	$(X = 1)$	$(X = 0)$

c **i** If two are accurate then $X = 2$.

ii If at least two are accurate then 2, 3 or 4 are accurate $\therefore$ $X = 2, 3$ or 4.

4 **a** If 3 coins are tossed then the number of heads X can be 0, 1, 2 or 3.

b Suppose H represents heads, T represents tails.

	HHT	TTH	
	HTH	THT	
HHH	THH	HTT	TTT
$(X = 3)$	$(X = 2)$	$(X = 1)$	$(X = 0)$

c $P(X = 0) = \frac{1}{8}$ $P(X = 1) = \frac{3}{8}$

$P(X = 2) = \frac{3}{8}$ $P(X = 3) = \frac{1}{8}$

d

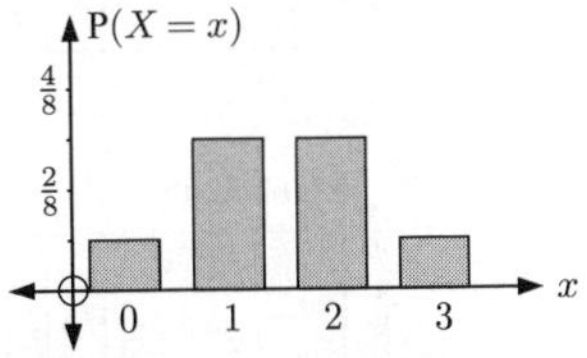

EXERCISE 23B

1 **a**
$$\sum_{x=0}^{2} P(x) = 1$$
$$\therefore \quad 0.3 + k + 0.5 = 1$$
$$\therefore \quad k = 0.2$$

b
$$\sum_{x=0}^{3} P(X = x) = 1$$
$$\therefore \quad k + 2k + 3k + k = 1$$
$$\therefore \quad 7k = 1$$
$$\therefore \quad k = \frac{1}{7}$$

2 **a** $P(2) = 0.1088$ (from table)

b Since this is a probability distribution, $\sum P(x_i) = 1$

$\therefore \quad a + 0.3333 + 0.1088 + 0.0084 + 0.0007 + 0.0000 = 1$

$\therefore \quad a + 0.4512 = 1$

$\therefore \quad a = 0.5488$

This is the probability that Jason does not hit a home run in a game.

c $P(1) + P(2) + P(3) + P(4) + P(5) = 0.3333 + 0.1088 + 0.0084 + 0.0007 + 0.0000$
$= 0.4512$

This represents the probability that Jason hits at least one home run in a game.

d

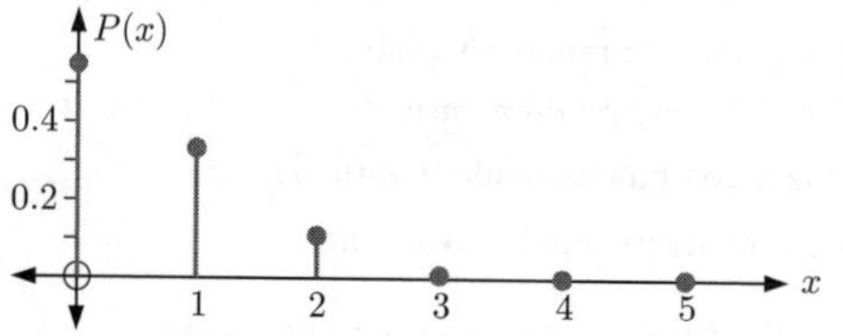

3 **a** Sum of probabilities $\sum P(x_i) = 0.2 + 0.3 + 0.4 + 0.2 = 1.1$

Since this sum $\neq 1$, this is not a valid probability distribution.

b $P(5) = -0.2$, so not all of the probabilities lie in $0 \leqslant P(x_i) \leqslant 1$.

$\therefore$ this is not a valid probability distribution.

4 **a** The random variable represents the number of hits that Sally has in each game.

b $0.07 + 0.14 + k + 0.46 + 0.08 + 0.02 = 1$ $\{$since $\sum P(x_i) = 1\}$

$\therefore \quad k + 0.77 = 1$

$\therefore \quad k = 0.23$

c **i** $\mathrm{P}(X \geqslant 2)$
$= \mathrm{P}(X = 2 \text{ or } X = 3 \text{ or } X = 4 \text{ or } X = 5)$
$= P(2) + P(3) + P(4) + P(5)$
$= 0.23 + 0.46 + 0.08 + 0.02$
$= 0.79$

ii $\mathrm{P}(1 \leqslant X \leqslant 3)$
$= P(1) + P(2) + P(3)$
$= 0.14 + 0.23 + 0.46$
$= 0.83$

5 **a** When rolling a die twice, the sample space is:

roll 1 \ roll 2	1	2	3	4	5	6
6	(6, 1)	(6, 2)	(6, 3)	(6, 4)	(6, 5)	(6, 6)
5	(5, 1)	(5, 2)	(5, 3)	(5, 4)	(5, 5)	(5, 6)
4	(4, 1)	(4, 2)	(4, 3)	(4, 4)	(4, 5)	(4, 6)
3	(3, 1)	(3, 2)	(3, 3)	(3, 4)	(3, 5)	(3, 6)
2	(2, 1)	(2, 2)	(2, 3)	(2, 4)	(2, 5)	(2, 6)
1	(1, 1)	(1, 2)	(1, 3)	(1, 4)	(1, 5)	(1, 6)

b $P(0) = 0$ $\quad P(1) = 0$
$P(2) = \frac{1}{36}$ $\quad P(3) = \frac{2}{36}$
$P(4) = \frac{3}{36}$ $\quad P(5) = \frac{4}{36}$
$P(6) = \frac{5}{36}$ $\quad P(7) = \frac{6}{36}$
$P(8) = \frac{5}{36}$ $\quad P(9) = \frac{4}{36}$
$P(10) = \frac{3}{36}$ $\quad P(11) = \frac{2}{36}$
$P(12) = \frac{1}{36}$

c

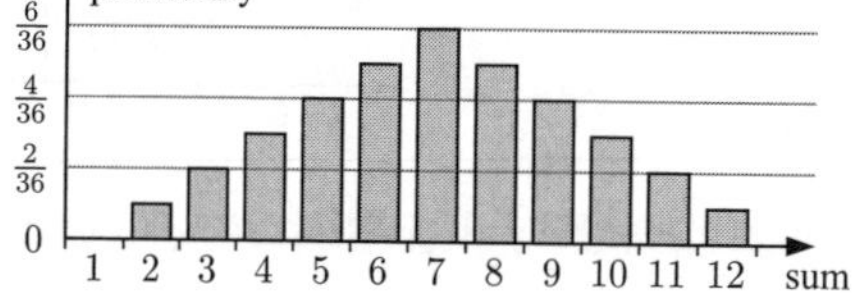

6 **a** $P(x) = k(x+2), \quad x = 1, 2, 3$

$\therefore \quad P(1) = 3k, \quad P(2) = 4k, \quad P(3) = 5k$

Since this is a probability distribution, $3k + 4k + 5k = 1$

$\therefore \quad 12k = 1$

$\therefore \quad k = \frac{1}{12}$

b $P(x) = \dfrac{k}{x+1}, \quad x = 0, 1, 2, 3$

$\therefore \; P(0) = k, \quad P(1) = \dfrac{k}{2},$

$P(2) = \dfrac{k}{3}, \quad P(3) = \dfrac{k}{4}.$

Since $\sum P(x_i) = 1, \quad k + \dfrac{k}{2} + \dfrac{k}{3} + \dfrac{k}{4} = 1$

$\therefore \; \dfrac{12k + 6k + 4k + 3k}{12} = 1$

$\therefore \; \dfrac{25k}{12} = 1$

$\therefore \; k = \frac{12}{25}$

7 **a** $\text{P}(X = x) = k\left(\frac{1}{3}\right)^x \left(\frac{2}{3}\right)^{4-x}, \quad x = 0, 1, 2, 3, 4$

$\text{P}(X = 0) = k\left(\frac{1}{3}\right)^0 \left(\frac{2}{3}\right)^4 = \dfrac{16k}{81}$ $\quad \text{P}(X = 1) = k\left(\frac{1}{3}\right)^1 \left(\frac{2}{3}\right)^3 = \dfrac{8k}{81}$

$\text{P}(X = 2) = k\left(\frac{1}{3}\right)^2 \left(\frac{2}{3}\right)^2 = \dfrac{4k}{81}$ $\quad \text{P}(X = 3) = k\left(\frac{1}{3}\right)^3 \left(\frac{2}{3}\right)^1 = \dfrac{2k}{81}$

$\text{P}(X = 4) = k\left(\frac{1}{3}\right)^4 \left(\frac{2}{3}\right)^0 = \dfrac{k}{81}$

b Since $\sum \text{P}(X = i) = 1,$

$\therefore \; \dfrac{16k}{81} + \dfrac{8k}{81} + \dfrac{4k}{81} + \dfrac{2k}{81} + \dfrac{k}{81} = 1$

$\therefore \; \dfrac{31k}{81} = 1$

$\therefore \; k = \frac{81}{31}$

$\therefore \; k \approx 2.61$

$\therefore \; \text{P}(X \geqslant 2) = P(2) + P(3) + P(4)$

$= \dfrac{4k}{81} + \dfrac{2k}{81} + \dfrac{k}{81}$

$= \dfrac{7k}{81} = \dfrac{7}{81} \times \dfrac{81}{31}$

$= \frac{7}{31} \quad (\approx 0.226)$

8 **a** P(no faulty component)

$= \text{P}(X = 0)$

$= P(0)$

$= \binom{10}{0}(0.04)^0(0.96)^{10-0}$

$= (0.96)^{10}$

≈ 0.665

b P(at least one faulty component)

$= \text{P}(X \geqslant 1)$

$= 1 - \text{P(none are faulty)}$

$\approx 1 - 0.6648$

≈ 0.335

9 **a**

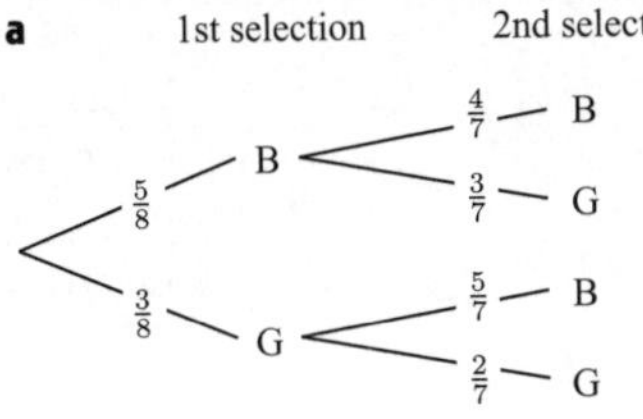

Event	X	Probability
BB	2	$\frac{5}{8} \times \frac{4}{7} = \frac{20}{56}$
BG	1	$\frac{5}{8} \times \frac{3}{7} = \frac{15}{56}$
GB	1	$\frac{3}{8} \times \frac{5}{7} = \frac{15}{56}$
GG	0	$\frac{3}{8} \times \frac{2}{7} = \frac{6}{56}$

x	0	1	2
$\text{P}(X = x)$	$\frac{3}{28}$	$\frac{15}{28}$	$\frac{10}{28}$

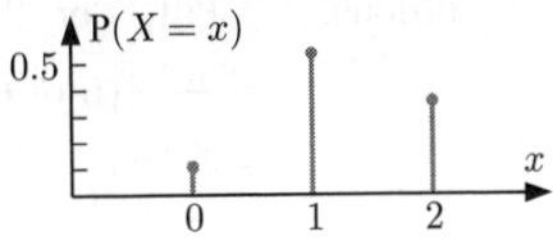

b

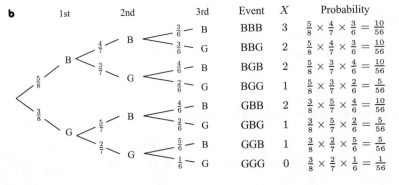

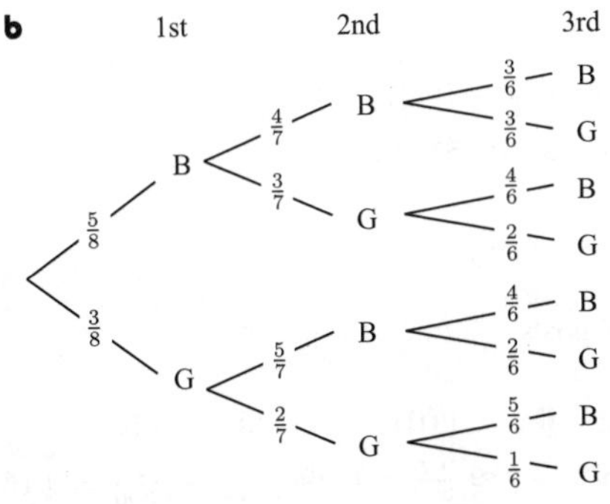

Event	X	Probability
BBB	3	$\frac{5}{8} \times \frac{4}{7} \times \frac{3}{6} = \frac{10}{56}$
BBG	2	$\frac{5}{8} \times \frac{4}{7} \times \frac{3}{6} = \frac{10}{56}$
BGB	2	$\frac{5}{8} \times \frac{3}{7} \times \frac{4}{6} = \frac{10}{56}$
BGG	1	$\frac{5}{8} \times \frac{3}{7} \times \frac{2}{6} = \frac{5}{56}$
GBB	2	$\frac{3}{8} \times \frac{5}{7} \times \frac{4}{6} = \frac{10}{56}$
GBG	1	$\frac{3}{8} \times \frac{5}{7} \times \frac{2}{6} = \frac{5}{56}$
GGB	1	$\frac{3}{8} \times \frac{2}{7} \times \frac{5}{6} = \frac{5}{56}$
GGG	0	$\frac{3}{8} \times \frac{2}{7} \times \frac{1}{6} = \frac{1}{56}$

x	0	1	2	3
$\text{P}(X = x)$	$\frac{1}{56}$	$\frac{15}{56}$	$\frac{30}{56}$	$\frac{10}{56}$

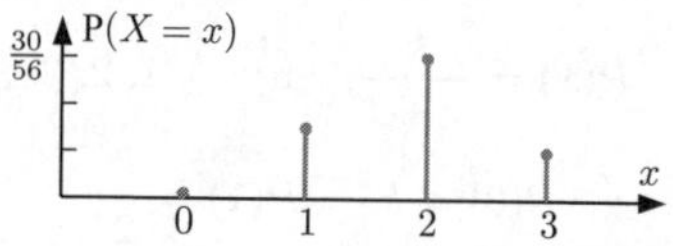

10 **a** Die 2

Die 1	1	2	3	4	5	6
1	2	3	4	5	6	7
2	3	4	5	6	7	8
3	4	5	6	7	8	9
4	5	6	7	8	9	10
5	6	7	8	9	10	11
6	7	8	9	10	11	12

36 possible results

b $\text{P}(D = 7) = \frac{6}{36} = \frac{1}{6}$

c

x	2	3	4	5	6	7	8	9	10	11	12
$P(x)$	$\frac{1}{36}$	$\frac{2}{36}$	$\frac{3}{36}$	$\frac{4}{36}$	$\frac{5}{36}$	$\frac{6}{36}$	$\frac{5}{36}$	$\frac{4}{36}$	$\frac{3}{36}$	$\frac{2}{36}$	$\frac{1}{36}$

d $\text{P}(D \geqslant 8 \mid D \geqslant 6) = \dfrac{\text{P}(D \geqslant 8 \cap D \geqslant 6)}{\text{P}(D \geqslant 6)}$

$= \dfrac{\text{P}(D \geqslant 8)}{\text{P}(D \geqslant 6)}$

$= \frac{15}{36} \div \frac{26}{36}$

$= \frac{15}{26}$

11 **a** Die 2

Die 1	1	2	3	4	5	6
1	0	1	2	3	4	5
2	1	0	1	2	3	4
3	2	1	0	1	2	3
4	3	2	1	0	1	2
5	4	3	2	1	0	1
6	5	4	3	2	1	0

b

n	0	1	2	3	4	5
$\text{P}(N = n)$	$\frac{6}{36}$	$\frac{10}{36}$	$\frac{8}{36}$	$\frac{6}{36}$	$\frac{4}{36}$	$\frac{2}{36}$

c $\text{P}(N = 3) = \frac{6}{36} = \frac{1}{6}$

d $\text{P}(N \geqslant 3 \mid N \geqslant 1)$

$= \dfrac{\text{P}(N \geqslant 3 \cap N \geqslant 1)}{\text{P}(N \geqslant 1)}$

$= \dfrac{\text{P}(N \geqslant 3)}{\text{P}(N \geqslant 1)}$

$= \frac{12}{36} \div \frac{30}{36}$

$= \frac{2}{5}$

EXERCISE 23C

1 P(rain) $= 0.28$ $\therefore$ we would expect rain on $0.28 \times 365.25 \approx 102$ days a year.

2 **a** P(HHH)

$= \frac{1}{2} \times \frac{1}{2} \times \frac{1}{2}$

$= \frac{1}{8}$

b For 200 tosses, we expect $200 \times \frac{1}{8} = 25$ to be '3 heads'.

3 P(double) $=$ P(1,1 or 2,2 or 3,3 or 4,4 or 5,5 or 6,6)

$= \frac{6}{36}$ {6 of the possible 36 outcomes}

$= \frac{1}{6}$

$\therefore$ when rolling the dice 180 times, we expect $180 \times \frac{1}{6} = 30$ doubles.

4

result	*win*
H	\$2
T	$-$\$1

For playing *once*,
we would expect to win $\frac{1}{2} \times \$2 + \frac{1}{2} \times (-\$1) = \$0.50$
$\therefore$ for 3 games we would expect to win \$1.50.

5 Udo could expect to see snow falling on $\frac{3}{7} \times 5 \times 7 = 15$ days.

6 The goalkeeper would expect to save $\frac{3}{10} \times 90 = 27$ goals.

7 **a** $165 + 87 + 48 = 300$

i P(A) $\approx \frac{165}{300} = 0.55$

ii P(B) $\approx \frac{87}{300} = 0.29$

iii P(C) $\approx \frac{48}{300} = 0.16$

b **i** We expect $7500 \times 0.55 = 4125$ to vote for A.
ii We expect $7500 \times 0.29 = 2175$ to vote for B.
iii We expect $7500 \times 0.16 = 1200$ to vote for C.

8 **a** Expect to win $\frac{1}{6} \times €1 + \frac{1}{6} \times €2 + \frac{1}{6} \times €3 + \frac{1}{6} \times €4 + \frac{1}{6} \times €5 + \frac{1}{6} \times €6$
$= \frac{1}{6} \times €21 = €3.50$

b The expected gain is $€3.50 - €4 = -€0.50$
$\therefore$ the player should not play several games, as on each occasion he would expect to lose an average of €0.50.

c **i** The game is fair when the expected gain is 0.
$\therefore\ 3.50 - k = 0$, so $k = 3.50$.

ii $k > 3.50$

9 **a** **i** P(wins \$10)
= P(rolls a 6)
$= \frac{1}{6}$

ii P(wins \$4)
= P(rolls 4 or 5)
$= \frac{2}{6}$ (or $\frac{1}{3}$)

iii P(wins \$1)
= P(rolls 1, 2 or 3)
$= \frac{3}{6}$ (or $\frac{1}{2}$)

b **i** Expectation
$= \frac{2}{6} \times \$4$
$\approx \$1.33$

ii Expectation
$= \frac{3}{6} \times \$1$
$= \$0.50$

iii Expectation
$= \frac{1}{6} \times \$10 + \frac{2}{6} \times \$4 + \frac{3}{6} \times \$1$
$= \frac{1}{6}(\$21) = \$3.50$

c It costs \$4 to play and the expected return is \$3.50.
$\therefore$ you expect to lose \$0.50 per game. (50 cents)

d Over 100 games you expect to lose $100 \times \$0.50 = \$50$.

10

result	*win*
HH	£10
HT or TH	£3
TT	−£5

a Expectation $= \frac{1}{4} \times £10 + \frac{2}{4} \times £3 + \frac{1}{4} \times (-£5) = £2.75$

b Expected win per game (payout) $= £2.75$
$\therefore$ the organiser would charge $£2.75 + £1.00 = £3.75$ to play each game.

11

x_i	0	1	2	3	4	5	> 5
$P(x_i)$	0.54	0.26	0.15	k	0.01	0.01	0.00

a $0.54 + 0.26 + 0.15 + k + 0.01 + 0.01 = 1$
$\therefore\ k + 0.97 = 1$
$\therefore\ k = 0.03$

b $\mu = \sum x_i p_i$
$= 0 \times 0.54 + 1 \times 0.26 + 2 \times 0.15 + 3 \times 0.03 + 4 \times 0.01 + 5 \times 0.01$
$= 0.26 + 0.30 + 0.09 + 0.04 + 0.05$
$= 0.74$ So, over a long period the mean number of deaths per dozen crayfish is 0.74.

12 $P(x) = \dfrac{x^2 + x}{20}$ for $x = 1, 2, 3$

x_i	1	2	3
$P(x_i)$	$\frac{2}{20} = 0.1$	$\frac{6}{20} = 0.3$	$\frac{12}{20} = 0.6$

$\mu = \sum x_i p_i$
$= 1 \times 0.1 + 2 \times 0.3 + 3 \times 0.6$
$= 2.5$

13

Die 2 (columns), Die 1 (rows)

	1	2	3	4	5	6
1	1	2	3	4	5	6
2	2	2	3	4	5	6
3	3	3	3	4	5	6
4	4	4	4	4	5	6
5	5	5	5	5	5	6
6	6	6	6	6	6	6

a

m	1	2	3	4	5	6
$P(M = m)$	$\frac{1}{36}$	$\frac{3}{36}$	$\frac{5}{36}$	$\frac{7}{36}$	$\frac{9}{36}$	$\frac{11}{36}$

b $\mu = \sum m_i p_i$

$= 1\left(\frac{1}{36}\right) + 2\left(\frac{3}{36}\right) + 3\left(\frac{5}{36}\right) + \ldots\ldots + 6\left(\frac{11}{36}\right)$

$= \frac{1}{36} + \frac{6}{36} + \frac{15}{36} + \frac{28}{36} + \frac{45}{36} + \frac{66}{36}$

$= \frac{161}{36}$

≈ 4.47

14 **a** **Exercise 23B Q 10** gives the probability of each individual value of X.

So, $P(X \leqslant 3) = \frac{1}{36} + \frac{2}{36} = \frac{1}{12}$

$P(4 \leqslant X \leqslant 6) = \frac{3}{36} + \frac{4}{36} + \frac{5}{36} = \frac{1}{3}$

$P(7 \leqslant X \leqslant 9) = \frac{6}{36} + \frac{5}{36} + \frac{4}{36} = \frac{5}{12}$

$P(X \geqslant 10) = \frac{3}{36} + \frac{2}{36} + \frac{1}{36} = \frac{1}{6}$

b The expected gain is $\left(\frac{1}{12} + \frac{5}{12}\right)\left(-\frac{a}{3}\right) + \frac{1}{3}(7) + \frac{1}{6}(21) - a$

$= -\frac{a}{6} + \frac{7}{3} + \frac{21}{6} - a$

$= -\frac{7a}{6} + \frac{35}{6} = \frac{1}{6}(35 - 7a)$ dollars, as required.

c The game is fair when the expected gain is 0.

$\therefore\ \frac{1}{6}(35 - 7a) = 0$

$\therefore\ 35 - 7a = 0$

$\therefore\ a = 5$

d If $a = 4$, expected gain $= \frac{1}{6}(35 - 7(4))$

$= \frac{7}{6}$ dollars

So, the people playing would expect to win about \$1.17 per game, which means the organisers expect to lose \$1.17 per game.

e If $a = 6$, expected gain $= \frac{1}{6}(35 - 7(6))$

$= -\frac{7}{6}$ dollars

Expectation from 2406 games is $-\frac{7}{6} \times 2406 = -2807$.

$\therefore$ the organisers would expect to gain \$2807.

EXERCISE 23D

1 **a** The binomial distribution applies, as tossing a coin has two possible outcomes (a head or a tail) and each toss is independent of every other toss.

b The binomial distribution applies, as this is equivalent to tossing one coin 100 times.

c The binomial distribution applies as we can draw out a red or a blue marble with the same chances each time.

d The binomial distribution does not apply as the result of each draw is dependent upon the results of previous draws.

e The binomial distribution does not apply, assuming that ten bolts are drawn without replacement, as we do not have a repetition of independent trials.

2 X is the random variable for the number working night-shift.
$\therefore \quad X = 0, 1, 2, 3, 4, 5, 6, 7 \quad \text{and} \quad X \sim \text{B}(7, 0.35)$.

a $\text{P}(X = 3)$
$= \binom{7}{3}(0.35)^3(0.65)^4$
≈ 0.268

b $\text{P}(X < 4)$
$= \text{P}(X \leqslant 3)$
≈ 0.800

c P(at least 4 work night-shift)
$= \text{P}(X \geqslant 4)$
$= 1 - \text{P}(X \leqslant 3)$
$\approx 1 - 0.800$
≈ 0.200

3 X is the number of faulty items.
$\therefore \quad X = 0, 1, 2, 3, ..., 12 \quad \text{and} \quad X \sim \text{B}(12, 0.06)$.

a $\text{P}(X = 0)$
$= \binom{12}{0}(0.06)^0(0.94)^{12}$
≈ 0.476

b P(at most one is faulty)
$= \text{P}(X \leqslant 1)$
≈ 0.840

c P(at least two are faulty)
$= \text{P}(X \geqslant 2)$
$= 1 - \text{P}(X \leqslant 1)$
≈ 0.160 {from **b**}

d P(less than four are faulty)
$= \text{P}(X < 4)$
$= \text{P}(X \leqslant 3)$
≈ 0.996

4 X is the random variable for the number of apples with a blemish.
$\therefore \quad X = 0, 1, 2, 3, ..., 25 \quad \text{and} \quad X \sim \text{B}(25, 0.05)$.

a $\text{P}(X = 2)$
$= \binom{25}{2}(0.05)^2(0.95)^{23}$
≈ 0.231

b $\text{P}(X \geqslant 1)$
$= 1 - \text{P}(X = 0)$
$= 1 - \binom{25}{0}(0.05)^0(0.95)^{25}$
≈ 0.723

c $\text{E}(X) = np$
$= 25 \times 0.05$
$= 1.25$ apples

5 X is the random variable for the number of times in a week that the bus is on time.
Since it is late 2 in every 5 days, and on time 3 in every 5 days,
$X = 0, 1, 2, 3, 4, 5, 6$ or $7 \quad \text{and} \quad X \sim \text{B}(7, 0.6)$.

a $\text{P}(X = 7)$
$= \binom{7}{7}(0.6)^7(0.4)^0$
≈ 0.0280

b P(on time only on Monday)
$= 0.6 \times (0.4)^6$
$\approx 0.002\,46$

c $\text{P}(X = 6)$
$= \binom{7}{6}(0.6)^6(0.4)$
≈ 0.131

d $\text{P}(X \geqslant 4)$
$= 1 - \text{P}(X \leqslant 3)$
$\approx 1 - 0.290$
≈ 0.710

6 X is the random variable for the number of students with the flu.
$\therefore \quad X = 0, 1, 2, 3, ..., 25 \quad \text{and} \quad X \sim \text{B}(25, 0.3)$.

a $\text{P}(X \geqslant 2)$
$= 1 - \text{P}(X \leqslant 1)$
$\approx 1 - 0.001\,57$
≈ 0.998

b P(test cancelled)
$= \text{P}(X \geqslant 6)$ {20% of 25 = 5}
$= 1 - \text{P}(X \leqslant 5)$
$\approx 1 - 0.193$
≈ 0.807

c Expected absentees from 350 students
$= 0.3 \times 350$
$= 105$ students

7 X is the random variable for the number of successful shots from the free throw line.

$\therefore\ X = 0, 1, 2, 3, ..., 20$ and $X \sim \text{B}(20, 0.94)$.

a **i** $\text{P}(X = 20) = \binom{20}{20}(0.94)^{20}(0.06)^0$
≈ 0.290

ii $\text{P}(X \geqslant 18) = 1 - \text{P}(X \leqslant 17)$
$\approx 1 - 0.115$
≈ 0.885

b $\text{E}(X) = np = 20 \times 0.94$
$= 18.8$ successful throws

REVIEW SET 23A

1 **a** $P(x) = f(x) = \dfrac{a}{x^2 + 1}$ for $x = 0, 1, 2, 3$

x_i	0	1	2	3
$P(x_i)$	a	$\frac{a}{2}$	$\frac{a}{5}$	$\frac{a}{10}$

Now $a + \frac{a}{2} + \frac{a}{5} + \frac{a}{10} = 1$ {as $\sum P(x_i) = 1$}

$\therefore\ 10a + 5a + 2a + a = 10$

$\therefore\ 18a = 10$

$\therefore\ a = \frac{5}{9}$

b $\text{P}(X \geqslant 1) = \text{P}(X = 1, 2 \text{ or } 3)$
$= \text{P}(X = 1) + \text{P}(X = 2) + \text{P}(X = 3)$
$= \frac{5}{18} + \frac{1}{9} + \frac{5}{90}$
$= \frac{4}{9}$

or $\text{P}(X \geqslant 1) = 1 - \text{P}(X < 1)$
$= 1 - \text{P}(X = 0)$
$= 1 - \frac{5}{9}$
$= \frac{4}{9}$

2 Let X be the number of defective toothbrushes.

$\therefore\ X \sim \text{B}(120, 0.04)$ $\mu = np$
$= 120 \times 0.04$
$= 4.8$ defectives

3

x_i	0	1	2	3	4
$P(x_i)$	0.10	0.30	0.45	0.10	k

a If this is a probability distribution then $\sum P(x_i) = 1$

$\therefore\ 0.1 + 0.3 + 0.45 + 0.1 + k = 1$

$\therefore\ 0.95 + k = 1$

$\therefore\ k = 0.05$

b $\mu = \sum x_i p_i$
$= 0(0.1) + 1(0.3) + 2(0.45) + 3(0.1) + 4(0.05)$
$= 0 + 0.3 + 0.9 + 0.3 + 0.2$
$= 1.7$

4 **a**

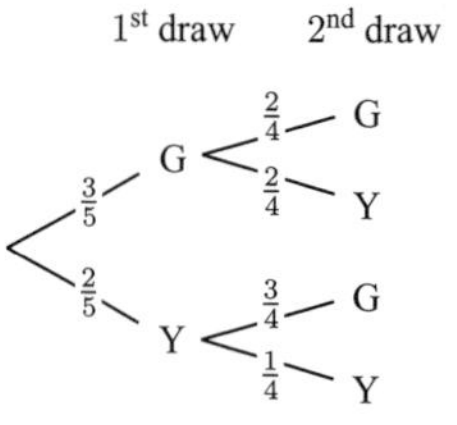

Event	X	Probability
GG	2	$\frac{3}{5} \times \frac{2}{4} = \frac{3}{10}$
GY	1	$\frac{3}{5} \times \frac{2}{4} = \frac{3}{10}$
YG	1	$\frac{2}{5} \times \frac{3}{4} = \frac{3}{10}$
YY	0	$\frac{2}{5} \times \frac{1}{4} = \frac{1}{10}$

i **ii** **iii**

x	0	1	2
$\text{P}(X = x)$	$\frac{1}{10}$	$\frac{3}{5}$	$\frac{3}{10}$

b $\text{E}(X) = 0 \times \frac{1}{10} + 1 \times \frac{3}{5} + 2 \times \frac{3}{10} = \frac{6}{5}$

5 **a**

Result	*Pays*
1	£2
2	£4
3	£6
4	£8
5	£10
6	£12

Expectation
$= \frac{1}{6} \times £2 + \frac{1}{6} \times £4 + \frac{1}{6} \times £6 + \frac{1}{6} \times £8 + \frac{1}{6} \times £10 + \frac{1}{6} \times £12$
$= \frac{1}{6} \times £42$
$= £7$

b Expected gain is $£7 - £8 = -£1$.
$\therefore$ advise Lakshmi against playing several games, as £1 is expected to be lost per game in the long run.

REVIEW SET 23B

1 **a** $P(x) = k\left(\frac{3}{4}\right)^x\left(\frac{1}{4}\right)^{3-x}$ for $x = 0, 1, 2, 3$

$P(0) = k\left(\frac{3}{4}\right)^0\left(\frac{1}{4}\right)^3 = \frac{k}{64}$

$P(1) = k\left(\frac{3}{4}\right)^1\left(\frac{1}{4}\right)^2 = \frac{3k}{64}$

$P(2) = k\left(\frac{3}{4}\right)^2\left(\frac{1}{4}\right)^1 = \frac{9k}{64}$

$P(3) = k\left(\frac{3}{4}\right)^3\left(\frac{1}{4}\right)^0 = \frac{27k}{64}$

x	0	1	2	3
$P(x)$	$\frac{k}{64}$	$\frac{3k}{64}$	$\frac{9k}{64}$	$\frac{27k}{64}$

Now $\frac{k}{64} + \frac{3k}{64} + \frac{9k}{64} + \frac{27k}{64} = 1$ {as $\sum P(x_i) = 1$}
$\therefore \frac{40k}{64} = 1$
$\therefore k = \frac{8}{5}$ or 1.6

b $P(X \geqslant 1) = 1 - P(X = 0)$
$= 1 - \frac{k}{64}$
$= 1 - \frac{1.6}{64}$
$= 0.975$

c $E(X) = \sum x_i p_i$
$= 0 \times \frac{1.6}{64} + 1 \times \frac{3 \times 1.6}{64} + 2 \times \frac{9 \times 1.6}{64} + 3 \times \frac{27 \times 1.6}{64}$
$= 2.55$

2 X is the number of defectives. Then $X \sim B(10, 0.18)$. $X = 0, 1, 2, 3, ..., 10$.

a $P(X = 1)$
$= \binom{10}{1}(0.18)^1(0.82)^9$
≈ 0.302

b $P(X = 2)$
$= \binom{10}{2}(0.18)^2(0.82)^8$
≈ 0.298

c $P(X \geqslant 2)$
$= 1 - P(X \leqslant 1)$
$\approx 1 - 0.439$
≈ 0.561

3 Expected number of major knee surgeries $= np$
$= 487 \times 0.0132$
≈ 6.43

4 If X is the number of X-rays which show the fracture, then $X = 0, 1, 2, 3, 4$ and $X \sim B(4, 0.96)$.

a $P(X = 4)$
$= \binom{4}{4}(0.96)^4(0.04)^0$
≈ 0.849

b $P(X = 0)$
$= \binom{4}{0}(0.96)^0(0.04)^4$
$\approx 2.56 \times 10^{-6}$

c $P(X \geqslant 3)$
$= 1 - P(X \leqslant 2)$
$\approx 1 - 0.009\,10$
≈ 0.991

d $P(X = 1)$
$= \binom{4}{1}(0.96)^1(0.04)^3$
$\approx 0.000\,246$

5 X is the number of visitors who make a voluntary donation upon entry.
Then $X = 0, 1, 2, 3, ..., 175$ and $X \sim \text{B}(175, 0.24)$.

a $\text{E}(X) = np$
$= 175 \times 0.24$
$= 42$

b $\text{P}(X < 40) = \text{P}(X \leqslant 39)$
≈ 0.334

6 X is the number of players who turn up to a game.
Then $X = 0, 1, 2, 3, ..., 8$ and $X \sim \text{B}(8, 0.75)$.

a **i** $\text{P}(X = 8) = \binom{8}{8}(0.75)^8(0.25)^0$
≈ 0.100

ii $\text{P(team has to forfeit)} = \text{P}(X \leqslant 4)$
≈ 0.114

b Expected number of games forfeited in $30 = np$
$\approx 30 \times 0.1138$ {from **a ii**}
≈ 3.41

REVIEW SET 23C

1 **a** $\sum \text{P}(x_i) = 1$

$\therefore \frac{k}{2 \times 1} + \frac{k}{2 \times 2} + \frac{k}{2 \times 3} = 1$

$\therefore \frac{k}{2} + \frac{k}{4} + \frac{k}{6} = 1$

$\therefore 6k + 3k + 2k = 12$

$\therefore 11k = 12$

$\therefore k = \frac{12}{11}$

b $\sum \text{P}(x_i) = 1$

$\therefore \frac{k}{2} + 0.2 + k^2 + 0.3 = 1$

$\therefore 2k^2 + k + 1 = 2$

$\therefore 2k^2 + k - 1 = 0$

$\therefore (2k - 1)(k + 1) = 0$

$\therefore k = -1, \frac{1}{2}$

If $k = -1$, then $P(0) = \frac{-1}{2} < 0$, so $P(x)$ would not be a valid probability distribution function.

$\therefore k = \frac{1}{2}$

2 **a** $\text{P}(X = x) = \binom{4}{x}\left(\frac{1}{2}\right)^x\left(\frac{1}{2}\right)^{4-x}$

$\therefore \text{P}(X = 0) = \binom{4}{0}\left(\frac{1}{2}\right)^0\left(\frac{1}{2}\right)^4 = 0.0625$

$\text{P}(X = 1) = \binom{4}{1}\left(\frac{1}{2}\right)^1\left(\frac{1}{2}\right)^3 = 0.25$

$\text{P}(X = 2) = \binom{4}{2}\left(\frac{1}{2}\right)^2\left(\frac{1}{2}\right)^2 = 0.375$

$\text{P}(X = 3) = \binom{4}{3}\left(\frac{1}{2}\right)^3\left(\frac{1}{2}\right)^1 = 0.25$

$\text{P}(X = 4) = \binom{4}{4}\left(\frac{1}{2}\right)^4\left(\frac{1}{2}\right)^0 = 0.0625$

x	0	1	2	3	4
$\text{P}(X = x)$	0.0625	0.25	0.375	0.25	0.0625

b $\mu = \sum x_i \text{P}(X = x_i)$
$= 0 \times 0.0625 + 1 \times 0.25 + 2 \times 0.375 + 3 \times 0.25 + 4 \times 0.0625$
$= 2$

3 $X \sim \text{B}(1200, 0.4)$ So, mean of $X = np$
$= 1200 \times 0.4$
$= 480$

4 X is the number of trees that survive the first year.
$\therefore X = 0, 1, 2, 3, 4, 5$ and $X \sim \text{B}(5, 0.4)$

a $\text{P}(X = 1)$
$= \binom{5}{1}(0.4)^1(0.6)^4$
≈ 0.259

b $\text{P}(X \leqslant 1)$
≈ 0.337

c $\text{P}(X \geqslant 1)$
$= 1 - \text{P}(X = 0)$
$= 1 - \binom{5}{0}(0.4)^0(0.6)^5$
≈ 0.922

5 **a** **i** In the numbers 1 to 20, there are 10 even numbers.
However, '4' and '16' are square numbers, so 8 of the numbers in the bag win \$3.
$\therefore$ P(player wins \$3) $= \frac{8}{20} = \frac{2}{5}$

ii 1, 4, 9 and 16 are the only square numbers in the bag.
But '4' and '16' are even, so 2 of the numbers in the bag win \$6.
$\therefore$ P(player wins \$6) $= \frac{2}{20} = \frac{1}{10}$

iii 2 numbers are both even and square (4 and 16).
$\therefore$ P(player wins \$9) $= \frac{2}{20} = \frac{1}{10}$

b Expected winnings $= \frac{2}{5} \times \$3 + \frac{1}{10} \times \$6 + \frac{1}{10} \times \$9$
$= \$2.70$

$\therefore$ for the game to be fair, players should be charged \$2.70 per game.

Chapter 24

STATISTICAL DISTRIBUTIONS OF CONTINUOUS RANDOM VARIABLES

EXERCISE 24A

1 **a**

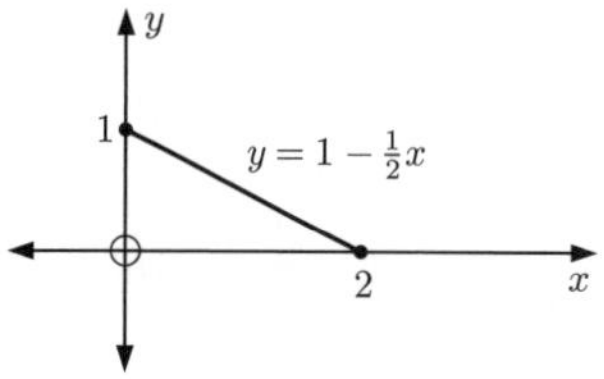

From the graph, we see that $f(x) \geqslant 0$ for $0 \leqslant x \leqslant 2$.

Area $= \frac{1}{2} \times 2 \times 1 = 1$ ✓

or $\int_0^2 (1 - \frac{1}{2}x)\, dx = \left[x - \frac{1}{4}x^2\right]_0^2 = \left(2 - \frac{1}{4}(4)\right)$
$= 1$ ✓

$\therefore$ $f(x)$ is a probability density function.

b $\mu = \int_0^2 x\, f(x)\, dx$
$= \int_0^2 \left(x - \frac{1}{2}x^2\right) dx$
$= \left[\frac{1}{2}x^2 - \frac{1}{6}x^3\right]_0^2$
$= \left(\frac{1}{2}(4) - \frac{1}{6}(8)\right) - (0 - 0)$
$= \frac{2}{3}$

2 **a** $\int_0^4 ax(x-4)\, dx = 1$
$\therefore\ a \int_0^4 (x^2 - 4x)\, dx = 1$
$\therefore\ a \left[\frac{x^3}{3} - \frac{4x^2}{2}\right]_0^4 = 1$
$\therefore\ a\left(\frac{64}{3} - 32\right) = 1$
$\therefore\ a\left(\frac{-32}{3}\right) = 1$
$\therefore\ a = -\frac{3}{32}$

b $f(x) = -\frac{3}{32}x(x-4), \quad 0 \leqslant x \leqslant 4$

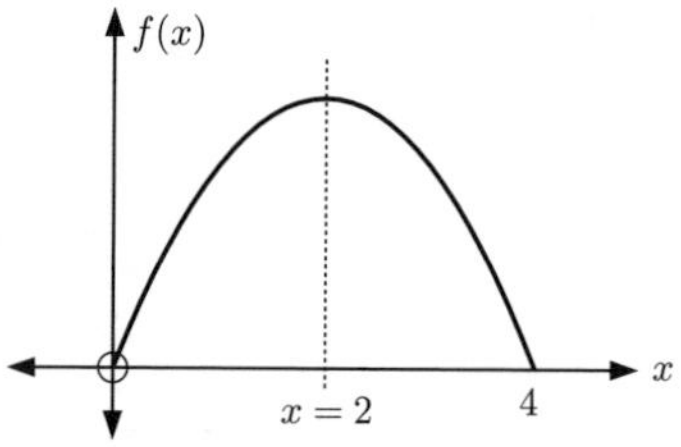

c $P(0 \leqslant X \leqslant 1)$
$= \int_0^1 -\frac{3}{32}x(x-4)\, dx$
$= \int_0^1 \left(-\frac{3}{32}x^2 + \frac{3}{8}x\right) dx$
$= \left[-\frac{1}{32}x^3 + \frac{3}{16}x^2\right]_0^1$
$= -\frac{1}{32} + \frac{3}{16}$
$= \frac{5}{32}$

d $\mu = \int_0^4 x\, f(x)\, dx$
$= \int_0^4 -\frac{3}{32}x^2(x-4)\, dx$
$= -\frac{3}{32} \int_0^4 (x^3 - 4x^2)\, dx$
$= -\frac{3}{32} \left[\frac{1}{4}x^4 - \frac{4}{3}x^3\right]_0^4$
$= -\frac{3}{32} \left(\frac{1}{4}(4)^4 - \frac{4}{3}(4)^3\right)$
$= -\frac{3}{32} \left(4^3 - \frac{4}{3} \times 4^3\right)$
$= -\frac{3}{32} \left(-\frac{64}{3}\right) = 2$

3 **a** $\int_0^5 kx^2(x-6)\, dx = 1$
$\therefore\ k \int_0^5 (x^3 - 6x^2)\, dx = 1$
$\therefore\ k \left[\frac{1}{4}x^4 - \frac{6}{3}x^3\right]_0^5 = 1$
$\therefore\ k\left(\frac{625}{4} - 250\right) = 1$
$\therefore\ k\left(\frac{-375}{4}\right) = 1$
$\therefore\ k = -\frac{4}{375}$

b $E(X) = \int_0^5 x\, f(x)\, dx$
$= \int_0^5 -\frac{4}{375}\, x^3(x-6)\, dx$
$= 3\frac{1}{3}$ {using technology}

c $P(3 \leqslant X \leqslant 5) = \int_3^5 -\frac{4}{375}x^2(x-6)\, dx$
$= 0.64$ {using technology}

4 $\text{P}\left(X \leqslant \frac{2}{3}\right) = \frac{1}{243}$

$\therefore \int_0^{\frac{2}{3}} ax^4\,dx = \frac{1}{243}$

$\therefore \left[\frac{1}{5}ax^5\right]_0^{\frac{2}{3}} = \frac{1}{243}$

$\therefore \frac{1}{5}a \times \left(\frac{2}{3}\right)^5 = \frac{1}{243}$

$\therefore \frac{1}{5}a \times \frac{32}{243} = \frac{1}{243}$

$\therefore a = \frac{5}{32}$

So, $f(x) = \frac{5}{32}x^4, \quad 0 \leqslant x \leqslant k$

$\therefore \int_0^k \frac{5}{32}x^4\,dx = 1$

$\therefore \left[\frac{1}{32}x^5\right]_0^k = 1$

$\therefore \frac{1}{32}k^5 = 1$

$\therefore k^5 = 32$

$\therefore k = 2$

5 **a** $\int_0^4 k\sqrt{x}\,dx = 1$

$\therefore k\int_0^4 x^{\frac{1}{2}}\,dx = 1$

$\therefore k\left[\frac{2}{3}x^{\frac{3}{2}}\right]_0^4 = 1$

$\therefore \frac{2}{3}k\left(4^{\frac{3}{2}} - 0\right) = 1$

$\therefore \frac{2}{3}k(8) = 1$

$\therefore k = \frac{3}{16}$

b $\mu = \int_0^4 x\,f(x)\,dx$

$= \int_0^4 \frac{3}{16}x^{\frac{3}{2}}\,dx$

$= \frac{3}{16}\left[\frac{2}{5}x^{\frac{5}{2}}\right]_0^4$

$= \frac{3}{40}\left(4^{\frac{5}{2}} - 0^{\frac{5}{2}}\right)$

$= \frac{12}{5} = 2.4$ hours or 2 hours 24 minutes

c $\text{P}(1 \leqslant X \leqslant 2.5) = \int_1^{2.5} \frac{3}{16}\sqrt{x}\,dx$

≈ 0.369

{using technology}

d $\text{P}(0 \leqslant X \leqslant a) = \text{P}(a \leqslant X \leqslant 4)$

Since $\text{P}(0 \leqslant X \leqslant 4) = 1$,

$\text{P}(0 \leqslant X \leqslant a) + \text{P}(a \leqslant X \leqslant 4) = 1$

$\therefore \text{P}(0 \leqslant X \leqslant a) = \frac{1}{2} = \text{P}(a \leqslant X \leqslant 4)$

$\therefore \int_0^a \frac{3}{16}x^{\frac{1}{2}}\,dx = \frac{1}{2}$

$\therefore \frac{3}{8}\left[\frac{2}{3}x^{\frac{3}{2}}\right]_0^a = 1$

$\therefore \frac{1}{4}\left(a^{\frac{3}{2}} - 0^{\frac{3}{2}}\right) = 1$

$\therefore a^{\frac{3}{2}} = 4$

$\therefore a = \sqrt[3]{16} \approx 2.52$ hours

EXERCISE 24B.1

1

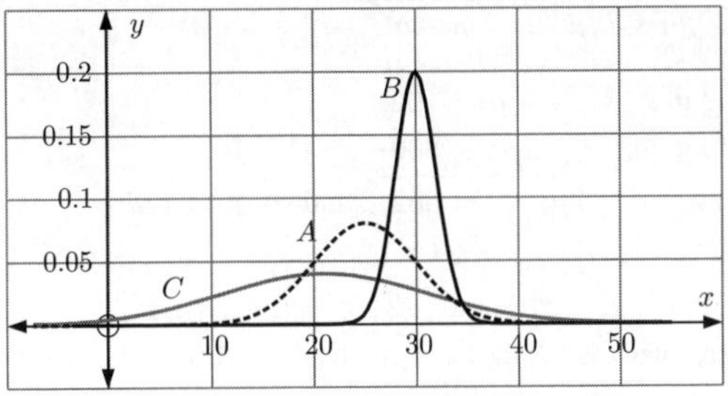

2 **a, b** The mean volume (or diameter) is likely to occur most often with variations around the mean occurring symmetrically as a result of random variations in the production process.

3

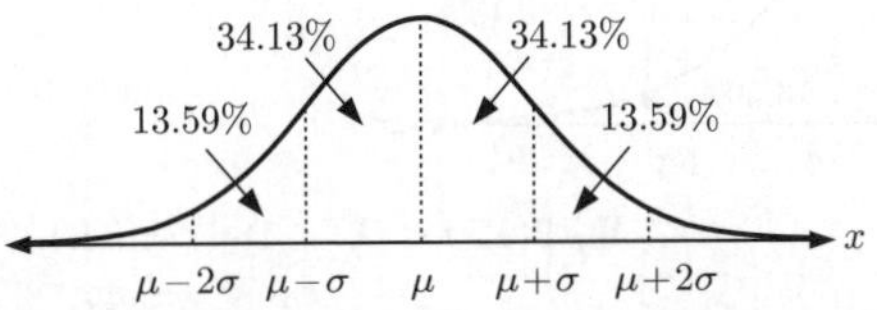

a P(value between $\mu - \sigma$ and $\mu + \sigma$)

$\approx 34.13\% + 34.13\%$

≈ 0.683

b P(value between μ and $\mu + 2\sigma$)

$\approx 34.13\% + 13.59\%$

≈ 0.477

4 Without fertiliser

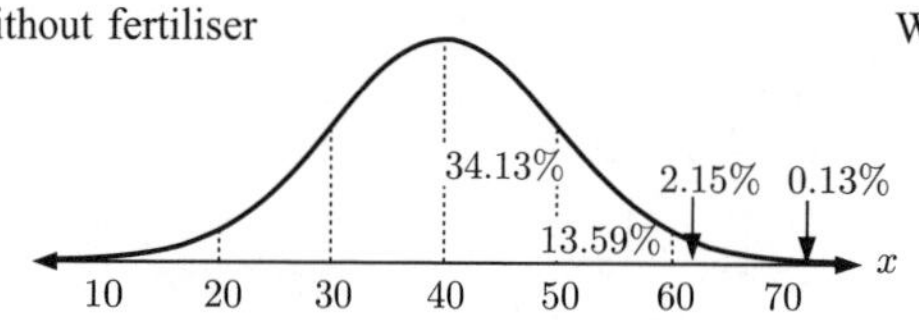

With fertiliser

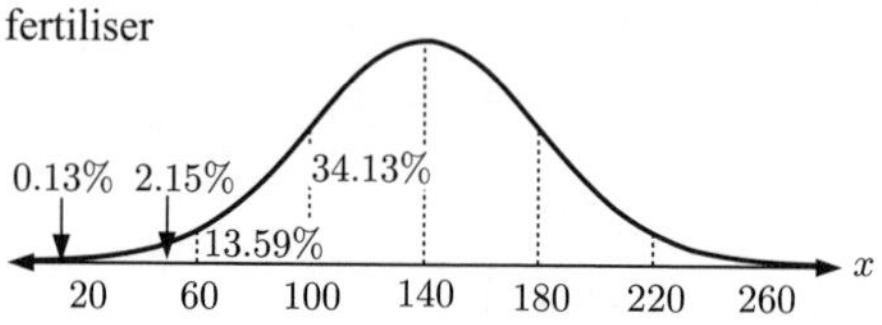

a P(without and < 50)
$\approx 50\% + 34.13\%$
$\approx 84.1\%$

b P(with and < 60)
$\approx 0.13\% + 2.15\%$
$\approx 2.28\%$

c **i** P(with and $20 \leqslant X \leqslant 60$)
$\approx 2.15\%$

ii P(without and $20 \leqslant X \leqslant 60$)
$\approx 2(34.13\% + 13.59\%)$
$\approx 95.4\%$

d **i** P(with and $X \geqslant 60$)
$\approx 13.59\% + 34.13\% + 50\%$
$\approx 97.7\%$

ii P(without and $X \geqslant 60$)
$\approx 2.15\% + 0.13\%$
$\approx 2.28\%$

5

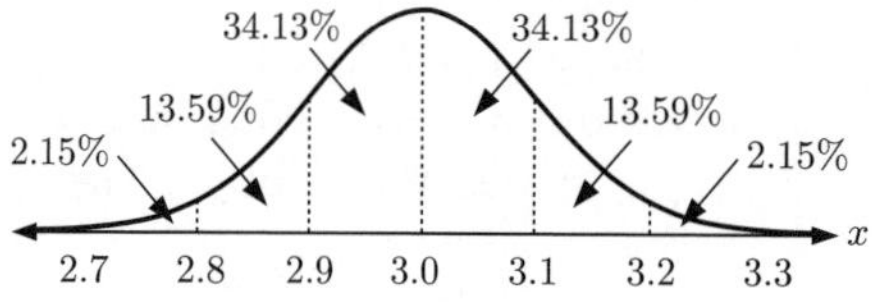

a P(value is within 2 standard deviations of the mean)
$= \text{P}(2.8 \leqslant X \leqslant 3.2)$
$\approx 13.59\% + 34.13\% + 34.13\% + 13.59\%$
≈ 0.954

b The value 1.7 standard deviations below the mean is $X = 3 - 1.7 \times 0.1 = 2.83$

6 **a**

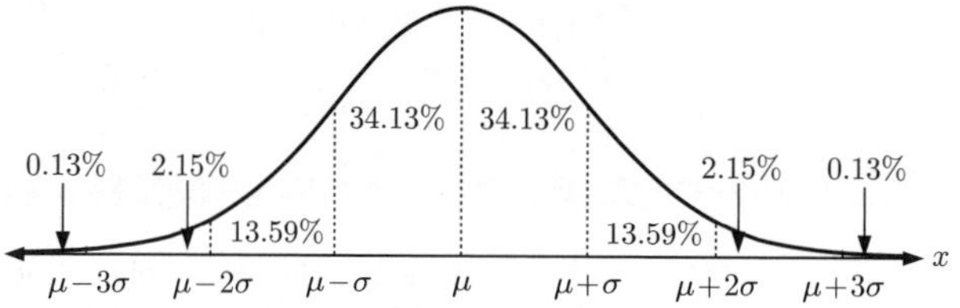

84% of the crop weigh more than 152 g $\therefore \quad \mu - \sigma = 152$
16% of the crop weigh more than 200 g $\therefore \quad \mu + \sigma = 200$ .... (1)
Adding: $2\mu = 352$, and so $\mu = 176$ g
Substituting $\mu = 176$ into (1) gives $\sigma = 200 - \mu = 24$ g.

b For $\mu = 176$ g and $\sigma = 24$ g, $152 \text{ g} = \mu - \sigma$, and $224 \text{ g} = \mu + 2\sigma$.
$\therefore$ between 152 g and 224 g, the percentage is $34.13\% + 34.13\% + 13.59\% \approx 81.9\%$

7

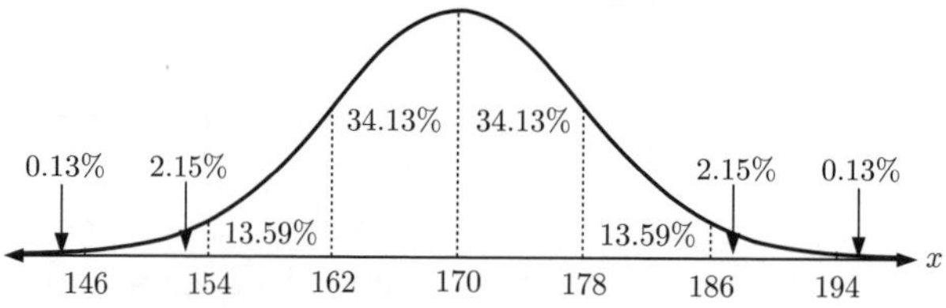

a **i** $\text{P}(162 < X < 170) \approx 34.1\%$

ii $\text{P}(170 < X < 186) \approx 34.13\% + 13.59\%$
$\approx 47.7\%$

b **i** $P(178 < X < 186)$
$\approx 13.59\%$
≈ 0.136

ii $P(X < 162)$
$\approx 1 - (0.5 + 0.3413)$
≈ 0.159

iii $P(X < 154)$
$\approx 0.0215 + 0.0013$
≈ 0.0228

iv $P(X > 162)$
$\approx 1 - 0.159$ {using **b ii**}
≈ 0.841

c 16% of students are taller than 178 cm $\{13.59\% + 2.15\% + 0.13\% \approx 16\%\}$
$\therefore \; k = 178$

8

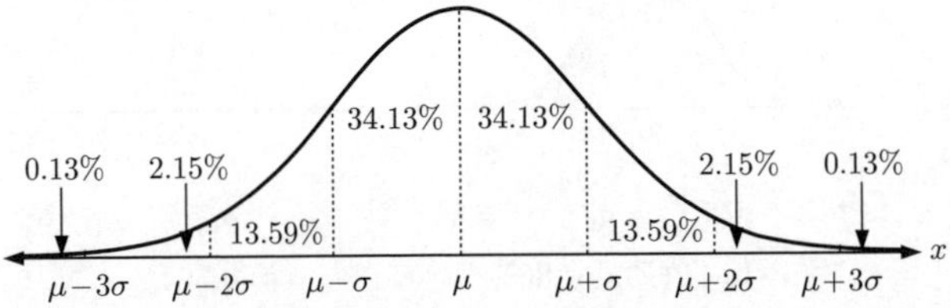

a 97.72% of 13 year old boys are taller than 131 cm $\therefore \; \mu - 2\sigma = 131$
2.28% of 13 year old boys are taller than 179 cm $\therefore \; \mu + 2\sigma = 179$ .... (1)
Adding: $2\mu = 310$, and so $\mu = 155$ cm

Substituting $\mu = 155$ cm into (1) gives $\sigma = \dfrac{179 - \mu}{2} = \dfrac{24}{2} = 12$ cm

b For $\mu = 155$ cm and $\sigma = 12$ cm, $143 \text{ cm} = \mu - \sigma$, and $191 \text{ cm} = \mu + 3\sigma$
$\therefore$ between 143 cm and 191 cm, the percentage is $34.13\% + 34.13\% + 13.59\% + 2.15\% \approx 84.0\%$
So the probability is 0.84.

9

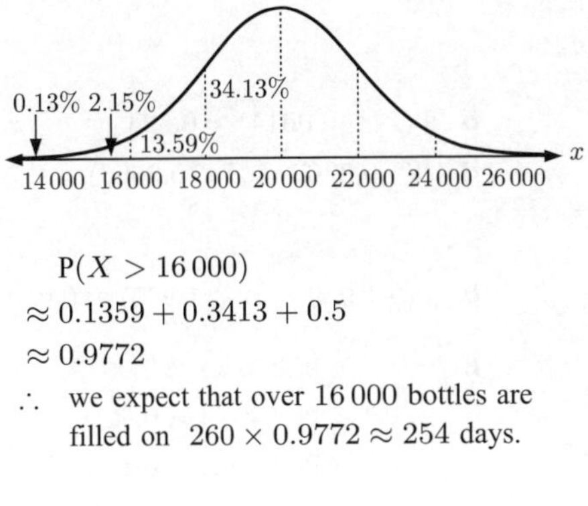

a $P(X < 18\,000)$
$\approx 1 - 0.5 - 0.3413$
≈ 0.1587
$\therefore$ we expect that less than 18 000 bottles are filled on $260 \times 0.1587 \approx 41$ days.

b $P(X > 16\,000)$
$\approx 0.1359 + 0.3413 + 0.5$
≈ 0.9772
$\therefore$ we expect that over 16 000 bottles are filled on $260 \times 0.9772 \approx 254$ days.

c $P(18\,000 \leqslant X \leqslant 24\,000)$
$\approx 0.3413 \times 2 + 0.1359$
≈ 0.8185
$\therefore$ we expect that between 18 000 and 24 000 bottles are filled on 260×0.8185
≈ 213 days.

EXERCISE 24B.2

1 **a** 0.341 **b** 0.383 **c** 0.106

2 **a** 0.341 **b** 0.264 **c** 0.212 **d** 0.945 **e** 0.579 **f** 0.383

3 **a** $P(X < a) = 0.378$
$\therefore \; a \approx 21.4$

b $P(X \geqslant a) = 0.592$
$\therefore \; P(X < a) = 1 - 0.592 = 0.408$
$\therefore \; a \approx 21.8$

c $P(23 - a < X < 23 + a) = 0.427$

$\therefore\ P(23 < X < 23 + a) = \frac{1}{2}(0.427) = 0.2135$

$\therefore\ P(X < 23 + a) = 0.5 + 0.2135 = 0.7135$

$\therefore\ 23 + a \approx 25.82$

$\therefore\ a \approx 2.82$

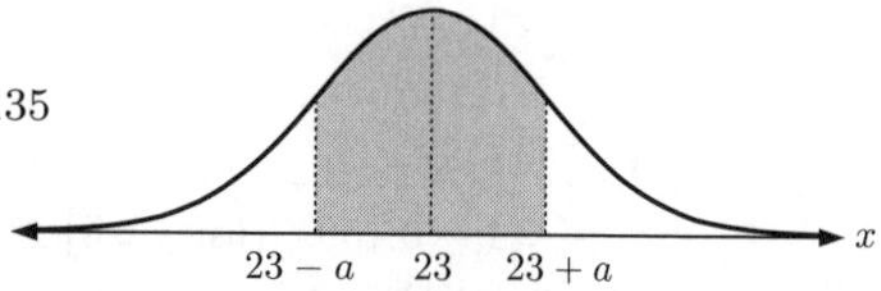

EXERCISE 24C.1

1 **a**

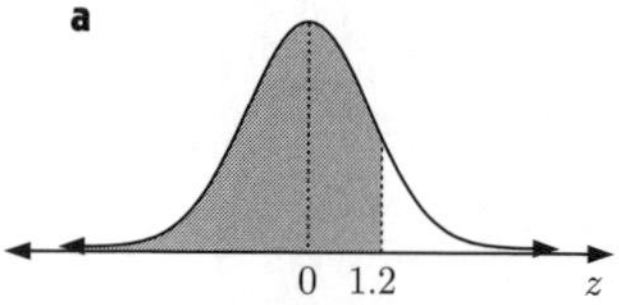

$P(Z \leqslant 1.2)$
≈ 0.8849
≈ 0.885

b

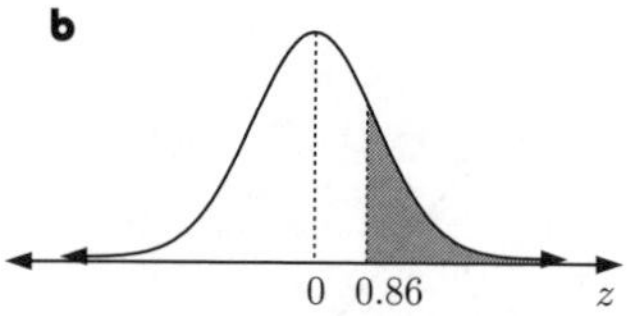

$P(Z \geqslant 0.86)$
$= 1 - P(Z < 0.86)$
$\approx 1 - 0.8051$
≈ 0.195

c

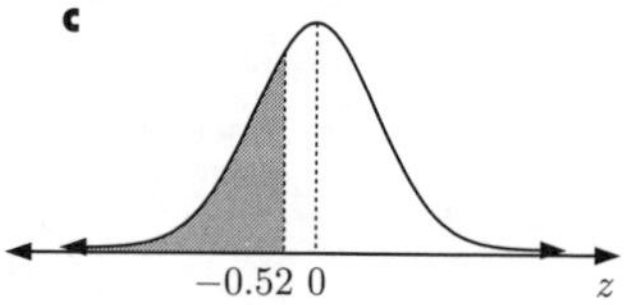

$P(Z \leqslant -0.52)$
≈ 0.3015
≈ 0.302

d

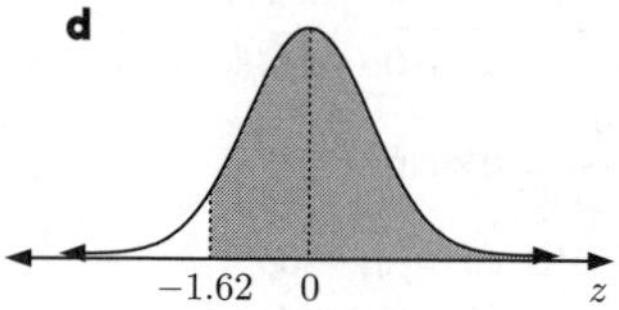

$P(Z \geqslant -1.62)$
$= 1 - P(Z < -1.62)$
$\approx 1 - 0.0526$
≈ 0.947

e

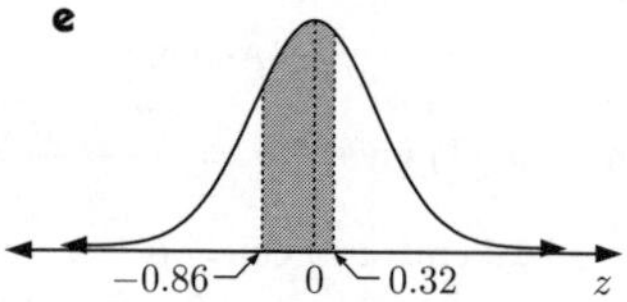

$P(-0.86 \leqslant Z \leqslant 0.32)$
$= P(Z \leqslant 0.32) - P(Z \leqslant -0.86)$
$\approx 0.6255 - 0.1949$
≈ 0.431

2 **a** $P(Z \geqslant 0.837) \approx 0.201$ **b** $P(Z \leqslant 0.0614) \approx 0.524$

c $P(Z \geqslant -0.876) \approx 0.809$ **d** $P(-0.3862 \leqslant Z \leqslant 0.2506) \approx 0.249$

e $P(-2.367 \leqslant Z \leqslant -0.6503) \approx 0.249$

3 **a** $P(-0.5 < Z < 0.5) \approx 0.383$ **b** $P(-1.960 < Z < 1.960) \approx 0.950$

4 **a** $P(Z \leqslant a) = 0.95$

$\therefore\ a \approx 1.645$ {searching in tables

$\therefore\ a \approx 1.64$ or using technology}

b $P(Z \geqslant a) = 0.90$

$\therefore\ 1 - P(Z < a) = 0.90$

$\therefore\ P(Z < a) = 0.1$

$\therefore\ a \approx -1.28 - \frac{3}{18}(0.01)$

$\therefore\ a \approx -1.282$

$\therefore\ a \approx -1.28$

5 **a** For Physics, $Z = \dfrac{83 - 78}{10.8} \approx 0.463$ For Chemistry, $Z = \dfrac{77 - 72}{11.6} \approx 0.431$

For Maths, $Z = \dfrac{84 - 74}{10.1} \approx 0.990$ For German, $Z = \dfrac{91 - 86}{9.6} \approx 0.521$

For Biology, $Z = \dfrac{72 - 62}{12.2} \approx 0.820$

b Maths, Biology, German, Physics, Chemistry

6 Z-score for algebra $= \dfrac{56-50.2}{15.8} \approx 0.3671$ $\qquad$ Z-score for geometry $= \dfrac{x-58.7}{18.7}$

$\therefore$ we need to solve $\dfrac{x-58.7}{18.7} = 0.3671$

$\therefore \quad x - 58.7 \approx 6.86$

$\therefore \quad x \approx 65.6$ $\qquad$ So, Pedro needs a result of 65.6%.

EXERCISE 24C.2

1 X is normal with mean 70, standard deviation 4.

a $\mathrm{P}(X \geqslant 74)$

$= \mathrm{P}\left(\dfrac{X-70}{4} \geqslant \dfrac{74-70}{4}\right)$

$= \mathrm{P}(Z \geqslant 1)$

$= 1 - \mathrm{P}(Z < 1)$

$\approx 1 - 0.8413$

≈ 0.159

b $\mathrm{P}(X \leqslant 68)$

$= \mathrm{P}\left(\dfrac{X-70}{4} \geqslant \dfrac{68-70}{4}\right)$

$= \mathrm{P}(Z \leqslant -\frac{1}{2})$

≈ 0.309

c $\mathrm{P}(60.6 \leqslant X \leqslant 68.4)$

$= \mathrm{P}\left(\dfrac{60.6-70}{4} \leqslant \dfrac{X-70}{4} \leqslant \dfrac{68.4-70}{4}\right)$

$= \mathrm{P}(-2.35 \leqslant Z \leqslant -0.4)$

$\approx 0.3446 - 0.0094$

≈ 0.335

2 X is normal with mean 58.3 and standard deviation 8.96.

a $\mathrm{P}(X \geqslant 61.8)$

$= \mathrm{P}\left(\dfrac{X-58.3}{8.96} \geqslant \dfrac{61.8-58.3}{8.96}\right)$

$= \mathrm{P}(Z \geqslant 0.390\,625)$

≈ 0.348

b $\mathrm{P}(X \leqslant 54.2)$

$= \mathrm{P}\left(\dfrac{X-58.3}{8.96} \leqslant \dfrac{54.2-58.3}{8.96}\right)$

$\approx \mathrm{P}(Z \leqslant -0.4576)$

≈ 0.324

c $\mathrm{P}(50.67 \leqslant X \leqslant 68.92)$

$= \mathrm{P}\left(\dfrac{50.67-58.3}{8.96} \leqslant \dfrac{X-58.3}{8.96} \leqslant \dfrac{68.92-58.3}{8.96}\right)$

$\approx \mathrm{P}(-0.851\,56 \leqslant Z \leqslant 1.1853)$

≈ 0.685

3 L is normal with mean 50.2 mm and standard deviation 0.93 mm.

a $\mathrm{P}(L \geqslant 50)$

$= \mathrm{P}\left(\dfrac{L-50.2}{0.93} \geqslant \dfrac{50-50.2}{0.93}\right)$

$\approx \mathrm{P}(Z \geqslant -0.2151)$

≈ 0.585

b $\mathrm{P}(L \leqslant 51)$

$= \mathrm{P}\left(\dfrac{L-50.2}{0.93} \leqslant \dfrac{51-50.2}{0.93}\right)$

$\approx \mathrm{P}\,(Z \leqslant 0.8602)$

≈ 0.805

c $\mathrm{P}(49 \leqslant L \leqslant 50.5)$

$= \mathrm{P}\left(\dfrac{49-50.2}{0.93} \leqslant \dfrac{L-50.2}{0.93} \leqslant \dfrac{50.5-50.2}{0.93}\right)$

$= \mathrm{P}\,(-1.2903 \leqslant Z \leqslant 0.3226)$

≈ 0.528

EXERCISE 24D

1 **a**

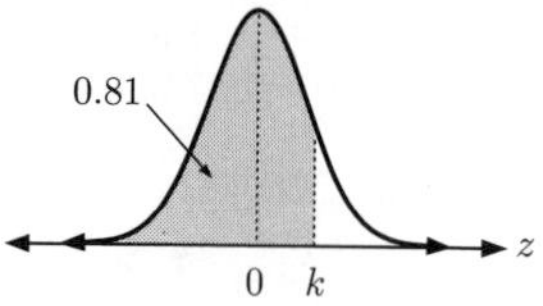

$P(Z \leqslant k) = 0.81$

$\therefore \quad k \approx 0.87 + \frac{22}{28}(0.01)$

$\therefore \quad k \approx 0.878$

b

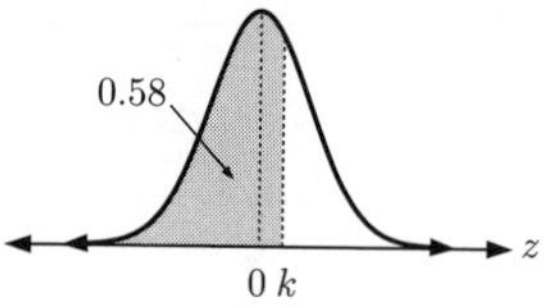

$P(Z \leqslant k) = 0.58$

$\therefore \quad k \approx 0.20 + \frac{7}{39}(0.01)$

$\therefore \quad k \approx 0.202$

c

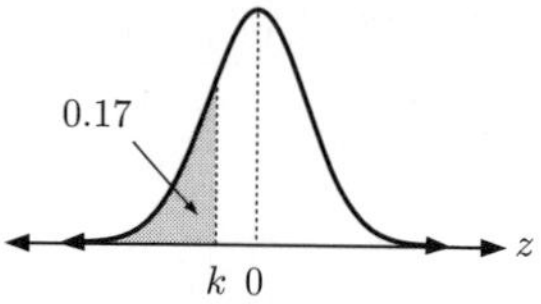

$P(Z \leqslant k) = 0.17$

$\therefore \quad k \approx -0.96 + \frac{15}{26}(0.01)$

$\therefore \quad k \approx -0.954$

2 **a**

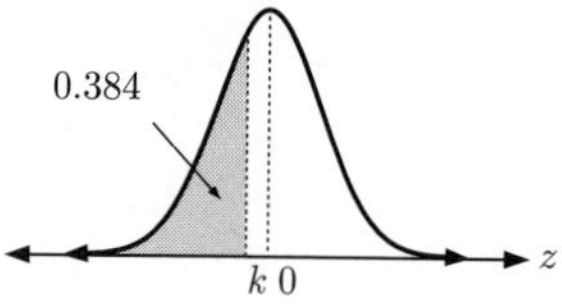

$P(Z \leqslant k) = 0.384$

$\therefore \quad k \approx -0.295$

b

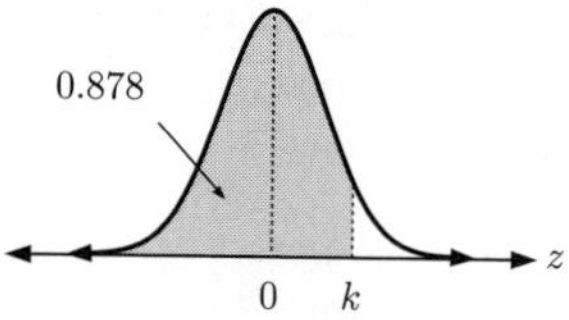

$P(Z \leqslant k) = 0.878$

$\therefore \quad k \approx 1.17$

c

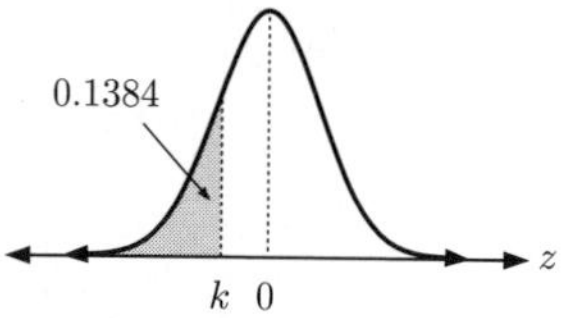

$P(Z \leqslant k) = 0.1384$

$\therefore \quad k \approx -1.09$

3 **a** $P(-k \leqslant Z \leqslant k)$

$= P(Z \leqslant k) - P(Z < -k)$

$= P(Z \leqslant k) - P(Z > k)$ {as area 1 = area 2}

$= P(Z \leqslant k) - [1 - P(Z \leqslant k)]$

$= P(Z \leqslant k) - 1 + P(Z \leqslant k)$

$= 2P(Z \leqslant k) - 1$

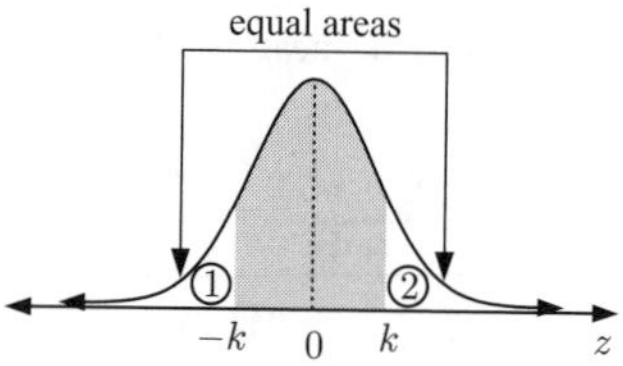

b **i** $P(-k \leqslant Z \leqslant k) = 0.238$

$\therefore \quad 2P(Z \leqslant k) - 1 = 0.238$

$\therefore \quad 2P(Z \leqslant k) = 1.238$

$\therefore \quad P(Z \leqslant k) = 0.619$

$\therefore \quad k \approx 0.303$

ii $P(-k \leqslant Z \leqslant k) = 0.7004$

$\therefore \quad 2P(Z \leqslant k) - 1 = 0.7004$

$\therefore \quad 2P(Z \leqslant k) = 1.7004$

$\therefore \quad P(Z \leqslant k) = 0.8502$

$\therefore \quad k \approx 1.04$

4 **a** $P(X \leqslant k) = 0.9$

$\therefore \quad k \approx 79.1$ {using technology}

b $P(X \geqslant k) = 0.8$

$\therefore \quad P(X < k) = 0.2$

$\therefore \quad k \approx 31.3$ {using technology}

EXERCISE 24E

1 Let X be the length of a bolt in cm.

Then X is normally distributed with $\mu = 19.8$ and $\sigma = 0.3$.

$\therefore \quad P(19.7 < X < 20) \approx 0.378$

2 Let X be the money collected in dollars.

Then X is normally distributed with $\mu = 40$ and $\sigma = 6$.

a $P(30.00 < X < 50.00) \approx 0.904$

$\approx 90.4\%$

b $P(X \geqslant 50) \approx 0.0478$

$\approx 4.78\%$

3 Let X be the result of the Physics test.

Then X is normally distributed with $\mu = 46$ and $\sigma = 25$.

We need to find k such that $P(X \geqslant k) = 0.07$

$\therefore \quad 1 - P(X < k) = 0.07$

$\therefore \quad P(X < k) = 0.93$

$\therefore \quad k \approx 82.894$

$\therefore \quad k \approx 83$ {assuming k is an integer}

So, the lowest score to get an A would be 83.

4 Let X be the length of an eel in cm.

Then X is normally distributed with $\mu = 41$ and $\sigma = \sqrt{11}$.

a $P(X \geqslant 50) \approx 0.003\,33$

b $P(40 \leqslant X \leqslant 50) \approx 0.615$

$\approx 61.5\%$

c $P(X \geqslant 45) \approx 0.114$

So, we would expect $200 \times 0.114 \approx 23$ eels to be at least 45 cm long.

5 $P(X \geqslant 35) = 0.32$

$\therefore \quad P(X < 35) = 0.68$

$\therefore \quad P\left(\dfrac{X-\mu}{\sigma} < \dfrac{35-\mu}{\sigma}\right) = 0.68$

$\therefore \quad P\left(Z < \dfrac{35-\mu}{\sigma}\right) = 0.68$

$\therefore \quad \dfrac{35-\mu}{\sigma} \approx 0.4677$

$\therefore \quad 35 - \mu \approx 0.4677\sigma \quad .... \ (1)$

and

$P(X \leqslant 8) = 0.26$

$\therefore \quad P\left(\dfrac{X-\mu}{\sigma} \leqslant \dfrac{8-\mu}{\sigma}\right) = 0.26$

$\therefore \quad P\left(Z \leqslant \dfrac{8-\mu}{\sigma}\right) = 0.26$

$\therefore \quad \dfrac{8-\mu}{\sigma} \approx -0.6433$

$\therefore \quad 8 - \mu \approx -0.6433\sigma \quad .... \ (2)$

Solving (1) and (2) simultaneously, $35 - 0.4677\sigma \approx 8 + 0.6433\sigma$

$\therefore \quad 27 \approx 1.111\sigma$

$\therefore \quad \sigma \approx 24.3$ and $\mu = 35 - 0.4677 \times 24.3$

$\therefore \quad \mu \approx 23.6$

So, $\mu \approx 23.6$ and $\sigma \approx 24.3$

6 **a** Let the mean be μ and standard deviation be σ.

Then $\text{P}(X \geqslant 80) = 0.1$

$\therefore \text{P}(X < 80) = 0.9$

$\therefore \text{P}\left(\frac{X-\mu}{\sigma} < \frac{80-\mu}{\sigma}\right) = 0.9$

$\therefore \text{P}\left(Z < \frac{80-\mu}{\sigma}\right) = 0.9$

$\therefore \frac{80-\mu}{\sigma} \approx 1.2816$

$\therefore 80 - \mu \approx 1.2816\sigma$ .... (1)

and $\text{P}(X \leqslant 30) = 0.15$

$\therefore \text{P}\left(\frac{X-\mu}{\sigma} \leqslant \frac{30-\mu}{\sigma}\right) = 0.15$

$\therefore \text{P}\left(Z \leqslant \frac{30-\mu}{\sigma}\right) = 0.15$

$\therefore \frac{30-\mu}{\sigma} \approx -1.0364$

$\therefore 30 - \mu \approx -1.0364\sigma$ .... (2)

Solving (1) and (2) simultaneously, $(80-\mu) - (30-\mu) \approx 1.2816\sigma + 1.0364\sigma$

$50 \approx 2.318\sigma$

$\therefore \sigma \approx \frac{50}{2.318} \approx 21.57$

Using (1), $80 - \mu \approx 1.2816 \times 21.57 \approx 27.6$

$\therefore \mu \approx 52.36$

$\therefore \mu \approx 52.4$ and $\sigma \approx 21.6$

b Let X be the result of the mathematics exam.

X is normally distributed with mean μ and standard deviation σ.

We know that $\text{P}(X \geqslant 80) = 0.1$ and $\text{P}(X \leqslant 30) = 0.15$.

So, from **a**, $\mu \approx 52.36$ and $\sigma \approx 21.57$.

If part marks can be given, $\text{P}(X > 50) \approx 0.544$

$\approx 54.4\%$

If only integer marks can be given, $\text{P}(X \geqslant 51) \approx 0.525$

$\approx 52.5\%$

7 Let X be the IQ of a student at the school.

X is normally distributed with mean μ and standard deviation 15.

Now, $\text{P}(X \geqslant 125) = 0.2$

$\therefore \text{P}\left(\frac{X-\mu}{15} \geqslant \frac{125-\mu}{15}\right) = 0.2$

$\therefore \text{P}\left(Z \geqslant \frac{125-\mu}{15}\right) = 0.2$

$\therefore \text{P}\left(Z < \frac{125-\mu}{15}\right) = 0.8$

$\therefore \frac{125-\mu}{15} \approx 0.8416$

$\therefore \mu \approx 112$

The mean IQ at the school is 112.

8 Let X be the distance jumped by the athlete.

X is normally distributed with mean 5.2 m and standard deviation σ.

Now, $\text{P}(X < 5) = 0.15$

$\therefore \text{P}\left(\frac{X-5.2}{\sigma} < \frac{5-5.2}{\sigma}\right) = 0.15$

$\therefore \text{P}\left(Z < -\frac{0.2}{\sigma}\right) = 0.15$

$\therefore -\frac{0.2}{\sigma} \approx -1.036$

$\therefore \sigma \approx 0.193$ m

9 **a** Let the mean be μ and standard deviation be σ and X be the diameter in cm.

$\therefore\ P(X < 1.94) = 0.02$ and $P(X > 2.06) = 0.03$

$\therefore\ P\left(\frac{X-\mu}{\sigma} < \frac{1.94-\mu}{\sigma}\right) = 0.02$ $\quad\therefore\ P\left(\frac{X-\mu}{\sigma} > \frac{2.06-\mu}{\sigma}\right) = 0.03$

$\therefore\ P\left(Z < \frac{1.94-\mu}{\sigma}\right) = 0.02$ $\quad\therefore\ P\left(Z > \frac{2.06-\mu}{\sigma}\right) = 0.03$

$\therefore\ \frac{1.94-\mu}{\sigma} \approx -2.054$ $\quad\therefore\ P\left(Z \leqslant \frac{2.06-\mu}{\sigma}\right) = 0.97$

$\therefore\ 1.94-\mu \approx -2.054\sigma$ .... (1) $\quad \frac{2.06-\mu}{\sigma} \approx 1.881$

$2.06-\mu \approx 1.881\sigma$ .... (2)

Solving (1) and (2) simultaneously, $(2.06-\mu)-(1.94-\mu) = 1.881\sigma + 2.054\sigma$

$\therefore\ 3.935\sigma = 0.12$

$\therefore\ \sigma \approx 0.0305$

Using (1), $1.94-\mu \approx -2.054 \times 0.0305 \approx -0.0626$

$\therefore\ \mu \approx 2.00$

$\therefore\ \mu \approx 2.00$ and $\sigma \approx 0.0305$

b Let Y be the number of tokens which will not operate the machine. This is a binomial situation with the probability $p = 0.02 + 0.03 = 0.05$ of failure to operate and $n = 20$. So, $Y \sim B(20, 0.05)$.

$\therefore$ P(at most one will not operate) $= P(Y \leqslant 1)$

≈ 0.736

REVIEW SET 24A

1 X is the contents of the container in mL.
X is normally distributed with $\mu = 377$ and $\sigma = 4.2$.

a **i** $P(X < 368.6)$
$\approx 2.15\% + 0.13\%$
$\approx 2.28\%$

ii $P(372.8 < X < 389.6)$
$\approx 2 \times 34.13\% + 13.59\% + 2.15\%$
$\approx 84.0\%$

b $P(377 < X < 381.2)$
≈ 0.341

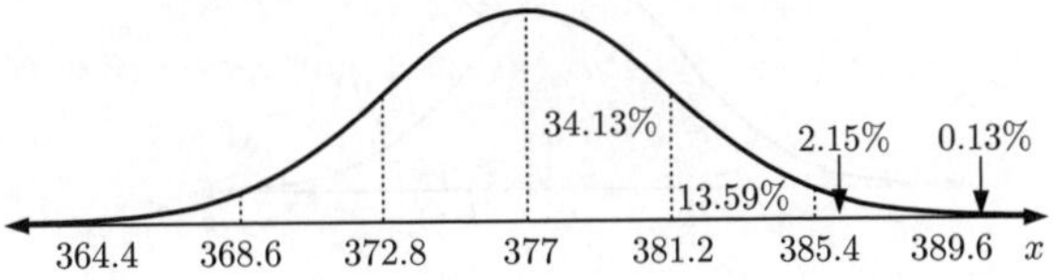

2 Let X denote the mass of a Coffin Bay Oyster. X is distributed normally with a mean of 38.6 and a standard deviation of 6.3.

a $P(38.6 - a \leqslant X \leqslant 38.6 + a) = 0.6826$

$\therefore\ P\left(\frac{38.6-a-38.6}{6.3} \leqslant \frac{X-38.6}{6.3} \leqslant \frac{38.6+a-38.6}{6.3}\right) = 0.6826$

$\therefore\ P\left(-\frac{a}{6.3} \leqslant Z \leqslant \frac{a}{6.3}\right) = 0.6826$

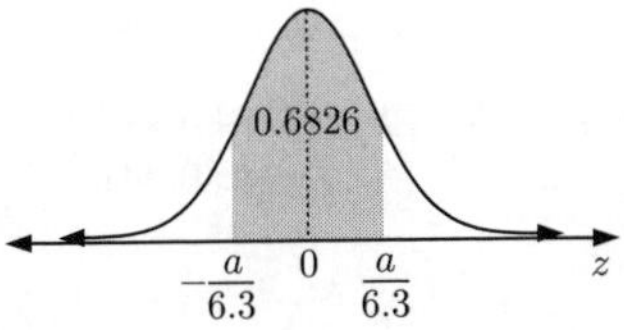

$\therefore$ by symmetry, $P\left(Z \leqslant -\frac{a}{6.3}\right) = \frac{1-0.6826}{2}$

$\therefore\ P\left(Z \leqslant -\frac{a}{6.3}\right) = 0.1587$ .... (*)

$\therefore\ -\frac{a}{6.3} \approx -1.00$

$\therefore\ a \approx 6.30$ g

b
$$\begin{aligned}
& P(X \geqslant b) = 0.8413 \\
\therefore\ & P(X < b) = 0.1587 \\
\therefore\ & P\left(\frac{X-38.6}{6.3} < \frac{b-38.6}{6.3}\right) = 0.1587 \\
\therefore\ & P\left(Z < \frac{b-38.6}{6.3}\right) = 0.1587
\end{aligned}$$

Comparing with $(*)$, $\dfrac{b-38.6}{6.3} = -\dfrac{a}{6.3}$

$\therefore\ b - 38.6 \approx -6.30$

$\therefore\ b \approx 32.3$ g

3 **a**
$$\begin{aligned}
& \int_0^2 ax(x-3)\,dx = 1 \\
\therefore\ & a\int_0^2 (x^2-3x)\,dx = 1 \\
\therefore\ & a\left[\tfrac{1}{3}x^3 - \tfrac{3}{2}x^2\right]_0^2 = 1 \\
\therefore\ & a\left[\tfrac{8}{3} - 6\right] = 1 \\
\therefore\ & a\left(-\tfrac{10}{3}\right) = 1 \\
\therefore\ & a = -\tfrac{3}{10}
\end{aligned}$$

b

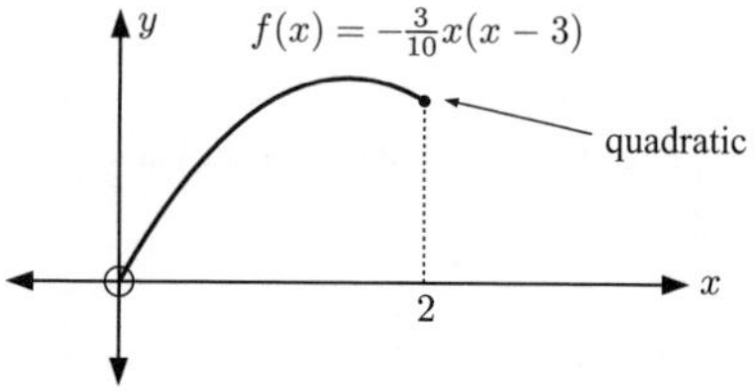

c
$$\begin{aligned}
\mu &= \int_0^2 x\,f(x)\,dx \\
&= \int_0^2 -\tfrac{3}{10}x^2(x-3)\,dx \\
&= -\tfrac{3}{10}\int_0^2 (x^3-3x^2)\,dx \\
&= -\tfrac{3}{10}\left[\tfrac{1}{4}x^4 - x^3\right]_0^2\,dx \\
&= -\tfrac{3}{10}\left(\tfrac{1}{4}(16) - 8\right)dx \\
&= -\tfrac{3}{10}(-4) \\
&= \tfrac{6}{5} = 1.2
\end{aligned}$$

d
$$\begin{aligned}
& P(1 \leqslant X \leqslant 2) \\
&= \int_1^2 -\tfrac{3}{10}x(x-3)\,dx \\
&= -\tfrac{3}{10}\int_1^2 (x^2-3x)\,dx \\
&= -\tfrac{3}{10}\left[\tfrac{1}{3}x^3 - \tfrac{3}{2}x^2\right]_1^2 \\
&= -\tfrac{3}{10}\left[\left(\tfrac{1}{3}(8) - \tfrac{3}{2}(4)\right) - \left(\tfrac{1}{3}(1) - \tfrac{3}{2}(1)\right)\right] \\
&= -\tfrac{3}{10}\left(-\tfrac{10}{3} - (-\tfrac{7}{6})\right) \\
&= \tfrac{13}{20}
\end{aligned}$$

4 $P(-k \leqslant Z \leqslant k) = 0.4$

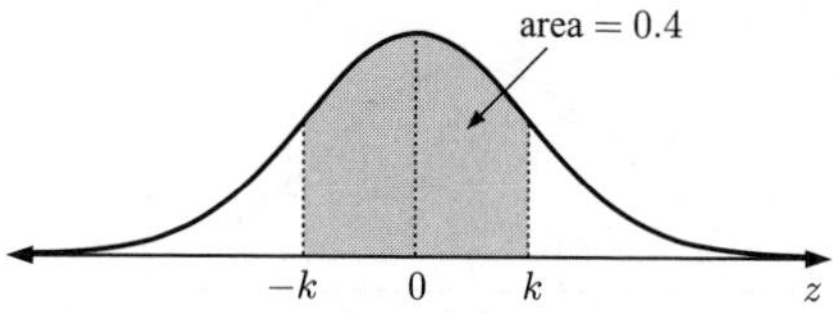

$$\begin{aligned}
\therefore\ & P(Z > k \text{ or } Z < -k) = 0.6 \\
\therefore\ & P(Z < -k) = \tfrac{1}{2}(0.6) = 0.3 \\
\therefore\ & -k \approx -0.524 \\
\therefore\ & k \approx 0.524
\end{aligned}$$

5 Jarrod's z-score is $\dfrac{41-35}{4} = 1.5$

$\therefore$ Paul needs x such that $\dfrac{x-25}{3} = 1.5$

$\therefore\ x = 25 + 4.5 = 29.5$

Paul needs to throw a tennis ball 29.5 m to perform as well as Jarrod.

6

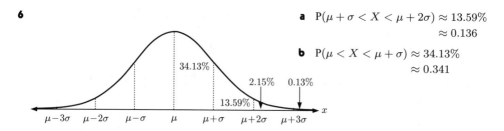

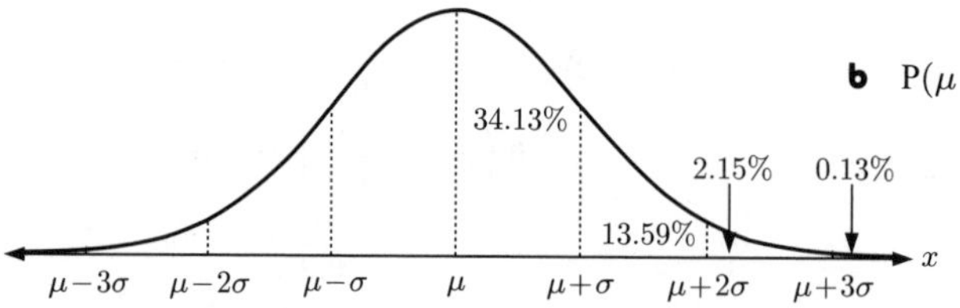

a $P(\mu + \sigma < X < \mu + 2\sigma) \approx 13.59\%$
≈ 0.136

b $P(\mu < X < \mu + \sigma) \approx 34.13\%$
≈ 0.341

REVIEW SET 24B

1 If random variable X is the arm length in cm then X is normally distributed with $\mu = 64$ and $\sigma = 4$.

a **i** $P(60 < X < 72)$
$\approx 2 \times 34.13\% + 13.59\%$
$\approx 81.9\%$

ii $P(X > 60)$
$\approx 50\% + 34.13\%$
$\approx 84.1\%$

b $P(56 < X < 64) \approx 0.3413 + 0.1359$
≈ 0.477

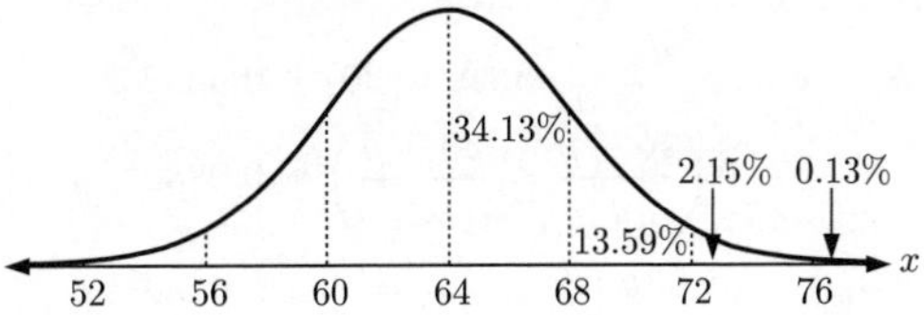

c $P(X > x) = 0.7$
$\therefore\ P(X \leqslant x) = 0.3$
$\therefore\ x \approx 61.9$ {using technology}

2 Let X be the rod length in mm.
X is normally distributed with mean μ and $\sigma = 3$.

Now $P(X < 25) = 0.02$

$\therefore\ P\left(\dfrac{X-\mu}{3} < \dfrac{25-\mu}{3}\right) = 0.02$

$\therefore\ P\left(Z < \dfrac{25-\mu}{3}\right) = 0.02$

$\therefore\ \dfrac{25-\mu}{3} \approx -2.0537$

$\therefore\ 25 - \mu \approx -6.161$

$\therefore\ \mu \approx 31.2$

$\therefore$ the mean rod length is 31.2 mm.

3 **a** Since Area A = Area B, 20 and 38 must be equal distances away from the mean μ, because of the symmetry of the normal distribution.

$\therefore\ \mu$ is halfway between 20 and 38, so $\mu = \dfrac{20+38}{2} = 29$.

Now $P(X \leqslant 20) = 0.2$

$\therefore\ P\left(\dfrac{X-29}{\sigma} \leqslant \dfrac{20-29}{\sigma}\right) = 0.2$

$\therefore\ P\left(Z \leqslant -\dfrac{9}{\sigma}\right) = 0.2$

$\therefore\ -\dfrac{9}{\sigma} \approx -0.8416$

$\therefore\ \sigma \approx 10.69$

$\therefore\ \mu = 29,\quad \sigma \approx 10.7$

b Using the values obtained for μ and σ in **a** and technology:

i $P(X \leqslant 35) \approx 0.713$

ii $P(23 \leqslant X \leqslant 30) \approx 0.250$

4 Let X be the marks in the examination. Then X is normally distributed with $\mu = 49$ and $\sigma = 15$.

a $P(X \geqslant 45) \approx 0.6051$
So, $2376 \times 0.6051 \approx 1438$ students passed the examination.

b Let k be the minimum mark required for a '7'.

$\therefore\ P(X \geqslant k) = 0.07$

$\therefore\ P(X < k) = 1 - 0.07 = 0.93$

$\therefore\ k \approx 71.1$

$\therefore\ k \approx 71$ (to the nearest integer)

So the minimum mark required to obtain a '7' is 71 marks.

5 X is the life of a battery in weeks.
X is normally distributed with $\mu = 33.2$ and $\sigma = 2.8$.

a $\text{P}(X \geqslant 35) \approx 0.260$

b We need to find k such that $\text{P}(X \leqslant k) = 0.08$
$\therefore \; k \approx 29.3$

So, the manufacturer can expect the batteries to last 29.3 weeks before 8% of them fail.

6 **a** $\text{P}(X \leqslant 30) = 0.0832$ and $\text{P}(X \geqslant 90) = 0.101$

$$\therefore \; \text{P}\left(\frac{X-\mu}{\sigma} \leqslant \frac{30-\mu}{\sigma}\right) \approx 0.0832$$
$$\therefore \; \text{P}\left(Z \leqslant \frac{30-\mu}{\sigma}\right) \approx 0.0832$$
$$\therefore \; \frac{30-\mu}{\sigma} \approx -1.383\,864$$
$$\therefore \; 30-\mu \approx -1.383\,864\sigma \quad .... (1)$$

$$\therefore \; \text{P}(X < 90) = 0.899$$
$$\therefore \; \text{P}\left(\frac{X-\mu}{\sigma} < \frac{90-\mu}{\sigma}\right) = 0.899$$
$$\therefore \; \text{P}\left(Z < \frac{90-\mu}{\sigma}\right) = 0.899$$
$$\therefore \; \frac{90-\mu}{\sigma} \approx 1.275\,874$$
$$\therefore \; 90-\mu = 1.275\,874\sigma \quad .... (2)$$

Solving (1) and (2) simultaneously,
$$(90-\mu)-(30-\mu) \approx 1.275\,874\sigma - (-1.383\,864\sigma)$$
$$\therefore \; 60 \approx 2.6597\sigma$$
$$\therefore \; \sigma \approx 22.559$$

Using (2), $90-\mu \approx 1.275\,874(22.559)$
$$\therefore \; \mu \approx 90 - 1.275\,874(22.559) \approx 61.218$$

b $\text{P}(-7 \leqslant X-\mu \leqslant 7) \approx \text{P}(-7 \leqslant X - 61.218 \leqslant 7)$
$\approx \text{P}(54.218 \leqslant X \leqslant 68.218)$
≈ 0.244

REVIEW SET 24C

1 Using technology:

a $\text{P}(X \geqslant 22) \approx 0.364$ **b** $\text{P}(18 \leqslant X \leqslant 22) \approx 0.356$ **c** $\text{P}(X \leqslant k) = 0.3$
$\therefore \; k \approx 18.2$

2 Let X be the volume of drink in mL.
Then X is normally distributed with $\mu = 376$.

Now $\text{P}(X < 375) = 0.023$
$$\therefore \; \text{P}\left(\frac{X-376}{\sigma} < \frac{375-376}{\sigma}\right) = 0.023$$
$$\therefore \; \text{P}\left(Z < \frac{-1}{\sigma}\right) = 0.023$$
$$\therefore \; -\frac{1}{\sigma} \approx -1.995$$
$$\therefore \; \sigma \approx 0.501$$

$\therefore$ the standard deviation is 0.501 mL

3 $\text{P}(-0.524 < X-\mu < 0.524) = \text{P}\left(\frac{-0.524}{2} < \frac{X-\mu}{2} < \frac{0.524}{2}\right)$
$= \text{P}(-0.262 < Z < 0.262)$
≈ 0.207 {using technology}

4 Let X be the length of the rods. X is normally distributed with $\sigma = 6$.

$$\text{Now} \quad \text{P}(X \geqslant 89.52) = 0.0563$$
$$\therefore \quad \text{P}(X < 89.52) = 1 - 0.0563$$
$$\therefore \quad \text{P}\left(\frac{X-\mu}{6} < \frac{89.52-\mu}{6}\right) = 0.9437$$
$$\therefore \quad \text{P}\left(Z < \frac{89.52-\mu}{6}\right) = 0.9437$$
$$\therefore \quad \frac{89.52-\mu}{6} \approx 1.5866$$
$$\therefore \quad 89.52 - \mu \approx 9.52$$
$$\therefore \quad \mu \approx 80.0$$

So, the mean is 80.0 cm.

5

$$\text{P}(X < 90) \approx 0.975$$
$$\therefore \quad \text{P}\left(\frac{X-50}{\sigma} < \frac{90-50}{\sigma}\right) \approx 0.975$$
$$\therefore \quad \text{P}\left(Z < \frac{40}{\sigma}\right) \approx 0.975$$
$$\therefore \quad \frac{40}{\sigma} \approx 1.959\,96$$
$$\therefore \quad \sigma \approx 20.409$$

So, $X \sim \text{N}(50, 20.409^2)$

Now, the shaded area $= \text{P}(X \geqslant 80)$
$\approx 0.0708 \text{ units}^2$

6 Let X be the heights of 18 year old boys. X is normally distributed with $\mu = 187$.

$$\text{Now} \quad \text{P}(X > 193) = 0.15$$
$$\therefore \quad \text{P}(X \leqslant 193) = 0.85$$
$$\therefore \quad \text{P}\left(\frac{X-187}{\sigma} \leqslant \frac{193-187}{\sigma}\right) = 0.85$$
$$\therefore \quad \text{P}\left(Z \leqslant \frac{6}{\sigma}\right) = 0.85$$
$$\therefore \quad \frac{6}{\sigma} \approx 1.0364$$
$$\therefore \quad \sigma \approx 5.789$$

So, $\text{P}(X > 185) \approx 0.635$

$\therefore$ the probability that two 18 year old boys are taller than 185 cm $\approx 0.635^2$
≈ 0.403

Chapter 25
MISCELLANEOUS QUESTIONS

EXERCISE 25A

1 **a** $S_1 = u_1 = 2$ and $S_2 = u_1 + u_2 = 8$
$\therefore\ u_1 = 2$ and $u_2 = 6$
But $u_2 = u_1 r$
$\therefore\ 6 = 2r$
$\therefore\ r = 3$

b $u_{20} = u_1 r^{19}$
$= 2 \times 3^{19}$

2 $\ln 2 + \ln 4 + \ln 8 + \ln 16 + ...$
$= \ln 2 + \ln 2^2 + \ln 2^3 + \ln 2^4 + ...$
$= \ln 2 + 2\ln 2 + 3\ln 2 + 4\ln 2 + ...$
which is arithmetic with $u_1 = \ln 2$ and $d = \ln 2$

Now $S_n = \frac{n}{2}(2u_1 + (n-1)d)$
$\therefore\ S_{40} = 20(2\ln 2 + 39\ln 2)$
$= 20 \times 41 \ln 2$
$= 820 \ln 2$

3 $f(x) = be^x$ and $g(x) = \ln(bx)$

a $(f \circ g)(x) = f(g(x))$
$= f(\ln(bx))$
$= be^{\ln(bx)}$
$= b(bx)$
$= b^2 x$

b $(g \circ f)(x) = g(f(x))$
$= g(be^x)$
$= \ln(bbe^x)$
$= \ln(b^2 e^x)$
$= \ln b^2 + \ln e^x$
$= 2\ln b + x$

c $(f \circ g)(x^*) = (g \circ f)(x^*)$
$\therefore\ b^2 x^* = 2\ln b + x^*$
$\therefore\ (b^2 - 1)x^* = 2\ln b$
$\therefore\ x^* = \dfrac{2\ln b}{b^2 - 1}$

4 **a** The vertex is $(b, 2)$.

b $f(0) = -2b^2 + 2$
$\therefore$ the y-intercept is $2 - 2b^2$
$f(x)$ cuts the x-axis when $f(x) = 0$
$\therefore\ -2(x-b)^2 + 2 = 0$
$\therefore\ (x-b)^2 = 1$
$\therefore\ x - b = \pm 1$
$\therefore\ x = b \pm 1$
$\therefore$ the x-intercepts are $b - 1$ and $b + 1$

c

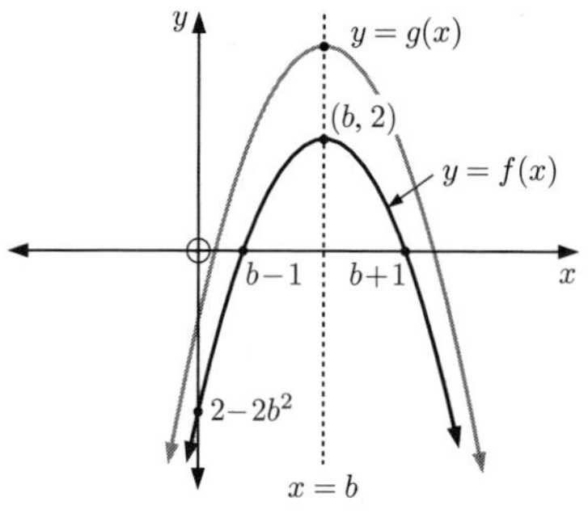

$g(x) = f(x) + b$
$= -2(x-b)^2 + 2 + b$
$= -2(x^2 - 2bx + b^2) + 2 + b$
$= -2x^2 + 4bx - 2b^2 + b + 2$
which has discriminant
$\Delta = (4b)^2 - 4(-2)(-2b^2 + b + 2)$
$= 16b^2 - 16b^2 + 8b + 16$
$= 8b + 16$

i g has exactly one x-intercept if $8b + 16 = 0$
$\therefore\ b = -2$

ii g has no x-intercepts if $8b + 16 < 0$
$\therefore\ 8b < -16$
$\therefore\ b < -2$

iii g passes through the origin if $g(0) = 0$
$\therefore\ -2b^2 + b + 2 = 0$
$\therefore\ b = \dfrac{-1 \pm \sqrt{1 - 4(-2)(2)}}{-4}$
$= \dfrac{1 \pm \sqrt{17}}{4}$

5 **a** $(x-2)^3$
$= x^3 + 3x^2(-2) + 3x(-2)^2 + (-2)^3$
$= x^3 - 6x^2 + 12x - 8$

b $(3x^2-7)(x-2)^3$
$= (3x^2-7)(x^3 - 6x^2 + 12x - 8)$

$\therefore$ coefficient of x^3 is $3 \times 12 + (-7) \times 1$
$= 29$

6 $f(x) = \sqrt{1-2x}$

a $f(0) = \sqrt{1} = 1$

b $f(-4) = \sqrt{1-2(-4)}$
$= \sqrt{9}$
$= 3$

c $f(x)$ is defined when $1 - 2x \geqslant 0$
$\therefore\ 1 \geqslant 2x$
$\therefore\ x \leqslant \frac{1}{2}$

So, the domain is $\{x \mid x \leqslant \frac{1}{2},\ x \in \mathbb{R}\}$

d As $f(x) = \sqrt{1-2x}$,
$f(x) \geqslant 0$ for all x in the domain.
$\therefore$ the range is $\{y \mid y \geqslant 0,\ y \in \mathbb{R}\}$

Check:

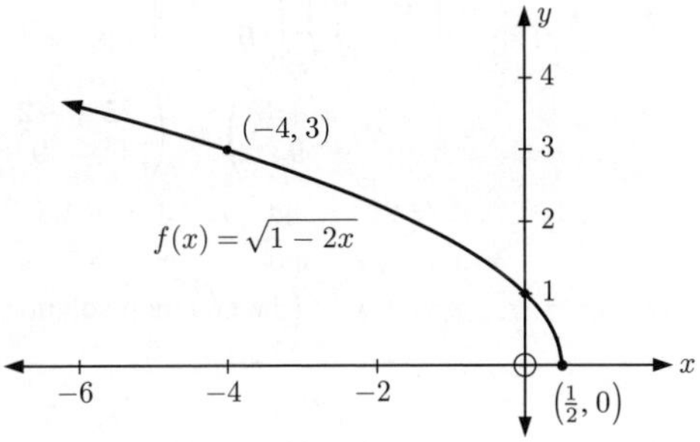

7 **a** $\sin 160^o$
$= \sin 20^o \quad \{\sin\theta = \sin(180-\theta)\}$
$= a$

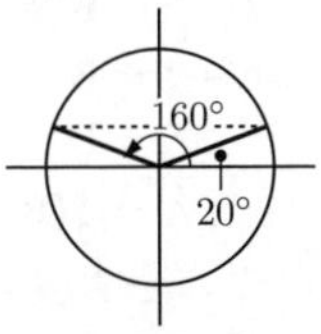

b $\tan(-50^o)$
$= -\tan 50^o \quad \{\tan(-\theta) = -\tan\theta\}$
$= -b$

c $\cos 70^o$
$= \sin(90-70)^o \quad \{\cos\theta = \sin(90-\theta)\}$
$= \sin 20^o$
$= a$

d $\tan 20^o = \dfrac{\sin 20^o}{\cos 20^o}$

Now $\cos^2\theta + \sin^2\theta = 1$, so $\cos\theta = \pm\sqrt{1-\sin^2\theta}$

But $\cos 20^o > 0$, so in this case $\cos 20^o = \sqrt{1-\sin^2 20^o} = \sqrt{1-a^2}$

So, $\tan 20^o = \dfrac{a}{\sqrt{1-a^2}}$

8 **a** $(f \circ g)(x) = 1$
$\therefore\ f(g(x)) = 1$
$\therefore\ f(2x) = 1$
$\therefore\ \cos 2x = 1$
$\therefore\ 2x = 0 + k2\pi,\ k \in \mathbb{Z}$
$\therefore\ x = k\pi,\ k \in \mathbb{Z}$
$\therefore\ x = 0, \pi$ or $2\pi \quad \{0 \leqslant x \leqslant 2\pi\}$

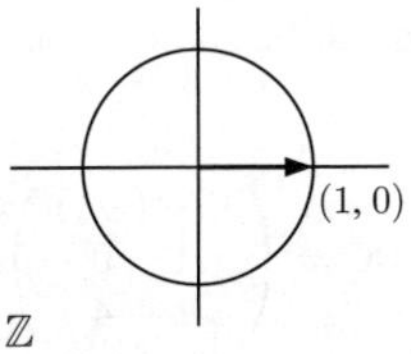

b $(g \circ f)(x) = 1$
$\therefore\ g(f(x)) = 1$
$\therefore\ g(\cos x) = 1$
$\therefore\ 2\cos x = 1$
$\therefore\ \cos x = \frac{1}{2}$
$\therefore\ x = \frac{\pi}{3}$ or $\frac{5\pi}{3}$

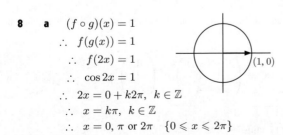

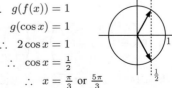

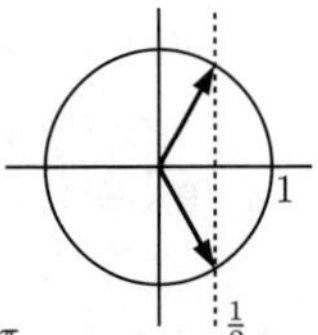

9 Consider $f(x) = ax^2 + bx + c$.
$f(x)$ is concave up, so $a > 0$.
The y-intercept of $f(x)$ is positive, so $c > 0$.
The axis of symmetry is to the right of the y-axis, so $\dfrac{-b}{2a} > 0$
$\therefore\ b < 0 \quad \{a > 0\}$
$f(x)$ does not cut the x-axis, so $\Delta < 0$.

Consider $g(x) = dx^2 + ex + h$.
$g(x)$ is concave down, so $d < 0$.
$g(x)$ passes through the origin, so $h = 0$.
The axis of symmetry is to the right of the y-axis, so $\dfrac{-e}{2d} > 0$
$\therefore\ e > 0 \quad \{d < 0\}$
$g(x)$ cuts the x-axis twice, so $\Delta > 0$.

Constant	a	b	c	d	e	h	Δ of $f(x)$	Δ of $g(x)$
Sign	>0	<0	>0	<0	>0	$=0$	<0	>0

10 **a** $\mathbf{A}^2 = \begin{pmatrix} x & 1 \\ 0 & 3 \end{pmatrix}\begin{pmatrix} x & 1 \\ 0 & 3 \end{pmatrix}$
$= \begin{pmatrix} x^2 & x+3 \\ 0 & 9 \end{pmatrix} = \begin{pmatrix} 25 & -2 \\ 0 & 9 \end{pmatrix}$
$\therefore\ x^2 = 25$ and $x + 3 = -2$
$\therefore\ x = \pm 5$ and $x = -5$
$\therefore\ x = -5$ {the common solution}

b $\begin{pmatrix} 2 & 4 \\ 1 & 1 \end{pmatrix}\begin{pmatrix} a & 2 \\ -a & -1 \end{pmatrix} = \begin{pmatrix} 1 & 0 \\ 0 & 1 \end{pmatrix}$
$\therefore\ \begin{pmatrix} 2a-4a & 4-4 \\ a-a & 2-1 \end{pmatrix} = \begin{pmatrix} 1 & 0 \\ 0 & 1 \end{pmatrix}$
$\therefore\ \begin{pmatrix} -2a & 0 \\ 0 & 1 \end{pmatrix} = \begin{pmatrix} 1 & 0 \\ 0 & 1 \end{pmatrix}$
$\therefore\ -2a = 1$ and so $a = -\frac{1}{2}$

11 **a** $(3\mathbf{A}+\mathbf{I})^{-1} = \mathbf{B}$
$\therefore\ \left((3\mathbf{A}+\mathbf{I})^{-1}\right)^{-1} = \mathbf{B}^{-1}$
$\therefore\ 3\mathbf{A}+\mathbf{I} = \mathbf{B}^{-1}$
$\therefore\ 3\mathbf{A} = \mathbf{B}^{-1} - \mathbf{I}$
$\therefore\ \mathbf{A} = \frac{1}{3}(\mathbf{B}^{-1} - \mathbf{I})$

b $\mathbf{A} = \frac{1}{3}\left(\frac{1}{-1}\begin{pmatrix} 5 & 4 \\ -1 & -1 \end{pmatrix} - \begin{pmatrix} 1 & 0 \\ 0 & 1 \end{pmatrix}\right)$
$= \frac{1}{3}\begin{pmatrix} -6 & -4 \\ 1 & 0 \end{pmatrix}$
$= \begin{pmatrix} -2 & -1\frac{1}{3} \\ \frac{1}{3} & 0 \end{pmatrix}$

c $\det \mathbf{A} = 0 - (-\frac{4}{9}) = \frac{4}{9}$

12 **a** **i** $\mathbf{u} \perp \mathbf{v}$
$\therefore\ \mathbf{u} \bullet \mathbf{v} = 0$
$\therefore\ (1)(3) + (-2)(p) + (1)(-1) = 0$
$\therefore\ 3 - 2p - 1 = 0$
$\therefore\ 2p = 2$
$\therefore\ p = 1$

ii $|\mathbf{u}|\,|\mathbf{v}| = \sqrt{1+4+1}\sqrt{9+p^2+1}$
$= \sqrt{6}\sqrt{p^2+10}$
$= \sqrt{6p^2+60}$

b $\mathbf{v} - \mathbf{u} = (3\mathbf{i} + p\mathbf{j} - \mathbf{k}) - (\mathbf{i} - 2\mathbf{j} + \mathbf{k})$
$= 2\mathbf{i} + (p+2)\mathbf{j} - 2\mathbf{k}$

For $\mathbf{v} - \mathbf{u}$ to be parallel to $\mathbf{u}$, there must exist some scalar s such that
$s(\mathbf{i} - 2\mathbf{j} + \mathbf{k}) = 2\mathbf{i} + (p+2)\mathbf{j} - 2\mathbf{k}$
$\therefore\ s = 2,\ -2s = p + 2$ and $s = -2$
There is no common solution to these equations, so no value of p exists such that $\mathbf{v} - \mathbf{u}$ and $\mathbf{u}$ are parallel.

13 **a** $\overrightarrow{BA} = \begin{pmatrix} 2-(-1) \\ 1-0 \\ 5-4 \end{pmatrix} = \begin{pmatrix} 3 \\ 1 \\ 1 \end{pmatrix}$ and $\overrightarrow{BC} = \begin{pmatrix} 0-(-1) \\ 1-0 \\ 1-4 \end{pmatrix} = \begin{pmatrix} 1 \\ 1 \\ -3 \end{pmatrix}$

b $|\overrightarrow{BA}| = \sqrt{9+1+1}$
$= \sqrt{11}$ units

$|\overrightarrow{BC}| = \sqrt{1+1+9}$
$= \sqrt{11}$ units

c ABCD is a parallelogram, so the opposite sides are equal in length, and $|\overrightarrow{BA}| = |\overrightarrow{BC}|$ from **b**.
$\therefore$ all the sides are equal in length, so ABCD is a rhombus.

d
$$\cos(C\widehat{B}A) = \frac{\overrightarrow{BC} \bullet \overrightarrow{BA}}{|\overrightarrow{BC}||\overrightarrow{BA}|}$$
$$= \frac{3+1-3}{\sqrt{11}\sqrt{11}}$$
$$= \tfrac{1}{11}$$

e
$$\sin^2(C\widehat{B}A) + \cos^2(C\widehat{B}A) = 1$$
$$\therefore \sin^2(C\widehat{B}A) + \tfrac{1}{121} = 1$$
$$\therefore \sin(C\widehat{B}A) = \tfrac{\sqrt{120}}{11}$$

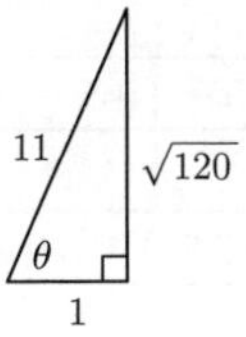

f
$$\text{area} = 2 \times \text{area of } \Delta ABC$$
$$= 2 \times \tfrac{1}{2}\,|\overrightarrow{BA}||\overrightarrow{BC}|\sin(C\widehat{B}A)$$
$$= \sqrt{11}\sqrt{11} \times \tfrac{\sqrt{120}}{11}$$
$$= \sqrt{120}$$
$$= 2\sqrt{30} \text{ units}^2$$

14 **a** There are 13 data values, and $\dfrac{13+1}{2} = 7$

$\therefore$ the median $=$ 7th data value $= g$

b **i** range $= m - a$

ii $\underbrace{a\ b\ c\ d\ e\ f}_{\text{lower}}\ \underset{\text{median}}{\underset{\uparrow}{g}}\ \underbrace{h\ i\ j\ k\ l\ m}_{\text{upper}}$

$Q_1 =$ median of lower half $= \dfrac{c+d}{2}$ $\qquad$ $Q_3 =$ median of upper half $= \dfrac{j+k}{2}$

$\therefore$ IQR $= Q_3 - Q_1 = \dfrac{j+k-c-d}{2}$

15 The data set $\{a, b, c\}$ has mean 17.5 and standard deviation 3.2

$$\therefore \frac{a+b+c}{3} = 17.5 \quad \text{and} \quad \sqrt{\frac{(a-17.5)^2 + (b-17.5)^2 + (c-17.5)^2}{3}} = 3.2$$

a For the data set $\{2a, 2b, 2c\}$,

$$\mu = \frac{2a+2b+2c}{3}$$
$$= 2\left(\frac{a+b+c}{3}\right)$$
$$= 2 \times 17.5 = 35$$

$$\sigma = \sqrt{\frac{(2a-35)^2 + (2b-35)^2 + (2c-35)^2}{3}}$$
$$= \sqrt{\frac{2^2(a-17.5)^2 + 2^2(b-17.5)^2 + 2^2(c-17.5)^2}{3}}$$
$$= 2\sqrt{\frac{(a-17.5)^2 + (b-17.5)^2 + (c-17.5)^2}{3}}$$
$$= 2 \times 3.2 = 6.4$$

b For the data set $\{a+2, b+2, c+2\}$,

$$\mu = \frac{(a+2)+(b+2)+(c+2)}{3}$$
$$= \frac{a+b+c}{3} + \frac{6}{3}$$
$$= 17.5 + 2 = 19.5$$

$$\sigma = \sqrt{\frac{(a+2-19.5)^2 + (b+2-19.5)^2 + (c+2-19.5)^2}{3}}$$
$$= \sqrt{\frac{(a-17.5)^2 + (b-17.5)^2 + (c-17.5)^2}{3}}$$
$$= 3.2$$

c For the data set $\{3a+5,\ 3b+5,\ 3c+5\}$,

$$\mu = \frac{(3a+5)+(3b+5)+(3c+5)}{3}$$
$$= \frac{3(a+b+c)}{3} + \frac{15}{3}$$
$$= 3 \times 17.5 + 5 = 57.5$$

$$\sigma = \sqrt{\frac{(3a+5-57.5)^2+(3b+5-57.5)^2+(3c+5-57.5)^2}{3}}$$
$$= \sqrt{\frac{(3a-52.5)^2+(3b-52.5)^2+(3c-52.5)^2}{3}}$$
$$= \sqrt{\frac{3^2(a-17.5)^2+3^2(b-17.5)^2+3^2(c-17.5)^2}{3}}$$
$$= 3\sqrt{\frac{(a-17.5)^2+(b-17.5)^2+(c-17.5)^2}{3}}$$
$$= 3 \times 3.2 = 9.6$$

16 Let the circle have radius r. The equal angles are $\frac{2\pi}{5}$ radians.

The area of each triangle $= \frac{1}{2} \times r \times r \times \sin(\frac{2\pi}{5})$
$= \frac{1}{2}r^2 \sin(\frac{2\pi}{5})$

$\therefore$ P(lands on shaded region | lands on board) $= \dfrac{\pi r^2 - \left(\frac{1}{2}r^2\sin(\frac{2\pi}{5})\right) \times 5}{\pi r^2}$
$= 1 - \frac{5}{2\pi}\sin(\frac{2\pi}{5})$

17 $f(x)$ is a quadratic, so $f'(x)$ is linear and $f''(x)$ is constant.

a $f(x)$ is concave down, so $f'(x)$ is decreasing.
$f'(x) = 0$ when $f(x)$ is maximised.

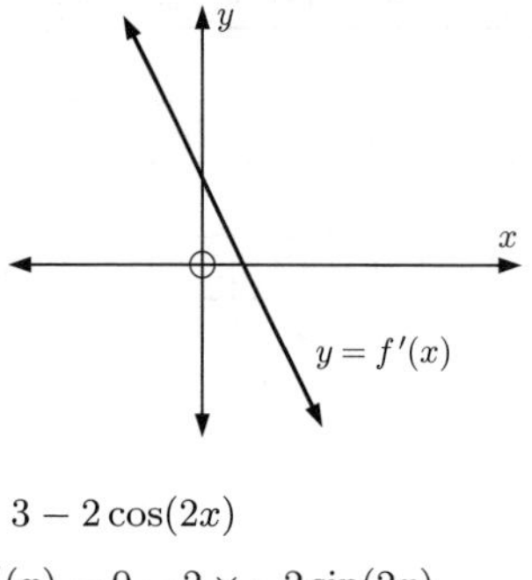

b $f'(x)$ is decreasing, so $f''(x)$ is negative.

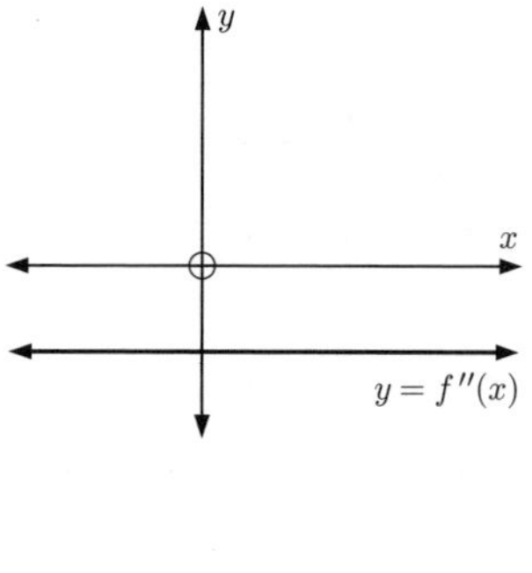

18 $g(x) = 3 - 2\cos(2x)$

a $g'(x) = 0 - 2 \times -2\sin(2x)$
$= 4\sin(2x)$

b

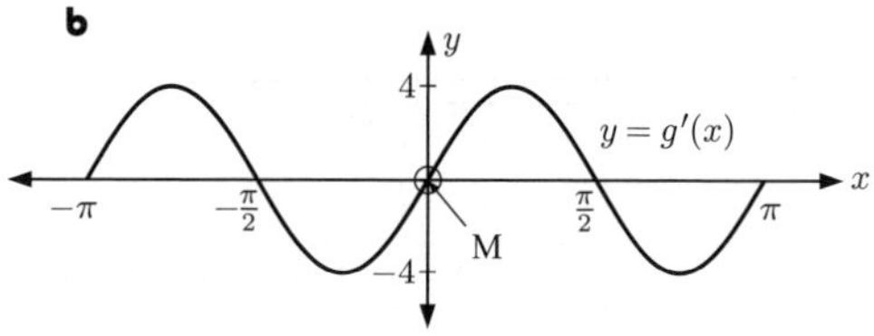

c $g'(x) = 0$ when $\sin(2x) = 0$

$\therefore\ x = -\pi,\ -\frac{\pi}{2},\ 0,\ \frac{\pi}{2},\ \pi$

$\therefore$ there are 5 solutions.

d If $g''(x) > 0$, $y = g'(x)$ is increasing

$\therefore$ M is at $(-\pi, 0)$, $(0, 0)$ or $(\pi, 0)$

19 **a** $P(B) = 1 - P(B') = 0.57$

As A and B are mutually exclusive,

$A \cap B = \varnothing$ and so $P(A \cap B) = 0$.

$\therefore\ P(A \cup B) = P(A) + P(B)$

$= x + 0.57$

b We need to solve $x + 0.57 = 0.73$

$\therefore\ x = 0.16$

20 $g'(0) = 0$ and $g''(0) = 0$

$\therefore$ there is a stationary inflection point at A(0, 2)

$g''(2) = 0$ and $g'(2) \neq 0$

$\therefore$ there is a non-stationary inflection point at B(2, 0)

$g'(4) = 0$ and $g''(4) > 0$

$\therefore$ there is a local minimum at C(4, −2)

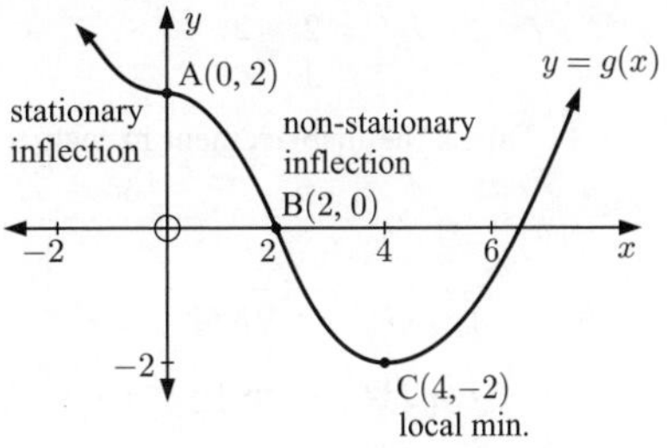

21 $f(x) = xe^{1-2x}$

a $f'(x) = 1e^{1-2x} + xe^{1-2x}(-2)$

$= e^{1-2x}(1 - 2x)$

b $f'(x) = 0$ when $1 - 2x = 0$

$\therefore\ x = \frac{1}{2}$

and $f(\frac{1}{2}) = \frac{1}{2}e^{1-2(\frac{1}{2})} = \frac{1}{2}$

$\therefore$ the tangent is horizontal at $(\frac{1}{2}, \frac{1}{2})$

c **i** $f(x) > 0$ when $xe^{1-2x} > 0$

But e^{1-2x} is always > 0

$\therefore\ f(x) > 0$ when $x > 0$

ii $f'(x) > 0$ when $1 - 2x > 0$

$\therefore\ 1 > 2x$

$\therefore\ x < \frac{1}{2}$

22 $s(t) = 3 - 4e^{2t} + kt$ metres

a $v(t) = 0 - 4e^{2t}(2) + k$

$= k - 8e^{2t}$ m s^{-1}

b $v(\ln 3) = 0$

$\therefore\ k - 8e^{2\ln 3} = 0$

$\therefore\ k - 8e^{\ln 3^2} = 0$

$\therefore\ k - 8 \times 9 = 0$

$\therefore\ k = 72$

23 **a** $\log_3 27 = x$

$\therefore\ 3^x = 27$

$\therefore\ 3^x = 3^3$

$\therefore\ x = 3$

b $\log_x 7 = 2$

$\therefore\ x^2 = 7$

$\therefore\ x = \sqrt{7}\quad \{x > 0\}$

c $e^{5-2x} = 8$

$\therefore\ 5 - 2x = \ln 8$

$\therefore\ 2x = 5 - \ln 8$

$\therefore\ x = \dfrac{5 - \ln 8}{2}$

d

$$\ln(x^2 - 3) - \ln(2x) = 0$$

$$\therefore\ \ln\left(\frac{x^2 - 3}{2x}\right) = 0$$

$$\therefore\ \frac{x^2 - 3}{2x} = 1$$

$$\therefore\ x^2 - 3 = 2x$$

$$\therefore\ x^2 - 2x - 3 = 0$$

$$\therefore\ (x - 3)(x + 1) = 0$$

$$\therefore\ x = 3 \text{ or } -1$$

But $x^2 - 3$ and $2x$ must be > 0

$\therefore\ x = 3$ is the only solution

24 **a** $v(0) = 1\ \text{m s}^{-1}$ and is the initial velocity.

b $v(t) = 2$, a constant for $1 \leqslant t \leqslant 3$

$\therefore\ v'(t) = 0$ on $1 \leqslant t \leqslant 3$

$\therefore\ v'(2) = 0$

This is the acceleration at $t = 2$.
The velocity is constant at this time.

c $\int_1^3 v(t)\,dt =$ area under $y = v(t)$ from $t = 1$ to $t = 3$

$= 2 \times 2$

$= 4$

This is the displacement in metres of the particle for $1 \leqslant t \leqslant 3$.

25 **a**

$$\begin{aligned}&\int_{-1}^{2} (f(x) - 6)\ dx\\ &= \int_{-1}^{2} f(x)\ dx - \int_{-1}^{2} 6\ dx\\ &= 10 - [6x]_{-1}^{2}\\ &= 10 - (12 - -6)\\ &= 10 - 18\\ &= -8\end{aligned}$$

b

$$\begin{aligned}&\int_{2}^{-1} k\,f(x)\ dx = -5\\ \therefore\ &\int_{-1}^{2} k\,f(x)\ dx = 5\\ \therefore\ &k\int_{-1}^{2} f(x)\ dx = 5\\ \therefore\ &k(10) = 5\\ \therefore\ &k = \tfrac{1}{2}\end{aligned}$$

26 **a**

$$\begin{aligned}&(\sin\theta - \cos\theta)^2\\ &= \sin^2\theta - 2\sin\theta\cos\theta + \cos^2\theta\\ &= \sin^2\theta + \cos^2\theta - \sin 2\theta\\ &= 1 - \sin 2\theta\end{aligned}$$

b

$$\begin{aligned}&\int_0^{\frac{\pi}{4}} (\sin\theta - \cos\theta)^2\ d\theta\\ &= \int_0^{\frac{\pi}{4}} (1 - \sin 2\theta)\ d\theta\\ &= \left[\theta + \tfrac{1}{2}\cos 2\theta\right]_0^{\frac{\pi}{4}}\\ &= \left(\tfrac{\pi}{4} + \tfrac{1}{2}\cancel{\cos\tfrac{\pi}{2}}^{\,0}\right) - \left(0 + \tfrac{1}{2}\right)\\ &= \tfrac{\pi}{4} - \tfrac{1}{2}\end{aligned}$$

27

$$\begin{aligned}\mathrm{E}(X) &= \textstyle\sum p_i x_i\\ &= -0.2 - a + 0 + b + 0.3\\ &= b - a + 0.1\end{aligned}$$

But $\mathrm{E}(X) = 0$

$\therefore\ a - b = 0.1$ .... (1)

Also, $\sum p_i = 1$

$\therefore\ 0.1 + a + 0.25 + b + 0.15 = 1$

$\therefore\ a + b = 0.5$ .... (2)

Adding (1) and (2) gives $2a = 0.6$

$\therefore\ a = 0.3$ and so $b = 0.2$

28 **a** The common ratio

$$r = \frac{\frac{1}{e}}{e} = \frac{1}{e^2}$$

b

$$\begin{aligned}u_{101} &= u_1 r^{100}\\ &= e \times \left(\frac{1}{e^2}\right)^{100}\\ &= e \times e^{-200}\\ &= e^{-199}\end{aligned}$$

c

$$\begin{aligned}S &= \frac{u_1}{1 - r}\\ &= \frac{e}{1 - \frac{1}{e^2}}\\ &= \left[\frac{e}{1 - \frac{1}{e^2}}\right]\frac{e^2}{e^2}\\ &= \frac{e^3}{e^2 - 1}\end{aligned}$$

29 **a**
$$\begin{vmatrix} 1 & 0 & 2 \\ -1 & x & 4 \\ 5 & -6 & 2 \end{vmatrix} = 12$$

$$\therefore\ 1\begin{vmatrix} x & 4 \\ -6 & 2 \end{vmatrix} - 0\begin{vmatrix} -1 & 4 \\ 5 & 2 \end{vmatrix} + 2\begin{vmatrix} -1 & x \\ 5 & -6 \end{vmatrix} = 12$$

$$\therefore\ 2x + 24 + 2(6-5x) = 12$$
$$\therefore\ 2x + 24 + 12 - 10x = 12$$
$$\therefore\ -8x = -24$$
$$\therefore\ x = 3$$

b
$$\begin{pmatrix} 1 & 0 & 2 \\ -1 & x & 4 \\ 5 & -6 & 2 \end{pmatrix}\begin{pmatrix} 6 \\ 3 \\ 1 \end{pmatrix} = \begin{pmatrix} 8 \\ 4 \\ 14 \end{pmatrix}$$

$$\therefore\ \begin{pmatrix} 8 \\ 3x-2 \\ 14 \end{pmatrix} = \begin{pmatrix} 8 \\ 4 \\ 14 \end{pmatrix}$$

$$\therefore\ 3x - 2 = 4$$
$$\therefore\ x = 2$$

30 **a, b**

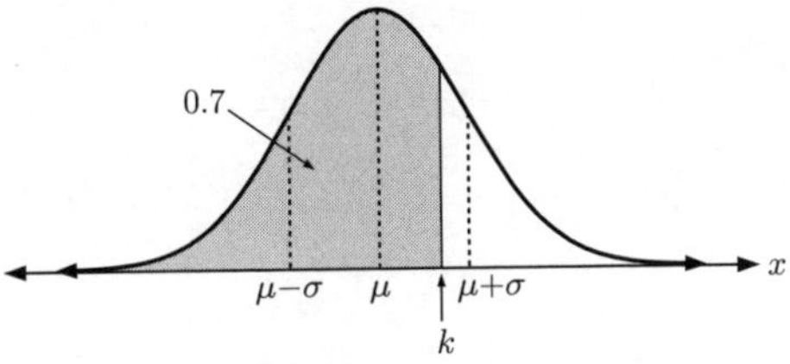

c **i** $\mathrm{P}(X > k) = 1 - 0.7 = 0.3$

ii $\mathrm{P}(\mu < X < k) = 0.7 - 0.5 = 0.2$

iii $\mathrm{P}(\mu - \sigma < X < k)$
$= \mathrm{P}(\mu - \sigma < X < \mu) + \mathrm{P}(\mu < X < k)$
$\approx 0.341 + 0.2 \approx 0.541$

d $\mathrm{P}(k \leqslant X \leqslant t)$
$= \mathrm{P}(X \leqslant t) - \mathrm{P}(X < k)$
$= 0.8 - 0.7$
$= 0.1$

31 **a** $200 - 60 = 140$ out of 200 students passed
$\therefore$ 70% passed

b **i** $m = \mathrm{Q}_1 \approx 27.5$

ii $n = \mathrm{Q}_2 \approx 35$

iii $p = \mathrm{Q}_3 \approx 42.5$

iv $q =$ maximum score $= 100$

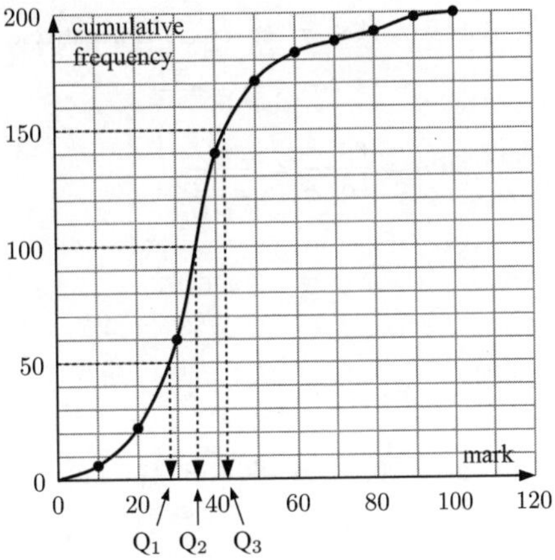

32 $y = (\tan 60^o)x$ is $y = \sqrt{3}x$

a $(10,\ p)$ lies on L if $p = \sqrt{3} \times 10$
$\therefore\ p = 10\sqrt{3}$

b The gradient of L is $\sqrt{3}$
$\therefore$ the gradient of the perpendicular line is $-\frac{1}{\sqrt{3}}$

$\therefore$ its equation is $\dfrac{y - 10\sqrt{3}}{x - 10} = -\dfrac{1}{\sqrt{3}}$

$$\therefore\ \sqrt{3}y - 30 = -x + 10$$
$$\therefore\ x + \sqrt{3}y = 40$$

33 **a**

$$a(t) = 1 - 3\cos(2t + \tfrac{\pi}{2})\ \text{cm s}^{-2}$$
$$v(t) = \int \left[1 - 3\cos(2t + \tfrac{\pi}{2})\right]\, dt$$
$$= t - 3(\tfrac{1}{2})\sin(2t + \tfrac{\pi}{2}) + c$$

But $v(0) = 5$

$$\therefore\ -\tfrac{3}{2}\sin(\tfrac{\pi}{2}) + c = 5$$
$$\therefore\ -\tfrac{3}{2}(1) + c = 5$$
$$\therefore\ c = 6\tfrac{1}{2}$$
$$\therefore\ v(t) = t - \tfrac{3}{2}\sin(2t + \tfrac{\pi}{2}) + 6\tfrac{1}{2}\ \text{cm s}^{-1}$$

b

$$v(\tfrac{\pi}{4}) = \tfrac{\pi}{4} - \tfrac{3}{2}\sin\pi + 6\tfrac{1}{2}$$
$$= \tfrac{\pi}{4} + \tfrac{13}{2}$$
$$= \tfrac{\pi+26}{4}\ \text{cm s}^{-1}$$

34 **a**

$$f(x) = e^{3x-4} + 1 \quad \text{has inverse}$$
$$x = e^{3y-4} + 1$$
$$\therefore\ e^{3y-4} = x - 1$$
$$\therefore\ 3y - 4 = \ln(x-1)$$
$$\therefore\ 3y = \ln(x-1) + 4$$
$$\therefore\ f^{-1}(x) = \frac{\ln(x-1)+4}{3}$$

b

$$f^{-1}(8) - f^{-1}(3)$$
$$= \frac{\ln 7 + 4}{3} - \frac{\ln 2 + 4}{3}$$
$$= \frac{\ln 7 + 4 - \ln 2 - 4}{3}$$
$$= \tfrac{1}{3}\ln(\tfrac{7}{2})$$

35

$$\sin A = \tfrac{2}{5}$$

Now $\cos^2 A + \sin^2 A = 1$

$$\therefore\ \cos^2 A + \tfrac{4}{25} = 1$$
$$\therefore\ \cos^2 A = \tfrac{21}{25}$$
$$\therefore\ \cos A = \pm\tfrac{\sqrt{21}}{5}$$

But $\frac{\pi}{2} \leqslant A \leqslant \pi$, so $\cos A$ is negative

$$\therefore\ \cos A = -\tfrac{\sqrt{21}}{5}$$

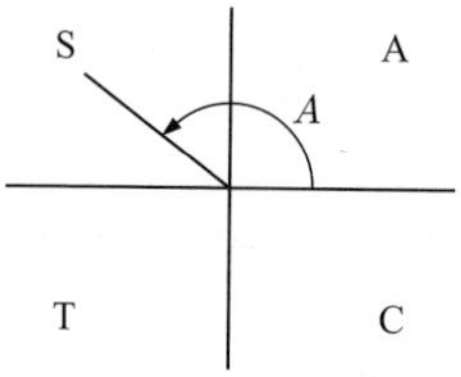

a

$$\tan A = \frac{\sin A}{\cos A} = \frac{\frac{2}{5}}{-\frac{\sqrt{21}}{5}}$$
$$= -\tfrac{2}{\sqrt{21}}$$

b

$$\sin 2A = 2\sin A\cos A$$
$$= 2(\tfrac{2}{5})(-\tfrac{\sqrt{21}}{5})$$
$$= -\tfrac{4\sqrt{21}}{25}$$

36 **a**

$$u_1 = 20, \quad r = \tfrac{10\sqrt{2}}{20} = \tfrac{1}{\sqrt{2}}$$
$$u_{12} = u_1 r^{11}$$
$$= 20 \times \left(\tfrac{1}{\sqrt{2}}\right)^{11}$$
$$= 20 \times \tfrac{1}{(\sqrt{2})^{11}}$$
$$= 20 \times \tfrac{1}{32\sqrt{2}}$$
$$= \tfrac{5}{8\sqrt{2}}$$
$$= \tfrac{5}{16}\sqrt{2}$$

b

$$S = \frac{u_1}{1-r} = \left(\frac{20}{1-\frac{1}{\sqrt{2}}}\right) \times \frac{\sqrt{2}}{\sqrt{2}}$$
$$= \left(\frac{20\sqrt{2}}{\sqrt{2}-1}\right) \times \left(\frac{\sqrt{2}+1}{\sqrt{2}+1}\right)$$
$$= \frac{40 + 20\sqrt{2}}{2-1}$$
$$= 40 + 20\sqrt{2}$$

37 **a** $\overrightarrow{PQ} = \begin{pmatrix} 1-3 \\ 3-1 \\ 4-(-2) \end{pmatrix} = \begin{pmatrix} -2 \\ 2 \\ 6 \end{pmatrix}$

b $|\overrightarrow{PQ}| = \sqrt{4+4+36}$
$= \sqrt{44}$ units
$\therefore$ speed $= \frac{\sqrt{44}}{2}$ units s^{-1}
$= \sqrt{11}$ units s^{-1}

c The line has equation

$$\mathbf{r} = \begin{pmatrix} 3 \\ 1 \\ -2 \end{pmatrix} + t\begin{pmatrix} -2 \\ 2 \\ 6 \end{pmatrix}, \quad t \geqslant 0$$

$\therefore\ x = 3-2t, \quad y = 1+2t, \quad z = -2+6t, \quad t \geqslant 0$

38 **a** $\mathbf{PQ} = \begin{pmatrix} -1 & 3 \\ a & -4 \end{pmatrix}\begin{pmatrix} -4 & -3 \\ -2 & b \end{pmatrix}$

$= \begin{pmatrix} 4-6 & 3+3b \\ -4a+8 & -3a-4b \end{pmatrix}$

$= \begin{pmatrix} -2 & 3b+3 \\ 8-4a & -3a-4b \end{pmatrix}$

b $\mathbf{PQ} = k\mathbf{I}$

$\therefore \begin{pmatrix} -2 & 3b+3 \\ 8-4a & -3a-4b \end{pmatrix} = \begin{pmatrix} k & 0 \\ 0 & k \end{pmatrix}$

$\therefore\ k=-2, \quad 3b+3=0 \quad \therefore\ b=-1$
$8-4a=0 \quad \therefore\ a=2$

Check: $-3a-4b = -3(2)-4(-1)$
$= -6+4$
$= -2$ ✓

Thus $a=2, \ b=-1, \ k=-2$.

c $\mathbf{PQ} = -2\mathbf{I}$
$\therefore\ \mathbf{P}(-\frac{1}{2}\mathbf{Q}) = \mathbf{I}$
$\therefore\ \mathbf{P}^{-1} = -\frac{1}{2}\mathbf{Q}$

39 **a, c**

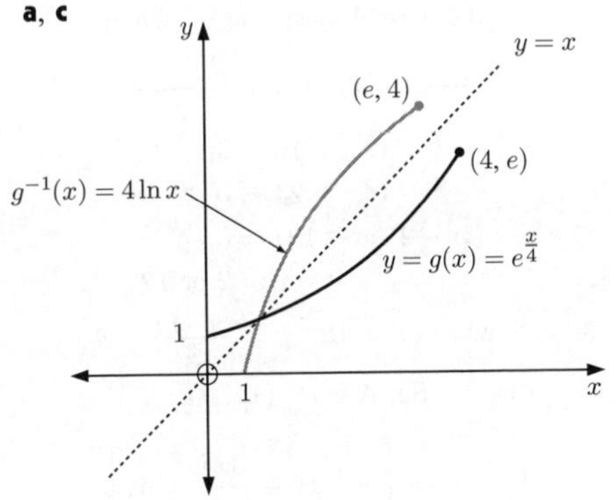

b The range of g is $\{y \mid 1 \leqslant y \leqslant e\}$

d The domain of g^{-1} is $\{x \mid 1 \leqslant x \leqslant e\}$
The range of g^{-1} is $\{y \mid 0 \leqslant y \leqslant 4\}$

e The inverse of $y = e^{\frac{x}{4}}$ is
$x = e^{\frac{y}{4}}$
$\therefore\ \ln x = \frac{y}{4}$
$\therefore\ y = 4\ln x$
So, $g^{-1}(x) = 4\ln x$

40 **a** By the cosine rule,
$(\text{QS})^2 = 3^2+5^2-2\times 3\times 5\times\cos\phi$
$= 9+25-30\cos\phi$
$= 34-30\cos\phi$
$\therefore\ \text{QS} = \sqrt{34-30\cos\phi}$ cm $\{\text{QS} > 0\}$

b If $\phi = 60^o$, $\text{QS} = \sqrt{34-30(\frac{1}{2})}$
$= \sqrt{34-15}$
$= \sqrt{19}$ cm

Now $\frac{\sin\theta}{7} = \frac{\sin 30^o}{\sqrt{19}}$ {sine rule}

$\therefore\ \sin\theta = \frac{7(\frac{1}{2})}{\sqrt{19}} = \frac{7}{2\sqrt{19}}$

c Let $\text{RS} = x$ cm
$\therefore\ \text{QS}^2 = 7^2+x^2-2\times 7\times x\times\cos 30^o$ {cosine rule}
$\therefore\ 19 = 49+x^2-14x(\frac{\sqrt{3}}{2})$
$\therefore\ x^2-7\sqrt{3}x+30=0$
$\therefore\ x = \frac{7\sqrt{3}\pm\sqrt{147-4(1)(30)}}{2}$

$\therefore\ x = \dfrac{7\sqrt{3} \pm \sqrt{27}}{2}$

$\therefore\ x = \dfrac{7\sqrt{3} \pm 3\sqrt{3}}{2}$

$\therefore\ x = \frac{10\sqrt{3}}{2}$ or $\frac{4\sqrt{3}}{2}$

$\therefore\ x = 5\sqrt{3}$ or $2\sqrt{3}$

$\therefore\ x = 5\sqrt{3}$ $\left\{\text{If } x = 2\sqrt{3},\ \cos\theta = \dfrac{19+12-49}{2\sqrt{19}\times 2\sqrt{3}} \text{ which is } < 0 \quad \therefore\ \theta \text{ is obtuse}\right\}$

So, the length of [RS] is $5\sqrt{3}$ cm.

d Perimeter $= 5 + 3 + 7 + 5\sqrt{3}$
$= 15 + 5\sqrt{3}$ cm

Area $= \frac{1}{2}(3)(5)\sin 60^o + \frac{1}{2}(7)(5\sqrt{3})\sin 30^o$
$= \frac{15}{2}(\frac{\sqrt{3}}{2}) + \frac{35}{4}\sqrt{3}$
$= \frac{50}{4}\sqrt{3}$
$= \frac{25}{2}\sqrt{3}$ cm^2

41 **a** **i** $f(x) = -\frac{1}{4}x^2 + 3x + 4$
$\therefore\ f'(x) = -\frac{1}{2}x + 3$

ii $f'(2) = -1 + 3 = 2$
$\therefore$ the normal has gradient $-\frac{1}{2}$,
and passes through (2, 9).

$\therefore$ its equation is $\dfrac{y-9}{x-2} = -\dfrac{1}{2}$
$\therefore\ 2y - 18 = -x + 2$
$\therefore\ x + 2y = 20$

iii The normal has equation $x + 2y = 20$
$\therefore\ y = \dfrac{20-x}{2}$

$\therefore$ the normal meets $f(x)$ where
$-\frac{1}{4}x^2 + 3x + 4 = \dfrac{20-x}{2}$
$\therefore\ -x^2 + 12x + 16 = 40 - 2x$
$\therefore\ x^2 - 14x + 24 = 0$
$\therefore\ (x-2)(x-12) = 0$
$\therefore\ x = 2$ or 12

when $x = 12$, $y = \frac{20-12}{2} = 4$

So, A is at (12, 4).

b **i** area $= \int_2^6 \left(-\frac{1}{4}x^2 + 3x + 4\right) dx$

ii area $= \left[-\frac{1}{12}x^3 + \dfrac{3x^2}{2} + 4x\right]_2^6$
$= (-18 + 54 + 24) - (-\frac{2}{3} + 6 + 8)$
$= 46\frac{2}{3}$ units^2

iii volume $= \pi \int_2^6 \left(-\frac{1}{4}x^2 + 3x + 4\right)^2 dx$

42 **a** **i** $r = \frac{-12}{4} = -3$

ii $u_{14} = u_1 r^{13}$
$= 4 \times (-3)^{13}$ or -4×3^{13}

b **i** $r = \dfrac{x-2}{x} = \dfrac{2x-7}{x-2}$
$\therefore\ (x-2)^2 = x(2x-7)$
$\therefore\ x^2 - 4x + 4 = 2x^2 - 7x$
$\therefore\ x^2 - 3x - 4 = 0$
$\therefore\ (x-4)(x+1) = 0$
$\therefore\ x = 4$ or -1

ii When $x = 4$, sequence is 4, 2, 1, ... with $r = \frac{1}{2}$
$\therefore\ S = \dfrac{4}{1-\frac{1}{2}} = 8$

When $x = -1$, sequence is $-1, -3, -9$, ...
with $r = 3$
$\therefore$ S does not exist $\{S$ only exists if $-1 < r < 1\}$

c The sequence is arithmetic if

$$u_2 - u_1 = u_3 - u_2$$
$$\therefore\ x - 2 - x = 2x - 7 - (x - 2)$$
$$\therefore\ -2 = 2x - 7 - x + 2$$
$$\therefore\ x - 5 = -2$$
$$\therefore\ x = 3$$

i When $x = 3$, the sequence is 3, 1, −1, ... which is arithmetic with $u_1 = 3$ and $d = -2$.

$$\therefore\ u_{30} = u_1 + 29d = 3 + 29(-2) = -55$$

ii
$$S_{50} = \tfrac{50}{2}\,[2(3) + 49(-2)] = 25\,[6 - 98] = 25 \times -92 = -2300$$

43 **a** $\overrightarrow{OA} = \begin{pmatrix} 3 \\ 2 \\ -1 \end{pmatrix}$, $\overrightarrow{OB} = \begin{pmatrix} 2 \\ -1 \\ -8 \end{pmatrix}$

i
$$\overrightarrow{AB} = \overrightarrow{AO} + \overrightarrow{OB} = \begin{pmatrix} -3 \\ -2 \\ 1 \end{pmatrix} + \begin{pmatrix} 2 \\ -1 \\ -8 \end{pmatrix} = \begin{pmatrix} -1 \\ -3 \\ -7 \end{pmatrix}$$

ii
$$\overrightarrow{BA} = \begin{pmatrix} 1 \\ 3 \\ 7 \end{pmatrix}$$
$$\therefore\ |\overrightarrow{BA}| = \sqrt{1 + 9 + 49} = \sqrt{59}$$
$$\therefore\ \text{the unit vector } \mathbf{u} = \tfrac{1}{\sqrt{59}}\begin{pmatrix} 1 \\ 3 \\ 7 \end{pmatrix}.$$

b
$$\mathbf{u} \bullet \overrightarrow{OA} = \tfrac{1}{\sqrt{59}}\begin{pmatrix} 1 \\ 3 \\ 7 \end{pmatrix} \bullet \begin{pmatrix} 3 \\ 2 \\ -1 \end{pmatrix} = \tfrac{1}{\sqrt{59}}(3 + 6 - 7) = \tfrac{2}{\sqrt{59}} \neq 0$$

$\therefore$ $\mathbf{u}$ and $\overrightarrow{OA}$ are not perpendicular.

c
$$\overrightarrow{OC} = \begin{pmatrix} 1 \\ 1 \\ a \end{pmatrix}$$
$$\therefore\ \begin{pmatrix} 1 \\ 1 \\ a \end{pmatrix} \bullet \begin{pmatrix} a \\ -1 \\ 4 \end{pmatrix} = 0$$
$$\therefore\ a - 1 + 4a = 0$$
$$\therefore\ 5a = 1$$
$$\therefore\ a = \tfrac{1}{5}$$

d M is $(\tfrac{5}{2}, \tfrac{1}{2}, -\tfrac{9}{2})$

$\therefore$ $\overrightarrow{OM} = \tfrac{1}{2}(5\mathbf{i} + \mathbf{j} - 9\mathbf{k})$

e $\mathbf{r}_1 = \tfrac{1}{2}\begin{pmatrix} 5 \\ 1 \\ -9 \end{pmatrix} + t\begin{pmatrix} 3 \\ 2 \\ -1 \end{pmatrix}$, $t \in \mathbb{R}$

f $\mathbf{r}_2 = \begin{pmatrix} m \\ 1 \\ -1 \end{pmatrix} + s\begin{pmatrix} 2 \\ -3 \\ 1 \end{pmatrix}$

i L_1 and L_2 are not parallel as $\begin{pmatrix} 3 \\ 2 \\ -1 \end{pmatrix} \neq k\begin{pmatrix} 2 \\ -3 \\ 1 \end{pmatrix}$ for any $k \in \mathbb{R}$.

ii If L_1 and L_2 intersect then $\tfrac{5}{2} + 3t = m + 2s$, $\tfrac{1}{2} + 2t = 1 - 3s$ and $-\tfrac{9}{2} - t = -1 + s$

$$\therefore\ \begin{cases} 3s + 2t = \tfrac{1}{2} \\ s + t = -\tfrac{7}{2} \end{cases} \qquad \therefore\ \begin{aligned} 3s + 2t &= \tfrac{1}{2} \\ -2s - 2t &= 7 \end{aligned}$$

adding: $s = 7\tfrac{1}{2}$ and so $t = -11$

$$\therefore\ \tfrac{5}{2} + (-33) = m + 15$$
$$\therefore\ -30\tfrac{1}{2} = m + 15$$
$$\therefore\ m = -45\tfrac{1}{2}$$

iii Using L_1, when $t = -11$,

$$\mathbf{r}_1 = \frac{1}{2}\begin{pmatrix} 5 \\ 1 \\ -9 \end{pmatrix} - 11\begin{pmatrix} 3 \\ 2 \\ -1 \end{pmatrix} = \begin{pmatrix} -30\frac{1}{2} \\ -21\frac{1}{2} \\ 6\frac{1}{2} \end{pmatrix}$$

So, P is at $(-30\frac{1}{2}, -21\frac{1}{2}, 6\frac{1}{2})$.

44 **a** **i** $\mathbf{A}^2 = \begin{pmatrix} 2 & 1 \\ 0 & 1 \end{pmatrix}\begin{pmatrix} 2 & 1 \\ 0 & 1 \end{pmatrix} = \begin{pmatrix} 4 & 3 \\ 0 & 1 \end{pmatrix}$ **ii** $\mathbf{A}^3 = \begin{pmatrix} 4 & 3 \\ 0 & 1 \end{pmatrix}\begin{pmatrix} 2 & 1 \\ 0 & 1 \end{pmatrix} = \begin{pmatrix} 8 & 7 \\ 0 & 1 \end{pmatrix}$

iii $\mathbf{A}^4 = \begin{pmatrix} 8 & 7 \\ 0 & 1 \end{pmatrix}\begin{pmatrix} 2 & 1 \\ 0 & 1 \end{pmatrix} = \begin{pmatrix} 16 & 15 \\ 0 & 1 \end{pmatrix}$

b $\mathbf{A}^1 = \begin{pmatrix} 2 & 1 \\ 0 & 1 \end{pmatrix} = \begin{pmatrix} 2^1 & 2^1 - 1 \\ 0 & 1 \end{pmatrix}$, $\mathbf{A}^2 = \begin{pmatrix} 4 & 3 \\ 0 & 1 \end{pmatrix} = \begin{pmatrix} 2^2 & 2^2 - 1 \\ 0 & 1 \end{pmatrix}$,

$\mathbf{A}^3 = \begin{pmatrix} 8 & 7 \\ 0 & 1 \end{pmatrix} = \begin{pmatrix} 2^3 & 2^3 - 1 \\ 0 & 1 \end{pmatrix}$, $\mathbf{A}^4 = \begin{pmatrix} 16 & 15 \\ 0 & 1 \end{pmatrix} = \begin{pmatrix} 2^4 & 2^4 - 1 \\ 0 & 1 \end{pmatrix}$

c $\mathbf{A}^{10} = \begin{pmatrix} 2^{10} & 2^{10} - 1 \\ 0 & 1 \end{pmatrix} = \begin{pmatrix} 1024 & 1023 \\ 0 & 1 \end{pmatrix}$

d **i** $S_n = \begin{pmatrix} 2 & 1 \\ 0 & 1 \end{pmatrix} + \begin{pmatrix} 4 & 3 \\ 0 & 1 \end{pmatrix} + \begin{pmatrix} 8 & 7 \\ 0 & 1 \end{pmatrix} + \begin{pmatrix} 16 & 15 \\ 0 & 1 \end{pmatrix} + ... + \begin{pmatrix} 2^n & 2^n - 1 \\ 0 & 1 \end{pmatrix}$

ii $r = \underbrace{0 + 0 + 0 + ... + 0}_{n \text{ of these}} = 0$ $\qquad s = \underbrace{1 + 1 + 1 + ... + 1}_{n \text{ of these}} = n$

iii $p = 2 + 4 + 8 + 16 + ... + 2^n$ which is a geometric series with $u_1 = 2$, $r = 2$

$$\therefore\ p = \frac{u_1(r^n - 1)}{r - 1} = \frac{2(2^n - 1)}{2 - 1} = \frac{2^{n+1} - 2}{1} = 2^{n+1} - 2$$

and $q = 2^1 - 1 + 2^2 - 1 + 2^3 - 1 + ... + 2^n - 1$

$= (2^1 + 2^2 + 2^3 + ... + 2^n) - n$

$\therefore\ q = 2^{n+1} - 2 - n$

iv $S_{14} = \begin{pmatrix} 2^{15} - 2 & 2^{15} - 2 - 14 \\ 0 & 14 \end{pmatrix} = \begin{pmatrix} 2^{15} - 2 & 2^{15} - 16 \\ 0 & 14 \end{pmatrix}$

45 **a** **i** $y = -x^2 + 12x - 20$ has $a = -1 < 0$

$\therefore$ its shape is , and so the quadratic has a maximum value.

ii y is maximised when $x = \frac{-b}{2a} = \frac{-12}{-2} = 6$

iii When $x = 6$, $y = -6^2 + 12(6) - 20 = -36 + 72 - 20 = 16$

$\therefore$ the maximum value of $y = -x^2 + 12x - 20$ is 16.

b **i** The perimeter is 20.

$\therefore\ x + y + 8 = 20$

$\therefore\ y = 12 - x$

ii $y^2 = x^2 + 8^2 - 2 \times x \times 8 \times \cos\theta$

$\therefore\ y^2 = x^2 + 64 - 16x\cos\theta$

iii Since $y = 12 - x$, $(12 - x)^2 = x^2 + 64 - 16x\cos\theta$

$$\therefore\ 144 - 24x + x^2 = x^2 + 64 - 16x\cos\theta$$
$$\therefore\ 16x\cos\theta = 24x - 80$$
$$\therefore\ \cos\theta = \frac{24x - 80}{16x} = \frac{3x - 10}{2x}$$

c Area $A = \frac{1}{2} \times x \times 8 \times \sin\theta$
$= 4x\sin\theta$
$\therefore\ A^2 = 16x^2\sin^2\theta$

d $A^2 = 16x^2(1-\cos^2\theta)$
$= 16x^2\left[1 - \left(\dfrac{3x-10}{2x}\right)^2\right]$
$= 16x^2\left[1 - \dfrac{9x^2-60x+100}{4x^2}\right]$
$= 16x^2 - 4(9x^2-60x+100)$
$= 16x^2 - 36x^2 + 240x - 400$
$= -20x^2 + 240x - 400$
$= 20(-x^2+12x-20)$

e A is maximised when A^2 is maximised since $A > 0$.
From **a**, $-x^2+12x-20$ has a maximum value of 16 when $x = 6$.
When $x=6$, $A^2 = 20(16)$
$= 320$
$\therefore\ A = \sqrt{320}$ $\{A > 0\}$
$= 8\sqrt{5}$
$\therefore$ the maximum area of the triangle is $8\sqrt{5}$ units^2.

f When $x = 6$, $y = 12 - 6$
$= 6$
$\therefore$ the triangle is isosceles.

46 **a** $f(x) = 4x - 3$ has inverse $x = 4y - 3$
$\therefore\ x + 3 = 4y$
$\therefore\ f^{-1}(x) = \dfrac{x+3}{4}$

$g(x) = x + 2$ has inverse $x = y + 2$
$\therefore\ y = x - 2$
$\therefore\ g^{-1}(x) = x - 2$

b $(f \circ g)^{-1}(x) = f(g^{-1}(x))$
$= f(x-2)$
$= 4(x-2) - 3$
$= 4x - 11$

c $(f \circ g^{-1})(x) = f^{-1}(x)$
$\therefore\ 4x - 11 = \dfrac{x+3}{4}$
$\therefore\ 16x - 44 = x + 3$
$\therefore\ 15x = 47$
$\therefore\ x = \frac{47}{15}$

d **i, iv** $H(x) = \dfrac{4x-3}{x+2}$
$\therefore\ H(0) = \dfrac{4(0)-3}{0+2} = -\frac{3}{2}$
So the y-intercept is $-1\frac{1}{2}$.
When $H(x) = 0$, $4x - 3 = 0$
$\therefore\ x = \frac{3}{4}$
$\therefore$ the x-intercept is $\frac{3}{4}$.
$g(x) = 0$ when $x = -2$, so $x = -2$ is a vertical asymptote.
$H(x) = \dfrac{4x-3}{x+2} = \dfrac{4 - \frac{3}{x}}{1 + \frac{2}{x}}$
which $\to \frac{4}{1} = 4$ as $x \to \pm\infty$,
so $y = 4$ is a horizontal asymptote.

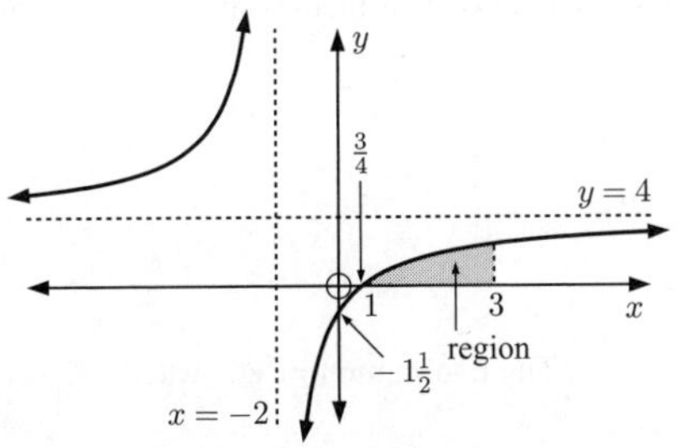

ii $\dfrac{4x-3}{x+2} = A + \dfrac{B}{x+2}$
$= \dfrac{A(x+2)+B}{x+2}$
$\therefore\ Ax + 2A + B = 4x - 3$ for all x
$\therefore\ A = 4$ and $2A + B = -3$
$\therefore\ 8 + B = -3$
$\therefore\ B = -11$

iii $\int_{-1}^{2} H(x)\,dx$
$= \int_{-1}^{2}\left(4 + \dfrac{-11}{x+2}\right)dx$
$= [4x - 11\ln(x+2)]_{-1}^{2}$
$\{x + 2 > 0$ for $-1 \leqslant x \leqslant 2\}$
$= (8 - 11\ln 4) - (-4 - \underbrace{11\ln 1}_{0})$
$= 8 - 11\ln 4 + 4$
$= 12 - 11\ln 4$

47 **a** Hannah is wrong as $0.1+0.3+0.3+0.2+0.2=1.1\neq 1$

b $0.2+a+0.3+b+0.2=1$

$\therefore\ a+b=0.3,\quad a,\ b\geqslant 0\quad\text{and}\quad a,\ b\leqslant 0.3$

c **i** $P(X=2)=\dfrac{2\times 4}{50}=0.16$

ii $P(X\neq 2)=1-P(X=2)$
$=0.84$

48 **a**

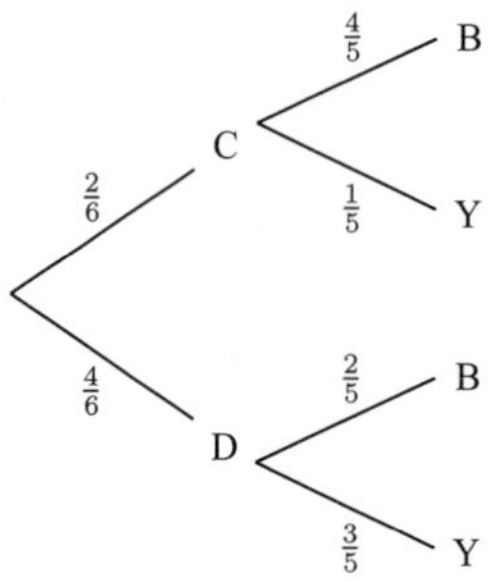

b P(a yellow drawn from D)
$=P(\text{D and Y})$
$=\frac{4}{6}\times\frac{3}{5}$
$=\frac{2}{5}$

c P(a yellow drawn from either bag)
$=P(C\cap Y\ \text{ or }\ D\cap Y)$
$=\frac{2}{6}\times\frac{1}{5}+\frac{4}{6}\times\frac{3}{5}$
$=\frac{14}{30}$
$=\frac{7}{15}$

d $P(D\mid B)$
$=\dfrac{P(D\cap B)}{P(B)}$
$=\dfrac{\frac{4}{6}\times\frac{2}{5}}{\frac{2}{6}\times\frac{4}{5}+\frac{4}{6}\times\frac{2}{5}}$
$=\dfrac{8}{8+8}$
$=\frac{1}{2}$

e $P(B)=\frac{2}{6}\times\frac{4}{5}+\frac{4}{6}\times\frac{2}{5}=\frac{16}{30}$

$P(Y)=\frac{2}{6}\times\frac{1}{5}+\frac{4}{6}\times\frac{3}{5}=\frac{14}{30}$

$\therefore$ expected return
$=\frac{16}{30}\times \$6+\frac{14}{30}\times \$9$
$=\frac{\$96}{30}+\frac{\$126}{30}$
$=\$3.20+\$4.20$
$=\$7.40$

49 **a**

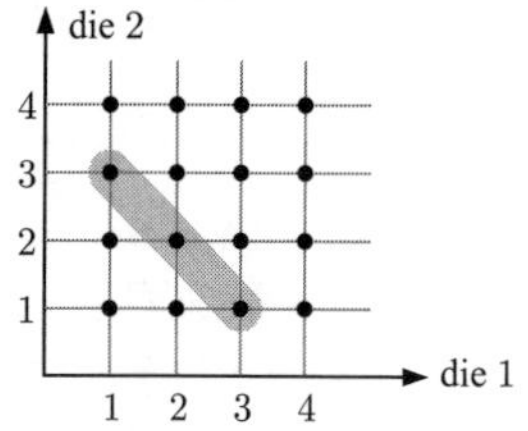

b The possible values of X range from $1+1=2$ to $4+4=8$.
$\therefore$ the possible values of X are 2, 3, 4, 5, 6, 7 and 8.

c **i** 3 of the 16 possible outcomes result in a sum of 4. {those shaded}
$\therefore\ P(X=4)=\frac{3}{16}$

ii 10 of the 16 possible outcomes result in a sum greater than 4.
$\therefore\ P(X>4)=\frac{10}{16}=\frac{5}{8}$

iii Of the 10 outcomes with a sum greater than 4, 3 of them result in a score of 6.
$\therefore\ P(X=6\mid X>4)=\frac{3}{10}$

d From **c**, $P(X<4)=1-\frac{3}{16}-\frac{10}{16}=\frac{3}{16}$

Expectation $=\frac{3}{16}\times 5+\frac{5}{8}\times 1-\frac{3}{16}\times d$
$=\frac{25}{16}-\dfrac{3d}{16}$
$=\left(\dfrac{25-3d}{16}\right)$ euros

The expectation $=$ €0 when $25-3d=0$
$\therefore\ d=8\frac{1}{3}$

50 $a(t) = 3t - \sin t \text{ cm s}^{-2}$

a $a(0) = 3(0) - \sin 0 = 0 \text{ cm s}^{-2}$

$a(\frac{\pi}{2}) = 3(\frac{\pi}{2}) - \sin(\frac{\pi}{2}) = \frac{3\pi}{2} - 1 \text{ cm s}^{-2}$

b $v(t) = \int a(t)\, dt$

$\therefore\ v(t) = \dfrac{3t^2}{2} + \cos t + c \text{ cm s}^{-1}$

But $v(0) = 3$

$\therefore\ 0 + \cos 0 + c = 3$

$\therefore\ c = 2$

Thus $v(t) = \dfrac{3t^2}{2} + \cos t + 2 \text{ cm s}^{-1}$

c $\int_0^{\frac{\pi}{2}} v(t)\, dt$

$= \int_0^{\frac{\pi}{2}} \left(\frac{3t^2}{2} + \cos t + 2\right) dt$

$= \left[\dfrac{t^3}{2} + \sin t + 2t\right]_0^{\frac{\pi}{2}}$

$= \left(\dfrac{\pi^3}{16} + 1 + \pi\right) - (0)$

$= \left(\dfrac{\pi^3}{16} + \pi + 1\right)$ cm

As $\pi \approx 3.14$, $\dfrac{\pi^3}{16} + \pi + 1 > 0$

$\therefore\ \int_0^{\frac{\pi}{2}} v(t)\, dt > 0$

d This integral represents the displacement of the particle in the first $\frac{\pi}{2}$ seconds of motion.

51 $f(t) = a \sin b(t - c) + d$

a $a = \text{amplitude} = \frac{17-3}{2} = 7$

$\text{period} = \dfrac{2\pi}{b} = 13 - -3 = 16$

$\therefore\ b = \frac{\pi}{8}$

$c =$ the x-coordinate of the point halfway between the first minimum and the following maximum

$= \dfrac{-3+5}{2} = 1$

The equation of the principal axis is $y = \frac{3+17}{2} = 10 \quad \therefore\ d = 10$

b **i** Under a translation of $\binom{2}{-3}$, A(5, 17) $\mapsto$ $A_1(7, 14)$

Under a vertical stretch with scale factor 2, $A_1(7, 14) \mapsto A'(7, 28)$

$\therefore$ A(5, 17) $\mapsto$ A′(7, 28)

ii Under a translation of $\binom{2}{-3}$, $y = 7\sin \frac{\pi}{8}(t-1) + 10$ becomes

$y = 7\sin \frac{\pi}{8}(t-1-2) + 10 - 3$

$\therefore\ y = 7\sin \frac{\pi}{8}(t-3) + 7$

and then under a vertical stretch of factor 2 it becomes $y = 2\left[7\sin \frac{\pi}{8}(t-3) + 7\right]$

$\therefore\ y = 14\sin \frac{\pi}{8}(t-3) + 14$

c To map g back to f, we perform the inverse transformations in the reverse order.

So, the transformation is a vertical stretch (x-axis invariant) of scale factor $\frac{1}{2}$ followed by a translation of $\binom{-2}{3}$.

52 **a** $4^x - 2^x - 20$

$= (2^x)^2 - 2^x - 20$

$= (2^x + 4)(2^x - 5)$

b $2^x(2^x - 1) = 20$

$\therefore\ 2^{2x} - 2^x - 20 = 0$

$\therefore\ (2^x + 4)(2^x - 5) = 0$

$\therefore\ 2^x = -4$ or 5

But 2^x is never negative

$\therefore\ 2^x = 5$

$\therefore\ \log 2^x = \log 5$

$\therefore\ x \log 2 = \log 5$

$\therefore\ x = \dfrac{\log 5}{\log 2}$

or $\log_2 5$

c **i** If $p = \log_5 2$ then

$$p = \frac{\log 2}{\log 5}$$

$$\therefore\ x = \frac{1}{p}$$

ii

$$8^x = 5^{1-x}$$

$$\therefore\ 2^{3x} = 5^{1-x}$$

$$\therefore\ 3x \log 2 = (1-x)\log 5$$

$$\therefore\ \frac{1-x}{3x} = \frac{\log 2}{\log 5} = p$$

$$\therefore\ 1 - x = 3px$$

$$\therefore\ x(3p+1) = 1$$

$$\therefore\ x = \frac{1}{3p+1}$$

53 **a**

$$f(x) = a\cos 2x + b\sin^2 x$$
$$= a\cos 2x + b[\sin x]^2$$
$$\therefore\ f'(x) = a(-2\sin 2x) + 2b[\sin x]^1 \cos x$$
$$= -2a\sin 2x + b\sin 2x$$
$$= (b - 2a)\sin 2x$$

b $b < 2a \ \therefore\ b - 2a < 0$

$\therefore$ the max value of $f'(x)$ is $2a - b$

when $\sin 2x = -1$

$$\therefore\ 2x = \tfrac{3\pi}{2} + k2\pi,\ k \in \mathbb{Z}$$

$$\therefore\ x = \tfrac{3\pi}{4} + k\pi$$

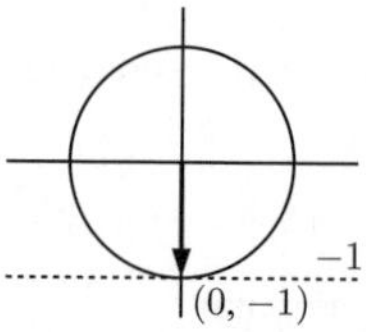

So, the maximum value of $f'(x)$ is $2a - b$ when $x = \frac{3\pi}{4}$ or $\frac{7\pi}{4}$ $\{0 \leqslant x \leqslant 2\pi\}$

c $f'(x) = 0$ when $\sin 2x = 0$

$$\therefore\ 2x = 0 + k\pi,\quad k \in \mathbb{Z}$$

$$\therefore\ x = \tfrac{k\pi}{2}$$

$$\therefore\ x = 0,\ \tfrac{\pi}{2},\ \pi,\ \tfrac{3\pi}{2},\ 2\pi$$

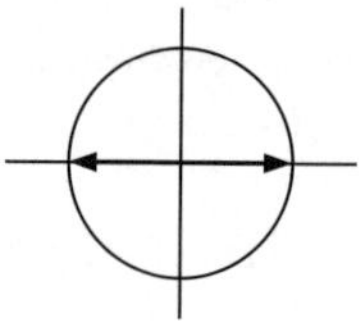

Sign diagram of $f'(x)$:

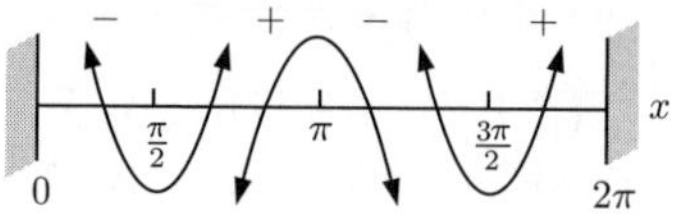

$$f(0) = a\cos 0 + b\sin^2 0 = a$$
$$f(\tfrac{\pi}{2}) = a\cos\pi + b\sin^2\tfrac{\pi}{2} = b - a$$
$$f(\pi) = a\cos 2\pi + b\sin^2\pi = a$$
$$f(\tfrac{3\pi}{2}) = a\cos 3\pi + b\sin^2\tfrac{3\pi}{2} = b - a$$
$$f(2\pi) = a\cos 4\pi + b\sin^2 2\pi = a$$

So, the maximum turning points are $(0, a)$, (π, a) and $(2\pi, a)$,
and the minimum turning points are $(\frac{\pi}{2}, b-a)$ and $(\frac{3\pi}{2}, b-a)$.

54 **a**

$$[C(x)]^2 - [S(x)]^2$$
$$= \left(\frac{e^x + e^{-x}}{2}\right)^2 - \left(\frac{e^x - e^{-x}}{2}\right)^2$$
$$= \left(\frac{e^x + e^{-x}}{2} + \frac{e^x - e^{-x}}{2}\right)\left(\frac{e^x + e^{-x}}{2} - \frac{e^x - e^{-x}}{2}\right)$$
$$= \left(\frac{2e^x}{2}\right)\left(\frac{2e^{-x}}{2}\right)$$
$$= e^0$$
$$= 1$$

b $\dfrac{d}{dx}[S(x)] = \frac{1}{2}\left(e^x - e^{-x}(-1)\right)$
$= \dfrac{e^x + e^{-x}}{2}$
$= C(x)$

c $\dfrac{d}{dx}[C(x)] = \frac{1}{2}\left(e^x + e^{-x}(-1)\right)$
$= \dfrac{e^x - e^{-x}}{2}$
$= S(x)$

d $T(x) = \dfrac{S(x)}{C(x)}$

$\therefore \dfrac{d}{dx}(T(x)) = \dfrac{S'(x)C(x) - S(x)C'(x)}{[C(x)]^2}$

$= \dfrac{[C(x)]^2 - [S(x)]^2}{[C(x)]^2}$ {using **b**, **c**}

$= \dfrac{1}{[C(x)]^2}$ {using **a**}

55 $P(t) = \dfrac{60\,000}{1+2e^{-\frac{t}{4}}}, \quad t \geqslant 0$

$= 60\,000\left(1+2e^{-\frac{t}{4}}\right)^{-1}$

a $P'(t) = -60\,000\left(1+2e^{-\frac{t}{4}}\right)^{-2} \times 2e^{-\frac{t}{4}}\left(-\frac{1}{4}\right)$

$= \dfrac{30\,000e^{-\frac{t}{4}}}{\left(1+2e^{-\frac{t}{4}}\right)^2}$

Since $e^{-\frac{t}{4}} > 0$ for all t and $\left(1+2e^{-\frac{t}{4}}\right)^2 > 0$ for all t,

$P'(t) > 0$ for all $t \geqslant 0$

b $P(t)$ is increasing for all $t \geqslant 0$

c $P''(t)$

$$= \frac{30\,000e^{-\frac{t}{4}}\left(-\frac{1}{4}\right)\left(1+2e^{-\frac{t}{4}}\right)^2 - 30\,000e^{-\frac{t}{4}} \times 2\left(1+2e^{-\frac{t}{4}}\right)^1 2e^{-\frac{t}{4}}\left(-\frac{1}{4}\right)}{\left(1+2e^{-\frac{t}{4}}\right)^4}$$

$$= \frac{-7500e^{-\frac{t}{4}}\left(1+2e^{-\frac{t}{4}}\right) + 30\,000e^{-\frac{t}{2}}}{\left(1+2e^{-\frac{t}{4}}\right)^3}$$

$$= \frac{-7500e^{-\frac{t}{4}} - 15\,000e^{-\frac{t}{2}} + 30\,000e^{-\frac{t}{2}}}{\left(1+2e^{-\frac{t}{4}}\right)^3}$$

$$= \frac{15\,000e^{-\frac{t}{2}} - 7500e^{-\frac{t}{4}}}{\left(1+2e^{-\frac{t}{4}}\right)^3}$$

$$= \frac{7500e^{-\frac{t}{4}}\left(2e^{-\frac{t}{4}} - 1\right)}{\left(1+2e^{-\frac{t}{4}}\right)^3}$$

d The growth is given by $P'(t)$ which is maximised when $P''(t) = 0$

$\therefore 2e^{-\frac{t}{4}} = 1$

$\therefore e^{-\frac{t}{4}} = \frac{1}{2}$

$\therefore -\dfrac{t}{4} = \ln(0.5)$

$\therefore t = -4\ln(0.5)$

$\therefore t = 4\ln 2$

At this time, $P'(t) = \dfrac{30\,000(\frac{1}{2})}{\left(1+2(\frac{1}{2})\right)^2} = \dfrac{15\,000}{4}$

$\therefore$ the maximum growth rate is 3750 per year at $t = 4\ln 2$ years.

e as $t \to \infty$, $e^{-\frac{t}{4}} \to 0$

$\therefore \; P(t) \to \dfrac{60\,000}{1+2(0)}$

$\therefore \; P(t) \to 60\,000$

and $P(0) = \dfrac{60\,000}{1+2} = 20\,000$

f

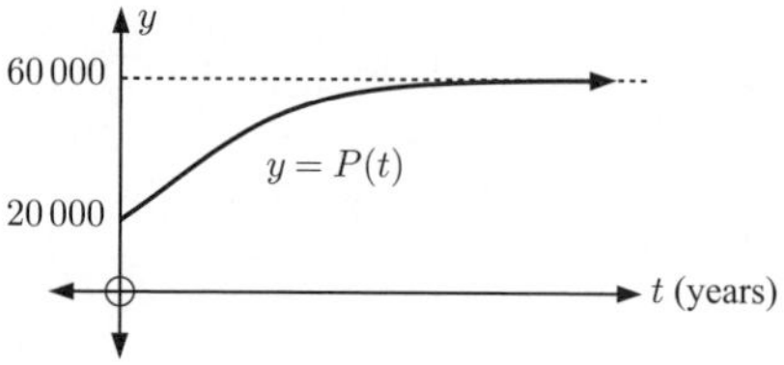

56

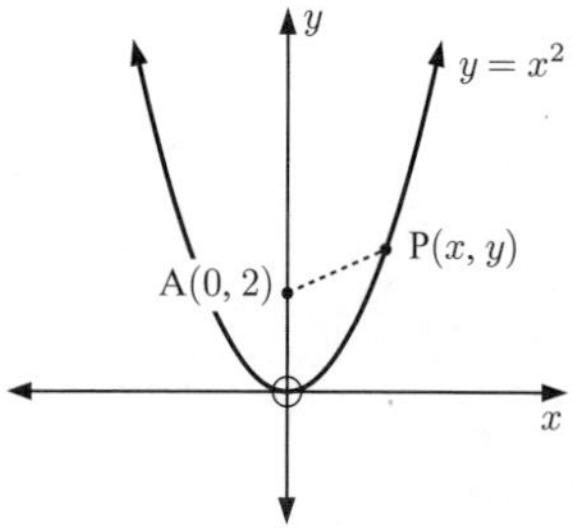

a $\text{AP} = \sqrt{(x-0)^2+(y-2)^2}$

$= \sqrt{x^2+(x^2-2)^2}$ {as $y = x^2$}

$= \sqrt{x^2+x^4-4x^2+4}$

$= \sqrt{x^4-3x^2+4}$

b Since $\text{AP} > 0$, AP will be minimised when AP^2 is minimised.

$\text{AP}^2 = x^4 - 3x^2 + 4$

$\therefore \; \dfrac{d(\text{AP}^2)}{dx} = 4x^3 - 6x$

$= 2x(2x^2-3)$

$= 2x(\sqrt{2}x+\sqrt{3})(\sqrt{2}x-\sqrt{3})$

Sign diagram of $\dfrac{d(\text{AP}^2)}{dx}$:

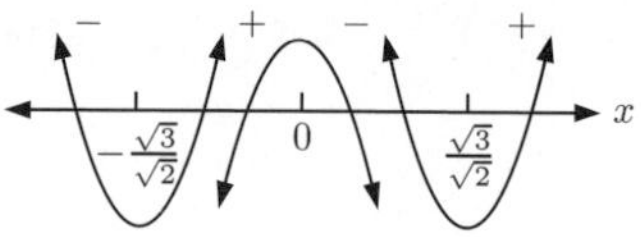

AP is a minimum when $x = \pm\frac{\sqrt{3}}{\sqrt{2}}$ and $y = x^2 = \frac{3}{2}$

So, AP is a minimum when P is at $\left(-\frac{\sqrt{3}}{\sqrt{2}}, \frac{3}{2}\right)$ or $\left(\frac{\sqrt{3}}{\sqrt{2}}, \frac{3}{2}\right)$.

The minimum distance $\text{AP}_{\min} = \sqrt{\frac{9}{4} - 3(\frac{3}{2}) + 4}$

$= \sqrt{\frac{7}{4}}$

$= \frac{1}{2}\sqrt{7}$ units

57 $f(x) = \cos^3 x = [\cos x]^3$

a As $-1 \leqslant \cos x \leqslant 1$ then $-1 \leqslant \cos^3 x \leqslant 1$

$\therefore$ the range is $\{y \mid -1 \leqslant y \leqslant 1\}$

b $8\cos^3 x = 1$

$\therefore \; \cos^3 x = \frac{1}{8}$

$\therefore \; \cos x = \frac{1}{2}$

$\therefore \; x = \frac{\pi}{3}, \frac{5\pi}{3}$

So, there are 2 solutions.

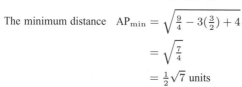

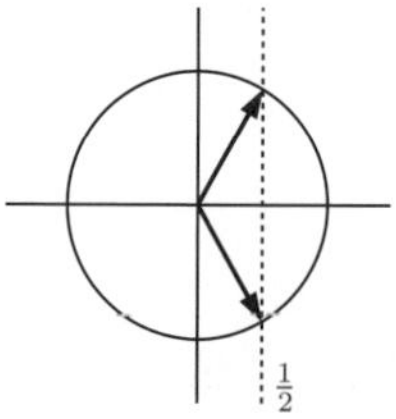

c $f'(x) = 3[\cos x]^2(-\sin x)$

$= -3\sin x \cos^2 x$

d $\text{Volume} = \pi \int_0^{\frac{\pi}{2}} \left[\sqrt{3}\cos x\sqrt{\sin x}\,\right]^2 dx$

$= -\pi \int_0^{\frac{\pi}{2}} (-3\sin x\cos^2 x)\, dx$

$= -\pi \left[(\cos x)^3\right]_0^{\frac{\pi}{2}}$

$= -\pi(0^3 - 1^3)$

$= \pi$ units^3

58

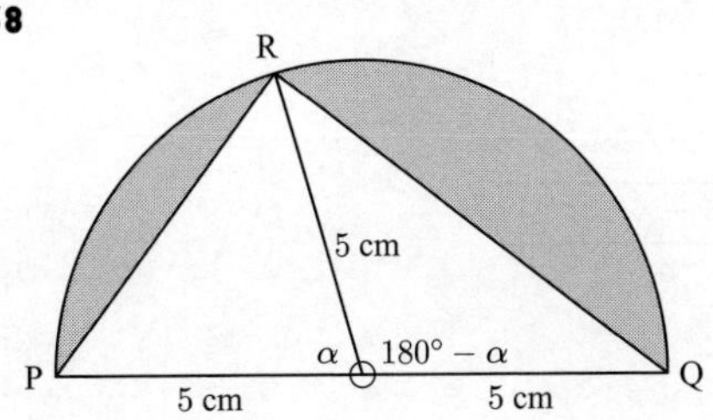

a Area of ΔPQR

$= \frac{1}{2}(5^2)\sin\alpha + \frac{1}{2}(5^2)\sin(180^o - \alpha)$

$= \frac{25}{2}(\sin\alpha + \sin\alpha)$

$= 25\sin\alpha \text{ cm}^2$

b $A = \frac{1}{2}\pi(5^2) - 25\sin\alpha$

$= \frac{25\pi}{2} - 25\sin\alpha \text{ cm}^2$

c Now $0 \leqslant \sin\alpha \leqslant 1$ $\{$as $0 \leqslant \alpha \leqslant \pi\}$

$\therefore$ the area is a minimum when $\sin\alpha = 1$

$\therefore$ $A_{\min} = \frac{25\pi}{2} - 25(1)$

$= 25(\frac{\pi}{2} - 1)$ when $\alpha = \frac{\pi}{2}$

The area is a maximum when $\sin\alpha = 0$.

$\therefore$ $A_{\max} = \frac{25\pi}{2}$ cm^2 when $\alpha = 0$ or π.

59 **a** The vertex is $\left(\frac{1+7}{2}, 18\right)$ which is $(4, 18)$.

Thus **i** $h = 4$ **ii** $k = 18$

$\therefore$ $f(x) = a(x-4)^2 + 18$

iii Now $f(1) = 0$

$\therefore$ $a(-3)^2 + 18 = 0$

$\therefore$ $9a = -18$

$\therefore$ $a = -2$

b $f(x) = -2(x-4)^2 + 18$

$= -2(x^2 - 8x + 16) + 18$

$= -2x^2 + 16x - 14$

$\therefore$ the shaded area $= \int_1^3 (-2x^2 + 16x - 14)\, dx$

$= \left[-\frac{2}{3}x^3 + 8x^2 - 14x\right]_1^3$

$= \left(-\frac{2}{3}(27) + 8(9) - 42\right) - \left(-\frac{2}{3} + 8 - 14\right)$

$= 12 + 6\frac{2}{3} = 18\frac{2}{3}$ units^2

60 **a** **i** $2^{1-2x} = 0.5 = \frac{1}{2}$

$\therefore$ $2^{1-2x} = 2^{-1}$

$\therefore$ $1 - 2x = -1$

$\therefore$ $x = 1$

ii $\log_x 7 = 5$

$\therefore$ $x^5 = 7$

$\therefore$ $x = \sqrt[5]{7}$

b $25^x - 6(5^x) + 5 = 0$

$\therefore$ $(5^x)^2 - 6(5^x) + 5 = 0$

$\therefore$ $(5^x - 1)(5^x - 5) = 0$

$\therefore$ $5^x = 1$ or 5

$\therefore$ $x = 0$ or 1

c $2^x = 3^{1-x}$

$\therefore$ $2^x = \dfrac{3^1}{3^x}$

$\therefore$ $6^x = 3$

$\therefore$ $x = \log_6 3$

61 **a** **i** $\dfrac{1-\cos 2\theta}{\sin 2\theta} = \sqrt{3}$, $0 < \theta < \frac{\pi}{2}$

$\therefore$ $\dfrac{1 - (1 - 2\sin^2\theta)}{2\sin\theta\cos\theta} = \sqrt{3}$

$\therefore$ $\dfrac{\not{2}\sin^{\not{2}}\theta}{\not{2}\not{\sin\theta}\cos\theta} = \sqrt{3}$

$\therefore$ $\dfrac{\sin\theta}{\cos\theta} = \sqrt{3}$ $\{\sin\theta \neq 0$ on $0 < \theta < \frac{\pi}{2}\}$

$\therefore$ $\tan\theta = \sqrt{3}$

ii As $0 < \theta < \frac{\pi}{2}$, $\theta = \frac{\pi}{3}$.

b If $\cos 2x = 2\cos x$ then $2\cos^2 x - 1 = 2\cos x$

$$\therefore\ 2\cos^2 x - 2\cos x - 1 = 0$$

$$\therefore\ \cos x = \frac{2 \pm \sqrt{4 - 4(2)(-1)}}{4} = \frac{2 \pm \sqrt{12}}{4} = \frac{2 \pm 2\sqrt{3}}{4} = \frac{1 \pm \sqrt{3}}{2}$$

$$\therefore\ \cos x = \frac{1 - \sqrt{3}}{2} \quad \{-1 \leqslant \cos x \leqslant 1\}$$

62 $S_n = n^3 + 2n - 1$

Now $u_n = S_n - S_{n-1},\quad n > 1$

$= n^3 + 2n - 1 - \left[(n-1)^3 + 2(n-1) - 1\right]$

$= n^3 + 2n - 1 - \left[n^3 - 3n^2 + 3n - 1 + 2n - 2 - 1\right]$

$= \cancel{n^3} + \cancel{2n} - \cancel{1} - \cancel{n^3} + 3n^2 - 3n + \cancel{1} - \cancel{2n} + 3$

$= 3n^2 - 3n + 3,\quad n > 1$

and $u_1 = S_1 = 2$

$\therefore\ u_1 = 2,\ u_n = 3n^2 - 3n + 3,\ n > 1$

63 $\sin^2 x + \sin x - 2 = 0,\quad -2\pi \leqslant x \leqslant 2\pi$

$\therefore\ (\sin x + 2)(\sin x - 1) = 0$

$\therefore\ \sin x = -2$ or 1

$\therefore\ \sin x = 1 \quad \{\text{as } -1 \leqslant \sin x \leqslant 1\}$

$\therefore\ x = \frac{\pi}{2} + k2\pi,\quad k$ an integer

$\therefore\ x = -\frac{3\pi}{2}$ or $\frac{\pi}{2}$

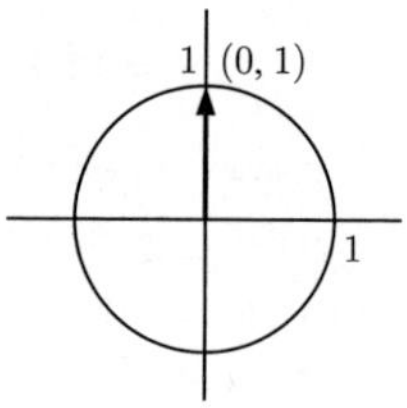

64 $f(x) = \ln x$ has inverse $f^{-1}(x) = e^x$.

$g(x) = 3 + x$ has inverse given by $x = 3 + y$

$\therefore\ y = x - 3$ so $g^{-1}(x) = x - 3$.

a $f^{-1}(2) \times g^{-1}(2)$

$= e^2 \times -1$

$= -e^2$

b $(f \circ g)(x) = f(g(x)) = f(3 + x)$

$= \ln(3 + x)$

$\therefore$ the inverse of $(f \circ g)(x)$ is $x = \ln(3 + y)$

$\therefore\ 3 + y = e^x$

$\therefore\ y = e^x - 3$

So, $(f \circ g)^{-1}(x) = e^x - 3$

and $(f \circ g)^{-1}(2) = e^2 - 3$

65

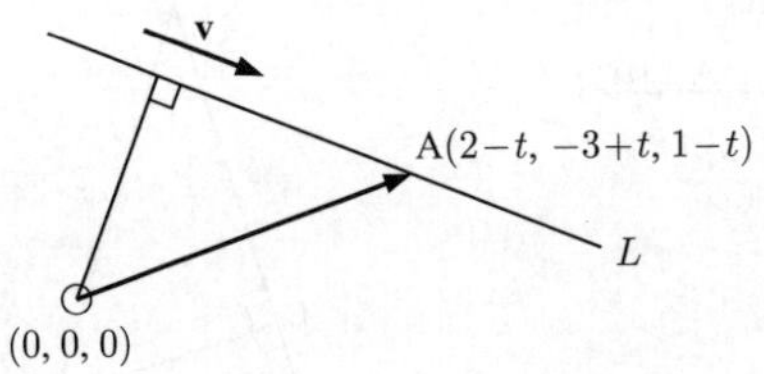

$\overrightarrow{OA} = \begin{pmatrix} 2-t \\ t-3 \\ 1-t \end{pmatrix}$ and $\mathbf{v} = \begin{pmatrix} -1 \\ 1 \\ -1 \end{pmatrix}$

The shortest distance occurs when $\overrightarrow{OA} \bullet \mathbf{v} = 0$

$\therefore \; -(2-t) + t - 3 + (-1)(1-t) = 0$

$\therefore \; t - 2 + t - 3 - 1 + t = 0$

$\therefore \; 3t = 6$

$\therefore \; t = 2$

So, the point on L that is nearest the origin is $(2-2, -3+2, 1-2)$, which is $(0, -1, -1)$.

66 $f'(x) > 0$ and $f''(x) < 0$ for all x

$\therefore$ $f(x)$ is increasing and concave downwards for all x.

a $f(2) = 1$ and $f'(2) = 2$

$\therefore$ $(2, 1)$ lies on the curve and the tangent at this point has gradient 2

$\therefore$ the equation of the tangent is $y = 2x + c$

and $1 = 2(2) + c$, so $c = -3$

$\therefore$ the tangent has equation $y = 2x - 3$.

b

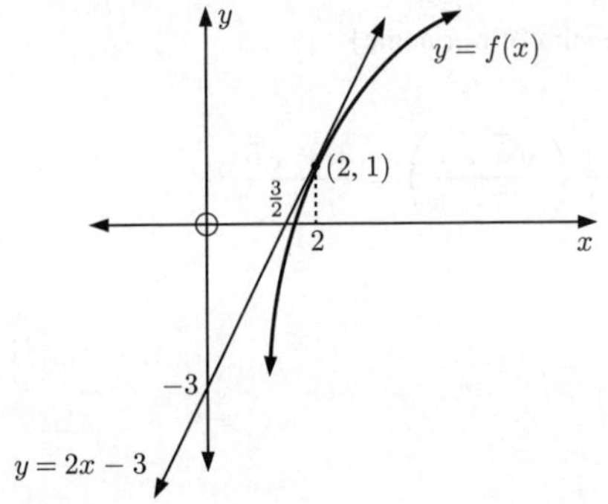

c As $f(x)$ is increasing it has *at most one* zero. But $f(x)$ is also concave downwards for all x, so it always lies below the tangent shown. So, for $x < \frac{3}{2}$, the tangent's y-values are negative and so $f(x)$ is also negative. Thus $f(x)$ has *exactly one* zero.

d From the graph, the x-intercept of $y = f(x)$ lies inside $\frac{3}{2} < x < 2$.

67

$\log_y x - \log_x y = \frac{8}{3}$

$\therefore \; \log_y 16y - \log_{16y} y = \frac{8}{3}$ {as $x = 16y$}

$\therefore \; \dfrac{\log 16y}{\log y} - \dfrac{\log y}{\log 16y} - \frac{8}{3} = 0$

$\therefore \; m - \dfrac{1}{m} - \frac{8}{3} = 0$ {letting $\dfrac{\log 16y}{\log y} = m$}

$\therefore \; 3m^2 - 8m - 3 = 0$ $\{\times 3m\}$

$\therefore \; (3m+1)(m-3) = 0$

$\therefore \; m = \dfrac{\log 16y}{\log y} = -\frac{1}{3}$ or 3

$\therefore \; \log 16y = -\frac{1}{3}\log y$ or $\log 16y = 3\log y$

$\therefore \; 16y = y^{-\frac{1}{3}}$ or $16y = y^3$

$\therefore \; y^{\frac{4}{3}} = \frac{1}{16}$ or $y(y^2 - 16) = 0$

$\therefore \; y = \left(\pm\frac{1}{2}\right)^3$ or $y = 0$ or ± 4

$\therefore \; y = \frac{1}{8}$ or 4 {as y is a base, $y > 0$}

$\therefore \; y = \frac{1}{8}, \; x = 2$ or $y = 4, \; x = 64$

68 After labelling the triangles,

ΔABC and ΔABD are isosceles {equal base angles}

and $\widehat{\text{BDC}} = 2\alpha$ {exterior angle of ΔABD}

$\therefore$ ΔBDC is isosceles {equal base angles}

Thus AD = BD = BC = x, say.

Now ΔADE and ΔBDE are congruent {AAcorS}

$\therefore$ AE = EB

If we let AE = EB = 1 unit, then AC = 2 and DC = $2 - x$.

Now ΔABC and ΔBCD are similar {equiangular}

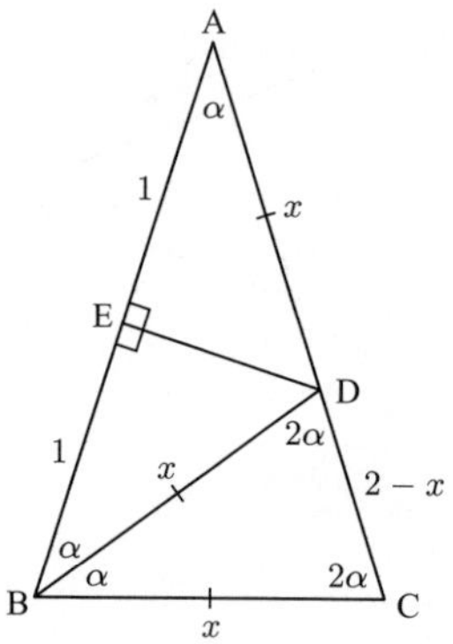

$$\therefore \frac{\text{AB}}{\text{BC}} = \frac{\text{BC}}{\text{CD}} \text{ and so } \frac{2}{x} = \frac{x}{2-x}$$

$$\therefore x^2 = 4 - 2x$$

$$\therefore x^2 + 2x - 4 = 0$$

$$\therefore x = \frac{-2 \pm \sqrt{4 - 4(-4)}}{2}$$

$$\therefore x = \frac{-2 \pm 2\sqrt{5}}{2}$$

$$\therefore x = -1 \pm \sqrt{5}$$

$$\therefore x = \sqrt{5} - 1 \quad \{\text{as } x \text{ must be} > 0\}$$

$$\text{Now } 5\alpha = 180^o \quad \{\text{angle sum of a triangle}\}$$

$$\therefore \alpha = 36^o$$

$$\text{so in } \Delta\text{BED}, \ \cos 36^o = \frac{1}{\sqrt{5}-1} = \frac{1}{\sqrt{5}-1}\left(\frac{\sqrt{5}+1}{\sqrt{5}+1}\right) = \frac{1+\sqrt{5}}{4}.$$

69

$$\text{Area } = \int_a^{a+2} x^2\,dx = \tfrac{31}{6}$$

$$\therefore \left[\frac{x^3}{3}\right]_a^{a+2} = \tfrac{31}{6}$$

$$\therefore \frac{(a+2)^3}{3} - \frac{a^3}{3} = \tfrac{31}{6}$$

$$\therefore \frac{\cancel{a^3} + 6a^2 + 12a + 8 - \cancel{a^3}}{3} = \tfrac{31}{6}$$

$$\therefore 12a^2 + 24a + 16 = 31$$

$$\therefore 12a^2 + 24a - 15 = 0$$

$$\therefore 4a^2 + 8a - 5 = 0$$

$$\therefore (2a-1)(2a+5) = 0$$

$$\therefore a = \tfrac{1}{2} \text{ or } -\tfrac{5}{2}$$

But $a > 0$, so $a = \frac{1}{2}$

70 If X and Y are independent events then $\text{P}(X \cap Y) = \text{P}(X)\text{P}(Y)$

Thus $\text{P}((A \cap B) \cap (A \cup B)) = \text{P}(A \cap B)\,\text{P}(A \cup B)$

$\therefore \text{P}(A \cap B) = \text{P}(A \cap B)\,\text{P}(A \cup B)$ {since $A \cap B \subseteq A \cup B$}

$\therefore \text{P}(A \cap B) = 0$ or $\text{P}(A \cup B) = 1$

$\therefore$ A and B are disjoint or either A or B must occur.

71 $\sin\theta\cos\theta = \frac{1}{4}$

$\therefore \frac{1}{2}\sin 2\theta = \frac{1}{4}$

$\therefore \sin 2\theta = \frac{1}{2}$

$\therefore 2\theta = \frac{\pi}{6} + k2\pi$ or $\frac{5\pi}{6} + k2\pi, \quad k \in \mathbb{Z}$

$\therefore 2\theta = \frac{\pi}{6}, -\frac{11\pi}{6}, \frac{5\pi}{6}, -\frac{7\pi}{6}$ $\quad$ {as $-2\pi \leqslant 2\theta \leqslant 2\pi$}

$\therefore \theta = -\frac{11\pi}{12}, -\frac{7\pi}{12}, \frac{\pi}{12}, \frac{5\pi}{12}$

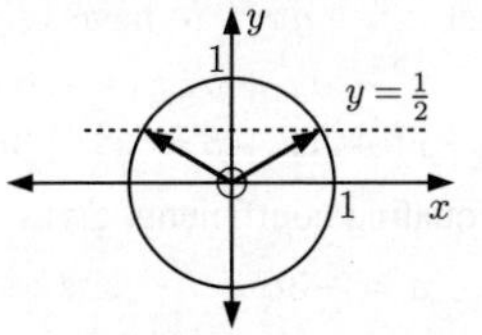

72 **a** $f(x) = \ln(x(x-2))$ is defined when $x(x-2) > 0$

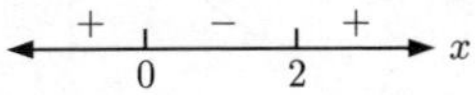

$\therefore x < 0$ or $x > 2$

So the domain is $\{x \mid x < 0$ or $x > 2\}$

b $f(x) = \ln x + \ln(x-2)$ {log law}

$\therefore f'(x) = \frac{1}{x} + \frac{1}{x-2}$

c $f'(3) = \frac{1}{3} + 1 = \frac{4}{3}$ at $(3, \ln 3)$

$\therefore$ the tangent has equation

$$\frac{y - \ln 3}{x - 3} = \frac{4}{3}$$

$\therefore 4x - 12 = 3y - 3\ln 3$

$\therefore 4x - 3y = 12 - 3\ln 3$

73 **a**

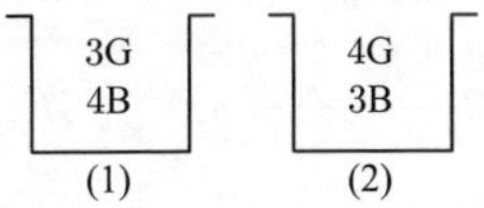

P(both same colour)

$= \mathrm{P}(\text{GG or BB})$

$= \frac{3}{7}\times\frac{4}{7} + \frac{4}{7}\times\frac{3}{7}$

$= \frac{24}{49}$

b P(G from (2) | both different)

$= \dfrac{\text{P(G from (2)} \cap \text{both different)}}{\text{P(both different)}}$

$= \dfrac{\text{P(G from (2) \textbf{and} B from (1))}}{1 - \frac{24}{49}}$

$= \dfrac{\frac{4}{7}\times\frac{4}{7}}{\frac{25}{49}}$

$= \frac{16}{25}$

74

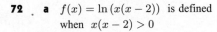

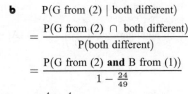

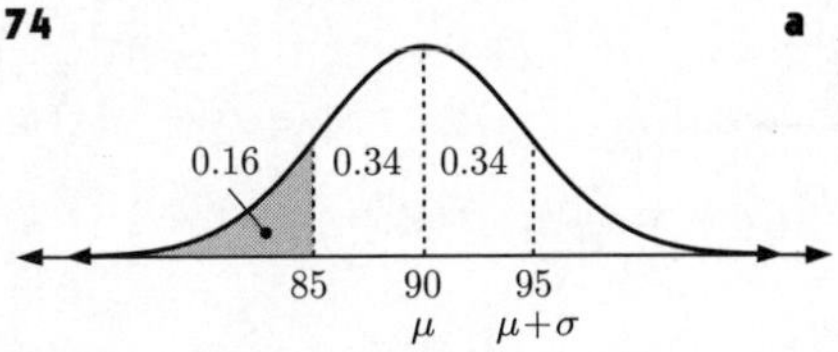

a $\mathrm{P}(X < 85) \approx 0.16$
From the diagram,
$\mathrm{P}(90 < X < 95) \approx 0.34$

b As roughly 34% of scores lie between μ and $\mu + \sigma$ for the normal distribution then $\sigma \approx 5$.

75 $\mathrm{P}(X = x) = a\left(\frac{2}{5}\right)^x$ where $x = 0, 1, 2, 3, 4, 5, ...$

$\therefore a\left(\frac{2}{5}\right)^0 + a\left(\frac{2}{5}\right)^1 + a\left(\frac{2}{5}\right)^2 + ... = 1$ $\quad$ $\{\sum \mathrm{P}(x) = 1\}$

$\therefore a\left(1 + \frac{2}{5} + \left(\frac{2}{5}\right)^2 + ...\right) = 1$

$\therefore a\left(\dfrac{1}{1-\frac{2}{5}}\right) = 1$ $\quad$ {infinite geometric series, $u_1 = 1, \ r = \frac{2}{5}$}

$\therefore \dfrac{a}{\frac{3}{5}} = 1$

$\therefore a = \frac{3}{5}$

76 $x = \log_3 y^2 \quad \therefore y^2 = 3^x = (81^{\frac{1}{4}})^x = 81^{\frac{x}{4}}$

$\therefore y = (81^{\frac{x}{4}})^{\frac{1}{2}} = 81^{\frac{x}{8}}$

$\therefore 81 = y^{\frac{8}{x}}$ and so $\log_y 81 = \dfrac{8}{x}$

77 Let $x^2 + ax + b$ have zeros α and 2α

$\therefore \quad x^2 + ax + b = (x - \alpha)(x - 2\alpha)$

$\therefore \quad x^2 + ax + b = x^2 - 3\alpha x + 2\alpha^2$

Equating coefficients gives

$\qquad a = -3\alpha \qquad$ and $\qquad b = 2\alpha^2$

$\therefore \quad \alpha = \dfrac{-a}{3} \qquad$ and $\qquad b = 2\alpha^2$

$\therefore \quad b = 2\left(\dfrac{a^2}{9}\right) \qquad$ and so $\quad 2a^2 = 9b$

78 **a** $y = f(x - 2) + 1$ is a translation of $y = f(x)$ through $\begin{pmatrix} 2 \\ 1 \end{pmatrix}$. $\quad \therefore \quad$ A$(-2, 3) \mapsto$ A$'(0, 4)$.

b $y = 2f(x - 2)$ is obtained from $y = f(x)$ by a translation through $\begin{pmatrix} 2 \\ 0 \end{pmatrix}$ followed by a vertical stretch with scale factor $p = 2$. $\therefore \quad$ A$(-2, 3) \mapsto$ A$'(0, 3) \mapsto$ A$''(0, 6)$.

c $y = f(2x - 3) = f\left(\dfrac{x - \frac{3}{2}}{\frac{1}{2}}\right)$

It is obtained from $y = f(x)$ by a horizontal stretch with scale factor $q = \frac{1}{2}$ followed by a translation through $\begin{pmatrix} \frac{3}{2} \\ 0 \end{pmatrix}$. $\therefore \quad$ A$(-2, 3) \mapsto$ A$'(-1, 3) \mapsto$ A$''(\frac{1}{2}, 3)$.

d Consider $y = \dfrac{1}{f(x)}$. When $x = -2$, $y = \dfrac{1}{f(-2)} = \frac{1}{3}$. $\therefore \quad$ A$(-2, 3) \mapsto$ A$'(-2, \frac{1}{3})$.

e Consider $y = f^{-1}(x)$. For an inverse function, the point is reflected in the line $y = x$.
$\therefore \quad$ A$(-2, 3) \mapsto$ A$'(3, -2)$.

79

$y = a\sin(b(x - c)) + d$

(0, −1)   (6, −1) $y = -1$

(3, −5)

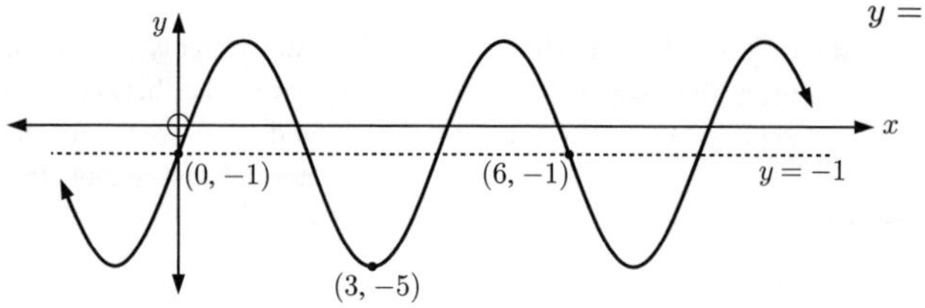

The amplitude $= a = 4$.   The period $= 4 = \dfrac{2\pi}{b}$ $\therefore \quad b = \frac{\pi}{2}$

The basic sine curve has been translated through $\begin{pmatrix} 0 \\ -1 \end{pmatrix}$. $\therefore \quad c = 0, \; d = -1$

Thus $y = 4\sin\left(\frac{\pi}{2}x\right) - 1$

Check: $y(3) = 4\sin\left(\frac{3\pi}{2}\right) - 1 = 4(-1) - 1 = -5$ ✓

$\qquad\quad y(6) = 4\sin(3\pi) - 1 \; = 4(0) - 1 = -1$ ✓

80 **a** A and B are mutually exclusive if $A \cap B = \varnothing$. In this case P$(A \cap B) = 0$, so $x = 0$.

b If A and B are independent, then $\qquad$ P$(A \cap B) =$ P(A)P(B)

$\therefore \quad x = (0.3 + x)(0.2 + x)$

$\therefore \quad x = 0.06 + 0.5x + x^2$

$\therefore \quad x^2 - 0.5x + 0.06 = 0$

$\therefore \quad (x - 0.2)(x - 0.3) = 0$

$\therefore \quad x = 0.2$ or 0.3

81 Since **A** is its own inverse, $\mathbf{A} = \mathbf{A}^{-1}$

$\therefore\ \mathbf{AA} = \mathbf{AA}^{-1}$ {premultiplying by **A**}

$\therefore\ \mathbf{A}^2 = \mathbf{I}$

$$\therefore \begin{pmatrix} a & -1 \\ b & 2 \end{pmatrix}\begin{pmatrix} a & -1 \\ b & 2 \end{pmatrix} = \begin{pmatrix} 1 & 0 \\ 0 & 1 \end{pmatrix}$$

$\therefore\ a^2 - b = 1,\quad -a - 2 = 0,\quad ab + 2b = 0,\quad -b + 4 = 1$

$\therefore\ a = -2$ and $b = 3$ *Check:* $a^2 - b = 4 - 3 = 1$ ✓
$ab + 2b = -6 + 6 = 0$ ✓

So, $\mathbf{A}^{11} = (\mathbf{A}^2)^5\mathbf{A} = \mathbf{I}^5\mathbf{A} = \mathbf{IA} = \mathbf{A} = \begin{pmatrix} -2 & -1 \\ 3 & 2 \end{pmatrix}$.

82 $f(x) = \dfrac{x^2+1}{(x+1)^2} = \dfrac{x^2+1}{x^2+2x+1} = \dfrac{1+\frac{1}{x^2}}{1+\frac{2}{x}+\frac{1}{x^2}}$

a Vertical asymptote is $x = -1$.
As $x \to \pm\infty$, $y \to \frac{1}{1} = 1$
So, the horizontal asymptote is $y = 1$.

b $f'(x) = \dfrac{2x(x+1)^2 - (x^2+1)2(x+1)^1}{(x+1)^4}$

$= \dfrac{2x(x+1) - 2(x^2+1)}{(x+1)^3}$

$= \dfrac{2(x-1)}{(x+1)^3}$

which has sign diagram:

+ ⋮ − | + (critical values -1 and 1 on the x-axis)

$\therefore$ there is a local minimum at $(1, \frac{1}{2})$.

c $f''(x) = \dfrac{2(x+1)^3 - 2(x-1)3(x+1)^2}{(x+1)^6}$

$= \dfrac{2(x+1) - 6(x-1)}{(x+1)^4}$

$= \dfrac{-4x+8}{(x+1)^4}$

$= \dfrac{-4(x-2)}{(x+1)^4}$

which has sign diagram:

+ ⋮ + | − (critical values -1 and 2 on the x-axis)

A change of sign about $x = 2$ indicates that $(2, \frac{5}{9})$ is a point of inflection.

d

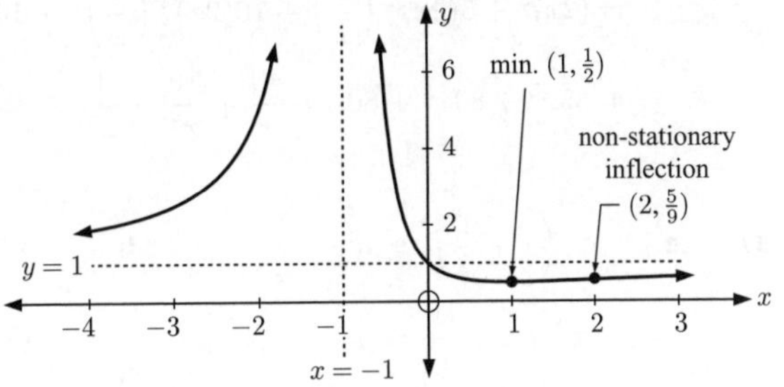

83 **a** $(f \circ g)(x) = f(g(x))$

$= f\left(\dfrac{x+1}{x-2}\right)$

$= 2\left(\dfrac{x+1}{x-2}\right) + 1$

$= \dfrac{2x+2+x-2}{x-2}$

$= \dfrac{3x}{x-2}$

b $y = \dfrac{x+1}{x-2}$ has inverse $x = \dfrac{y+1}{y-2}$

$\therefore\ xy - 2x = y + 1$

$\therefore\ y(x-1) = 2x + 1$

$\therefore\ y = \dfrac{2x+1}{x-1}$

$\therefore\ f^{-1}(x) = \dfrac{2x+1}{x-1}$

84 **a** **i** If A and B are mutually exclusive then $\text{P}(A \cup B) = \text{P}(A) + \text{P}(B) = \frac{1}{3} + \frac{2}{7} = \frac{13}{21}$

ii $$\begin{aligned}\text{P}(A \cup B) &= \text{P}(A) + \text{P}(B) - \text{P}(A \cap B)\\ &= \text{P}(A) + \text{P}(B) - \text{P}(A)\text{P}(B) \quad \{A \text{ and } B \text{ independent}\}\\ &= \tfrac{1}{3} + \tfrac{2}{7} - \tfrac{1}{3} \times \tfrac{2}{7} = \tfrac{11}{21}\end{aligned}$$

b $$\text{P}(A \mid B) = \frac{\text{P}(A \cap B)}{\text{P}(B)} = \frac{\text{P}(A) + \text{P}(B) - \text{P}(A \cup B)}{\text{P}(B)} = \left(\frac{\frac{13}{21} - \frac{3}{7}}{\frac{2}{7}}\right)\frac{21}{21} = \frac{4}{6} = \frac{2}{3}$$

85 $$\begin{aligned}\log_a(x+2) &= \log_a x + 2\\ \therefore\ \log_a(x+2) - \log_a x &= 2\\ \therefore\ \log_a\left(\frac{x+2}{x}\right) &= 2\\ \therefore\ \frac{x+2}{x} &= a^2\\ \therefore\ 1 + \frac{2}{x} &= a^2\\ \therefore\ \frac{2}{x} &= a^2 - 1\\ \therefore\ \text{since } a > 1,\ x &= \frac{2}{a^2 - 1}\end{aligned}$$

86 **a** $(a-b)^5 = a^5 - 5a^4b + 10a^3b^2 - 10a^2b^3 + 5ab^4 - b^5$

b This expression is the binomial expansion of $(0.4 + 0.6)^5 = 1^5$
$= 1$

c $$\begin{aligned}&\left(2x + \frac{1}{x}\right)^5\\ &= (2x)^5 + 5(2x)^4\left(\frac{1}{x}\right) + 10(2x)^3\left(\frac{1}{x}\right)^2 + 10(2x)^2\left(\frac{1}{x}\right)^3 + 5(2x)\left(\frac{1}{x}\right)^4 + \left(\frac{1}{x}\right)^5\\ &= 32x^5 + 80x^3 + 80x + \frac{40}{x} + \frac{10}{x^3} + \frac{1}{x^5}\end{aligned}$$

87 **a** $$\begin{aligned}\left(x + \frac{1}{x}\right)^2 &= a^2\\ \therefore\ x^2 + 2 + \frac{1}{x^2} &= a^2\\ \therefore\ x^2 + \frac{1}{x^2} &= a^2 - 2\end{aligned}$$

b $$\begin{aligned}\left(x + \frac{1}{x}\right)^3 &= a^3\\ \therefore\ x^3 + 3x^2\left(\frac{1}{x}\right) + 3x\left(\frac{1}{x}\right)^2 + \left(\frac{1}{x}\right)^3 &= a^3\\ \therefore\ x^3 + 3x + \frac{3}{x} + \frac{1}{x^3} &= a^3\\ \therefore\ x^3 + \frac{1}{x^3} + 3\left(x + \frac{1}{x}\right) &= a^3\\ \therefore\ x^3 + \frac{1}{x^3} &= a^3 - 3a\end{aligned}$$

88 **a** **i** The ellipse cuts the x-axis when $y = 0$

$$\therefore\ \frac{x^2}{16} = 1 \qquad \therefore\ x^2 = 16 \qquad \therefore\ x = \pm 4$$

So, A is $(4, 0)$ and B is $(-4, 0)$.

ii The ellipse cuts the y-axis when $x = 0$

$$\therefore\ \frac{y^2}{4} = 1 \qquad \therefore\ y^2 = 4 \qquad \therefore\ y = \pm 2$$

So, C is $(0, 2)$ and D is $(0, -2)$.

b Since $\dfrac{y^2}{4} = 1 - \dfrac{x^2}{16}$

then $y^2 = 4 - \dfrac{x^2}{4}$

$\therefore\ y = \pm\sqrt{4 - \dfrac{x^2}{4}}$

But $y > 0$, so $y = \sqrt{4 - \dfrac{x^2}{4}}$.

c $\displaystyle\int_0^4 \sqrt{4 - \frac{x^2}{4}}\, dx$ is the area of one quarter of the ellipse

$\therefore$ area of the ellipse $= 4\displaystyle\int_0^4 \sqrt{4 - \frac{x^2}{4}}\, dx$

d Volume $= \pi \int_{-4}^{4} y^2\, dx$

$= 2\pi \int_0^4 y^2\, dx$

$= 2\pi \int_0^4 (4 - \frac{1}{4}x^2)\, dx$

$= 2\pi \left[4x - \frac{1}{4}\frac{x^3}{3}\right]_0^4$

$= 2\pi \left\{\left(16 - \frac{16}{3}\right) - (0)\right\}$

$= 2\pi \times \frac{32}{3}$

$= \frac{64\pi}{3}$ units^3

89 **a**

x	0	$\frac{\pi}{4}$	$\frac{\pi}{2}$	$\frac{3\pi}{4}$	π	$\frac{5\pi}{4}$	$\frac{3\pi}{2}$	$\frac{7\pi}{4}$	2π
$f(x)$	0	$\frac{1}{2}$	1	$\frac{1}{2}$	0	$\frac{1}{2}$	1	$\frac{1}{2}$	0

b

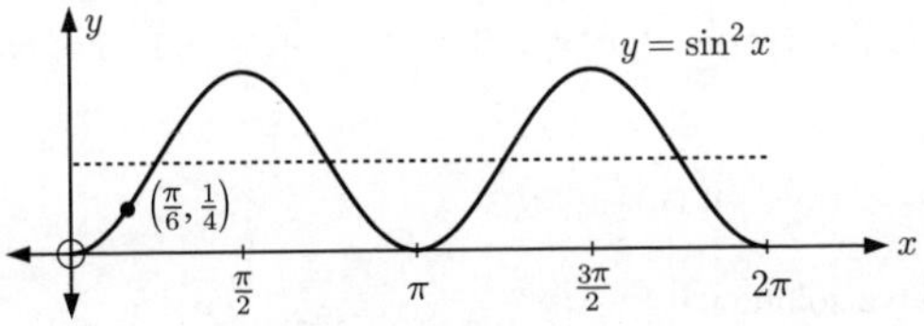

c When $x = \frac{\pi}{6}$, $\sin x = \frac{1}{2}$

$\therefore\ y = \left(\frac{1}{2}\right)^2 = \frac{1}{4}$ ✓

d The range is $\{y \mid 0 \leqslant y \leqslant 1\}$

e Area $= \int_0^{\pi} \sin^2 x\, dx$

$= \int_0^{\pi} \left(\frac{1}{2} - \frac{1}{2}\cos 2x\right)\, dx$

$= \left[\frac{1}{2}x - \frac{1}{2}(\frac{1}{2})\sin 2x\right]_0^{\pi}$

$= \left(\frac{\pi}{2} - \frac{1}{4}(0)\right) - (0 - 0)$

$= \frac{\pi}{2}$ units^2

f $f'(x) = 2(\sin x)^1 \cos x$

$= \sin 2x$

$\therefore\ f'(\frac{\pi}{4}) = \sin\frac{\pi}{2} = 1$

$\therefore$ the slope of the tangent is $\frac{1}{1}$ at $(\frac{\pi}{4}, \frac{1}{2})$

$\therefore$ the equation of the tangent is $x - y = \frac{\pi}{4} - \frac{1}{2}$

90 $f(x) = 2e^{-x},\ \ 0 \leqslant x \leqslant k$

a $f(0) = 2e^0 = 2$

b

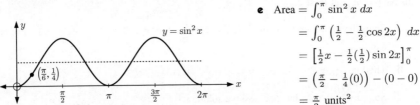

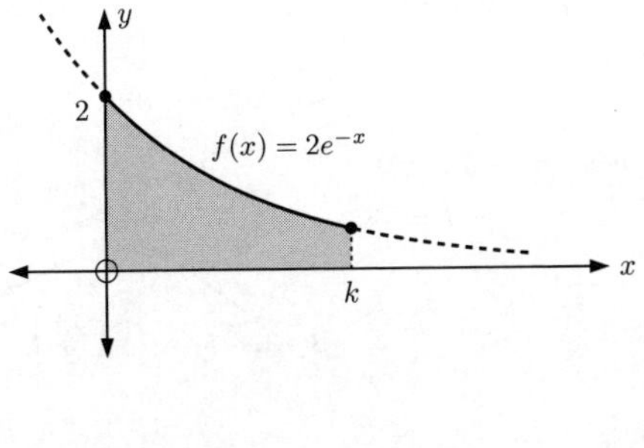

c The shaded region has area 1

$\therefore\ \int_0^k 2e^{-x}\, dx = 1$

$\therefore\ \int_0^k e^{-x}\, dx = \frac{1}{2}$

$\therefore\ \left[-e^{-x}\right]_0^k = \frac{1}{2}$

$\therefore\ -e^{-k} - (-e^0) = \frac{1}{2}$

$\therefore\ e^{-k} = \frac{1}{2}$

$\therefore\ e^k = 2$ {reciprocals}

$\therefore\ k = \ln 2$

d $\mathrm{P}(0 \leqslant X \leqslant \frac{1}{2})$
$= \int_0^{\frac{1}{2}} f(x)\,dx$
$= \int_0^{\frac{1}{2}} 2e^{-x}\,dx$
$= \left[-2e^{-x}\right]_0^{\frac{1}{2}}$
$= (-2e^{-\frac{1}{2}}) - (-2e^0)$
$= -\dfrac{2}{\sqrt{e}} + 2$
$= 2 - \dfrac{2}{\sqrt{e}}$

e $\mathrm{P}(X \geqslant \frac{1}{2})$
$= 1 - \mathrm{P}(0 \leqslant X \leqslant \frac{1}{2})$
$= 1 - \left(2 - \dfrac{2}{\sqrt{e}}\right)$
$= \dfrac{2}{\sqrt{e}} - 1$

91 $f(x) = x + x^{-1}$

a $f'(x) = 1 - x^{-2}$, and $f'(x) = 0$
when $1 - \dfrac{1}{x^2} = 0$
$\therefore\ 1 = \dfrac{1}{x^2}$
$\therefore\ x^2 = 1$
$\therefore\ x = 1$ {as $x > 0$}

b Sign diagram of $f'(x)$:

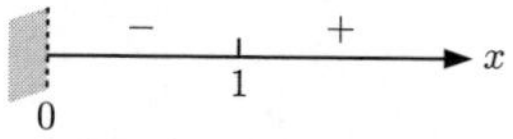

$\therefore$ f has a minimum at $x = 1$.
$f(1) = 1 + \frac{1}{1} = 2$
$\therefore$ A is at $(1, 2)$

c Since the minimum value of $f(x) = x + \dfrac{1}{x}$ for $x > 0$ is 2, the sum of a positive number and its reciprocal is at least 2.

d **i** Since the minimum value of $f(x) = x + \dfrac{1}{x}$ for $x > 0$ is 2, $x + \dfrac{1}{x} = 1$ has no positive solutions.

ii The line $y = 2$ touches $y = x + \dfrac{1}{x}$ at A.
So, $x + \dfrac{1}{x} = 2$ has one positive solution.

iii The line $y = 3$ cuts $y = x + \dfrac{1}{x}$ in two places. So, $x + \dfrac{1}{x} = 3$ has two positive solutions.

92 **a** $\mathbf{r} = \begin{pmatrix} 2 \\ 0 \\ -3 \end{pmatrix} + t\begin{pmatrix} 1 \\ -1 \\ 2 \end{pmatrix}$, $t \in \mathbb{R}$

b $x = 2 + t$, $y = -t$, $z = -3 + 2t$, $t \in \mathbb{R}$

c $(2 + t, -t, -3 + 2t)$ represents any point on the line.

d $\overrightarrow{\mathrm{BP}} = \begin{pmatrix} 2 + t - -1 \\ -t - 3 \\ -3 + 2t - 5 \end{pmatrix} = \begin{pmatrix} t + 3 \\ -t - 3 \\ 2t - 8 \end{pmatrix}$

e $\overrightarrow{\mathrm{BP}} \bullet (\mathbf{i} - \mathbf{j} + 2\mathbf{k}) = (t + 3)1 + (-t - 3)(-1) + (2t - 8)2$
$= t + 3 + t + 3 + 4t - 16$
$= 6t - 10$

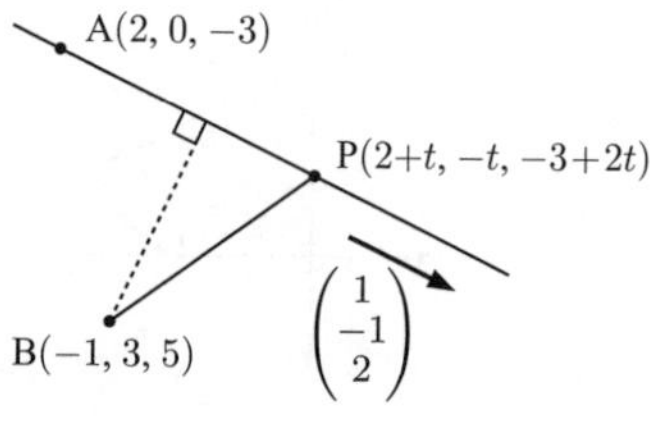

f [BP] is perpendicular to the original line when $6t - 10 = 0$
$\therefore\ t = \frac{5}{3}$

g When $t = \frac{5}{3}$, P is at $(2 + \frac{5}{3}, -\frac{5}{3}, -3 + \frac{10}{3})$
$\therefore$ $(\frac{11}{3}, -\frac{5}{3}, \frac{1}{3})$ is closest to B.

93 **a** **i** A is the minimum value of X.
ii B is Q_1, the lower quartile.
iii C is Q_2, the median.
iv D is Q_3, the upper quartile.
v E is the maximum value of X.

b **i** maximum value − minimum value $= E - A$ is the range.
ii $Q_3 - Q_1 = D - B$ is the interquartile range or IQR.

c **i** 75% of the scores are less than D, and 25% of the scores are less than B.
$\therefore \; P(B \leqslant X \leqslant D) = 0.75 - 0.25 = 0.5$
ii 75% of the scores are greater than B.
$\therefore \; P(X \geqslant B) = 0.75$

d

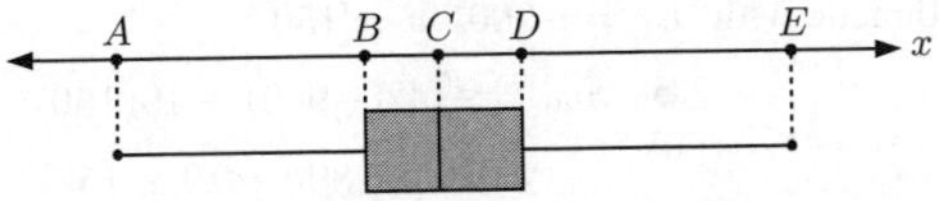

EXERCISE 25B

1 $\sum_{k=1}^{n}(2k-31) = 0$

$\therefore \; (-29) + (-27) + (-25) + ... + (2n-31) = 0$

The LHS is arithmetic with $u_1 = -29$, $d = 2$ and "n" $= n$.

$\therefore \; \frac{n}{2}(-58 + (n-1)2) = 0$

$\therefore \; \frac{n}{2}(2n-60) = 0$

$\therefore \; n(n-30) = 0$

$\therefore \; n = 30$ as $n \neq 0$

2 $f(x) = 5\ln(x-4) + 2$

a, c

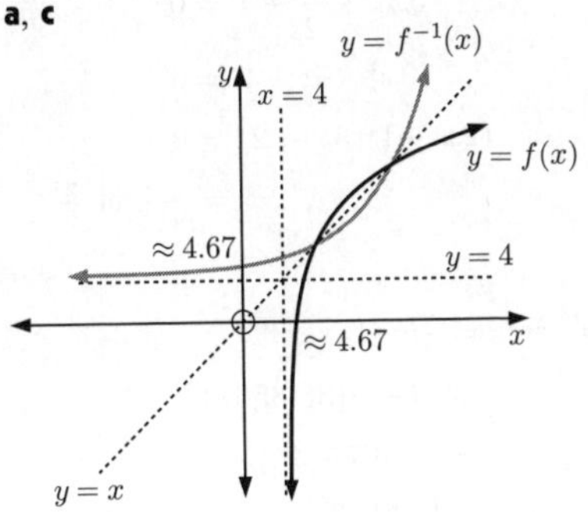

b $f(x) = 1$ when $x \approx 4.82$ {technology}

d $f'(x) = 5\left(\frac{1}{x-4}\right) + 0$

$\therefore \; f'(5) = 5$

$\therefore$ the normal has gradient $-\frac{1}{5}$ when $x = 5$.

$f(5) = 5\ln(1) + 2 = 2$

$\therefore$ at $(5, 2)$ the normal is $\frac{y-2}{x-5} = -\frac{1}{5}$

$\therefore \; 5y - 10 = -x + 5$

$\therefore \; x + 5y = 15$

3 The general term is $T_{r+1} = \binom{9}{r}x^{9-r}\left(-\frac{1}{5x^2}\right)^r$

$= \binom{9}{r}x^{9-r}\left(-\frac{1}{5}\right)^r x^{-2r}$

$= \binom{9}{r}\left(-\frac{1}{5}\right)^r x^{9-3r}$

If we let $9 - 3r = 0$ then $r = 3$

$\therefore \; T_4 = \binom{9}{3}\left(-\frac{1}{5}\right)^3 x^0$

$= 84 \times -\frac{1}{125}$

$= -\frac{84}{125}$

So, the constant term is $-\frac{84}{125}$.

4 **a** The vertex is $(1, 3)$.

$\therefore\ f(x) = (x-1)^2 + 3$ {coefficient of x^2 is 1}

$= x^2 - 2x + 4$

Thus $m = -2$, $n = 4$.

b $f(3) = k \ \therefore\ 9 - 2(3) + 4 = k$

$\therefore\ k = 7$

c $g(x) = f(x-1) + 2$ is a translation of $f(x)$ through $\binom{1}{2}$ and $\binom{1}{3} + \binom{1}{2} = \binom{2}{5}$

$\therefore$ the vertex of $g(x)$ is $(2, 5)$

d Domain of $f(x)$ is $\{x \mid x \in \mathbb{R}\}$
Range of $f(x)$ is $\{y \mid y \geqslant 3\}$

Domain of $g(x)$ is $\{x \mid x \in \mathbb{R}\}$
Range of $g(x)$ is $\{y \mid y \geqslant 5\}$

5 $-900, -750, -600, -450, ...$ is arithmetic with $u_1 = -900$, $d = 150$

a $u_{20} = u_1 + 19d$

$= -900 + 19(150)$

$= 1950$

b $S_{20} = \frac{20}{2}[2(-900) + 19(150)]$

$= 10[-1800 + 19 \times 150]$

$= 10\,500$

6 **a** $(f \circ g)(x) = -8$

$\therefore\ f(g(x)) = -8$

$\therefore\ f(1 - 5x^2) = -8$

$\therefore\ 2(1 - 5x^2) = -8$

$\therefore\ 2 - 10x^2 = -8$

$\therefore\ 10x^2 = 10$

$\therefore\ x = \pm 1$

b $(g \circ f)(x) = -8$

$\therefore\ g(f(x)) = -8$

$\therefore\ g(2x) = -8$

$\therefore\ 1 - 5(2x)^2 = -8$

$\therefore\ 1 - 20x^2 = -8$

$\therefore\ 20x^2 = 9$

$\therefore\ x^2 = \frac{9}{20}$

$\therefore\ x \approx \pm 0.671$

c $f(x) = 2x$ and $g(x) = 1 - 5x^2$

$f'(x) = g'(x)$ implies that

$2 = -10x$

$\therefore\ x = -0.2$

d $f(x)$ is $y = 2x$

$\therefore\ f^{-1}(x)$ is $x = 2y$

$\therefore\ y = \dfrac{x}{2}$

$\therefore\ f^{-1}(x) = \dfrac{x}{2}$

Thus $\dfrac{x}{2} = 1 - 5x^2$

$\therefore\ 5x^2 + \dfrac{x}{2} - 1 = 0$

$\therefore\ 10x^2 + x - 2 = 0$

$\therefore\ (2x+1)(5x-2) = 0$

$\therefore\ x = -\frac{1}{2}$ or $\frac{2}{5}$

7 **a** $P = 50\pi$

$\therefore\ 2r + r(\frac{7\pi}{9}) = 50\pi$

$\therefore\ r(2 + \frac{7\pi}{9}) = 50\pi$

$\therefore\ r = \dfrac{50\pi}{2 + \frac{7\pi}{9}} \approx 35.3507$

$\therefore\ r \approx 35.4$ cm

b area $= \frac{1}{2}\theta r^2$

$\approx \frac{1}{2}(\frac{7\pi}{9})(35.3507)^2$

≈ 1526.77

$\approx 1530\ \text{cm}^2$

c

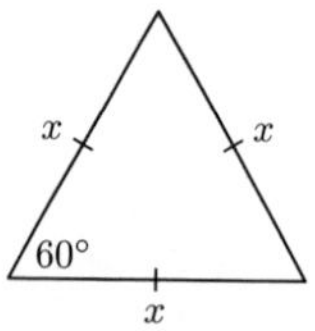

area of triangle $= \frac{1}{2} \times x \times x \times \sin 60^\circ$

$\therefore\ \frac{1}{2}x^2(\frac{\sqrt{3}}{2}) \approx 1526.77$

$\therefore\ \frac{\sqrt{3}}{4}x^2 \approx 1526.77$

$\therefore\ x^2 \approx 3525.9$

$\therefore\ x \approx 59.4$

$\therefore$ the sides are 59.4 cm long.

8 $f(x) = x\sin x - 3\cos x, \quad g(x) = \ln x$

a, c

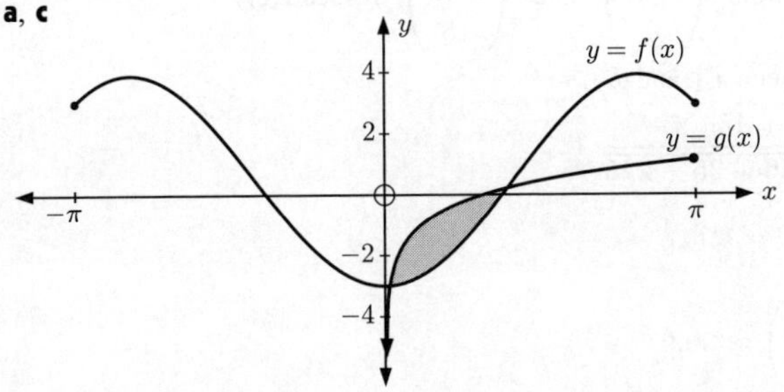

b $x \approx \pm 1.68$ {using technology}

d **i** The curves meet at $x \approx 0.0501$ and $x \approx 1.245$

$\therefore \ A \approx \int_{0.0501}^{1.245} (\ln x - (x\sin x - 3\cos x))\, dx$

$\therefore \ A \approx \int_{0.0501}^{1.245} (\ln x - x\sin x + 3\cos x)\, dx$

ii $A \approx 1.37$ units^2 {using technology}

9 **a** $\mathbf{A}^{-1} = \begin{pmatrix} -2 & -3 & -2 \\ -\frac{1}{2} & -\frac{1}{2} & -\frac{1}{2} \\ -\frac{7}{10} & -\frac{9}{10} & -\frac{1}{2} \end{pmatrix}$ {using technology}

b In matrix form, the equations are

$$\begin{pmatrix} 2 & -3 & -5 \\ -1 & 4 & 0 \\ -1 & -3 & 5 \end{pmatrix}\begin{pmatrix} x \\ y \\ z \end{pmatrix} = \begin{pmatrix} 0 \\ 0 \\ 5 \end{pmatrix}$$

$$\therefore \ \begin{pmatrix} x \\ y \\ z \end{pmatrix} = \begin{pmatrix} 2 & -3 & -5 \\ -1 & 4 & 0 \\ -1 & -3 & 5 \end{pmatrix}^{-1}\begin{pmatrix} 0 \\ 0 \\ 5 \end{pmatrix}$$

$$= \begin{pmatrix} -10 \\ -2.5 \\ -2.5 \end{pmatrix}$$

$\therefore \ x = -10, \ y = -2.5, \ z = -2.5$

10 $f(t) = m\cos n(t - p) + r$

a The amplitude $m = \dfrac{12 - 4}{2} = 4$

period $= 2(5 - 1) = 8$

$\therefore \ \dfrac{2\pi}{n} = 8$

$\therefore \ n = \frac{\pi}{4}$

maximum value $= m + r = 12$

$\therefore \ 4 + r = 12$

$\therefore \ r = 8$

The maximum value occurs when $t = 1$.

$\therefore$ when $t = 1$, $\cos n(t - p) = 1$

$\therefore \ \frac{\pi}{4}(1 - p) = 0$

$\therefore \ p = 1$

Thus, $m = 4, \ n = \frac{\pi}{4}, \ p = 1, \ r = 8$.

b $f(t) = 4\cos\frac{\pi}{4}(t - 1) + 8$

i $f(6) = 4\cos(\frac{\pi}{4} \times 5) + 8$

$= 4\cos(\frac{5\pi}{4}) + 8$

$= 8 - 2\sqrt{2} \approx 5.17$

ii $f(t) = 10$

$\therefore \ 4\cos\frac{\pi}{4}(t - 1) + 8 = 10$

$\therefore \ \cos\frac{\pi}{4}(t - 1) = \frac{1}{2}$

$\therefore \ \frac{\pi}{4}(t - 1) = \pm\frac{\pi}{3} + k2\pi$

$\therefore \ t - 1 = \pm\frac{4}{3} + 8k$

$\therefore \ t = 2\frac{1}{3}$ is the smallest positive t

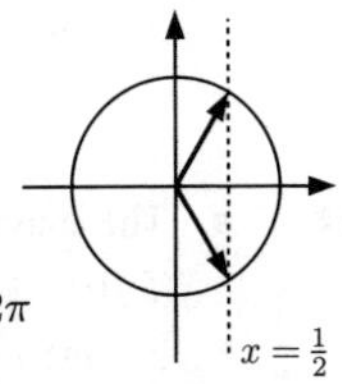

11 **a** L_1 and L_2 have direction vectors $\begin{pmatrix} 4 \\ -1 \\ 3 \end{pmatrix}$ and $\begin{pmatrix} 6 \\ -5 \\ 15 \end{pmatrix}$ respectively.

Let θ be the acute angle between L_1 and L_2.

$$\therefore \quad \cos\theta = \frac{|24+5+45|}{\sqrt{16+1+9}\sqrt{36+25+225}}$$

$$= \frac{74}{\sqrt{26}\sqrt{286}}$$

$$\therefore \quad \theta = \cos^{-1}\left(\frac{74}{\sqrt{7436}}\right) \approx 30.9^{\circ}$$

b **i** When $t = -2$, $\mathbf{r}_1 = \begin{pmatrix} -2 \\ 1 \\ 4 \end{pmatrix} - 2\begin{pmatrix} 4 \\ -1 \\ 3 \end{pmatrix} = \begin{pmatrix} -10 \\ 3 \\ -2 \end{pmatrix}$

$\therefore$ $(-10, 3, -2)$ lies on L_1.

ii We require $\begin{pmatrix} -10 \\ 3 \\ -2 \end{pmatrix} = \begin{pmatrix} -5+6s \\ -5s \\ 7+15s \end{pmatrix}$ for some $s \in \mathbb{R}$

$\therefore$ $3 = -5s$ and so $s = -\frac{3}{5}$

But $-5+6s = -5+6(-\frac{3}{5})$

$= -8\frac{3}{5}$

$\neq -10$

$\therefore$ P does not lie on L_2.

c If L_1 and L_2 meet then
$$\begin{cases} -2+4t = -5+6s & \text{.... (1)} \\ 1-t = -5s & \text{.... (2)} \\ 4+3t = 7+15s & \text{.... (3)} \end{cases}$$

From (1) and (2), $\begin{cases} 6s-4t = 3 \\ t = 1+5s \end{cases}$

$\therefore$ $6s - 4(1+5s) = 3$

$\therefore$ $6s - 4 - 20s = 3$

$\therefore$ $-14s = 7$

$\therefore$ $s = -\frac{1}{2}$

and $t = 1 + 5(-\frac{1}{2}) = -\frac{3}{2}$

Check in (3): LHS $= 4 + 3t = 4 - \frac{9}{2} = -\frac{1}{2}$

RHS $= 7 + 15s = 7 - \frac{15}{2} = -\frac{1}{2}$ ✓

$\therefore$ L_1 and L_2 meet where $s = -\frac{1}{2}$, $t = -\frac{3}{2}$

$\therefore$ they meet at $(-8, 2\frac{1}{2}, -\frac{1}{2})$

d $\begin{pmatrix} 4 \\ -1 \\ 3 \end{pmatrix} \bullet \begin{pmatrix} a \\ 2 \\ 8 \end{pmatrix} = 0$

$\therefore$ $4a - 2 + 24 = 0$

$\therefore$ $4a = -22$

$\therefore$ $a = -5\frac{1}{2}$

12 **a** The x-axis intercepts are -1 and β.

$f(0) = (1)(-\beta) = -\beta$

$\therefore$ the y-intercept is $-\beta$

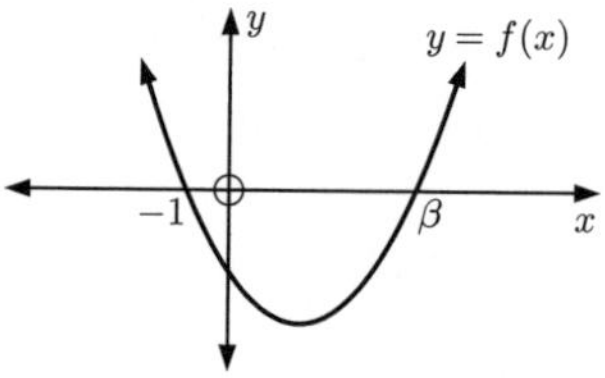

b $g(x) = -f(x-1)$ is obtained by translating $f(x)$ 1 unit to the right, then reflecting the result in the x-axis.

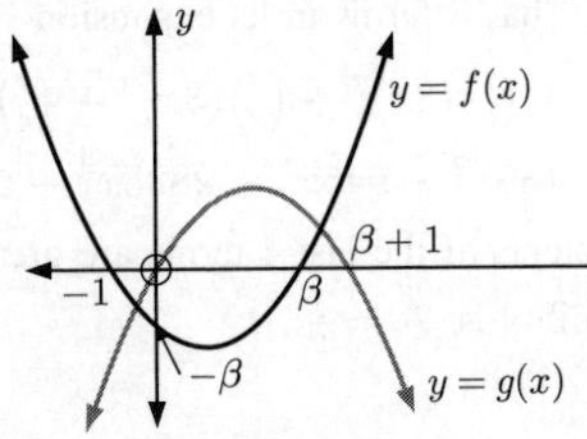

c The x-intercepts of $g(x)$ are 1 unit to the right of the x-intercepts of $f(x)$.

$\therefore$ the x-intercepts of $g(x)$ are 0 and $\beta + 1$.

The y-intercept is 0.

13 **a**

Time	*f*	*Cumulative freq.*
$0 < t \leqslant 10$	15	15
$10 < t \leqslant 20$	10	25
$20 < t \leqslant 30$	25	50
$30 < t \leqslant 40$	50	100
$40 < t \leqslant 50$	30	130
$50 < t \leqslant 60$	20	150

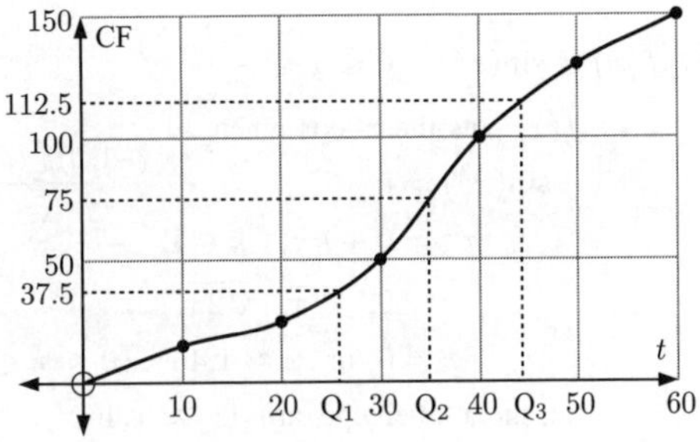

b **i** median ≈ 35 min $(= Q_2)$

ii $\text{IQR} = Q_3 - Q_1$

$\approx 44 - 25$

≈ 19

iii $\approx 50\%$ of the data is less than 35 min

$\therefore$ the probability ≈ 0.5

14 Total number of seats $= 16 + 18 + 20 + 22 + \ldots$ $(n = 30)$

which is arithmetic with $u_1 = 16$, $d = 2$ and $n = 30$

$\therefore$ $S_{30} = \frac{30}{2}(2 \times 16 + 29(2))$ and $u_{30} = u_1 + 29d$

$= 15(32 + 58)$ $\qquad = 16 + 29(2)$

$= 15 \times 90$ $\qquad = 74$

$= 1350$

$\therefore$ P(seated in last row) $= \frac{74}{1350} \approx 0.0548$

15 Let $X =$ the number of successful shots

$X \sim \text{B}(5, 0.86)$

a $\text{P}(X = 5) = (0.86)^5 \approx 0.470$

b $\text{P}(X = 3) = \binom{5}{3}(0.86)^3(0.14)^2$ as there are $\binom{5}{3} = 10$ different ways of scoring 3 from 5.

These are: SSSMM SMSSM

SSMSM MSSSM

SSMMS MSSMS $\{S = \text{score},\ M = \text{miss}\}$

SMMSS MSMSS

SMSMS MMSSS

Max used only one of these.

16 $(1+3x)^7$ has 8 terms in its expansion

$= 1 + \binom{7}{1}3x + \binom{7}{2}(3x)^2 + \binom{7}{3}(3x)^3 + \binom{7}{4}(3x)^4 + \binom{7}{5}(3x)^5 + \binom{7}{6}(3x)^6 + \binom{7}{7}(3x)^7$

$= 1 + 21x + 189x^2 + 945x^3 + 2835x^4 + 5103x^5 + 5103x^6 + 2187x^7$

So, the coefficients of the last 4 terms are greater than 1000.

$\therefore$ the probability is $\frac{4}{8} = \frac{1}{2}$.

17 **a** Carl's z-score for the 100 m $= \dfrac{9.99 - 10.20}{0.113}$
≈ -1.86

His z-score for the 200 m $= \dfrac{17.30 - 18.50}{0.706}$
≈ -1.70

b His z-score is further from the mean 0 for the 100 m, indicating that his performance is better in that event.

18 $f(x) = \sin(x^3)$, $0 \leqslant x \leqslant \frac{\pi}{2}$

a $f(x)$ cuts the x-axis when

$\sin(x^3) = 0$

$\therefore\ x^3 = 0 + k\pi,\quad k \in \mathbb{Z}$

$\therefore\ x = 0,\ \sqrt[3]{\pi},\ \sqrt[3]{2\pi},\ \ldots.$

$\therefore\ x = 0$ or $x \approx 1.46$ $\{0 \leqslant x \leqslant \frac{\pi}{2}\}$

So, the x-intercepts are 0, ≈ 1.46.

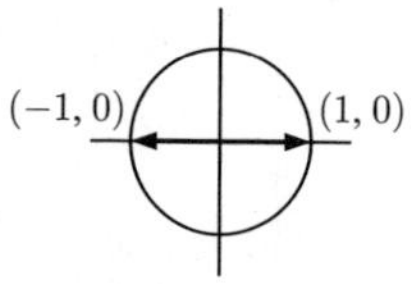

b

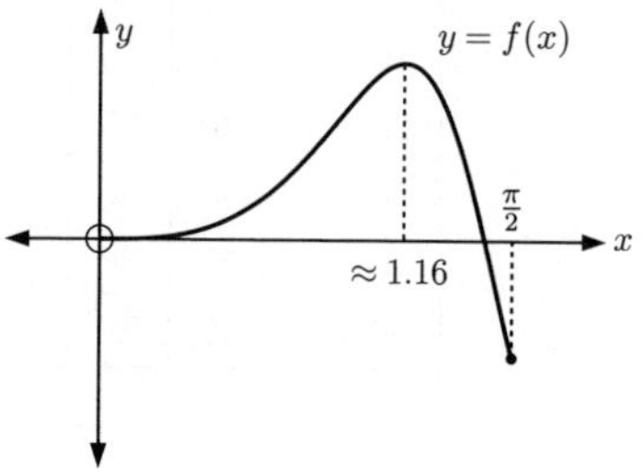

c **i** $f'(x) = \cos(x^3) \times 3x^2$

$\therefore\ f'(\frac{\pi}{4}) = \cos\left(\frac{\pi^3}{64}\right) \times 3\left(\frac{\pi^2}{16}\right)$

≈ 1.64

ii So, at $\left(\frac{\pi}{4}, f(\frac{\pi}{4})\right)$ the tangent has gradient ≈ 1.64

and since $f(\frac{\pi}{4}) = \sin\left(\frac{\pi^3}{64}\right) \approx 0.466$ the tangent has equation

$\dfrac{y - 0.466}{x - \frac{\pi}{4}} \approx 1.64$

$\therefore\ y - 0.466 \approx 1.64x - 1.286$

$\therefore\ y \approx 1.64x - 0.820$

d $f''(x) = -\sin(x^3)3x^2(3x^2) + \cos(x^3)6x$

$= 6x\cos(x^3) - 9x^4\sin(x^3)$

$= 3x\left[2\cos(x^3) - 3x^3\sin(x^3)\right]$

So, $f''(x) = 0$ when $x = 0$ or when $2\cos(x^3) = 3x^3\sin(x^3)$

$\therefore\ x \approx 0.903$ {technology}

From the graph in **b**, $x \approx 0.903$ is the only solution where $f(x) > 0$ and $f'(x) > 0$.

$\therefore$ P is (0.903, 0.671)

19

	Under 35	Over 35	
Successful	951	257	1208
Unsuccessful	174	415	589
	1125	672	1797

a **i** P(successfully treated) $= \dfrac{1208}{1797}$
≈ 0.672

ii P(over 35 | unsuccessful) $= \dfrac{415}{589}$
≈ 0.705

b This is not strictly a binomial situation as the probability of selecting a successfully treated person changes with each selection.
However, as the population is very large, the binomial model provides a *very good approximation.*
So, Harry's method is valid to obtain a very good approximation.
Note: If $X \sim \text{B}(10, 0.672)$, $\text{P}(X = 8) \approx 0.202$

20 $f(x) = 3e^{1-4x}$

a $f'(x) = 3e^{1-4x}(-4)$
$= -12e^{1-4x}$

b $\int f(x)\,dx$
$= \int 3e^{1-4x}\,dx$
$= 3(\frac{1}{-4})e^{1-4x} + c$
$= -\frac{3}{4}e^{1-4x} + c$

c $\int_0^2 f(x)\,dx$
$= \left[-\frac{3}{4}e^{1-4x}\right]_0^2$
$= \left(-\frac{3}{4}e^{-7}\right) - \left(-\frac{3}{4}e^{1}\right)$
$= \frac{3}{4}(e - e^{-7})$
≈ 2.04

21 **a** $y = (\ln x)\sin x$ cuts the x-axis when $y = 0$
$\therefore$ $\ln x = 0$ or $\sin x = 0$

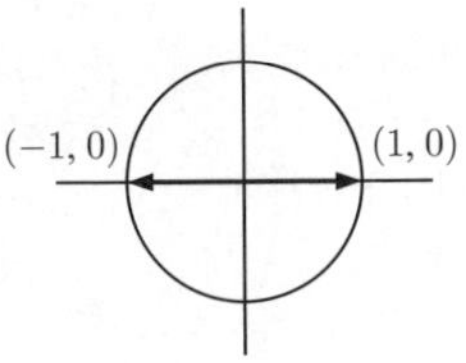

$\therefore$ $x = 1$ or $x = 0 + k\pi$, $k \in \mathbb{Z}$
$\therefore$ A is $(1, 0)$ and B is $(\pi, 0)$

b C is $\approx (2.128, 0.641)$ {using technology}

c $\dfrac{dy}{dx} = \left(\dfrac{1}{x}\right)\sin x + (\ln x)\cos x$

From the graph, the point we are seeking is the point where $\dfrac{dy}{dx}$ is a maximum, between $x = 1$ and $x \approx 2.128$.
This is when $x \approx 1.101$ {technology}
$\therefore$ the point is $(1.101, 0.086)$

d A non-stationary inflection.

22 **a**
$$r = 1 + \frac{0.045}{12} = 1.003\,75$$
$$u_1 = 2000$$
$$\therefore\ u_{n+1} = u_1 \times r^n = 2000 \times (1.003\,75)^n$$

The investment doubles when $u_{n+1} = 2000(1.003\,75)^n = 4000$
$\therefore$ $(1.003\,75)^n = 2$
$\therefore$ $n \log 1.003\,75 = \log 2$
$\therefore$ $n = \dfrac{\log 2}{\log 1.003\,75}$
$\therefore$ $n \approx 185.19$ months

So, it will take 186 months for the investment value to double.

b The investment value will quadruple when $(1.003\,75)^n = 4$
$\therefore$ $n \log(1.003\,75) = \log 4$
$\therefore$ $n \approx 370.37$

So, it will take 371 months for the investment to quadruple in value.

23 **a** $(x^2+2)^5$ has

$T_{r+1} = \binom{5}{r}\left(x^2\right)^{5-r} 2^r$

$\therefore\ T_4 = \binom{5}{3}\left(x^2\right)^2 2^3 = 80x^4$

$T_5 = \binom{5}{4}\left(x^2\right)^1 2^4 = 80x^2$

$T_6 = 2^5 = 32$

b $\int (x^2+2)^5\ dx$

$= \int (x^{10} + 10x^8 + 40x^6 + 80x^4 + 80x^2 + 32)\ dx$

$= \frac{1}{11}x^{11} + \frac{10}{9}x^9 + \frac{40}{7}x^7 + 16x^5 + \frac{80}{3}x^3 + 32x + c$

24 **a** $f(x) = a(x-1)^2 + 4$

But $f(0) = 0$ $\quad\therefore\ a(-1)^2 + 4 = 0$

$\therefore\ a = -4$

So, $f(x) = -4(x-1)^2 + 4$

b **i** The curves meet where $(x-1)^2 = -4(x-1)^2 + 4$

$\therefore\ 5(x-1)^2 = 4$

$\therefore\ x - 1 = \pm\sqrt{\frac{4}{5}}$

$\therefore\ x = 1 \pm \sqrt{\frac{4}{5}}$

$\therefore\ x \approx 0.106,\ 1.89$

$\therefore\ A \approx \int_{0.106}^{1.89} [f(x) - g(x)]\ dx$

ii $A \approx \int_{0.106}^{1.89} [-4(x-1)^2 + 4 - (x-1)^2]\ dx$

≈ 4.77 units^2

25 **a** $\mathrm{P}(A \cup B) = \mathrm{P}(A) + \mathrm{P}(B) - \mathrm{P}(A \cap B)$

But A and B are independent $\quad\therefore\ \mathrm{P}(A \cap B) = \mathrm{P}(A)\,\mathrm{P}(B)$

$\therefore\ 0.68 = \dfrac{\mathrm{P}(B)}{3} + \mathrm{P}(B) - \dfrac{\mathrm{P}(B)}{3}\mathrm{P}(B)$

$\therefore\ 2.04 = \mathrm{P}(B) + 3\,\mathrm{P}(B) - [\mathrm{P}(B)]^2$

$\therefore\ [\mathrm{P}(B)]^2 - 4\,\mathrm{P}(B) + 2.04 = 0$

b The solutions are $\mathrm{P}(B) = 3.4$ or 0.6 {technology}

But $0 \leqslant \mathrm{P}(B) \leqslant 1$

$\therefore\ \mathrm{P}(B) = 0.6$ and $\mathrm{P}(A) = 0.2$

26 **a** $\overline{x} = \dfrac{63 + 76 + 99 + ... + 83}{12} = 70.5$ kg

b Total weight of the students $= 12 \times 70.5 = 846$ kg.

Let x be the weight of the student who left $\quad\therefore\ \dfrac{846 - x}{11} = 70$

$\therefore\ 846 - x = 770$

$\therefore\ x = 76$

$\therefore$ the student weighed 76 kg.

c $s \approx 15.1$ {technology}

d The heaviest student is 99 kg.

The z-score $\approx \dfrac{99 - 70}{15.1}$

≈ 1.92

$\therefore$ the heaviest student is about 1.92 standard deviations above the mean.

27 Let $X =$ the number of correct answers.
$X \sim \text{B}(30, \frac{1}{5})$.

a $\text{P}(X = 10) \approx 0.0355$ {technology}

b $\text{P}(\text{no more than 10 correct}) = \text{P}(X \leqslant 10)$
≈ 0.974 {technology}

28 **a** Let $D =$ the number of defective batteries.
$D \sim \text{B}(20, 0.03)$

i $\text{P}(D = 0) \approx 0.544$

ii $\text{P}(\text{at least one is defective}) = \text{P}(D \geqslant 1)$
$= 1 - \text{P}(D = 0)$
≈ 0.456

b $X \sim \text{B}(n, 0.03)$

i $\text{P}(X = r) = \binom{n}{r}(0.03)^r(0.97)^{n-r}$
$\therefore \text{P}(X = 0) = \binom{n}{0}(0.03)^0(0.97)^n$
$= (0.97)^n$

ii $\text{P}(X \geqslant 1) \geqslant 0.3$
$\therefore 1 - \text{P}(X = 0) \geqslant 0.3$
$\therefore 0.7 \geqslant \text{P}(X = 0)$
$\therefore (0.97)^n \leqslant 0.7$
$\therefore n\log(0.97) \leqslant \log(0.7)$
$\therefore n \geqslant \dfrac{\log(0.7)}{\log(0.97)}$ $\{\log(0.97) < 0\}$
$\therefore n \geqslant 11.709$
$\therefore$ the smallest n is 12.

29 **a** **i**

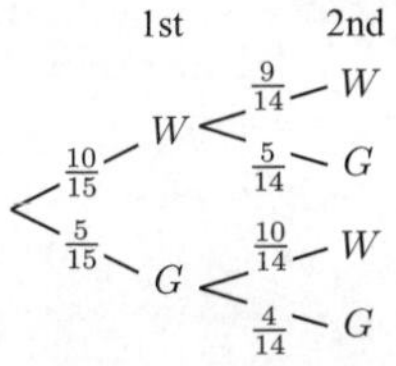

ii P(same colour)
$= \text{P}(WW \text{ or } GG)$
$= \left(\frac{10}{15}\right)\left(\frac{9}{14}\right) + \left(\frac{5}{15}\right)\left(\frac{4}{14}\right)$
$= \frac{110}{210}$
$= \frac{11}{21}$

b

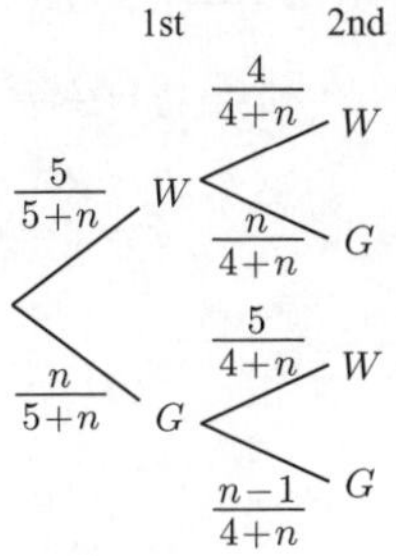

If $\text{P}(WW) = \frac{2}{11}$, then
$\left(\dfrac{5}{5+n}\right)\left(\dfrac{4}{4+n}\right) = \frac{2}{11}$
$\therefore (n+4)(n+5) = 110$
$\therefore n^2 + 9n - 90 = 0$
$\therefore (n-6)(n+15) = 0$
$\therefore n = 6 \text{ or } -15$
As $n > 0$, the only solution is $n = 6$.

30 Let $X =$ mass of a sea lion.
$X \sim \text{N}(\mu, \sigma^2)$

$\text{P}(X > 900) = 0.1$
$\therefore \text{P}(X \leqslant 900) = 0.9$
$\therefore \text{P}\left(\dfrac{X-\mu}{\sigma} \leqslant \dfrac{900-\mu}{\sigma}\right) = 0.9$
$\therefore \text{P}\left(Z \leqslant \dfrac{900-\mu}{\sigma}\right) = 0.9$
$\therefore \dfrac{900-\mu}{\sigma} \approx 1.2816$
$\therefore 900 - \mu \approx 1.2816\sigma$ .... (1)

Also $\text{P}(X < 500) = 0.15$
$\therefore \text{P}\left(Z < \dfrac{500-\mu}{\sigma}\right) = 0.15$
$\therefore \dfrac{500-\mu}{\sigma} \approx -1.0364$
$\therefore 500 - \mu \approx -1.0364\sigma$ .... (2)

Solving (1) and (2) simultaneously:

$$\begin{aligned} 900 - \mu &\approx 1.2816\sigma \\ -500 + \mu &\approx 1.0364\sigma \\ \hline 400 &\approx 2.318\sigma \quad \therefore \ \sigma \approx 172.6 \end{aligned}$$

and $\mu \approx 500 + 1.0364 \times 172.6$

$\therefore \ \mu \approx 678.9$

$\therefore \ \mu \approx 679$ kg, $\sigma \approx 173$ kg

31 **a** $\theta = (360 - 70 - 184)^o$

$\therefore \ \theta = 106^o$

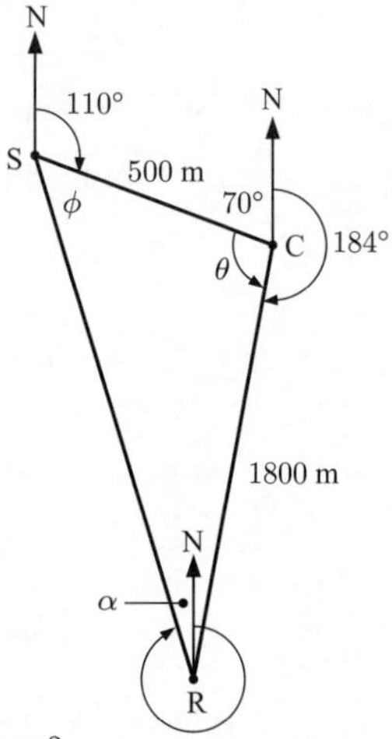

$\therefore$ $\text{RS}^2 = 500^2 + 1800^2 - 2 \times 500 \times 1800 \times \cos 106^o$

$\therefore$ RS ≈ 1996.53

$\therefore$ RS ≈ 2000 m

b First we find ϕ

$$\frac{\sin\phi}{1800} \approx \frac{\sin 106^o}{1996.53}$$

$$\therefore \ \sin\phi \approx \frac{1800 \times \sin 106^o}{1996.53}$$

$\therefore \ \sin\phi \approx 0.866\,64$

$\therefore \ \phi \approx 60.07^o$

Thus $\alpha \approx (180 - 110 - 60.07)^o$

$\therefore \ \alpha \approx 9.93^o$

and $360^o - \alpha \approx 350.07^o$

$\therefore$ the bearing from R to S is 350°.

32 **a** $\text{P}(A') = 0.2 \quad \therefore \ \text{P}(A) = 0.8$

$\therefore \ x + y = 0.8$ .... (1)

$\text{P}(A \cup B) = x + y + z = 0.9$

$\therefore \ 0.8 + z = 0.9$ {using (1)}

$\therefore \ z = 0.1$

b $\text{P}(A \mid B) = \dfrac{y}{y+z} = 0.5$

$\therefore \ y = \frac{1}{2}(y+z)$

$\therefore \ 2y = y + z$

$\therefore \ y = z = 0.1$

c From (1), $x + 0.1 = 0.8$

$\therefore \ x = 0.7$

33 **a** $y = \dfrac{1}{x}$ meets $y = x + 2$ where $x + 2 = \dfrac{1}{x}$

$\therefore \ x^2 + 2x = 1$

$\therefore \ x^2 + 2x - 1 = 0$

$$\therefore \ x = \frac{-2 \pm \sqrt{4 - 4(1)(-1)}}{2}$$

$$\therefore \ x = \frac{-2 \pm \sqrt{8}}{2}$$

$$\therefore \ x = \frac{-2 \pm 2\sqrt{2}}{2}$$

$\therefore \ x = -1 \pm \sqrt{2}$

$\therefore \ m = -1, \quad n = 2$

b **i** $y = \frac{1}{x}$ under a translation of $\binom{-2}{0}$ becomes $y = \frac{1}{x+2}$ which under a reflection in the x-axis becomes $y = -\frac{1}{x+2}$

$\therefore \; g(x) = -\frac{1}{x+2}$

ii $g(x)$ is undefined when $x + 2 = 0$

$\therefore \; x = -2$ is a vertical asymptote.

As $x \to \pm\infty$, $g(x) \to 0$

$\therefore \; y = 0$ is a horizontal asymptote.

iii $g(0) = -\frac{1}{2}$

$\therefore$ the y-intercept is $-\frac{1}{2}$.

iv

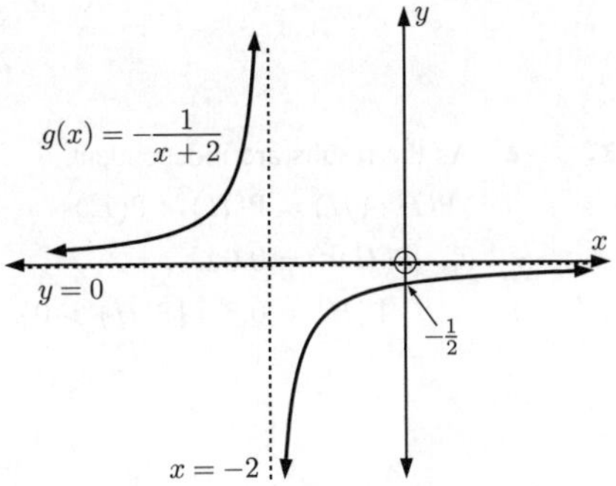

34 **a** **i** When $t = 0$, $\mathbf{r}_1 = \binom{-1}{3}$

So, the initial position of the first object is $(-1, 3)$.

ii When $t = 10$, $\mathbf{r}_1 = \binom{-1}{3} + \binom{40}{20}$

$= \binom{39}{23}$

So, the first object is at $(39, 23)$ after 10 seconds.

b distance $= \sqrt{(39 - -1)^2 + (23 - 3)^2}$

$= \sqrt{40^2 + 20^2}$

≈ 44.7 m

c When $t = 1$, $\mathbf{r}_2 = \binom{0}{4} + \binom{9}{4}$

$= \binom{9}{8}$

$\therefore$ the second object passes through $(9, 8)$.

d We consider $\binom{9}{8} = \binom{-1}{3} + t\binom{4}{2}$

$\therefore \; 9 = -1 + 4t$ and

$8 = 3 + 2t$

$\therefore \; t = 2.5$ is the common solution

So, both objects pass through $(9, 8)$, but they do not collide since they pass through $(9, 8)$ at different times.

35

$$\text{area of } A = \text{area of } B$$

$$\therefore \; -\int_{-2}^{k} f(x)\,dx = \int_1^3 f(x)\,dx$$

$$\therefore \; \int_k^{-2} f(x)\,dx = \int_1^3 f(x)\,dx$$

$$\therefore \; \left[-\frac{x^4}{4} + \frac{2x^3}{3} + \frac{5x^2}{2} - 6x\right]_k^{-2} = 5\tfrac{1}{3} \quad \{\text{technology}\}$$

$$\therefore \; (-4 - \tfrac{16}{3} + 10 + 12) - \left(-\frac{k^4}{4} + \frac{2k^3}{3} + \frac{5k^2}{2} - 6k\right) = 5\tfrac{1}{3}$$

$$\therefore \; 12\tfrac{2}{3} + \frac{k^4}{4} - \frac{2k^3}{3} - \frac{5k^2}{2} + 6k = 5\tfrac{1}{3}$$

$$\therefore \; \frac{k^4}{4} - \frac{2k^3}{3} - \frac{5k^2}{2} + 6k + 7\tfrac{1}{3} = 0$$

$$\therefore \; 3k^4 - 8k^3 - 30k^2 + 72k + 88 = 0$$

$$\therefore \; k \approx -0.969 \quad \{\text{technology}\}$$

36 $y = e^{-x^2}$

a When $x = 0$, $y = e^0 = 1$ and when $x = 2$, $y = e^{-4}$

$\therefore$ A is $(0, 1)$ and B is $\left(2, \dfrac{1}{e^4}\right)$.

b Area $= \int_0^2 e^{-x^2}\,dx$
≈ 0.882 units^2

37 **a** As the results are independent,

$P(H \cap H) = P(H) \times P(H)$

$\therefore [P(H)]^2 = 0.64$

$\therefore P(H) = 0.8 \quad \{P(H) > 0\}$

b Let X = the number of heads obtained.
$X \sim B(10, 0.8)$

i $P(X = 6) \approx 0.0881$ {technology}

ii $P(X \geqslant 6) = 1 - P(X \leqslant 5)$
$\approx 1 - 0.0328$
≈ 0.967

38 Let X = height of a maize plant.
$X \sim N(\mu, 6.8^2)$

a
$$P(X < 45) = 0.75$$
$$\therefore P\left(\frac{X-\mu}{6.8} < \frac{45-\mu}{6.8}\right) = 0.75$$
$$\therefore P\left(Z < \frac{45-\mu}{6.8}\right) = 0.75$$
$$\therefore \frac{45-\mu}{6.8} \approx 0.6745$$
$$\therefore 45 - \mu \approx 4.59$$
$$\therefore \mu \approx 40.4$$

b $X \sim N(40.41, 6.8^2)$

$\therefore P(X < 25) \approx 0.0117$

c

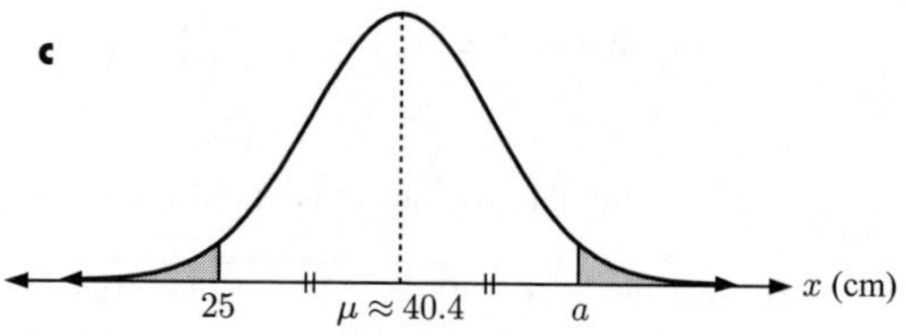

If $P(X < 25) = P(X > a)$ then
$$\mu - 25 = a - \mu$$
$$\therefore a = 2\mu - 25$$
$$\approx 55.8$$

39 $\mathbf{A} = \begin{pmatrix} 3 & 1 \\ -2 & 2 \end{pmatrix}$

a
$$\mathbf{A}^2 - 5\mathbf{A} = \begin{pmatrix} 3 & 1 \\ -2 & 2 \end{pmatrix}\begin{pmatrix} 3 & 1 \\ -2 & 2 \end{pmatrix} - 5\begin{pmatrix} 3 & 1 \\ -2 & 2 \end{pmatrix}$$
$$= \begin{pmatrix} 7 & 5 \\ -10 & 2 \end{pmatrix} - \begin{pmatrix} 15 & 5 \\ -10 & 10 \end{pmatrix}$$
$$= \begin{pmatrix} -8 & 0 \\ 0 & -8 \end{pmatrix}$$

b $|\mathbf{A}| = 6 - (-2) = 8 \neq 0$

$\therefore \mathbf{A}^{-1}$ exists

c As $\mathbf{A}^2 - 5\mathbf{A} = -8\mathbf{I}$ {from **a**}
$$\mathbf{A}^2 = 5\mathbf{A} - 8\mathbf{I}$$
$$\therefore \mathbf{A}^2\mathbf{A} = 5\mathbf{A}^2 - 8\mathbf{A}$$
$$\therefore \mathbf{A}^3 = 5(5\mathbf{A} - 8\mathbf{I}) - 8\mathbf{A}$$
$$\therefore \mathbf{A}^3 = 17\mathbf{A} - 40\mathbf{I}$$

d As $\mathbf{A}^2 = 5\mathbf{A} - 8\mathbf{I}$
$$\mathbf{A}^2\mathbf{A}^{-1} = 5\mathbf{A}\mathbf{A}^{-1} - 8\mathbf{I}\mathbf{A}^{-1}$$
$$\therefore \mathbf{A} = 5\mathbf{I} - 8\mathbf{A}^{-1}$$
$$\therefore 8\mathbf{A}^{-1} = -\mathbf{A} + 5\mathbf{I}$$
$$\therefore \mathbf{A}^{-1} = \tfrac{1}{8}(-\mathbf{A} + 5\mathbf{I})$$
$$\therefore \mathbf{A}^{-1} = -\tfrac{1}{8}\mathbf{A} + \tfrac{5}{8}\mathbf{I}$$
$$\therefore a = -\tfrac{1}{8}, \quad b = \tfrac{5}{8}$$

40 $f(x) = x\sin(2x), \quad 0 < x < 3$

a

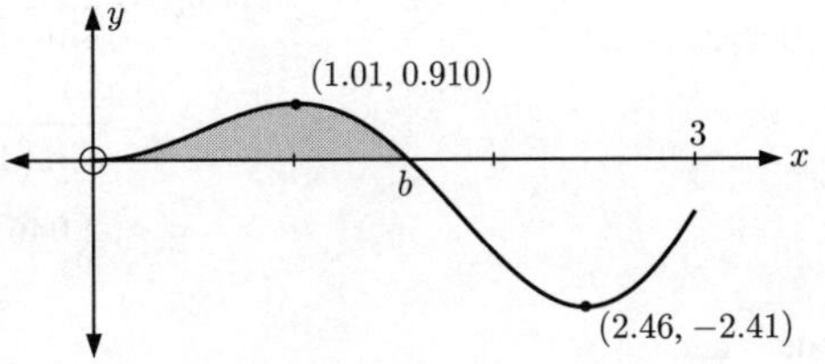

b Using technology, the maximum value of $f(x)$ is 0.91, and the minimum value is -2.41.

$\therefore$ the range is $\{y \mid -2.41 \leqslant y \leqslant 0.91\}$

c $f(x) = 0$ when $\sin(2x) = 0 \quad \{0 < x < 3\}$

$\therefore\ 2x = 0 + k\pi, \quad k \in \mathbb{Z}$

$\therefore\ x = \dfrac{k\pi}{2}$

$\therefore\ b = \frac{\pi}{2}$

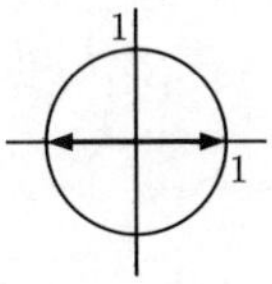

d $A = \int_0^{\frac{\pi}{2}} x\sin(2x)\,dx$

≈ 0.785 units^2 {technology}

e Volume $= \pi \int_0^{\frac{\pi}{2}} [x\sin(2x)]^2\,dx$

≈ 1.721 units^3

41 $f(x) = e^{-3x}\sin x, \quad -\frac{1}{2} \leqslant x \leqslant 3$

a $f'(x) = e^{-3x}(-3)\sin x + e^{-3x}\cos x$

$= e^{-3x}(\cos x - 3\sin x)$

b $f'(\frac{\pi}{2}) = e^{-\frac{3\pi}{2}}\left(\cos\frac{\pi}{2} - 3\sin\frac{\pi}{2}\right)$

$= e^{-\frac{3\pi}{2}}(0 - 3(1))$

$= -\dfrac{3}{e^{\frac{3\pi}{2}}}$

$\therefore$ the tangent has equation

$y - e^{-\frac{3\pi}{2}} = -\dfrac{3}{e^{\frac{3\pi}{2}}}\left(x - \frac{\pi}{2}\right)$

$\therefore\ e^{\frac{3\pi}{2}}y - e^{-\frac{3\pi}{2}}\left(e^{\frac{3\pi}{2}}\right) = -3x - 3(-\frac{\pi}{2})$

$\therefore\ e^{\frac{3\pi}{2}}y - 1 = -3x + \frac{3\pi}{2}$

$\therefore\ 3x + e^{\frac{3\pi}{2}}y = \frac{3\pi}{2} + 1$

c

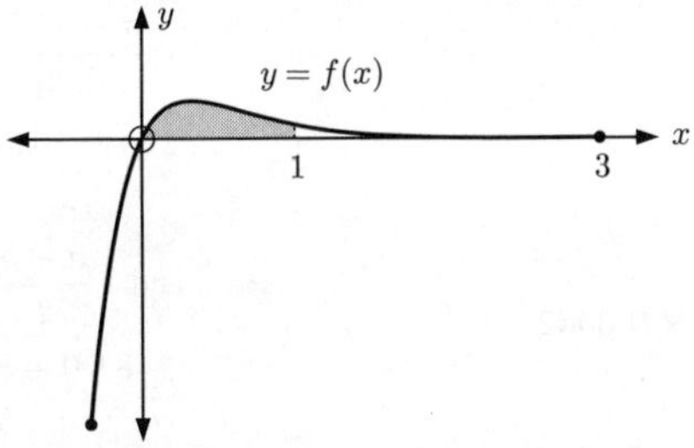

area $= \int_0^1 e^{-3x}\sin x\,dx$

≈ 0.0847 units^2

42 $f(x) = \dfrac{\sin x}{\cos x} + \dfrac{\cos x}{\sin x}, \quad 0 \leqslant x \leqslant \frac{\pi}{2}$

a $f(x) = \dfrac{\sin^2 x + \cos^2 x}{\cos x \sin x}$

$= \dfrac{1}{\sin x \cos x} \times \left(\dfrac{2}{2}\right)$

$= \dfrac{2}{\sin 2x}$

b $\sin 2x = 0$

$\therefore\ 2x = 0 + k\pi, \quad k \in \mathbb{Z}$

$\therefore\ x = k\frac{\pi}{2}, \quad k \in \mathbb{Z}$

$\therefore\ x = 0$ or $\frac{\pi}{2}$

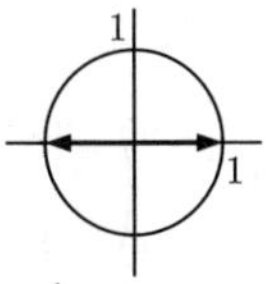

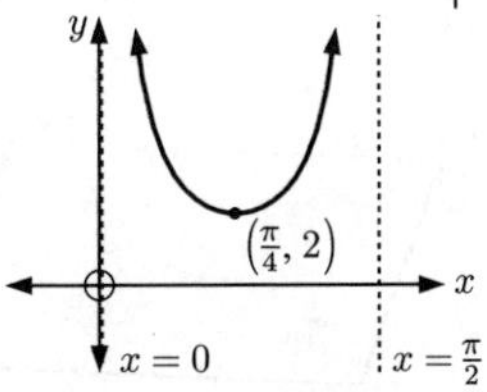

$\therefore\ x = 0$ and $x = \frac{\pi}{2}$ are vertical asymptotes of $f(x)$

c The least value of $f(x)$ is $\frac{2}{1} = 2$

when $\sin 2x = 1$

$\therefore\ 2x = \frac{\pi}{2} \quad \{0 \leqslant 2x \leqslant \pi\}$

$\therefore\ x = \frac{\pi}{4}$

d $\sin a = \frac{1}{3}$ so $a = \sin^{-1}(\frac{1}{3})$

$$\therefore\ f(2a) = \frac{2}{\sin(4a)} = \frac{2}{\sin\left(4\sin^{-1}(\frac{1}{3})\right)} \approx 2.046$$

43 Let F = weight of a female, and
M = weight of a male

$\therefore\ F \sim \text{N}(78.6, 5.03^2)$ and $M \sim \text{N}(91.3, 6.29^2)$

a **i** $\text{P}(M < 80) \approx 0.0362$ **ii** $\text{P}(F < 80) \approx 0.610$ **iii** $\text{P}(70 < F < 80) \approx 0.566$

b
$$\text{P}(F < k) = 0.2$$
$$\therefore\ \text{P}\left(\frac{F - 78.6}{5.03} < \frac{k - 78.6}{5.03}\right) = 0.2$$
$$\therefore\ \text{P}\left(Z < \frac{k - 78.6}{5.03}\right) = 0.2$$
$$\therefore\ \frac{k - 78.6}{5.03} \approx -0.841\,62$$
$$\therefore\ k - 78.6 \approx -4.23$$
$$\therefore\ k \approx 74.4$$

c

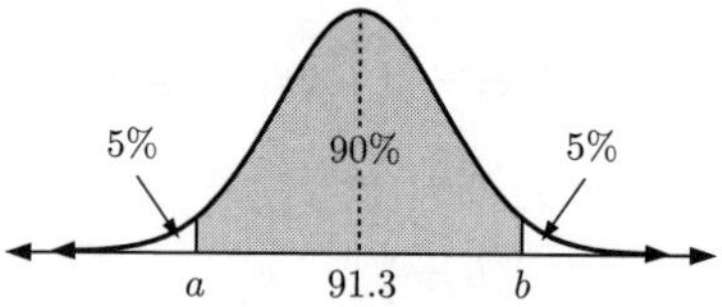

$$\text{P}(M < a) = 0.05$$
$$\therefore\ \text{P}\left(\frac{M - 91.3}{6.29} < \frac{a - 91.3}{6.29}\right) = 0.05$$
$$\therefore\ \text{P}\left(Z < \frac{a - 91.3}{6.29}\right) = 0.05$$
$$\therefore\ \frac{a - 91.3}{6.29} \approx -1.645$$
$$\therefore\ a - 91.3 \approx -10.35$$
$$\therefore\ a \approx 81.0$$

and $\frac{a + b}{2} = 91.3$

$$\therefore\ 81.0 + b \approx 182.6$$
$$\therefore\ b \approx 101.6$$

d **i**

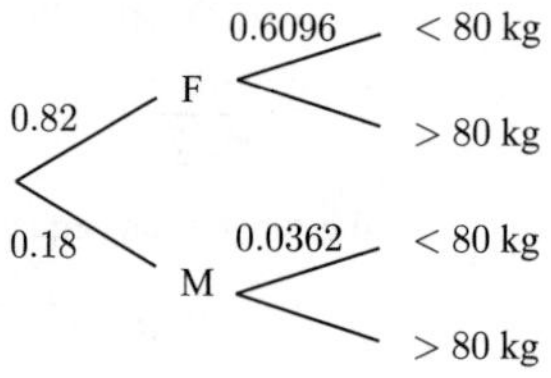

$$\text{P}(< 80\text{ kg}) \approx 0.82 \times 0.6096 + 0.18 \times 0.0362 \approx 0.506$$

ii
$$\text{P}(M < 100) \approx 0.916\,69$$
$$\text{P}(F < 100) \approx 0.999\,99$$
$$\therefore\ \text{P}(M \mid < 100) = \frac{\text{P}(M \cap < 100)}{\text{P}(< 100)}$$
$$\approx \frac{0.18 \times 0.916\,69}{0.18 \times 0.916\,69 + 0.82 \times 0.999\,99}$$
$$\approx 0.168$$

44 **a** $\overrightarrow{\text{BA}} = \begin{pmatrix} 3 - 1 \\ 2 - -1 \\ -1 - 4 \end{pmatrix} = \begin{pmatrix} 2 \\ 3 \\ -5 \end{pmatrix}$ $\quad \therefore\ |\overrightarrow{\text{BA}}| = \sqrt{4 + 9 + 25} = \sqrt{38}$ units

b

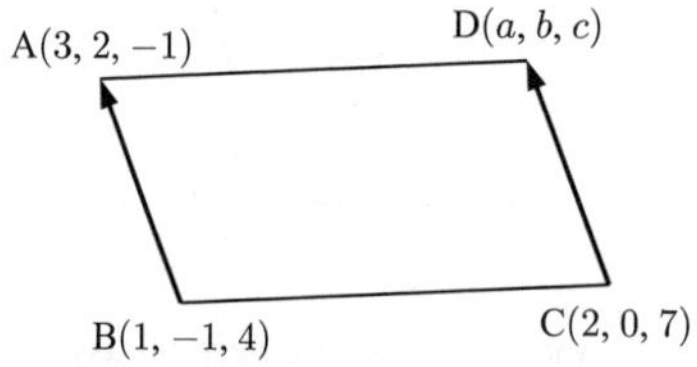

$$\overrightarrow{\text{CD}} = \overrightarrow{\text{BA}}$$
$$\therefore\ \begin{pmatrix} a - 2 \\ b - 0 \\ c - 7 \end{pmatrix} = \begin{pmatrix} 2 \\ 3 \\ -5 \end{pmatrix}$$
$$\therefore\ a = 4,\ b = 3,\ c = 2$$

$\therefore$ D is at $(4, 3, 2)$.

c Let F be (p, q, r).

$$\therefore\ \overrightarrow{BF} = \begin{pmatrix} p-1 \\ q+1 \\ r-4 \end{pmatrix} = \begin{pmatrix} 6 \\ -3 \\ -6 \end{pmatrix}$$

$\therefore\ p = 7,\quad q = -4,\quad r = -2$

So, F is at $(7, -4, -2)$.

d
$$\overrightarrow{BA} \bullet \overrightarrow{BF} = \begin{pmatrix} 2 \\ 3 \\ -5 \end{pmatrix} \bullet \begin{pmatrix} 6 \\ -3 \\ -6 \end{pmatrix}$$
$$= 12 - 9 + 30$$
$$= 33$$

$$\cos(A\widehat{B}F) = \frac{\overrightarrow{BA} \bullet \overrightarrow{BF}}{|\overrightarrow{BA}||\overrightarrow{BF}|}$$
$$= \frac{33}{\sqrt{38}\sqrt{36+9+36}}$$
$$= \frac{33}{\sqrt{38} \times 9}$$
$$= \frac{11}{3\sqrt{38}}$$

e Area ABFE $= 2 \times$ area ΔABF
$$= 2 \times \tfrac{1}{2}\ |\overrightarrow{BA}||\overrightarrow{BF}|\sin(A\widehat{B}F)$$
$$= \sqrt{38} \times 9 \times \sqrt{1 - \cos^2(A\widehat{B}F)}$$
$$= \sqrt{38} \times 9 \times \sqrt{1 - \frac{121}{9 \times 38}}$$
$$= \sqrt{38} \times 9 \times \sqrt{\frac{9 \times 38 - 121}{9 \times 38}}$$
$$= \sqrt{38} \times 9 \times \frac{\sqrt{221}}{3\sqrt{38}}$$
$$= 3\sqrt{221} \text{ units}^2$$

45 **a**
$$\sum p_i = 1$$
$$\therefore\ \tfrac{1}{10} + 2t + \tfrac{3}{20} + 2t^2 + \frac{t}{2} = 1$$
$$\therefore\ 2t^2 + \tfrac{5}{2}t - \tfrac{3}{4} = 0$$
$$\therefore\ 8t^2 + 10t - 3 = 0$$
$$\therefore\ (4t-1)(2t+3) = 0$$
$$\therefore\ t = \tfrac{1}{4} \text{ or } -\tfrac{3}{2}$$
But $t \geqslant 0 \quad \therefore\ t = \tfrac{1}{4}$

b
$$E(Y) = \sum y_i p_i$$
$$= 1(\tfrac{1}{10}) + 2(\tfrac{1}{2}) + 3(\tfrac{3}{20}) + 4(\tfrac{1}{8}) + 5(\tfrac{1}{8})$$
$$= 2.675$$

c 2.675 is the mean of the Y distribution.

46 $f(x) = 5x + e^{1-x^2} - 2, \quad -1 \leqslant x \leqslant 2$

a $f(0) = 0 + e^1 - 2$

$\therefore$ the y-intercept is $e - 2 \approx 0.718$

b

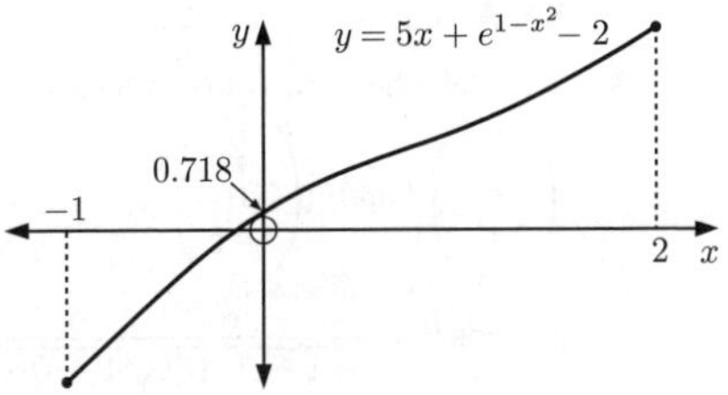

c x-intercept ≈ -0.134 {technology}

d
$$f'(x) = 5 + e^{1-x^2}(-2x)$$
$$\therefore\ f'(1) = 5 + e^0(-2) = 3$$
So the tangent has gradient 3 when $x = 1$.

47 Let X = the length of a zucchini.

$X \sim \text{N}(24.3, 6.83^2)$

a

$$\text{P}(X < a) = 0.15$$
$$\therefore\ \text{P}\left(\frac{X-24.3}{6.83} < \frac{a-24.3}{6.83}\right) = 0.15$$
$$\therefore\ \text{P}\left(Z < \frac{a-24.3}{6.83}\right) = 0.15$$
$$\therefore\ \frac{a-24.3}{6.83} \approx -1.0364$$
$$\therefore\ a - 24.3 \approx -7.08$$
$$\therefore\ a \approx 17.2$$

and $\text{P}(X > b) = 0.2$

$$\therefore\ \text{P}(X \leqslant b) = 0.8$$
$$\therefore\ \text{P}\left(Z \leqslant \frac{b-24.3}{6.83}\right) = 0.8$$
$$\therefore\ \frac{b-24.3}{6.83} \approx 0.8416$$
$$\therefore\ b - 24.3 \approx 5.75$$
$$\therefore\ b \approx 30.0$$

b $100\% - 15\% - 20\% = 65\%$, so 65% of zucchinis are of saleable length.

i

$$\text{P}(20 < X < 26 \mid \text{OK}) = \frac{\text{P}(20 < X < 26 \cap \text{OK})}{\text{P}(\text{OK})} = \frac{\text{P}(20 < X < 26)}{0.65} \approx \frac{0.3338}{0.65} \approx 0.514$$

ii

$$\text{P}(X < 24.3 \mid \text{OK}) = \frac{\text{P}(X < 24.3 \cap \text{OK})}{\text{P}(\text{OK})} = \frac{0.35}{0.65} \approx 0.538$$

{50% are < 24.3, 15% are too small}

48 **a** Let X = number of days of more than 5 mm rainfall.

$$\text{E}(X) = 7 \times 0.37 = 2.59 \text{ days}$$

b $X \sim \text{B}(7, 0.37)$

i $\text{P}(X = 3) \approx 0.279$

ii $\text{P}(X \geqslant 2) = 1 - \text{P}(X \leqslant 1) \approx 1 - 0.201 \approx 0.799$

49 **a** $3 + s = -2 \quad \therefore\ s = -5$

$$\therefore\ \mathbf{r}_1 = \begin{pmatrix} 2 \\ -1 \\ 3 \end{pmatrix} + (-5)\begin{pmatrix} 1 \\ -1 \\ 1 \end{pmatrix} = \begin{pmatrix} -3 \\ 4 \\ -2 \end{pmatrix}$$

So, A is at $(-3, 4, -2)$.

b They meet if
$2 + s = 6 + 2t$ .... (1)
$-1 - s = -3$ .... (2)
$3 + s = 4 - t$ .... (3)

$\therefore\ s = 2$ and in (3), $5 = 4 - t$

$\therefore\ t = -1$

So in (1), $2 + s = 4$ and $6 + 2t = 4$ ✓

$\therefore\ L_1$ and L_2 meet at $(4, -3, 5)$.

c The lines have direction vectors

$\begin{pmatrix} 1 \\ -1 \\ 1 \end{pmatrix}$ and $\begin{pmatrix} 2 \\ 0 \\ -1 \end{pmatrix}$.

$$\therefore\ \cos\theta = \frac{|2 + 0 - 1|}{\sqrt{1+1+1}\sqrt{4+0+1}} = \frac{1}{\sqrt{3}\sqrt{5}}$$

$$\therefore\ \theta = \cos^{-1}\left(\tfrac{1}{\sqrt{15}}\right) \approx 75.0^o$$

50 $f(x) = e^x(x^2 - 3x + 2)$

a $y = f(x)$ cuts the x-axis when $y = 0$

$\therefore\ x^2 - 3x + 2 = 0$ {as $e^x > 0$}

$\therefore\ (x-1)(x-2) = 0$

$\therefore\ x = 1$ or 2

So, A is at $(1, 0)$ and B is at $(2, 0)$.

$y = f(x)$ cuts the y-axis when $x = 0$

$\therefore\ y = e^0(2) = 2$

$\therefore$ C is at $(0, 2)$.

b As $x \to -\infty$, $f(x) \to 0$ (above)

$\therefore\ y = 0$ is a horizontal asymptote.

c $f'(x) = e^x(x^2 - 3x + 2) + e^x(2x - 3)$

$= e^x(x^2 - 3x + 2 + 2x - 3)$

$= e^x(x^2 - x - 1)$

d $f'(x) = 0$ when $x^2 - x - 1 = 0$

$\therefore\ x \approx -0.618$ or 1.618

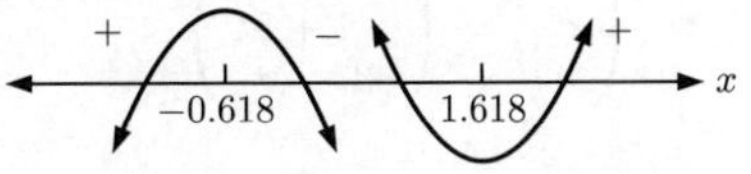

$\therefore$ the local maximum has x-coordinate ≈ -0.618 and the local minimum has x-coordinate ≈ 1.62.

e $f'(1) = e(1 - 1 - 1) = -e$

$\therefore$ the normal at A(1, 0) has gradient $\frac{1}{e}$,

and the equation of the normal is

$\frac{y - 0}{x - 1} = \frac{1}{e}$

$\therefore\ ey = x - 1$

$\therefore\ x - ey = 1$

f $x - ey = 1$ meets $y = e^x(x^2 - 3x + 2)$

where $\frac{x-1}{e} = e^x(x^2 - 3x + 2)$

$\therefore\ x - 1 = e^{x+1}(x-1)(x-2)$

$\therefore\ (x-1)\left[1 - e^{x+1}(x-2)\right] = 0$

$\therefore\ x = 1$ or $x \approx 2.0475$ {technology}

$\therefore$ the x-coordinate of D ≈ 2.05

g Area $\approx \int_1^{2.0475} \left[\frac{x-1}{e} - e^x(x^2 - 3x + 2)\right] dx$

≈ 0.959 units^2

51 $f(x) = e^{-x}x^3 - 1$

a $f(0) = e^0(0) - 1 = -1$

$\therefore$ the y-intercept is -1

b At P, $x \approx 1.857$ {technology}

At Q, $x \approx 4.536$ {technology}

c $f'(x) = e^{-x}(-1)x^3 + e^{-x}(3x^2)$

$= x^2e^{-x}(3 - x)$

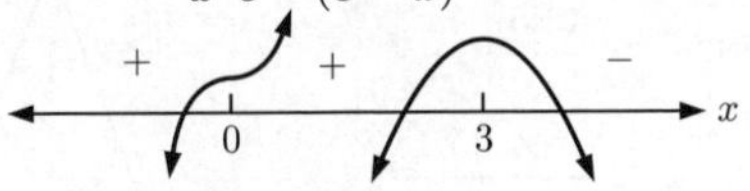

$\therefore$ B is at $(3, 0.344)$.

d $f'(x) = e^{-x}(-x^3 + 3x^2)$

$\therefore\ f''(x) = e^{-x}(-1)(-x^3 + 3x^2) + e^{-x}(-3x^2 + 6x)$

$= e^{-x}(x^3 - 3x^2 - 3x^2 + 6x)$

$= e^{-x}(x^3 - 6x^2 + 6x)$

$= xe^{-x}(x^2 - 6x + 6)$

which is 0 when $x^2 - 6x + 6 = 0$

$\therefore\ x = \frac{6 \pm \sqrt{36 - 4(1)(6)}}{2}$

$= \frac{6 \pm 2\sqrt{3}}{2}$

$= 3 \pm \sqrt{3}$

$\therefore$ the x-coordinate of A is $3 - \sqrt{3}$ and the x-coordinate of C is $3 + \sqrt{3}$

e area $\approx \int_{1.857}^{4.536} (e^{-x}x^3 - 1)\ dx$

≈ 0.595 units^2

52 **a**

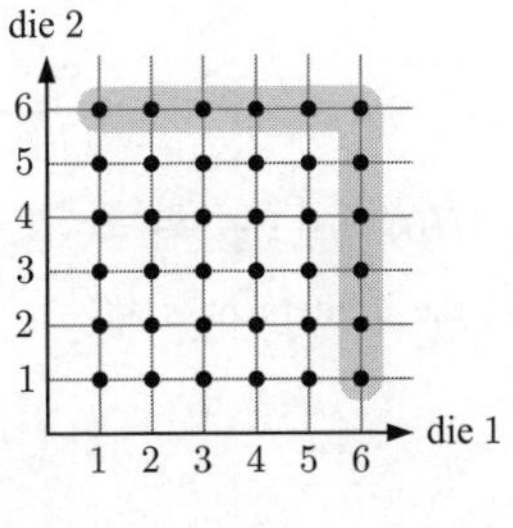

b $P(E) = \frac{11}{36}$ {enclosed points}

c Let $X =$ the number of times E occurs.

$X \sim B(10, \frac{11}{36})$

i $P(X = 2) \approx 0.227$

ii P(at most 3 times) $= P(X \leqslant 3)$

≈ 0.635

53 **a** $\overrightarrow{AB} = \begin{pmatrix} -2-2 \\ 4--1 \\ -1-3 \end{pmatrix} = \begin{pmatrix} -4 \\ 5 \\ -4 \end{pmatrix}$

$\therefore$ L_1 has equation

$$\mathbf{r}_1 = \begin{pmatrix} 2 \\ -1 \\ 3 \end{pmatrix} + t\begin{pmatrix} -4 \\ 5 \\ -4 \end{pmatrix} = \begin{pmatrix} 2-4t \\ -1+5t \\ 3-4t \end{pmatrix}$$

$\therefore$ $\mathbf{r}_1 = (2-4t)\mathbf{i} + (5t-1)\mathbf{j} + (3-4t)\mathbf{k}$

b $\begin{pmatrix} 4 \\ a \\ b \end{pmatrix} = \begin{pmatrix} 2-4t \\ -1+5t \\ 3-4t \end{pmatrix}$

$\therefore$ $2-4t=4$

$\therefore$ $-4t=2$

$\therefore$ $t=-\frac{1}{2}$

So, $a=-1+5(-\frac{1}{2})=-3\frac{1}{2}$

and $b=3-4(-\frac{1}{2})=5.$

c They meet if $2-4t=3s-5$ .... (1)

$5t-1=s-16$ .... (2)

$3-4t=16-s$ .... (3)

From (2) and (3), $5t-1=-(3-4t)$

$\therefore$ $5t-1=-3+4t$

$\therefore$ $t=-2$

and $5(-2)-1=s-16$

$\therefore$ $-11=s-16$

$\therefore$ $s=5$

In (1), LHS $=2-4t=2+8=10$

RHS $=3s-5=15-5=10$ ✓

$\therefore$ all 3 equations are satisfied by $t=-2,\ s=5$

So, the lines meet at

$(2-4(-2),\ 5(-2)-1,\ 3-4(-2))$

$\therefore$ D is at $(10, -11, 11)$.

54 **a** $DB^2 = 7.6^2 + 8.1^2 - 2\times 7.6\times 8.1\times \cos 30^o$

$\therefore$ $DB \approx 4.09$ m

$BC^2 = 16^2 + 8.1^2 - 2\times 16\times 8.1\times\cos 30^o$ $\{AC = 7.6 + 8.4 = 16\}$

$\therefore$ $BC \approx 9.86$ m

b Let $A\widehat{B}E = \theta$

In ΔABD, $\dfrac{\sin\theta}{7.6} \approx \dfrac{\sin 30^o}{4.092}$

$\therefore$ $\sin\theta \approx \dfrac{3.8}{4.092}$

$\therefore$ $\theta \approx 68.2^o$

$\therefore$ $A\widehat{B}E \approx 68.2^o$

Let $D\widehat{B}C = \phi$

$\therefore$ $\dfrac{\sin\phi}{8.4} \approx \dfrac{\sin(30^o + 68.22^o)}{9.856}$ {exterior angle of a Δ}

$\therefore$ $\sin\phi \approx \dfrac{8.4\times\sin(98.22^o)}{9.856}$

$\therefore$ $\sin\phi \approx 0.8435$

$\therefore$ $\phi \approx 57.5^o$

$\therefore$ $D\widehat{B}C \approx 57.5^o$

c Area BCD

$= \frac{1}{2}\times BC\times DB\times\sin(D\widehat{B}C)$

$\approx \frac{1}{2}\times 9.86\times 4.09\times\sin(57.5^o)$

$\approx 17.0\ m^2$

d In ΔAME,

$\cos 68.2^o \approx \dfrac{4.05}{x}$

$\therefore$ $x \approx \dfrac{4.05}{\cos(68.2^o)}$

$\therefore$ $x \approx 10.9$

$\therefore$ AE ≈ 10.9 m

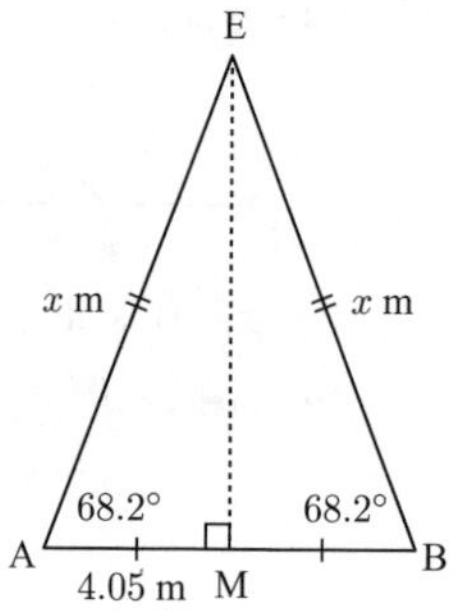

55 **a** As $x\to\infty$, $\dfrac{3}{(x-1)(x+b)} \to 0$

$\therefore$ $y \to a$

But $y=-1$ is the H.A.

$\therefore$ $a=-1$

The vertical asymptotes are $x=1,\ x=-2$

$\therefore$ $b=2$

b $f(x) = -1 + \dfrac{3}{(x-1)(x+2)}$

$\therefore$ $f(0) = -1 + \frac{3}{-2} = -2\frac{1}{2}$

$\therefore$ the y-intercept is $-2\frac{1}{2}$.

c The function cuts the x-axis when $y = 0$.

$$\therefore \frac{3}{(x-1)(x+2)} = 1$$
$$\therefore (x-1)(x+2) = 3$$
$$\therefore x^2 + x - 2 = 3$$
$$\therefore x^2 + x - 5 = 0$$
$$\therefore x = \frac{-1 \pm \sqrt{1 - 4(1)(-5)}}{2}$$
$$\therefore x = \frac{-1 \pm \sqrt{21}}{2}$$

$\therefore$ x-intercepts are $\frac{-1-\sqrt{21}}{2}$ and $\frac{-1+\sqrt{21}}{2}$.

d
$$f(x) = -1 + 3(x^2 + x - 2)^{-1}$$
$$\therefore f'(x) = 0 - 3(x^2 + x - 2)^{-2}(2x+1)$$
$$= \frac{-3(2x+1)}{(x^2 + x - 2)^2}$$
$$\therefore f'(x) = 0 \text{ when } -3(2x+1) = 0$$
$$\therefore x = -\tfrac{1}{2}$$
$$\text{and } f(-\tfrac{1}{2}) = -1 + \frac{3}{(-\frac{3}{2})(\frac{3}{2})}$$
$$= -1 + \frac{3}{-\frac{9}{4}}$$
$$= -1 - \tfrac{4}{3}$$
$$= -2\tfrac{1}{3}$$

$\therefore$ D is at $(-\frac{1}{2}, -2\frac{1}{3})$.

e **i** $A = -\displaystyle\int_{\frac{\sqrt{21}-1}{2}}^{k} \left(-1 + \frac{3}{x^2 + x - 2}\right) dx$

ii $A = \displaystyle\int_{3}^{\frac{\sqrt{21}-1}{2}} \left(-1 + \frac{3}{x^2 + x - 2}\right) dx$

≈ 0.558 units^2

56 **a** **i**
$$a + b = 0.3$$
$$b + c = 0.4$$
$$a + b + c = 0.58$$

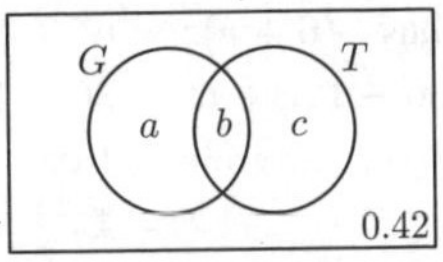

$$\therefore a = 0.18$$
$$b = 0.12$$
$$c = 0.28$$

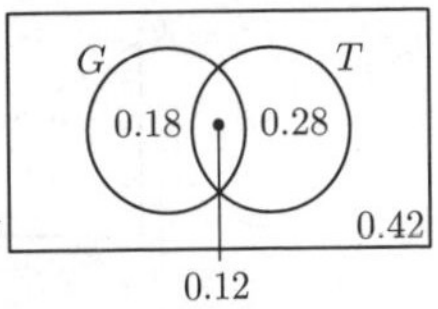

ii $P(G \cap T) = b$
$= 0.12$

The chance of a randomly selected customer buying both a goldfish and a tortoise is 0.12.

iii $P(G)\,P(T) = 0.3 \times 0.4$
$= 0.12$
$= P(G \cap T)$

$\therefore$ G and T are independent events.

b **i** Since $P(T) = 2\,P(G)$,
$$b + c = 2(a+b)$$
$$\therefore b + c = 2a + 2b$$
$$\therefore c = 2a + b$$

ii Since G and T are independent,
$$P(G \cap T) = P(G)\,P(T)$$
$$\therefore b = (a+b)(b+c)$$
$$\therefore b = (a+b)(b + 2a + b) \quad \{\text{from } \mathbf{i}\}$$
$$\therefore b = (a+b)(2a+2b)$$
$$\therefore b = 2(a+b)^2$$
$$\therefore (a+b)^2 = \frac{b}{2}$$
$$\therefore a + b = \sqrt{\frac{b}{2}} \quad \{\text{as } a + b > 0\}$$
$$\therefore a = \sqrt{\frac{b}{2}} - b$$

iii As $a + b + c = 0.58$,
$$a + b + 2a + b = 0.58$$
$$\therefore 3a + 2b = 0.58$$
$$\therefore 3\left(\sqrt{\frac{b}{2}} - b\right) + 2b = 0.58$$
$$\therefore 3\sqrt{\frac{b}{2}} - b = 0.58$$

So, $b \approx 0.104$ $\{0 < b < 1\}$
and $a \approx 0.124$

iv $P(G) = a + b \approx 0.228$

57 $f(x) = ax^3 + bx^2 + cx - 6$

a $f(1) = -10$

$\therefore\ a + b + c - 6 = -10$

$\therefore\ a + b + c = -4$ .... (1)

b $f(3) = 36$

$\therefore\ 27a + 9b + 3c - 6 = 36$

$\therefore\ 27a + 9b + 3c = 42$

$\therefore\ 9a + 3b + c = 14$ .... (2)

and $f(-2) = -4$

$\therefore\ -8a + 4b - 2c - 6 = -4$

$\therefore\ -8a + 4b - 2c = 2$

$\therefore\ -4a + 2b - c = 1$ .... (3)

c In matrix form, equations (1), (2) and (3) can be written as

$$\begin{pmatrix} 1 & 1 & 1 \\ 9 & 3 & 1 \\ -4 & 2 & -1 \end{pmatrix}\begin{pmatrix} a \\ b \\ c \end{pmatrix} = \begin{pmatrix} -4 \\ 14 \\ 1 \end{pmatrix}$$

d $$\begin{pmatrix} a \\ b \\ c \end{pmatrix} = \begin{pmatrix} 1 & 1 & 1 \\ 9 & 3 & 1 \\ -4 & 2 & -1 \end{pmatrix}^{-1}\begin{pmatrix} -4 \\ 14 \\ 1 \end{pmatrix}$$

$$\therefore\ \begin{pmatrix} a \\ b \\ c \end{pmatrix} = \begin{pmatrix} 2 \\ 1 \\ -7 \end{pmatrix} \quad \{\text{technology}\}$$

$\therefore\ a = 2,\ b = 1,\ c = -7$

e $f(x) = 2x^3 + x^2 - 7x - 6 = (x^2 - x - 2)(px + q)$

Equating coefficients of x^3 gives $2 = p$

Equating constant terms gives $-6 = -2q$

$\therefore\ q = 3$

So, $p = 2$ and $q = 3$.

58 **a** Let the smaller semi-circle have radius r cm.

In ΔPTR, we have:

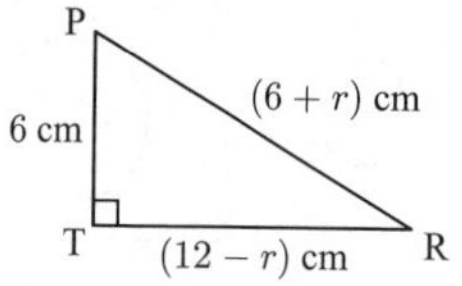

Thus $(6 + r)^2 = 6^2 + (12 - r)^2$

$\therefore\ 36 + 12r + r^2 = 36 + 144 - 24r + r^2$

$\therefore\ 36r = 144$

$\therefore\ r = 4$

b **i** $\cos(\widehat{\text{TPR}}) = \frac{6}{10} = 0.6$

$\therefore\ \widehat{\text{TPR}} \approx 53.1^o \approx 0.927^c$

ii $\widehat{\text{PRT}} \approx 90^o - 53.1^o$

$\approx 36.9^o \approx 0.644^c$

c **i** Area of A = area ΔPTR − (area sector PQT + area sector RQS)

$\approx \frac{1}{2}(8 \times 6) - \left(\frac{1}{2}(0.927)(6^2) + \frac{1}{2}(0.644)(4^2)\right)$

$\approx 24 - 16.69 - 5.15$

$\approx 2.16\ \text{cm}^2$

ii Area of B = area of quarter circle − area of semi-circles − area of A

$\approx \frac{1}{4}\pi(12^2) - \frac{1}{2}\pi(6^2) - \frac{1}{2}\pi(4^2) - 2.16$

$\approx 36\pi - 18\pi - 8\pi - 2.16$

$\approx 10\pi - 2.16$

$\approx 29.3\ \text{cm}^2$

59 **a** $\log_2(x^2 - 2x + 1) = 1 + \log_2(x - 1)$

$\therefore\ \log_2(x - 1)^2 - \log_2(x - 1) = 1$

$\therefore\ 2\log_2(x - 1) - \log_2(x - 1) = 1$

$\therefore\ \log_2(x - 1) = 1$

$\therefore\ x - 1 = 2^1$

$\therefore\ x = 3$

b $3^{2x+1} = 5(3^x) + 2$

$\therefore\ 3(3^x)^2 - 5(3^x) - 2 = 0$

$\therefore\ 3m^2 - 5m - 2 = 0 \quad \{m = 3^x\}$

$\therefore\ (3m+1)(m-2) = 0$

$\therefore\ m = -\frac{1}{3}$ or 2

$\therefore\ 3^x = -\frac{1}{3}$ or $3^x = 2$

The first equation is impossible as $3^x > 0$ for all x.

$\therefore\ 3^x = 2$

$\therefore\ x = \dfrac{\ln 2}{\ln 3}$ or $\log_3 2$

60 **a** $L = \int_0^1 \sqrt{1+(2x)^2}\, dx$

≈ 1.48 units {using technology}

b $y = \sin x$, so $\dfrac{dy}{dx} = \cos x$

$\therefore\ L = \int_0^{\pi} \sqrt{1+\cos^2 x}\, dx$

≈ 3.82 units {using technology}

61 **a**

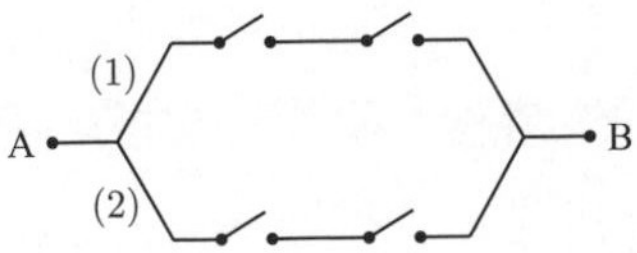

P(current flows)

$=$ P((1) closed $\cup$ (2) closed)

$=$ P((1) closed) $+$ P((2) closed) $-$ P((1) **and** (2) closed)

$= p^2 + p^2 - p^4$

$= 2p^2 - p^4$

b We need to solve $2p^2 - p^4 \geqslant \frac{1}{2}$

We graph $y = 2x^2 - x^4$ for $0 \leqslant x \leqslant 1$.

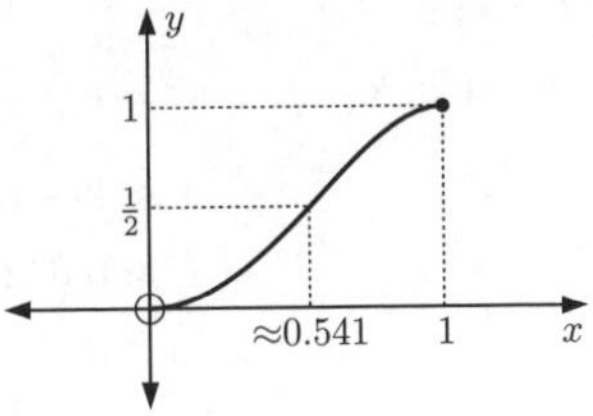

So, for $2p^2 - p^4 \geqslant \frac{1}{2}$, $p \geqslant 0.541$

$\therefore$ the least value of p is ≈ 0.541

62 **a**

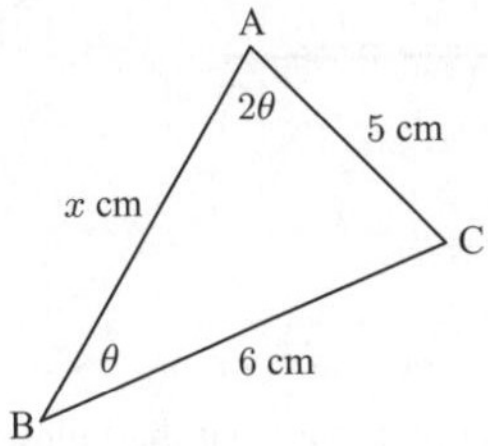

Let the angle at B be θ and at A be 2θ

By the sine rule: $\dfrac{\sin 2\theta}{6} = \dfrac{\sin\theta}{5}$

$\therefore\ \dfrac{2\sin\theta\cos\theta}{\sin\theta} = \dfrac{6}{5}$

$\therefore\ \cos\theta = \frac{3}{5}$ .... $(*)$ $\{$as $\sin\theta \neq 0\}$

b Let AB $= x$ cm.

Using the cosine rule,

$5^2 = x^2 + 6^2 - 2x(6)\cos\theta$

$\therefore\ 25 = x^2 + 36 - 12x\left(\frac{3}{5}\right)$

$\therefore\ x^2 - \frac{36}{5}x + 11 = 0$

$\therefore\ 5x^2 - 36x + 55 = 0$

$\therefore\ (x-5)(5x-11) = 0$

$\therefore\ x = 5$ or $\frac{11}{5}$

$\therefore$ AB $= 5$ cm or 2.2 cm

c If AB $= 5$ we have an isosceles triangle

$\therefore\ 4\theta = 180^o$

$\therefore\ \theta = 45^o$

which contradicts $(*)$

as $\cos 45^o = \frac{1}{\sqrt{2}} \neq \frac{3}{5}$.

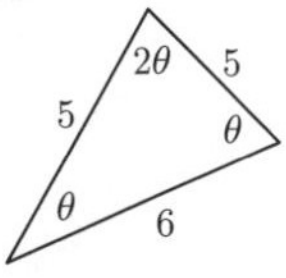

If AB $= 2.2$, no such problem occurs.

$\therefore$ AB $= 2.2$ is the only valid solution.

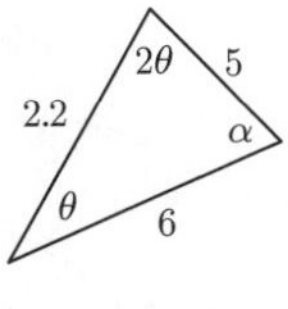

63

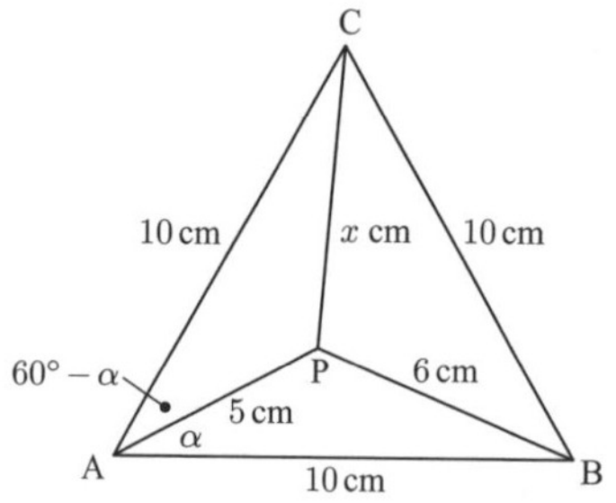

Using the cosine rule in ΔABP,

$$\cos\alpha = \frac{5^2 + 10^2 - 6^2}{2(5)(10)} = \tfrac{89}{100} = 0.89$$

$\therefore\ \alpha \approx 27.127^o$

$\therefore\ 60^o - \alpha \approx 32.873^o$

So, in ΔAPC, $x^2 = 10^2 + 5^2 - 2(10)(5)\cos 32.873^o$

$\therefore\ x^2 \approx 41.0127$

$\therefore\ x \approx 6.40$

Thus P is about 6.40 cm from C.

64 Let F be the event of a faulty chip.

$\therefore$ $\text{P}(F) = 0.03$ and $\text{P}(F') = 0.97$

If X is the number which are faulty then $X \sim \text{B}(500,\ 0.03)$

So, $\text{P}(5 \leqslant X \leqslant 10) = \text{P}(X \leqslant 10) - \text{P}(X \leqslant 4)$ $\quad\{1\%$ is 5, 2% is $10\}$

$\approx 0.114\,787 - 0.000\,754$

≈ 0.114

65 $X \sim \text{N}(\mu,\ 2.83^2)$

$$\therefore\quad \text{P}(-4 < X - \mu < 4) = \text{P}\left(\frac{-4}{2.83} < \frac{X-\mu}{2.83} < \frac{4}{2.83}\right)$$

$\approx \text{P}(-1.4134 < Z < 1.4134)$

≈ 0.842

66 **a**

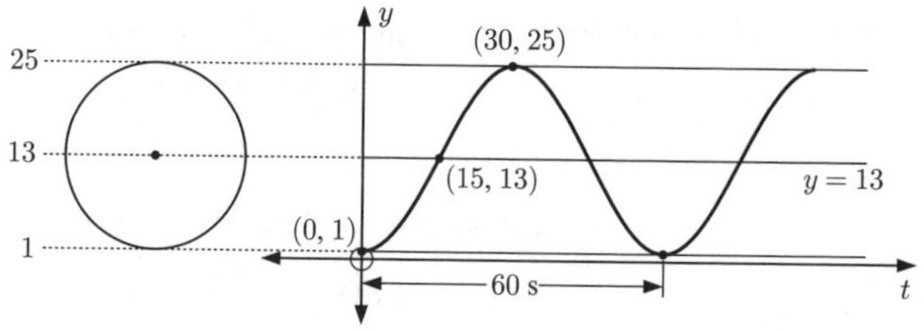

We model the Ferris wheel using $h(t) = a + b\sin(c(t - d))$.

The amplitude $= b = 12$. The period $= \dfrac{2\pi}{c} = 60 \quad \therefore\ c = \frac{\pi}{30}$

$d =$ the t-coordinate of the point halfway between the first minimum and the following maximum

$= \dfrac{0 + 30}{2} = 15$

The equation of the principal axis is $\quad y = \dfrac{1 + 25}{2} = 13$

$\therefore\ a = 13$

Thus $a = 13,\ \ b = 12,\ \ c = \frac{\pi}{30},\ \ d = 15.$

So, $h(t) = 12\sin\left(\frac{\pi}{30}(t - 15)\right) + 13$

Check: $h(0) = 12\sin\left(\frac{-\pi}{2}\right) + 13 = 12(-1) + 13 = 1$ ✓

$h(30) = 12\sin\left(\frac{\pi}{2}\right) + 13 = 12(1) + 13 = 25$ ✓

b When $t = 91$, $h(91) = 12\sin\left(\dfrac{\pi \times 76}{30}\right) + 13 \approx 24.9$ m

67 $\mathbf{AB} = \begin{pmatrix} 2 & 1 & -1 \\ -1 & 2 & 1 \\ 0 & 6 & 1 \end{pmatrix} \begin{pmatrix} 4 & 7 & -3 \\ -1 & -2 & 1 \\ 6 & 12 & -5 \end{pmatrix} = \begin{pmatrix} 1 & 0 & 0 \\ 0 & 1 & 0 \\ 0 & 0 & 1 \end{pmatrix} = \mathbf{I}$

Now $\begin{pmatrix} 4 & 7 & -3 \\ -1 & -2 & 1 \\ 6 & 12 & -5 \end{pmatrix} \begin{pmatrix} a \\ b \\ c \end{pmatrix} = \begin{pmatrix} -8 \\ 3 \\ -15 \end{pmatrix}$

$$\therefore \begin{pmatrix} a \\ b \\ c \end{pmatrix} = \begin{pmatrix} 4 & 7 & -3 \\ -1 & -2 & 1 \\ 6 & 12 & -5 \end{pmatrix}^{-1} \begin{pmatrix} -8 \\ 3 \\ -15 \end{pmatrix}$$

$$= \begin{pmatrix} 2 & 1 & -1 \\ -1 & 2 & 1 \\ 0 & 6 & 1 \end{pmatrix} \begin{pmatrix} -8 \\ 3 \\ -15 \end{pmatrix} \quad \{\text{since } \mathbf{B}^{-1} = \mathbf{A}\}$$

$$= \begin{pmatrix} 2 \\ -1 \\ 3 \end{pmatrix}$$

$\therefore\ a = 2,\ b = -1,\ c = 3$

68 $f(x) = e^{\sin^2 x}, \quad 0 \leqslant x \leqslant \pi$

a $f'(x) = e^{\sin^2 x} \times 2\sin x \cos x$

$= e^{\sin^2 x} \sin 2x$

which is 0 when $\sin 2x = 0$

$\therefore\ 2x = 0 + k\pi, \quad k \in \mathbb{Z}$

$\therefore\ x = 0 + \dfrac{k\pi}{2}$

$\therefore\ x = 0,\ \frac{\pi}{2},\ \pi$

Sign diagram:

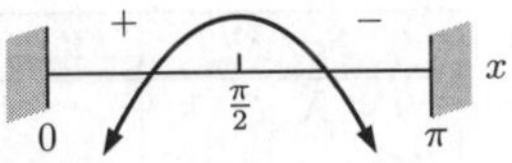

$\therefore\ f(x)$ has maximum value when $x = \frac{\pi}{2}$, and this maximum value is e.

b $f''(x) = e^{\sin^2 x} \times \sin 2x \sin 2x + e^{\sin^2 x} 2\cos 2x$

$= e^{\sin^2 x}(\sin^2 2x + 2\cos 2x)$

$= e^{\sin^2 x}\left[(2\sin x \cos x)^2 + 2(1 - 2\sin^2 x)\right]$

$= e^{\sin^2 x}\left(4\sin^2 x \cos^2 x + 2 - 4\sin^2 x\right)$

$= e^{\sin^2 x}\left(2 - 4\sin^2 x(1 - \cos^2 x)\right)$

$= e^{\sin^2 x}(2 - 4\sin^4 x)$

$\therefore\ f''(x) = 0$ when $2 - 4\sin^4 x = 0$

$\therefore\ \sin^4 x = \frac{1}{2}$

$\therefore\ \sin^2 x = \frac{1}{\sqrt{2}}$

So, $\sin^2 x = \frac{1}{\sqrt{2}}$ needs to be solved.

c $\sin^2 x = \frac{1}{\sqrt{2}}$

$\therefore\ \sin x = \frac{1}{\sqrt[4]{2}} \quad \{0 \leqslant x \leqslant \pi\}$

$\therefore\ x \approx 0.999$ or 2.14

$\therefore$ the points of inflection are $\left(0.999,\ e^{\frac{1}{\sqrt{2}}}\right)$ and $\left(2.14,\ e^{\frac{1}{\sqrt{2}}}\right)$.

69 $\dfrac{u_1}{1-r} = 49$ and $u_1 r = 10$

$\therefore\ \dfrac{10}{r} = 49(1-r)$

$\therefore\ 10 = 49r - 49r^2$

$\therefore\ 49r^2 - 49r + 10 = 0$

$\therefore\ (7r-2)(7r-5) = 0$

$\therefore\ r = \frac{2}{7}$ or $\frac{5}{7}$

When $r = \frac{2}{7},\ u_1 = 35,\ u_2 = 10,\ u_3 = \frac{20}{7} = 2\frac{6}{7}$

When $r = \frac{5}{7},\ u_1 = 14,\ u_2 = 10,\ u_3 = \frac{50}{7} = 7\frac{1}{7}$

Thus $S_3 = 35 + 10 + 2\frac{6}{7} = 47\frac{6}{7}$

or $S_3 = 14 + 10 + 7\frac{1}{7} = 31\frac{1}{7}$

70 $f(x) = xe^{1-2x^2}$

a $f'(x) = 1e^{1-2x^2} + xe^{1-2x^2}(-4x)$
$= e^{1-2x^2}(1-4x^2)$
$f''(x) = -4xe^{1-2x^2}(1-4x^2) + e^{1-2x^2}(-8x)$
$= e^{1-2x^2}(-4x+16x^3-8x)$
$= e^{1-2x^2}(16x^3-12x)$

b $f'(x) = e^{1-2x^2}(1+2x)(1-2x)$

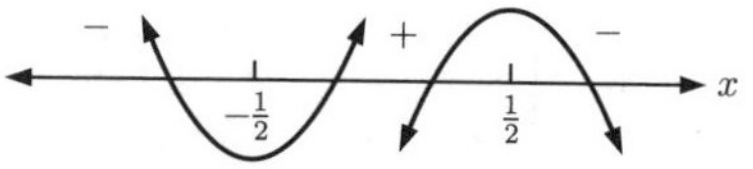

$\therefore$ there is a local minimum at $(-\frac{1}{2}, -\frac{\sqrt{e}}{2})$ and a local maximum at $(\frac{1}{2}, \frac{\sqrt{e}}{2})$.

c $f''(x) = 0$ when $16x^3 - 12x = 0$
$\therefore\ 4x(4x^2-3) = 0$
$\therefore\ x = 0$ or $\pm\frac{\sqrt{3}}{2}$

d As $x \to \infty$, $f(x) \to 0$ (above).
As $x \to -\infty$, $f(x) \to 0$ (below).

e

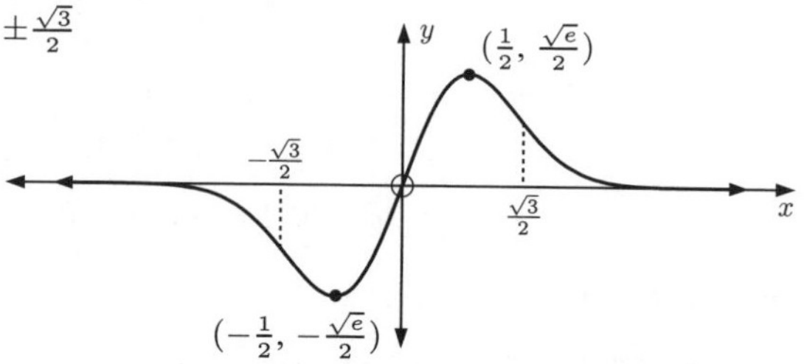

71

Score	*Midpoint* (x)	f
$9 < X \leqslant 11$	10	2
$11 < X \leqslant 13$	12	7
$13 < X \leqslant 15$	14	6
$15 < X \leqslant 17$	16	21
$17 < X \leqslant 19$	18	17
$19 < X \leqslant 21$	20	5
	Total	58

a mean $\approx \dfrac{10\times 2 + 12\times 7 + ... + 20\times 5}{58}$
≈ 16.0

b standard deviation ≈ 2.48 {technology}

72 **a** $|\mathbf{a}| = \sqrt{\cos^2\theta + \sin^2\theta + \cos^2\theta}$
$= \sqrt{1+\cos^2\theta}$

b As $-1 \leqslant \cos\theta \leqslant 1$ for all θ,
then $0 \leqslant \cos^2\theta \leqslant 1$
$\therefore\ 1 \leqslant 1+\cos^2\theta \leqslant 2$
$\therefore\ 1 \leqslant \sqrt{1+\cos^2\theta} \leqslant \sqrt{2}$
$\therefore\ 1 \leqslant |\mathbf{a}| \leqslant \sqrt{2}$

c $\mathbf{a}\bullet\mathbf{b} = 0$ $\{\mathbf{a}\perp\mathbf{b}\}$
$\therefore\ \cos\theta\sin\theta - \sin^2\theta + \cos^2\theta = 0$
$\therefore\ \frac{1}{2}\sin 2\theta + \cos^2\theta - \sin^2\theta = 0$
$\therefore\ \frac{1}{2}\sin 2\theta + \cos 2\theta = 0$
$\therefore\ \frac{1}{2}\sin 2\theta = -\cos 2\theta$
$\therefore\ \tan 2\theta = -2$
$\therefore\ 2\theta \approx -1.1071 + k\pi,\quad k\in\mathbb{Z}$
$\therefore\ \theta \approx -0.5536 + \dfrac{k\pi}{2}$
$\therefore\ \theta \approx 1.02,\ 2.59,\ 4.16$ or 5.73

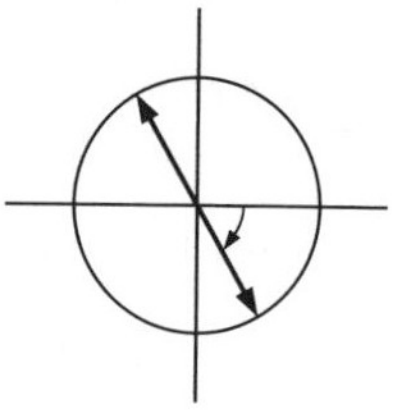

73 **a** $3 + 2\sin x = 0$
$\therefore\ \sin x = -\frac{3}{2}$ which is impossible as $-1 \leqslant \sin x \leqslant 1$
$\therefore$ no solutions exist

b $3\cos\left(\frac{x}{2}\right)+1=0$

$\therefore\ \cos\left(\frac{x}{2}\right)=-\frac{1}{3}$

$\therefore\ \frac{x}{2}\approx\pi\pm1.231+k2\pi,\quad k\in\mathbb{Z}$

$\therefore\ x\approx2\pi\pm2.462+k4\pi$

$\therefore\ x\approx3.82$

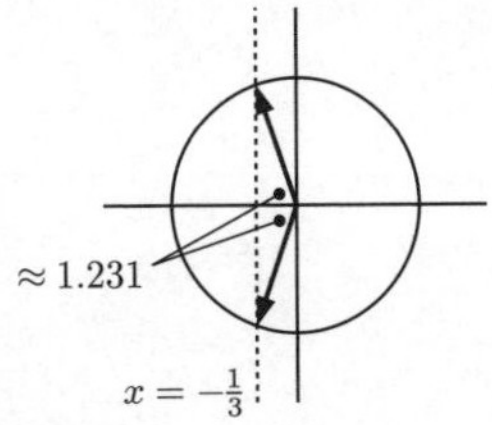

74 **a** As $\sum_{x=0}^{4} f(x)=1,$

$\frac{1+k}{50}+\frac{4+2k}{50}+\frac{9+3k}{50}+\frac{16+4k}{50}=1$

$\therefore\ \frac{30+10k}{50}=1$

$\therefore\ 30+10k=50$

$\therefore\ k=2$

b

x	0	1	2	3	4
$f(x)$	0	$\frac{3}{50}$	$\frac{8}{50}$	$\frac{15}{50}$	$\frac{24}{50}$

$\mu=\sum x_i f_i$

$=0+\frac{3}{50}+\frac{16}{50}+\frac{45}{50}+\frac{96}{50}$

$=\frac{160}{50}$

$=3.2$

c $P(X\geqslant2)=\frac{8}{50}+\frac{15}{50}+\frac{24}{50}$

$=\frac{47}{50}$

75 $f(x)=\sin x\cos(2x),\quad 0\leqslant x\leqslant\pi$

a $f'(x)=\cos x\cos(2x)+\sin x\,(-2\sin(2x))$

$=\cos x(2\cos^2x-1)-2\sin x(2\sin x\cos x)$

$=2\cos^3x-\cos x-4\cos x(1-\cos^2x)$

$=2\cos^3x-\cos x-4\cos x+4\cos^3x$

$=6\cos^3x-5\cos x$

b $f'(x)=0\quad\text{if}\quad\cos x(6\cos^2x-5)=0$

$\therefore\ \cos x=0\quad\text{or}\quad\cos^2x=\frac{5}{6}$

$\therefore\ \cos x=0\quad\text{or}\quad\pm\sqrt{\frac{5}{6}}$

c $\cos x=0\quad\text{or}\quad\pm0.912\,87$

$\therefore\ x=\frac{\pi}{2}\quad\text{or}\quad x\approx0.421,\ 2.72$

Sign diagram of $f'(x)$:

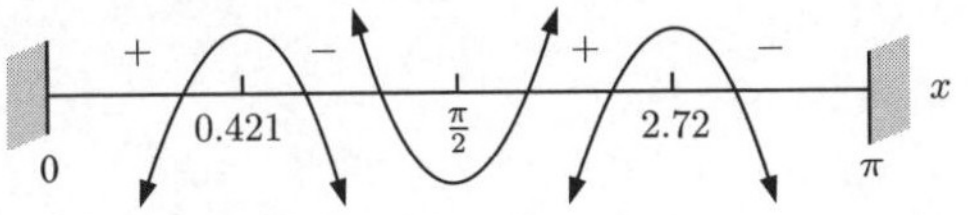

$\therefore$ there is a local maximum at $(0.421,\ 0.272)$ and $(2.72,\ 0.272)$ and a local minimum at $(\frac{\pi}{2},\ -1)$.

d

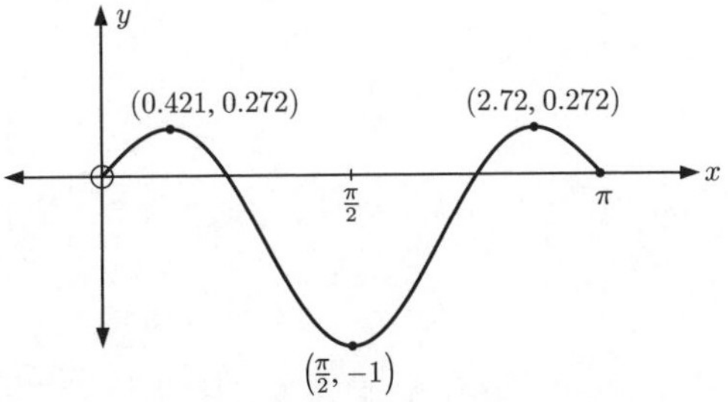